Get connected to the power of the Internet

McDougal Littell's online resources for teachers provide time-saving planning, instruction, and assessment support.

classzone.com

- Links correlated to the text
- Web Research Guide
- Demographic data updates
- Self-scoring quizzes
- Interactive games and activities
- Links to current events
- Test practice
- Teacher Center

You have immediate access to *ClassZone's* teacher resources.

MCDYJQ49FDTCF

Use this code to create your own username and password.

Visit classzone.com for purchasing information and free demos

easyPlanner Plus ONLINE

- Customizable lesson plans
- Trackable standards correlated to each lesson

eEdition Plus ONLINE

- Online version of the text
- Interactive features to explore geography

eTest Plus ONLINE

- Customizable assessment tool
- Automatically grades tests
- Generates reports correlated to standards

Now it all clicks!™

McDougal Littell

TEACHER'S EDITION

World Cultures AND GEOGRAPHY

Western Hemisphere and Europe

McDOUGAL LITTELL

Acknowledgments begin on page R53.

ISBN 0-618-37759-X

Printed in the United States of America

2 3 4 5 6 7 8-VJM-07 06 05 04

Contents

Motivation

McDougal Littell's ***World Cultures and Geography*** offers a vibrant, visual approach that motivates students to gain a greater appreciation of the diversity of peoples around the world.

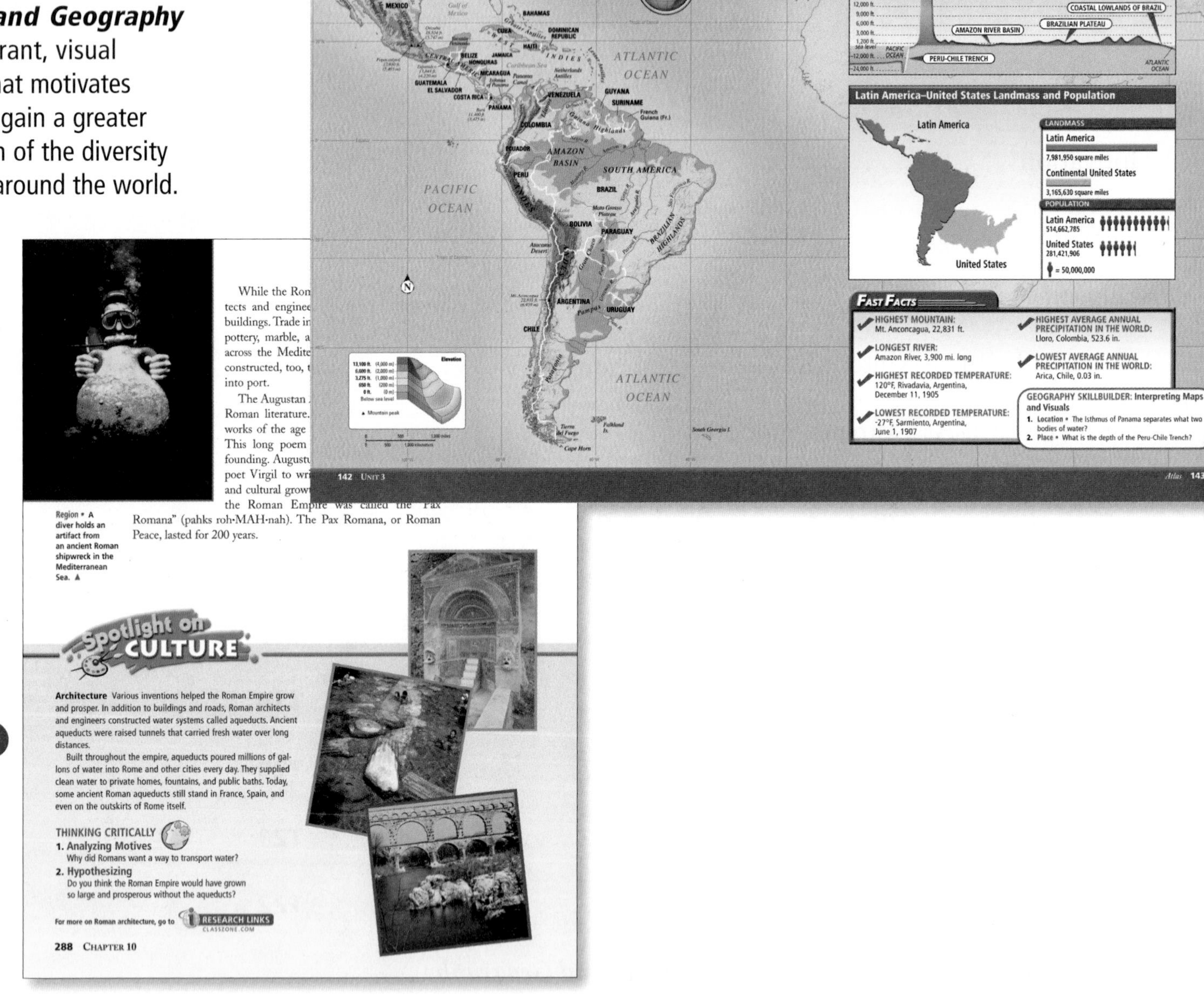

1

UNIT Atlas 3 Physical Geography — Latin America

Latin America: Physical

Elevation Profile: South America

ANDES MOUNTAINS · COASTAL LOWLANDS OF BRAZIL · BRAZILIAN PLATEAU · AMAZON RIVER BASIN · PERU-CHILE TRENCH · PACIFIC OCEAN · ATLANTIC OCEAN

Latin America–United States Landmass and Population

Latin America · United States

LANDMASS: Latin America 7,981,950 square miles; Continental United States 3,165,630 square miles

POPULATION: Latin America 514,662,785; United States 281,421,906; = 50,000,000

FAST FACTS

HIGHEST MOUNTAIN: Mt. Aconcagua, 22,831 ft.

LONGEST RIVER: Amazon River, 3,900 mi. long

HIGHEST RECORDED TEMPERATURE: 120°F, Rivadavia, Argentina, December 11, 1905

LOWEST RECORDED TEMPERATURE: -27°F, Sarmiento, Argentina, June 1, 1907

HIGHEST AVERAGE ANNUAL PRECIPITATION IN THE WORLD: Lloro, Colombia, 523.6 in.

LOWEST AVERAGE ANNUAL PRECIPITATION IN THE WORLD: Arica, Chile, 0.03 in.

GEOGRAPHY SKILLBUILDER: Interpreting Maps and Visuals

1. Location • The Isthmus of Panama separates what two bodies of water?
2. Place • What is the depth of the Peru-Chile Trench?

142 UNIT 3 · Atlas 143

While the Ron… tects and enginee… buildings. Trade in… pottery, marble, a… across the Medite… constructed, too, t… into port.

The Augustan… Roman literature… works of the age… This long poem… founding. Augustu… poet Virgil to wri… and cultural grow… the Roman Empire was called the "Pax Romana" (pahks roh·MAH·nah). The Pax Romana, or Roman Peace, lasted for 200 years.

Region • A diver holds an artifact from an ancient Roman shipwreck in the Mediterranean Sea. ▲

2

Spotlight on CULTURE

Architecture Various inventions helped the Roman Empire grow and prosper. In addition to buildings and roads, Roman architects and engineers constructed water systems called aqueducts. Ancient aqueducts were raised tunnels that carried fresh water over long distances.

Built throughout the empire, aqueducts poured millions of gallons of water into Rome and other cities every day. They supplied clean water to private homes, fountains, and public baths. Today, some ancient Roman aqueducts still stand in France, Spain, and even on the outskirts of Rome itself.

THINKING CRITICALLY

1. **Analyzing Motives** Why did Romans want a way to transport water?
2. **Hypothesizing** Do you think the Roman Empire would have grown so large and prosperous without the aqueducts?

For more on Roman architecture, go to RESEARCH LINKS CLASSZONE.COM

288 CHAPTER 10

1 Unit Atlas

A **Unit Atlas** introduces each unit with state-of-the-art maps and up-to-date charts and graphs—giving a colorful overview of each region. Every **Unit Atlas** includes a full-page physical map, a full-page political map, and several thematic maps.

2 Spotlight on Culture

High-interest features like **Spotlight on Culture** provide a comprehensive look at cultural activities around the world, including dance, food, music, and art.

3

4

Technology: 1400

Andean Agriculture

Long before the rise of the Inca Empire, people living in the Andes had learned to farm the steep valley walls by building terraces into the sides of the mountains. They had also learned to build canals, many of them lined with stone, to carry water to their crops. The Inca improved and expanded the existing terraces and canals until they could feed 15 million people, with enough food left over to put away stores for three to seven years.

In the Andes, valley walls rise as high as 10,000 feet and temperatures can span a 55-degree range.

Inca canals stretched for miles. They were often lined and covered with stones. Some were cut through solid rock.

The Inca grew maize, hundreds of kinds of potatoes, and many other crops. Farmers had to plant crops adapted to many different climates because of the great variations in altitude and temperature.

The Inca had few farm tools. The most widely used was the *taclla*, or digging stick. It consisted of a pointed hardwood pole with a footrest for pushing the tool into the ground. Some *tacllas* had metal tips. The other main tools were hoes and clubs.

Workers directed by royal architects built stone retaining walls. Inside the walls, they placed layers of stone, clay, gravel, and topsoil. This combination allowed water to slowly work its way to lower terraces.

THINKING Critically

1. Analyzing Motives
Why did people living in the Andes need to build terraces and canals?

2. Recognizing Effects
What role did agriculture play in the building and maintenance of the Inca Empire?

164 UNIT 3 *Latin America*

CHAPTER 7 ASSESSMENT

TERMS & NAMES

Explain the significance of each of the following:

1. criollo
2. mestizo
3. Father Miguel Hidalgo
4. Treaty of Guadalupe Hidalgo
5. Benito Juárez
6. hacienda
7. maquiladora
8. PEMEX
9. Diego Rivera
10. fiesta

REVIEW QUESTIONS

The Roots of Modern Mexico (pages 173—178)

1. Why did Cortés need the help of some Native Americans to defeat Montezuma II?
2. How was Mexico's War of Independence connected to events elsewhere in the world at that time?

Government in Mexico: Revolution and Reform (pages 179—184)

3. What concerns did Mexicans in the 1800s and 1900s have about land ownership?
4. What was unique about Mexico's presidential election in 2000?

Mexico's Changing Economy (pages 185—189)

5. Why has the *ejido* system been unsuccessful at ending poverty for many villagers?
6. What is the role of oil in Mexico's economy?

Mexico's Culture Today (pages 190—194)

7. What are some of the problems caused by Mexico City's rapid growth?
8. What event do Mexicans celebrate every September 16?

CRITICAL THINKING

Sequencing Events

1. Using your completed time line from Reading Social Studies, p. 172, list the points in Mexico's history when the nation's government changed drastically.

Forming and Supporting Opinions

2. Do you think the Mexican government's decision in the 1990s to privatize farming and business was a good one? Explain.

Hypothesizing

3. Modern artists, such as Diego Rivera and Octavio Paz, have used images from both Mexico's past and its present in their works. What do you think this says about the way Mexicans today view their nation's history?

Visual Summary

1 The Roots of Modern Mexico
- From 1810 to 1821, Mexico fought for and gained independence from Spain.

2 Government in Mexico: Revolution and Reform
- The Mexican Revolution (1910–1920) established a new government.
- Today, the government of Mexico is a federal republic.

3 Mexico's Changing Economy
- Farming is still an important part of Mexico's economy, but today its top businesses are oil production and tourism.

4 Mexico's Culture Today
- Mexico's culture combines its Native American and Spanish pasts with new elements.
- Mexico's holidays and arts show the influence of its history.

198 CHAPTER 7

3 Technology

Cross-curricular features such as **Technology** involve students in learning about how the peoples of South America were able to develop agriculture high in the Andes Mountains, China's Three Gorges Dam, and The Great Zimbabwe. Students learn about the impact of technology on culture through the combination of pictures and words in these engaging and interactive features.

4 Visual Summary

A **Visual Summary** at the end of each chapter summarizes the key ideas of that chapter. The key ideas are organized in easy-to-read charts that give visual learners and others another path to understanding.

Relevance

World Cultures and Geography offers interactive lessons that help students make the connection between geography and current events. Students see geography as more relevant when they see how it connects to other subjects they are studying. ***World Cultures and Geography*** makes numerous interdisciplinary connections through activities and features that show how geography and culture relate to literature, art, science, math, and other subjects.

❶

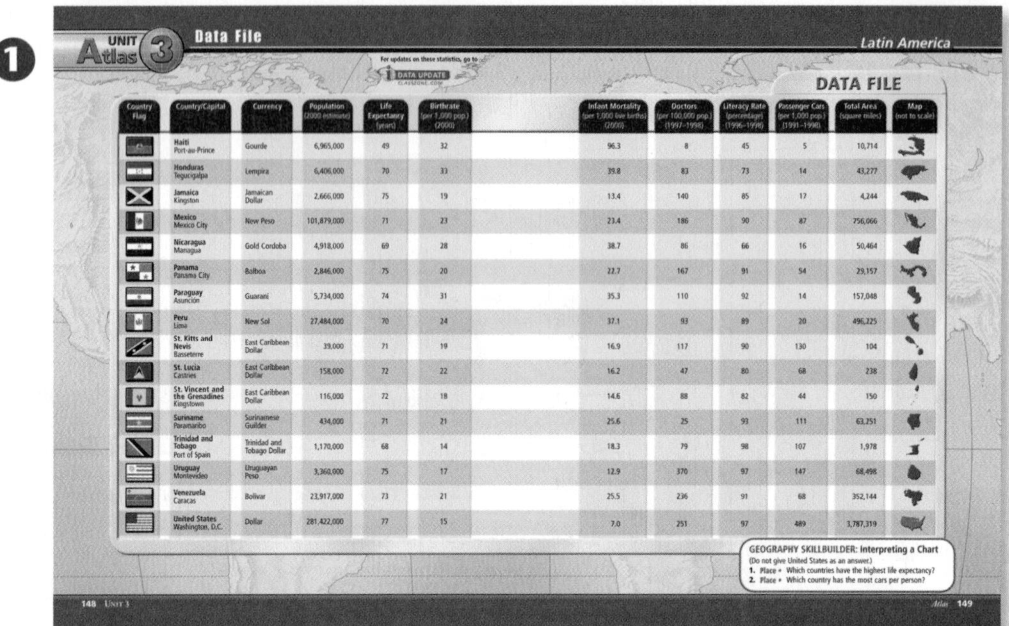

Unit Atlas 3 Data File — Latin America

For updates on these statistics, go to DATA UPDATE CLASSZONE.COM

DATA FILE

Country Flag	Country/Capital	Currency	Population (2000 estimate)	Life Expectancy (years)	Birthrate (per 1,000 pop.) (2000)	Infant Mortality (per 1,000 live births) (2000)	Doctors (per 100,000 pop.) (1997–1998)	Literacy Rate (percentage) (1996–1998)	Passenger Cars (per 1,000 pop.) (1991–1998)	Total Area (square miles)	Map (not to scale)
	Haiti Port-au-Prince	Gourde	6,965,000	49	32	96.3	8	45	5	10,714	
	Honduras Tegucigalpa	Lempira	6,406,000	70	33	39.8	83	73	14	43,277	
	Jamaica Kingston	Jamaican Dollar	2,666,000	75	19	13.4	140	85	17	4,244	
	Mexico Mexico City	New Peso	101,879,000	71	23	23.4	186	90	87	756,066	
	Nicaragua Managua	Gold Cordoba	4,918,000	69	28	38.7	86	66	16	50,464	
	Panama Panama City	Balboa	2,846,000	75	20	22.7	167	91	54	29,157	
	Paraguay Asunción	Guarani	5,734,000	74	31	35.3	110	92	14	157,048	
	Peru Lima	New Sol	27,484,000	70	24	37.1	93	89	20	496,225	
	St. Kitts and Nevis Basseterre	East Caribbean Dollar	39,000	71	19	16.9	117	90	130	104	
	St. Lucia Castries	East Caribbean Dollar	158,000	72	22	16.2	47	80	68	238	
	St. Vincent and the Grenadines Kingstown	East Caribbean Dollar	116,000	72	18	14.6	88	82	44	150	
	Suriname Paramaribo	Surinamese Guilder	434,000	71	21	25.6	25	93	111	63,251	
	Trinidad and Tobago Port of Spain	Trinidad and Tobago Dollar	1,170,000	68	14	18.3	79	98	107	1,978	
	Uruguay Montevideo	Uruguayan Peso	3,360,000	75	17	12.9	370	97	147	68,498	
	Venezuela Caracas	Bolivar	23,917,000	73	21	25.5	236	91	68	352,144	
	United States Washington, D.C.	Dollar	281,422,000	77	15	7.0	251	97	489	3,787,319	

GEOGRAPHY SKILLBUILDER: Interpreting a Chart
(Do not give United States as an answer.)
1. Place • Which countries have the highest life expectancy?
2. Place • Which country has the most cars per person?

148 UNIT 3 Atlas 149

❷

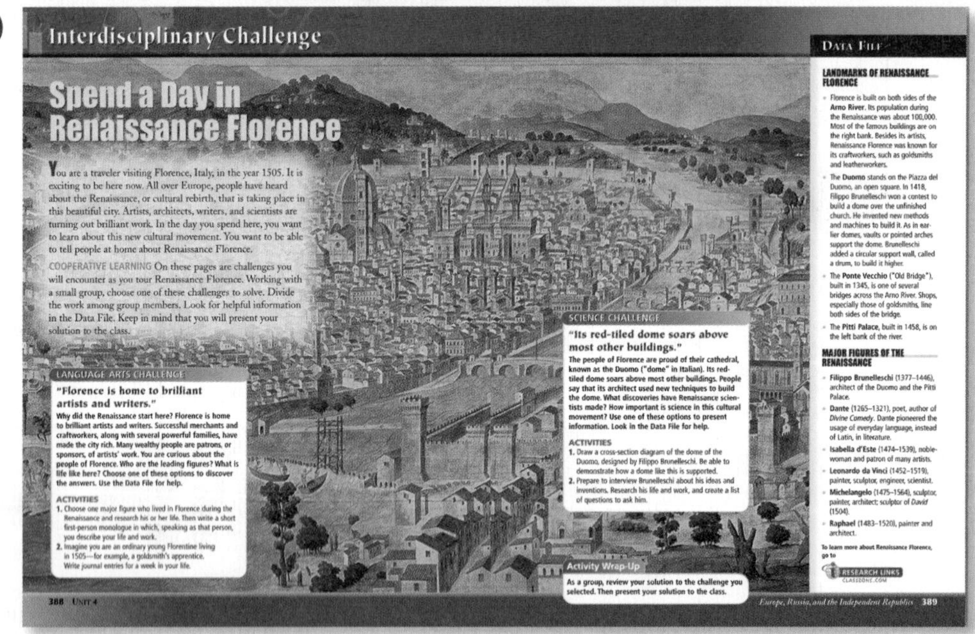

Interdisciplinary Challenge

Spend a Day in Renaissance Florence

You are a traveler visiting Florence, Italy, in the year 1505. It is exciting to be here now. All over Europe, people have heard about the Renaissance, or cultural rebirth, that is taking place in this beautiful city. Artists, architects, writers, and scientists are turning out brilliant work. In the day you spend here, you want to learn about this new cultural movement. You want to be able to tell people at home about Renaissance Florence.

COOPERATIVE LEARNING On these pages are challenges you will encounter as you tour Renaissance Florence. Working with a small group, choose one of these challenges to solve. Divide the work among group members. Look for helpful information in the Data File. Keep in mind that you will present your solution to the class.

LANGUAGE ARTS CHALLENGE

"Florence is home to brilliant artists and writers."

Why did the Renaissance start here? Florence is home to brilliant artists and writers. Successful merchants and craftworkers, along with several powerful families, have made the city rich. Many wealthy people are patrons, or sponsors, of artists' work. You are curious about the people of Florence. Who are the leading figures? What is life like here? Choose one of these options to discover the answers. Use the Data File for help.

ACTIVITIES
1. Choose one major figure who lived in Florence during the Renaissance and research his or her life. Then write a short first-person monologue in which, speaking as that person, you describe your life and work.
2. Imagine you are an ordinary young Florentine living in 1505—for example, a goldsmith's apprentice. Write journal entries for a week in your life.

SCIENCE CHALLENGE

"Its red-tiled dome soars above most other buildings."

The people of Florence are proud of their cathedral, known as the Duomo ("dome" in Italian). Its red-tiled dome soars above most other buildings. People say that its architect used new techniques to build the dome. What discoveries have Renaissance scientists made? How important is science in this cultural movement? Use one of these options to present information. Look in the Data File for help.

ACTIVITIES
1. Draw a cross-section diagram of the dome of the Duomo, designed by Filippo Brunelleschi. Be able to demonstrate how a dome like this is supported.
2. Prepare to interview Brunelleschi about his ideas and inventions. Research his life and work, and create a list of questions to ask him.

Activity Wrap-Up

As a group, review your solution to the challenge you selected. Then present your solution to the class.

DATA FILE

LANDMARKS OF RENAISSANCE FLORENCE

- Florence is built on both sides of the **Arno River**. Its population during the Renaissance was about 100,000. Most of the famous buildings are on the right bank. Besides its artists, Renaissance Florence was known for its craftworkers, such as goldsmiths and leatherworkers.
- The **Duomo** stands on the Piazza del Duomo, an open square. In 1418, Filippo Brunelleschi won a contest to build a dome over the unfinished church. He invented new methods and machines to build it. As in earlier domes, vaults or pointed arches support the dome. Brunelleschi added a circular support wall, called a drum, to build it higher.
- The **Ponte Vecchio** ("Old Bridge"), built in 1345, is one of several bridges across the Arno River. Shops, especially those of goldsmiths, line both sides of the bridge.
- The **Pitti Palace**, built in 1458, is on the left bank of the river.

MAJOR FIGURES OF THE RENAISSANCE

- **Filippo Brunelleschi** (1377–1446), architect of the Duomo and the Pitti Palace.
- **Dante** (1265–1321), poet, author of *Divine Comedy*. Dante pioneered the usage of everyday language, instead of Latin, in literature.
- **Isabella d'Este** (1474–1539), noblewoman and patron of many artists.
- **Leonardo da Vinci** (1452–1519), painter, sculptor, engineer, scientist.
- **Michelangelo** (1475–1564), sculptor, painter, architect; sculptor of *David* (1504).
- **Raphael** (1483–1520), painter and architect.

To learn more about Renaissance Florence, go to RESEARCH LINKS CLASSZONE.COM

388 UNIT 4 Europe, Russia, and the Independent Republics 389

❶ Data File

Every unit contains a **Data File** as part of the **Unit Atlas.** This comprehensive, easy-to-read chart contains detailed demographic data on each region to help students make comparisons and connections. For regular data updates, go to **classzone.com**, the companion Web site for ***World Cultures and Geography.***

❷ Interdisciplinary Challenge

These interactive features place students in situations that challenge them to solve problems by making connections to other disciplines. Every **Interdisciplinary Challenge** includes a **Data File** of information that students use in solving the problems.

Section 3
Mexico's Changing Economy

TERMS & NAMES
Carlos Salinas de Gortari
privatization
distribution
maquiladora
nationalize
PEMEX

3

MAIN IDEA
In the mid-1900s, the basis of Mexico's economy changed from farming to industry and tourism.

WHY IT MATTERS NOW
Mexico's successful expansion of its economy has helped the nati[on] to prosper.

DATELINE

MEXICO CITY, MEXICO, MARCH 18, 1938

President Lázaro Cárdenas's radio address today established a new course for Mexico's economy. Speaking to the nation, the president announced that foreigners will no longer be allowed to control petroleum companies in Mexico. The oil industry will now be run by the Mexican government itself. It is hoped that this change will boost both Mexico's economy and its national identity.

Human-Environment Interaction • The Mexican government will now own all oil-producing equipment, such as these oil wells in Veracruz. ▶

Farming in a Time of Change

The 1938 decision that the government would own [...] industry was made in an effort to expand Mexico[...] The expansion was necessary because, from ancient [...] the mid-1900s, most Mexicans worked in just one [...] farming. Since the 1950s, great numbers of Mexica[...] farming for other kinds of work. However, farming is sti[...] to Mexico's economy.

Section 3
The European Union

TERMS & NAMES
European Union
currency
euro
tariff
standard of living
Court of Human Rights

MAIN IDEA
Europeans want to maintain a high quality of life for all citizens while preserving their unique cultures.

WHY IT MATTERS NOW
A prosperous and culturally diverse Europe provides goods and markets for the rest of the world.

4

DATELINE

WESTERN EUROPE, DECEMBER 2001—Starting next month, people in many Western European nations will begin trading their old bills and coins for euros—the new money of the European Union (EU). The design of the bills, below, is the same for all EU members.

The design of the euro coins, however, will be different. Individual countries are minting their own. As shown here, one side has a standard euro design. The other side has national symbols that relate to each country. In 1996, artists and sculptors from all over Europe entered a contest to design the coins. The winner was Luc Luycx (lewk lowx) from Belgium.

Western Europe Today

Today, in Western Europe, all national leaders share their pow[er] with elected lawmakers. Citizens take part in government by v[ot]ing and through membership in a variety of political parties. T[he] Unit Atlas on pages 260–269 shows modern Europe.

Location • Draw a picture on the entire surface of an orange and then peel the orange in one continuous piece. After you lay the peel flat, your image will be distorted. ◀

5

Differences Between Maps and Globes Both maps and globes represent Earth and its features. A globe is an accurate model of the world because it has three dimensions and can show its actual shape. Globes are difficult to carry around, however. Maps are more practical. They can be folded, carried, hung on a wall, or printed in a book or magazine. However, because maps show the world in only two dimensions, they are not perfectly accurate. Look at the pictures above to see why. When the orange peel is flattened out, the picture on the orange is distorted, or twisted out of shape. Cartographers have the same problem with maps.

Reading Social Studies
A. Clarifying Why does a globe represent Earth better than a map?

Three Kinds of Maps General reference maps, which show natural and human-made features, are used to locate a place. **Thematic maps** focus on one specific idea or theme. The population map on page 48 is an example of a thematic map. Pilots and sailors use nautical maps to find their way through air and over water. A nautical map is sometimes called a chart.

Location • A road map is a reference map that shows how to get from one place to another. ▼

Connections to Math

Measuring Earth In 230 B.C., the Greek scientist Eratosthenes used basic geometry to measure the circumference of Earth. Eratosthenes knew that at noon on June 21, the sun cast no shadow in the Egyptian city of Syene (now Aswan). (See the diagram below.) At the same time, the sun cast a shadow of 7°12′ in Alexandria, about 500 miles from Syene.

The circumference of a circle is 360°; 7°12′ is about 2 percent, or 1/50, of 360°. Therefore, he concluded, 500 miles must be about 2 percent of the distance around Earth, which at the equator would be about 25,000 miles.

7°12′ = 1/50 circumference
N
Syene
512 miles
Sun
S

46 CHAPTER 2

3 Main Idea and Why It Matters Now

By answering the age-old question "Why does it matter now?" students begin to understand that events from around the world have an impact on their own lives. Every section of ***World Cultures and Geography*** clearly tells students what the main idea is and why it is important.

4 Dateline

The lively **Dateline** re-creates a moment in history through a news account or brief story. This feature acts as a warm-up to interest students in the section and promote their understanding of the relationship of the past, present, and future.

5 Connections Features

The **Connections** features link culture and geography to other disciplines, including math, history, science, language arts, and other subjects.

Skills

McDougal Littell's ***World Cultures and Geography*** places special emphasis on critical thinking, content-area reading, and map skills to provide continual social studies skill development.

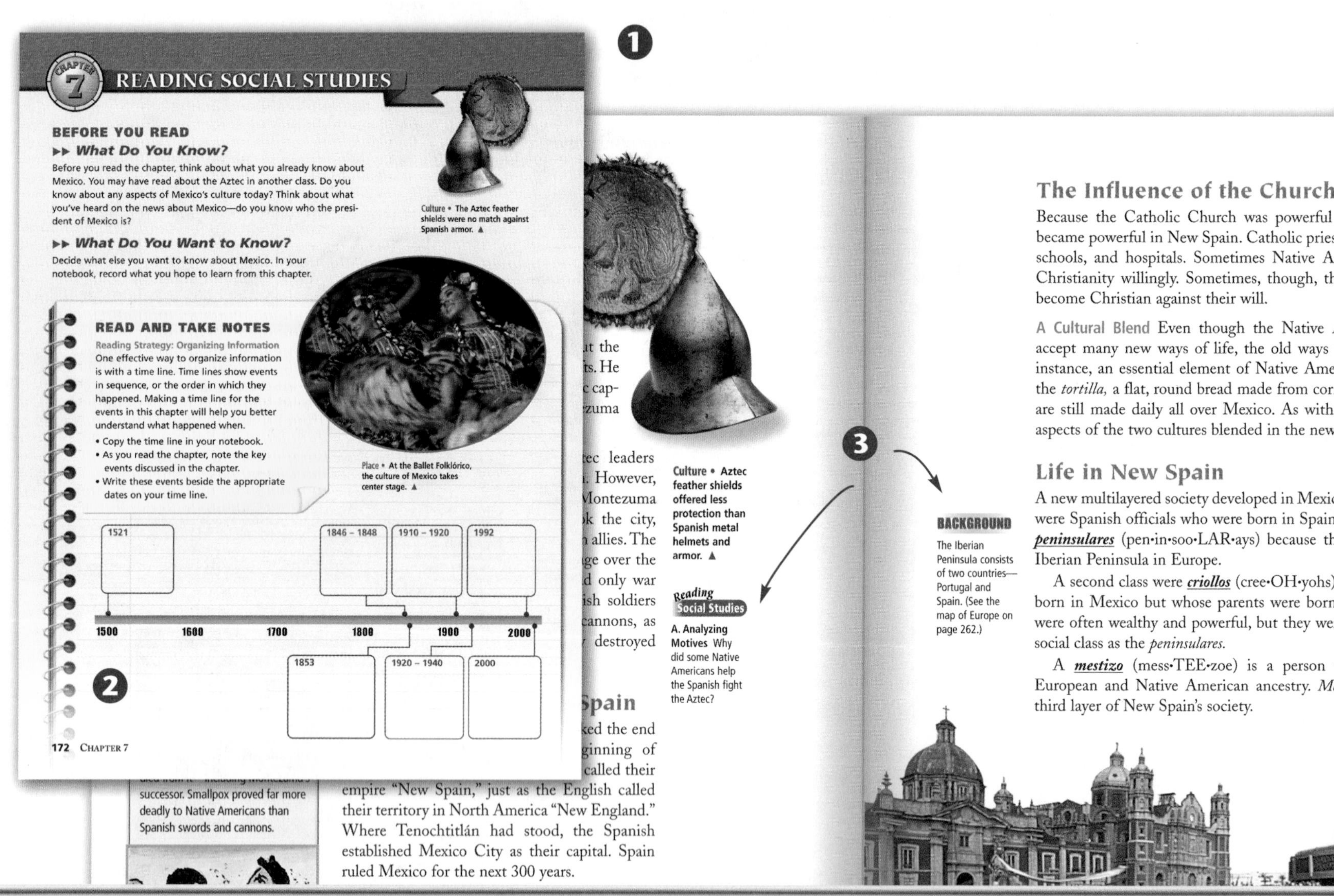

1 Reading Social Studies

Every chapter begins with **Reading Social Studies.** The page features the widely used **K-W-L** strategy, asking students what they **know,** and **what** they would like to know. The Chapter Assessment then asks students what they **learned.**

2 Graphic Organizers

The "Read and Take Notes" graphic organizer helps students determine the main ideas and other important information as they work their way through the text.

3 Reading Social Studies, Vocabulary, and Background Notes

World Cultures and Geography provides reading support for students as they read. Side-column notes on every page support student comprehension. **Reading Social Studies** notes ask student-comprehension and critical-thinking questions. **Vocabulary** notes define important words at point of use. **Background** notes provide vital historical information that helps to clarify events for students.

4

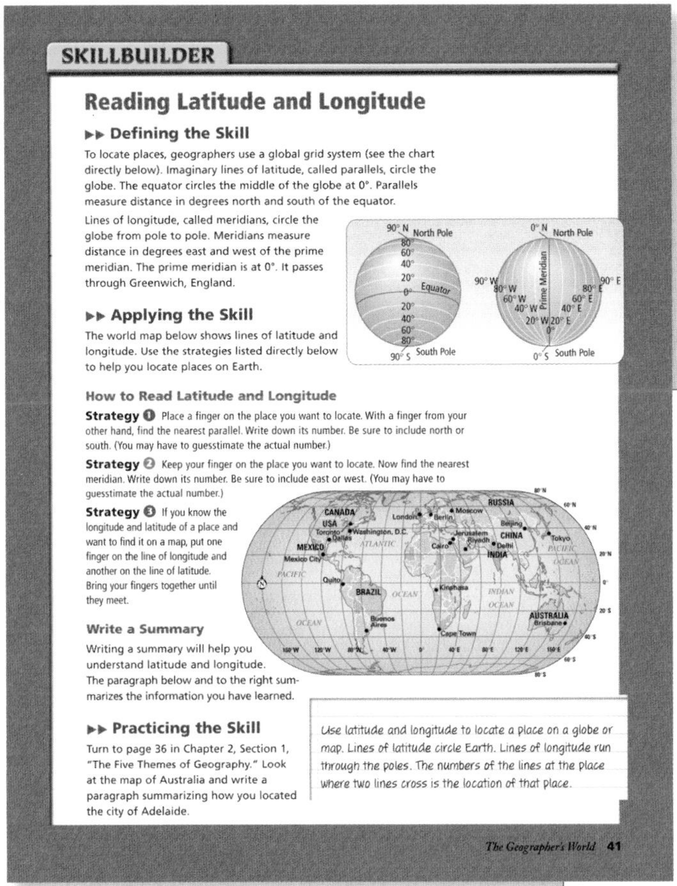

SKILLBUILDER

Reading Latitude and Longitude

▸▸ Defining the Skill

To locate places, geographers use a global grid system (see the chart directly below). Imaginary lines of latitude, called parallels, circle the globe. The equator circles the middle of the globe at 0°. Parallels measure distance in degrees north and south of the equator.

Lines of longitude, called meridians, circle the globe from pole to pole. Meridians measure distance in degrees east and west of the prime meridian. The prime meridian is at 0°. It passes through Greenwich, England.

▸▸ Applying the Skill

The world map below shows lines of latitude and longitude. Use the strategies listed directly below to help you locate places on Earth.

How to Read Latitude and Longitude

Strategy 1 Place a finger on the place you want to locate. With a finger from your other hand, find the nearest parallel. Write down its number. Be sure to include north or south. (You may have to guesstimate the actual number.)

Strategy 2 Keep your finger on the place you want to locate. Now find the nearest meridian. Write down its number. Be sure to include east or west. (You may have to guesstimate the actual number.)

Strategy 3 If you know the longitude and latitude of a place and want to find it on a map, put one finger on the line of longitude and another on the line of latitude. Bring your fingers together until they meet.

Write a Summary

Writing a summary will help you understand latitude and longitude. The paragraph below and to the right summarizes the information you have learned.

Use latitude and longitude to locate a place on a globe or map. Lines of latitude circle Earth. Lines of longitude run through the poles. The numbers of the lines at the place where two lines cross is the location of that place.

▸▸ Practicing the Skill

Turn to page 36 in Chapter 2, Section 1, "The Five Themes of Geography." Look at the map of Australia and write a paragraph summarizing how you located the city of Adelaide.

The Geographer's World 41

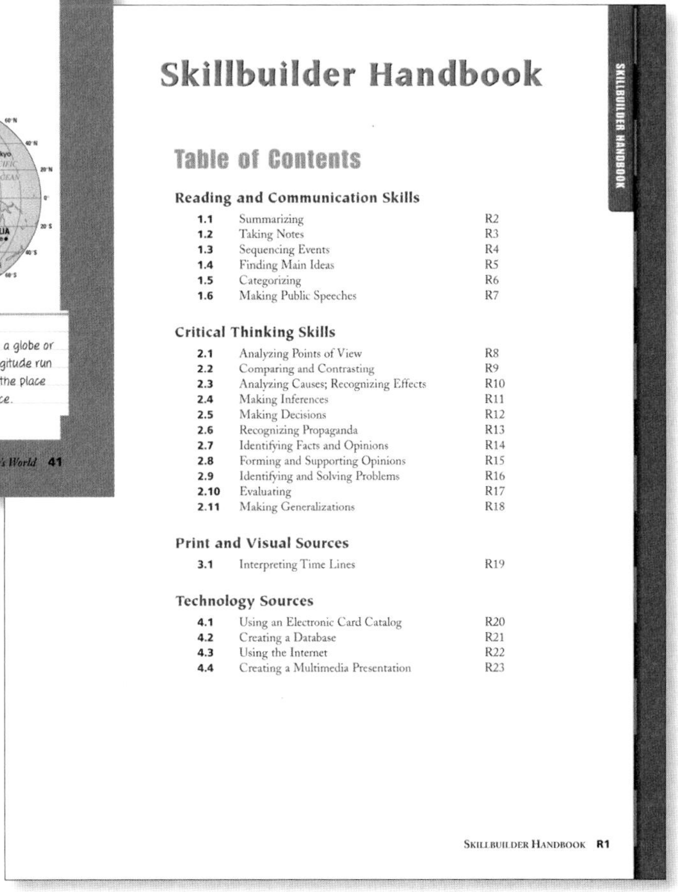

SKILLBUILDER HANDBOOK

Skillbuilder Handbook

Table of Contents

Reading and Communication Skills

Critical Thinking Skills

Print and Visual Sources

Technology Sources

Skillbuilder Handbook R1

5

4 Social Studies Skillbuilder

A one-page **Skillbuilder** feature in each chapter focuses on a specific skill related to social studies, such as sequencing events, reading a timeline, or identifying cause and effect. Students work through a set of strategies in order to master each skill. There is also one **Skillbuilder Mini-Lesson** in each chapter, which provides additional practice with the skill from the previous chapter.

5 Skillbuilder Handbook

The **Skillbuilder Handbook** provides access to skills support right in the textbook. This convenient reference section gives students direct and clear explanations of essential social study skills and helps them develop their abilities in such areas as critical thinking, researching, and map reading.

Teacher's Resources

World Cultures and Geography offers a wide variety of resources to help teachers manage their classroom and support students as they interact with the world.

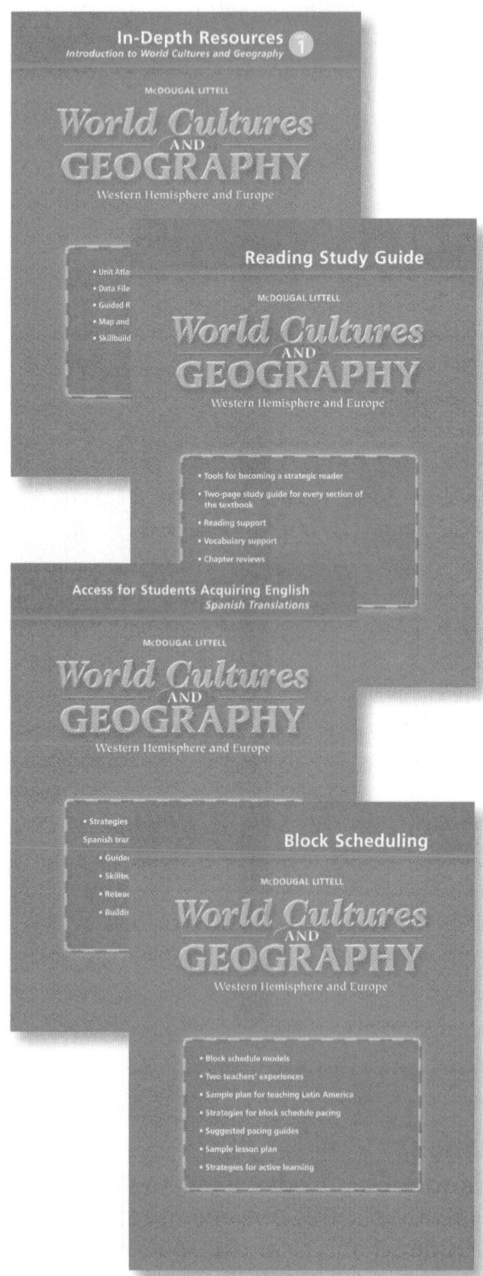

Teacher's Resource Package

In-Depth Resources
Resources organized by unit, chapter, and section include:

- Unit Atlas Activities
- Data File Activities
- Guided Reading Worksheets
- Map Skills
- Skillbuilder Practice
- Vocabulary Worksheets
- Geography Workshops
- Reteaching Activities

Outline Map Activities
Outline maps and blackline masters for geography skills development

Reading Study Guide Workbook
Section summaries written below level, main idea questions, and vocabulary activities to support less-proficient readers (English and Spanish)

Reading Study Guide Workbook Answer Key

Access for Students Acquiring English: Spanish Translations
Strategies for teaching ESL students and Spanish translations of Chapter Summaries and selected In-Depth Resources

Interdisciplinary Activities

Issues in World Geography: Central Asia

World Literature

Block Scheduling Strategies
A pacing guide, chapter teaching models, organization charts, and suggestions for addressing multiple learning styles

Formal Assessment
Three levels of tests for each chapter, section quizzes, and rubrics for assessing writing

Integrated Assessment
Provides explanations and forms for a variety of assessment options, including cooperative learning, group discussion, role-playing, oral presentations, peer assessment, self-assessment, and portfolio assessment

Strategies for Test Preparation
Strategies and exercises to help prepare students for a variety of test-taking experiences

Strategies for Test Preparation Teacher's Edition

McDougal Littell Student Atlas of the World
Over 100 pages of information and physical and political maps, divided by region and topic

Document-Based Questions Strategies and Practice
Provides a set of document-based, constructed-response, and multiple-choice questions for each chapter, allowing students to practice test-taking strategies

Modified Lesson Plans for English Learners
Includes tips for teachers on adapting lessons and activities for English learners

Multi-Language Glossary of Social Studies Terms
Includes key terms from the text and other commonly-used social studies terms defined in English with Spanish, Cambodian, Vietnamese, Hmong, Cantonese, Portuguese, Russian, Arabic, and Haitian Creole translations

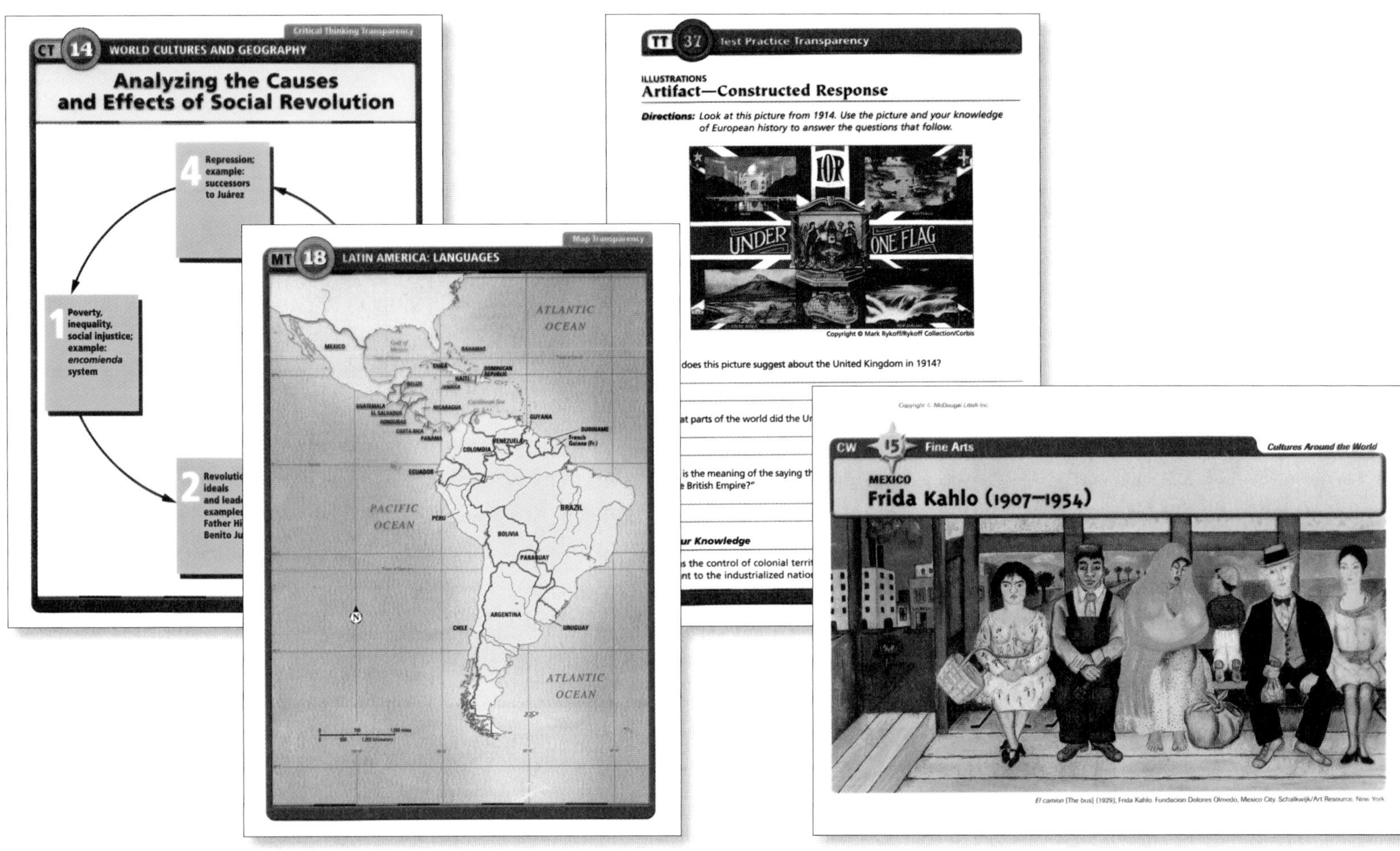

Additional Resources

World Cultures and Geography Workbook
Features notetaking strategies and graphic organizers for enhancing reading comprehension

Geography Posters
Two colorful posters for each unit

Writing for Social Studies
Provides extra content-area writing support for writing research projects, historical narratives, essays, interviews, oral histories, book reviews, and short reports

Reading Toolkit for Social Studies
Collection of reading support resources that includes graphic organizers, vocabulary practice and support, plus Skillbuilder transparencies

Transparencies

Map Transparencies
Overlays and transparencies provide more than 70 maps for in-depth coverage of regions and locations

Critical Thinking Transparencies
Transparencies provide graphic organizers, cause-and-effect charts, and visual summaries

Cultures Around the World Transparencies
60 articles and transparencies on a variety of cultural topics such as food, music, dance, art, literature, architecture, and cultural artifacts

Test Practice Transparencies
One transparency for each section of the textbook reviews content and familiarizes students with a variety of testing items

Comprehensive Assessment Support

Comprehensive assessment materials in print, transparency, CD-ROM, and Internet formats offer a variety of testing options for teachers and test practice opportunities for students.

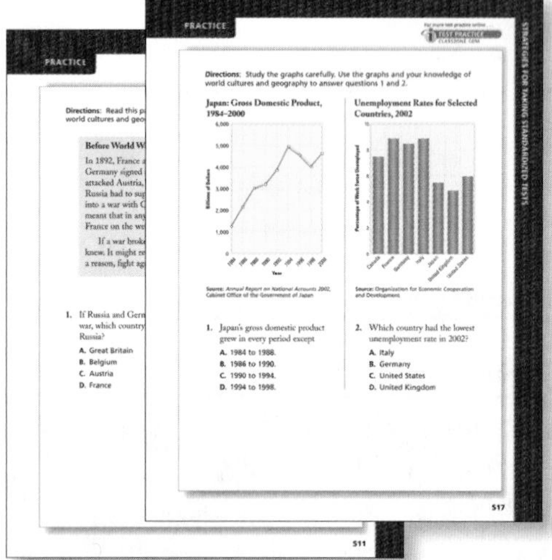

Strategies for Taking Standardized Tests

This innovative section of the *Pupil's Edition* includes strategies for answering multiple-choice, constructed-response, extended-response, and document-based questions, and guidelines for analyzing primary and secondary sources. Each strategy is followed by a set of practice items.

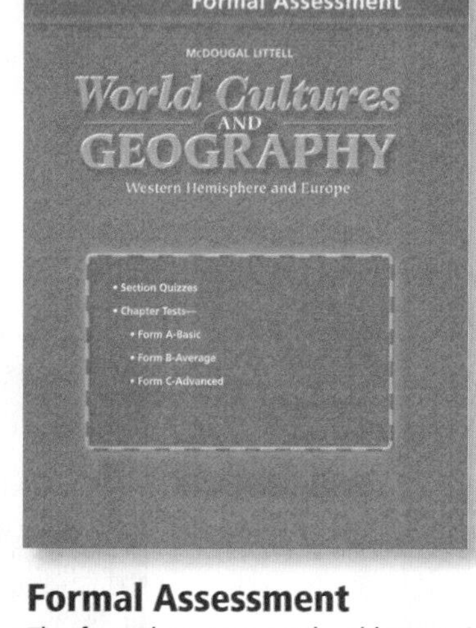

Formal Assessment

The formal assessment booklet provides section quizzes and three levels of tests for each chapter.

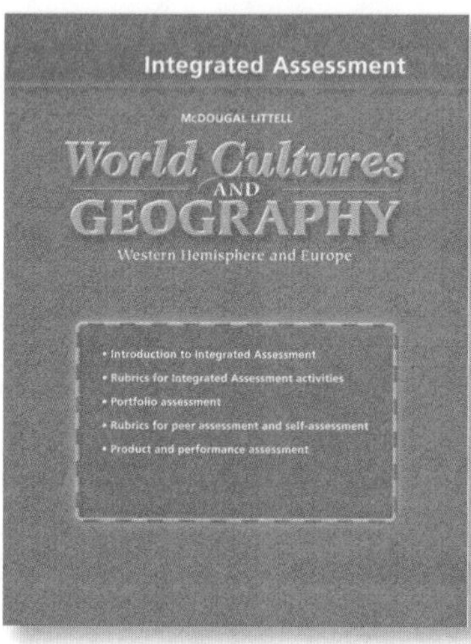

Integrated Assessment

This booklet includes rubrics for evaluating alternative assessments, including portfolios, cooperative learning activities, group discussions, and presentations.

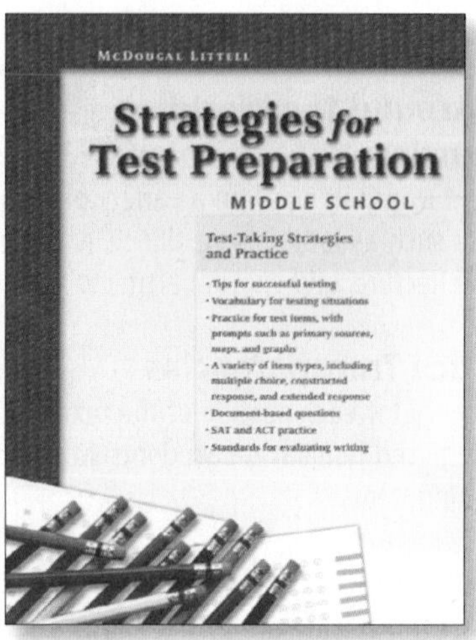

Strategies for Test Preparation

The strategies and exercises in this booklet help prepare students for a variety of test-taking experiences.

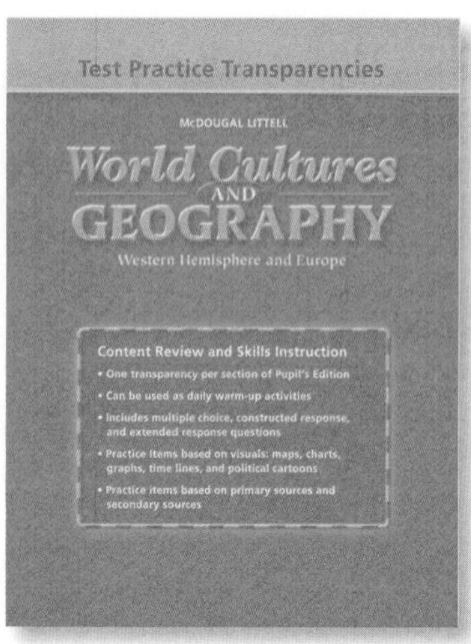

Test Practice Transparencies

This booklet provides one transparency for each section of the textbook that covers social studies content and familiarizes students with a variety of testing items.

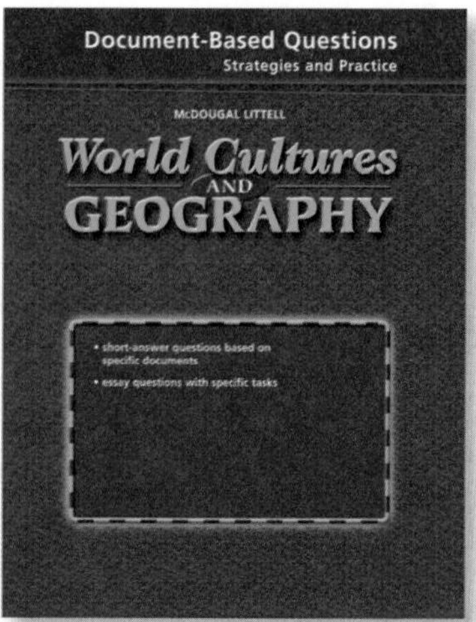

Document-Based Questions Strategies and Practice

This set of document-based, constructed-response, and multiple-choice questions provides students with practice and test-taking strategies for each chapter.

Test Generator CD-ROM

This CD-ROM contains a variety of pre-made tests and a test bank of items for creating customized tests. Questions are provided in three levels: basic, average, and advanced. Tools walk the user through the searching and editing steps and help teachers correlate tests to national and state standards. Test A is translated into Spanish.

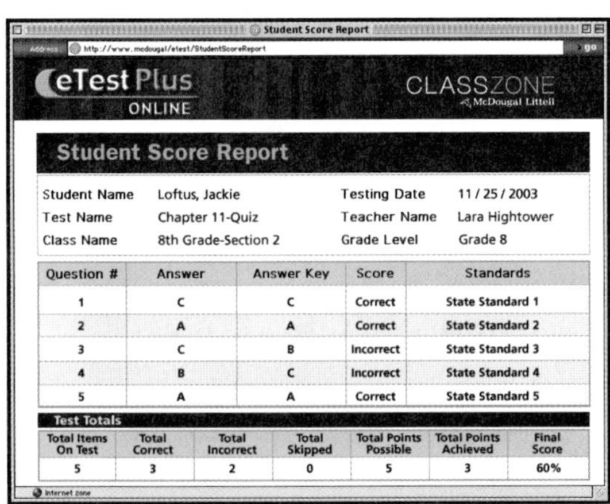
Student Score Report

eTest Plus ONLINE

CLASSZONE McDougal Littell

Student Score Report

Student Name	Loftus, Jackie	Testing Date	11 / 25 / 2003
Test Name	Chapter 11-Quiz	Teacher Name	Lara Hightower
Class Name	8th Grade-Section 2	Grade Level	Grade 8

Question #	Answer	Answer Key	Score	Standards
1	C	C	Correct	State Standard 1
2	A	A	Correct	State Standard 2
3	C	B	Incorrect	State Standard 3
4	B	C	Incorrect	State Standard 4
5	A	A	Correct	State Standard 5

Test Totals

Total Items On Test	Total Correct	Total Incorrect	Total Skipped	Total Points Possible	Total Points Achieved	Final Score
5	3	2	0	5	3	60%

eTest Plus Online

This customizable assessment tool allows teachers to publish tests from the *Test Generator CD-ROM*. Students take the tests online, with results graded automatically. Individual and group score reports can be viewed, printed, and correlated to national and state standards.

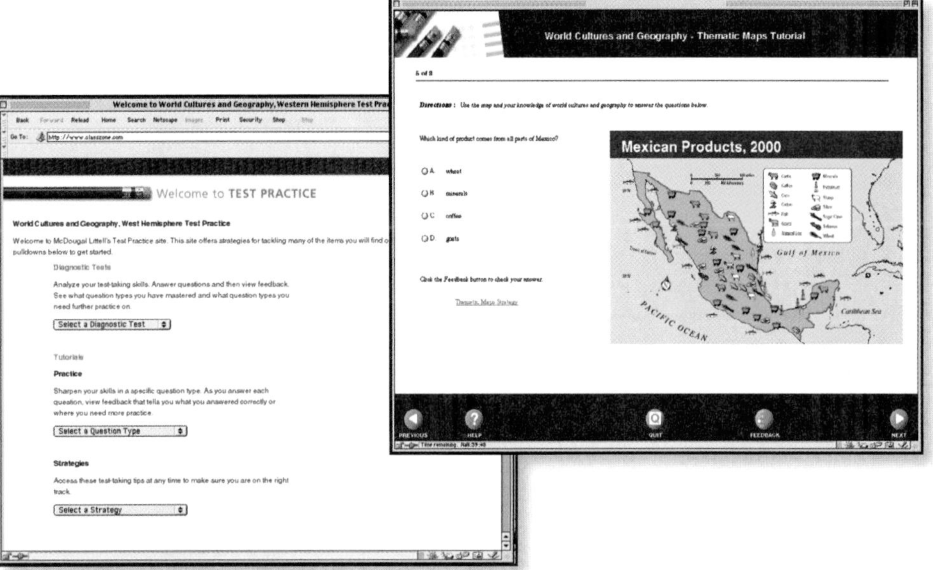

Online Test Preparation

This online, student test practice feature can be accessed through the *ClassZone* Web site.The test practice includes test-taking tips, diagnostic tests, skill-based tutorials, and skills and strategies help.

English Learner Support

These print and audio resources focus on reading comprehension and offer specific support to help English learners better understand the content as they study geography.

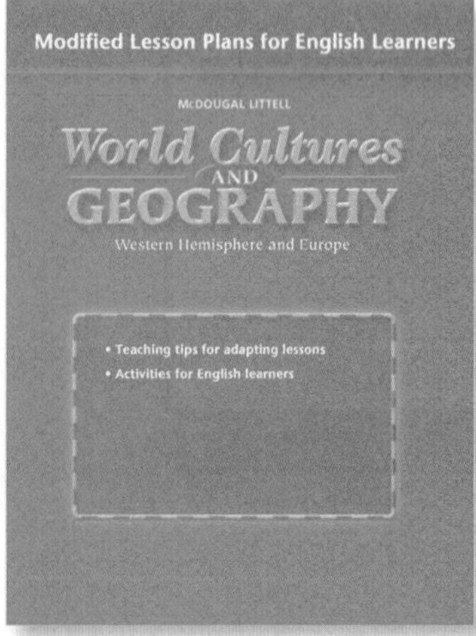

Modified Lesson Plans for English Learners

This lesson plan booklet includes tips for teachers on adapting lessons and activities for English learners.

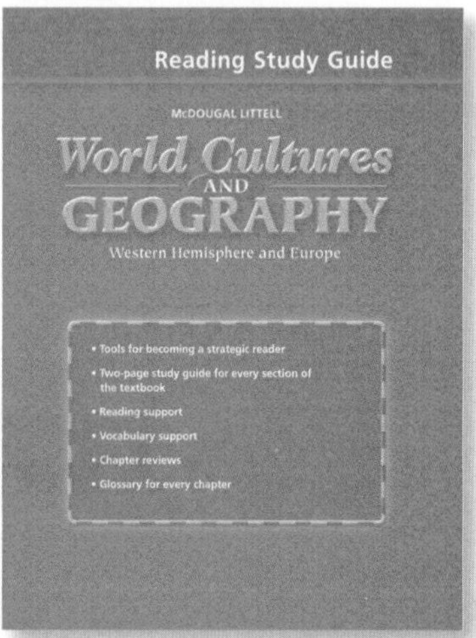

Reading Study Guide Workbook

This workbook contains chapter summaries and reading comprehension questions written at the 6–7 grade level. (English and Spanish)

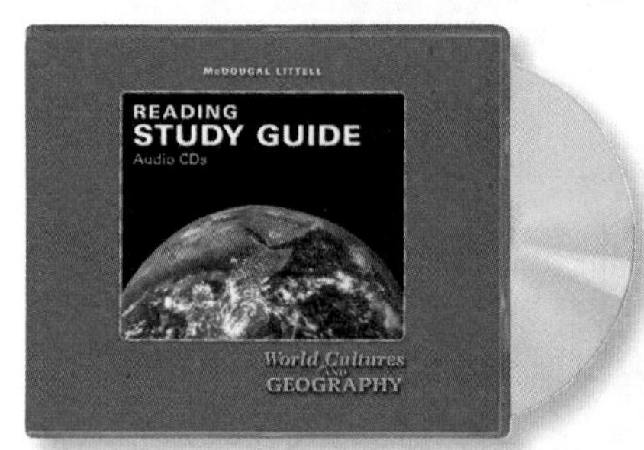

Reading Study Guide Audio CDs

These audio versions of below grade-level section summaries support struggling readers. (English and Spanish).

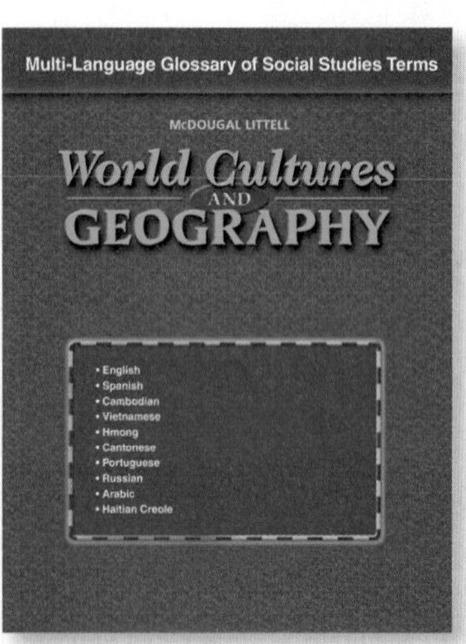

Multi-Language Glossary of Social Studies Terms

This glossary contains key terms from the text and other commonly-used social studies terms defined in English with Spanish, Cambodian, Vietnamese, Hmong, Cantonese, Portuguese, Russian, Arabic, and Haitian Creole translations.

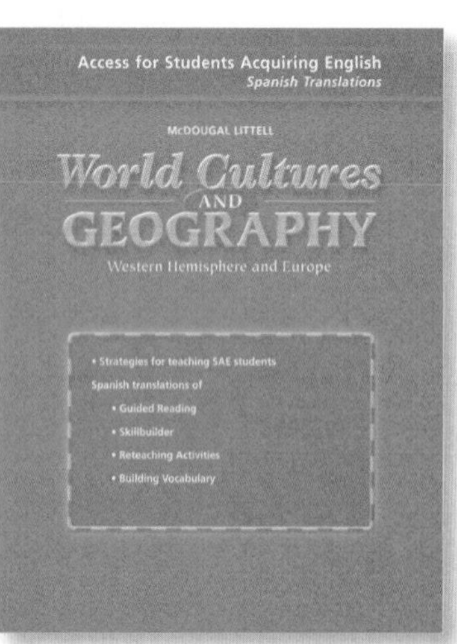

Access for Students Acquiring English: Spanish Translations

This English learner support book offers strategies for teaching ESL students and Spanish translations of selected ancillaries.

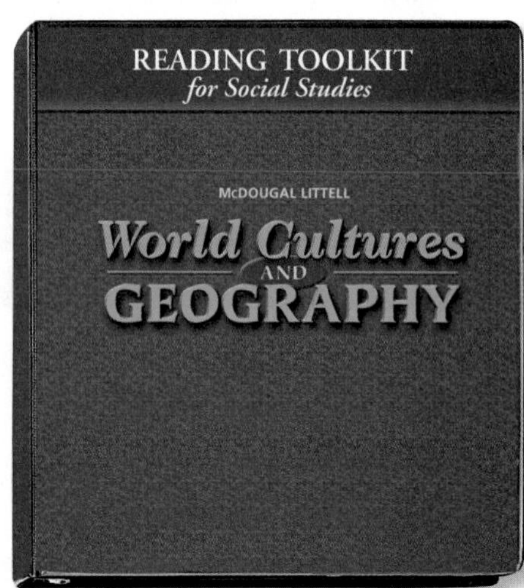

Reading Toolkit for Social Studies

This collection of resources provides reading support for the social studies classroom. Included are graphic organizers, vocabulary practice and support, plus Skillbuilder transparencies.

Distribution of Bilingual Speakers, 2001

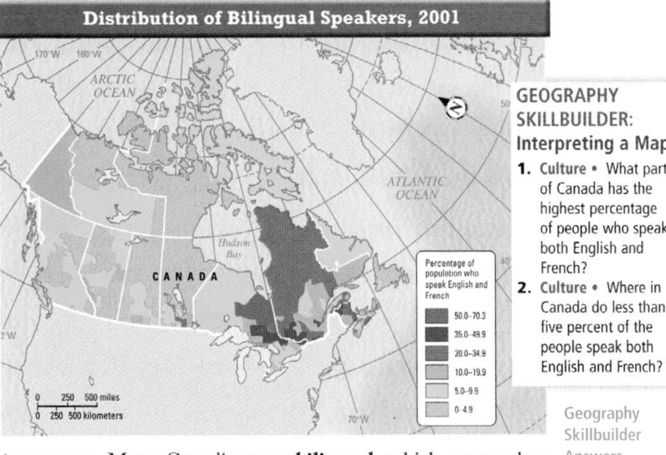

GEOGRAPHY SKILLBUILDER: Interpreting a Map

1. **Culture** • What part of Canada has the highest percentage of people who speak both English and French?
2. **Culture** • Where in Canada do less than five percent of the people speak both English and French?

Geography Skillbuilder Answers

1. eastern Canada, also north of Great Lakes
2. far northwest (Yukon)

Languages Many Canadians are **bilingual,** which means they speak two languages. Look at this map to see where bilingual Canadians live. Canada has two official languages, English and French. Literature, official documents, road signs, newspapers, and television broadcasts are in both languages. The two languages are not exactly like those spoken in England, the United States, and France. **Francophones** are French-speaking people. Canadian French, based on the French of the 1800s, is pronounced differently from the French spoken in modern France.

Culture • Business signs on a street in Quebec City reflect the strong influence of French culture. ▶

FOCUS ON VISUALS

Interpreting the Map Have students read the text in the paragraph "Languages" in connection with this map. Direct them in using the map key to find the major areas of bilingualism in eastern Canada and the smaller areas elsewhere. Ask them to speculate about the small areas of bilingualism in the prairie provinces. Why do they exist?

Possible Responses They are near large cities.

Extension Ask students to take a survey of second languages spoken by class members and make a bar graph to record the number of speakers of each.

CRITICAL THINKING ACTIVITY

Forming and Supporting Opinions Ask students to infer what language the United States might adopt as its second official language if it was to follow Canada's example. Discuss students' suggestions and their arguments in support of them.

Class Time 10 minutes

Activity Options

Differentiating Instruction: Students Acquiring English/ESL

Using Prefixes Have students read "Languages" as a group. Point out the word *bilingual,* and ask a volunteer to identify context clues that explain the word's meaning ("speaking two languages"). Tell students that the prefix *bi-* means "two." Then ask students to use the word *bilingual* in a sentence describing themselves.

Create a chart such as the one shown, and list other words containing the prefix *bi-*. Have students use a dictionary to identify the meaning of each word.

bicycle	
binoculars	
bipartisan	
biped	
bivalve	

Differentiated Instruction

Every chapter provides teaching suggestions, practice, and activities that specifically address the needs of English learners.

Nancy Siddens
Consulting Editor
English for Speakers of Other Languages
McDougal Littell

Teaching English Learners

McDougal Littell recognizes the challenges that English learners and their teachers face in the mainstream classroom. The trend to use English instruction in content classes (with less support in students' primary languages) and the federal legislation (No Child Left Behind) requiring instruction in content simultaneously with instruction in English, have driven the need for comprehensive, flexible support materials for teaching in English to English learners.

McDougal Littell ***World Cultures and Geography*** provides a wide range of support for these students and their teachers. This support gives teachers the tools needed to employ research-based principles for teaching standards-based curriculum to English learners while they learn the English language in meaningful historical contexts.

Research on facilitating the learning of content by students acquiring English has determined that effective instruction is built on these three principles:

- **Increase comprehension**
- **Enhance student interaction**
- **Improve thinking and study skills**

These principles, identified as effective through research by the Center for Applied Linguistics (Jameson), have guided McDougal Littell's approach to developing materials for teaching social studies and language to English learners.

Principle 1: Increase comprehension
Providing understandable information to the English learner is crucial for accessing both language and content. ***World Cultures and Geography*** provides materials that enable teachers to employ practical strategies for increasing student comprehension. Verbal as well as nonverbal support is provided to make history more accessible to the English learner. The role of visual and experiential clues is especially important to beginning and intermediate language learners. Examples of nonverbal instruction include:

- A strong visual component in the Pupil's Edition, including vivid images, large maps, a variety of graphic organizers, charts, graphs, and timelines.
- Complete audiovisual support through audio chapter summaries, extensive online services, and the video series *There is No Food Like My Food.*
- A transparency program that includes *Map Transparencies, Critical Thinking Transparencies*, and *Cultures Around the World Transparencies*.

Examples of verbal and print support include:

- The Reading Study Guide in both English and Spanish, chapter summaries written below grade level, accompanied by comprehension questions.
- The Multi-Language Glossary of Social Studies Terms, a compilation of translations of the key terms from the Pupil's Edition in nine languages.
- Access for Students Acquiring English, Spanish translations of selected ancillary support pages.

Principle 2: Enhance student interaction
Research shows that language is learned through communication with others—negotiating meaning to accomplish real purposes (Long and Porter). Participants in a discussion restate, question, explain, and clarify in order to come to a common understanding. This process helps students learn social studies as well as language. ***World Cultures and Geography***—and *Modified Lesson Plans for*

English Learners, which accompanies it—provide opportunities to increase interaction in the classroom in a variety of ways, including:

- Pair activities, in which students work in pairs to compare their own experiences relating to a particular lesson or topic
- Group work such as "jigsaw" reading, in which each student takes on responsibility for presenting one part of a lesson to a small group
- Class activities such as debates, unit projects, or games
- Opportunities for family involvement

Principle 3: Improve thinking and study skills

Explicit teaching of academic skills helps develop thinking skills and "thinking language" for English learners (Chamot and O'Malley). ***World Cultures and Geography*** provides materials that focus instruction on performing higher-order thinking tasks, asking critical thinking questions, assessing learning in a manner and language consistent with instruction, and reinforcing study and test-taking skills.

In the Pupil's Edition of ***World Cultures and Geography***, support for developing thinking and study skills includes:

- An emphasis on the main idea of each section before reading begins
- Graphic organizers for notetaking practice in each section
- Highlighted vocabulary terms
- Main Idea comprehension questions that reinforce important content
- Skillbuilder questions accompanying large visuals such as maps and charts
- Section and chapter assessments that provide leveled questions from Main Idea to Critical Thinking
- A skills strand woven throughout the section and chapter assessments and reinforced in the Skillbuilder Handbook
- Test practice opportunities such as the 32-page "Strategies for Taking Standardized Tests" in the front of the book and Standards-Based Assessment practice in the chapter assessments

The Teacher's Edition and ancillaries for ***World Cultures and Geography*** provide additional support for developing students' skills through activities for differentiated instruction and Guided Reading and Skillbuilder Practice pages.

World Cultures and Geography provides teachers with the comprehensive support they need for teaching English learners. The program provides verbal and nonverbal support to increase understanding, promotes increased interaction and opportunities for communication, and encourages the development of academic thinking and study skills. Each of these areas has been identified, through research, as crucial for promoting the success of the English learner in acquiring content knowledge as well as language.

For more information:

Chamot, A.U. and O'Malley, J.M. (1994). *The CALLA Handbook: Implementing the Cognitive Academic Language Learning Approach*. Addison-Wesley: Reading, MA.

Jameson, J. (1998). *Three Principles for Success: English Language Learners in Mainstream Content Classes, From Theory to Practice 6*, Center for Applied Linguistics: Region XIV Comprehensive Center.

Long, M. and Porter, P.A. (1985). *Group Work, Interlanguage Talk, and Second Language Acquisition*, in TESOL Quarterly 19: 207–227.

Integrated Technology

Audio, video, online, and computer resources provide students with additional opportunities to explore graphic information.

eEdition CD-ROM
The electronic version of the textbook allows students and teachers to access all the same features of the printed text plus interactive maps, infographics, and links to *ClassZone.*

EasyPlanner CD-ROM
This convenient CD-ROM contains all of the teacher resources in the print ancillaries. Teachers can view, search, and print the ancillaries, which are organized by resource and chapter.

Power Presentations CD-ROM
This multimedia presentation tool was designed to augment teachers' lectures and provide resources for students to create their own PowerPoint® presentations. Detailed, multimedia lecture notes are enhanced by an image gallery of photographs, maps, art, charts, and graphs. The CD-ROM also includes an interactive review game for every chapter.

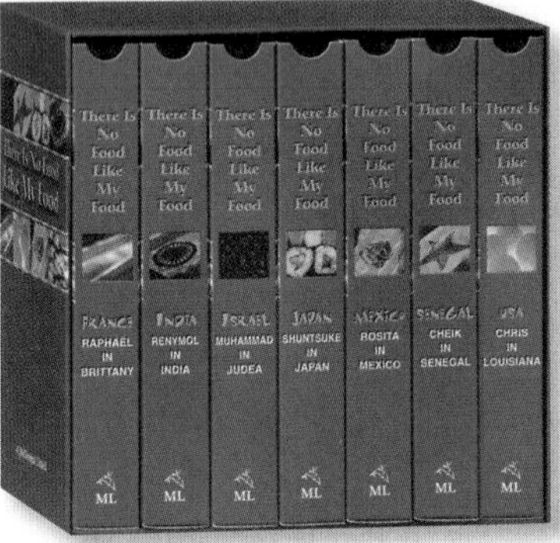

World Cultures and Geography Video Series: There Is No Food Like My Food
This series of seven videos introduces foods enjoyed in a variety of cultural settings by pre-teens and their families.
- USA: Chris in Louisiana
- Mexico: Rosita in Mexico
- France: Raphaël in Brittany
- Israel: Muhammad in Judea
- Senegal: Cheik in Senegal
- India: Renymol in India
- Japan: Shuntsuke in Japan

Test Generator CD-ROM
Pre-made, customizable tests with three levels of questions, including document-based questions, are correlated to national and state standards. Form A is available in Spanish. The *Test Generator* interfaces with *eTest Plus Online.*

Reading Study Guide Audio CDs
Below grade-level section summaries on audio CD support struggling readers. (English and Spanish)

The World's Music Audio CD
A variety of recordings gives students the opportunity to hear music from different cultures.

The online guide to *World Cultures and Geography*

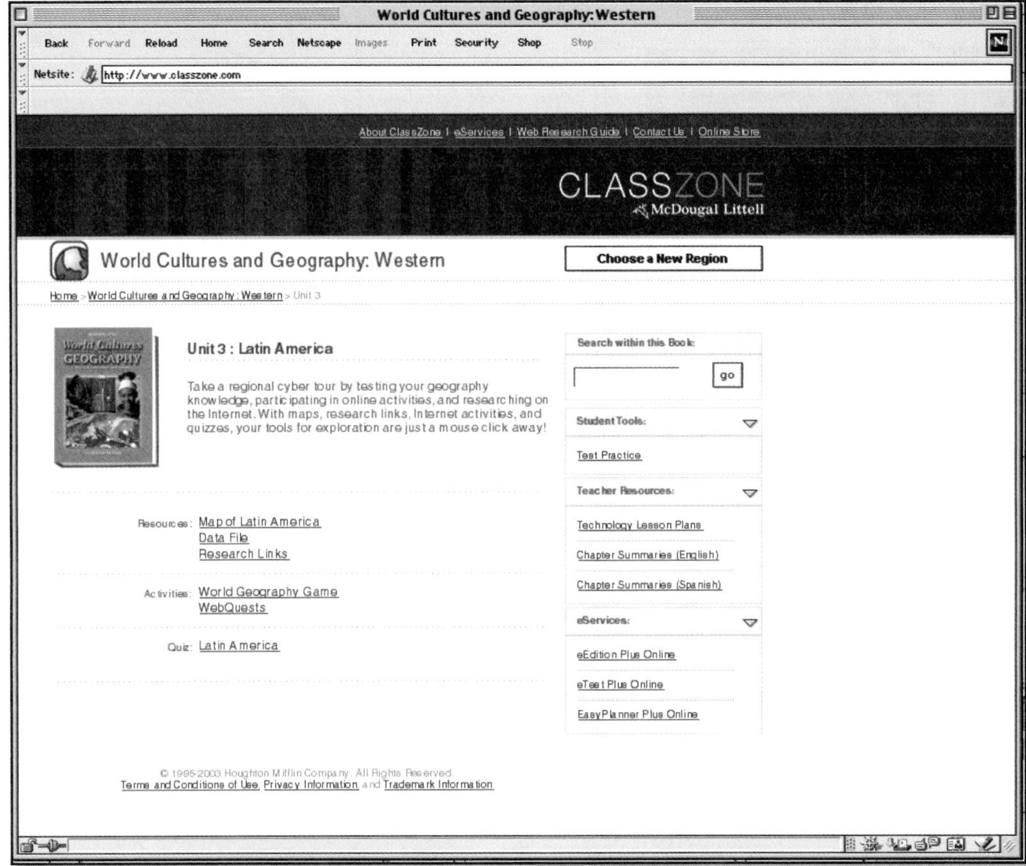

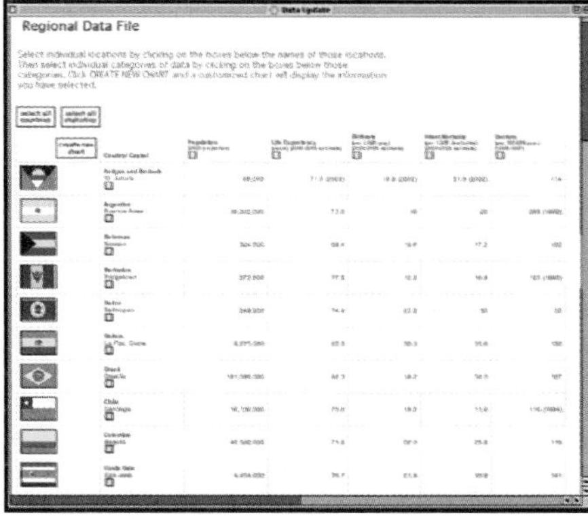

Data Update

Compare countries around the globe with regularly updated demographic data.

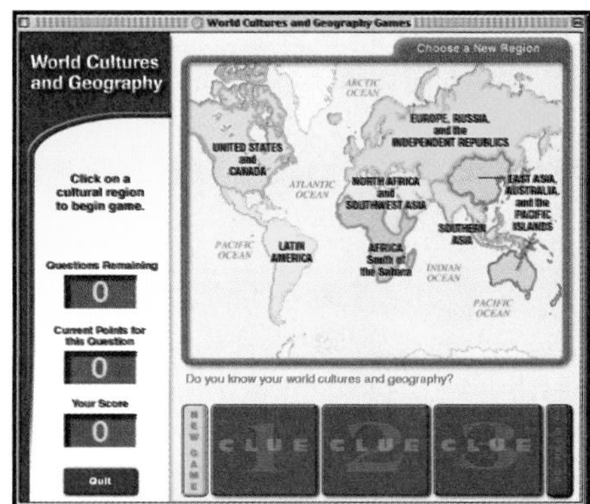

Activities

Encourage student interactivity through online games and activities.

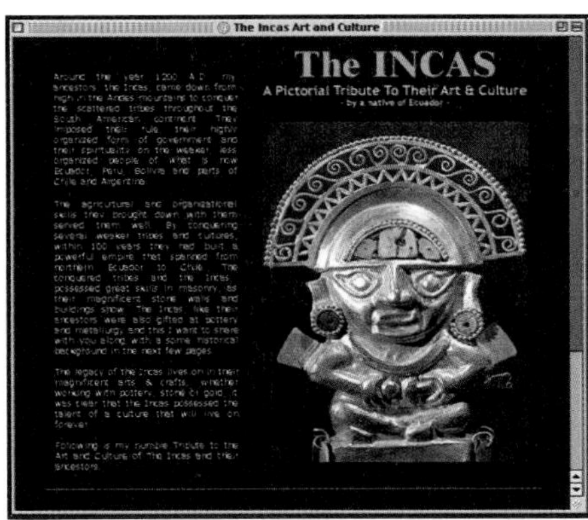

Research Links

Explore the topics of each chapter with links correlated to the textbook.

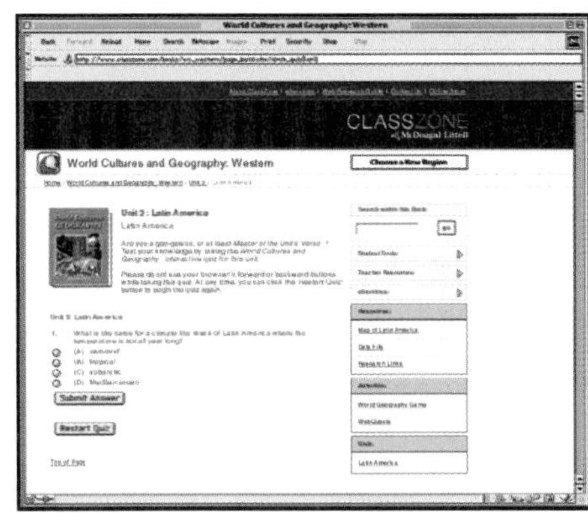

Quizzes

Check students' history knowledge with interactive quizzes.

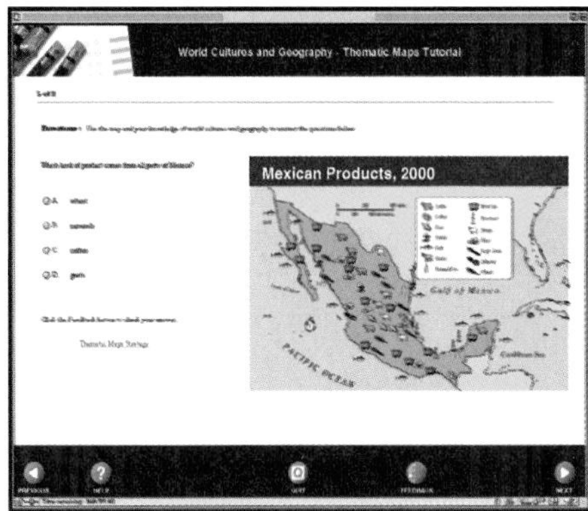

Test Practice

Support students with test-taking tips and skill-based tutorials.

McDougal Littell eServices

Plan it

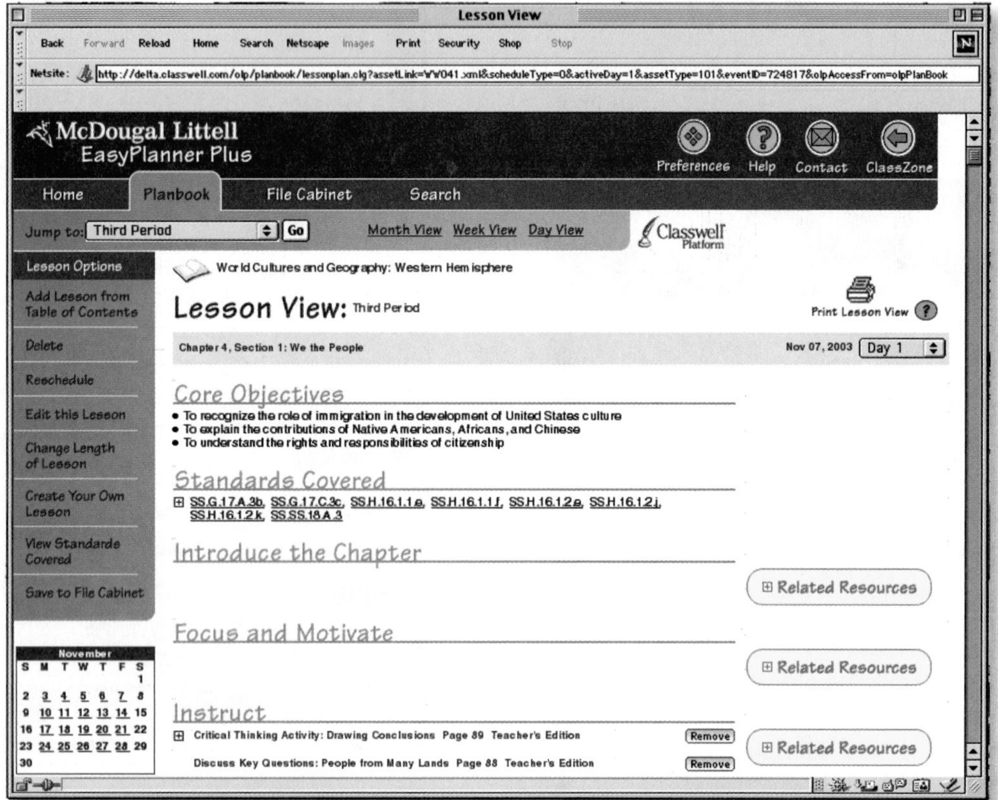

easyPlanner Plus ONLINE

Teach it

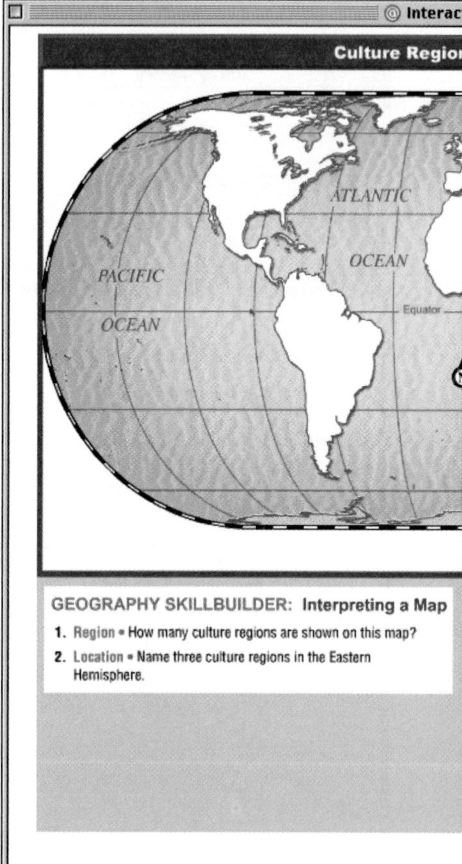

EasyPlanner Plus Online

Lesson Planning When and Where You Want It

- Customize lesson plans using all teacher resources.
- Plan a lesson, or plan the whole year.
- Access, view, and print all teacher resources.
- Print plans in daily, weekly, or monthly views.
- Track state standards correlated to each lesson.

eEdition Plus Online

A Textbook for the Internet Generation

- Access interactive textbook online.
- View animated maps and infographics.
- Take notes onscreen.
- Send and receive e-mail to build student-teacher communication.
- Post worksheets and assignments right to student book.

eEdition Plus ONLINE

Assess it

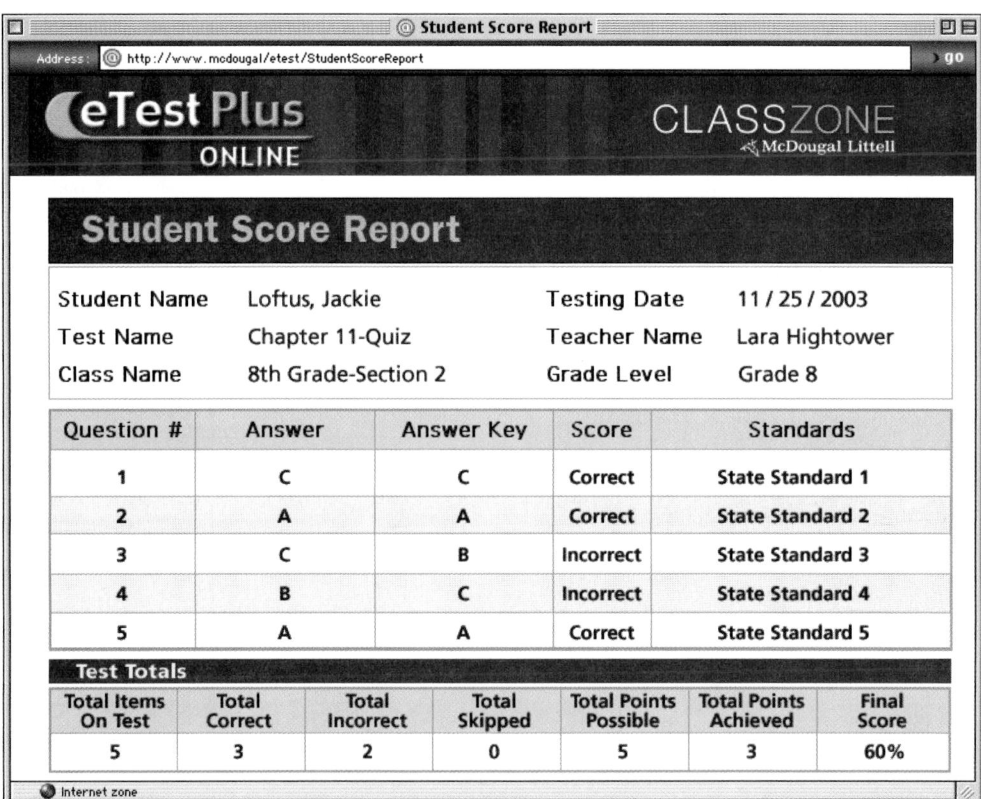

Student Score Report

Student Name	Loftus, Jackie	Testing Date	11 / 25 / 2003
Test Name	Chapter 11-Quiz	Teacher Name	Lara Hightower
Class Name	8th Grade-Section 2	Grade Level	Grade 8

Question #	Answer	Answer Key	Score	Standards
1	C	C	Correct	State Standard 1
2	A	A	Correct	State Standard 2
3	C	B	Incorrect	State Standard 3
4	B	C	Incorrect	State Standard 4
5	A	A	Correct	State Standard 5

Test Totals

Total Items On Test	Total Correct	Total Incorrect	Total Skipped	Total Points Possible	Total Points Achieved	Final Score
5	3	2	0	5	3	60%

eTest Plus Online

Customized Assessment Online

- Create tests with the *Test Generator* and publish them to the *eTest Plus Online* service.
- Have students take tests online, with results graded automatically.
- View and print individual and class score reports.
- Generate reports correlated to national and state standards.

Focus on the World

A two-page **Planning Guide** at the beginning of every chapter of the ***World Cultures and Geography*** Teacher's Edition provides a chapter overview at a glance. It also outlines program resources section by section.

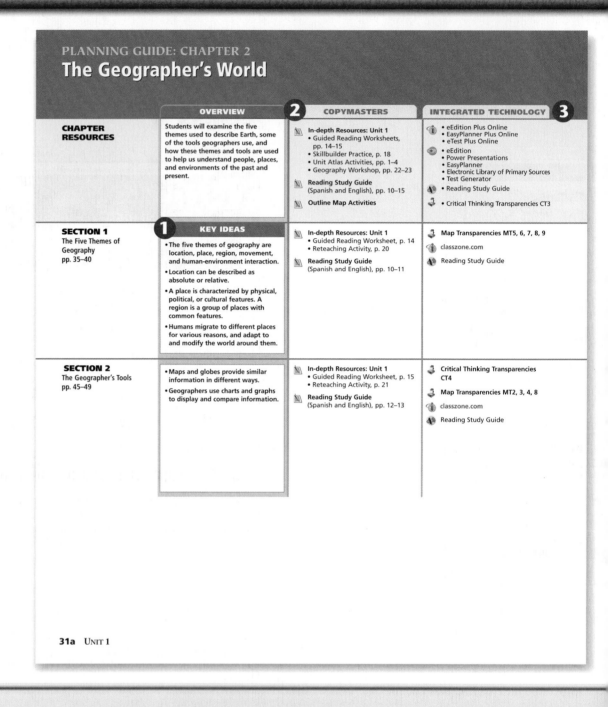

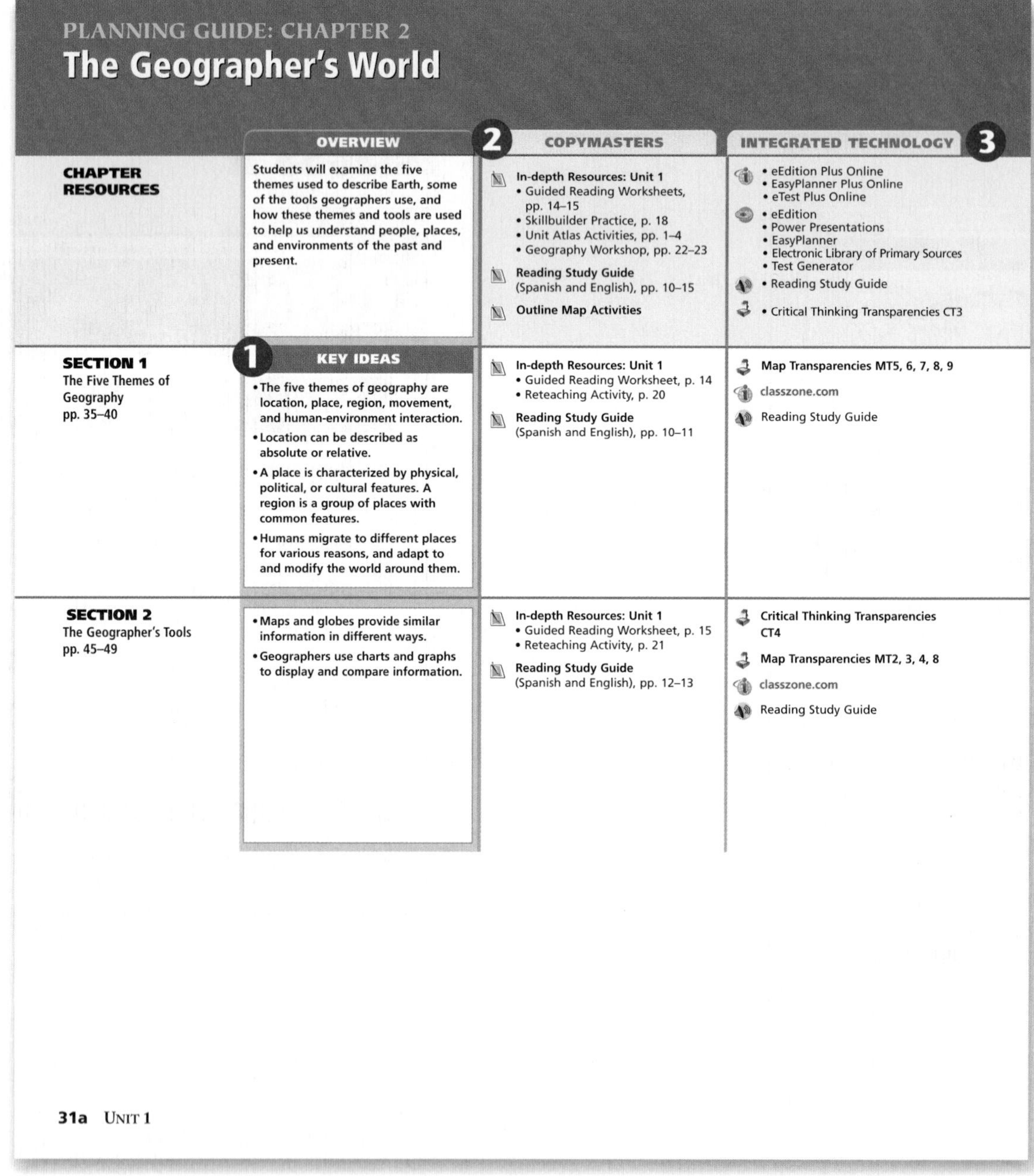

PLANNING GUIDE: CHAPTER 2

The Geographer's World

	OVERVIEW	2 COPYMASTERS	INTEGRATED TECHNOLOGY 3
CHAPTER RESOURCES	Students will examine the five themes used to describe Earth, some of the tools geographers use, and how these themes and tools are used to help us understand people, places, and environments of the past and present.	**In-depth Resources: Unit 1** • Guided Reading Worksheets, pp. 14–15 • Skillbuilder Practice, p. 18 • Unit Atlas Activities, pp. 1–4 • Geography Workshop, pp. 22–23 **Reading Study Guide** (Spanish and English), pp. 10–15 **Outline Map Activities**	• eEdition Plus Online • EasyPlanner Plus Online • eTest Plus Online • eEdition • Power Presentations • EasyPlanner • Electronic Library of Primary Sources • Test Generator • Reading Study Guide • Critical Thinking Transparencies CT3
	1 KEY IDEAS		
SECTION 1 The Five Themes of Geography pp. 35–40	• The five themes of geography are location, place, region, movement, and human-environment interaction. • Location can be described as absolute or relative. • A place is characterized by physical, political, or cultural features. A region is a group of places with common features. • Humans migrate to different places for various reasons, and adapt to and modify the world around them.	**In-depth Resources: Unit 1** • Guided Reading Worksheet, p. 14 • Reteaching Activity, p. 20 **Reading Study Guide** (Spanish and English), pp. 10–11	**Map Transparencies MT5, 6, 7, 8, 9** classzone.com Reading Study Guide
SECTION 2 The Geographer's Tools pp. 45–49	• Maps and globes provide similar information in different ways. • Geographers use charts and graphs to display and compare information.	**In-depth Resources: Unit 1** • Guided Reading Worksheet, p. 15 • Reteaching Activity, p. 21 **Reading Study Guide** (Spanish and English), pp. 12–13	**Critical Thinking Transparencies CT4** **Map Transparencies MT2, 3, 4, 8** classzone.com Reading Study Guide

31a UNIT 1

1 Key Ideas and Chapter Overview

A brief summary of each section and a **Chapter Overview** show how the content is structured to present a unifying chapter theme.

2 CopyMasters

A complete listing of reproducible materials for each section reveals the depth of resource material that is available.

3 Integrated Technology

Technology is listed for each section and includes resources for audio and visual learners, electronic and online teacher tools, and Internet resources. Selected chapters also include the video resource **There Is No Food Like My Food,** which features foods from different cultures.

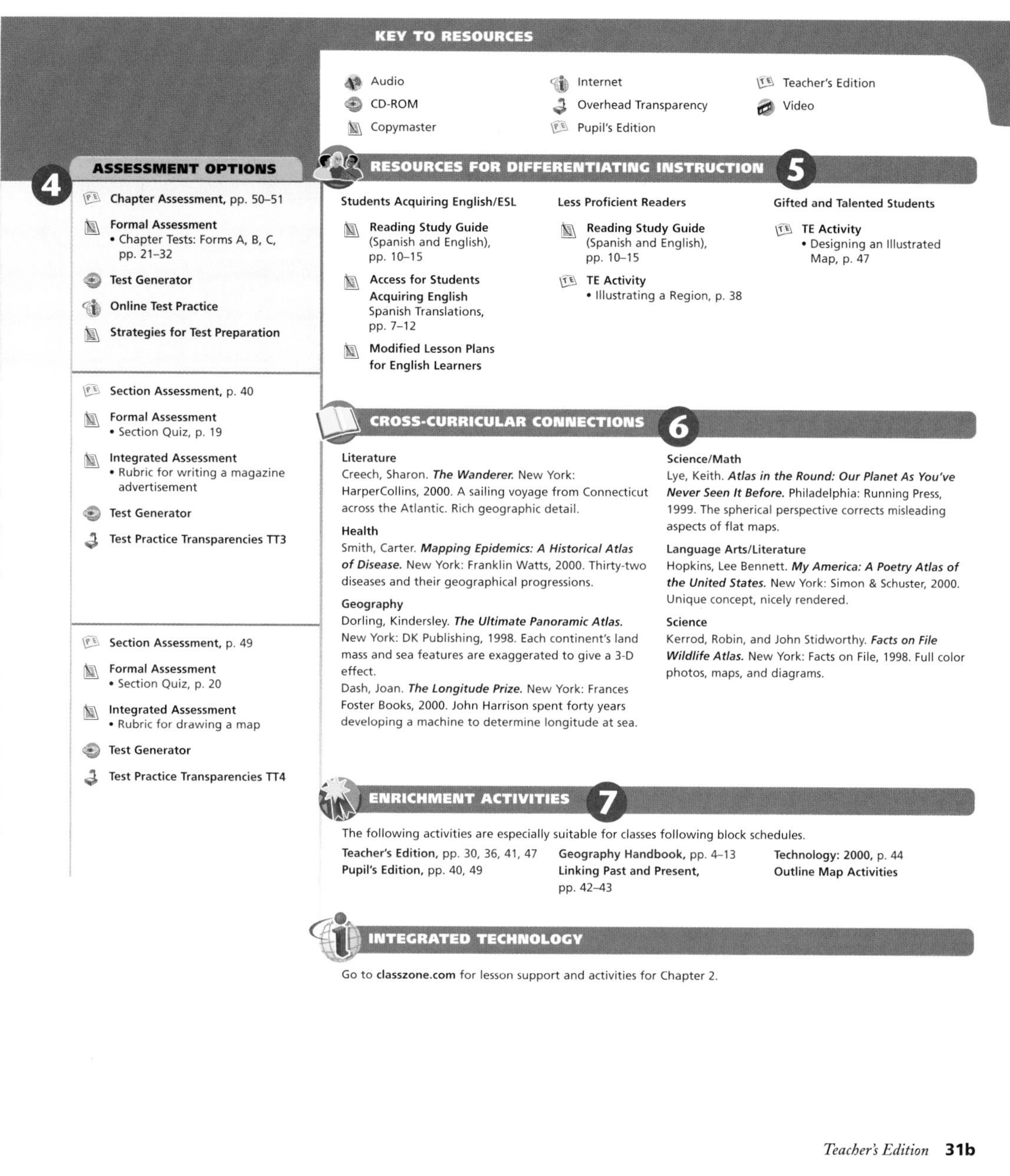

KEY TO RESOURCES

Audio
CD-ROM
Copymaster
Internet
Overhead Transparency
Pupil's Edition
Teacher's Edition
Video

ASSESSMENT OPTIONS 4

Chapter Assessment, pp. 50–51

Formal Assessment
- Chapter Tests: Forms A, B, C, pp. 21–32

Test Generator

Online Test Practice

Strategies for Test Preparation

Section Assessment, p. 40

Formal Assessment
- Section Quiz, p. 19

Integrated Assessment
- Rubric for writing a magazine advertisement

Test Generator

Test Practice Transparencies TT3

Section Assessment, p. 49

Formal Assessment
- Section Quiz, p. 20

Integrated Assessment
- Rubric for drawing a map

Test Generator

Test Practice Transparencies TT4

RESOURCES FOR DIFFERENTIATING INSTRUCTION 5

Students Acquiring English/ESL

Reading Study Guide (Spanish and English), pp. 10–15

Access for Students Acquiring English Spanish Translations, pp. 7–12

Modified Lesson Plans for English Learners

Less Proficient Readers

Reading Study Guide (Spanish and English), pp. 10–15

TE Activity
- Illustrating a Region, p. 38

Gifted and Talented Students

TE Activity
- Designing an Illustrated Map, p. 47

CROSS-CURRICULAR CONNECTIONS 6

Literature
Creech, Sharon. ***The Wanderer.*** New York: HarperCollins, 2000. A sailing voyage from Connecticut across the Atlantic. Rich geographic detail.

Health
Smith, Carter. ***Mapping Epidemics: A Historical Atlas of Disease.*** New York: Franklin Watts, 2000. Thirty-two diseases and their geographical progressions.

Geography
Dorling, Kindersley. ***The Ultimate Panoramic Atlas.*** New York: DK Publishing, 1998. Each continent's land mass and sea features are exaggerated to give a 3-D effect.
Dash, Joan. ***The Longitude Prize.*** New York: Frances Foster Books, 2000. John Harrison spent forty years developing a machine to determine longitude at sea.

Science/Math
Lye, Keith. ***Atlas in the Round: Our Planet As You've Never Seen It Before.*** Philadelphia: Running Press, 1999. The spherical perspective corrects misleading aspects of flat maps.

Language Arts/Literature
Hopkins, Lee Bennett. ***My America: A Poetry Atlas of the United States.*** New York: Simon & Schuster, 2000. Unique concept, nicely rendered.

Science
Kerrod, Robin, and John Stidworthy. ***Facts on File Wildlife Atlas.*** New York: Facts on File, 1998. Full color photos, maps, and diagrams.

ENRICHMENT ACTIVITIES 7

The following activities are especially suitable for classes following block schedules.

Teacher's Edition, pp. 30, 36, 41, 47
Pupil's Edition, pp. 40, 49
Geography Handbook, pp. 4–13
Linking Past and Present, pp. 42–43
Technology: 2000, p. 44
Outline Map Activities

INTEGRATED TECHNOLOGY

Go to **classzone.com** for lesson support and activities for Chapter 2.

Teacher's Edition **31b**

4 Assessment Options

This column lists the variety of **Assessment Options;** page numbers make chapter and section assessments quick to find.

5 Resources for Differentiating Instruction

Page references for the Teacher's Edition and ancillary activities for teaching students acquiring English, less proficient readers, and gifted and talented students are provided, as are Spanish language resources.

6 Cross-Curricular Connections

Resources for interdisciplinary teaching are listed. Books are categorized for literature, popular culture, history, science, and other disciplines.

7 Enrichment Activities

This listing highlights appropriate activities for block scheduling.

Chapter Pacing for Block Schedules

Teachers who teach in blocks will appreciate a special **Pacing Guide.** This easy-to-use guide is a valuable aid in planning lessons.

PACING GUIDE: CHAPTER 2

1

BLOCK SCHEDULE LESSON PLAN OPTIONS: 90-MINUTE PERIOD

DAY 1

CHAPTER PREVIEW, pp. 32–33
Class Time 20 minutes

- **Hypothesize** Use "What do you think?" questions in Focus on Geography on PE p. 33 to help students hypothesize about the uses of geography in exploring and understanding the world.

SECTION 1, pp. 35–40
Class Time 70 minutes

- **Summarizing** Have students write a brief description of a place they know well. Ask them to imagine they are describing it to someone who has never seen it. Write the five geography themes (location, place, region, movement, human-environment interaction) on the chalkboard. Have students share their descriptions, and ask them to decide into which geographical theme their description best fits. Review the themes as you discuss. **Class Time** 30 minutes
- **Comparing** To compare the ideas of absolute location and relative location, have a student stand somewhere in the room. Ask students to describe his/her location first in an absolute way, then in a relative way. Have students explain the differences. **Class Time** 10 minutes
- **Internet** Divide students into groups and have them visit **classzone.com** to learn more about Pangaea and changes in the locations of Earth's continents. **Class Time** 30 minutes

DAY 2

SECTION 1, continued
Class Time 60 minutes

- **Reading a Map** Divide students into teams. Have them use the Rand McNally World Physical map on PE pp. A2–3 to find two location examples of each region shown in the Natural Regions of the World chart, on PE p. 38. Have groups present their examples to the class; they can point them out on a map. **Class Time** 30 minutes
- **Analyzing Causes** Have students review the Human Migration map on PE p. 39. Lead a discussion about what natural features may have influenced or affected these movement patterns. **Class Time** 10 minutes
- **Peer Review** To practice using longitude and latitude, ask students to choose a place in the world they would like to visit. Have students use the Skillbuilder on PE p. 41 to plot the longitude and latitude of their choice, keeping the location a secret. Then have students exchange coordinates, figure out the location, and check their answer with their partner. **Class Time** 20 minutes

SECTION 2, pp. 45–49
Class Time 30 minutes

- **Making a Map** Review the concept of thematic maps, then ask students to make a thematic map of the school. They should clearly indicate the theme in the map's legend. Some possible themes: the locations of student activities, or the places where one can get food in the school.

DAY 3

SECTION 2, continued
Class Time 35 minutes

- **Cartographer** Have students use PE pp. 42–44 to review the skills cartographers use to construct maps. Encourage them to make an illustrated map of their town or neighborhood. Display the finished maps.

CHAPTER 2 REVIEW AND ASSESSMENT, pp. 50–51
Class Time 55 minutes

- **Review** Have students use the charts they created for Reading Social Studies on PE p. 34 to review the five geography themes. **Class Time** 20 minutes
- **Assessment** Have students complete the Chapter 2 Assessment. **Class Time** 35 minutes

31c UNIT 1

1 Comprehensive Planning

From the **Chapter Preview** to **Review and Assessment,** suggestions for content and pacing are all here.

2 Estimated Times

Estimated times needed for each activity are provided to help teachers make efficient use of the block period.

3 Teaching Options

Numerous teaching options are presented so teachers can vary the pacing of the class as well as the types of activities in which students are engaged.

TECHNOLOGY IN THE CLASSROOM

DESIGNING WEB SITES

Students can design their own Web sites to organize information and to share their knowledge with other students at school and around the world. By creating a Web site, students gain experience in organizing information in a nonlinear fashion and get a behind-the-scenes look at what goes into developing materials for the Internet. Their Web sites can be kept on the classroom computer, uploaded to the school district's server, or uploaded to the Internet for students at other schools to view. The Web site at **classzone.com** is a helpful resource for students to learn about designing their own Web pages.

ACTIVITY OUTLINE

Objective Students will design Web pages to teach other students about the Five Themes of Geography.

Task Have students work in groups to create a Web site that provides examples of the Five Themes of Geography and that teaches other students about each of the themes.

Class Time Three class periods

DIRECTIONS

1. Hold a class discussion reviewing the Five Themes of Geography, as discussed in Chapter 2. Ask students to provide examples to explain each theme, and ask them to describe the reasons why it is important to be familiar with the concepts covered by each of these themes. For example, why is it important to identify your location (Theme 1) or to understand the impact of migration and transportation (Theme 4)?
2. Divide the class into small groups of about four students each. Assign one of the Five Themes. It may be necessary to assign some themes to more than one group.
3. Ask groups to create Web pages that will teach students in their grade or the grade below them about their assigned theme. They will need to provide text and images to help students learn the concepts of the theme. For example, for place (Theme 2), they might provide a brief description of what *place* means and then create a Web page that shows pictures of the landscape and people in their home town. They could then ask the audience to think about the special characteristics of their own home and to compare and contrast those features to the things they see on the Web page.
4. Suggest that students use the following criteria for completing their Web pages:
   - They must provide a two- to four-sentence description of what their theme is about.
   - They must include at least one example of how this theme relates to things they already know about (e.g., their home town, transportation, world regions).
   - They must include at least one image, and they must cite the source of the image if they have taken it from a book or another Web site.
5. Help several students design an overall home page that will link to each group's Web page. This main home page should contain a list of the Five Themes. As an option, upload this page and each group's site onto the school district's server, and register it with a search engine.

Teacher's Edition **31d**

Technology in the Classroom

Every chapter in the ***World Cultures and Geography*** Teacher's Edition includes a page providing an innovative strategy for using the Internet and other technologies in the classroom.

Activity Outline

An overview introduces the strategy, presents an instructional rationale for using it with students, states the objective, and summarizes the task in direct, easy-to-understand terms. Recommended class time is also provided.

Directions

The directions spell out, step-by-step, what students need to do to complete the activity. These straightforward directions ensure that students can incorporate technology into their daily lives.

Open Up the World

Teacher's Edition Lesson Support

The ***World Cultures and Geography*** Teacher's Edition provides a wealth of information and practical teaching suggestions to meet the needs of each student, link geography and culture to other subjects, develop critical thinking skills, and more.

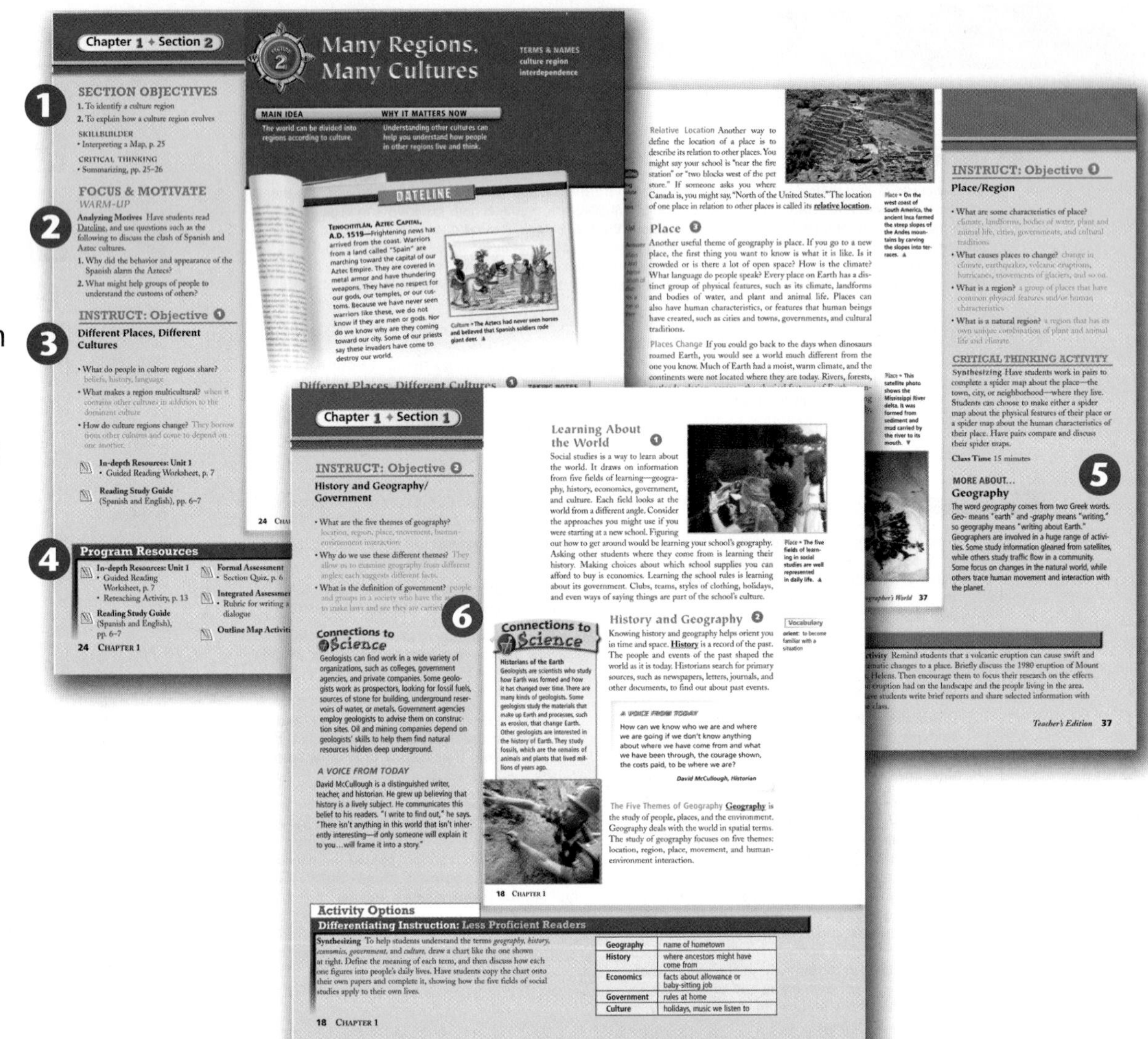

1 Section Objectives
The **Objectives** for the section are clearly spelled out. **Skillbuilder** and **Critical Thinking** skills covered in the section are also listed.

2 Focus & Motivate
Focus questions to stimulate students' thinking about the content of the section begin every lesson plan.

3 Instruct
Questions for every boldface heading in the body of the text help teachers review content and determine comprehension. Answers appear in blue type.

4 Program Resource References at Point of Use
Throughout the Teacher's Edition, references to the **Program Resources** at their point of use will help teachers plan and teach every lesson.

5 More About . . .
These short features provide additional information, often little-known facts, to supplement the text.

6 Enrichment
Additional content is often provided for these Pupil's Edition features: **A Voice from** (the country or region), **Biography, Connections to Citizenship** (and content area subjects), **Spotlight on Culture, Strange but True,** and **Citizenship in Action.**

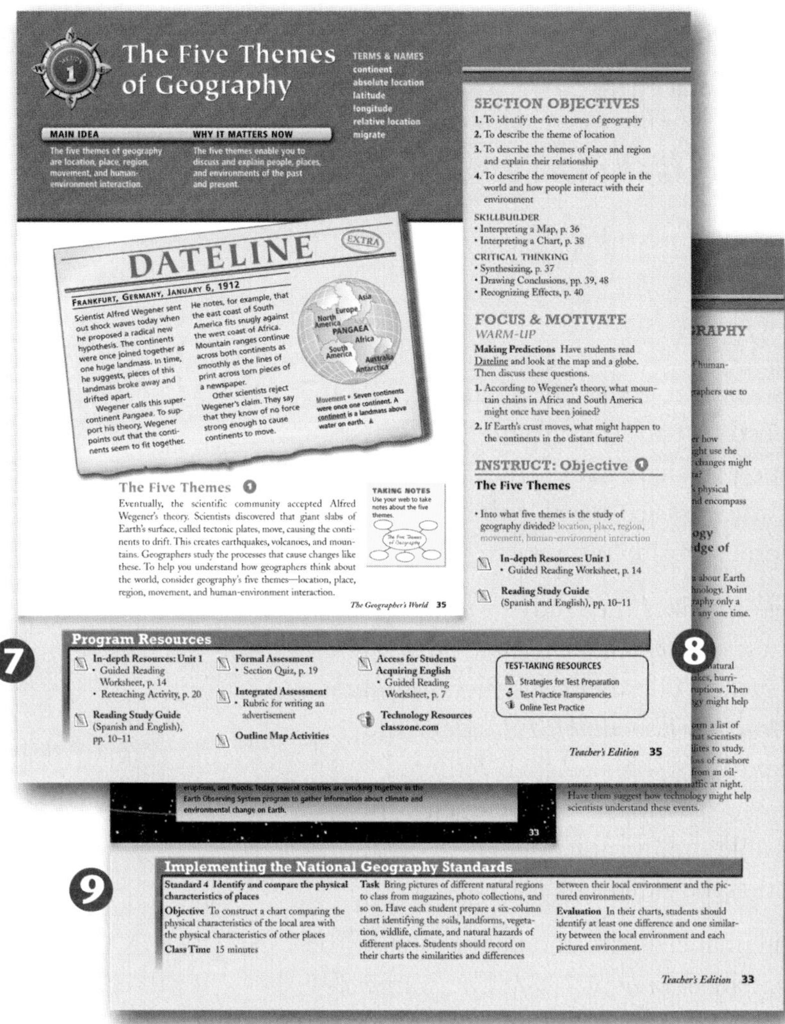

The Five Themes of Geography

TERMS & NAMES: continent, absolute location, latitude, longitude, relative location, migrate

MAIN IDEA
The five themes of geography are location, place, region, movement, and human-environment interaction.

WHY IT MATTERS NOW
The five themes enable you to discuss and explain people, places, and environments of the past and present.

DATELINE
FRANKFURT, GERMANY, JANUARY 6, 1912

Scientist Alfred Wegener sent out shock waves today when he proposed a radical new hypothesis. The continents were once joined together as one huge landmass. In time, he suggests, pieces of this landmass broke away and drifted apart.

Wegener calls this super-continent Pangaea. To support his theory, Wegener points out that the continents seem to fit together. He notes, for example, that the east coast of South America fits snugly against the west coast of Africa. Mountain ranges continue smoothly as the lines of print across torn pieces of a newspaper.

Other scientists reject Wegener's claim. They say that they know of no force strong enough to cause continents to move.

Movement • Seven continents were once one continent. A continent is a landmass above water on Earth.

The Five Themes

Eventually, the scientific community accepted Alfred Wegener's theory. Scientists discovered that giant slabs of Earth's surface, called tectonic plates, move, causing the continents to drift. This creates earthquakes, volcanoes, and mountains. Geographers study the processes that cause changes like these. To help you understand how geographers think about the world, consider geography's five themes—location, place, region, movement, and human-environment interaction.

TAKING NOTES: Use your web to take notes about the five themes.

The Geographer's World 35

SECTION OBJECTIVES
1. To identify the five themes of geography
2. To describe the theme of location
3. To describe the themes of place and region and explain their relationship
4. To describe the movement of people in the world and how people interact with their environment

SKILLBUILDER
• Interpreting a Map, p. 36
• Interpreting a Chart, p. 38

CRITICAL THINKING
• Synthesizing, p. 37
• Drawing Conclusions, pp. 39, 48
• Recognizing Effects, p. 40

FOCUS & MOTIVATE
WARM-UP
Making Predictions Have students read Dateline and look at the map and a globe. Then discuss these questions.
1. According to Wegener's theory, what mountain chains in Africa and South America might once have been joined?
2. If Earth's crust moves, what might happen to the continents in the distant future?

INSTRUCT: Objective 1
The Five Themes
• Into what five themes is the study of geography divided? location, place, region, movement, human-environment interaction

In-depth Resources: Unit 1
• Guided Reading Worksheet, p. 14

Reading Study Guide (Spanish and English), pp. 10–11

Program Resources
In-depth Resources: Unit 1
• Guided Reading Worksheet, p. 14
• Reteaching Activity, p. 20
Reading Study Guide (Spanish and English), pp. 10–11
Formal Assessment
• Section Quiz, p. 19
Integrated Assessment
• Rubric for writing an advertisement
Outline Map Activities
Access for Students Acquiring English
• Guided Reading Worksheet, p. 7
Technology Resources classzone.com
TEST-TAKING RESOURCES
Strategies for Test Preparation
Test Practice Transparencies
Online Test Practice

Teacher's Edition 35

Implementing the National Geography Standards
Standard 4 Identify and compare the physical characteristics of places
Objective To construct a chart comparing the physical characteristics of the local area with the physical characteristics of other places
Class Time 15 minutes
Task Bring pictures of different natural regions to class from magazines, photo collections, and so on. Have each student prepare a six-column chart identifying the soils, landforms, vegetation, wildlife, climate, and natural hazards of different places. Students should record on their charts the similarities and differences between their local environment and the pictured environments.
Evaluation In their charts, students should identify at least one difference and one similarity between the local environment and each pictured environment.

Teacher's Edition 33

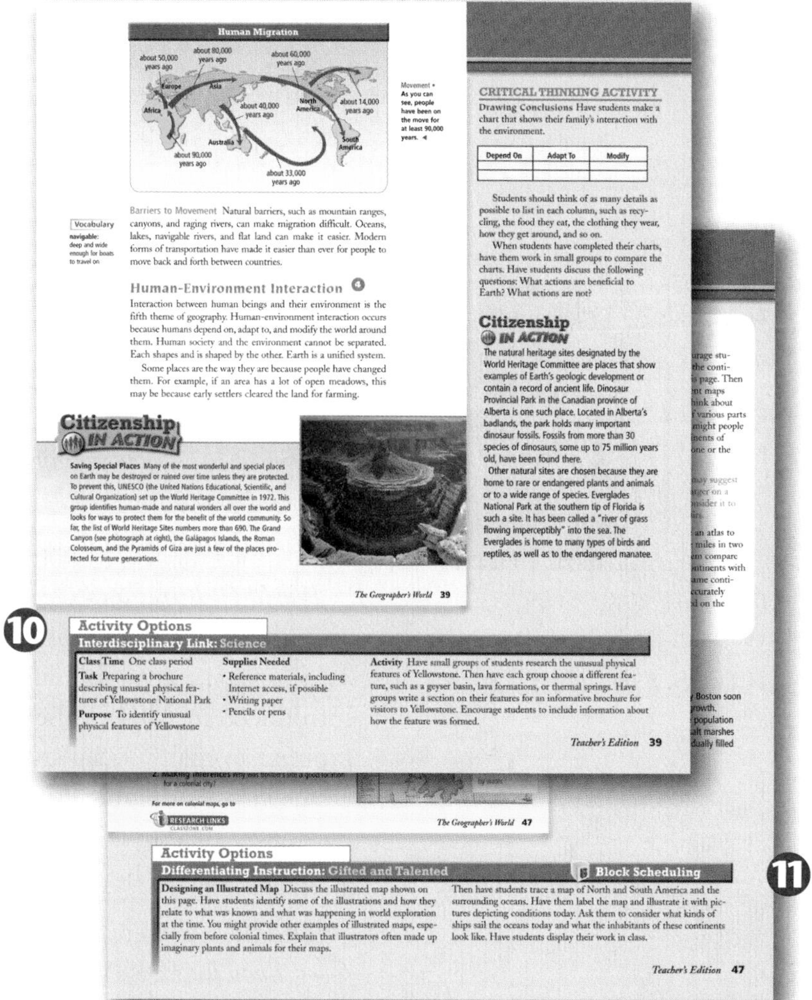

Human Migration

Movement • As you can see, people have been on the move for at least 90,000 years.

Barriers to Movement Natural barriers, such as mountain ranges, canyons, and raging rivers, can make migration difficult. Oceans, lakes, navigable rivers, and flat land can make it easier. Modern forms of transportation have made it easier than ever for people to move back and forth between countries.

Vocabulary: navigable: deep and wide enough for boats to travel on

Human-Environment Interaction

Interaction between human beings and their environment is the fifth theme of geography. Human-environment interaction occurs because humans depend on, adapt to, and modify the world around them. Human society and the environment cannot be separated. Each shapes and is shaped by the other. Earth is a unified system.

Some places are the way they are because people have changed them. For example, if an area has a lot of open meadows, this may be because early settlers cleared the land for farming.

Citizenship IN ACTION

Saving Special Places Many of the most wonderful and special places on Earth may be destroyed or ruined over time unless they are protected. To prevent this, UNESCO (the United Nations Educational, Scientific, and Cultural Organization) set up the World Heritage Committee in 1972. This group identifies human-made and natural wonders all over the world and looks for ways to protect them for the benefit of the world community. So far, the list of World Heritage Sites numbers more than 690. The Grand Canyon (see photograph at right), the Galápagos Islands, the Roman Colosseum, and the Pyramids of Giza are just a few of the places protected for future generations.

The Geographer's World 39

CRITICAL THINKING ACTIVITY
Drawing Conclusions Have students make a chart that shows their family's interaction with the environment.

Depend On	Adapt To	Modify

Students should think of as many details as possible to list in each column, such as recycling, the food they eat, the clothing they wear, how they get around, and so on.

When students have completed their charts, have them work in small groups to compare the charts. Have students discuss the following questions: What actions are beneficial to Earth? What actions are not?

Citizenship IN ACTION
The natural heritage sites designated by the World Heritage Committee are places that show examples of Earth's geologic development or contain a record of ancient life. Dinosaur Provincial Park in the Canadian province of Alberta is one such place. Located in Alberta's badlands, the park holds many important dinosaur fossils. Fossils from more than 30 species of dinosaurs, some up to 75 million years old, have been found there.

Other natural sites are chosen because they are home to rare or endangered plants and animals or to a wide range of species. Everglades National Park at the southern tip of Florida is such a site. It has been called a "river of grass flowing imperceptibly" into the sea. The Everglades is home to many types of birds and reptiles, as well as to the endangered manatee.

Activity Options
Interdisciplinary Link: Science
Class Time One class period
Task Preparing a brochure describing unusual physical features of Yellowstone National Park
Purpose To identify unusual physical features of Yellowstone
Supplies Needed
• Reference materials, including Internet access, if possible
• Writing paper
• Pencils or pens
Activity Have small groups of students research the unusual physical features of Yellowstone. Then have each group choose a different feature, such as a geyser basin, lava formations, or thermal springs. Have groups write a section on their features for an informative brochure for visitors to Yellowstone. Encourage students to include information about how the feature was formed.

Teacher's Edition 39

Activity Options
Differentiating Instruction: Gifted and Talented — Block Scheduling
Designing an Illustrated Map Discuss the illustrated map shown on this page. Have students identify some of the illustrations and how they relate to what was known and what was happening in world exploration at the time. You might provide other examples of illustrated maps, especially from before colonial times. Explain that illustrators often made up imaginary plants and animals for their maps.

Then have students trace a map of North and South America and the surrounding oceans. Have them label the map and illustrate it with pictures depicting conditions today. Ask them to consider what kinds of ships sail the oceans today and what the inhabitants of these continents look like. Have students display their work in class.

Teacher's Edition 47

7 Program Resources

This listing of specific ancillaries for ***World Cultures and Geography*** will help teachers select appropriate support items to reinforce lessons.

8 Test-Taking Resources

Test-Taking Resources include Strategies for Test Preparation, Test Practice Transparencies, and Online Test Practice.

9 Implementing the National Geography Standards

World Cultures and Geography has 4 Unit Activities and 14 Chapter Activities that correlate to the *National Geography Standards*.

10 Activity Options

The bottom column of the Teacher's Edition pages contains a variety of activities for enriching teaching, including critical thinking activities, links to other subjects, activities for students acquiring English, plus activities for less proficient readers and gifted and talented students.

11 Block Scheduling

The **Block Scheduling** logo appears wherever an Activity Option is appropriate for block scheduling situations.

And there is more, including:

- **Recommended Resources**
- **Focus on Visuals**
- **Critical Thinking Activities**
- **Assess & Reteach**
- **Reteaching Activities**

The National Standards and *World Cultures and Geography*

Dear Educator:

As a geographer and former social studies teacher, I am pleased to welcome you to the exciting opportunity to teach young learners about the world in which we live. I hope you share my enthusiasm for teaching geography and helping students see the role that geography plays in understanding contemporary world cultures.

World Cultures and Geography organizes the study of today's world around five key geographic themes: *location, place, region, human-environment interaction,* and *movement*. The themes, which were first introduced in 1984 in the *Guidelines for Geographic Education,* are familiar to many social studies educators. This teacher-centered framework to organize geography instruction was developed two decades ago as an easy and accessible tool for introducing geographic concepts that go beyond map and globe skills and place names.

The themes were updated and enhanced with the publication of the *Geography for Life: National Geography Standards 1994.* The *National Geography Standards* represent the next big step toward improving the quality of geographic education in American schools. While the five themes help teachers plan opportunities for students to learn geography, the *National Geography Standards,* arranged in six essential elements, are student-centered. The standards present what students should know and be able to do in order to understand our world's people, places, and environments.

The essential elements of the national standards expand upon and deepen the five themes. For example, the theme *location* is the pivotal point of geography. If history is about time and chronology (when), geography is about space and location (where). The parallel element in the national standards is *Seeing the World in Spatial Terms.* For Grades 5–8, Standards 1, 2, and 3 implement this essential element by identifying specific knowledge and expectations that will help students to view the world in spatial terms.

The themes of *place* and *region* are also expanded upon and made more explicit in the national standards. The essential element *Places and Regions* introduces the theme of *place* in Standard 4 and *region* in Standard 5. Standard 6 examines how culture and experience influence people's perceptions of both place and region. The physical characteristics of place, and the processes that produce those characteristics, are explored in the essential element *Physical Systems* (Standards 7 and 8). The human characteristics of places, such as population, culture, settlement patterns, and migration, are described in greater detail in *Human Systems* (Standards 9, 10, 11, 12, and 13).

The theme *human-environment interaction* is elaborated upon in the essential element *Environment and Society,* with three related standards. Standard 14 looks at how human actions modify the physical environment, Standard 15 examines how physical systems affect human systems, and Standard 16 focuses on changes in resources.

The theme of *movement,* again a fundamental of geography, is a component of all the essential elements. It includes the movement of weather fronts, landslides, and other natural phenomena (*Physical Systems*) and the movement of people, goods, and ideas (*Human Systems*).

However, the five themes do not address one of the most important and beloved topics of the social studies—the link between history and geography. The essential element *The Uses of Geography* and Standards 17 and 18 give teachers clear guidelines and suggestions for teaching about the geography-history relationship. In summary, the *National Geography Standards* encompass the five themes, expand upon them, and transform them into student expectations.

As you will see in the pages of ***World Cultures and Geography*** and in suggested student activities, the program incorporates key aspects of the *National Geography Standards* to promote innovative practices in geography and to reflect current understandings about effective learning and teaching strategies. We hope that by presenting both the themes and the national standards, we will support teachers making the transition from the themes to the standards, benefiting students in the process.

Sarah W. Bednarz, Ph.D.
Texas A&M University
College Station, Texas

Correlation to the *National Geography Standards*

World Cultures and Geography correlates to the *National Geography Standards* through the Pupil's Edition and the Teacher's Edition. The following page references are representative of the many ways the textbook meets the national standards.

	Pupil's Edition	Teacher's Edition
GEOGRAPHY STANDARD 1 • THE WORLD IN SPATIAL TERMS		
How to use maps and other geographic representations, tools, and technologies to acquire, process, and report information from a spatial perspective	A1–A25, 19, 25, 27, 30–31, 35–39, 41–43, 45–51, 54–65, 77, 80, 83, 97, 107, 113, 115, 122, 125, 129, 133, 139, 142–149, 154–155, 156–159, 169, 176–178, 183, 188, 192, 199, 204, 206, 210, 214, 219, 227, 232, 240, 246, 251, 257, 260–269, 273, 277, 279, 286–287, 297, 302, 308, 311, 320, 325, 330, 332, 335, 338, 344–345, 349, 355, 361, 363, 365, 375, 380, 385, 391, 395, 400, 405	22, 23, 27, 32, 36, 41, 44, 45, 46, 47, 49, 50, 51, 54, 57, 70, 80, 106, 142, 143, 147, 155, 169, 183, 201, 214, 218, 219, 232, 240, 251, 260, 262, 269, 274, 277, 279, 286, 287, 330, 344, 365, 370, 380, 385, 392, 400
GEOGRAPHY STANDARD 2 • THE WORLD IN SPATIAL TERMS		
How to use mental maps to organize information about people, places, and environments in a spatial context	29, 77, 80, 100–101, 152, 172, 189, 191, 194, 198–199, 216, 227, 236–237, 239, 257, 271, 281, 286, 289, 297, 325, 347, 349, 370, 405	28, 29, 47, 54, 118, 139, 165, 167, 233, 236, 237, 260, 276, 309, 331, 334, 350, 378, 395, 397, 401

	Pupil's Edition	Teacher's Edition
GEOGRAPHY STANDARD 3 • THE WORLD IN SPATIAL TERMS		
How to analyze the spatial organization of people, places, and environments on Earth's surface	19, 38, 39, 48, 78, 87–91, 99, 111, 122, 125, 137, 154–155, 156–159, 174, 188, 198, 199, 203, 204, 214, 240, 246, 251, 253, 274, 277, 299, 310, 325, 365, 375	24, 25, 48, 49, 50, 51, 56, 66, 77, 107, 121, 122, 132, 135, 136, 139, 158, 170, 176, 178, 195, 196, 197, 204, 205, 206, 243, 246, 251, 252, 255, 261, 265, 271, 275, 278, 279, 280, 289, 291, 294, 295, 296, 315, 332, 335, 337, 338, 340, 355, 361, 363, 391, 396, 401
GEOGRAPHY STANDARD 4 • PLACES AND REGIONS		
The physical and human characteristics of places	4–13, 19, 21, 24–26, 31, 35, 38, 48, 112–113, 115, 120, 127, 132, 133, 154–155, 156–163, 166, 168–169, 171, 191–192, 198, 201, 204, 208, 211–213, 220, 224–227, 229, 232, 235, 240–241, 246, 251–253, 256, 274–275, 278, 284, 313–314, 375	24, 25, 33, 37, 42, 43, 72, 92, 93, 137, 150, 168, 263, 289, 298, 308, 310, 311, 320, 321, 322, 325, 379, 394
GEOGRAPHY STANDARD 5 • PLACES AND REGIONS		
That people create regions to interpret Earth's complexity	19, 24–25, 72, 77, 80, 124, 133, 153, 158, 159, 167, 168, 183, 203, 302, 344–345, 355, 361, 368, 379–380	15, 65, 72, 108, 109, 181, 201, 268, 378, 388, 389
GEOGRAPHY STANDARD 6 • PLACES AND REGIONS		
How culture and experience influence people's perception of places and regions	52–53, 108–109, 133, 140–141, 164, 236–237, 248–249, 258–259, 372–373, 380–381	84, 100, 101, 144, 190, 248, 249, 327, 330, 388, 389
GEOGRAPHY STANDARD 7 • PHYSICAL SYSTEMS		
The physical processes that shape the patterns of Earth's surface	33, 35, 73–77, 115, 151, 154, 156–159, 163, 201, 271, 275, 277, 297	29, 35, 37, 39, 55, 67, 73, 74, 75, 76, 77, 79, 82, 122, 154, 156, 157, 168, 201, 250, 271, 276, 277, 341, 386
GEOGRAPHY STANDARD 8 • PHYSICAL SYSTEMS		
The characteristics and spatial distribution of ecosystems on Earth's surface	19, 33, 38, 139, 151, 154, 156, 158, 159, 163, 168, 169, 201	

	Pupil's Edition	Teacher's Edition
GEOGRAPHY STANDARD 9 • HUMAN SYSTEMS		
The characteristics, distribution, and migration of human populations on Earth's surface	19, 25, 38, 39, 48, 49, 127, 133, 136, 158, 162, 163, 166, 173, 174, 176, 177, 190, 191, 193, 204, 211–213, 220, 224–226, 229, 232, 235, 241, 246, 247, 253, 256, 257, 315, 325, 349, 375	14, 38, 62, 74, 84, 85, 87, 88, 91, 120, 123, 145, 212, 238, 246, 247, 253, 267, 345
GEOGRAPHY STANDARD 10 • HUMAN SYSTEMS		
The characteristics, distribution, and complexity of Earth's cultural mosaics	21, 24–26, 30–31, 78, 85, 87–91, 111–112, 114, 132–136, 138, 166, 190–194, 198, 211–213, 224–225, 242, 246–247, 253–254, 256, 257, 302–304, 306, 310–311, 324, 364, 370–371, 382–383, 386–387	190, 212, 227, 370, 373
GEOGRAPHY STANDARD 11 • HUMAN SYSTEMS		
The patterns and networks of economic interdependence on Earth's surface	20, 26, 30, 79, 102–107, 114, 128–131, 138, 157, 166, 169, 186–189, 198, 206, 209–211, 213, 217, 218, 220, 222–226, 232, 234, 235, 238–242, 244, 247, 250, 252, 254, 256, 294, 308–310, 325, 330, 351, 355–356, 362, 365–366, 369, 377, 383, 385–386, 401–402, 405	14, 72, 75, 102, 103, 104, 105, 106, 107, 109, 111, 114, 128, 129, 130, 131, 148, 208, 209, 210, 213, 218, 219, 223, 226, 228, 241, 244, 250, 254, 276, 294, 299, 304, 306, 307, 313, 314, 316, 317, 324, 325, 362, 364, 367, 368, 369, 371, 372, 374, 375, 377, 382, 383, 384, 387, 391, 392, 404
GEOGRAPHY STANDARD 12 • HUMAN SYSTEMS		
The processes, patterns, and functions of human settlement	12, 15, 40, 163, 168, 174, 193, 207, 225, 226, 229, 241, 245–247, 253, 256–257, 288, 294	68, 69, 78, 110, 171, 192, 193, 194, 198, 224, 229, 256, 294, 296, 315, 319, 381
GEOGRAPHY STANDARD 13 • HUMAN SYSTEMS		
How the forces of cooperation and conflict among people influence the division and control of Earth's surface	24, 136, 162, 166, 175, 176, 178, 180–182, 204–206, 216–218, 222, 226, 232, 263, 297, 327, 394–397	20, 31, 124, 138, 166, 173, 174, 175, 177, 179, 180, 181, 184, 203, 204, 207, 211, 213, 215, 216, 217, 220, 221, 222, 225, 226, 231, 232, 235, 241, 299, 327, 331, 334, 342, 348, 357, 380, 381, 383, 398, 399, 403, 404

	Pupil's Edition	Teacher's Edition
GEOGRAPHY STANDARD 14 • ENVIRONMENT AND SOCIETY		
How human actions modify the physical environment	19, 33, 40, 113, 151, 157–159, 162, 163–164, 166, 169, 187, 192, 206, 245, 323, 363, 386	40, 74, 157, 163
GEOGRAPHY STANDARD 15 • ENVIRONMENT AND SOCIETY		
How physical systems affect human systems	12, 19, 51, 117, 151, 153–154, 156, 158–159, 162, 163, 165–166, 168, 171, 177, 188, 201, 208–209, 238–239, 242, 250, 252, 254, 387	19, 39, 67, 68, 69, 70, 71, 117, 119, 150, 154, 164, 199, 216, 228, 270, 271, 273, 332
GEOGRAPHY STANDARD 16 • ENVIRONMENT AND SOCIETY		
The changes that occur in the meaning, use, distribution, and importance of resources	19, 44, 78–79, 129, 130, 151, 157, 158, 185, 210, 240, 242, 351, 375, 392–393	26, 38, 72, 75, 78, 79, 82, 185, 186, 187, 188, 189, 198, 248, 252, 269, 351, 365, 377, 386
GEOGRAPHY STANDARD 17 • THE USES OF GEOGRAPHY		
How to apply geography to interpret the past	39–40, 42, 45, 47, 76, 82, 87–91, 115, 120–121, 123, 131, 136, 151–152, 156, 160–163, 165–166, 168, 177, 204–205, 207, 229, 232, 235, 239, 250–251, 253–254, 256, 275, 284–290, 294, 296, 297, 311, 315, 317, 325, 351, 354–356, 370–371	44, 59, 60, 61, 64, 83, 88, 89, 91, 99, 115, 120, 121, 123, 125, 126, 127, 133, 134, 138, 144, 151, 153, 158, 159, 160, 161, 162, 166, 168, 172, 182, 183, 191, 200, 202, 203, 231, 233, 234, 239, 242, 265, 266, 267, 268, 270, 272, 278, 281, 282, 284, 285, 286, 288, 289, 290, 291, 292, 293, 300, 301, 302, 303, 305, 306, 311, 318, 319, 320, 322, 324, 328, 329, 333, 335, 336, 337, 338, 339, 343, 344, 346, 347, 348, 349, 352, 353, 354, 355, 356, 357, 358, 360, 364, 366, 369, 374, 376, 388, 389, 390, 391, 393, 400, 402, 403, 404
GEOGRAPHY STANDARD 18 • THE USES OF GEOGRAPHY		
How to apply geography to interpret the present and plan for the future	44, 48, 49, 51, 79, 113, 124, 127, 131, 151, 154, 157–159, 171, 192, 194, 229, 242, 245, 273, 363, 369, 377, 386, 392, 404	34, 48, 58, 63, 83, 90, 94, 95, 96, 97, 98, 112, 113, 114, 139, 146, 148, 149, 219, 229, 230, 234, 245, 246, 257, 263, 264, 283, 326, 368, 375

World Cultures AND GEOGRAPHY

Western Hemisphere and Europe

McDOUGAL LITTELL

McDOUGAL LITTELL

World Cultures AND GEOGRAPHY

Western Hemisphere and Europe

Sarah Witham Bednarz

Inés M. Miyares

Mark C. Schug

Charles S. White

McDougal Littell

A DIVISION OF HOUGHTON MIFFLIN COMPANY

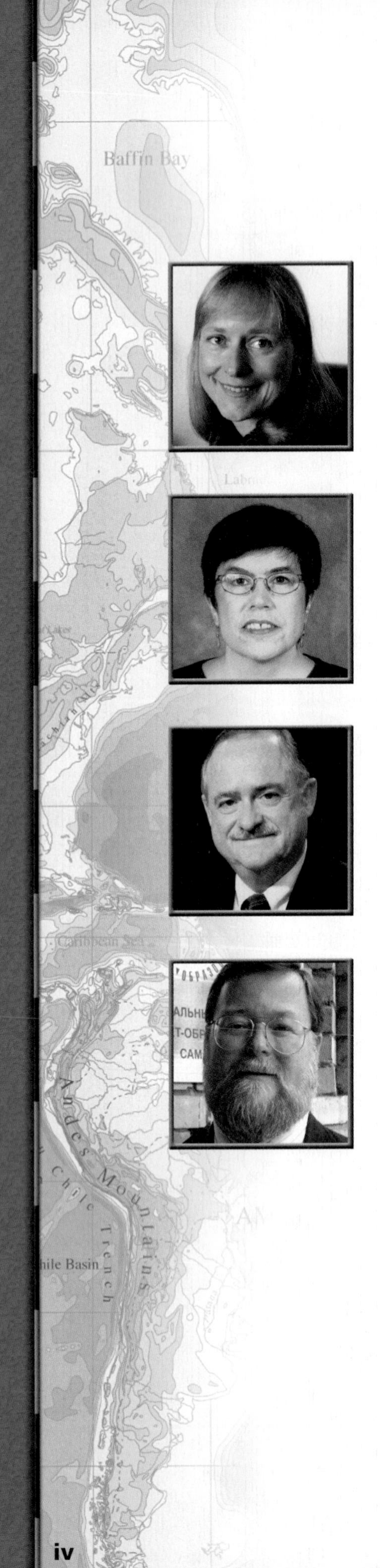

Senior Consultants

Sarah Witham Bednarz is associate professor of geography at Texas A&M University, where she has taught since 1988. She earned a Ph.D. in educational curriculum and instruction in 1992 from Texas A&M University and has written extensively about geography literacy and education. Dr. Bednarz was an author of *Geography for Life: National Geography Standards,* 1994. In 1997 she received the International Excellence Award from the Texas A&M University International Programs Office.

Inés M. Miyares is associate professor of geography at Hunter College–City University of New York. Born in Havana, Cuba, and fluent in Spanish, Dr. Miyares has focused much of her scholarship on Latin America, immigration and refugee policy, and urban ethnic geography. She holds a Ph.D. in geography from Arizona State University. In 1999 Dr. Miyares was the recipient of the Hunter College Performance Excellence Award for excellence in teaching, research, scholarly writing, and service.

Mark C. Schug is director of the University of Wisconsin–Milwaukee Center for Economic Education. A 30-year veteran of middle school, high school, and university classrooms, Dr. Schug has been cited for excellence in teaching by the University of Wisconsin–Milwaukee and the Minnesota Council on Economic Education. In addition to coauthoring eight national economics curriculum programs, Dr. Schug has spoken on economic issues to audiences throughout the world. Dr. Schug edited *The Senior Economist* for the National Council for Economics Education from 1986 to 1996.

Charles S. White is associate professor in the School of Education at Boston University, where he teaches methods of instruction in social studies. Dr. White has written and spoken extensively on the role of technology in social studies education. He has received numerous awards for his scholarship, including the 1995 Federal Design Achievement Award from the National Endowment for the Arts, for the Teaching with Historic Places project. In 1997, Dr. White taught his Models of Teaching doctoral course at the Universidad San Francisco de Quito, Ecuador.

Acknowledgments begin on page R53.

ISBN 0-618-37756-5

Printed in the United States of America
1 2 3 4 5 6 7 8 9 – VJM – 07 06 05 04 03

iv

Consultants and Reviewers

Content Consultants

Charmarie Blaisdell
Department of History
Northeastern University
Boston, Massachusetts

David Buck, Ph.D.
Department of History
University of Wisconsin–Milwaukee
Milwaukee, Wisconsin

Erich Gruen, Ph.D.
Departments of Classics and History
University of California, Berkeley
Berkeley, California

Charles Haynes, Ph.D.
Senior Scholar for Religious Freedom
The Freedom Forum First Amendment Center
Arlington, Virginia

Alusine Jalloh, Ph.D.
The Africa Program
University of Texas at Arlington
Arlington, Texas

Shabbir Mansuri
Council on Islamic Education
Fountain Valley, California

Michelle Maskiell, Ph.D.
Department of History
Montana State University
Bozeman, Montana

Vasudha Narayanan, Ph.D.
Department of Religion
University of Florida
Gainesville, Florida

Amanda Porterfield, Ph.D.
Department of Religious Studies
University of Wyoming
Laramie, Wyoming

Mark Wasserman, Ph.D.
Department of History
Rutgers University
New Brunswick, New Jersey

Multicultural Advisory Board

Dr. Munir Bashshur
Education Department
American University of Beirut
Beirut, Lebanon

Stephen Fugita
Ethnic Studies Program
Santa Clara University
Santa Clara, California

Sharon Harley
Afro-American Studies Program
University of Maryland at College Park
College Park, Maryland

Doug Monroy
Department of Southwest Studies
Colorado College
Colorado Springs, Colorado

Cliff Trafzer
Departments of History and Ethnic Studies
University of California, Riverside
Riverside, California

v

UNIT 1

Introduction to World Cultures and Geography

Some scientists believe the continents were once joined, page 35.

A variety of people inhabit the world, page 14.

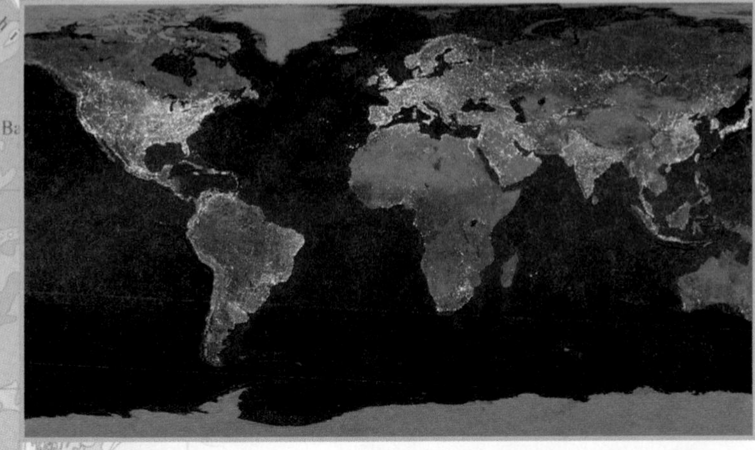

Satellite photographs of Earth, pages 2–3

vi

The United States and Canada

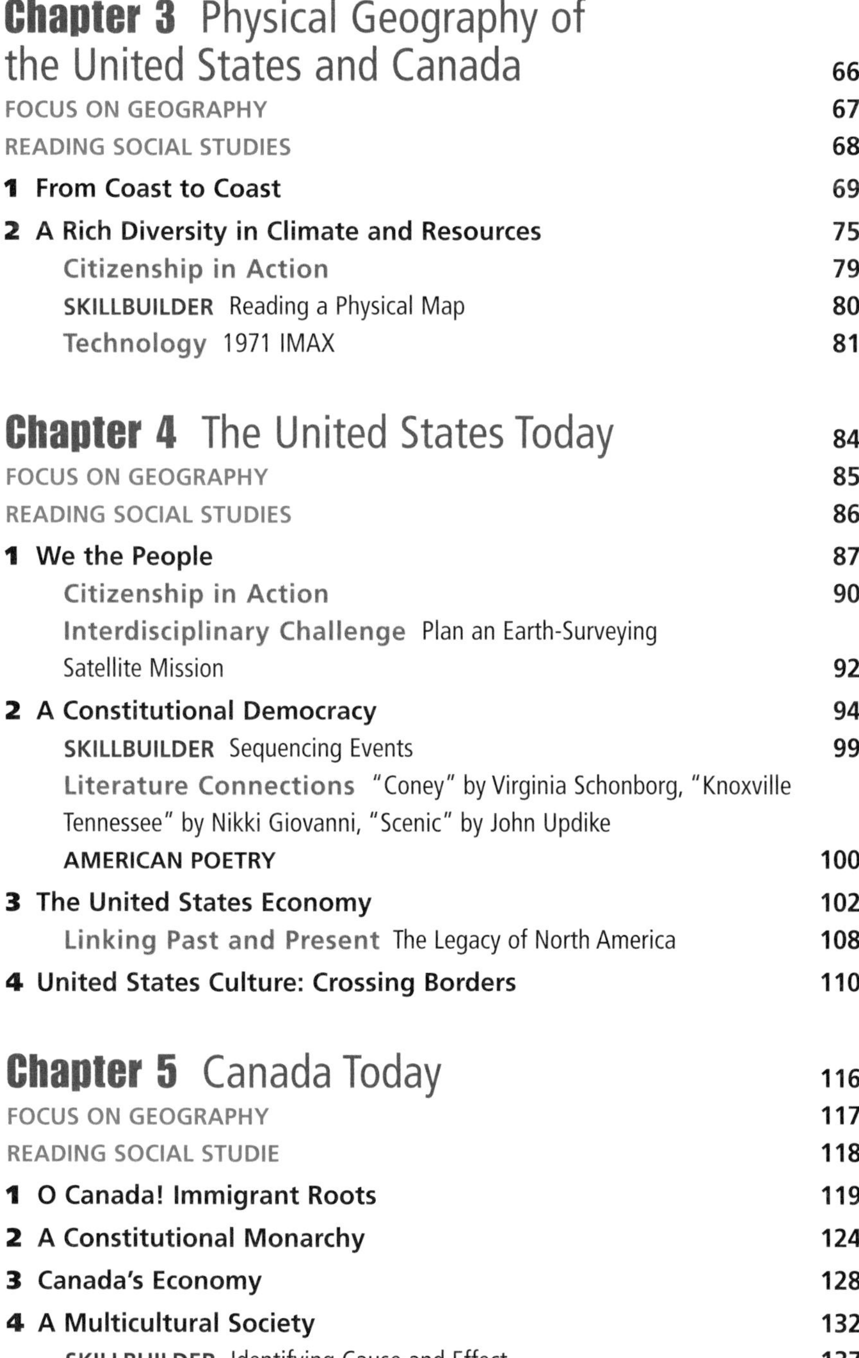

A bald eagle, page 71

Baseball is often called America's pastime, page 111.

Inuit build an igloo, page 124.

vii

Latin America

An Aztec stone carving, page 162

Mount Popocatépetl in Mexico is a volcano, page 153.

A Quechua mother and daughter, page 254

viii

Europe, Russia, and the Independent Republics

A stained-glass window from the Middle Ages, page 294

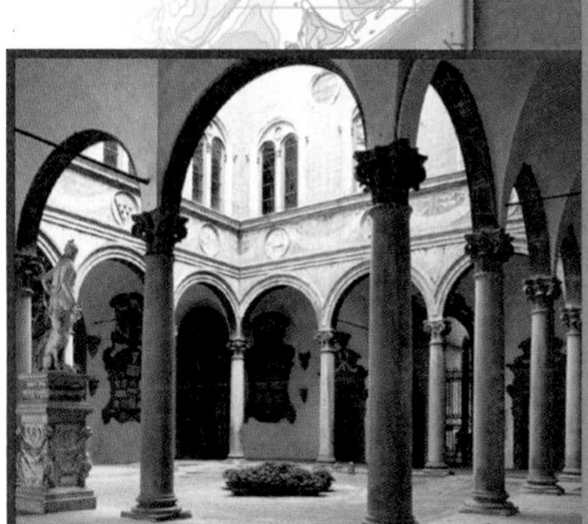

A wealthy merchant family built this Renaissance palace, page 303.

These dogs helped fight World War I, page 335.

J.K. Rowling wrote the popular Harry Potter books, page 383.

The Brandenburg Gate in Berlin, page 347

continued from page ix

A high-speed train rushes across France, pages 350–351.

x

Features

Interdisciplinary Challenge

Infographics, Time Lines, and Cartoons

Literature Connections

Linking Past and Present

DATELINE

xi

SKILLBUILDER

Citizenship IN ACTION

Strange but TRUE

Connections To...

xii

Biography

FOCUS ON GEOGRAPHY

The WORLD'S HERITAGE

VOICES FROM . . .

xiii

MAPS

UNIT 1

UNIT 2

UNIT 3

UNIT 4

ATLAS

SPECIAL REPORT: TERRORISM AND THE WAR IN IRAQ

xiv

CHARTS, DIAGRAMS, AND GRAPHS

xv

This section will help you develop and practice the skills you need to study social studies and to take standardized tests. Part 1, **Strategies for Studying Social Studies,** shows you the features of this book. It also shows you how to improve your reading and study skills.

Part 2, **Test-Taking Strategies and Practice,** gives you strategies to help you answer the different kinds of questions that appear on standardized tests. Each strategy is followed by a set of questions you can use for practice.

CONTENTS

S1

Part 1: Strategies for Studying Social Studies

Reading is important in the study of social studies or any other subject. You can improve your reading skills by practicing certain strategies. Good reading skills help you remember more when you read. The next four pages show how some of the features of *World Cultures and Geography: Western Hemisphere* can help you learn and understand social studies.

Preview Chapters Before You Read

Each chapter begins with a two-page chapter opener. Study these pages to help you get ready to read.

1. Read the chapter title. Read the section titles. These tell what topics will be covered in the chapter.
2. Look at the art and photographs. Use the illustrations to help you identify themes or messages of the chapter.
3. Study the **Focus on Geography** feature. Use the questions to help you think about the information you might find in the chapter.

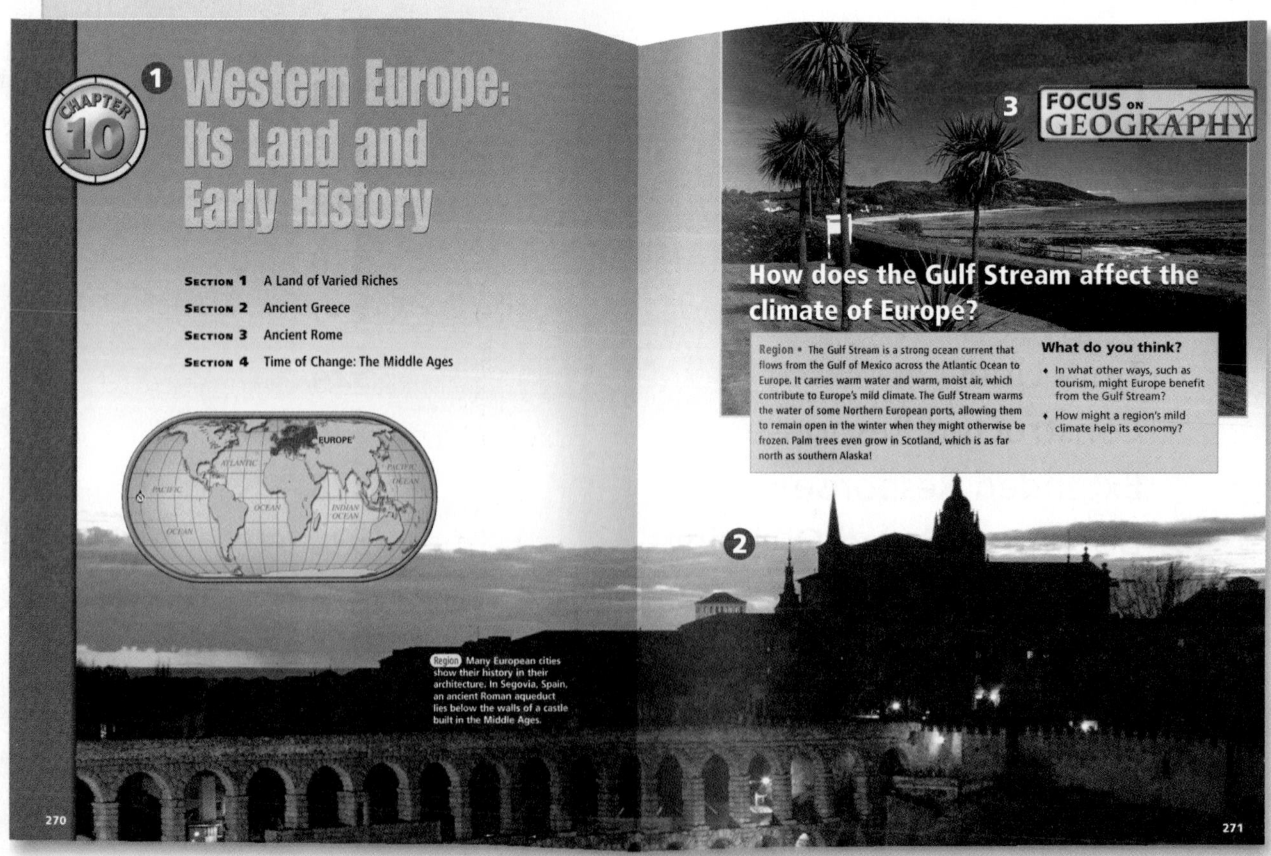

CHAPTER 10
1
Western Europe: Its Land and Early History
SECTION 1 A Land of Varied Riches
SECTION 2 Ancient Greece
SECTION 3 Ancient Rome
SECTION 4 Time of Change: The Middle Ages
EUROPE
Region Many European cities show their history in their architecture. In Segovia, Spain, an ancient Roman aqueduct lies below the walls of a castle built in the Middle Ages.
270
3
FOCUS ON GEOGRAPHY
How does the Gulf Stream affect the climate of Europe?
Region • The Gulf Stream is a strong ocean current that flows from the Gulf of Mexico across the Atlantic Ocean to Europe. It carries warm water and warm, moist air, which contribute to Europe's mild climate. The Gulf Stream warms the water of some Northern European ports, allowing them to remain open in the winter when they might otherwise be frozen. Palm trees even grow in Scotland, which is as far north as southern Alaska!
What do you think?
• In what other ways, such as tourism, might Europe benefit from the Gulf Stream?
• How might a region's mild climate help its economy?
2
271

S2

Preview Sections Before You Read

Each chapter has three, four, or five sections. These sections cover shorter time periods or certain themes.

1. Study the sentences under the headings **Main Idea** and **Why It Matters Now.** These headings tell what's important in the material you're about to read.
2. Look at the **Terms & Names** list. This list tells you what people and issues will be covered in the section.
3. Read the feature titled **Dateline.** It tells about a historical event as if it were happening today.
4. Skim the pages to see how the section is organized. Red headings are major topics. Blue headings are smaller topics or subtopics. The headings provide an outline of the section.
5. Skim the pages of the section to find key words. These words will often be in **boldface** type. Use the **Vocabulary** notes in the margin to help you with unfamiliar terms.

TERMS & NAMES
Mediterranean Sea
peninsula
fjord
Ural Mountains
plain

SECTION 1

A Land of Varied Riches

2 TERMS & NAMES
Mediterranean Sea
peninsula
fjord
Ural Mountains
plain

1 MAIN IDEA

Europe is a continent with varied geographic features, abundant natural resources, and a climate that can support agriculture.

WHY IT MATTERS NOW

The development of Europe's diverse cultures has been shaped by the continent's diverse geography.

3 DATELINE EXTRA

LONDON, ENGLAND, MAY 6, 1994

Rough waters have always made the English Channel, which separates England and France, difficult to cross. Now, however, you can travel under the water! Today, a tunnel nicknamed "the Chunnel" opens, allowing high-speed trains to travel between London and Paris in about three hours. The Chunnel—short for Channel Tunnel—was carved through chalky earth under the sea floor and took seven years to build. It is the largest European construction project of the 20th century.

Movement • Eurostar trains make the 31-mile trip under the English Channel in only 20 minutes. ▲

Location • The Channel Tunnel connects England and France. ▲

The Geography of Europe 4

Today, cars, airplanes, and trains are common forms of high-speed transportation across Europe. Before the 19th century, however, the fastest form of transportation was to travel by water—on top of it, rather than under it.

TAKING NOTES
Use your chart to take notes about Western Europe.

Time Period	Social Structure	Art/Architecture
Ancient Greece		
Ancient Rome		

Western Europe: Its Land and Early History 273

Place • The Alps remain snowcapped year-round. ▶

Mountain ranges, including the towering Alps, also stretch across much of the continent. Along Europe's eastern border, the **Ural Mountains** (YUR•uhl) divide the continent from Asia. The many mountain ranges of Europe separated groups of people from one another as they settled the land thousands of years ago. This is one of the reasons why different cultures developed across the continent.

Reading Social Studies
B. Clarifying What natural landform separates Europe from Asia?

The Great European Plain Not all of Europe is mountainous. A vast region called the Great European Plain stretches from the coast of France to the Ural Mountains. A **plain** is a large, flat area of land, usually without many trees. The Great European Plain is the location of some of the world's richest farmland. Ancient trading centers attracted many people to this area, which today includes some of the largest cities in Europe—Paris, Berlin, Warsaw, and Moscow.

Climate

5 Vocabulary
Gulf Stream: a warm ocean current that flows northeast from the Gulf of Mexico through the Atlantic Ocean.

Although the Gulf Stream brings warm air and water to Europe, the winters are still severe in the mountains and in the far north. In some of these areas, cold winds blow southward from the Arctic Circle and make the average temperature fall below 0°F in January. The Alps and the Pyrenees, however, protect the European countries along the Mediterranean Sea from these chilling winds. In these warmer parts of southern Europe, the average temperature in January stays above 50°F.

Western Europe: Its Land and Early History 275

S3

Use Active Reading Strategies As You Read

Now you're ready to read the chapter. Read one section at a time, from beginning to end.

1. Begin by looking at the **Reading Social Studies** page. Consider the questions under the **Before You Read** heading. Think about what you know already about the chapter topic and what you'd like to learn.
2. Review the suggestions in the **Read and Take Notes** section. These will help you understand and remember the information in the chapter.
3. Ask and answer questions as you read. Look for the **Reading Social Studies** questions in the margin. Answering these questions will show whether you understand what you've just read.
4. Study the **Background** notes in the margin for additional information on people, places, events, or ideas discussed in the chapter.

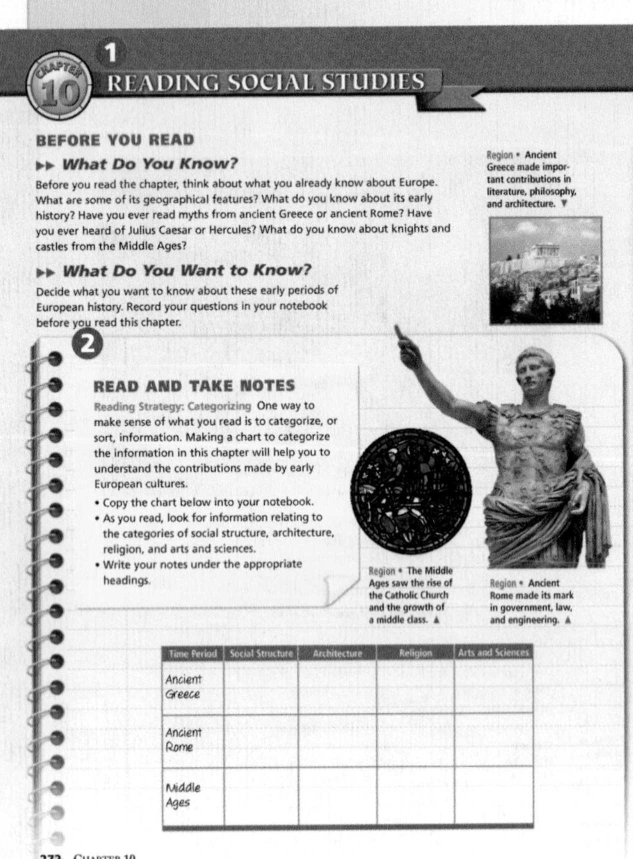

1

CHAPTER 10 READING SOCIAL STUDIES

BEFORE YOU READ

What Do You Know?

Before you read the chapter, think about what you already know about Europe. What are some of its geographical features? What do you know about its early history? Have you ever read myths from ancient Greece or ancient Rome? Have you ever heard of Julius Caesar or Hercules? What do you know about knights and castles from the Middle Ages?

What Do You Want to Know?

Decide what you want to know about these early periods of European history. Record your questions in your notebook before you read this chapter.

Region • Ancient Greece made important contributions in literature, philosophy, and architecture. ▼

2

READ AND TAKE NOTES

Reading Strategy: Categorizing One way to make sense of what you read is to categorize, or sort, information. Making a chart to categorize the information in this chapter will help you to understand the contributions made by early European cultures.

- Copy the chart below into your notebook.
- As you read, look for information relating to the categories of social structure, architecture, religion, and arts and sciences.
- Write your notes under the appropriate headings.

Region • The Middle Ages saw the rise of the Catholic Church and the growth of a middle class. ▲

Region • Ancient Rome made its mark in government, law, and engineering. ▲

Time Period	Social Structure	Architecture	Religion	Arts and Sciences
Ancient Greece				
Ancient Rome				
Middle Ages				

272 CHAPTER 10

Reading Social Studies

A. Clarifying Why were waterways important for the movement of people and goods?

Waterways Look at the map of Europe on page 277. Water surrounds the continent to the north, south, and west. The southern coast of Europe borders the warm waters of the **Mediterranean Sea.** Europe also has many rivers. The highly traveled Rhine and Danube rivers are two of the most important. The Volga, which flows nearly 2,200 miles through western Russia, is the continent's longest. For hundreds of years, these and other waterways have been home to boats and barges carrying people and goods inland across great distances.

Reading Social Studies 3

A. Clarifying Why were waterways important for the movement of people and goods?

Landforms Several large **peninsulas,** or bodies of land surrounded by water on three sides, form the European continent. In Northern Europe, the Scandinavian Peninsula is home to Norway and Sweden. Along the jagged shoreline of this peninsula are beautiful fjords (fyawrdz). A **fjord** is a long, narrow, deep inlet of the sea located between steep cliffs. In Western Europe, the Iberian Peninsula includes Portugal and Spain. The Iberian Peninsula is separated from the rest of the continent by a mountain range called the Pyrenees (PEER•uh•NEEZ). The entire continent of Europe, itself surrounded by water on three sides, is a giant peninsula.

BACKGROUND 4

Europe can be divided into four areas: Western Europe, Northern Europe, Eastern Europe, and Russia and its neighboring countries.

Place • The Scandinavian Peninsula is the location of many spectacular fjords, such as this one in Norway. ▶

274 CHAPTER 10

BACKGROUND

Europe can be divided into four areas: Western Europe, Northern Europe, Eastern Europe, and Russia and its neighboring countries.

S4

Review and Summarize What You Have Read

When you finish reading a section, review and summarize the information you have learned. Reread any information that is still unclear.

1. Look again at the red and blue headings for a quick summary of the section.
2. Study the photographs, maps, charts, graphs, and illustrated features in the section. Think about how these visuals relate to the information you've learned.
3. Answer the questions in the **Assessment** section. This will help you think critically about what you've just read.

The summers in the south are usually hot and dry, with an average July temperature around 80°F. This makes the Mediterranean coast a popular vacation spot. Elsewhere in Europe, in all but the coldest areas of the mountains and the far north, the average July temperature ranges from 50°F to 70°F.

1 Natural Resources

Europe has a large variety of natural resources, including minerals. The rich coal deposits of Germany's Ruhr (rur) Valley region have helped to make that area one of the world's major industrial centers. Russia and Ukraine have large deposits of iron ore, which is used to make iron for automobiles and countless other products.

Region • Western Europe benefits from a varied landscape rich in natural resources. ▼

2

276 CHAPTER 10

2

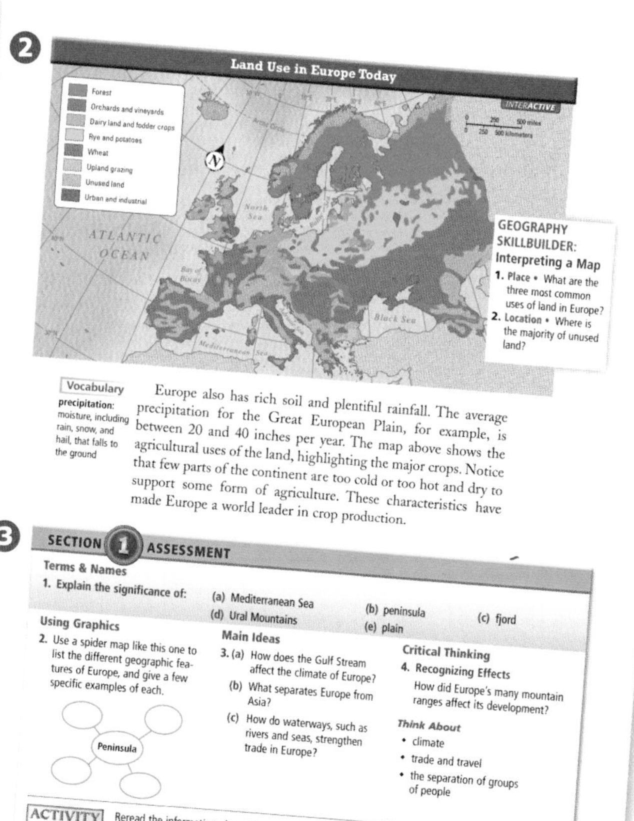

GEOGRAPHY SKILLBUILDER: Interpreting a Map
1. Place • What are the three most common uses of land in Europe?
2. Location • Where is the majority of unused land?

Vocabulary
precipitation: moisture, including rain, snow, and hail, that falls to the ground

Europe also has rich soil and plentiful rainfall. The average precipitation for the Great European Plain, for example, is between 20 and 40 inches per year. The map above shows the agricultural uses of the land, highlighting the major crops. Notice that few parts of the continent are too cold or too hot and dry to support some form of agriculture. These characteristics have made Europe a world leader in crop production.

3

SECTION 1 ASSESSMENT

Terms & Names
1. Explain the significance of: (a) Mediterranean Sea (b) peninsula (c) fjord (d) Ural Mountains (e) plain

Using Graphics
2. Use a spider map like this one to list the different geographic features of Europe, and give a few specific examples of each.

Peninsula

Main Ideas
3. (a) How does the Gulf Stream affect the climate of Europe?
(b) What separates Europe from Asia?
(c) How do waterways, such as rivers and seas, strengthen trade in Europe?

Critical Thinking
4. Recognizing Effects
How did Europe's many mountain ranges affect its development?

Think About
- climate
- trade and travel
- the separation of groups of people

ACTIVITY -OPTION- Reread the information about the Channel. Write a **short story** in which you imagine what it might have been like to work on the Channel's construction.

Western Europe: Its Land and Early History 277

STRATEGIES FOR TAKING STANDARDIZED TESTS

S5

USING STRATEGIES FOR . . .

Multiple Choice Explain to students that they will do best on test questions by thinking them through carefully and by applying test-taking strategies, such as the following.

1. Read the question carefully and evaluate each choice to find the correct answer. As the hint points out in question 1, (C) should be eliminated since Cuba is an island. (A) and (D) should be eliminated since these are countries in South America, which does not share a border with the United States. The correct answer is (B).
2. *Most* is a key word. All the languages listed in the choices are spoken in Latin America and the Caribbean, but Spanish is the language that *most* people in the region speak. The correct answer is (D).
3. In question 3 the last choice is *all of the above.* Make sure that the other choices are all correct if you pick this answer. Since (A), (B), and (C) are all correct, (D) is the correct answer.
4. Be careful when reading questions that include the word *not.* First rule out all the islands that are in the Caribbean—(A), (B), and (D). The one that remains, Hawaii, is in the Pacific Ocean, not the Caribbean, so (C) is the correct choice.

GENERAL TEST-TAKING TIPS

Share these tips with your students.

- The night before a test, make sure you get at least eight hours of sleep.
- Have a healthy breakfast or lunch before taking your test.
- Wear clothes that make you comfortable.
- Relax and enjoy the challenge!

Part 2: Test-Taking Strategies and Practice

Use the strategies in this section to improve your test-taking skills. First read the tips on the left page. Then use them to help you with the practice items on the right page.

Multiple Choice

A multiple-choice question is a question or incomplete sentence and a set of choices. One of the choices correctly answers the question or completes the sentence.

1. Read the question or incomplete sentence carefully. Try to answer the question before looking at the choices.
2. Read each choice with the question. Don't decide on your final answer until you have read all the choices.
3. Rule out any choices that you know are wrong.
4. Look for key words in the question. They may help you figure out the correct answer.
5. Sometimes the last choice is *All of the above.* Make sure that the other choices are all correct if you pick this answer.
6. Be careful with questions that include the word *not.*

answers: 1 (B); 2 (D); 3 (D); 4 (C)

❶ 1. Which of the following countries shares a border with the United States?

❷ choices
❸ Since Cuba is an island, you know that it cannot share a border with the United States.

A. Brazil
B. Canada
C. Cuba
D. Venezuela

2. Which language is spoken by most of the people who live in Latin America and the Caribbean?

❹ The word *most* is key here. All of the languages listed are spoken in Latin America and the Caribbean, but most people there speak Spanish.

A. English
B. French
C. Portuguese
D. Spanish

3. The Andes mountains run through

A. Colombia.
B. Ecuador.
C. Peru.
D. all of the above

❺ Before selecting *All of the above,* make sure that all of the choices are, indeed, correct.

4. Which of the following is *not* an island in the Caribbean?

❻ First rule out all the choices that name islands in the Caribbean. The choice that remains is the correct answer.

A. Barbados
B. Cuba
C. Hawaii
D. Jamaica

Activity Options

Individual Needs: Students Acquiring English

Word Meaning Make sure students understand the following terms and concepts on these pages.

Strategy Page

Question 2 *Latin America:* the region made up of Mexico, Central America, and South America
Caribbean: the region made up of the island countries in the Caribbean Sea
Question 3 *Andes mountains:* world's longest mountain chain

Practice Page

Question 1 *conquered:* defeated
Inca: early Native American civilization whose empire stretched along the Andes mountains in South America

For more test practice online . . .
TEST PRACTICE
CLASSZONE.COM

Directions: Read each question carefully. Choose the *best* answer from the four choices.

1. Which of the following conquered the Inca?

A. Francisco Pizarro
B. Miguel Hidalgo
C. Hernán Cortés
D. Simón Bolívar

2. Until recent years, many Central American and Caribbean countries had single-product economies, depending on such crops as

A. bananas.
B. coffee.
C. sugar cane.
D. all of the above

3. Many Canadians are bilingual, which means that they speak

A. French at home and English at work.
B. English and their native language.
C. two languages.
D. several languages.

4. During the 1700s, England controlled which of the following?

A. the sugar trade
B. the Atlantic slave trade
C. the cotton trade
D. the coconut trade

PRACTICE QUESTIONS

Thinking It Through Share the following explanations with students as they discuss the strategies they used to answer the practice questions.

1. Read the question and try to answer the question before looking at the choices. Rule out any choices that you know are wrong. Rule out (B) and (D) since these people led independence movements against Spain in Mexico and Latin America during the early 1800s. Hernán Cortés conquered the Aztecs, so rule out (C). The correct answer is (A).
2. In this question, the last choice is *all of the above*. So make sure that the other choices are all correct if you pick this answer. The correct answer is (D), since all of the choices are correct.
3. *Bilingual* is the key word. Look for the definition of bilingual in the choices. The correct answer is (C).
4. *1700s* and *England* are key words in this question. In 1700 the English controlled the Atlantic slave trade, so the correct answer is (B).

Skills Tested in the Items

STRATEGY ITEMS		PRACTICE ITEMS	
Item Number	**Skill Tested**	**Item Number**	**Skill Tested**
1.	Clarifying	1.	Clarifying
2.	Clarifying	2.	Summarizing
3.	Summarizing	3.	Clarifying
4.	Categorizing	4.	Clarifying

USING STRATEGIES FOR . . .

Primary Sources Explain to students that they will do best on test questions by thinking them through carefully and by applying test-taking strategies, such as the following.

1. Review the question. Then reread the primary source to answer the question. Look for key words such as *Indians, now,* and *different.* The correct answer is (B).
2. Review the question. Then skim the article to find information to help identify the main idea. Then review the choices to find the sentence that *best* expresses the main idea. The correct answer is (C).

GENERAL TEST-TAKING TIPS

Share these tips with your students.

- Read the directions carefully before you begin to answer the questions.
- Plan the time you are given to take the test.
- Check your answers.
- Believe in yourself.

STRATEGIES

Primary Sources

Sometimes you will need to look at a document to answer a question. Some documents are primary sources. Primary sources are written or made by people who either saw an event or were actually part of the event. A primary source can be a photograph, letter, diary, speech, or autobiography.

1. Look at the source line to learn about the document and its author. Consider how reliable the information might be.
2. Skim the article to get an idea of what it is about. As you read, look for the main idea. The main idea is the writer's most important point. Sometimes it is not directly stated.
3. Note any special punctuation. For example, ellipses (...) mean that words and sentences have been left out.
4. Ask yourself questions about the document as you read.
5. Review the questions. This will give your reading a purpose and also help you find the answers more easily. Then reread the document.

answers: 1 (B); 2 (C)

A Native American View of Nature

Plants are of different families… It is the same with animals… It is the same with human beings; there is some place which is best adapted to each. The seeds of the plants are blown about by the wind until they reach the place where they will grow best… and there they take ❸ root and grow… In the early days the animals probably roamed over a very wide country until they found a proper place. An animal depends on the natural conditions around it. If the buffalo were here today, I think they would be different from the buffalo of the old days because all the natural conditions have changed. They would not find the same food, nor the same surroundings. We see the change in our ponies… Now… they have less endurance and must have constant care [unlike in the past]. It is the same with the Indians; they have less freedom and they fall an easy prey to disease.

—Okute, a Teton Sioux (1911)

❶ The Sioux were a Native American people who lived on the Great Plains of North America.

1. Okute thinks that horses and Indians now are different from those of the early days because

A. the buffalo are gone.
B. natural conditions have changed.
C. they cannot roam to find the best place to live.
D. there is no water.

2. Which sentence *best* expresses the main idea of this passage?

A. "Animals and humans must always roam."
B. "Each plant, animal, and human is different."
C. "Each plant, animal, and human thrives in a particular place."
D. "Plants and animals were created for humans to eat."

S8

Activity Options

Individual Needs: Students Acquiring English

Word Meaning Make sure students understand the following terms and concepts on these pages.

Strategy Page: Primary Source

roamed: wandered
ponies: horses
endurance: toughness, hardiness
Question 2 *thrives:* grows or lives very well

Practice Page: Primary Source

declaration: statement
lords: rulers; leaders
commons: ordinary people
suspending: stopping
execution: carrying out
parliament: group of people chosen to make the laws
subjects: people living under the rule of a king or queen
petition: ask; question

For more test practice online . . .
TEST PRACTICE
CLASSZONE.COM

Directions: Read this passage from the English Bill of Rights. Use this document and your knowledge of world cultures and geography to answer questions 1 and 2.

A Declaration of English Rights

The… Lords… and Commons,… being now assembled in a full and free representative of this nation,… declare:

1. That the pretended power of suspending of laws, or the execution of laws, by [the King], without consent of Parliament, is illegal…
4. That [raising] money for or to the use of the [King], without grant of Parliament,… is illegal.
5. That it is the right of the subjects to petition the King, and all commitments and prosecutions for such petitioning are illegal.
6. That the raising or keeping a standing army within the kingdom in time of peace, unless it be with consent of parliament, is against the law…

—*Declaration of Rights,* 1689
(English Bill of Rights)

1. Which of the articles of the *Declaration of Rights* is concerned with taxation?

A. Article 1
B. Article 4
C. Article 5
D. Article 6

2. The *Declaration of Rights* was aimed at limiting the power of the

A. Parliament.
B. nobles and clergy.
C. rising middle class.
D. King.

PRACTICE QUESTIONS

Thinking It Through Share the following explanations with students as they discuss the strategies they used to answer the practice questions.

1. *Taxation* is a key word. Skim the articles to find information about taxation. Since Article 4 refers to raising money, which is taxation, the correct answer is (B).
2. You need to skim the article to get an idea of whose powers are limited by the *Declaration of Rights.* The correct answer is (D).

Skills Tested in the Items

STRATEGY ITEMS		PRACTICE ITEMS	
Item Number	**Skill Tested**	**Item Number**	**Skill Tested**
1.	Analyzing Causes; Recognizing Effects	1.	Clarifying
2.	Analyzing Primary Sources: Determining Main Ideas	2.	Analyzing Primary Sources: Analyzing Motives

USING STRATEGIES FOR . . .

Secondary Sources Explain to students that they will do best on test questions by thinking them through carefully and by applying test-taking strategies, such as the following.

1. Question 1 asks you to identify a statement about Malinche that is an opinion. (A), (B), and (D) are facts, which can be proved. (C) is the correct answer, since it is a statement that cannot be proved.
2. According to the account, Malinche was a Native American who helped the Spanish conquistador Cortés conquer the Aztecs. As a result, the soldiers and officers in Cortés's army would view Malinche as a heroine. So the correct answer is (B).

GENERAL TEST-TAKING TIPS

Share these tips with your students.

- Glance over the test to determine the types and numbers of questions.
- Estimate the amount of time you have to spend on each type of question.

STRATEGIES

Secondary Sources

A secondary source is an account of events by a person who did not actually experience them. The author often uses information from several primary sources to write about a person or event. Biographies, many newspaper articles, and history books are examples of secondary sources.

1. Read the title to get an idea of what the passage is about. (The title here indicates that the passage is about a person named Malinche about whom people have different opinions.)
2. Skim the paragraphs to find the main idea of the passage.
3. Look for words that help you understand the order in which events happen.
4. Ask yourself questions as you read. (You might ask yourself: Why did people's opinions about Malinche change over time?)
5. Review the questions to see what information you will need to find. Then reread the passage.

answers: 1 (C); 2 (B)

1 **Malinche—Heroine or Traitor?**

No one knows much about Malinche's early life. People do know that in 1519 she met Hernán Cortés. The Spanish conquistador had landed in Mexico earlier that year. Malinche was only 15 years old. Even though she was very
2 young, Malinche helped Cortés conquer the Aztecs. She spoke the languages of the Aztec and the Maya. Over time, she learned Spanish. She translated for Cortés and advised him on Native American politics.

The Spanish conquistadors admired Malinche, calling
3 her Doña Marina. For many centuries, the Spanish people regarded her as a heroine. In the 1800s, however, Mexico won its independence from Spain. People rejected their Spanish rulers. Writers and artists started calling Malinche a traitor to her people. Today, however, she is seen as a heroine again. **4**

5

1. Which of the following statements about Malinche is an opinion?
   - A. She was very young when she met Cortés.
   - B. She became a translator for Cortés.
   - C. She was a traitor to her own people.
   - D. She understood Native American politics.

Remember that an opinion is a statement that cannot be proved. A fact is a statement that can be proved.

2. Based on this source, which person or group would view Malinche as a heroine?
   - A. a fighter for Mexican independence from Spain
   - B. the soldiers and officers in Cortés's army
   - C. the Aztec ruler and his court in Mexico
   - D. a historian writing about Mexico in the 1800s

S10

Activity Options

Individual Needs: Students Acquiring English

Word Meaning Make sure students understand the following terms and concepts on these pages.

Strategy Page: Secondary Source
Explain that women were treated like possessions in the 1500s in most of the world.

conquistador: Spanish adventurer in sixteenth-century Americas
Question 2 *heroine:* a woman admired for her achievements and qualities

Practice Page: Secondary Source

allies: countries that agree to help one another
neutral: not siding with one country or another

For more test practice online . . .
TEST PRACTICE
CLASSZONE.COM

Directions: Read this passage. Use the passage and your knowledge of world cultures and geography to answer questions 1 and 2.

Before World War I

In 1892, France and Russia had become military allies. Later, Germany signed an agreement to protect Austria. If any nation attacked Austria, Germany would fight on its side. France and Russia had to support each other as well. For instance, if France got into a war with Germany, Russia had to fight Germany, too. This meant that in any war, Germany would have to fight on two fronts: France on the west and Russia on the east.

If a war broke out, what part would Great Britain play? No one knew. It might remain neutral, like Belgium. It might, if given a reason, fight against Germany.

1. If Russia and Germany went to war, which country had to help Russia?

A. Great Britain
B. Belgium
C. Austria
D. France

2. When World War I broke out, what part did Great Britain play?

A. It remained neutral, like Belgium.
B. It sided with Germany and Austria.
C. It joined France in fighting Germany.
D. It fought Russia after its revolution.

PRACTICE QUESTIONS

Thinking It Through Share the following explanations with students as they discuss the strategies they used to answer the practice questions.

1. You may recall the answer to this question without needing to reread the passage. If not, you should skim the passage to identify the country which had to help Russia in a war against Germany. The first paragraph identifies (D) as the correct answer.
2. The secondary source gives information about military allies before World War I. So you need to use your knowledge of world cultures and geography to answer this question. The correct answer is (C).

Skills Tested in the Items

STRATEGY ITEMS		PRACTICE ITEMS	
Item Number	**Skill Tested**	**Item Number**	**Skill Tested**
1.	Distinguishing Fact from Opinion	1.	Recognizing Effects
2.	Making Inferences	2.	Summarizing

USING STRATEGIES FOR . . .

Political Cartoons Political cartoons are one kind of primary source. Remind students to analyze the political cartoon before reading the questions. They should identify the subject, note important symbols and details, interpret the message, and analyze the point of view. Then they will read the questions to identify the information they need to find.

Explain to students that they will do best on test questions by thinking them through carefully and by applying test-taking strategies, such as the following.

1. Since the swastika was the symbol of Nazi Germany, the correct answer for question 1 is (B).
2. To answer question 2, analyze the cartoonist's message. Look at how the cartoonist exaggerates the main object in the cartoon. The swastika, which represents Nazi Germany, is huge and appears to be turning like a wheel, about to roll down onto Poland and crush it. The label "Poland" tells what country is the subject of the cartoon's title, "Next!" Therefore, (A) is the correct answer.

GENERAL TEST-TAKING TIPS

Share these tips with your students.

- Ask questions before the test begins.
- Know how to fill in the answer form.
- Read and listen to directions carefully.

Political Cartoons

Cartoonists who draw political cartoons use both words and art to express opinions about political issues.

1. Try to figure out what the cartoon is about. Titles and captions may give clues.
2. Use labels to help identify the people, places, and events represented in the cartoon.
3. Note when and where the cartoon was published.
4. Look for symbols—that is, people, places, or objects that stand for something else.
5. The cartoonist often exaggerates the physical features of people and objects. This technique will give you clues as to how the cartoonist feels about the subject.
6. Try to figure out the cartoonist's message and summarize it in a sentence.

answers: 1 (B); 2 (A)

(1) NEXT!

(4) The cartoonist uses the swastika, a symbol used during World War II.

(5) The swastika looks like a huge, frightening machine. It can easily crush Poland.

(2) The label "Poland" tells which country is the subject of the cartoon's title.

Daniel Fitzpatrick / *St. Louis Post-Dispatch*, August 24, 1939.

(3) The date is a clue that the cartoon refers to the beginning of World War II.

1. What does the swastika in the cartoon stand for?
   - **A.** the Soviet Union
   - **B.** Nazi Germany
   - **C.** the Polish army
   - **D.** Great Britain

(6) 2. Which sentence *best* summarizes the cartoonist's message?
   - **A.** Germany will attack Poland next.
   - **B.** Poland should stop Germany.
   - **C.** Germany will lose this battle.
   - **D.** Poland will fight a civil war.

S12

Activity Options

Individual Needs: Students Acquiring English

Word Meaning Make sure students understand the following terms and concepts on these pages.

Strategy Page

swastika: cross with the ends at right angles; symbol of Nazi Germany

Practice Page

Question 2 *twain:* two; pair

tyranny: a government whose ruler has absolute power

For more test practice online . . .
TEST PRACTICE
CLASSZONE.COM

Directions: Study this cartoon. Use the cartoon and your knowledge of world cultures and geography to answer questions 1 and 2.

The Granger Collection, New York

Benjamin Franklin (1754)

1. What do the sections of the snake in the cartoon represent?

A. army units
B. states
C. Native American groups
D. colonies

2. Which phrase *best* states the message of the cartoon?

A. "East is East, and West is West, and never the twain shall meet."
B. "Taxation without representation is tyranny."
C. "United we stand, divided we fall."
D. "Out of many, one."

PRACTICE QUESTIONS

Thinking It Through Share the following explanations with students as they discuss the strategies they used to answer the practice questions.

1. To correctly answer this question, try to figure out what the cartoon is about by reading the title, the labels, and noting by whom and when the cartoon was published. Note that the date 1754 is the year the French and Indian War began. Since the cartoon was published before the American Revolution, the sections could not represent states. The sections in the cartoon are labeled using the initials of colonies. (N. E. stands for New England.) The correct answer is (D).
2. To answer this question, you need to figure out what the cartoon is about. Look at the symbols and use the caption to help you figure out the cartoonist's message. Then decide which phrase best states the message of the cartoon from the choices given. Franklin's message is that unless the colonies join together in the fight against the French and Indians, the colonies will die. Therefore, the correct answer is (C).

Skills Tested in the Items

STRATEGY ITEMS		PRACTICE ITEMS	
Item Number	**Skill Tested**	**Item Number**	**Skill Tested**
1.	Analyzing Political Cartoons: Clarifying	**1.**	Analyzing Political Cartoons: Clarifying
2.	Analyzing Political Cartoons: Determining Main Ideas	**2.**	Analyzing Political Cartoons: Determining Main Ideas

USING STRATEGIES FOR . . .

Charts Explain to students that they will do best on test questions that they think through carefully by applying test-taking strategies, such as the following.

1. Question 1 asks you to compare the number of immigrants in the column with the heading "Number of Immigrants." Look down this column to find the largest number. Then move your eye left across this row to find the name of the country that received the most immigrants. The correct answer is (C) because the United States had the greatest number of immigrants.
2. To answer question 2, you should look at the column with the heading "Years." Look down this column to find the three earliest years. Then move your eye left across these rows to find the names of the countries that received immigrants the earliest. The correct answer is (B) because all three of these countries received immigrants starting in 1821—the earliest date on the chart.

GENERAL TEST-TAKING TIPS

Share these tips with your students.

- Use practice tests, such as the one you are taking now, to learn about your test-taking habits and weaknesses.
- Use this information to practice strategies that will help you be a successful test-taker.

Charts

Charts present facts in a visual form. History textbooks use several different types of charts. The chart that is most often found on standardized tests is the table. A table organizes information in columns and rows.

1. Read the title of the chart to find out what information is represented.
2. Read the headings at the top of each column. Then read the headings at the left of each row.
3. Notice how the information in the chart is organized.
4. Compare the information from column to column and row to row.
5. Try to draw conclusions from the information in the chart.
6. Read the questions and then study the chart again.

(1) This chart is about the number of people who immigrated to different countries.

(4) Notice that different years are used for different countries.

Immigration to Selected Countries

Country	Years	Number of Immigrants
Argentina	1856-1932	6,405,000
Australia	1861-1932	2,913,000
Brazil	1821-1932	4,431,000
British West Indies	1836-1932	1,587,000
Canada	1821-1932	5,206,000
Cuba	1901-1932	857,000
Mexico	1911-1931	226,000
New Zealand	1851-1932	594,000
South Africa	1881-1932	852,000
United States	1821-1932	34,244,000
Uruguay	1836-1932	713,000

Source: Alfred W. Crosby, Jr. *The Columbian Exchange: Biological and Cultural Consequences of 1492*

(3) This chart lists countries in alphabetical order. Other charts organize information by years or by numbers.

(5) Of all the countries listed, six received the most immigrants. Think about what these countries have in common.

(6)

1. The country that received the most immigrants was

A. Canada.
B. the British West Indies.
C. the United States.
D. Brazil.

2. Different countries received immigrants in different years. Which countries received immigrants the earliest?

A. Argentina, New Zealand, and Canada
B. Canada, Brazil, and United States
C. Mexico, United States, and British West Indies
D. Brazil, South Africa, and Cuba

answers: 1 (C); 2 (B)

Activity Options

Individual Needs: Students Acquiring English

Word Meaning Make sure students understand the following terms and concepts on these pages.

Strategy Page Chart

immigrants: people who move to another country to live

Practice Page Chart

per capita: per person

gross domestic product: the value of goods and services created within a country in a year

For more test practice online . . .
TEST PRACTICE
CLASSZONE.COM

Directions: Read the chart carefully. Use the chart and your knowledge of world cultures and geography to answer questions 1 and 2.

Mexico and the Nations of Central America				
Country	Capital	Area (sq. miles)	Population	Per Capita Gross Domestic Product (in U.S. dollars)
Belize	Belmopan	8,900	249,183	3,000
Costa Rica	San José	19,700	3,710,558	6,700
El Salvador	San Salvador	8,100	6,122,515	3,000
Guatemala	Guatemala City	42,000	12,639,939	3,800
Honduras	Tegucigalpa	43,300	6,249,598	2,400
Mexico	Mexico City	761,600	100,349,766	8,300
Nicaragua	Managua	49,998	4,812,569	2,500
Panama	Panama City	32,200	2,808,268	7,300

Source: *World Almanac and Book of Facts* (2001)

1. The largest country in terms of both area and population is

A. Guatemala.
B. Honduras.
C. Mexico.
D. Nicaragua.

2. Which correctly states the countries' rank from high to low in terms of per capita gross domestic product?

A. Honduras, Nicaragua, Belize, Guatemala
B. Mexico, Panama, El Salvador, Honduras
C. Panama, Costa Rica, Mexico, Belize
D. Mexico, Costa Rica, Panama, Guatemala

PRACTICE QUESTIONS

Thinking It Through the following explanations with students as they discuss the strategies they used to answer the practice questions.

1. The key words in this question are *largest, area,* and *population.* Read down the columns with the headings "Area" and "Population" to find the largest country in terms of both area and population. The correct answer is (C).
2. *Per capita gross domestic product* is the key term. Look at the column with the heading "Per Capita Gross Domestic Product." Find the choice that correctly ranks the countries from high to low in terms of per capita gross domestic product. You can eliminate (A) and (C) since Mexico has the highest per capita gross domestic product. The correct answer is (B), since Panama has a higher per capita gross domestic product than Costa Rica has.

Skills Tested in the Items

STRATEGY ITEMS		PRACTICE ITEMS	
Item Number	Skill Tested	Item Number	Skill Tested
1.	Interpreting Charts: Comparing	1.	Interpreting Charts: Comparing; Contrasting
2.	Interpreting Charts: Clarifying	2.	Interpreting Charts: Comparing; Contrasting

USING STRATEGIES FOR . . .

Line and Bar Graphs Remind students that the vertical axis goes up and down and is on the left side of the graph. The horizontal axis runs from left to right across the bottom of the graph.

Explain to students that they will do best on test questions by thinking them through carefully and by applying test-taking strategies, such as the following.

1. To answer question 1, you need to read each answer choice and compare the statement to the information in the chart to find the true statement. Notice that the trend of both total and Atlantic exports is to grow over time, so (A), (B), and (D) are false statements. The correct answer is (C).
2. To answer question 2, you need to study the information in the bar graph to find the country with the tallest bar, which represents the nation with the largest foreign debt. The correct answer is (A).

GENERAL TEST-TAKING TIPS

Share these tips with your students.

- Do not spend too much time on one question.
- Skip a question you are having problems with. Go back to it later, if you have time.
- If you skip a question, be sure to skip the same number on your answer sheet.

Line and Bar Graphs

Graphs are often used to show numbers. Line graphs often show changes over time. Bar graphs make it easy to compare numbers.

1. Read the title of the graph to find out what information is represented.
2. Study the labels on the graph.
3. Look at the source line that tells where the graph is from. Decide whether you can depend on the source to provide reliable information.
4. See if you can make any generalizations about the information in the graph. Note whether the numbers change over time.
5. Read the questions carefully and then study the graph again.

1 Exports of English Manufactured Goods, 1699–1774

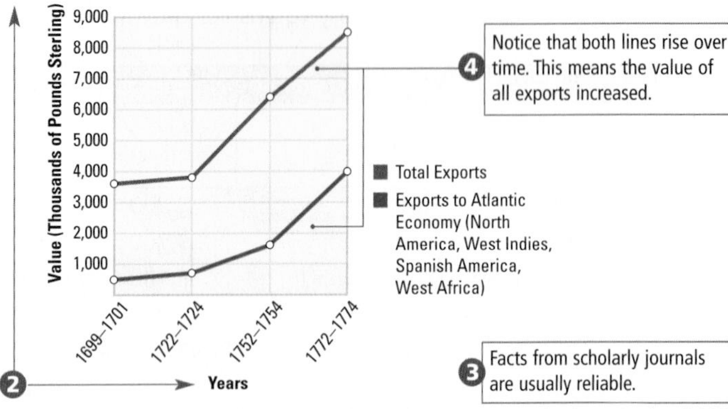

Source: R. Davis, "English Foreign Trade, 1700–1774," *Economic History Review* (1962)

5 **1.** Which of the following is a true statement?

A. Exports to the New World declined over time.
B. Total exports stayed the same over time.
C. Total exports rose sharply after 1724.
D. Exports to the New World fell sharply after 1754.

1 Nations with High Foreign Debt, 2000

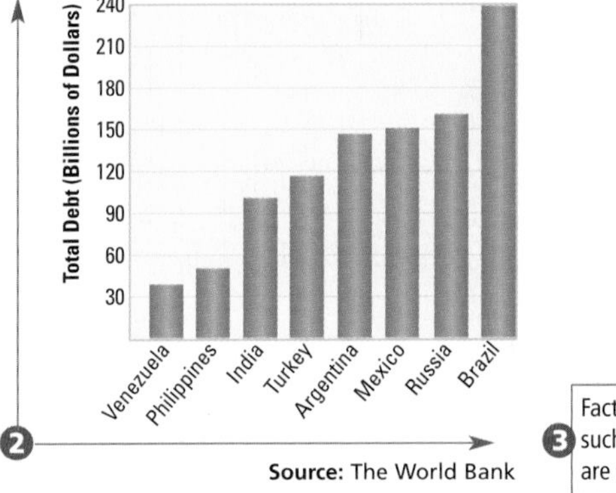

Source: The World Bank

3 Facts from major organizations, such as the World Bank, usually are reliable.

5 **2.** The nation with the largest foreign debt is

A. Brazil.
B. Argentina.
C. Mexico.
D. Venezuela.

answers: 1 (C); 2 (A)

Activity Options

Individual Needs: Students Acquiring English

Word Meaning Make sure students understand the following terms and concepts on these pages.

Strategy Page: Line Graph

Point out that the line graph covers years of discovery and exploration (1699–1774), which explains why one land is called "Spanish America," a phrase not used now.

Question 1. *decline:* downward movement

Bar Graph

foreign debt: money owed to other countries

Practice Page: Line Graph

imports: resources or goods brought into a country

exports: resources or goods sent from one country to another country

Bar Graph

unemployment rate: the number of people who are able to work, but do not have a job

For more test practice online . . .
TEST PRACTICE CLASSZONE.COM

Directions: Study the graphs carefully. Use the graphs and your knowledge of world cultures and geography to answer questions 1 and 2.

Canada: Imports and Exports, 1995–2000

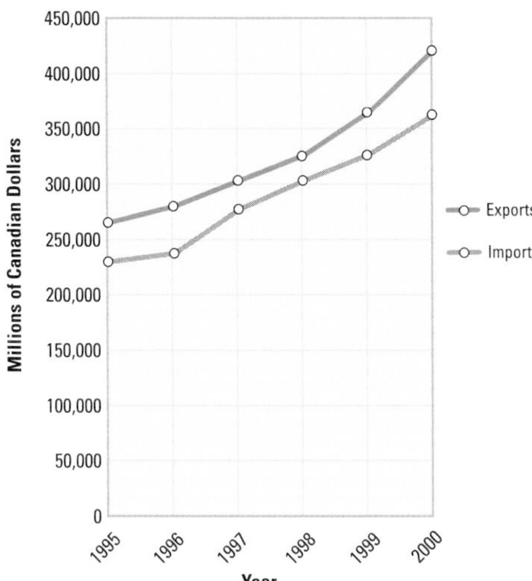

Source: *Statistics Canada*

Unemployment Rates for Selected Countries, 2002

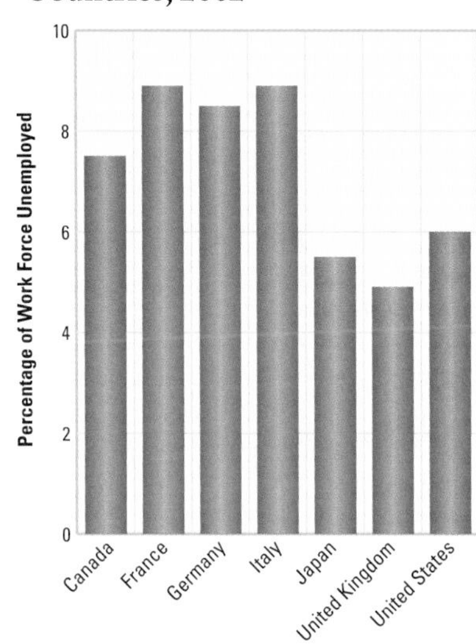

Source: Organization for Economic Cooperation and Development

1. In which year was the difference in the value of imports and exports the greatest?

A. 1995
B. 1996
C. 1999
D. 2000

2. Which country had the lowest unemployment rate in 2002?

A. Canada
B. Japan
C. United Kingdom
D. United States

PRACTICE QUESTIONS

Thinking It Through Share the following explanations with students as they discuss the strategies they used to answer the practice questions.

1. *Greatest* is a key word. This is an example of an item in which you need to note the intervals between amounts for each date shown on the graph. Since the difference in the value of imports and exports was greatest in 2000, the correct answer is (D).
3. *Lowest* is a key word. To answer this item, you need to study the information in the bar graph to find the country with the shortest bar, or lowest unemployment rate. The correct answer is (C).

Skills Tested in the Items

STRATEGY ITEMS		PRACTICE ITEMS	
Item Number	**Skill Tested**	**Item Number**	**Skill Tested**
1.	Interpreting Charts: Clarifying	1.	Interpreting Graphs: Comparing; Contrasting
2.	Interpreting Charts: Comparing	2.	Interpreting Graphs: Comparing; Contrasting

USING STRATEGIES FOR . . .

Pie Graphs Explain to students that they will do best on test questions by thinking them through carefully and by applying test-taking strategies, such as the following.

1. The key phrase is *nearly 5 percent.* Look at the pie graph and find the slice of the pie marked 5 percent. Then look at the legend to see which region is represented by that slice. The correct answer is (B).
2. *About one-twelfth* is the key phrase. Remember, one-twelfth of something is *not* the same as 12 percent of something. To answer this question you must first convert one-twelfth into a percentage number. To do this, divide 100 by 12, which is 8.33. Now find the slice on the pie graph closest to 8.33 percent, and read the legend to see what that slice represents. The correct answer is (C).

GENERAL TEST-TAKING TIPS

Share these tips with your students.

- Read the question and each answer choice before answering.
- Many items include choices that may seem right at first glance, but are actually wrong.

Pie Graphs

A pie, or circle, graph shows the relationship among parts of a whole. These parts look like slices of a pie. Each slice is shown as a percentage of the whole pie.

1. Read the title of the chart to find out what information is represented.
2. The graph may provide a legend, or key, that tells you what different slices represent.
3. The size of the slice is related to the percentage. The larger the percentage, the larger the slice.
4. Look at the source line that tells where the graph is from. Ask yourself if you can depend on this source to provide reliable information.
5. Read the questions carefully, and study the graph again.

answers: 1 (B); 2 (C)

❶ World Population by Region, 2002

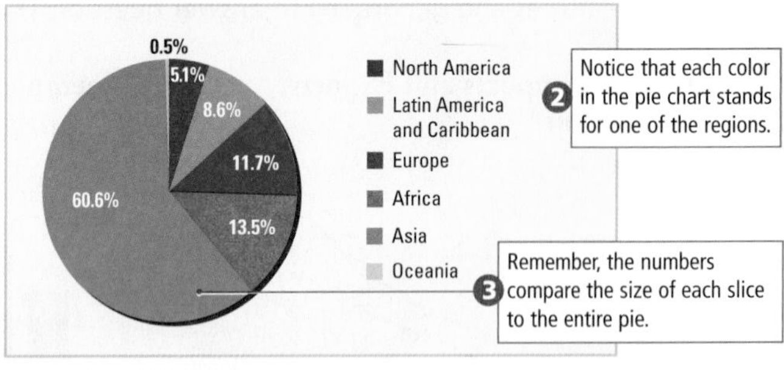

❷ Notice that each color in the pie chart stands for one of the regions.

❸ Remember, the numbers compare the size of each slice to the entire pie.

❹ The Population Reference Bureau studies population data for the United States and other countries.

Source: Population Reference Bureau

❺

1. Which region accounts for nearly 5 percent of the world's population?

A. Africa
B. North America
C. Europe
D. Asia

2. About one-twelfth of the world's population lives in

A. Africa.
B. Europe.
C. Latin America and Caribbean.
D. North America.

Activity Options

Individual Needs: Students Acquiring English

Word Meaning Make sure students understand the following terms and concepts on these pages.

Strategy Page
Question 1 *accounts for:* has

Practice Page: Pie Graph
energy consumption: energy use
Developing Asia: countries in Asia that are in the process of becoming industrialized

For more test practice online . . .
TEST PRACTICE
CLASSZONE.COM

Directions: Study the pie graph. Use the graph and your knowledge of world cultures and geography to answer questions 1 and 2.

World Energy Consumption by Region

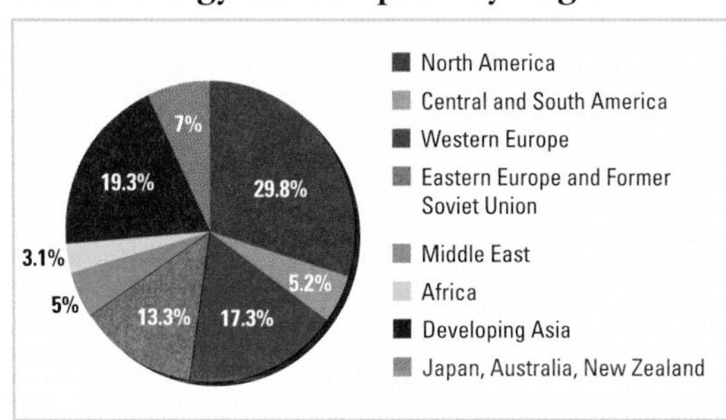

Source: "Earth Pulse," *National Geographic* (March 2001)

1. Which region uses the most energy?
   - **A.** Western Europe
   - **B.** Developing Asia
   - **C.** North America
   - **D.** Eastern Europe and Former Soviet Union

2. Two regions consumed nearly the same amount of the world's energy. They are Central and South America and
   - **A.** Middle East.
   - **B.** Western Europe.
   - **C.** Developing Asia.
   - **D.** North America.

PRACTICE QUESTIONS

Thinking It Through Share the following explanations with students as they discuss the strategies they used to answer the practice questions.

1. The key word is *most.* Compare the slices of the pie. At 29.8 percent, North America uses the most energy. The correct answer is (C).
2. Use the legend to find the color that represents Central and South America. Then find the amount of energy consumed by this region on the pie graph (5.2 percent). Next, use the pie graph and legend to find another region that consumes nearly the same amount of the world's energy as Central and South America. Since the Middle East consumes about 5 percent of the world's energy, the correct answer is (A).

Skills Tested in the Items

STRATEGY ITEMS		PRACTICE ITEMS	
Item Number	**Skill Tested**	**Item Number**	**Skill Tested**
1.	Interpreting Graphs: Comparing	**1.**	Interpreting Graphs: Comparing; Contrasting
2.	Interpreting Graphs: Comparing	**2.**	Interpreting Graphs: Comparing; Contrasting

USING STRATEGIES FOR . . .

Political Maps Explain to students that they will do best on test questions by thinking them through carefully and by applying test-taking strategies, such as the following.

1. *West* is the key word in this question. Use the North arrow to figure out the direction of places on the map. Then find the province or territory that is furthest west. The correct answer is (B).
2. To answer question 2, use the scale to estimate the distance of the United States-Canada border from the Great Lakes to the Pacific Ocean. If you are allowed to use scratch paper during your test, place the paper next to the United States-Canada border. Mark the length of the border from the Great Lakes to the Pacific Ocean on your paper. Then mark off how many times the map scale fits on the border length on your paper. Since the scale fits about three times, the correct answer is (B).

GENERAL TEST-TAKING TIPS

Share these tips with your students.

- Try to answer every question on the test.
- If you are not sure of an answer, make an educated guess.
- First eliminate the choices you are sure are *not* correct. Then choose from the choices that remain.

Political Maps

Political maps show the divisions within countries. A country may be divided into states, provinces, etc. The maps also show where major cities are. They may also show mountains, oceans, seas, lakes, and rivers.

1. Read the title of the map. This will give you the subject and purpose of the map.
2. Read the labels on the map. They also give information about the map's subject and purpose.
3. Study the key or legend to help you understand the symbols in the map.
4. Use the scale to estimate distances between places shown on the map. Maps usually show the distance in both miles and kilometers.
5. Use the North arrow to figure out the direction of places on the map.
6. Read the questions. Carefully study the map to find the answers.

1 Canada and Its Provinces

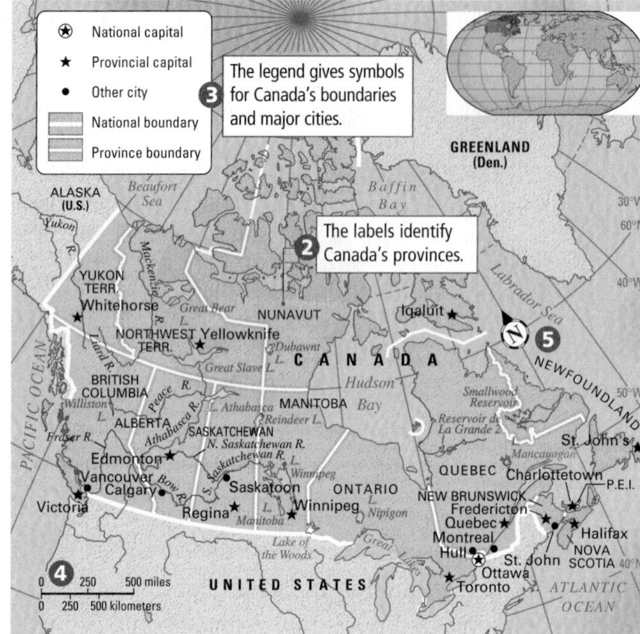

1. Which province or territory is the furthest west?

A. Northwest Territories
B. Yukon Territory
C. British Columbia
D. Alberta

6

2. About how many miles is the United States-Canada border from the Great Lakes west to Vancouver on the Pacific Ocean?

A. 1,000
B. 1,500
C. 2,000
D. 2,500

answers: 1 (B); 2 (B)

Activity Options

Individual Needs: Students Acquiring English

Word Meaning Make sure students understand the following terms and concepts on this page.

Strategy Page: Political Map

political: showing borders of countries, states, and the like
N, S, E, W: north, south, east, and west
province: a political division of a country
territory: a geographical area belonging to a government

For more test practice online . . .
TEST PRACTICE
CLASSZONE.COM

Directions: Study the map carefully. Use the map and your knowledge of world cultures and geography to answer questions 1 and 2.

The Roman Empire, A.D. 400

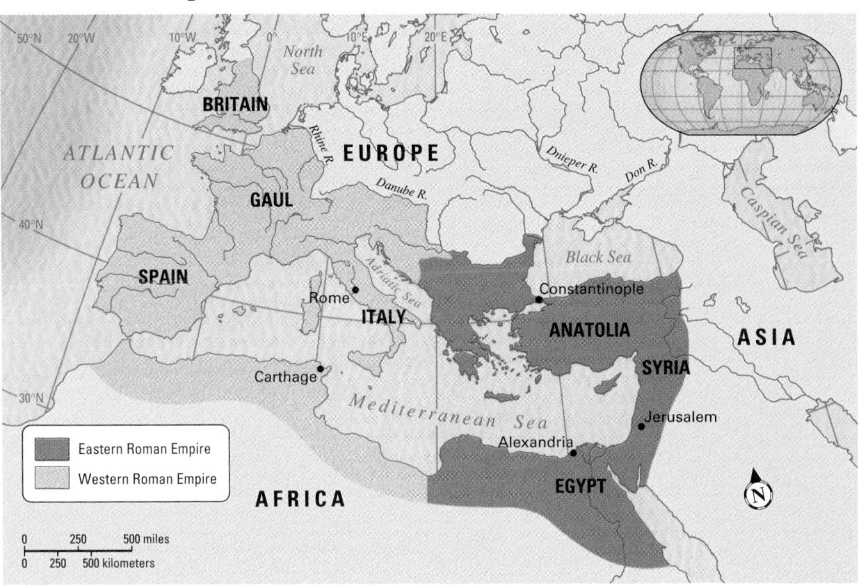

1. Which area was part of the Eastern Roman Empire?
   - **A.** Spain
   - **B.** Gaul
   - **C.** Anatolia
   - **D.** all of the above

2. The most northern country in the Western Roman Empire was
   - **A.** Syria.
   - **B.** Gaul.
   - **C.** Spain.
   - **D.** Britain.

PRACTICE QUESTIONS

Thinking It Through Share the following explanations with students as they discuss the strategies they used to answer the practice questions.

1. *Eastern Roman Empire* are key words. Study the legend and the map to find the Eastern Roman Empire. Compare the answer choices with this area. The correct answer is (C).
2. *Most northern* are key words. To answer this question, use the legend and the map to find the Western Roman Empire. Next, use the North arrow to help you figure out which of the choices is the most northern country in the Western Roman Empire. The correct answer is (D).

Skills Tested in the Items

STRATEGY ITEMS		PRACTICE ITEMS	
Item Number	**Skill Tested**	**Item Number**	**Skill Tested**
1.	Interpreting Maps	**1.**	Interpreting Maps
2.	Interpreting Maps	**2.**	Interpreting Maps

Strategies for Taking Standardized Tests

USING STRATEGIES FOR . . .

Thematic Maps Explain to students that they will do best on test questions by thinking them through carefully and by applying test-taking strategies, such as the following.

1. Use the map labels to find the Great Lakes. Then use the map legend to find the meaning of the color around the Great Lakes. The correct answer is (B).
2. Be careful with questions that contain *not*. This question asks you to find the statement that is *not* true. To do this, check each answer choice with the information shown on the map and eliminate true statements. Since (A), (B), and (C) are all true statements, the correct answer is (D) because this statement is not true.

GENERAL TEST-TAKING TIPS

Share these tips with your students.

- Think positively.
- Tell yourself that you can do it!
- If you have studied for the test, you are prepared to succeed.

STRATEGIES

Thematic Maps

Thematic maps focus on special topics. For example, a thematic map might show a country's natural resources or major battles in a war.

1. Read the title of the map. This will give you the subject and purpose of the map.
2. Read the labels on the map. They give information about the map's subject and purpose. (The labels identify the three European empires.)
3. Study the key or legend to help you understand the symbols and/or colors on the map. (The legend shows the colors that indicate the three European Empires.)
4. Try to make generalizations about information shown on the map.
5. Read the questions. Carefully study the map to find the answers.

answers: 1 (B); 2 (D)

1 **European Empires in North America, 1700**

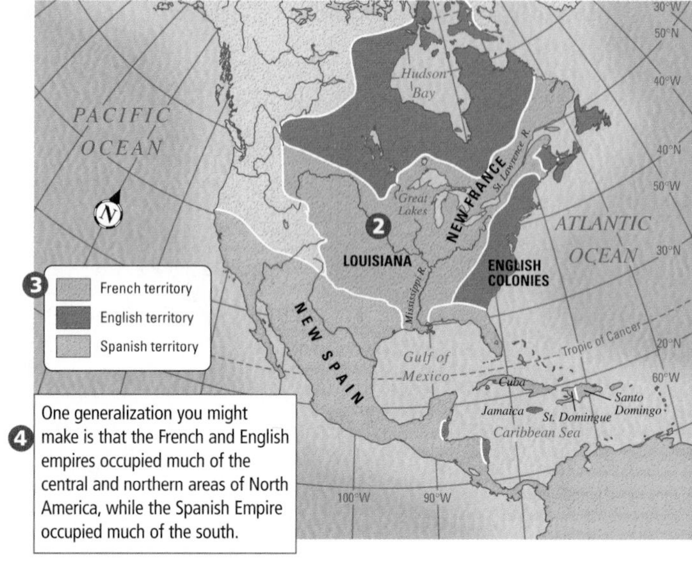

4 One generalization you might make is that the French and English empires occupied much of the central and northern areas of North America, while the Spanish Empire occupied much of the south.

5

1. The land around the Great Lakes was controlled by
   - **A.** England.
   - **B.** France.
   - **C.** Spain.
   - **D.** all of the above

2. Which of the following statements about the English colonial empire in North America is *not* true?
   - **A.** English lands shared borders with both French and Spanish lands.
   - **B.** English lands did not stretch as far west as those held by Spain.
   - **C.** England held no land on the Pacific Coast of North America.
   - **D.** England held no lands east of 60° W longitude.

Activity Options

Individual Needs: Students Acquiring English

Word Meaning Make sure students understand the following terms and concepts on these pages.

Practice Page: Map

conquest: the conquering; the capture of
Christian: referring to one of the major religions of the world
Muslim: referring to one of the major religions of the world

For more test practice online . . .
TEST PRACTICE
CLASSZONE.COM

Directions: Study the map carefully. Use the map and your knowledge of world cultures and geography to answer questions 1 and 2.

The Christian Conquest of Muslim Spain

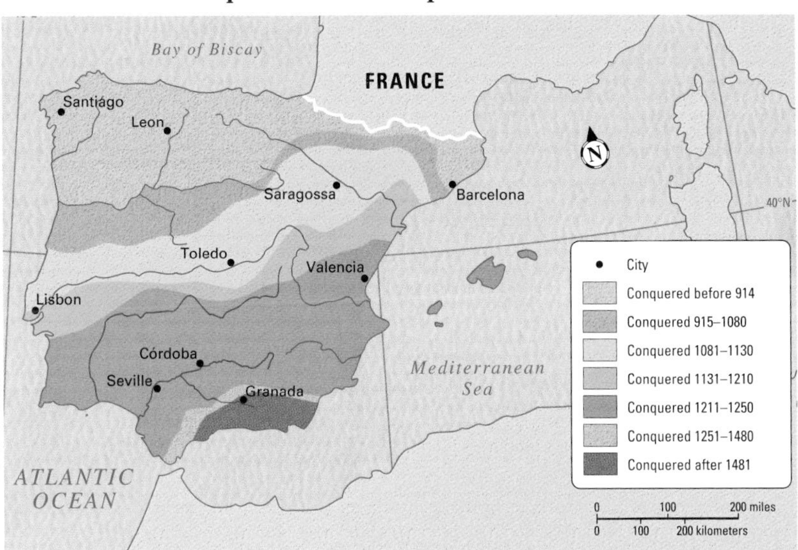

1. By A.D. 1250, how much of Spain did Christians control?

A. only a small portion
B. about one third
C. about one half
D. almost all the land

2. When did Spain recover Granada?

A. 1000
B. 1150
C. 1450
D. 1492

PRACTICE QUESTIONS

Thinking It Through Share the following explanations with students as they discuss the strategies they used to answer the practice questions.

1. *A.D. 1250* are key words in the question. Use the map labels and the legend to determine how much of Spain the Christians controlled by A.D. 1250. The conquered areas cover all but a small part around Granada, so the answer is (D).
2. According to the legend, Granada is located within the area that was not conquered until after 1480, so the correct answer is (D).

Skills Tested in the Items

STRATEGY ITEMS		PRACTICE ITEMS	
Item Number	**Skill Tested**	**Item Number**	**Skill Tested**
1.	Interpreting Maps	**1.**	Interpreting Maps: Comparing
2.	Interpreting Maps: Drawing Conclusions	**2.**	Interpreting Maps

USING STRATEGIES FOR . . .

Time Lines Explain to students that they will do best on test questions by thinking them through carefully and by applying test-taking strategies, such as the following.

1. To answer question 1, you need to read the question and then study the time line to find the answer. Skim the time line for information about Father José María Morelos. The correct answer is (B).
2. To answer question 2, skim the time line for information about each of the leaders listed in the choices. Find the leader who switched sides during the struggle for Mexican independence. The correct answer is (C).

GENERAL TEST-TAKING TIPS

Share these tips with your students.

- Relax during the test.
- Several times during the test, take a few seconds to relax and breathe deeply.
- Occasional deep breaths will help relieve anxiety and keep you focused.

Time Lines

A time line is a chart that lists events in the order in which they occurred. Time lines can be vertical or horizontal.

1. Read the title to learn what subject the time line covers.
2. Note the dates when the time line begins and ends.
3. Read the events in the order they occurred.
4. Think about what else was going on in the world on these dates. Try to make connections.
5. Read the questions. Then carefully study the time line to find the answers.

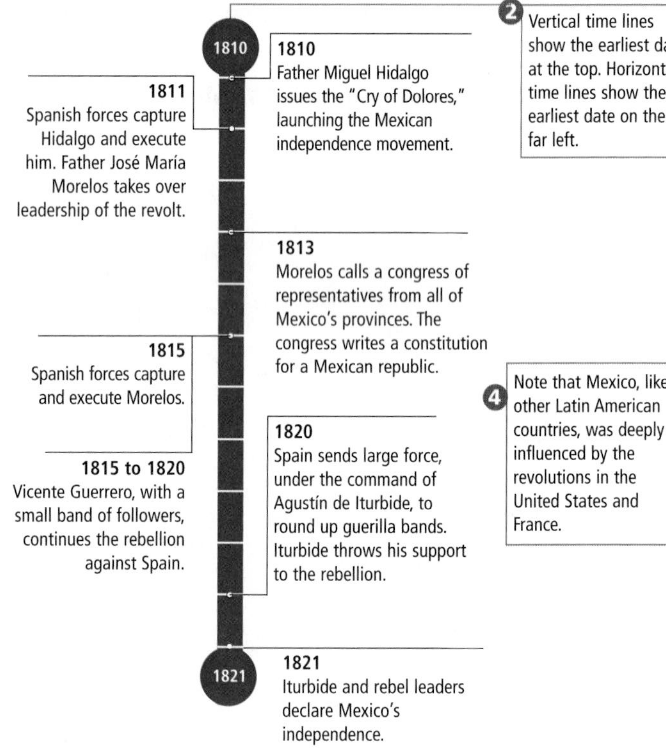

2 Vertical time lines show the earliest date at the top. Horizontal time lines show the earliest date on the far left.
1810
1810 Father Miguel Hidalgo issues the "Cry of Dolores," launching the Mexican independence movement.
1811 Spanish forces capture Hidalgo and execute him. Father José María Morelos takes over leadership of the revolt.
1813 Morelos calls a congress of representatives from all of Mexico's provinces. The congress writes a constitution for a Mexican republic.
1815 Spanish forces capture and execute Morelos.
4 Note that Mexico, like other Latin American countries, was deeply influenced by the revolutions in the United States and France.
1815 to 1820 Vicente Guerrero, with a small band of followers, continues the rebellion against Spain.
1820 Spain sends large force, under the command of Agustín de Iturbide, to round up guerilla bands. Iturbide throws his support to the rebellion.
1821
1821 Iturbide and rebel leaders declare Mexico's independence.

5

1. In which year did Father José María Morelos become the leader of the Mexican independence movement?
   - **A.** 1810
   - **B.** 1811
   - **C.** 1813
   - **D.** 1815

2. Which of the following switched sides during the struggle for Mexican independence?
   - **A.** Vicente Guerrero
   - **B.** Father Miguel Hidalgo
   - **C.** Augustín de Iturbide
   - **D.** José María Morelos

answers: 1 (B); 2 (C)

Activity Options

Individual Needs: Students Acquiring English

Word Meaning Make sure students understand the following terms and concepts on these pages.

Strategy Page: Time Line

congress: meeting
province: a division of a country
constitution: plan of government

guerrilla bands: loosely organized fighting forces that make surprise attacks on enemy troops

Practice Page: Time Line

reforms: changes
constitution: plan of government
hardliners: people who take an uncompromising approach

For more test practice online . . .
TEST PRACTICE
CLASSZONE.COM

Directions: Study the time line. Use the information shown and your knowledge of world cultures and geography to answer questions 1 and 2.

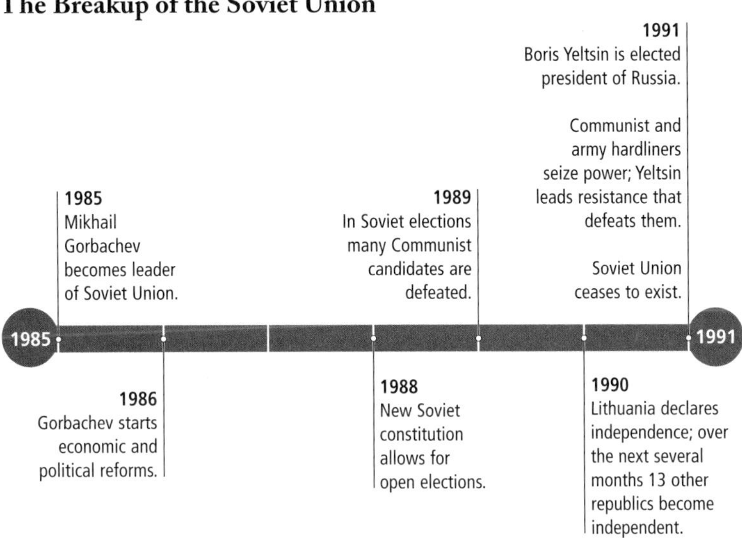
The Breakup of the Soviet Union
1985
1985 Mikhail Gorbachev becomes leader of Soviet Union.
1986 Gorbachev starts economic and political reforms.
1988 New Soviet constitution allows for open elections.
1989 In Soviet elections many Communist candidates are defeated.
1990 Lithuania declares independence; over the next several months 13 other republics become independent.
1991 Boris Yeltsin is elected president of Russia.
Communist and army hardliners seize power; Yeltsin leads resistance that defeats them.
Soviet Union ceases to exist.
1991

1. What happened after Lithuania became independent?
   - **A.** Gorbachev started economic and political reforms.
   - **B.** Many other republics became independent.
   - **C.** A new constitution allowed for open elections.
   - **D.** Gorbachev defeated Yeltsin in a new election.

2. In which year did Communist and army hardliners try to seize power?
   - **A.** 1985
   - **B.** 1988
   - **C.** 1990
   - **D.** 1991

PRACTICE QUESTIONS

Thinking It Through Share the following explanations with students as they discuss the strategies they used to answer the practice questions.

1. *After* is a key word. Find the name *Lithuania* on the time line, and read what happened after Lithuania became independent. The correct answer is (B).
2. To answer question 2, skim the time line to find the date when Communist and army hardliners tried to seize power. The correct answer is (D).

Skills Tested in the Items

STRATEGY ITEMS		PRACTICE ITEMS	
Item Number	**Skill Tested**	**Item Number**	**Skill Tested**
1.	Clarifying	1.	Recognizing Effects
2.	Contrasting	2.	Following Chronological Order

USING STRATEGIES FOR . . .

Constructed Response Remind students of the following:

- To answer the constructed-response questions on this page, you need to use the document and your knowledge of world cultures and geography.
- Some constructed-response questions do not include a document. Instead, all the questions may require you to use only your knowledge of world cultures and geography to answer the questions.
- Sometimes constructed-response questions start with short answer questions and build up to a short essay. The short answers may help you write the short essay, so try to answer the questions in the order they are asked. When your responses are scored, each part will be worth some points, but the short essay will probably be worth more than the short-answer questions.
- Useful information may be found in a title, a caption, or a source line as well as in the document itself.

GENERAL TEST-TAKING TIPS

Share these tips with your students.

- Be sure to answer all parts of constructed-response questions or as many parts as you can. Each part is worth points.
- As you answer each question, make sure that the number of the answer and the number of the question are the same.

STRATEGIES

Constructed Response

Constructed-response questions focus on a document, such as a photograph, cartoon, chart, graph, or time line. Instead of picking one answer from a set of choices, you write a short response. Sometimes, you can find the answer in the document. Other times, you will use what you already know about a subject to answer the question.

1. Read the title of the document to get an idea of what it is about.
2. Study the document.
3. Read the questions carefully. Study the document again to find the answers.
4. Write your answers. You don't need to use complete sentences unless the directions say so.

❶ The Temple of Inscriptions, Palenque, Mexico

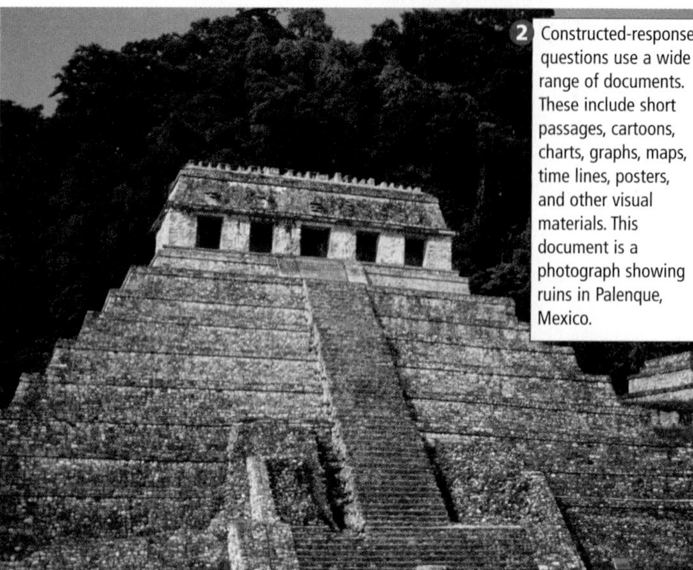

❷ Constructed-response questions use a wide range of documents. These include short passages, cartoons, charts, graphs, maps, time lines, posters, and other visual materials. This document is a photograph showing ruins in Palenque, Mexico.

The Temple of Inscriptions stands over the tomb of Pakal, a great Maya king. The Maya used the temple for religious purposes. Such flat-topped, pyramid-like buildings are typical of Maya architecture.

❸ 1. Palenque, with its massive pyramid structures, was one of the city-states of what civilization?

❹ *Maya*

2. For what purpose was this building used?

religious purposes

3. This civilization went into decline around A.D. 900. Which great civilization arose in its place?

Aztec civilization

S26

Activity Options

Individual Needs: Students Acquiring English

Word Meaning Make sure students understand the following terms and concepts on these pages.

Strategy Page: Document

inscriptions: engravings
Maya: a highly complex civilization in Central America during the early centuries A.D.

Question 1 *city-state:* a city and its surrounding lands that function as an independent political unit

Practice Page: Document

Zlata Filipovic: Make sure students know that Zlata Filipovic was a young girl when she wrote this diary entry
Sarajevo: the capital city of Bosnia and Herzegovina.
Question 1 *Balkans:* the countries occupying the Balkan Peninsula

For more test practice online . . .
TEST PRACTICE
CLASSZONE.COM

Directions: Read the following passage from *Zlata's Diary.* Use the passage and your knowledge of world cultures and geography to answer questions 1 through 3. You do not need to use complete sentences.

Saturday, May 2, 1992

Dear Mimmy,

Today was truly, absolutely the worst day ever in Sarajevo. The shooting started around noon. Mommy and I moved into the hall. Daddy was in his office, under our apartment, at the time. We told him on the intercom to run quickly to the downstairs lobby where we'd meet him… The gunfire was getting worse, and we couldn't get over the wall to the Bobars', so we ran down to our own cellar.

The cellar is ugly, dark, smelly. Mommy, who's terrified of mice, had two fears to cope with. The three of us were in the same corner as the other day. We listened to the pounding shells, the shooting, the thundering noise overhead. We even heard planes. At one moment I realized that this awful cellar was the only place that could save our lives. Suddenly, it started to look almost warm and nice. It was the only way we could defend ourselves against all this terrible shooting. We heard glass shattering in our street. Horrible. I put my fingers in my ears to block out the terrible sounds. I was worried about Cicko. We had left him behind in the lobby. Would he catch cold there? Would something hit him? I was terribly hungry and thirsty. We had left our half-cooked lunch in the kitchen.

—Zlata Filipovic, *Zlata's Diary: A Child's Life in Sarajevo* (1994)

1. In the early 1990s, war broke out in the Balkans. Why were people fighting in Bosnia and Herzegovina?

2. What does Zlata say is happening in the city of Sarajevo?

3. How does the war affect Zlata and her family?

PRACTICE QUESTIONS

Thinking It Through Share the following explanations with students as they discuss the strategies they used to answer the practice questions.

1. Use your knowledge of world cultures and geography to answer the question. The answer is that Serbs in Bosnia wanted to rid the country of Muslims.
2. Read the question and then study the document to find the answer. The answer is that Zlata says the city is being bombed and there is shooting in the streets.
3. You need to study the document to answer this question. The answer is that the family lives in fear of being killed. They must be ready to drop everything to find shelter when the bombing and shooting starts. Things are so bad that even the ugly, dark, smelly cellar starts to look almost warm and nice because hiding in it is the only way to save themselves.

Scoring Constructed-Response Questions

Constructed-response questions usually are scored using a rubric, or scoring guide. The questions on this page might be scored by giving 1 point for each question—a total score of 3 points. Another way of scoring these questions might be to give 1 point for each correct answer for questions 1 and 2, and 2 points for question 3 (1 point for knowing that the family lives in fear of being killed and 1 point for knowing that the family must be ready to drop everything to seek shelter in the cellar)—a total score of 4 points.

Skills Tested in the Items

STRATEGY ITEMS		PRACTICE ITEMS	
Item Number	**Skill Tested**	**Item Number**	**Skill Tested**
1.	Clarifying	**1.**	Analyzing Causes
2.	Analyzing Motives	**2.**	Analyzing Primary Sources
3.	Following Chronological Order	**3.**	Recognizing Effects

USING STRATEGIES FOR . . .

Extended Response Remind students of the following:

- Read the extended-response questions that go with one document before beginning to answer any questions. Look for words that tell you how to organize your answer.
- In question 1, you are asked to complete a chart. You need to apply your knowledge of world cultures to complete the chart.
- In question 2, you need to apply your knowledge of the Industrial Revolution to write the essay. Key words are *changed people's lives.* Jot down your ideas and create an outline on a separate piece of paper. Use this outline to write a short essay to answer the question. Support your main ideas with details and examples.

GENERAL TEST-TAKING TIPS

Share these tips with your students.

- Write in complete sentences whenever appropriate. Extended-response essays require complete sentences.
- Remember, neatness counts! If the scorer cannot read your answer, you will not get credit for it.
- Use correct grammar, punctuation, and spelling to help the scorer understand your answer.

STRATEGIES

Extended Response

Extended-response questions, like constructed-response questions, focus on a document of some kind. However, they are more complicated and require more time to complete. Some extended-response questions ask you to present the information in the document in a different form. You might be asked to present the information in a chart in graph form, for example. Other questions ask you to complete a document such as a chart or graph. Still others require you to apply your knowledge to information in the document to write an essay.

1. Read the title of the document to get an idea of what it is about.
2. Carefully read directions and questions.
3. Study the document.
4. Sometimes the question may give you part of the answer. (The answer given tells how inventions were used and what effects they had on society. Your answers should have the same kind of information.)
5. The question may require you to write an essay. Write down some ideas to use in an outline. Then use your outline to write the essay. (A good essay will contain the ideas shown in the rubric to the right.)

❸ Read the column heads carefully. They offer important clues about the subject of the chart. For instance, the column head "Impact" is a cl about why these inventions were so importan

❶ **Inventions of the Industrial Revolution**

Invention	Impact
Flying shuttle, spinning jenny, water frame, spinning mule, power loom	Spun thread and wove cloth faster; more factories were built and more people were hired ❹
Cotton gin	Cleaned seeds faster from cotton; companies produced more cotton
Macadam road, steamboat, locomotive	Made travel over land and water faster; could carry larger, heavier loads; railroads needed more coal and iron
Mechanical reaper	Made harvesting easier; increased wheat production

❷ **1.** Read the list of inventions in the left-hand column. Then in the right-hand column briefly state what the inventions meant to industry. The first item has been filled in for you.

2. The chart shows how some inventions helped create the Industrial Revolution. Write a short essay describing how the Industrial Revolution changed people's lives.

❺ **Sample Response** The best essays will point out that progress in agriculture meant that fewer people were needed to work the farms. As a result, many farm workers went to the city looking for work in factories. As cities grew, poor sanitation and poor housing made them unhealthy and dangerous places to live. Life for factory workers was hard. They worked long hours under very bad conditions. At first, the Industrial Revolution produced three classes of people: an upper class of landowners and aristocrats; a middle class of merchants and factory owners; and a large lower class of poor people. Over the long term, though, working and living conditions improved, even for the lower class. This was partly because factory goods could be sold at a lower cost. In time, even lower classes could afford to buy many goods and services.

S28

Activity Options

Individual Needs: Students Acquiring English

Word Meaning Make sure students understand the following terms and concepts on these pages.

Strategy Page: Document

Industrial Revolution: radical changes needed to mechanize the making of products and farming

impact: effect

Practice Page: Document

smallpox: a disease that is highly contagious

Aztecs: early Native American civilization of Central America

Question 1 *Inca:* early Native American civilization of South America

For more test practice online . . .
TEST PRACTICE
CLASSZONE.COM

Directions: Use the drawing and passage below and your knowledge of world cultures and geography to answer question 1.

Smallpox Spreads Among the Aztecs

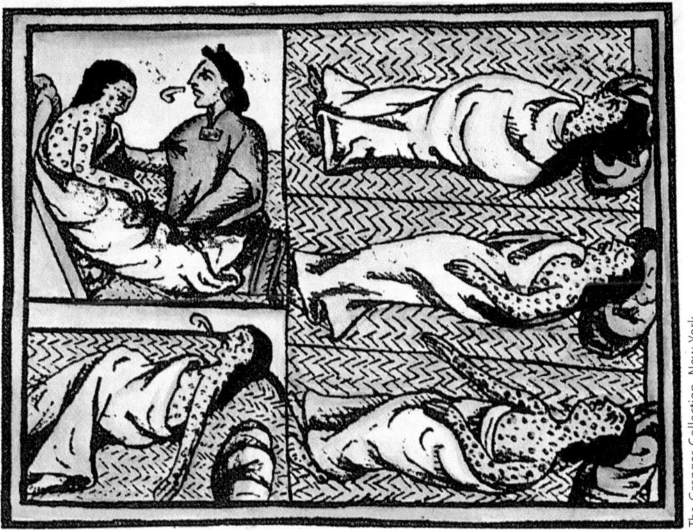

The Granger Collection, New York.

European diseases were like a second "army" of conquerors. Native people had no way to treat diseases like smallpox, typhoid fever, or measles. This "army" was more deadly than swords or guns.

— Based on P. M. Ashburn, *The Ranks of Death* (1947)

1. What role did disease play in the Spanish conquest of the Aztecs and Inca?

PRACTICE QUESTIONS

Thinking It Through Share the following explanations with students as they discuss the strategies they used to answer the practice question.

- Study the illustration and the extended-response question. Note that the question asks you to explain the role of disease in the Spanish conquest of the Aztec and Inca.
- Apply your knowledge of world cultures and geography to write a short essay to answer this question.

The best essays should include the following information:

The role of disease in the Spanish conquest of the Aztec and Inca

- The Europeans brought diseases with them to the Americas. For the most part, the Europeans had become immune to them. The Aztec and Inca, however, had no immunity since their people had never been exposed to these diseases before.
- The Aztec and Inca could do little to stop the spread of European diseases. Many people died. The Aztec and Inca forces were greatly reduced by these diseases.
- The reduced forces of the Aztecs and Inca weakened their resistance. This made it much easier for the Spanish to conquer the Aztec and Inca.

Scoring Extended-Response Questions

Extended-response questions usually are scored using a rubric, or scoring guide. The question on this page might be scored by giving each part 2 points, for a total score of 6 points.

Skills Tested in the Items

STRATEGY ITEMS		PRACTICE ITEMS	
Item Number	**Skill Tested**	**Item Number**	**Skill Tested**
1.	Analyzing Causes; Recognizing Effects	1.	Making Inferences; Recognizing Effects
2.	Recognizing Effects		

USING STRATEGIES FOR . . .

Document-Based Questions Remind students of the following:

- Document-based questions are designed to help you work like a historian. You are given several documents from a variety of sources that you must analyze, evaluate, and synthesize in order to write an essay, much the way a historian would proceed.
- Use the "Introduction" to help you organize your essay. The "Context" gives you the focus of the document-based questions. The document-based questions shown here focus on the common features including the ways of life and the challenges of the countries in South America.
- Use the "Task" section to help you make a graphic organizer such as an outline, chart, or concept web to organize the information for your essay. This "Task" section explains that the essay must: (1) identify the common features that make the countries of South America a culture region; (2) discuss those common features. Make a two-column chart on a piece of scratch paper. Label the columns "Common Features of South American Countries" and "Common Features That Make South America a Culture Region."
- As you answer the "Part 1: Short Answers" questions, also complete the chart.
- To answer the "Part 2: Essay" question, use the documents, the answers to the short-answer questions, the notes in your graphic organizer, and your knowledge of world cultures and geography to help you write the essay.

GENERAL TEST-TAKING TIPS

Share these tips with your students.

- Write legibly.
- Write dark enough for electronic scanners to read.

Document-Based Questions

To answer a document-based question, you have to study more than one document. First you answer questions about each document. Then you use those answers and information from the documents as well as your own knowledge of world cultures and geography to write an essay.

1. Read the "Context" section. It will give you an idea of the topic that will be covered in the question.
2. Read the "Task" section carefully. It tells you what you will need to write about in your essay.
3. Study each document. Think about the connection the documents have to the topic in the "Task" section.
4. Read and answer the questions about each document. Think about how your answers connect to the "Task" section.

Introduction

1 **Context:** South America is made up of 12 very different countries. However, these countries have many features in common. The ways of living in these countries are quite similar. The challenges they face are similar, too.

Task: Identify and discuss the common features that make the countries of South America a culture region.

2 **Part 1: Short Answers**

Study each document carefully. Answer the questions that follow.

3 **Document 1: The Countries of South America**

Country	Main Language	Major Religion	Colonized by	Date Became Independent
Argentina	Spanish	Catholic	Spain	1816
Bolivia	Spanish	Catholic	Spain	1825
Brazil	Portuguese	Catholic	Portugal	1822
Chile	Spanish	Catholic	Spain	1818
Colombia	Spanish	Catholic	Spain	1819
Ecuador	Spanish	Catholic	Spain	1830
Guyana	English	Christian	Great Britain	1966
Paraguay	Spanish, Guarani	Catholic	Spain	1811
Peru	Spanish	Catholic	Spain	1821
Suriname	Dutch	Christian	Netherlands	1975
Uruguay	Spanish	Catholic	Spain	1828
Venezuela	Spanish	Catholic	Spain	1830

4 **What are the dominant language and religion in South America?**

Spanish, Catholicism

Activity Options

Individual Needs: Students Acquiring English

Word Meaning Make sure students understand the following terms and concepts on these pages.

Strategy Pages

Task *culture region:* division of the earth based on a variety of factors, including language, government, social groups, economic systems, and religion

Question 1 *dominant:* main

Document 2 *colonialism:* control by one power over a dependent area or people

Document 2: Government in South America

Oligarchy (government by the few) and military rule have characterized the governments of many of the countries of South America since they won their independence…

Although many South American nations gained freedom in the 1800s, hundreds of years of colonialism had its effects. Strong militaries, underdeveloped economies, and social class divisions still exist in the region today.

—Daniel D. Arreola, et al., *World Geography*

What kinds of governments have ruled in South America for much of the time since independence?

Oligarchies and military governments

Document 3: Crowded São Paulo

Over the last 60 years, many South Americans have moved to the cities. Today, nearly 80 percent of the people live in cities. Close to 18 million people live in the Brazilian city of São Paulo.

What is one of the major challenges facing the countries of South America today?

rapid urbanization and overcrowding of cities

Part 2: Essay

Using information from the documents, your answers to the questions in Part 1, and your knowledge of world cultures and geography, write an essay identifying and discussing the features that make South America a culture region. 6

5 Read the essay question carefully. Then write a brief outline for your essay.

6 Write your essay. The first paragraph should introduce your topic. The middle paragraphs should explain it. The closing paragraph should restate the topic and your conclusion. Support your ideas with quotations or details from the documents. Add other supporting facts or details from your knowledge of world history.

7 A good essay will contain the ideas in the sample response below.

Sample Response The best essays will begin by noting that a culture region is a large geographic area marked by a common culture. Essays will go on to discuss the common historical and cultural background of the countries of South America (Document 1). Essays also will note that most South American countries share similar experiences in political and economic development (Document 2). Finally, essays will point out that the major challenges that South American countries face today are alike (Document 3).

Rubric for DBQ Essay

This sample rubric might be used to score a DBQ essay.

To score a 5, the DBQ essay:
- thoroughly answers all parts of the Task.
- uses data from all documents.
- is supported with relevant facts.
- has outside knowledge.
- is well developed and organized.
- has a strong introduction and conclusion.

To score a 4, the DBQ essay:
- answers all parts of the Task.
- uses data from most documents.
- is supported with relevant facts.
- has outside knowledge.
- is well developed and organized.
- has a good introduction and conclusion.

To score a 3, the DBQ essay:
- answers most parts of the Task.
- uses data from some documents.
- is supported by some relevant facts.
- has little outside knowledge.
- is satisfactorily developed and organized.
- restates the essay theme.

To score a 2, the DBQ essay:
- answers some parts of the Task or all parts in a limited way.
- uses limited data from documents.
- uses few facts to support the essay.
- has little or no outside knowledge.
- is poorly organized.
- has limited or missing intro or conclusion.

To score a 1, the DBQ essay:
- shows limited understanding of the Task.
- uses limited data from documents.
- uses few or no facts or details to support essay.
- has no outside knowledge.
- is poorly organized.
- has limited or missing intro or conclusion.

To score a 0, the DBQ essay:
- does not answer the Task.
- is illegible.
- is blank or missing.

Skills Tested in the Items

STRATEGY ITEMS	
Item Number	**Skill Tested**
Documents 1	Interpreting Charts; Summarizing
Documents 2	Analyzing Secondary Sources
Documents 3	Analyzing Issues
Part 2: Essay	Synthesizing

PRACTICE QUESTIONS

Thinking It Through Share the following explanations with students as they discuss the strategies they used to answer the practice questions.

Part 1: Short Answer

Document 1. Analyze the picture and the caption (*abus:* "injustice"). French poor paid most of the taxes and, therefore, "carried" the nobles and clergy.

Document 2. Skim the source: Natural rights, such as liberty, property, security, and resistance to oppression.

Document 3. Read the timeline: Except for the Bastille, early events were fairly moderate. Nobles lost feudal rights and priests lost land. Later, the king's power was limited. Events became more violent after a mob captured the king and arrested the royal family. After Robespierre gained power, thousands were put to death. This Reign of Terror lasted from 1793–1794.

Part 2: Essay. Share the sample rubric on page S31 with students so they know the criteria they must meet to earn the maximum number of points for this essay. Tell students the following:

- Use the "Introduction" to help you organize your essay. Jot down things you know about the time period or question theme. Use the "Task" to help you make a graphic organizer, such as an outline with three sections: "I. Social Classes," "II. Declaration of Rights," and "III. Revolution Turns Violent." As you answer the short-answer questions, also complete this outline.
- Use the documents, the answers to the short-answer questions, the notes in your graphic organizer, and your knowledge of world cultures and geography to help you write the essay.

PRACTICE

Introduction

Historical Context: For many centuries, kings and queens ruled the countries of Europe. Their power was supported by nobles and armies. European society began to change. In the late 1700s, those changes produced a violent revolution in France.

Task: Discuss how social conflict and new ideas contributed to the French Revolution and why the Revolution turned radical.

Part 1: Short Answers

Study each document carefully. Answer the questions that follow.

Document 1: Social Classes in Pre-Revolutionary France

This cartoon shows a peasant woman carrying women of nobility and the Church. What does the cartoon say about the lives of the poor before the revolution?

S32 Engraving: *Le Grand Abus.* Engraving of a cartoon held in the collection of M. de baron de Vinck d'Orp of Brussels/Mary Evans Picture Library, London.

Activity Options

Individual Needs: Students Acquiring English

Word Meaning Make sure students understand the following terms and concepts on these pages.

Practice Pages

Document 1 *social classes:* social levels

peasant: person in the lowest social level 200 years ago in France

Document 2 *preservation:* protection

resistance to oppression: fighting against the loss of freedoms

Document 2: A Declaration of Rights

> 1. Men are born and remain free and equal in rights…
> 2. The aim of all political association is the preservation of the natural and [unlimited] rights of man. These rights are liberty, property, security, and resistance to oppression…
>
> — *Declaration of the Rights of Man and of the Citizen* (1789)

According to this document, which rights belong to all people?

Document 3: The French Revolution — Major Events

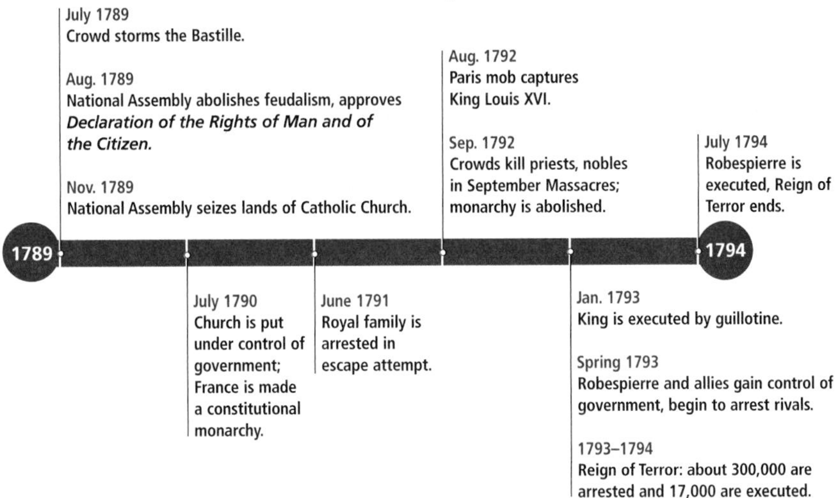

1789
July 1789
Crowd storms the Bastille.
Aug. 1789
National Assembly abolishes feudalism, approves *Declaration of the Rights of Man and of the Citizen.*
Nov. 1789
National Assembly seizes lands of Catholic Church.
July 1790
Church is put under control of government; France is made a constitutional monarchy.
June 1791
Royal family is arrested in escape attempt.
Aug. 1792
Paris mob captures King Louis XVI.
Sep. 1792
Crowds kill priests, nobles in September Massacres; monarchy is abolished.
Jan. 1793
King is executed by guillotine.
Spring 1793
Robespierre and allies gain control of government, begin to arrest rivals.
1793–1794
Reign of Terror: about 300,000 are arrested and 17,000 are executed.
July 1794
Robespierre is executed, Reign of Terror ends.
1794

Over time, the revolution became more violent. How does the information in the timeline show this?

Part 2: Essay

Write an essay discussing how social conflict and new ideas led to the French Revolution and why it became so violent. Use information from the documents, your short answers, and your knowledge of world cultures and geography.

Rubric for Essay

The best essays will address all three parts of the question—how social conflict contributed to the French Revolution, how new ideas led to the French Revolution, and why the Revolution became so violent.

Social Conflict

- Food shortages (outside knowledge)
- Resentment felt by the peasants and members of the Third Estate
    - toward the high taxes they paid
    - toward the privileges of the nobles, the clergy, and the king.

(Documents 1 and 3)

New Ideas

The spread of new ideas promoted rights for all people such as

- Equality
- Natural rights
- Liberty

(Document 2)

Reasons for Increased Violence

- Rumors of conservative reaction against revolutionary advances
- France's war with Austria and Prussia
- The radicalism of the Paris *sansculottes*
- Internal struggles among the revolutionaries

(outside knowledge)

Skills Tested in the Items

PRACTICE ITEMS	
Item Number	**Skill Tested**
Documents 1. 2. 3.	Analyzing Political Cartoons: Making Inferences Analyzing Primary Sources Analyzing Causes; Recognizing Effects
Part 2: Essay	Analyzing Issues Synthesizing

RAND McNALLY

World Atlas

Ab

Contents

Complete Legend for Physical and Political Maps

Symbols

 Lake

 Salt Lake

 Seasonal Lake

 River

\ Waterfall

 Canal

△ Mountain Peak

 Highest Mountain Peak

Cities

■ **Los Angeles** — City over 1,000,000 population

▣ **Calgary** — City of 250,000 to 1,000,000 population

• Haifa — City under 250,000 population

⊛ *Paris* — National Capital

★ **Vancouver** — Secondary Capital (State, Province, or Territory)

Type Styles Used to Name Features

CHINA — Country

O N T A R I O — State, Province, or Territory

PUERTO RICO (U.S.) — Possession

A T L A N T I C O C E A N — Ocean or Sea

A l p s — Physical Feature

Borneo — Island

Boundaries

 International Boundary

Secondary Boundary

Land Elevation and Water Depths

Land Elevation

Meters	Feet
3,000 and over	9,840 and over
2,000 - 3,000	6,560 - 9,840
500 - 2,000	1,640 - 6,560
200 - 500	656 - 1,640
0 - 200	0 - 656

Water Depth

Meters	Feet
Less than 200	Less than 656
200 - 2,000	656 - 6,560
Over 2,000	Over 6,560

ATLAS

A1

ARCTIC OCEAN
Greenland
Jan Mayen
Baffin Island
Baffin Bay
Arctic Circle
Iceland
Faroe Is.
British Isles
London
Mt. McKinley
20,320 Ft.
6,194m
Yukon
Mackenzie
Hudson Bay
Canadian Shield
NORTH AMERICA
Rocky Mountains
Great Plains
Aleutian Islands
Vancouver
Newfoundland
St. Lawrence
Appalachian Mts.
Washington D.C.
Cape Hatteras
Mississippi
Colorado
Los Angeles
Azores
Iberian Peninsula
Atlas Mts.
Canary Islands
Midway Is.
Baja California
Gulf of Mexico
Tropic of Cancer
Hawaiian Islands
ATLANTIC
Yucatan Peninsula
Cuba
Hispaniola
Puerto Rico
Jamaica
Caribbean Sea
Cape Verde Islands
Cape Verde
Niger
PACIFIC
Trinidad
Orinoco
OCEAN
Palmyra
Equator
Galapagos Islands
Amazon
Amazon Basin
SOUTH AMERICA
Kiribati
OCEAN
Marquesas Is.
Andes
Samoa Islands
Mato Grosso Plateau
St. Helena
Tonga Is.
Cook Islands
Tahiti
Tropic of Capricorn
Rio de Janeiro
Easter Island
Paraná
Mt. Aconcagua
22,831 Ft.
6,959m
Buenos Aires
Archipiélago Juan Fernández
N
Chatham Is.
Patagonia
0 1000 2000 Miles
0 1000 2000 3000 Kilometers
Copyright by Rand McNally & Co.
Robinson Projection
Falkland Is.
South Georgia
Tierra del Fuego
Cape Horn
South Sandwich Is.
South Orkney Is.
South Shetland Is.
Antarctic Circle
Antarctic Peninsula
Weddell Sea
Ross Sea
Marie Byrd Land
Vinson Massif
16,066 Ft.
4,897m

A2

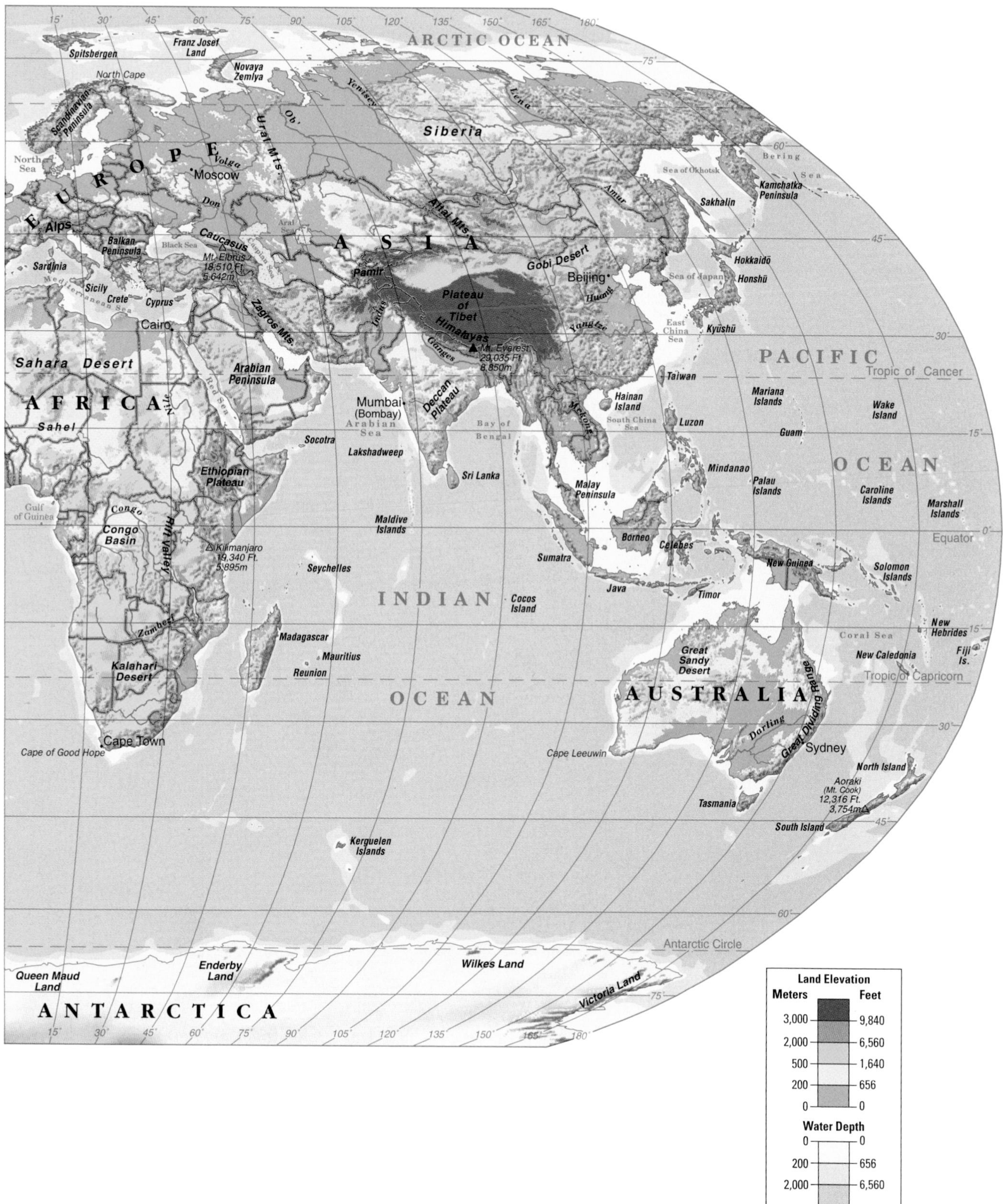
ARCTIC OCEAN
Spitsbergen
Franz Josef Land
Novaya Zemlya
North Cape
Scandinavian Peninsula
North Sea
EUROPE
Moscow
Volga
Don
Ural Mts.
Ob'
Yenisey
Lena
Siberia
Bering Sea
Sea of Okhotsk
Kamchatka Peninsula
Sakhalin
Amur
Altai Mts.
Aral Sea
Caspian Sea
Alps
Balkan Peninsula
Black Sea
Caucasus
Mt. Elbrus 18,510 Ft. 5,642m
ASIA
Pamir
Gobi Desert
Beijing
Hokkaidō
Honshū
Sea of Japan
Huang
Sardinia
Sicily
Crete
Cyprus
Mediterranean Sea
Cairo
Zagros Mts.
Indus
Plateau of Tibet
Himalayas
Ganges
Mt. Everest 29,035 Ft. 8,850m
Yangtze
East China Sea
Kyūshū
PACIFIC
Tropic of Cancer
Sahara Desert
Arabian Peninsula
Red Sea
Nile
AFRICA
Sahel
Mumbai (Bombay)
Arabian Sea
Deccan Plateau
Bay of Bengal
Mekong
Taiwan
Hainan Island
South China Sea
Luzon
Mariana Islands
Wake Island
Guam
Socotra
Lakshadweep
Sri Lanka
Mindanao
Palau Islands
OCEAN
Caroline Islands
Marshall Islands
Ethiopian Plateau
Gulf of Guinea
Congo
Congo Basin
Rift Valley
Kilimanjaro 19,340 Ft. 5,895m
Maldive Islands
Malay Peninsula
Borneo
Celebes
Equator
Seychelles
Sumatra
New Guinea
Solomon Islands
Java
Timor
INDIAN
Cocos Island
Zambezi
Madagascar
Mauritius
Reunion
Coral Sea
New Hebrides
New Caledonia
Fiji Is.
Great Sandy Desert
Tropic of Capricorn
Kalahari Desert
OCEAN
AUSTRALIA
Great Dividing Range
Darling
Sydney
Cape Town
Cape of Good Hope
Cape Leeuwin
North Island
Aoraki (Mt. Cook) 12,316 Ft. 3,754m
Tasmania
South Island
Kerguelen Islands
Antarctic Circle
Queen Maud Land
Enderby Land
Wilkes Land
Victoria Land
ANTARCTICA
Land Elevation
Meters Feet
3,000 9,840
2,000 6,560
500 1,640
200 656
0 0
Water Depth
0 0
200 656
2,000 6,560

ATLAS

RAND McNALLY

A3

ARCTIC OCEAN
RUSSIA
ALASKA
Yukon (U.S.)
Anchorage
Aleutian Islands
CANADA
Baffin Bay
Hudson Bay
GREENLAND (Den.)
Arctic Circle
ICELAND
FAROE IS. (Den.)
UNITED KINGDOM
IRELAND
London
FRANCE
Newfoundland
Vancouver
Missouri
Montreal
Ottawa
Chicago
New York
Washington D.C.
UNITED STATES
Colorado
Mississippi
Los Angeles
Houston
Azores (Port.)
PORTUGAL
Madrid
SPAIN
Casablanca
MOROCCO
Canary Islands (Sp.)
W. SAHARA
MIDWAY IS. (U.S.)
Gulf of Mexico
MEXICO
ATLANTIC
Tropic of Cancer
BAHAMAS
Hawaiian Islands (U.S)
CUBA
HAITI
DOM. REP.
PUERTO RICO (U.S.)
Mexico City
BELIZE
JAMAICA
Caribbean Sea
GUAT.
HOND.
EL SAL.
NIC.
CAPE VERDE
MAURITANIA
MALI
SENEGAL
GAMBIA
GUINEA-BISSAU
GUINEA
Niger
BURK. FASO
SIERRA LEONE
COTE D'IVOIRE
GHANA
LIBERIA
PACIFIC
Caracas
TRINIDAD AND TOBAGO
COSTA RICA
PANAMA
VENEZUELA
GUYANA
SURINAME
FRENCH GUIANA
COLOMBIA
Equator
KIRIBATI
Galapagos Islands (Ecuador)
ECUADOR
Amazon
BRAZIL
PERU
Lima
OCEAN
OCEAN
SAMOA
AMERICAN SAMOA
COOK ISLANDS (N.Z.)
TONGA
FRENCH POLYNESIA
ST. HELENA (U.K.)
BOLIVIA
PARAGUAY
Rio de Janeiro
Tropic of Capricorn
Easter Island (Chile)
ARGENTINA
URUGUAY
Santiago
Buenos Aires
CHILE
N
0 1000 2000 Miles
0 1000 2000 3000 Kilometers
Copyright by Rand McNally & Co.
Robinson Projection
FALKLAND IS. (U.K.)
South Georgia (U.K.)
South Orkney Is. (U.K.)
South Shetland Is. (U.K.)
Antarctic Circle
Weddell Sea
180° 165° 150° 135° 120° 105° 90° 75° 60° 45° 30° 15°
75° 60° 45° 30° 15° 0° 15° 30° 45° 60° 75°

 A4

⊛ National Capital
• Major Cities

A5

180° 150° 120° 90° 60° 30°

75° 60° 45° 30° 15° 0° 15° 30° 75°

Arctic

Tropic of Cancer

Equator

Tropic of Capricorn

Antarctic Circle

Vancouver

Montréal

Chicago

New York

Los Angeles

Houston

Mexico City

Caracas

Rio de Janeiro

Buenos Aires

London

Madrid

PACIFIC OCEAN

ATLANTIC OCEAN

N

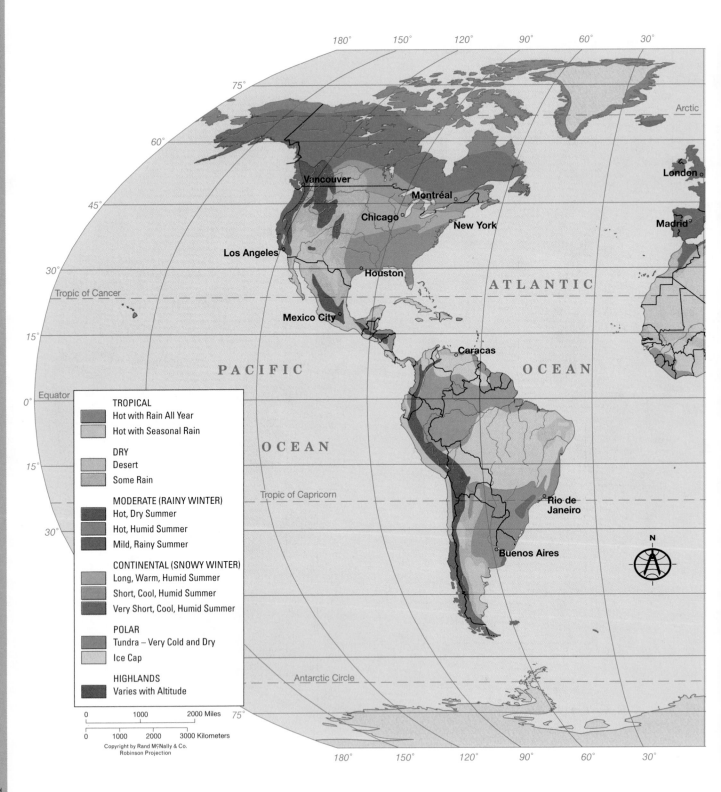

TROPICAL
- Hot with Rain All Year
- Hot with Seasonal Rain

DRY
- Desert
- Some Rain

MODERATE (RAINY WINTER)
- Hot, Dry Summer
- Hot, Humid Summer
- Mild, Rainy Summer

CONTINENTAL (SNOWY WINTER)
- Long, Warm, Humid Summer
- Short, Cool, Humid Summer
- Very Short, Cool, Humid Summer

POLAR
- Tundra – Very Cold and Dry
- Ice Cap

HIGHLANDS
- Varies with Altitude

0 1000 2000 Miles

0 1000 2000 3000 Kilometers

Robinson Projection

180° 150° 120° 90° 60° 30°

A6

30° 60° 90° 120° 150° 180°

ARCTIC OCEAN

75°

Circle

Stockholm

60°

Moscow

Paris

45°

Rome

Beijing

Algiers

Tehrān

Tōkyō

Cairo

PACIFIC

30°

Tropic of Cancer

Mumbai (Bombay)

15°

Bangkok

OCEAN

Lagos

Equator

0°

Nairobi

Jakarta

INDIAN

15°

Tropic of Capricorn

Johannesburg

OCEAN

30°

Sydney

Melbourne

45°

60°

Antarctic Circle

75°

30° 60° 90° 120° 150° 180°

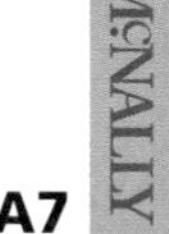

A7

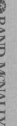

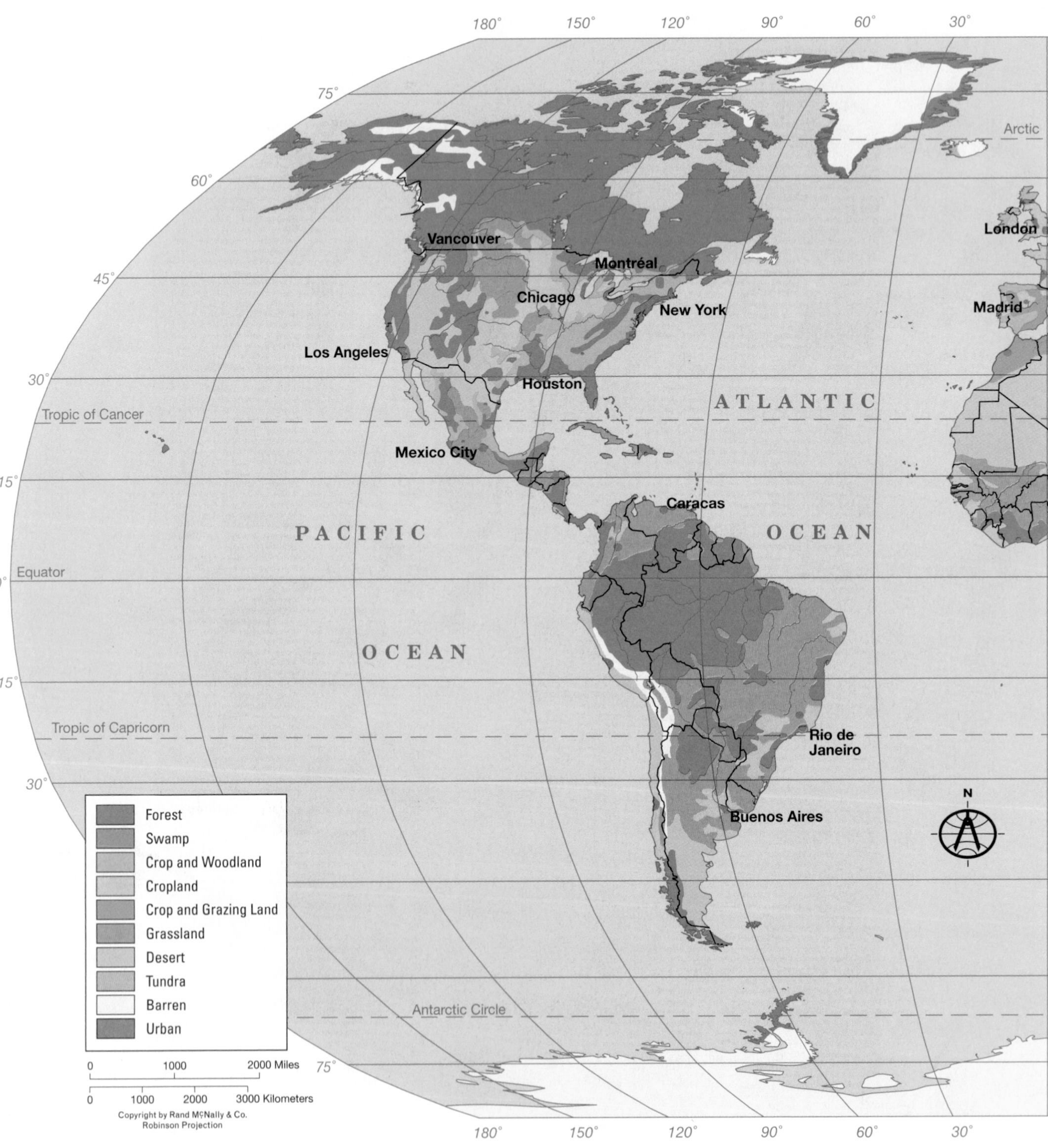
180°
150°
120°
90°
60°
30°
75°
60°
45°
30°
15°
0°
15°
30°
75°
Arctic
Tropic of Cancer
Equator
Tropic of Capricorn
Antarctic Circle
Vancouver
Montréal
Chicago
New York
London
Madrid
Los Angeles
Houston
Mexico City
Caracas
Rio de Janeiro
Buenos Aires
ATLANTIC
OCEAN
PACIFIC
OCEAN
N
Forest
Swamp
Crop and Woodland
Cropland
Crop and Grazing Land
Grassland
Desert
Tundra
Barren
Urban
0
1000
2000 Miles
0
1000
2000
3000 Kilometers
Robinson Projection

A8

30°
60°
90°
120°
150°
180°
ARCTIC OCEAN
75°
Circle
Stockholm
60°
Moscow
Paris
45°
Rome
Beijing
Algiers
Tehrān
Tōkyō
Cairo
PACIFIC
30°
Tropic of Cancer
Mumbai
(Bombay)
15°
Bangkok
OCEAN
Lagos
Equator
0°
Nairobi
Jakarta
INDIAN
15°
Tropic of Capricorn
OCEAN
Johannesburg
30°
Sydney
Melbourne
45°
60°
Antarctic Circle
75°
30°
60°
90°
120°
150°
180°

A9

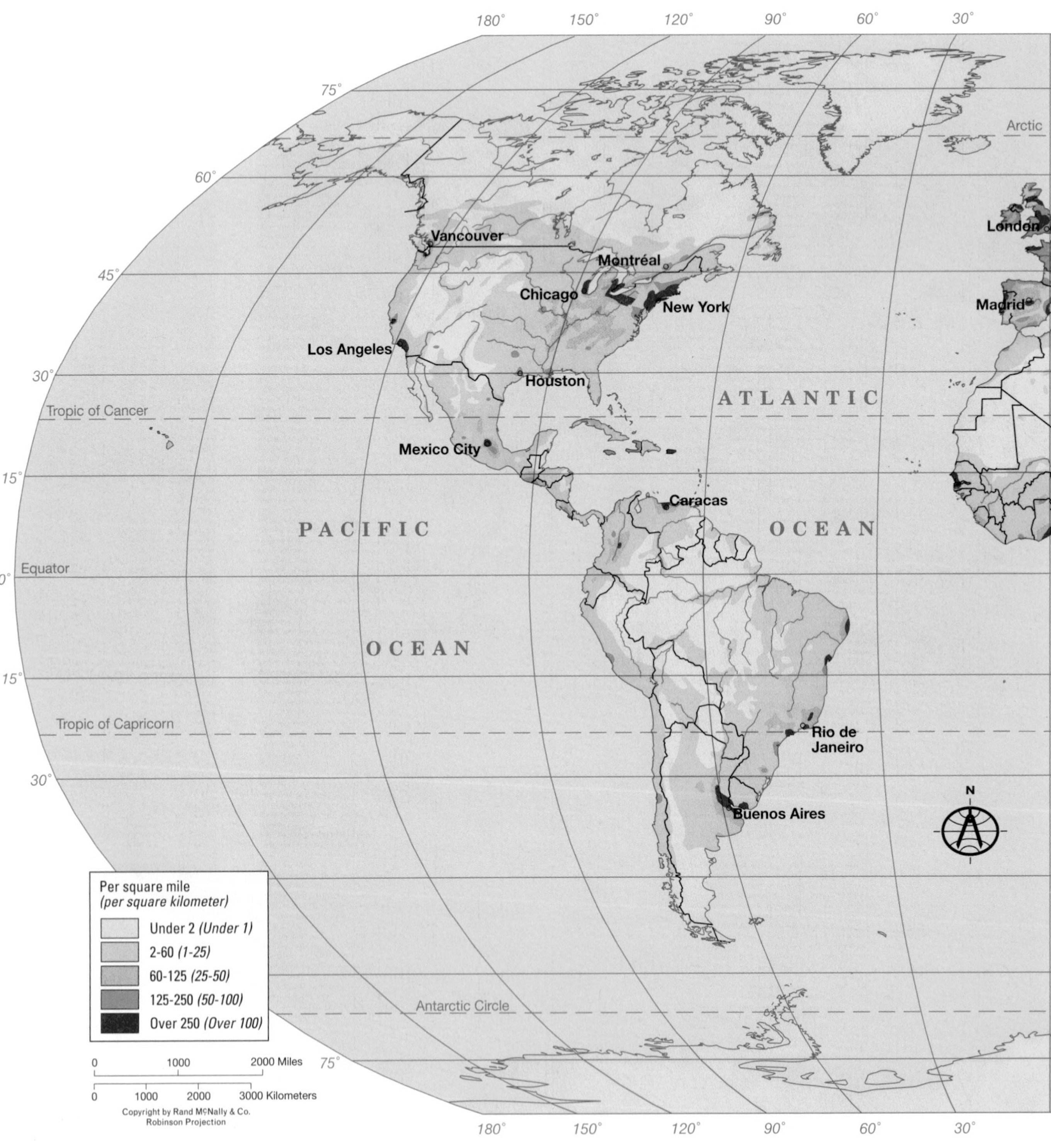
180°
150°
120°
90°
60°
30°
75°
60°
45°
30°
15°
0°
15°
30°
75°
Arctic
Tropic of Cancer
Equator
Tropic of Capricorn
Antarctic Circle
Vancouver
Montréal
Chicago
New York
Los Angeles
Houston
Mexico City
Caracas
Rio de Janeiro
Buenos Aires
London
Madrid
ATLANTIC
OCEAN
PACIFIC
OCEAN
N
Per square mile
(per square kilometer)
Under 2 (Under 1)
2-60 (1-25)
60-125 (25-50)
125-250 (50-100)
Over 250 (Over 100)
0 1000 2000 Miles
0 1000 2000 3000 Kilometers
Copyright by Rand McNally & Co.
Robinson Projection

A10

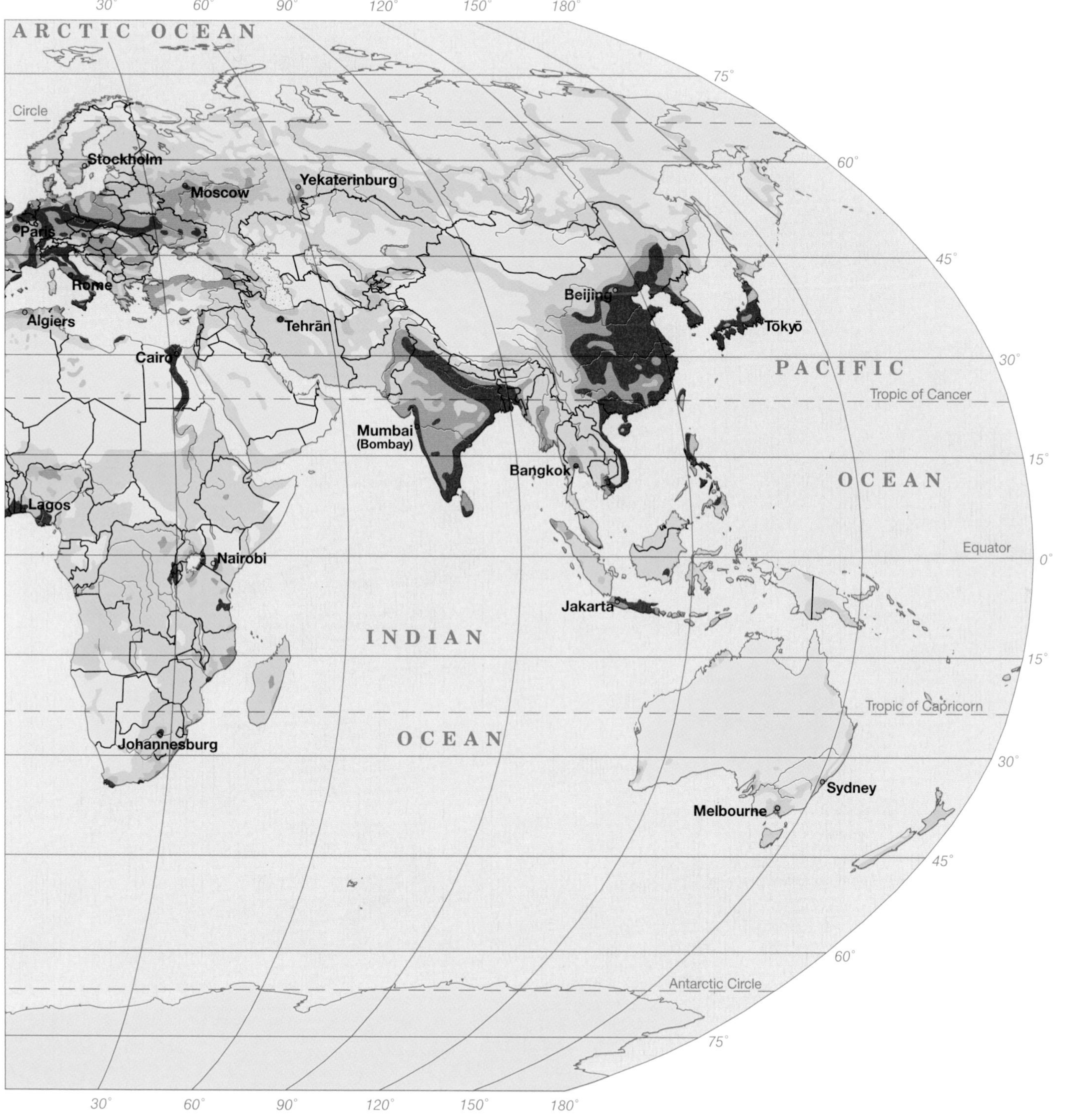
ARCTIC OCEAN
Circle
Stockholm
Moscow
Yekaterinburg
Paris
Rome
Algiers
Tehrān
Cairo
Beijing
Tōkyō
PACIFIC
Tropic of Cancer
Mumbai
(Bombay)
Bangkok
OCEAN
Lagos
Equator
Nairobi
Jakarta
INDIAN
Tropic of Capricorn
OCEAN
Johannesburg
Sydney
Melbourne
Antarctic Circle
30°
60°
90°
120°
150°
180°
75°
60°
45°
30°
15°
0°
15°
30°
45°
60°
75°

A11

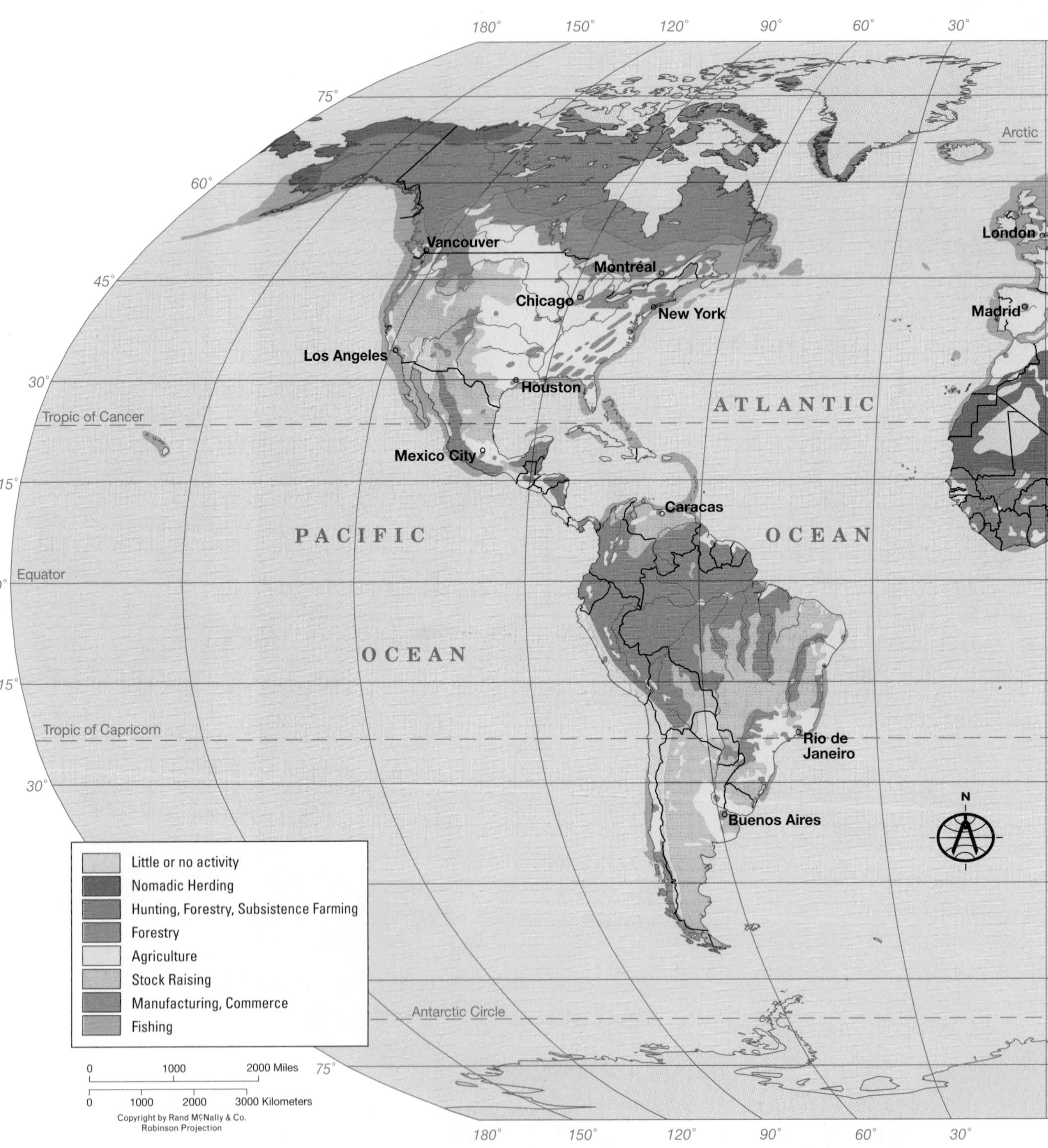
180°
150°
120°
90°
60°
30°
75°
Arctic
60°
London
Vancouver
Montréal
45°
Chicago
New York
Madrid
Los Angeles
30°
Houston
ATLANTIC
Tropic of Cancer
Mexico City
15°
Caracas
PACIFIC
OCEAN
0°
Equator
OCEAN
15°
Tropic of Capricorn
Rio de Janeiro
30°
Buenos Aires
N
Little or no activity
Nomadic Herding
Hunting, Forestry, Subsistence Farming
Forestry
Agriculture
Stock Raising
Manufacturing, Commerce
Fishing
Antarctic Circle
0
1000
2000 Miles
75°
0
1000
2000
3000 Kilometers
Copyright by Rand McNally & Co.
Robinson Projection

A12

30° 60° 90° 120° 150° 180°
ARCTIC OCEAN
Circle
75°
60°
45°
30°
15°
0°
Stockholm
Moscow
Paris
Rome
Algiers
Tehrān
Cairo
Beijing
Tōkyō
PACIFIC
Tropic of Cancer
Mumbai
(Bombay)
Bangkok
OCEAN
Lagos
Nairobi
Equator
Jakarta
INDIAN
OCEAN
Tropic of Capricorn
Johannesburg
Sydney
Melbourne
Antarctic Circle

A13

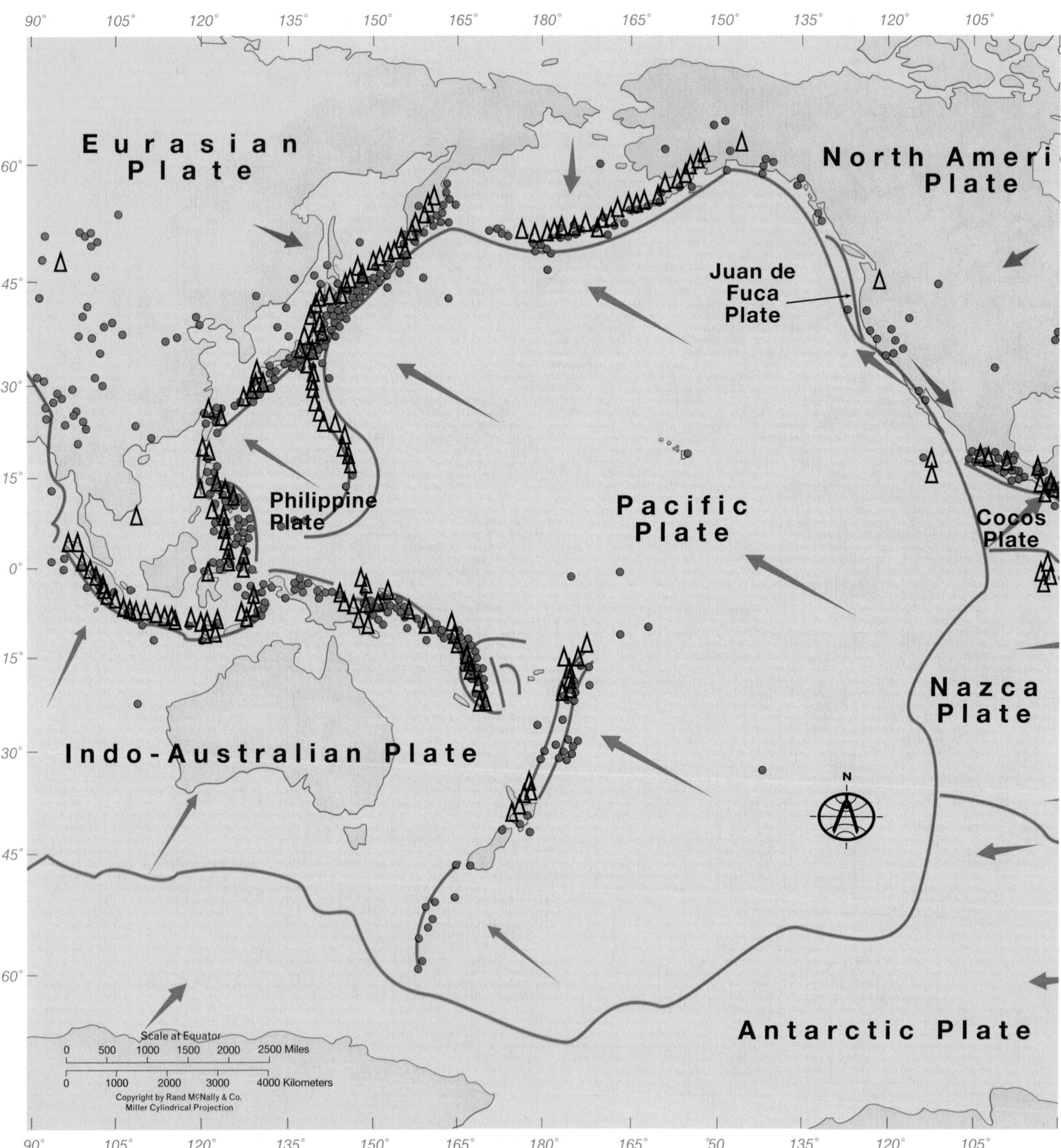
Eurasian Plate
North American Plate
Juan de Fuca Plate
Philippine Plate
Pacific Plate
Cocos Plate
Nazca Plate
Indo-Australian Plate
Antarctic Plate
Scale at Equator
0 500 1000 1500 2000 2500 Miles
0 1000 2000 3000 4000 Kilometers
Copyright by Rand McNally & Co.
Miller Cylindrical Projection

75° 60° 45° 30° 15° 0° 15° 30° 45° 60° 75° 90°

can

Eurasian Plate

Caribbean Plate

Arabian Plate

African Plate

Indo-Australian Plate

South American Plate

Antarctic Plate

Scotia Plate

60° 45° 30° 15° 0° 15° 30° 45° 60°

Volcanic eruptions since 1900

Earthquakes of 7.7 magnitude and above since 10 A.D.

Directions of plate movement

75° 60° 45° 30° 15° 0° 15° 30° 45° 60° 75° 90°

A15

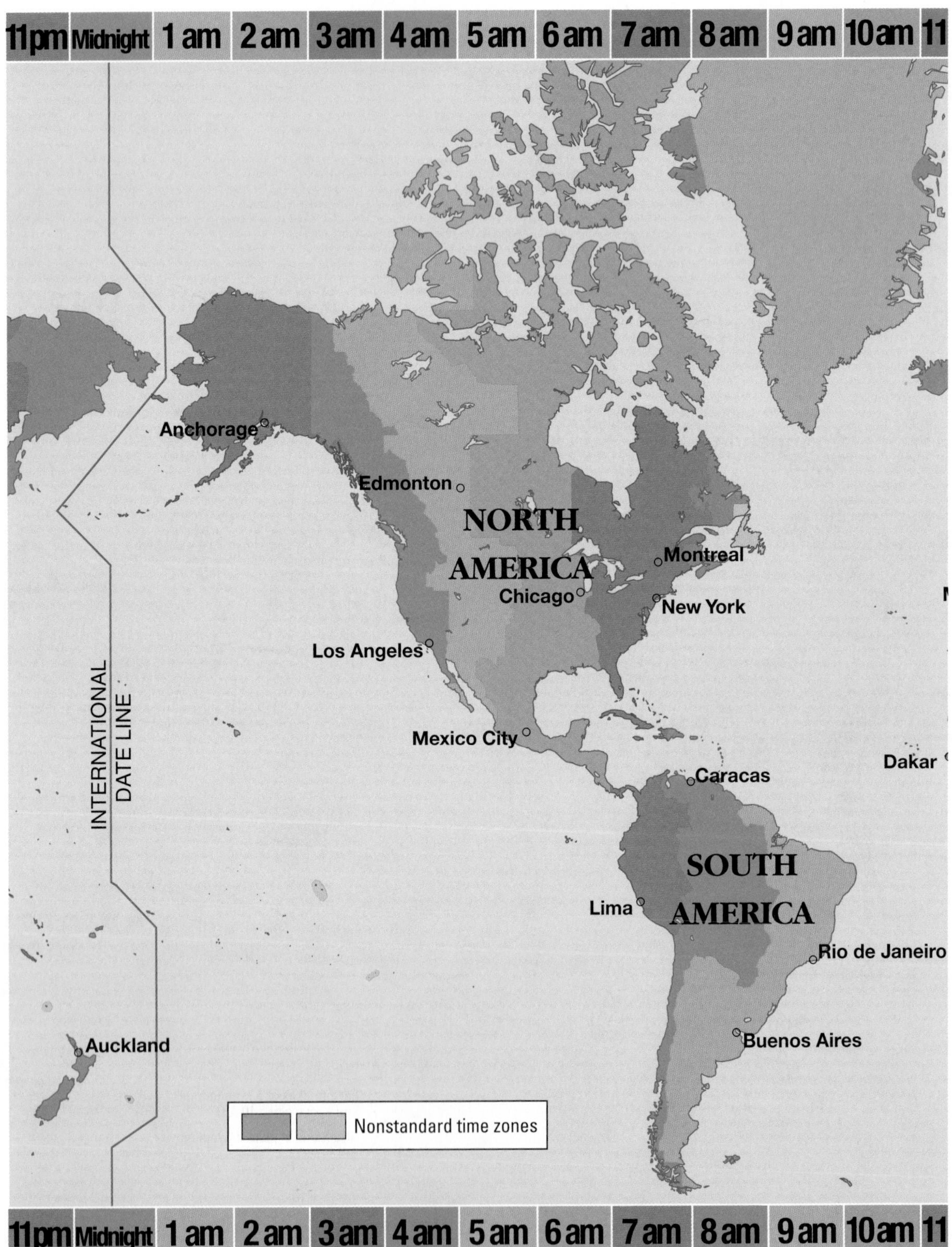
11pm
Midnight
1 am
2 am
3 am
4 am
5 am
6 am
7 am
8 am
9 am
10 am
11
INTERNATIONAL DATE LINE
Anchorage
Edmonton
NORTH AMERICA
Montreal
Chicago
New York
Los Angeles
Mexico City
Caracas
Dakar
SOUTH AMERICA
Lima
Rio de Janeiro
Buenos Aires
Auckland
Nonstandard time zones

A16

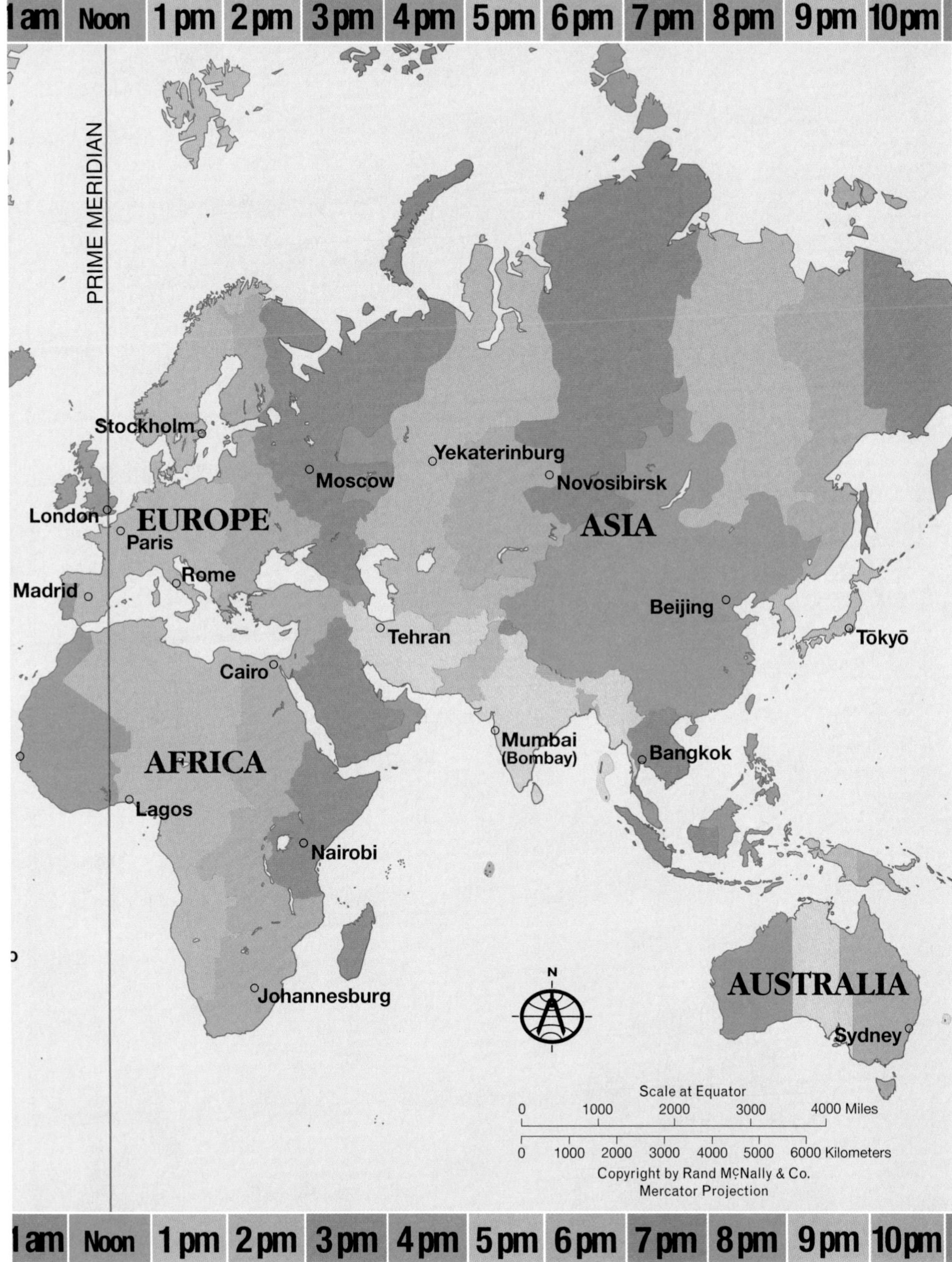
1 am
Noon
1 pm
2 pm
3 pm
4 pm
5 pm
6 pm
7 pm
8 pm
9 pm
10 pm
PRIME MERIDIAN
Stockholm
Moscow
Yekaterinburg
Novosibirsk
London
EUROPE
Paris
ASIA
Madrid
Rome
Beijing
Tehran
Tōkyō
Cairo
Mumbai
(Bombay)
Bangkok
AFRICA
Lagos
Nairobi
Johannesburg
N
AUSTRALIA
Sydney
Scale at Equator
0 1000 2000 3000 4000 Miles
0 1000 2000 3000 4000 5000 6000 Kilometers
Copyright by Rand McNally & Co.
Mercator Projection
1 am
Noon
1 pm
2 pm
3 pm
4 pm
5 pm
6 pm
7 pm
8 pm
9 pm
10 pm

A17

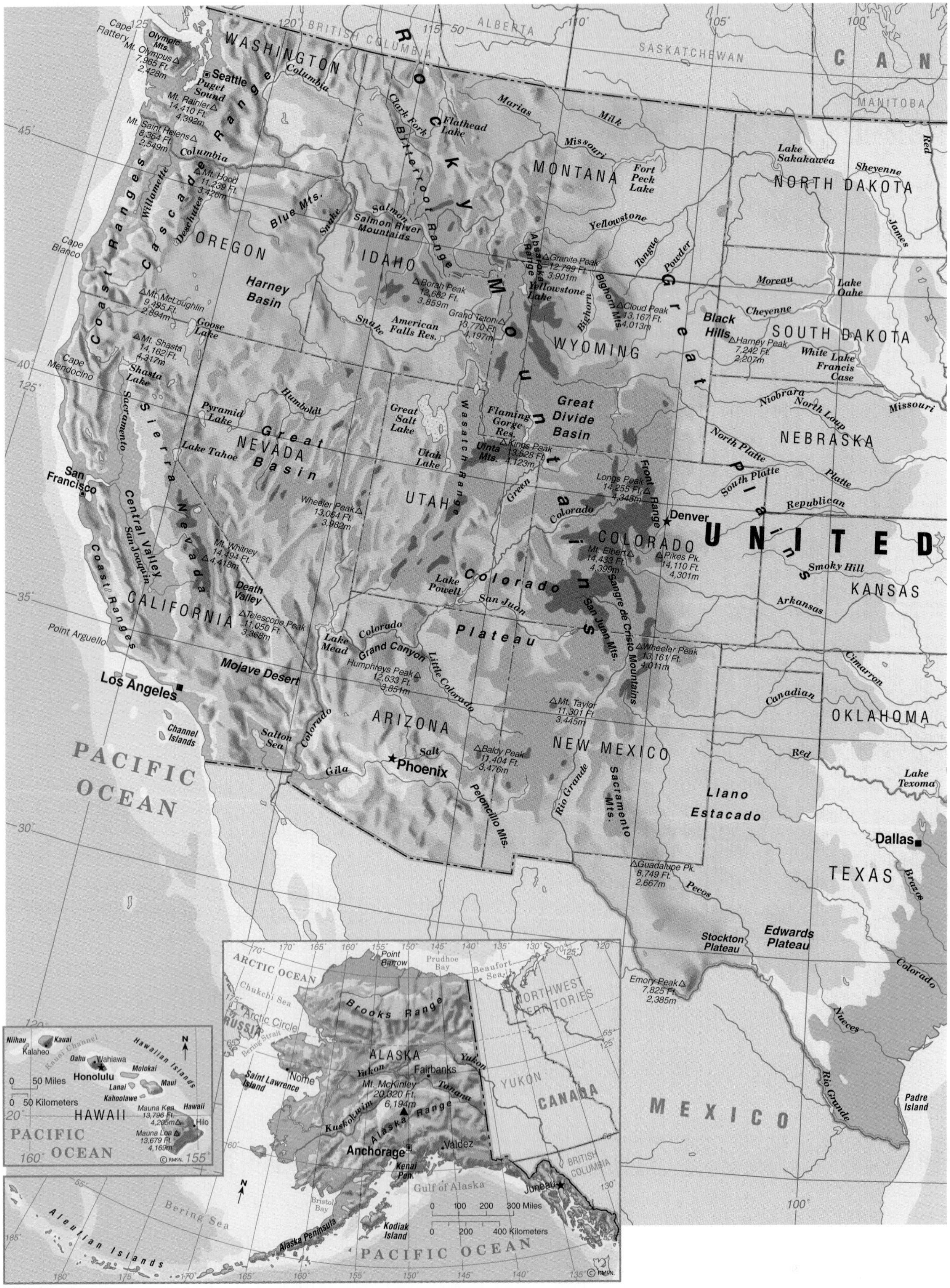

WASHINGTON
OREGON
IDAHO
MONTANA
WYOMING
NORTH DAKOTA
SOUTH DAKOTA
NEBRASKA
KANSAS
COLORADO
UTAH
NEVADA
CALIFORNIA
ARIZONA
NEW MEXICO
OKLAHOMA
TEXAS
UNITED
MEXICO
CAN
BRITISH COLUMBIA
ALBERTA
SASKATCHEWAN
MANITOBA
PACIFIC OCEAN
Rocky Mountains
Great Plains
Great Basin
Colorado Plateau
Great Divide Basin
Cascade Range
Coast Ranges
Sierra Nevada
Olympic Mts.
Mt. Olympus 7,965 Ft. 2,428m
Cape Flattery
Seattle
Puget Sound
Mt. Rainier 14,410 Ft. 4,392m
Mt. Saint Helens 8,364 Ft. 2,549m
Mt. Hood 11,239 Ft. 3,426m
Columbia
Willamette
Deschutes
Blue Mts.
Snake
Salmon
Salmon River Mountains
Bitterroot Range
Clark Fork
Flathead Lake
Marias
Milk
Missouri
Fort Peck Lake
Yellowstone
Yellowstone Lake
Absaroka Range
Granite Peak 12,799 Ft. 3,901m
Borah Peak 12,662 Ft. 3,859m
Grand Teton 13,770 Ft. 4,197m
Cloud Peak 13,167 Ft. 4,013m
Bighorn
Bighorn Mts.
Tongue
Powder
American Falls Res.
Harney Basin
Cape Blanco
Mt. McLoughlin 9,495 Ft. 2,894m
Goose Lake
Mt. Shasta 14,162 Ft. 4,317m
Shasta Lake
Cape Mendocino
Sacramento
Pyramid Lake
Humboldt
Lake Tahoe
Great Salt Lake
Utah Lake
Wasatch Range
Uinta Mts.
Flaming Gorge Res.
Kings Peak 13,528 Ft. 4,123m
Green
Wheeler Peak 13,064 Ft. 3,982m
San Francisco
Central Valley
San Joaquin
Mt. Whitney 14,494 Ft. 4,418m
Death Valley
Telescope Peak 11,050 Ft. 3,368m
Point Arguello
Lake Mead
Lake Powell
San Juan
Colorado
Grand Canyon
Little Colorado
Humphreys Peak 12,633 Ft. 3,851m
Mojave Desert
Los Angeles
Channel Islands
Salton Sea
Gila
Salt
Phoenix
Peloncillo Mts.
Baldy Peak 11,404 Ft. 3,476m
Mt. Taylor 11,301 Ft. 3,445m
Rio Grande
Sacramento Mts.
Sangre de Cristo Mountains
San Juan Mts.
Wheeler Peak 13,161 Ft. 4,011m
Front Range
Longs Peak 14,255 Ft. 4,345m
Denver
Mt. Elbert 14,433 Ft. 4,399m
Pikes Pk. 14,110 Ft. 4,301m
Guadalupe Pk. 8,749 Ft. 2,667m
Emory Peak 7,825 Ft. 2,385m
Pecos
Stockton Plateau
Edwards Plateau
Llano Estacado
Dallas
Brazos
Colorado
Nueces
Padre Island
Red
Lake Texoma
Canadian
Cimarron
Arkansas
Smoky Hill
Republican
Platte
North Platte
South Platte
North Loup
Niobrara
Missouri
Black Hills
Harney Peak 7,242 Ft. 2,207m
White
Lake Francis Case
Cheyenne
Moreau
Lake Oahe
James
Sheyenne
Lake Sakakawea

Niihau
Kauai
Kalaheo
Kauai Channel
Oahu
Wahiawa
Honolulu
Molokai
Lanai
Maui
Kahoolawe
Hawaiian Islands
HAWAII
Hawaii
Mauna Kea 13,796 Ft. 4,205m
Mauna Loa 13,679 Ft. 4,169m
Hilo
PACIFIC OCEAN
0 50 Miles
0 50 Kilometers

ARCTIC OCEAN
Chukchi Sea
Arctic Circle
RUSSIA
Bering Strait
Point Barrow
Prudhoe Bay
Beaufort Sea
NORTHWEST TERRITORIES
Brooks Range
ALASKA
Yukon
Fairbanks
Nome
Saint Lawrence Island
Mt. McKinley 20,320 Ft. 6,194m
Tanana
Kuskokwim
Alaska Range
Anchorage
Valdez
Kenai Pen.
Gulf of Alaska
Juneau
BRITISH COLUMBIA
YUKON
CANADA
Bering Sea
Bristol Bay
Aleutian Islands
Alaska Peninsula
Kodiak Island
PACIFIC OCEAN
0 100 200 300 Miles
0 200 400 Kilometers
© RMcN.

A18

ATLAS

RAND McNALLY

A19

ONTARIO
QUÉBEC
NEW BRUNSWICK
Lake Winnipeg
Lake of the Woods
Lake Nipigon
International Falls
Isle Royale
Lake Superior
Duluth
MINNESOTA
St. Cloud
Minneapolis
St. Paul
Mankato
Rochester
Marquette
Sault Ste. Marie
MICHIGAN
Georgian Bay
Lake Huron
Ottawa
Montréal
Lake Champlain
VERMONT
Burlington
Montpelier
Connecticut
NEW HAMPSHIRE
Concord
Manchester
MAINE
Moosehead Lake
Bangor
Augusta
Portland
Gulf of Maine
Eau Claire
WISCONSIN
Green Bay
Appleton
OshKosh
Sheboygan
Milwaukee
Madison
Racine
Lake Michigan
Traverse City
Saginaw
Flint
Grand Rapids
Lansing
Detroit
Kalamazoo
Ann Arbor
Toronto
Lake Ontario
Rochester
Syracuse
Watertown
NEW YORK
Albany
Hudson
Buffalo
Binghamton
MASSACHUSETTS
Boston
Worcester
CONNECTICUT
Hartford
R.I.
Providence
Bridgeport
Long Island
Nantucket Island
New York
Newark
Lake Erie
Erie
Des Moines
Sioux City
Waterloo
Dubuque
Cedar Rapids
IOWA
Des Moines
Davenport
Omaha
Rockford
Aurora
Chicago
Gary
South Bend
Toledo
Cleveland
Oil City
PENNSYLVANIA
Scranton
Allentown
Trenton
Philadelphia
Akron
Youngstown
Fort Wayne
Lima
OHIO
Columbus
Pittsburgh
Harrisburg
Susquehanna
Wilmington
NEW JERSEY
Moline
Peoria
Illinois
Bloomington
INDIANA
Muncie
Indianapolis
Springfield
Dayton
Baltimore
Dover
Delaware Bay
DELAWARE
Mississippi
Missouri
St. Joseph
Kansas City
Topeka
Kansas City
Springfield
Decatur
ILLINOIS
Terre Haute
Bloomington
Cincinnati
Ohio
WEST VIRGINIA
Washington D.C.
Annapolis
MARYLAND
ATLANTIC OCEAN
Columbia
Jefferson City
Missouri
St. Louis
Emporia
Wabash
Evansville
Louisville
Frankfort
Lexington
Huntington
Charleston
VIRGINIA
Richmond
Newport News
Norfolk
Virginia Beach
Owensboro
KENTUCKY
Roanoke
Albemarle Sound
MISSOURI
Cape Girardeau
Springfield
Clarksville
Cumberland
Johnson City
Winston-Salem
Greensboro
Durham
Raleigh
Pamlico Sound
Tulsa
Fayetteville
Muskogee
Kentucky Lake
Nashville
Knoxville
Asheville
NORTH CAROLINA
Charlotte
Jonesboro
TENNESSEE
Memphis
Chattanooga
Greenville
SOUTH CAROLINA
Columbia
Wilmington
Fort Smith
Arkansas
Little Rock
ARKANSAS
Pine Bluff
Huntsville
Athens
Atlanta
Birmingham
Augusta
Savannah
Charleston
Texarkana
Ouachita
MISSISSIPPI
Tombigbee
Tuscaloosa
ALABAMA
Macon
GEORGIA
Columbus
Savannah
Shreveport
Monroe
Tyler
LOUISIANA
Jackson
Montgomery
Chattahoochee
Albany
Altamaha
Sam Rayburn Res.
Toledo Bend Res.
Red
Hattiesburg
Dothan
Jacksonville
Trinity
Lake Charles
Baton Rouge
Lafayette
Mobile
Pensacola
Tallahassee
Gulfport
Beaumont
Houston
Galveston
New Orleans
Gainesville
Daytona Beach
Orlando
Tampa
Lakeland
St. Petersburg
FLORIDA
Lake Okeechobee
West Palm Beach
Fort Myers
Fort Lauderdale
Miami
Key West
GULF OF MEXICO
National Capital
Secondary Capital (State, Province, or Territory)
City over 1,000,000 population
City of 250,000 to 1,000,000 population
City under 250,000 population
N
0 100 200 300 Miles
0 100 200 300 400 Kilometers
Copyright by Rand McNally & Co.
Alber's Conic Equal Area Projection
ATLANTIC OCEAN
Arecibo
San Juan
Mayagüez
Caguas
Ponce
PUERTO RICO (U.S.)
0 25 50 Miles
0 25 50 Kilometers
Caribbean Sea
© RMcN.

A21

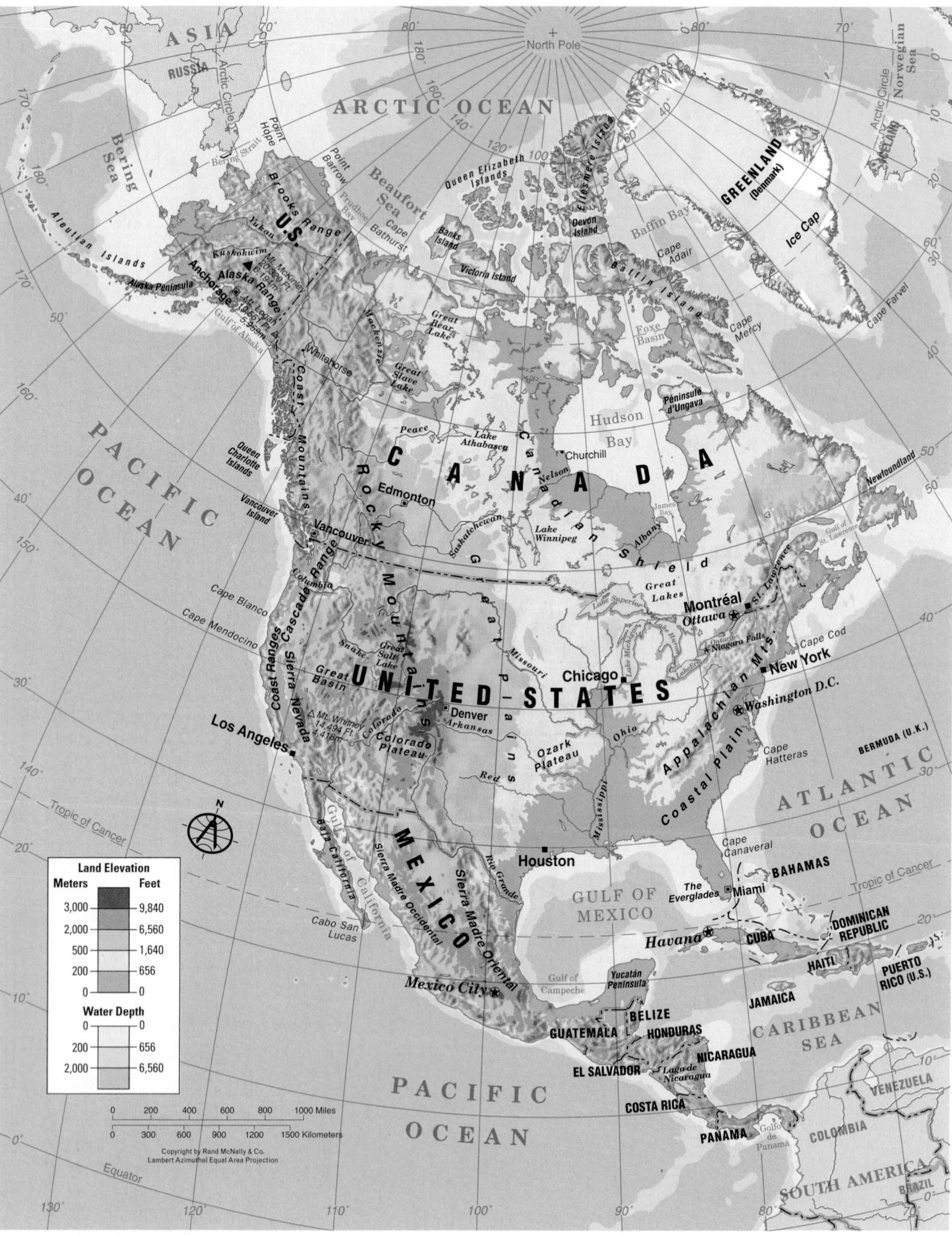
North Pole
ASIA
RUSSIA
Arctic Circle
ARCTIC OCEAN
Bering Sea
Bering Strait
Point Hope
Point Barrow
Prudhoe Bay
Beaufort Sea
Brooks Range
U.S.
Yukon
Kuskokwim
Mt. McKinley 20,320 Ft. 6,194m
Alaska Range
Anchorage
Mt. Logan 19,551 Ft. 5,959m
Gulf of Alaska
Aleutian Islands
Alaska Peninsula
Cape Bathurst
Queen Elizabeth Islands
Ellesmere Island
Devon Island
Banks Island
Victoria Island
Baffin Bay
Cape Adair
Baffin Island
Cape Mercy
Foxe Basin
GREENLAND (Denmark)
Ice Cap
Cape Farvel
Norwegian Sea
ICELAND
Mackenzie
Great Bear Lake
Great Slave Lake
Whitehorse
Coast Mountains
Queen Charlotte Islands
Vancouver Island
Vancouver
Peace
Lake Athabasca
Hudson Bay
Churchill
Péninsule d'Ungava
CANADA
Canadian Shield
Nelson
Edmonton
Rocky Mountains
Saskatchewan
Lake Winnipeg
Albany
James Bay
Newfoundland
Gulf of St. Lawrence
PACIFIC OCEAN
Columbia
Cascade Range
Cape Blanco
Cape Mendocino
Coast Ranges
Sierra Nevada
Snake
Great Salt Lake
Great Basin
Great Plains
Missouri
Great Lakes
Lake Superior
Lake Michigan
Lake Huron
Lake Erie
Lake Ontario
St. Lawrence
Montréal
Ottawa
Niagara Falls
Cape Cod
New York
Chicago
UNITED STATES
Denver
Arkansas
Mt. Whitney 14,494 Ft. 4,418m
Colorado
Colorado Plateau
Los Angeles
Ozark Plateau
Ohio
Red
Mississippi
Appalachian Mts.
Washington D.C.
Coastal Plain
Cape Hatteras
BERMUDA (U.K.)
ATLANTIC OCEAN
Tropic of Cancer
Baja California
Gulf of California
Sierra Madre Occidental
Sierra Madre Oriental
MEXICO
Rio Grande
Houston
Cape Canaveral
The Everglades
Miami
BAHAMAS
GULF OF MEXICO
Cabo San Lucas
Havana
CUBA
DOMINICAN REPUBLIC
HAITI
PUERTO RICO (U.S.)
JAMAICA
Mexico City
Gulf of Campeche
Yucatán Peninsula
BELIZE
GUATEMALA
HONDURAS
CARIBBEAN SEA
NICARAGUA
EL SALVADOR
Lago de Nicaragua
COSTA RICA
PANAMA
Golfo de Panamá
VENEZUELA
COLOMBIA
SOUTH AMERICA
BRAZIL
Equator
Land Elevation
Meters
Feet
3,000 — 9,840
2,000 — 6,560
500 — 1,640
200 — 656
0 — 0
Water Depth
0 — 0
200 — 656
2,000 — 6,560
0 200 400 600 800 1000 Miles
0 300 600 900 1200 1500 Kilometers
Copyright by Rand McNally & Co.
Lambert Azimuthal Equal Area Projection

A22

North America: Political

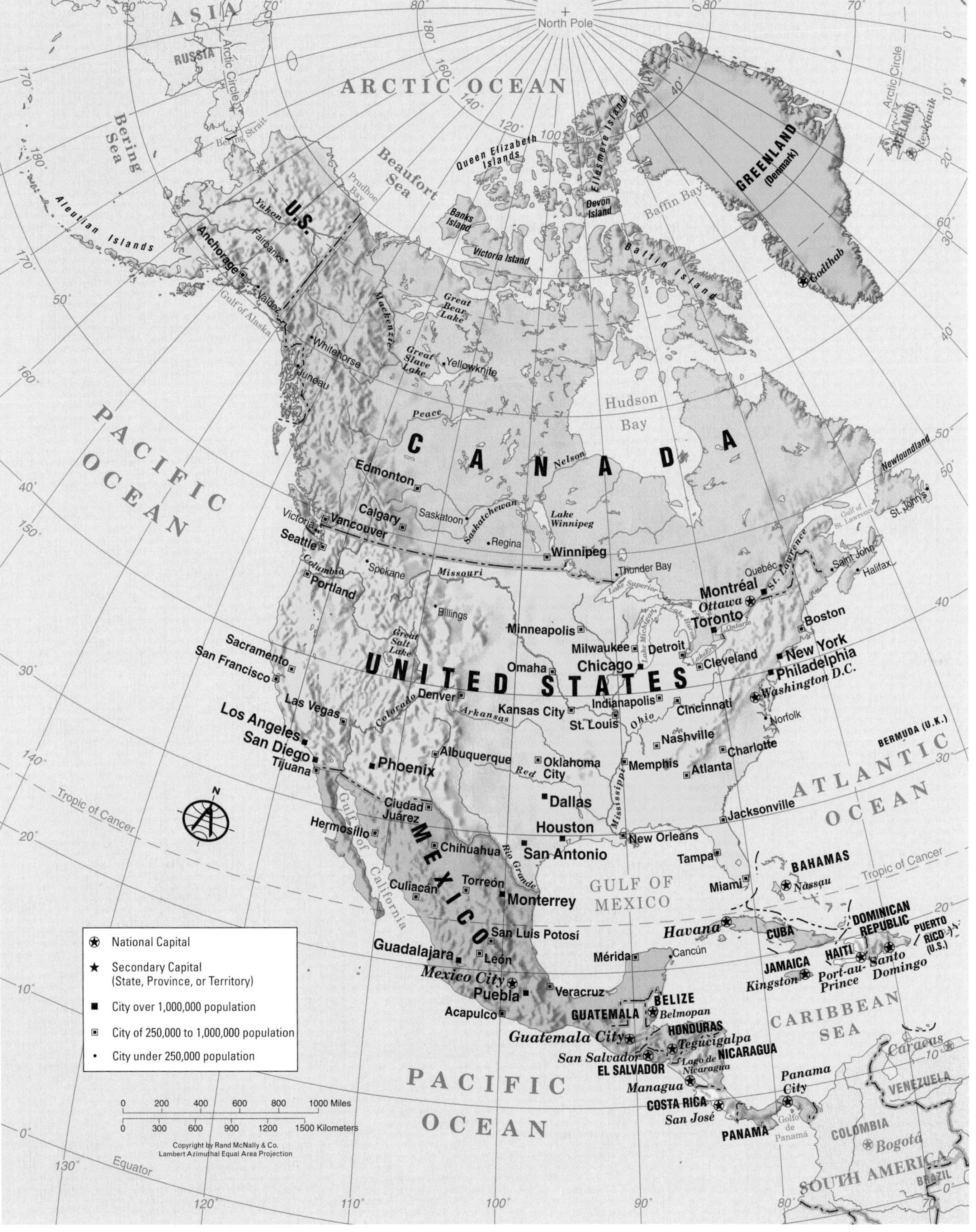

North Pole
ASIA
RUSSIA
Arctic Circle
ARCTIC OCEAN
Bering Sea
Bering Strait
Aleutian Islands
Beaufort Sea
Prudhoe Bay
Queen Elizabeth Islands
Ellesmere Island
Devon Island
Banks Island
Victoria Island
Baffin Bay
Baffin Island
GREENLAND (Denmark)
Godthab
ICELAND
Reykjavik
U.S.
Yukon
Fairbanks
Anchorage
Valdez
Gulf of Alaska
Mackenzie
Great Bear Lake
Great Slave Lake
Yellowknife
Whitehorse
Juneau
Hudson Bay
Peace
PACIFIC OCEAN
CANADA
Nelson
Edmonton
Calgary
Saskatoon
Saskatchewan
Lake Winnipeg
Regina
Winnipeg
Victoria
Vancouver
Seattle
Newfoundland
St. John's
Gulf of St. Lawrence
St. Lawrence
Québec
Saint John
Halifax
Thunder Bay
Lake Superior
Montréal
Ottawa
Toronto
Lake Ontario
Lake Huron
Lake Michigan
Lake Erie
Columbia
Spokane
Missouri
Portland
Billings
Minneapolis
Great Salt Lake
Sacramento
San Francisco
Milwaukee
Detroit
Chicago
Cleveland
Boston
New York
Philadelphia
Washington D.C.
UNITED STATES
Omaha
Denver
Colorado
Las Vegas
Kansas City
Arkansas
Indianapolis
Cincinnati
Ohio
St. Louis
Norfolk
Los Angeles
San Diego
Tijuana
Nashville
Charlotte
Albuquerque
Phoenix
Oklahoma City
Red
Memphis
Atlanta
Mississippi
BERMUDA (U.K.)
ATLANTIC OCEAN
Tropic of Cancer
Ciudad Juárez
Dallas
Jacksonville
Hermosillo
Gulf of California
Houston
New Orleans
MEXICO
Chihuahua
Rio Grande
San Antonio
Tampa
BAHAMAS
Nassau
Culiacán
Torreón
Monterrey
GULF OF MEXICO
Miami
Havana
CUBA
DOMINICAN REPUBLIC
PUERTO RICO (U.S.)
San Luis Potosí
Guadalajara
León
Mérida
Cancún
JAMAICA
Kingston
HAITI
Port-au-Prince
Santo Domingo
Mexico City
Puebla
Veracruz
BELIZE
Belmopan
CARIBBEAN SEA
Acapulco
GUATEMALA
HONDURAS
Tegucigalpa
Guatemala City
San Salvador
EL SALVADOR
Lago de Nicaragua
NICARAGUA
Managua
Caracas
VENEZUELA
Panama City
COSTA RICA
San José
PANAMA
Golfo de Panamá
COLOMBIA
Bogotá
SOUTH AMERICA
BRAZIL
Equator
National Capital
Secondary Capital (State, Province, or Territory)
City over 1,000,000 population
City of 250,000 to 1,000,000 population
City under 250,000 population
0 200 400 600 800 1000 Miles
0 300 600 900 1200 1500 Kilometers
Copyright by Rand McNally & Co.
Lambert Azimuthal Equal Area Projection
N

ATLAS

RAND McNALLY

A23

GULF OF MEXICO
NORTH AMERICA
CUBA
Greater Antilles
DOMINICAN REPUBLIC
HAITI
JAMAICA
PUERTO RICO (U.S.)
Lesser Antilles
BELIZE
Gulf of Honduras
MEXICO
HONDURAS
GUATEMALA
CARIBBEAN SEA
ATLANTIC OCEAN
EL SALVADOR
NICARAGUA
Cristóbal Colón Peak 18,948 Ft. 5,775m
Caracas
TRINIDAD AND TOBAGO
COSTA RICA
PANAMA
Gulf of Panama
Orinoco
Llanos
VENEZUELA
GUYANA
Magdalena
Bogotá
COLOMBIA
SURINAME
FRENCH GUIANA
Cape Orange
Galapagos Islands (Ec.)
ECUADOR
Chimborazo 20,703 Ft. 6,310m
Putumayo
Japurá
Negro
Manaus
Amazon
Ilha de Marajó
Belém
Equator
Amazon Basin
Tocantins
Tapajós
Juruá
Madeira
BRAZIL
Selvas
Ucayali
PERU
Andes
Recife
Mt. Huascarán 22,133 Ft. 6,746m
Lima
Mato Grosso Plateau
Mt. Illampu 21,066 Ft. 6,421m
Lake Titicaca
BOLIVIA
Cordillera Oriental
Brasília
São Francisco
Serra do Espinhaço
Mt. Sajama 21,463 Ft. 6,542m
Gran Chaco
Atacama Desert
PARAGUAY
Paraná
São Paulo
Rio de Janeiro
Tropic of Capricorn
Isla San Ambrosio (Chile)
Mt. Ojos del Salado 22,615 Ft. 6,893m
Isla San Felix (Chile)
ARGENTINA
PACIFIC OCEAN
Archipiélago Juan Fernández (Chile)
URUGUAY
Mt. Aconcagua 22,831 Ft. 6,959m
Santiago
Buenos Aires
Río de la Plata
Pampas
N
CHILE
Land Elevation
Meters
Feet
3,000 9,840
2,000 6,560
500 1,640
200 656
0 0
Water Depth
0 0
200 656
2,000 6,560
San Matías Gulf
Península Valdés
Chiloé
Patagonia
San Jorge Gulf
Point Medanoso
Grand Bay
West Falkland
FALKLAND ISLANDS (U.K.)
East Falkland
Strait of Magellan
Tierra del Fuego
Cape Horn
0 200 400 600 800 1000 Miles
0 300 600 900 1200 1500 Kilometers
Copyright by Rand McNally & Co.
Lambert Azimuthal Equal Area Projection
South Georgia (U.K.)
Drake Passage
South Shetland Islands (U.K.)
South Orkney Islands (U.K.)
South Sandwich Islands (U.K.)

A24

GULF OF MEXICO
Havana
CUBA
NORTH AMERICA
MEXICO
BELIZE
HONDURAS
GUATEMALA
EL SALVADOR
NICARAGUA
COSTA RICA
PANAMA
JAMAICA
HAITI
DOMINICAN REPUBLIC
PUERTO RICO (U.S.)
Lesser Antilles
CARIBBEAN SEA
ATLANTIC OCEAN
TRINIDAD AND TOBAGO
Barranquilla
Cartagena
Maracaibo
Caracas
Barquisimeto
Valencia
Cúcuta
Bucaramanga
Orinoco
Ciudad Guayana
VENEZUELA
Georgetown
GUYANA
Paramaribo
Cayenne
SURINAME
FRENCH GUIANA
Medellín
Magdalena
Bogotá
COLOMBIA
Cali
Macapá
Galapagos Islands (Ec.)
Quito
ECUADOR
Guayaquil
Putumayo
Japurá
Negro
Manaus
Amazon
Santarém
Belém
São Luís
Fortaleza
Tocantins
Imperatriz
Teresina
Iquitos
Juruá
Madeira
Tapajós
BRAZIL
Natal
Recife
Maceió
Aracaju
Chiclayo
Trujillo
PERU
Ucayali
Pôrto Velho
Lima
Feira de Santana
Salvador
Cusco
Lake Titicaca
Arequipa
BOLIVIA
La Paz
Cochabamba
Santa Cruz
Sucre
Cuiabá
Goiânia
Brasília
Montes Claros
Uberlândia
Belo Horizonte
Vitória
Campo Grande
Paraná
Campinas
Rio de Janeiro
São Paulo
PARAGUAY
Antofagasta
Tropic of Capricorn
Isla San Ambrosio (Chile)
Isla San Felix (Chile)
Salta
San Miguel de Tucumán
Asunción
Curitiba
Caxias do Sul
Pôrto Alegre
Córdoba
Santa Fe
Rosario
URUGUAY
Archipiélago Juan Fernández (Chile)
Valparaíso
Santiago
Mendoza
Buenos Aires
La Plata
Montevideo
Río de la Plata
PACIFIC OCEAN
Concepción
Bahía Blanca
Mar del Plata
CHILE
ARGENTINA
Chiloé
Archipiélago de los Chonos
Comodoro Rivadavia
ATLANTIC OCEAN
FALKLAND ISLANDS (U.K.)
West Falkland
East Falkland
Strait of Magellan
Punta Arenas
Tierra del Fuego
South Georgia (U.K.)
South Sandwich Islands (U.K.)
Drake Passage
South Shetland Islands (U.K.)
South Orkney Islands (U.K.)
Equator
N
National Capital
Secondary Capital (State, Province, or Territory)
City over 1,000,000 population
City of 250,000 to 1,000,000 population
City under 250,000 population
0 200 400 600 800 1000 Miles
0 300 600 900 1200 1500 Kilometers
Copyright by Rand McNally & Co.
Lambert Azimuthal Equal Area Projection

ATLAS

A25

ICELAND
Horn
Fontur
Surtsey
Arctic Circle
NORWEGIAN SEA
Lofoten Islands
Kebnekaise 6,926 Ft. 2,111m
Torneälven
Luleälven
Lap
Umeälven
Scandinavian Peninsula
Gulf of Bothnia
ATLANTIC OCEAN
FAROE ISLANDS (Den.)
NORWAY
SWEDEN
Galdhøpiggen 8,100 Ft. 2,469m
Glåma
Klarälven
Dalälven
Stockholm
Hebrides
Orkney Islands
Grampian Mts.
UNITED KINGDOM
Cheviot Hills
NORTH SEA
Skagerrak
Vänern
Vättern
Öland
BALTIC SEA
DENMARK
IRELAND
Irish Sea
St. George's Channel
Great Britain
Thames
London
Bornholm (Den.)
RUSSIA
NETHERLANDS
Elbe
Northern Europ
Berlin
Oder
GERMANY
POLAND
Wisła
English Channel
Strait of Dover
BELGIUM
Rhine
LUX.
Paris
Paris Basin
Seine
Loire
CZECH REPUBLIC
Bohemian Forest
Black Forest
SLOVAKIA
Danube
FRANCE
Jura
Saone
SWITZERLAND
LIECH.
AUSTRIA
HUNGARY
Great Hungarian Plain
Bay of Biscay
Mt. Blanc 15,771 Ft. 4,808m
Alps
Dordogne
Massif Central
Rhône
SLOVENIA
Drava
CROATIA
Po
Cantabrian Mts.
Pyrenees
Douro
Duero
Iberian Mts.
Ebro
ANDORRA
MONACO
SAN MARINO
BOSNIA AND HERZEGOVINA
Dinaric Alps
Apennines
ADRIATIC SEA
Balkan
YUGOSLAVIA
Lisbon
PORTUGAL
Iberian Peninsula
Tagus
SPAIN
Corsica (Fr.)
Rome
ITALY
ALBANIA
MACEDONIA
Sierra Morena
Balearic Islands
Minorca
Majorca
Ibiza
Sardinia (It.)
Vesuvius 4,190 Ft. 1,277m
Pindus Mts.
Strait of Gibraltar
GIBRALTAR (U.K.)
TYRRHENIAN SEA
MEDITERRANEAN
Mt. Etna 10,902 Ft. 3,323m
Sicily
IONIAN SEA
Algiers
AFRICA
MOROCCO
ALGERIA
TUNISIA
MALTA
Land Elevation
Meters
Feet
3,000 — 9,840
2,000 — 6,560
500 — 1,640
200 — 656
0 — 0
Water Depth
0 — 0
200 — 656
2,000 — 6,560
0 100 200 300 400 Miles
0 200 400 600 Kilometers
Copyright by Rand McNally & Co.
Lambert Conformal Conic Projection

A26

ATLAS

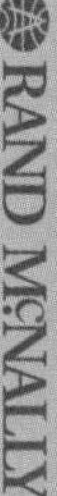

RAND McNALLY

A27

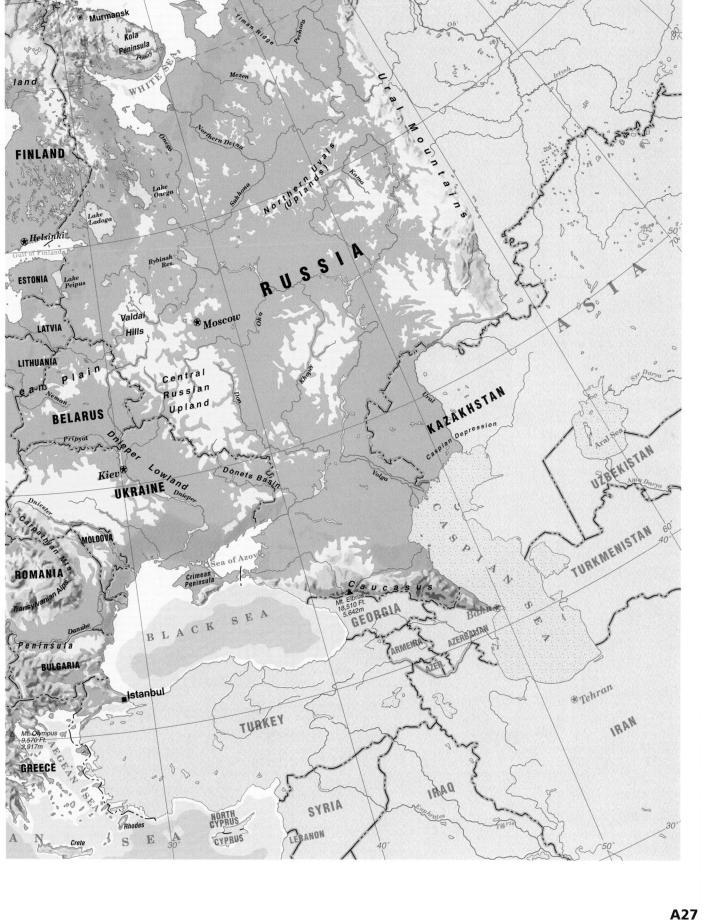

- ✪ National Capital
- ★ Secondary Capital (State, Province, or Territory)
- ■ City over 1,000,000 population
- ▣ City of 250,000 to 1,000,000 population
- • City under 250,000 population

0 100 200 300 400 Miles

0 200 400 600 Kilometers

Lambert Conformal Conic Projection

ATLANTIC OCEAN
NORWEGIAN SEA
NORTH SEA
BALTIC SEA
Gulf of Bothnia
Skagerrak
Irish Sea
St. George's Channel
English Channel
Strait of Dover
Bay of Biscay
Strait of Gibraltar
MEDITERRANEAN
TYRRHENIAN SEA
ADRIATIC SEA
IONIAN SEA
Arctic Circle

ICELAND
Reykjavík
FAROE ISLANDS (Den.)
NORWAY
Hammerfest
Trondheim
Bergen
Oslo
SWEDEN
Umeå
Stockholm
Vänern
Vättern
Göteborg
Tampere
DENMARK
Copenhagen
LITHUANIA
RUSSIA
Kaliningrad
Gdańsk
POLAND
Szczecin
Warsaw
Wisła
Oder
Wroclaw
Kraków
UNITED KINGDOM
SCOTLAND
Aberdeen
Glasgow
Edinburgh
NORTHERN IRELAND
Belfast
IRELAND
Dublin
Cork
Liverpool
Manchester
WALES
Birmingham
ENGLAND
Cardiff
Thames
London
Plymouth
NETHERLANDS
Amsterdam
The Hague
Brussels
BELGIUM
Luxembourg
LUX.
GERMANY
Hamburg
Elbe
Berlin
Rhine
Cologne
Bonn
Frankfurt
Dresden
Stuttgart
Munich
CZECH REPUBLIC
Prague
Danube
SLOVAKIA
Bratislava
AUSTRIA
Vienna
HUNGARY
Budapest
FRANCE
Le Havre
Paris
Seine
Loire
Nantes
Strasbourg
Bordeaux
Lyon
Rhône
Toulouse
Marseille
Nice
MONACO
SWITZERLAND
Zürich
Bern
Geneva
LIECH.
SLOVENIA
Ljubljana
CROATIA
Zagreb
Belgrade
BOSNIA AND HERZEGOVINA
Sarajevo
YUGOSLAVIA
Skopje
ALBANIA
Tiranë
MACEDONIA
ITALY
Milan
Turin
Genoa
Venice
Po
SAN MARINO
Florence
Corsica (Fr.)
Rome
VATICAN CITY
Sardinia (It.)
Cagliari
Naples
Bari
Palermo
Sicily
Catania
Valletta
MALTA
SPAIN
A Coruña
Gijón
Bilbao
Valladolid
Ebro
Zaragoza
ANDORRA
Barcelona
Madrid
Tagus
Valencia
Palma
Córdoba
Seville
Málaga
GIBRALTAR (U.K.)
PORTUGAL
Porto
Lisbon
AFRICA
MOROCCO
Rabat
ALGERIA
Algiers
TUNISIA
Tunis

A28

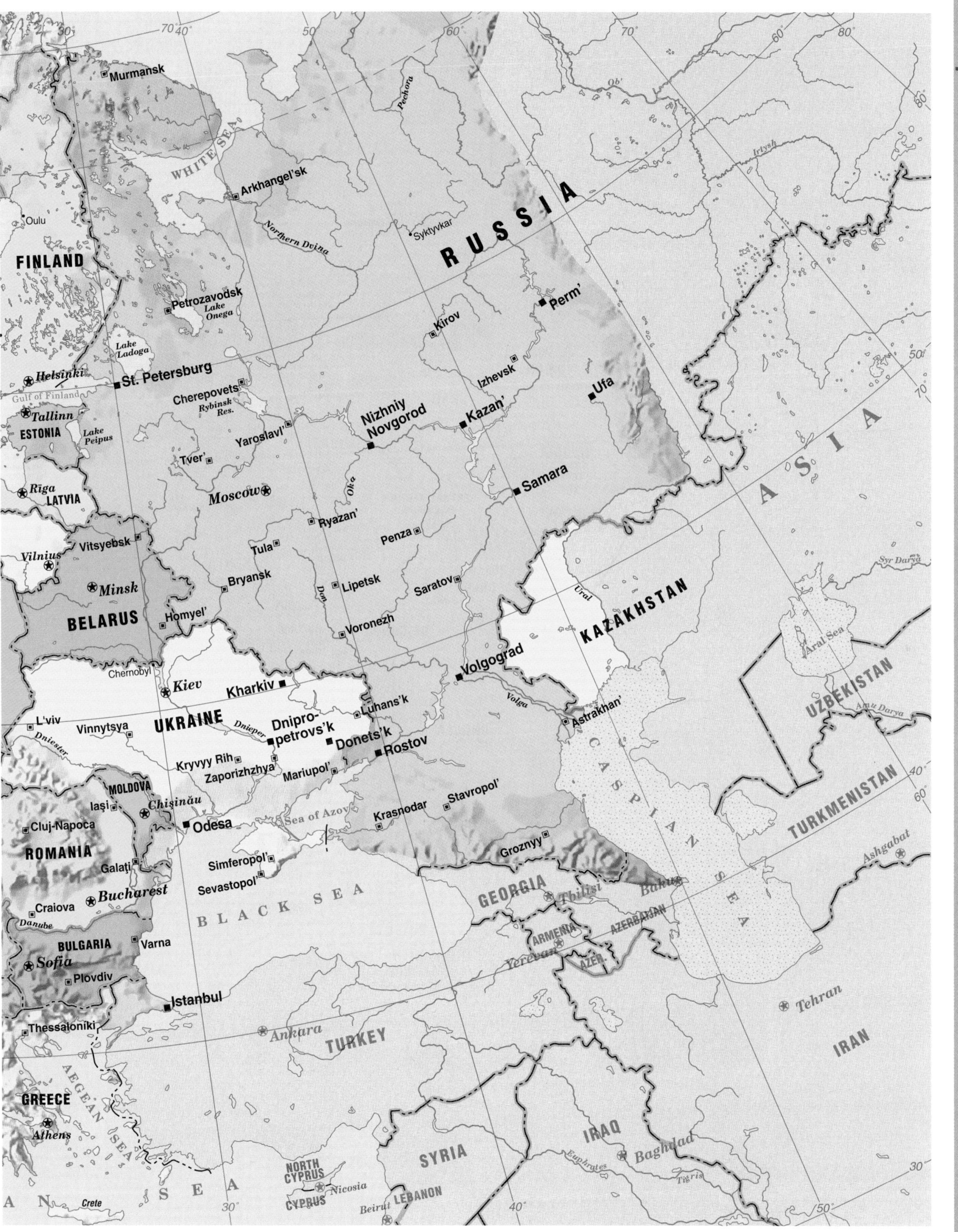

ATLAS

RAND McNALLY

ATLANTIC OCEAN
EUROPE
FRANCE
AUS.
HUNG.
ROMANIA
UKRAINE
RUSSIA
KAZ.
Aral Sea
UZBEKISTAN
PORTUGAL
SPAIN
ITALY
BOS.
YUGO.
BUL.
ALB.
GREECE
Black Sea
GEORGIA
ARM.
AZER.
Caspian Sea
TURKMENISTAN
TURKEY
Azores (Port.)
Mediterranean Sea
Algiers
MALTA
CYPRUS
SYRIA
ASIA
IRAN
Strait of Gibraltar
Madeira Islands (Port.)
MOROCCO
Atlas Mountains
TUNISIA
LEBANON
IRAQ
ISRAEL
Canary Islands (Spain)
Great Western Desert
Great Eastern Desert
Gulf of Sidra
Qattara Depression
Cairo
JORDAN
KUWAIT
Persian Gulf
ALGERIA
LIBYA
EGYPT
QATAR
WESTERN SAHARA (MOROCCO)
Tropic of Cancer
Tahat 9,541 Ft. 2,908m
Ahaggar Mts.
Libyan Desert
Lake Nasser
Red Sea
U.A.E.
SAUDI ARABIA
Ijafene
Sahara Desert
Tibesti Massif
Nubian Desert
MAURITANIA
Mt. Koussi 11,204 Ft. 3,415m
Aïr (Mts.)
Nile
OMAN
CAPE VERDE
Sénégal
MALI
NIGER
Ennedi
ERITREA
YEMEN
Cape Verde
SENEGAL
Dakar
GAMBIA
Sahel
Lake Chad
CHAD
Khartoum
Socotra (Yem.)
Gulf of Aden
Niger
Blue Nile
White Nile
Lake Tana
DJIBOUTI
Cape Gwardafuy
GUINEA-BISSAU
GUINEA
BURKINA FASO
SUDAN
Ethiopian Plateau
SIERRA LEONE
GHANA
Lake Volta
TOGO
BENIN
NIGERIA
Jos Plateau
Benue
As Sudd
Mountain Nile
Great Rift Valley
COTE D'IVOIRE
LIBERIA
Lagos
CAMEROON
CENTRAL AFRICAN REPUBLIC
ETHIOPIA
SOMALIA
Mt. Cameroon 13,451 Ft. 4,100m
Gulf of Guinea
Bioko
Ubangi
Congo
Uele
UGANDA
Lake Turkana
KENYA
EQUATORIAL GUINEA
REP. OF CONGO
Mt. Kenya 17,058 Ft. 5,199m
Equator
SAO TOME AND PRINCIPE
GABON
Congo Basin
Lake Victoria
RWANDA
Nairobi
INDIAN OCEAN
DEM. REP. OF CONGO
Serengeti Plain
Kilimanjaro 19,340 Ft. 5,895m
BURUNDI
Kinshasa
Lake Tanganyika
Masai Steppe
Zanzibar
SEYCHELLES
TANZANIA
Ascension (St. Helena)
Kwango
Kasai
Cuanza
ANGOLA
MALAWI
Lake Nyasa
COMOROS
Mayotte (Fr.)
Cape Ambre
ATLANTIC OCEAN
ZAMBIA
MOZAMBIQUE
St. Helena (U.K.)
Victoria Falls
Cunene
Okavango
Lake Kariba
Zambezi
Mozambique Channel
ZIMBABWE
MAURITIUS
NAMIBIA
BOTSWANA
MADAGASCAR
Namib Desert
Kalahari Desert
Limpopo
Réunion (Fr.)
Tropic of Capricorn
Barra Point
Johannesburg
Vaal
SWAZILAND
Cape Sainte-Marie
Orange
LESOTHO
Drakensberg
SOUTH AFRICA
Cape of Good Hope
Cape Agulhas
Tristan da Cunha Group (St. Helena)
Prince Edward Islands (S. Af.)
Crozet Islands (Fr.)

Land Elevation

Meters	Feet
3,000	9,840
2,000	6,560
500	1,640
200	656
0	0

Water Depth

Meters	Feet
0	0
200	656
2,000	6,560

0 200 400 600 800 1000 Miles
0 300 600 900 1200 1500 Kilometers

Lambert Azimuthal Equal Area Projection

A30

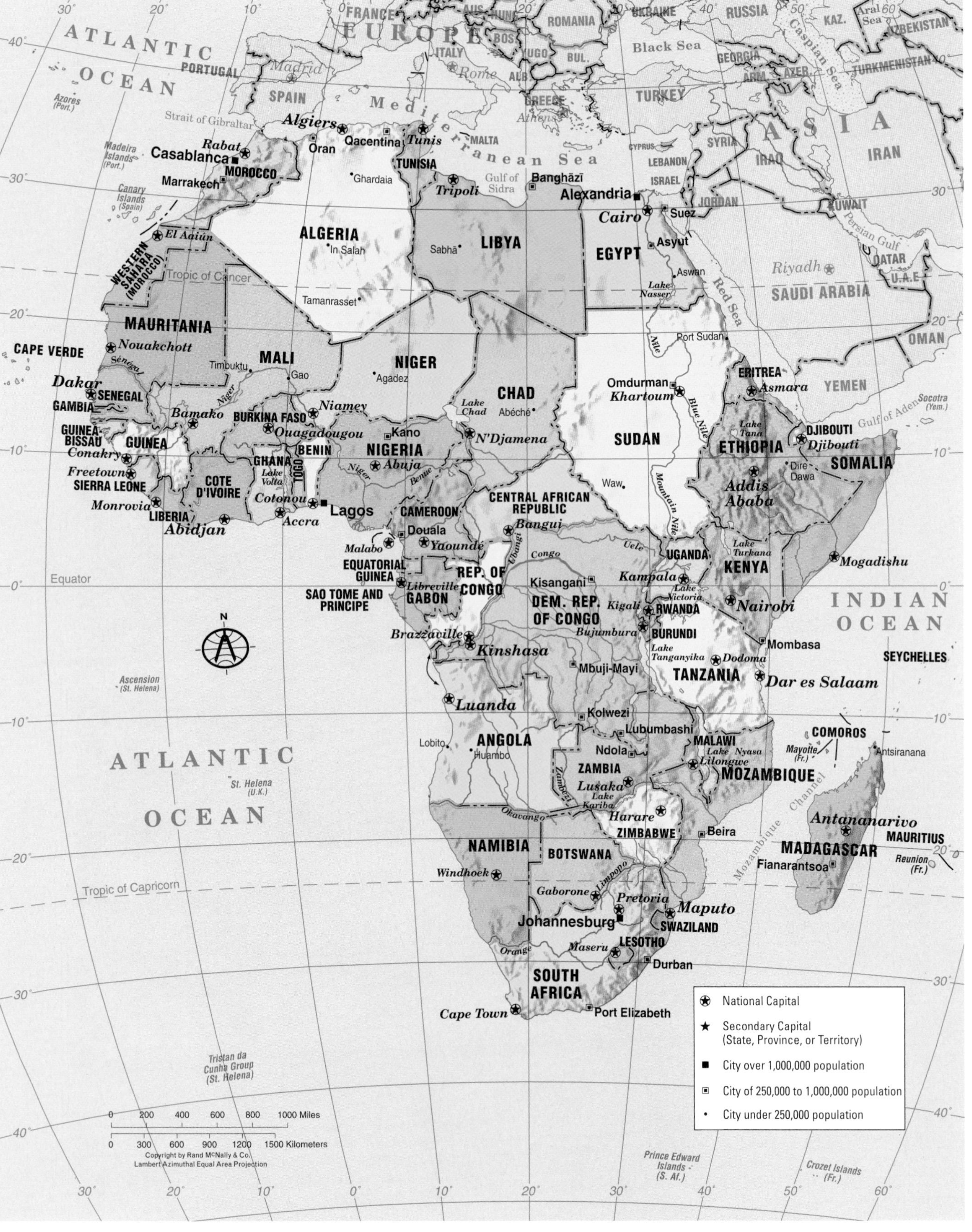
ATLANTIC OCEAN
PORTUGAL
SPAIN
FRANCE
EUROPE
ITALY
Rome
ROMANIA
UKRAINE
RUSSIA
KAZ.
Aral Sea
UZBEKISTAN
Black Sea
Caspian Sea
BUL.
YUGO.
BOS.
ALB.
GREECE
Athens
GEORGIA
ARM.
AZER.
TURKMENISTAN
TURKEY
ASIA
SYRIA
IRAQ
IRAN
CYPRUS
LEBANON
ISRAEL
JORDAN
KUWAIT
Persian Gulf
QATAR
U.A.E.
Riyadh
SAUDI ARABIA
OMAN
YEMEN
Socotra (Yem.)
Gulf of Aden
Azores (Port.)
Madrid
Strait of Gibraltar
Mediterranean Sea
MALTA
Madeira Islands (Port.)
Rabat
Casablanca
MOROCCO
Marrakech
Canary Islands (Spain)
Algiers
Oran
Qacentina
Tunis
TUNISIA
Ghardaia
Tripoli
Gulf of Sidra
Banghāzī
Alexandria
Cairo
Suez
Asyut
EGYPT
Aswan
Lake Nasser
Red Sea
ALGERIA
In Salah
Sabhā
LIBYA
El Aaiún
WESTERN SAHARA (MOROCCO)
Tropic of Cancer
Tamanrasset
MAURITANIA
Nouakchott
CAPE VERDE
MALI
Timbuktu
Gao
NIGER
Agadez
CHAD
Lake Chad
Abéché
Port Sudan
Nile
Omdurman
Khartoum
Blue Nile
ERITREA
Asmara
Dakar
SENEGAL
GAMBIA
GUINEA-BISSAU
GUINEA
Conakry
Bamako
BURKINA FASO
Ouagadougou
Niamey
Kano
NIGERIA
Abuja
N'Djamena
SUDAN
Lake Tana
ETHIOPIA
DJIBOUTI
Djibouti
Dire Dawa
SOMALIA
Freetown
SIERRA LEONE
Monrovia
LIBERIA
COTE D'IVOIRE
Abidjan
GHANA
Lake Volta
TOGO
BENIN
Cotonou
Accra
Lagos
Niger
Benue
CAMEROON
Douala
Yaoundé
Malabo
EQUATORIAL GUINEA
CENTRAL AFRICAN REPUBLIC
Bangui
Waw
Mountain Nile
Addis Ababa
Lake Turkana
KENYA
Mogadishu
Equator
SAO TOME AND PRINCIPE
Libreville
GABON
REP. OF CONGO
Ubangi
Congo
Uele
UGANDA
Kampala
Lake Victoria
Nairobi
Kisangani
DEM. REP. OF CONGO
Kigali
RWANDA
Bujumbura
BURUNDI
INDIAN OCEAN
Brazzaville
Kinshasa
Mombasa
Lake Tanganyika
Dodoma
TANZANIA
Dar es Salaam
SEYCHELLES
Mbuji-Mayi
Ascension (St. Helena)
Luanda
Kolwezi
Lubumbashi
Lobito
Huambo
ANGOLA
Ndola
MALAWI
Lake Nyasa
Lilongwe
COMOROS
Mayotte (Fr.)
Antsiranana
ZAMBIA
Zambezi
Lusaka
Lake Kariba
MOZAMBIQUE
Mozambique Channel
St. Helena (U.K.)
Okavango
Harare
ZIMBABWE
Beira
Antananarivo
MAURITIUS
MADAGASCAR
Fianarantsoa
Reunion (Fr.)
NAMIBIA
BOTSWANA
Windhoek
Limpopo
Tropic of Capricorn
Gaborone
Pretoria
Maputo
Johannesburg
SWAZILAND
Orange
Maseru
LESOTHO
Durban
SOUTH AFRICA
Cape Town
Port Elizabeth
National Capital
Secondary Capital (State, Province, or Territory)
City over 1,000,000 population
City of 250,000 to 1,000,000 population
City under 250,000 population
Tristan da Cunha Group (St. Helena)
0 200 400 600 800 1000 Miles
0 300 600 900 1200 1500 Kilometers
Copyright by Rand McNally & Co.
Lambert Azimuthal Equal Area Projection
Prince Edward Islands (S. Af.)
Crozet Islands (Fr.)
N
40° 30° 20° 10° 0° 10° 20° 30° 40° 50° 60°

ATLAS

RAND McNALLY

A31

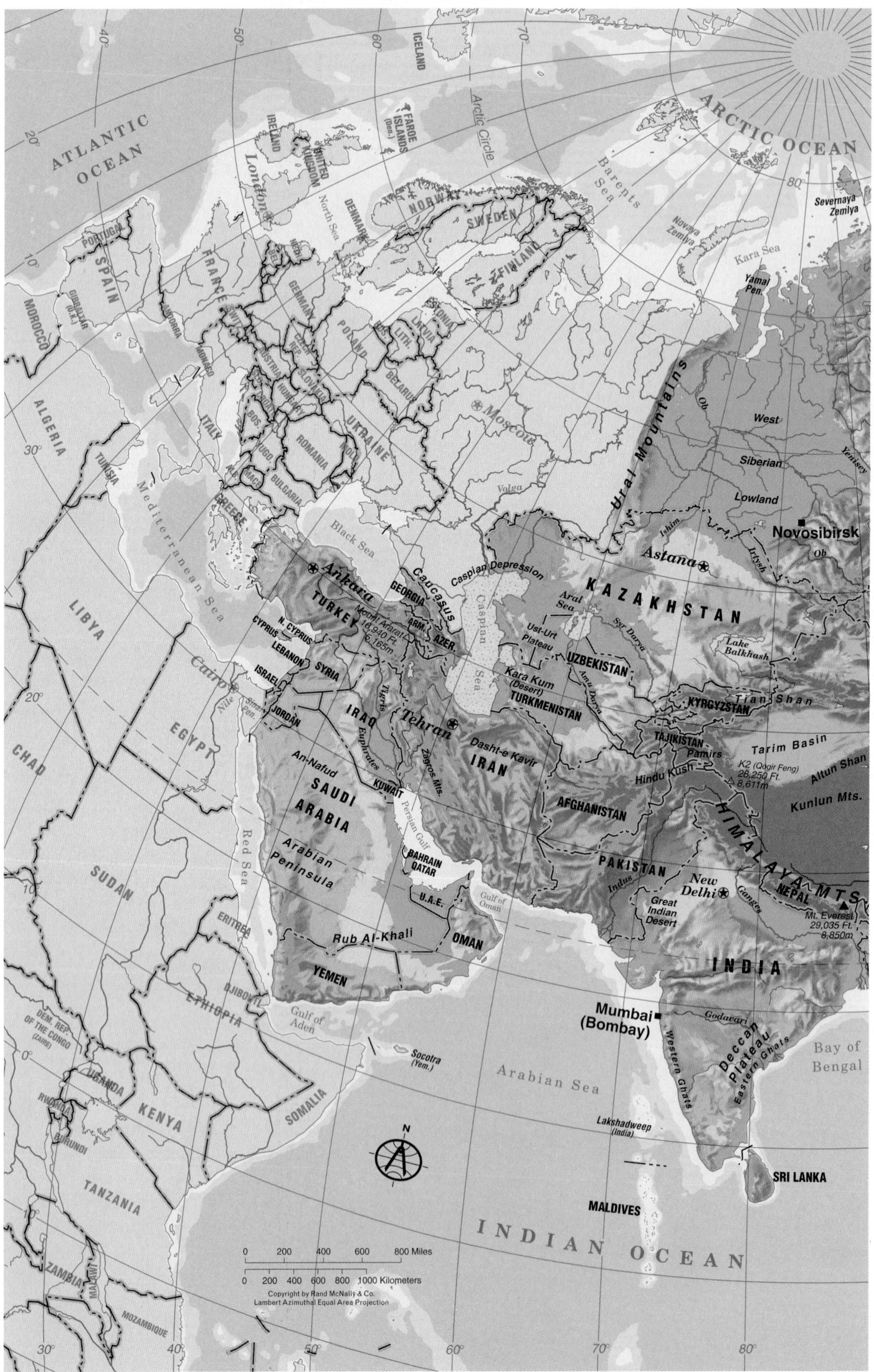
ATLANTIC OCEAN
ARCTIC OCEAN
INDIAN OCEAN
ICELAND
FAROE ISLANDS (Den.)
IRELAND
UNITED KINGDOM
London
North Sea
DENMARK
NORWAY
SWEDEN
FINLAND
Barents Sea
Arctic Circle
Novaya Zemlya
Severnaya Zemlya
Kara Sea
Yamal Pen.
PORTUGAL
SPAIN
GIBRALTAR (U.K.)
MOROCCO
ANDORRA
MONACO
FRANCE
GERMANY
POLAND
ESTONIA
LATVIA
LITH.
BELARUS
CZECH REP.
SLOVAKIA
AUSTRIA
HUNGARY
ITALY
BOS.
YUGO.
ROMANIA
BULGARIA
GREECE
UKRAINE
Moscow
Volga
Ural Mountains
Ob
West Siberian Lowland
Yenisey
Novosibirsk
Ishim
Irtysh
Astana
ALGERIA
TUNISIA
Mediterranean Sea
Black Sea
LIBYA
Caspian Depression
Caucasus
GEORGIA
ARM.
AZER.
Caspian Sea
Ankara
TURKEY
Mount Ararat 16,940 Ft. 5,165m
N. CYPRUS
CYPRUS
LEBANON
SYRIA
ISRAEL
JORDAN
Cairo
Nile
Sinai Pen.
EGYPT
IRAQ
Tigris
Euphrates
Tehran
Zagros Mts.
Dasht-e Kavir
IRAN
KAZAKHSTAN
Aral Sea
Syr Darya
Ust-Urt Plateau
UZBEKISTAN
Amu Darya
Kara Kum (Desert)
TURKMENISTAN
Lake Balkhash
Tian Shan
KYRGYZSTAN
TAJIKISTAN
Pamirs
Tarim Basin
K2 (Qogir Feng) 28,250 Ft. 8,611m
Altun Shan
Kunlun Mts.
Hindu Kush
AFGHANISTAN
HIMALAYA MTS.
PAKISTAN
Indus
New Delhi
Ganges
NEPAL
Mt. Everest 29,035 Ft. 8,850m
Great Indian Desert
INDIA
CHAD
SUDAN
An-Nafud
SAUDI ARABIA
KUWAIT
Persian Gulf
BAHRAIN
QATAR
U.A.E.
Gulf of Oman
OMAN
Arabian Peninsula
Red Sea
Rub Al-Khali
YEMEN
ERITREA
DJIBOUTI
ETHIOPIA
Gulf of Aden
Socotra (Yem.)
Mumbai (Bombay)
Godavari
Western Ghats
Deccan Plateau
Eastern Ghats
Bay of Bengal
Arabian Sea
Lakshadweep (India)
SRI LANKA
MALDIVES
DEM. REP. OF THE CONGO (ZAIRE)
UGANDA
KENYA
SOMALIA
RWANDA
BURUNDI
TANZANIA
ZAMBIA
MALAWI
MOZAMBIQUE
N
0 200 400 600 800 Miles
0 200 400 600 800 1000 Kilometers
Copyright by Rand McNally & Co.
Lambert Azimuthal Equal Area Projection

A32

Land Elevation

Meters	Feet
3,000	9,840
2,000	6,560
500	1,640
200	656
0	0

Water Depth

Meters	Feet
0	0
200	656
2,000	6,560

RUSSIA
Siberia
Central Siberian Uplands
Taymyr Peninsula
Laptev Sea
New Siberian Islands
East Siberian Sea
Arctic Circle
Indigirka
Kolyma
Verkhoyansk Mts.
Lena
Bering Sea
Aleutian Islands (U.S.)
Kamchatka Peninsula
Sea of Okhotsk
Sakhalin
Kuril Islands
Stanovoy Range
Angara
Lake Baikal
Amur
Greater Khingan Range
Sikhote-Alin Mts.
Tatar Strait
Hokkaido
Honshu
Tokyo
Sea of Japan
Sayan Mountains
MONGOLIA
Altai Mts.
Gobi Desert
Beijing
NORTH KOREA
SOUTH KOREA
Mt. Fuji 12,388 ft. 3,776m
JAPAN
Shikoku
Kyushu
Yellow Sea
Qilian Shan
Huang
Shanghai
East China Sea
CHINA
Qinling Shandi
Chang (Yangtze)
PACIFIC OCEAN
Tropic of Cancer
NORTHERN MARIANA ISLANDS (U.S.)
TAIWAN
Taiwan Strait
Philippine Sea
GUAM (U.S.)
FEDERATED STATES OF MICRONESIA
BHUTAN
Brahmaputra
Irrawaddy
Salween
Xi
Luzon Strait
Luzon
Gulf of Tonkin
Hainan Island
Red
PHILIPPINES
BNGL.
MYANMAR
LAOS
Mekong
South China Sea
Manila
THAILAND
Bangkok
VIETNAM
CAMBODIA
Sulu Sea
Mindanao
PALAU
Equator
Andaman Islands (India)
Andaman Sea
Gulf of Thailand
Nicobar Islands (India)
MALAY PENINSULA
MALAYSIA
BRUNEI
Celebes Sea
Moluccas
New Guinea
PAPUA NEW GUINEA
Str. of Malacca
Singapore
Borneo
Celebes
Ceram
Banda Sea
Sumatra
Greater Sunda Islands
INDONESIA
Arafura Sea
Coral Sea
Gulf of Carpentaria
EAST TIMOR
Timor
Java Sea
Jakarta
Java
Timor Sea
AUSTRALIA
90° 100° 110° 120° 130° 140° 150°
70° 60° 50° 40° 30° 20° 10° 0° 10° 20°
170° 180°

RAND McNALLY

A33

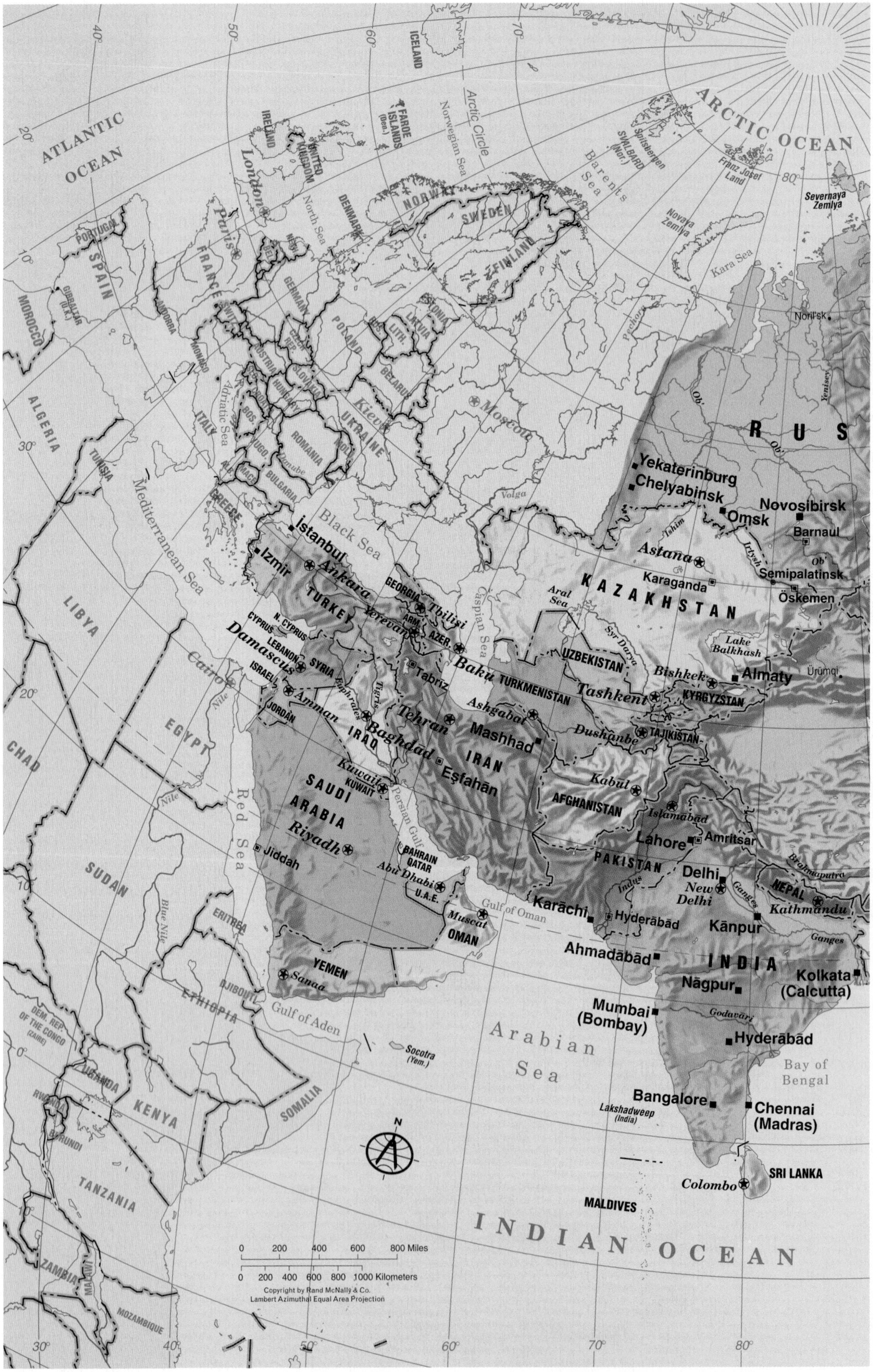

ARCTIC OCEAN

ATLANTIC OCEAN

ICELAND

FAROE ISLANDS (Den.)

IRELAND

UNITED KINGDOM

London

Arctic Circle

Norwegian Sea

North Sea

SVALBARD (Nor.)

Spitsbergen

Franz Josef Land

Severnaya Zemlya

Barents Sea

Novaya Zemlya

Kara Sea

NORWAY

SWEDEN

FINLAND

DENMARK

PORTUGAL

SPAIN

FRANCE

Paris

GERMANY

POLAND

ESTONIA

LATVIA

LITH.

BELARUS

UKRAINE

Kiev

Moscow

ROMANIA

BULGARIA

GREECE

ITALY

Adriatic Sea

MOROCCO

ALGERIA

TUNISIA

LIBYA

EGYPT

Cairo

CHAD

SUDAN

ERITREA

ETHIOPIA

DJIBOUTI

SOMALIA

KENYA

UGANDA

RWANDA

BURUNDI

TANZANIA

ZAMBIA

MALAWI

MOZAMBIQUE

DEM. REP. OF THE CONGO (ZAIRE)

Mediterranean Sea

Black Sea

Istanbul

Izmir

Ankara

TURKEY

GEORGIA

Tbilisi

ARM.

AZER.

Yerevan

Baku

Caspian Sea

N. CYPRUS

CYPRUS

LEBANON

SYRIA

Damascus

ISRAEL

JORDAN

Amman

IRAQ

Baghdad

Tabrīz

Tehrān

IRAN

Eşfahān

Mashhad

KUWAIT

Kuwait

SAUDI ARABIA

Riyadh

Jiddah

Red Sea

Persian Gulf

BAHRAIN

QATAR

Abu Dhabi

U.A.E.

Gulf of Oman

Muscat

OMAN

YEMEN

Sanaa

Gulf of Aden

Socotra (Yem.)

Arabian Sea

RUSSIA

Noril'sk

Ob'

Yenisey

Yekaterinburg

Chelyabinsk

Omsk

Novosibirsk

Barnaul

Volga

Astana

Karaganda

Semipalatinsk

Öskemen

KAZAKHSTAN

Aral Sea

Syr Darya

Lake Balkhash

Almaty

Ürümqi

UZBEKISTAN

TURKMENISTAN

Ashgabat

Tashkent

Bishkek

KYRGYZSTAN

Dushanbe

TAJIKISTAN

Kabul

AFGHANISTAN

Islamabad

Lahore

Amritsar

PAKISTAN

Karāchi

Hyderābād

Indus

Delhi

New Delhi

NEPAL

Kathmandu

Brahmaputra

Ganges

Kānpur

Ahmadābād

INDIA

Kolkata (Calcutta)

Nāgpur

Mumbai (Bombay)

Godavari

Hyderābād

Bay of Bengal

Bangalore

Chennai (Madras)

Lakshadweep (India)

SRI LANKA

Colombo

MALDIVES

INDIAN OCEAN

0 200 400 600 800 Miles

0 200 400 600 800 1000 Kilometers

Lambert Azimuthal Equal Area Projection

RAND McNALLY

A34

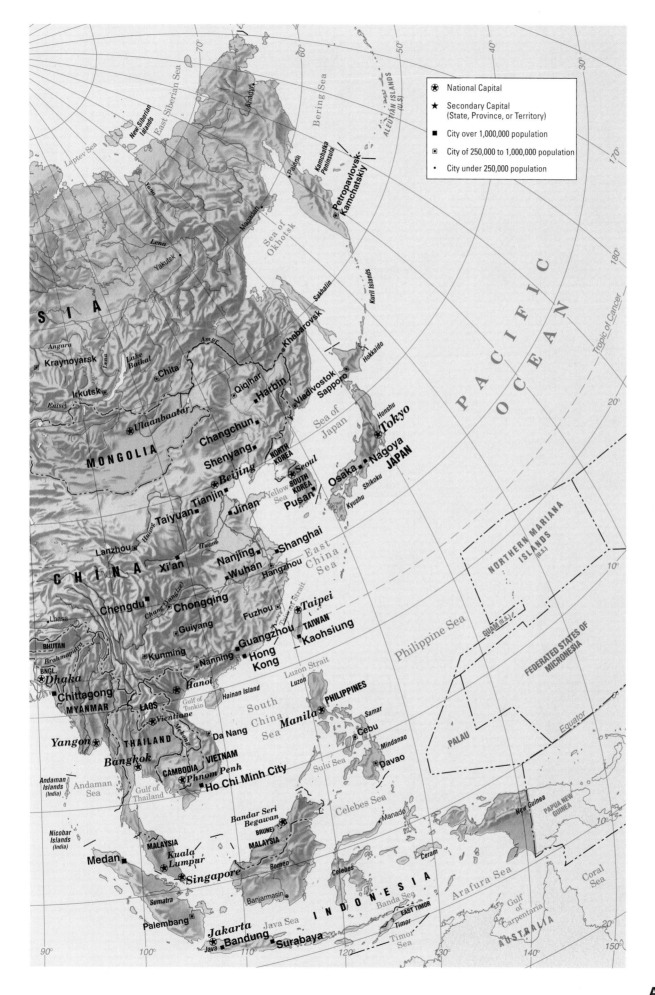

National Capital
Secondary Capital (State, Province, or Territory)
City over 1,000,000 population
City of 250,000 to 1,000,000 population
City under 250,000 population
PACIFIC OCEAN
East Siberian Sea
New Siberian Islands
Laptev Sea
Bering Sea
ALEUTIAN ISLANDS (U.S.)
Anadyr
Palana
Kamchatka Peninsula
Petropavlovsk-Kamchatskiy
Magadan
Sea of Okhotsk
Yana
Lena
Yakutsk
Kuril Islands
Sakhalin
Khabarovsk
Amur
Angara
Kraynoyarsk
Lake Baikal
Chita
Irkutsk
Enisej
Qiqihar
Harbin
Vladivostok
Sapporo
Hokkaido
Honshu
Tokyo
Ulaanbaatar
MONGOLIA
Changchun
Sea of Japan
Shenyang
NORTH KOREA
Seoul
SOUTH KOREA
Beijing
Tianjin
Yellow Sea
Pusan
Osaka
Nagoya
JAPAN
Shikoku
Kyushu
Jinan
Taiyuan
Huang
Lanzhou
Shanghai
Nanjing
East China Sea
Xi'an
CHINA
Wuhan
Hangzhou
Chengdu
Chang (Yangtze)
Chongqing
Fuzhou
Taiwan Strait
Taipei
TAIWAN
Kaohsiung
Lhasa
Guiyang
Guangzhou
BHUTAN
Brahmaputra
Kunming
Nanning
Hong Kong
BNGL.
Dhaka
Chittagong
Hanoi
Hainan Island
Gulf of Tonkin
MYANMAR
LAOS
Vientiane
Mekong
South China Sea
Luzon Strait
Luzon
PHILIPPINES
Manila
Samar
Cebu
Mindanao
Davao
Da Nang
Yangon
THAILAND
Bangkok
VIETNAM
CAMBODIA
Phnom Penh
Ho Chi Minh City
Andaman Islands (India)
Andaman Sea
Gulf of Thailand
Sulu Sea
Celebes Sea
Bandar Seri Begawan
BRUNEI
MALAYSIA
Nicobar Islands (India)
Medan
Kuala Lumpur
Singapore
Borneo
Celebes
Manado
Ceram
New Guinea
PAPUA NEW GUINEA
Philippine Sea
NORTHERN MARIANA ISLANDS (U.S.)
GUAM (U.S.)
FEDERATED STATES OF MICRONESIA
PALAU
Equator
Tropic of Cancer
Banjarmasin
Sumatra
Palembang
INDONESIA
Banda Sea
EAST TIMOR
Timor
Timor Sea
Arafura Sea
Gulf of Carpentaria
AUSTRALIA
Coral Sea
Jakarta
Java Sea
Java
Bandung
Surabaya
90°
100°
110°
120°
130°
140°
150°
170°
180°
70°
60°
50°
40°
30°
20°
10°

A35

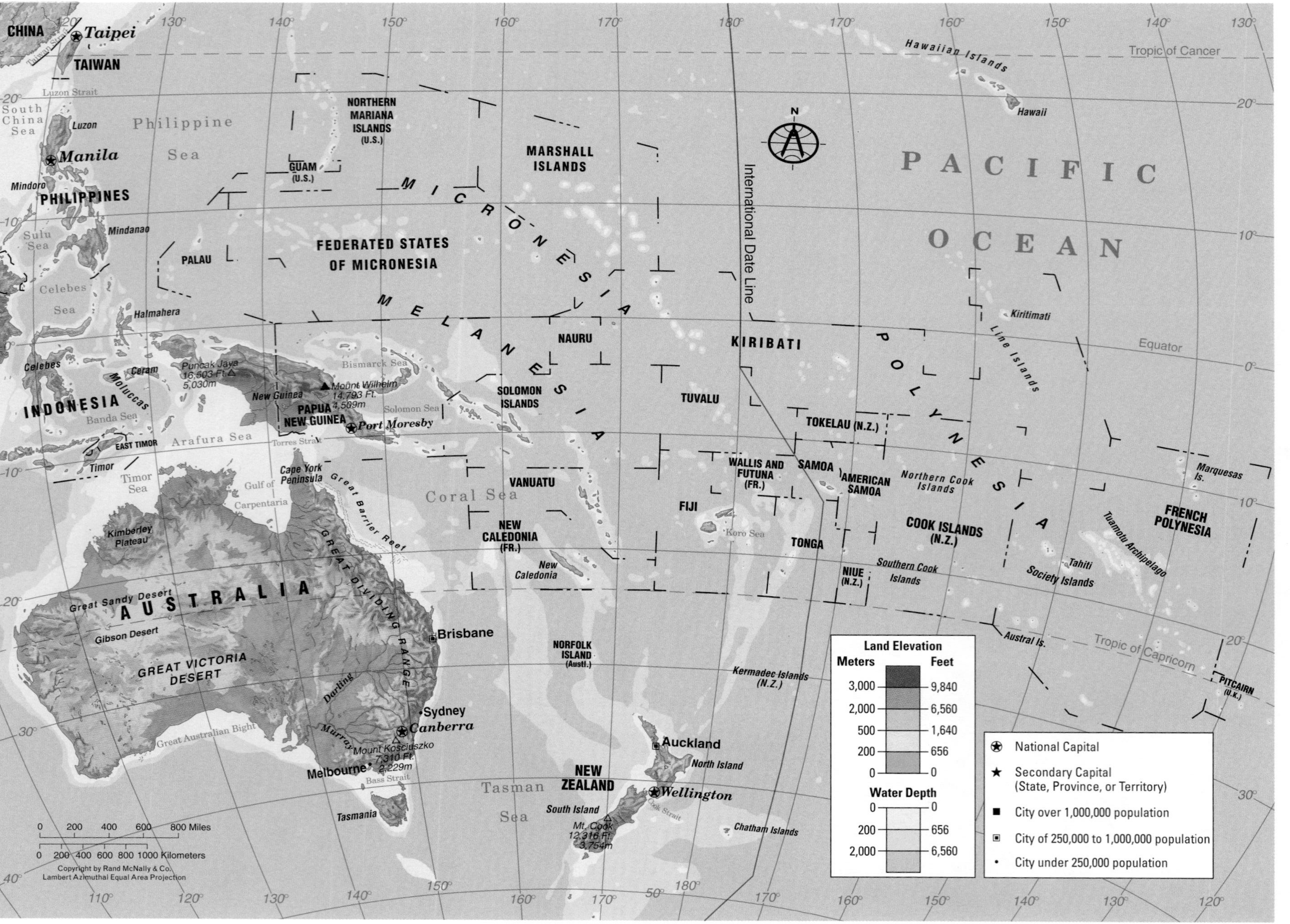
PACIFIC OCEAN
MICRONESIA
MELANESIA
POLYNESIA
AUSTRALIA
CHINA
Taipei
TAIWAN
Taiwan Strait
Luzon Strait
South China Sea
Luzon
Philippine Sea
Manila
Mindoro
PHILIPPINES
Mindanao
Sulu Sea
Celebes Sea
Celebes
Halmahera
Ceram
Moluccas
INDONESIA
Banda Sea
EAST TIMOR
Timor
Timor Sea
Arafura Sea
Torres Strait
Puncak Jaya 16,503 Ft. 5,030m
New Guinea
Mount Wilhelm 14,793 Ft. 4,509m
PAPUA NEW GUINEA
Port Moresby
Bismarck Sea
Solomon Sea
PALAU
NORTHERN MARIANA ISLANDS (U.S.)
GUAM (U.S.)
FEDERATED STATES OF MICRONESIA
MARSHALL ISLANDS
NAURU
SOLOMON ISLANDS
KIRIBATI
TUVALU
TOKELAU (N.Z.)
WALLIS AND FUTUNA (FR.)
SAMOA
AMERICAN SAMOA
FIJI
Koro Sea
TONGA
NIUE (N.Z.)
VANUATU
NEW CALEDONIA (FR.)
New Caledonia
Coral Sea
Cape York Peninsula
Gulf of Carpentaria
Great Barrier Reef
GREAT DIVIDING RANGE
Kimberley Plateau
Great Sandy Desert
Gibson Desert
GREAT VICTORIA DESERT
Great Australian Bight
Darling
Murray
Brisbane
Sydney
Canberra
Mount Kosciuszko 7,310 Ft. 2,229m
Melbourne
Bass Strait
Tasmania
Tasman Sea
NORFOLK ISLAND (Austl.)
NEW ZEALAND
Auckland
North Island
Wellington
Cook Strait
South Island
Mt. Cook 12,316 Ft. 3,754m
Chatham Islands
Kermadec Islands (N.Z.)
International Date Line
Hawaiian Islands
Hawaii
Kiritimati
Line Islands
Northern Cook Islands
COOK ISLANDS (N.Z.)
Southern Cook Islands
Society Islands
Tahiti
Tuamotu Archipelago
FRENCH POLYNESIA
Marquesas Is.
Austral Is.
PITCAIRN (U.K.)
Tropic of Cancer
Tropic of Capricorn
Equator
N
Land Elevation
Meters
Feet
3,000 — 9,840
2,000 — 6,560
500 — 1,640
200 — 656
0 — 0
Water Depth
0 — 0
200 — 656
2,000 — 6,560
National Capital
Secondary Capital (State, Province, or Territory)
City over 1,000,000 population
City of 250,000 to 1,000,000 population
City under 250,000 population
0 200 400 600 800 Miles
0 200 400 600 800 1000 Kilometers
Copyright by Rand McNally & Co.
Lambert Azimuthal Equal Area Projection

North Pole

ALASKA (U.S.)
CANADA
Brooks Range
Barrow
Point Barrow
Chukchi Sea
Wrangell I.
East Siberian Sea
Srednekolymsk
Indigirka
New Siberian I.
New Siberian Islands
Kotelny I.
Verkhoyansk
Verkhoyansk Mts.
Lena
Tiksi
Olenek
RUSSIA
Anabar
Laptev Sea
Taymyr Peninsula
Lake Taymyr
Khatanga
Kotui
Severnaya Zemlya
Dikson
Kara Sea
Franz Josef Land
Novaya Zemlya
Vorkuta
SVALBARD (Nor.)
SPITSBERGEN
Barents Sea
Greenland Sea
Pechora
Murmansk
Kola Peninsula
Arhangel'sk
North Cape
Hammerfest
NORWAY
FINLAND
Jan Mayen (Nor.)
Beaufort Sea
Inuvik
Norman Wells
Great Bear Lake
Amundsen Gulf
Banks I.
QUEEN ELIZABETH ISLANDS
Prince Patrick Island
Melville I.
VICTORIA I.
Viscount Melville Sound
Iqaluktuutiak
Prince of Wales
Somerset
Gulf of Boothia
Devon I.
Ellef Ringnes I.
North Magnetic Pole
Axel Heiberg
ELLESMERE I.
Alert
ARCTIC OCEAN
North Pole
Bylot I.
Etah
Thule
BAFFIN I.
Baffin Bay
Peary Land
GREENLAND (Den.)
Davis Strait
Disko
Godhavn
Godthab
Gunnbjorn Fjeld 12,139 Ft. 3,700m
Angmagssalik
Yukon

0 200 400 600 Miles
0 200 400 600 800 1000 Kilometers
Copyright by Rand McNally & Co.
Azimuthal Equidistant Projection

Land Elevation	
Meters	Feet
3,000	9,840
2,000	6,560
500	1,640
200	656
0	0

Water Depth	
0	0
200	656
2,000	6,560

South Pole

Strait of Magellan
FALKLAND ISLANDS (U.K.)
Cape Horn
Drake Passage
South Shetland Islands (U.K.)
Graham Land
Adelaide I.
Alexander I.
Antarctic Peninsula
Larsen Ice Shelf
South Orkney Islands (U.K.)
Scotia Sea
South Georgia (U.K.)
South Sandwich Islands (U.K.)
ATLANTIC OCEAN
Thurston I.
Bellingshausen Sea
Amundsen Sea
Ellsworth Land
Mt. Sidley 13,717 Ft. 4,181m
Ellsworth Mts.
Vinson Massif 16,066 Ft. 4,897m
Ronne Ice Shelf
Weddell Sea
Berkner I.
Filchner Ice Shelf
Marie Byrd Land
Rockefeller Plateau
Pensacola Mts.
Coats Land
Cape Norvegia
PACIFIC OCEAN
Antarctic Circle
Roosevelt I.
Mt. Kirkpatrick 14,856 Ft. 4,528m
Ross Sea
Ross Ice Shelf
Transantarctic Mountains
South Pole
Queen Maud Land
Mühlig Hofmann Mts.
Cape Adare
Mt. Erebus 12,451 Ft. 3,795m
Victoria Land
ANTARCTICA
Sør Rondane Mts.
George V Coast
Macquarie Island (Austl.)
South Magnetic Pole
Wilkes Land
Lambert Glacier
American Highland
Enderby Land
Napier Mts.
Amery Ice Shelf
Cape Ann
Cape Darnley
Cape Poinsett
Prince Edward Is. (S. Afr.)
Crozet Archipelago (Fr.)
INDIAN OCEAN

0 200 400 600 800 1000 Miles
0 300 600 900 1200 1500 Kilometres
Copyright by Rand McNally & Co.
Polar Stereographic Projection

ATLAS

RAND McNALLY

A37

Introduction to World Cultures and Geography

Chapter 1
Welcome to the World

Chapter 2
The Geographer's World

Before You Read

Previewing Unit 1

Unit 1 introduces the term *social studies* and identifies and describes the five fields of learning that contribute to social studies—geography, history, economics, government, and culture. The unit also explains the five themes of geography. It concludes with an explanation of how culture regions are grouped, how they change, and how they evolve. The importance of social studies as a way to understand Earth and its people is examined throughout the unit.

UNIT 1

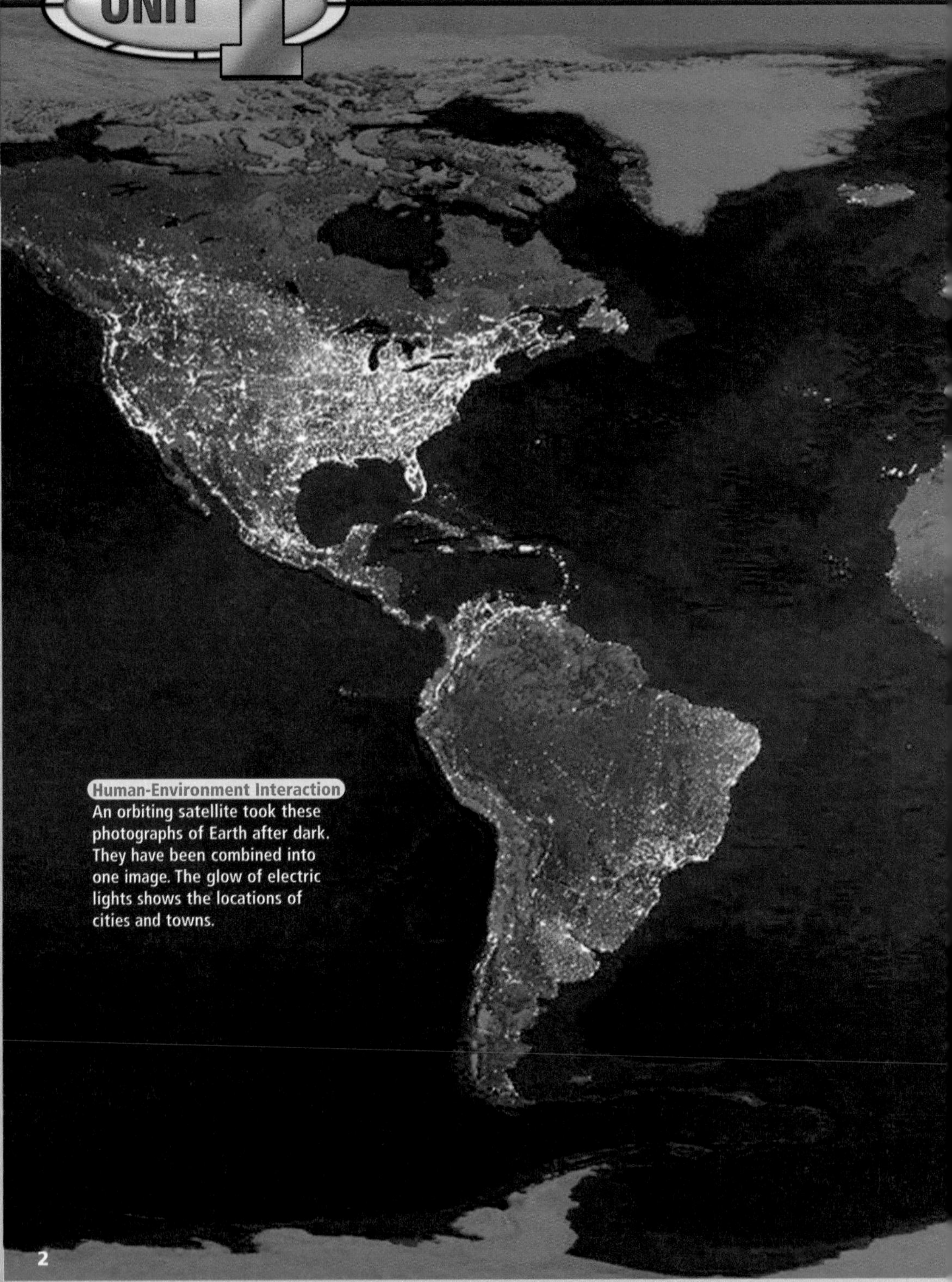

Human-Environment Interaction An orbiting satellite took these photographs of Earth after dark. They have been combined into one image. The glow of electric lights shows the locations of cities and towns.

2

Unit Level Activities

1. Write an Editorial

After students read about culture regions, ask them to imagine that a historic site in their area is about to be destroyed due to highway construction. Have students decide whether they support or oppose the highway, citing reasons why the history of a people is a vital part of their culture, why the current needs of a people are important, and so on.

2. Write Interview Questions

Have students write questions they might ask a visitor from another country. Remind them to include questions about the five fields of social studies—geography, history, economics, government, and culture. Then have students trade questions with a partner and choose a foreign country about which they will answer the questions they have received. Have students interview each other using the questions and answers.

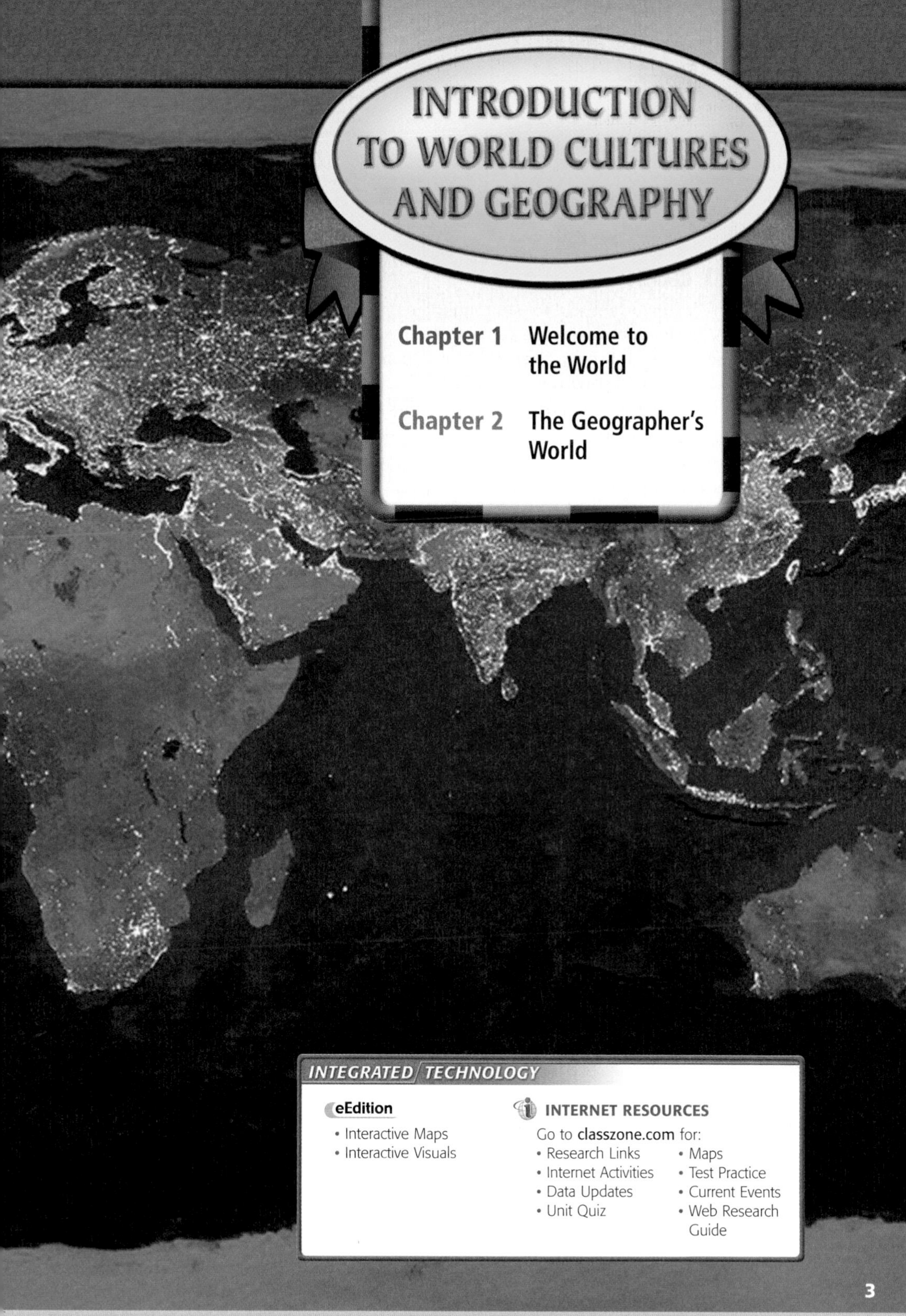

INTRODUCTION TO WORLD CULTURES AND GEOGRAPHY
Chapter 1 Welcome to the World
Chapter 2 The Geographer's World
INTEGRATED TECHNOLOGY
eEdition
• Interactive Maps
• Interactive Visuals
INTERNET RESOURCES
Go to classzone.com for:
• Research Links
• Internet Activities
• Data Updates
• Unit Quiz
• Maps
• Test Practice
• Current Events
• Web Research Guide
3

FOCUS ON VISUALS

Interpreting the Photograph Have students examine the photograph. Ask a volunteer to read the caption aloud. Explain that satellites are human-made objects that orbit Earth. Ask students to identify the continents and major bodies of water in the photograph. Then ask them to locate the area in which they live. Guide students to think about what insights seeing Earth from the perspective of outer space can provide.

Possible Responses Only a camera many miles above Earth can capture these photographs. From this perspective, the continents seem close together despite the great distances between them. Combining photographs taken from this perspective creates an image of how Earth looks as a whole.

Extension Ask students to write a short radio broadcast in which a passenger in a manned satellite describes what Earth looks like from outer space.

Implementing the National Geography Standards

Standard 10 Describe visible cultural elements in the student's own community

Objective To identify various cultural elements in the student's environment

Class Time 25 minutes

Task Students work in groups to find examples of the different cultures represented in their communities. Assign each group one of these topics: religious institutions, restaurants, specialty stores, distinctive buildings, and media. Each group should list local examples of their topic. Have the groups share their examples.

Evaluation Groups should identify at least two foreign and two domestic examples of cultural elements in the community.

OBJECTIVES

1. To understand the methods for determining location
2. To interpret the different elements of a map
3. To understand the elements and purpose of the geographic grid
4. To examine different types of projections
5. To understand and interpret the different types of maps

FOCUS & MOTIVATE

WARM-UP

Using Maps Ask students to create a class list of the types of information that can be found on a map. Use the following questions to help them understand what is on a map.

1. Why would you look at a map?
2. What information would you expect to find on a map?

INSTRUCT: Objective 1

Map Basics

- Where can you find the subject and basic information contained on a map? in its title
- What is the function of a compass rose? to show direction
- What is a legend? the explanation of symbols and colors used on a map
- What map features would you use to determine the distance between two places? scale

MORE ABOUT...

Magnetic Compasses

A compass consists of a small, lightweight magnet balanced atop a nearly frictionless pivot point. The magnet is sometimes called a needle. In response to Earth's magnetic field, the needle points toward the north. In the photograph on page 4 the needle is in the shape of an arrow.

Map Basics

Maps are an important tool for studying the use of space on Earth. This handbook covers the basic map skills and information that geographers rely on as they investigate the world—and the skills you will need as you study geography.

Mapmaking depends on surveying, or measuring and recording the features of Earth's surface. Until recently, this could be undertaken only on land or sea. Today, aerial photography and satellite imaging are the most popular ways to gather data.

Location • Magnetic compasses, introduced by the Chinese in the 1100s, help people accurately determine directions. ▲

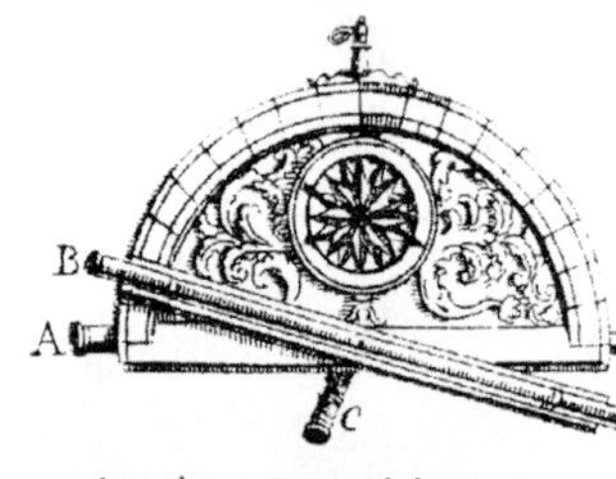

Location • Determining a ship's location at sea was the purpose of this 1750 instrument, called a sextant. ▲

Human-Environment Interaction • Surveyors use a theodolite, which measures angles and distances on Earth. ▲

Location • An early example of a three-dimensional geographic grid. ▼

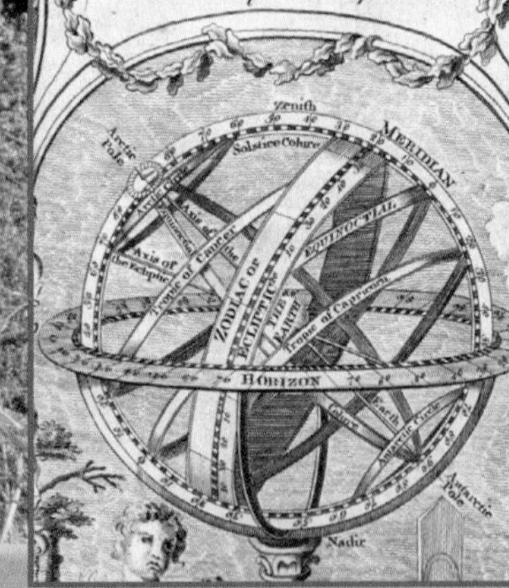

Recommended Resources

BOOKS FOR THE TEACHER

The National Geographic Desk Reference. Washington, D.C., 1999. A geographic reference with hundreds of maps, charts, and graphs.

Erickson, Jon. ***Making of the Earth: Geologic Forces That Shape Our Planet.*** Facts on File, Inc., NY, 2000. Discussion of formation of landforms.

VanCleave, Janice. ***Janice VanCleave's Geography for Every Kid.*** John Wiley and Sons, Inc., NY, 1993. Book of activities to help children understand geographic concepts such as how early explorers used maps, latitude and longitude, scale, grid, geographic versus magnetic north, compass rose, and time zones.

INTERNET

For more about geography skills, visit **classzone.com.**

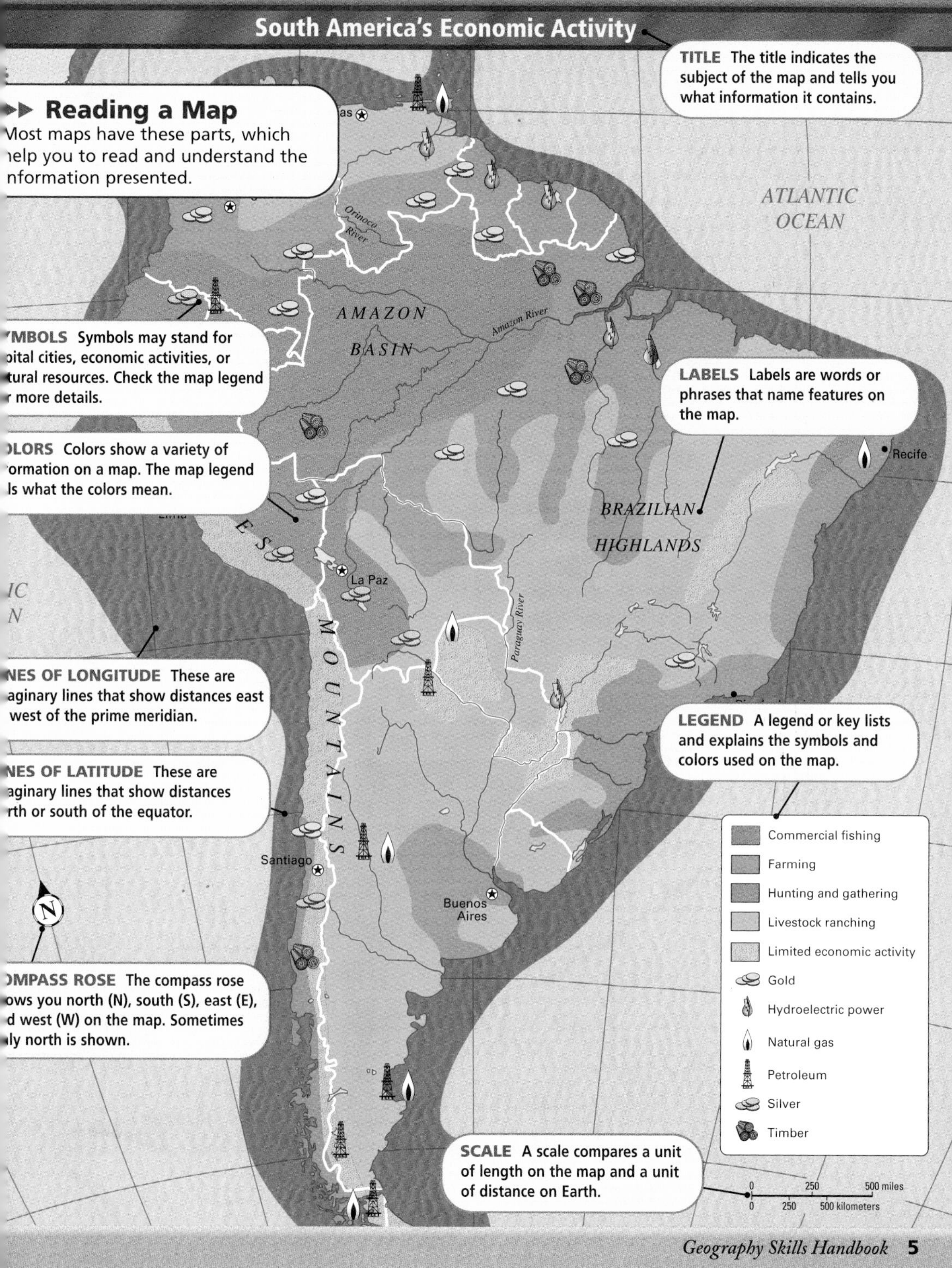

South America's Economic Activity
TITLE The title indicates the subject of the map and tells you what information it contains.
Reading a Map
Most maps have these parts, which help you to read and understand the information presented.
ATLANTIC OCEAN
Orinoco River
AMAZON BASIN
Amazon River
MBOLS Symbols may stand for pital cities, economic activities, or tural resources. Check the map legend r more details.
LABELS Labels are words or phrases that name features on the map.
Recife
LORS Colors show a variety of ormation on a map. The map legend ls what the colors mean.
BRAZILIAN HIGHLANDS
La Paz
Paraguay River
NES OF LONGITUDE These are aginary lines that show distances east west of the prime meridian.
LEGEND A legend or key lists and explains the symbols and colors used on the map.
NES OF LATITUDE These are aginary lines that show distances rth or south of the equator.
Commercial fishing
Farming
Hunting and gathering
Livestock ranching
Limited economic activity
Gold
Hydroelectric power
Natural gas
Petroleum
Silver
Timber
Santiago
Buenos Aires
MOUNTAINS
OMPASS ROSE The compass rose ows you north (N), south (S), east (E), d west (W) on the map. Sometimes ly north is shown.
SCALE A scale compares a unit of length on the map and a unit of distance on Earth.
0 250 500 miles
0 250 500 kilometers

Geography Skills Handbook 5

MORE ABOUT...

Compass Rose

Ship charts and old-fashioned maps often use a compass rose—an ornamental directional arrow. The rose is usually drawn from a circle divided into 360 degrees and is used to tell directions from magnetic north, or 0 degrees, on the compass.

Activity Options

Multiple Learning Styles: Spatial

Time One class period

Task Creating a map to use the basic features of mapmaking

Purpose To explore how the basic features of a map work

Supplies Needed

- Paper
- Colored pens and markers
- Large sheets of paper

Activity Have students create maps of the areas around their homes or school. Encourage them to make imaginative use of the basic features examined above. Students might create a scale using the length of their footsteps or design symbols to designate the location of favorite stores or homes of friends and relatives. With colors, they might show areas where they play. Display and discuss the maps in class.

Map Basics, cont.

INSTRUCT: Objective 2

Longitude and Latitude Lines/Hemisphere/Scale

- What features of a map will help you find absolute locations? lines of longitude and latitude
- What is the term for half of the globe? hemisphere
- What determines the scale of a map? how much detail is to be shown on the map

MORE ABOUT...

Scale

Ratio scales, also called representative fraction scales or fractional scales, are the most accurate of all scale statements. Since they are presented numerically, they can be understood in any language.

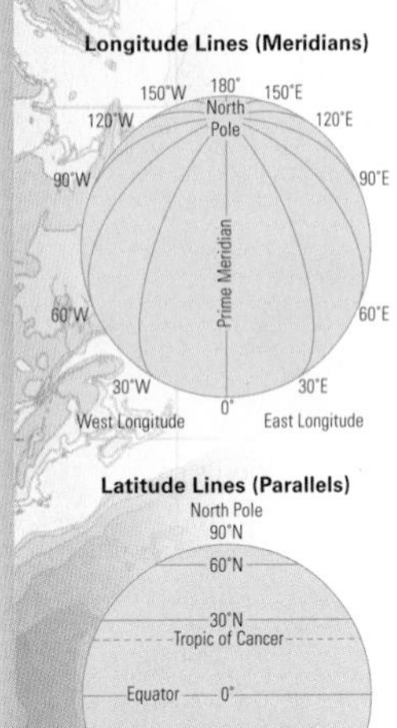

▶▶ Longitude and Latitude Lines

Longitude and latitude lines appear together on a map and allow you to pinpoint the absolute locations of cities and other geographic features. You express these locations as coordinates of intersecting lines. These are measured in degrees.

Longitude lines are imaginary lines that run north and south; they are also known as meridians. They show distances in degrees east or west of the prime meridian. The prime meridian is a longitude line that runs from the North Pole to the South Pole through Greenwich, England. It marks 0° longitude.

Latitude lines are imaginary lines that run east to west around the globe; they are also known as parallels. They show distances in degrees north or south of the equator. The equator is a latitude line that circles Earth halfway between the north and south poles. It marks 0° latitude. The tropics of Cancer and Capricorn are parallels that form the boundaries of the tropical zone, a region that stays warm all year.

▶▶ Hemisphere

Hemisphere is a term for half the globe. The globe can be divided into northern and southern hemispheres (separated by the equator) or into eastern and western hemispheres. The United States is located in the northern and western hemispheres.

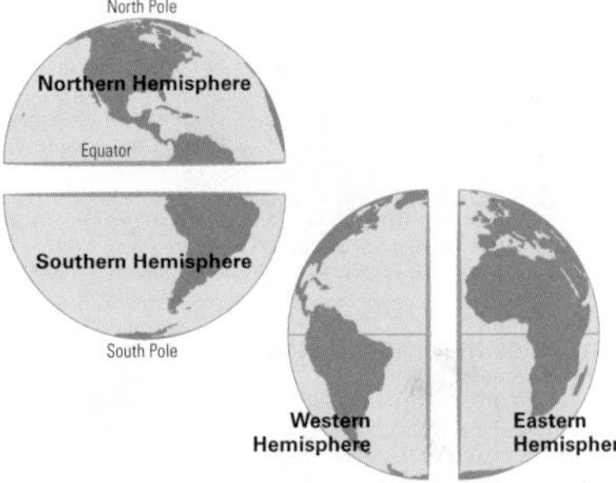

▶▶ Scale

A geographer decides what scale to use by determining how much detail to show. If many details are needed, a large scale is used. If fewer details are needed, a small scale is used.

Small scale used, without a lot of detail. ▼

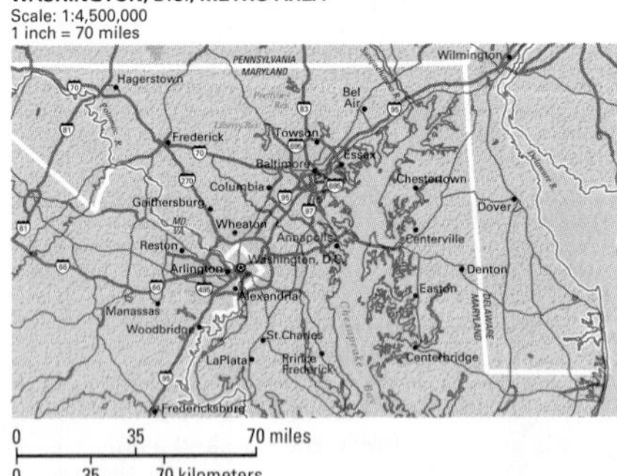

Larger scale used, with a lot of detail. ▼

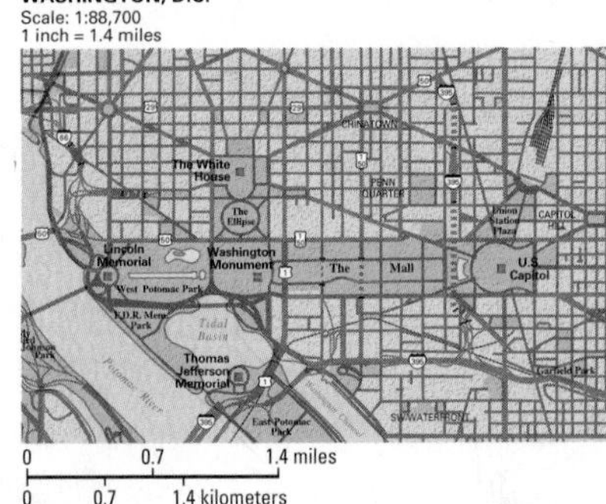

Activity Options

Interdisciplinary Link: Mathematics

Time 30 minutes

Task Creating a map and scale for the classroom

Purpose To understand how to use scale to represent distance and detail on a map

Supplies Needed

- Tape measures
- Rulers
- Grid paper
- Colored pencils

Activity Divide students into groups. Have one group measure the dimensions of the classroom. Have another group measure the sizes of the room's larger furnishings. A third group could measure the distance of these furnishings from the walls. After students finish collecting this data, have them draw a map that shows the classroom and its dimensions on the chalkboard. Next, provide students with a ratio (for example, 5 inches = 1 foot) and have them redraw the map of the room on their grid paper using this ratio.

▶▶ Projections

A projection is a way of showing the curved surface of Earth on a flat map. Flat maps cannot show sizes, shapes, and directions with total accuracy. As a result, all projections distort some aspect of Earth's surface. Below are four projections.

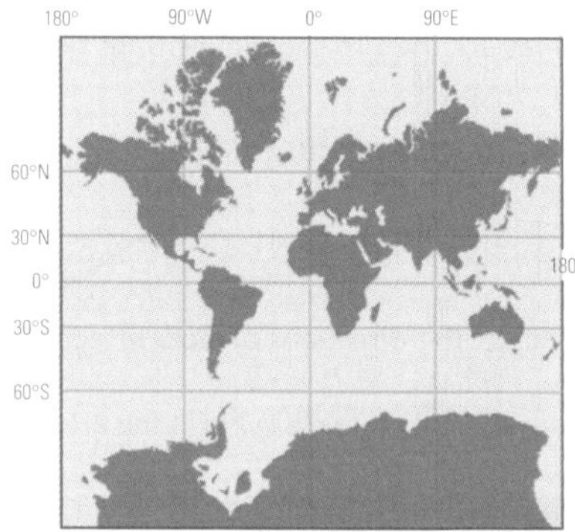

Mercator Projection • **The Mercator projection shows most of the continents as they look on a globe. However, the projection stretches out the lands near the north and south poles. The Mercator projection is used for all kinds of navigation. ▲**

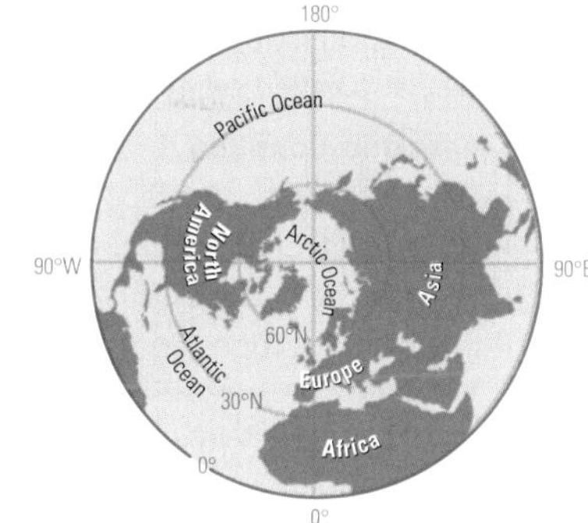

Azimuthal Projection • **An azimuthal projection shows Earth so that a straight line from the central point to any other point on the map corresponds to the shortest distance between the two points. Sizes and shapes of the continents are distorted. ▲**

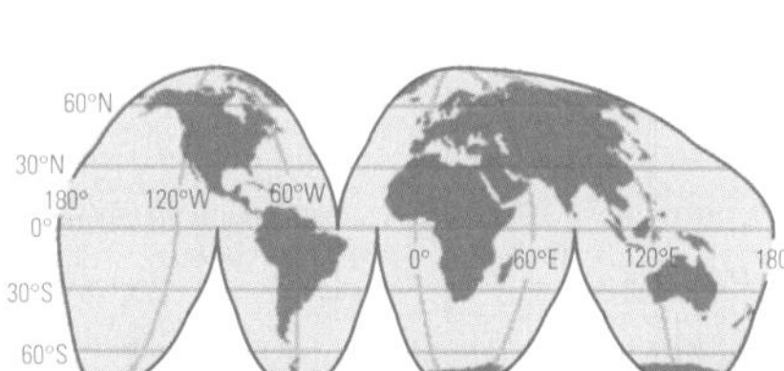

Homolosine Projection • **This projection shows landmasses' shapes and sizes accurately, but distances are not correct. ▲**

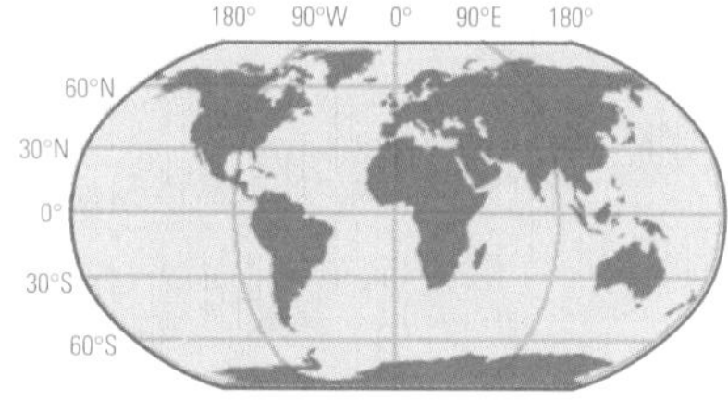

Robinson Projection • **For textbook maps, the Robinson projection is commonly used. It shows the entire Earth, with continents and oceans having nearly their true sizes and shapes. However, the landmasses near the poles appear flattened. ▲**

MAP PRACTICE

MAIN IDEAS

1. (a) What are the longitude and latitude of your city or town?

(b) What information is provided by the legend in the map on page 5?

(c) What is a projection? Compare and contrast the depictions of Antarctica in the Mercator and Robinson projections.

CRITICAL THINKING

2. Making Inferences Why do you think latitude and longitude are important to sailors?

Think About

- the landmarks you use to find your way around
- the landmarks available to sailors on the ocean

INSTRUCT: Objective 3

Projections

- What is a projection? a way of showing the curved surface of Earth on a flat map
- How might a projection be misleading? Because Earth is a sphere, projecting it onto a flat surface will cause some distortion.

MORE ABOUT...

Projections

Projections have been known to involve heated debates. In 1974, a German cartographer named Arno Peters introduced a different kind of projection. He claimed that his new world map showed more accurate proportions than maps with traditional projections—which he considered biased toward developed countries. These claims were hotly contested by other cartographers, a controversy that continues today.

CRITICAL THINKING ACTIVITY

Drawing Conclusions Have students examine the azimuthal projection on this page. Ask them why a bar scale might be misleading if they were to measure the distance between San Francisco and Rome and try to convert the measurement using the bar scale. The bar scale might be misleading because the size and shape of Earth have been distorted by the projection.

Class Time 10 minutes

Map Practice

1. Main Ideas

a. Responses will vary.
b. economic activities and resources
c. A projection shows the curved surfaces of Earth on a flat map. On the Mercator projection Antarctica looks stretched out and large; on the Robinson projection Antarctica looks thin and narrow.

2. Critical Thinking

Possible Response Students may note that sailors rely on longitude and latitude because they have no visual landmarks to use.

INSTRUCT: Objective 4

Different Types of Maps

- What is the purpose of a physical map? to show landforms and bodies of water in a specific area
- How does a physical map represent relief? with color, shading, or contour lines
- What is the purpose of a political map? to show features on Earth's surface that are created by humans, such as countries, states, cities, and other political entities

MORE ABOUT...

Sea Level

Elevations are based on a landform's distance above or below sea level. But what if one sea is higher than another? And what happens when the level of the sea changes, for example, with the change of the tides? Because of these questions, geographers measure elevation using global mean sea level—the height of the surface of the sea averaged over all tide stages and over long periods of time.

Different Types of Maps

▸▸ Physical Maps

Physical maps help you see the landforms and bodies of water in specific areas. By studying a physical map, you can learn the relative locations and characteristics of places in a region.

On a physical map, color, shading, or contour lines are used to show elevations or altitudes, also called relief.

Ask these questions about the physical features shown on a physical map:

- Where on Earth's surface is this area located?
- What is its relative location?
- What is the shape of the region?
- In which directions do the rivers flow? How might the directions of flow affect travel and transportation in the region?
- Are there mountains or deserts? How might they affect the people living in the area?

Mexico and Central America: Physical

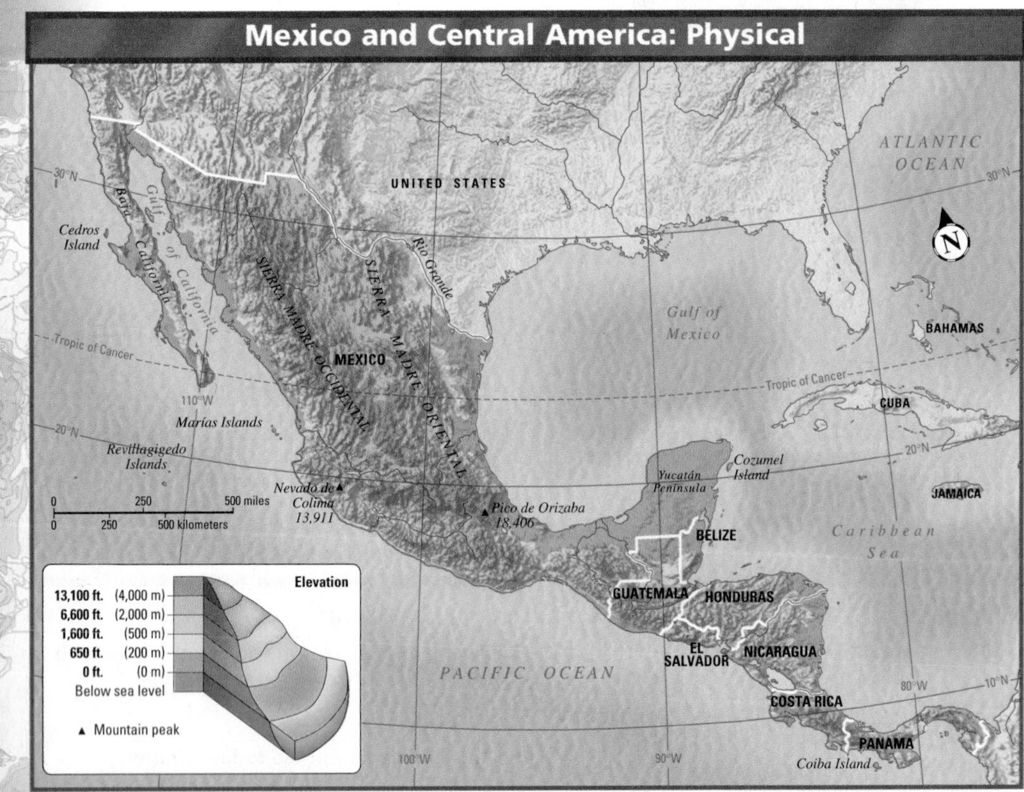

8 UNIT 1

Activity Options

Differentiating Instruction: Less Proficient Readers

Class Time 10 minutes

Task Using mnemonic devices to help remember key terms

Purpose To help students differentiate between and remember longitude and latitude

Activity Some students may have difficulty remembering the difference between latitude and longitude. Have them use the following mnemonic device to differentiate the two terms and to remember them.

Latitude goes **a**round Earth. **L**ongitude goes **o**ver Earth.

Help students devise a simple quiz game in which the answers to the questions are latitude or longitude.

▸▸ Political Maps

Political maps show features that humans have created on Earth's surface. Included on a political map may be cities, states, provinces, territories, and countries. Boarders

Ask these questions about the political features shown on a political map:

- Where on Earth's surface is this area located?
- What is its relative location? How might a country's location affect its economy and its relationships with other countries?
- What is the shape and size of the country? How might its shape and size affect the people living in the country?
- Who are the region's, country's, state's, or city's neighbors?
- How populated does the area seem to be? How might that affect activities there?

Mexico and Central America: Political

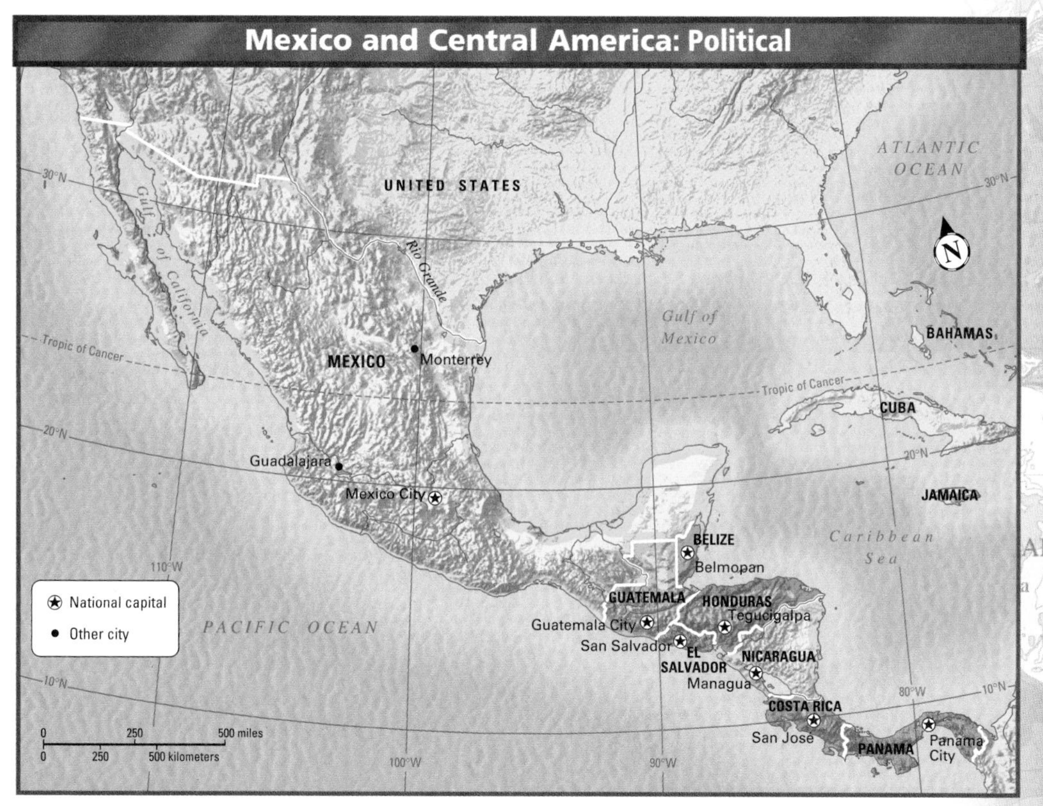

MORE ABOUT...

Political Boundaries

The boundaries we see on political maps are not always as fixed as they seem. Boundary disputes occur even in the United States. Recently the U.S. Supreme Court had to settle a dispute between New York and New Jersey. Leaders from each state argued that Ellis Island, the historic entry point for millions of immigrants to the United States, fell within their borders. In May 1998, the Supreme Court decided in favor of New Jersey.

Population

A population is the total number of people living in a defined area. That area may be as small as a neighborhood school or as large as the world. Births, deaths, and migration—the movement of people—determine the size of a population.

The study of population is called demography. Demographics are critical in making plans for the future. Demographers ask questions about where people are being born, how long they will live, and whether their basic needs can be met.

Activity Options

Multiple Learning Styles: Intrapersonal

Class Time 30 minutes

Task Creating a chart with maps and descriptions of border disputes

Purpose To have students examine contested political boundaries

Supplies Needed

- Magazines, newspapers
- Internet access

Activity Divide students into small groups. Ask them to use newspapers, magazines, or the Internet to find stories about disputed political borders. Encourage students to collect articles discussing various conflicts. Then have them create a large chart summarizing the conflicts.

map showing labeled contested border	origin of dispute and current status

Different Types of Maps, cont.

INSTRUCT: Objective 5

Thematic Maps

- What are thematic maps? maps that focus on specific themes or types of information
- What are some examples of types of thematic maps? maps that show climate, population density, vegetation
- What are some of the ways in which thematic maps are presented? as qualitative and flow-line maps and as cartograms

MORE ABOUT...

Cartographers

Cartography is the art of making maps or charts. People who draw maps are called cartographers. The word comes from the French *carte*, meaning "map" and the Greek *graph* meaning "writing."

▸▸ Thematic Maps

Geographers also rely on thematic maps, which focus on specific ideas. For example, in this textbook you will see thematic maps that show climates, types of vegetation, natural resources, population densities, and economic activities. Some thematic maps show historical trends; others may focus on movements of people or ideas. Thematic maps may be presented in a variety of ways.

Major Rain Forests of Latin America

Oil Reserves in the Western Hemisphere Cartogram

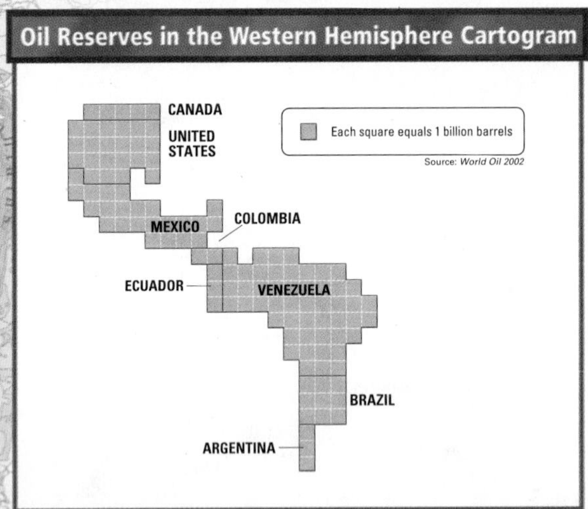

Qualitative Maps On a qualitative map, colors, symbols, dots, or lines are used to help you see patterns related to a specific idea. The map shown here depicts the influence of the Roman Empire on Europe, North Africa, and Southwest Asia.

Use the suggestions below to help you interpret the map.

- Check the title to identify the theme and the data being presented.
- Carefully study the legend to understand the theme and the information presented.
- Look at the physical or political features of the area. How might the theme of the map affect them?
- What are the relationships among the data?

Cartograms A cartogram presents information about countries other than their shapes and sizes. The size of each country is determined by the data being presented, and not by its actual land size. On the cartogram shown here, the countries' sizes show the amounts of their oil reserves.

Use the suggestions below to help you interpret the map.

- Check the title and the legend to identify the data being presented.
- Look at the relative sizes of the countries shown. Which is the largest?
- Which countries are smallest?
- How do the sizes of these countries on a physical map differ from their sizes in the cartogram?
- What are the relationships among the data?

Activity Options

Multiple Learning Styles: Logical

Class Time 30 minutes

Task Locating different thematic maps

Purpose To familiarize students with different thematic maps

Supplies Needed

- Atlases
- Internet access
- Road maps, satellite maps, etc.

Activity Divide students into groups of four or more students. Provide each group with a list of different types of thematic maps. Within each group, have individual students pick one type of map from the list.

Students will be responsible for finding maps that focus on the theme they have chosen. When they have located their maps, have them interpret the information given on the map and prepare a presentation to the class that explains the information and its uses.

Flow-Line Maps Flow-line maps illustrate movements of people, goods, or ideas. The movements are usually shown by a series of arrows. Locations, directions, and scopes of movement can be seen. The width of an arrow may show how extensive a flow is. Often the information is related to a period of time. The map shown here portrays immigration to the United States during 2000.

Use the suggestions below to help you interpret the map.

- Check the title and the legend to identify the data being presented.
- Over what period of time did the movement occur?
- In what directions did the movement occur?
- How extensive was the movement?

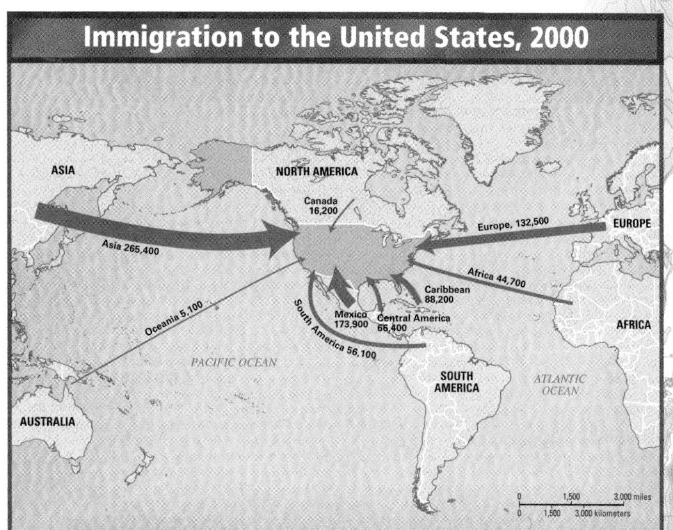

Map Practice

Use pages 8–11 to help you answer these questions. Use the maps on pages 8–9 to answer questions 1–3.

1. What areas in Mexico are of lowest elevation?
2. Belmopan is the capital of which nation?
3. Which is the highest mountain on the physical map?
4. Why are only a few nations shown in the cartogram?
5. Which kind of thematic map would be best for showing the locations of climate zones?

Exploring Local Geography Obtain a physical-political map of your state. Use the data on it to create **two separate maps.** One should show physical features only, and the other should show political features only.

CRITICAL THINKING ACTIVITY

Interpreting Maps Have students examine the immigration map. Use the following questions to help them interpret the map: What period of American history does this map cover? the 1900s During this period, from what part of the world did most immigrants come? the Americas

Class Time 10 minutes

Map Practice

Responses

1. coastal areas
2. Belize
3. Pico de Orizaba
4. The cartogram shows countries that have oil reserves in the Western Hemisphere. All other countries are left off the cartogram.
5. A qualitative map would be best for showing the location of climate zones.

GeoActivity

Integrated Assessment

- Rubric for creating a map

MORE ABOUT...

Deserts

Deserts are regions where the land is covered in sand or bare soil. Precipitation totals are always small, usually less than ten inches a year. Deserts are not always hot though. Temperatures can vary from extremely hot days to cold, even freezing, nights. The plants and animals that survive in the desert do so because they have adapted to the environment. Succulents, for example, store water in their leaves and stems. Many desert animals are nocturnal, taking advantage of the cool evenings to be active, while burrowing in the earth during the day.

Volcanoes

Most volcanoes form when molten rock from deep inside the earth rises to the surface at a fault line or soft spot in a tectonic plate. The molten rock that spurts out of the top of the volcano is known as lava.

Geographic Dictionary

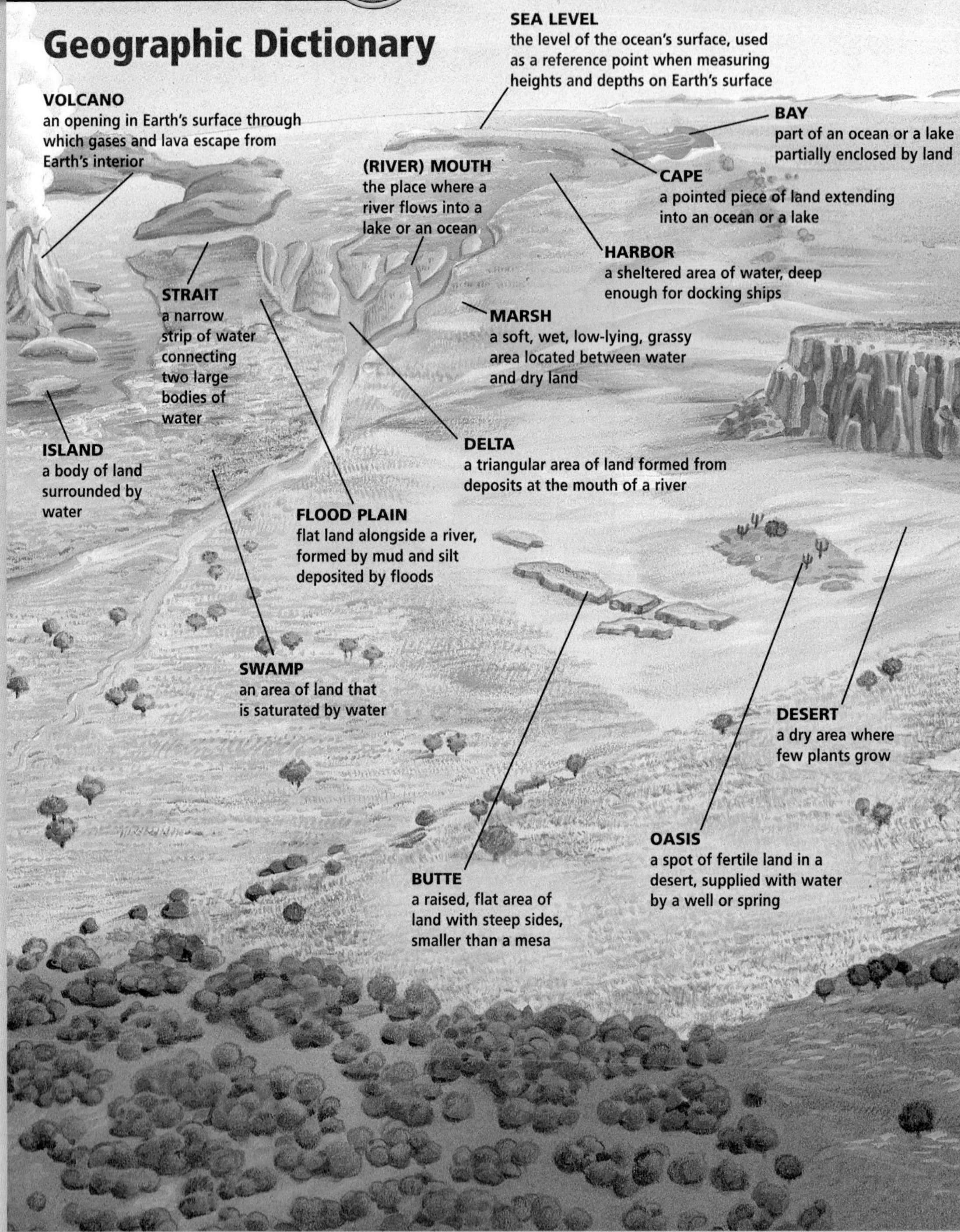
SEA LEVEL
the level of the ocean's surface, used as a reference point when measuring heights and depths on Earth's surface
VOLCANO
an opening in Earth's surface through which gases and lava escape from Earth's interior
BAY
part of an ocean or a lake partially enclosed by land
(RIVER) MOUTH
the place where a river flows into a lake or an ocean
CAPE
a pointed piece of land extending into an ocean or a lake
HARBOR
a sheltered area of water, deep enough for docking ships
STRAIT
a narrow strip of water connecting two large bodies of water
MARSH
a soft, wet, low-lying, grassy area located between water and dry land
ISLAND
a body of land surrounded by water
DELTA
a triangular area of land formed from deposits at the mouth of a river
FLOOD PLAIN
flat land alongside a river, formed by mud and silt deposited by floods
SWAMP
an area of land that is saturated by water
DESERT
a dry area where few plants grow
OASIS
a spot of fertile land in a desert, supplied with water by a well or spring
BUTTE
a raised, flat area of land with steep sides, smaller than a mesa

Activity Options

Differentiating Instruction: Gifted and Talented

Class Time one hour

Task Writing a report on the measurement of sea level using information gathered on the Internet or from library resources

Purpose To learn more about the use of sea level as a basis for determining elevation

Supplies Needed
- Internet access
- Library resource material

Activity Challenge gifted and interested students to use the Internet or library to investigate sea level as a basis for determining elevation.

They might examine the history of sea level measurement, the causes of fluctuations in sea levels, how these fluctuations are measured, and how global mean sea level is defined and measured.

Ask students to use their research to write a short report that includes a visual component to clarify difficult ideas and concepts. They should use this visual in a presentation to the class.

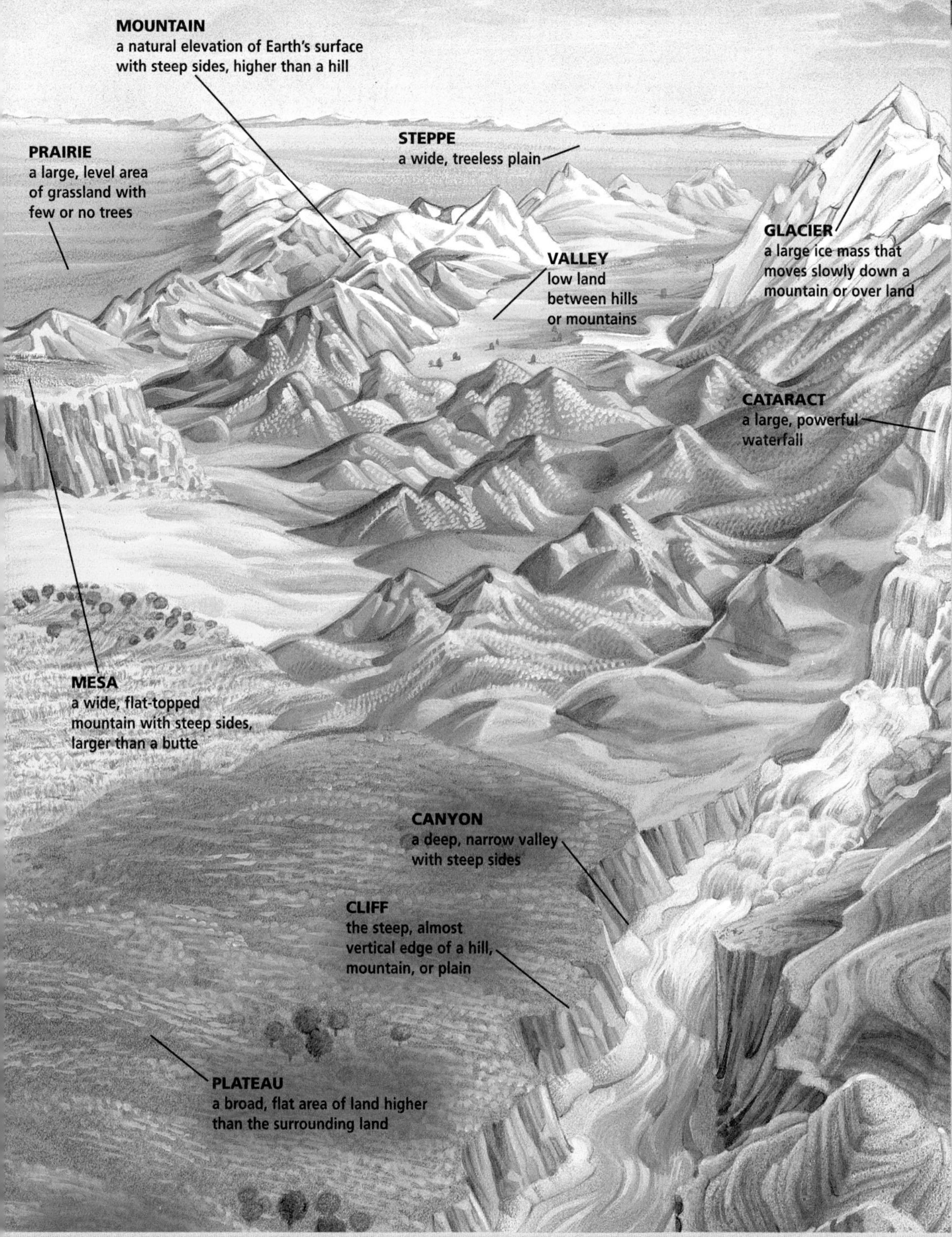
MOUNTAIN
a natural elevation of Earth's surface with steep sides, higher than a hill
PRAIRIE
a large, level area of grassland with few or no trees
STEPPE
a wide, treeless plain
GLACIER
a large ice mass that moves slowly down a mountain or over land
VALLEY
low land between hills or mountains
CATARACT
a large, powerful waterfall
MESA
a wide, flat-topped mountain with steep sides, larger than a butte
CANYON
a deep, narrow valley with steep sides
CLIFF
the steep, almost vertical edge of a hill, mountain, or plain
PLATEAU
a broad, flat area of land higher than the surrounding land

MORE ABOUT...

Glaciers

Glaciers cover about 6 million square miles or 3 percent of Earth's surface. They form at high elevations where it is cold enough for more snow to fall each year than melts. Over the years the snow gets deeper and deeper, and pressure from the weight of the snow turns it into huge sheets of ice. These sheets of ice flow, like slow-moving rivers, down the mountainside until they reach the warmer air along the ocean. When they break off, they form floating icebergs.

Steppes

A steppe is an area mainly covered with grassland. A steppe usually receives about 10 to 20 inches of rain a year. That is enough rain to support short grasses, but not enough for tall grass or trees. The Eurasian steppe, which extends from Hungary to China, is the largest grassland in the world.

Canyons

A deep, narrow valley with steep sides is called a canyon, from the Spanish word *cañon*, which means "tube" or "pipe." Most canyons develop in areas with arid climates, hard rocks, and streams that cascade down steep slopes. Scientists estimate that it took millions of years for the Grand Canyon in Arizona to form. Carved by the Colorado River the canyon is 18 miles wide in some places; at one point it is 1 mile deep from rim to river.

Activity Options

Differentiating Instruction: Students Acquiring English/ESL

Students who are acquiring English may have difficulty understanding that when we ask the question, "What is it like?" we are referring to place, one of the five themes of geography. Remind them that when we describe a place, we are describing physical features, such as climate, or cultural characteristics, such as religion or ethnicity.

Tell students that cartograms are useful for showing these characteristics. Have them brainstorm a list of other features of place that can be represented on a cartogram. water resources, deserts, forests, roads, home ownership Then ask for a list of what political units such as states can be represented. counties, cities, towns, neighborhoods

PLANNING GUIDE: CHAPTER 1

Welcome to the World

	OVERVIEW	COPYMASTERS	INTEGRATED TECHNOLOGY
CHAPTER RESOURCES	The student will examine the fields of learning that contribute to social studies and develop an understanding of culture and culture traits.	**In-depth Resources: Unit 1** • Guided Reading Worksheets, pp. 6–7 • Skillbuilder Practice, p. 10 • Unit Atlas Activities, pp. 1–4 • Geography Workshop, pp. 22–23 **Reading Study Guide** (Spanish and English), pp. 4–9 **Outline Map Activities**	• eEdition Plus Online • EasyPlanner Plus Online • eTest Plus Online • eEdition • Power Presentations • EasyPlanner • Electronic Library of Primary Sources • Test Generator • Reading Study Guide • Critical Thinking Transparencies CT1
SECTION 1 The World at Your Fingertips pp. 17–21	**KEY IDEAS** • Social studies draws from five fields of learning: geography, history, economics, government, and culture. • The five themes of geography are location, region, place, movement, and human-environment interaction. • Culture is the shared beliefs, customs, laws, and ways of living of a people. • A culture trait is any food, clothing, technology, language, tool, or belief shared by a cultural group.	**In-depth Resources: Unit 1** • Guided Reading Worksheet, p. 6 • Reteaching Activity, p. 12 **Reading Study Guide** (Spanish and English), pp. 4–5	**Critical Thinking Transparencies CT2** **Map Transparencies MT10** **classzone.com** Reading Study Guide
SECTION 2 Many Regions, Many Cultures pp. 24–26	• People in culture regions share beliefs, history, and language. • Culture regions borrow from other cultures and depend on one another; many regions are multicultural.	**In-depth Resources: Unit 1** • Guided Reading Worksheet, p. 7 • Reteaching Activity, p. 13 **Reading Study Guide** (Spanish and English), pp. 6–7	**Map Transparencies MT1** **classzone.com** Reading Study Guide

KEY TO RESOURCES

 Audio
 CD-ROM
Copymaster
 Internet
 Overhead Transparency
Pupil's Edition
Teacher's Edition
Video

ASSESSMENT OPTIONS

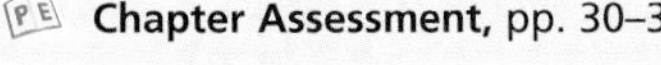

Chapter Assessment, pp. 30–31

Formal Assessment
- Chapter Tests: Forms A, B, C, pp. 17–18

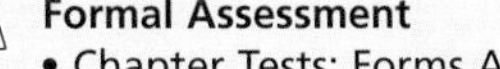

Test Generator

Online Test Practice

Strategies for Test Preparation

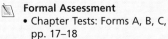

Section Assessment, p. 21

Formal Assessment
- Section Quiz, p. 5

Integrated Assessment
- Rubric for making a poster

Test Generator

Test Practice Transparencies TT1

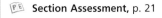

Section Assessment, p. 26

Formal Assessment
- Section Quiz, p. 6

Integrated Assessment
- Rubric for writing a dialogue

Test Generator

Test Practice Transparencies TT2

RESOURCES FOR DIFFERENTIATING INSTRUCTION

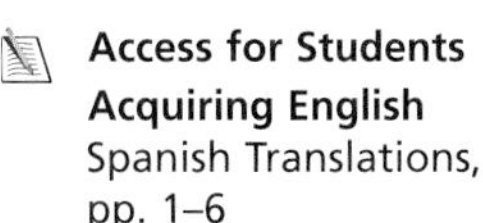

Students Acquiring English/ESL

Reading Study Guide (Spanish and English), pp. 4–9

Access for Students Acquiring English Spanish Translations, pp. 1–6

Modified Lesson Plans for English Learners

Less Proficient Readers

Reading Study Guide (Spanish and English), pp. 4–9

TE Activity
- Synthesizing, p. 18

Gifted and Talented Students

TE Activity
- Making an Oral Report, p. 25

CROSS-CURRICULAR CONNECTIONS

Humanities
Markel, Michelle. ***Cornhusk, Silk, and Wishbones: A Book of Dolls From Around the World.*** New York: Houghton Mifflin, 2000. Museum-quality dolls from 26 cultures.

Literature
Koch, Kenneth. ***Talking to the Sun: An Illustrated Anthology of Poems for Young People.*** New York: Henry Holt, 1985. International collection illustrated with museum art.

Popular Culture
Braman, Arlette N. ***Kids Around the World Cook!: The Best Foods and Recipes from Many Lands.*** New York: John Wiley & Sons, 2000. Inclusive collection and fun facts.

Health
Ichord, Loretta. ***Toothworms and Spider Juice: An Illustrated History of Dentistry.*** Brookfield, CT: Millbrook Press, 2000. Dental details from many cultures and times.

Storring, Rod. ***A Doctor's Life.*** New York: NAL, 1998. Medicine from A.D. 50 to present.

Science/Math
D'Amico, Joan. ***The Science Chef Travels Around the World: Fun Food Experiments and Recipes for Kids.*** New York: John Wiley & Sons, 1996. Combines math and science with multicultural viewpoint.

Mathematics
Bruno, Leonard C. ***Math and Mathematicians: The History of Math Discoveries Around the World.*** Farmington Hills, MI: Gale Group, 1999. Multicultural context for mathematics.

Science
Branley, Franklyn. ***Keeping Time: From the Beginning and into the 21st Century.*** New York: Houghton Mifflin, 1993. Clear explanations of humanity's attempt to keep time.

Economics
Young, Robert. ***Money.*** Minneapolis, MN: Carolrhoda Books, 1998. History of money and types of economies.

ENRICHMENT ACTIVITIES

The following activities are especially suitable for classes following block schedules.

Teacher's Edition, pp. 19, 20, 25, 27
Pupil's Edition, pp. 21, 26
Geography Handbook, pp. 4–13
Interdisciplinary Challenge, pp. 22–23
Literature Connections, pp. 28–29
Outline Map Activities

INTEGRATED TECHNOLOGY

Go to **classzone.com** for lesson support and activities for Chapter 1.

BLOCK SCHEDULE LESSON PLAN OPTIONS: 90-MINUTE PERIOD

DAY 1

UNIT PREVIEW, pp. 14–15
Class Time 20 minutes

- **Discussion** Lead a class discussion about why people settle where they do, using Focus on Geography on PE p. 15. Encourage students to identify what needs people are meeting when they choose a location in which to settle.

GEOGRAPHY HANDBOOK, pp. 4–13
Class Time 35 minutes

- **Peer Teaching** Divide students into pairs and have them alternate quizzing each other about content on the Geography Handbook pages.

SECTION 1, pp. 17–21
Class Time 35 minutes

- **Making Connections** Have students reread Learning About the World, PE p. 18. Divide the class into five groups. Assign each group a field of social studies. Have each group develop one specific detail about the different grade levels in your school that is an example of the assigned social studies theme. Reconvene as a whole class for discussion.

DAY 2

SECTION 1, continued
Class Time 50 minutes

- **Applying Understanding** Write the five themes of geography on the chalkboard and review them as a class. Choose a country and have students write five questions about it, one question for each geography theme. Review and compare questions as a group.
Class Time 30 minutes
- **Clarifying Economics** Lead a discussion about economics, using the question prompts under Objective 3, on TE p. 18. Be sure that students understand the important terms and can identify examples of each.
Class Time 20 minutes

SECTION 2, pp. 24–26
Class Time 40 minutes

- **Reading Maps** Divide the class into seven groups. Assign each a culture region from the map Culture Regions of the World, PE p. 25. Have groups use the map to create a brief profile of their region. Each profile should include the region's size and location, its bordering regions, and one generalization. Have teams present their profiles, and discuss each concluding generalization as a class.

DAY 3

SECTION 2, continued
Class Time 20 minutes

- **Understanding Connections** Have students reread the section Culture Regions Change, PE p. 26. Ask students to share examples of items or events from their daily lives that are examples of interdependence upon other regions. Write their examples on the board.

CHAPTER 1 REVIEW AND ASSESSMENT, pp. 30–31
Class Time 70 minutes

- **Review** Have students work in teams to prepare a brief oral summary of the chapter by reviewing the Main Ideas and Why It Matters Now features in each section in Chapter 1. Have one student from each group deliver the summary.
Class Time 35 minutes
- **Assessment** Have students complete the Chapter 1 Assessment.
Class Time 35 minutes

TECHNOLOGY IN THE CLASSROOM

DIGGING FOR INFORMATION ON THE WEB

This type of activity asks students to dig through a few Web sites to answer specific questions. This exercise requires them to practice their reading and skimming skills, to pay attention to detail, and to be patient enough to look for the correct information. It can serve as a warm-up for larger Internet research projects. Students can do this activity individually or with partners. To make the project shorter, simply omit one or more questions.

ACTIVITY OUTLINE

Objective Students will answer eight questions about world cultures, based on information they find at specified Web sites.

Task Discuss the meaning of *culture,* and have students dig through the Web sites to find the answers to the eight questions listed below. As an option, have students choose one culture to research in more depth and write a two-page report on the computer about this culture.

Class Time 1–2 class periods (not including optional research project)

DIRECTIONS

1. Ask students what they think *culture* means, and write their ideas on the chalkboard. Explain that culture includes such things as customs, daily activities, language, religion, art, and music. Can they think of other things that are part of a person's culture? What are some elements of their own culture?
2. Have students go to the Web sites listed at **classzone.com** to find the answers to the following questions about world cultures. The answers are all available at the Web sites, but students will have to do some digging to find them.
    - What do Jewish children in Israel (and other parts of the world) do on Purim?
    - How do you play "sick cat," a Brazilian game?
    - Describe three traditions, stories, songs, or dances of the Tuareg people of the Sahara Desert.
    - What is Komba, and what is its importance to the Baka people of the African rain forest?
    - How do most teenagers in France get around town?
    - Why might you experience culture shock on a trip to Kathmandu, Nepal?
    - What is Talavera, a Mexican tradition?
3. Discuss students' answers as a group, and have them point out the places they have learned about on a world map.

CHAPTER 1 OBJECTIVE

Students will examine the fields of learning that contribute to social studies and develop an understanding of culture and culture traits.

FOCUS ON VISUALS

Interpreting the Photograph Direct students' attention to the photograph. Read the caption aloud, and then ask students to explain how the photograph supports the caption. Have students discuss how the photograph compares to a group photograph taken in their own school.

Possible Responses Students may say that the diversity shown in the photograph is comparable to the diversity found in their own school.

Extension Have students create a collage of class photos to demonstrate the differences and similarities among classmates.

CRITICAL THINKING ACTIVITY

Recognizing Effects Prompt a discussion about how people adapt to where they live. Brainstorm a list of factors that affect people's lifestyles, such as climate and geography. Have students list these factors and make notes about the ways that people might adapt to each one.

Class Time 15 minutes

CHAPTER 1

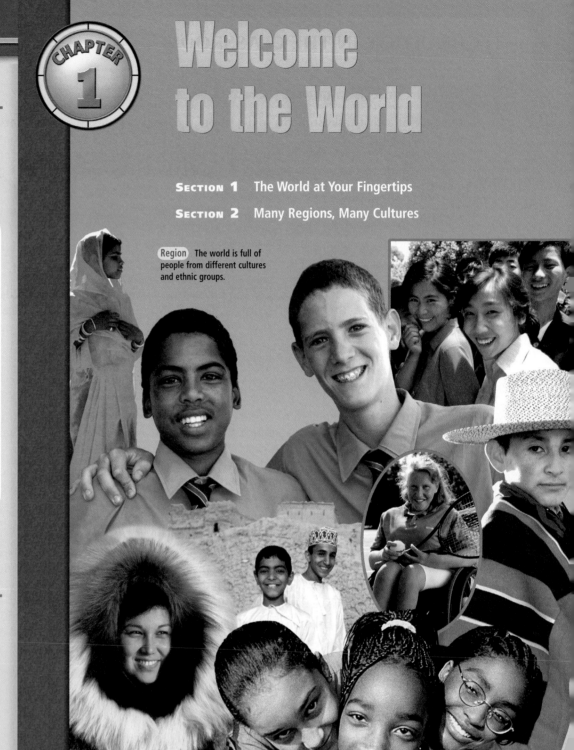

Welcome to the World

SECTION 1 The World at Your Fingertips

SECTION 2 Many Regions, Many Cultures

Region The world is full of people from different cultures and ethnic groups.

14

Recommended Resources

BOOKS FOR THE TEACHER

Angeloni, Elvio. ***Anthropology 01/02.*** Guilford, CT: McGraw-Hill, 2001. Forty articles covering cultures, structures, and perspectives.

Geertz, Clifford. ***The Interpretation of Cultures.*** New York: Basic Books, 2000. What culture is, what role it plays in social life, and how to study it.

Mark, Joan T. ***Margaret Mead: Coming of Age in America.*** New York: Oxford University Press, 1999. Traces Mead's life and groundbreaking work.

Moehn, Heather. ***World Holidays.*** Danbury, CT: Franklin Watts, 2000. Alphabetical and chronological guide.

VIDEOS

People. New York: Lightyear Entertainment, 1995. Cultures around the world.

INTERNET

For more information about world cultures, visit **classzone.com**

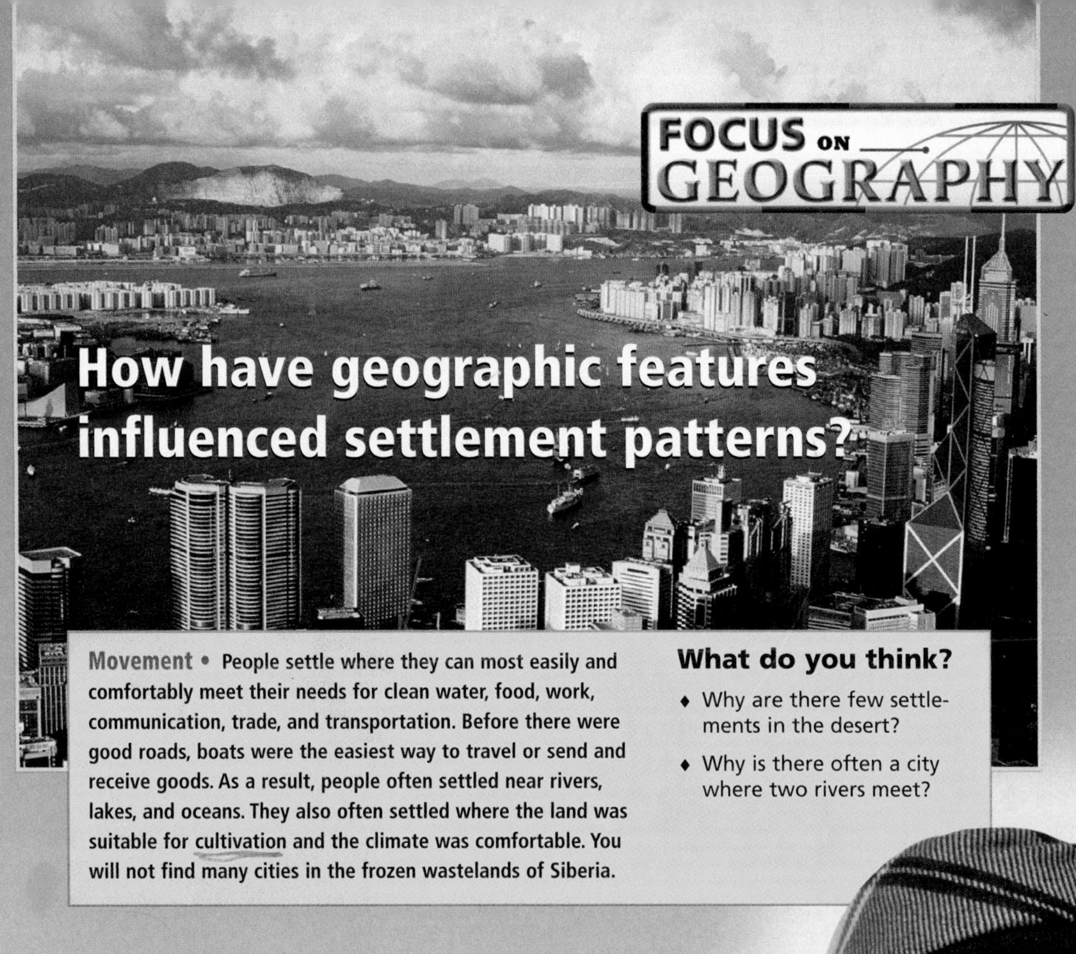

How have geographic features influenced settlement patterns?

Movement • People settle where they can most easily and comfortably meet their needs for clean water, food, work, communication, trade, and transportation. Before there were good roads, boats were the easiest way to travel or send and receive goods. As a result, people often settled near rivers, lakes, and oceans. They also often settled where the land was suitable for cultivation and the climate was comfortable. You will not find many cities in the frozen wastelands of Siberia.

What do you think?

- Why are there few settlements in the desert?
- Why is there often a city where two rivers meet?

FOCUS ON GEOGRAPHY

Objectives

- To help students identify the relationship between geography and the way people meet their needs
- To explain how geographic factors present global challenges

What Do You Think?

1. Make sure students know that a desert is a region so dry that relatively few animals and plants exist there. Then guide a discussion of why the lack of resources discourages settlement in most deserts.
2. Guide students to think about trade, transportation, communication, and other opportunities provided by rivers.

How have geographic features influenced settlement patterns?

Ask students to consider the immediate needs of people who settled in places before transportation and communication systems were established. Have them think about what they might have brought with them, what they would need to make a living, and how they would get food and water. Ask students to think about how people might have changed the land once they settled there.

MAKING GEOGRAPHIC CONNECTIONS

Ask students to make a list of geographic features that might have presented problems to settlers. Then ask them to list the specific problem next to each feature.

Implementing the National Geography Standards

Standard 5 Give examples of regions at different spatial scales

Objective To have students form a mental map of where their community is in the world

Class Time 15 minutes

Task Students list the regions they live in, going from the largest region (e.g., Earth) to the smallest region (e.g., their street). Show the class a map of the world. Have the students identify what hemisphere they live in. Then ask students to identify what continent, country, state, city, and neighborhood they live in. Tell students to identify other regions to which their community belongs.

Evaluation Students should correctly order the regions by size.

BEFORE YOU READ

What Do You Know?

Prompt a discussion about the term *social studies.* Have students discuss the kinds of information they have learned in social studies classes in the past. Have small groups of students work together to come up with a definition of social studies. Then have groups share and compare their definitions, looking for common words and ideas. Compare their definitions with dictionary and encyclopedia entries.

What Do You Want to Know?

Write the headings *history, geography, economics, government,* and *culture* on the chalkboard. Have students take turns listing what they already know about the terms. Then ask what more they would need to know in order to write definitions for the terms. Suggest that students list questions that they would like the chapter to answer. Students should look for answers to their questions as they read and record facts they find particularly interesting.

READ AND TAKE NOTES

Reading Strategy: Categorizing Remind students that a chart is an excellent format for organizing information, such as words and their meanings. Have them read the terms listed in the chart. Then explain that they will find these **boldfaced** terms in the chapter. As students locate each one, have them write its definition and any other important information in the chart. Tell them that when the chart is completed, they will be able to use it as a study resource.

In-depth Resources: Unit 1
- Guided Reading Worksheets, pp. 6–7

CHAPTER 1 READING SOCIAL STUDIES

BEFORE YOU READ

▸▸ What Do You Know?

You live in the world, but how much do you know about it? The best way to find out is through social studies. *Social studies* is an umbrella term. It covers history, geography, economics, government, and culture. History, as you probably know, is the study of the past. How clearly can you define the other terms? How do you think they can help you to learn about the world?

▸▸ What Do You Want to Know?

Think about what else you need to know before you can come up with clear, complete, and accurate definitions. Record any questions you have in a notebook before you read this chapter.

Region • Citizens have more rights under some governments than others. ▼

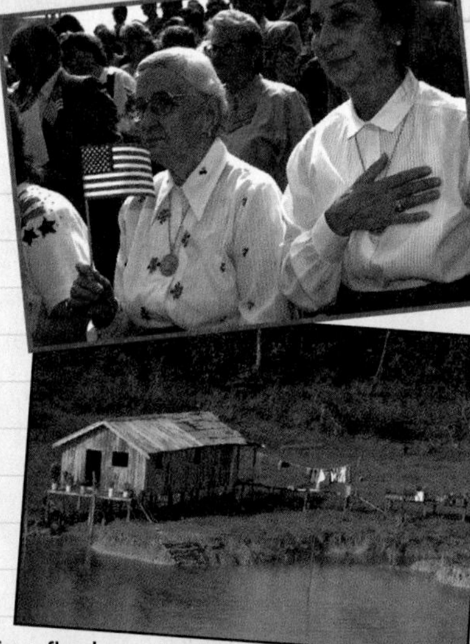

Region • Prepared for a flood, this house was built to suit its environment. ▲

READ AND TAKE NOTES

Reading Strategy: Categorizing One way to make sense of information is to organize it in a chart. Writing your notes in a chart with categories can help you remember the most important parts of what you have read.

- Copy the chart below into your notebook.
- As you read the chapter, note the definition of each term listed on the chart.
- Write these definitions next to the appropriate heading.

Term	Definition
history	The study of the past with the help of written records
geography	The study of people, places, and the environment
economics	The study of the ways in which people produce and exchange goods
government	The people and groups of people that have the power to make laws and see that they are obeyed
culture	The beliefs, customs, and ways of living that a group of people share

Teaching Strategy

Reading the Chapter This is a thematic chapter that defines social studies and the five areas of learning that contribute to the field of social studies: geography, history, economics, government, and culture. Encourage students to use prior knowledge and the photographs in the chapter to help them understand how each area contributes to the study of the world and its cultures.

Integrated Assessment The Chapter Assessment on page 30 describes several activities that may be used for integrated assessment. You may wish to have students work on these activities during the course of the chapter and then present them at the end.

The World at Your Fingertips

TERMS & NAMES
history
geography
government
citizen
economics
scarcity
culture
culture trait

MAIN IDEA

Social studies includes information from five fields of learning to provide a well-rounded picture of the world and its peoples.

WHY IT MATTERS NOW

Understanding your world is essential if you are to be an informed citizen of a global society.

DATELINE

EXTRA

SAN FRANCISCO, USA, JUNE 26, 1945

Fifty nations signed a charter today to establish a new organization called the United Nations. The organization will go into effect October 24.

The United Nations is a successor to the old League of Nations, founded after World War I to prevent another world war, which it failed to do.

The purpose of the new organization is to maintain peace and develop friendly relations among nations.

The member nations hope to cooperate to solve economic, social, cultural, and humanitarian problems and to promote respect for human rights and freedom.

Region • Flags of member countries fly in front of the United Nations headquarters in New York City. ▲

The Peoples of the World ❶

For centuries, people in different parts of the world have been trying to get along with one another, not always with success. Part of the problem is a lack of understanding of other people's ways of life. Certain advances in communication and transportation, such as the Internet and high-speed planes, have brought people closer together. So have increased international trade and immigration. Knowledge of other societies can be a key to understanding them.

TAKING NOTES
Use your chart to take notes about these terms.

Term	Definition
history	
geography	

SECTION OBJECTIVES

1. To define social studies and identify the five fields of learning that it draws from
2. To identify the impact of geography and government on people
3. To explain economics and related terms
4. To define culture and culture traits

CRITICAL THINKING
- Drawing Conclusions, p. 19
- Clarifying, p. 20

FOCUS & MOTIVATE

WARM-UP

Identifying Problems Have students read Dateline and discuss the following questions to help them understand the United Nations.

1. What organization did the United Nations replace?
2. What are the purpose and goals of the United Nations?

INSTRUCT: Objective ❶

The Peoples of the World/ Learning About the World

- What is the key to understanding other societies? knowledge of the other societies
- What five fields of learning contribute to the field of social studies? geography, history, economics, government, culture

In-depth Resources: Unit 1
- Guided Reading Worksheet, p. 6

Reading Study Guide
(Spanish and English), pp. 4–5

Program Resources

In-depth Resources: Unit 1
- Guided Reading Worksheet, p. 6
- Reteaching Activity, p. 12

Reading Study Guide
(Spanish and English), pp. 4–5

Formal Assessment
- Section Quiz, p. 5

Integrated Assessment
- Rubric for making a poster

Outline Map Activities

Access for Students Acquiring English
- Guided Reading Worksheet, p. 1

Technology Resources
classzone.com

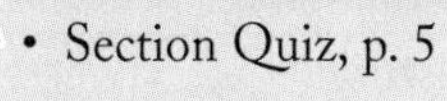

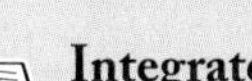

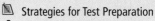

TEST-TAKING RESOURCES
- Strategies for Test Preparation
- Test Practice Transparencies
- Online Test Practice

INSTRUCT: Objective 2

History and Geography/ Government

SS.C.1.3.2

- What are the five themes of geography? location, region, place, movement, human-environment interaction
- Why do we use these different themes? They allow us to examine geography from different angles; each suggests different facts.
- What is the definition of government? people and groups in a society who have the authority to make laws and see they are carried out

Connections to Science

Geologists can find work in a wide variety of organizations, such as colleges, government agencies, and private companies. Some geologists work as prospectors, looking for fossil fuels, sources of stone for building, underground reservoirs of water, or metals. Government agencies employ geologists to advise them on construction sites. Oil and mining companies depend on geologists' skills to help them find natural resources hidden deep underground.

A VOICE FROM TODAY

David McCullough is a distinguished writer, teacher, and historian. He grew up believing that history is a lively subject. He communicates this belief to his readers. "I write to find out," he says. "There isn't anything in this world that isn't inherently interesting—if only someone will explain it to you…will frame it into a story."

Learning About the World 1

Social studies is a way to learn about the world. It draws on information from five fields of learning—geography, history, economics, government, and culture. Each field looks at the world from a different angle. Consider the approaches you might use if you were starting at a new school. Figuring out how to get around would be learning your school's geography. Asking other students where they come from is learning their history. Making choices about which school supplies you can afford to buy is economics. Learning the school rules is learning about its government. Clubs, teams, styles of clothing, holidays, and even ways of saying things are part of the school's culture.

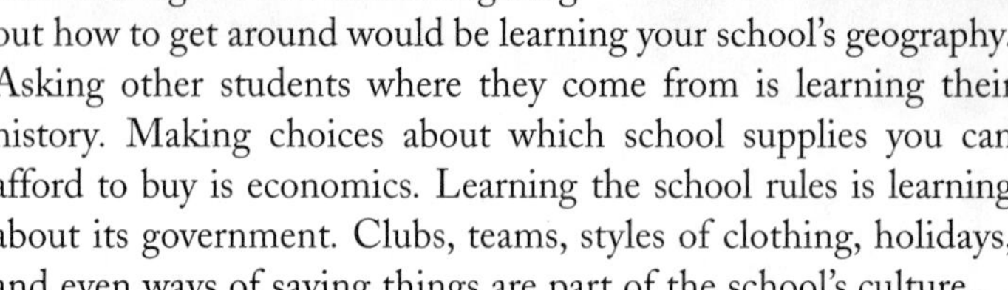

Place • **The five fields of learning in social studies are well represented in daily life.** ▲

History and Geography 2

Knowing history and geography helps orient you in time and space. **History** is a record of the past. The people and events of the past shaped the world as it is today. Historians search for primary sources, such as newspapers, letters, journals, and other documents, to find out about past events.

Vocabulary

orient: to become familiar with a situation

A VOICE FROM TODAY

How can we know who we are and where we are going if we don't know anything about where we have come from and what we have been through, the courage shown, the costs paid, to be where we are?

David McCullough, Historian

The Five Themes of Geography **Geography** is the study of people, places, and the environment. Geography deals with the world in spatial terms. The study of geography focuses on five themes: location, region, place, movement, and human-environment interaction.

Connections to Science

Historians of the Earth
Geologists are scientists who study how Earth was formed and how it has changed over time. There are many kinds of geologists. Some geologists study the materials that make up Earth and processes, such as erosion, that change Earth. Other geologists are interested in the history of Earth. They study fossils, which are the remains of animals and plants that lived millions of years ago.

Activity Options

Differentiating Instruction: Less Proficient Readers

Synthesizing To help students understand the terms *geography, history, economics, government,* and *culture,* draw a chart like the one shown at right. Define the meaning of each term, and then discuss how each one figures into people's daily lives. Have students copy the chart onto their own papers and complete it, showing how the five fields of social studies apply to their own lives.

Geography	name of hometown
History	where ancestors might have come from
Economics	facts about allowance or baby-sitting job
Government	rules at home
Culture	holidays, music we listen to

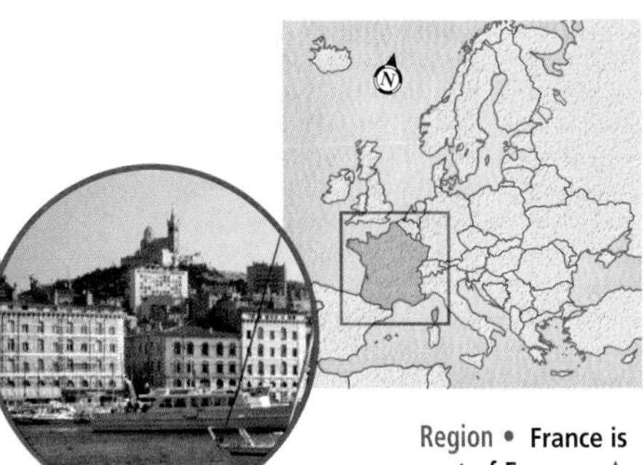

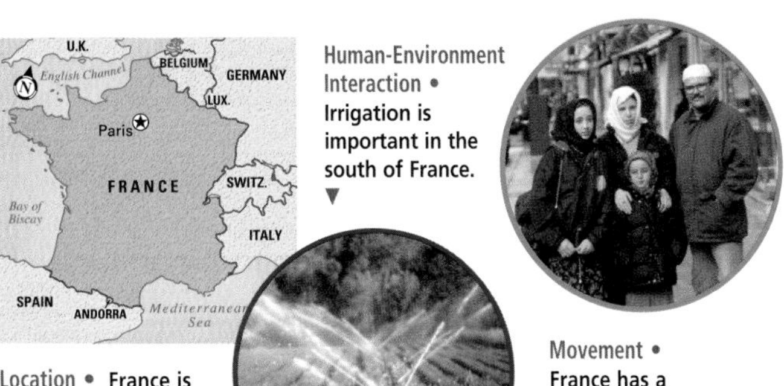

Region • France is part of Europe. ▲

Location • France is located in western Europe. ▲

Human-Environment Interaction • Irrigation is important in the south of France. ▼

Movement • France has a large immigrant population. ▲

Place • France has a dry climate in the south, and a wetter climate in the north, with prosperous farms and thriving cities. ▲

Location tells where a place is. Several countries that have features in common form a region. Place considers an area's distinguishing characteristics. Movement is a study of the migrations of people, animals, and even plants. Human-environment interaction considers how people change and are changed by the natural features of Earth.

Government ❷

Every country has laws and a way to govern itself. Laws are the rules by which people live. **Government** is the people and groups within a society that have the authority to make laws, to make sure they are carried out, and to settle disagreements about them. The kind of government determines who has the authority to make the laws and see that they are carried out.

Limited and Unlimited Governments In a limited government, everyone, including those in charge, must obey the laws. Some of the laws tell the government what it cannot do. Democracies and republics are two forms of limited government. In a democracy, the people have the authority to make laws directly. In a republic, the people make laws through elected representatives. The governments of the United States, Mexico, and India are examples of republics.

Rulers in an unlimited government can do whatever they want without regard to the law. Totalitarianism is a form of unlimited government. In a totalitarian government the people have no say. Rulers have total control.

BACKGROUND

Local, state, and national governments provide needed services, such as schools, parks, electricity, and roads.

Vocabulary

totalitarian government: a government in which the rulers have total control

CRITICAL THINKING ACTIVITY

Drawing Conclusions Review with students the characteristics of a limited government (everyone, including those in charge, must obey the laws) and an unlimited government (rulers do not have to obey the laws). Have students work in small groups to discuss why a country might have an unlimited government. Help them draw conclusions about the advantages and disadvantages of an unlimited government. Have groups share their ideas.

Class Time 15 minutes

FOCUS ON VISUALS

Interpreting Photographs and Maps Have students study the maps and photographs that represent the themes of geography. Ask the following questions to help students understand how the maps and photographs relate to the themes. What nation is represented? What characteristics of France are seen in the left-hand photo? What does the middle photo show? What does it tell you about the climate and about how the French interact with the environment? What countries border France? Have students look at a map to find other countries in the region.

Possible Responses France is represented. Characteristics shown are a modern city on the sea. The middle photograph shows an irrigation system. The climate in the south is dry, but farmers have found ways to water their crops. Spain, Italy, Switzerland, Germany, Luxembourg, and Belgium all border France. Other counttries in the region are the United Kingdom and the Netherlands.

Extension Have students write descriptions of maps and photographs that could represent the five themes as they pertain to your state.

Activity Options

Interdisciplinary Link: Art

Class Time One class period

Task Creating posters showing the five themes of geography

Purpose To understand the five themes of geography

Supplies Needed
- Poster board
- Glue
- Scissors
- Markers
- Old magazines

 Block Scheduling

Activity Review the five themes of geography. Explain that students will make collages of photographs that pertain to each of the five themes. Brainstorm the subject matter of photographs that could be used for each theme. For example, for the theme of human-environment interaction, students might find photographs of igloos or houses with solar panels. Display the finished posters.

MORE ABOUT…

Naturalization

Nearly half a million people were naturalized as citizens in the United States in 1998. Applicants for citizenship must be 18 years or older and be permanent residents for at least five years. They must also be able to speak, understand, read, and write simple English and pass a test about basic American history.

INSTRUCT: Objective 3

Economics/Kinds of Economies

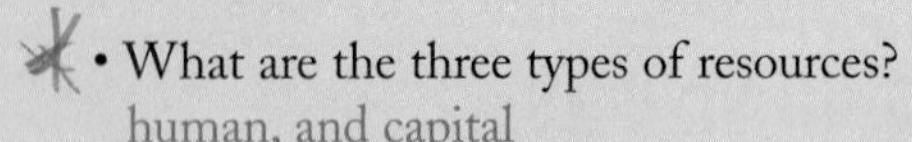

- What are the three types of resources? natural, human, and capital
- What is the difference between a command economy and a market economy? command: government decides quantity and cost of product; market: individual businesses decide

MORE ABOUT…

Kofi Annan

Kofi Annan's philosophy of conflict resolution is evident in the following quote: "Each of us has the right to take pride in our particular faith or heritage. But the notion that what is ours is necessarily in conflict with what is theirs is both false and dangerous. . . . It need not be so. People of different religions live side by side in almost every part of the world. . . . We *can* love what we are, without hating what—and who—we are *not*."

CRITICAL THINKING ACTIVITY

Clarifying For students who may not understand the two economies, ask these questions:

- In what type of economy does the government decide how many videos to produce?
- In what type of economy do video producers decide how many videos will be produced?

Citizenship A **citizen** is a legal member of a country. Citizens have rights, such as the right to vote in elections, and duties, such as paying taxes. Being born in a country can make you a citizen. Another way is to move to a country, complete certain requirements, and take part in a naturalization ceremony.

Vocabulary

naturalization: the process of becoming a citizen

Movement • One way people become American citizens is by participating in a naturalization ceremony. ▲

Economics 3

Looking at the long list of flavors at the ice cream store, you have a decision to make. You have only enough money for one cone. Will it be mint chip or bubble gum flavor? You will have to choose. **Economics** is the study of how people manage their resources by producing, exchanging, and using goods and services. Economics is about choice.

Some economists claim that people's desires are unlimited. Resources to satisfy these desires, however, are limited. These economists refer to the conflict between people's desires and their limited resources as **scarcity.**

Resources Economists identify three types of resources: natural, human, and capital. Natural resources are gifts of nature, such as forests, fertile soil, and water. Human resources are skills people have to produce goods and services. Capital resources are the things people make, such as machines and equipment, to produce goods and services.

Biography

Kofi Annan (b. 1938)
Kofi Annan (KOH•fee AN•uhn) was born in Ghana, West Africa. He is the seventh Secretary-General of the United Nations (UN). Annan studied economics in the United States and Switzerland. In 1972, he earned a graduate degree in management from the Massachusetts Institute of Technology. Annan is a passionate reformer. He cares deeply about HIV/AIDS funding, conflict resolution, educational reform, and ending poverty. In 2001, Annan and the UN jointly received the Nobel Peace Prize.

Kinds of Economies 3

Blue jeans are a product. Who decides whether to make them and how many to make and what price to charge? In a command economy, the government decides. In a market economy, individual businesses decide, based on what they think consumers want.

Levels of Development Different countries and regions have different levels of economic development. In a country with a high level of development, most people are well educated, have good health, and earn decent salaries. Services such as clean running water, electricity, and transportation are plentiful. Technology is advanced, and businesses flourish.

Reading Social Studies
A. Possible Answer
In a market economy, businesses make decisions about how much of a product to make and how much to charge. These decisions are made by the government in a command economy.

A. Contrasting How does a market economy differ from a command economy?

Activity Options

Skillbuilder Mini-Lesson: Reading a Textbook

Block Scheduling

Explaining the Skill Explain to students that when they read a text book, they should look for important information and answers to specific questions.

Applying the Skill Have students scan a chapter in this textbook and use these strategies to find out what information it contains.

1. Look at the chapter opener to find out what is covered in the chapter.
2. Read the headings of each section and subsection in the chapter. These headings indicate what the section will cover.
3. Look for maps, graphs, tables, and photographs. Point out that these usually provide additional information that supports the text.
4. Scan the chapter for terms that are boldfaced, underlined, or italicized. These terms are vocabulary that is key to the subject matter.

Vocabulary

literacy: ability to read and write

life expectancy: average number of years people live

A country with a low level of development is marked by few jobs in industry, poor services, and low literacy rates. Life expectancy is low. These countries are often called developing countries.

Culture 

Some people wear saris. Others wear T-shirts. Some people eat cereal and milk for breakfast. Others eat pickled fish. Some people go to church on Sunday morning. Others kneel and pray to Allah five times a day. All these differences are expressions of **culture.** Culture consists of the beliefs, customs, laws, art, and ways of living that a group of people share.

Religion is part of most cultures; so is a shared language. The ways people express themselves through music, dance, literature, and the visual arts are important parts of every culture; so are the technology and tools they use to accomplish various tasks. Each kind of food, clothing, or technology, each belief, language, or tool shared by a culture is called a **culture trait.** Taken together, the culture traits of a people shape their way of life.

Reading **Social Studies**

B. Recognizing Important Details What are three characteristics that can define a culture?

Reading Social Studies B. Possible Answer Students should name any three of the following: beliefs, customs, laws, art, language, technology, tools, clothing, ways of living.

Citizenship IN ACTION

Habitat for Humanity International Since 1984 former President Jimmy Carter, below, has worked with Habitat for Humanity International. Habitat is a charity that builds and restores homes for people in need. The houses have a low cost, and the new homeowners pay for them over time.

The homeowners also help build their own houses. Carter explains that taking part gives people self-respect: "The people who will live in the homes work side-by-side with the volunteers, so they feel very much that they are on an equal level."

SECTION 1 ASSESSMENT

Terms & Names

1. Explain the significance of: (a) history (b) geography (c) government (d) citizen (e) economics (f) scarcity (g) culture (h) culture trait

Using Graphics

2. Use a chart like this one to list the five themes of geography and their characteristics.

Theme	Characteristics

Main Ideas

3. (a) What five areas of learning does social studies include?

(b) What are the three main kinds of resources, and how is each one defined?

(c) What is the difference between limited and unlimited government?

Critical Thinking

4. Making Inferences

Does the United States have a shared, or common, culture?

Think About

- what you eat and wear, where you live, how you spend your free time
- who else shares these activities with you

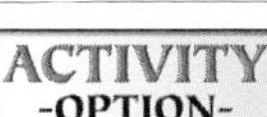

ACTIVITY -OPTION- Reread the section on citizenship. Make a **poster** showing the rights and responsibilities of a citizen.

Citizenship IN ACTION

On May 31, 2003, Habitat for Humanity founder Millard Fuller opened a new six-acre educational center at Habitat's headquarters in Americus, Georgia. A dedication by former President Jimmy Carter followed on June 7. The center includes life-size models of Habitat houses from around the world. Other highlights include hands-on classes in making compressed-earth blocks or roof tile.

INSTRUCT: Objective 4

Culture

SS.B.1.3.4

- What shared factors make up a culture? beliefs, customs, laws, art, ways of living
- What is a culture trait? food, clothing, tools, technology, language, or belief shared by a culture

ASSESS & RETEACH

Reading Social Studies Have students fill in definitions in the chart on page 16.

Formal Assessment
- Section Quiz, p. 5

RETEACHING ACTIVITY

Pair students, and have one explain the importance of history and geography. Have the partner ask questions to clarify the explanation. Then have students switch roles to define government and economics.

In-depth Resources: Unit 1
- Reteaching Activity, p. 12

Access for Students Acquiring English
- Reteaching Activity, p. 5

Section 1 Assessment

1. Terms & Names

a. history, p. 18
b. geography, p. 18
c. government, p. 19
d. citizen, p. 20
e. economics, p. 20
f. scarcity, p. 20
g. culture, p. 21
h. culture trait, p. 21

2. Using Graphics

Location	where it is
Region	what it is similar to
Place	characteristics such as climate, soil, land use
Movement	who and what have come and gone
Human-environment interaction	how people have changed and been changed by the natural world

3. Main Ideas

a. Social studies includes geography, history, economics, government, and culture.
b. The three kinds of resources are natural (gifts of nature), human (skills people have), and capital (tools people make).
c. In a limited government, everyone must obey the laws; in an unlimited government, rulers do not have to obey the laws.

4. Critical Thinking

Students might mention the music they listen to, the food they eat, or the religion they practice.

ACTIVITY OPTION

Integrated Assessment
- Rubric for making a poster

Interdisciplinary Challenge

OBJECTIVE

Students work alone or in groups to create a programmed, interactive globe and related materials.

 **Block Scheduling**

PROCEDURE

Provide materials such as paper, tracing paper, poster board, and pencils. Have students form groups of four or five and divide the work among the groups. Then ask each group to plan a strategy for completing the challenge and assign tasks to each group member.

HISTORY/ECONOMICS CHALLENGE

Class Time 50 minutes

Students will need to do research in order to complete this challenge. Discuss with them possible sources of information and search terms for using the Internet.

Possible Solutions

Time lines should focus on your community and include

- a broad scope of major events starting with Native Americans and ending at the present.
- clear but brief explanations of the events.

The thematic maps should

- clearly show the products and industries important to your state and where they are located.

Investigate Your World

Suppose that someone has given you a globe as a gift. What a great present! Unlike a flat map, your globe gives you a more accurate view of the world. Best of all, this new globe is programmable. You can input new information about different features and places on Earth. In fact, the manufacturer has set up a contest—the Global Game—giving prizes for the best and most creative approaches to programming the globe. Good luck!

COOPERATIVE LEARNING On these pages are challenges you will meet in trying to win the Global Game. Working with a small group, choose which one you want to solve. Divide the work among group members. Look for helpful information in the Data File. Keep in mind that you will present your solution to the class.

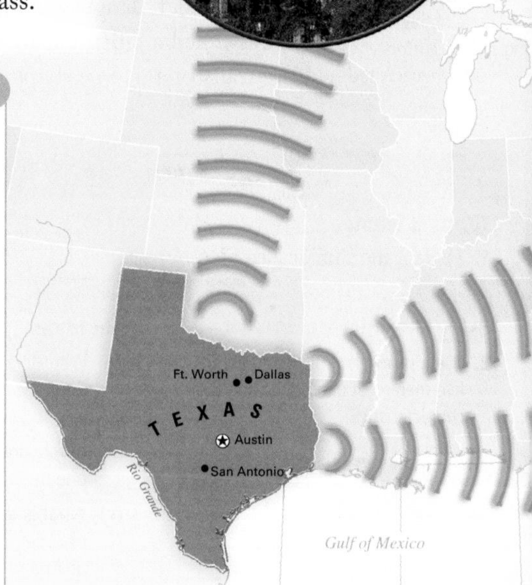

HISTORY/ECONOMICS CHALLENGE

". . . you want to know more about the world closer to home."

Now that your globe has shown you the worldwide picture, you want to know more about the world closer to home. How has geography influenced the history of your community? What features or resources brought settlers there? Choose one of these options. Look in the Data File for information.

ACTIVITIES

1. Make a time line of major events in the growth of your community. If possible, begin with the Native Americans who originally inhabited the area. Include the arrival of immigrants from various places.
2. Draw or trace an outline of your state. Then make a thematic map of its major products and industries. Use words or symbols (such as a cow, a factory, a computer) and create a map key to identify each product.

Standards for Evaluation

HISTORY/ECONOMICS CHALLENGE

Option 1 Time lines should
- highlight key events in your community's history.
- begin with Native Americans and end in the present.

Option 2 Thematic maps should
- show the location of major industries and products of your state.

LANGUAGE ARTS CHALLENGE

Option 1 The script should
- present key information about the geography and culture of the continent.
- be informative and appealing.

Option 2 The geography game should
- require use of a globe.
- be fun and challenging.

GEOGRAPHY CHALLENGE

Option 1 The list should
- focus on geographical records.
- include records for every continent.

Option 2 The collector's cards should
- be factually true.
- include information from all aspects of geography.

LANGUAGE ARTS CHALLENGE

"... you are taken on an audio journey to new places."

Your new globe has built-in sensors activated by a laser wand. When you point the wand at a spot on the globe, you are taken on an audio journey to new places. The sound clip introduces you to a place's culture—the things that make it unique.

As part of the Global Game, the manufacturer is looking for new ways to present this information. What will you include in your approach? How can you add to the globe's popular appeal? Choose one of these options. Look in the Data File for help.

ACTIVITIES

1. Choose one continent and write a script about it for a seven-minute "audio journey." Remember to include information about geographic features as well as aspects of culture.
2. Design another geography game that the manufacturer can use to market its globe. The game should appeal to students of your age. Write a brief description of the game and its rules.

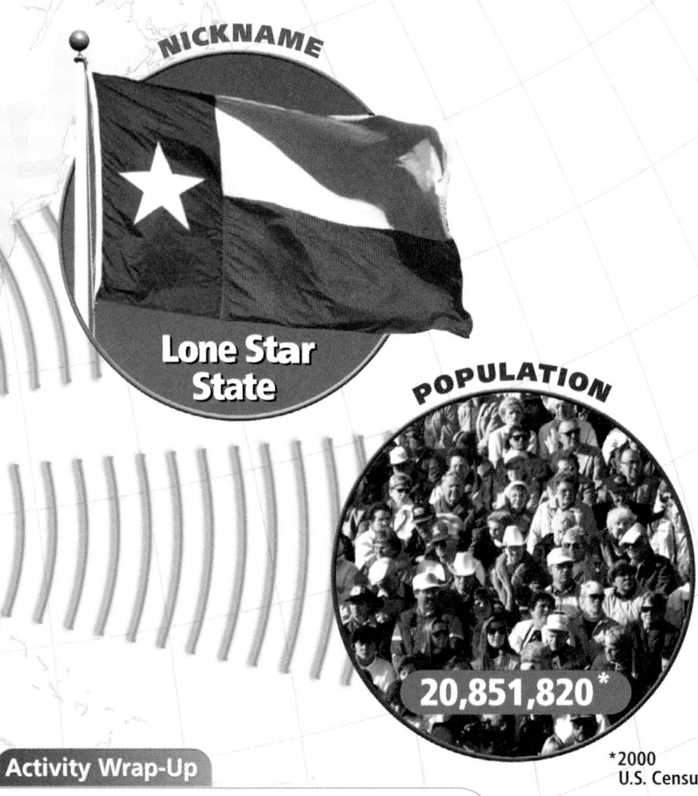

*2000 U.S. Census

Activity Wrap-Up

As a group, review your solution to the challenge you selected. Then present your solution to the class.

DATA FILE

WORLD STATISTICS

- **Circumference** at Equator: 24,902 mi.
- **Earth's speed** of orbit: 18.5 mi./sec.
- **Total area:** 197,000,000 sq. mi.; land area: 57,900,000 sq. mi.
- **Highest point:** 29,035 ft.—Mt. Everest.
- **Lowest point:** 35,800 ft. below sea level—Marianas Trench, Pacific Ocean.
- **Lowest point on land:** 1,312 ft. below sea level, Dead Sea, Israel, and Jordan.

HIGHEST ELEVATIONS BY CONTINENT

- **Asia:** Mt. Everest, Nepal–Tibet, 29,035 ft.
- **South America:** Mt. Aconcagua, Argentina–Chile, 22,834 ft.
- **North America:** Mt. McKinley (Denali), Alaska, 20,320 ft.
- **Africa:** Mt. Kilimanjaro, Tanzania, 19,340 ft.
- **Europe:** Mt. Elbrus, Russia, 18,510 ft.
- **Antarctica:** Vinson Massif, 16,066 ft.
- **Western Europe:** Mont Blanc, France, 15,771 ft.
- **Australia:** Mt. Kosciusko, 7,310 ft.

SOME MAJOR RIVER SYSTEMS

- **Nile,** Africa: 4,160 mi.
- **Amazon,** South America: 4,080 mi.
- **Mississippi**–Missouri, North America (U.S.): 3,740 mi.
- **Chang Jiang** (Yangtze), China: 3,915 mi.
- **Yenisey,** Russia: 2,566 mi.
- **Plata,** South America: 3,030 mi.
- **Huang He** (Yellow), China: 3,010 mi.
- **Congo** (Zaire), Africa: 2,880 mi.

To learn more about Earth's geography, go to

LANGUAGE ARTS CHALLENGE

Class Time 50 minutes

For the "audio journey," students might consider background music unique to the continent. For the second option, have students use familiar games as models.

Possible Solutions

For the first option, the scripts will

- highlight the most important, unique features of the continent's geography and cultures.

Student games will

- include the use of the globe.
- have simple, easy-to-follow directions.

ALTERNATIVE CHALLENGE... GEOGRAPHY CHALLENGE

On which continent do you find the longest river system? the highest mountain? the biggest rock? the tallest tree? On which continent are the most languages spoken? On which continent is the country with the most telephones per person? the fastest train? the tallest building? All of these are questions about Global Geography Records.

- Imagine that you can touch any continent on your programmable globe and hear about that continent's records. Use information from the Data File and research in other sources to list the records that you would hear for each continent.
- Create a set of collector's cards of Global Geography Records.

Activity Wrap-Up

To help students evaluate the creativity of their challenge solutions, have them make a grid with criteria like the one shown. Then have them rate each solution on a scale from 1 to 5.

• Originality	1	2	3	4	5
• Creativity	1	2	3	4	5
• Accuracy	1	2	3	4	5
• Overall effectiveness	1	2	3	4	5

Chapter 1 ✦ Section 2

SECTION OBJECTIVES

1. To identify a culture region
2. To explain how a culture region evolves

SKILLBUILDER
- Interpreting a Map, p. 25

CRITICAL THINKING
- Summarizing, pp. 25–26

FOCUS & MOTIVATE

WARM-UP

Analyzing Motives Have students read Dateline, and use questions such as the following to discuss the clash of Spanish and Aztec cultures.

1. Why did the behavior and appearance of the Spanish alarm the Aztecs?
2. What might help groups of people to understand the customs of others?

INSTRUCT: Objective 1

Different Places, Different Cultures

- What do people in culture regions share? beliefs, history, language
- What makes a region multicultural? when it contains other cultures in addition to the dominant culture
- How do culture regions change? They borrow from other cultures and come to depend on one another.

In-depth Resources: Unit 1
- Guided Reading Worksheet, p. 7

Reading Study Guide
(Spanish and English), pp. 6–7

SECTION 2 Many Regions, Many Cultures

TERMS & NAMES
culture region
interdependence

MAIN IDEA

The world can be divided into regions according to culture.

WHY IT MATTERS NOW

Understanding other cultures can help you understand how people in other regions live and think.

DATELINE

TENOCHTITLÁN, AZTEC CAPITAL, A.D. 1519—Frightening news has arrived from the coast. Warriors from a land called "Spain" are marching toward the capital of our Aztec Empire. They are covered in metal armor and have thundering weapons. They have no respect for our gods, our temples, or our customs. Because we have never seen warriors like these, we do not know if they are men or gods. Nor do we know why are they coming toward our city. Some of our priests say these invaders have come to destroy our world.

Culture • The Aztecs had never seen horses and believed that Spanish soldiers rode giant deer. ▲

Different Places, Different Cultures 1

In 1519, one of the greatest culture clashes in history took place. That year, the Spanish invaded the powerful Aztec Empire, in what is now Mexico. The Spanish and the Aztecs were confused by each other's customs because they came from different culture regions. A **culture region** is an area of the world in which many people share similar beliefs, history, and languages. The people in a culture region may have religion, technology, and ways of earning a living in common as well. They may grow and eat similar foods, wear similar kinds of clothes, and build houses in similar styles.

TAKING NOTES
Use your chart to take notes about these terms.

Term	Definition
history	
geography	

Program Resources

In-depth Resources: Unit 1
- Guided Reading Worksheet, p. 7
- Reteaching Activity, p. 13

Reading Study Guide
(Spanish and English), pp. 6–7

Formal Assessment
- Section Quiz, p. 6

Integrated Assessment
- Rubric for writing a dialogue

Outline Map Activities

Access for Students Acquiring English
- Guided Reading Worksheet, p. 2

Technology Resources
classzone.com

TEST-TAKING RESOURCES
- Strategies for Test Preparation
- Test Practice Transparencies
- Online Test Practice

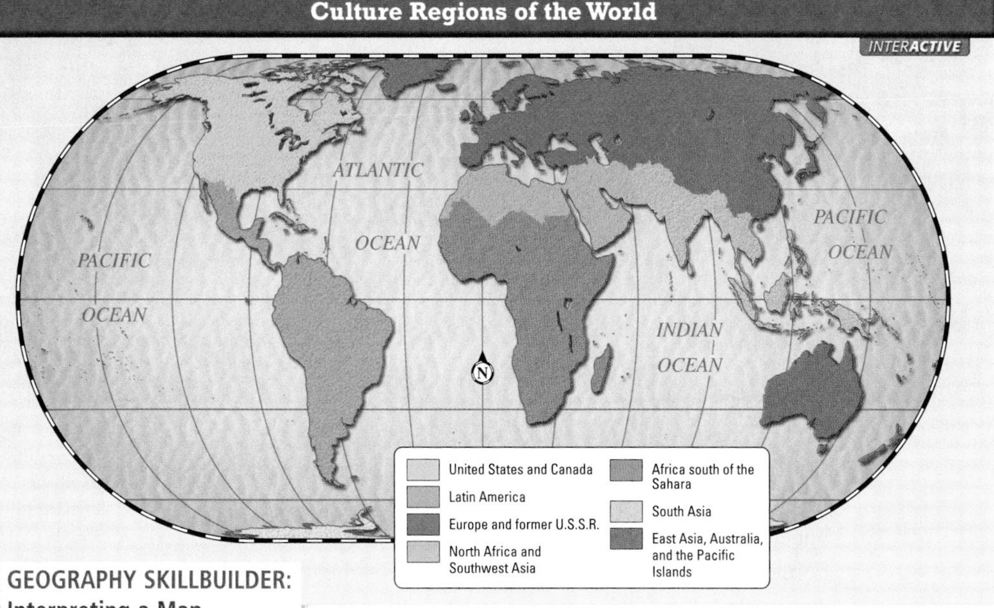

GEOGRAPHY SKILLBUILDER: Interpreting a Map

1. **Region •** How many culture regions are shown on this map?
2. **Location •** Name three culture regions in the Eastern Hemisphere.

Geography Skillbuilder Answers

7

any three of Europe and the former U.S.S.R.; North Africa and Southwest Asia; Africa south of the Sahara; South Asia; East Asia, Australia, and the Pacific Islands; United States and Canada; Latin America

The World's Culture Regions

The map above shows the major culture regions of the world. Latin America is one culture region. The Spanish and Portuguese languages help to tie its people together. So does its common history. Southwest Asia and North Africa is another culture region. Most countries in this region share a common desert climate and landscape. People have adapted to the desert in similar ways, thus creating a common culture. Islam, which is the major religion in this region, also helps shape a common culture.

Usually, not every person in a region belongs to the dominant, or mainstream, culture. Some regions are multicultural. For example, the United States and Canada contain other cultures besides the dominant one. Although most people in this region speak English, many people in eastern Canada speak French. Many people in the United States speak Spanish, especially in the Southwest. In both countries, Catholics, Protestants, Jews, Muslims, Buddhists, and members of other religions are free to worship.

Reading Social Studies

A. Recognizing Important Details Name two characteristics that make the United States multicultural.

Reading Social Studies A. Possible Answer Many people speak Spanish; many religions flourish.

Children Invent Language In the late 1700s, most people in Hawaii spoke English or native Hawaiian. As people began immigrating to Hawaii to work on the sugar cane plantations, they brought their native languages with them: Japanese, Chinese, Korean, Spanish, and Portuguese.

At school, the children of these immigrants began speaking a form of English that was a blend of native Hawaiian, Pidgin English, Pidgin Hawaiian, and their native languages, especially Portuguese. Eventually, children began to learn this form of English as their native language. This was the beginning of Hawaii Creole English, and it soon became the language of the majority of Hawaiians.

Da kaet ste in da haus.

FOCUS ON VISUALS

Interpreting a Map Ask students to find the United States on the map. Ask students what other country shares the same culture region (Canada). Then ask what country has a border with the United States but belongs to a different culture region (Mexico). Ask students to think about the culture regions of these three countries. What do the countries have in common? What differences are there? What kinds of things are shared among the three nations and the two culture regions? What kinds of things remain different?

Possible Responses Students may mention that foods and forms of entertainment might be shared; religious beliefs, art, and ways of living might remain different.

Extension Ask students to discuss what might cause changes between two culture groups and whether those changes are always positive.

Strange but TRUE

Wherever human society exists, language exists. According to linguists, approximately 6,000 languages are spoken in the world today. Some are spoken by fewer than one thousand people; others by a million or more. Twenty-three languages, including English and Arabic, have 50 million or more speakers.

CRITICAL THINKING ACTIVITY

Summarizing Ask students to review the two paragraphs on the page and summarize the factors that unite people in a culture region. Remind them that a summary includes only the main ideas. Have volunteers present their summaries to the group.

Class Time 10 minutes

Activity Options

Differentiating Instruction: Gifted and Talented

Task Making an oral report about varieties of one musical instrument family found around the world

Purpose To understand similarities and differences of world music and the way environment affects musical instruments

Supplies Needed
- Record player or tape deck

Block Scheduling

Activity There are several different families of musical instruments: percussion, winds, strings, and horns. Have students choose one instrument family and research instruments found in that family in different parts of the world. Their oral report should include the materials needed to make the instruments and how the materials relate to the environment where those instruments are played.

FOCUS ON VISUALS

Interpreting the Photographs Have students study the three photographs and draw conclusions about the cultures shown. Have students use the chart headings below to organize their ideas.

Where Live	Information from Photo	Conclusions About Culture

Have students find photographs in newspapers or magazines that reveal information about the culture shown. Ask students to share the photographs and discuss what aspects of culture are demonstrated in each.

CRITICAL THINKING ACTIVITY

Summarizing Have students work in pairs to make a list of ways their community is dependent on or affected by other cultures. Ask pairs to share lists and discuss similarities and differences among them.

Class Time 15 minutes

ASSESS & RETEACH

Reading Social Studies Have students add terms and definitions to the chart on page 16.

Formal Assessment
- Section Quiz, p. 6

RETEACHING ACTIVITY

Divide the class into two groups. Have one group of students work together to explain cultural interdependence, and have the other group explain how cultures change.

In-depth Resources: Unit 1
- Reteaching Activity, p. 13

Access for Students Acquiring English
- Reteaching Activity, p. 6

Region • Home life can differ greatly in different culture regions, sometimes depending on a region's climate or natural resources. ▲

Culture Regions Change For thousands of years, culture regions have changed and evolved as they have borrowed culture traits from one another. They have also come to depend upon each other economically. Decisions and events in one part of the world affect other parts. Advances in transportation and communication have increased this **interdependence.** When oil-producing nations in the Middle East raise the price of oil, for example, the price of gasoline at the neighborhood gas station is likely to rise. If there is an especially abundant banana crop in parts of Latin America, the price of bananas may drop at the local grocery store. More and more, people of different countries are becoming part of one world.

SECTION 2 ASSESSMENT

Terms & Names

1. **Explain the significance of:** (a) culture region (b) interdependence

Using Graphics

2. Use a chart to list the major culture regions of the world.

Major Culture Regions of the World
1.
2.
3.
4.
5.
6.
7.

Main Ideas

3. (a) List at least three things people in a culture region may have in common.
   (b) Which continents have more than one culture region?
   (c) What is one cause of cultural change?

Critical Thinking

4. **Clarifying**
   Why might Brazilian coffee at your local supermarket suddenly cost more?

   Think About
   - price setting
   - coffee supplies

ACTIVITY -OPTION- Write a **dialogue** between you and a visitor from another country in which you explain what makes the culture in your region different from others.

Section 2 Assessment

1. Terms & Names
- a. culture region, p. 24
- b. interdependence, p. 26

2. Using Graphics

1. United States/Canada
2. Latin America
3. Europe
4. North Africa/Southwest Asia
5. Africa south of the Sahara
6. South Asia
7. East Asia/Australia/Pacific Islands

3. Main Ideas
- a. The people in a culture region may have religion, history, and language in common.
- b. North America, Africa, Europe, Asia
- c. One cause of cultural change is interdependence.

4. Critical Thinking

The coffee harvest may have been poor.

ACTIVITY OPTION

Integrated Assessment
- Rubric for writing a dialogue

SKILLBUILDER

Reading a Time Zone Map

▶▶ Defining the Skill

A time zone map shows the 24 time zones of the world. The prime meridian runs through Greenwich (GREHN•ich), England. Each zone east of Greenwich is one hour later than the zone before. Each zone west of Greenwich is one hour earlier. The International Date Line runs through the Pacific Ocean. It is the location where each day begins. If it is Saturday to the east of the International Date Line, then it is Sunday to the west of it.

▶▶ Applying the Skill

Use the strategies listed below to help you find times and time differences on a time zone map.

How to Read a Time Zone Map

Strategy ❶ Read the title. It tells you what the map is intended to show.

Strategy ❷ Read the labels at the top of the map. They show the hours across the world when it is noon in Greenwich. The labels at the bottom show the number of hours earlier or later than the time in Greenwich.

Strategy ❸ Locate a place whose time you know. Locate the place where you want to know the time. Count the number of time zones between them. Then add or subtract that number of hours.

For example, if it is noon time on the west coast of Africa, you can see that it is 7:00 A.M. on the east coast of the United States. That is a difference of five hours.

❶ **WORLD TIME ZONES**

▶▶ Practicing the Skill

Practice determining the difference in hours between various time zones. For example, if you select the yellow zone in western Asia and you live in Texas, you will have a time difference of 11 hours. Now select one time zone in Africa, one in Europe, and one in Australia. For each location, determine the number of hours' difference with the time zone in which you live.

SKILLBUILDER

Reading a Time Zone Map

Defining the Skill

Display a globe and point out your location. Ask students to note the time. Then point to a location in another time zone and ask students if they think the time is the same or different. Explain that the time changes by one hour as you move from one time zone to the next. Ask students to think about why it might be important to understand time differences in the world. Discuss information needed to plan a trip or make a long-distance call.

Applying the Skill

How to Read a Time Zone Map Point out the three strategies for reading a time zone map. Ask a volunteer to read the title and labels on the map. Explain that the labels on the top of the map show hours across the world in one-hour increments, and that the labels at the bottom show time differences.

Make a Chart

Discuss the information on the map with students. Ask questions such as: How many time zones are there? (24) What time is it on the East Coast of the United States when it is noon in London? (7 A.M.) What time is it on the West Coast of the United States when it is noon on the East Coast? (9 A.M.)

Practicing the Skill

Have students turn to page 25 and read the title of the map. Explain that since this map shows culture regions, the strategy for reading it is slightly different. Color is used to indicate each culture region. Name each continent; have students tell which culture region(s) can be found there.

In-depth Resources: Unit 1
- Skillbuilder Practice, p. 10

Career Connection: Cultural Geographer

Encourage students who enjoy reading the time zone map and thinking about differences in time around the world to find out about careers that analyze the relationship between people and their environment. For example, within the broad field of geography, some geographers study the cultural characteristics of different regions of the United States or the world. These geographers are called cultural geographers.

1. Suggest that small groups of students look for information about specific jobs that cultural geographers hold. Examples include a community or urban planner and a university professor. Have students explain how each job they identify relates to cultural geography.
2. With the different directions that cultural geography can take, what education is needed to become a cultural geographer? Help students learn what schooling cultural geographers get before embarking on their careers.
3. Have each group of students share an oral summary of its findings.

 **Block Scheduling**

OBJECTIVE

Students analyze a Native American myth, which explains how the Pacific Ocean was formed.

FOCUS & MOTIVATE

Making Inferences To help students understand the origin and purpose of this myth, have them study the illustration and answer the following questions:

1. What predictions can you make about the story from the illustrations?
2. Why would water be falling from seashells?

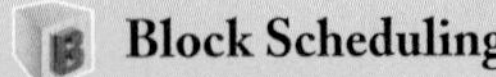
Block Scheduling

MORE ABOUT...

"How Thunder and Earthquake Made Ocean"

This story comes from the Yurok, a Native American people who lived along the Pacific coast of northwestern California. The Yurok depended on coastal waters not only for food, but for money as well—they used the shell of a mollusk as currency. This story reveals the importance of the ocean in Yurok culture.

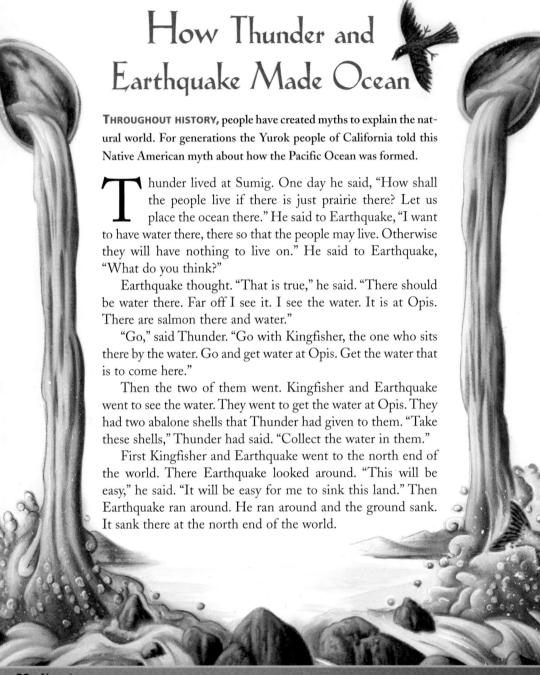

How Thunder and Earthquake Made Ocean

THROUGHOUT HISTORY, people have created myths to explain the natural world. For generations the Yurok people of California told this Native American myth about how the Pacific Ocean was formed.

Thunder lived at Sumig. One day he said, "How shall the people live if there is just prairie there? Let us place the ocean there." He said to Earthquake, "I want to have water there, there so that the people may live. Otherwise they will have nothing to live on." He said to Earthquake, "What do you think?"

Earthquake thought. "That is true," he said. "There should be water there. Far off I see it. I see the water. It is at Opis. There are salmon there and water."

"Go," said Thunder. "Go with Kingfisher, the one who sits there by the water. Go and get water at Opis. Get the water that is to come here."

Then the two of them went. Kingfisher and Earthquake went to see the water. They went to get the water at Opis. They had two abalone shells that Thunder had given to them. "Take these shells," Thunder had said. "Collect the water in them."

First Kingfisher and Earthquake went to the north end of the world. There Earthquake looked around. "This will be easy," he said. "It will be easy for me to sink this land." Then Earthquake ran around. He ran around and the ground sank. It sank there at the north end of the world.

Activity Options

Differentiating Instruction: Less Proficient Readers

Building Language Skills Have students work in pairs or small groups, matching less proficient readers with more fluent readers. Have pairs take turns reading the story aloud together, noting and discussing any words that are unfamiliar or difficult to pronounce. Then ask students to create a two-column chart with the heads *Land Creatures* and *Ocean Creatures.* Students can list the animals mentioned in the story in the appropriate column. Have pairs check their work with other pairs.

Then Kingfisher and Earthquake started for Opis. They went to the place at the end of the water. They made the ground sink behind them as they went. At Opis they saw all kinds of seals and salmon. They saw all the kinds of animals and fish that could be eaten there in the water at Opis. Then they took water in the abalone shells.

"Now we will go to the south end of the world," said Earthquake. "We will go there and look at the water. Thunder, who is at Sumig, will help us by breaking down the trees. The water will extend all the way to the south end of the world. There will be salmon and fish of all kinds and seals in the water."

Now Kingfisher and Earthquake came back to Sumig. They saw that Thunder had broken down the trees. Together the three of them went north. As they went together they kept sinking the ground. The Earth quaked and quaked and water flowed over it as Kingfisher and Earthquake poured it from their abalone shells. Kingfisher emptied his shell and it filled the ocean halfway to the north end of the world. Earthquake emptied his shell and it filled the ocean the rest of the way.

As they filled in the ocean, the creatures which would be food swarmed into the water. The seals came as if they were thrown in in handfuls. Into the water they came, swimming toward shore. Earthquake sank the land deeper to make gullies and the whales came swimming through the gullies where that water was deep enough for them to travel. The salmon came running through the water.

Now all the land animals, the deer and elk, the foxes and mink, the bear and others had gone inland. Now the water creatures were there. Now Thunder and Kingfisher and Earthquake looked at the ocean. "This is enough," they said. "Now the people will have enough to live on. Everything that is needed is in the water."

So it is that the prairie became ocean. It is so because Thunder wished it so. It is so because Earthquake wished it so. All kinds of creatures are in the ocean before us because Thunder and Earthquake wished the people to live.

Reading THE LITERATURE

Before you read, examine the title. What kind of story might this be? What is the purpose of the story?

Thinking About THE LITERATURE

What does this myth suggest about the importance of the ocean in the lives of Native Americans? What does the myth reveal about Native American attitudes to the natural world?

Writing About THE LITERATURE

Write a continuation of the story, showing how the people reacted when they saw that an ocean had replaced the prairie.

Further Reading

Legends of Landforms by Carole G. Vogel explores Native American legends about the origins of many places in the United States.

Keepers of the Earth by Michael J. Caduto and Joseph Bruchac retells Native American stories.

INSTRUCT

Reading the Literature

Possible Responses

The story is a myth or legend. Its purpose is to explain how the ocean was formed.

Thinking About the Literature

Possible Responses

Native Americans depended on the ocean for food. The myth reveals that Native Americans thought that the world provides an abundance of food for human beings.

Writing About the Literature

Possible Responses

Students' stories should describe the people's surprise and delight when they discover the ocean and its supply of food.

MORE ABOUT...

Floods and Earthquakes

Northwest California is one of the most earthquake-prone regions in the United States. The force of an earthquake along coastal waters can produce a disastrous tidal surge or tsunami that can easily flood low-lying areas. The memory of such an awesome event is reflected in the story, which links the natural forces of flood and earthquake.

More to Think About

Making Personal Connections Ask students what this story explains. What other stories have they read that explain natural phenomena? Have students brainstorm a list of questions people might have about landforms or weather in your area. Have students come up with a myth and a scientific explanation for each question. This can be a class activity, or students can choose a question to address.

Vocabulary Activity List the news questions *Who? Where? When? Why? What? How?* on the chalkboard. Tell students to answer each question with information from the story. Then have them use their answers, in any order, to write a news flash for this headline: "New Ocean Forms."

ASSESSMENT

TERMS & NAMES

1. history, p. 18
2. geography, p. 18
3. government, p. 19
4. citizen, p. 20
5. economics, p. 20
6. scarcity, p. 20
7. culture, p. 21
8. culture trait, p. 21
9. culture region, p. 24
10. interdependence, p. 26

REVIEW QUESTIONS

Possible Responses

1. The five themes of geography are location, region, place, movement, and human-environment interaction.
2. In a limited government, everyone must obey the laws; in an unlimited government, the rulers do not have to obey the laws.
3. To become a citizen of a country, you are either born there or you move there, complete certain requirements, and then take part in a naturalization ceremony.
4. In a command economy, the government makes decisions about what is produced and the price of those goods. In a market economy, businesses make these decisions.
5. Characteristics of a culture include religion, language, customs, art, laws, and ways of living.
6. Answers may vary. For example, if a country decides to produce less of a good, that good will be more expensive in other countries.
7. Shared aspects of daily life might include clothing, food, language, and religious rituals.
8. Most people in the United States and Canada speak English, but some people in Canada speak French and some in the United States speak Spanish. Many religions are practiced in both countries.

ASSESSMENT

TERMS & NAMES

Explain the significance of each of the following:

1. history
2. geography
3. government
4. citizen
5. economics
6. scarcity
7. culture
8. culture trait
9. culture region
10. interdependence

REVIEW QUESTIONS

The World at Your Fingertips *(pages 17–21)*

1. What are the five themes of geography?
2. What are the main differences between a limited and an unlimited government?
3. How can someone become a citizen of a country?
4. What is the difference between a command economy and a market economy?
5. What are some characteristics of a culture?

Many Regions, Many Cultures *(pages 24–26)*

6. How can decisions made in one part of the world affect people in another part of the world?
7. What aspects of daily life might people in the same culture region share?
8. What makes the United States and Canada a multicultural region?

CRITICAL THINKING

Remembering Definitions

1. Using your completed chart from Reading Social Studies, p. 16, tell how understanding the culture could help you make friends in a new country.

Making Inferences

2. Why might someone's life expectancy be low in a region with a low level of development?

Identifying Problems

3. If countries in the Middle East stopped producing oil, how might that affect the economy of the United States?

Visual Summary

1 The World at Your Fingertips

- History, geography, government, economics, and culture are five ways to understand Earth and its peoples.

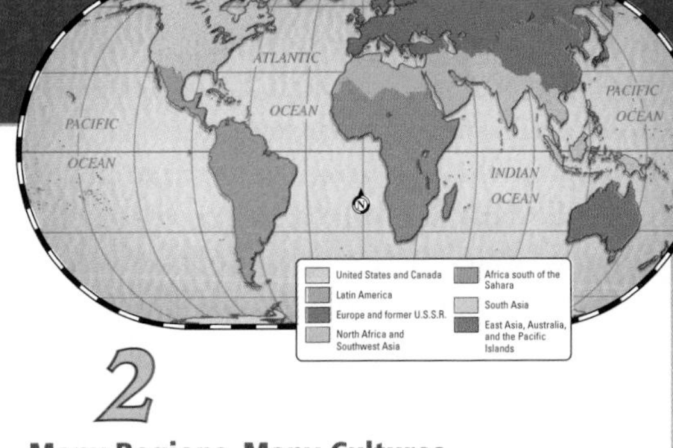

2 Many Regions, Many Cultures

- People live, dress, and think differently in each of the world's culture regions.

CRITICAL THINKING: Possible Responses

1. Remembering Definitions

When you can identify the language someone speaks, what is valued in his or her culture, and what is shared between your cultures, you have a better chance of becoming friends.

2. Making Inferences

Life expectancy in a developing economy is low because of poor health care services, low literacy rates, lack of clean running water, and low salaries.

3. Identifying Problems

If the countries in the Middle East stopped producing oil, the cost of oil would go up. The increased cost of oil would affect transportation and production in the United States.

se the map and your knowledge of world cultures and eography to answer questions 1 and 2.

dditional Test Practice, pp. S1–S33

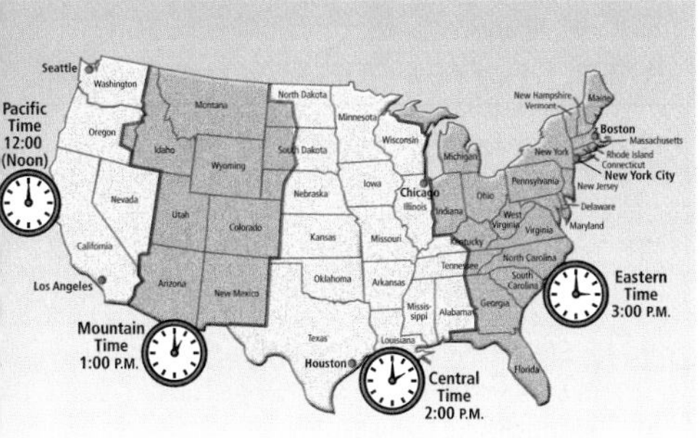

What is the time difference between Los Angeles and New York City?

A. one hour
B. two hours
C. three hours
D. four hours

If it is 2 P.M. in Houston, what time would it be in Chicago?

A. 1 P.M.
B. 2 P.M.
C. 3 P.M.
D. 4 P.M.

In the following quotation, writer Aime Cesair talks about culture. Use the quotation and your knowledge of world cultures and geography to answer question 3.

PRIMARY SOURCE

Culture is everything. Culture is the way we dress, the way we carry our heads, the way we walk, the way we tie our ties—it is not only the fact of writing books or building houses.

AIME CESAIR, speech to the World Congress of Black Writers and Artists in Paris

3. Which of the following statements would Cesair agree with most?

A. Culture refers to fine art, drama, and classical music.
B. Culture includes all the things that characterize a group.
C. Most people in the world do not take part in culture.
D. Some parts of culture are more important than others.

LTERNATIVE ASSESSMENT

WRITING ABOUT HISTORY

The United States and Canada share a culture region. Think about the elements that contribute to this culture region. Then write a script for a one-act play in which a person from another culture region experiences our culture for the first time. Focus on unique aspects of our language, technology, food, and clothing that a person might find surprising and strange.

COOPERATIVE LEARNING

With a group of three to five students, set up a peace conference to help two warring groups make peace. Choose two specific groups and a specific issue that caused the hostility, such as a conflict over the scarcity of water. Do research to understand each side's point of view in the conflict. Group members should take on specific roles, such as conference moderator and spokesperson for each side. Present your peace conference to the class as a skit.

INTEGRATED TECHNOLOGY

Doing Internet Research

Use the Internet to research a culture, such as the people of Lebanon or Hong Kong. Write a report of your findings. List the Web sites you used to prepare your report.

- Specifically find out about the daily life of the people, including what jobs they have, what foods they eat, what clothes they wear, and what their homes are like.
- Also, research what most people in the culture value, what governments they live under, and what common difficulties they face.

For Internet links to support this activity, go to

STANDARDS-BASED ASSESSMENT

1. **Answer C** is the correct answer because New York is in the Eastern time zone, while Los Angeles is in the Pacific time zone. The map shows that when it is three o'clock Eastern Time, it is noon Pacific Time. Answers A, B, and D are incorrect because they do not give the correct difference between those two time zones.
2. **Answer B** is the correct answer because Chicago and Houston are in the same time zone. Answers A, C, and D are incorrect because they give the time for other time zones.
3. **Answer B** is the correct answer because it echoes the idea that culture includes everything. Answer A is incorrect because it contradicts Cesair's main point; answer C is incorrect because everyone has the traits that Cesair mentions; answer D is incorrect because Cesair does not rank aspects of culture.

INTEGRATED TECHNOLOGY

Discuss how students might find information using the Internet. Brainstorm possible key words, Web sites, and search engines.

Discuss other issues, such as language, education, food, or women in government.

Suggest that students make a chart listing their areas of interest to be filled in with information they locate.

Alternative Assessment

1. Rubric

The script should

- reflect the student's understanding of the culture of the United States and Canada.
- focus on the unique aspects of the culture.
- exhibit creativity.
- be logically organized.

2. Cooperative Learning Activity

Discuss the purpose of a peace conference and how one might be arranged. Then brainstorm locations where tension exists, such as the Middle East, Ireland, or India. Help students understand the causes of the tension. Provide source materials and information for a Web search.

Guide groups to focus on a single issue. Assign the roles of intermediary and spokesperson.

PLANNING GUIDE: CHAPTER 2
The Geographer's World

	OVERVIEW	COPYMASTERS	INTEGRATED TECHNOLOGY
CHAPTER RESOURCES	**Students will examine the five themes used to describe Earth, some of the tools geographers use, and how these themes and tools are used to help us understand people, places, and environments of the past and present.**	**In-depth Resources: Unit 1** • Guided Reading Worksheets, pp. 14–15 • Skillbuilder Practice, p. 18 • Unit Atlas Activities, pp. 1–4 • Geography Workshop, pp. 22–23 **Reading Study Guide** (Spanish and English), pp. 10–15 **Outline Map Activities**	• eEdition Plus Online • EasyPlanner Plus Online • eTest Plus Online • eEdition • Power Presentations • EasyPlanner • Electronic Library of Primary Sources • Test Generator • Reading Study Guide • Critical Thinking Transparencies CT3
	KEY IDEAS		
SECTION 1 The Five Themes of Geography pp. 35–40	• **The five themes of geography are location, place, region, movement, and human-environment interaction.** • **Location can be described as absolute or relative.** • **A place is characterized by physical, political, or cultural features. A region is a group of places with common features.** • **Humans migrate to different places for various reasons, and adapt to and modify the world around them.**	**In-depth Resources: Unit 1** • Guided Reading Worksheet, p. 14 • Reteaching Activity, p. 20 **Reading Study Guide** (Spanish and English), pp. 10–11	**Map Transparencies MT5, 6, 7, 8, 9** **classzone.com** Reading Study Guide
SECTION 2 The Geographer's Tools pp. 45–49	• **Maps and globes provide similar information in different ways.** • **Geographers use charts and graphs to display and compare information.**	**In-depth Resources: Unit 1** • Guided Reading Worksheet, p. 15 • Reteaching Activity, p. 21 **Reading Study Guide** (Spanish and English), pp. 12–13	**Critical Thinking Transparencies CT4** **Map Transparencies MT2, 3, 4, 8** **classzone.com** Reading Study Guide

KEY TO RESOURCES

 Audio
 CD-ROM
Copymaster
 Internet
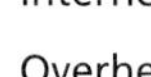 Overhead Transparency
Pupil's Edition
 Teacher's Edition
 Video

ASSESSMENT OPTIONS

Chapter Assessment, pp. 50–51

 **Formal Assessment**
- Chapter Tests: Forms A, B, C, pp. 21–32

 **Test Generator**

 **Online Test Practice**

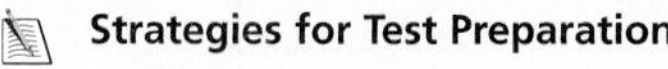 **Strategies for Test Preparation**

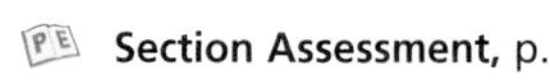 **Section Assessment,** p. 40

 **Formal Assessment**
- Section Quiz, p. 19

 **Integrated Assessment**
- Rubric for writing a magazine advertisement

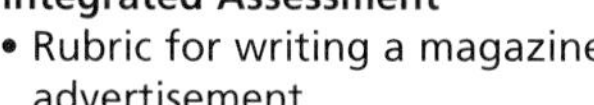 **Test Generator**

Test Practice Transparencies TT3

Section Assessment, p. 49

Formal Assessment
- Section Quiz, p. 20

 **Integrated Assessment**
- Rubric for drawing a map

 **Test Generator**

 **Test Practice Transparencies TT4**

RESOURCES FOR DIFFERENTIATING INSTRUCTION

Students Acquiring English/ESL

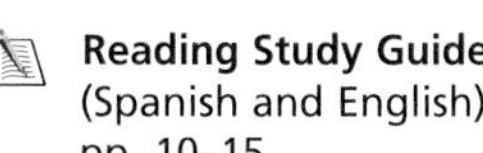 **Reading Study Guide** (Spanish and English), pp. 10–15

Access for Students Acquiring English Spanish Translations, pp. 7–12

Modified Lesson Plans for English Learners

Less Proficient Readers

Reading Study Guide (Spanish and English), pp. 10–15

TE Activity
- Illustrating a Region, p. 38

Gifted and Talented Students

 **TE Activity**
- Designing an Illustrated Map, p. 47

CROSS-CURRICULAR CONNECTIONS

Literature
Creech, Sharon. ***The Wanderer.*** New York: HarperCollins, 2000. A sailing voyage from Connecticut across the Atlantic. Rich geographic detail.

Health
Smith, Carter. ***Mapping Epidemics: A Historical Atlas of Disease.*** New York: Franklin Watts, 2000. Thirty-two diseases and their geographical progressions.

Geography
Dorling, Kindersley. ***The Ultimate Panoramic Atlas.*** New York: DK Publishing, 1998. Each continent's land mass and sea features are exaggerated to give a 3-D effect.
Dash, Joan. ***The Longitude Prize.*** New York: Frances Foster Books, 2000. John Harrison spent forty years developing a machine to determine longitude at sea.

Science/Math
Lye, Keith. ***Atlas in the Round: Our Planet As You've Never Seen It Before.*** Philadelphia: Running Press, 1999. The spherical perspective corrects misleading aspects of flat maps.

Language Arts/Literature
Hopkins, Lee Bennett. ***My America: A Poetry Atlas of the United States.*** New York: Simon & Schuster, 2000. Unique concept, nicely rendered.

Science
Kerrod, Robin, and John Stidworthy. ***Facts on File Wildlife Atlas.*** New York: Facts on File, 1998. Full color photos, maps, and diagrams.

ENRICHMENT ACTIVITIES

The following activities are especially suitable for classes following block schedules.

Teacher's Edition, pp. 30, 36, 41, 47
Pupil's Edition, pp. 40, 49
Geography Handbook, pp. 4–13
Linking Past and Present, pp. 42–43
Technology: 2000, p. 44
Outline Map Activities

INTEGRATED TECHNOLOGY

Go to **classzone.com** for lesson support and activities for Chapter 2.

BLOCK SCHEDULE LESSON PLAN OPTIONS: 90-MINUTE PERIOD

DAY 1

CHAPTER PREVIEW, pp. 32–33
Class Time 20 minutes

- **Hypothesize** Use "What do you think?" questions in Focus on Geography on PE p. 33 to help students hypothesize about the uses of geography in exploring and understanding the world.

SECTION 1, pp. 35–40
Class Time 70 minutes

- **Summarizing** Have students write a brief description of a place they know well. Ask them to imagine they are describing it to someone who has never seen it. Write the five geography themes (location, place, region, movement, human-environment interaction) on the chalkboard. Have students share their descriptions, and ask them to decide into which geographical theme their description best fits. Review the themes as you discuss.
  Class Time 30 minutes
- **Comparing** To compare the ideas of absolute location and relative location, have a student stand somewhere in the room. Ask students to describe his/her location first in an absolute way, then in a relative way. Have students explain the differences.
  Class Time 10 minutes
- **Internet** Divide students into groups and have them visit **classzone.com** to learn more about Pangaea and changes in the locations of Earth's continents.
  Class Time 30 minutes

DAY 2

SECTION 1, continued
Class Time 60 minutes

- **Reading a Map** Divide students into teams. Have them use the Rand McNally World Physical map on PE pp. A2–3 to find two location examples of each region shown in the Natural Regions of the World chart, on PE p. 38. Have groups present their examples to the class; they can point them out on a map.
  Class Time 30 minutes
- **Analyzing Causes** Have students review the Human Migration map on PE p. 39. Lead a discussion about what natural features may have influenced or affected these movement patterns.
  Class Time 10 minutes
- **Peer Review** To practice using longitude and latitude, ask students to choose a place in the world they would like to visit. Have students use the Skillbuilder on PE p. 41 to plot the longitude and latitude of their choice, keeping the location a secret. Then have students exchange coordinates, figure out the location, and check their answer with their partner.
  Class Time 20 minutes

SECTION 2, pp. 45–49
Class Time 30 minutes

- **Making a Map** Review the concept of thematic maps, then ask students to make a thematic map of the school. They should clearly indicate the theme in the map's legend. Some possible themes: the locations of student activities, or the places where one can get food in the school.

DAY 3

SECTION 2, continued
Class Time 35 minutes

- **Cartographer** Have students use PE pp. 42–44 to review the skills cartographers use to construct maps. Encourage them to make an illustrated map of their town or neighborhood. Display the finished maps.

CHAPTER 2 REVIEW AND ASSESSMENT, pp. 50–51
Class Time 55 minutes

- **Review** Have students use the charts they created for Reading Social Studies on PE p. 34 to review the five geography themes.
  Class Time 20 minutes
- **Assessment** Have students complete the Chapter 2 Assessment.
  Class Time 35 minutes

TECHNOLOGY IN THE CLASSROOM

DESIGNING WEB SITES

Students can design their own Web sites to organize information and to share their knowledge with other students at school and around the world. By creating a Web site, students gain experience in organizing information in a nonlinear fashion and get a behind-the-scenes look at what goes into developing materials for the Internet. Their Web sites can be kept on the classroom computer, uploaded to the school district's server, or uploaded to the Internet for students at other schools to view. The Web site at **classzone.com** is a helpful resource for students to learn about designing their own Web pages.

ACTIVITY OUTLINE

Objective Students will design Web pages to teach other students about the Five Themes of Geography.

Task Have students work in groups to create a Web site that provides examples of the Five Themes of Geography and that teaches other students about each of the themes.

Class Time Three class periods

DIRECTIONS

1. Hold a class discussion reviewing the Five Themes of Geography, as discussed in Chapter 2. Ask students to provide examples to explain each theme, and ask them to describe the reasons why it is important to be familiar with the concepts covered by each of these themes. For example, why is it important to identify your location (Theme 1) or to understand the impact of migration and transportation (Theme 4)?
2. Divide the class into small groups of about four students each. Assign one of the Five Themes. It may be necessary to assign some themes to more than one group.
3. Ask groups to create Web pages that will teach students in their grade or the grade below them about their assigned theme. They will need to provide text and images to help students learn the concepts of the theme. For example, for place (Theme 2), they might provide a brief description of what *place* means and then create a Web page that shows pictures of the landscape and people in their home town. They could then ask the audience to think about the special characteristics of their own home and to compare and contrast those features to the things they see on the Web page.
4. Suggest that students use the following criteria for completing their Web pages:
   - They must provide a two- to four-sentence description of what their theme is about.
   - They must include at least one example of how this theme relates to things they already know about (e.g., their home town, transportation, world regions).
   - They must include at least one image, and they must cite the source of the image if they have taken it from a book or another Web site.
5. Help several students design an overall home page that will link to each group's Web page. This main home page should contain a list of the Five Themes. As an option, upload this page and each group's site onto the school district's server, and register it with a search engine.

CHAPTER 2 OBJECTIVE

Students will examine the five themes used to describe Earth, some of the tools geographers use, and how these themes and tools help us understand people, places, and environments of the past and present.

FOCUS ON VISUALS

Interpreting the Photograph Have students observe the model of the solar system and compare it to the illustration of the Ptolemaic idea of the solar system shown on p. 33. Ask them to describe the difference between the two conceptions of the solar system.

Possible Response The modern understanding of the solar system places the sun at the center, with planets revolving around it. The Ptolemaic idea of the solar system imagined Earth at the center, surrounded by the rotating planets.

Extension Using an encyclopedia or the Internet, have students research Copernicus and his challenge to traditional ideas.

CRITICAL THINKING ACTIVITY

Recognizing Effects Discuss with students why people resisted the idea that the Earth was not at the center of the solar system. Have students imagine an idea that might challenge our modern understanding of the universe.

Class Time 10 minutes

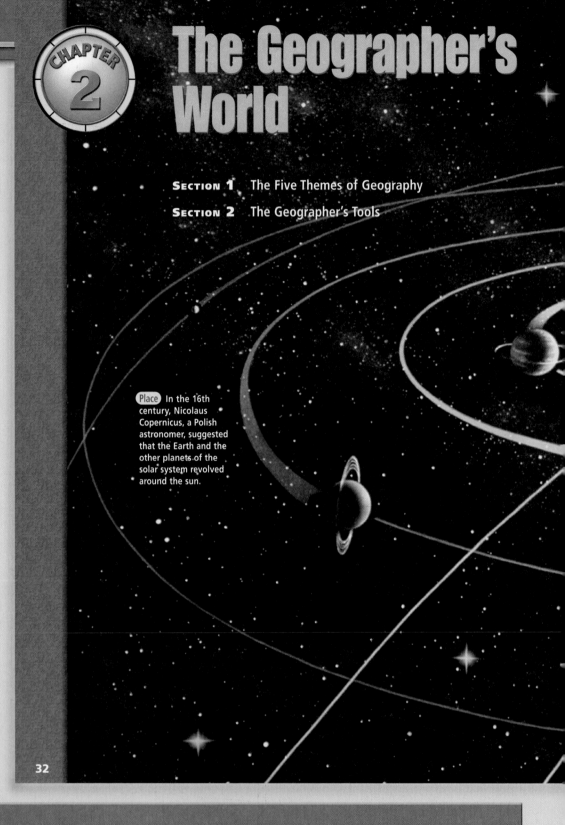

CHAPTER 2

The Geographer's World

Place In the 16th century, Nicolaus Copernicus, a Polish astronomer, suggested that the Earth and the other planets of the solar system revolved around the sun.

Recommended Resources

BOOKS FOR THE TEACHER

Elsom, Derek. ***Planet Earth.*** Detroit, MI: Macmillan Reference USA, 2000. Geography and geology.

The National Geographic Desk Reference: A Geographical Reference with Hundreds of Photographs, Maps, Charts, and Graphs. Washington, D.C.: National Geographic Society, 1999. Mapmaking and evolution of Earth's people.

VIDEOS

Geography Tutor, Volume 2: Types of Maps and Map Projections. Venice, CA: TMW Media Group. Explanation of mapping Earth.

SOFTWARE

Planet Earth: Explore the Worlds Within. Library Video Company, 2000. Lands, habitats, peoples, and cultures of the world.

INTERNET

For more information about Earth, visit **classzone.com.**

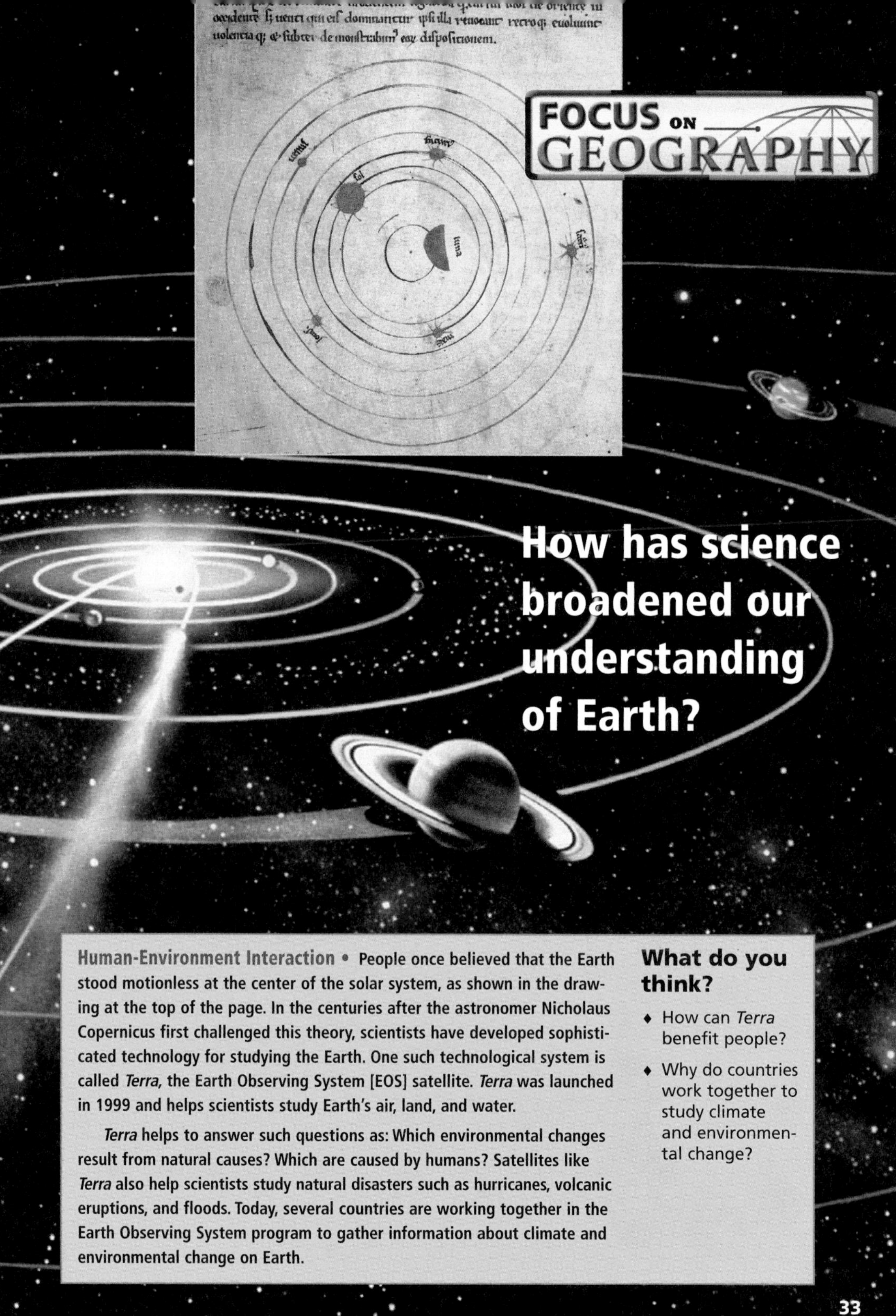

FOCUS ON GEOGRAPHY

How has science broadened our understanding of Earth?

Human-Environment Interaction • People once believed that the Earth stood motionless at the center of the solar system, as shown in the drawing at the top of the page. In the centuries after the astronomer Nicholaus Copernicus first challenged this theory, scientists have developed sophisticated technology for studying the Earth. One such technological system is called *Terra,* the Earth Observing System [EOS] satellite. *Terra* was launched in 1999 and helps scientists study Earth's air, land, and water.

Terra helps to answer such questions as: Which environmental changes result from natural causes? Which are caused by humans? Satellites like *Terra* also help scientists study natural disasters such as hurricanes, volcanic eruptions, and floods. Today, several countries are working together in the Earth Observing System program to gather information about climate and environmental change on Earth.

What do you think?

- How can *Terra* benefit people?
- Why do countries work together to study climate and environmental change?

33

FOCUS ON GEOGRAPHY

Objectives

- To recognize the importance of human-environment interaction
- To describe the tools that geographers use to learn about Earth's features

What Do You Think?

1. Encourage students to consider how scientists and governments might use the data collected by *Terra.* What changes might people make based on this data?
2. Point out that many of Earth's physical features cover a large region and encompass many countries.

How has new technology increased our knowledge of Earth?

Have students consider how data about Earth was collected before satellite technology. Point out that even with aerial photography only a limited area could be observed at any one time.

MAKING GEOGRAPHIC CONNECTIONS

Ask students to brainstorm a list of natural disasters, such as floods, earthquakes, hurricanes, tornadoes, and volcanic eruptions. Then have them suggest how technology might help track and predict these events.

Then ask students to brainstorm a list of other events on Earth's surface that scientists could use information from satellites to study. These events could include the loss of seashore or rain forests, the spread of oil from an oil-tanker spill, or the increase in traffic at night. Have them suggest how technology might help scientists understand these events.

Implementing the National Geography Standards

Standard 4 Identify and compare the physical characteristics of places

Objective To construct a chart comparing the physical characteristics of the local area with the physical characteristics of other places

Class Time 15 minutes

Task Bring pictures of different natural regions to class from magazines, photo collections, and so on. Have each student prepare a six-column chart identifying the soils, landforms, vegetation, wildlife, climate, and natural hazards of different places. Students should record on their charts the similarities and differences between their local environment and the pictured environments.

Evaluation In their charts, students should identify at least one difference and one similarity between the local environment and each pictured environment.

BEFORE YOU READ

What Do You Know?

Ask students if they have ever given someone directions to their house. What kind of information do they use? If they had to give directions to the continent of North America, what would they say? Ask students to tell about what makes their neighborhood unique.

Ask students to pick a region in the United States that they have visited. Ask how the climate, landforms, and rainfall or other water resources differ from the region in which they live. Then ask what the region in which they live was like 100 years ago. How has it changed? What changes did humans make, and why? What changes are humans making today, and why?

What Do You Want to Know?

Have students work in pairs to generate questions about what they want to know about the field of geography and record the questions in their notebooks. Have students record the answers they find as they read this chapter. Students can compare their answers when they have completed the chapter.

READ AND TAKE NOTES

Reading Strategy: Identifying Main Ideas
Point out that each of the five themes is a major heading in Section 1. Explain that each heading is a main idea and that the text under each heading contains supporting details. Encourage students to find several supporting details for each main idea and expand their webs to include these details.

In-depth Resources: Unit 1
- Guided Reading Worksheets, pp. 14–15

CHAPTER 2 READING SOCIAL STUDIES

BEFORE YOU READ

▸▸ What Do You Know?

Do you know how to find important places in your town? Have you visited cities, towns, or rural areas and noticed what made these places special? Do you ever use terms like "up north" or "back east"? Have you ever moved from one neighborhood, town, or country to another? Do you know about the harmful effects of pollution on wildlife habitats? If you answered yes, then you know something about each of geography's five big themes—location, place, region, movement, and human-environment interaction.

▸▸ What Do You Want to Know?

Decide what more you want to learn about geography's five themes. Write your ideas, and any questions you may have, in your notebook before you read this chapter.

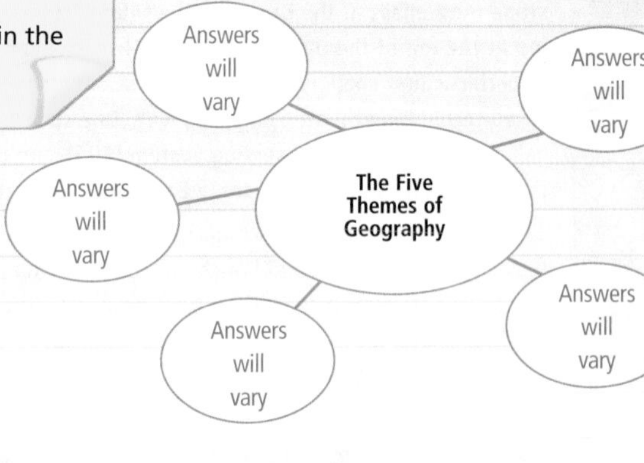

Place • Physical and human characteristics reveal patterns in places. ▲

READ AND TAKE NOTES

Reading Strategy: Identifying Main Ideas One way to make sense of what you read is to look for main ideas and supporting details. Each paragraph, topic heading, and section in a chapter usually has a main idea. Supporting details help to explain the main idea. Use this spider map to write a main idea and its supporting details from this chapter.

- Copy the spider map in your notebook.
- As you read, look for information about the five themes of geography.
- Write a main idea in the center circle.
- Write details supporting the main idea in the other circles.

The Five Themes of Geography

Answers will vary

Answers will vary

Answers will vary

Answers will vary

Answers will vary

Teaching Strategy

Reading the Chapter This is a thematic chapter focusing on the five themes of geography and on the variety of maps, globes, charts, and other tools used to depict information about Earth and human societies. Encourage students to focus on understanding the five themes and explaining how tools are used to describe and learn about Earth.

Integrated Assessment The Chapter Assessment on page 50 describes several activities for integrated assessment. You may wish to have students work on these activities during the course of the chapter and then present them at the end.

The Five Themes of Geography

SECTION 1

TERMS & NAMES
continent
absolute location
latitude
longitude
relative location
migrate

MAIN IDEA

The five themes of geography are location, place, region, movement, and human-environment interaction.

WHY IT MATTERS NOW

The five themes enable you to discuss and explain people, places, and environments of the past and present.

DATELINE

EXTRA

FRANKFURT, GERMANY, JANUARY 6, 1912

Scientist Alfred Wegener sent out shock waves today when he proposed a radical new hypothesis. The continents were once joined together as one huge landmass. In time, he suggests, pieces of this landmass broke away and drifted apart.

Wegener calls this supercontinent *Pangaea*. To support his theory, Wegener points out that the continents seem to fit together. He notes, for example, that the east coast of South America fits snugly against the west coast of Africa. Mountain ranges continue across both continents as smoothly as the lines of print across torn pieces of a newspaper.

Other scientists reject Wegener's claim. They say that they know of no force strong enough to cause continents to move.

Movement • Seven continents were once one continent. A continent is a landmass above water on earth. ▲

The Five Themes ❶

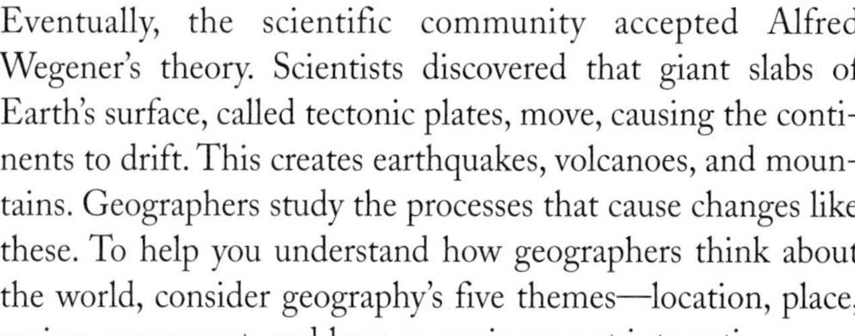

Eventually, the scientific community accepted Alfred Wegener's theory. Scientists discovered that giant slabs of Earth's surface, called tectonic plates, move, causing the continents to drift. This creates earthquakes, volcanoes, and mountains. Geographers study the processes that cause changes like these. To help you understand how geographers think about the world, consider geography's five themes—location, place, region, movement, and human-environment interaction.

TAKING NOTES
Use your web to take notes about the five themes.

The Five Themes of Geography

SECTION OBJECTIVES

1. To identify the five themes of geography
2. To describe the theme of location
3. To describe the themes of place and region and explain their relationship
4. To describe the movement of people in the world and how people interact with their environment

SKILLBUILDER
- Interpreting a Map, p. 36
- Interpreting a Chart, p. 38

CRITICAL THINKING
- Synthesizing, p. 37
- Drawing Conclusions, pp. 39, 48
- Recognizing Effects, p. 40

FOCUS & MOTIVATE

WARM-UP

Making Predictions Have students read Dateline and look at the map and a globe. Then discuss these questions.

1. According to Wegener's theory, what mountain chains in Africa and South America might once have been joined?
2. If Earth's crust moves, what might happen to the continents in the distant future?

INSTRUCT: Objective ❶

The Five Themes

- Into what five themes is the study of geography divided? location, place, region, movement, human-environment interaction

 **In-depth Resources: Unit 1**
- Guided Reading Worksheet, p. 14

 **Reading Study Guide**
(Spanish and English), pp. 10–11

Program Resources

 **In-depth Resources: Unit 1**
- Guided Reading Worksheet, p. 14
- Reteaching Activity, p. 20

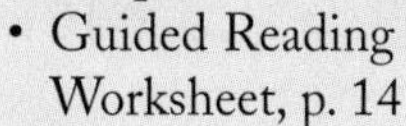

 **Reading Study Guide**
(Spanish and English), pp. 10–11

 **Formal Assessment**
- Section Quiz, p. 19

 **Integrated Assessment**
- Rubric for writing an advertisement

 **Outline Map Activities**

 **Access for Students Acquiring English**
- Guided Reading Worksheet, p. 7

 **Technology Resources**
classzone.com

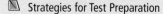

TEST-TAKING RESOURCES
Strategies for Test Preparation
Test Practice Transparencies

Online Test Practice

INSTRUCT: Objective 2

Location

- How is absolute location determined? by using latitude and longitude lines
- How is relative location determined? by describing it in relation to other places

FOCUS ON VISUALS

Interpreting the Map Guide students to name each of the lines of latitude shown on the map. Then have them identify the states that are located north of latitude 36°N.

Possible Response Students should identify Missouri, Illinois, Kentucky, and Virginia.

Extension Have students use a map to locate their community or the closest city and identify the latitude at which their community is located.

The World's Heritage

In 1835, British naturalist Charles Darwin visited the Galápagos Islands. He was intrigued with the many variations in a type of bird—the finch—that he observed there. He noticed that some finches ate mainly plants, while others fed only on insects. The shapes of their bills varied depending on their diet. Their bills were adapted for grasping, biting, crushing, or probing. Darwin's observations in the Galápagos and elsewhere in South America helped him form the basis for his theory of natural selection. In part, the theory states that animals change over time to adapt to their specialized environments.

Geography Skillbuilder Answers

1. Tennessee, Mississippi, Alabama, Georgia, North Carolina
2. Florida

GEOGRAPHY SKILLBUILDER: Interpreting a Map

1. **Location** • Which states have borders that run along latitude 35°N?
2. **Location** • Which state extends south to longitude 80°W and latitude 25°N?

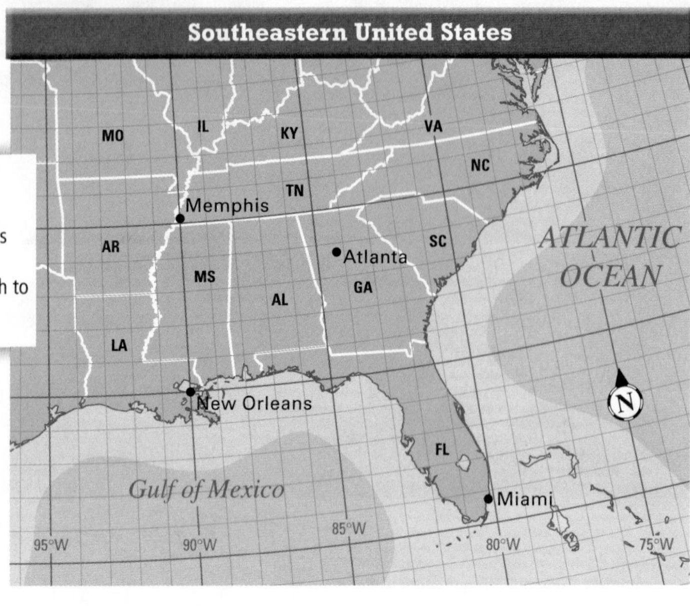

The Galápagos Islands The Galápagos Islands are an archipelago, or group of islands, 600 miles off the coast of South America. These islands, which contain many forms of plant and animal life found nowhere else in the world, are a unique "living museum."

Scientists and tourists are fascinated by the islands' creatures, such as Galápagos hawks, land iguanas, waved albatrosses, and blue-footed boobies. The islands are also home to giant tortoises, which can weigh up to 650 pounds and can live to be 200 years old. One of the 11 subspecies of giant tortoises has only one member left. Lonesome George, shown below, is about 80 years old.

Location 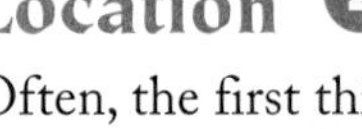

Often, the first thing you want to know about a place is where it is located in space. Geography helps you think about things spatially—where they are located and how they got there. Location allows you to discuss places in the world in terms everyone can understand.

Absolute Location If someone asks you where your school is, you might say, "At the corner of Fifth Street and Second Avenue." Ask a geographer where Melbourne, Australia, is located, and you may get the answer "38° south latitude, 145° east longitude." This is the absolute location of the city of Melbourne. **Absolute location** is the exact spot on Earth where a place can be found.

Using a system of imaginary lines drawn on its surface, geographers can locate any place on Earth. Lines that run parallel to the equator are called **latitude** lines. They show distance north and south of the equator. Lines that run between the North and South Poles are called **longitude** lines. They show distance east and west of the prime meridian.

Activity Options

Skillbuilder Mini-Lesson: Reading a Time Zone Map

Block Scheduling

Explaining the Skill Tell students that a time zone map shows 24 time zones. The day begins at the International Date Line, 180° longitude.

Applying the Skill Refer students to the Time Zone Map on page 27. Have them use the following strategies to read the map.

1. Read the title of the map to find out what it is about.
2. Read the labels across the top of the map that show the hours across the world. Read the labels at the bottom of the map that show the time differences from the prime meridian.
3. Locate the time zone of your community. Compare your time with the time at the prime meridian.

Relative Location Another way to define the location of a place is to describe its relation to other places. You might say your school is "near the fire station" or "two blocks west of the pet store." If someone asks you where Canada is, you might say, "North of the United States." The location of one place in relation to other places is called its **relative location.**

Reading Social Studies

A. Contrasting Contrast absolute location with relative location.

Place • On the west coast of South America, the ancient Inca farmed the steep slopes of the Andes mountains by carving the slopes into terraces. ▲

Place ❸

Another useful theme of geography is place. If you go to a new place, the first thing you want to know is what it is like. Is it crowded or is there a lot of open space? How is the climate? What language do people speak? Every place on Earth has a distinct group of physical features, such as its climate, landforms and bodies of water, and plant and animal life. Places can also have human characteristics, or features that human beings have created, such as cities and towns, governments, and cultural traditions.

Reading Social Studies

A. Possible Answer Absolute location uses latitude and longitude to name the exact location of a place. Relative location names a place's location in relation to other places.

Places Change If you could go back to the days when dinosaurs roamed Earth, you would see a world much different from the one you know. Much of Earth had a moist, warm climate, and the continents were not located where they are today. Rivers, forests, wetlands, glaciers, oceans—the physical features of Earth—continue to change. Some changes are dramatic, caused by erupting volcanoes, earthquakes, and hurricanes. Others happen slowly, such as the movement of glaciers or the formation of a delta.

Region ❸

Geographers group places into regions. A region is a group of places that have physical features or human characteristics, or both, in common. A geographer interested in languages, for example, might divide the world into language regions. All the countries where Spanish is the major language would form one Spanish-speaking language region. Geographers compare regions to understand the differences and similarities among them.

Place • This satellite photo shows the Mississippi River delta. It was formed from sediment and mud carried by the river to its mouth. ▼

INSTRUCT: Objective ❸

Place/Region

- What are some characteristics of place? climate, landforms, bodies of water, plant and animal life, cities, governments, and cultural traditions
- What causes places to change? change in climate, earthquakes, volcanic eruptions, hurricanes, movements of glaciers, and so on.
- What is a region? a group of places that have common physical features and/or human characteristics
- What is a natural region? a region that has its own unique combination of plant and animal life and climate

CRITICAL THINKING ACTIVITY

Synthesizing Have students work in pairs to complete a spider map about the place—the town, city, or neighborhood—where they live. Students can choose to make either a spider map about the physical features of their place or a spider map about the human characteristics of their place. Have pairs compare and discuss their spider maps.

Class Time 15 minutes

MORE ABOUT... Geography

The word *geography* comes from two Greek words. *Geo-* means "earth" and *-graphy* means "writing," so geography means "writing about Earth." Geographers are involved in a huge range of activities. Some study information gleaned from satellites, while others study traffic flow in a community. Some focus on changes in the natural world, while others trace human movement and interaction with the planet.

Activity Options

Interdisciplinary Links: Language Arts/History

Class Time One class period

Task Writing a report on the 1980 eruption of Mount St. Helens

Purpose To learn about the effects of a volcanic eruption on the environment and humans

Supplies Needed

- Reference materials about the 1980 eruption of Mount St. Helens
- Pencils or pens
- Writing paper

Activity Remind students that a volcanic eruption can cause swift and dramatic changes to a place. Briefly discuss the 1980 eruption of Mount St. Helens. Then encourage them to focus their research on the effects the eruption had on the landscape and the people living in the area. Have students write brief reports and share selected information with the class.

INSTRUCT: Objective 4

Movement/Human-Environment Interaction

- What are some reasons that people migrate to other areas? to leave difficult circumstances; to find a better life
- What are some natural barriers to human movement? mountains, canyons, raging rivers
- How do humans and the environment interact? Humans depend on, adapt to, and modify the world around them. The environment and human society are shaped by each other.

FOCUS ON VISUALS

Interpreting the Chart Have students notice the relationship between the climate and plant life shown for each region. Ask them to identify the natural region in which they live and what kind of vegetation is in the area.

Possible Response If students have trouble identifying the natural region in which they live, have them identify the climate of their surroundings.

Extension Have students use an encyclopedia to find out more about one of these types of regions. Then have them describe a specific place on Earth that typifies that region.

INTERACTIVE

Natural Regions of the World

Region	Climate	Plant Life
Tropical Rain Forest	Hot and wet all year	Thick trees, broad leaves Trees stay green all year
Tropical Grassland	Hot all year Wet and dry seasons	Tall grasses Some trees
Mediterranean	Hot, dry summers Cool-to-mild winters	Open forests Some clumps of trees Many shrubs, herbs, grasses
Temperate Forest	Warm summers Cold-to-cool winters	Mixed forests; some trees lose leaves in winter, others stay green all year
Cool Forest	Cool-to-mild summers Long, cold winters	Mostly trees with needles; stay green all year; some trees lose leaves in winter
Cool Grassland	Warm summers Cool winters Drier than forest regions	Prairies: Tall, thick grass Higher lands: Shorter grass
Desert	Hot all year Very little rain	Sand or bare soil, few plants May have cactus, some grass and bushes
Tundra	Short, cool summers Long, cold winters Little rain or snow	Rolling plains: No trees Some patches of moss, short grass, flowering plants
Arctic	Very cold Covered in ice all year	None
High Mountain	Varies, depending on altitude	Varies, depending on altitude

SKILLBUILDER: Interpreting a Chart
1. **Region •** How are desert regions and tropical grasslands alike and how are they different?
2. **Region •** In which type of climate are trees most likely to stay green all year?

Skillbuilder Possible Answers
1. They are both hot all year. Deserts are mostly dry all year and have few plants. Tropical grasslands have dry and wet seasons and some plants.
2. tropical rain forest and temperate forest

Region • The tropical rain forest is one of the ten natural regions of the world. ▲

Natural Regions The world can be divided into ten natural regions. A natural region has its own unique combination of plant and animal life and climate. Tropical rain forest regions are in Central and South America, Africa south of the Sahara, Southeast Asia, Australia, and the Pacific Islands. Where are desert regions located?

Movement 4

People, goods, and ideas move from one place to another. So do animals, plants, and other physical features of Earth. Movement is the fourth geographic theme. The Internet is a good tool for the movement of ideas. Sometimes people move within a country. For example, vast numbers of people have migrated from farms to cities. **Migrate** means to move from one area to settle in another. You may have ancestors who immigrated to the United States—perhaps from Africa, Europe, Latin America, or Asia. When people emigrate, they take their ideas and customs with them. They may also adopt new ideas from their new home.

Reading Social Studies
B. Possible Answer
Problems push people out of one place. The advantages of another place pull them in.

Vocabulary
immigrate: to move to an area
emigrate: to move away from an area

Reasons for Moving Migration is a result of push and pull factors. Problems in one place push people out. Advantages in another place pull people in. Poverty, overcrowding, lack of jobs and schooling, prejudice, war, and political oppression are push factors. Pull factors include a higher standard of living, employment and educational opportunities, rights, freedom, peace, and safety.

B. Synthesizing How do push and pull factors work together?

Activity Options

Differentiating Instruction: Less Proficient Readers

Illustrating a Region To help students visualize what types of plants and animals might be found in the regions described in the chart, display photos depicting these features for some of the regions. Discuss the plants and animals they see, and have students choose one of the natural regions to illustrate. Ask them to draw illustrations of vegetation and animals that might be found in the region. Encourage students to label each illustration.

When students have completed their illustrations, have them present the drawings to the class and tell about what the drawings show and why they chose those particular plants and animals. After each student presents his or her drawing, ask the following questions: Would you like to live here? Why or why not? What would you have to do to live comfortably in this region?

Human Migration

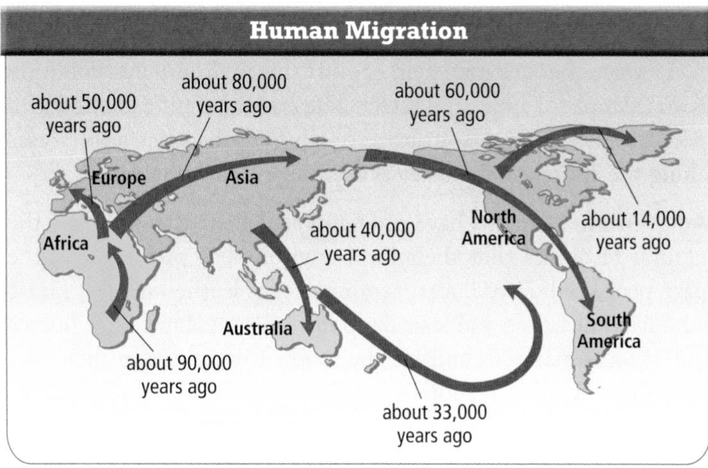

Movement • As you can see, people have been on the move for at least 90,000 years. 

Barriers to Movement Natural barriers, such as mountain ranges, canyons, and raging rivers, can make migration difficult. Oceans, lakes, navigable rivers, and flat land can make it easier. Modern forms of transportation have made it easier than ever for people to move back and forth between countries.

Vocabulary
navigable: deep and wide enough for boats to travel on

Human-Environment Interaction 4

Interaction between human beings and their environment is the fifth theme of geography. Human-environment interaction occurs because humans depend on, adapt to, and modify the world around them. Human society and the environment cannot be separated. Each shapes and is shaped by the other. Earth is a unified system.

Some places are the way they are because people have changed them. For example, if an area has a lot of open meadows, this may be because early settlers cleared the land for farming.

Citizenship IN ACTION

Saving Special Places Many of the most wonderful and special places on Earth may be destroyed or ruined over time unless they are protected. To prevent this, UNESCO (the United Nations Educational, Scientific, and Cultural Organization) set up the World Heritage Committee in 1972. This group identifies human-made and natural wonders all over the world and looks for ways to protect them for the benefit of the world community. So far, the list of World Heritage Sites numbers more than 690. The Grand Canyon (see photograph at right), the Galápagos Islands, the Roman Colosseum, and the Pyramids of Giza are just a few of the places protected for future generations.

CRITICAL THINKING ACTIVITY

Drawing Conclusions Have students make a chart that shows their family's interaction with the environment.

Depend On	Adapt To	Modify

Students should think of as many details as possible to list in each column, such as recycling, the food they eat, the clothing they wear, how they get around, and so on.

When students have completed their charts, have them work in small groups to compare the charts. Have students discuss the following questions: What actions are beneficial to Earth? What actions are not?

Citizenship IN ACTION

The natural heritage sites designated by the World Heritage Committee are places that show examples of Earth's geologic development or contain a record of ancient life. Dinosaur Provincial Park in the Canadian province of Alberta is one such place. Located in Alberta's badlands, the park holds many important dinosaur fossils. Fossils from more than 30 species of dinosaurs, some up to 75 million years old, have been found there.

Other natural sites are chosen because they are home to rare or endangered plants and animals or to a wide range of species. Everglades National Park at the southern tip of Florida is such a site. It has been called a "river of grass flowing imperceptibly" into the sea. The Everglades is home to many types of birds and reptiles, as well as to the endangered manatee.

Activity Options

Interdisciplinary Link: Science

Class Time One class period

Task Preparing a brochure describing unusual physical features of Yellowstone National Park

Purpose To identify unusual physical features of Yellowstone

Supplies Needed
- Reference materials, including Internet access, if possible
- Writing paper
- Pencils or pens

Activity Have small groups of students research the unusual physical features of Yellowstone. Then have each group choose a different feature, such as a geyser basin, lava formations, or thermal springs. Have groups write a section on their features for an informative brochure for visitors to Yellowstone. Encourage students to include information about how the feature was formed.

CRITICAL THINKING ACTIVITY

Recognizing Effects Review the ways in which the environment affects how people live. Choose one region from the chart on page 38 and work with students to complete a web. Write the name of the region in the center and topics such as food, clothing, and shelter in surrounding circles. Then have students add details about how people might adapt to that environment. For example, people who live near oceans might make fish an important part of their diet.

Class Time 15 minutes

ASSESS & RETEACH

Reading Social Studies Have students complete the web diagram on page 34.

Formal Assessment
- Section Quiz, p. 19

RETEACHING ACTIVITY

Have pairs of students create a three-column chart of the five themes of geography. Charts should include theme names, definitions, and examples.

In-depth Resources: Unit 1
- Reteaching Activity, p. 20

Access for Students Acquiring English
- Reteaching Activity, p. 11

Human changes may help or hurt the environment. Pollution is an example of a harmful effect. The environment can also harm people. For example, hurricanes wash away beaches and houses along the shore; earthquakes cause fire and destruction.

Adaptation Humans have often adapted their way of life to the natural resources that their local environment provided. In the past, people who lived near teeming oceans learned to fish. Those who lived near rich soil learned to farm. People built their homes out of local materials and ate the food easily grown in their surroundings. Cultural choices, such as what clothes to wear or which sports to participate in, often reflected the environment.

Because of technology, this close adaptation to the environment is not as common as it once was. Airplanes, for example, can quickly fly frozen fish from the coast to towns far inland. Even so, there are many more ice skaters in Canada and surfers in California than the other way around.

Interaction People and the environment continually interact. For example, when thousands of people in a city choose to use public transportation or ride bicycles rather than drive, less gasoline is burned. When less gasoline is burned, there is less air pollution. In other words, when the environment is healthy, the people who live in it are able to lead healthier lives.

SECTION 1 ASSESSMENT

Terms & Names

1. **Explain the significance of:** (a) continent (b) absolute location (c) latitude (d) longitude (e) relative location (f) migrate

Using Graphics

2. Use a chart like this one to list and explain the five themes of geography.

Theme	Explanation

Main Ideas

3. (a) What physical processes can cause places to change over time?
   (b) How do push and pull factors cause migration?
   (c) What are some ways people have adapted to their environment?

Critical Thinking

4. **Making Inferences**
   What factors make your part of the United States a region?

   Think About
   - similar human geography
   - similar physical geography

ACTIVITY -OPTION- Write and illustrate a **magazine advertisement** to persuade people to move to a new place. Include several pull factors for the place you are advertising.

40 CHAPTER 2

Section 1 Assessment

1. Terms & Names

a. absolute location, p. 36
b. latitude, p. 36
c. longitude, p. 36
d. relative location, p. 37
e. migrate, p. 38

2. Using Graphics

Location	where place is located in space
Place	area with distinct group of physical features
Region	places with common physical/human characteristics
Movement	people, animals, plants, and ideas go from one place to another
Human-environment interaction	when humans depend on, adapt to, and modify the world around them

3. Main Ideas

a. Volcanic eruptions, earthquakes, hurricanes, and glacial movement cause change.
b. Difficult conditions push people to leave a place; the promise of better conditions pulls people to a new place.
c. Adaptation may involve building homes from local material, learning new work skills, or eating locally grown food.

4. Critical Thinking

Students should describe human and physical characteristics that make up a region.

ACTIVITY OPTION

Integrated Assessment
- Rubric for writing an advertisement

SKILLBUILDER

Reading Latitude and Longitude

▶▶ Defining the Skill

To locate places, geographers use a global grid system (see the chart directly below). Imaginary lines of latitude, called parallels, circle the globe. The equator circles the middle of the globe at 0°. Parallels measure distance in degrees north and south of the equator.

Lines of longitude, called meridians, circle the globe from pole to pole. Meridians measure distance in degrees east and west of the prime meridian. The prime meridian is at 0°. It passes through Greenwich, England.

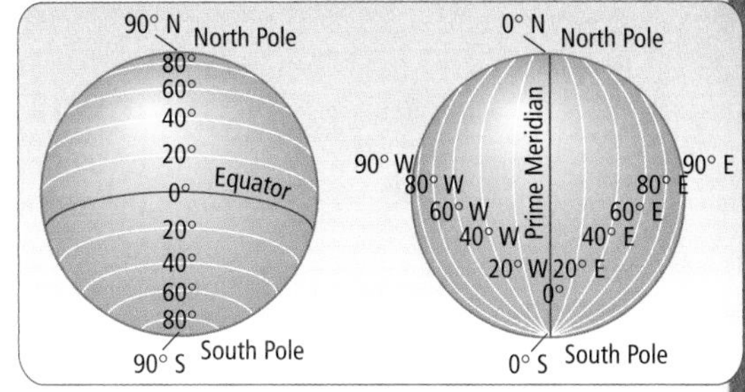

▶▶ Applying the Skill

The world map below shows lines of latitude and longitude. Use the strategies listed directly below to help you locate places on Earth.

How to Read Latitude and Longitude

Strategy ❶ Place a finger on the place you want to locate. With a finger from your other hand, find the nearest parallel. Write down its number. Be sure to include north or south. (You may have to guesstimate the actual number.)

Strategy ❷ Keep your finger on the place you want to locate. Now find the nearest meridian. Write down its number. Be sure to include east or west. (You may have to guesstimate the actual number.)

Strategy ❸ If you know the longitude and latitude of a place and want to find it on a map, put one finger on the line of longitude and another on the line of latitude. Bring your fingers together until they meet.

Write a Summary

Writing a summary will help you understand latitude and longitude. The paragraph below and to the right summarizes the information you have learned.

Use latitude and longitude to locate a place on a globe or map. Lines of latitude circle Earth. Lines of longitude run through the poles. The numbers of the lines at the place where two lines cross is the location of that place.

▶▶ Practicing the Skill

Turn to page 36 in Chapter 2, Section 1, "The Five Themes of Geography." Look at the map of Australia and write a paragraph summarizing how you located the city of Adelaide.

SKILLBUILDER

Reading Latitude and Longitude

Defining the Skill

Tell students that globes and world maps usually show longitude and latitude lines. Point out that if they know the longitude and latitude coordinates of a place, they can find the location of that place on any map or globe, whether for a classroom assignment or for planning a trip.

Applying the Skill

How to Read Latitude and Longitude

Suggest that students work through each strategy. Remind students that meridians indicate distance east or west of the prime meridian and are marked E or W. Point out that they must identify the north/south and east/west lines of longitude and latitude.

Write a Summary

Discuss the summary in detail with students before they begin to write. Encourage them to note the importance of identifying the longitude and latitude lines as accurately as they can.

Practicing the Skill

Have students locate two cities on the map and estimate their exact locations. If students need more practice, ask a group to determine the longitude and latitude for two cities. Then suggest the group give the coordinates to another group and ask them to identify the cities.

In-depth Resources: Unit 1
- Skillbuilder Practice, p. 18

Career Connection: Ship Captain

Encourage students who enjoy reading longitude and latitude to learn about careers that use navigational skills. For example, a ship captain uses navigation to determine the ship's position and course on the open sea, the Great Lakes, or major rivers.

Block Scheduling

1. Suggest that students find out how ship captains use longitude and latitude in their navigation. Encourage them to learn how changing technology is affecting navigation.
2. Help students find out what education and training they would need to embark on this career.
3. Have students present an oral summary of their findings to the class.

OBJECTIVE

Students learn about the long-term effects of the Age of Exploration, which began in the 1400s.

FOCUS & MOTIVATE

Synthesizing and Summarizing Ask students to study the pictures and read each paragraph. Then have them answer these questions:

1. What are some of the positive effects of the Age of Exploration?
2. What are some of the negative effects?

 **Block Scheduling**

MORE ABOUT...

Movement of People

Europeans emigrated to what is now the United States in increasing numbers. By 1700, about 250,000 colonists lived in America. Between 1700 and 1775, about 450,000 European immigrants arrived. More than 7 million Europeans immigrated to the United States between 1820 and 1870. The number of enslaved Africans that came to America was much smaller. By the time Congress outlawed the importation of slaves in 1803, about 375,000 Africans had been brought to this country.

MORE ABOUT...

Cultural Exchange

Almost every aspect of Western culture has been influenced to some extent by other cultures. For example, African music influenced many forms of modern music, while African art inspired 20th-century painters such as Picasso. Chinese and Japanese art and architecture have also had an impact on Western design for centuries.

Linking Past and Present

The Legacy of World Exploration

Movement of People

The Age of Exploration that began in the 1400s led to massive movements of peoples. Millions of Europeans settled in the Americas. Many Native Americans were forced to move as settlers took their lands. Europeans also brought large numbers of enslaved Africans to the Americas. The racial and ethnic mix of modern societies is a legacy of this period of world exploration.

Global Trade

The European demand for Eastern goods such as porcelain, silk, and spices encouraged explorers to seek a direct sea route to East Asia. Global trade increased—Europeans bought products in Asia, using gold and silver from Spanish colonies in the Americas to buy goods in India and China. In this way global trade encouraged contact between different cultures. It also helped prepare the way for today's multicultural societies.

Cultural Exchange

Explorers and settlers helped spread European languages, religions, and ideas around the globe. At the same time, Western culture was enriched by contact with foreign cultures. For example, as English speakers traveled, their vocabulary grew. English now includes words from India, such as *shampoo* and *pajamas,* as well as Native American words, such as *toboggan* and *barbecue.*

Activity Option

Interdisciplinary Link: Science/World History

Class Time 60 minutes

Task Researching and reporting on inventions that reflect the contributions of more than one culture

Supplies Needed

- Encyclopedia and other research materials on technology and scientific inventions

Activity Have students work in groups. Ask each group to research the inventions or discoveries made by one of the following cultures: ancient Greek and Roman, Chinese, Arab. Tell each group to research the history of the invention, investigating how the invention or discovery influenced developments in other cultures. Let students share their findings with the class.

Exchange of Plants and Animals

Exploration encouraged not only the movement of people and goods, but also the exchange of plants and animals. From the Americas, Europeans transported tomatoes, corn, and potatoes to other continents. They brought wheat, rice, and sugarcane to the Americas. This exchange of plants and animals provided new food sources and improved many aspects of life. However, introduced species sometimes threatened and destroyed native species.

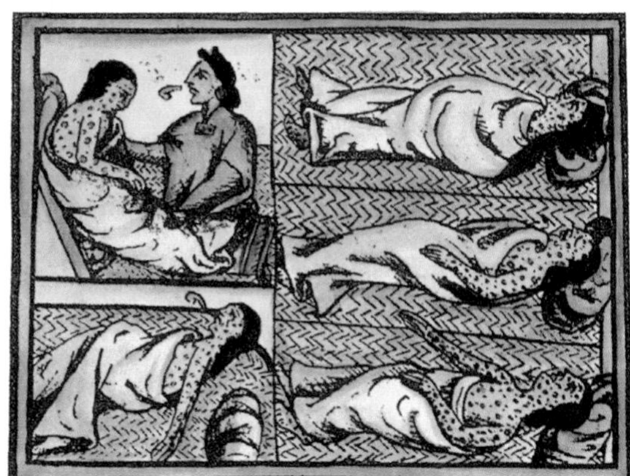

Spread of Disease

Europeans unknowingly brought a host of deadly new diseases to the Americas, including smallpox, malaria, and bubonic plague. With no natural defenses against these diseases, Native Americans died in large numbers. The spread of disease as a result of global trade and travel continues today. Medical organizations around the world are struggling to control the spread of diseases such as SARS.

Science and Technology

World exploration helped spread Western technology around the globe. However, Western technology had itself developed from the inventions of many different cultures. China had invented gunpowder and paper; Muslims had made many important mathematical and scientific discoveries. One particularly important invention was the astrolabe, which measured the sun's and stars' angles above the horizon. This instrument became essential to world explorers as it helped them to determine their ships' position at sea.

Find Out More About It!

Study the text and photos on these pages to learn about world exploration. Then choose the topic that interests you the most and research it in the library or on the Internet to learn more about it. Use the information you gather to write a short play that you and your classmates can perform.

INSTRUCT

- What initiated the Age of Exploration?
- What foods from the Americas were brought to Europe?
- Why did so many Native Americans perish from European diseases?

MORE ABOUT...

Exchange of Plants and Animals

In the United States, Hawaii has been especially affected by the introduction of foreign plants and animals. Except for upper mountain slopes and a few protected lands, Hawaii is now dominated by foreign species.

Newly introduced plant diseases can also change ecosystems dramatically. One example is the chestnut blight fungus. Introduced into New York City from Asia in the late 1800s, this fungus spread over 225 million acres of land in the eastern United States. It killed almost all the chestnut trees.

MORE ABOUT...

Spread of Disease

Smallpox played a major role in Cortes's conquest of the Aztecs in Mexico in 1521. In fact, throughout the Americas, more Native Americans died from European diseases than from warfare. Before 1492, an estimated 40 to 50 million people inhabited the Americas. The majority of this population was wiped out after contact with the Europeans.

More to Think About

Making Personal Connections Ask students to think about the origins of foods that they eat everyday. From what lands or cultures do these foods come? How does the clothing that students wear reflect the importance of global trade? Where were their clothes made?

Vocabulary Activity Have students use a dictionary to find the origin of these words: *algebra, almanac, sauna, tote,* and *tycoon.* Ask students to use the dictionary to make up their own list off English words that have come from foreign languages.

OBJECTIVES

1. To explain how technology has changed the way maps are made and used
2. To explain the usefulness of data collected from maps of Earth's surface
3. To use information from a photograph and captions to support the text

INSTRUCT

- How was radar used to produce 3-D images during the SRTM mission?
- How might maps of Earth's surface be useful to the general public?

Block Scheduling

MORE ABOUT...

Gathering Information for Maps

Although photographs of Earth were taken from balloons in the 1800s, it was not until after World War I that they were routinely used in making maps. In the 1970s, the United States used satellites called Landsats to gather data. Much of this information is now collected by radar, which can penetrate obstructions in ways that photography cannot.

Connect to History

Strategic Thinking As a mapmaker in the 15th century, how might you have gathered data?

Connect to Today

Strategic Thinking How have photography and flight changed the way maps are made?

Technology: 2004

INTERACTIVE

A Map of Earth in 3-D

On February 11, 2000, the space shuttle *Endeavour* was launched into space on an 11-day mission to complete the most in-depth mapping project in history. The Shuttle Radar Topography Mission (SRTM) collected data on 80 percent of Earth's surface. This information was gathered by beaming radar waves at Earth and converting the echoes into images through a process known as interferometry (IHN•tuhr•fuh•RAHM•ih•tree).

With the aid of computers, the resulting information can be used to produce almost limitless numbers of three-dimensional (3-D) maps. These maps show the topography—rivers, forests, mountains, and valleys—of Earth's surface. It took one year to process the data into 3-D maps. These maps, the most accurate topographical maps ever, will help scientists to better study Earth's surface. The data will also be useful to the general public; for example, it can be used to find new locations for cellular-phone towers and to create maps for hikers.

The data collected on the 11-day SRTM mission can be used by many people—such as the military, the science community, and civic groups—and can be tailored to their needs.

The 200-foot mast is the longest structure used in space today.

Radar interferometry uses radar images taken from two different angles to produce a single 3-D image.

1. Drawing Conclusions How will new, sophisticated tools such as radar interferometry and computers change the study of Earth and the environment?

2. Making Predictions How will these topographical maps help the world?

Thinking Critically

1. Drawing Conclusions Possible Responses Scientists will be able to look at Earth in ways never available before. They will be able to see physical characteristics more clearly, and the data they receive will be more accurate. They will be able to study concepts such as population groupings and changes in climate.

2. Making Predictions Possible Responses The data from these maps will be available to a variety of groups all over the world, including scientists, the military, politicians, civic groups, industrialists, and the general public. Greater knowledge may help deal with environmental issues, military strategies, and overcrowding of cities.

The Geographer's Tools

TERMS & NAMES
cartographer
thematic map
map projection

MAIN IDEA

Geographers use maps, globes, charts, graphs, and new technology to learn about and display the features of Earth.

WHY IT MATTERS NOW

Knowing how to use the tools of geography adds to your ability to understand the world.

DATELINE

PARIS, 1650—The world looks very different today. A new map of North America has just been published by Nicolas Sanson, the famous French mapmaker. Sanson based his map on reports from explorers, missionaries, and other travelers. The map shows five great lakes. Two of these lakes, Lake Ontario and Lake Superior, have been named for the first time.

Certain regions to the north and west remain unknown. Could there be a northwest passage that will allow us to sail to China and the Far East? Only one thing is certain—our knowledge of this mysterious continent will grow with time.

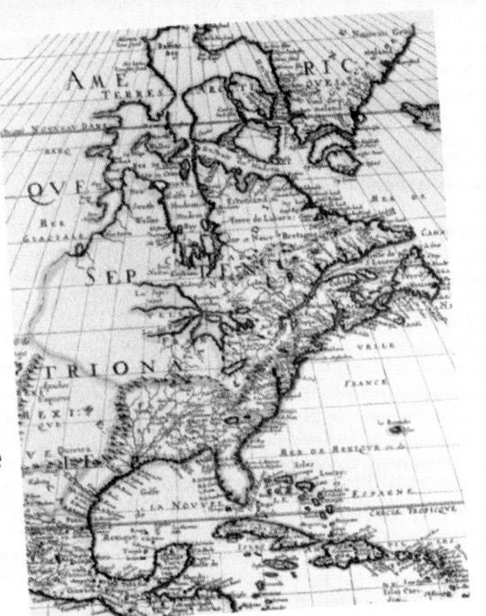

Location • Nicolas Sanson created one of the most important 17th-century maps of North America. ▲

Maps and Globes 1

People have been drawing maps of their world for thousands of years. Geographers today have many tools, such as remote sensing and the Global Positioning System, to help them represent Earth. Increased knowledge and technology allows a **cartographer,** or mapmaker, to construct maps that give a much more detailed and accurate picture of the world. The "Linking Past and Present" and "Technology: 2004" features on pages 42–44 provide more information on modern mapmaking technology.

TAKING NOTES
Use your web to take notes about the five themes.

The Five Themes of Geography

SECTION OBJECTIVES

1. To explain the differences between maps and globes, and to identify types of maps and map projections
2. To describe how geographers use charts and graphs to display and compare information

SKILLBUILDER
- Interpreting a Map, pp. 47, 48
- Reading a Graph, p. 49

CRITICAL THINKING
- Comparing, p. 46

FOCUS & MOTIVATE

WARM-UP

Making Inferences Have students read Dateline and examine the map and the diagram of the map. Then discuss these questions.

1. Why does the map of North America appear incomplete?
2. Which areas seem inaccurately drawn, compared to a modern map of North America?

INSTRUCT: Objective 1 

Maps and Globes

- How do maps and globes differ? globe—accurate, three-dimensional; maps—two-dimensional, not perfectly accurate

 **In-depth Resources: Unit 1**
- Guided Reading Worksheet, p. 15

 **Reading Study Guide**
(Spanish and English), pp. 12–13

Program Resources

 **In-depth Resources: Unit 1**
- Guided Reading Worksheet, p. 15
- Reteaching Activity, p. 21

Reading Study Guide
(Spanish and English), pp. 12–13

 **Formal Assessment**
- Section Quiz, p. 20

 **Integrated Assessment**
- Rubric for making a map

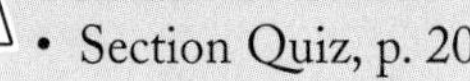

 **Outline Map Activities**

 **Access for Students Acquiring English**
- Guided Reading Worksheet, p. 8

 **Technology Resources**
classzone.com

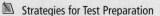

TEST-TAKING RESOURCES
- Strategies for Test Preparation
- Test Practice Transparencies
- Online Test Practice

CRITICAL THINKING ACTIVITY

Comparing Review the differences and similarities between maps and globes and the advantages and disadvantages of each. Have students decide if a map, a globe, or both would be the preferred tool for each of the following tasks: to convey a specific theme, to locate latitude and longitude, to take on a trip, to determine a continent's size in relation to another, and to find a route. Have students give reasons for their answers.

Class Time 15 minutes

Connections to Math

Eratosthenes used geometry to calculate the circumference of Earth. Geometry is also the basis for determining longitude lines. The Prime Meridian is an imaginary line drawn halfway around the center of Earth from pole to pole. Directly opposite the prime meridian is the longitude line designated 180°. The lines of longitude are numbered from 0° to 180° east or west. Combined, these lines equal 360°, which is the circumference of a circle.

FOCUS ON VISUALS

Interpreting the Map Have students study the map of Prince Edward Island. Ask what different kinds of information can be found on the map.

Possible Responses Information on the map includes cities, three types of highways, national parks, and the names of bodies of water.

Extension Have students name three other kinds of information that could be included on the map, such as camping areas or railways.

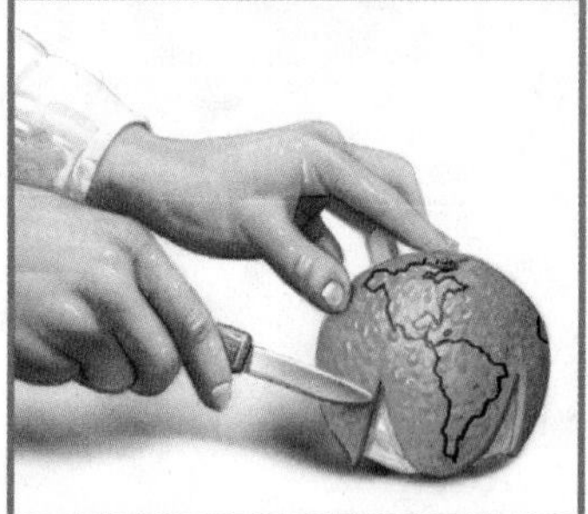

Location • Draw a picture on the entire surface of an orange and then peel the orange in one continuous piece. After you lay the peel flat, your image will be distorted.

Differences Between Maps and Globes Both maps and globes represent Earth and its features. A globe is an accurate model of the world because it has three dimensions and can show its actual shape. Globes are difficult to carry around, however. Maps are more practical. They can be folded, carried, hung on a wall, or printed in a book or magazine. However, because maps show the world in only two dimensions, they are not perfectly accurate. Look at the pictures above to see why. When the orange peel is flattened out, the picture on the orange is distorted, or twisted out of shape. Cartographers have the same problem with maps.

Reading Social Studies

A. Clarifying Why does a globe represent Earth better than a map?

Reading Social Studies A. Possible Response Since Earth is round and a globe is round, a globe is a more accurate representation of Earth.

Connections to Math

Measuring Earth In 230 B.C., the Greek scientist Eratosthenes used basic geometry to measure the circumference of Earth. Eratosthenes knew that at noon on June 21, the sun cast no shadow in the Egyptian city of Syene (now Aswan). (See the diagram below.) At the same time, the sun cast a shadow of 7°12′ in Alexandria, about 500 miles from Syene.

The circumference of a circle is 360°; 7°12′ is about 2 percent, or 1/50, of 360°. Therefore, he concluded, 500 miles must be about 2 percent of the distance around Earth, which at the equator would be about 25,000 miles.

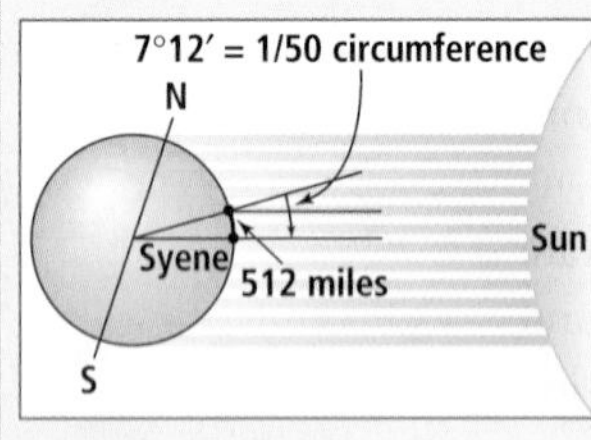

Three Kinds of Maps General reference maps, which show natural and human-made features, are used to locate a place. **Thematic maps** focus on one specific idea or theme. The population map on page 48 is an example of a thematic map. Pilots and sailors use nautical maps to find their way through air and over water. A nautical map is sometimes called a chart.

Location • A road map is a reference map that shows how to get from one place to another. ▼

Road Map of Prince Edward Island, 2002

Activity Options

Differentiating Instruction: Students Acquiring English/ESL

Reading in Pairs Students may have difficulty with the details of this section. Pair fluent readers with students acquiring English, and have them read this section together, listing any difficult passages or concepts. When students have completed the section, have the pairs meet in groups to discuss and clarify the problem passages with one another. You may wish to help students clarify any remaining questions.

To help students understand the three kinds of maps, provide samples of each kind. Guide students as they study the maps. Point out what makes the map a general reference map, a reference map, or a nautical chart. Then have students make a chart that lists the three kinds of maps and describes the function and properties of each kind.

Mercator Projection

NORTH AMERICA
EUROPE
ASIA
AFRICA
SOUTH AMERICA
AUSTRALIA
ANTARCTICA

Robinson Projection

NORTH AMERICA
EUROPE
ASIA
AFRICA
SOUTH AMERICA
AUSTRALIA
ANTARCTICA

GEOGRAPHY SKILLBUILDER: Interpreting a Map

1. Location • Compare the size of Africa in relation to other continents on the two projections. How do they differ?
2. Location • What other differences do you notice between the Mercator projection and the Robinson projection?

Reading Social Studies

B. Identifying Problems What are the main problems faced by cartographers?

Reading Social Studies
B. Possible Answer
All map projections distort Earth.

Map Projections The different ways of showing Earth's curved surface on a flat map are called **map projections.** All projections distort Earth, but different projections distort it in different ways. Some make places look bigger or smaller than they really are in relation to other places. Other projections distort shapes. For more than 400 years, the Mercator projection was most often shown on maps of the world. Recently, the Robinson projection has come into common use because it gives a fairer and more accurate picture of the world.

Geography Skillbuilder Possible Answers

1. On Robinson, Africa appears larger in relation to Europe and Asia.
2. On Robinson, South America is larger in relation to North America; Europe is smaller in relation to Asia.

Colonial Map This map of Boston was drawn by Jacques Nicolas Bellin and published in Paris around 1764. It shows the harbor and the bays surrounding Boston in the mid-18th century.

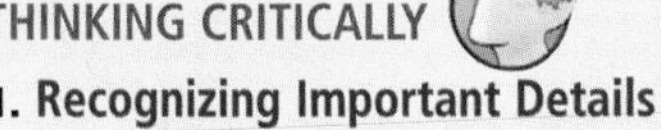

1. **Recognizing Important Details** How was it possible for travelers to enter Boston by land?
2. **Making Inferences** Why was Boston's site a good location for a colonial city?

For more on colonial maps, go to

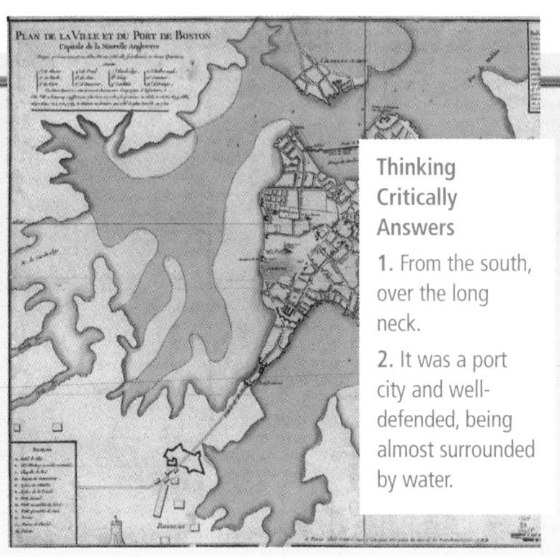

Thinking Critically Answers

1. From the south, over the long neck.
2. It was a port city and well-defended, being almost surrounded by water.

FOCUS ON VISUALS

Interpreting the Maps Encourage students to compare the size of all the continents shown on the maps on this page. Then ask what effect these two different maps might have on the way people think about the world and the importance of various parts of the world. For example, how might people think differently about the continents of Africa or South America using one or the other of these projections?

Possible Response Students may suggest that when a continent appears larger on a map projection, people might consider it to be more important in world affairs.

Extension Ask students to use an atlas to determine the number of square miles in two of the continents. Then have them compare the square mileage of the two continents with the Mercator projection of the same continents. Is the difference in size accurately depicted? Is it accurately depicted on the Robinson projection?

Spotlight on CULTURE

The peninsula that protected early Boston soon became an obstacle to the city's growth. Demand for land increased as the population grew, and many of the bays and salt marshes that surrounded the city were gradually filled in.

Activity Options

Differentiating Instruction: Gifted and Talented

Designing an Illustrated Map Discuss the illustrated map shown on this page. Have students identify some of the illustrations and how they relate to what was known and what was happening in world exploration at the time. You might provide other examples of illustrated maps, especially from before colonial times. Explain that illustrators often made up imaginary plants and animals for their maps. Then have students trace a map of North and South America and the surrounding oceans. Have them label the map and illustrate it with pictures depicting conditions today. Ask them to consider what kinds of ships sail the oceans today and what the inhabitants of these continents look like. Have students display their work in class.

FOCUS ON VISUALS

Interpreting the Maps Guide students to name the continents with the densest and least dense populations. Guide them to name the continents where people can expect to live longest and where their lives are much shorter.

Possible Responses The continents with the densest populations are Asia and Europe, and Australia has the least dense. People live longer in North America, Europe, and Australia. People in Africa have the shortest lives.

Extension Have students find a population map for the United States. Have them compare the way the information is shown with the population map on this page. Ask them to identify the states with the densest and the least dense populations.

CRITICAL THINKING ACTIVITY

Drawing Conclusions Have students compare the two maps on this page. Have them look for areas with dense population and high life expectancy. Then have them look for areas with dense population and low life expectancy. What conclusions can they draw from making these comparisons, if any?

Class Time 15 minutes

MORE ABOUT...

Life Expectancy

Many factors determine the life expectancy of a population. These include education, health care, clean running water, low pollution, and employment.

World Population and Life Expectancy, 2000

Population

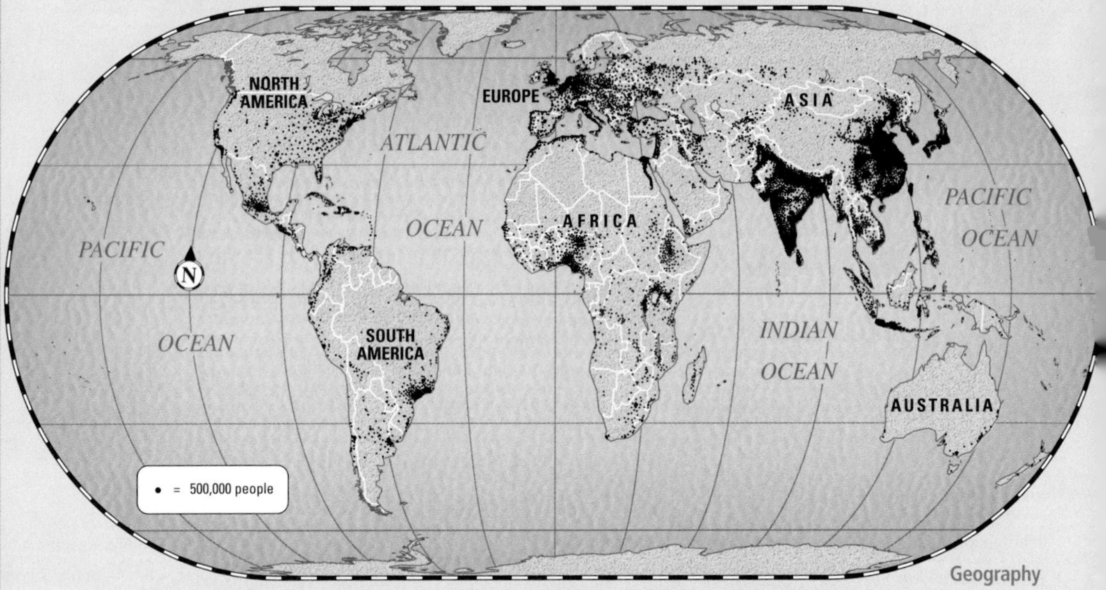

Life Expectancy

Geography Skillbuilder Possible Answers

1. Asia; Australia
2. 75 years or more

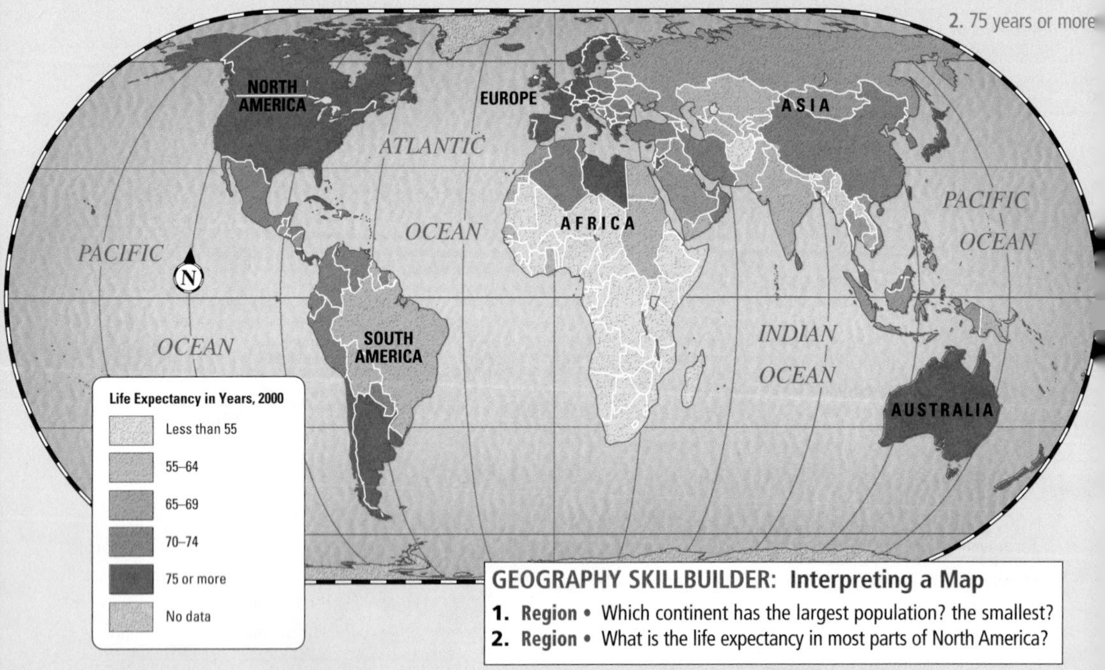

GEOGRAPHY SKILLBUILDER: Interpreting a Map

1. **Region** • Which continent has the largest population? the smallest?
2. **Region** • What is the life expectancy in most parts of North America?

48 CHAPTER 2

Activity Options

Interdisciplinary Links: Visual/Geography

Explaining the Skill Tell students that the population map uses dots to show the number of people and where they live. The life expectancy map uses color to show how long people in different regions of the world can expect to live.

Applying the Skill Have students use the following strategies to read the maps.

1. Read the titles of the maps to find out what they show.
2. Read the map key. The population map uses a dot to represent 500,000 people. Read the key of the life expectancy map to see what life expectancy each color represents.
3. Study each map. Where is the population densest? Where in the world can people expect to live longest?

Comparing Maps, Charts, and Graphs ❷

Along with maps, geographers use charts and graphs to display and compare information. The graphs on this page and the maps on page 48 contain related information about the world's population. Notice how each quickly and clearly presents facts that would otherwise take up many paragraphs of text.

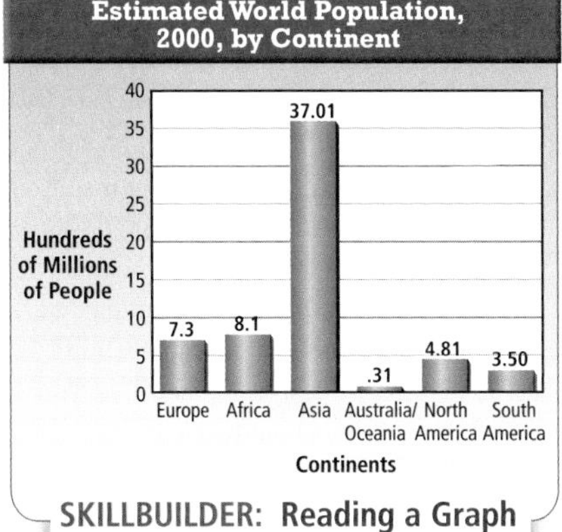

SKILLBUILDER: Reading a Graph

1. How many people live in Europe?
2. Which continent has the smallest population?

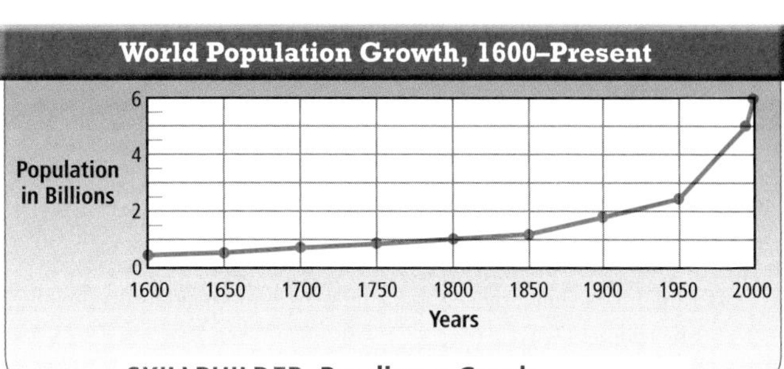

SKILLBUILDER: Reading a Graph

1. How many people lived in the world in 1900?
2. How much did the world's population increase between 1600 and 1900? between 1900 and 2000?

Skillbuilder Answers

1. 729 million
2. Oceania

1. 1.9 billion
2. 1.5 billion; 3.3 billion

SECTION 2 ASSESSMENT

Terms & Names

1. **Explain the significance of:** (a) cartographer (b) thematic map (c) map projection

Using Graphics

2. Use a chart like the one below to compare the advantages and disadvantages of maps and globes.

	Maps	Globes
Advantages		
Disadvantages		

Main Ideas

3. (a) What are the differences among the three main kinds of maps?
   (b) How have new tools and knowledge helped cartographers?
   (c) What kinds of information can be displayed in maps and graphs?

Critical Thinking

4. **Using Maps**
   What kind of map would show how many students are in each school in your district?

Think About

- the three kinds of maps
- what information different kinds of population maps show

ACTIVITY -OPTION- Draw a **map** of the route you take to and from school or some other familiar destination. Include the names of streets, landmarks such as shops and other buildings, and any other useful information.

INSTRUCT: Objective ❷

Comparing Maps, Charts, and Graphs

- What is an advantage of using maps, charts, and graphs instead of text? They quickly and clearly present information.

FOCUS ON VISUALS

Interpreting the Graphs Guide students as they read the graphs. What information is along the x-axis and the y-axis? Be sure students understand that in the right-hand graph, the numbers are in billions, and in the left-hand graph, they are in hundreds of millions.

Possible Response Right graph: x-axis is years, y-axis is population; left graph: x-axis is continents, y-axis is population.

Extension Ask students to think of other information about population that could be displayed in charts.

ASSESS & RETEACH

Reading Social Studies Have students create a web diagram with a center circle labeled "Geographer's Tools." Then have them add details for each category.

Formal Assessment

- Section Quiz, p. 20

RETEACHING ACTIVITY

Have students list ways in which each of the geographer's tools might be used.

In-depth Resources: Unit 1

- Reteaching Activity, p. 21

Access for Students Acquiring English

- Reteaching Activity, p. 12

Section 2 Assessment

1. Terms & Names

a. cartographer, p. 45
b. thematic map, p. 46
c. map projection, p. 47

2. Using Graphics

	Maps	Globes
Advantages	portable; can show thematic material	show physical features accurately
Disadvantages	distort physical features	cannot be easily carried

3. Main Ideas

a. general reference maps: natural and human-made features; thematic maps: one specific idea; nautical maps: used for navigation
b. They allow cartographers to make more detailed and accurate maps.
c. facts, data, and statistics

4. Critical Thinking

Students may say a thematic map that shows population density.

ACTIVITY OPTION

Integrated Assessment

- Rubric for making a map

ASSESSMENT

TERMS & NAMES

1. absolute location, p. 36
2. latitude, p. 36
3. longitude, p. 36
4. relative location, p. 37
5. cartographer, p. 45
6. thematic map, p. 46
7. map projection, p. 47
8. migrate, p. 38
9. continent, p. 35

REVIEW QUESTIONS

Possible Responses

1. Latitude and longitude are used to determine absolute location.
2. Relative location tells where a place is in relation to another place, whereas absolute location uses latitude and longitude.
3. Mountain ranges, canyons, and rivers made migration difficult in the past.
4. Because of technology, humans have not had to adapt as closely to their environments.
5. Because a globe is round like Earth, there is less distortion.
6. A pilot would use a nautical map to find his or her way through air and over water.
7. Modern technology has helped cartographers to be more detailed and accurate.
8. The Robinson projection more accurately shows the size of continents in relation to one another.

ASSESSMENT

TERMS & NAMES

Explain the significance of each of the following:

1. absolute location
2. latitude
3. longitude
4. relative location
5. cartographer
6. thematic map
7. map projection
8. migrate
9. continent

REVIEW QUESTIONS

The Five Themes of Geography *(pages 35–40)*

1. What system do geographers use to determine absolute location?
2. How is relative location different from absolute location?
3. What are some of the natural barriers that made migration difficult in the past?
4. How has technology changed the way humans adapt to their environment?

The Geographer's Tools *(pages 45–49)*

5. Why is a globe an accurate representation of the world?
6. Why would a pilot use a nautical map?
7. How has modern technology helped cartography?
8. Why do most modern cartographers prefer the Robinson projection to the Mercator projection?

CRITICAL THINKING

Drawing Conclusions

1. Using your completed spider map from Reading Social Studies, p. 34, draw a conclusion about which theme of geography is most useful in familiarizing you with an area of the world. Which details in your chart help you understand a country or region?

Contrasting

2. Maps and globes both represent Earth and its features. Contrast the advantages of a map with the advantages of a globe.

Clarifying

3. Why would the leaders of a country find a population density map of their country useful?

Visual Summary

The Five Themes of Geography 1

- The five themes of geography are location, place, region, movement, and human-environment interaction.
- These themes are the keys to understanding the geography of the world.

The Geographer's Tools 2

- Maps, globes, charts, graphs, and other tools are available to geographers to help them understand the features of Earth.
- Geographers use these tools to organize and explain Earth's features.

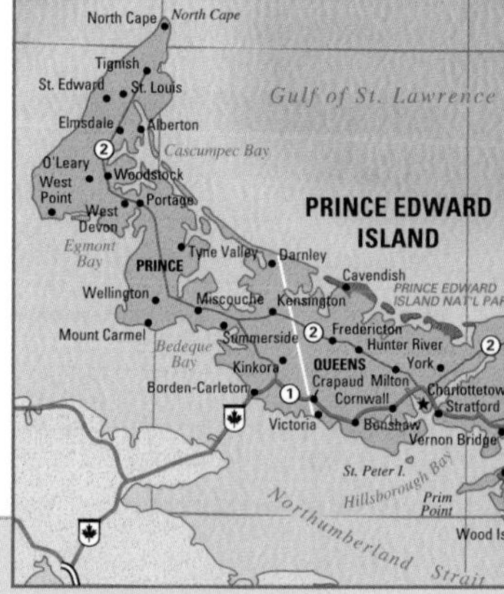

CRITICAL THINKING: Possible Responses

1. Drawing Conclusions

Answers may vary. Students might say that place is the most useful theme because it gives information about the climate, landforms, and bodies of water and tells about human characteristics such as cultural traditions. Accept any answer that students can justify.

2. Contrasting

Students may note that the advantages of a map are that it can be readily carried around and referred to, is easy to read, and can depict a small and specific area. The advantages of a globe are that it is a more accurate picture of the whole world, there is no distortion, and you can see parts of the world in relation to other parts.

3. Clarifying

Students may note that leaders could use population density maps to apportion natural resources where they are needed and to anticipate needs in the future.

STANDARDS-BASED ASSESSMENT

se the map and your knowledge of world cultures and ography to answer questions 1 and 2.

dditional Test Practice, pp. S1–S33

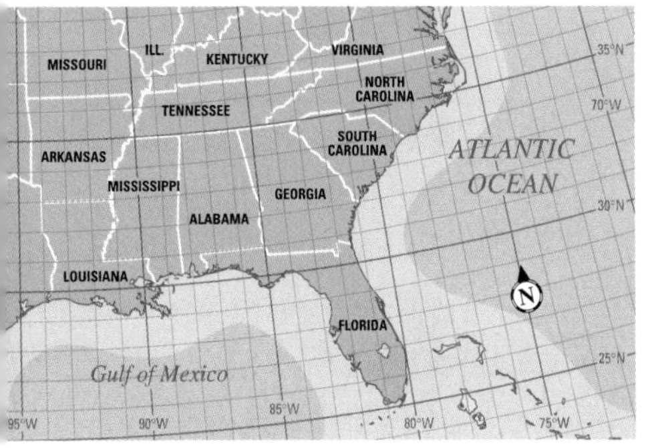

Which states have their northern border near latitude 35°N?

A. Arkansas, Tennessee, North Carolina
B. Louisiana, Mississippi, Alabama
C. Mississippi, Alabama, Georgia
D. Missouri, Kentucky, Virginia

Which state extends to about latitude 25°N and longitude 81°W?

A. Florida
B. Mississippi
C. North Carolina
D. South Carolina

In this quotation from a press release, a UNESCO spokesperson describes the archaeological site Mapungubwe. It was the center of the largest kingdom to exist in Africa south of the Sahara before the 14th century. Use the quotation and your knowledge of world cultures and geography to answer question 3.

PRIMARY SOURCE

What survives are the almost untouched remains of the palace sites and also the entire settlement area dependent upon them, as well as two earlier capital sites, the whole presenting an unrivalled picture of the development of social and political structures over some 400 years.

JASMINI SOPOVA, UNESCO spokesperson, *allAfrica.com*

3. Why is it so important that the archaeological remains are untouched?

A. They will be clean, not damaged by dirt and mud.
B. They will accurately show how people once lived.
C. They will make the person who found them famous.
D. They can be rearranged however the archaeologists want.

STANDARDS-BASED ASSESSMENT

1. **Answer C** is the correct answer because the northern borders of Mississippi, Alabama, and Georgia are close to latitude 35°N. Answer A is incorrect because latitude 35°N runs through Arkansas and North Carolina. Answer B is incorrect because all of Louisiana lies south of 35°N. Answer D is incorrect because Missouri, Kentucky, Virginia are all north of 35°N.
2. **Answer A** is the correct answer because the southern tip of Florida is close to latitude 25°N and longitude 81°W. Answers B, C, and D are all incorrect because none of the states named extend further south than latitude 30°N.
3. **Answer B** is correct because it echoes the idea that the site presents an unrivalled picture of the past. Answer A is incorrect because "untouched" means that humans have not tampered with it, not that it is clean; Answer C is incorrect because the passage does not speak of fame; Answer D is incorrect because it misrepresents what archaeologists do.

LTERNATIVE ASSESSMENT

WRITING ABOUT HISTORY

The world can be divided into ten natural regions. Find out what these ten regions are. Then choose one region and write a poem about it. In the poem, provide information about where the region is located, what plant life it has, what animal life it has, and what the climate is like in that particular region. Share your poem with the class.

COOPERATIVE LEARNING

With a small group of classmates, choose any country in the world and use the five themes of geography to describe it. Divide the tasks of answering these questions: Where is the country located? (location) What are its physical and human characteristics? (place) How can you classify the region? (region) What movement has occurred in the country? (movement) How have the people who live there adapted to the environment? (human-environment interaction) Arrange your findings on a poster board.

INTEGRATED TECHNOLOGY

Doing Internet Research

Use the Internet to research how technology can be used to study natural events. Write a report of your findings, including a list of the Web sites you used.

- Focus on one natural event, such as tropical rainfall or a hurricane. Look into how technology is helping us understand it.
- Include a prediction about the future uses of technology that involve these kinds of events.

For Internet links to support this activity, go to

INTEGRATED TECHNOLOGY

Brainstorm with students possible key words they might explore, such as "global warming" and "technology." Point out that reports should have a main idea statement identifying the natural event and give details about the specific technology and how it is used. Suggest that students conclude with an explanation of how the technology might be used in the future.

Alternative Assessment

1. Rubric

The poem should

- show an understanding of the natural region selected.
- utilize appropriate elements of poetry, such as images and symbolism.
- use appropriate lyrics possibly containing rhythm and rhyme and connecting to musical phrasing.

2. Cooperative Learning

Remind students that their book or poster should describe the location, climate, and landforms of their country. Encourage them to describe the people who live there, the type of region, the patterns of human migration, and the way the inhabitants adapt to their environment. Remind them that they can classify the region according to physical features, climate, or human characteristics.

The United States and Canada

Chapter 3
The Physical Geography of the United States and Canada

Chapter 4
The United States Today

Chapter 5
Canada Today

Before You Read

Previewing Unit 2

Unit 2 introduces the physical geography of the United States and Canada and identifies major landforms, mountain ranges, river systems, climates, and vegetation zones. Then the focus shifts to provide students with an understanding of the diverse societies of the United States and Canada. The unit examines and compares the history, government, economy, and culture of these two countries.

UNIT 2

Place Mount Rushmore is a monument located in the Black Hills of South Dakota. The 60-foot likenesses of (from left to right) George Washington, Thomas Jefferson, Theodore Roosevelt, and Abraham Lincoln took 14 years to carve into the granite cliff.

52

Unit Level Activities

1. Write a Letter

Ask students to pick a city in the United States they are familiar with, and a Canadian city they might like to visit. Tell them to imagine the Canadian city and writing a letter home to family or friends. In the letter, they should compare the two cities, discussing climate, transportation, industry, and cultural origins of the population.

2. Teach a Lesson

Explain to students that one way to really know about a subject is to teach someone else about it. Have students choose one region in North America and plan a lesson that discusses the region's geographic features and economic activities. Students should organize their ideas in a lesson plan with these headings: geographic features, economic activities, visuals to use (graphic organizer, video, Internet reference, and so on), questions to ask, and assignments to give.

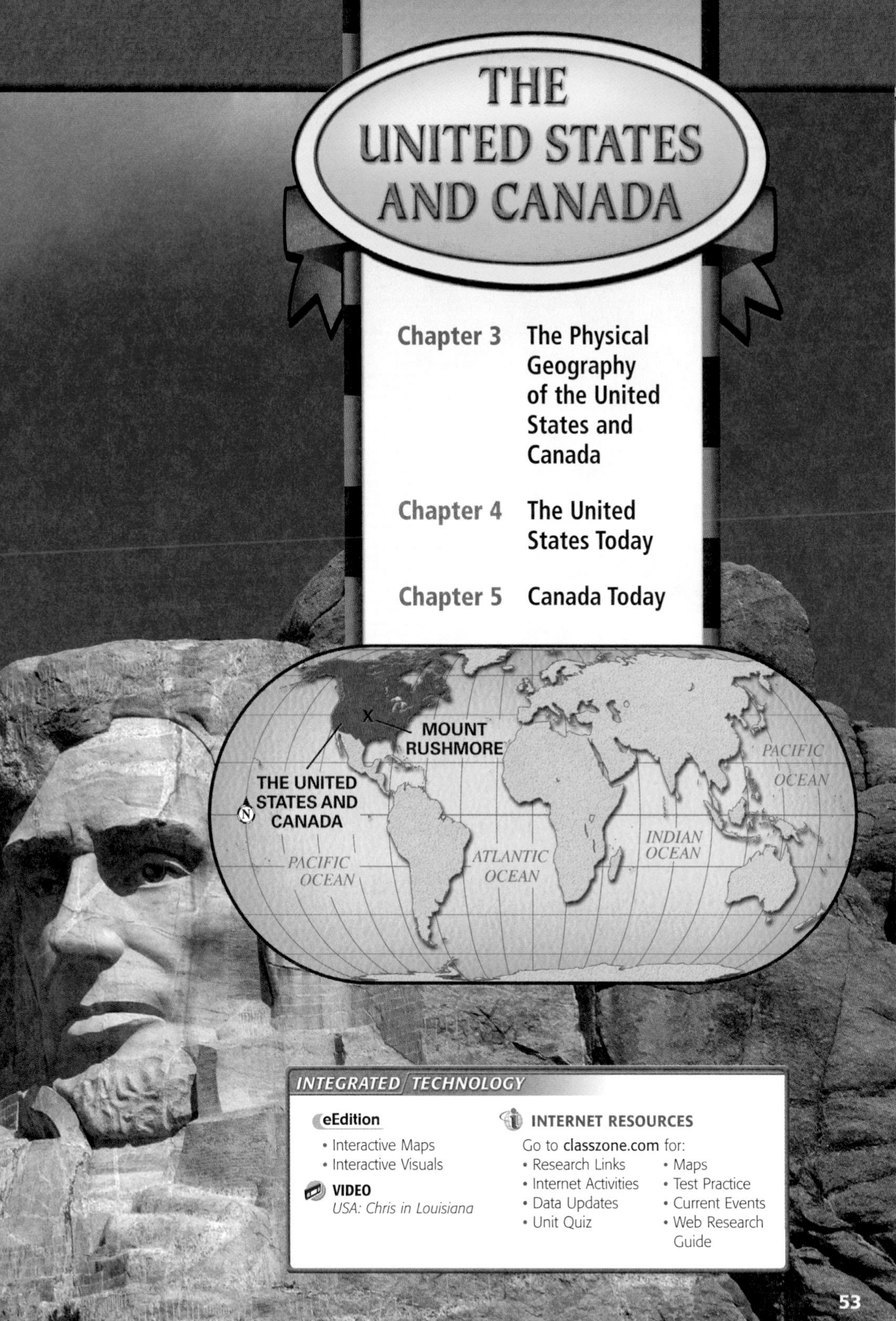

INTEGRATED TECHNOLOGY

eEdition
- Interactive Maps
- Interactive Visuals

VIDEO
USA: Chris in Louisiana

INTERNET RESOURCES
Go to **classzone.com** for:
- Research Links
- Internet Activities
- Data Updates
- Unit Quiz
- Maps
- Test Practice
- Current Events
- Web Research Guide

FOCUS ON VISUALS

Interpreting the Photograph Have students examine the photograph of Mount Rushmore and ask a volunteer to read the caption aloud. Provide students with information about each of the presidents whose likenesses are carved there. Help them to match faces and names. Encourage students to imagine the impact of viewing Mount Rushmore from the air for the first time. Do they know of any place that might make a similar impression? Have them suggest places with a similar impact. Are these places man-made or created by nature?

Possible Responses Niagara Falls or the Rocky Mountains are examples of impressive sights made by nature; skyscrapers and large suspension bridges like the Golden Gate Bridge are examples of impressive sights made by man.

Extension Have students write a paragraph describing Mount Rushmore as it might be seen from an airplane for the first time.

Implementing the National Geography Standards

Standard 11 Describe the effects of the gradual disappearance of small-scale retail facilities

Objective To interview a senior citizen about the changes American stores have undergone in the past 50 years

Class Time 15 minutes

Task Students tape-record interviews with senior citizens about the effects of the gradual disappearance of small-scale retail facilities, such as corner general stores, in American towns. Students should ask senior citizens questions about where they shopped 50 years ago versus where they shop today. Have students present a transcript of their interviews.

Evaluation Students should ask the senior citizens at least five questions about the disappearance of small-scale retail facilities.

ATLAS The United States and Canada

ATLAS OBJECTIVES

1. Describe and locate physical features of the United States and Canada
2. Compare the landmass and population of the United States and Canada
3. Identify political features of the United States and Canada
4. Analyze the population of the United States and Canada today and in 1600
5. Compare states, provinces, and territories

FOCUS & MOTIVATE

Have students describe the geography of the United States and Canada. Ask them what information a physical map provides. After students have identified major landforms, tell them a physical map identifies features of the land.

INSTRUCT: Objective 1

Physical Map of the United States and Canada

- What three major rivers run through the central United States? Mississippi, Missouri, Ohio rivers
- What mountain ranges are found in Alaska? Brooks and Alaska ranges
- What large body of water cuts deep into Canada? Hudson Bay
- What is the elevation of the Great Plains? between 1,600 and 6,600 feet

In-depth Resources: Unit 2
- Unit Atlas Activity, p. 1

UNIT Atlas 2 **Physical Geography**

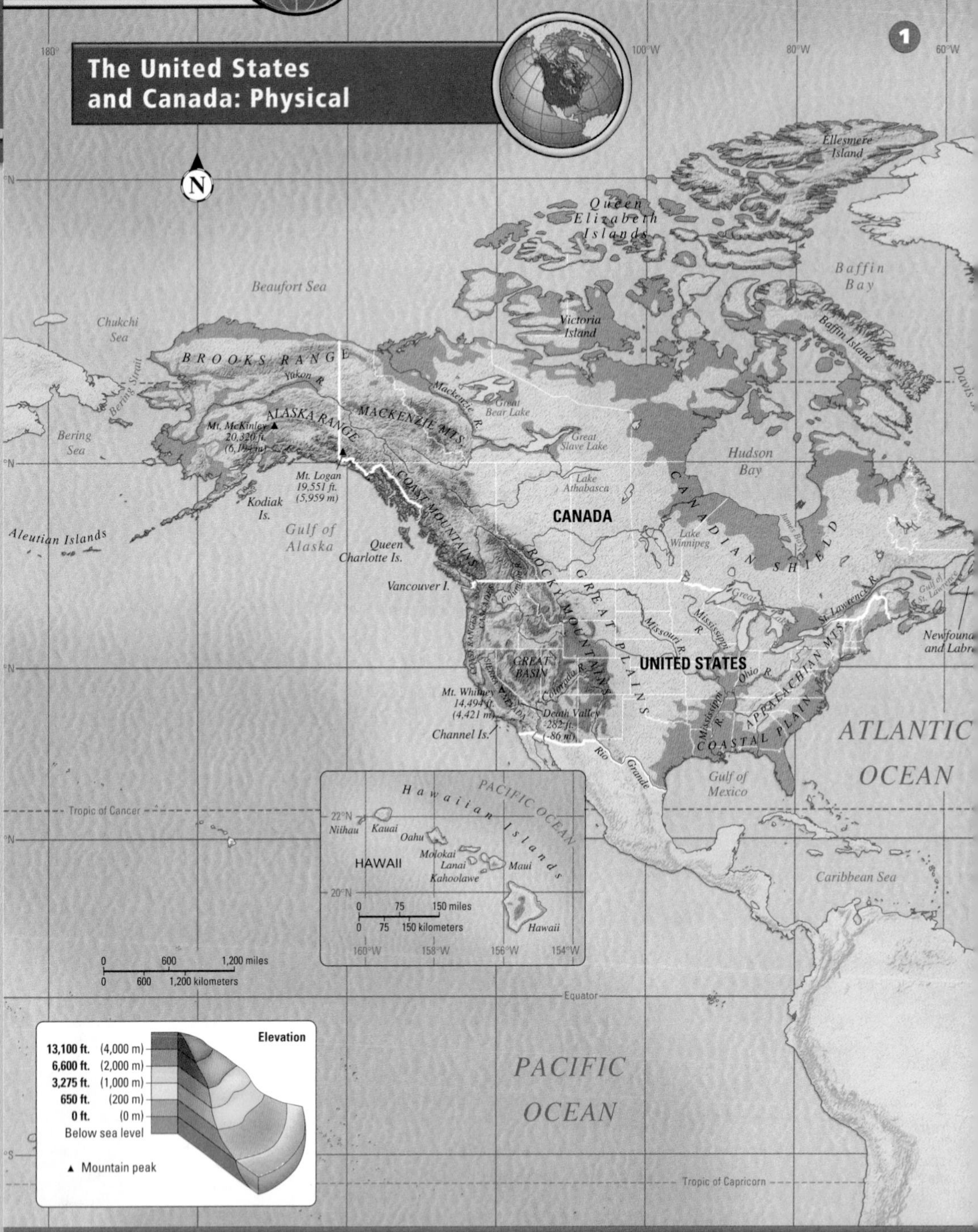

The United States and Canada: Physical

Elevation
13,100 ft. (4,000 m)
6,600 ft. (2,000 m)
3,275 ft. (1,000 m)
650 ft. (200 m)
0 ft. (0 m)
Below sea level
▲ Mountain peak

0 600 1,200 miles
0 600 1,200 kilometers

54 UNIT 2

Activity Options

Interdisciplinary Link: Writing

Explaining the Skill Students will demonstrate an understanding of how to read a physical map.

Applying the Skill Tell students to find the approximate location of their community on the map of the United States and Canada. Then tell them to plan a trip that will take them 1,000 miles in any direction. Have students write a description of the trip they take. It should include the direction they travel and the landforms they pass through. Tell them to identify the highest and lowest elevations they reach.

Block Scheduling

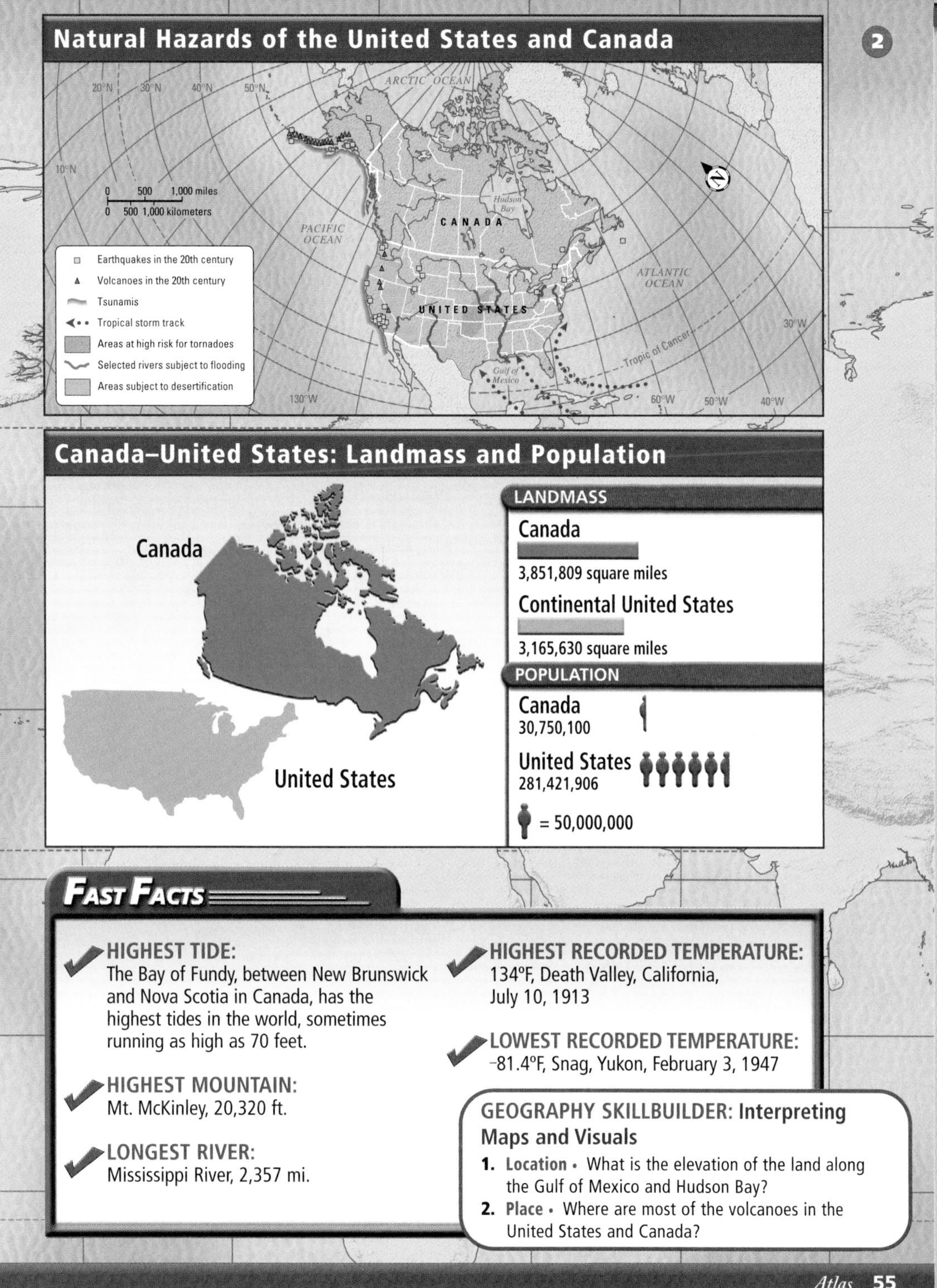

FAST FACTS

HIGHEST TIDE:
The Bay of Fundy, between New Brunswick and Nova Scotia in Canada, has the highest tides in the world, sometimes running as high as 70 feet.

HIGHEST MOUNTAIN:
Mt. McKinley, 20,320 ft.

LONGEST RIVER:
Mississippi River, 2,357 mi.

HIGHEST RECORDED TEMPERATURE:
134°F, Death Valley, California, July 10, 1913

LOWEST RECORDED TEMPERATURE:
−81.4°F, Snag, Yukon, February 3, 1947

GEOGRAPHY SKILLBUILDER: Interpreting Maps and Visuals

1. **Location** • What is the elevation of the land along the Gulf of Mexico and Hudson Bay?
2. **Place** • Where are most of the volcanoes in the United States and Canada?

INSTRUCT: Objective 2

Natural Hazards of the U.S. and Canada

- Where do most earthquakes occur? in southern California

Landmass of the U.S. and Canada

- About how much larger is Canada than the United States? 686,179 square miles

Population of the U.S. and Canada

- What is the population of Canada? 30,750,100 people

Fast Facts

- What is the lowest temperature ever recorded in the region? −81.4°F

MORE ABOUT...

Tsunamis

Tsunamis, destructive waves, are produced when an undersea volcanic eruption, earthquake, or landslide creates gigantic waves that can travel at 600 miles per hour and cross vast distances. In 1946, a tsunami traveled 2,500 miles from Alaska's Aleutian Islands and crashed into the island of Hilo, killing 159 people. In 1958, an Alaska earthquake started a landslide that created a tsunami. The wave raced across a bay and destroyed trees 1,700 feet up a mountain.

Fast Facts

Review the Fast Facts with students. Then instruct them to develop their own Fast Facts file. Almanacs, atlases, encyclopedias, and the Internet are good places to begin.

GEOGRAPHY SKILLBUILDER

Answers

1. between 0 ft. and 650 ft.
2. along the Pacific coast

Country Profiles

Quebec Province In many ways, Quebec is the most unusual of the Canadian provinces. It is the largest province and has a greater population than any but Ontario. What makes Quebec most unusual, however, is the French ancestry of the majority of its people and its pride in its French traditions. About 80 percent of Quebec's population are descendants of French settlers that arrived in Canada during the 1600s and 1700s. Although the colony was taken over by Great Britain in 1763, the French settlers remained separate and maintained their French traditions. Even today, 60 percent of the people speak only French and another 30 percent speak both French and English. There are many French-style homes and most schools are taught in French.

INSTRUCT: Objective 3

Political Map of the United States and Canada

- How is this map similar to and different from the map of Selected Native Peoples on page 57? Both are political maps, showing the different nations of North America at different times. The Native Peoples map also shows food sources; the map on this page shows national capitals and major cities.
- What is the capital of Canada? Ottawa
- What U.S. state borders Canada's Yukon Territory? Alaska
- What three oceans border the United States and Canada? Arctic, Pacific, Atlantic

MORE ABOUT...

The St. Lawrence Seaway

The St. Lawrence Seaway was built between 1954 and 1959 in a joint effort by the United States and Canada. Its purpose was to link the Great Lakes with the Atlantic Ocean so that ocean-going vessels could reach far inland to pick up and deliver goods. The seaway runs 186 miles between Montreal and Lake Ontario. Because the Great Lakes are over 200 feet higher than the Atlantic, a series of locks and dams were built to raise and lower ships as they pass through the seaway. It takes a ship about one and one-half days to make the entire trip.

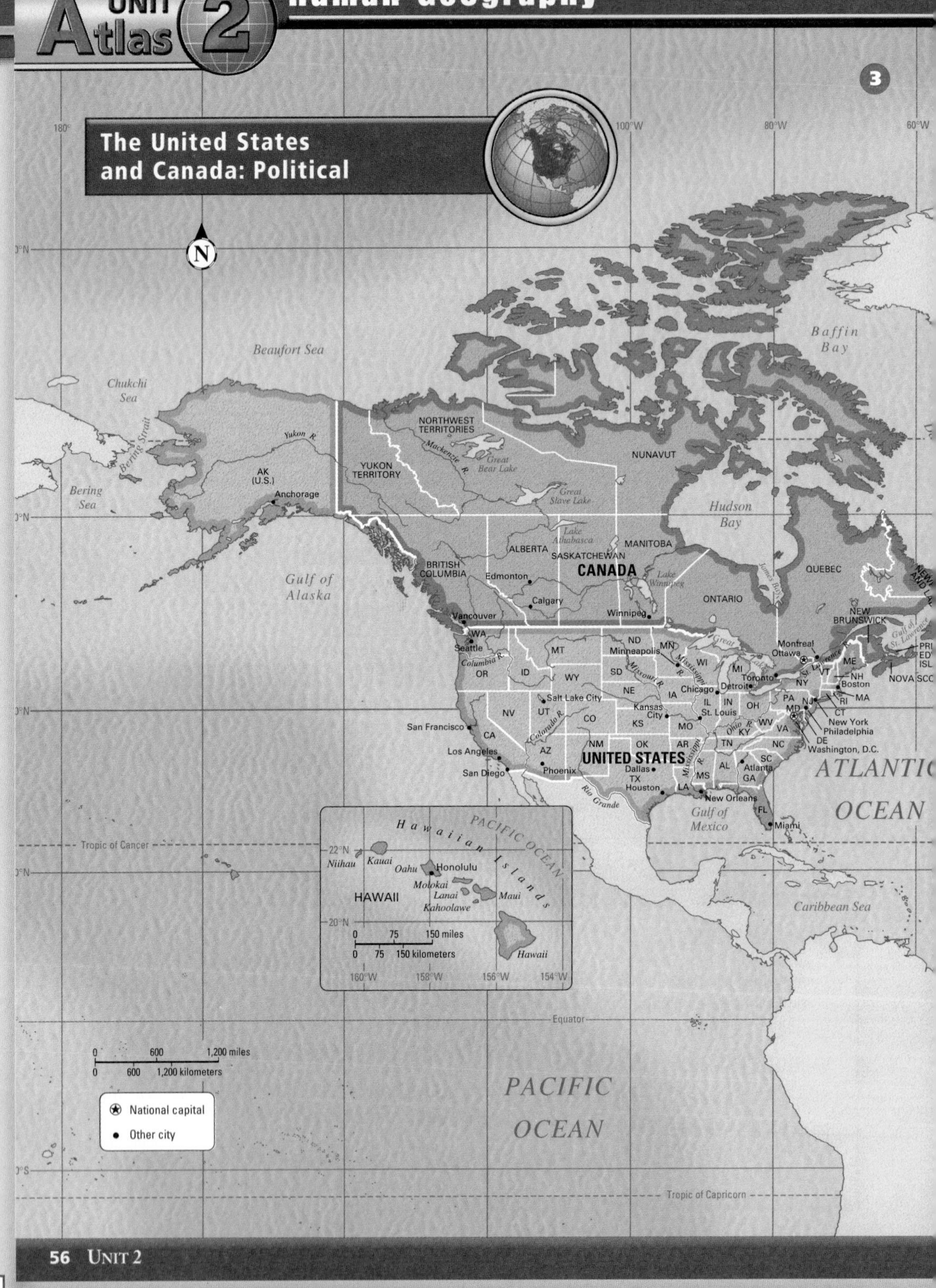

Activity Options

Interdisciplinary Link: Current Events

Class Time Two class periods

Task Comparing the United States and Canada

Purpose To learn about the relationships between Canada and the United States

Supplies Needed

- Encyclopedias, Internet resources
- Pens or pencils
- Paper

Activity Have students work in pairs to investigate relationships between the United States and Canada. Encourage them to choose a specific point of comparison, such as international relations, trade, entertainment, or tourism. Have pairs divide up the research and then work together to compile their findings in a brief report. Ask them to share what they learn in class discussion.

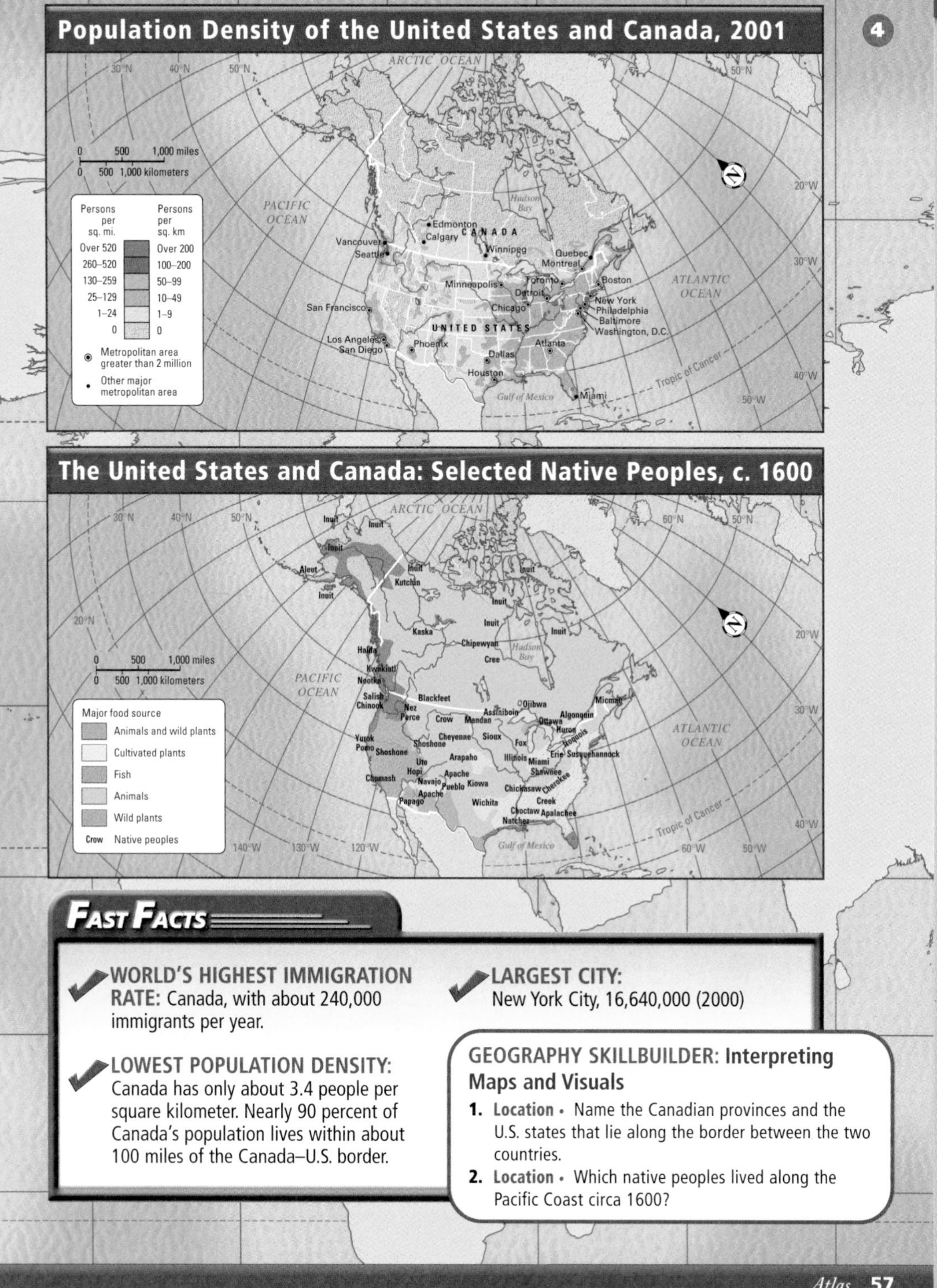

Fast Facts

- **WORLD'S HIGHEST IMMIGRATION RATE:** Canada, with about 240,000 immigrants per year.
- **LOWEST POPULATION DENSITY:** Canada has only about 3.4 people per square kilometer. Nearly 90 percent of Canada's population lives within about 100 miles of the Canada–U.S. border.
- **LARGEST CITY:** New York City, 16,640,000 (2000)

GEOGRAPHY SKILLBUILDER: Interpreting Maps and Visuals

1. **Location** • Name the Canadian provinces and the U.S. states that lie along the border between the two countries.
2. **Location** • Which native peoples lived along the Pacific Coast circa 1600?

INSTRUCT: Objective 4

Population Density of the U.S. and Canada

- What part of the United States and Canada is most heavily populated? The least populated? the middle portion of the United States' east coast; the northern portion of Canada

Native Peoples of North America in 1600

- What are some Native American groups that lived in the southeastern part of what is today the United States? Chickasaw, Cherokee, Creek, Choctaw, Natchez, Apalachee

Fast Facts

- What country has the highest immigration rate in the world? Canada

MORE ABOUT...

Native Americans Today

Approximately 2.5 million Native Americans live in the United States today. About one-third of them live on reservations or traditional lands. Most of the rest live in urban areas. Canada's population includes 550,000 Native Americans. Many live in Nunavut.

Fast Facts

Urge students to add to their Fast Facts file by looking for facts and statistics in almanacs and encyclopedias, and on the Internet.

GEOGRAPHY SKILLBUILDER

Answers

1. Canada: British Columbia, Alberta, Saskatchewan, Manitoba, Ontario, Quebec, New Brunswick; United States: Washington, Idaho, Montana, North Dakota, Minnesota, Michigan, New York, Vermont, New Hampshire, Maine
2. Aleut, Inuit, Haida, Kwakiutl, Nootka, Salish, Chinook, Yurok, Pano, Chumask

Country Profiles

Hawaii Hawaii is the newest state in the United States. It was admitted to the union in 1959. It is also the only state that is not part of the North American mainland. It lies about 2,400 miles west of the California coast. Hawaii is made up of 132 islands that extend for 1,523 miles, a distance equal to the distance from New York City to Denver. All of the islands were formed by volcanoes on the ocean floor that slowly built up until they rose above the ocean. Most of Hawaii's population live on seven islands, and of these, 72 percent live on the island of Oahu, which is the location of Honolulu, Hawaii's capital and largest city. The people of Hawaii are of many ethnic groups. About 15 percent are of Polynesian descent. These were the first inhabitants of Hawaii. Other groups include those of European ancestry (35 percent) and Japanese ancestry (25 percent). Other people are of Chinese, Filipino, Korean, and Samoan ancestry.

DATA FILE OBJECTIVE

Examine and compare data on the states of the United States and the provinces and territories of Canada

FOCUS & MOTIVATE

Ask students to read the headings on the Data File. Then ask them to find the state that covers the largest area. How large is it? Next, have them turn to pages 64 and 65. Ask them which province or territory has the greatest number of high school graduates. Alaska; 615,230 square miles; British Columbia

INSTRUCT: Objective 5

Data File

- Which of the states has the highest population density? New Jersey (Explain that Washington, D.C., is not a state but a special political unit because it is the capital of the United States. Likewise, Puerto Rico is a territory of the United States.)
- Which state has the lowest infant mortality rate? New Hampshire
- Which Canadian province or territory has the highest per capita income? Yukon Territory

In-depth Resources: Unit 2
- Data File Activity, p. 2

UNIT 2 Atlas Data File

For updates on these statistics, go to

DATA UPDATE CLASSZONE.COM

5

State Flag	State/ Capital	Population (2000)	Population Rank (2000)	Infant Mortality (per 1,000 live births) (1998)	Doctors (per 100,000 pop.) (1998–1999)
	Alabama (AL) Montgomery	4,447,100	23	10.2	198
	Alaska (AK) Juneau	626,900	48	5.9	167
	Arizona (AZ) Phoenix	5,130,600	20	7.5	202
	Arkansas (AR) Little Rock	2,673,400	33	8.9	190
	California (CA) Sacramento	33,871,600	1	5.8	247
	Colorado (CO) Denver	4,301,300	24	6.7	238
	Connecticut (CT) Hartford	3,405,600	29	7.0	354
	Delaware (DE) Dover	783,600	45	9.6	234
	*District of Columbia (DC)	572,100	—	12.5	737
	Florida (FL) Tallahassee	15,982,400	4	7.2	238
	Georgia (GA) Atlanta	8,186,500	10	8.5	211
	Hawaii (HI) Honolulu	1,211,500	42	6.9	265
	Idaho (ID) Boise	1,294,000	39	7.2	154
	Illinois (IL) Springfield	12,419,300	5	8.4	260
	Indiana (IN) Indianapolis	6,080,500	14	7.6	195
	Iowa (IA) Des Moines	2,926,300	30	6.6	173
	Kansas (KS) Topeka	2,688,400	32	7.0	203

*The federal district of Washington, D.C., is the capital city of the United States.

Activity Options

Interdisciplinary Links: Logical/Mathematical

Class Time 20 minutes

Task Drawing conclusions based on generalizations

Purpose To make a generalization based on per capita income and percentage of high school graduates

Supplies Needed
- Data File

Activity Have students work independently to randomly select ten states or provinces. They should write the per capita income of each and next to it the percentage of high school graduates. Have them make a generalization based on the data. Point out that every piece of data may not support their generalization, but most data should. Have students share what they learn in class discussion and draw conclusions based on their generalizations.

5

DATA FILE

Population Density (per square mile)	Per Capita Income ($) (1999)	High School Graduates (%) (1998)	Area Rank (2000)	Total Area (square miles)	Map (not to scale)
85.1	21,941	78.8	30	52,237	
1.0	27,274	90.6	1	615,230	
45.0	24,199	81.9	6	114,006	
50.3	21,146	76.8	28	53,182	
213.2	28,513	80.1	3	158,869	
41.3	30,291	89.6	8	104,100	
614.3	37,452	83.7	48	5,544	
327.0	29,341	85.2	49	2,396	
8,412.6	36,554	83.8	51	68	
266.7	26,796	81.9	23	59,928	
138.8	26,007	80.0	24	58,977	
187.6	26,623	84.6	47	6,459	
15.5	22,418	82.7	14	83,574	
214.4	29,908	84.2	25	57,918	
167.0	24,949	83.5	38	36,420	
51.9	24,600	87.7	26	56,276	
32.7	25,467	89.2	15	82,282	

MORE ABOUT...

District of Columbia

The District of Columbia is the site of our national capital. The location was chosen in 1791 by George Washington, and Maryland and Virginia donated the land. The U.S. Constitution gave Congress authority to govern the new capital. Although city council members were elected by citizens of the city, the president chose the mayor. By 1820, citizens were given the right to choose their own mayor. However, they did not get the right to vote in national elections until 1964, when Congress permitted them to vote for president. In 1970, citizens of Washington were allowed to elect a representative to the House of Representatives. The representative could vote in committees but was not allowed to participate in full House votes. Many citizens felt their political rights were unfairly limited. Beginning in the 1970s, many began to push for making the District of Columbia into a state. There were several near successes in this effort, but all have failed. Interest in statehood for the District of Columbia continues, however.

Activity Options

Differentiating Instruction: Less Proficient Readers

Class Time One class period

Task To use Data File information in an educational game format

Purpose To collect and master information given in the Data File

Supplies Needed
- Pens or pencils
- Index cards

Activity Have students work in groups. Give each group ten index cards. The group should write the name of one of the states, territories, or provinces at the top of each card and two clues to its identity. Have groups exchange their completed cards, placing the cards face down. Students should take turns drawing the top card and reading the clues. Other players will race to identify the state, province, or territory.

MORE ABOUT...

U.S. Census

The United States Constitution requires that a census be taken every ten years. The first took place in 1790. Many different types of information are gathered, including numbers of people, how many live in each household, age, education, race, sex, marital status, employment, and income. In fact, each ten-year period, the amount of information gathered increases. In 1790, the census report filled a 56-page book. In 1990, the report was 500,000 pages long. The information is used in many ways. The government uses it to decide on necessary public programs and how much money to spend on different needs. One of the most important uses is the report's effect on a state's representation in the House of Representatives. The greater the population, the more representatives a state has. State legislatures are also based on the population of different areas of the states.

UNIT 2 Atlas

Data File

For updates on these statistics, go to DATA UPDATE CLASSZONE.COM

5

State Flag	State/ Capital	Population (2000)	Population Rank (2000)	Infant Mortality (per 1,000 live births) (1998)	Doctors (per 100,000 pop.) (1998–1999)
	Kentucky (KY) Frankfort	4,041,800	25	7.5	209
	Louisiana (LA) Baton Rouge	4,469,000	22	9.1	246
	Maine (ME) Augusta	1,274,900	40	6.3	223
	Maryland (MD) Annapolis	5,296,500	19	8.6	374
	Massachusetts (MA) Boston	6,349,100	13	5.1	412
	Michigan (MI) Lansing	9,938,400	8	8.2	224
	Minnesota (MN) St. Paul	4,919,500	21	5.9	249
	Mississippi (MS) Jackson	2,844,700	31	10.1	163
	Missouri (MO) Jefferson City	5,595,200	17	7.7	230
	Montana (MT) Helena	902,200	44	7.4	190
	Nebraska (NE) Lincoln	1,711,300	38	7.3	218
	Nevada (NV) Carson City	1,998,300	35	7.0	173
	New Hampshire (NH) Concord	1,235,800	41	4.4	237
	New Jersey (NJ) Trenton	8,414,400	9	6.4	295
	New Mexico (NM) Santa Fe	1,819,000	36	7.2	212
	New York (NY) Albany	18,976,500	3	6.3	387
	North Carolina (NC) Raleigh	8,049,300	11	9.3	232

60 UNIT 2

Activity Options

Multiple Learning Styles: Interpersonal

Class Time 30 minutes

Task Determining a population pattern from a population map

Purpose To recognize population patterns and to speculate about reasons behind the patterns

Supplies Needed
- Art paper
- Pencils

Activity Have students work in pairs and draw an outline map of the United States. Then have them write the population of each state on the map. Ask them what patterns they see in where people live. Invite students to suggest reasons for population patterns. Have them share their ideas in class discussion.

DATA FILE

Population Density (per square mile)	Per Capita Income ($) (1999)	High School Graduates (%) (1998)	Area Rank (2000)	Total Area (square miles)	Map (not to scale)
100.0	22,147	77.9	37	40,411	
90.0	21,794	78.6	31	49,651	
37.8	23,867	86.7	39	33,741	
430.7	30,757	84.7	42	12,297	
687.1	34,168	85.6	45	9,241	
102.8	26,625	85.4	11	96,705	
56.6	29,281	89.4	12	86,943	
58.9	19,608	77.3	32	48,286	
80.3	25,040	82.9	21	69,709	
6.1	21,337	89.1	4	147,046	
22.1	26,235	87.7	16	77,358	
18.1	29,022	89.1	7	110,567	
133.1	29,552	84.0	44	9,283	
1,024.3	34,525	86.5	46	8,215	
15.0	21,097	79.6	5	121,598	
351.5	32,459	81.5	27	53,989	
153.0	25,072	81.4	29	52,672	

MORE ABOUT...

Statistical Data

Point out that the statistics given in the Regional Data File are presented in different ways. The population, for example, is the total number of people living in an area. The number of doctors is given as a number per 100,000 people. To find the actual number of doctors, students would have to divide 100,000 into the population of a state or province and then multiply that number by the number of doctors per 100,000 people. Likewise, the number of high school graduates is given as a percentage of the population. To find the total number of high school graduates in a state, students would multiply the total population by the percentage of graduates. You may wish to model both of these processes for students.

Activity Options

Interdisciplinary Links: Government/History

Class Time One class period

Task Learning about the flag of students' home states

Purpose To learn the history and symbolism of a state's flag

Supplies Needed

- Library or Internet resources
- Pencils or pens
- Paper

Activity Have students work independently or in pairs to learn about their state's flag. They should look for the meaning of the flag's colors and of any symbols or state seal that may appear. Have them investigate the flag's history. When was it designed? Has it changed during the history of the state? If so, how and why? When students finish, guide a class discussion of the flag.

MORE ABOUT...
American Samoa

American Samoa is a group of seven islands located about 2,300 miles southwest of Hawaii. The islands had been occupied for about 2,000 years before they were settled by Polynesians who traveled there from islands far to the west. The people are considered nationals by the U.S. government. This means they can come to the United States at any time. However, since they are not citizens, they lack certain rights, such as voting in U.S. elections. The primary language remains Samoan, although many people speak English as well.

UNIT 2 Atlas Data File

For updates on these statistics, go to

5

State Flag	State/ Capital	Population (2000)	Population Rank (2000)	Infant Mortality (per 1,000 live births) (1998)	Doctors (per 100,000 pop.) (1998–1999)
	North Dakota (ND) Bismarck	642,200	47	8.6	222
	Ohio (OH) Columbus	11,353,100	7	8.0	235
	Oklahoma (OK) Oklahoma City	3,450,700	27	8.5	169
	Oregon (OR) Salem	3,421,400	28	5.4	225
	Pennsylvania (PA) Harrisburg	12,281,100	6	7.1	291
	Rhode Island (RI) Providence	1,048,300	43	7.0	338
	South Carolina (SC) Columbia	4,012,000	26	9.6	207
	South Dakota (SD) Pierre	754,800	46	9.1	184
	Tennessee (TN) Nashville	5,689,300	16	8.2	246
	Texas (TX) Austin	20,851,800	2	6.4	203
	Utah (UT) Salt Lake City	2,233,200	34	5.6	200
	Vermont (VT) Montpelier	608,800	49	7.0	305
	Virginia (VA) Richmond	7,078,500	12	7.7	241
	Washington (WA) Olympia	5,894,100	15	5.7	235
	West Virginia (WV) Charleston	1,808,300	37	8.0	215
	Wisconsin (WI) Madison	5,363,700	18	7.2	227
	Wyoming (WY) Cheyenne	493,800	50	7.2	171
	United States Washington, D.C.	281,422,000	3	7.0	251

62 UNIT 2

Activity Options

Multiple Learning Styles: Intrapersonal

Class Time 90 minutes

Task Creating a travel brochure for a state or province

Purpose To use various sources to gain information about one state or province of the United States or Canada

Supplies Needed
- Library or Internet resources
- Art supplies
- Pencils and paper

Activity Ask students to work independently and choose a state or province they would like to visit. Then have them do research to learn about the area. Next, have them create a travel brochure to promote the region. They might highlight physical features, wildlife, cities, museums, or other attractions. Have them post their brochures on the bulletin board.

DATA FILE

Population Density (per square mile)	Per Capita Income ($) (1999)	High School Graduates (%) (1998)	Area Rank (2000)	Total Area (square miles)	Map (not to scale)
9.1	22,488	84.3	18	70,704	
253.3	25,895	86.2	34	44,828	
49.4	21,802	84.6	20	69,903	
35.2	25,947	85.5	10	97,132	
266.6	27,420	84.1	33	46,058	
851.6	24,418	80.7	50	1,231	
128.6	22,467	78.6	40	31,189	
9.8	24,007	86.3	17	77,121	
135.0	24,461	76.9	36	42,146	
78.0	25,363	78.3	2	267,277	
26.3	22,333	89.3	13	84,904	
63.3	24,758	86.7	43	9,615	
167.2	28,193	82.6	35	42,326	
83.4	28,968	92.0	19	70,637	
74.6	19,973	76.4	41	24,231	
82.0	26,212	88.0	22	65,499	
5.0	24,864	90.0	9	97,818	
74.3	33,900	83.0	4	3,787,319	

5

MORE ABOUT...
Puerto Rico

Puerto Rico lies in the Caribbean Sea about 1,000 miles south of Florida. It is a commonwealth rather than a territory of the United States. The Congress of the United States officially governs the island, and U.S. federal laws apply. Puerto Rico also has considerable powers of self-government. The people are U.S. citizens and can move to the United States at any time. While on the island, they do not have to pay federal income taxes, but they also cannot vote in elections for U.S. presidents. Puerto Rico has two official languages, Spanish and English. Puerto Rico has often been spoken of as a possible 51st state. When given the choice, however, Puerto Ricans have consistently voted to remain a commonwealth.

Activity Options

Multiple Learning Styles: Visual

Class Time One class period

Task Creating a political poster

Purpose To express a view for or against statehood for Puerto Rico

Supplies Needed
- Poster board
- Colored markers
- Internet or library resources (optional)

B Block Scheduling

Activity Tell students to imagine that they are Puerto Ricans. The government has asked citizens to vote for or against having Puerto Rico become one of the United States. Have students work independently and create a political poster expressing their view. (You may wish to have them first do research to learn more about the arguments for and against statehood.) Have students post their work and share their opinions in class discussion.

Unit 2 Atlas

Data File

For updates on these statistics, go to DATA UPDATE CLASSZONE.COM

Province or Territory Flag	Province or Territory/ Capital	Population (2000)	Population Rank (2000)	Infant Mortality (per 1,000 live births) (1998)	Doctors (per 100,000 pop.) (1998–1999)
	Alberta (AB) Edmonton	2,997,200	4	4.8	162
	British Columbia (BC) Victoria	4,063,800	3	4.7	193
	Manitoba (MB) Winnipeg	1,147,900	5	7.5	177
	New Brunswick (NB) Fredericton	756,600	8	5.7	153
	Newfoundland and Labrador (NF) St. John's	538,800	9	5.2	171
	Northwest Territories (NT) Yellowknife	42,100	11	10.9	92
	Nova Scotia (NS) Halifax	941,000	7	4.4	196
	Nunavut (NU) Iqaluit	27,700	13	N/A	N/A
	Ontario (ON) Toronto	11,669,300	1	5.5	178
	Prince Edward Island (PE) Charlottetown	138,900	10	4.4	128
	Quebec (QC) Quebec City	7,372,400	2	5.6	211
	Saskatchewan (SK) Regina	1,023,600	6	8.9	149
	Yukon Territory (YT) Whitehorse	30,700	12	8.4	149
	Canada Ottawa, Ontario	30,750,000	36	5.5	185

64 UNIT 2

MORE ABOUT...

The Canadian Flag

Although Canada became independent of the United Kingdom in 1931, Canada continued to fly the British flag. Some changes were made to the flag over the years, such as adding a special Canadian symbol to it. However, none of these flags were adopted as the official flag of Canada. That did not happen until 1964 when the Canadian Parliament had the present Canadian flag with the red maple leaf created.

Provincial Flags

Many Canadians remain proud of their British heritage, and although the country now has its own distinct national flag, several provinces—British Columbia, Manitoba, and Ontario—continue to include the British flag as an element in their own flags. Quebec, which has a strong French tradition and many French-speaking citizens, includes four fleurs-de-lis on its provincial flag. The fleurs-de-lis is a symbol of France.

Activity Options

Interdisciplinary Link: Citizenship

Class Time One class period

Task Illustrating rules for proper use of the United States flag

Purpose To learn proper ways of displaying and caring for the flag of the United States

Supplies
- Colored markers, colored pencils, or crayons
- Art paper
- Encyclopedia, Internet, or other information sources

Activity Explain to students that there are specific rules for displaying, handling, and caring for the United States flag. Have students do research to learn about these rules. Then have students work in pairs and draw a picture demonstrating one of the rules. Ask them to display their pictures before the class, describe the rule, and explain the reason for it.

DATA FILE

Population Density (per square mile)	Per Capita Income ($) (1999)	High School Graduates (%) (1998)	Area Rank (2000)	Total Area (square miles)	Map (not to scale)
11.7	30,038	86	6	255,285	
11.1	31,592	87	5	366,255	
4.6	26,829	79	8	250,934	
26.7	26,607	78	11	28,345	
3.4	27,692	71	10	156,649	
0.08	33,738 (1994)	64 (1996)	3	503,951	
44.0	25,712	78	12	21,425	
0.03	27,421 (1994)	N/A	1	818,959	
28.3	32,537	84	4	412,582	
49.4	25,534	74	13	2,814	
12.4	28,826	78	2	594,860	
4.1	26,463	82	7	251,700	
0.2	36,130	67 (1996)	9	186,661	
8.0	23,000	82	3	3,850,420	

GEOGRAPHY SKILLBUILDER: Interpreting a Chart

1. **Place** • Which province or territory in Canada has the fewest people? How is it ranked in area? How do these two facts explain the population density?
2. **Place** • What is the highest percentage of high school graduates in a state? What is the lowest percentage?

MORE ABOUT...
Nunavut

Nunavut is Canada's newest territory. It was formed from part of what was formerly the Northwest Territories. Nunavut was created on April 1, 1999, in order to allow the Inuit a higher level of self-government. The Inuit are a native people and make up most of the population of Nunavut. This new territory covers a vast area and has a harsh climate, and the population is widely dispersed in small villages. There are few roads and no railroads. People depend on airplanes for travel between towns. New communications technologies have helped overcome some problems, but many villages remain isolated. As a result, many details about the population, such as infant mortality rates, numbers of doctors, and high school graduates, have yet to be gathered.

GEOGRAPHY SKILLBUILDER
Answers

1. Nunavut has the fewest people; it is ranked first in area; the population density is close to zero because there are so few people in relation to the vast area of land.
2. The highest percentage of high school graduates is 87 percent in British Columbia; the lowest is 64 percent in the Northwest Territories.

Country Profiles

Greenland Greenland is the largest island in the world, located northeast of Canada between the Arctic and North Atlantic oceans. The island covers about 840,000 square miles, 84 percent of which is covered by ice. This ice cap is nearly two miles deep at points. Land that is not covered with ice is generally rough, rocky terrain that does not yield to crops. The population of Greenland is just over 56,000, made up primarily of Inuit people native to this region, Canada, and Siberia. Greenland gained self-rule in 1979, but the country still remains a part of the Kingdom of Denmark. The new government is a parliamentary republic. Greenland's economy is almost wholly dependent upon the fishing industry.

The Physical Geography of the United States and Canada

	OVERVIEW	COPYMASTERS	INTEGRATED TECHNOLOGY
UNIT ATLAS AND CHAPTER RESOURCES	Students will learn about the physical geography of the United States and Canada and how it influences the economic development of the regions within those countries.	**In-depth Resources: Unit 2** • Guided Reading Worksheets, pp. 3–4 • Skillbuilder Practice, p. 7 • Unit Atlas Activities, pp. 1–2 • Geography Workshop, pp. 35–36 **Reading Study Guide** (Spanish and English), pp. 16–21 **Outline Map Activities**	• eEdition Plus Online • EasyPlanner Plus Online • eTest Plus Online • eEdition • Power Presentations • EasyPlanner • Electronic Library of Primary Sources • Test Generator • Reading Study Guide • Critical Thinking Transparencies CT5
	KEY IDEAS		
SECTION 1 From Coast to Coast pp. 69–74	• North America's location in the middle latitudes features a productive farming area that helps feed large populations. • The United States and Canada contain many different agricultural and climate regions. • Glaciers change the physical geography of Earth by smoothing out rough surfaces, carving depressions and deep trenches, and piling up rocks and dirt.	**In-depth Resources: Unit 2** • Guided Reading Worksheet, p. 3 • Reteaching Activity, p. 9 **Reading Study Guide** (Spanish and English), pp. 16–17	**Critical Thinking Transparencies CT6** **Map Transparencies MT13, 15** **classzone.com** Reading Study Guide
SECTION 2 A Rich Natural Diversity pp. 75–79	• Weather is the state of the atmosphere near Earth at a given time and place; climate is the typical weather in a region over a long period of time. • North America's five vegetation zones are polar, tundra, forest, grassland, and desert. • The natural wealth of the United States and Canada positions them as leaders in the global economy.	**In-depth Resources: Unit 2** • Guided Reading Worksheet, p. 4 • Reteaching Activity, p. 10 **Reading Study Guide** (Spanish and English), pp. 18–19	**Map Transparencies MT 15** **classzone.com** Reading Study Guide

KEY TO RESOURCES

 Audio
 CD-ROM
Copymaster
 Internet
Overhead Transparency
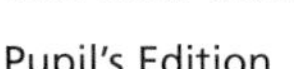 Pupil's Edition
 Teacher's Edition
 Video

ASSESSMENT OPTIONS

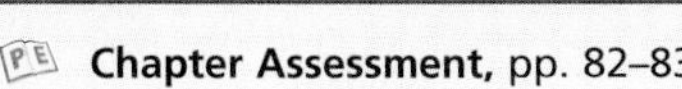 **Chapter Assessment,** pp. 82–83

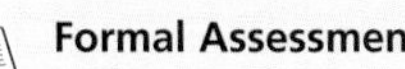 **Formal Assessment**
- Chapter Tests: Forms A, B, C, pp. 35–46

Test Generator

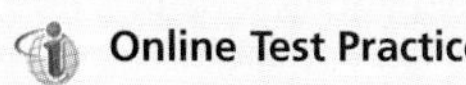 **Online Test Practice**

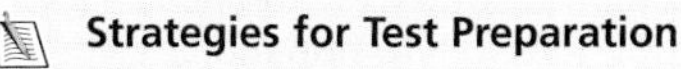 **Strategies for Test Preparation**

 **Section Assessment,** p. 74

 **Formal Assessment**
- Section Quiz, p. 33

 **Integrated Assessment**
- Rubric for describing features of a region

Test Generator

Test Practice Transparencies TT5

 **Section Assessment,** p. 79

Formal Assessment
- Section Quiz, p. 34

 **Integrated Assessment**
- Rubric for drawing physical features

Test Generator

Test Practice Transparencies TT6

RESOURCES FOR DIFFERENTIATING INSTRUCTION

Students Acquiring English/ESL

Reading Study Guide (Spanish and English), pp. 16–21

Access for Students Acquiring English Spanish Translations, pp. 13–18

Modified Lesson Plans for English Learners

Less Proficient Readers

Reading Study Guide (Spanish and English), pp. 16–21

TE Activity
- Organizing information, p. 72

Gifted and Talented Students

TE Activity
- Categorizing, p. 76

CROSS-CURRICULAR CONNECTIONS

Popular Culture

Peterson, Cris. ***Century Farm: One Hundred Years on a Family Farm.*** Honesdale, PA: Boyds Mills Press, 1999. Wisconsin farm over a century of change.

Ricciuti, Edward R. ***America's Top 100.*** Woodbridge, CT: Blackbirch Press, 2001. Includes national parks and geographical wonders.

Geography

Grupper, Jonathan. ***Destination: Rocky Mountains.*** Washington, D.C.: National Geographic Society, 2001. Stunning photographs and humorous narrative.

Language Arts/Literature

DeSpain, Pleasant. ***Sweet Land of Story: Thirty-Six American Tales to Tell.*** Little Rock, AR: August House, 2000. Stories reflective of region and geography.

Science

Vogel, Carole Garbuny. ***Legends of Landforms: Native American Lore and the Geology of the Land.*** Brookfield, CT: Millbrook Press, 1999. Unique coupling of folklore and geological history.

Murphy, Jim. ***Blizzard! The Storm That Changed America.*** New York: Scholastic Press, 2000. Blizzard of 1888 that devastated the Atlantic seaboard.

History

Maurer, Richard. ***The Wild Colorado: The True Adventures of Fred Dellenbaugh, Age 17, on the Second Powell Expedition into the Grand Canyon.*** New York: Crown Publishers, 1999. Photo documentary of the 1869 expedition.

ENRICHMENT ACTIVITIES

The following activities are especially suitable for classes following block schedules.

Teacher's Edition, pp. 71, 73, 76, 77, 78
Pupil's Edition, pp. 74, 79

Unit Atlas, pp. 54–65
Technology: 1971, p. 81

Outline Map Activities

INTEGRATED TECHNOLOGY

Go to **classzone.com** for lesson support and activities for Chapter 3.

BLOCK SCHEDULE LESSON PLAN OPTIONS: 90-MINUTE PERIOD

DAY 1

UNIT PREVIEW, pp. 52–53
Class Time 10 minutes

- **Discussion** Discuss the Unit Introduction, using the discussion prompts on TE pp. 52–53.

UNIT ATLAS, pp. 54–65
Class Time 20 minutes

- **Small Groups** Divide the class into four groups and have each group answer Making Connections questions for one section of the Unit Atlas: Physical Geography, Human Geography, Regional Patterns, and Regional Data File.

SECTION 1, pp. 69–74
Class Time 60 minutes

- **Recognizing Effects** Use the Critical Thinking Activity on TE p. 70 to help students recognize the link between where a place is located and economic and other human activities.
Class Time 10 minutes
- **Travel Poster** Have students work in small groups to create a travel poster that features one of the seven geographic regions that Canada and the United States share.
Class Time 40 minutes
- **Newspaper Reporters** Have students reread Physical Processes Shape the Land. Guide them to think about the landforms and bodies of water in their region and to speculate about what natural processes shaped the area in which they live. Tell them to pretend that they are newspaper reporters and the event just happened. Have them write a headline for the local newspaper announcing the earth-changing news.
Class Time 10 minutes

DAY 2

SECTION 1, continued
Class Time 35 minutes

- **Peer Competition** Divide the class into pairs. Assign each pair one of the Terms & Names for this section. Have pairs make up five questions that can be answered with the term or name. Have groups take turns asking the class their questions.

SECTION 2, pp. 75–79
Class Time 55 minutes

- **Oral Report** Have students summarize the reports they prepared on climate change for the Interdisciplinary Link: Science on TE p. 78. Ask them to present their summaries to the class.

DAY 3

SECTION 2, continued
Class Time 35 minutes

- **Peer Teaching** Have pairs of students review the Main Idea for the section on PE p. 75 and find three details to support it. Then have each pair trade lists with another group and find two additional details to support the Main Idea.
Class Time 10 minutes
- **Internet** Extend students' background knowledge of the climate and vegetation of North America by visiting **classzone.com.**
Class Time 25 minutes

CHAPTER 3 REVIEW AND ASSESSMENT, pp. 82–83
Class Time 55 minutes

- **Review** Have students prepare a summary of the chapter by reviewing the Main Idea and Why It Matters Now features of each section in Chapter 3.
Class Time 20 minutes
- **Assessment** Have students complete the Chapter 3 Assessment.
Class Time 35 minutes

TECHNOLOGY IN THE CLASSROOM

VIRTUAL TRAVEL ON THE INTERNET

The Internet will never be an ideal substitute for a real vacation, but it can help students see what it would be like to visit different places. By looking at photographs and reading descriptions, they can learn about geography, geology, and other features of a place and perhaps even plan a future vacation.

ACTIVITY OUTLINE

Objective Students will take a virtual trip to several national parks in the United States and Canada to find out what the natural landscape looks like in different parts of these countries.

Task Have students visit Web sites to see pictures of U.S. and Canadian national parks. Then ask them to create multimedia presentations that showcase four of the national parks.

Class Time 2–4 class periods

DIRECTIONS

1. Ask the class if anyone has ever taken a road trip across the United States or Canada. Where did they go? What was the scenery and weather like? What did they do?
2. Have students look at a physical map of both the United States and Canada, and describe some of the major landscape features they see. What are the big mountain ranges, rivers, and lakes?
3. Ask students to imagine that their class has been given some money, a bus and driver, and two months to explore the continental United States and Canada next summer. They want to focus on natural scenery, so they have chosen to visit some national parks.
4. Have students make a computer-generated chart or one on paper with rows for the following 14 national parks: United States: Acadia, Everglades, Great Smoky Mountains, Rocky Mountain, Yellowstone, Grand Canyon, Yosemite, Olympic; Canada: Gros Morne, Jasper, Nahanni, Auyuittuq, Aulavik, and Grasslands.
5. Have students, either individually or in small groups, visit the park Web sites at **classzone.com.** They should look at the pictures of the parks and read information about the plant and animal life and geology. Ask them to list in each row of their charts three distinctive features of the natural landscape they see at each park. For example, for Yellowstone, they might say "geysers," "the Grand Canyon of the Yellowstone," and "bison." If there is not enough time for students to visit every park, divide the class into groups and assign each group to several parks.
6. As they travel on the field trip, students should label the national parks on blank outline maps of the United States and Canada (available at the last Web site at **classzone.com**).
7. Finally, have each student or group choose two national parks from the United States and two from Canada. The parks should be located in different parts of the countries. Ask them to create multimedia presentations to serve as "electronic brochures" for the parks they have visited over the summer. The presentations should focus on the scenery and physical geography of the parks.

CHAPTER 3 OBJECTIVE

Students will learn about the physical geography of the United States and Canada, and how it influences the economic development of the regions within those countries.

FOCUS ON VISUALS

Interpreting the Photograph Invite students to examine the photograph of the aurora borealis (northern lights) and read the caption. Ask them to describe the colors and images they see. Point out the globe. Ask students to describe the location of the United States and Canada.

Possible Responses Students may say north of the equator, between the equator and the North Pole, surrounded by oceans, and so on.

Extension Ask students to use the Data File in the Unit Atlas to compare the land area of Canada with the land area of the United States.

CRITICAL THINKING ACTIVITY

Hypothesizing Direct students' attention to the map of Canada and the United States on page 56. Point out Ottawa and Washington, D.C., the capital cities. Have students speculate as to why these locations might have been chosen as capital cities.

Class Time 10 minutes

The Physical Geography of the United States and Canada

Place The aurora borealis, or northern lights, is a natural phenomenon most commonly seen in the winter skies over Alaska and northern Canada.

66

Recommended Resources

BOOKS FOR THE TEACHER

America: A Celebration of the United States. Skokie, IL: Rand McNally, 1999. Geography, history, and cultural traditions of U.S. regions.

Schwartz, Seymour. ***This Land Is Your Land: The Geographic Evolution of the United States.*** New York: Harry N. Abrams, 2000. Growth of the United States, with historic maps and text.

SOFTWARE

Where in the U.S.A. is Carmen Sandiego? Novato, CA: Broderbund, 1997. Geographical clues to track down the fictional Carmen Sandiego.

INTERNET

For more information about the United States and Canada, visit **classzone.com.**

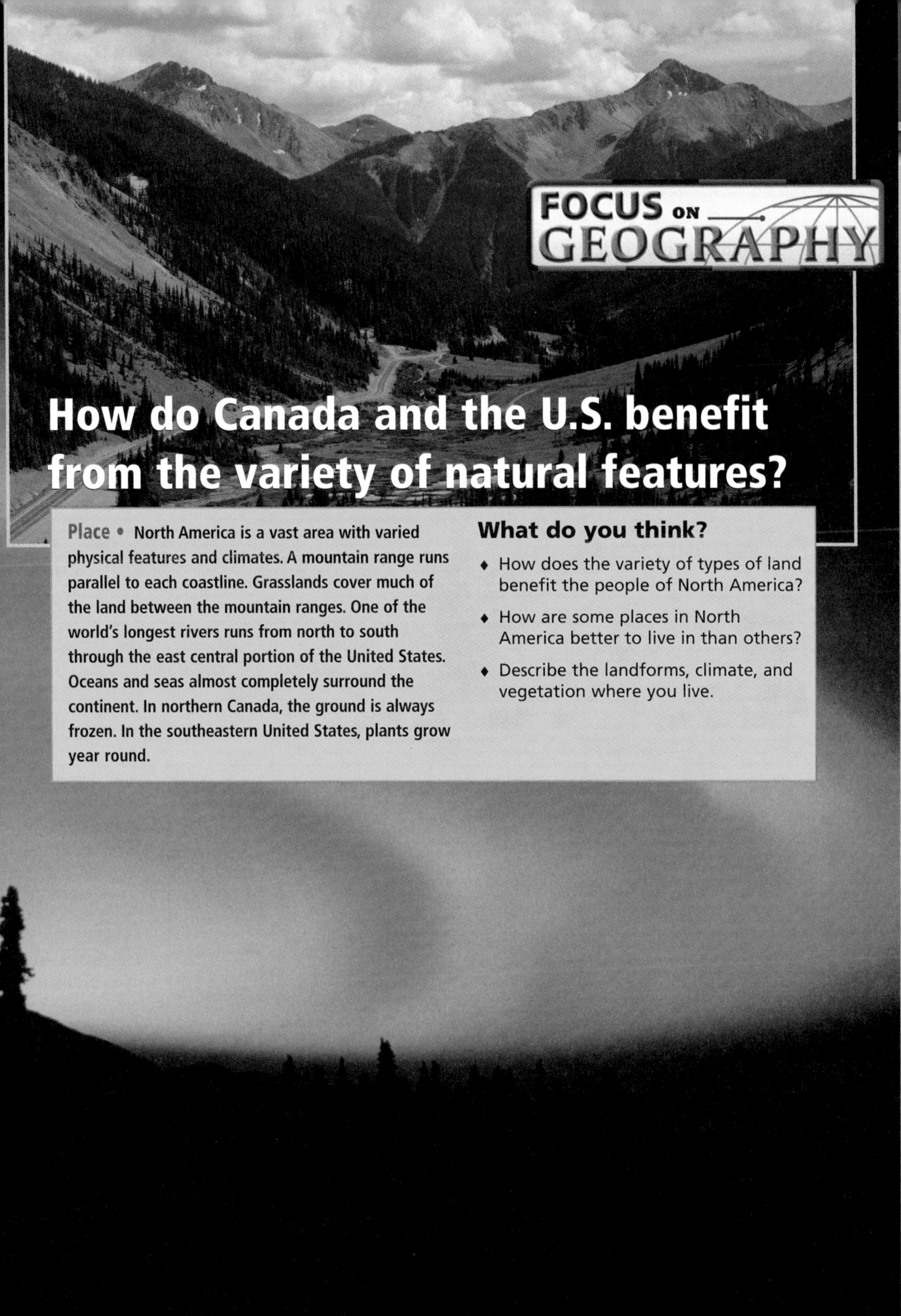

How do Canada and the U.S. benefit from the variety of natural features?

Place • North America is a vast area with varied physical features and climates. A mountain range runs parallel to each coastline. Grasslands cover much of the land between the mountain ranges. One of the world's longest rivers runs from north to south through the east central portion of the United States. Oceans and seas almost completely surround the continent. In northern Canada, the ground is always frozen. In the southeastern United States, plants grow year round.

What do you think?

- How does the variety of types of land benefit the people of North America?
- How are some places in North America better to live in than others?
- Describe the landforms, climate, and vegetation where you live.

FOCUS ON GEOGRAPHY

Objectives

- To introduce students to the geographical features of Canada and the United States, two countries with similar cultures
- To help students understand the relationship between geography and economy in Canada and the United States

What Do You Think?

1. Help students understand that the varied types of land provide the opportunity for a variety of economic and other activities.
2. Guide students to consider why some places are less suitable for human living than others and that livability can be related to factors such as landforms, waterways, and climate.
3. Be sure students understand the terms landforms, climate, and vegetation. Students might record their answers in their notebooks and share them with classmates.

How do Canada and the U.S. benefit from the variety of natural features?

Have students think about the resources that might be available in each country and how they differ. Ask them to speculate on how the two countries are able to share the benefits of those resources.

MAKING GEOGRAPHIC CONNECTIONS

Ask students to make a list of resources they have used in the past 24 hours, such as specific foods, clothing, school supplies, and entertainment items. Ask them to identify the origin (grasslands, forests, rivers, etc.) of as many resources as they can.

Implementing the National Geography Standards

Standard 7 Construct and analyze climate graphs

Objective To create and analyze bar graphs comparing the climate of different U.S. states

Class Time 20 minutes

Task Have students create and analyze bar graphs, using a map and the following information: Average Annual Temperature (°F)—Juneau, AK, 40.6; Denver, CO, 50.3; Houston, TX, 67.9; Boston, MA, 51.3. Average Annual Precipitation (inches)—Juneau, AK, 54.3; Denver, CO, 15.4; Houston, TX, 46.1; Boston, MA, 41.5. Have students make a list of some factors affecting climate.

Evaluation Students should include in their lists the following three factors: latitude, nearness to the ocean, and position in relation to mountains.

BEFORE YOU READ

What Do You Know?

Divide the class in half. Designate one half as Canada and the other as the United States. Have students in both groups write three facts about their countries. Each fact should relate to the physical features or natural resources of the country. After five minutes, collect the papers and randomly read statements aloud. Invite students to tell which facts they think apply to the United States, which to Canada, and which to both countries.

What Do You Want to Know?

Suggest that students organize their questions in the form of a K-W-L chart. Have them make three columns, titled What I Know, What I Want to Know, and What I Learned. As they read, students should record information in the third column.

READ AND TAKE NOTES

Reading Strategy: Drawing Conclusions
Explain to students that as they read, they will learn how various geographic features affect economic activities. Tell students that when they complete the chart, they will have a summary of the cause-effect relationships between geographic features and economic strengths and benefits.

In-depth Resources: Unit 2
- Guided Reading Worksheets, pp. 3–4

CHAPTER 3 **READING SOCIAL STUDIES**

BEFORE YOU READ

▸▸ What Do You Know?

Before you read the chapter, reflect on what you already know about the United States and Canada. On which continent will you find these countries? Remember what you have read, seen, or learned in other classes about their physical features and natural resources. They are two of the wealthiest and most powerful countries in the modern world. What makes nations powerful in today's world?

▸▸ What Do You Want to Know?

Decide what you want to know about the United States and Canada. In your notebook, record what you hope to learn from this chapter.

Region • **Water resources, such as the Mississippi River, contribute to the wealth of the United States and Canada and to their positions of leadership in the world.** ▲

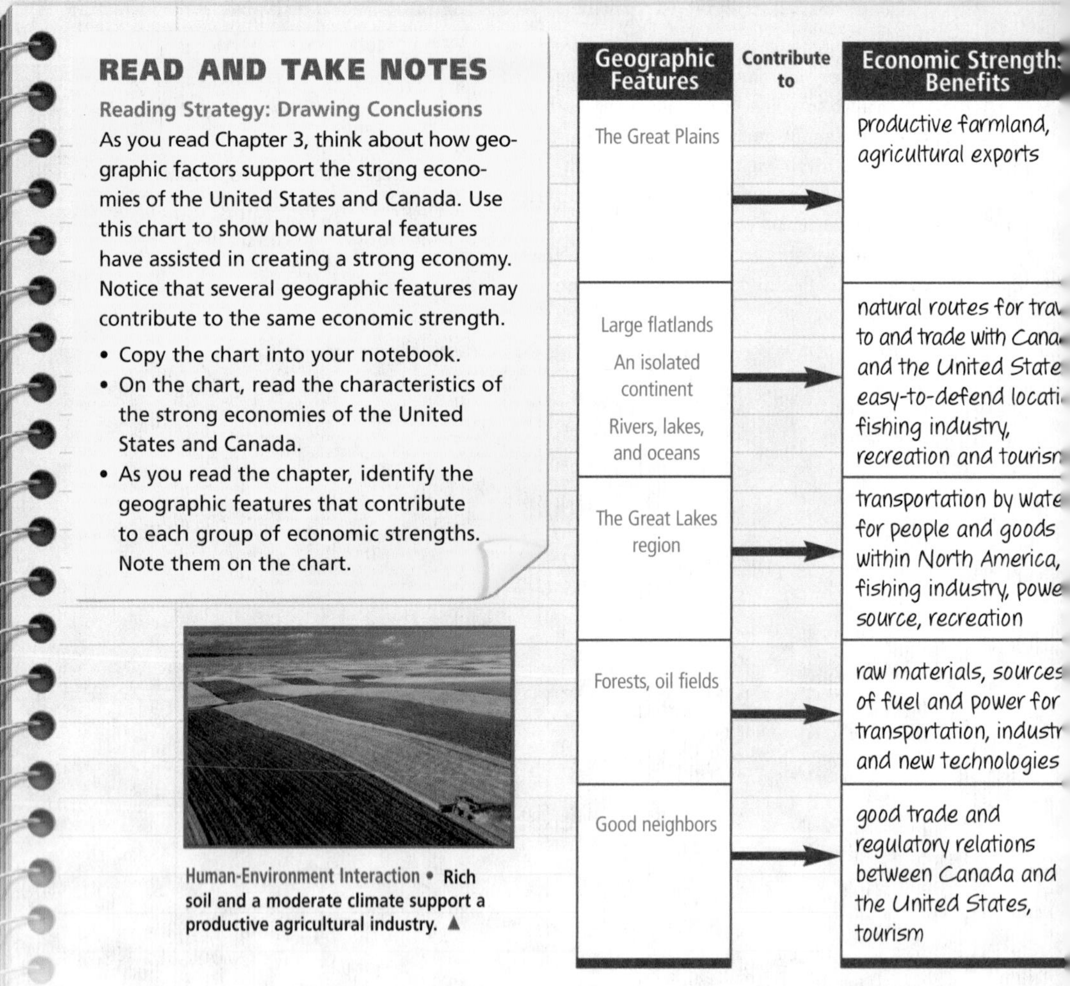

READ AND TAKE NOTES

Reading Strategy: Drawing Conclusions
As you read Chapter 3, think about how geographic factors support the strong economies of the United States and Canada. Use this chart to show how natural features have assisted in creating a strong economy. Notice that several geographic features may contribute to the same economic strength.

- Copy the chart into your notebook.
- On the chart, read the characteristics of the strong economies of the United States and Canada.
- As you read the chapter, identify the geographic features that contribute to each group of economic strengths. Note them on the chart.

Geographic Features	Contribute to	Economic Strengths Benefits
The Great Plains	→	productive farmland, agricultural exports
Large flatlands An isolated continent Rivers, lakes, and oceans	→	natural routes for trav to and trade with Cana and the United State easy-to-defend locati fishing industry, recreation and tourisn
The Great Lakes region	→	transportation by wate for people and goods within North America, fishing industry, powe source, recreation
Forests, oil fields	→	raw materials, sources of fuel and power for transportation, industr and new technologies
Good neighbors	→	good trade and regulatory relations between Canada and the United States, tourism

Human-Environment Interaction • **Rich soil and a moderate climate support a productive agricultural industry.** ▲

Teaching Strategy

Reading the Chapter This is a thematic chapter focusing on how the physical geography of North America influenced settlement patterns. It also analyzes how North America's location affects its relationship with the rest of the world. As they read, ask students to identify how geographic factors are responsible for the location of particular economic activities in regions of North America.

Integrated Assessment The Chapter Assessment on page 83 describes several activities for integrated assessment. You may wish to have students work on these activities during the course of the chapter and then present them at the end.

From Coast to Coast

TERMS & NAMES
Sacagawea
landform
glacier
erosion
river system

AIN IDEA

orth America has varied regions nd landforms.

WHY IT MATTERS NOW

North America's geography contributes to the prosperity of the people who live there.

DATELINE EXTRA

MOUNT ST. HELENS, WASHINGTON, MAY 18, 1980

Mount St. Helens, an ancient volcano in Washington's Cascade Mountains, erupted this morning, killing 60 people and thousands of animals. An earthquake caused the mountain's north face to fall away. The debris from this landslide filled Spirit Lake. Hot air blasts traveling at 300 miles per hour threw both gas and volcanic ash 12 miles high and destroyed 10 million trees.

The avalanche and the mudslide that followed buried parts of the Toutle River Valley to a depth of almost 500 feet. The hot ash and rock started forest fires, ruined crops, and covered cities. Fortunately, the land around the volcano has begun to recover from the eruptions.

Place • Gas and volcanic ash from the eruption of Mount St. Helens vaporized trees and caused widespread devastation as far as 19 miles from the volcano. ▲

North America ❶

Earth's geography changes continually. Sometimes change happens violently, as in the eruption of a volcano such as Mount St. Helens, or the jolt of an earthquake. At other times, change occurs very slowly, as when rain washes away soil, or weather wears down a mountain. All these natural processes affect the physical geography of North America.

TAKING NOTES
Use your chart to take notes about the United States and Canada.

Geographic Features	Economic Strengths/ Benefits
	productive farmland

SECTION OBJECTIVES

1. To define North America as a region and explain how its location affects its relationships with other nations
2. To identify regions of the U.S. and Canada
3. To explain physical processes that shape the land

SKILLBUILDER
- Interpreting a Map, p. 70

CRITICAL THINKING
- Recognizing Effects, p. 70
- Finding Causes, p. 72

FOCUS & MOTIVATE

WARM-UP

Making Inferences Have students read Dateline and discuss changes in the area.

1. How might the changes affect vegetation and animal life in the area?
2. What effect could these changes have on the human population?

INSTRUCT: Objective ❶

North America/An Isolated Continent

- Why do geographers study the United States and Canada separately from Mexico, even though they are all in North America? Canada and the United States share a cultural heritage.
- How does location in the middle latitudes help a country's economic health? Moderate climate allows productive farming.

 **In-depth Resources: Unit 2**
- Guided Reading Worksheet, p. 3

 **Reading Study Guide**
(Spanish and English), pp. 16–17

Program Resources

 **In-depth Resources: Unit 2**
- Guided Reading Worksheet, p. 3
- Reteaching Activity, p. 9

Reading Study Guide
(Spanish and English), pp. 16–17

 **Formal Assessment**
- Section Quiz, p. 33

 **Integrated Assessment**
- Rubric for writing a description

 **Outline Map Activities**

 **Access for Students Acquiring English**
- Guided Reading Worksheet, p. 13

 **Technology Resources**
classzone.com

TEST-TAKING RESOURCES
- Strategies for Test Preparation
- Test Practice Transparencies
- Online Test Practice

CRITICAL THINKING ACTIVITY

Recognizing Effects After students read the paragraph "Middle Latitudes," point out that productive farming is possible in this area because of the temperate climate. Have students brainstorm a list of other activities that are possible or likely because of this climate. Encourage them to consider aspects of economics, transportation, and recreation.

Class Time 10 minutes

FOCUS ON VISUALS

Interpreting the Map Help students use the illustrated globe to orient themselves and describe the location of the North American continent. Ask students to point out the equator and the Arctic Circle and then describe how Canada and the United States are located on the globe relative to these major reference points. Are any parts of the United States north of the Arctic Circle?

Possible Responses All of Canada and the United States are north of the equator; northern Canada and part of Alaska are above the Arctic Circle.

Extension Ask students to locate the tropic of Cancer. Point out that the warm tropics, or tropical zone, lie between the tropic of Cancer (23°27' N) and the tropic of Capricorn (23°27' S). Ask them to locate the area of the United States closest to the tropic of Cancer and explain how its climate might differ from other parts of the country.

The Middle Latitudes

GEOGRAPHY SKILLBUILDER: Interpreting a Map

1. **Region** • Do you think Canada or the United States has the colder climate? Why?
2. **Location** • Where in Canada would you expect to find the most people living? Explain.

Countries on the Continent North America's huge landmass is home to several large countries and many smaller ones. Find Canada and the United States on the Unit Atlas map on page 56. Canada is the second largest country in area in the world. The United States is almost as great in area. North America also includes the Danish dependency Greenland, which is the world's largest island.

Mexico, the Central American countries, and the Caribbean island nations, such as Cuba, the Dominican Republic, and Haiti, are part of the continent of North America. These countries, along with the South American nations, make up what is considered Latin America. Find these countries on the Unit 3 Atlas map on page 144. Their heritage differs from that of the United States and Canada. Historically, Latin America owes much of its culture to Spain and Portugal. The United States and Canada were greatly influenced by the British and French. Because of these different cultural heritages, geographers study the United States and Canada separately from Latin America.

Middle Latitudes Most of the United States and Canada is located in the middle latitudes of the northern hemisphere of Earth. This area between the Arctic Circle and the Tropic of Cancer has a temperate climate. It is not as hot as land closer to the Equator. It is not as cold as regions near the North or South Poles. Many plants and animals thrive in this climate. Productive farming enables countries in the middle latitudes to feed large populations.

Human-Environment Interaction • Crops raised in the United States and Canada feed the people of these countries and are exported to other countries all over the world. ▼

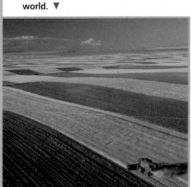

An Isolated Continent ❶

North America is almost completely surrounded by water. Its landmass stretches from the Arctic Ocean to the Gulf of Mexico and from the Pacific Ocean to the Atlantic Ocean. Find these bodies of water on page 54 of the Unit Atlas.

Reading Social Studies
A. Possible Answer
They are relatively close to the equator and far from the North or South Poles.

Geography Skillbuilder Answers
1. Canada: It is farther from the equator, and part of it lies within the Arctic Circle.
2. along its southern borders and the coasts, because the climate is milder

A. Analyzing Effects Why do the middle latitudes have a moderate climate?

Activity Options

Multiple Learning Styles: Bodily-Kinesthetic

Class Time 15 minutes

Task Integrating stretching exercises and geography concepts

Purpose To review the vocabulary of geography: latitude, longitude, meridian, parallel, Arctic Circle, tropic of Cancer, Antarctic Circle, tropic of Capricorn, North Pole, South Pole, equator

Supplies Needed
• World globe or map

Activity Use the globe or map to quickly review the terms listed to the left. Then have students stand and orient themselves by associating their waistlines with the equator, heads with the North Pole, shoulders with the tropic of Cancer, knees with the tropic of Capricorn, and feet with the South Pole. Call out the terms randomly as students stretch and gently reach for the correct "region." For latitude, longitude, and so forth, students should stretch either north-south or east-west.

At one time, these waters isolated North America, or kept it separate from the rest of the world. Unique plants, such as the giant sequoia and the saguaro cactus, and animals, such as the bald eagle and the American alligator, developed in North America.

Region • The bald eagle has been the national bird of the United States since 1782. ▲

The oceans and seas were also a barrier to people. The earliest settlers arrived 12,000 to 35,000 years ago. No other people reached this continent until thousands of years later.

Crossing the Barriers ❷

As people learned more about shipbuilding and navigation, the oceans became a hazardous but passable travel route. Settlers arrived in North America with plants and animals from their home countries. Many of these plants and animals were new to the continent. In some places, these replaced the native plants and animals.

In the 20th century, the distance from other countries helped protect Canada and the United States mainland from attack during the two World Wars. Today, satellites, the Internet, modern transportation, and other technologies link people everywhere.

BACKGROUND

Hawaii, the 50th state of the United States, is a group of volcanic islands in the central Pacific Ocean.

Place • Sailors have used the sextant to navigate their ships since its invention in 1731. ▲

Biography

Sacagawea Sacagawea (SAK•uh•guh•WEE•uh) was a Shoshone woman who had a vital role in the exploration of what is now the northwestern United States. She guided explorers Meriwether Lewis and William Clark from what is now North Dakota into the Pacific Northwest. They had been sent to explore the newly purchased Louisiana Territory. Sacagawea's husband, French Canadian trapper Toussaint Charbonneau, and their baby son were also on the journey, which lasted from 1804 to 1806.

Sacagawea identified fruits and vegetables for the group to eat and helped the explorers communicate with the Native Americans whom they met along the trail. Historians believe that she was born around 1786 and probably died in 1812.

INSTRUCT: Objective ❷

Crossing the Barriers/Regions of the United States and Canada

- How did knowledge of shipbuilding and navigation affect North America? made reaching the continent possible; brought settlers with new plants and animals
- What features characterize the Atlantic Coastal Plain? rich farmland, swamps, wetlands
- In which regions do ranchers raise cattle? Great Plains and Intermontane Region
- What activity takes place in the Canadian Shield? mining

Biography

Some facts about the fabled guide Sacagawea are verifiable, while others are not. What is certain is that she joined Lewis and Clark's expedition in what is now North Dakota, after her husband was hired as an interpreter. Her name is probably from the Hidatsa words *sacaga* (bird) and *wea* (woman).

In August 1805 the explorers met a band of Shoshone Indians whose chief was Sacagawea's brother, so the expedition was able to obtain horses from him.

One report has Sacagawea dying from a fever in 1812. Another account, though, says she survived until 1884.

In 2000, the United States Mint issued a new "Golden Dollar" coin featuring a picture of Sacagawea with her baby. However, there are no actual known portraits of Sacagawea.

Activity Options

Interdisciplinary Link: Science

Block Scheduling

Class Time One class period

Task Making annotated drawings of new animal or plant species

Purpose To record accurately new species in nature as a scientist would

Supplies Needed
- Drawing paper
- Drawing tools
- Reference sources

Activity Explain that many explorers took naturalists with them on their journeys of discovery to keep records of new species of plants and animals. Have students draw one of the species mentioned in the text: the giant sequoia, saguaro cactus, bald eagle, or American alligator. Encourage them to refer to illustrations in books and to include as many details as they can. Tell them to add labels that would help future settlers identify that species.

CRITICAL THINKING ACTIVITY

Finding Causes Explain that the present time is referred to as the information age because of the technological tools that allow people to communicate regarding a variety of economic activities to and from almost anywhere in the world. Point out that before this was possible, people's economic activity depended mainly on nearby resources. Ask them to imagine seeing any one of the described regions for the very first time. Have volunteers tell what economic activity they would pursue in that region and explain why.

Class Time 15 minutes

MORE ABOUT...

The Canadian Shield

The broad, flat, rocky plateau of the Canadian Shield is the ancient heart of the North American continent. Each continent has at least one similar region of stable, ancient Precambrian crystalline rock, called a "continental shield." The oldest of these very old shield rocks are 2 to 3 billion years old. On other continents, most of the ancient bedrock is covered with layers of soil and sediment, but in North America, glaciers scraped the ancient rock bare in many places. Like the Canadian Shield, other shield areas are rich in mineral deposits.

Regions of the United States and Canada ❷

The United States and Canada share many geographic regions. Find these regions on the Unit Atlas map on page 54.

Atlantic Coastal Plain This region runs along the Gulf of Mexico and the east coast of North America. The region has much rich farmland and some swamps and wetlands.

Appalachian Mountains This 400-million-year-old mountain range lies west of the Atlantic Coastal Plain. These forest-covered mountains have weathered, or worn down, over time.

Central Lowlands West of the Appalachians are the Central Lowlands. They extend west to the Great Plains and are generally flat. The soil is rich, and many farms are located here.

Great Plains The Great Plains have grasslands and few trees. The land gradually rises from the Central Lowlands to the Rocky Mountains. Farmers grow crops and ranchers raise cattle in some areas.

The Rocky Mountains and Coastal Ranges North America's highest mountain ranges lie in the west. They include the Rocky Mountains, the Sierra Nevada and the Cascade ranges of the United States, and the Coast Mountains of Canada. These high, rugged, and heavily forested mountain ranges run along the western part of the continent from Mexico to Alaska.

Intermountain Region Located between the Rocky Mountains and the western coastal mountains, this region is dry and contains plateaus, basins, and deserts. Ranchers raise cattle and sheep in some areas. The Grand Canyon is found here.

Region • The Canadian Rocky Mountains are part of the rugged range that reaches from Mexico to Alaska in western North America. ▼

Region • The Grand Canyon in Arizona is one of the natural wonders of the world. ▼

Activity Options

Differentiating Instruction: Less Proficient Readers

Organizing Information To help students organize and retain the information on pages 72–73, have them work with partners to learn about one region. Ask each pair to summarize what they learned in a chart like the one shown. After they have completed their charts, have each partner pair off with a student who has studied another region. They can then share what they learned. Have them continue until every student has reviewed each region.

APPALACHIAN MOUNTAINS	
Physical Geography	**Resources**
weathered mountains, covered by forests	forests for lumber

Canadian Shield or Laurentian Plateau The Canadian Shield covers most of Greenland, curves around the Hudson Bay, and reaches into the United States along the Great Lakes. The central and northwestern part of this huge rocky region has flat plains with hills and lakes. The northeast has high mountains and the south is covered with forests. The shield is rich in minerals, such as iron, gold, copper, and uranium. Most of the land is not farmable and is sparsely populated.

Reading Social Studies

B. Clarifying Which regions of Canada and the United States have productive farmland?

Reading Social Studies B. Answer Atlantic Coastal Plain, Central Lowlands, and Great Plains

Physical Processes That Shaped the Land ❸

Natural processes have shaped North America. Some of the continent's most dramatic landforms were created by the action of wind, water, ice, and moving slabs of Earth's crust. **Landforms** are features of Earth's surface, such as mountains, valleys, and plateaus.

A **glacier** is a thick sheet of ice that moves slowly across land. Thousands of years ago, when Earth was much colder, glaciers covered much of North America. As they flowed across the land, they smoothed out rough surfaces, carved depressions and deep trenches, and piled up rock and dirt. When the ice melted, North America had new valleys, lakes, and hills.

Quakes Shake Central U.S. Lands Usually, earthquakes shake up California and other parts of the western United States. But from December 1811 to February 1812, Mississippi River Valley residents were shocked by several powerful quakes.

The earthquakes changed the Mississippi River's path. Islands disappeared, riverbanks collapsed, and waves capsized boats and drowned people. Farmland was flooded, new lakes appeared, and forests were destroyed. Trees such as this one were uprooted.

Eyewitnesses saw "houses, gardens, and fields . . . swallowed up" in New Madrid, Missouri, in the final quake.

Formation of the Rocky Mountains

INTERACTIVE

Plates collide over millions of years

Main Ranges

Collision causes uplifting of rock layers

Front Ranges

Movement of plate under Pacific Ocean

Movement of North American plate

Region • The Rocky Mountains were formed 40 to 70 million years ago as a result of a collision between the tectonic plate under the Pacific Ocean and the North American plate. ◄

Strange but TRUE

The Missouri earthquakes may have been some of the strongest ever to hit the United States. If seismographs had been available, the quakes might have measured 8 or higher on the Richter scale.

Charles F. Richter was an American seismologist who in 1835 developed the numbering system to measure earthquakes. Each whole-number step up on the scale means an increase of 32 times the force of the number before. The Missouri quakes would probably have measured higher on the scale than the devastating San Francisco earthquake of 1906. However, fewer people lived in the Missouri region, so property damage and the death toll were relatively small.

INSTRUCT: Objective ❸

Physical Processes Shape the Land/Waterways

- How did glaciers change the physical geography of Earth? smoothed out rough surfaces, carved depressions and deep trenches, piled up rock and dirt
- How was the Grand Canyon formed? at least partly from erosion by the Colorado River
- What are the three longest rivers in North America? Mississippi, Missouri, Mackenzie

Activity Options

Interdisciplinary Link: Art

Block Scheduling

Class Time One class period

Task Creating before-and-after posters

Purpose To illustrate changes in the physical features of the land caused by glaciers

Supplies Needed
- Reference materials
- Poster board or large sheets of drawing paper
- Poster paint and brushes

Activity Recall with students the kinds of changes caused by the movement and melting of glaciers. Have students create a two-part landscape, the first showing an area as it might have been before the Ice Age, the second showing an area after the glaciers gouged out river valleys and left behind mounds of soil and rocks as well as lakes filled with water. Provide a space for students to display their posters.

Though the Atlantic Ocean and the Great Lakes are at different elevations, ships are able to navigate the seaway that links them because of a system of canals and locks. Each of the 19 locks is 80 feet (24 meters) wide, 766 feet (233 meters) long, and 30 feet (9 meters) deep. It takes a ship about 45 minutes to pass through each lock.

ASSESS & RETEACH

Reading Social Studies Have students add details to the first column of the graphic organizer on page 68.

Formal Assessment
- Section Quiz, p. 33

RETEACHING ACTIVITY

Students can work in small groups to create a section review. Have individual students write one- or two-sentence summaries of each section. Then have each student share his or her summaries with the group.

In-depth Resources: Unit 2
- Reteaching Activity, p. 9

Access for Students Acquiring English
- Reteaching Activity, p. 17

The St. Lawrence Seaway This seaway, completed in 1959 by Canada and the United States, is one of the largest civil engineering projects ever built. It enables ships to travel 2,340 miles inland from the Atlantic Ocean through locks such as this one, to the Great Lakes. As a result, trade between the United States and Canada and between North America and other continents has improved. Grain, iron ore, and coal are three vital goods shipped on the seaway.

Wind, rivers, and rain wear away soil and stone in a process called **erosion.** Erosion can create magnificent landforms. The Grand Canyon is at least partly the result of millions of years of erosion by the Colorado River. Volcanoes, such as Mount St. Helens, and earthquakes are other natural forces that change the land. All these mighty forces have created landforms across North America.

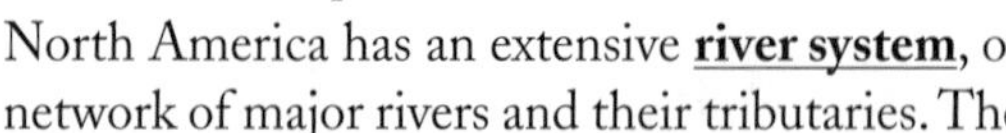

Waterways 3

North America has an extensive **river system**, or network of major rivers and their tributaries. The longest rivers are the Mississippi and Missouri rivers in the United States and the Mackenzie River in Canada. Find these rivers on the Unit Atlas map on page 54. When snow melts and rain falls, the water runs down into creeks that collect more water, becoming rivers. North America's rivers empty into bays, oceans, seas, gulfs, lakes, and other rivers.

Vocabulary

gulf: large area of a sea or ocean partially enclosed by land

SECTION 1 ASSESSMENT

Terms & Names

1. **Explain the significance of:** (a) Sacagawea (b) landform (c) glacier (d) erosion (e) river system

Using Graphics

2. Use a Venn diagram like the one shown, to compare two geographic regions of North America.

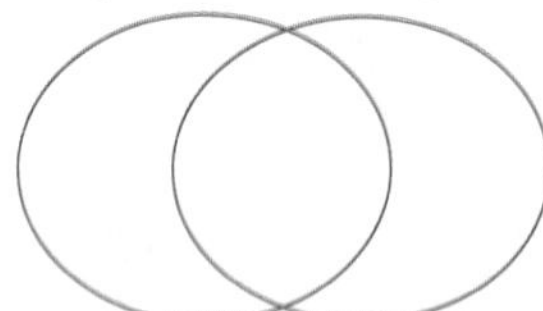

Main Ideas

3. (a) Describe North America's location on Earth and tell how this affects its climate and plant and animal life.
(b) What barriers prevented plants, animals, and people from reaching North America? How were the barriers overcome?
(c) What processes of nature help to shape the land?

Critical Thinking

4. **Making Inferences**
What natural features of North America attracted people from other lands? Support your conclusion with details from the text.

Think About
- the natural resources
- the climate

ACTIVITY -OPTION- **List** two regions of North America. Describe in a few words the kinds of plants, animals, and jobs that are found in each.

Section 1 Assessment

1. Terms & Names
- **a.** Sacagawea, p. 71
- **b.** landform, p. 73
- **c.** glacier, p. 73
- **d.** erosion, p. 74
- **e.** river system, p. 74

2. Using Graphics

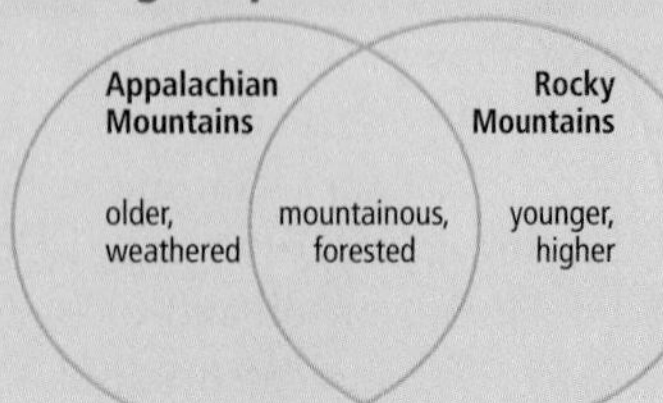

3. Main Ideas
- **a.** Most of North America is in the middle latitudes, giving it a moderate climate and allowing a variety of plants and animals to live there.
- **b.** Seas and oceans were barriers. Better ships and navigation helped overcome barriers.
- **c.** Wind, rain, rivers, glaciers, volcanoes, and earthquakes help shape the land.

4. Critical Thinking

The continent had thick forests, a large amount of rich soil suitable for farming, a moderate climate, an abundance of plant and animal life, minerals, and waterways.

ACTIVITY OPTION

Integrated Assessment
- Rubric for writing a description

A Rich Diversity in Climate and Resources

TERMS & NAMES
weather
precipitation
climate
vegetation
economy

MAIN IDEA

A region's climate, vegetation, and natural resources are contributing factors to economic activities.

WHY IT MATTERS NOW

The prosperity of people living in the United States and Canada affects the prosperity of the modern global economy.

DATELINE

ST. LOUIS, MISSOURI, U.S.A., OCTOBER 7, 1993—The worst flood in United States history is finally over. Heavy, almost continuous rain fell on much of the upper Midwest. In St. Louis, where the Mississippi and Missouri rivers meet, the flooding began in April and continued for six months.

The disaster affected nine states. Tens of thousands of people had to leave their homes. More than 10,000 homes were destroyed, and 50 people lost their lives.

Human-Environment Interaction • At least 75 towns are under water as a result of the flooding. ▲

Climate and Vegetation 1

The flood of 1993 was caused by unusual **weather.** Weather is the state of the atmosphere near Earth at a given time and place. It includes temperature, wind, and **precipitation,** or moisture such as rain or snow that falls to Earth. **Climate** is the typical weather in a region over a long period of time. A region's climate helps determine what types of **vegetation**—trees, shrubs, grasses, and other plants—will grow there.

TAKING NOTES
Use your chart to take notes about the United States and Canada.

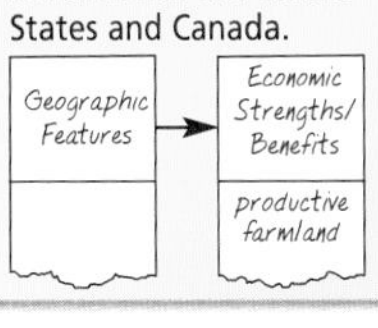

SECTION OBJECTIVES

1. To define weather, climate, and vegetation
2. To identify the vegetation zones in North America
3. To explain how the natural wealth of the United States and Canada positions them as leaders in the global economy

CRITICAL THINKING
- Hypothesizing, p. 76

FOCUS & MOTIVATE

WARM-UP

Finding Causes Have students read Dateline and discuss questions about the 1993 flood.

1. What do you know about large rivers that suggests that smaller rivers and streams in the area were also flooding?
2. By studying the photograph, how high do you think the water was in this town?

INSTRUCT: Objective 1

Climate and Vegetation

- What is the difference between weather and climate? Weather is the state of the atmosphere near Earth at a given time and place; climate is the typical weather in a region over a long period of time.
- What determines the type of vegetation that can grow in a region? climate

 **In-depth Resources: Unit 2**
- Guided Reading Worksheet, p. 4

 **Reading Study Guide**
(Spanish and English), pp. 18–19

Program Resources

 **In-depth Resources: Unit 2**
- Guided Reading Worksheet, p. 4
- Reteaching Activity, p. 10

 **Reading Study Guide**
(Spanish and English), pp. 18–19

 **Formal Assessment**
- Section Quiz, p. 34

 **Integrated Assessment**
- Rubric for drawing a picture

 **Outline Map Activities**

 **Access for Students Acquiring English**
- Guided Reading Worksheet, p. 14

 **Technology Resources**
classzone.com

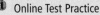

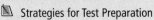

TEST-TAKING RESOURCES
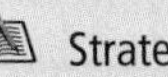 Strategies for Test Preparation
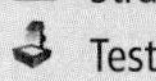 Test Practice Transparencies
 Online Test Practice

INSTRUCT: Objective 2

Vegetation Zones

- What vegetation zones are found in North America? polar and tundra, forest, rain forest, grassland, desert
- In which vegetation zone does the temperature rise above freezing for only two months a year? polar and tundra
- Where is North America's rain forest vegetation zone located? along the Pacific Coast

Connections to History

Established by the federal government in 1935, the Soil Conservation Service taught farmers in the area various methods to protect the soil and slow the erosion process that caused dust storms. More than 18,500 miles of trees were planted to break the winds that blew across the plains. However, after the crisis passed, many farmers stopped practicing these protective measures, and dust storms again damaged the region in the 1950s and the 1970s.

CRITICAL THINKING ACTIVITY

Hypothesizing Remind students that farmers in the southern Great Plains destroyed native grasses by overgrazing and planting crops. Point out that it was only after those changes took place that the dust storms began. Ask students to hypothesize as to how the native grasses might have differed from the crops that farmers planted, and why major dust storms did not occur before farmers changed the vegetation there.

Class Time 10 minutes

Connections to History

The Dust Bowl Disaster From 1931 to 1939, the southern Great Plains suffered one of the worst droughts in U.S. history. In the 1920s, farmers used up the soil's natural nutrients, and cattle and sheep ranchers destroyed native grasses by overgrazing. When the drought began, crops died, and there were no plant root systems to hold the soil.

The southern Great Plains became known as the Dust Bowl. By 1934, dust storms of topsoil were causing serious damage in 27 states. Ships 300 miles off the Atlantic Coast were covered by blowing dirt. Thousands of people, like this boy, had to leave their farms.

Vegetation Zones 2

North America's vegetation zones are determined by the climate and physical geography of each area. It is usually warmer in the south and colder in the north, but physical features such as mountains and oceans also affect the climate. Find these vegetation zones on the map on page 77.

Polar and Tundra Northern Canada and Alaska have cool summers and very cold winters. It is usually above freezing (32°F/0°C) there for only two months each year. Precipitation varies from 4 to 20 inches a year. Much of the ground is frozen all year except for the surface, which thaws in summer.

Forest Forests of conifer (evergreen) and broadleaf trees cover much of Canada and the northwest, northeast, and southeast United States. Precipitation averages between 10 and 80 inches annually. Temperatures vary from mild to cold in different forested areas.

Reading Social Studies
A. Possible Answer Climate and geography help determine economic activities in an area, which may attract people.

Rain Forest Along the Pacific Coast, precipitation can reach 167 inches each year. Rain forests with trees 300 feet tall grow in these areas. The ground is covered in bushes, small trees, and other plants. Moss and lichen are the smallest vegetation. One acre of rain forest might have 6,000 pounds of these tiny plants. The temperature is moderate even in the north, seldom falling below 32°F in winter.

Vocabulary
lichen (LY•kuhn): organism that grows with algae on rocks or tree trunks

Grassland The center of North America is covered by grasslands. The prairie in the Mississippi Valley may get 30 inches or more of precipitation each year. Grasses are tall and thick. Farther west, the land gets less rainfall—as little as 15 inches in Alberta, Canada—and the grass is shorter. People grow grain and raise cattle in these areas.

Reading Social Studies
A. Recognizing Effects How do climate and geography influence the attraction of people to an area?

Desert The deserts of the American Southwest get less than 10 inches of precipitation a year. Plants in the deserts must be able to endure harsh sun, high temperatures, and little rain. Only the hardiest bushes, shrubs, a few small trees, and cacti survive there.

Activity Options

Differentiating Instruction: Gifted and Talented

B Block Scheduling

Categorizing Challenge advanced students to create a board game based on information about the vegetation zones of North America. Explain that the purpose of the games is to reinforce players' knowledge and understanding of the characteristics of each vegetation zone. For example, a game might focus on trivia about each zone, with playing pieces advancing for each correct answer; or it might represent a journey through one or more zones, with challenges and obstacles along the way. ("A snowstorm traps you in the Rockies. Lose one turn.")

Suggest to students that they use the map on page 77 in devising their games. Each student should draw the board for his or her game, make or describe the playing pieces, and write a brief book of rules.

Vegetation Zones of Canada and the United States

INTERPRETING MAPS

Vegetation Zones of North America

Emphasize to students the importance of understanding the map key in reading a map like this one. Review the vegetation zones one by one, asking students to, first, locate the area or areas of the continent in which each zone is found and, then, supply descriptions of the kinds of vegetation there. Point out that the rain forest of the Pacific Coast is a temperate rain forest, not a tropical rain forest. Remind students that these zones are based on natural vegetation and that in many places people have introduced new types of plants. Ask students to look more closely at the map and answer the following questions:

- Since climate determines vegetation, in what zones would you find the most extreme climates? tundra and ice cap (northern Canada, Alaska); tropical rain forest (Hawaii)
- Leaving out Alaska and Hawaii, what is the latitude of most of the United States? between about 30° N and 50° N
- What vegetation zones are around the Great Lakes? coniferous forest, deciduous and mixed forest, and a small area of temperate grassland

Activity Options

Interdisciplinary Links: Geography/Art

Block Scheduling

Class Time One class period

Task Designing and making a vegetation poster

Purpose To identify and illustrate native vegetation of students' own environment

Supplies Needed
- Poster paper or board
- Poster paint or colored markers
- Information on local plants and trees

Activity Help the class identify the vegetation zone in which their community is located. Then have students work in small groups to design and illustrate posters showing examples of the typical vegetation (trees, shrubs, grasses; needles, cones, leaves) of this zone. They can use drawings or photographs as illustrations. Provide space for students to display their finished posters.

INSTRUCT: Objective 3

Natural Wealth/Neighbors and Leaders

- What power resources do the United States and Canada have? oil fields, coal, rivers
- Why did the early settlers tend to live near rivers? to be near fresh water and good soil for farming and raising cattle
- How do Canada and the United States cooperate? trade, tourism, defense, environmental regulation, air traffic, fishing regulations

Spotlight on CULTURE

Food is another important and unique ingredient of Cajun culture. Once the Cajuns arrived in Louisiana, their cuisine mixed with the Creole styles of New Orleans. Today, people all over the United States enjoy spicy Cajun specialties such as blackened catfish, crayfish étouffée, and red beans and rice. Rich seafood stews like jambalaya and gumbo also reflect the tradition. Cajun cooking features pork, oysters, crayfish, crab, and even alligator!

Natural Wealth 3

The United States and Canada are rich in natural resources. This wealth has influenced their economic development.

Land and Power Resources The farmlands of the midwestern United States and the prairies in the central provinces of Canada have rich soil. Forests are found in western Canada and the northwestern, northeastern, and southeastern United States. There are oil fields in Alberta, Canada; in Texas, California, Louisiana, Oklahoma, and Alaska; and in the Gulf of Mexico. Coal is in Canada's western provinces; in the Appalachian Mountains; and in Illinois, Indiana, and Wyoming.

Region • The Mississippi River has been an important commercial shipping route for more than a century. ▼

Water Resources Water routes affect where people and industry are located. Settlers in North America followed rivers to areas where fresh water and good soil permitted farming and raising cattle. Businesses grew in new communities. People still use rivers to ship natural resources, such as timber and coal, and as trade and travel routes. Fishing is a food source and an industry. Rivers and lakes supply water and power, and offer recreational activities.

Thinking Critically Answers
1. It forced most of them to move away, mainly to French-held territory.
2. a distinct language, a unique style of music

BACKGROUND

The Mississippi River is the largest river in North America. From its source, Lake Itasca in Minnesota, to its mouth in the Gulf of Mexico, it flows for 2,350 miles. The name *Mississippi* is from the Native American Algonquian language and means "Big River."

The Cajuns: Americans with Canadian Roots The French settle in Acadia, which is now Nova Scotia, Canada, in 1604. The British gained control of much of Nova Scotia, and in 1755, they expelled most French Acadians. Many of the displaced settlers relocated to southern Louisiana, which was under the rule of France at that time

Known as Cajuns, the descendants of those French Canadians share a special cultural heritage. Their language has French, English Spanish, German, and Native American influences. Their unique music is played with fiddles, accordions, and guitars. The man show at left plays the accordion at a Cajun music festival in Louisiana.

THINKING CRITICALLY 

1. **Recognizing Effects** What effect did Britain's rule over Nova Scotia have on the French in Acadia?
2. **Recognizing Important Details** What are some features of the Cajun culture?

For more on the Cajuns, go to 

Activity Options

Interdisciplinary Link: Science

Class Time One class period

Task Writing a report on climate change

Purpose To research climate change based on human activity

Supplies Needed
- Reference materials, including the Internet
- Writing tools

Block Scheduling

Activity Explain that human activity can cause both short- and long-term changes in climate. Have students write reports on how activities such as the building of cities, the clearing of forests, and the burning of fossil fuels can change how much solar radiation reaches Earth's surface, thus affecting climate. Allow time for students to share their research with the class.

Neighbors and Leaders 3

More than 200 million people cross the U.S.-Canadian border every year. Trade between the two countries exceeds $1 billion a day. They cooperate on issues as diverse as national security and defense, the environment, air traffic, and fishing regulations. A United States president described the relationship between these countries.

Reading Social Studies
B. Possible Answer
trade, economics, shared history

> *A VOICE FROM THE UNITED STATES*
>
> Geography has made us neighbors, history has made us friends, economics has made us partners, and necessity has made us allies.
>
> *John F. Kennedy*

Reading Social Studies

B. Finding Causes What common interests make the United States and Canada allies and partners?

Both countries have strong economies and are leaders in world trade. An **economy** is the way that business owners use resources to provide the goods and services that people want.

Citizenship IN ACTION

The Nature Conservancy The Conservancy works with communities to protect natural areas, plants, and animals. It has safeguarded 12 million acres and 1,400 land preserves, such as this one, in the United States. In its Great Lakes Program, and in Minnesota, the Conservancy is working with Canadian and U.S. groups to protect wildlife and 10,000 acres of the last tallgrass prairie on the U.S.-Canadian border.

SECTION 2 ASSESSMENT

Terms & Names

1. **Explain the significance of:** (a) weather (b) precipitation (c) climate (d) vegetation (e) economy

Using Graphics

2. Make a chart such as this one to list details about vegetation zones of the United States and Canada.

Polar and Tundra	Forest	Rain Forest	Grassland	Desert

Main Ideas

3. (a) How do climate and geography affect vegetation?
   (b) What natural resources are found in North America?
   (c) How have waterways affected settlement and development in the United States and Canada?

Critical Thinking

4. **Drawing Conclusions**
   How does the variety of vegetation zones affect the economies of the United States and Canada?

Think About
- crops and resources found in each vegetation zone
- world economy

ACTIVITY -OPTION- Choose one of the vegetation zones discussed in the section. Draw a **picture** showing what the land looks like.

Citizenship IN ACTION

The Nature Conservancy is an international conservation organization. It has helped preserve millions of acres in Latin America alone, including tropical habitats. In the United States, its Pine Butte Swamp Preserve in western Montana is the site of the first discovery of a nest of dinosaur eggs containing petrified baby dinosaurs.

Incorporated in 1951, the organization has a membership of more than 1 million people. Its journal, *Nature Conservancy,* is published six times a year.

ASSESS & RETEACH

Reading Social Studies Have students complete the graphic organizer on page 68.

Formal Assessment
- Section Quiz, p. 34

RETEACHING ACTIVITY

Have students work in pairs to create a section review. Instruct partners to work together to create a list of important concepts and vocabulary from the section. Then have students divide the list and write a definition or explanation for each item. Have partners come back together to share their definitions.

In-depth Resources: Unit 2
- Reteaching Activity, p. 10

Access for Students Acquiring English
- Reteaching Activity, p. 18

Section 2 Assessment

1. Terms & Names

a. weather, p. 75
b. precipitation, p. 75
c. climate, p. 75
d. vegetation, p. 75
e. economy, p. 79

2. Using Graphics

Polar and Tundra	Forest	Rain forest	Grassland	Desert
4–20 in. precipitation/year ground frozen all year	conifer and broadleaf trees temperatures from mild to cold	warm, wet up to 167 in. precipitation/year	15–30 in. precipitation/year grass-covered	high temperatures fewer than 10 in. precipitation/year

3. Main Ideas

a. Factors of climate and geography, such as temperature, rainfall, and landforms, determine what kinds of vegetation grow in a region.
b. Soil, trees, water, oil, and coal are natural resources found in North America.
c. People settled along rivers. Waterways are routes for trade and travel, supply fresh water and fish, and can be used to provide power and for recreation.

4. Critical Thinking

Both countries use the varied resources of the different zones to provide goods and services that are marketable in the global economy.

ACTIVITY OPTION

Integrated Assessment
- Rubric for drawing a picture

SKILLBUILDER

Reading a Physical Map

Defining the Skill

Ask students to suggest situations in which people might need to consult a physical map. Specifically, ask them who would find it necessary to know land elevations in an area; who would need to know about the location of mountains; who would need detailed information about rivers, lakes, and oceans; and who might need to know about national borders. Ask them to consider what information might appear on a physical map of their state.

Applying the Skill

How to Read a Physical Map Point out the three strategies for reading a physical map and guide students in working through each one. For example, ask students to locate on the map at least one of each type of physical feature represented. Then have volunteers locate an example of each elevation. Ask students to use the scale to work through the steps for calculating the distance between two landforms, latitudes, or political boundaries.

Make a Chart

Point out to students that they will find it helpful to use a chart to organize the information they learned from the map.

Practicing the Skill

Have students turn to the map on page 54 and notice the similar features. If students need additional practice with this skill, you might have them study a physical map of your state and make a chart that organizes the information.

In-depth Resources: Unit 2
- Skillbuilder Practice, p. 7

SKILLBUILDER

Reading a Physical Map

▶▶ Defining the Skill

Physical maps show the natural features of Earth's surface. These include landforms such as mountains, hills, and plains; and bodies of water such as rivers, lakes, bays, and oceans. Land elevation may be shown in a map key. National boundaries and major cities may also be included.

▶▶ Applying the Skill

The physical map below shows the natural features of Canada. Use these strategies to identify the information shown on the map.

How to Read a Physical Map

Strategy ❶ Read the title. It tells you which region's physical features are being represented.

Strategy ❷ Read the key. It tells you the elevation of land each color represents. A map key may also show boundaries between nations, national capitals, and other cities.

Strategy ❸ Read the scale. It tells you how many miles or kilometers each inch on the map represents.

Make a Chart

A chart can help you organize information given on maps. The chart below organizes information from the map on this page.

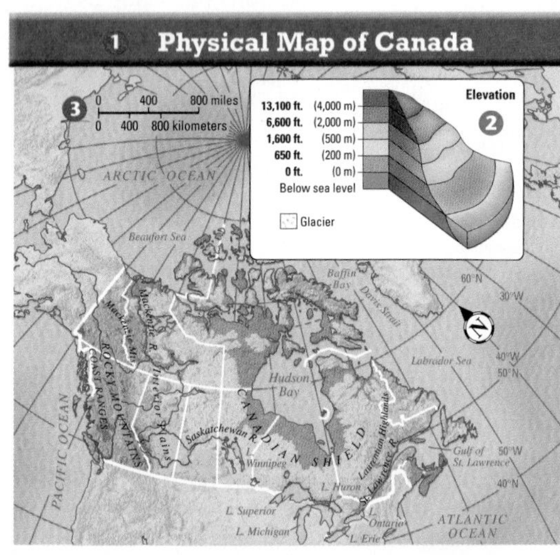

Canada	
Bodies of Water	Atlantic Ocean, Pacific Ocean, Arctic Ocean, Hudson Bay, Davis Strait, Baffin Bay, Labrador Sea, Beaufort Sea, Mackenzie River, Lake Winnipeg, Saskatchewan River, Lake Superior, Lake Huron, Lake Erie, Lake Ontario, St. Lawrence River, Gulf of St. Lawrence
Landforms	Rocky Mountains, Coast Ranges, Mackenzie Mountains, Canadian Shield, Interior Plains, Laurentian Highlands

▶▶ Practicing the Skill

Turn to page 54 in the Unit 2 Atlas. Study the physical map of North America. Make a chart listing major physical landforms of North America that you see on the map. You may also include a section in your chart labeled "Countries."

Career Connection: Geologist

Encourage students who enjoy reading physical maps to find out about careers that use this skill. For example, geologists are scientists who study the composition of Earth. Geologists may both use and create physical maps in an effort to learn about Earth's past and present and to make predictions about its future.

Block Scheduling

1. Suggest that students investigate how much field, laboratory, and office work is involved in geology and how those areas are balanced.
2. Help students learn about the aptitudes and training needed to become a geologist.
3. Have students contribute their findings to a class booklet.

Technology: 1971

TIVE

IMAX

Although the IMAX system was first successfully demonstrated during Expo '70 (a world's fair) in Osaka, Japan, it has its roots in Expo '67 in Montreal, Canada. There people flocked to special motion-picture theaters, where they stood in the middle of circular auditoriums to watch images projected on the surrounding walls. What they saw was incredible. Hockey players seemed to skate right through them, and cars seemed to drive over them!

The first permanent IMAX theater was built in Toronto in 1971. Two years later, the first IMAX Dome (also called OMNIMAX) was built in San Diego. Since then, IMAX has developed three-dimensional and high-definition systems and incorporated the technology into amusement-park rides. Because of the size of its frames, an IMAX movie generally has a shorter running time than a standard movie. However, IMAX movies require more film. A typical 45-minute IMAX feature requires nearly 16,000 feet of film!

1 Dome Early IMAX theaters had flat screens up to eight stories high, but most are now domes up to 100 feet in diameter. When large images are projected, they extend beyond normal peripheral vision—what a person can see without moving his or her eyes—and create the sensation of being in the middle of the action.

2 Film IMAX movies use what is called the 15/70 format, in which the film has 15 perforations—little holes along the side—per frame and is 70 millimeters wide. This makes each frame ten times larger than conventional 35-millimeter film and accounts for the exceptional size and sharpness of IMAX images.

3 Projector Because of its size, IMAX film is fed through the projector horizontally in a wavelike motion, or "rolling loop" movement. The newest IMAX systems project film at 48 frames per second, twice the rate of standard movie film, which results in even sharper images.

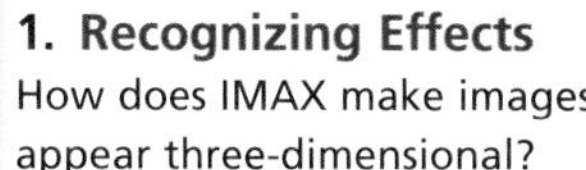

1. Recognizing Effects How does IMAX make images appear three-dimensional?

2. Drawing Conclusions Despite the system's technical advantages, IMAX theaters have not been as financially successful as standard movie theaters. Why do you think this is so?

OBJECTIVES

1. To explain how IMAX works
2. To explain the differences between large format movies and standard movie formats

INSTRUCT

- Where did the idea for IMAX begin?
- Why are IMAX images larger and sharper than standard film images?

Block Scheduling

MORE ABOUT...

IMAX Cameras

IMAX film is so large that the IMAX camera can hold only three minutes of film at a time. After shooting a scene for three minutes, the camera has to be reloaded, which takes about half an hour. A loaded IMAX camera weighs about 228 pounds! Yet it is very versatile. It can endure extreme heat, has been strapped to the tail of a jet, and has performed perfectly more than two miles beneath the ocean.

Connect to History

Strategic Thinking Why were the first movies called magic lantern shows?

Connect to Today

Strategic Thinking How has the invention of motion pictures changed society?

Thinking Critically

1. **Recognizing Effects Possible Response** The screens are so big that they extend beyond normal peripheral vision, making the images appear to pop out in front of you. This gives the sensation of being in the middle of the action.
2. **Drawing Conclusions Possible Response** It is difficult to produce traditional feature films using large format technology. Most IMAX films are educational in nature. Expensive IMAX technology results in higher ticket prices.

ASSESSMENT

TERMS & NAMES

1. Sacagawea, p. 71
2. landform, p. 73
3. glacier, p. 73
4. erosion, p. 74
5. river system, p. 74
6. weather, p. 75
7. precipitation, p. 75
8. climate, p. 75
9. vegetation, p. 75
10. economy, p. 79

REVIEW QUESTIONS

Possible Responses

1. The continent of North America is located north of the equator, in the middle latitudes.
2. At first, oceans isolated North America; later, they became travel passages.
3. farming and ranching: Atlantic Coastal Plain, Central Lowlands, Great Plains; mining: Canadian Shield; lumber: Rocky Mountains, Coastal Ranges, Canadian Shield
4. These include erosion by wind and water, glaciation, and movement of tectonic plates.
5. There are different vegetation zones because varying climate conditions, such as temperature and rainfall, allow different types of vegetation to grow.
6. grasslands
7. The variety of natural resources—good soil, forests, minerals, water—allows the development of strong economies and a high standard of living.
8. Waterways were important in bringing new settlers and establishing cities; today they provide routes for trade and travel, waterpower, fresh water, fishing, and recreation.

CHAPTER 3 ASSESSMENT

TERMS & NAMES

Explain the significance of each of the following:

1. Sacagawea
2. landform
3. glacier
4. erosion
5. river system
6. weather
7. precipitation
8. climate
9. vegetation
10. economy

REVIEW QUESTIONS

From Coast to Coast *(pages 69–74)*

1. Describe the location of North America.
2. What effect did the oceans have on the settlement of North America?
3. What regions are best for farming and ranching? for mining? for lumber?
4. What natural processes have changed the geography of North America?

A Rich Diversity in Climate and Resources *(pages 75–79)*

5. Why are there different vegetation zones?
6. Which vegetation zones permit farming and ranching?
7. How do natural resources affect the economies of the United States and Canada?
8. How have waterways helped people in the past, and how do they help people today?

CRITICAL THINKING

Synthesizing

1. Using your completed chart from Reading Social Studies p. 68, list the geographic features that contribute to productive farmland and agricultural exports.

Drawing Conclusions

2. Choose one of the regions of the United States and Canada. Think about what you know about the region's natural resources and economics. What do you think are the economic advantages and disadvantages of the region?

Making Inferences

3. What industries that depend on natural resources might flourish in the United States and Canada? Why?

Visual Summary

1 From Coast to Coast

- North America is located in the middle latitudes and is surrounded on most sides by oceans.
- Various regions, each with its own characteristics, make up the United States and Canada.

2 A Rich Diversity in Climate and Resources

- The vegetation of North America is affected by both climate and geography.
- Natural resources contribute to the prosperous economies of the United States and Canada.

CRITICAL THINKING: Possible Responses

1. Synthesizing

These include rich soil, temperate climate, adequate rainfall, and a good growing season.

2. Drawing Conclusions

Possible Responses The Appalachian Mountains have large coal deposits and extensive forests. The coal can provide energy for industry, and the forests can provide lumber and wood products. Transportation might be difficult in the mountains.

3. Making Inferences

Students may mention wood products, fuel, minerals, food products, and clothing; because the natural resources are plentiful

STANDARDS-BASED ASSESSMENT

se the map and your knowledge of world cultures and eography to answer questions 1 and 2.

dditional Test Practice, pp. S1–S33

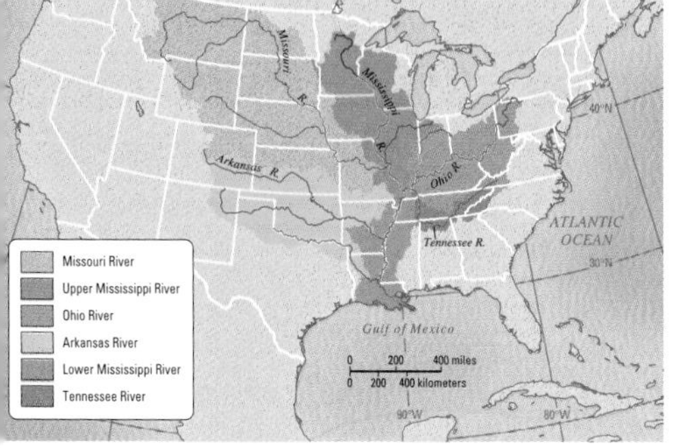

Into which body of water or bodies of water does the Mississippi River flow?

A. Atlantic Ocean
B. Great Lakes
C. Gulf of Mexico
D. Ohio River

Which of the following is a major tributary of the Mississippi River?

A. Atlantic Ocean
B. Gulf of Mexico
C. Hudson River
D. Missouri River

The following passage describes farmers' reactions to the Dust Bowl. Use the quotation and your knowledge of world cultures and geography to answer question 3.

PRIMARY SOURCE

The story of the dust bowl was the story of people—folks who had the courage and fortitude to endure the stress and strain . . . and to emerge scarred but victorious. . . . The vast majority of the residents had come when the land first opened for homesteading. . . . The pioneering days on subsistence farms provided the folks with the know-how to survive depressions and droughts.

PAUL BONNIFIELD, excerpt from *The Dust Bowl: Men, Dirt, and Depression*

3. This passage supports which of the following points of view?

A. The pioneers should not have tried so hard to farm this land.
B. The farmers were too weak to continue and would have to give up.
C. The farmers needed government help to survive the hard times.
D. The farmers' past experience enabled them to deal with the Dust Bowl.

ALTERNATIVE ASSESSMENT

WRITING ABOUT HISTORY

Jedediah Smith, John Jacob Astor, Alexander Mackenzie, and James McGill were fur traders who traveled across the United States and Canada. Write a journal entry that one of these men might have written about his experience.

- Use the Internet or other library resources to learn about the journey of one of the fur traders.
- Write your journal entry from the point of view of the fur trader. You might want to include a map illustrating an important trade route. Show and describe important landmarks along the way.

COOPERATIVE LEARNING

With a small group of classmates, prepare and present a television news report on the Lewis and Clark Expedition. One group member can be the news anchor. Other group members can take the roles of reporters, Lewis, Clark, Sacagawea, or Native Americans. As a group, make a list of interview questions. Then share the responsibility of researching to find answers to the questions.

INTEGRATED TECHNOLOGY

Doing Internet Research

The climate and vegetation differ in the several regions of North America. Choose one region and do research to learn about its climate and vegetation. Then use your research to write a report.

- Research online or print encyclopedias, atlases, and almanacs. You might also find articles in periodicals or newspapers.
- Focus your research on how climate and vegetation affect people's lives, the economy, and the history of the region.

For Internet links to support this activity, go to

STANDARDS-BASED ASSESSMENT

1. **Answer C** is the correct answer because the map shows the river flowing into the Gulf of Mexico. Answer A is incorrect because the Atlantic Ocean is east of the Mississippi; Answer B is incorrect because the Great Lakes are in the north, near the source of the Mississippi; Answer D is incorrect because the Ohio River branches off from the Mississippi.
2. **Answer D** is the correct answer because the four major tributaries of the Mississippi shown on the map are the Missouri, Ohio, Tennessee, and Arkansas rivers. Answers A, B, and C are incorrect because they name other bodies of water.
3. **Answer D** is the correct answer because it focuses on the know-how based on experience described in the passage. Answer A is incorrect because the passage speaks respectfully of farmers' hard work; answer B is incorrect because it states a view that contradicts the passage; answer C is incorrect because the passage refers to the farmers' own know-how.

INTEGRATED TECHNOLOGY

Discuss with students the keywords they might use to find information about the climate and vegetation of the regions of North America. Remind them that not all sources are trustworthy. Students should use only well-known sites, such as government databases, encyclopedias, and reliable online magazines. They should keep a record of sources used.

Alternative Assessment

1. Rubric

The journal entry should

- accurately reflect the thoughts and experiences of the fur trader selected.
- cover the fur trader's travels across the United States or Canada.
- reflect the student's understanding of the history of the fur trade.
- use correct grammar, spelling, and punctuation.

2. Cooperative Learning

Briefly discuss the style of television news reports that students have seen. Suggest that interviewers follow up yes or no answers with more probing questions, such as asking the interviewee to explain his or her answer further. Allow time for role-players to rehearse presentations.

The United States Today

	OVERVIEW	COPYMASTERS	INTEGRATED TECHNOLOGY
UNIT ATLAS AND CHAPTER RESOURCES	The students will examine the values and beliefs that have influenced the history, government, economics, and culture of the United States.	**In-depth Resources: Unit 2** • Guided Reading Worksheets, pp. 11–14 • Skillbuilder Practice, p. 17 • Unit Atlas Activities, pp. 1–2 • Geography Workshop, pp. 35–36 **Reading Study Guide** (Spanish and English), pp. 22–31 **Outline Map Activities**	• eEdition Plus Online • EasyPlanner Plus Online • eTest Plus Online • eEdition • Power Presentations • EasyPlanner • Electronic Library of Primary Sources • Test Generator • Reading Study Guide • The World's Music • Critical Thinking Transparencies CT7 • There Is No Food Like My Food
	KEY IDEAS		
SECTION 1 We the People pp. 87–91	• Immigration has played an important role in the development of culture in the United States. • U.S. citizens have a responsibility to participate in the political process.	**In-depth Resources: Unit 2** • Guided Reading Worksheet, p. 11 • Reteaching Activity, p. 19 **Reading Study Guide** (Spanish and English), pp. 22–23	**Critical Thinking Transparencies CT8** **Map Transparencies MT14** **classzone.com** Reading Study Guide
SECTION 2 A Constitutional Democracy pp. 94–98	• The Constitution is the framework of the U.S. government. • The federal government can declare war, make laws, and levy taxes. State governments control most other functions. • The U.S. government is divided into an executive branch, a legislative branch, and a judicial branch.	**In-depth Resources: Unit 2** • Guided Reading Worksheet, p. 12 • Reteaching Activity, p. 20 **Reading Study Guide** (Spanish and English), pp. 24–25	**classzone.com** Reading Study Guide
SECTION 3 The United States Economy pp. 102–107	• The four factors of production are natural resources, labor resources, capital resources, and entrepreneurs. • In a free enterprise system, citizens and businesses make economic decisions. Government plays a limited role.	**In-depth Resources: Unit 2** • Guided Reading Worksheet, p. 13 • Reteaching Activity, p. 21 **Reading Study Guide** (Spanish and English), pp. 26–27	**classzone.com** Reading Study Guide
SECTION 4 United States Culture: Crossing Borders pp. 110–113	• Values are the principles and ideals by which people live, such as individual freedoms and equal opportunity. • Globalization is the spreading of cultural influences around the world. • Modern technology allows U.S. scientists to work and share ideas with other people around the world.	**In-depth Resources: Unit 2** • Guided Reading Worksheet, p. 14 • Reteaching Activity, p. 22 **Reading Study Guide** (Spanish and English), pp. 28–29	**classzone.com** Reading Study Guide

KEY TO RESOURCES

 Audio
 CD-ROM
Copymaster
 Internet
 Overhead Transparency
Pupil's Edition
 Teacher's Edition
Video

ASSESSMENT OPTIONS

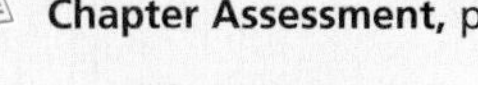

Chapter Assessment, pp. 114–115

Formal Assessment
- Chapter Tests: Forms A, B, C, pp. 51–62

Test Generator

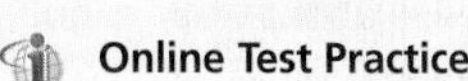

Online Test Practice

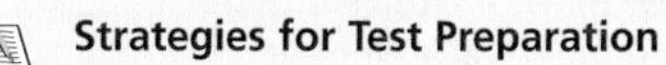

Strategies for Test Preparation

Section Assessment, p. 91

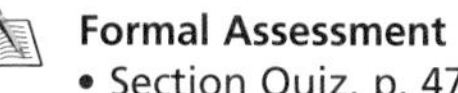

Formal Assessment
- Section Quiz, p. 47

Integrated Assessment
- Rubric for writing a magazine article

Test Generator

Test Practice Transparencies TT7

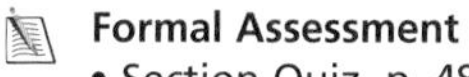

Section Assessment, p. 98

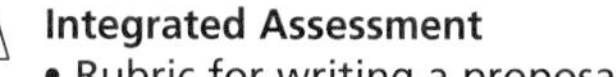

Formal Assessment
- Section Quiz, p. 48

Integrated Assessment
- Rubric for writing a proposal

Test Generator

Test Practice Transparencies TT8

Section Assessment, p. 107

Formal Assessment
- Section Quiz, p. 49

Integrated Assessment
- Rubric for writing an advertisement

Test Generator

Test Practice Transparencies TT9

Section Assessment, p. 113

Formal Assessment
- Section Quiz, p. 50

Integrated Assessment
- Rubric for writing an essay

Test Generator

Test Practice Transparencies TT10

RESOURCES FOR DIFFERENTIATING INSTRUCTION

Students Acquiring English/ESL

Reading Study Guide (Spanish and English), pp. 22–31

Access for Students Acquiring English Spanish Translations, pp. 19–28

TE Activity
- Word Meanings, p. 103

Modified Lesson Plans for English Learners

Less Proficient Readers

Reading Study Guide (Spanish and English), pp. 22–31

TE Activities
- Recognizing Important Details, p. 95
- Finding Main Ideas, p. 105
- Understanding Sequences, p. 97

Gifted and Talented Students

TE Activity
- Creating a Mural, p. 88

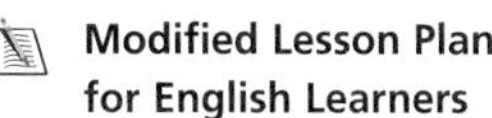

CROSS-CURRICULAR CONNECTIONS

Humanities
West, Delno C. ***Uncle Sam and Old Glory: Symbols of America.*** New York: Atheneum, 2000. Liberty Bell, the log cabin, the bald eagle, Uncle Sam, the Mayflower, the peace pipe, the Minuteman, and Smokey the Bear.

Popular Culture
Miller, Millie. ***The United States of America: A State-by-State Guide.*** New York: Scholastic Reference, 1999. Alphabetical collection of one-page articles provides basic information on each state.

Science/Math
Curlee, Lynn. ***Liberty.*** New York: Atheneum, 2000. How the statue was designed and built.

Primary Sources
Lehman, Jeffrey. ***Gale Encyclopedia of Multicultural America: Primary Documents.*** Farmington Hills, MI: Gale Group, 2000. Letters, journals, interviews, poems, songs, legislation, speeches, and photographs.

Language Arts/Literature
Knight, Margy Burns. ***Who Belongs Here? An American Story.*** Gardiner, ME: Tilbury House, 1996. Provocative story about prejudice in the United States.

Science
Duncan, Beverly. ***Explore the Wild: A Nature Search-and-Find Book.*** New York: HarperCollins, 1996. Seven North American habitats: alpine tundra, kelp forest, desert, prairie, salt marsh, swamp, and the Arctic.

Government
Krull, Kathleen. ***A Kids' Guide to the American Bill of Rights: Curfews, Censorship, and the 100-Pound Giant.*** New York: Avon, 1999. Directly related to children's issues.

Economics
Parker, Nancy Winslow. ***Money, Money, Money: The Meaning of the Art and Symbols on United States Paper Currency.*** New York: HarperCollins, 1995. History unfolds on our currency.

ENRICHMENT ACTIVITIES

The following activities are especially suitable for classes following block schedules.

Teacher's Edition, pp. 88, 90, 99, 111
Pupil's Edition, pp. 91, 98, 107, 113
Unit Atlas, pp. 54–65
Interdisciplinary Challenge, pp. 92–93
Literature Connections, pp. 100–101
Linking Past and Present, pp. 108–109
Outline Map Activities

INTEGRATED TECHNOLOGY

Go to **classzone.com** for lesson support and activities for Chapter 4.

BLOCK SCHEDULE LESSON PLAN OPTIONS: 90-MINUTE PERIOD

DAY 1

CHAPTER PREVIEW, pp. 84–85
Class Time 20 minutes

- **Small Groups** Form four small groups of students. Assign one of these topics (democracy, history, economy, or culture) to each group. Have each group use a word web to brainstorm what they already know about the topic. Reconvene as a whole class for discussion.

SECTION 1, pp. 87–91
Class Time 70 minutes

- **Outline Maps** In preparation for discussing Section 1, have students complete the physical map of the United States in Outline Map Activities. Students should label the states and physical features such as mountains and rivers. They should color the map, using different colors for different landforms.
Class Time 35 minutes
- **Press Conference** Have a group of four students field questions about the landforms and resources of the United States. The rest of the class should prepare questions to ask the group of four either in class or as homework.
Class Time 35 minutes

DAY 2

SECTION 2, pp. 94–98
Class Time 45 minutes

- **Speech** Use the Multiple Learning Styles: Linguistic to help students understand the content and purpose of the Bill of Rights by presenting speeches in support of the first eight amendments.

SECTION 3, pp. 102–107
Class Time 45 minutes

- **Seeking Elective Office** Ask students to work in pairs. Have the pairs select one branch of government (executive, legislative, or judicial) and develop a platform for running for office within that branch. Have students present their platforms to the whole class.

DAY 3

SECTION 4, pp. 110–113
Class Time 35 minutes

- **Hypothesizing** Have students work in small groups and hypothesize about what might happen to the United States if most Americans did not exercise their right to vote. Reconvene as a whole class for discussion.

CHAPTER 4 REVIEW AND ASSESSMENT, pp. 114–115
Class Time 55 minutes

- **Review** Have students prepare a summary of the chapter, using the Terms & Names listed on the first page of each section.
Class Time 20 minutes
- **Assessment** Have students complete the Chapter 4 Assessment.
Class Time 35 minutes

TECHNOLOGY IN THE CLASSROOM

INTERNET RESEARCH, SEARCHING THE WEB, AND DIGITAL CAMERAS

The Web lends itself to student research because of the availability of sites with valuable, up-to-date information, including text and pictures. By searching child-friendly Web sites, students can find information appropriate for their grade level and relevant to the subjects they are studying. In the process, they gain experience in focusing their searches and sifting through search results to find what they are looking for.

ACTIVITY OUTLINE

Digital cameras are becoming increasingly popular. Students can take pictures of people, buildings, scenery, or other subjects and upload the pictures directly onto a computer for storage and sharing. Students will commonly incorporate these photographs into multimedia presentations or Web sites. The use of digital cameras is optional in this activity.

Objective Students will search the Web for examples of activities related to democracy and Constitutional rights. They will write paragraphs on the computer describing how they might participate in one of these activities.

Task Have students conduct an Internet search to find some examples of how Americans practice democracy and exercise their Constitutional rights. Suggest they write paragraphs explaining how they would participate in one of these activities. As an option, ask them to take digital photographs of people in their school participating in such activities.

Class Time Two class periods (not including the optional digital camera activity)

DIRECTIONS

1. Review the topics covered in Section 2 of Chapter 4 ("A Constitutional Democracy"), and ask students to list examples of how people in the United States practice democracy and exercise their Constitutional rights (e.g., free speech and freedom of religion). For example, they might list voting, volunteering for environmental groups, writing letters to politicians, and calling radio talk shows. Discuss the lists as a class. Ask if any students have ever taken part in these activities. Have their parents or other people they know?
2. Suggest that students, either individually or in small groups, search the directory listed at **classzone.com** to find Web sites providing examples of children or adults practicing democracy- and rights-related activities. Three good keywords to use are *volunteering, democracy,* and *news* (the latter term will allow them to browse through news sources such as magazines and newspapers).
3. Ask students to take notes on the examples they find. Each student or group should find at least four examples of Americans involved in activities related to democracy and Constitutional rights. Discuss students' findings as a class.
4. Have students choose an example of democratic participation or another rights-related activity and write a paragraph explaining how they could get involved in a similar activity. Ask them to consider what they would like to do and what their goals would be.
5. (optional step) If you have access to digital cameras, ask groups of students to photograph people at the school who are participating in democratic activities or exercising their Constitutional rights. They might photograph other students campaigning or voting in a school election, or they could take pictures of a guest speaker. Have students upload their photographs onto a computer and create a multimedia presentation that includes text descriptions of the activities they observed.

CHAPTER 4 OBJECTIVE

Students will examine the values and beliefs that have influenced the history, government, economics, and culture of the United States.

FOCUS ON VISUALS

Interpreting the Photograph Direct students' attention to the photograph of the New York harbor and identify Ellis Island. Point out that immigrants brought to this country their own ideas about government, economics, traditions, and cultural values. Ask students to look at the photograph as they think about people who immigrate to the United States, what they hope to find, and how they contribute to the development of the country.

Possible Responses People from all over the world come to the United States, often in search of a better way of life. They bring their own ideas about what life should be like, as well as cultural traditions for music, holidays, art, food, and so on.

Extension Using an encyclopedia or the Internet, have students research Ellis Island and its importance in our history.

CRITICAL THINKING ACTIVITY

Analyzing Motives Prompt a discussion about the arrival of immigrants to the United States. Ask questions such as these: Why did they come? What skills did they have? What ideas did they contribute? What did they leave behind?

Class Time 15 minutes

CHAPTER 4

The United States Today

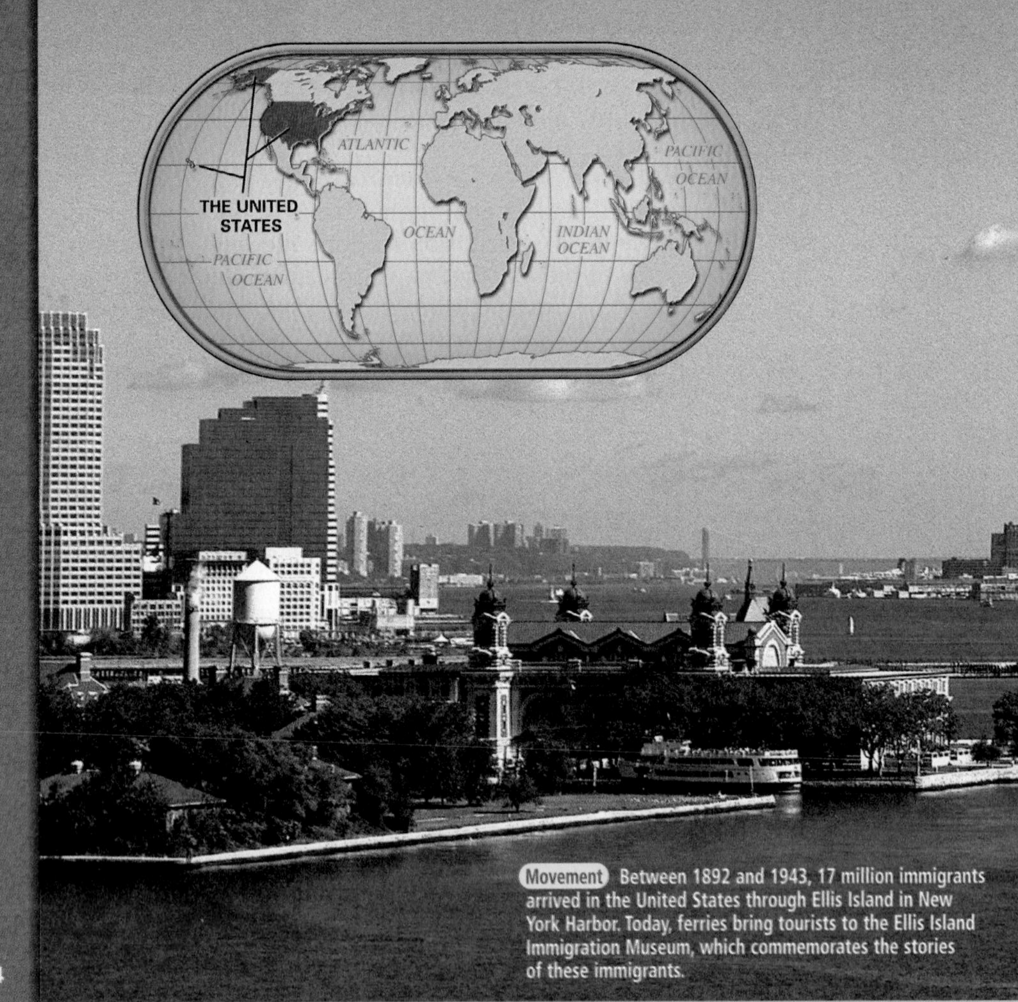

Movement Between 1892 and 1943, 17 million immigrants arrived in the United States through Ellis Island in New York Harbor. Today, ferries bring tourists to the Ellis Island Immigration Museum, which commemorates the stories of these immigrants.

Recommended Resources

BOOKS FOR THE TEACHER

Gordon, Patricia. ***Kids Learn America!: Bringing Geography to Life with People, Places, & History.*** Charlotte, VT: Williamson Publishing, 1999. Suggests projects to go along with state study.

Takaki, Ronald, T. ***A Different Mirror: A History of Multicultural America.*** New York: Little, Brown, 1994. Groundbreaking look at the cultures of America.

VIDEOS

We the People: The Story of the Constitution of the United States. Las Vegas: Distributed by Educational Distributors of America, 1997. How the government was formed and modified.

INTERNET

For more information about the United States, visit **classzone.com.**

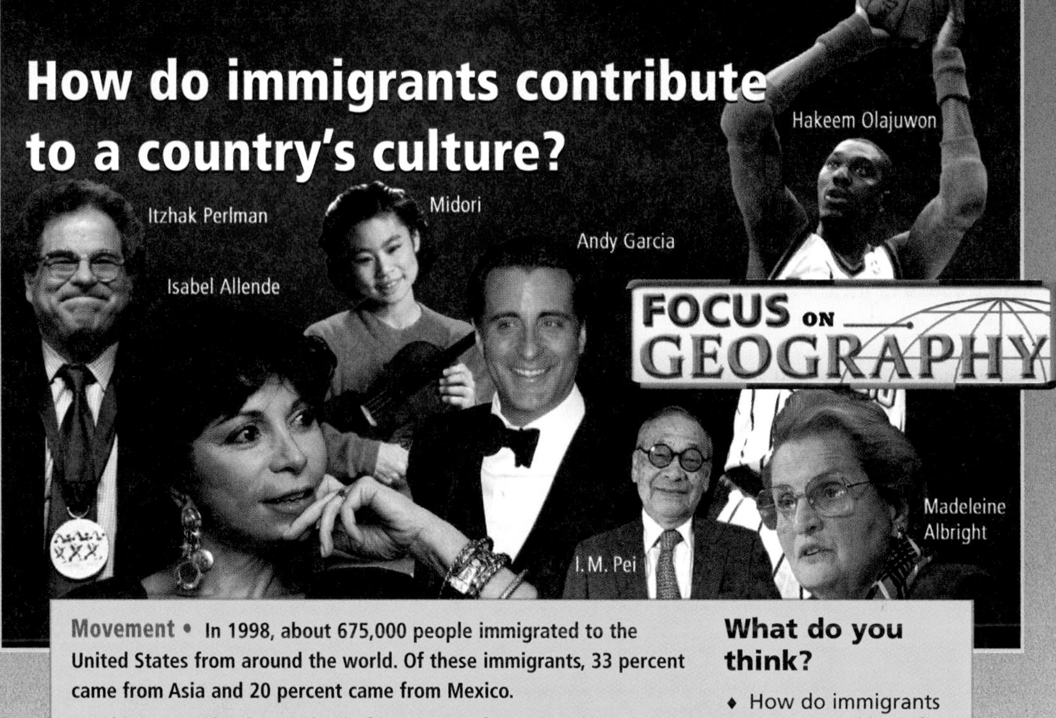

Movement • In 1998, about 675,000 people immigrated to the United States from around the world. Of these immigrants, 33 percent came from Asia and 20 percent came from Mexico.

This country has been a land of immigrants from its earliest days. People have contributed customs, art, music, languages, foods, and ideas from their homelands to enrich U.S. culture. Violinist Itzhak Perlman, writer Isabel Allende, violinist Midori, actor Andy Garcia, architect I. M. Pei, basketball player Hakeem Olajuwon, and former Secretary of State Madeleine Albright are just a few famous U.S. immigrants.

What do you think?

- How do immigrants change and benefit the United States?
- What contributions by immigrants have influenced your life?

FOCUS ON GEOGRAPHY

Objectives

- To help students identify how immigration affected the culture of the United States
- To identify and discuss the contributions made by immigrants to the culture of the United States

What Do You Think?

1. Guide students to consider what ideas, talents, and customs immigrants bring with them to another country.
2. Encourage students to think about medical advances, music, fashion, and other aspects of their lives that are derived from other cultures.

How do immigrants contribute to a country's culture?

Challenge students to add to the list of immigrants who have enriched life in the United States. Ask them to name other musicians, sports figures, actors, and writers they know of who have brought their talents to the United States. What about scientists, chefs, and business owners? As students respond, you may want to record names, contributions, and countries of origin on a chart.

MAKING GEOGRAPHIC CONNECTIONS

Ask students which countries their parents, grandparents, and great-grandparents came from. Then place pushpins with label flags on a world map to show all the countries that students' ancestors came from.

Implementing the National Geography Standards

Standard 9 Explain how the movement of people can alter the character of a place

Objective To make a collage illustrating immigrants' contributions to the character of the United States

Class Time 40 minutes

Task Have students create a collage of pictures cut from magazines and newspapers showing how immigrants have affected the character of the United States. Before students start their collages, have them brainstorm cultural contributions that were brought here by immigrants. Write the students' ideas on the board.

Evaluation Students should show contributions from at least three cultural groups in their collages.

BEFORE YOU READ

What Do You Know?

Point out to students that they probably know that the United States is a democracy. Encourage a discussion about what that means by asking how democracy was formed in the United States and having students identify the benefits and obligations of a democracy. Repeat a similar exercise about the history, the economy, and the culture of the United States.

What Do You Want to Know?

Suggest that students make a list with these headings: history, economics, government, and culture. Then have them write the specific questions they hope to have answered about each. As they read they can record on the list facts they find in the text.

READ AND TAKE NOTES

Reading Strategy: Identifying Main Ideas Direct students' attention to the chart. Read aloud the topics in the first column and explain that each of these topics is the focus of a section of this chapter. Remind students to use headings and topic sentences in paragraphs to identify main ideas. Have them record main ideas in the chart. Point out that their completed charts will become valuable study resources.

In-depth Resources: Unit 2

- Guided Reading Worksheets, pp. 11–14

READING SOCIAL STUDIES

BEFORE YOU READ

▸▸ What Do You Know?

Before you read the chapter, think about how you would describe the United States to a person from another country. What do you and your classmates do for fun? What information could you give about U.S. history, government, economy, and culture? Reflect on your experience and what you have already read or learned about these aspects of the United States. How accurate and complete do you think your knowledge is?

▸▸ What Do You Want to Know?

Decide what you want to know about the characteristics of history, government, economy, and culture that make the United States unique. In your notebook, record what you hope to learn from this chapter.

Place • After years of campaigning, women won the right to vote in 1920 when the 19th Amendment to the Constitution was adopted. ▲

READ AND TAKE NOTES

Reading Strategy: Identifying Main Ideas Look for main ideas as you read to help you identify and remember important information. Record the main ideas of this chapter in the chart below.

- Copy the chart into your notebook.
- Notice that each section of Chapter 4 covers a different topic: history, government, economy, and culture.
- As you read each section, look for the main ideas. Use section titles and headings, and find the main idea of each paragraph.
- After you read each section, record what you think are the main ideas about aspects of the United States.

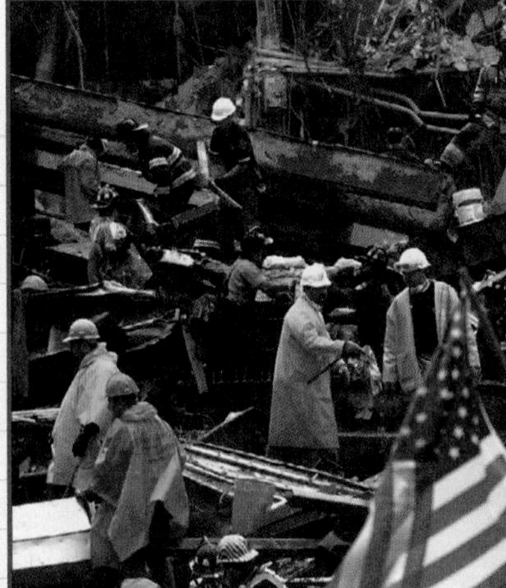

Place • Firefighters and rescue personnel rushed t the scene of the terrorist attacks on New York Cit World Trade Center, on September 11, 2001. ▲

The United States Today	
History	country of immigrants; fight for freedom
Government	constitutional democracy; limited power; three branches of government
Economy	free enterprise; market economy; competition
Culture	values; private property; individual freedom; science and technology

86 CHAPTER 4

Teaching Strategy

Reading the Chapter This is a thematic chapter that focuses on how the history, government, economy, and culture of the United States developed and how they exist today. Encourage students to use their personal knowledge as well as the visuals in each section to augment what they learn from the text.

Integrated Assessment The Chapter Assessment on page 115 describes several activities for integrated assessment. You may wish to have students work on these activities during the course of the chapter and then present them at the end.

We the People

TERMS & NAMES
immigrant
Anasazi
equal opportunity
citizenship
democracy
republic
political process
patriotism

MAIN IDEA

Citizens of the United States come from many cultures and share the same rights and responsibilities.

WHY IT MATTERS NOW

The ideas and values of U.S. immigrants have helped shape the success of the country in the world today.

DATELINE

EXTRA

NEW YORK CITY, 1849

Ships docked today, bringing more people from their Irish homeland to live in the United States. These immigrants have suffered through four years of a deadly potato famine. A blight, or disease, has almost completely destroyed the potato crop every year since 1845. The Irish, especially those who are poor, depend on this crop to survive.

The failure of the potato crop has resulted in the deaths of more than 1 million people—12% of the population of Ireland—from starvation and disease. Today's arrivals join more than 1 million Irish people who are already in the United States.

Movement • Irish immigrants wait to board a ship that will bring them to the United States. ▲

One Country, Many Cultures 1

Immigrants, such as the Irish, have brought unique contributions to the United States from their homelands all over the world. The United States is sometimes called a "melting pot," a "salad bowl," or a "patchwork quilt" to illustrate how U.S. society combines aspects of many cultures. Some features may blend into the culture of the United States, while others retain their original characteristics.

TAKING NOTES
Use your chart to take notes about the United States today.

The United States Today	
History	
Government	

SECTION OBJECTIVES

1. To recognize the role of immigration in the development of United States culture
2. To explain the contributions of Native Americans, Africans, and Chinese
3. To understand the rights and responsibilities of citizenship

SKILLBUILDER
- Interpreting a Map, p. 89

CRITICAL THINKING
- Drawing Conclusions, p. 89

FOCUS & MOTIVATE

WARM-UP

Drawing Conclusions Have students read Dateline and answer the questions.

1. What conclusions can you draw about the economy of Ireland in the mid-1800s?
2. What skills do you think these Irish immigrants brought with them?

INSTRUCT: Objective 1

One Country, Many Cultures

- Why is the term "salad bowl" used to describe the United States? Some immigrant features blend together, while others retain their original characteristics, like vegetables in a salad.
- What are some of the reasons that people immigrate? to escape discrimination, persecution, war, or natural disaster; for educational and economic opportunities

In-depth Resources: Unit 2
- Guided Reading Worksheet, p. 11

Reading Study Guide
(Spanish and English), pp. 22–23

Program Resources

In-depth Resources: Unit 2
- Guided Reading Worksheet, p. 11
- Reteaching Activity, p. 19

Reading Study Guide
(Spanish and English), pp. 22–23

Formal Assessment
- Section Quiz, p. 47

Integrated Assessment
- Rubric for writing a magazine article

Outline Map Activities

Access for Students Acquiring English
- Guided Reading Worksheet, p. 19

Technology Resources
classzone.com

TEST-TAKING RESOURCES
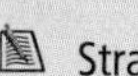 Strategies for Test Preparation
 Test Practice Transparencies
Online Test Practice

INSTRUCT: Objective 2

People from Many Lands

- What accomplishments were the Anasazi known for? irrigation systems, cliff dwellings
- What was a constant source of conflict between the Europeans and the Native Americans? competition for land
- Why were Africans forced to come to America? They were enslaved to provide cheap labor.
- Why did many Chinese immigrate to the United States? to work in mines and build the transcontinental railroad

MORE ABOUT...

The Great Seal

In July 1776 the Second Continental Congress set up a committee to design a national seal that would reflect the ideals of the new nation. Early design proposals varied, and the final design was not approved until June 1782. The eagle, which symbolizes self-reliance, supports a shield with 13 vertical white and red stripes on it. In the eagle's right talon is an olive branch; in its left is a bundle of 13 arrows. This symbolizes that while the United States prefers peace, it is capable of waging war if necessary. In the eagle's beak is a scroll inscribed with the motto *E pluribus unum.* The seal is used to authenticate important documents.

BACKGROUND

The motto *E Pluribus Unum,* Latin for "out of many, one," is on the Great Seal of the United States and many U.S. coins.

For example, settlers from Great Britain brought English, the most widely spoken language in the United States. Spanish is often spoken in the Southeast and the Southwest, where people from Spain and Mexico settled. French is heard in Louisiana, which was once held by France. People in the United States enjoy the influence of different groups on their foods, music, sports, and other areas of their lives.

Why People Immigrate An **immigrant** is someone who chooses to move to a new country. They come to the United States for different reasons. Some are escaping from discrimination, persecution, or war. Others leave their homelands because of drought, earthquake, or other natural disasters. Often, people come hoping to improve their economic or educational opportunities.

Reading Social Studies

A. Summarizing Why have people immigrated to the United States?

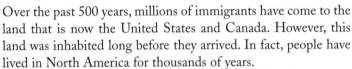

Reading Social Studies
A. Possible Answe[r]
People immigrated to the United State[s] to escape discrimination, persecution, or war; to improve their opportunities in life; or to escape problems caused by natural disasters in their homelands.

People from Many Lands 2

Over the past 500 years, millions of immigrants have come to the land that is now the United States and Canada. However, this land was inhabited long before they arrived. In fact, people have lived in North America for thousands of years.

The First Americans Native Americans were the first people to inhabit the Western Hemisphere. They came to North America from Eastern Asia, 12,000 to 35,000 years ago. Some groups, such as the Mississippians and **Anasazi** (Navajo for "Ancient Ones"), developed complex civilizations.

The Anasazi civilization developed around A.D. 100 and reached its height in the 11th to 13th centuries. The Anasazi were experts at irrigation. They built homes called cliff dwellings that had from 20 to 1,000 rooms. Remains of these structures survive in the Mesa Verde National Park in Colorado and in other places in the Southwestern United States.

Human-Environment Interaction • The remains of Anasazi cliff dwellings, such as Cliff Palace in Mesa Verde National Park, are found in the Southwestern United States. ◄

Activity Options

Differentiating Instruction: Gifted and Talented

Block Scheduling

Class Time One class period

Task Creating a mural depicting information about the Anasazis

Purpose To learn more about the first people who immigrated to the Western Hemisphere

Activity Invite students to work as a group to create a mural showing aspects of Anasazi culture. Have students divide the responsibilities of researching Anasazi agriculture, dwellings, clothing, and religious ceremonies. Then have students draw pictures on a mural and write captions to explain their illustrations.

North American Settlement, 1750

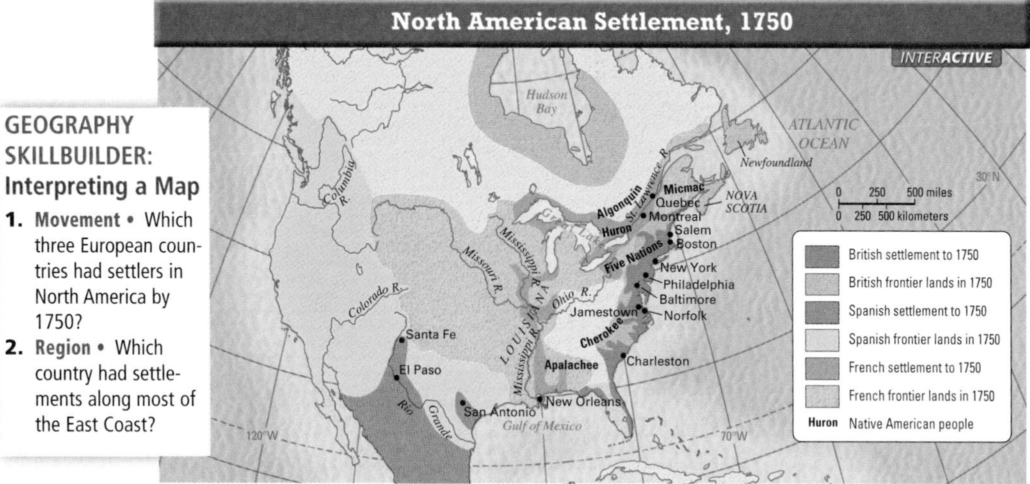

GEOGRAPHY SKILLBUILDER: Interpreting a Map

1. **Movement** • Which three European countries had settlers in North America by 1750?
2. **Region** • Which country had settlements along most of the East Coast?

Geography Skillbuilder Answers

1. Britain, Spain, France
2. Britain

The Europeans Arrive European exploration of the Americas began in the late 1400s. Colonists soon followed the explorers. The British settled along the Atlantic coast, in what is now southeastern Canada and the Northeastern United States. Spaniards settled in Florida and came north from Mexico to build towns in the Southwest. Often, the settlers' ways of life and needs for resources conflicted with those of the Native Americans. As the European population grew, competition for land intensified. Europeans often took land from Native Americans. Cultural differences and land disputes led to distrust and war.

Slaves in the Colonies European settlers began to plant and harvest crops and started businesses and towns. This created a demand for cheap labor, so Europeans forced some people to migrate to America.

They had been buying people from slave traders in Africa since the 1500s. Beginning in 1619, enslaved Africans were shipped to the American colonies under such harsh conditions that many died during the journey. Those who survived were bought and sold as property and forced to work for free all their lives. Their children were born into slavery. Although these Africans did not arrive by choice, their labor helped build the country, and their influence is seen in our culture today.

Connections to History

Indentured Servants Indentured servants were immigrants who agreed to work in the colonies for a certain number of years in exchange for passage to America. Their indentures, or contracts, could be bought and sold by employers.

Indentured servants were often forced to work long hours and were sometimes treated very badly. Many did not live long enough to gain their freedom. Most, however, settled in the colonies after completing their contracts.

Connections to History

Indentured servitude first appeared approximately ten years after the 1607 founding of Jamestown. Land was plentiful, but labor was hard to find. Those without money to travel to the New World could indenture themselves to someone willing to pay the passage. The difficult trip could take anywhere from 8 to 12 weeks. Many died along the way. At the end of the journey they faced three to six years of servitude. The very young were often indentured until their 21st birthdays.

CRITICAL THINKING ACTIVITY

Drawing Conclusions Have students reread the paragraphs about indentured servants in "Connections to History." Ask students to draw conclusions about the living conditions in Europe and the economic opportunities available there. Remind them to consider that people were willing to act as indentured servants for several years in exchange for passage to North America.

Class Time 10 minutes

Activity Options

Interdisciplinary Links: Language Arts/Writing

Class Time One class period

Task Writing a letter from the point of view of an indentured servant

Purpose To gain an understanding of the experiences of indentured servants

Supplies Needed
- Paper
- Pens or pencils

Activity Have each student write a letter as if he or she were an indentured servant corresponding with a relative back in Europe. Suggest that students describe what the journey to America was like, the tasks they do all day, and how their expectations have turned out to be different from the reality of their lives.

INSTRUCT: Objective 3

Rights of Citizens/ Responsibilities of Citizenship

- What are the duties of citizens? to pay taxes, to vote, to serve on juries, to serve in the military, to volunteer, to obey laws
- What is included in the political process? legal activities through which citizens influence decisions on public policy

Citizenship IN ACTION

Americans from many different walks of life came together to volunteer their time and skills in rescue efforts at the World Trade Center. Ironworkers, carpenters, and electricians worked to clear the rubble. Health-care workers, such as doctors and nurses, provided 24-hour emergency service. Teachers developed materials for New York City schoolchildren. Longshoremen and teamsters arranged and loaded supplies of food and water. Tobacco workers sent paper face masks for rescue workers. Aid poured in from places far from New York City such as Ohio, West Virginia, Michigan, Tennessee, Illinois, and California.

From Far and Near In the second half of the 1800s, many Chinese immigrants entered the United States. Some worked in mines, while others helped build the transcontinental railroad. In the 1880s and the 1920s, new laws limited the number of U.S. immigrants from various countries. In 1952, legislation again allowed immigrants of all nationalities to become citizens.

Reading Social Studies
B. Possible Answer
Students may say that citizens of a democracy are fortunate to have rights that may be denied to others, so they have a responsibility to take steps to protect those rights.

Rights of Citizens 3

Although the United States is among the world's leaders in protecting individual freedom, many U.S. citizens have struggled for their rights. Even after African Americans were freed from slavery in 1865 by the 13th Amendment to the Constitution, they were denied their rights. Women could not vote in the United States until 1920. Native Americans, as well as Hispanics, Asians, the Irish, and other immigrants, have fought against discrimination.

The guarantee of **equal opportunity** in education, employment, and other areas of life has expanded over the years. Today, it is illegal for the government or private institutions to discriminate because of race, gender, religion, age, or disability.

Reading **Social Studies**

B. Making Inferences Do you think citizens of a democracy have greater responsibilities because they have more rights? Explain.

Responsibilities of Citizenship 3

U.S. citizens' rights come with responsibilities. Citizens should help decide who will run their government and what actions it will take.

Americans Join Together On September 11, 2001, terrorists hijacked four U.S. planes. They flew two into the twin towers of the World Trade Center in New York City and one into the Pentagon in Washington, D.C. Both towers collapsed, and one wing of the Pentagon was damaged. Thousands of people were killed, and hundreds more were injured or trapped under debris. The fourth plane crashed in Pennsylvania after passengers struggled with the hijackers.

During the crisis, Americans like these rescue workers showed their patriotism and heroism. Hundreds of firefighters, police officers, and medical personnel worked tirelessly, risking their own lives to save others. Citizens and companies across the country donated time, blood, supplies, and millions of dollars to the victims and their families through organizations such as the Red Cross and the United Way. Americans came together in response to the attack on their nation.

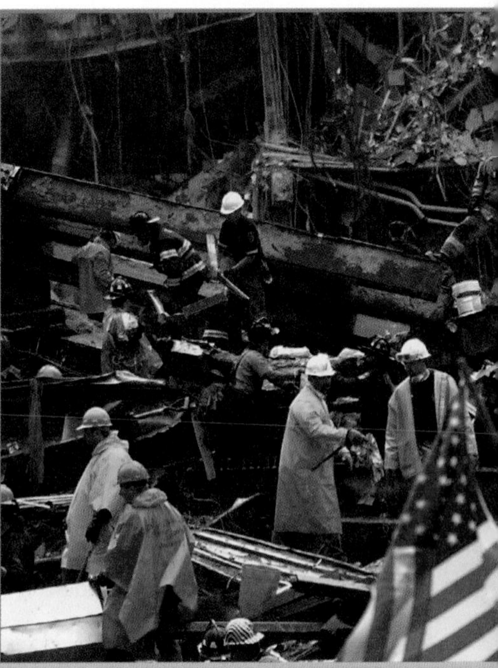

Activity Options

Interdisciplinary Link: Government

Block Scheduling

Class Time One class period

Task Preparing a time line showing when specific citizens' rights were formalized in the Constitution

Purpose To recognize how long the struggle for certain rights has taken

Supplies Needed
- A copy of the Constitution, with amendments

Activity Have students reread the section "Rights of Citizens." Working as a class, make a list of the rights identified in the section. Divide students into groups, and have each group find the Constitutional amendment that guarantees each right on the list. Have groups make a time line showing when each amendment was passed. When they have finished, review and discuss when the various rights were guaranteed. How long did it take, and how many years it was from the ratification of the Constitution?

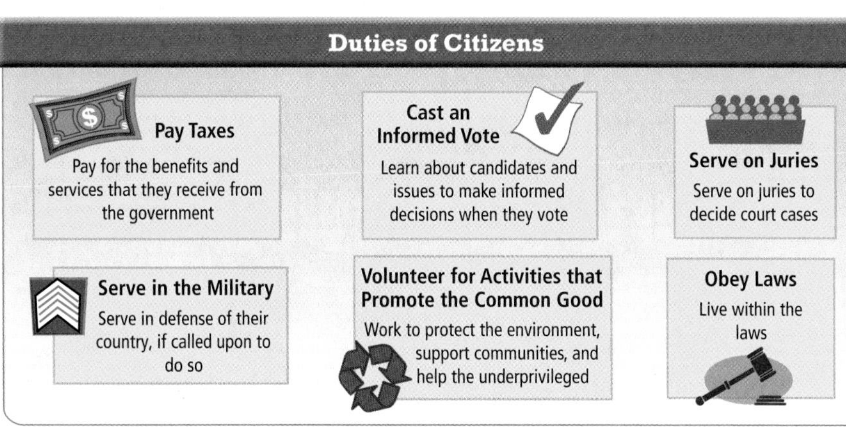

Duties of Citizens
Pay Taxes
Pay for the benefits and services that they receive from the government
Cast an Informed Vote
Learn about candidates and issues to make informed decisions when they vote
Serve on Juries
Serve on juries to decide court cases
Serve in the Military
Serve in defense of their country, if called upon to do so
Volunteer for Activities that Promote the Common Good
Work to protect the environment, support communities, and help the underprivileged
Obey Laws
Live within the laws

Citizenship is a combination of the duties and rights of a citizen. Good citizenship means doing more than the minimum required by law to secure the good of the people.

The Political Process in a Democracy and a Republic In a democracy, government receives its power from the people. **Democracy** is a Greek word that means "rule of the people." In a **republic**, the people also hold power, but they rule through elected representatives. The United States is a republic. The citizens of a democracy or a republic have the responsibility to take part in the political process. The **political process** refers to those legal activities through which citizens can change public policy. By becoming involved, citizens demonstrate their **patriotism,** or love for their country.

> **Vocabulary**
> **public policy:** Actions that a government takes to carry out its responsibilities, such as making laws and creating rules and regulations.

SECTION 1 ASSESSMENT

Terms & Names

1. Explain the significance of: (a) immigrant (b) Anasazi (c) equal opportunity (d) citizenship (e) democracy (f) republic (g) political process (h) patriotism

Using Graphics

2. Use a chart like this one to list five groups that immigrated to America. List the approximate date they began arriving.

Group	First Arrived

Main Ideas

3. (a) Why do people immigrate to the United States?

(b) What are some of the rights guaranteed to U.S. citizens by the Constitution?

(c) What are some responsibilities of U.S. citizens?

Critical Thinking

4. Draw Conclusions

In what ways do the rights and duties of U.S. citizens reflect the ideas of the writers of the Constitution?

Think About

- reasons people immigrated to the colonies
- the way a democracy works

ACTIVITY -OPTION- Think about one of the immigrant populations described. Write and illustrate a **magazine article** about the contributions this group has made to the culture of the United States.

MORE ABOUT...

Taxes

Although today over 90 percent of all federal income is collected from income and payroll taxes, Americans did not always pay income tax. Income taxes were levied in 1862 to help pay for the Civil War, but were dropped about 10 years later. When a second income tax was attempted in 1894, the Supreme Court declared it unconstitutional. It was not until 1913 that the 16th Amendment to the U.S. Constitution allowed Congress to collect income taxes.

ASSESS & RETEACH

Reading Social Studies Have students fill in the main ideas of the history of the United States covered in this section on the chart on page 86.

Formal Assessment
- Section Quiz, p. 47

RETEACHING ACTIVITY

Ask students to work in small groups to create an oral section review. Assign each group a heading from the section. Ask each group to prepare oral reports and present them to another group.

In-depth Resources: Unit 2
- Reteaching Activity, p. 19

Access for Students Acquiring English
- Reteaching Activity, p. 25

Section 1 Assessment

1. Terms & Names

a. immigrant, p. 88
b. Anasazi, p. 88
c. equal opportunity, p. 90
d. citizenship, p. 91
e. democracy, p. 91
f. political process, p. 91
g. patriotism, p. 91

2. Using Graphics

Group	First Arrived
Native American	12,000–35,000 years ago
English and Northern Europeans	late 1400s
African Americans	1619
Irish	1840s
Chinese	1850s

3. Main Ideas

a. People immigrate to the United States to escape persecution, war, and natural disaster; for employment and educational opportunities.
b. Some rights guaranteed are equal opportunity in education and employment.
c. Responsibilities include paying taxes, voting, serving on juries, and obeying laws.

4. Critical Thinking

The writers of the Constitution believed in democracy, in which each person takes part in the political process. The rights and duties of U.S. citizens are part of that process.

ACTIVITY OPTION

Integrated Assessment
- Rubric for writing a magazine article

OBJECTIVE

Students work alone or in groups to plan an Earth-surveying satellite mission.

 Block Scheduling

PROCEDURE

Have students form groups of three or four. For the Language Arts Challenge, provide students with books about careers in the sciences, plus newspaper "help wanted" pages. For the Geography Challenge, students will need poster board, colored markers, and an atlas.

LANGUAGE ARTS CHALLENGE

Class Time 50 minutes

Suggest that students use the newspaper want-ads as a "source" model for finding new team members. A scientist from the area might be invited to talk to the class about the variety of jobs that are part of a satellite mission.

Possible Solutions

Students' recruitment speeches might be supplemented with a video or another kind of graphic presentation. A question-and-answer session regarding the job could follow. After reading the "help wanted" ads, students might prepare short resumes or letters to show how their experience and personality traits could fulfill the job requirements.

Plan an Earth-Surveying Satellite Mission

Satellites supply scientists with information such as data about Earth's surface, the atmosphere, and the ocean floor. They carry instruments to analyze weather conditions and survey the ozone layer. Satellites measure heat radiation, plant growth, ocean conditions, and other features of our environment. They relay phone calls and TV programs. You and a team of scientists are getting ready to launch a new Earth-surveying satellite.

COOPERATIVE LEARNING On these pages are challenges you will face as you plan your mission. Working in a small group, choose one of the challenges. Divide the work among group members. You will find helpful information in the Data File. Keep in mind that you will present your solution to the class.

LANGUAGE ARTS CHALLENGE

"How can you find these talented people?"

As you plan, you find that you need new team members for several jobs. You need computer scientists and engineers who design instruments to measure data. Experts in forestry, geology, weather, and other fields will interpret the satellite data. How can you find these talented people? Choose one of these options. Use the Data File for help.

ACTIVITIES

1. As part of the satellite team, you give recruitment speeches to science students. Write and deliver a speech explaining what you do, what background and education you needed, and why your job is interesting.
2. Write help-wanted ads for three different jobs. Describe the education, technical background, and personality traits needed for each position.

Engineers working on the satellite. ▲

Standards for Evaluation

LANGUAGE ARTS CHALLENGE

Option 1 Speeches should
- describe the job and its requirements.
- explain in detail the nature of the job.

Option 2 "Help wanted" ads should
- explain education and experience needed.
- describe the desirable candidate.

GEOGRAPHY CHALLENGE

Option 1 The diagram should
- show Earth and a satellite orbiting it from the North to the South Pole.

Option 2 The orbit should
- show all countries and bodies of water along the equator.

SCIENCE CHALLENGE

Option 1 The chart should
- list at least five specific jobs that can be accomplished with satellites.
- list the orbit necessary for each job.

Option 2 The paragraph should explain that satellites help to
- plan a ship's course and follow weather.

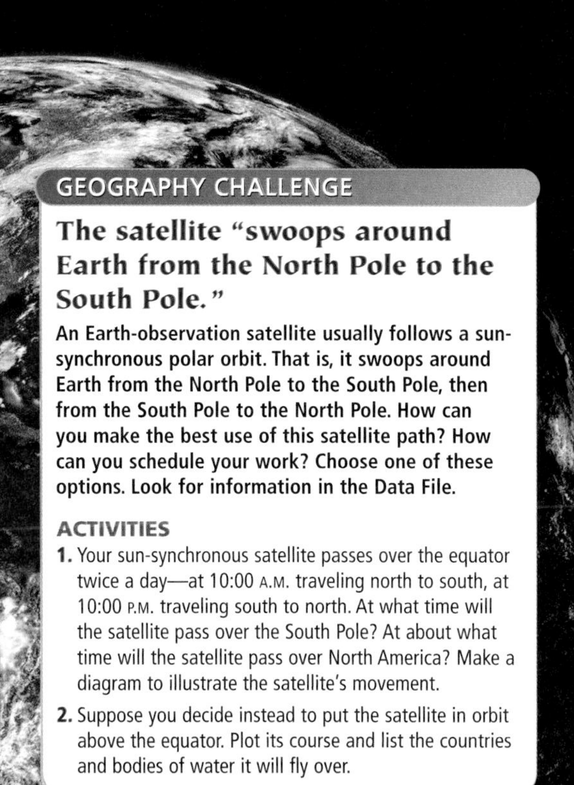

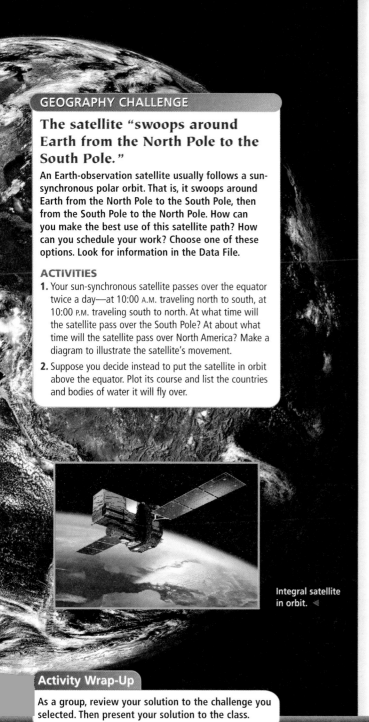

GEOGRAPHY CHALLENGE

The satellite "swoops around Earth from the North Pole to the South Pole."

An Earth-observation satellite usually follows a sun-synchronous polar orbit. That is, it swoops around Earth from the North Pole to the South Pole, then from the South Pole to the North Pole. How can you make the best use of this satellite path? How can you schedule your work? Choose one of these options. Look for information in the Data File.

ACTIVITIES

1. Your sun-synchronous satellite passes over the equator twice a day—at 10:00 A.M. traveling north to south, at 10:00 P.M. traveling south to north. At what time will the satellite pass over the South Pole? At about what time will the satellite pass over North America? Make a diagram to illustrate the satellite's movement.
2. Suppose you decide instead to put the satellite in orbit above the equator. Plot its course and list the countries and bodies of water it will fly over.

Integral satellite in orbit. ◄

Activity Wrap-Up

As a group, review your solution to the challenge you selected. Then present your solution to the class.

DATA FILE

SATELLITE ORBITS

- **High-altitude geosynchronous orbit:** The satellite travels above the equator at a height of about 22,300 miles. It moves at the same speed and in the same direction as Earth is turning. Because the satellite's orbit coordinates with Earth's movement, to an observer on the ground it seems to hover overhead at the same place in the sky.
- **Sun-synchronous polar orbit:** The satellite passes over the north and south poles, circling the globe north to south and south to north in a 24-hour day. As a result, the satellite always crosses the equator at the same local times—say, 2:00 A.M. and 2:00 P.M. As it circles Earth, it makes observations of lands at every latitude.
- **Low-altitude orbit:** The satellite travels within Earth's atmosphere, but in the highest layer, where friction is less because the air is thin.

TYPES OF ARTIFICIAL SATELLITES

- **Earth observation:** map and monitor physical changes on Earth.
- **Scientific research:** gather data for scientists to analyze.
- **Weather:** make local or worldwide observations, depending on orbit; measure cloud patterns, air pressure, air pollution, and air chemistry.
- **Communications:** relay radio, TV, and telephone signals.
- **Navigation:** provide location information.
- **Military:** gather information for military use.

To learn more about Earth-surveying satellites, go to

GEOGRAPHY CHALLENGE

Class Time 50 minutes

Students can work individually or in pairs to do this activity. They may want to devise a way to record information. An atlas will be helpful for the second option.

Possible Solutions

Students should use the text and the Data File to make an accurate diagram. They may also need an outside source, such as a library book or a Web site. The diagram should show Earth and an orbiting satellite on a path that crosses both poles. If the satellite crosses the equator at 10 A. M. going south, then it will cross the South Pole at 4 P. M., and it will cross North America during the night.

ALTERNATIVE CHALLENGE...
SCIENCE CHALLENGE

The word satellite comes from Latin and means attendant, or follower. Meteorologists use satellites to follow weather. They can track hurricanes or tornadoes. Scientists use satellites to track the change of shorelines. Ocean ships track their path using information from satellites. A satellite in a polar orbit will eventually circle all of Earth, so scientists can gather information about any place on the planet.

- Using information from the Data File, make a chart of at least five specific jobs that can be accomplished using satellites. Then list which kind of satellite orbit would be most suitable for each job.
- Imagine that you are the captain of an ocean-going ship. Write a paragraph that describes how important satellites are to you.

Activity Wrap-Up

Solutions to the challenges should

- clearly identify the problems in each challenge.
- provide effective, accurate definitions for each challenge.
- demonstrate an understanding of satellites and their importance to us.

Chapter 4 ✦ Section 2

SECTION 2: A Constitutional Democracy

SECTION OBJECTIVES

1. To recognize that the Constitution is the framework of the U.S. government
2. To identify the rights of citizenship granted by the Bill of Rights
3. To delineate the powers of the federal and state governments
4. To identify the three branches of government

SKILLBUILDER
- Interpreting a Map, p. 97

CRITICAL THINKING
- Drawing Conclusions, p. 97

FOCUS & MOTIVATE

WARM-UP

Analyzing Motives Have students read Dateline and answer questions about the U.S. Supreme Court's ruling.

1. Why might Oliver Brown have wanted his daughter to attend the all-white school?
2. Do you think this ruling impacted people outside Topeka, Kansas? Explain your answer.

INSTRUCT: Objective 1

The Law of the Land/Limited and Unlimited Governments

- Under the U.S. Constitution, from whom does the government receive power? the people
- What goals did the writers of the Constitution hope to achieve? to protect individual rights and establish a stable government

In-depth Resources: Unit 2
- Guided Reading Worksheet, p. 12

Reading Study Guide
(Spanish and English), pp. 24–25

TERMS & NAMES
- **United States Constitution**
- **limited government**
- **unlimited government**
- **constitutional amendment**
- **Bill of Rights**
- **federal government**

MAIN IDEA
The founders of the United States drafted a constitution that protected the rights of citizens.

WHY IT MATTERS NOW
After more than 200 years, the Constitution continues to protect the freedoms of U.S. citizens.

DATELINE

U.S. SUPREME COURT, MAY 17, 1954—Oliver Brown, an African American, wanted his daughter to attend a nearby all-white school. When the Board of Education in Topeka, Kansas, refused to admit her, Brown went to court. Today, the U.S. Supreme Court ruled that separate schools for different races violate the equal protection guaranteed by the 14th Amendment to the U.S. Constitution.

In 1896, the Supreme Court had allowed states to provide "separate but equal" facilities for blacks and whites. The words in the Constitution remain the same, but changes in society have led to a new interpretation of those words.

Place • Soldiers and police escort students into a formerly all-white school following this Supreme Court ruling. ▲

The Law of the Land 1

The basis for U.S. law is the **United States Constitution,** written by the country's first leaders. Amazingly, this document remains the foundation for all laws and the framework for the U.S. government more than 200 years after its creation. The Supreme Court decides whether the actions of states, businesses, and individuals are in accordance with the ideas in the Constitution, as it did in this 1954 case.

TAKING NOTES
Use your chart to take notes about the United States today.

The United States Today	
History	
Government	

Program Resources

In-depth Resources: Unit 2
- Guided Reading Worksheet, p. 12
- Reteaching Activity, p. 20

Reading Study Guide
(Spanish and English), pp. 24–25

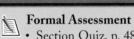

Formal Assessment
- Section Quiz, p. 48

Integrated Assessment
- Rubric for writing a proposal

Outline Map Activities

Access for Students Acquiring English
- Guided Reading Worksheet, p. 20

Technology Resources
classzone.com

TEST-TAKING RESOURCES

Strategies for Test Preparation

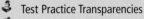

Test Practice Transparencies

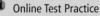

Online Test Practice

Forming a New Government American colonists living under British rule did not have the rights and the protections they wanted. After gaining independence from Great Britain in 1783, they established a nation called the United States of America. The writers of the Constitution designed a government that received its power from the people.

The founders, or early American leaders, wanted to protect people's individual rights and freedoms from government interference. They also knew that a society needs strong laws and a stable government to ensure the common good. They wrote a constitution that achieved both goals. The U.S. Constitution describes and limits the power of the government and its leaders. It also defines the rights of citizens and their role in governing their country. In 1902, President Theodore Roosevelt explained the relationship between U.S. citizens and their government.

Reading Social Studies

Making Inferences What were two goals that the writers of the Constitution wanted to achieve?

Reading Social Studies Possible Answer

They wanted to protect individual rights and establish a stable government.

A VOICE FROM THE UNITED STATES

The government is us; we are the government, you and I.

Theodore Roosevelt

Limited and Unlimited Governments 

The constitutional republic of the United States is one example of a **limited government.** In other types of government, called **unlimited governments,** the leaders have almost total power. For instance, dictators control their countries' laws and people.

The United States Constitution The U.S. Constitution is the oldest national constitution still in use. It was written in 1787 at the Constitutional Convention in Philadelphia and ratified in 1789. Ideas that shaped the U.S. Constitution came from many places and times, including Great Britain, France, and ancient Rome. Native American nations, such as the Iroquois Confederacy, may also have influenced political ideas at the time that the Constitution was drafted. This painting, *Scene at the Signing of the Constitution of the United States* by H. C. Christy, hangs in the Capitol in Washington, D.C.

A VOICE FROM THE UNITED STATES

Provide students with some background about Theodore "Teddy" Roosevelt. Explain that during his administration, from 1901 to 1909, Congress passed laws to regulate railroads, conserve the nation's forests, and protect the public from harmful foods and drugs. Then ask students to explain how this quote supports evidence that the founders were truly successful in designing a government that receives its power from the people.

The United States Constitution is one of many important historic documents that are preserved and made available for public viewing at the National Archives in Washington, D.C. The Archives, established in 1934, hold a collection of records, maps, sound recordings, photographs, and motion pictures from 1774 to the present. More than 1 million people visit the Exhibition Hall every year. The Constitution, along with documents such as the Declaration of Independence, are on display in sealed bronze and glass cases filled with helium for their protection.

Activity Options

Differentiating Instruction: Less Proficient Readers

Recognizing Important Details Some students may have difficulty identifying the goals of the writers of the Constitution. Pair a less proficient reader with a more skilled reader and ask them to reread the paragraphs under "Forming a New Government." Ask them to list the goals of the writers of the Constitution as they read.

Goals of the Writers of the Constitution
individuals' rights
power from the people
freedom from government interference
strong laws
stable government

INSTRUCT: Objective 2

The Constitution Grows and Changes

- What are some freedoms granted in the Bill of Rights? freedom of speech and religion, right to a fair trial, right to gather peaceably
- What are some amendments that have been made to adapt to the country's changing needs? ending slavery, giving women the right to vote, limiting the president to two terms

As one of the world's best-known advocates of nonviolent social change, Reverend Martin Luther King, Jr., was also a powerful and effective speaker. In 1955 Reverend King was asked to be president of the organization planning the boycott of city buses in Montgomery, Alabama. In his first speech as leader he said, "First and foremost, we are American citizens. . . . We are not here advocating violence. . . . The only weapon that we have . . . is the weapon of protest. . . . The great glory of American democracy is the right to protest for right."

INSTRUCT: Objective 3

Limiting Powers of Government

- What did leaders do to limit government power? divided power between federal and state governments
- What powers does the Constitution give the federal government? power to establish an army, wage war, collect taxes, make laws

Place • Women began the fight for suffrage, or the right to vote, in the early 1800s and continued until they succeeded in 1920. ▲

The Constitution Grows and Changes 

The Constitution went into effect in 1789. A condition of ratifying, or approving, it in many states was the promise of a bill of rights. In 1791, the states adopted ten constitutional amendments proposed by Congress. A **constitutional amendment** is a change or addition to the Constitution. This **Bill of Rights** lists specific freedoms guaranteed to every U.S. citizen. Among them are freedom of speech and religion, the right to a fair trial, and the right to gather peaceably. In all, 27 amendments have adapted the Constitution to the country's changing needs. Some amendments passed after the Bill of Rights include ones that ended slavery, gave women the right to vote, and limited a president's terms to two.

Limiting Powers of Government 3

Leaders of the new country wanted to limit government power and to preserve each state's right to govern itself. To accomplish these goals, they created a federal system in which power is divided between the **federal government,** or national government, and the state governments. The federal government is a republic headed by the President.

Martin Luther King, Jr. (1929–1968) Reverend King was a civil rights leader in the 1950s and 1960s. A gifted speaker, he argued for voting rights, equal opportunities in education and jobs, and justice not based on the color of people's skin. He expressed these hopes in a famous 1963 speech in Washington, D.C., shown at right.

Influenced by his study of Christianity and his admiration for India's civil rights leader Mahatma Gandhi, Reverend King used nonviolent protest. He received the Nobel Peace Prize in 1964.

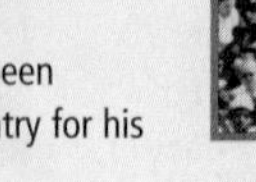

On April 4, 1968, he was assassinated. Since 1986, a U.S. holiday has been observed in January to honor him, and he is remembered around the country for his leadership in the civil rights movement.

Activity Options

Interdisciplinary Link: Speech

Class Time One class period

Task Memorizing and/or reciting lines from the speech "I Have a Dream" by Martin Luther King, Jr.

Purpose To familiarize students with one of history's most famous speeches

Supplies Needed
- Copies of the speech "I Have a Dream"

Activity Review with students the events of August 1963 when Dr. Martin Luther King, Jr., delivered this famous speech. Assign a few sentences of the speech to each student, and ask them to memorize their assigned sentences. Then have students deliver their sentences in order, reciting the entire speech. After the presentation, discuss how hearing King present his speech might have affected the listeners in 1963.

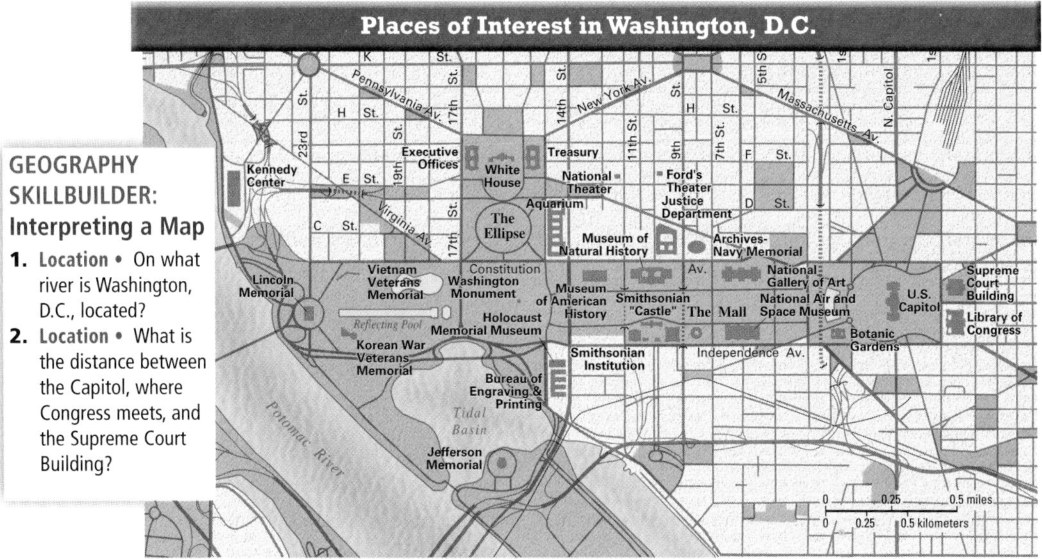

GEOGRAPHY SKILLBUILDER: Interpreting a Map

1. **Location** • On what river is Washington, D.C., located?
2. **Location** • What is the distance between the Capitol, where Congress meets, and the Supreme Court Building?

Geography Skillbuilder Answers

1. Potomac River
2. 0.25 of a mile, a little less than 0.5 kilometers

Federal and State Government The Constitution gives the federal government specific powers, including establishing an army, waging war, raising money through taxes, and making laws to carry out its duties. All other powers are held by the states. The Constitution does not refer to local government, so each state determines the form of town or county rule.

Checks and Balances Three branches share the powers of the U.S. government. Each branch checks the power of the other branches. The process by which a bill, or a proposal for a new law, becomes a law shows how this balance of power works.

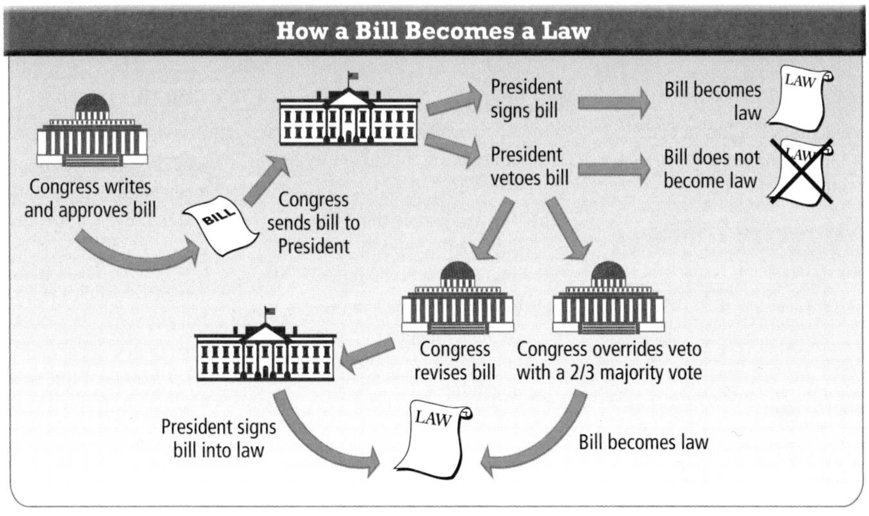

FOCUS ON VISUALS

Interpreting the Map Tell students that the three branches of government are represented by three different buildings shown on this map. Lead students to identify the White House with the executive branch, the Capitol with the legislative branch, and the Supreme Court building with the judicial branch. Ask students to notice the distances between these three important buildings. Discuss why the architect might have placed them in these relative positions.

Possible Responses Students might speculate that the architect wanted to physically represent the governmental separation of the branches.

Extension Have students work in groups to design a new arrangement of the three government buildings. Ask each group to show their arrangement by drawing a new map of Washington.

CRITICAL THINKING ACTIVITY

Drawing Conclusions Ask students to reread "How a Bill Becomes a Law." Point out and discuss the ways in which the President's power balances Congress' in the passing of a law. Ask students to identify the ways in which Congress's power balances the President's in the same process. Discuss.

Class Time 15 minutes

Activity Options

Differentiating Instruction: Less Proficient Readers

Understanding Sequences Less proficient readers may need to review the sequence of how a bill does or does not become a law. Divide students into groups and assign each group one of the following possible sequences: Congress approves the bill, the President signs the bill, the bill becomes law. Congress approves the bill, the President vetoes the bill, the bill does not become law. Congress approves the bill, the President vetoes the bill, Congress revises the bill, the President signs the bill, the bill becomes law. Congress approves the bill, the President vetoes the bill, Congress overrides the veto, the bill becomes law.

Have each group create a sequence chart for the assigned sequence. When they have finished, ask them to present their sequences orally and discuss.

INSTRUCT: Objective 4

Three Branches of Government

- What are the three branches of the government? executive, legislative, judicial
- Who leads the executive branch? President
- What makes up the legislative branch? Senate and House of Representatives
- What is the process for becoming a Supreme Court judge? nominated by President, approved by Senate

ASSESS & RETEACH

Reading Social Studies Have students fill in the main ideas of the government on the chart on page 86.

Formal Assessment
- Section Quiz, p. 48

RETEACHING ACTIVITY

Have students work in small groups to create a section review. Have one student in each group explain the Constitution, while the rest of the group asks questions to clarify the explanation. Then repeat with another group explaining and clarifying the Bill of Rights, and another the three branches of government.

In-depth Resources: Unit 2
- Reteaching Activity, p. 20

Access for Students Acquiring English
- Reteaching Activity, p. 26

Three Branches of Government 4

The Constitution separates powers of government into the executive, legislative, and judicial branches. Each branch has its own job. All are located in the U.S. capital, Washington, D.C.

Executive Branch
Enforces the laws

The Executive Branch The President is elected to head the executive branch. He enforces the laws, serves as commander in chief of the armed forces, and conducts foreign affairs. The Vice-President is elected with the President. The President's cabinet includes the secretaries of the 14 executive departments and other key members of the executive branch.

Legislative Branch
Makes the laws and controls taxes and spending

The Legislative Branch Congress is made up of two houses—the Senate and the House of Representatives—and makes national laws. Two senators are elected from each of the 50 states. The House of Representatives has 435 members, elected from each state according to its population. The two houses have some shared responsibilities and some separate ones.

Judicial Branch
Decides if laws agree with the U.S. Constitution

The Judicial Branch The judicial branch is the system of federal courts that makes sure all laws and treaties are constitutional, or agree with the U.S. Constitution. The highest federal court, the Supreme Court, has nine justices, or judges, nominated by the President and approved by the Senate.

SECTION 2 ASSESSMENT

Terms & Names

1. Explain the significance of: (a) United States Constitution (b) limited government (c) unlimited government (d) constitutional amendment (e) Bill of Rights (f) federal government

Using Graphics

2. Use a diagram like this one to record the organization of the government.

Branches of Government		
Executive	Legislative	Judicial

Main Ideas

3. (a) What two goals did the writers of the U.S. Constitution try to achieve?

(b) Why did people think it was important to add the Bill of Rights to the Constitution?

(c) Why does the Constitution create a balance of powers among the three branches of government?

Critical Thinking

4. Drawing Inferences
Why do you think Congress is made up of two parts, the Senate and the House of Representatives?

Think About
- the wish to limit the power of government
- the number of senators and representatives from each state

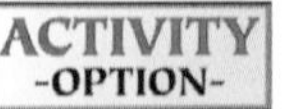

ACTIVITY -OPTION- If you were going to add an amendment to the Constitution, what would it be? Write a **proposal** for it. Tell what the law would do, whom it would affect, and why people should support it.

Section 2 Assessment

1. Terms & Names
- **a.** United States Constitution, p. 94
- **b.** limited government, p. 95
- **c.** unlimited government, p. 95
- **d.** constitutional amendment, p. 96
- **e.** Bill of Rights, p. 96
- **f.** federal government, p. 96

2. Using Graphics

Branches of Government		
Executive	**Legislative**	**Judicial**
President Enforces laws Commander in chief Conducts foreign affairs	Senate House of Representatives Makes national laws	Federal courts Supreme Court Nine judges Decides if laws are constitutional

3. Main Ideas
- **a.** They tried to create a stable government and to protect individuals' rights.
- **b.** They wanted to list the specific freedoms granted to every citizen.
- **c.** The balance prevents any one branch from becoming too powerful.

4. Critical Thinking

Each part can serve as a check on the other's actions; the different methods of determining representation ensure fair treatment among states.

ACTIVITY OPTION

Integrated Assessment
- Rubric for writing a proposal

SKILLBUILDER

Sequencing Events

▸▸ Defining the Skill

Sequence is the order in which events follow one another. If you learn to follow the sequence of events through history, you can better understand how events relate to each other.

▸▸ Applying the Skill

This passage at the right describes the sequence of events that improved the transportation of natural resources and goods in the United States. Use the strategies listed below to understand how transportation has improved.

How to Sequence Events

Strategy ❶ Look for time periods of events or discoveries. Some dates may be exact, and others may be indicated only by decades or centuries. Words such as *day, month, year,* or *century* may help you to sequence the events or discoveries.

Strategy ❷ Look for clues about time that allow you to order events according to sequence. Words such as *first, next, later,* and *finally* may help you order events that are not dated.

RESOURCES, GOODS, AND PEOPLE

The United States could not have developed without new ways of transporting natural resources and goods. In the early 19th century, water provided the fastest way to transport goods. Canals linked some lakes and rivers, so places became more accessible. A little later, steamboats allowed river traffic to go upstream and downstream easily.

By the middle of the century, steam-powered railroads had replaced steamboats because they could reach more places and carry larger loads. At the beginning of the 20th century, people and goods in many cities and towns were linked by railroad.

In the early 1900s, automobiles and trucks began to move raw materials and goods. Airplanes began to transport goods in the 1930s. By the 1950s, paved roads connected much of the country. Today, fast jets also bring goods and people to many parts of the United States.

Make a Time Line

Making a time line can help you sequence events. This time line shows the sequence of events in the passage you just read.

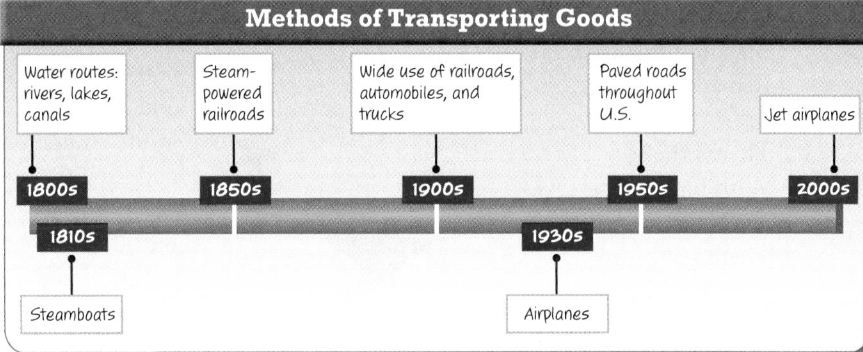

▸▸ Practicing the Skill

Turn to Chapter 4, Section 1, "We the People." Read about the different groups of people who settled the land that became the United States. Make a list showing the sequence of the arrival of different immigrant groups.

SKILLBUILDER

Sequencing Events

Defining the Skill

Ask students to recall the sequence of their day so far. Discuss milestones in their lives and in what order they occurred. Why might it be important to understand and record events in sequential order? Have students give examples of how using sequential order would be useful in planning a trip or completing a project.

Applying the Skill

How to Sequence Events Point out strategies for sequencing events to students. Guide them through each strategy, and be sure they understand the importance of dates and general time periods Review word clues that will help them recognize sequential order.

Make a Time Line

Review and discuss the time line "Methods of Transporting Goods." Have students connect the sequence of events in the passage they read with the sequence of the events on the time line.

Practicing the Skill

Point out to students that the information about different groups of immigrants who settled in the United States covers a long time span. Have students share and compare information about the groups and corresponding arrivals that they included on their lists. Provide students with the following information about waves of immigration to the United States: Italians (1880s to 1920s), Cubans (1960s to 1980s). Discuss where this information should be inserted sequentially into the list.

In-depth Resources: Unit 2
- Skillbuilder Practice, p. 17

Activity Options

Career Connection: Archaeologist

Encourage students who enjoy sequencing events to find out about careers that use this skill. For example, archaeologists study the past by excavating, or digging up, the remains of bones and the remains of things that earlier people left behind. They use their knowledge and findings to figure out what happened in the past, including the sequence of important events.

Block Scheduling

1. Encourage students to find out about the specific tasks an archaeologist performs, as well as about different branches of archaeology.
2. Help students find out what education and training a person needs to become an archaeologist.
3. Have students present their findings on a large poster.

OBJECTIVE

Students analyze three poems that describe places in different parts of the United States: Coney Island, New York; Knoxville, Tennessee; and San Francisco, California.

FOCUS & MOTIVATE

Drawing Conclusions To help students picture the places described in the poems and understand the purpose of the selections, have them study the photographs on pages 100–101 and answer the following questions:

1. What can you tell about each place from the photograph?
2. What do the photographs together show you about the United States?

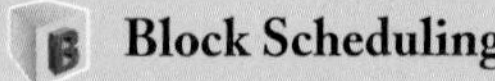

Block Scheduling

MORE ABOUT...

Coney Island

"Coney" describes Coney Island, an oceanfront amusement area in Brooklyn, New York. At the turn of the 20th century, Coney Island was the most extravagant playground in the country. Between 1902 and 1911 it was the site of three of the world's greatest amusement parks: Steeplechase Park, Dreamland, and Luna Park. One of the first amusement rides on Coney Island was the famous "Wonder Wheel"—a Ferris wheel patterned after the "Observation Wheel" introduced at the 1893 Chicago World's Fair.

LITERATURE CONNECTIONS

WITH A LAND AREA of over 3.7 million square miles and a population of more than 285 million, the United States is an amazingly varied country. From small towns, forests, and farmland to big cities like San Francisco and New York, the United States has a wealth of both human-made and natural features. The poems here celebrate this variety.

CONEY

By Virginia Schonborg

There's hot corn
And franks.
There's the boardwalk
With lots of games,
With chances
To win or lose.
There's the sun.
Underneath the boardwalk
It's cool,
And the sand is salty.
The beach is
Like a fruitstand of people,
Big and little,
Red and white,
Brown and yellow.
There's the sea
With high green waves.
And after,
There's hot corn
And franks.

KNOXVILLE, TENNESSEE

By Nikki Giovanni

I always like summer
best
you can eat fresh corn
from daddy's garden
and okra
and greens
and cabbage
and lots of
barbecue
and buttermilk
and homemade ice-cream
at the church picnic
and listen to
gospel music
outside
at the church
homecoming
and go to the mountains with
your grandmother
and go barefooted
and be warm
all the time
not only when you go to bed
and sleep

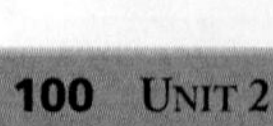

Activity Options

Differentiating Instruction: Less Proficient Readers

Building Language Skills Have students work in pairs, matching less proficient readers with more fluent readers. Duplicate copies of the Literature Connection. Give each pair of students the three poems and a set of colored pencils. Have partners read the poems, determine where sentences begin and end, and underline each sentence in a different color. Help groups decide where to separate ideas in "Knoxville, Tennessee." Tell students to circle any unfamiliar words and look them up in a dictionary. Then have each student recite his or her favorite of the three poems.

Scenic

By John Updike

O when in San Francisco do
As natives do: they sit and stare
And smile and stare again. The view
Is visible from anywhere.

Here hills are white with houses whence,
Across a multitude of sills,
The owners, lucky residents,
See other houses, other hills.

The meanest[1] San Franciscan knows,
No matter what his past has been,
There are a thousand patios
Whose view he is included in.

The Golden Gate, the cable cars,
Twin Peaks, the Spreckles habitat,[2]
The local ocean, sun, and stars—
When fog falls, one admires *that.*

Here homes are stacked in such a way
That every picture window has
An unmarred prospect of the Bay
And, in its center, Alcatraz.[3]

1. Poorest.
2. In the late 1950s, Twin Peaks was a well-to-do residential neighborhood where a very wealthy family, the Spreckles, lived.
3. An island in San Francisco Bay, site of a former federal prison.

Reading THE LITERATURE

Pick out the words each poet uses to give a sense of place. How can you tell that some places are rural, some urban? What aspects of geography has each poet highlighted?

Thinking About THE LITERATURE

Imagine that you live in one of these places. What kinds of jobs might people have there? What might they do for fun? How does the place where you live influence your daily activities?

Writing About THE LITERATURE

Which of the poems creates the strongest image in your mind? Using examples from that poem, write a short paragraph explaining why.

About the Authors

Virginia Schonborg (b. 1913) was born in Rhode Island and is the author of *The Salt Marsh* and *Subway Swinger.*

Nikki Giovanni (b. 1943) was born in Tennessee and became well-known as a poet in the 1960s.

John Updike (b. 1932), born in Shillington, Pennsylvania, is the author of many short stories, novels, and poems.

Further Reading For more outstanding poems about the United States, read *My America: A Poetry Atlas of the United States*, edited by Lee Bennett Hopkins.

INSTRUCT

Reading the Literature
Possible Responses

"Coney": boardwalk, beach, sea, sand; "Knoxville, Tennessee": daddy's garden, church picnic, mountains; "Scenic": San Francisco, hills, Golden Gate, cable cars, Bay, Ocean, fog, Alcatraz

Thinking About the Literature
Possible Responses

Near Coney Island, people might work as vendors or lifeguards. Visitors can swim, eat special food, and play arcade games for fun. Geography, climate, and local industries influence what people do for employment and recreation.

Writing About the Literature
Possible Responses

Students' paragraphs should name the poem that creates the strongest image for them and explain why. For example, details in "Coney" may remind students of vacations. Details in "Knoxville, Tennessee" may remind them of summer. In "Scenic," students may identify with people who see their city's landmarks and feel lucky to live there.

MORE ABOUT...

Poetic Technique

Explain that a simile is a figure of speech that compares one thing with another, using the word *like.* Point out to students that "Coney" includes a simile: "The beach is like a fruitstand of people...." Discuss the simile and how effectively it describes the scene. Challenge students to brainstorm other similes that could describe a mix of colors and sizes.

 **World Literature**

More to Think About

Making Personal Connections Point out that writers often write about places they know well. Ask students to think of stories set in different parts of the United States. Talk about how each setting affects the characters' lives. Then ask students to name stories, movies, art, or music about their home state. How is their state portrayed? How does where they live shape their lives?

Vocabulary Activity Have students work in pairs or small groups. Instruct one student to read aloud images from a poem until listeners can name the title. Continue until all students have had a turn. Alternatively, have partners write their favorite images from each poem on slips of paper, shuffle them, and sort them by title.

SECTION OBJECTIVES

1. To differentiate between goods and services
2. To identify the four factors of production
3. To explain the free enterprise system
4. To recognize the workings of a global economy

SKILLBUILDER
- Interpreting a Map, p. 107

CRITICAL THINKING
- Making Inferences, p. 103
- Hypothesizing, p. 105

FOCUS & MOTIVATE

WARM-UP

Making Inferences Have students read Dateline, and ask these questions to help them understand the rise and fall of dot-coms.

1. What was the attraction of start-ups?
2. What long-term effects might dot-coms have on the general economy?

INSTRUCT: Objective 1

The Study of Economics

- How are goods and services defined? Goods are objects you buy to satisfy wants; services are actions that satisfy wants.
- What goods and services do governments need to buy? goods: schools, roads; services: fire protection, military forces

In-depth Resources: Unit 2
- Guided Reading Worksheet, p. 13

Reading Study Guide (Spanish and English), pp. 26–27

SECTION 3 The United States Economy

TERMS & NAMES
- factors of production
- GDP
- free enterprise/market economy
- consumer
- profit
- competition

MAIN IDEA	WHY IT MATTERS NOW
The United States has an economy based on free enterprise. Consumers and business owners decide what goods and services to produce.	This economic system has made it possible for the United States to become a leader in the worldwide economy.

DATELINE

JUNE 2001, THE UNITED STATES—The collapse of many dot-coms has shaken consumer confidence. Internet-based companies, nicknamed "dot-coms" for the last part of their Web addresses, have been failing by the hundreds.

In the mid-1990s, investors backed dot-com companies in hopes of making a huge return on their investments. This ready money encouraged people with limited management experience, untested technologies, and experimental Web sales strategies to start companies. In 1999, more than 1,700 Internet start-ups were born.

When investors realized that many start-ups were losing money, they began to withdraw their support. Between January 2000 and May 2001, 435 Internet companies shut down. Over 31,000 people have lost their jobs.

Place • When companies fail, employees are often left without jobs and office equipment is sold. ▲

The Study of Economics 1

Business start-ups and shutdowns, the rise and fall of investor and consumer confidence, the increase and decrease in the number of people without jobs—all of these changes are part of the market economy of the United States. Investors, service providers, manufacturers, and consumers make choices each day, and these choices affect the state of the economy.

TAKING NOTES
Use your chart to take notes about the United States today.

The United States Today	
History	
Government	

Program Resources

In-depth Resources: Unit 2
- Guided Reading Worksheet, p. 13
- Reteaching Activity, p. 21

Reading Study Guide (Spanish and English), pp. 26–27

Formal Assessment
- Section Quiz, p. 49

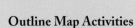

Integrated Assessment
- Rubric for writing an advertisement

Outline Map Activities

Access for Students Acquiring English
- Guided Reading Worksheet, p. 21

Technology Resources
classzone.com

TEST-TAKING RESOURCES
- Strategies for Test Preparation
- Test Practice Transparencies
- Online Test Practice

102 CHAPTER 4

Goods and Services Suppose you want a CD recording just released by your favorite music group. To earn the money to pay for it, you might rake your neighbor's leaves or care for his or her child for a few hours. The CD is a good. A good is any object you can buy to satisfy a want. Raking leaves or baby-sitting is a service you provide. A service is an action that meets a want. Your neighbor buys your service to meet his or her want.

What to Buy People constantly decide which goods and services to buy. They usually satisfy basic needs such as food, clothing, housing, transportation, childcare, and medical treatment first. If there is money left over, they might choose to spend it on music CDs, in-line skates, a computer game, or a vacation.

A government must also make decisions. Tax dollars must be set aside to pay for police and fire protection, schools, roads, and military forces. Once these expenses have been determined, other choices can be made.

Reading Social Studies

A. Synthesizing After a government has set aside money for basic wants, what additional goods or services might it pay for?

Reading Social Studies A. Answer
It might pay for environmental protection, medical research, or arts development.

A Growing Economy ❷

A nation must produce goods and provide services to support a growing economy. In an expanding economy, citizens have better-paying jobs, so the government collects more tax money. Then people and the government can satisfy more wants.

To sustain a growing economy, business owners must keep production at a high level. Production is the making of goods and services. The four **factors of production** are the ingredients, or elements, needed for production to occur.

Culture • **Christina Aguilera's critical contribution in the production of a CD is as one of the labor resources. ▼**

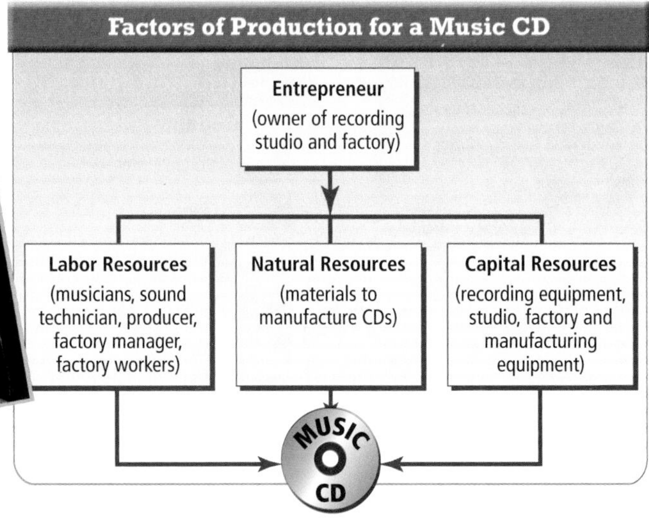

CRITICAL THINKING ACTIVITY

Making Inferences Ask students to study the chart "Factors of Production for a Music CD." Lead a discussion about how changes in any of the four factors of production might change the price of the final product. What happens to the CD's price if the cost of natural resources goes up, or the supply goes down? What about changes in labor cost or supply? What if the entrepreneur runs out of money? Encourage students to understand the relationship between the four factors of production.

Class Time 15 minutes

INSTRUCT: Objective ❷

A Growing Economy

- What are the four factors of production? natural resources, labor resources, capital resources, entrepreneurs
- What are entrepreneurs? people who take the risk and initiative to bring resources together to produce goods and services

Activity Options

Differentiating Instruction: Students Acquiring English/ESL

Word Meanings Students who are acquiring English may be confused about how the word *good* is used in this section. Tell students that the word *good* has more than one meaning. Say that they are probably familiar with *good* as an adjective meaning well-behaved, kind, or in sound condition. Explain that *good* can also mean "a thing that is bought and sold."

To help students understand the meaning of the word *good* or *goods* as it is used in "The Study of Economics," ask the following questions:

- What is an example of a good in this section?
- What is an example of a good that you or your family bought this week?
- What good might you buy with money received as a birthday present?

INSTRUCT: Objective 3

The United States Economy

- What does the GDP indicate? the total value of the goods and services that a country produces each year
- What are the qualities of a free enterprise system? Citizens and businesses make economic decisions; business controls production; government has a limited role.
- What is the law of supply and demand? Price and availability are affected by how much consumers are willing to pay for an item, how much sellers decide to charge for it, and how much of the item is available.
- What choices does a company have if it wants to sell its goods? offer a better product, sell a product at a lower price, or make fewer

Strange but TRUE

What happens to torn or worn-out money? Banks send damaged bills to a Federal Reserve Bank to be exchanged for new bills. The average life span of a 1-dollar bill is about 18 months; the 10-dollar bill has the same life span. A 5-dollar bill usually lasts 15 months; a 20-dollar bill lasts 2 years. Even longer lasting are 50-dollar and 100-dollar bills.

How much cash is in circulation? The amount increased dramatically in the 20th century. In 1910 the figure was $3,148,700,000. In 1990 it was up to $266,902,367,798.

Natural Resources Raw materials are used to make goods. Examples include land, water, forests, minerals, soil, and climate.

Labor Resources Workers are needed with the appropriate knowledge, skills, and experience to make goods or provide services.

Capital Resources Machines, factories, and supplies are needed.

Entrepreneurs These are the people who bring natural resources, labor resources, and capital resources together to produce goods and services.

The Value of Money The U.S. Department of the Treasury first issued paper money in 1861. During the Civil War, people hoarded coins. This caused a severe shortage of coins in circulation.

The Bureau of Engraving and Printing printed "fractional currency" bills of 3 cents, 5 cents, 10 cents, 25 cents, and 50 cents. These were the smallest bills ever printed in this country. The largest note was the $100,000 Gold Certificate printed in 1934 and 1935. These notes were used for bank transactions and were not available for the public to use.

The United States Economy 3

The U.S. economy is one of the wealthiest in the world. One way to measure a country's economy is to look at its **GDP**, or gross domestic product. This tells the total value of the goods and services that a country produces each year. The GDP is also a way to compare the economies of different countries.

U.S. industries include services, such as health care and legal services; communications, such as publishing, television and radio, telephone, and mail delivery; finance, such as banks and stock markets; manufacturing, such as food products, automobiles, and clothing; and electronics, such as televisions, computers, and sound equipment. The success of these industries helps make many Americans wealthier than people in other parts of the world.

BACKGROUND

In 2000, 2.8 billion of the world's 6 billion people lived on less than $2 a day.

Estimated 2000 Per Capita* GDP

Country	U.S. Dollars
United States	36,200
China	3,600
India	2,200
Saudi Arabia	10,500
France	24,400
Zimbabwe	2,500

**Per capita* means for each person.
Source: *CIA World Factbook*

Activity Options

Interdisciplinary Link: Economics

Class Time One class period

Task Applying the four factors of production to a familiar product

Purpose To identify and apply the four factors of production

Supplies Needed
- Drawing paper
- Pencils or markers

Activity Brainstorm with students a list of products they purchase on a regular basis, such as particular types of clothing, food, or electronics. Have students select a product from the list and work in small groups to discuss the four factors of production that are necessary to create that product. Then have them create a graphic such as the one on page 103 for the product they chose.

The Free Enterprise or Market Economy U.S. citizens and businesses make most economic decisions. Business owners control the factors of production. The government plays a limited role. It does not decide which or how many goods are produced. It does not set prices or tell people where to work. These are the qualities of a **free enterprise/market economy.** A market is a setting for exchanging goods and services. In a free enterprise economy, business owners compete in the market with little government interference. Other nations, such as Canada, many countries in Western Europe, Japan, and some Latin American countries, also have market economies.

Culture • The machines that make CDs are a capital resource, one of the four factors of production. ▲

Supply and Demand In a free enterprise or market economy, **consumers**—the people who use goods and services—help decide what will be produced. Prices affect how products are distributed to consumers. For example, suppose a music company produces 1,000 CDs priced at $16.95 each, but 1,100 people want to buy them. There are not enough CDs to satisfy the wants of these consumers. Because demand for the good is greater than the supply, the seller can increase the price to $17.95. He or she sells all the CDs and makes an extra $1.00 profit on each. The seller, like all entrepreneurs, wants to increase profit. **Profit** is the money that remains after all the costs of producing a product have been paid.

Now suppose the seller offers 1,000 more CDs at the original $16.95 price. One hundred sell right away, leaving 900 CDs that no one wants at this price. The seller then reduces the price to $15.95. Consumers who didn't want the CD at $16.95 may want it at this lower price. Because the supply of the good is greater than the demand, the seller must reduce the price.

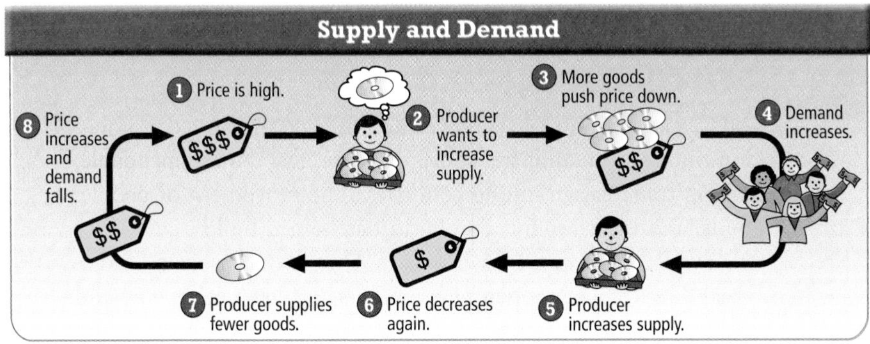

MORE ABOUT...
Consumerism

Consumerism is a movement that promotes the interests of consumers of goods and services by working to protect them from unsafe products, fraudulent advertising, and unfair business activities. Consumer groups believe that consumers are entitled to four basic rights: the right to quality, the right to safety, the right to information, and the right to choose. Consumer groups are particularly active when it comes to products for babies and young children, electrical products, and automobiles.

CRITICAL THINKING ACTIVITY

Hypothesizing Ask students to predict what steps the music company described on this page will take next. Ask the following: Based on the actions the music company has taken so far with the CDs, what do you think will happen to the CDs that were reduced to $15.95? Will they sell? What will the company do if they do not sell? How long do you think the company will offer them at a reduced price? What might the CD company learn from this experience?

Class Time 10 minutes

Activity Options

Differentiating Instruction: Less Proficient Readers

Finding the Main Idea Some students may have difficulty understanding the various concepts and terms introduced in the section. Write the following chart on the chalkboard. Have students copy it onto their own papers, then scan the section to find the terms. Then have them restate the meaning of each term in their own words and write their definitions on the chart. Encourage them to include examples where appropriate.

goods	
services	
capital resources	
GDP	
profit	
demand	
free enterprise	

MORE ABOUT...

Competition

In a free market economy, businesses that produce the same or similar products must compete with each other. As a result of this competition, prices are kept at a reasonable level. For example, if one store raises the price of an article, shoppers are free to look around for another store with a lower price for the item. This competition is so central to the economy of the United States that there are laws to protect it. Sellers cannot agree together to limit competition. Most monopolies, or companies that control all the production of a single product, are illegal.

INSTRUCT: Objective 4

Other Economic Systems/ The Global Economy

- In a command economy, who sets the quantity and prices of goods? the government
- In a traditional economy, what determines how goods and services are produced and how much they cost? social roles and culture
- What factors have led to the development of a global economy? the movement of people, goods, and ideas around the world
- Which countries are involved in NAFTA? United States, Mexico, Canada

Culture • Stores compete in malls and online to attract consumers. They use advertising and appealing store displays to catch the attention of potential customers. ▲

Supply and demand explain how price and availability are affected by how much consumers are willing to pay for an item and how much the seller decides to charge for it. The number of CDs offered at each price is the supply. The number of CDs that people will buy at each price is the demand.

Competition In a free enterprise or market economy, many businesses produce similar goods or services. There is competition to attract consumers. **Competition** is the rivalry among businesses to sell goods to consumers and make the greatest profit. To achieve these goals, a company may offer an improved product, manufacture it more cheaply, or sell it at a better price.

Other Economic Systems 

Most countries combine features from three types of economic systems: market, command, and traditional economies.

Command Economy In this system, the government decides how many of which goods are produced and sets the prices. Countries, such as North Korea and Cuba with Communist governments, have command economies. China has elements of market and command economies.

Traditional Economy In this system, social roles and culture determine how goods and services are produced, what prices and individual incomes are, and which consumers are allowed to buy certain goods. For instance, a family's status may determine whether they can own a tractor. Farmers may give much of their produce to community leaders. India has features of both market and traditional economies.

Reading Social Studies

B. Making Inferences Why might countries combine features of different economic systems?

Reading Social Studies
B. Possible Answer
Countries might combine features of different economic systems to further economic goals by combining the best of different systems, or to satisfy certain political or social constraints.

The Global Economy 

Today, more countries than ever before have market economies. Communication and transportation are fast and dependable, making trade easier among countries. The movement of people, goods, and ideas around the world has helped build a global, or worldwide, economy in which the United States is a leader. Expanding trade can open new markets and keep prices low and quality high for consumers. U.S. citizens buy many cars and clothes from other countries that take part in the global economy.

Activity Options

Skillbuilder Mini-Lesson: Reading a Physical Map

Explaining the Skill Review with students that physical maps show the natural features of Earth's surface. Remind them that understanding the landforms of a region can help us understand how people live and work there.

Applying the Skill To practice this, students will identify bodies of water in the United States that can be used to transport goods from one region to another.

1. Have students study the physical map of North America in the Unit 2 Atlas. On an outline map of the U.S., have them find, draw, and label waterways that can be used to transport goods both within a region and to other regions. Remind them to look at lakes as well as rivers.
2. When finished, have students share their maps and identify which regions are linked by the waterways they identified. Encourage them to see how the whole country can be linked by waterways.

World Trade Partners of the United States, 2000

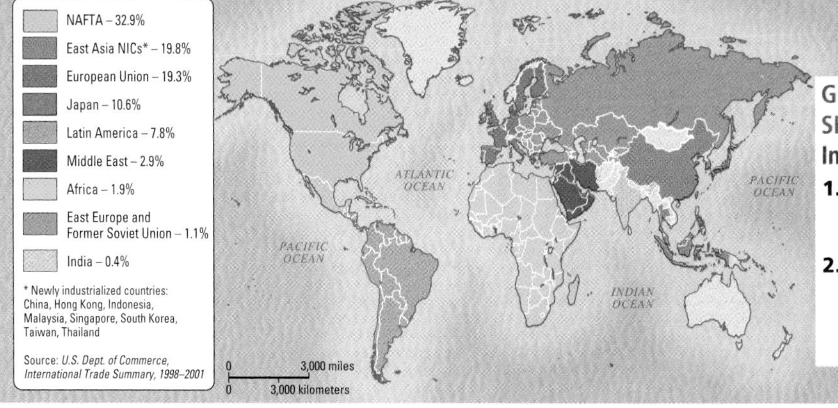

GEOGRAPHY SKILLBUILDER: Interpreting a Map

1. **Movement •** Who are the largest U.S. trade partners?
2. **Movement •** With which countries or regions does the United States trade least?

Geography Skillbuilder Answers

1. NAFTA, East Asia NICS, European Union
2. Africa, East Europe and former Soviet Union, India

Trade Barriers Sometimes countries establish barriers to restrict trade because they prefer to produce their own goods or services. Tariffs, or taxes on imported goods, raise the price to the consumer and make it more difficult for other countries to compete.

In 1994, NAFTA, the North American Free Trade Agreement, reduced trade barriers among the United States, Canada, and Mexico. There was concern that companies would move factories to Mexico, where workers earn less, and U.S. workers would lose jobs. However, many economists believe free trade benefits all three countries.

SECTION 3 ASSESSMENT

Terms & Names

1. **Explain the significance of:** (a) factors of production (b) GDP (c) free enterprise/market economy (d) consumer (e) profit (f) competition

Using Graphics

2. Use a chart like this one to take notes on the characteristics of free enterprise/market, command, and traditional economies.

Free Enterprise/ Market Economy	Command Economy	Traditional Economy

Main Ideas

3. (a) How is the price of goods decided in the U.S. economic system?
   (b) What are the four factors of production?
   (c) What role does government have in the U.S. market economy?

Critical Thinking

4. **Making Inferences**
   What effect will reducing trade barriers between countries have on the price of goods? Explain.

Think About

- the characteristics of a free enterprise/market economy
- competition

ACTIVITY -OPTION- Write an **advertisement** for a new business showing the goods or services it will produce. Make a chart showing the type of materials, labor, and capital you will need to produce the product.

FOCUS ON VISUALS

Interpreting the Map Direct students' attention to the map key. Ask them which countries are included in NAFTA, and what might be one reason why the other two countries have become such large trade partners with the United States.

Possible Responses Students should be able to identify Canada and Mexico as the non-U.S. NAFTA partners. Encourage students to see the sharing of borders as one important factor in forming trading partnerships.

Extension Ask students to study the map and come up with two or three countries that might be included in NAFTA-like organizations in Europe and in Africa.

ASSESS & RETEACH

Reading Social Studies Have students fill in the main ideas of the economy covered in this section on the chart on page 86.

Formal Assessment
- Section Quiz, p. 49

RETEACHING ACTIVITY

In small groups, have students complete the graphic in the section assessment and discuss the characteristics of the different economies.

In-depth Resources: Unit 2
- Reteaching Activity, p. 21

Access for Students Acquiring English
- Reteaching Activity, p. 27

Section 3 Assessment

1. Terms & Names

a. factors of production, p. 103
b. GDP, p. 104
c. free enterprise/market economy, p. 105
d. consumer, p. 105
e. profit, p. 105
f. competition, p. 106

2. Using Graphics

Free Enterprise/ Market Economy	Command Economy	Traditional Economy
Price and quantity of goods determined by supply and demand	The government decides what goods are produced	Cultural traditions determine economic decisions
Competition ensures the best product at the lowest price	The government sets prices	Status in society determines economic success

3. Main Ideas

a. The price of goods is determined by the supply of goods, the demand for them, and competition among sellers.
b. The four factors of production are natural resources, labor resources, capital resources, and entrepreneurs.
c. Government has a minor role.

4. Critical Thinking

The price of goods may go down because of increased competition and reduced production costs.

ACTIVITY OPTION

Integrated Assessment
- Rubric for writing an advertisement

The Legacy of North America

OBJECTIVE

Students learn about the contributions of North America to transportation, communication, and recreation.

FOCUS & MOTIVATE

Drawing Conclusions Ask students to study the pictures and read each paragraph. Then have them answer the following questions:

1. What inventions discussed in this feature help people contact each other?
2. What inventions discussed in this feature are used both for work and recreation?

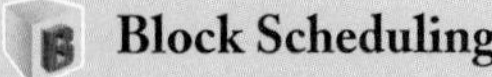

MORE ABOUT...

Telephones

Alexander Graham Bell applied for his first telephone-related patent in 1876. Then, on March 10, 1876, he and his partner, Thomas Watson, sent the human voice over wire. They were working in different rooms and about to try a new type of transmitter when Watson heard Bell say, "Mr. Watson, come here. I want you!" Bell had spilled battery acid on his clothes, and was calling for help. When both men realized that they had transmitted the human voice electronically, the accident was quickly forgotten.

Lacrosse

Lacrosse is the oldest continuously played sport in North America. It is based on a game American Indian tribes played hundreds of years ago. George Beers, a Canadian lacrosse player, standardized the game in 1867. He set field dimensions, limited the number of players, and provided a set of rules.

Telephones

In 1876, the Scottish-born American Alexander Graham Bell (1847–1922) invented the telephon Telephones have come a long way since Bell's da Today, caller identification, automatic redialing, and call waiting are common telephone services Cordless phones, mobile phones, and the Intern allow people throughout the world to communicate over great distances.

Lacrosse

Native peoples of North America played an early version of lacrosse long before Columbus landed. Players used a racket, now known as a crosse, to throw a hard ball down the field and into the opposing team's goal. Today, lacrosse is a popular sport not only in North America but also in Ireland, Australia, and South Africa. Men's teams play for the ILF World Championship, and women's teams compete for the World Cup.

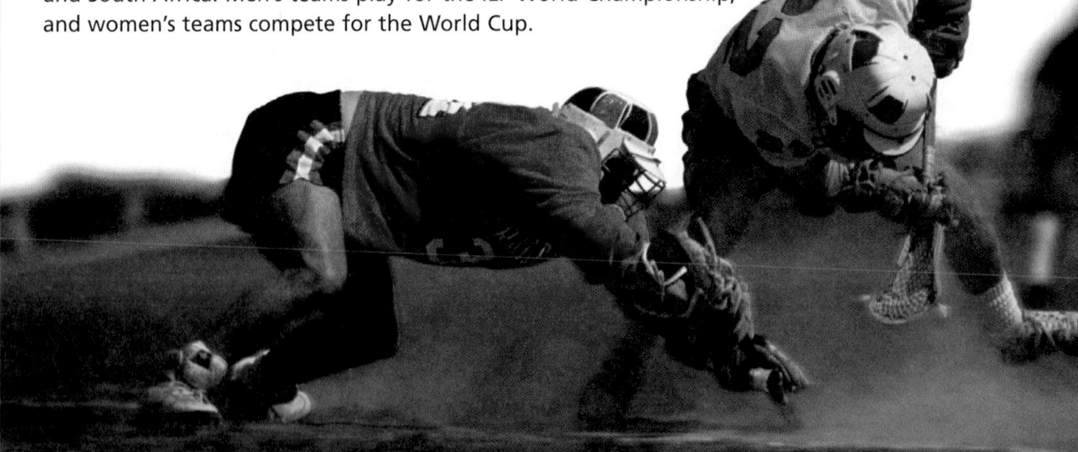

Activity Option

Interdisciplinary Link: Speech

Block Scheduling

Class Time 60 minutes

Task Developing and delivering a sales presentation for an invention/item to make skyscrapers practical

Supplies Needed

- Information about the development of skyscrapers

Activity Innovative inventions often need additional inventions to become practical for widespread use. Brainstorm what was needed to take skyscrapers from a creative idea to a practical one. Then break students into groups and assign each group one of the features discussed. Have each group prepare a sales presentation to convince a builder that his new building cannot be built without this item. Tell students to use information about the development of skyscrapers to support their arguments. Encourage them to develop creative ways of making the sales speech.

irplanes

1903, Wilbur (1867–1912) d Orville (1871–1948) ight achieved the first wered, sustained flight ar Kitty Hawk, N.C. The ght lasted 12 seconds and ached a height of 120 et. By the 1920s, aviation d become a booming dustry. Today, jumbo and personic jets allow people travel around the world just a day or two. More an 1.5 billion passengers nd and take off at the orld's 50 busiest airports ery year.

Find Out More About It!

Study the text and photos on these pages to learn about inventions, creations, and contributions that have come from North America. Then choose the item that interests you the most and do research in the library or on the Internet to learn more about it. Use the information you gather to construct a model of the item you choose.

RESEARCH LINKS
CLASSZONE.COM

Skyscrapers

In the late 1800s, U.S. architects and engineers worked together to build the first skyscrapers in Chicago and New York City. Today, these enormous structures rise in cities all over the world. The tallest skyscrapers in the world are the Petronas Towers in Kuala Lumpur, Malaysia, which reach a height of 1,483 feet.

Snowmobiles

In the winter of 1922, 15-year-old Joseph-Armand Bombardier of Canada invented the first crude snowmobile by attaching a car engine to sleigh runners and adding a wooden propeller to the back. In 1935, he refined his invention after his young son died from appendicitis because they couldn't make it through the snow to the hospital in time. Today, people around the world use snowmobiles for winter fun, and people in the remote north depend on them for transportation when the snows are deep.

INSTRUCT

- Where were the first skyscrapers built?
- Why did the inventor of the snowmobile refine his first invention?
- Who played lacrosse in North America before Columbus landed?

MORE ABOUT...

Airplanes

The Wright Brothers' historic flight on December 17, 1903, was witnessed by four men and one boy, and one man took a picture just as Orville Wright left the ground. Still, it received little public attention. Only a few newspapers mentioned it, and their stories were inaccurate. The Wrights continued flying, but despite some accurate reports, their achievement remained basically unknown for five years. It was not until they signed a contract with the U.S. Department of War in 1908 that newspapers began to report on their flights.

From this quiet beginning, aviation quickly became big business. By 1919, the first commercial airlines began service in Europe. Airlines began operating in other parts of the world during the 1920s. By the late 1930s, there were three and a half million air passengers a year.

Skyscrapers

The skyscraper developed in stages during the second half of the 1800s. Safety elevators, metal frames, and other new technologies made the modern skyscraper possible. The elevator, for example, made it possible to construct buildings taller than the number of stairs people could climb. The metal frame supported more stories than masonry walls.

More to Think About

Making Personal Connections Ask students to think about the widespread daily use of telephones in their lives. Have them consider what aspects of their lives would be changed if there were no telephones. What transactions, events, and arrangements would be altered? What might be used in place of the telephone? Ask students to come up with examples and ideas from their own lives. Discuss these as a class.

Vocabulary Activity Ask students to use the title word from each section to make a word-find puzzle. The words should be placed in rows with other letters. Tell them the words can appear left to right, right to left, up or down, reversed, or on a diagonal, but all of the letters of the words must appear in the correct order to form the word. Have them exchange their puzzles and solve them.

SECTION OBJECTIVES

1. To understand the values that define culture
2. To recognize the influences of other cultures on United States society
3. To understand how globalization affects the United States
4. To identify how discoveries made by United States scientists affect the world

SKILLBUILDER

• Interpreting a Map, p. 113

FOCUS & MOTIVATE

WARM-UP

Making Inferences Have students read Dateline and answer these questions.

1. Why do you think the opening of a pizzeria might have been newsworthy in 1905?
2. What would people today think about the last sentence?

INSTRUCT: Objective 1

American Way of Life

• What are values? the principles and ideals by which people live
• What are some examples of values? individual freedom, equal opportunity for jobs and education, fair treatment for all, private ownership of property
• Why is education considered to be so important to U.S. citizens? They believe they can improve their lives through education.

In-depth Resources: Unit 2
• Guided Practice Worksheet, p. 14

Reading Study Guide (Spanish and English), pp. 28–29

United States Culture: Crossing Borders

SECTION 4

TERMS & NAMES
value
globalization
technology

MAIN IDEA	WHY IT MATTERS NOW
The American way of life reflects the cultures of people from many countries around the world.	People around the world are more closely connected than ever before.

DATELINE

NEW YORK CITY, 1905—Visit Gennaro Lombardi's pizzeria—the first in the United States—to try a pizza. This baked, flat pie has a bread crust covered with cheese, tomato sauce, and seasonings.

Lombardi's pizzas are not exactly like the pizzas made in Naples, Italy, because different seasonings, flour, and cheese are available here. Also, instead of using sliced tomatoes like those put on Italian pizzas, Lombardi adds a spicy homemade tomato sauce. Customers say this pizza is as good as any from Naples. Judging by the warm reception it has received so far, pizza may become a popular U.S. food.

Place • Pizza, brought here by Italian immigrants, is becoming a popular food. ▲

American Way of Life 1

People in the United States have brought diverse customs, traditions, and foods, like pizza, from their homelands, but they share many of the same values. **Values** are the principles and ideals by which people live. U.S. citizens care about individual freedoms; equal opportunities for jobs and education; fair treatment of people regardless of race, religion, or gender; and private ownership of property. Many of these values are part of the U.S. Constitution and help define American culture.

TAKING NOTES
Use your chart to take notes about the United States today.

The United States Today	
History	
Government	

Program Resources

In-depth Resources: Unit 2
• Guided Reading Worksheet, p. 14
• Reteaching Activity, p. 22

Reading Study Guide (Spanish and English), pp. 28–29

Formal Assessment
• Section Quiz, p. 50

Integrated Assessment
• Rubric for writing an essay

Outline Map Activities

Access for Students Acquiring English
• Guided Reading Worksheet, p. 22

Technology Resources
classzone.com

TEST-TAKING RESOURCES
Strategies for Test Preparation
Test Practice Transparencies
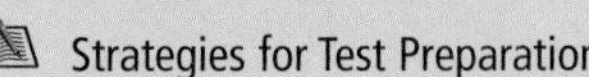
Online Test Practice

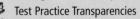

Education U.S. citizens believe they can improve their lives through education. In 1647, Massachusetts established the first colonial public school system. Today, state laws require that all children attend school or be taught at home until they are at least 16. More than 99 percent of U.S. children finish elementary school, and more than 85 percent complete high school.

U.S. Religions About 70 percent of all U.S. citizens are members of religious groups. Many colonists, such as British Protestants and Catholics, settled in America so that they could worship as they wished. Since then, people with many different religious beliefs have come to the United States. Most Spanish, French, and Italian immigrants were Catholic. In the 1900s, many European Jews settled in the United States. Asian immigrants practice Buddhism and Hinduism. North Africans and Southwest Asians brought Islam. Many Native Americans continue to practice their ancestors' religions.

Reading Social Studies

Drawing Conclusions Why do you think values such as freedom of religion were written into the U.S. Constitution?

Reading Social Studies Possible Answer

The writers of the Constitution felt these values were important and should be specifically protected.

The Arts and Entertainment ❷

Leisure activities in the United States reflect the influence of other cultures. For example, sports such as tennis, golf, soccer, and even baseball originated in other countries. Tennis came from France, golf from Scotland, and soccer from England. Baseball is probably based on rounders, a game played in Great Britain in the late 1700s. Basketball was invented in the United States by a Canadian and later spread to other countries. Football is played chiefly in the United States and Canada.

Culture • Baseball, often called America's national pastime, was probably adapted from a British game played in the 1700s. ▲

The movie and television industries and certain musical forms, such as rock 'n' roll, developed in the United States, although they were affected by other cultures. Jazz was greatly influenced by the blues, which is rooted in spirituals once sung by enslaved Africans. Today, artists and audiences around the world enjoy these American musical styles.

Place • Americans eat foods from the traditions of many countries, such as Japanese sushi and sashimi. ▼

Globalization of Culture The international popularity of U.S. music is an example of the globalization of culture. **Globalization** means spreading around the world. Today, cultural influences often cross national boundaries. People around the world enjoy blue jeans, sodas, and fast food from the United States. McDonald's serves about 45 million people a day in 121 countries.

INSTRUCT: Objective ❷

The Arts and Entertainment

- What are some leisure activities played in the United States that reflect the influence of other countries? soccer, tennis, golf, baseball
- What is meant by the globalization of culture? the spreading around the world of cultural influences

MORE ABOUT...

Basketball

James Naismith, the man who invented basketball, was working as a physical-education instructor in Springfield, Massachusetts, when the head of his department asked him to create a team sport that could be played indoors during the winter. His idea was to have teams score points by throwing a ball into boxes placed at opposite sides of the gym. He used a soccer ball because it was large enough to catch. He asked for two boxes to use as goals for his new game, but was given two peach baskets instead. These baskets were attached to the balcony railing of the indoor gym. This new sport caught on immediately. Soon it was being played throughout the United States and Canada.

Activity Options

Multiple Learning Styles: Logical

Class Time 15 minutes

Task Naming examples of globalization of culture in daily life

Purpose To realize how other cultures are commonly reflected in our daily lives

Supplies Needed
- Beanbag

Block Scheduling

Activity Have small groups of students sit in circles. Give a beanbag to one student in each group and ask him or her to name an example of how he or she recently has been influenced by a person or product from another culture. Examples might be a music group, a car, a food, a movie actor, or an author from another country. Then tell him or her to toss the beanbag to someone else. Continue until students run out of examples.

INSTRUCT: Objective 3

U.S. Science and Technology

- What does modern technology enable United States scientists to do? work with other scientists around the world
- What are two examples of the negative impact of technology? pollution, loss of unique cultural features, poorer nations may not benefit fully from technology

Spotlight on CULTURE

The 102-story Empire State Building, measuring 1,250 feet from the sidewalk to the roof, was completed in 1931. The building rises in a series of step-like shapes to a tower topped by a thin spire. The architecture is an example of art deco, a geometric style popular in the 1920s and 1930s. A steel framework is covered with panels of limestone and an alloy of chrome, nickel, and steel. This construction is so strong that only two floors suffered serious damage when a plane accidentally crashed into it in 1945.

U.S. citizens eat Japanese sushi, listen to Italian operas, and drive South Korean cars. Literature from many nations is translated into different languages. Print and electronic communication, television, movies, and the Internet provide speedy ways to share the products and creations of different cultures.

Thinking Critically Answers

1. Two factors were limited space and new technology.
2. Building height is measured from the ground level main entrance to the structural top, not including antennas or flagpoles.

U.S. Science and Technology 3

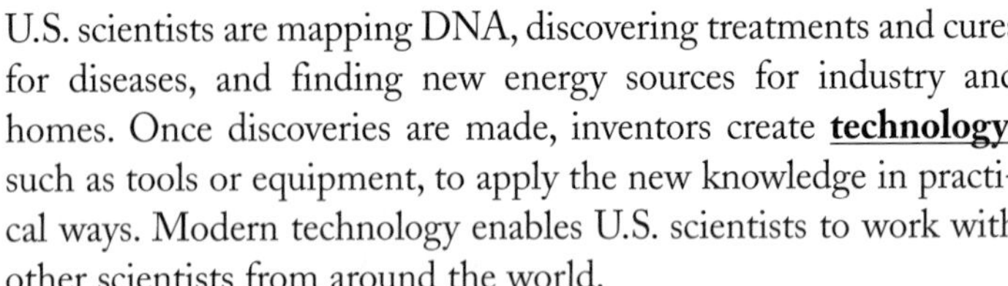

U.S. scientists are mapping DNA, discovering treatments and cures for diseases, and finding new energy sources for industry and homes. Once discoveries are made, inventors create **technology,** such as tools or equipment, to apply the new knowledge in practical ways. Modern technology enables U.S. scientists to work with other scientists from around the world.

BACKGROUND

DNA is the molecule that carries the information that determines the characteristics of every living thing. It was first identified by Francis Crick, James Watson, and others in 1953.

Science and Technology Change the World Discoveries by U.S. scientists help people throughout the world. Polio, a disease that usually affects children, was widespread in the 1940s and 1950s.

Spotlight on CULTURE

Skyscrapers The skyscraper originated in the United States in the 1870s and has caught on in other countries. Limited space in cities, plus materials and technology, such as steel and elevators, have inspired architects to design taller buildings.

The height of each building is measured from the ground-level main entrance to the structural top, including spires but not antennas or flagpoles. Today, many of the tallest buildings are located outside the United States in places such as Kuala Lumpur, Malaysia, and Hong Kong and Shanghai in China.

World's Tallest Skyscrapers

Height in Feet: 1,800; 1,600; 1,400; 1,200; 1,000; 800; 600; 400; 200

Name	Petronas Towers	Sears Tower	Jin Mao Building	Plaza Rakyat	Empire State Building	Central Plaza	Bank of China
City	Kuala Lumpur	Chicago	Shanghai	Kuala Lumpur	New York	Hong Kong	Hong Kong
Year Built	1998	1974	1998	1999	1931	1992	1989

THINKING CRITICALLY

1. **Finding Causes** What were two factors that led architects to design skyscrapers?
2. **Recognizing Important Details** How is building height measured?

For more on skyscrapers RESEARCH LINKS CLASSZONE.COM

Activity Options

Interdisciplinary Link: Art

Class Time One class period

Task Creating a collage that illustrates the cultural mix of the United States

Purpose To explore the effects of cultural influences in the United States

Supplies Needed
- Old magazines, newspapers
- Paint, markers, pencils
- Poster board, paste

Activity Have students use words and pictures cut out of old magazines and newspapers, markers, and paint to create a collage that illustrates the mix of cultures that exists in the United States. Encourage students to write sentences that express how this mix of cultures has enriched their own lives.

112 CHAPTER 4

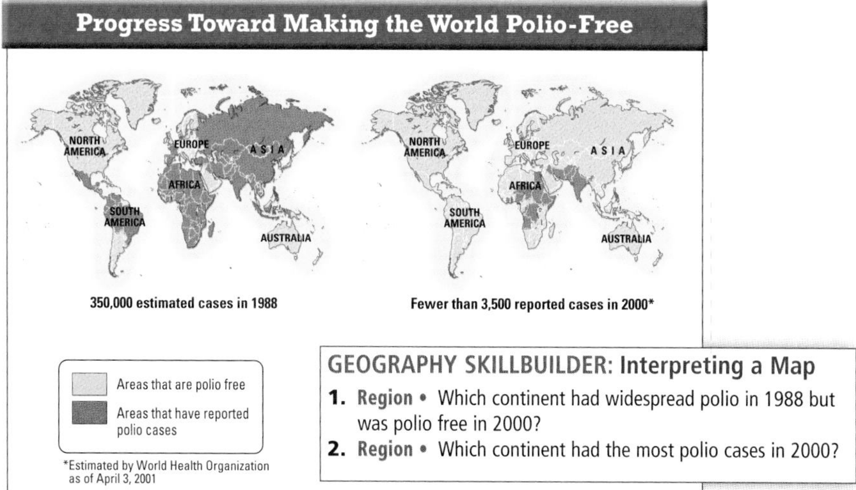

GEOGRAPHY SKILLBUILDER: Interpreting a Map

1. **Region** • Which continent had widespread polio in 1988 but was polio free in 2000?
2. **Region** • Which continent had the most polio cases in 2000?

Then, U.S. doctors Jonas Salk and Albert Sabin each developed a different vaccine. As a result, great progress has been made toward making the world free of polio.

Negative effects of technology include increased pollution of the environment and the loss of unique cultural features as countries share languages, foods, and customs. Poorer nations may lack the money and skilled labor needed to benefit from new applications of science.

Geography Skillbuilder Answers
1. South America
2. Africa

SECTION 4 ASSESSMENT

Terms & Names

1. **Explain the significance of:** (a) value (b) globalization (c) technology

Using Graphics

2. Make a chart like the one below to show features of U.S. culture and science that have spread to other parts of the world.

U.S. Culture and Science Around the World	
Culture	Science and Technology

Main Ideas

3. (a) What are some values shared by people in the United States?
   (b) What are some examples of contributions from other cultures to the American way of life?
   (c) In what ways does American culture influence people in other countries?

Critical Thinking

4. **Forming and Supporting Opinions**
   Do you think globalization has a positive or a negative effect on the world? Support your opinion.

Think About

- the effects of globalization
- the changes caused by international trade and communication
- the effects of scientific discoveries worldwide

ACTIVITY -OPTION- Reread "The Arts and Entertainment" on page 111. Think about the sports, music, and movies you enjoy. Write an **essay** describing how the cultures of other countries influence your activities.

MORE ABOUT...
Jonas Salk

In 1953 Jonas Edward Salk announced the development of a trial vaccine to prevent poliomyelitis. Salk, his wife, and their three sons were among the first to receive the vaccine. It was administered to 1,830,000 schoolchildren. Salk received many honors including a Congressional Medal of Honor for "great achievement in the field of medicine." He refused all cash awards.

ASSESS & RETEACH

Reading Social Studies Have students fill in the chart on page 86 with the main ideas about our culture covered in this section.

Formal Assessment
- Section Quiz, p. 50

RETEACHING ACTIVITY

Ask students to work in small groups to present a section review. One group might stage a panel discussion about the U.S. contributions to technology and science. Another group can present a panel on arts and entertainment; another on American values. Panel members should be prepared to answer questions posed by the class.

In-depth Resources: Unit 2
- Reteaching Activity, p. 22

Access for Students Acquiring English
- Reteaching Activity, p. 28

Section 4 Assessment

1. Terms & Names

a. value, p. 110
b. globalization, p. 111
c. technology, p. 112

2. Using Graphics

Culture	Science and Technology
clothing	medicine
television shows	communication advances
food	energy sources
music	

3. Main Ideas

a. a belief in equal opportunity, education, private ownership of property
b. different religious beliefs, foods, types of music, sports
c. American culture influences people as globalization helps spread ideas around the world. People in other countries wear blue jeans, drink soda, and eat fast food.

4. Critical Thinking

Possible Response The mixing of cultures opens up new ideas and spreads advances in medicine and technology. However, there may be a loss of unique cultural traits.

ACTIVITY OPTION

Integrated Assessment
- Rubric for writing an essay

TERMS & NAMES

1. immigrant, p. 88
2. citizenship, p. 91
3. democracy, p. 91
4. political process, p. 91
5. United States Constitution, p. 94
6. Bill of Rights, p. 96
7. factors of production, p. 103
8. free enterprise/market economy, p. 105
9. value, p. 110
10. globalization, p. 111

REVIEW QUESTIONS

Possible Responses

1. People have immigrated to the U.S. for better jobs and educational opportunities; to escape oppression, war, or natural disaster.
2. They must pay taxes, vote, serve on juries, and serve in the military.
3. The early leaders wanted to create a stable government that protected the rights and freedoms of the people.
4. They wanted to create a structure in which each branch of government would serve as a check on the power of the other branches.
5. Government must fund fire protection, schools, roads, and the military.
6. Natural resources provide the raw materials; labor resources provide the workers; capital resources provide money for machines, factories, and supplies; entrepreneurs bring the other factors of production together.
7. The writers of the Constitution wanted to support and protect these values and ensure that they would not be challenged at a later time.
8. Globalization may blur the differences among cultures of the world.

CHAPTER 4 ASSESSMENT

TERMS & NAMES

Explain the significance of each of the following:

1. immigrant
2. citizenship
3. democracy
4. political process
5. United States Constitution
6. Bill of Rights
7. factors of production
8. free enterprise/market economy
9. value
10. globalization

REVIEW QUESTIONS

We the People *(pages 87–91)*

1. What are some reasons that people have immigrated to the United States?
2. What are some responsibilities of citizens of the United States?

A Constitutional Democracy *(pages 94–98)*

3. What were the goals of the early leaders when they wrote the United States Constitution?
4. What were the leaders attempting to do when they set up a system of checks and balances within the federal government?

The United States Economy *(pages 102–107)*

5. What are examples of wants that a government must fund?
6. How does each factor of production contribute to the making of a product?

United States Culture: Crossing Borders *(pages 110–113)*

7. What can you conclude from the fact that cultural values were directly stated in the U.S. Constitution?
8. If globalization continues, how will it affect the cultures of the world?

CRITICAL THINKING

Synthesizing

1. Using your completed chart from Reading Social Studies, p. 86, explain why the economy, form of government, and cultural values of the United States continue to attract so many immigrants each year.

Forming and Supporting Opinions

2. Martin Luther King, Jr., fought against laws he thought were unjust. How was this good citizenship? Explain.

Making Inferences

3. What were the major concerns of the leaders who wrote the U.S. Constitution and how did they address them?

Visual Summary

1 We the People

- The United States is made up of immigrants from many countries.
- U.S. citizens share many rights and responsibilities.

2 A Constitutional Democracy

- After the American colonists gained independence, they wrote a Constitution that created a limited government and protected their rights and freedoms.

3 The United States Economy

- The United States has a free enterprise/market economy.
- Because more countries have market economies, a global economy is developing with expanded trade among nations.

4 United States Culture: Crossing Borders

- Americans believe in strong values, many of which are protected by the U.S. Constitution.
- American culture has been shaped by many groups and now influences the cultures of other nations.

CRITICAL THINKING: Possible Responses

1. Synthesizing
Students might note that the United States offers more political, economic, and cultural freedoms than some other countries.

2. Forming and Supporting Opinions
Students should recognize that good citizenship involves doing what is necessary to improve society. In a democracy, working to change unfair laws is a right that Martin Luther King, Jr., exercised.

3. Making Inferences
They were committed to both democracy and stability. Their careful organization of the government and their emphasis on protecting the rights of individuals indicates this.

STANDARDS-BASED ASSESSMENT

the map and your knowledge of world cultures and graphy to answer questions 1 and 2.

Additional Test Practice, pp. S1–S33

Percentage of Foreign-Born Population

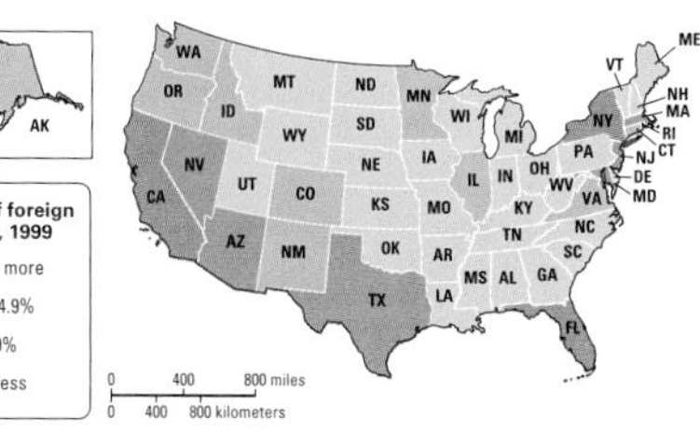

1. Which states have the highest percentage of foreign-born people?
   A. Alaska, Colorado, Oregon, Washington
   B. Arizona, Maryland, Nevada, Texas
   C. Arkansas, Iowa, Louisiana, Mississippi
   D. California, Florida, Hawaii, New York

2. Where are most states with a high percentage of foreign-born people located?
   A. On the coasts or borders
   B. In the southeast region
   C. In the central region
   D. Only by the northern border

This quotation is from a presentation made to high school and middle school students. Use the quotation and your knowledge of world cultures and geography to answer question 3.

PRIMARY SOURCE

[Martin Luther King day] is America's celebration.
It is each of our individual's day of celebration.
It is a moment of rejoicing for all Americans.
It is a moment of reflection for all Americans.
It is a moment of renewal on our long diversity journey.
It is a moment of challenge for each of us.

FRANK A. BLETHEN, Publisher, *The Seattle Times,* from a presentation on January 17, 2003

3. Which of the following best expresses the point of view of the passage?
   A. The holiday can be both an individual and a group holiday.
   B. The holiday contradicts the basic values of the U.S. Constitution.
   C. The holiday is a time set aside only for celebration.
   D. It is not necessary for all Americans to celebrate or observe the day.

ALTERNATIVE ASSESSMENT

WRITING ABOUT HISTORY

The music of other cultures has influenced many of the forms of music enjoyed in the United States and many American musicians. Write a biography of a favorite musician of today or from the past.

Research the musician to learn what musical influences shaped his or her work. Find out if these influences were from other cultures or times.

If possible, include copies of photographs with captions to accompany your biography.

COOPERATIVE LEARNING

With a group of four to six classmates, think of a natural event such as a flood, drought, or storm that affected your community or your region of the country. Research the event and interview people who remember it. Create a display about the event, explaining its causes and its effects on the area. Share the responsibilities of researching, interviewing, and creating drawings, models, or diagrams.

INTEGRATED TECHNOLOGY

Doing Internet Research

Use the Internet, as well as other sources in the library, to learn about the Constitutional Convention. Write a report of your findings.

- You might research historic documents or biographies of the founding fathers to find information about process of creating the Constitution.
- In your research, find out what the specific concerns of the small and large states were and how they differed. Explore how a compromise was reached that allowed the convention to draft the Constitution.
- You might use a chart or graph to present information about the different proposals and the final compromise.

For Internet links to support this activity, go to

STANDARDS-BASED ASSESSMENT

1. **Answer D** is the correct answer because it includes the states with 15.00 percent or more of foreign-born people as indicated by the map. Answers A, B, and C are incorrect because they include states with lower percentages of foreign-born people.
2. **Answer A** is correct because most of the states indicated by purple and red are located on coasts. Answer B is incorrect because several states with a high percentage of foreign-born people are not in the southeast; answers C and D are incorrect because such states are not in the central region or near the northern border.
3. **Answer A** is the correct answer because it restates the point that the holiday is both for individuals and the group. Answer B is incorrect because the holiday honors the values of freedom and equality; answer C is incorrect because the holiday is a time for reflection as well as celebration; answer D is incorrect because, according to the passage, the essence of the holiday is important to all Americans.

INTEGRATED TECHNOLOGY

Discuss how students might find information about the Constitutional Convention using the Internet. Brainstorm possible key words and Web sites to explore.

Suggest that students make a chart to organize the concerns of large and small states. They can use the information to write their reports.

Alternative Assessment

1. Rubric

The biography should
- accurately convey information about the musician's life.
- be logically organized.
- be written in an interesting style.
- use correct grammar, spelling, and punctuation.

2. Cooperative Learning

Have students look for specific facts about the event, such as the date it occurred and its effects in terms of personal and economic losses. Suggest that students find out if residents can do anything to prepare for a similar disaster.

You might want to make available a variety of art supplies, such as watercolor paints, cardboard, clay, or papier-mâché.

Canada Today

	OVERVIEW	COPYMASTERS	INTEGRATED TECHNOLOGY
UNIT ATLAS AND CHAPTER RESOURCES	The students will explore Canada today and examine the ways in which its geography and history have shaped aspects of Canadian life.	**In-depth Resources: Unit 2** • Guided Reading Worksheets, pp. 23–26 • Skillbuilder Practice, p. 29 • Unit Atlas Activities, pp. 1–2 • Geography Workshop, pp. 35–36 **Reading Study Guide** (Spanish and English), pp. 32–41 **Outline Map Activities**	• eEdition Plus Online • EasyPlanner Plus Online • eTest Plus Online • eEdition • Power Presentations • EasyPlanner • Electronic Library of Primary Sources • Test Generator • Reading Study Guide • The World's Music • Critical Thinking Transparencies CT9
	KEY IDEAS		
SECTION 1 O Canada! Immigrant Roots pp. 119–123	• Canada supports its different citizen groups with an official policy of multiculturalism. • Geographic features and economic opportunities lead most Canadians to settle in the southern part of the country.	**In-depth Resources: Unit 2** • Guided Reading Worksheet, p. 23 • Reteaching Activity, p. 31 **Reading Study Guide** (Spanish and English), pp. 32–33	**Critical Thinking Transparencies CT10** **Map Transparencies MT14** **classzone.com** Reading Study Guide
SECTION 2 A Constitutional Monarchy pp. 124–127	• The prime minister leads Canada's government. • The Charter of Rights guarantees many freedoms for all Canadians, including freedom of speech and religion. • Separatists want Quebec to become an independent nation.	**In-depth Resources: Unit 2** • Guided Reading Worksheet, p. 24 • Reteaching Activity, p. 32 **Reading Study Guide** (Spanish and English), pp. 34–35	**classzone.com** Reading Study Guide
SECTION 3 Canada's Economy pp. 128–131	• Timber, agriculture, and mining are important industries in Canada. • Much of Canada's work force is well-educated — a boon to Canada's economy. • Transportation is a vital part of Canada's economy.	**In-depth Resources: Unit 2** • Guided Reading Worksheet, p. 25 • Reteaching Activity, p. 33 **Reading Study Guide** (Spanish and English), pp. 36–37	**classzone.com** Reading Study Guide
SECTION 4 A Multicultural Society pp. 132–136	• Many Canadians speak both English and French. • The Canada Council for the Arts provides grants and support for arts organizations. • Canada has many culture regions, areas whose residents share common languages, customs, or lifestyles.	**In-depth Resources: Unit 2** • Guided Reading Worksheet, p. 26 • Reteaching Activity, p. 34 **Reading Study Guide** (Spanish and English), pp. 38–39	**classzone.com** Reading Study Guide

KEY TO RESOURCES

 Audio
 CD-ROM
Copymaster
 Internet
Overhead Transparency
Pupil's Edition
 Teacher's Edition
 Video

ASSESSMENT OPTIONS

Chapter Assessment, pp. 138–139

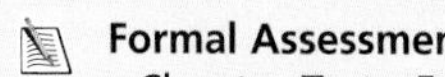

Formal Assessment
- Chapter Tests: Forms A, B, C, pp. 67–68

Test Generator

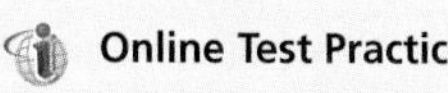

Online Test Practice

Strategies for Test Preparation

Section Assessment, p. 123

Formal Assessment
- Section Quiz, p. 63

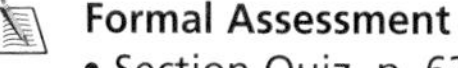

Integrated Assessment
- Rubric for writing a magazine article

Test Generator

Test Practice Transparencies TT11

Section Assessment, p. 127

Formal Assessment
- Section Quiz, p. 64

Integrated Assessment
- Rubric for creating a travel poster

Test Generator

Test Practice Transparencies TT12

Section Assessment, p. 131

Formal Assessment
- Section Quiz, p. 65

Integrated Assessment
- Rubric for writing a newspaper article

Test Generator

Test Practice Transparencies TT13

Section Assessment, p. 136

Formal Assessment
- Section Quiz, p. 66

Integrated Assessment
- Rubric for creating a mural or collage

Test Generator

Test Practice Transparencies TT14

RESOURCES FOR DIFFERENTIATING INSTRUCTION

Students Acquiring English/ESL

Reading Study Guide (Spanish and English), pp. 32–41

Access for Students Acquiring English Spanish Translations, pp. 29–38

TE Activity
- Using Prefixes, p. 133

Modified Lesson Plans for English Learners

Less Proficient Readers

Reading Study Guide (Spanish and English), pp. 32–41

TE Activity
- Organizing Information, p. 125

CROSS-CURRICULAR CONNECTIONS

Literature
Granfield, Linda. ***High Flight: A Story of World War II.*** Toronto: Tundra Books, 1999. Stirring story of the fighter pilot John Magee who wrote "High Flight," a legendary poem.

Popular Culture
Kalman, Bobbie. ***Canada Celebrates Multiculturalism.*** New York: Crabtree, 1993. A look at the many cultures within the borders.

Geography
Bowers, Vivien. ***Wow, Canada!: Exploring This Land from Coast to Coast to Coast.*** Toronto: Owl Communications, 1999. A family's lighthearted look at traveling through Canada.

Science/Math
Shemie, Bonnie. ***Building Canada.*** Toronto: Tundra Books, 2001. Architecture defines Canadian culture.

Language Arts/Literature
Mowat, Farley. ***The Dog Who Wouldn't Be.*** New York: Bantam, 1983. Mowat recalls growing up with his hilarious dog, Mutt, on the Canadian prairie.

Science
Shell, Barry. ***Great Canadian Scientists.*** Vancouver: Polestar Press, 2000. Biographies and projects for students.

Government
LeVert, Suzanne. ***Dominion of Canada.*** Philadelphia: Chelsea House, 2001. French nationalism in Quebec, foreign relations, native rights.

History
Junior Worldmark Encyclopedia of the Canadian Provinces. Detroit: U.X.L., 1999. Geography, economy, culture, politics, and social demographics.

ENRICHMENT ACTIVITIES

The following activities are especially suitable for classes following block schedules.

Teacher's Edition, pp. 126, 129, 130, 134

Pupil's Edition, pp. 123, 127, 131, 136

Unit Atlas, pp. 54–65

Outline Map Activities

INTEGRATED TECHNOLOGY

Go to **classzone.com** for lesson support and activities for Chapter 5.

BLOCK SCHEDULE LESSON PLAN OPTIONS: 90-MINUTE PERIOD

DAY 1

CHAPTER PREVIEW, pp. 116–117
Class Time 20 minutes

- **Hypothesize** Use the "What do you think?" questions in Focus on Geography on PE p. 117 to help students hypothesize about how the geography of Canada influences its culture.

SECTION 1, pp. 119–123
Class Time 70 minutes

- **Peer Teaching** Have students work in pairs to come up with five questions about immigration to Canada. Then have groups exchange questions and answer them.
Class Time 25 minutes
- **Making Generalizations** Have groups present the results of the research they did for Activity Options, Interdisciplinary Link: Science on TE p. 122 to the entire class. After all the groups have made their presentations, lead the class in making generalizations about climate in Canada.
Class Time 35 minutes
- **Geography Skills** Before students complete Geography Skillbuilder on PE p. 122, remind them that a population density map is an example of a thematic map. Discuss with students the purpose of a thematic map. Ask students what symbol is used to identify cities on the map. What do the different colors on the map represent?
Class Time 10 minutes

DAY 2

SECTION 2, pp. 124–127
Class Time 45 minutes

- **Small Groups** Divide the class into small groups. Have each group use Terms & Names on PE p. 124 to write one or two paragraphs supporting the main idea of this section: Canada is a democracy that protects the rights of individuals and of different cultures. Reconvene as a class and have groups read their paragraphs.
Class Time 30 minutes
- **Compare and Contrast** Use the questions in Skillbuilder: Reading a Chart on PE p. 126 to lead a discussion comparing and contrasting the governments of the United States and Canada.
Class Time 15 minutes

SECTION 3, pp. 128–131
Class Time 45 minutes

- **Poster** Have students work in pairs to create a poster to illustrate the goods produced in each of Canada's regions. Display student posters in the classroom.

DAY 3

SECTION 4, pp. 132–136
Class Time 35 minutes

- **Press Conference** As an interesting way to review the section, have the class conduct a press conference about Canadian culture. Select a group of four students to field questions about Canadian culture and society. The rest of the class should prepare questions to ask the group.

CHAPTER 5 REVIEW AND ASSESSMENT, pp. 138–139
Class Time 55 minutes

- **Review** Have students use the charts they created for Reading Social Studies on PE p. 118 to review how geographic, historical, and cultural factors have influenced Canada's development.
Class Time 20 minutes
- **Assessment** Have students complete the Chapter 5 Assessment.
Class Time 35 minutes

TECHNOLOGY IN THE CLASSROOM

KEYPALS

Keypals are pen pals who correspond via e-mail. Exchanging e-mail with keypals is one of the most popular ways for classes to use the Internet. When using keypals with your class, have a goal in mind—a project to collaborate on or set of questions to ask the other class. It is a good idea to avoid assigning individual students to their own keypals with the goal of completing a class project. If the other student does not cooperate, is out of school, etc., your student will not get the information necessary to finish the project. It is better to have groups become keypals with other groups or to correspond as an entire class.

ACTIVITY OUTLINE

Objective Students will correspond with keypals in different parts of Canada to learn what it is like to live there.

Task Have students hypothesize what it might be like to live in Canada. Then have them find out by e-mailing questions to students in different parts of the country.

Class Time 2–5 class periods, over a period of time

DIRECTIONS

1. Ask students if any of them have ever been to Canada. Where did they go? What was it like? What do they think it would be like to live there? If no one has been to Canada, ask students what they think it might be like to live there.
2. Locate a few Canadian classes with which to correspond, using the suggested Web sites at **classzone.com** or others you know about. Try to find at least one class from Eastern Canada, one from Central Canada, one from Western Canada, and one from the North.
3. Have students label the locations of their partner schools on a wall map of Canada, and display the map in the classroom until the end of the project.
4. Divide the class into small groups, and have each group write questions to ask their keypals. They should include a few questions about daily life and a few about things they have learned in Chapter 5, such as information about Canada's diversity of culture or its economy.
5. Have groups send their questions to their keypals. Specify a time by which they hope to receive a reply. As they receive their responses, suggest that they summarize the answers in a notebook or an e-file.
6. Once students have answers from their keypals, ask them to compare the responses they received to their original ideas about what life in Canada might be like. Have them write reports explaining whether their original ideas were accurate and describing what life is like for students their age living in Canada. They should conclude by answering the questions, "How is life in Canada different from or similar to your own life? What are some differences between life in different parts of Canada, according to what you have learned from your keypals?"

CHAPTER 5 OBJECTIVE

Students will explore the people, government, and economy of present-day Canada and examine the ways in which its geography and history have shaped aspects of Canadian life.

FOCUS ON VISUALS

Interpreting the Photograph Have students study the photograph of the icy waters of Hudson Bay. Ask them to locate Canada on the inset map and notice the country's size and far northern location. Ask students to think of words or phrases that describe the setting of the photo. Explain that this chapter focuses on the influence of geography on Canada's people and history. What might they assume about Canada from this picture?

Possible Responses cold, freezing, icy, wintry, dangerous; that at least some parts of the country are very cold and wild, without many people

Extension Ask students to write a descriptive paragraph or poem about the scene in the photograph.

CRITICAL THINKING ACTIVITY

Contrasting Have students look at the buildings in the inset photograph. Ask them to contrast the aspects of Canada they see there with the impression given by the photograph of Hudson Bay. Ask them to point out details in the inset photo that add to the contrasts.

Class Time 5 minutes

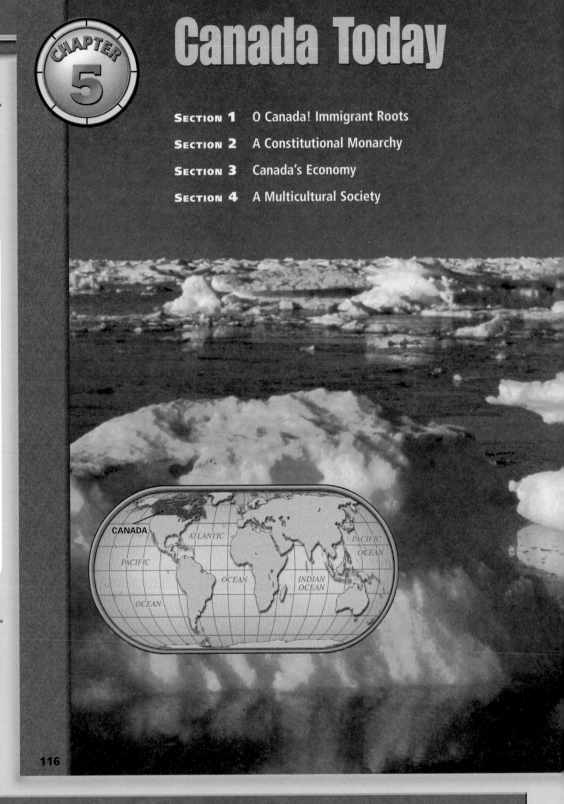

CHAPTER 5

Canada Today

CANADA
ATLANTIC
PACIFIC
OCEAN
PACIFIC OCEAN
INDIAN OCEAN
OCEAN

116

Recommended Resources

BOOKS FOR THE TEACHER

Canadian Almanac & Directory 2001: Over 47,000 Facts and Figures About Canada. Toronto, ON: Micromedia, 2000. Almanac and miscellany about Canada.

The Canadian Encyclopedia. Toronto, ON: McClelland & Stewart, 2000. People, places, and events.

Ivory, Michael. ***The National Geographic Traveler: Canada.*** Washington, D.C.: National Geographic Society, 1999. Essays on the history, culture, and contemporary life of the country.

SOFTWARE

Crosscountry Canada Platinum. Blaine, WA: Ingenuity Works, Inc. Interactive game reinforces math and map-reading skills.

INTERNET

For more information about Canada today, visit **classzone.com.**

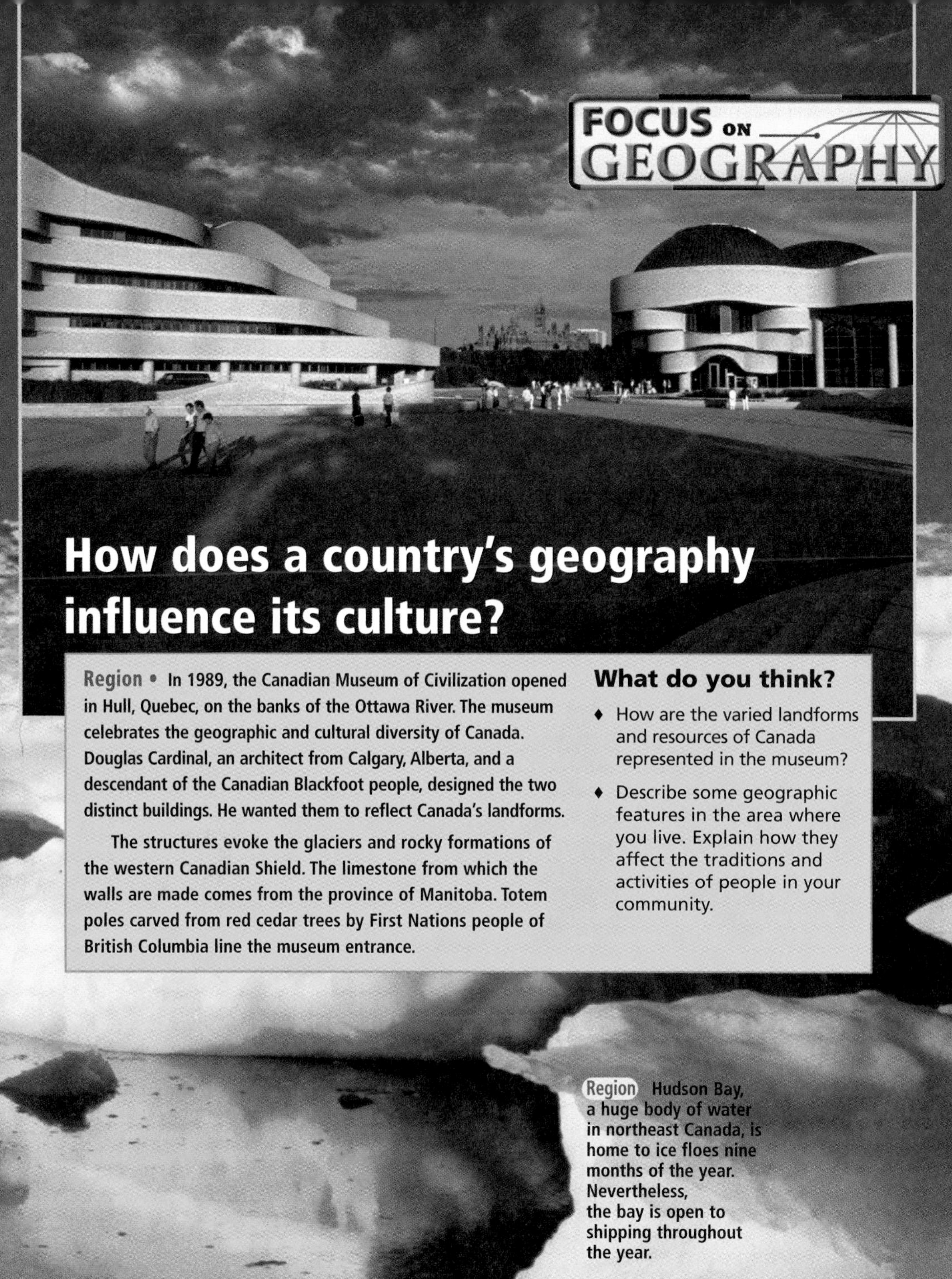

How does a country's geography influence its culture?

Region • In 1989, the Canadian Museum of Civilization opened in Hull, Quebec, on the banks of the Ottawa River. The museum celebrates the geographic and cultural diversity of Canada. Douglas Cardinal, an architect from Calgary, Alberta, and a descendant of the Canadian Blackfoot people, designed the two distinct buildings. He wanted them to reflect Canada's landforms.

The structures evoke the glaciers and rocky formations of the western Canadian Shield. The limestone from which the walls are made comes from the province of Manitoba. Totem poles carved from red cedar trees by First Nations people of British Columbia line the museum entrance.

What do you think?

- How are the varied landforms and resources of Canada represented in the museum?
- Describe some geographic features in the area where you live. Explain how they affect the traditions and activities of people in your community.

Region Hudson Bay, a huge body of water in northeast Canada, is home to ice floes nine months of the year. Nevertheless, the bay is open to shipping throughout the year.

117

FOCUS ON GEOGRAPHY

Objectives

- To introduce students to the range and variety of Canada's landforms
- To demonstrate the influence of geography on Canada's history and culture

What Do You Think?

1. Ask students to identify the different cultures and landforms the museum's exterior represents. Guide a discussion about how the museum's building is an expression of its content.
2. Encourage students to identify events, traditions, or activities in their own community that are influenced by its location and geography—for example, a winter ice-skating race or a museum of desert plants and animals.

How does a country's geography influence its culture?

Ask students to think about the diversity shown in these two photographs of different parts of Canada. Help them understand that Canada, like the United States, is a land with great variety in landforms and climate. Encourage them to see how that diversity affects other aspects of Canadian life, such as trade, economy, and settlement patterns.

MAKING GEOGRAPHIC CONNECTIONS

Ask students to imagine a museum in their own community that would reflect local geography and landforms in the same way the Canadian Museum of Civilization does. Invite them to suggest design elements and building materials, such as local stone and wood from native trees.

Implementing the National Geography Standards

Standard 15 Give examples of ways people take the environment into account when deciding on locations for human activities

Objective To write a report on the location of human settlements in Canada

Class Time 25 minutes

Task Students use a map showing the population density of Canada to write a brief report describing where most Canadians live. The reports should identify which aspects of their environment Canadians took into account when deciding where to build cities and towns.

Evaluation The reports should show that most Canadians live in warmer areas near bodies of water.

BEFORE YOU READ

What Do You Know?

Write the words *Past* and *Present* on the board. Ask students to brainstorm things they know about, or associate with, Canada. As students share their ideas, ask the class to identify whether the item belongs to Canada's past or its present. List each item in the appropriate column. Tell students that while reading this chapter they will learn about how events in Canada's past have shaped the present-day country.

What Do You Want to Know?

Have students work in pairs to develop questions about aspects of life in Canada. As they read the chapter, have them look for and write answers to their questions.

READ AND TAKE NOTES

Reading Strategy: Analyzing Causes and Effects Explain to students that the chart will help them organize information to show cause-and-effect relationships. Tell them that when the chart is completed, they will have a visual guide connecting factors in Canada's history and geography to aspects of its government, economy, and culture.

In-depth Resources: Unit 2
- Guided Reading Worksheets, pp. 23–26

READING SOCIAL STUDIES

BEFORE YOU READ

▸▸ What Do You Know?

Before you read this chapter, consider what you know about Canada. Have you ever read that Canada has the world's longest coastline? Did you know that Canada is an independent democracy with a constitution, yet it pledges loyalty to the British monarch? Perhaps you know that 60 percent of the National Hockey League's players are Canadian? Recall what you have learned about Canada from personal experience, television, and other classes. Think about how Canada is similar to the United States and how it is different.

▸▸ What Do You Want to Know?

Decide what you would like to know about Canada's history, government, economy, and culture. In your notebook, record what you hope to learn from this chapter.

Culture • The art of totem pole carving almost disappeared in the mid-1800s. Museums began to preserve totem poles in the 1850s. ▼

Culture • The 1893 Montreal Amateur Athletic Association was the first hockey team to win the Stanley Cup. ▲

READ AND TAKE NOTES

Reading Strategy: Analyzing Causes and Effects To help you understand how geographic, historical, and cultural factors have influenced Canada's development, pay attention to causes and effects as you read Chapter 5. Notice that several factors may cause the same effect. Use the chart below to make connections between statements about geography, history, and culture and statements describing Canada today.

- Copy the chart into your notebook.
- On the chart, read statements describing issues and conditions in Canada today (effects).
- As you read the chapter, identify the effects of the listed causes and note them on the chart.

Causes		Effects
Early settlers were from France and Britain, two nations that had conflicts. French-speaking Canadians have kept their own language and culture separate from the rest of the nation. The people of First Nations and other culture groups want to preserve their traditions.	→	Canada is a multilingual and multicultural country.
Cold climate, geographic barriers, and poor soil exist in northern parts of Canada.	→	few people live in the north; not very significant economically
The Arctic Ocean and Hudson Bay are frozen for most of the year. Landforms such as the Rocky Mountains create transportation barriers.	→	Barriers make transportation and communication difficult and generate isolation.

118 CHAPTER 5

Teaching Strategy

Reading the Chapter This is a thematic chapter that explores the historical, geographic, and human elements that have shaped present-day Canada. Encourage students, as they read the chapter, to look for examples of how geographic features and historical events have influenced aspects of life in Canada.

Integrated Assessment The Chapter Assessment on page 139 describes several activities for integrated assessment. You may wish to have students work on these activities during the course of the chapter and then present them at the end.

O Canada! Immigrant Roots

TERMS & NAMES
First Nation
multiculturalism
refugee

MAIN IDEA

Canada's population includes many groups of people from different lands who retain their cultural identities.

WHY IT MATTERS NOW

Knowing the history of the people of Canada helps in understanding Canada's policy of multiculturalism.

DATELINE EXTRA

QUEBEC, NEW FRANCE, JUNE 1609

New colonists have just arrived to join the first settlers of Quebec. Only French explorer Samuel de Champlain and 8 of the 32 men he led here survived their first winter in the new colony. Champlain chose this location that the Algonquins call Quebec, or the Narrows, after much searching.

After arriving last July, he and his men built houses, planted grain, and worked to encourage fur trading and friendly relations with the native people. Champlain has spent years traveling around New France and mapping and recording information about the seacoast and rivers.

Place • Champlain has great hopes for the future of Quebec. ▲

Who Are the Canadians?

The people of Canada come from many countries—not just France. More than 50 ethnic groups make up the population. More than two-thirds of Canadians have European ancestry. About 40 percent have British roots and 27 percent share a French heritage. Other Canadians trace their families back to Germany, Italy, and Ukraine, as well as to nations in Africa and Asia. Less than 5 percent of all Canadians are people of the First Nations.

TAKING NOTES
Use your chart to take notes about Canada.

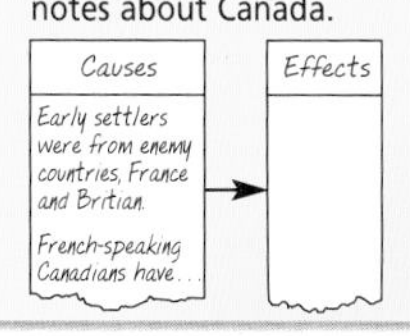
Causes
Effects
Early settlers were from enemy countries, France and Britian
French-speaking Canadians have...

SECTION OBJECTIVES

1. To identify the groups that make up Canada's population and their origins
2. To trace Canada's immigration history and policies regarding citizenship
3. To describe settlement patterns in Canada

SKILLBUILDER
- Interpreting a Map, p. 122

CRITICAL THINKING
- Making Inferences, p. 121

FOCUS & MOTIVATE

WARM-UP

Drawing Conclusions Have students read Dateline and discuss the following questions.

1. What does the survival rate of Champlain's group suggest about the living conditions in this region?
2. How did Champlain prepare for building the new settlement?

INSTRUCT: Objective 1

Who Are the Canadians?/ The First Nations

- What ancestry do more than two-thirds of Canadians share? European
- Who are the First Nations people? descendants of the first settlers from Asia

In-depth Resources: Unit 2
- Guided Reading Worksheet, p. 23

Reading Study Guide
(Spanish and English), pp. 32–33

Program Resources

In-depth Resources: Unit 2
- Guided Reading Worksheet, p. 23
- Reteaching Activity, p. 31

Reading Study Guide
(Spanish and English), pp. 32–33

Formal Assessment
- Section Quiz, p. 63

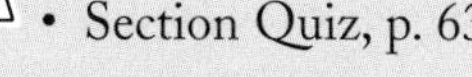

Integrated Assessment
- Rubric for writing a magazine article

Outline Map Activities

Access for Students Acquiring English
- Guided Reading Worksheet, p. 29

Technology Resources
classzone.com

TEST-TAKING RESOURCES
- Strategies for Test Preparation
- Test Practice Transparencies
- Online Test Practice

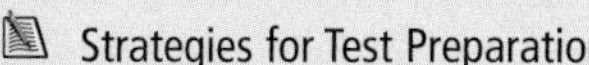

INSTRUCT: Objective 2

European Immigrants/Canadian Citizens and Citizenship

- What two European countries first struggled to control Canada? Britain and France
- Which province, or area of Canada, remained largely French under British control? Quebec
- Which provinces made up the original Dominion of Canada? Ontario, Quebec, Nova Scotia, New Brunswick
- How does Canada publicly support its different citizen groups? with an official policy of multiculturalism
- What freedoms are guaranteed in Canada? freedom of religion, speech, and assembly; equal protection under the law

Spotlight on CULTURE

The brightly carved totems on a totem pole are symbols of a tribe, clan, or family. Clan totems can be a variety of animal, plant, or other natural object. Some groups think of the totem as an ancestor of the clan; clans may also have taboos against killing or eating the species to which the totem belongs. Clan members are often known by the name of the clan's totem. It was the Chippewa, or Ojibwa, Indians who first used the term *totem* for the animals or birds associated with their clans.

The First Nations

People have lived in North America for at least 12,000 years. At times in the past, the levels of the oceans were as much as 300 feet lower than they are today. Then the narrow water passage between Asia and North America—the Bering Strait—became dry land. Small bands of people crossed this land bridge into North America and settled throughout North America and South America.

BACKGROUND

The name *Canada* comes from *Kanata,* a First Nations Huron-Iroquois word that means "village."

The Canadians of the **First Nations** are descendants of those first settlers from Asia. In the Arctic north, Inuit and other native people make up more than half the population. Large numbers of First Nations people, including Cree, Micmac, Abenaki, and Ojibwa, live in southern Canada near the United States border.

Thinking Critically Possible Answer

1. The First Nations people have a deep, complex relationship with nature.
2. Totem poles tell stories, preserve history, celebrate important events, and mark graves and important sites.

European Immigrants

The first major wave of European settlement began in the 1600s. Both Britain and France established colonies in what is now Canada. These two countries had a long history of conflict, and they continued their rivalry on the North American continent. Between 1754 and 1763, they fought the French and Indian War for control of North America.

Totem Poles—Carving History The Haida people in Canada's Queen Charlotte Islands and the Kwakiutl in central British Columbia have been skilled totem carvers for centuries. Early craftspeople believed that red cedar was a gift from the Great Spirit. They used simple tools to carve beautiful, detailed totem poles from these trees.

Totem poles, such as these in Stanley Park, Vancouver, display brightly painted animal figures, or totems. These include eagles, whales, grizzly bears, wolves, ravens, frogs, and halibut. Totems are symbols that tell stories, celebrate important events, and preserve the history of native clans. Totem poles have also been used as grave markers and monuments.

THINKING CRITICALLY 

1. **Making Inferences**
   What do the totem poles tell you about the First Nations people's relationship with nature?
2. **Drawing Conclusions**
   What roles do totem poles play in native culture?

For more on totem poles, go to 

120 CHAPTER 5

Activity Options

Interdisciplinary Link: Geography

Class Time One class period

Task Preparing and presenting a broadcast interview about human migration from Asia

Purpose To recognize the impact of geographical change on human migration

Supplies Needed
- Reference materials about the Bering Strait land bridge and human migration from Asia
- Writing paper
- Pens or pencils

 **Block Scheduling**

Activity Invite groups of students to prepare and present an interview focused on the role of the land bridge in human migration from Asia to North America. Have students share the roles of interviewer and panel of experts. Suggest that the "experts" focus on narrow topics, such as sea level changes, human group movements, or size and length of the land bridge. Review the interview format before students make their presentations.

France lost the war and surrendered most of its Canadian territory to Great Britain. However, many French settlers remained, and disputes continued between them and the fast-growing population of British settlers.

Culture • British General James Wolfe's troops defeated the French and captured Quebec in 1759 during the French and Indian War. Benjamin West's painting *The Death of General Wolfe* shows Wolfe's death at the end of the battle. ▲

Canada and the United Kingdom In 1791, the British government established itself in two areas in Canada. Upper Canada, now Ontario, had mostly British settlers. Lower Canada, now Quebec, remained largely French. Although hostilities continued between the two populations, in 1867 they were united as the Dominion of Canada, along with Nova Scotia and New Brunswick. Canada became a self-governing nation, although the British monarch remained its head of state.

In 1869, the Hudson's Bay Company sold land to Canada that later became the provinces of Manitoba, Alberta, and Saskatchewan. In 1871, British Columbia joined the Dominion, and Canada now reached to the Pacific Ocean. In 1931, with the enactment of the Statute of Westminster, Canada gained equal status with the United Kingdom and joined the Commonwealth of Nations. In 1982, the last legal connection between Canada and the British Parliament ended, although Canada remains a member of the Commonwealth.

Later Immigrants Most of Canada's early immigrants were English, Scottish, Irish, and French. After World War I, other Europeans arrived from countries such as Italy, Poland, and Ukraine. Most Italian immigrants settled in Toronto and Montreal. Most Ukrainians moved to the prairies of central Canada. After World War II, Germans and Dutch entered the country, settling primarily in Ontario and British Columbia. In the 1960s, new immigration laws allowed people to migrate from Africa, Latin America, Asia, and the Pacific Islands.

Reading Social Studies
A. Possible Answer The policy of multiculturalism might attract more immigrants because it would assure them of tolerance toward their culture along with economic opportunities.

Canadian Citizens and Citizenship ❷

As Canadian citizens, those of English or French descent have retained their separate languages and identities. Other groups have also kept the traditions of their homelands after settling in Canada. To support these citizen groups, Canada has adopted an official policy of **multiculturalism**—an acceptance of many cultures instead of just one.

Reading Social Studies

A. Drawing Inferences How might Canada's policy of multiculturalism lead to increased immigration?

CRITICAL THINKING ACTIVITY

Making Inferences Remind students that the French and British populations in Canada were united under one government about 100 years after the defeat of France in the French and Indian War. Until then, the two groups were separate and distinct from each other. Ask students to make inferences about how smoothly unification took place. How might the two groups have viewed each other and themselves? What problems might have surfaced? How might they have worked for or against unification? In what ways might those experiences have influenced Canada's present policy of multiculturalism?

Class Time 15 minutes

MORE ABOUT...

The French and Indian War

The French and Indian War (1754–1763) was fought between Britain and France for control of northern and eastern lands in North America. It was closely related to, and sometimes referred to as a theater of, the Seven Years' War—a complex global war between France and Britain and their allies. The name "French and Indian War" is derived from the alliances between France and the majority of Native American tribes. On February 10, 1763, France signed the Treaty of Paris, which ceded territory on mainland North America to Great Britain.

Activity Options

Skillbuilder Mini-Lesson: Sequencing Events

Explaining the Skill Remind students that sequence is the order of events in time, and that identifying sequence helps us understand events in relation to one other.

Applying the Skill Have students create a set of historical maps showing Canada's boundary changes from 1791 to the present.

1. Give pairs of students an outline map of Canada. Assign each pair a date and boundary distinction, and have them create a map to show this information.
2. When students have finished, have them line up their maps in chronological order to show the sequence of events in Canada's expansion. Discuss any confusions in the sequence of events.

FOCUS ON VISUALS

Interpreting the Map Explain to students that a population distribution map shows the density of people living in a given area. Ask students to use the map and key to determine where the population distribution in Canada is generally high and where it is generally low.

Possible Response The population is generally high in cities and low in the northern areas of Canada.

Extension Have students use a current reference source to identify and locate the five largest cities in Canada and find the population of each.

INSTRUCT: Objective 3

Where Do Most Canadians Live?

- What factors influence where Canadians live? geographic features and economic opportunities
- Why do few Canadians live in the northern regions? rugged terrain and cold climate
- Where do one-quarter of Canada's immigrants live? Toronto, Ontario
- What has made Toronto a center of industry and trade? access to the Atlantic Ocean and the United States

Population Distribution of Canada, 2000

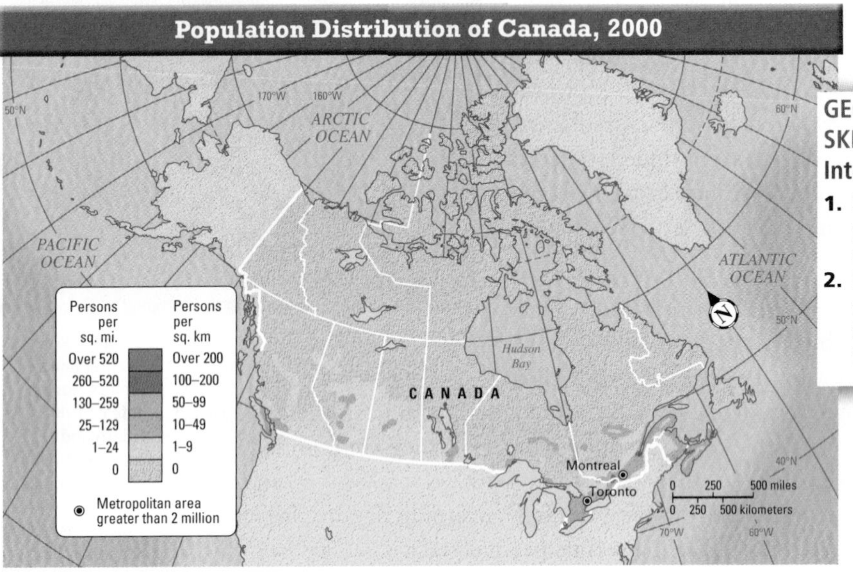

GEOGRAPHY SKILLBUILDER: Interpreting a Map

1. **Region** • In what part of Canada do most Canadians live?
2. **Place** • What two cities have a population of more than two million?

Geography Skillbuilder Answers

1. in the southeast, along the Great Lakes, the St. Lawrence River, and the Atlantic Coast
2. Montreal, Toronto

Canadian citizens have many of the same rights and responsibilities as U.S. citizens. They must obey Canada's laws. They have the option of voting and participating in the political system. They are guaranteed freedom of religion, speech, and assembly, as well as equal protection and treatment for all under Canadian law.

Where Do Most Canadians Live? 3

While Canada's land area is second only to Russia's, its population is a relatively small 31 million people. Canadians often live where they find a favorable combination of geographic features and economic opportunities. Three-fourths of the population live in the cities and towns of southern Canada. In this region, the Great Lakes, the St. Lawrence Seaway, numerous rivers, and an excellent railway system provide convenient transportation for people and goods. Some Canadians live on farms in the central prairies and in port cities along the coasts. The northern regions of Canada are rugged and very cold. Few people live in those remote areas.

Reading Social Studies

B. Analyzing Motives What geographic and economic features attract people to settle in some parts of Canada?

Reading Social Studies B. Possible Answer Fertile soil, transportation, moderate climate, and job opportunities make some parts of Canada more attractive to people than others.

Vancouver, Gateway to the Pacific Vancouver, British Columbia, is called Canada's "Gateway to the Pacific." As Canada's largest port, it trades heavily with Asian countries.

Thousands of Chinese from Hong Kong and many Japanese arrived in Canada at the end of the 20th century. Recent refugees have come from Vietnam, Laos, and Cambodia. **Refugees** are people who flee a country because of war, disaster, or persecution.

Activity Options

Interdisciplinary Link: Science

Class Time One class period

Task Comparing the climate and seasons of two Canadian communities

Purpose To learn about the differences in weather and climate in northern and southern Canada

Supplies Needed
- Reference sources
- Political and physical maps of Canada
- Paper
- Pens and pencils

Activity Have students work in groups to choose a community or location in northern Canada and another in southern Canada. Ask students to research climate and seasonal information for each community and then compare and contrast the communities based on what they learned. Encourage students to present the results of their research in an interesting and informative way, such as a climate graph.

Place • Toronto is on the shore of Lake Ontario, the easternmost Great Lake. Toronto's skyline is highlighted by the CN Tower. ▲

Toronto, City of Immigrants Toronto, Ontario's capital, is home to one-twelfth of Canada's population but contains one-fourth of its immigrants. More than 70,000 immigrants arrive each year from more than 100 countries in Asia, Europe, the West Indies, and North America. More than 40 percent of Toronto's population is foreign-born, and 10 percent arrived after 1991. Toronto's location, with access to the Atlantic Ocean and the United States, has helped it become a center of industry and international trade.

SECTION 1 ASSESSMENT

Terms & Names

1. **Explain the significance of:** (a) First Nation (b) multiculturalism (c) refugee

Using Graphics

2. Make a spider map like this one to record details about the people who settled in Canada.

Canadian People | First Nations | Canadians | European Immigration | Recent Immigration

Main Ideas

3. (a) How did the first people reach North America? Who are their descendants?
   (b) Describe the relationship between the British and Canada in the 1700s and the 1800s.
   (c) What are some of the rights and responsibilities of Canadian citizens?

Critical Thinking

4. **Synthesizing**
   How has the policy of multiculturalism benefited recent immigrants to Canada?

Think About

- the historic relations between the French and the English
- the many groups of immigrants and refugees in Canada

ACTIVITY -OPTION- Choose one place in Canada where you might like to live. Look at the information in the Unit Atlas and in this section. Write and illustrate a **magazine article** about this location.

ASSESS & RETEACH

Reading Social Studies Have students add details to the "Effects" column of the first section of the chart on page 118.

Formal Assessment
- Section Quiz, p. 63

RETEACHING ACTIVITY

Ask pairs of students to write six true-or-false statements about Section 1. Encourage them to write at least one statement about the text under each heading. Have pairs of students exchange statements, read them, and determine whether each is true or false.

In-depth Resources: Unit 2
- Reteaching Activity, p. 31

Access for Students Acquiring English
- Reteaching Activity, p. 35

Section 1 Assessment

1. Terms & Names

a. First Nation, p. 120
b. multiculturalism, p. 121
c. refugee, p. 122

2. Using Graphics

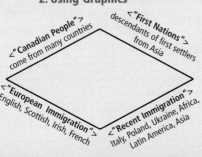

"Canadian People" come from many countries
"First Nations" descendants of first settlers from Asia
"European Immigration" English, Scottish, Irish, French
"Recent Immigration" Italy, Poland, Ukraine, Africa, Latin America, Asia

3. Main Ideas

a. by crossing a land bridge in the Bering Strait; people of the First Nations
b. British colonies united with Nova Scotia and New Brunswick as self-governing Dominion of Canada in 1867, with British monarch as head of state.
c. should obey laws and vote; have freedom of speech, religion, and assembly; equal protection of laws

4. Critical Thinking

Possible Response The policy gives recent immigrants opportunities equal to those for established residents from other backgrounds.

ACTIVITY OPTION

Integrated Assessment
- Rubric for writing a magazine article

SECTION OBJECTIVES

1. To examine the structure of Canada's government
2. To describe the role of government in protecting civil rights and preserving multiculturalism
3. To explore the issue of independence for Quebec

SKILLBUILDER

- Interpreting a Map, p. 125
- Interpreting a Chart, p. 126

CRITICAL THINKING

- Contrasting, p. 125

FOCUS & MOTIVATE

WARM-UP

Making Inferences Have students read Dateline and discuss the new Inuit territory.

1. What does this article suggest about Inuit political status before April 1, 1999?
2. Which values of the Inuit do you think might be evident in their land policies?

INSTRUCT: Objective 1

A Nation of Provinces and Territories/Organization of Canada's Government

- How is the country of Canada divided? into ten provinces and three territories
- What is modern Canada's relationship with Great Britain? British monarch is the head of state.

In-depth Resources: Unit 2
- Guided Reading Worksheet, p. 24

Reading Study Guide
(Spanish and English), pp. 34–35

A Constitutional Monarchy

TERMS & NAMES
constitutional monar
Parliament
prime minister
Pierre Trudeau
separatist

MAIN IDEA

Canada is a democracy that protects the rights of individuals and of different cultures.

WHY IT MATTERS NOW

Canada's form of government has enabled the country to remain united despite conflicts among different groups of citizens.

DATELINE

NUNAVUT, CANADA, APRIL 1, 1999—The Inuit, the native people of Arctic Canada, have today been granted a separate territory in northern Canada. This historic day comes as a result of almost 25 years of negotiations with Canada's government. The Inuit argued that as a First Nations people, they have lived here for at least 4,000 years and have the right to govern their own land.

Today the Canadian government turned over 733,600 square miles of land that had been the eastern part of the Northwest Territories. Nunavut (NOO•nuh•voot)—which means "our land" in Inuktitut, the language of the Inuit—becomes the third territory of Canada. The capital is Iqaluit on Baffin Island.

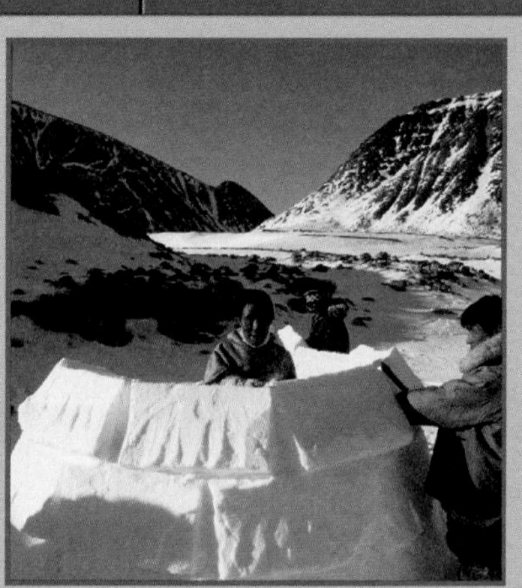

Place • The Inuit build igloos out of blocks of snow to use as temporary winter homes. ▲

1 A Nation of Provinces and Territories

Other First Nations people also seek the self-government that the Inuit won. At this time, Canada remains a nation of ten provinces and three territories. The responsibilities of the central government include national defense, trade and banking, immigration, criminal law, and postal service. The provincial governments administer education, property rights, local government, hospitals, and provincial taxes. Territorial governments have fewer responsibilities but still enjoy limited self-government.

TAKING NOT
Use your chart to notes about Cana

Causes
Early settlers were from enemy countries, France and Britian
French-speaking Canadians have...

Program Resources

In-depth Resources: Unit 2
- Guided Reading Worksheet, p. 24
- Reteaching Activity, p. 32

Reading Study Guide
(Spanish and English), pp. 34–35

Formal Assessment
- Section Quiz, p. 64

Integrated Assessment
- Rubric for creating a travel poster

Outline Map Activities

Access for Students Acquiring English
- Guided Reading Worksheet, p. 30

Technology Resources
classzone.com

TEST-TAKING RESOURCES

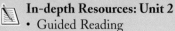

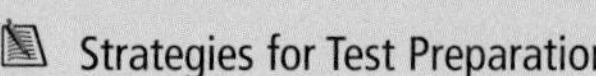
- Strategies for Test Preparation
- Test Practice Transparencies
- Online Test Practice

The First Nations of Canada, 2001

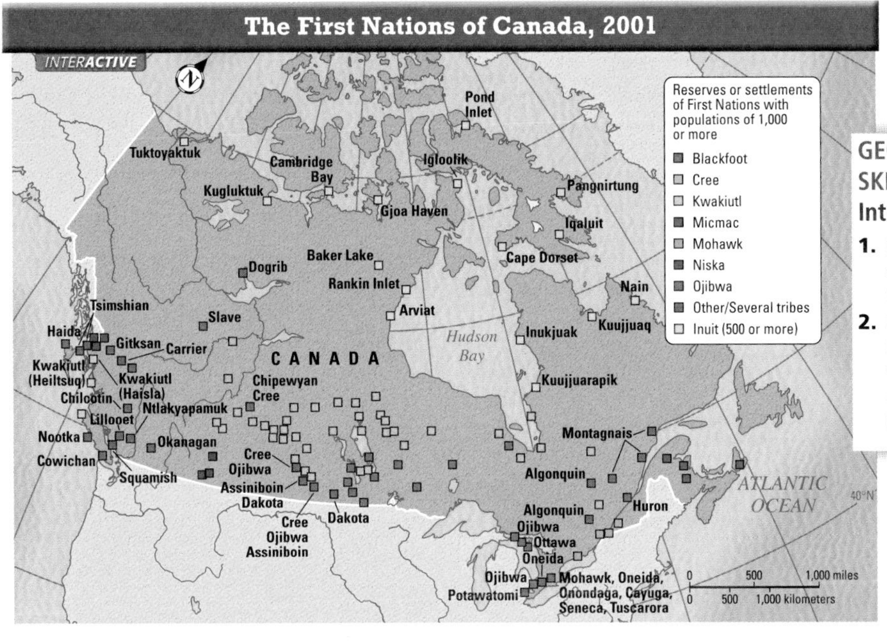

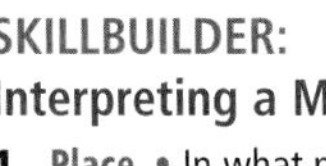

GEOGRAPHY SKILLBUILDER: Interpreting a Map

1. **Place** • In what part of Canada do most Inuits live?
2. **Place** • Name some First Nations peoples who live along Canada's southern border.

Geography Skillbuilder Answers

1. in the far north and the Arctic, around Hudson Bay and Baffin Island
2. Ojibwa, Oneida, Mohawk, Cree, Dakota, Squamish, Assiniboin, Potawatomi

Organization of Canada's Government ❶

BACKGROUND

The prime minister and his cabinet are accountable to the members of the House of Commons. If they lose the support of the majority of members, they must resign, or ask the governor-general to dissolve Parliament and call an election.

Canada is a **constitutional monarchy.** It has a constitution to explain the powers of the government and owes allegiance to a monarch, a king or a queen. The Canadian government consists of the legislative and the judicial branches. Executive duties are within the legislature.

Head of State The British monarch is Canada's head of state. Since the queen or king does not live in Canada, she or he selects a governor-general as a representative. The monarch and the governor-general have little genuine power in Canadian government. They represent the historical traditions of Canada.

Reading Social Studies A. Possible Answer Both are heads of government, set policies, make treaties, propose legislation. The president also heads executive branch.

Reading Social Studies

A. Comparing What are some of the powers of the prime minister, and how do these differ from those of the U.S. president?

Legislature Canada's legislature, called **Parliament,** has two bodies, the House of Commons and the Senate. Together they determine Canadian laws and policies. Citizens elect members of the House of Commons. The leader of the party with the most members becomes the head of government, or **prime minister,** who runs the executive branch within the legislature. Senators are chosen by the prime minister from each of the ten provinces and three territories.

Location • The Parliament Buildings in Ottawa, Canada's capital, house the legislature of the central government. ▼

FOCUS ON VISUALS

Interpreting the Map Review the map key with students. Ask them to use the key to point out at least one locale on the map for each of the First Nations peoples listed. Ask them to point out names of groups that European settlers adopted for cities and geographical locations.

Possible Responses Ottawa, Oneida (cities); Dakota (state); Huron, Cayuga (lakes)

Extension Have students research which Native American groups live or once lived in their own state and determine whether any are the same as Canada's First Nations. Do any Native American tribes or nations still have land and communities in the area?

CRITICAL THINKING ACTIVITY

Contrasting Lead a discussion contrasting the way the head of state and senators are chosen in Canada—which has a parliamentary government—with the way they are chosen in the United States—a presidential government. Ask in what ways Canada's method affects the prime minister's power. How does this differ from what is done in the United States, and from the President's power? What might be the advantages and disadvantages of the Canadian system?

Class Time 15 minutes

Activity Options

Differentiating Instruction: Less Proficient Readers

Organizing Information To help students understand the structure and components of government in Canada, have them write each of the following terms on index cards: legislative branch, Parliament, House of Commons, Senate, prime minister, judicial branch, Supreme Court, British monarch, governor-general. Have students shuffle their cards. Then have students arrange the cards to show the organization of the Canadian government. Ask volunteers to explain the responsibilities of each component, as well as the relationships between them. Be sure students understand that the position of prime minister is part of the legislative branch, unlike the executive position held by the U.S. President.

FOCUS ON VISUALS

Interpreting the Chart Focus students' attention on the chart, and have them read the title and labels. Ask students to identify one way in which the two governments differ.

Possible Responses Head of state and head of government are two distinct positions in Canada; monarchy versus republic

Extension Ask students to do research to compare term limits and requirements for the senators and members of Canada's House of Commons with those for senators and representatives in the United States Congress.

INSTRUCT: Objective 2

Equality and Justice

- What freedoms does the Charter of Rights guarantee? speech, religion; equal rights regardless of race, religion, gender, age, national origin

Biography

Pierre Trudeau was born and raised in Montreal. He attended Jean-de-Brébeuf, a Jesuit college, and went on to pursue a law degree at the University of Montreal. In 1945, Trudeau received his master's degree in political economy at Harvard University. He then traveled to Europe to attend the School of Political Sciences in Paris and the London School of Economics and Political Science in Britain.

Comparing the Canadian and U.S. Governments

Aspects of Government	Canada	United States
Type	Constitutional Monarchy (limited power)	Constitutional Republic (limited power)
Head of State	Monarch	President
Head of Government	Prime Minister	President
Legislature	Parliament	Congress
System	Federal (central and provinces)	Federal (central and states)

SKILLBUILDER: Reading a Chart

1. **Place •** Name three ways in which the government of Canada differs from that of the United States.
2. **Place •** How are the governments of Canada and the United States alike?

Judiciary Canada has both federal and provincial courts. The highest court is the federal Supreme Court. It is made up of the chief justice of Canada and eight other judges.

Vocabulary

judiciary: the judicial branch of government, the court system

Equality and Justice 2

Canada is a democracy. Its government is responsible for protecting people's rights.

Civil Rights Prime Minister Pierre Trudeau led an effort to add a Charter of Rights and Freedoms to the Canadian Constitution in 1982. The Charter is similar to the U.S. Constitution's Bill of Rights. Among other rights, the Charter guarantees freedom of speech and freedom of religion. It protects every citizen's right to vote and to be assisted by a lawyer if arrested. It says that Canadians are free to live and work anywhere in Canada. The Charter also says that people have equal rights regardless of their race, religion, gender, age, or national origin.

Biography

Pierre Elliott Trudeau (1919–2000) From 1968 to 1979 and from 1980 to 1984, Pierre Trudeau was Canada's prime minister. Born in Montreal, Quebec, of French and Scottish ancestry, he grew up speaking both French and English. Despite his French-Canadian background, Trudeau successfully opposed Quebec's attempts to separate from Canada. He considered keeping Quebec a part of Canada one of his great achievements.

In 1982, Trudeau also helped enact a new Canadian constitution. At right, British Queen Elizabeth II signs a proclamation in 1982, making the new Canadian Constitution law, while Trudeau, seated, looks on. He worked to establish diplomatic relations with China and achieved Canada's complete independence from the British Parliament.

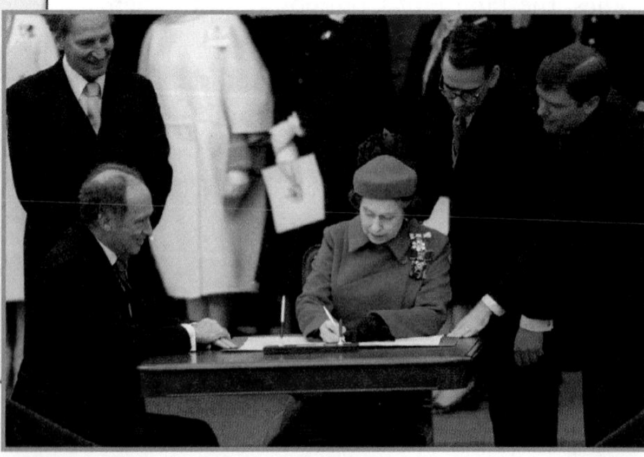

Reading Social Studies

B. Possible Answer The charter would strengthen all Canadians' civil rights and might help convince Quebec to remain a part of Canada.

Reading Social Studies

B. Analyzing Motives Why did Prime Minister Trudeau want a special document stating the rights of all Canadians?

Skillbuilder Answers

1. constitutional monarchy instead of republic; prime minister as head of government, not a president; legislature called Parliament, not Congress
2. Both have a federal system, two-chamber legislature, and a government with powers limited by a constitution.

Activity Options

Interdisciplinary Links: Government/History

Class Time One class period

Task Comparing Canada's Charter of Rights and Freedoms to the U.S. Constitution's Bill of Rights

Purpose To explore similarities and differences regarding citizens' rights

Supplies Needed

- Copies of the Bill of Rights
- Writing paper
- Pens or pencils

Block Scheduling

Activity Have students list the rights and freedoms guaranteed in Canada's Charter of Rights and Freedoms and then list the rights protected by the Bill of Rights of the U.S. Constitution. Invite students to compare the items on the lists and then discuss them as a class. Prompt discussion with questions such as: Which rights are the same, and which are different? Do the documents differ in style or language?

3 Many Cultures, Many Needs

Place • Quebec City, overlooking the St. Lawrence River, is the capital of the province of Quebec and the center of French-Canadian culture. ▲

Canada's people come from different cultures, and many wish to safeguard their special language and customs. Some French-speaking Canadians are **separatists,** or people who want the province of Quebec to become an independent country. In 1980 and in 1995, separatists asked for a vote on whether Quebec should become independent. Both times the issue was defeated, but the separatists promised to try again.

Reading Social Studies

C. Possible Answer The federal government was trying to satisfy the Quebec separatists and keep them in the country.

Quebec's Importance The federal government wants Quebec to remain part of Canada. Quebec is a major contributor to Canada's economy. Quebec is responsible for half of Canada's aerospace production, half of its information technology, and 38 percent of its high-tech industry. French culture is important in Canada's history and modern-day identity.

Reading Social Studies

C. Finding Causes What led to the passage of the Multicultural Act of 1988?

Laws Protecting Multiculturalism The Quebec provincial government has passed laws to preserve its citizens' French heritage. In an attempt to satisfy the separatists, Canada's federal government passed the Canadian Multicultural Act in 1988. This act guarantees the right of all Canadians to preserve their cultural heritage. Finding ways to maintain a unified country remains a critical issue in Canada today.

SECTION 2 ASSESSMENT

Terms & Names

1. **Explain the significance of:** (a) constitutional monarchy (b) Parliament (c) prime minister (d) Pierre Trudeau (e) separatist

Using Graphics

2. Make a diagram like the one shown below. Add details to show how the Canadian government is organized.

Canada's Government	
Head of state	
Legislature	
Judiciary	

Main Ideas

3. (a) What are some similarities and differences between Canadian and U.S. governments?
   (b) Describe some of the rights guaranteed in the Charter of Rights and Freedoms.
   (c) What is the purpose of the Multicultural Act?

Critical Thinking

4. **Predict**
   What might happen if Quebec became a separate country?

Think About

- Quebec's location
- Canada's multiculturalism
- economic impacts on Quebec and Canada

ACTIVITY -OPTION- Conduct research and create a **travel poster** to attract tourists to French-speaking Canada. Highlight the culture of Quebec.

INSTRUCT: Objective 3

Many Cultures, Many Needs

- What do Quebec separatists want? to make the province an independent country

ASSESS & RETEACH

Reading Social Studies Have students identify and write two specific details from the section that support the main idea that Canada is a democracy that protects the rights of individuals and different cultures. Then have them add relevant details to the first section of the chart on page 118.

Formal Assessment
- Section Quiz, p. 64

RETEACHING ACTIVITY

Have students work in pairs to develop arguments for or against Quebec's independence. Ask students to consider the impact of independence on Quebec and Canada in their arguments. Have pairs meet to share their ideas in a group discussion.

In-depth Resources: Unit 2
- Reteaching Activity, p. 32

Access for Students Acquiring English
- Reteaching Activity, p. 36

Section 2 Assessment

1. Terms & Names

a. constitutional monarchy, p. 125
b. Parliament, p. 125
c. prime minister, p. 125
d. Pierre Trudeau, p. 126
e. separatist, p. 127

2. Using Graphics

Head of state	the British monarch, governor-general, little real power
Legislature	Parliament, House of Commons, Senate, prime minister
Judiciary	federal and provincial courts, Supreme Court

3. Main Ideas

a. Both have a constitution, a two-house legislature, and a federal system. Canada has a prime minister and ties to the British monarch.
b. freedom of speech and religion; equal rights, voting rights, legal protection, freedom to travel
c. guarantees the rights of all Canadians to preserve their cultural heritage

4. Critical Thinking

Possible Response Losing Quebec's high-tech and aerospace industries and tourism would hurt the Canadian economy.

ACTIVITY OPTION

Integrated Assessment
- Rubric for creating a travel poster

Canada's Economy

TERMS & NAMES
industry
export
import
transportation corridor
transportation barrier

MAIN IDEA

Canada has a strong economy built on natural resources, a variety of industries, and good transportation.

WHY IT MATTERS NOW

Canada is a leader in the global economy.

SECTION OBJECTIVES

1. To identify Canada's natural resources and the importance of trade
2. To examine the connection between industry and the economy
3. To show the importance of transportation in the economy

SKILLBUILDER
- Interpreting a Map, p. 129

CRITICAL THINKING
- Making Inferences, p. 129

FOCUS & MOTIVATE

WARM-UP

Drawing Conclusions After students read Dateline, discuss the Gold Rush in Canada.

1. How has the discovery of gold changed Dawson?
2. What benefits and problems might the Gold Rush bring to the region?

INSTRUCT: Objective 1

Contributors to the Economy

- Why is Canada's economy strong? skilled labor, natural resources, international trade
- What are important industries in Canada? timber, agriculture, mining, fishing
- How has Canada's attitude regarding trade affected its economy? Openness to trade has promoted economic growth.

In-depth Resources: Unit 2
- Guided Reading Worksheet, p. 25

Reading Study Guide (Spanish and English), pp. 36–37

DATELINE

DAWSON, YUKON TERRITORY, AUGUST 1898— Gold! That's the cry on the street here in Dawson. Just two years ago, this was a small, unknown town of 5,000 people. Then gold was discovered nearby on a branch of the Klondike River. The Klondike Gold Rush was on. People came here from all over the world hoping to strike it rich.

Now Dawson has more than 30,000 people crowding its streets. Most live in tents. Every day, thousands head up the creeks looking for gold. Most of these people haven't found any yet and never will. However, a few lucky prospectors have already made their fortunes.

Human-Environment Interaction • Prospectors use gold dust to pay for merchandise in a store in the Yukon Territory. ▲

Contributors to the Economy 1

Canada is rich in natural resources, including gold. Europeans were first drawn to Canada by the abundant fishing and fur trading. In the 1800s, gold and other minerals were discovered. Today, most Canadians work in the service and manufacturing industries. Canada's skilled labor force, natural resources, and international trade all contribute to the country's economy.

TAKING NOTES

Use your chart to take notes about Canada.

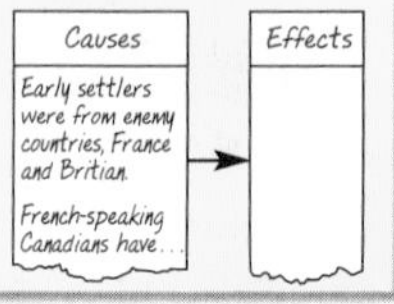
Causes
Effects
Early settlers were from enemy countries, France and Britian
French-speaking Canadians have...

Program Resources

In-depth Resources: Unit 2
- Guided Reading Worksheet, p. 25
- Reteaching Activity, p. 33

Reading Study Guide (Spanish and English), pp. 36–37

Formal Assessment
- Section Quiz, p. 65

Integrated Assessment
- Rubric for writing a newspaper article

Outline Map Activities

Access for Students Acquiring English
- Guided Reading Worksheet, p. 31

Technology Resources
classzone.com

TEST-TAKING RESOURCES
- Strategies for Test Preparation
- Test Practice Transparencies
- Online Test Practice

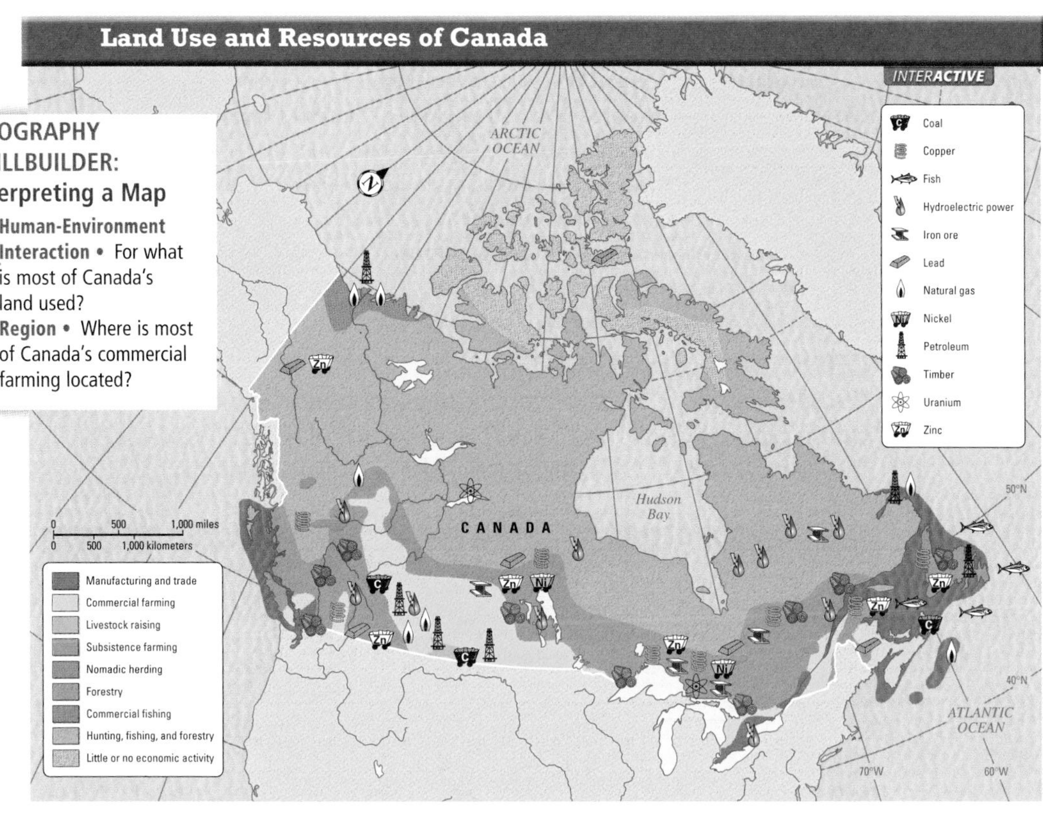

EOGRAPHY
KILLBUILDER:
nterpreting a Map

. **Human-Environment Interaction** • For what is most of Canada's land used?
. **Region** • Where is most of Canada's commercial farming located?

Industry Based on Natural Resources A nation's resources are a source of wealth. The prairie provinces of central Canada have extensive grasslands and good soil, making this area an ideal place to raise beef cattle and grow wheat. On the rich farmlands along the St. Lawrence River, farmers harvest grains, vegetables, and fruit. People plant potatoes and raise dairy cattle on the east coast. The Grand Banks, located off the coast of Newfoundland, is one of the world's most abundant fisheries. The salmon caught off Canada's Pacific coast enrich that area's economy.

Much of Canada is covered in forests, making the timber industry important, especially in British Columbia. **Industry** refers to any area of economic activity. Mining in the northern territories yields iron ore, gold, silver, copper, and other metals.

Trade Canada's openness to trade has contributed to the growth of its economy. Today almost 80 percent of Canada's raw materials are shipped as exports. **Exports** are goods traded to other countries. Canada's main exports are wood and paper products, fuel, minerals, aluminum, wheat, and oil. These and manufactured goods are sold around the world.

BACKGROUND

The Grand Banks, first noted by explorer John Cabot in 1498, extends 350 miles north to south and 420 miles east to west.

A. Recognizing Important Details What are Canada's main exports?

Geography Skillbuilder Answers

1. hunting, fishing, and forestry
2. in the west (prairie provinces) and along the St. Lawrence River

Reading Social Studies
A. Possible Answer
wood and paper products, minerals (aluminum), fuel (oil), wheat

CRITICAL THINKING ACTIVITY

Making Inferences Ask students to infer the economic effects on Canada if the demand for its exports was reduced. Encourage students to think about how the surplus of resources and the loss of income might affect Canada in terms of production, employment, and inflation.

Class Time 15 minutes

FOCUS ON VISUALS

Interpreting the Map Have students read the title of the map and notice that there are two keys: one showing land use and the other showing resources. Ask students to explain the Land Use key and then use it to locate areas for each type of economic use. Similarly, have them interpret the Resources key and locate at least one place where each type of resource is found. Ask them to point out the mineral resources found in western Canada.

Possible Responses oil, coal, natural gas, zinc, nickel, lead, uranium, copper

Extension Have students choose one province of Canada and research its resources and major economic activities.

Activity Options

Interdisciplinary Links: Language Arts/Writing

B Block Scheduling

Class Time One class period

Task Writing a journal entry during a trip across Canada

Purpose To develop an appreciation for Canada's natural resources

Supplies Needed
- Information on Canada's natural resources
- Physical and political maps of Canada
- Writing paper
- Pens or pencils

Activity Invite students to imagine that they are taking a trip across southern Canada. Have them write a journal entry describing their impression of the natural resources and industries they see and learn about. Suggest that they choose one area of Canada and a resource or industry they are interested in to write about. Encourage them to use the map on page 129 as well as other materials as a guide to the industries.

INSTRUCT: Objective 2

Industry and the Economy

- What are Canada's main exports? wood, paper products, fuel, minerals, aluminum, wheat, and oil
- What aspect of Canada's work force is a boon to its economy? Much of the work force is well-educated.
- In what type of industry does two-thirds of Canada's labor force work? tertiary, or service, industries

FOCUS ON VISUALS

Interpreting the Chart Ask individual students to read each of the four sections in the chart and explain it in his or her own words. For each type of industry, ask the class to suggest the kinds of jobs someone in that industry might hold.

Possible Responses primary: farmer; secondary: construction worker; tertiary: doctor, coach; quaternary: radio announcer

Extension Ask students to suggest jobs or careers they might be interested in following in the future. List their suggestions on the chalkboard, and then have students classify each job as part of a primary, secondary, tertiary, or quaternary industry.

Canada and the United States share a valuable trade partnership. Most of Canada's exports go to the United States. Most of its **imports**, or goods brought into the country, are from the United States. In 1994, Canada, the United States, and Mexico signed the North American Free Trade Agreement, or NAFTA, which lowered trade barriers among the three countries.

Industry and the Economy 2

Canada's well-educated work force is important to its economy. Canadians work in all four types of industry seen in the chart shown below. Since World War II, Canada has shifted from a mostly rural economy to a major industrial and urban economy.

Types of Industry	
Primary Industries	Prepare and process raw materials, such as timber, wheat, and iron ore, so other companies or consumers can use them *Examples:* **farms; mining companies; logging companies**
Secondary Industries	Manufacturing—turn raw materials into products that consumers or other businesses can use *Examples:* **bakeries; car manufacturers; furniture makers**
Tertiary Industries	Service industries—do not make goods or consume goods; distributors—move goods from the manufacturer to another business or to consumers *Examples:* **wholesalers; transportation companies (truck, train, airplane, or ship); retailers of food, clothing, and other goods; health care; education; recreation; banking**
Quaternary Industries	Pass on information *Examples:* **communication companies, such as Internet service providers and cable companies; financial, research, and other companies that gather and pass on information**

Tertiary, or service, industries, such as health care, recreation, education, transportation, banking, and the government, occupy about two-thirds of Canada's work force. About 30 percent of Canadians work in secondary, or manufacturing, industries. One of Canada's main products is transportation equipment, including automobiles, trucks, subway cars, and airplanes. Food processing, especially meat and poultry processing, is an important industry in Canada as well. Canada also makes chemicals, medicines, machinery, metal products, steel, and paper.

Transportation 3

Transportation is a major Canadian industry. The ability to import and export goods and move them from place to place across Canada's vast land area affects many consumers and businesses.

Activity Options

Multiple Learning Styles: Logical

Class Time One class period

Task Following one product through four different types of industry

Purpose To understand primary, secondary, tertiary, and quaternary industries as applied to one product

Supplies Needed

- Writing paper
- Pencils or pens

Block Scheduling

Activity Organize students into groups, and assign each a raw material, such as milk, iron ore, timber, wool, cotton, or wheat. Point out on the chart on page 130 that the production of these products is the role of primary industries. Have groups list as many secondary, tertiary, and quaternary industries associated with their assigned product as they can. Encourage them to find examples in addition to those on the chart.

Reading Social Studies

B. Analyzing Effects Why are transportation corridors important to the development of industry?

Vocabulary

transcontinental: spanning or crossing a continent

Reading Social Studies
B. Possible Answer
They are necessary to move resources and finished products.

Canada's geography both helps and hinders transportation. Canada has natural **transportation corridors**, or paths that make transportation easier. Rivers and coastal waters, sometimes combined with human-made canals and locks, provide convenient travel routes. The St. Lawrence Seaway, for example, allows oceangoing ships to travel between the Atlantic Ocean and the Great Lakes. Another important route is Canada's transcontinental railway system, which crosses the continent from coast to coast.

Canada also has **transportation barriers**, or geographic features that prevent or slow down transportation. In much of the north, snow and ice block travel by land or water. The Rocky Mountains in the west are another major obstacle. Industry develops slowly in regions where transportation is difficult.

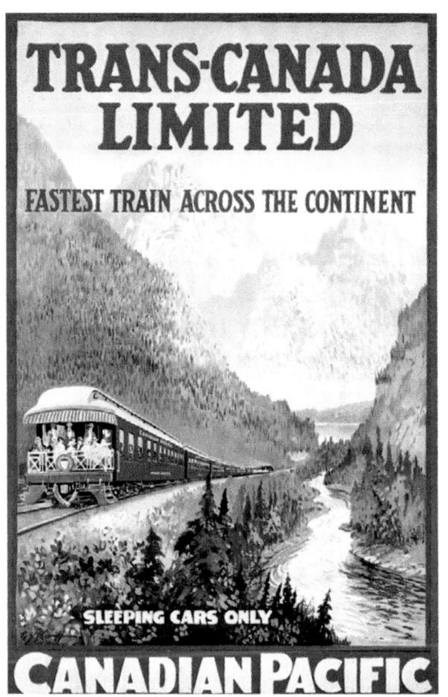

Region • The Canadian Pacific Railway Company completed a transcontinental line from Montreal to a Vancouver suburb in 1885. ▲

SECTION 3 ASSESSMENT

Terms & Names

1. **Explain the significance of:** (a) industry (b) export (c) import (d) transportation corridor (e) transportation barrier

Using Graphics

2. Make a chart like the one shown below to list goods that might be produced in each area.

Prairie Provinces	St. Lawrence River Valley	East Coast	British Columbia	Northern Territories

Main Ideas

3. (a) What important factors have helped build Canada's economy?
   (b) Give an example of a primary, a secondary, a tertiary, and a quaternary industry.
   (c) What are some transportation corridors and barriers in Canada?

Critical Thinking

4. **Drawing Conclusions**
   Why do you think Canada and the United States have become such good trade partners?

Think About
- geographic location
- their languages
- their governments

ACTIVITY -OPTION- Imagine that you are prospecting for gold in Dawson during the Klondike Gold Rush. Write a **newspaper article** describing what you have brought with you, how you traveled there, and what the town is like.

INSTRUCT: Objective 3

Transportation

- In what way is transportation a major industry? It is necessary to move goods across Canada's vast land area.
- How does Canada's geography help transportation? Rivers and coastal waters help provide travel routes.
- What are some transportation barriers in Canada? snow and ice, mountains

ASSESS & RETEACH

Reading Social Studies Have students add relevant details to the second and third sections of the chart on page 118.

Formal Assessment
- Section Quiz, p. 65

RETEACHING ACTIVITY

Give each student an outline map of Canada. Ask them to label the map with Canada's natural resources and industries. Allow time for students to share their maps.

In-depth Resources: Unit 2
- Reteaching Activity, p. 33

Access for Students Acquiring English
- Reteaching Activity, p. 37

Section 3 Assessment

1. Terms & Names

a. industry, p. 129
b. export, p. 129
c. import, p. 130
d. transportation corridor, p.131
e. transportation barrier, p. 131

2. Using Graphics

Prairie Provinces	beef cattle, wheat
St. Lawrence River Valley	grains, vegetables, fruit
East Coast	potatoes, dairy cattle
British Columbia	timber products
Northern Territories	iron ore, gold, silver, copper, other metals

3. Main Ideas

a. skilled labor force, natural resources, and international trade
b. primary industry: farming; secondary: auto manufacturing; tertiary: banking; quaternary: cable companies
c. transportation corridors: waterways and railway; transportation barriers: snow, ice, and Rocky Mountains

4. Critical Thinking

Possible Response They have a shared border, common languages, and similar governments.

ACTIVITY OPTION

Integrated Assessment
- Rubric for writing a newspaper article

A Multicultural Society

TERMS & NAMES
national identity
bilingual
Francophone

SECTION OBJECTIVES

1. To describe Canada's national identity
2. To explain how Canadians value the arts
3. To examine Canada's commitment to a multicultural society

SKILLBUILDER
- Interpreting a Map, p. 133

CRITICAL THINKING
- Forming and Supporting Opinions, p. 133
- Contrasting, p. 134

FOCUS & MOTIVATE

WARM-UP

Making Inferences Have students read Dateline and discuss Canada's love of hockey.

1. What does Lord Stanley's gift show about the importance of hockey to Canadians?
2. How does hockey reflect the contributions of different cultures?

INSTRUCT: Objective 1

Canadian Identity

- Why is it important for Canadians to have a national identity? to unite their many immigrant cultures
- Which two languages do many Canadians speak? French and English

In-depth Resources: Unit 2
- Guided Reading Worksheet, p. 26

Reading Study Guide (Spanish and English), pp. 38–39

MAIN IDEA

Many immigrant groups have contributed to Canadian culture while preserving their own identities.

WHY IT MATTERS NOW

Canada's desire to safeguard its cultural diversity is one of its most serious challenges.

DATELINE

MONTREAL, CANADA, 1893—Score! The Montreal Amateur Athletic Association team is the best hockey team in Canada, and it has a silver trophy cup to prove it. Canada's governor-general, Sir Frederick Arthur, Lord Stanley of Preston, presented the award to "the championship hockey club of the Dominion of Canada."

Many Canadians love this sport. First played by the Micmac, a First Nations people in Nova Scotia, ice hockey has spread across Canada and south to the United States. Competition for Stanley's Cup will make the sport even more exciting.

Culture • The winning team poses with Lord Stanley's Cup, which was purchased for about $50. ▲

Canadian Identity 1

Hockey is one of many good things about living in Canada. From 1994 to 2000, the United Nations rated Canada the best of 175 countries in a survey that examines the health, education, and wealth of each country's citizens. Yet, Canadians still seek a **national identity,** or sense of belonging to a nation, to unite its many immigrant cultures.

TAKING NOTES
Use your chart to take notes about Canada.

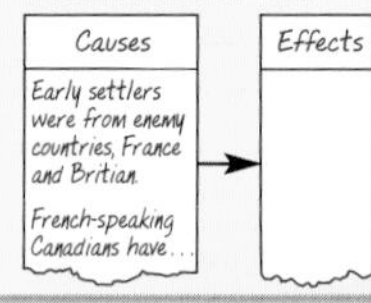
Causes
Effects
Early settlers were from enemy countries, France and Britian
French-speaking Canadians have...

Program Resources

In-depth Resources: Unit 2
- Guided Reading Worksheet, p. 26
- Reteaching Activity, p. 34

Reading Study Guide (Spanish and English), pp. 38–39

Formal Assessment
- Section Quiz, p. 66

Integrated Assessment
- Rubric for creating a mural or a collage

Outline Map Activities

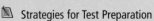

Access for Students Acquiring English
- Guided Reading Worksheet, p. 32

Technology Resources
classzone.com

TEST-TAKING RESOURCES

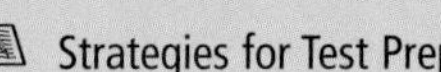

Strategies for Test Preparation

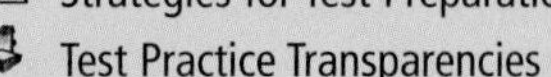

Test Practice Transparencies

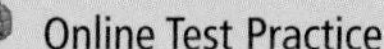

Online Test Practice

Distribution of Bilingual Speakers, 2001

ARCTIC OCEAN

ATLANTIC OCEAN

Hudson Bay

CANADA

Percentage of population who speak English and French

50.0–70.3
35.0–49.9
20.0–34.9
10.0–19.9
5.0–9.9
0–4.9

0 250 500 miles
0 250 500 kilometers

GEOGRAPHY SKILLBUILDER: Interpreting a Map

1. **Culture •** What part of Canada has the highest percentage of people who speak both English and French?
2. **Culture •** Where in Canada do less than five percent of the people speak both English and French?

Geography Skillbuilder Answers

1. eastern Canada, also north of Great Lakes
2. far northwest (Yukon)

Languages Many Canadians are **bilingual,** which means they speak two languages. Look at this map to see where bilingual Canadians live. Canada has two official languages, English and French. Literature, official documents, road signs, newspapers, and television broadcasts are in both languages. The two languages are not exactly like those spoken in England, the United States, and France. **Francophones** are French-speaking people. Canadian French, based on the French of the 1800s, is pronounced differently from the French spoken in modern France.

Culture • Business signs on a street in Quebec City reflect the strong influence of French culture. ▶

FOCUS ON VISUALS

Interpreting the Map Have students read the text in the paragraph "Languages" in connection with this map. Direct them in using the map key to find the major areas of bilingualism in eastern Canada and the smaller areas elsewhere. Ask them to speculate about the small areas of bilingualism in the prairie provinces. Why do they exist?

Possible Responses They are near large cities.

Extension Ask students to take a survey of second languages spoken by class members and make a bar graph to record the number of speakers of each.

CRITICAL THINKING ACTIVITY

Forming and Supporting Opinions Ask students to infer what language the United States might adopt as its second official language if it was to follow Canada's example. Discuss students' suggestions and their arguments in support of them.

Class Time 10 minutes

Activity Options

Differentiating Instruction: Students Acquiring English/ESL

Using Prefixes Have students read "Languages" as a group. Point out the word *bilingual,* and ask a volunteer to identify context clues that explain the word's meaning ("speaking two languages"). Tell students that the prefix *bi-* means "two." Then ask students to use the word *bilingual* in a sentence describing themselves.

Create a chart such as the one shown, and list other words containing the prefix *bi-*. Have students use a dictionary to identify the meaning of each word.

bicycle	
binoculars	
bipartisan	
biped	
bivalve	

INSTRUCT: Objective 2

Arts and Entertainment

- What is the purpose of the Canada Council for the Arts? to support Canadian artists and art organizations
- What do Canadians and Americans share in terms of the arts and entertainment? They enjoy many of the same newspapers, magazines, television shows, and movies.

CRITICAL THINKING ACTIVITY

Contrasting Point out to students that while many Canadian entertainers are popular in both Canada and the United States, many more Canadian performers seek to work in the United States than the reverse. Ask students to contrast the opportunities for performers in each country and to speculate why they are different. Encourage students to consider the opportunities available in New York and California.

Class Time 10 minutes

Canadian English uses some words, pronunciations, and spellings that differ from those used in the United States. For example, Canadians say *taps* and *serviettes* when people in the United States say *faucets* and *napkins*. For *about* and *house*, Canadians might say *aboot* and *hoos*. Many Canadians write *colour* for *color*, *theatre* for *theater*, and *cheque* for *check*. The nation's first prime minister, Sir John A. Macdonald, ordered that all official Canadian documents be written using standards set by dictionaries written in England.

Reading Social Studies

A. Synthesizing Why does Canada have two official languages?

Arts and Entertainment 2

Canada has rich traditions in the arts, actively supported by government funding. For example, the Canada Council for the Arts gives money to more than 8,400 artists and art organizations each year. Provincial governments also support regional arts programs.

Vocabulary

provincial: of, or relating to, a province

Canadians read many of the same newspapers and magazines, and watch many of the same television shows and movies as do people in the United States. Canadian musicians, such as Neil Young, Joni Mitchell, Buffy Ste. Marie, Céline Dion, and Shania Twain, are popular in both countries. Comedian-actors Dan Aykroyd and Jim Carrey are also from Canada.

Reading Social Studies

A. Possible Answer Both British and French communities and traditions have always been important in Canada's history.

Culture • The National Gallery of Canada in Ottawa is a visual arts museum that exhibits works by both Canadian and international artists. ◄

Activity Options

Interdisciplinary Link: The Arts

Class Time 30 minutes

Task Writing a letter supporting or opposing public funding of the arts

Purpose To consider the pros and cons of Canada's extensive public support of the arts

Supplies Needed
- Writing paper
- Pens or pencils

Block Scheduling

Activity Point out to students that public funding of the arts, such as that provided through the Canada Council for the Arts, is possible because of money collected from taxpayers. Ask students to take a position in support of, or in opposition to, public funding of the arts. Have them write a brief letter to the editor of a newspaper, expressing their opinion. Encourage students to support their opinions with specific reasons and examples.

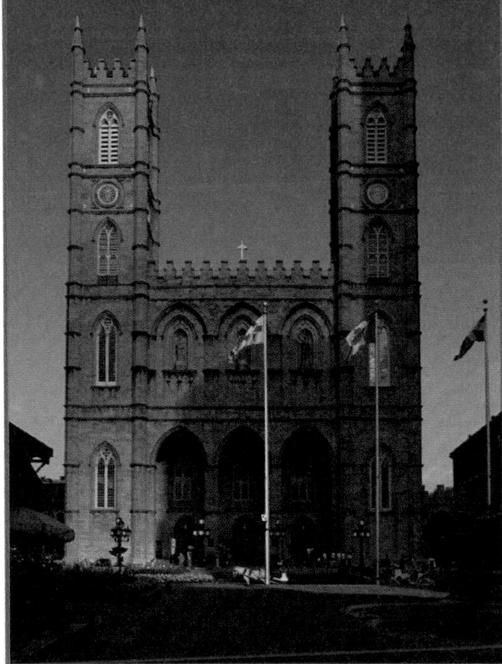

Culture • Notre-Dame was built in Montreal between 1824 and 1829. The architecture of the church—as well as the paintings, sculptures, and stained-glass windows inside—attracts many thousands of visitors each year. ◀

Religion 3

Christianity is widely practiced in Canada, but many other religions are followed as well, including Buddhism, Hinduism, Islam, and Judaism. Some religions are grounded in a spirituality based on respect for Earth and all forms of life. People of every cultural group are free to worship as they choose.

Culture Regions 

Most Canadian immigrants during the 1600s, 1700s, and 1800s were European. Recently, more people have arrived from Asia and South America. People who share the same language and background often settle in the same area. As a result, Canada has various culture regions, or areas where many people belonging to one cultural group live together.

Culture regions exist in different parts of Canada. Quebec is home to many French-speaking Canadians. In Nunavut more than 50 percent of the people are Inuit. Almost 16 percent of the population of Vancouver are Chinese, mostly from Hong Kong.

BACKGROUND

Inuit culture has been greatly affected by contact with the modern world. However, some Inuit traditions have been preserved in their arts and crafts, dialects, and ways of living in Arctic Canada.

Connections to History

aising the Maple Leaf A country's ag is an important national symbol. fter 1763, when the United Kingdom von the French and Indian War, the ritish Royal Union Flag, or Union ack, became Canada's flag. Efforts o design a new flag for Canada egan in 1925. The Red Ensign, which ad the Union Jack in its upper left corner and the Canadian coat of arms on its right side, was raised 20 years later.

In 1965, the Houses of Parliament adopted the Maple Leaf, which remains the flag of Canada today. The red background is a connection to the Red Ensign, and the maple leaf is Canada's national symbol.

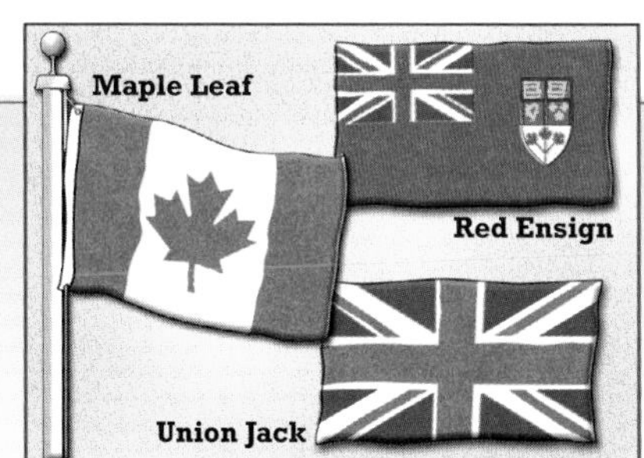

INSTRUCT: Objective 3

Religion/Culture Regions

- What religions do Canadians follow? Christianity, Islam, Judaism, Buddhism, Hinduism, and a variety of First Nations religions
- What are two examples of a culture region? Quebec (French-speaking Canadians), Nunavut (Inuit)
- How do Canada's cultural groups differ? languages, customs, and lifestyles

Connections to History

Historian George F. G. Stanley chose symbols significant to Canada's history when he designed the flag. The maple leaf had been a symbol of Canada since the early 1800s, and red and white had been Canada's official colors since the creation of the country's coat of arms in 1921.

Activity Options

Multiple Learning Styles: Visual

Class Time 45 minutes

Task Create a map of Canada's culture regions

Purpose To understand the variety and location of Canada's culture regions

Supplies Needed
- Information on Canada's population from the Unit Atlas
- Outline maps
- Writing paper
- Colored pencils or pens

Activity Distribute the outline maps, and have students use the information to create maps of Canada's different cultural groups. Tell students to create a key for their maps and use the key to differentiate culture regions. Point out that some places, like cities in southern Canada, are likely to have many more different groups than the northern rural areas.

A VOICE FROM CANADA

Direct students' attention to Speaker Bourget's comment about Canada's flag. Ask students to identify words in the quotation that explain why Bourget believes that the flag is a symbol of unity.

ASSESS & RETEACH

Reading Social Studies Ask students to write two sentences, one giving an example of the way Canada's many cultural groups strengthen the country, and the other giving an example of the way they provide challenges for the nation. Use the cause-and-effect model to discuss their examples.

Formal Assessment
- Section Quiz, p. 66

RETEACHING ACTIVITY

Organize students into groups of four. Have each student in the group write a summary of one of the four headings in the section. Then have each student read his or her summary aloud and answer any questions from the group.

In-depth Resources: Unit 2
- Reteaching Activity, p. 34

Access for Students Acquiring English
- Reteaching Activity, p. 38

Getting different culture regions to agree on national issues is sometimes difficult. The adoption of the Maple Leaf as Canada's flag in 1965 was one successful effort to unite all Canadians.

> *A VOICE FROM CANADA*
>
> The flag is the symbol of the nation's unity, for it, beyond any doubt, represents all the citizens of Canada without distinction of race, language, belief or opinion.
>
> *—Speaker of the Senate Maurice Bourget*

Reading Social Studies
B. Possible Answer
Language and lifestyle differences can lead to conflicts between different cultural groups.

Conflict and Cooperation Languages, customs, and lifestyles differ among the cultural groups of Canada. Sometimes these differences lead to conflict. For example, in the second half of the 20th century, some Canadians thought that the thousands of Chinese immigrants settling in the Vancouver area would change Canadian culture through their language and customs.

In 1975, the government began reviewing immigration policy. Chinese groups in Vancouver organized a Chinese-Canadian conference. They asked for continued support of multiculturalism and that immigration laws remain open for all people. The concerned groups solved the problem through human rights laws.

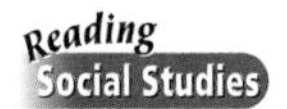

B. Analyzing Causes How do cultural differences cause conflicts among people?

SECTION 4 ASSESSMENT

Terms & Names

1. Explain the significance of: (a) national identity (b) bilingual (c) Francophone

Using Graphics

2. Make a spider map like the one shown below to illustrate how various groups contribute to making a unique Canadian culture.

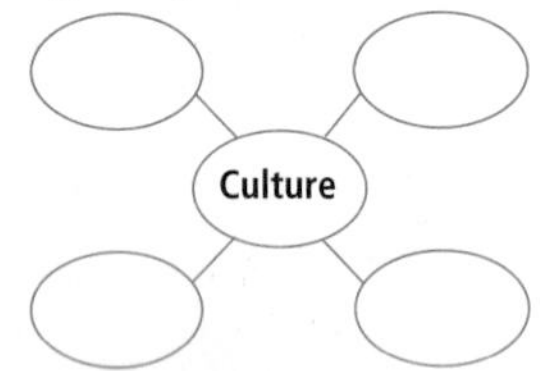

Main Ideas

3. (a) What are the two main languages spoken in Canada?
(b) How does Canada support its own arts and entertainment?
(c) What do people living in culture regions have in common?

Critical Thinking

4. **Forming and Supporting Opinions**
Do culture regions create more benefits or more disadvantages for Canada as a whole?

Think About
- existing and future conflicts
- how different groups contribute to Canadian culture

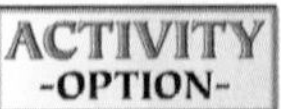

With a partner, choose a culture region of Canada. Create a **mural** or **collage** to show characteristics of the culture.

Section 4 Assessment

1. Terms & Names
- **a.** national identity, p. 132
- **b.** bilingual, p. 133
- **c.** Francophone, p. 133

2. Using Graphics

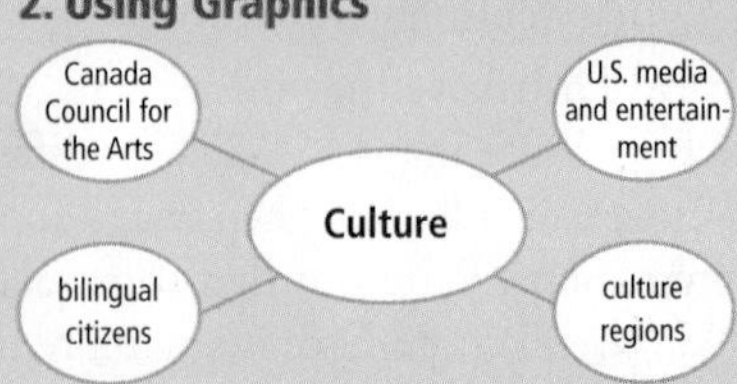

3. Main Ideas
- **a.** French and English
- **b.** Canada supports arts and entertainment through federal and provincial grants to arts organizations.
- **c.** People living in culture regions share languages and backgrounds.

4. Critical Thinking

Students may say that culture regions benefit Canada because they provide new immigrants with a familiar cultural environment.

ACTIVITY OPTION

Integrated Assessment
- Rubric for creating a mural or collage

SKILLBUILDER

Identifying Cause and Effect

▸▸ Defining the Skill

A cause is an event, a person, or an idea that brings about a result, or an effect. An effect is something that results from a cause. Understanding the relationship between cause and effect is key to understanding the world and its cultures.

▸▸ Applying the Skill

The following paragraph describes where most Canadians live. Use the strategies listed below to help you identify why Canadians live there.

How to Identify a Cause-and-Effect Relationship

Strategy ❶ Look for the cause of, or reason for, the cause-and-effect relationship. It might be suggested in the title and topic sentence. Ask yourself what happened and why it happened. Writers may indicate a cause-and-effect relationship by using words such as *thus, therefore, so,* and *as a result.* Use those words as clues.

Strategy ❷ Look for the results of the event or action. Ask yourself what happened as a result of the action. You have found the effect.

Strategy ❸ Remember that several causes can combine to create one event. Also remember that one cause can have several effects. Ask yourself if anything else helped to bring about the event. Ask yourself if there are any other results.

❶ WHERE CANADIANS LIVE

❷ Most of Canada's people live within 100 miles of the United States border. This heavily populated area covers only about 10 percent of the country. ❶ The mild climate in that part of the country makes living there more pleasant than living in the colder northern regions. ❸ People can find jobs more easily in the large cities located near the border. ❸ Many Americans who live in Canada can cross the border easily to visit family and friends.

Make a Diagram

Using a diagram can help you understand causes and effects. The diagram to the right shows what causes people in Canada to live close to the U.S. border and what effect is created.

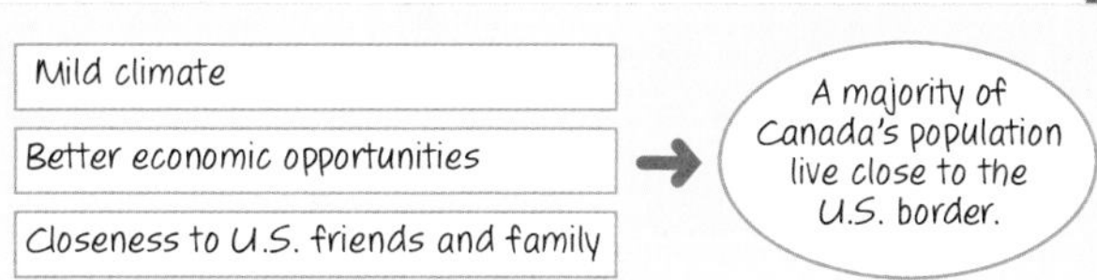

▸▸ Practicing the Skill

Turn to Chapter 5, "Canada Today," Section 3, "Canada's Economy," and make a diagram of the causes that have resulted in Canada's strong economy.

SKILLBUILDER

Identifying Cause and Effect

Defining the Skill

Tell students that understanding cause and effect helps us make connections between events. Present the following example of a cause-and-effect relationship: The prediction of a severe storm led to shortages at grocery stores and hardware stores. Ask students to identify other examples of cause-and-effect relationships in their own lives.

Applying the Skill

How to Identify a Cause-and-Effect Relationship Explain to students that they should carefully read and work through each strategy, in sequence, to identify a cause-and-effect relationship. Remind them that several causes can lead to one effect and that one cause can result in several effects.

Make a Diagram

Review and discuss the diagram. Help students identify the causes that resulted in a majority of Canada's population living near the U.S. border.

Practicing the Skill

Suggest that students create a diagram like the one on page 137. Invite volunteers to share the causes they identified. If students need additional practice, work with them to create a diagram that shows the effects of an event such as a storm or a decisive soccer game.

In-depth Resources: Unit 2

- Skillbuilder Practice, p. 29

Career Connection: Sociologist

Encourage students who enjoy identifying causes and effects to find out about careers that use this skill. For example, sociologists study how people are affected by each other and by the groups to which they belong. They conduct research that is used by educators, lawmakers, and other professionals.

1. Suggest that students look for information about the variety of topics that sociologists study. Help them see that any aspect of society is a possible topic for study.
2. Help students learn about other qualifications and education that a person needs to become a sociologist.
3. Have students share what they learn in an "experts' panel" format.

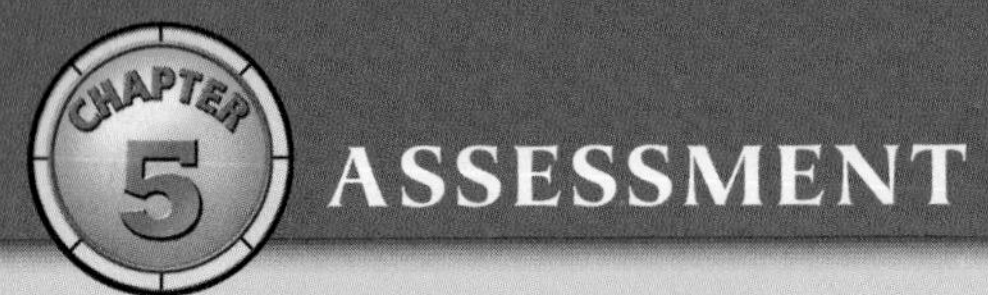

ASSESSMENT

TERMS & NAMES

1. multiculturalism, p. 121
2. refugee, p. 122
3. Parliament, p. 125
4. prime minister, p. 125
5. export, p. 129
6. import, p. 130
7. transportation corridor, p. 131
8. transportation barrier, p. 131
9. bilingual, p. 133
10. Francophone, p. 133

REVIEW QUESTIONS

Possible Responses

1. defeating France in the French and Indian War
2. Canada's policy of multiculturalism makes it a diverse country with distinct culture groups and two official languages.
3. into legislative and judicial branches, with a governor-general representing the British monarch as head of state
4. Canada has passed laws guaranteeing the preservation of different cultural heritages.
5. Canada's natural resources support farming, mining, logging, and fishing.
6. Canada's chief transportation corridors are waterways and railways. Its chief transportation barriers are the Rocky Mountains, snow, and ice.
7. Many Canadian French words are spelled and pronounced differently from standard French. Many English words differ from those used in the United States.
8. French Canadians, Inuit, Chinese, and South Americans

CHAPTER 5 ASSESSMENT

TERMS & NAMES

Explain the significance of each of the following:

1. multiculturalism
2. refugee
3. Parliament
4. prime minister
5. export
6. import
7. transportation corridor
8. transportation barrier
9. bilingual
10. Francophone

REVIEW QUESTIONS

O Canada! Immigrant Roots *(pages 119–123)*

1. How did the United Kingdom gain control of Canada?
2. What effects does the policy of multiculturalism have on Canada?

A Constitutional Monarchy *(pages 124–127)*

3. How is Canada's federal government organized?
4. How has Canada tried to satisfy the needs of its many culture groups?

Canada's Economy *(pages 128–131)*

5. How does Canada's wealth of natural resources contribute to its economy?
6. What are Canada's chief transportation corridors and barriers?

A Multicultural Society *(pages 132–136)*

7. How have the native languages of England and France changed in Canada?
8. What are some major culture groups of Canada?

CRITICAL THINKING

Analyzing Causes and Effects

1. Using your completed chart from Reading Social Studies, p. 118, describe some factors that influence where people settle in Canada.

Drawing Conclusions

2. Many people in Quebec wanted the province to separate from the rest of Canada. What in Prime Minister Pierre Trudeau's background made him effective in keeping Quebec part of Canada?

Making Inferences

3. How do you think the policy of multiculturalism affects the way Canadians of different cultures respond to one another?

Visual Summary

1 O Canada! Immigrant Roots

- The Canadian people have come from many countries.
- Canada is a very large country with a small population; most people settled in the southern part of the country.

2 A Constitutional Monarchy

- The Canadian government is a constitutional monarchy that is made up of two main branches: a legislative branch and a judicial branch.
- Canada has passed laws to protect the civil rights of its people and to support multiculturalism.

3 Canada's Economy

- Canada has many natural resources that contribute to its economy.
- Canada's most important trade partner is the United States.

4 A Multicultural Society

- Many cultural groups have helped to build the unique Canadian culture of today.

- The diverse cultures of the population create challenges in unifying Canada.

CRITICAL THINKING: Possible Responses

1. Analyzing Causes and Effects
Students may say that the extreme climate, transportation barriers, and poor soil of northern Canada have led most people to settle in southern Canada.

2. Drawing Conclusions
Pierre Trudeau was born in Montreal, of French and Scottish ancestry, and grew up speaking French and English. These factors may have made him more effective in keeping Quebec as part of Canada.

3. Making Inferences
Some students may point out that Canada's policy of multiculturalism ensures that different cultures are tolerant of each other. Others may argue that the policy encourages a separation of cultures that can lead to misunderstandings and conflicts.

e the map and your knowledge of world cultures and ography to answer questions 1 and 2.

dditional Test Practice, pp. S1–S33

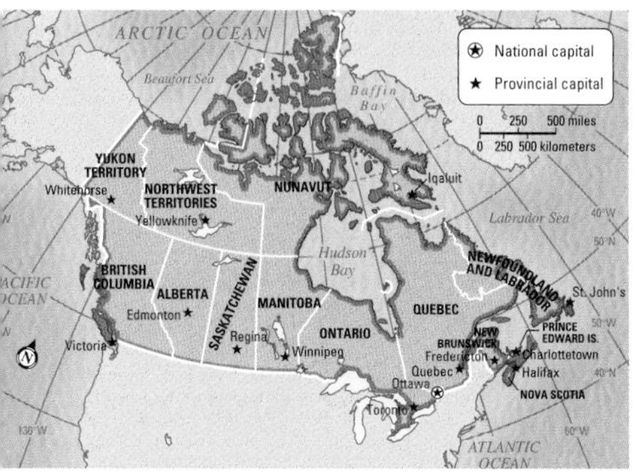

In which province is Canada's capital city located?

A. British Columbia
B. Nova Scotia
C. Ontario
D. Quebec

Which province is located on the Pacific Coast?

A. British Columbia
B. Manitoba
C. New Brunswick
D. Newfoundland and Labrador

This excerpt is from an article about the hockey competition of the 2002 winter Olympics in Utah. Use the quotation and your knowledge of world cultures and geography to answer question 3.

PRIMARY SOURCE

Winning its first Olympic gold medal in 50 years in the sport it invented, Team Canada overwhelmed the United States, 5–2, with a crushing third period Sunday to complete the finest hockey tournament the world has ever seen. . . . There was a major difference that inspired Canada's killing-blow two-goal period: The Americans *wanted* to win but the Canadians *had* to win.

MICHAEL HUNT, *Wisconsin Journal Sentinel*

3. The passage supports which of the following observations?

A. The Canadian team's success was due to the poor performance of the Americans.
B. The Canadian team's national pride helped them overpower the U.S. team.
C. The Canadian team was bound to win because Canada had invented hockey.
D. The Canadian team's success was due to better physical training of its players.

STANDARDS-BASED ASSESSMENT

1. **Answer C** is the correct answer because the capital city is Ottawa, which is located in southern Ontario. Answers A, B, and D are incorrect because although each province mentioned has its own provincial capital, the capital of the nation is in Ontario.
2. **Answer A** is the correct answer because the province of British Columbia is on the Pacific coast. Answer B is incorrect because Manitoba borders Hudson Bay; answer C is incorrect because New Brunswick borders an unlabeled river; answer D is incorrect because Newfoundland and Labrador is on the Atlantic coast.
3. **Answer B** is the correct answer because, according to the passage, the team believed it had to win the game because hockey is Canada's national sport. Answer A is incorrect because the article doesn't mention the American team's performance; answer C is incorrect because even though the Canadians invented hockey, they hadn't won in 50 years; answer D is incorrect because the article doesn't mention physical training.

LTERNATIVE ASSESSMENT

 **WRITING ABOUT HISTORY**

Should the Canadian government provide support for the arts? Research this topic and write an editorial based on your findings.

- Using the Internet and the library, research government support for the arts, currently and in the past. You might also contact Canadian national museums and arts associations for information.
- Based on your research, decide whether government support is necessary, and if so, whether it should be increased or decreased. State your opinion and support it with facts from your research.

COOPERATIVE LEARNING

Work with a small group to prepare an interview for a public television broadcast on the topic of self-government for cultural groups. One member of your group can take the role of news moderator, and other members can each represent a different group. Together, brainstorm a list of cultural groups and prepare a set of interview questions. Individuals can do research to find out how the group views self-government.

INTEGRATED TECHNOLOGY

Doing Internet Research

Canada is a large country with diverse geographic regions. Choose one region, such as the Yukon Territory or the Maritime Provinces, and research what it is like to live there. Prepare a presentation of your findings.

- Use the Internet or other resources in the library to learn about the geography of your chosen region.
- Other sources of information might be Canadian museums and tourist bureaus.
- Focus your research on topics such as the region's weather, types of plants that grow there, kinds of jobs residents have, how tourists would travel around the region, and what tourists might see and do.

For Internet links to support this activity, go to

INTEGRATED TECHNOLOGY

Students' reports should discuss the climate, natural resources, plants and animals, and human settlements and transportation of the chosen region. Students should include photographs or illustrations, as well as a list of Web sites used to research the information.

Alternative Assessment

1. Rubric

The editorial should

- clearly state a position on government support for the arts in Canada.
- present supporting reasons for the position.
- clearly rebut other viewpoints.
- use correct grammar, spelling, and punctuation.

2. Cooperative Learning

Review the procedure for presenting a news interview on television. Student groups should identify the cultural groups chosen. The student news moderator should ask questions clearly related to issues of self-government. Cultural group representatives should formulate answers based on their research.

Latin America

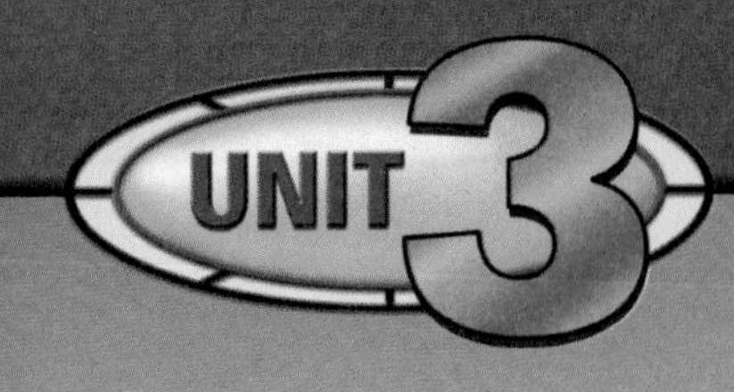

Before You Read

Previewing Unit 3

Unit 3 introduces the physical geography of Latin America and identifies major landforms, mountain ranges, river systems, and climate. Then the focus shifts to provide students with an understanding of the diverse societies of Mexico and the individual countries that make up Central America, the Caribbean, and South America. The history, government, economy, and culture of these countries are examined and compared throughout the unit.

140

Unit Level Activities

1. Create a Mural

Display a photograph of a Diego Rivera mural. Explain to students that Rivera was a well-known Mexican artist who painted murals depicting aspects of Mexico's culture and history. These murals can be seen today in Mexico City. Hang a large sheet of drawing paper in the classroom. As students read the chapter, invite them to illustrate events and famous people in a mural.

2. Conduct a Debate

After students read about the *ejido* and privatization systems, have them conduct a debate over the pros and cons of privatization. Divide students into two groups and assign each group the task of defending or criticizing the system. Allow time for students to prepare their arguments. Familiarize students with the rules of debate. You might want to act as the moderator.

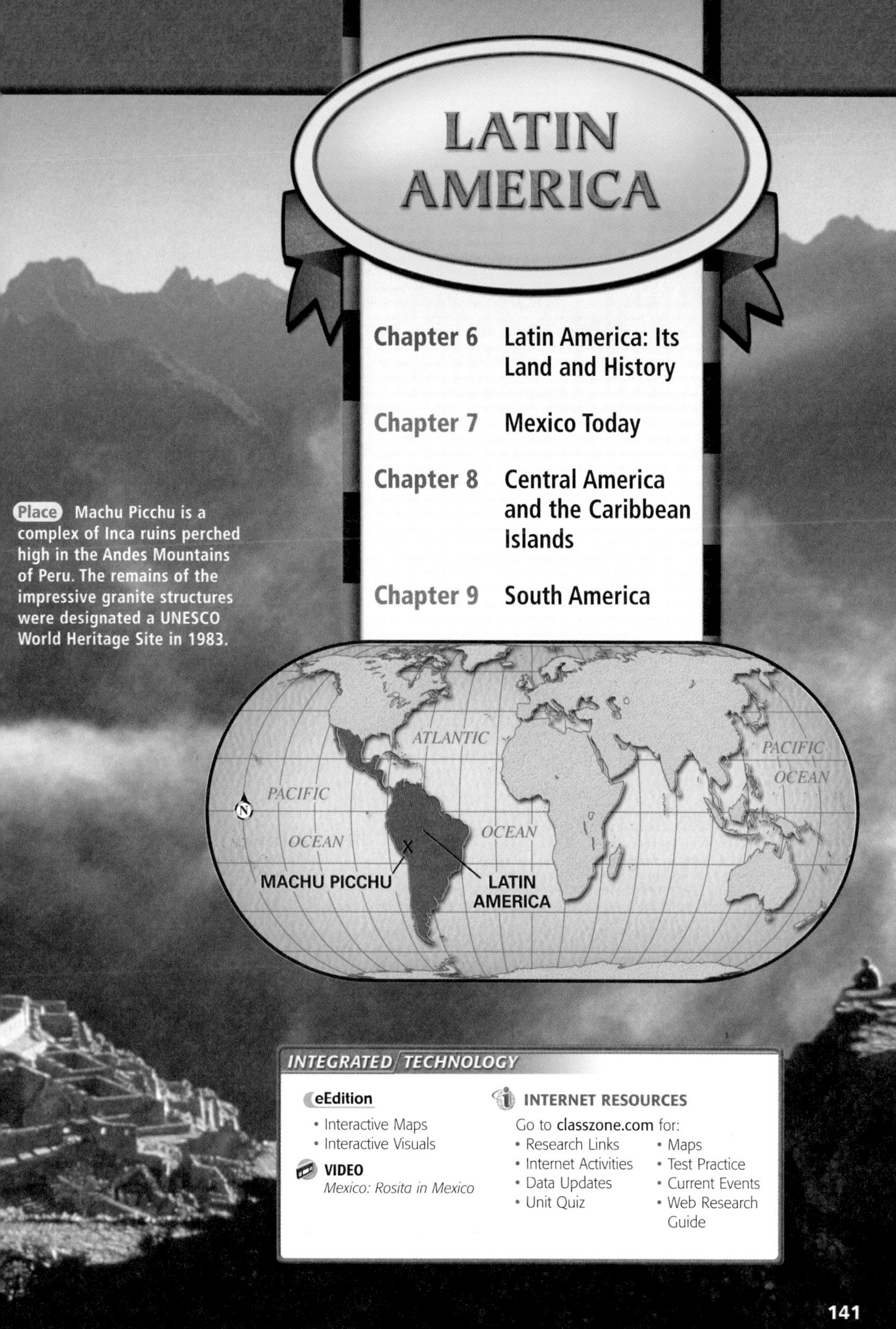

Place Machu Picchu is a complex of Inca ruins perched high in the Andes Mountains of Peru. The remains of the impressive granite structures were designated a UNESCO World Heritage Site in 1983.

INTEGRATED TECHNOLOGY

eEdition
- Interactive Maps
- Interactive Visuals

VIDEO
Mexico: Rosita in Mexico

INTERNET RESOURCES
Go to **classzone.com** for:
- Research Links
- Internet Activities
- Data Updates
- Unit Quiz
- Maps
- Test Practice
- Current Events
- Web Research Guide

141

FOCUS ON VISUALS

Interpreting the Photograph Have students examine the photograph, and ask a volunteer to read the caption aloud. Provide students with the background information that Machu Picchu was built by the Inca at approximately the beginning of the 16th century. Ask students why the Inca might have built Machu Picchu in this location. Ask why the Inca might have chosen this layout for the city. Finally, ask students to speculate as to what the large open area in the center of Machu Picchu might have been used for.

Possible Responses In this location, mountains protected the community from intruders. The most important people probably lived in higher areas and the others lived below. The large, open area might have been used for religious ceremonies, assemblies, or recreational purposes.

Extension Ask students to research the Andes Mountains and write three facts about them.

Implementing the National Geography Standards

Standard 6 Assess a region from the points of view of different types of people

Objective To show how points of view about a region may differ

Class Time 30 minutes

Task Have each student prepare a two-minute talk about the same region, speaking as a tourist, truck driver, teacher, shopkeeper, taxi driver, and so on. Students may choose to speak about the region's situation, scenery, cultural attractions, shopping, or other topic. Encourage speakers to represent the point of view of the person they have chosen. Compare and contrast the speakers' viewpoints.

Evaluation Students should include two examples of ways a person's prior experiences can affect how he or she views a region.

ATLAS Latin America

ATLAS OBJECTIVES

1. Describe and locate physical features of Latin America
2. Compare data on the physical geography of Latin American countries
3. Identify political features of Latin America
4. Compare city populations and languages spoken in Latin America

FOCUS & MOTIVATE

Ask students what they know about Latin America. Invite them to identify its location in the world and describe its physical geography. Ask them how maps can provide this specific information.

INSTRUCT: Objective 1

Physical Map of Latin America

- What do the colors on the map tell you? elevation of the land
- What is the largest mountain range in Latin America? Andes Mountains
- What sea lies north of South America? Caribbean Sea

In-depth Resources: Unit 3
- Unit Atlas Activity, p. 1

UNIT Atlas 3

Physical Geography

Latin America: Physical

MEXICO
Baja California
Gulf of California
Sierra Madre Occidental
Sierra Madre Oriental
Rio Grande
Gulf of Mexico
Orizaba 18,854 ft. (5,747 m)
Popocatépetl 17,930 ft. (5,465 m)
Yucatán Peninsula
Tajumulco 13,844 ft. (4,220 m)
CENTRAL AMERICA
BELIZE
HONDURAS
GUATEMALA
EL SALVADOR
NICARAGUA
COSTA RICA
PANAMA
Isthmus of Panama
Panama Canal
Barú 11,400 ft. (3,475 m)
BAHAMAS
CUBA
JAMAICA
HAITI
DOMINICAN REPUBLIC
Greater Antilles
WEST INDIES
Lesser Antilles
Netherlands Antilles
Caribbean Sea
Tropic of Cancer
ATLANTIC OCEAN
GUYANA
SURINAME
French Guiana (Fr.)
VENEZUELA
Llanos
Orinoco R.
Guiana Highlands
COLOMBIA
Negro R.
Equator
ECUADOR
AMAZON BASIN
Amazon R.
SOUTH AMERICA
PERU
Madeira R.
BRAZIL
Xingu R.
Araguaia R.
São Francisco R.
PACIFIC OCEAN
ANDES
Lake Titicaca
Mato Grosso Plateau
BOLIVIA
BRAZILIAN HIGHLANDS
PARAGUAY
Atacama Desert
Gran Chaco
Paraguay R.
Paraná R.
Tropic of Capricorn
Uruguay R.
Mt. Aconcagua 22,831 ft. (6,959 m)
ARGENTINA
Pampas
URUGUAY
Plata R.
CHILE
Patagonia
ATLANTIC OCEAN
Tierra del Fuego
Falkland Is.
Cape Horn
South Georgia

Elevation
13,100 ft. (4,000 m)
6,600 ft. (2,000 m)
3,275 ft. (1,000 m)
650 ft. (200 m)
0 ft. (0 m)
Below sea level
▲ Mountain peak

0 500 1,000 miles
0 500 1,000 kilometers

100°W 80°W 60°W 40°W

142 UNIT 3

Activity Options

Cooperative Learning: Creating an Elevation Profile

Explaining the Skill Direct students' attention to the Elevation Profile on page 143. Explain that this is one way to show the elevation of a place. Tell students that they will use the physical map of Latin America to create an elevation profile.

Applying the Skill Have students work in pairs to choose one of the larger countries of Latin America and trace its outline on a sheet of paper. Then have them draw a line from one side to the other. Next tell them to imagine that they are looking at the country from the side. Have them use the information from the map and key to create an elevation profile for the line they have drawn across the country. Have students meet in groups to share their elevation profiles.

INSTRUCT: Objective 2

Elevation Profile

- What is the highest elevation reached by the Brazilian Plateau? 3,000 feet above sea level

Land Area

- Which has a larger land area, Latin America or the United States? By about how much? Latin America; almost 5 million square miles

Fast Facts

- What is the highest mountain in Latin America? Mt. Aconcagua

MORE ABOUT...

The Amazon River

Although the Nile is longer, in many ways the Amazon River is the biggest river in the world. For most of its length, it ranges from 1.5 to 6 miles wide. At its mouth, it is up to 90 miles across. On average, it is about 40 feet deep, but it exceeds 300 feet deep in places. The Amazon carries more water than the Nile, Mississippi, and Yangtze rivers combined.

Fast Facts

Encourage students to do additional research to build their own Fast Facts file. Explain that atlases, almanacs, and encyclopedias are good sources of information.

GEOGRAPHY SKILLBUILDER

Answers

1. Atlantic Ocean, Pacific Ocean
2. -25,000 feet

Country Profiles

Argentina Argentina is the second largest country in South America. Only Brazil is larger. Argentina has a population of about 38 million people, about 90 percent of whom live in urban areas. Only Brazil and Colombia have greater populations. Buenos Aires is the capital and largest city. About one-third of all Argentineans live there. Native Americans make up just 15 percent of the population, a much lower proportion than in most Latin American countries. Most Argentinians—about 85 percent—are of European descent. Mt. Aconcagua, the highest mountain in South America and in the Western Hemisphere, is located in the Andes Mountains along Argentina's western border with Chile. The Pampa, which makes up about one-fifth of Argentina, is a great plain with some of the richest soil in the world. Argentina is one of the world's leading agricultural nations, exporting beef, corn, wheat, and flaxseed. Most Argentineans live on the Pampa.

INSTRUCT: Objective 3

Political Map of Latin America

- How is this map different from the physical map on page 142? It shows country names, major cities, and capitals.
- What two countries do not have a coastline? Paraguay and Bolivia
- What is the capital of Venezuela? Caracas
- Where is São Paulo located? on the southeast coast of Brazil

MORE ABOUT...

European Settlement

Although most of Latin America was colonized by Spain and Portugal, some of the islands have different histories. The Virgin Islands, for example, were claimed and settled by France, Denmark, and the United Kingdom. Denmark later sold its islands to the United States, and they are now called the U.S. Virgin Islands. The British gained control of Jamaica in 1670. The Netherlands established themselves in the Antilles. In the far south, the United Kingdom discovered the Falkland Islands in 1592.

Activity Options

Interdisciplinary Link: Popular Culture

Block Scheduling

Class Time One class period

Task Creating a travel brochure for a Latin American nation

Purpose To learn about the cultural attractions of a Latin American city

Supplies Needed

- Encyclopedias, almanacs, Internet access
- Writing paper
- Pencils or pens
- Art supplies

Activity Have students choose one capital city identified on the map and learn about its culture. Suggest that students explore museums, historic sites or monuments, festivals, and ethnic neighborhoods. Challenge them to create a travel brochure describing these places and encouraging tourists to visit. Have students share their brochures.

City Populations of Latin America

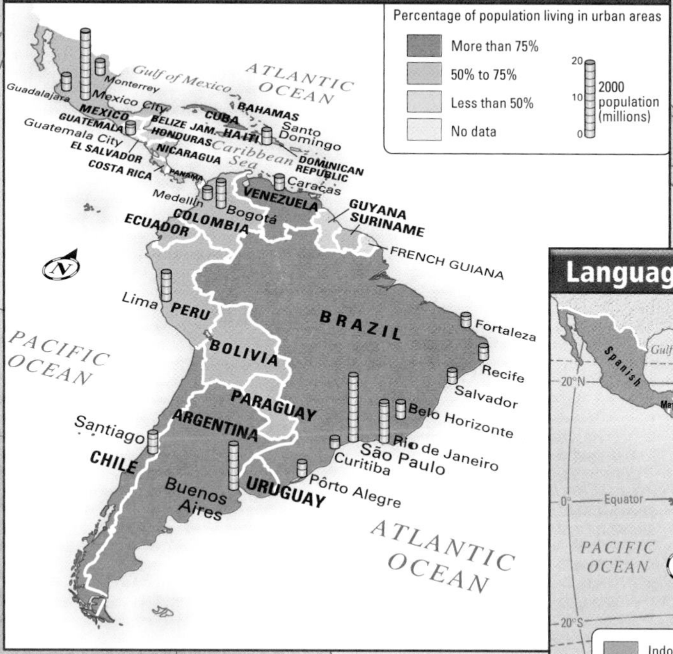

Languages of Latin America

FAST FACTS

- **LARGEST COUNTRY (in land area):** Brazil, 3,300,171 sq. mi.
- **SMALLEST COUNTRY (in land area):** Grenada, 120 sq. mi.
- **LARGEST CITY (population):** Mexico City, 18,131,000 (2000)
- **HIGHEST POPULATION DENSITY:** Barbados, 1,612 people per sq. mi.
- **LOWEST POPULATION DENSITY:** French Guiana, 5 people per sq. mi.

GEOGRAPHY SKILLBUILDER: Interpreting Maps and Visuals

1. **Location** • Which countries have no coastline?
2. **Region** • What language do most people in Brazil speak?

INSTRUCT: Objective 4

City Populations

- What percentage of Mexico's population lives in urban areas? between 50 and 75 percent

Languages of Latin America

- In what country is Portuguese the primary language? Brazil

Fast Facts

- What is the difference in population density between Barbados and French Guiana? 1,607 people per square mile

MORE ABOUT...

Native American Languages

When Columbus arrived, about 1,700 Native American languages were spoken in what was to become Latin America. Today, only a few of the original languages are still widely spoken. About 4 million people speak Maya, approximately 4 million people speak one of the Guarani languages, and about 2 million people speak an Aymara language. About 12 million people in Peru, Bolivia, Ecuador, and Argentina speak Quechua.

Fast Facts

Invite students to choose one country and research a Fast Fact about that country to add to the list. Students can share and compare their facts.

GEOGRAPHY SKILLBUILDER

Answers

1. Bolivia, Paraguay
2. Portuguese

Country Profiles

Chile Chile is a country of extremes. Its 2,640-mile length is ten times its width. The Atacama Desert, which occupies much of northern Chile, is one of the driest places on Earth. Rain may occur only once in 20 years in parts of this desert. The snowcapped Andes Mountains form Chile's eastern border. Along the Pacific coast lies another, lower range of mountains. In between the mountain ranges is the central valley. Most of Chile's 15 million people live here. The soil is fertile, and the climate is mild. In all, about 85 percent of Chileans live in cities. Chile's capital and largest city, Santiago, has about 4 million people, more than one-fourth of the country's population. The southern part of Chile is made up of the archipelago. It extends for 1,000 miles and is made up of thousands of islands. In the far south is Horn Island. At its tip is Cape Horn, the southernmost point of South America. Chile's economy is based on copper. In fact, Chile is the world's largest producer of this mineral.

DATA FILE OBJECTIVE

Examine and compare data on Latin American countries

FOCUS & MOTIVATE

Have students look at the Data File. To help them use the chart, ask questions such as, "In which Latin American country is life expectancy the longest? The shortest?"

INSTRUCT: Objective 5

Data File

- Which country has the fewest doctors per 100,000 people? Haiti; 8 per 100,000 people
- Which country has the highest birthrate? Guatemala; 35 per 1,000 people
- What is the official currency of Brazil? real
- What is the difference in literacy rates between El Salvador and Venezuela? The literacy rate in Venezuela is 20 percent higher.

If students have difficulty understanding the statistics on doctors or birthrate, discuss these categories as students review the Data File.

In-depth Resources: Unit 3
- Data File Activity, p. 2

UNIT 3 Atlas

Data File

For updates on these statistics, go to

DATA UPDATE CLASSZONE.COM

Country Flag	Country/Capital	Currency	Population (2001 estimate)	Life Expectancy (years)	Birthrate (per 1,000 pop.) (2000)
	Antigua and Barbuda St. John's	East Caribbean Dollar	67,000	70	20
	Argentina Buenos Aires	Peso	37,385,000	75	19
	Bahamas Nassau	Bahamian Dollar	298,000	71	20
	Barbados Bridgetown	Barbadian Dollar	275,000	73	14
	Belize Belmopan	Belizean Dollar	256,000	71	32
	Bolivia La Paz, Sucre	Boliviano	8,300,000	64	28
	Brazil Brasília	Real	174,469,000	63	19
	Chile Santiago	Chilean Peso	15,328,000	76	17
	Colombia Bogotá	Colombian Peso	40,349,000	70	23
	Costa Rica San José	Costa Rican Colon	3,773,000	76	21
	Cuba Havana	Peso	11,184,000	76	13
	Dominica Roseau	East Caribbean Dollar	71,000	73	18
	Dominican Republic Santo Domingo	Dominican Peso	8,581,000	73	25
	Ecuador Quito	U.S. Dollar and Sucre	13,184,000	71	27
	El Salvador San Salvador	Salvadoran Colon	6,238,000	70	29
	Grenada St. George's	East Caribbean Dollar	89,000	65	23
	Guatemala Guatemala City	Quetzal	12,974,000	66	35
	Guyana Georgetown	Guyanese Dollar	697,000	64	18

Activity Options

Interdisciplinary Link: Mathematics

Class Time 15 minutes

Task Determining the birthrate per year in Latin American countries

Purpose To use Data File information to develop new information

Supplies Needed
- Textbook
- Paper and pencils

Activity Tell students that they can use information from a chart to get additional information. Point out the columns for birthrate and population in the Data File. If students divide a population by 1,000 and then multiply that number by birthrate, they can determine about how many births occur per year in a country. Have students calculate births per year for Brazil, Colombia, and Mexico.

5

DATA FILE

Infant Mortality (per 1,000 live births) (2000)	Doctors (per 100,000 pop.) (1997–1998)	Literacy Rate (percentage) (1996–1998)	Passenger Cars (per 1,000 pop.) (1991–1998)	Total Area (square miles)	Map (not to scale)
20.0	114	90	207	171	
17.8	268	96	136	1,073,399	
17.8	152	98	245	5,386	
16.2	125	97	167	166	
30.8	55	93	10	8,867	
60.2	130	83	26	424,164	
33.8	127	85	84	3,300,171	
9.6	110	95	62	292,135	
23.2	116	91	31	440,831	
12.7	141	95	14	19,730	
7.7	530	96	2	42,804	
8.5	49	90	104	290	
40.8	216	82	14	18,704	
29.3	170	90	22	103,930	
27.2	107	71	6	8,124	
10.9	50	96	94	120	
45.0	93	56	9	42,042	
48.6	18	98	34	83,000	

MORE ABOUT...

Cars in Cuba

Only about two Cubans in 1,000 own a passenger car. These cars are likely to be old American cars from the 1950s or earlier. There are political and economic explanations for this. Prior to the Communist Revolution, the United States was Cuba's primary trade partner. Then, in 1961, the United States established a trade embargo against Cuba. Because Cubans could no longer import American cars, they kept their old ones. Also, the cost of gasoline in Cuba is prohibitive. At about four dollars per gallon, most Cubans cannot afford to own a car.

Activity Options

Interdisciplinary Links: Science/Health

Class Time 30 minutes

Task Creating a bar graph to represent infant mortality rates

Objective To gather information from a chart and present it in a bar graph

Supplies Needed
- Ruler
- Graph paper
- Pencils or pens

Activity Point out the column for infant mortality rate in the Data File. Tell students that they are going to create bar graphs comparing the infant mortality rates of five of the countries listed. Review the features of bar graphs and remind students to include a title and labels for the vertical and horizontal axes. Suggest that they number the vertical axis in increments of ten. Have students compare and discuss their graphs.

MORE ABOUT...
Panama

Although Panama is a small country, it is a vital transportation center. In 1914, after more than ten years of construction, the United States opened the Panama Canal. It was built to eliminate the need for ships to go around the tip of South America when traveling from the Atlantic Ocean to the Pacific Ocean. Many countries use the canal, which is the major source of economic activity in Panama. Ships pay tolls to use the canal, and the canal provides jobs for many Panamanians. Some work on the canal and others are employed by businesses that provide goods and services to the canal, its clients, and the people who live and work there. The United States transferred ownership of the canal to Panama in 1999.

UNIT 3 Atlas

Data File

For updates on these statistics, go t

DATA UPDATE
CLASSZONE.COM

Country Flag	Country/Capital	Currency	Population (2000 estimate)	Life Expectancy (years)	Birthrate (per 1,000 pop.) (2000)
	Haiti Port-au-Prince	Gourde	6,965,000	49	32
	Honduras Tegucigalpa	Lempira	6,406,000	70	33
	Jamaica Kingston	Jamaican Dollar	2,666,000	75	19
	Mexico Mexico City	New Peso	101,879,000	71	23
	Nicaragua Managua	Gold Cordoba	4,918,000	69	28
	Panama Panama City	Balboa	2,846,000	75	20
	Paraguay Asunción	Guarani	5,734,000	74	31
	Peru Lima	New Sol	27,484,000	70	24
	St. Kitts and Nevis Basseterre	East Caribbean Dollar	39,000	71	19
	St. Lucia Castries	East Caribbean Dollar	158,000	72	22
	St. Vincent and the Grenadines Kingstown	East Caribbean Dollar	116,000	72	18
	Suriname Paramaribo	Surinamese Guilder	434,000	71	21
	Trinidad and Tobago Port of Spain	Trinidad and Tobago Dollar	1,170,000	68	14
	Uruguay Montevideo	Uruguayan Peso	3,360,000	75	17
	Venezuela Caracas	Bolivar	23,917,000	73	21
	United States Washington, D.C.	Dollar	281,422,000	77	15

148 UNIT 3

Activity Options

Differentiating Instruction: Less Proficient Readers

Comparing and Contrasting Explain that Haiti and the Dominican Republic are both located on the island of Hispaniola. Have students make a chart like the one shown to compare these nations. After they complete their charts, have students meet in groups to summarize their findings.

Issues	Dominican Republic	Haiti
Currency		
Life Expectancy		
Birthrate		
Infant Mortality		
Doctors		
Literacy Rate		
Passenger Cars		

DATA FILE

Infant Mortality (per 1,000 live births) (2000)	Doctors (per 100,000 pop.) (1997–1998)	Literacy Rate (percentage) (1996–1998)	Passenger Cars (per 1,000 pop.) (1991–1998)	Total Area (square miles)	Map (not to scale)
96.3	8	45	5	10,714	
39.8	83	73	14	43,277	
13.4	140	85	17	4,244	
23.4	186	90	87	756,066	
38.7	86	66	16	50,464	
22.7	167	91	54	29,157	
35.3	110	92	14	157,048	
37.1	93	89	20	496,225	
16.9	117	90	130	104	
16.2	47	80	68	238	
14.6	88	82	44	150	
25.6	25	93	111	63,251	
18.3	79	98	107	1,978	
12.9	370	97	147	68,498	
25.5	236	91	68	352,144	
7.0	251	97	489	3,787,319	

GEOGRAPHY SKILLBUILDER: Interpreting a Chart

(Do not give United States as an answer.)

1. **Place** • Which countries have the highest life expectancy?
2. **Place** • Which country has the most cars per person?

MORE ABOUT...

Literacy

Explain to students that literacy is the ability to read and write. Standards for measuring literacy include the completion of a certain number of years of school or the ability to write a few sentences about oneself. The literacy of a population is important because it affects people's ability to prepare for and hold jobs. Literacy may also be an indicator of the health of a country's economy and the physical well-being and life expectancy of its people.

GEOGRAPHY SKILLBUILDER

Answers

1. Chile, Costa Rica, and Cuba (76 years)
2. Bahamas (245 cars)

Country Profiles

Cuba Cuba is the largest island in the Caribbean. It covers about 43,000 square miles (110,861 square kilometers), which is about the size of Tennessee. Cuba is located just 90 miles south of Key West, Florida. In 1492, Columbus became the first European to visit Cuba. Today, Cuba is the only Communist nation in Latin America. It has been ruled by Fidel Castro since 1959. Cuba has a population of about 11 million people, 75 percent of whom live in cities. Cuba's capital and largest city is Havana. The nation of Cuba is made up of the main island plus about 1,600 small islands. The main island has several mountain ranges. The highest mountain is Pico Turquino, which is 6,542 feet (1,994 meters) above sea level. Other parts of Cuba are covered by large grasslands and gently rolling hills. Between 1900 and 1960, the United States was Cuba's leading trade partner. When Castro brought communism to Cuba, however, trade ties between the countries ended.

PLANNING GUIDE: UNIT 3 ATLAS AND CHAPTER 6

Latin America: Its Land and History

	OVERVIEW	COPYMASTERS	INTEGRATED TECHNOLOGY
UNIT ATLAS AND CHAPTER RESOURCES	This chapter focuses on the geographic and ancient cultures of Latin America. The variety of landforms, waterways, and climates and the way in which these variations affected the development of ancient civilizations are discussed.	**In-depth Resources: Unit 3** • Guided Reading Worksheets, pp. 3–4 • Skillbuilder Practice, p. 7 • Unit Atlas Activities, pp. 1–2 • Geography Workshop, pp. 47–48 **Reading Study Guide** (Spanish and English), pp. 42–47 **Outline Map Activities**	• eEdition Plus Online • EasyPlanner Plus Online • eTest Plus Online • eEdition • Power Presentations • EasyPlanner • Electronic Library of Primary Sources • Test Generator • Reading Study Guide • Critical Thinking Transparencies CT11
SECTION 1 Physical Geography pp. 153–159	**KEY IDEAS** • Latin America has a wide range of environments and resources. • Physical geography influences Latin America's cultures, offering resources and obstacles. • Language is the cultural connection in Latin America. • Mexico City faces such problems as air pollution and earthquakes due to its location.	**In-depth Resources: Unit 3** • Guided Reading Worksheet, p. 3 • Reteaching Activity, p. 9 **Reading Study Guide** (Spanish and English), pp. 42–43	**Critical Thinking Transparencies CT12** **Map Transparencies MT19** **classzone.com** Reading Study Guide
SECTION 2 Ancient Latin America pp. 160–166	• The ancient cultures of Latin America thrived in challenging geographic settings. • These cultures serve as models for how successful civilizations develop. • The concept of zero, a 365-day solar calendar, and written language are some intellectual advances of the Maya. • Aztec managed to farm in the swamps by constructing floating gardens.	**In-depth Resources: Unit 3** • Guided Reading Worksheet, p. 4 • Reteaching Activity, p. 10 **Reading Study Guide** (Spanish and English), pp. 44–45	**classzone.com** Reading Study Guide

KEY TO RESOURCES

Audio
CD-ROM
Copymaster
Internet
Overhead Transparency
Pupil's Edition
Teacher's Edition
Video

ASSESSMENT OPTIONS

Chapter Assessment, pp. 168–169

Formal Assessment
- Chapter Tests: Forms A, B, C, pp. 81–89

Test Generator

Online Test Practice

Strategies for Test Preparation

Section Assessment, p. 159

Formal Assessment
- Section Quiz, p. 79

Integrated Assessment
- Rubric for writing a letter

Test Generator

Test Practice Transparencies TT15

Section Assessment, p. 166

Formal Assessment
- Section Quiz, p. 80

Integrated Assessment
- Rubric for making a list

Test Generator

Test Practice Transparencies TT16

RESOURCES FOR DIFFERENTIATING INSTRUCTION

Students Acquiring English/ESL

Reading Study Guide (Spanish and English), pp. 42–47

Access for Students Acquiring English Spanish Translations, pp. 39–44

TE Activity
- Compound Words, p. 156

Modified Lesson Plans for English Learners

Less Proficient Readers

Reading Study Guide (Spanish and English), pp. 42–47

TE Activity
- Recogizing Important Details, p. 161

Gifted and Talented Students

TE Activity
- Design a Travel Poster, p. 163

CROSS-CURRICULAR CONNECTIONS

Humanities
Fisher, Leonard Everett. ***Gods and Goddesses of the Ancient Maya.*** New York: Holiday House, 1999. Introduction to culture and pantheon of deities.

Literature
Suarez-Rivas, Maite. ***Latino Read-Aloud Stories.*** New York: Black Dog & Leventhal, 2000. Legends, fairy tales, and fables from ancient times.

Popular Culture
Dawson, Imogen. ***Clothes and Crafts in Aztec Times.*** Milwaukee, WI: Gareth Stevens Publishing, 2000. Information and projects.

Language Arts/Literature
Steele, Philip. ***The Aztec News.*** Cambridge, MA: Candlewick Press, 1997, Milwaukee, WI: Gareth Stevens Pub., 2001. Fictional newspaper from Aztec times.

Science
Getz, David. ***Frozen Girl.*** New York: H. Holt, 1998. Up-close look at the discovery of the Inca mummy alternates fictional and true stories.

ENRICHMENT ACTIVITIES

The following activities are especially suitable for classes following block schedules.

Teacher's Edition, pp. 154, 155, 158, 162, 167
Pupil's Edition, pp. 159, 166

Unit Atlas, pp. 142–149

Technology: 1400, p. 164
Outline Map Activities

INTEGRATED TECHNOLOGY

Go to **classzone.com** for lesson support and activities for Chapter 6.

BLOCK SCHEDULE LESSON PLAN OPTIONS: 90-MINUTE PERIOD

DAY 1

UNIT PREVIEW, pp. 140–141
Class Time 20 minutes

- **Discussion** Discuss the Unit Introduction, using the discussion prompts on TE p. 141.
  Class Time 10 minutes
- **Speculate** Use the discussion prompts in Analyzing the Photograph to help students speculate as to what the large open area in the center of Machu Picchu might have been used for.
  Class Time 10 minutes

UNIT ATLAS, pp. 142–149
Class Time 30 minutes

- **Small Groups** Divide the class into three groups and have each group prepare three key ideas for one section of the Unit Atlas: Physical Geography, Human Geography, and Regional Data File. Reconvene as a whole class for discussion.

SECTION 1, pp. 153–159
Class Time 40 minutes

- **Understanding Elevation** Use the discussion prompts in Applying the Skill to help students demonstrate an understanding about elevation shown on a physical map.

DAY 2

SECTION 2, pp. 160–166
Class Time 90 minutes

- **Summarize** To review Section 1 divide students into small groups. Have them study the chart on p. 152 and create a different way to present the information to the class.
  Class Time 30 minutes
- **Interview** Ask students to imagine that they are newspaper reporters who have been assigned to interview Maya, Aztec, and Inca citizens. Have each student make a list of questions to ask about the lives of the citizens. Conduct the interviews as a whole class.
  Class Time 30 minutes
- **Discussion** Assign the content under each heading in Section 2 to small groups of students. Each group is responsible for "teaching" the information to the class.
  Class Time 30 minutes

DAY 3

SECTION 2, continued
Class Time 35 Minutes

- **Main Idea** Have pairs of students review the Main Idea for each section and find three details to support it. Then have each pair list two additional important ideas and trade lists with another group to find details.

CHAPTER 6 REVIEW AND ASSESSMENT, pp. 168–169
Class Time 55 minutes

- **Review** Have students prepare a summary of the chapter, using Terms & Names listed on the first page of each section.
  Class Time 20 minutes
- **Assessment** Have students complete the Chapter 6 Assessment.
  Class Time 35 minutes

TECHNOLOGY IN THE CLASSROOM

CREATING WEB SITES

Students can design their own Web sites to organize information and to share their knowledge with other young people at school and around the world. By creating a Web site, students gain experience in organizing information in a nonlinear fashion and get a "behind-the-scenes" look at what goes into developing materials for the Internet. Their Web sites can be kept on the classroom computer, uploaded to the school's internal server, or uploaded to the Internet for students at other schools to view.

ACTIVITY OUTLINE

Objective Students will research ancient Latin American civilizations and create Web sites for people considering visiting the areas where these civilizations existed.

Task Have students conduct research on the Maya, Aztec, and Inca civilizations and take notes on these civilizations' cultures and geographic locations. Divide the class into groups, and have the groups design Web sites to educate potential visitors to Latin America about these civilizations and about what they can expect to see when visiting ancient ruins.

Class Time Three class periods

DIRECTIONS

1. Ask students to make charts with three columns and five rows. They should label the columns "Maya," "Aztec," and "Inca," and label the rows with these categories: geography and location, religion, daily life, arts and music, ruins to visit
2. Have students use their textbooks, library materials, and the Web sites listed at **classzone.com** to find out about the physical geography and location, religions, daily life, and arts and music of the ancient Maya, Aztec, and Inca civilizations. They should also look for evidence of the ruins that still remain from each civilization. Have them record their notes on these topics in the appropriate sections of their charts.
3. Divide the class into groups of approximately four students each, and assign each group to one of the three ancient Latin American civilizations they have studied.
4. Ask groups to pretend they have been hired by travel agencies to design Web sites that will inform potential visitors to Latin America about the ancient ruins they can visit and the civilizations that once lived there. Have groups create these Web sites, providing information from the charts they have made and additional information they think would be important to share with potential tourists. Their sites should contain text, images, and hyperlinks to Web sites with additional information. If students use photographs, they must make sure to cite the sources and the photographers (when known).
5. Have groups share their Web sites with the rest of the class.
6. As an option, have several students design an overall home page that will link to each group's Web site. This main home page may include a map or a time line showing the different Latin American civilizations the class has studied. Upload this page and each group's site onto the school district's server, and register it with your favorite search engines.

CHAPTER 6 OBJECTIVE

Students will explore the physical geography of Latin America and learn about the ancient civilizations of this culture region.

FOCUS ON VISUALS

Interpreting the Photograph Have students look at the photograph of Chichén Itzá, which shows El Castillo, the temple-pyramid of the serpent-god Kukulcan, as well as a Chac-Mool (at right of pyramid), a statue of the Maya rain god. Ask them to speculate about the significance of the huge pyramid. Point out that this chapter focuses on the land and history of Latin America. Ask them to describe what the photograph shows about the land and climate there.

Possible Responses The pyramid might be a burial place or a place for public ceremonies or religious rituals. The site looks hot and dry. There is thick jungle or forest.

Extension Have students look up *pyramid* in an encyclopedia. Ask them to find out which cultures have built pyramid-like structures and how their use and importance have varied from place to place.

CRITICAL THINKING ACTIVITY

Recognizing Effects Encourage students to think about how climate and physical geography influence life. Ask them to suggest the kinds of shelter, clothing, agriculture, and architecture that people would be likely to develop in the types of locations shown in these photographs.

Class Time 15 minutes

CHAPTER 6

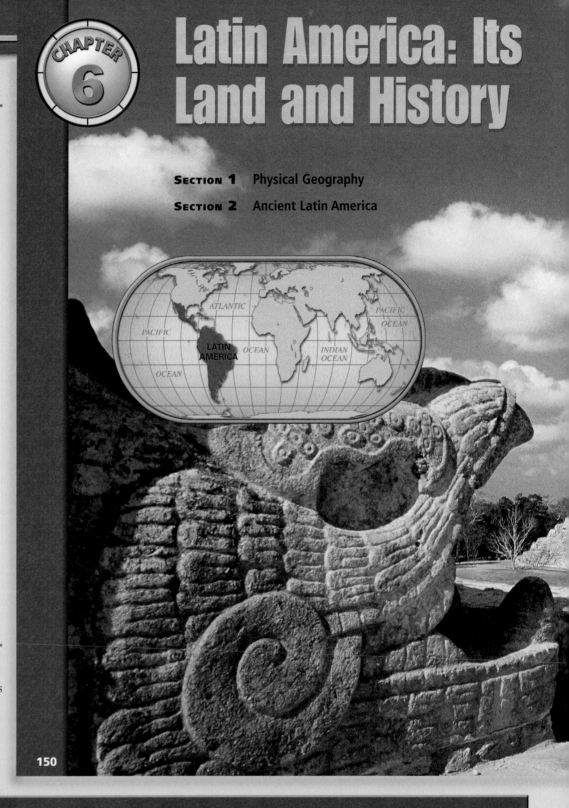

Latin America: Its Land and History

ATLANTIC
PACIFIC
PACIFIC
OCEAN
LATIN AMERICA
OCEAN
INDIAN OCEAN
OCEAN

150

Recommended Resources

BOOKS FOR THE TEACHER

Laughton, Timothy. ***The Maya: Life, Myth, and Art.*** New York: Stewart, Tabori & Chang, 1998. Appraisal of the cultural and artistic legacy of the Maya.

Muller, Karin. ***Along the Inca Road.*** Washington, D.C.: National Geographic Society, 2000. A woman's solo trip through ancient lands.

Townsend, Richard F. ***The Aztecs.*** New York: Thames & Hudson, 2000. Aztec history, everyday life, deities, calendars, ceremonies, and more.

VIDEOS

Aztec. Wynnewood, PA: Schlessinger Video, 1993. Emphasizes engineering skills, trade practices, religion, government, and social structure.

SOFTWARE

Aztecs. World Book, 1998. Explore a market, survive an interactive adventure, learn to play an ancient Aztec game, and make authentic pancakes.

INTERNET

For more information about Latin America, visit **classzone.com.**

FOCUS ON GEOGRAPHY

What conflicts arise when there is competition for a natural resource?

Human-Environment Interaction • About 50 percent of the world's rain forests are in Latin America, with 30 percent in Brazil alone. In Latin America, the rain forest is a natural resource that plays many different roles. It serves as a habitat for exotic plants and animals and a home to Native Americans. It is a moderator of the world climate and a destination for tourists. It also acts as a source of land that can be cleared and put to other uses, such as cattle ranching or logging.

What do you think?

- How might different groups want to use a rain forest as a resource? Think about the rain forest's inhabitants, tourists, environmentalists, and businesses.
- How might these different uses conflict with one another?

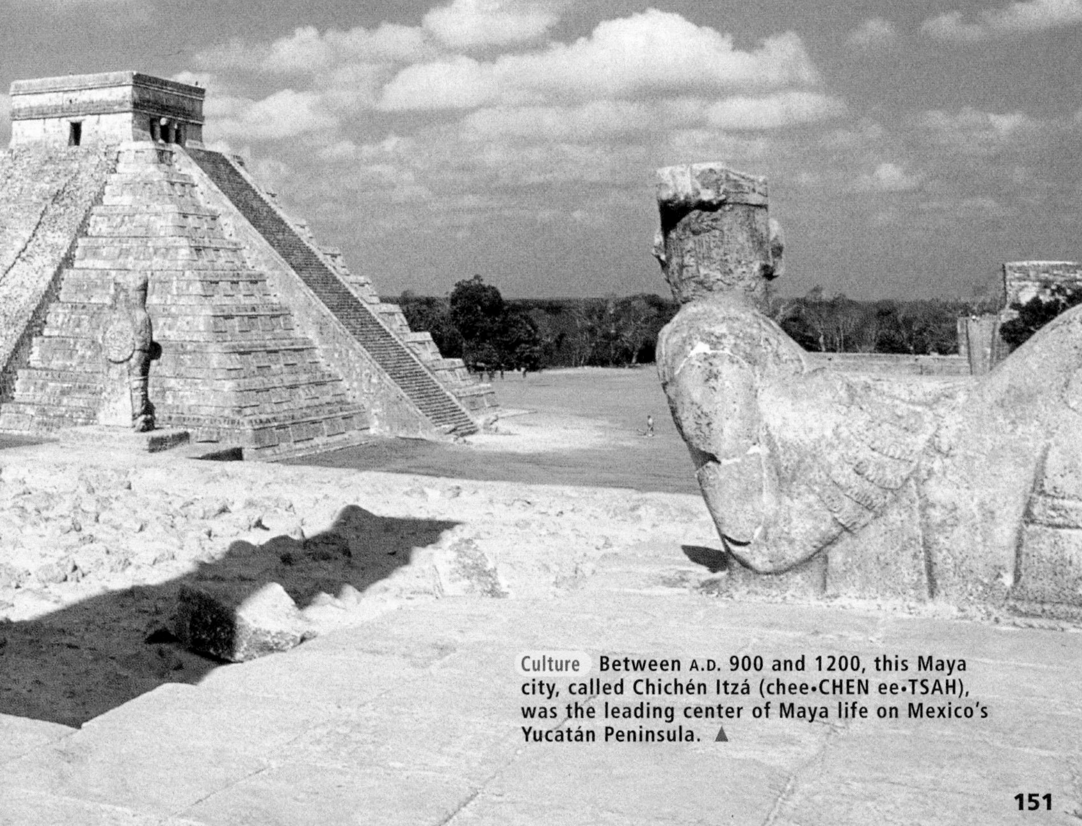

Culture Between A.D. 900 and 1200, this Maya city, called Chichén Itzá (chee•CHEN ee•TSAH), was the leading center of Maya life on Mexico's Yucatán Peninsula. ▲

FOCUS ON GEOGRAPHY

Objectives

- To identify the important uses of the Amazon rain forest
- To explain the conflicts surrounding the uses of natural resources in the Amazon rain forest

What Do You Think?

1. Create a two-column chart on the chalkboard to help students identify the competing groups and the ways in which each group might use the rain forest. Guide them to consider sites tourists might want to visit, products businesses might manufacture, and so on.
2. Guide students to understand that people who live in the rain forest have different needs from people who want to use the forest's resources for agricultural or commercial purposes. Environmentalists want to protect the resources that other groups may want to use.

What conflicts arise when there is competition for a natural resource?

Ask students to consider the conflicts that might arise over a resource such as trees. Ask what businesses might want to cut down trees, and why environmentalists might want to protect trees.

MAKING GEOGRAPHIC CONNECTIONS

Have students think about a threatened resource in your area, such as conservation land that a contractor wants to develop into a condominium community. Encourage them to consider groups that might be for or against such a use for the land. Ask them to think of reasons they would offer for their position. They might think about the need for lumber, space for agriculture, and areas for housing and recreation.

Implementing the National Geography Standards

Standard 17 Explain how differing perceptions of local, regional, national, and global resources have stimulated competition for natural resources

Objective To conduct a court hearing between competitors for natural resources in the rain forests of Central America

Class Time 30 minutes

Task Students will hold a hearing to determine how to allocate the resources of Central America's rain forests. Have students represent the farmers who need more land for growing crops, the industries that need raw timber, and the tribes that have lived in the forests for centuries. After each case has been argued in front of the judge (teacher), students should write a summary of the hearing.

Evaluation The summary should include an explanation of how the groups' different points of view stimulated competition for the forests' natural resources.

BEFORE YOU READ

What Do You Know?

Have students share what they know about Latin America from books they have read or movies they have seen. Then, ask any students who have come from a Latin American country, or who have family members or ancestors who did, to share aspects of their culture with the group. Encourage them to identify the languages spoken and to describe the climate, landforms and waterways, types of food, celebrations, and so on. Finally, encourage any students who have visited places in Latin America to describe the places, as well as events they participated in.

What Do You Want to Know?

Have small groups of students brainstorm lists of questions they hope to have answered in the chapter. You might model the process with questions such as the following: What types of animals (plants) live in the rain forest? What ancient peoples lived in this area? What is the highest mountain peak?

READ AND TAKE NOTES

Reading Strategy: Comparing and Contrasting Explain to students that by recording facts and details in a chart such as the one to the right, they will be able to compare and contrast the ancient civilizations of Latin America more easily and quickly. Seeing the similarities and differences among the three civilizations will help them understand and remember what they read.

In-depth Resources: Unit 3
- Guided Reading Worksheets, pp. 3–4

CHAPTER 6 READING SOCIAL STUDIES

BEFORE YOU READ

▸▸ What Do You Know?

Before you read the chapter, consider what you already know about the history and geography of Latin America. What countries make up Latin America? Does the region have well-known geographic features, such as mountains or rivers? You may have read folk tales from the ancient Maya, Aztec, and Inca civilizations. Think about what happened to these cultures and their people. Have you heard of the Aztec ruler Montezuma II?

▸▸ What Do You Want to Know?

Decide what else you want to know about Latin America. In your notebook, record what you hope to learn from this chapter.

Culture • T Maya carv many ston monumen

Culture • Montezuma II became the Aztec ruler in 1502. ▼

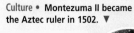

READ AND TAKE NOTES

Reading Strategy: Comparing and Contrasting Comparing and contrasting can be a useful strategy for studying cultures. When you compare, you look for similarities. When you contrast, you look for differences. Use the chart shown here to compare and contrast the ancient civilizations of Latin America.

- Copy the chart into your notebook.
- As you read Section 2, take notes on the Maya, Aztec, and Inca.
- Write your notes under the appropriate headings.

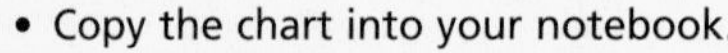

	Maya	Aztec	Inca
Location	Mexico and Central America	Valley of Mexico	Peru and Bolivia
Dates	1600 B.C.–A.D. 900	A.D. 1200–A.D. 1500	A.D. 1400–A.D. 1530
Achievements	Calendar	Chinampas	Terracing agriculture
Reasons for Decline	Unknown	Spanish conquest	Spanish conquest

Teaching Strategy

Reading the Chapter This is a thematic chapter focusing on the geographic features and ancient cultures of Latin America. As students read, encourage them to note the variety of landforms, waterways, and climates, and the way in which these variations affected the development of ancient civilizations.

Integrated Assessment The Chapter Assessment on page 169 describes several activities for integrated assessment. You may wish to have students work on these activities during the course of the chapter and then present them at the end.

Latin America: Physical Geography

TERMS & NAMES
tributary
deforestation
Tropical Zone
El Niño

AIN IDEA

atin America's landforms, odies of water, and climate offer wide range of environments nd resources.

WHY IT MATTERS NOW

Physical geography influences Latin America's cultures, offering them both resources and obstacles.

DATELINE

EXTRA

PUEBLA, MEXICO, DECEMBER 19, 2000

Yesterday, Mount Popocatépetl erupted, spewing out glowing five-foot-long rocks for miles in one of its largest eruptions in a thousand years. More than 30 million people live within sight of the volcano, and tens of thousands live close enough to be at risk when it erupts.

Government trucks drove through villages sounding warnings on speakers while church bells rang out the danger. In spite of the resulting damage, no one has been killed. However, authorities warn, a larger eruption could occur at any time.

Place • The Aztec named the volcano Popocatépetl, which means "smoking mountain." ▲

Defining Latin America ❶

Latin America includes Mexico, Central America, the Caribbean, and South America. Because the languages of most of its colonizers—Spanish and Portuguese—are derived from Latin, Europeans later referred to the region's colonies as *Latin America.* Because the region is defined by a cultural connection, in this case language, it is called a culture region.

TAKING NOTES
Use your chart to take notes about Latin America.

	Maya	Aztec
Location		
Dates		

SECTION OBJECTIVES

1. To identify Latin America as a culture region and show the influence of physical geography on Mexico
2. To describe Central America and the formation of the Caribbean Islands
3. To describe important geographic features of South America
4. To explain climate variations in the region

SKILLBUILDER
- Interpreting a Map, p. 155, p. 158
- Interpreting a Diagram, p. 159

CRITICAL THINKING
- Using Maps, p. 157
- Forming and Supporting Opinions, p. 158

FOCUS & MOTIVATE

WARM-UP

Making Inferences Have students read Dateline and discuss these questions.

1. What might have happened when many people in the region tried to escape?
2. How might people feel about returning to their hometown near Mt. Popocatépetl?

INSTRUCT: Objective ❶

Defining Latin America/Mexico

- What is the cultural connection in this region? language
- What problems does Mexico City face due to its location? air pollution, earthquakes

In-depth Resources: Unit 3
- Guided Reading Worksheet, p. 3

Reading Study Guide
(Spanish and English), pp. 42–43

Program Resources

In-depth Resources: Unit 3
- Guided Reading Worksheet, p. 3
- Reteaching Activity, p. 9

Reading Study Guide
(Spanish and English), pp. 42–43

Formal Assessment
- Section Quiz, p. 79

Integrated Assessment
- Rubric for making a list

Outline Map Activities

Access for Students Acquiring English
- Guided Reading Worksheet, p. 39

Technology Resources
classzone.com

TEST-TAKING RESOURCES
Strategies for Test Preparation

Test Practice Transparencies
Online Test Practice

Paricutín is an example of a cinder-cone volcano. This name comes from the words *cinder,* the type of volcanic material that spews from the vent, and *cone,* the shape that forms. When a cinder-cone volcano erupts, fragments of rock, known as tephra, burst forth from a vent in the earth. The tephra falls back to the earth around the vent, forming a cone-shaped mountain.

Paricutín has been classified as a dormant, or "sleeping," volcano because it has stopped erupting.

MORE ABOUT...

Earthquakes

Students may be surprised to learn that earthquakes are happening all the time along major fault lines. Many of these quakes are too low on the Richter scale to cause any major damage. For example, Southern California has been known to have as many as 45 quakes in a 24-hour period.

Mexico 1

Look at the map on page 155. Mexico is the farthest north of the Latin American countries. You can see that Mexico's major physical features include mountains, plateaus, and plains.

A Varied Landscape Mexico's two major mountain ranges share the name Sierra Madre (see•EHR•uh MAH•dray). Notice that in between the two ranges—the Sierra Madre Occidental and the Sierra Madre Oriental—sits Mexico's large central plateau. The vast northern stretches of the central plateau are desert.

Vocabulary

occidental: western

oriental: eastern

Now look just south of the central plateau. There you will see Mexico's two highest mountain peaks, Orizaba (or•ih•ZAH•buh) and Popocatépetl (POH•puh•KAT•uh•PEHT•uhl). Both are volcanoes. Volcanic activity and earthquakes frequently plague Mexico—and many other parts of Latin America too. They are caused by the movement of five tectonic plates.

A Problem of Place At the southern end of the central plateau sits Mexico City, the world's second most populated city. Air pollution is severe there, and the city's location has contributed to this problem. The mountains surrounding the city to the east, south, and west trap automobile exhaust and other pollutants that the city's huge population generates.

An added problem of location for Mexico City is the ground on which it was built—a drained lakebed. The vibrations that earthquakes send through Earth grow much stronger and more damaging when they pass through the soft, loose soils of a lakebed, thus making Mexico City highly vulnerable to the effects of earthquakes.

Eruption Disruption On February 20, 1943, something very strange happened in a cornfield in west-central Mexico. Before the eyes of a startled farmer, the land violently split open. Within 24 hours, a small, smoking cone had appeared—the tiny beginning of a mighty volcano called Paricutín (pah•REE•koo•TEEN). Scientists rushed to the area to watch the volcano being born. Within a year, a mountain stood in place of what had been farmland and a village.

While it was active, Paricutín rose to 10,400 feet above sea level. It poured lava over about a 10-square-mile area, burying streets and buildings, such as this nearby church. In 1952, it stopped erupting completely and became dormant. But the volcano that sprouted in a cornfield had given scientists a rare opportunity to study the life cycle of one of nature's most dramatic and dangerous features.

Activity Options

Interdisciplinary Link: Citizenship

Preparing a Speech Recall with students that Mexico City's location, between mountain ranges, is partially responsible for the problem of air pollution. Point out that the people living in the city also contribute to the air pollution problem. Ask students to brainstorm ways in which citizens in Mexico City might work to solve this problem. Then have students write brief speeches to encourage their fellow citizens to help reduce air pollution. Encourage them to use persuasive words and phrases in their speeches. Ask volunteers to deliver their speeches.

hysical Features of Mexico, Central America, and the Caribbean

INTERPRETING A MAP

Physical Features of Mexico, Central America, and the Caribbean

Guide a discussion of the map. Ask students what type of map it is. Have them give examples of the kind of information typically provided on a physical map. Point out that this map also includes information, such as capital cities and national boundaries, that is ordinarily provided on political maps. Ask students to look more closely at the map and answer the following questions:

- Describe the landscape where Mexico City is located. It is high, over 6,600 feet, and surrounded by mountains.
- What body of water lies north of the Yucatán Peninsula? Gulf of Mexico
- What is the average elevation of most of Cuba? 0–650 feet

GEOGRAPHY SKILLBUILDER

Answers

1. Tajumulco
2. the Yucatán Peninsula

Activity Options

Multiple Learning Styles: Logical

Class Time 40 minutes

Task Creating and completing a crossword puzzle

Purpose To use geographic clues to identify places on a map

Supplies Needed

- Pencils
- Graph paper
- Ruler

Block Scheduling

Activity Have students work in pairs to create crossword puzzles using information from the map. Ask them to choose names of places shown on the map and fit them together in a crossword puzzle. Then have them number the words and write clues taken from the map. They should then create blank crossword puzzles to exchange with other pairs of students. Have students complete the puzzles.

INSTRUCT: Objective 2

Central America and the Caribbean

- Where is Central America located? between Mexico and South America
- How would you describe the Central American landscape? mountainous or hilly, covered with forests—rain forests and deciduous trees
- Where are the Caribbean Islands located in relation to Central America? to the east
- How were the Caribbean Islands formed? some islands as volcanoes, other islands as coral reefs

MORE ABOUT...

Atolls and Coral Islands

An atoll is a circular band of coral that sometimes forms on the crater of a volcano that has sunk below the surface of the sea. A layer of soil attaches to the reef, and tropical plants grow in the soil. Eventually the atoll becomes a coral island.

Place • Seven islands make up the Caribbean island group that is called Guadeloupe (GWAHD•uhl•OOP). ▲

Central America and the Caribbean 2

Look at the physical map in the Unit Atlas. Central America is the mountainous landmass that forms a bridge between Mexico and South America. Now look to the east, and you will see island nations scattered throughout the Caribbean Sea.

Central America About 80 percent of Central America is hilly or mountainous, and most of it is covered with forests. Rain forests cover much of the lowlands. In the higher regions, deciduous trees cloak many of the slopes.

A string of more than 40 volcanoes lines 900 miles of Central America's Pacific coast, where two tectonic plates crash against each other. This is the most active group of volcanoes in North or South America. Earthquakes also occur frequently. They can completely destroy buildings, towns, and cities. They can also set landslides and mudslides in motion, sending land, houses, and people hurtling down the slopes.

Vocabulary

deciduous: a tree that loses its leaves each year

A. Clarifying What causes so many volcanoes and earthquakes in this region?

Reading Social Studies A. Answer Two tectonic plates crash into each other.

The Caribbean Islands As you can see on the map (page 155), the Caribbean Islands lie to the east of Central America. Some of these islands, such as St. Kitts and Grenada, are actually the peaks of volcanic mountains rising from the ocean floor. Over thousands of years, the volcanoes erupted, spewing lava that cooled, hardened, and added to the mountains' height.

Other islands, such as the Bahamas, began as coral reefs. Coral is made of organisms that shed hard skeletons when they die. The skeletons pile up, and a reef, or ridge, develops. A coral reef that becomes an island usually encircles a volcanic island and then grows over it.

Place • Coral reefs form in lots of different colors. ▼

Activity Options

Differentiating Instruction: Students Acquiring English/ESL

Compound Words Write the following words from the section on the chalkboard: *landmass, rain forest, earthquakes, landslides, mudslides, coral reefs.* Ask a volunteer to read the words aloud. Review with students that in a compound word, two words together change in meaning to form a single idea or identify a single thing. Point out that compound words may be written as one solid word, a hyphenated word, or two separate words. Explain that students can use the meaning of the individual words to figure out the meaning of the compound word. Students can also use context clues from the text to help them figure out the meaning of the words on the board. Work with students to define the words. Then have students take turns using the compound words in sentences.

South America ❸

Look at the map on page 142 of the Unit Atlas. You can see that the equator runs through Ecuador, Colombia, and Brazil. You can also see that only the Isthmus of Panama links South America to North America.

Vocabulary
isthmus: a narrow strip of land that connects two landmasses

The Andes On the map, you can see the Andes mountain range, which stretches over 5,000 miles along South America's west coast. It is the longest continuous mountain range on Earth's surface. Mount Aconcagua (AK•uhn•KAH•gwuh) in Argentina is the highest peak in the Western Hemisphere.

Beyond the Andes Notice the central plains east of the Andes. The plains in southern South America are called the *Pampas.* South America's largest rivers begin in the Andes, drain the central plains, and then flow into the Atlantic Ocean. They include the Orinoco (AWR•uh•NOH•koh), the Paraná-Paraguay-Plata (PAR•uh• NAH-PAR•uh•gwy-PLAH•tuh), and the Amazon.

BACKGROUND

The word *pampa,* meaning "flat surface," comes from the language of an Andean group of Native Americans called the Quechua (KEHCH•wuh).

The Amazon In 2000, 22 people explored the Andes's rivers to confirm the source of the Amazon River, which had been discovered in 1971. The mighty river begins in the Peruvian Andes as a trickle of water. It then flows for nearly 4,000 miles to the Atlantic Ocean. No other river carries as much water to the sea. Along with more than 1,000 **tributaries,** which are rivers or streams that flow into a larger body of water, the Amazon drains water from Peru, Ecuador, Colombia, Bolivia, Venezuela, and Brazil.

The Amazon at Risk **Deforestation,** or the process of cutting and clearing away trees from a forest, has greatly affected the Amazon rain forest. In recent years, Amazon deforestation has provided timber and cleared land for cattle ranches.

Vocabulary
timber: wood used as building material

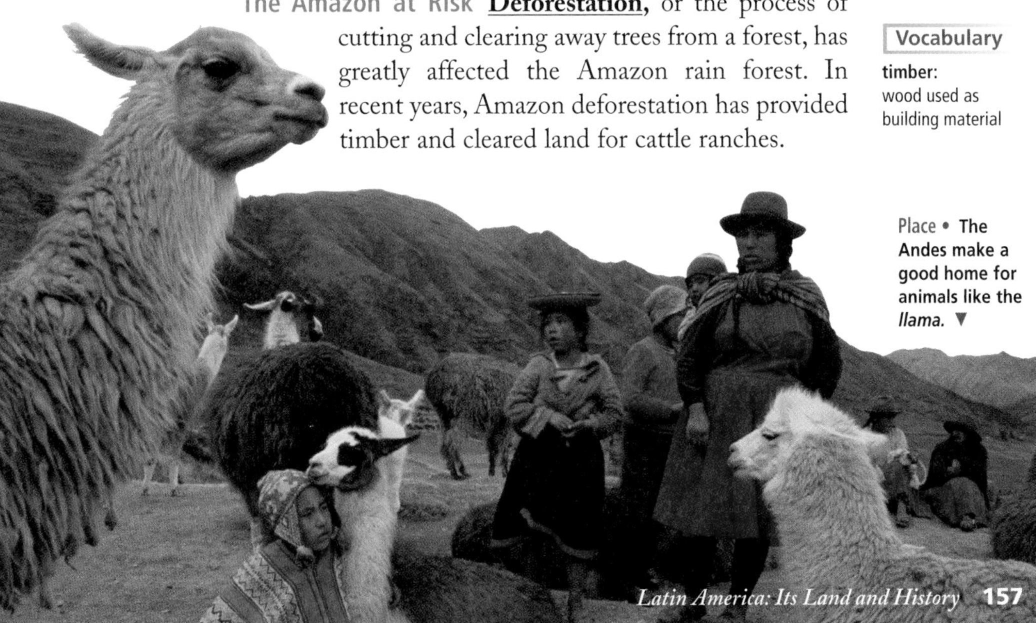

Place • **The Andes make a good home for animals like the *llama.*** ▼

INSTRUCT: Objective ❸

South America

- What is significant about the Andes Mountain range? longest mountain range on Earth's surface, has the highest peak in the Western Hemisphere
- What is the source of the Amazon River, and where does it flow? begins in Peruvian Andes, flows through Brazil to the Atlantic Ocean
- What is deforestation? the process of cutting and clearing away trees in a forest
- What have been some major causes and effects of deforestation in the rain forest? causes: cutting timber, clearing land for cattle ranches; effects: increased carbon dioxide in the air, possible global warming, displacement of animals and native peoples

CRITICAL THINKING ACTIVITY

Using Maps Refer students to the map on page 155. Ask them to consider how the climate of South America might vary due to features such as the equator, the Andes, vast plains, and the Amazon River. Have students describe how each feature might affect the climate.

Class Time 10 minutes

Activity Options

Multiple Learning Styles: Visual/Spatial

Class Time 30 minutes

Task Creating a diagram of the Andes mountain range

Purpose To visualize the formation of the Andes from the collision of tectonic plates

Supplies Needed
- Drawing paper
- Pencils or black markers

Activity Ask students to reread the text on pages 156–157 describing the Andes. As they read, encourage them to imagine how the collision of tectonic plates could have caused such high mountains to form. Have students work with partners to illustrate the process. Then have students compare their illustrations. Show students a diagram of this process when they have finished their drawings.

CRITICAL THINKING ACTIVITY

Forming and Supporting Opinions Ask students to list the causes of deforestation of the Amazon rain forest, suggesting causes not specified in the text, such as growing populations. Then have them list the effects of increased carbon dioxide in the environment. Finally, ask students to form an opinion about activities that result in deforestation. Encourage students to support their opinions with facts from the text.

Class Time 10 minutes

MORE ABOUT...
Rain Forests

According to the United States National Cancer Institute, the world's rain forests have about 2,000 plants that can fight cancer cells. Today, cancer-fighting drugs are made from plants grown in the rain forest. For example, vincristine, a drug used to fight leukemia in children, is made from periwinkle, a plant grown in the rain forest of Madigascar.

INSTRUCT: Objective 4

Climate

- What factors influence the climate of Latin America? elevation, distance from the equator, wind patterns, ocean currents
- What is *El Niño?* climate event in which a current of warm surface water in the Pacific flows eastward toward the coast of Latin America instead of being driven westward by the trade winds; it often causes heavy rains and flooding in Latin America and droughts in the western Pacific.

Most plants release oxygen into the air and absorb carbon dioxide. By reducing the number of trees and plants, deforestation increases the amount of carbon dioxide in the air. Some scientists think the increase in carbon dioxide contributes to global warming because carbon dioxide traps warm air at Earth's surface.

Others worry that animals who find food in the rain forest may need to move—or may die out—when large areas of forest are cut down. The Native American tribes that inhabit the rain forest are also at risk of being squeezed out of the land they live on.

Climate 4

Latin America's climate varies greatly from area to area. It is influenced by elevation, location, wind patterns, and ocean currents.

The Tropical Zone A large portion of Latin America lies in the **Tropical Zone,** which, as you can see on the map below, is between the latitudes 23°27' north and 23°27' south. The Tropical Zone may be rainy or dry, but it is typically hot. Also, temperature is always lower at higher elevations, but in the Tropical Zone, all elevations are warmer than they are elsewhere.

Wind and Water The waters in the Caribbean Sea stay warm most of the year and heat the air over them. A warm wind then blows across the islands, keeping the climate warm even in the winter.

Geography Skillbuilder Answers

1. Venezuela, Colombia, Ecuador, Peru, Bolivia, Brazil, Suriname, Guyana; parts of Chile and Paraguay. Students may also include French Guiana, which is a colony.
2. in the center

Reading Social Studies

B. Drawing Conclusions What aspect of the Tropical Zone's location on Earth causes it to be warmer than the other areas?

Reading Social Studies B. Answer

It is nearer the equator.

The Tropical Zone

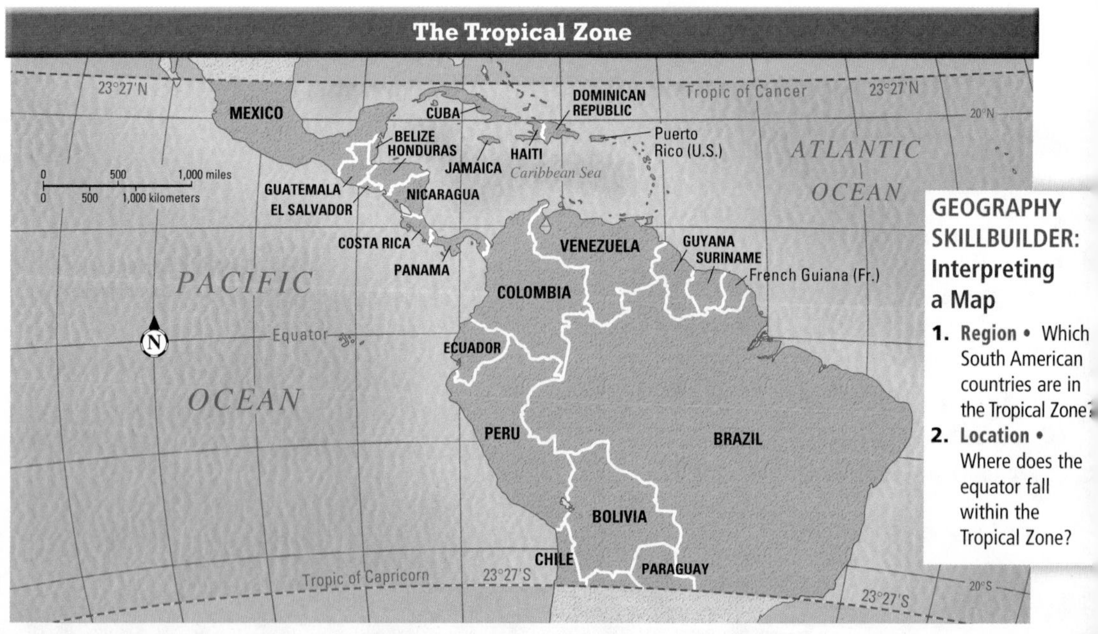

GEOGRAPHY SKILLBUILDER: Interpreting a Map

1. **Region •** Which South American countries are in the Tropical Zone?
2. **Location •** Where does the equator fall within the Tropical Zone?

Activity Options

Skillbuilder Mini-Lesson: Identifying Cause and Effect

Block Scheduling

Explaining the Skill Review with students that a cause is the event, person, or thought that makes something happen. The resulting event is the effect. Explain that geography and culture often have a cause-and-effect relationship.

Applying the Skill Have students reread "The Amazon at Risk" and use these strategies to identify the cause-and-effect relationship between the rain forest environment and the culture.

1. Look for the cause, or reason, that few towns or cities developed in the Amazon rain forest. The rain forest is too dense for large settlements of people.
2. Think about the effects, immediate results, of this cause. Fewer people in the rain forest means less development of the land.

Normal Year

Westerly trade winds

Warm surface water

El Niño Year

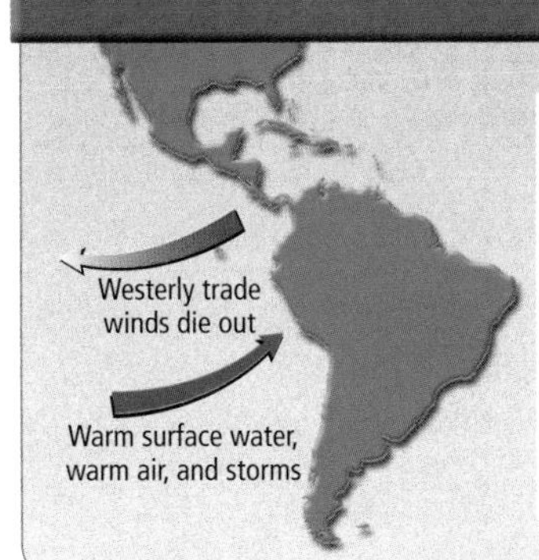

SKILLBUILDER: Interpreting a Diagram

1. What effect does the increased air pressure of *El Niño* have on the warm surface water of the Pacific?
2. What are the patterns of wind and water in a normal year?

Skillbuilder Answers

1. It forces warm, moist air to flow eastward, not westward, bringing rain to the coast of Latin America.
2. Trade winds and surface water flow to the west.

El Niño At times, unusually high air pressure in the south Pacific causes certain winds over the ocean, called trade winds, to die out. Without the trade winds, the ocean's sun-warmed surface water flows eastward, toward North and South America. Because this typically occurs around Christmastime, people call the current ***El Niño*** (ehl NEE•nyaw). This term is Spanish for "the Christ child" because Christmas celebrates the birth of Jesus.

The warmer water of *El Niño* warms the air. Warmer air holds more water and so releases more rain when it cools. This added precipitation often causes heavy rains and flooding in Latin America and other parts of the eastern Pacific. At the same time, areas in the western Pacific, such as Indonesia, may have less rain than usual.

SECTION 1 ASSESSMENT

Terms & Names

1. Explain the significance of: (a) tributary (b) deforestation (c) Tropical Zone (d) *El Niño*

Using Graphics

2. Use a chart like the one below to list the key physical features of each country or region.

Key Physical Features			
Mexico	Central America	The Caribbean Islands	South America

Main Ideas

3. (a) Name two types of natural disasters that occur frequently in Latin America.

(b) Describe two ways in which the Caribbean Islands formed.

(c) What happens when increased air pressure causes trade winds to lessen?

Critical Thinking

4. Forming and Supporting Opinions

Do you favor limiting the deforestation of the Amazon rain forest? Why? Why not?

Think About

- the economic benefits of timber production and cattle ranching
- the destruction of habitats
- the effects on the global environment

ACTIVITY -OPTION- Look at the map on page 158 that shows the Tropical Zone. Make a **list** of all the Latin American countries that are wholly or partly in the Tropical Zone.

ASSESS & RETEACH

Reading Social Studies Have students create a chart in which they compare and contrast Mexico, Central America, the Caribbean, and South America in terms of location, climate, landforms, and waterways. Such a chart will help students visualize each culture region more vividly.

Formal Assessment

- Section Quiz, p. 79

RETEACHING ACTIVITY

Have students work with partners to list ways in which the nations in this region are threatened. Ask volunteers to share their findings with the class.

In-depth Resources: Unit 3

- Reteaching Activity, p. 9

Access for Students Acquiring English

- Reteaching Activity, p. 43

Section 1 Assessment

1. Terms & Names

a. tributary, p. 157
b. deforestation, p. 157
c. Tropical Zone, p. 158
d. *El Niño*, p. 159

2. Using Graphics

Mexico	Central America	The Caribbean Islands	South America
mountains plateaus plains	mountains rain forests	volcanoes coral reefs	Andes Mountains Amazon River

3. Main Ideas

a. Two natural disasters are earthquakes and volcanoes.
b. Volcanoes and coral reefs formed the Caribbean Islands.
c. Without the trade winds (which blow from the east), warm surface water in the Pacific Ocean flows eastward, not westward, bringing moist air and heavy rains to the coast of Latin America.

4. Critical Thinking

Answers may vary. Encourage students to support their responses.

ACTIVITY OPTION

Integrated Assessment

- Rubric for making a list

SECTION OBJECTIVES

1. To identify major ancient civilizations in Latin America and explain the civilization and accomplishments of the ancient Maya
2. To describe the Aztec culture and religion
3. To identify features of the Inca Empire
4. To explain the effects of Spanish rule in Latin America

CRITICAL THINKING
- Drawing Conclusions, p. 163
- Using Maps, p. 165

FOCUS & MOTIVATE

WARM-UP

Making Inferences Have students read Dateline and discuss these questions.

1. How do you think workers moved large stone slabs or lifted heavy objects?
2. What purpose might the pyramids serve?

INSTRUCT: Objective 1

Ancient Civilizations of Latin America/The Maya

- Where did the major ancient civilizations of Latin America develop? Maya in Mexico and Central America; Aztec in Mexico; Inca in the Andes
- What are some intellectual advances of the Maya? the concept of zero, a 365-day solar calendar, written language
- What are the steps of the slash-and-burn method used by the Maya? cut down and burn trees, plant crops, let forest grow back, repeat

In-depth Resources: Unit 3
- Guided Reading Worksheet, p. 4

Reading Study Guide
(Spanish and English), pp. 44–45

SECTION 2 Ancient Latin America

MAIN IDEA

The ancient cultures of Latin America established civilizations in challenging geographic settings.

WHY IT MATTERS NOW

These cultures serve as models for how successful civilizations develop.

TERMS & NAMES
- hieroglyph
- *chinampa*
- Machu Picchu
- Hernán Cortés
- Montezuma II
- Francisco Pizarro
- Atahualpa
- Columbian Exchange

DATELINE

EL MIRADOR, GUATEMALA, 200 B.C.—In El Mirador today, a council of the city's leaders made a major announcement. Next month, construction will begin on a massive building complex for the city's center.

The plans include an enormous pyramid made of three smaller pyramids sitting atop a large stepped platform. The council expects to employ thousands of people to cut and carry the stone slabs that will be used to build the structure. The project is expected to take many months to complete.

Place • An artist made this drawing to show what El Mirador's three-part pyramid will look like. ▼

Ancient Civilizations of Latin America 1

Many ancient civilizations, such as the Egyptian, developed in river valleys and thrived there. The rivers provided water for both irrigation and transportation. In Latin America, however, some ancient civilizations flourished far from rivers. For example, the Maya of Mexico and Central America built cities in dense jungles. The Aztec of Mexico constructed their capital on a swampy island. The Inca of South America built cities high up in the Andes.

TAKING NOTES
Use your chart to take notes about Latin America.

	Maya	Aztec
Location		
Dates		

Program Resources

In-depth Resources: Unit 3
- Guided Reading Worksheet, p. 4
- Reteaching Activity, p. 10

Reading Study Guide
(Spanish and English), pp. 44–45

Formal Assessment
- Section Quiz, p. 80

Integrated Assessment
- Rubric for writing a letter

Outline Map Activities

Access for Students Acquiring English
- Guided Reading Worksheet, p. 40

Technology Resources
classzone.com

TEST-TAKING RESOURCES
- Strategies for Test Preparation
- Test Practice Transparencies
- Online Test Practice

The Maya ❶

In the areas that are today southern and eastern Mexico, western Honduras, Guatemala, El Salvador, and Belize, the ancient Maya built a widespread civilization. Small Maya communities existed as early as 1600 B.C. From A.D. 250 to A.D. 900, the Maya established one of Latin America's most important civilizations.

Maya Intellectual Advances The ancient Maya studied math and astronomy extensively. The Maya were among the first civilizations in the world known to understand the advanced mathematical concept of zero. They also had an intricate calendar system that included a 260-day calendar of sacred days, a 365-day calendar based on the sun's movement, and a calendar that measured the number of days that had passed since a fixed starting point.

The Maya established the best-developed written language in ancient Latin America. The basic units of the writing system were symbols called **hieroglyphs,** or glyphs. Each glyph represented a word or a syllable. The U.S. lawyer John Lloyd Stephens, while traveling through the Maya area in the 1800s, described his awe at seeing the glyphs and not being able to read them because no one had yet deciphered them.

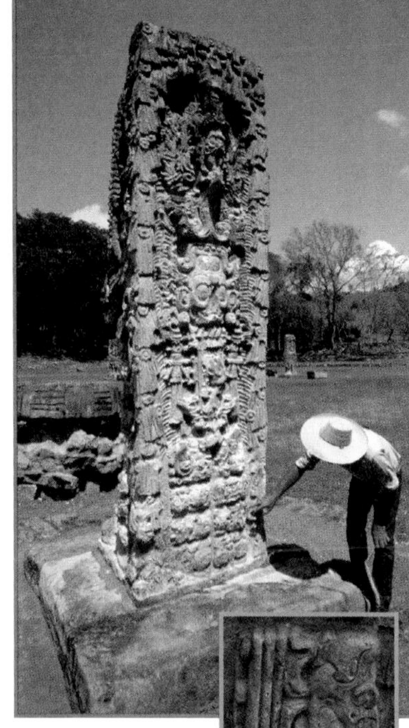

Culture • The Maya often carved hieroglyphs on stone monuments. ▲

BACKGROUND

Many of the Maya hieroglyphs have now been deciphered.

A VOICE FROM LATIN AMERICA

These structures . . . these stones . . . standing as they do in the depths of a tropical forest, silent and solemn, strange in design, excellent in sculpture, rich in ornament . . . their whole history so entirely unknown, with hieroglyphics explaining all, but perfectly unintelligible.

John Lloyd Stephens

Reading Social Studies

A. Drawing Conclusions How does having a system of writing help a civilization survive?

Reading Social Studies
A. Possible Answer
It aids communication and records past events.

Maya Agriculture Farming was essential to Maya life. Using a method called slash-and-burn agriculture, the Maya cut down and burned trees, planting crops in their place. After a few years, they let the forest grow back, so the soil could regain its nutrients. Later the area could again be cut, burned, and farmed. The Maya also built up ridges of farming land on floodplains. The floodplains were rich with nutrients, and the ridges kept the crops from getting too wet.

MORE ABOUT...
The Maya Calendar

The Maya based their 365-day calendar on the orbit of the sun around the Earth. They divided each year into 18 months of 20 days each. At the end of each year, there were five days that were left over. These five days were considered to be extremely unlucky. During this time, the Maya fasted and made many sacrifices to their gods.

A VOICE FROM LATIN AMERICA

Read aloud the passage and ask students to identify the descriptive words that John Lloyd Stephens used. Discuss the clever wording of his description ". . . explaining all, but perfectly unintelligible." Make sure that students understand that unless a person can translate the symbols, the symbols have no meaning for that person.

MORE ABOUT...
John Lloyd Stephens

Stephens (1805–1852) was a curious amateur archaeologist who, bored with the legal profession, began to travel to exotic places for his health. In about 1839, having heard about ancient ruins in Central America, he began to explore the jungles of Mexico, the Yucatán, and Central America with the English artist and archaeologist Frederick Catherwood. The accounts and pictures they published inspired the first wave of popular interest in Maya archaeology.

Activity Options

Differentiating Instruction: Less Proficient Readers

Recognizing Important Details Some students may need more help in identifying and reviewing the many intellectual advances made by the Maya, particularly those in mathematics and astronomy. Work with these students to create a spider map like the one shown here. Then encourage them to discuss the importance of discoveries such as these. What might such discoveries show about the Maya civilization? What tools and techniques would be needed to make them?

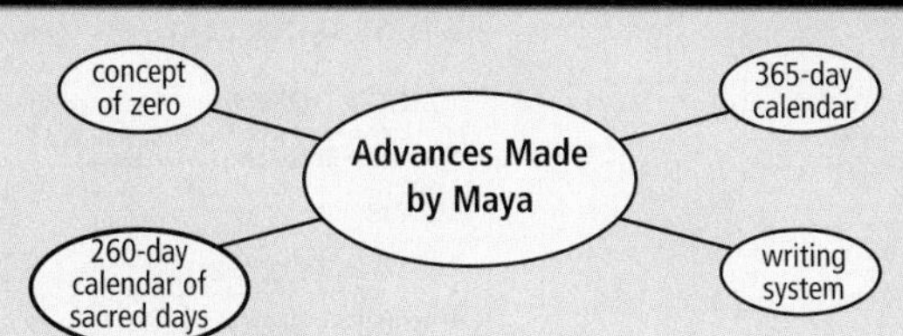

The Olmec are often considered the "parent culture" in Mesoamerica. Later peoples in the region adapted and developed Olmec cultural elements such as writing and number systems, the calendar, ritual ball games, and architectural styles.

INSTRUCT: Objective 2

The Aztec

- What was the name of the capital city of the Aztec Empire? Tenochtitlán
- For what two reasons did men join the Aztec army? to maintain a powerful empire; to die for Huitzilopochtli, the god of war
- How did the Aztec manage to farm in the swamps? constructed floating gardens

MORE ABOUT...

Aztec Gods

The Aztec worshipped many gods in addition to Huitzilopochtli. Each god ruled one or more aspects of nature or human activity. Since the Aztec economy was based on farming, the Aztec had many agricultural divinities. Among them were Centeocihuatl, the goddess of corn; Tlaloc, the god of rain and fertility; and Xipe Totec, the god of springtime and regrowth.

Colossal Olmec Heads Mexico's oldest known civilization is called the Olmec, which flourished from about 1200 to 600 B.C. The Olmec are famous for the colossal heads (like the one shown below) that they carved from a type of stone called basalt.

Thought to be portraits of Olmec rulers, some of these heads stand over nine feet tall. Each weighs thousands of pounds. All of Mexico's later Native American cultures were influenced by the Olmec. However, only the Olmec produced these giant stone monuments.

Decline of Maya Civilization Around A.D. 900, the Maya way of life began to change. For unknown reasons, the construction of massive temples and stone monuments stopped. Cities were abandoned. However, the Maya people did not disappear—they just spread out. More than 6 million Maya people still live in Guatemala, Belize, and southern Mexico and speak dialects based on the languages of their Maya ancestors.

The Aztec 2

Where modern Mexico City now stands, the waters of Lake Texcoco once lapped the shores of an island city called Tenochtitlán (teh•NOHCH•tee•TLAHN). With as many as 200,000 inhabitants, Tenochtitlán served as the capital of the Aztec Empire.

Aztec Origins The Aztec were composed of a number of tribes of wandering warriors. Of these, the Mexica (MEH•hee•KAH) were dominant. Mexico took its name from the Mexica. During the 1200s, the Aztec gradually grew in numbers and military strength until they controlled the region. They dominated until the early 1500s, when the Spanish conquered them.

Aztec Warfare and Religion The Aztec Empire centered on warfare. All able men, including priests, were expected to join the Aztec army, for two reasons. The first was to maintain a powerful empire, but the second was religious. The Aztec believed that anyone who died in battle had the great honor of dying for Huitzilopochtli (WEE•tsuh•loh•POHCH•tlee), the Aztec god of war.

Culture • This stone carving honors the Aztec sun god, whose face is shown in the center. ▼

Aztec Agriculture The Aztec held great power over their empire. One reason for their success was that the island location of their capital protected them from attack. However, much of the island was marsh, posing a major challenge to farming. The resourceful Aztec built floating gardens, called ***chinampas*** (chee•NAHM•pahs), on which they grew crops. First, they piled up plants from the water. Then they anchored these rafts between trunks of willow trees.

Activity Options

Multiple Learning Styles: Visual/Kinesthetic

Block Scheduling

Class Time One class period

Task Creating dioramas that portray aspects of Aztec daily life

Purpose To learn more about the Aztec culture

Supplies Needed
- Reference books
- Shoe boxes
- Construction paper
- Scissors
- Modeling clay
- Sticks, stones, sand

Activity Have small groups of students work together to research aspects of Aztec daily life, such as clothing, shelter, food, customs, and crafts. Then have students use art supplies and natural materials to create dioramas that portray these aspects of life.

Aztec *Chinampas*

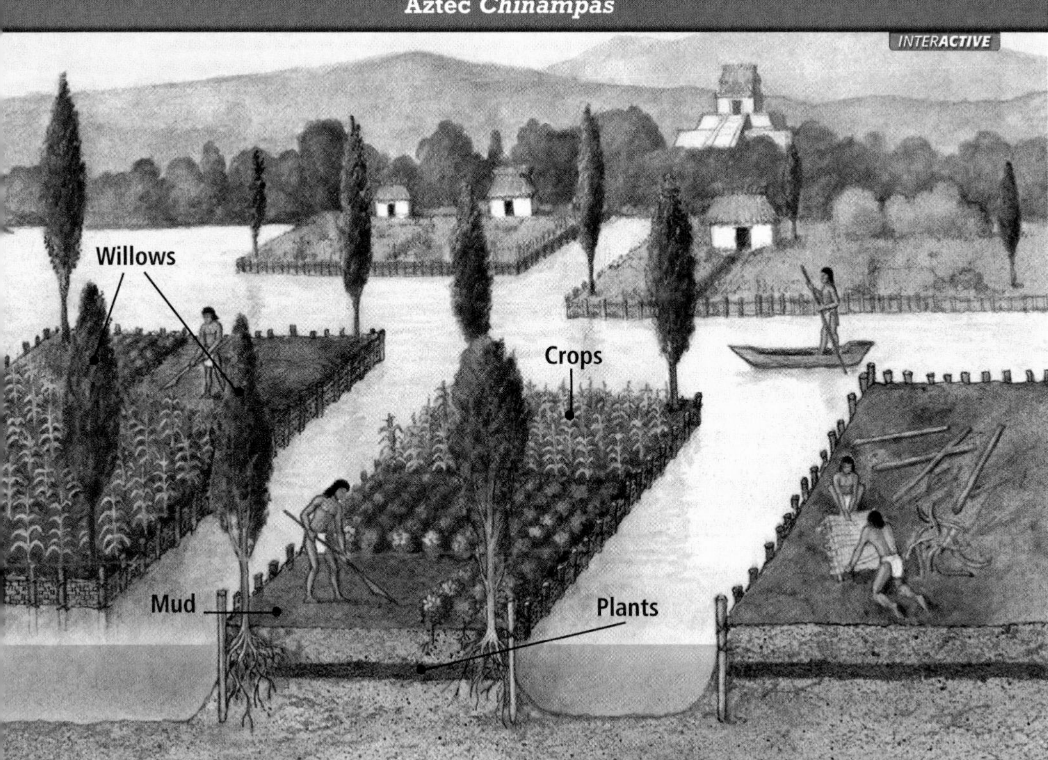

man-
vironment
eraction •
is infographic
ows how the
tec built and
rmed the
inampas. ▲

Finally, they heaped the lake's fertile mud on the piles to create plots for farming. The Aztec grew many crops, such as maize, beans, squash, avocados, tomatoes, peppers, and flowers. They also raised turkeys, ducks, geese, and dogs for food.

The Inca ❸

Around 1400, high in the Andes of Peru, a group of people called the Inca rose up to conquer the people of the surrounding areas. From their capital, Cuzco, the Inca soon ruled a huge empire that included parts of what are now Colombia, Ecuador, Bolivia, northern Chile, and northwestern Argentina.

Inca Agriculture To farm on the steep mountainsides, the Inca built stone terraces. These gave the Inca large areas of flat land to farm. The terraces also helped prevent erosion of the soil. In the desert lands to the west, the Inca built irrigation canals to water their crops. Some of these canals spanned entire valleys. Because of terracing and irrigation, Inca farmers were able to grow crops such as potatoes, maize, and a grain called *quinoa.*

INSTRUCT: Objective ❸

The Inca

- In which present-day countries did the Inca live? Peru, Colombia, Ecuador, Bolivia, Chile, Argentina
- How were the Inca able to farm on the mountainsides? They built stone terraces and irrigation canals.
- How did the Inca communicate throughout their empire without having a written language? Relay teams of messengers carrying verbal messages ran along roads.

CRITICAL THINKING ACTIVITY

Drawing Conclusions Ask students to reconsider what they have read about the Inca and their empire. Then ask them to draw conclusions about the type of government that could have formed and ruled such a vast empire. Have students discuss whether such an empire would be likely to have elements of democracy or whether the government might have rested on the power and influence of a ruling family.

Class Time 15 minutes

Activity Options

Differentiating Instruction: Gifted and Talented

Design a Travel Poster Tell students that Machu Picchu is the most popular tourist site in Peru. Each year, thousands of tourists hike along the Inca Trail on their way to visit Machu Picchu. On their way to the mysterious city, they pass other stone ruins as well as colorful local markets. The trail hike takes about four days, involving camping out and some strenuous climbing. Other travelers can reach Machu Picchu by train.

Have students research the Inca Trail and Machu Picchu, using the Internet, travel guides, and other resources. Ask them to work with a small group to create travel posters that will draw visitors to the site. Posters should use visuals and brief text to show major points of interest along the Inca Trail and in Machu Picchu itself.

Unit 3 ✦ Feature

OBJECTIVES

1. To demonstrate how Andean people overcame unfavorable climate and terrain to become extremely productive farmers
2. To show the influence of Andean farming on today's agriculture

INSTRUCT

- Why did the Inca use layers of soil and stone when building their terraces?
- Why did they grow many kinds of crops?

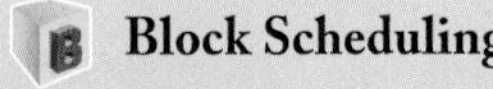

Block Scheduling

MORE ABOUT...

Inca Foods

More than half the products we eat today were developed by Andean farmers. These products include many varieties of corn and beans, squash, peppers, peanuts, manioc (used to make farina and tapioca), and a grain called quinoa. The most important crop was the potato. It could endure heavy frosts and could be planted as high as 15,000 feet. Long before the Inca, Andean people learned that potatoes could be preserved by drying them in the sun during the day and freezing them at night. The result was a light flour called *chuno*.

Connect to History

Drawing Conclusions What does the graphic say about the technological level of Inca tools? What tools would have made their jobs easier?

Connect to Today

Researching Andean farmers devised an effective method of freeze-drying food. What foods are freeze-dried today?

Technology: 1400

Andean Agriculture

Long before the rise of the Inca Empire, people living in the Andes had learned to farm the steep valley walls by building terraces into the sides of the mountains. They had also learned to build canals, many of them lined with stone, to carry water to their crops. The Inca improved and expanded the existing terraces and canals until they could feed 15 million people, with enough food left over to put away stores for three to seven years.

In the Andes, valley walls rise as high as 10,000 feet and temperatures can spa a 55-degree range.

Inca canals stretched for miles. They were often lined and covered with stones. Some were cut through solid rock.

The Inca grew maize, hundreds of kinds of potatoe and many other crops. Farmers had to plant crops adapted to many different climates because of the great variations in altitude and temperature.

The Inca had few farm tools. The most widely used was the *taclla,* or digging stick. It consisted of a pointed hardwood pole with a footrest for pushing the tool into the ground. Some *tacllas* had metal tips. The other main tools were hoes and clubs.

Workers directed by royal architects built stone retaining walls. Inside the walls, they placed layers of stone, clay, gravel, and topsoil. This combination allowed water to slowly work its way to lower terraces.

Critically THINKING

1. **Analyzing Motives** Why did people living in the Andes need to build terraces and canals?
2. **Recognizing Effects** What role did agriculture play in the building and maintenance of the Inca Empire?

Thinking Critically

1. **Analyzing Motives Possible Responses** Terraces were built to prevent erosion on steep hillsides. Terrace walls lined with layers of stone, soil, and gravel allowed water to seep from terrace to terrace without wearing away the soil. Canals were built to carry water to crops during dry spells.
2. **Recognizing Effects Possible Response** Because the Inca were able to grow their own food, they were able to stay in one location and focus on building their community.

eading Social
tudies
. Answer
unners could carry
erbal messages to
stant places.

Reading Social Studies

B. Recognizing Effects How did building stone roads improve the ability of the Inca rulers to control a large region?

hinking Critically
ossible Answers
. green and ripe
orn plants; sun,
noon, stars; storm
louds; newborn
nimals
. Cloth does not
ast as long as
tone; records
ould eventually
e lost.

Communicating Across the Inca Empire Stone roads were a major technological feat of the Inca. These roads are still in use today. Having no written language or knowledge of the wheel, Inca rulers ordered roads built on which runners carried verbal messages to distant places. The runners worked in relay teams stationed along the roads. One runner told the message to the next. Messages could travel 150 miles a day along the stone roads. This system of communication was important to the Inca because their empire spread out over thousands of miles.

Culture • The Inca kept records by tying knots in a series of strings called *quipu* (KEE•poo). ◄

Inca Stonework The Inca are known for their stonework. They erected many massive buildings, some with stones weighing as much as 200 tons. Wooden rollers were used to move these heavy stone blocks. The most remarkable of Inca stonework is the city of **Machu Picchu** (MAH•choo PEEK•choo), which still stands almost 8,000 feet above sea level. The walls of Machu Picchu were constructed so that they appear to emerge from the mountainsides. Around them, terraces connected by stairways run down the steep slopes. (See photograph on pages 140–141.)

potlight on CULTURE

ca Weaving The Inca had no formal written language, but they ed weaving as a means of representing ideas. Using wool sheared om llamas and alpacas, as well as many colorful plant dyes, the Inca ove images into the fabrics they wore and traded. Concepts related the passing of seasons, agricultural practices, and history were all presented in the weavings. In Peru today, Edwin Sulca Lagos is mous for his Inca-inspired weavings. This one is covered in designs om the Inca calendar.

HINKING CRITICALLY

Hypothesizing
What sorts of images might the Inca have used to convey concepts such as time or seasons?

Identifying Problems
What risks did the Inca face by recording ideas only on fabric?

more on Inca weaving, go to

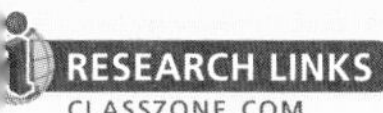
RESEARCH LINKS
CLASSZONE.COM

Spotlight on CULTURE

When the Spanish first came to Latin America, they were amazed at the large textile industry of the Inca. At that time there was a manufacturing center with hundreds of weavers, and storehouses full of textiles used for many purposes. Soldiers were paid in cloth, and elaborate cloth was woven for special rituals.

CRITICAL THINKING ACTIVITY

Using Maps Remind students that runners carried messages up to 150 miles a day along stone roads. Provide students with road maps of the area around the school. Have groups of students use the map scale to locate a destination that is approximately 150 miles away from the town the school is located in. Then have them write directions for a messenger to follow to that destination. While deciding on the best route, have them consider major obstacles such as rivers, swamps, and major highways.

Class Time 15 minutes

Activity Options

Interdisciplinary Link: Math

Class Time 15 minutes

Task Solving word problems

Purpose To find out how many people it took to deliver one message within the Inca Empire

Supplies Needed
- Paper
- Pencils

Activity Ask students to solve these math problems.

1. A message needs to be carried 250 miles. If each runner along the route runs 20 miles, but no more, how many runners will it take to carry the message? 13
2. If each runner takes about 6 minutes to run every mile, approximately how many hours will it take to deliver the message? 25

INSTRUCT: Objective 4

The Spanish in Latin America

- What was the name of the Aztec ruler conquered by Hernán Cortés? Montezuma II
- Who was the Spanish soldier that defeated the Inca ruler, Atahualpa? Francisco Pizarro
- What changes occurred in Latin America after the Spanish took control? Native Americans were enslaved and converted to Christianity; new products were shipped between Latin America and Spain.
- What kinds of products were part of the Columbian Exchange? Latin American crops, such as corn and tomatoes; Spanish textiles, foods, and animals

ASSESS & RETEACH

Reading Social Studies Have students complete the chart on page 152.

Formal Assessment
- Section Quiz, p. 80

RETEACHING ACTIVITY

Write the terms *Maya, Inca,* and *Aztec* on individual index cards. Print enough cards so each student will have one. Ask each student to draw a card from the pile and then to think of one fact to relate about that civilization. Have students form groups according to the civilizations and present their facts.

In-depth Resources: Unit 3
- Reteaching Activity, p. 10

Access for Students Acquiring English
- Reteaching Activity, p. 44

Culture • Montezuma II was a great warrior who was feared throughout the Aztec Empire. ▲

The Spanish in Latin America 4

Until about 500 years ago, Latin America was populated solely by Native Americans. In the 1500s, the Spanish arrived in the region. One famous Spanish soldier, **Hernán Cortés** (ehr•NAHN kawr•TEHS), captured the Aztec ruler, **Montezuma II** (MAHN•tih•ZOO•muh), in 1519. He claimed the Aztec Empire for Spain in 1521 and renamed it New Spain. A decade later, another Spanish soldier, **Francisco Pizarro,** defeated the Inca ruler, **Atahualpa** (AH•tuh•WAHL•puh), and claimed Atahualpa's empire for Spain.

Time of Change Once in control of Latin America, the Spanish enslaved many Native Americans and forced them to do labor, such as mining silver. The Spanish also worked hard to convert the Native Americans to Christianity.

Latin America and Spain also exchanged culture. Ships carrying Latin American goods sailed to Spain. The Spanish soon began growing corn, peppers, and tomatoes—crops they had never seen before. Manufactured products from Spain, especially textiles, were also shipped to Latin America. So were foods and animals, such as peaches and pigs. This trade was part of the **Columbian Exchange,** or the exchange of goods and ideas between European countries and their colonies in North and South America.

SECTION 2 ASSESSMENT

Terms & Names

1. Explain the signifcance of: (a) hieroglyph (b) *chinampa* (c) Machu Picchu (d) Hernán Cortés (e) Montezuma II (f) Francisco Pizarro (g) Atahualpa (h) Columbian Exchange

Using Graphics

2. Use a chart like the one below to list effects of the Spanish arriving in Latin America.

Effects of Spanish Arrival in Latin America	
1.	
2.	
3.	
4.	

Main Ideas

3. (a) Describe the writing system that the Maya developed.
   (b) How did the Inca pass important messages across great distances?
   (c) What was the Columbian Exchange and how did it work?

Critical Thinking

4. **Recognizing Effects**
   How did the Maya, Aztec, and Inca develop agricultural methods that responded to the environments in which they each lived?

   Think About
   - physical surroundings
   - available resources

ACTIVITY -OPTION- Imagine you live in Tenochtitlán and have spent the day constructing *chinampas.* Write a **letter** to a friend describing the process.

Section 2 Assessment

1. Terms & Names
- a. hieroglyph, p. 161
- b. *chinampa*, p. 162
- c. Machu Picchu, p. 165
- d. Hernán Cortés, p. 166
- e. Montezuma II, p. 166
- f. Francisco Pizarro, p. 166
- g. Atahualpa, p. 166
- h. Columbian Exchange, p. 166

2. Using Graphics

1. Many Native Americans were enslaved.
2. New foods and animals were introduced, such as peaches, horses, and pigs.
3. Spanish products, such as textiles, were shipped to Latin America.
4. The Spanish tried to convert people to Christianity.

3. Main Ideas
- a. The system involved hieroglyphs—symbols for words and word parts.
- b. Runners carried the messages by foot along stone roads.
- c. It was the exchange of goods between Europe and its American colonies. New products were introduced to both regions through trade.

4. Critical Thinking

The Maya used slash-and-burn farming to best utilize the forests. The Aztec built floating gardens. The Inca built terraces so they could farm on the mountain slopes.

ACTIVITY OPTION

Integrated Assessment
- Rubric for writing a letter

SKILLBUILDER

Finding and Summarizing the Main Idea

▸▸ Defining the Skill

When you find and summarize the main idea, you restate the subject of a written passage in fewer words. You include only the main idea and the most important details. It is important to use your own words when summarizing.

▸▸ Applying the Skill

The passage to the right tells about the climate in Latin America. Use the strategies listed below to help you find the main idea and summarize a passage.

How to Summarize

Strategy ❶ Look for a topic sentence stating the main idea. This is often at the beginning of a section or paragraph. Briefly restate the main idea in your own words.

Strategy ❷ Include key facts, numbers, dates, amounts, or percentages from the text.

Strategy ❸ After writing your summary, review it to see that you have included the main idea and the most important details.

The Tropical Zone

❶ A large portion of Latin America lies within the Tropical Zone, ❷ which is the region between the latitudes 23°27' north and 23°27' south. ❷ The Tropical Zone may be rainy or dry, but it is typically hot. ❷ Also, temperature is always lower at higher elevations, but in the Tropical Zone, all elevations are warmer than they are elsewhere.

Write a Summary

You can write your summary in a paragraph. The paragraph below summarizes the passage you just read.

❸ Much of Latin America is in the Tropical Zone. This zone is between latitudes 23°27' north and 23°27' south. Usually, the Tropical Zone is hot. Also, the elevations in the Tropical Zone are warmer than the same elevations in other places.

▸▸ Practicing the Skill

Turn to page 161 in Chapter 6, Section 2, "Ancient Latin America." Read "Maya Agriculture," find the main idea, and write a paragraph summarizing the passage.

SKILLBUILDER

Finding and Summarizing the Main Idea

Defining the Skill

Ask students to think of situations in which they might want to summarize what they read, such as when writing a book report or relating a current event from a newspaper article.

Applying the Skill

How to Summarize Point out the three strategies given, and guide students as they work through them. Emphasize the importance of following each step in order. Read the paragraph titled "The Tropical Zone" aloud. Then ask a volunteer to read the topic sentence aloud. Invite other students to identify important details that belong in a summary of the paragraph.

Write a Summary

Read the practice summary with students. Identify the information that was excluded from the original passage and discuss why it was not included. Discuss examples of how information was restated.

Practicing the Skill

Have students turn to page 161 and reread the paragraph titled "Maya Agriculture." Then allow students time to work individually to write a summary of the paragraph. Remind them to restate the main idea and then to include only the most important details. If students need additional practice, have them summarize paragraphs of newspaper articles.

In-depth Resources: Unit 3
- Skillbuilder Practice, p. 7

Career Connection: Political Reporter

Encourage students who enjoy identifying and summarizing main ideas to find out about careers that utilize this skill. For example, political reporters must read and understand long and detailed documents, such as current legislation (newly proposed laws) that is being debated in state legislatures or in Congress. They must then be able to identify the main idea of the documents and summarize it clearly for newspaper readers or television viewers.

Block Scheduling

1. Encourage students to find the text of a current or recently passed legislative bill from your state legislature or Congress. Discuss the length and difficulty of the text.
2. What other skills does a person need in order to become a political reporter? Help students think of and find information about the aptitudes and education needed to become a political reporter.
3. Have students develop a flow chart showing the education and possible career path of a political reporter.

ASSESSMENT

TERMS & NAMES

1. tributary, p. 157
2. deforestation, p. 157
3. Tropical Zone, p. 158
4. *El Niño,* p. 159
5. hieroglyph, p. 161
6. *chinampa,* p. 162
7. Machu Picchu, p. 165
8. Hernán Cortés, p. 166
9. Montezuma II, p. 166
10. Columbian Exchange, p. 166

REVIEW QUESTIONS

Possible Responses

1. Mountains surrounding Mexico City trap severe air pollution; its location on an unstable drained lakebed leads to earthquakes.
2. Caribbean islands were formed by either volcanoes or coral reefs.
3. The climate in the Tropical Zone is generally hot because of its location near the equator.
4. The movement of tectonic plates causes frequent earthquakes; volcanoes erupt along the Pacific coast of Central America, where two plates collide.
5. The Maya had a 365-day solar calendar, a 260-day calendar of sacred days, and another that measured the number of days from a fixed starting point.
6. Warfare helped maintain the Aztec Empire; soldiers believed it was an honor to die for the god of war.
7. Teams of runners carried verbal messages throughout the Inca empire; knotted strings *(quipu)* were used for record-keeping; weaving patterns symbolized ideas and history.
8. The Spanish enslaved Native Americans and tried to convert them to Christianity; new foods, animals, and manufactured goods were introduced.

ASSESSMENT

TERMS & NAMES

Explain the significance of each of the following:

1. tributary
2. deforestation
3. Tropical Zone
4. *El Niño*
5. hieroglyph
6. *chinampa*
7. Machu Picchu
8. Hernán Cortés
9. Montezuma II
10. Columbian Exchange

REVIEW QUESTIONS

Physical Geography *(pages 153–159)*

1. Describe two problems that Mexico City faces because of its location.
2. What two types of islands are found in the Caribbean?
3. Describe the climate in the Tropical Zone.
4. How have tectonic plates affected Latin America's physical geography? Give two examples.

Ancient Latin America *(pages 160–166)*

5. Explain the use of each of the three Maya calendars.
6. Why was warfare so important to the Aztec?
7. Describe two means of communication used by the Inca.
8. What types of changes took place in Latin America once the Spanish took control?

CRITICAL THINKING

Comparing

1. Using your completed chart from Reading Social Studies, p. 152, compare two ancient Latin American civilizations. List the similarities in the locations, dates, characteristics, achievements, and declines of the two civilizations.

Synthesizing

2. How are the farming methods of the Maya, the Aztec, and the Inca examples of people adapting to their environment?

Hypothesizing

3. Around A.D. 900, the Maya civilization—though not the people—began to disappear. What might have contributed to this decline?

Visual Summary

1 Physical Geography

- A variety of physical features, such as mountains and rivers, have impacted life in Latin America. Also, natural disasters, such as earthquakes and volcanic eruptions, occur frequently because of the region's location on five tectonic plates.
- The variations in Latin America's climate are caused by factors such as elevation, location, wind, and water.

2 Ancient Latin America

- The Maya, the Aztec, and the Inca all built major civilizations in Latin America.
- Each civilization developed creative solutions to problems presented by the physical geography around them.

168 CHAPTER 6

CRITICAL THINKING: Possible Responses

1. Comparing

Possible Responses The Maya and Aztec were both located in southern Mexico and made great achievements in agriculture. The Maya civilization flourished from A.D. 250 to 900. The Aztec ruled from the A.D. 1200s. Maya decline is unknown, and the Aztecs were conquered by the Spanish.

2. Synthesizing

Possible Responses The Maya used the slash-and-burn method to clear forests for farming. The Aztec built floating gardens on the marshes. The Inca built stone terraces so they could farm on mountainsides.

3. Hypothesizing

Possible Responses Climate change may have hurt crops and made it hard to support large cities. The population may have grown too large. Another group of people may have invaded and overthrown the political organization.

the map and your knowledge of world cultures and graphy to answer questions 1 and 2.

dditional Test Practice, pp. S1–S33

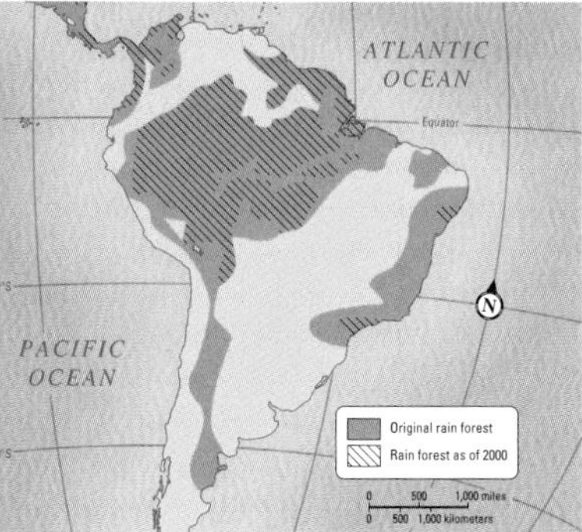

1. n what part of South America did no rain forest exist in 2000?
   A. East
   B. North
   C. Northwest
   D. South
2. What major trend does this map show?
   A. The rain forests of South America are decreasing.
   B. The rain forests of South America are increasing.
   C. The rain forests of South America have not changed.
   D. The rain forests of South America have disappeared.

This passage is from a speech that Prince Charles of the United Kingdom gave about tropical rain forests. Use the quotation and your knowledge of world cultures and geography to answer question 3.

PRIMARY SOURCE

It is not just those who depend directly on the tropical forests who suffer from deforestation, but the entire population of tropical forest countries. The forests assist in the regulation of local climate patterns, protecting watersheds, preventing floods, guaranteeing and controlling huge flows of life-giving water. Strip away the forests and there is, first, too much water . . . and then too little.

CHARLES, PRINCE OF WALES, from a speech given on February 6, 1990

3. Which of the following observations is supported by the passage?
   A. Those who inhabit the tropical forest suffer the most from the effects of deforestation.
   B. Efforts to save the tropical forest should be the sole responsibility of those people who live there.
   C. The preservation of tropical forests will positively affect the entire country in which they exist.
   D. Deforestation of the tropical forest has little effect on the climate of countries in the region.

STANDARDS-BASED ASSESSMENT

1. **Answer D** is correct because, as shown in the map, the original rain forest in the south was gone by 2000. Answer A is incorrect because the rain forest in the east was greatly reduced but still existed; answers B and C are incorrect because rain forest did exist in those regions.
2. **Answer A** is correct because the map shows the amount of rain forest was smaller in 2000 than originally; answer B is incorrect because the amount has not grown; answer C is incorrect because the amount has changed; answer D is incorrect because some rain forest still exists.
3. **Answer C** is correct because it restates the point that the entire population of tropical forest countries suffer the effects of deforestation; answer A is incorrect because it contradicts what the passage states; answer B is incorrect because the effects are widespread, so the responsibility should be shared; answer D is incorrect because deforestation affects climate.

TERNATIVE ASSESSMENT

WRITING ABOUT HISTORY

he Columbian Exchange brought a sharing of both goods and deas between European countries and the colonies in Latin merica. Write a magazine article about the Columbian xchange.

Use the Internet and library resources to learn about the crops, goods, animals, and ideas that were sent in each direction across the Atlantic Ocean.

Explain which trade items came from Latin America and which came from Europe. Describe how the new items were used in each region.

OOPERATIVE LEARNING

oth the Aztec and the Maya used two calendars—one with 60 days and one with 365 days. With a small group of lassmates, research one of the Aztec or Mayan calendars. Find ut how many months were in a year, how many days were in month, and what the names of the months or days were. hen, together, create a calendar based on your findings.

INTEGRATED TECHNOLOGY

Doing Internet Research

The ancient cultures of Latin America were among the earliest civilizations. Research one of these ancient cultures—Aztec, Maya, or Inca—in order to prepare a report.

- Use the Internet or reference sources in the library to learn about one aspect of daily life, such as religion, trade, or warfare, in the culture you chose.
- You might find information about ancient ruins on Web sites.
- If possible, include illustrations with your report.

For Internet links to support this activity, go to

INTEGRATED TECHNOLOGY

Discuss how students might use the Internet to find information about life in the Aztec, Maya, or Inca cultures. Brainstorm possible key words, Web sites, and search engines to explore. Have students take notes in the form of an outline. Suggest that they print pages containing interesting visuals to include with their reports.

Alternative Assessment

1. Rubric

The magazine article should
- use a journalistic style.
- cover the Columbian Exchange adequately including trade routes, goods, and ideas.
- present information in an unbiased way.
- use correct grammar, spelling, and punctuation.

2. Cooperative Learning

Before beginning this activity, briefly discuss different calendars students have seen in terms of format and style. Ask students to create their own calendars, possibly using symbols or icons to indicate holidays. Have students share the responsibilities of research, design, and layout.

Mexico Today

	OVERVIEW	COPYMASTERS	INTEGRATED TECHNOLOGY
UNIT ATLAS AND CHAPTER RESOURCES	This chapter examines how historical events influenced the development of Mexico's government, economy, and culture.	**In-depth Resources: Unit 3** • Guided Reading Worksheets, pp. 11–14 • Skillbuilder Practice, p. 17 • Unit Atlas Activities, pp. 1–2 • Geography Workshop, pp. 47–48 **Reading Study Guide** (Spanish and English), pp. 48–57 **Outline Map Activities**	• eEdition Plus Online • EasyPlanner Plus Online • eTest Plus Online • eEdition • Power Presentations • EasyPlanner • Electronic Library of Primary Sources • Test Generator • Reading Study Guide • The World's Music • Critical Thinking Transparencies CT13 • There Is No Food Like My Food
	KEY IDEAS		
SECTION 1 The Roots of Modern Mexico pp. 173–178	• Spanish settlers arrived in Mexico and conquered the Aztec. • The founding of New Spain brought about many changes. • Modern Mexico arose from conflict and cooperation among Native American, African, and Spanish settlers.	**In-depth Resources: Unit 3** • Guided Reading Worksheet, p. 11 • Reteaching Activity, p. 19 **Reading Study Guide** (Spanish and English), pp. 48–49	classzone.com Reading Study Guide
SECTION 2 Government in Mexico: Revolution and Reform pp. 179–184	• Land ownership, the power of national government, and the role of religion were concerns of reformers. • Distribution of land changed after the revolution. • The three levels of modern Mexico's government are as follows: national, state, and local.	**In-depth Resources: Unit 3** • Guided Reading Worksheet, p. 12 • Reteaching Activity, p. 20 **Reading Study Guide** (Spanish and English), pp. 50–51	**Critical Thinking Transparencies CT14** classzone.com Reading Study Guide
SECTION 3 Mexico's Changing Economy pp. 185–189	• Because farmers did not own land, they could not use land as security for loans. • Privatization changed farming. • In the mid-1900s the basis of Mexico's economy changed from farming to industry and tourism.	**In-depth Resources: Unit 3** • Guided Reading Worksheet, p. 13 • Reteaching Activity, p. 21 **Reading Study Guide** (Spanish and English), pp. 52–53	classzone.com Reading Study Guide
SECTION 4 Mexico's Culture Today pp. 190–194	• Native American, Spanish, and modern Mexican traditions have blended to form Mexico's culture. • Job opportunities, education, and cultural events have contributed to the growth of Mexico City. • Holidays play a significant role in Mexico's culture.	**In-depth Resources: Unit 3** • Guided Reading Worksheet, p. 14 • Reteaching Activity, p. 22 **Reading Study Guide** (Spanish and English), pp. 54–55	classzone.com Reading Study Guide

KEY TO RESOURCES

Audio	Internet	Teacher's Edition
CD-ROM	Overhead Transparency	Video
Copymaster	Pupil's Edition	

ASSESSMENT OPTIONS

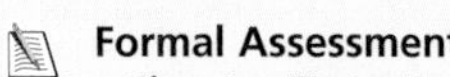

Chapter Assessment, pp. 198–199

Formal Assessment
- Chapter Tests: Forms A, B, C, pp. 97–108

Test Generator

Online Test Practice

Strategies for Test Preparation

Section Assessment, p. 178

Formal Assessment
- Section Quiz, p. 93

Integrated Assessment
- Rubric for writing a short story

Test Generator

Test Practice Transparencies TT17

Section Assessment, p.184

Formal Assessment
- Section Quiz, p. 94

Integrated Assessment
- Rubric for writing a speech

Test Generator

Test Practice Transparencies TT18

Section Assessment, p. 189

Formal Assessment
- Section Quiz, p. 95

Integrated Assessment
- Rubric for making a flow chart

Test Generator

Test Practice Transparencies TT19

Section Assessment, p. 194

Formal Assessment
- Section Quiz, p. 96

Integrated Assessment
- Rubric for writing a journal

Test Generator

Test Practice Transparencies TT20

RESOURCES FOR DIFFERENTIATING INSTRUCTION

Students Acquiring English/ESL

Reading Study Guide
(Spanish and English), pp. 48–57

Access for Students Acquiring English
Spanish Translations, pp. 45–54

TE Activity
- Word Usage, p.18

Modified Lesson Plans for English Learners

Less Proficient Readers

Reading Study Guide
(Spanish and English), pp. 48–57

TE Activities
- Comparing and Contrasting, p. 174
- Cause and Effect, p.183
- Comparing City and Country Life, p. 193

Gifted and Talented Students

TE Activity
- Presenting a Product, p. 187

CROSS-CURRICULAR CONNECTIONS

Humanities
Street-Porter, Tim. ***Casa Mexicana.*** New York: Workman Publishing, 1989. More than 350 Mexican houses.

Popular Culture
King, Elizabeth. ***Quinceañera: Celebrating Fifteen.*** New York: Dutton Children's Books, 1998. Mexican and Salvadoran coming-of-age ceremonies.
Coronado, Rosa. ***Cooking the Mexican Way.*** Minneapolis, MN: Lerner Publishing Group, 2001. Recipes and explanations of special ingredients.

Literature
Madrigal, Antonio Hernandez. ***The Eagle and the Rainbow: Timeless Tales from Mexico.*** Golden, CO: Fulcrum Kids, 1997. Five stories from five indigenous cultures.

Science
Pringle, Laurence. ***An Extraordinary Life: The Story of a Monarch Butterfly.*** New York: Orchard, 2000. Butterfly migration from Massachusetts to a winter refuge in Mexico.

Economics
Laufer, Peter. ***Made in Mexico.*** Washington, D.C.: National Geographic Society, 2000. Describes Paracho, the center of the Mexican guitar industry

ENRICHMENT ACTIVITIES

The following activities are especially suitable for classes following block schedules.

Teacher's Edition, pp. 177, 181, 182, 187, 195
Pupil's Edition, pp. 178, 184, 189, 194

Unit Atlas, pp. 142–149
Literature Connections, pp. 196–197

Outline Map Activities

INTEGRATED TECHNOLOGY

Go to **classzone.com** for lesson support and activities for Chapter 7.

BLOCK SCHEDULE LESSON PLAN OPTIONS: 90-MINUTE PERIOD

DAY 1

CHAPTER PREVIEW, pp. 170–171
Class Time 20 minutes

- **Hypothesize** Use "What do you think?" questions in Focus on Geography on PE p. 171 to help students hypothesize about the challenges presented by the land and climate.

SECTION 1, pp. 173–178
Class Time 70 minutes

- **Small Groups** Divide the class into four groups. Have each group select one section objective from TE p. 165 to help them prepare a summary of this section. Remind students that when they summarize, they should include the main ideas and most important details in their own words. Reconvene as a whole class for discussion.
Class Time 35 minutes
- **Peer Competition** Divide the class into pairs. Assign each pair one of the Terms & Names from this section. Have the pairs make up five questions that can be answered with the Term or Name that they have been assigned. Then have the pairs take turns asking the class their questions.
Class Time 35 minutes

DAY 2

SECTION 2, pp. 179–184
Class Time 45 minutes

- **Time Line** In preparation for presenting Section 2, have students use the time line to review key events presented in Section 1. As they read Section 2, ask students to note the key events on the time line.
Class Time 20 minutes
- **Historical Figure Description** Have students work in pairs. Ask each pair to select one of the historical figures from Terms & Names and prepare a description of that figure. Have one student from each pair present the description to the class. Class members should guess the name of the figure.
Class Time 25 minutes

SECTION 3, pp. 185–189
Class Time 45 minutes

- **News Article** Refer students to Data File, Unit Atlas PE pp. 142–149. Have each student choose a country and write a headline for it. Then have each student exchange his/her headline with another student and write a short news article to go with the headline received.

DAY 3

SECTION 4, pp. 190–194
Class Time 35 minutes

- **Travel Poster** Have students work in pairs to prepare a travel poster that reflects one feature of Mexico's rich culture. For example, students may select features such as art, holidays, architecture, music, or dance.

CHAPTER 7 REVIEW AND ASSESSMENT, pp. 198–199
Class Time 55 minutes

- **Review** Have students prepare a summary of the chapter by reviewing the Main Idea and Why It Matters Now features of each section in Chapter 7.
Class Time 20 minutes
- **Assessment** Have students complete the Chapter 7 Assessment.
Class Time 35 minutes

TECHNOLOGY IN THE CLASSROOM

GRAPHIC DESIGN

Computers can really come in handy when designing posters or other graphic displays. Computer drawing programs allow students to create original drawings right on the computer, while page layout and multimedia presentation programs let students combine their own drawings with images and text they find on the Internet, on CD-ROMs, or in print. Students should always be sure to cite the sources of these images.

ACTIVITY OUTLINE

aObjective Students will visit Web sites to gather information about important Mexican holidays. They will use the computer to create posters inviting people to attend celebrations of these holidays.

Task Have students go to the Web sites and take notes on the historical and religious backgrounds of Mexican Independence Day, Cinco de Mayo, and the Day of the Dead. Then have them pretend these holidays will be celebrated in their town. Students should use a computer drawing, page layout, or multimedia presentation program to create posters inviting people to join the festivities and informing them of the reasons for the celebrations.

Class Time Two class periods

DIRECTIONS

1. Review these basic questions with the class: From what country did Mexico gain independence? To what religion do the majority of Mexicans belong?
2. Inform students that Mexicans and many Mexican Americans celebrate several important holidays related to Mexico's history and to Mexican religious beliefs. Have them visit the Web sites listed at **classzone.com** to find out about Mexican Independence Day, Cinco de Mayo, and the Day of the Dead.
3. Ask students to make charts with three rows and three columns. They should label each column with one of the three Mexican holidays and each row with one of the following questions: When is this holiday celebrated? Why is this holiday celebrated (what event or group of people does this holiday commemorate)? What do Mexicans and Mexican Americans do to celebrate this holiday? As they go through the Web sites, have them take notes in the chart to answer the questions. Ask students to pay close attention to the holidays' historical and religious backgrounds.
4. If you have Mexican American students in your class, discuss the ways they celebrate these holidays. If any of them have lived in Mexico, have them share with the class the ways the holidays are celebrated differently in Mexico and in the United States.
5. Assign each student or group of students to one of the three holidays. Ask them to pretend that there will be a big celebration of this holiday in their town and that they have been hired to create the publicity posters. Have them use a computer drawing, page layout, or multimedia presentation program to design posters inviting people to the festivities. The posters should inform people of the holiday's historical or religious background and give them a preview of the things they can expect to see and do at the celebration.

CHAPTER 7 OBJECTIVE

Students will examine how historical events influenced the development of Mexico's government, economy, and culture.

FOCUS ON VISUALS

Interpreting the Photograph Direct students' attention to the photograph of the Zócalo. Ask students what they observe about the architecture of the buildings around the plaza. Point out that the chapter describes Mexico in terms of its past and present. Ask students to describe how this photograph shows elements belonging to both the past and the present.

Possible Responses The architecture is quite ornate, with decorative features. This older style of architecture stands in contrast to aspects of modern life, such as the use of electricity and the modern vehicles.

Extension Have students use an encyclopedia or almanac to compare the population of Mexico City with the largest city in their own state as well as New York City, the largest city in the United States.

CRITICAL THINKING ACTIVITY

Hypothesizing Prompt a discussion about why many major cities were originally established as settlements along or near the coastline of a country. Have students brainstorm a list of cities that fit this description, such as New York, Miami, Houston, and Seattle. Then direct students' attention to the map of Mexico and to Mexico City, in particular. Ask them to suggest possible reasons why this major city is located far inland, hundreds of miles from either coast.

Class Time 15 minutes

Mexico Today

Section 1 The Roots of Modern Mexico

Section 2 Government in Mexico: Revolution and Reform

Section 3 Mexico's Changing Economy

Section 4 Mexico's Culture Today

Place Mexico City's main plaza, the Zócalo, is the center of one of the world's largest cities.

170

Recommended Resources

BOOKS FOR THE TEACHER

Green, Jen. *Mexico.* Austin: Raintree Steck-Vaughn, 2000. Geography, history, economy, culture, and daily life of Mexico.

Hamnett, Brian R. *A Concise History of Mexico.* Cambridge: Cambridge University Press, 1999. Includes contemporary issues.

Harvey, Miles. *Look What Came from Mexico.* Danbury, CT: Franklin Watts, 1999. Inventions, sports, food, holidays, and customs.

Nye, Naomi Shihab, ed. *The Tree Is Older Than You Are: A Bilingual Gathering of Poems and Stories from Mexico with Paintings by Mexican Artists.* New York: Simon & Schuster, 1995. Rich contemporary cultural content.

VIDEOS

Mexico Close-Up. Maryknoll World Productions, 2001. Profiles two 14-year-olds: one in the city and one in a troubled rural area.

INTERNET

For more information about Mexico, visit **classzone.com.**

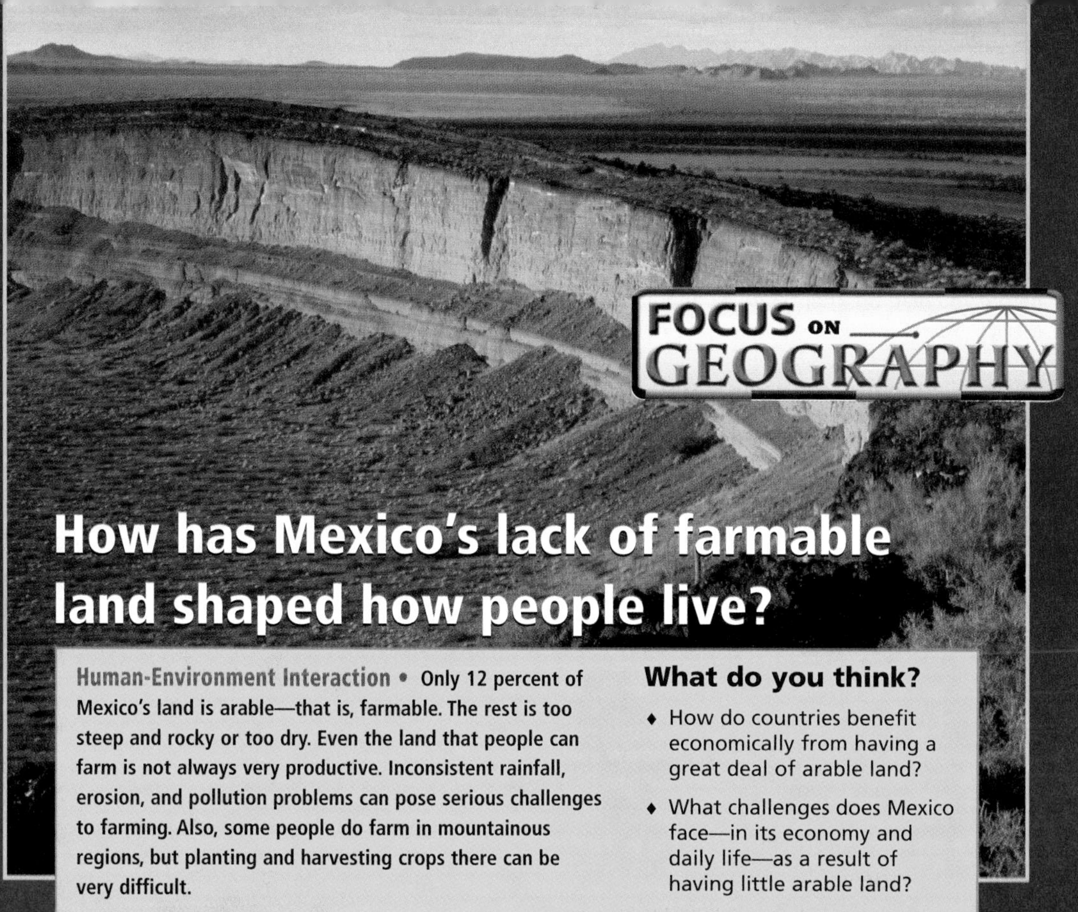

How has Mexico's lack of farmable land shaped how people live?

Human-Environment Interaction • Only 12 percent of Mexico's land is arable—that is, farmable. The rest is too steep and rocky or too dry. Even the land that people can farm is not always very productive. Inconsistent rainfall, erosion, and pollution problems can pose serious challenges to farming. Also, some people do farm in mountainous regions, but planting and harvesting crops there can be very difficult.

What do you think?

- How do countries benefit economically from having a great deal of arable land?
- What challenges does Mexico face—in its economy and daily life—as a result of having little arable land?

171

FOCUS ON GEOGRAPHY

Objectives

- To help students identify the relationship between geography and economy
- To identify and discuss challenges presented by the land and climate

What Do You Think?

1. Make sure that students understand that arable means "suitable for plowing." Then guide them to consider the relationship between arable land and the production of crops for consumption and trade purposes.
2. Guide students to think about how Mexico can compensate, economically, for its lack of arable land.

How has Mexico's lack of farmable land shaped how people live?

Have students consider how the lack of farmable land has influenced where people live, what they eat, and what the majority of people do for a living. Ask them whether they think people tried to adapt their lives to suit the land, or whether they ever attempted to make the land suitable for farming.

MAKING GEOGRAPHIC CONNECTIONS

Ask students to describe the area where they live in terms of the land and climate, and whether it is suitable for farming. Then have them compare their area to the one shown in the picture.

Implementing the National Geography Standards

Standard 12 Identify causes and consequences of urbanization

Objective To show why urban centers attract people and how people and urban centers subsequently are affected

Class Time 30 minutes

Task Have students prepare a two-column chart listing cultural, economic, and other issues that often are the results of urbanization. On one side ask them to list reasons why people move to urban areas. On the other side have them list problems resulting from urbanization. Ask students to compare the columns by asking questions such as: Are all items of the same importance? Do some factors outweigh other ones? What conclusions can we draw?

Evaluation Students should include at least two positive and two negative factors that they find are at work in urbanization.

BEFORE YOU READ

What Do You Know?

Poll students to see how many of them play soccer. Tell them that the most popular sport in Mexico is soccer, or *fútbol.* Almost every village has a soccer field. Aztec Stadium in Mexico City can hold up to 100,000 people.

Ask students who have eaten Mexican food to describe their favorite dishes. Then ask students if they know what a piñata is, and encourage them to describe how it is used. Finally, ask students to tell about any folk tales or other stories they have read that were set in Mexico or the region that is now Mexico.

What Do You Want to Know?

Suggest that students make a chart with the headings What I Want to Know and What I Learned in their notebooks. They can record questions they hope to have answered or topics they are curious about in the first column. As they read the chapter, they can record facts and ideas that they have learned.

READ AND TAKE NOTES

Reading Strategy: Organizing Information
Explain to students that this time line begins in the year 1500 and continues through the year 2000. Ask students how many years are covered in that period of time. (500) Point out that most of the events they will learn about happened after 1800. Tell students that when the time line is completed, they will have a list of the important events in the order in which they happened.

In-depth Resources: Unit 3
- Guided Reading Worksheets, pp. 11–14

CHAPTER 7 READING SOCIAL STUDIES

BEFORE YOU READ

▶▶ What Do You Know?

Before you read the chapter, think about what you already know about Mexico. You may have read about the Aztec in another class. Do you know about any aspects of Mexico's culture today? Think about what you've heard on the news about Mexico—do you know who the president of Mexico is?

Culture • **The Aztec feather shields were no match for Spanish armor.** ▲

▶▶ What Do You Want to Know?

Decide what else you want to know about Mexico. In your notebook, record what you hope to learn from this chapter.

READ AND TAKE NOTES

Reading Strategy: Organizing Information
One effective way to organize information is with a time line. Time lines show events in sequence, or the order in which they happened. Making a time line for the events in this chapter will help you better understand what happened when.

- Copy the time line in your notebook.
- As you read the chapter, note the key events discussed in it.
- Write these events beside the appropriate dates on your time line.

Place • **At the Ballet Folklórico, the culture of Mexico takes center stage.** ▲

1500 — 1600 — 1700 — 1800 — 1900 — 20[00]

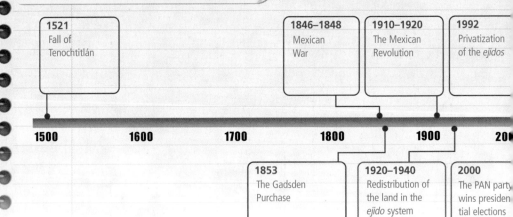

- **1521** Fall of Tenochtitlán
- **1846–1848** Mexican War
- **1853** The Gadsden Purchase
- **1910–1920** The Mexican Revolution
- **1920–1940** Redistribution of the land in the *ejido* system
- **1992** Privatization of the *ejidos*
- **2000** The PAN party wins presidential elections after 70 years of PRI rule

Teaching Strategy

Reading the Chapter This is a thematic chapter that focuses on how the people, government, economy, and culture of Mexico today reflect the influence of its early settlers—the Spanish, Native Americans, and Africans. Encourage students to use the visuals in each section to help them understand how the rich history of Mexico is still reflected in its daily life.

Integrated Assessment The Chapter Assessment on page 199 describes several activities that may be used for integrated assessment. You may wish to have students work on these activities during the course of the chapter and then present them at the end.

The Roots of Modern Mexico

MAIN IDEA

Modern Mexico arose from conflict and cooperation among Native American, African, and Spanish Mexicans.

WHY IT MATTERS NOW

The culture of Mexico today reflects the influences of and interactions among these groups.

TERMS & NAMES

peninsular
criollo
mestizo
encomienda
Father Miguel Hidalgo
Treaty of Guadalupe Hidalgo
Gadsden Purchase

DATELINE

EXTRA

TENOCHTITLÁN, AZTEC EMPIRE, 1519

Today, in the palace of the Aztec emperor Montezuma II, a Spanish explorer got a taste of something he is sure to remember. A treat enjoyed by both the Aztec and Maya, the drink, called *chocolatl*, is thick, smooth, and decidedly bitter. With no sugar added to sweeten it, this unusual drink is spiced with flavors such as chili pepper and vanilla.

Always served in liquid form, the drink is consumed by the Native Americans at a high rate. Emperor Montezuma II drinks up to 50 cups a day. The Spaniards hope to take the chocolate, as they call it, back to Europe—although they plan to find some way to help ease the bitter taste.

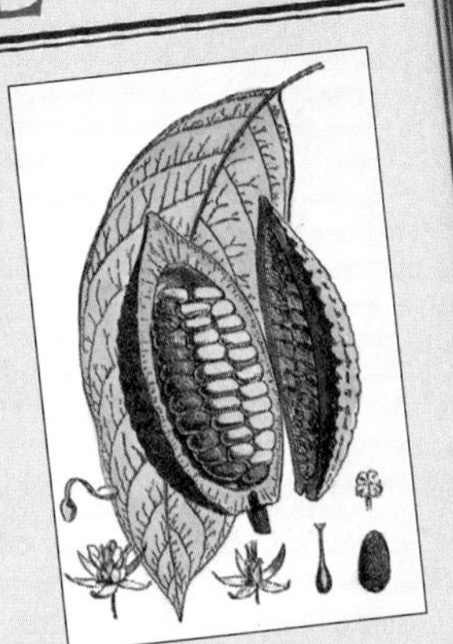

Region • Cacao seeds grow inside pods like this. ▲

The Arrival of the Spanish ❶

The leader of the Spanish army that first landed on the shores of Mexico was Hernán Cortés. He hoped to win new lands for Spain, as well as gold and glory for himself.

Cortés reached the east coast of Mexico in 1519 with about 500 soldiers. He claimed the land for the king and queen of Spain. Quickly, however, he learned that the land was ruled by the powerful Aztec emperor Montezuma II.

TAKING NOTES

Use your time line to take notes about Mexico.

1500 — 1700

SECTION OBJECTIVES

1. To explain when and how Spanish settlers arrived in Mexico and defeated the Aztecs
2. To describe the changes in New Spain
3. To identify the *mestizos* and other classes of people living in New Spain
4. To explain how Father Hidalgo began a revolution for independence

SKILLBUILDER

- Interpreting a Graph, p. 176
- Interpreting a Map, p. 178

CRITICAL THINKING

- Hypothesizing, p. 175
- Comparing, p. 177

FOCUS & MOTIVATE

WARM-UP

Making Inferences Have students read Dateline. Discuss these questions.

1. Why do you think that Montezuma offered *chocolatl* to the Spanish explorer when he visited the palace?
2. How do we know that the Spaniards really did take the chocolate back to Europe and add sugar to it?

INSTRUCT: Objective ❶

The Arrival of the Spanish/ A Clash of Cultures

- What motivated Cortés to conquer Mexico? acquiring new lands for Spain, personal glory
- What factors helped the Spanish conquer the Aztecs? better weapons, horses

In-depth Resources: Unit 3
- Guided Reading Worksheet, p. 11

Reading Study Guide (Spanish and English), pp. 48–49

Program Resources

In-depth Resources: Unit 3
- Guided Reading Worksheet, p. 11
- Reteaching Activity, p. 19

Reading Study Guide (Spanish and English), pp. 48–49

Formal Assessment
- Section Quiz, p. 93

Integrated Assessment
- Rubric for writing a short story

Outline Map Activities

Access for Students Acquiring English
- Guided Reading Worksheet, p. 45

Technology Resources classzone.com

TEST-TAKING RESOURCES

Strategies for Test Preparation

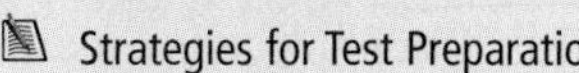

Test Practice Transparencies

Online Test Practice

INSTRUCT: Objective 2

The Founding of New Spain/ The Influence of the Church

- What is the historical significance of the fall of Tenochtitlán in 1521? marks beginning of Spanish rule
- What things did the Spanish introduce that changed life in the region? animals, trades, skills, religion
- What is one element of a Native American lifestyle that has been blended into Mexican culture? foods

Connections to Science

Smallpox, once one of the world's most feared diseases, was the first disease to be conquered by means of vaccination. In 1796, Edward Jenner, an English physician, developed a vaccine, which prevented the spread of smallpox. During the 1800s many countries passed laws requiring vaccination. Although the number of smallpox-infected countries steadily decreased, the disease continued to exist worldwide until the 1940s, when it was finally eliminated in the United States and Europe. In 1971, U.S. government health officials ended routine vaccinations for smallpox, except for people traveling to countries where the disease is still present.

A Clash of Cultures 1

Montezuma II ruled an empire of between 5 and 6 million people. However, many of his Native American subjects wanted to be free. They helped the Spanish conquer the Aztec king. They did not expect that the Spanish would become their new rulers.

The First Encounter Montezuma II heard about the arrival of the Spanish, and soon he welcomed Cortés with gifts. He even allowed Cortés to stay in a royal palace in the Aztec capital, Tenochtitlán. Within a week, Cortés took Montezuma II prisoner—and took control of the Aztec Empire.

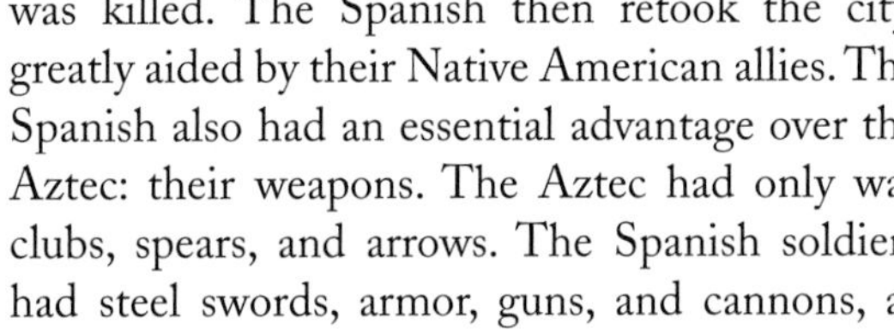

Culture • Aztec feather shields offered less protection than Spanish metal helmets and armor. ▲

The Spanish Takeover Other Aztec leaders drove the Spanish from Tenochtitlán. However, during the fighting that followed, Montezuma II was killed. The Spanish then retook the city, greatly aided by their Native American allies. The Spanish also had an essential advantage over the Aztec: their weapons. The Aztec had only war clubs, spears, and arrows. The Spanish soldiers had steel swords, armor, guns, and cannons, as well as horses. The invading army destroyed Tenochtitlán street by street.

Connections to Science

Invisible Weapons Smallpox and other diseases from Europe killed millions of Native Americans between 1500 and 1900. Smallpox (germ cell shown below right) had long been widespread in Europe, and most Europeans were at least partly immune. Native Americans, however, had no immunity to it because it had never before existed in the Americas.

Within months of the Spanish soldiers' arrival in Mexico, many thousands of Native Americans got sick with smallpox (shown below) and died from it—including Montezuma II's successor. Smallpox proved far more deadly to Native Americans than Spanish swords and cannons.

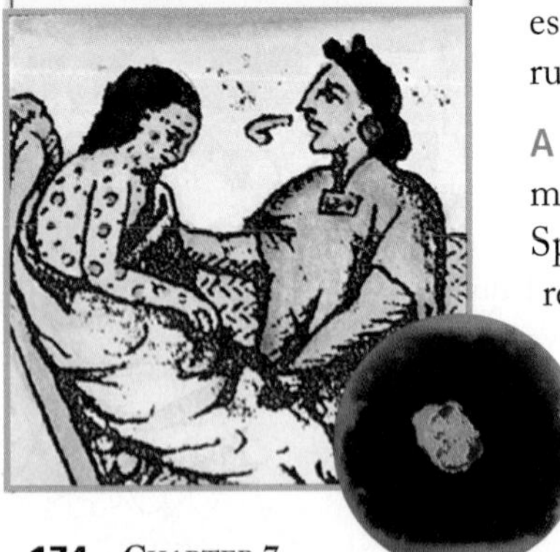

Reading Social Studies

A. Analyzing Motives Why did some Native Americans help the Spanish fight the Aztec?

Reading Social Studies
A. Possible Answer
They no longer wanted to be under Aztec rule. They wanted to be free and did not expect that the Spanish would become the new rulers.

The Founding of New Spain 2

The fall of Tenochtitlán in 1521 marked the end of the Aztec Empire and the beginning of Spanish rule in Mexico. The Spanish called their empire "New Spain," just as the English called their territory in North America "New England." Where Tenochtitlán had stood, the Spanish established Mexico City as their capital. Spain ruled Mexico for the next 300 years.

A New Way of Life The Spanish victory caused more than a change of rulers in Mexico. The Spanish introduced a different way of life to the region. They brought new animals, such as horses, cattle, sheep, and pigs. They also brought new trades, such as ironsmithing and shipbuilding. They brought a new religion as well—Christianity.

Vocabulary

ironsmithing: making items out of iron

Activity Options

Differentiating Instruction: Less Proficient Readers

Comparing and Contrasting Remind students that the fall of Tenochtitlán marked the end of Aztec rule and the beginning of Spanish rule in Mexico. Have students create a chart like the one shown that compares the empires that existed before and after 1521.

Empire Before 1521	Empire After 1521
Aztec rule	Spanish rule
Capital was Tenochtitlán	Empire renamed New Spain
	Capital was Mexico City
	New animals
	New trades
	Christianity

The Influence of the Church ❷

Because the Catholic Church was powerful in Spain, it soon became powerful in New Spain. Catholic priests set up churches, schools, and hospitals. Sometimes Native Americans accepted Christianity willingly. Sometimes, though, they were forced to become Christian against their will.

A Cultural Blend Even though the Native Americans had to accept many new ways of life, the old ways were not lost entirely. For instance, an essential element of Native American cooking was the tortilla, a flat, round bread made from corn or flour. Tortillas are still made daily all over Mexico. As with food, many other aspects of the two cultures blended in the new Mexican culture.

Life in New Spain ❸

A new multilayered society developed in Mexico. The ruling class were Spanish officials who were born in Spain. They were called ***peninsulares*** (peh•neen•soo•LAH•rehs) because they were from the Iberian Peninsula in Europe.

BACKGROUND

The Iberian Peninsula consists of two countries—Portugal and Spain. (See the map of Europe on page 262.)

A second class were ***criollos*** (kree•AW•yaws), people who were born in Mexico but whose parents were born in Spain. *Criollos* were often wealthy and powerful, but they were not in as high a social class as the *peninsulares.*

A ***mestizo*** (mehs•TEE•saw) is a person who is of Spanish and Native American ancestry. *Mestizos* formed the third layer of New Spain's society.

Movement • Mexicans today celebrate Catholic holy days that the Spanish established. This festival honors Our Lady of Guadalupe. ▼

INSTRUCT: Objective ❸

Life in New Spain

- What name was given to the class of people of both Spanish and Native American ancestry? *mestizo*
- How did the *criollos* differ from the *peninsulares? peninsulares:* born in Spain; *criollos:* parents born in Spain
- What was the system of *encomienda?* Native Americans had to pay tributes of labor or goods to Spanish overseers.

CRITICAL THINKING ACTIVITY

Hypothesizing Reread the sentence "A new multilayered society developed in Mexico." Ask students to discuss how they think layers of a society develop. To lead the discussion, ask questions such as these.

- Do new layers of society develop all at once?
- How do people in a society know what the new layers are?
- What do people in a society do—both knowingly and unknowingly—to develop these layers?
- How might layers in a society change or disappear?

Class Time 15 minutes

Activity Options

Multiple Learning Styles: Visual/Interpersonal

Class Time 15 minutes

Task Explaining the ancestry of *peninsulares, criollos, mestizos,* and Africans, and locating the ancestors of each on a world map

Purpose To point out the origin of the various social groups in Mexico

Supplies Needed

- World map

Activity Divide students into groups of four, and assign one student in each group the role of a *peninsulare*, a *criollo*, a *mestizo*, or a person from Africa. One at a time, have students tell what social group they belong to, explain what it is about their ancestry that makes them a member of that group, and identify the place(s) where they and their parents were born. As they identify these countries, have them point to each location on the world map.

INSTRUCT: Objective 4

The War of Independence/War with the United States

- Which groups of people benefited least from Mexico's independence from Spain? Native Americans and *mestizos*
- Which states in the current United States were once part of Mexico? California, Texas, New Mexico, Arizona, Nevada, Utah, part of Colorado, Wyoming, Oklahoma
- What was the result of the war between Mexico and the United States? the Treaty of Guadalupe Hidalgo; United States won control of northern Mexico

FOCUS ON VISUALS

Interpreting the Graph Have students compare the combined percentages of *criollos, mestizos,* and Native Americans with the combined percentages of *peninsulares* and Africans. Ask students if this difference is surprising. Ask why or why not. Point out that the *peninsulares* were the ruling group.

Possible Responses Neither *peninsulares* nor Africans were born in Mexico. They had to travel there, willingly or not. All the people in the other population groups were natives of the area.

Extension Ask students to make a graph to show how the percentages might have changed by 1860. Suggest that they think about which groups might have decreased in number, which groups might have increased in number, and why. Have students share and compare their completed graphs.

Population of Mexico in 1810

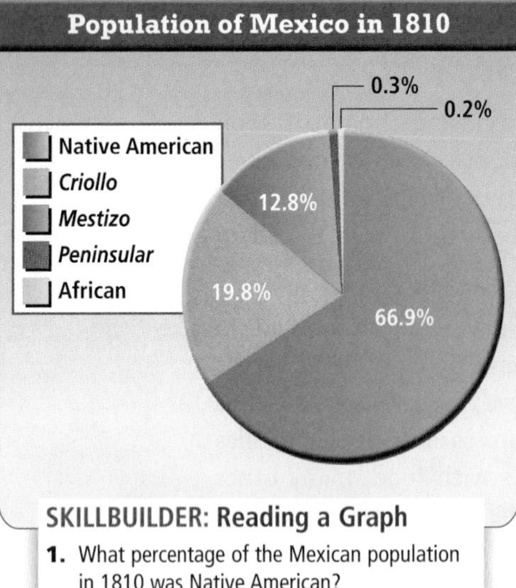

SKILLBUILDER: Reading a Graph

1. What percentage of the Mexican population in 1810 was Native American?
2. Were there more *peninsulares* or *criollos* in Mexico in 1810?

Skillbuilder Answers
1. 66.9%
2. *criollos*

A fourth group of people arrived in Mexico unwillingly—the enslaved Africans brought by European slave ships. African farming techniques, musical traditions, and crafts soon blended into the Mexican culture.

Encomienda New Spain's largest group was the Native Americans. They made up the bottom layer of society. The rulers of Spain set up in Mexico a system called ***encomienda*** (ehn•kaw•MYEHN•dah). Under this system, Spanish men were each given a Native American village to oversee. The villagers had to pay tribute—in goods, money, or labor—to this Spaniard. They were essentially enslaved. The results of their labor helped to make Spain rich. However, the villagers lived in poverty and hardship.

Vocabulary
tribute: a forced payment

The War of Independence 4

Based on earlier European and American political writers, many Mexican religious and political leaders in the early 1800s were saying that Mexicans should be free to choose their own government. They argued that Mexico should be independent from Spain. The demand for Mexican independence grew stronger after 1808, when France conquered Spain.

Reading Social Studies
B. Clarifying What ideas from other parts of the world did Mexicans agree with?

Reading Social Studies
B. Possible Answer
Mexicans agreed with the idea that people should be free to choose their own governments.

A Cry for Freedom Then, before dawn on September 16, 1810, the farmers in the mountain village of Dolores heard their church bells ringing. At the church, their priest, **Father Miguel Hidalgo,** gave a fiery speech urging them to throw off Spanish rule. No one knows the exact text of the speech, but it is known as the *Grito de Dolores* (Cry of Dolores). Urged on by his words, a small army of Native Americans and *mestizos* marched with Father Hidalgo toward Mexico City. Along the way, thousands more joined them.

A Difficult Challenge Father Hidalgo's army had few weapons. Mostly, his men carried clubs and farm tools, such as sickles and axes. When they faced the government soldiers, the farmers were soon defeated. Father Hidalgo was captured and executed, but the revolution he had sparked did not die.

Vocabulary
sickle: a blade used for cutting tall grass or grain

Activity Options

Skillbuilder Mini-Lesson: Finding and Summarizing the Main Idea

Explaining the Skill Explain to students that when you summarize, you include the main ideas and most important details in your own words.

Applying the Skill Have students reread the section entitled "The War of Independence" and use these strategies to summarize the section:

1. Briefly restate each main idea in your own words. Many people believed that Mexico should be independent from Spain. Father Hidalgo inspired *mestizos* and Native Americans to revolt. Mexico finally became independent.
2. Include key facts and figures from the section. On September 16, 1810, Father Hidalgo made a fiery speech.
3. Review your summary and delete nonessential details.

BACKGROUND

Mexico's independence was based on "three guarantees": that it would be independent from Spain, that it would be Catholic, and that *criollos* and *peninsulares* would be equal.

Independence at Last New leaders took Father Hidalgo's place. A few wealthy Spanish nobles and many *criollos* joined the fight for independence. The struggle lasted for 11 years. In 1821, the rebels finally overthrew the Spanish government, and Mexico became independent. However, the *peninsulares* and *criollos* still ruled the country. Native Americans and *mestizos* benefited little from independence from Spain.

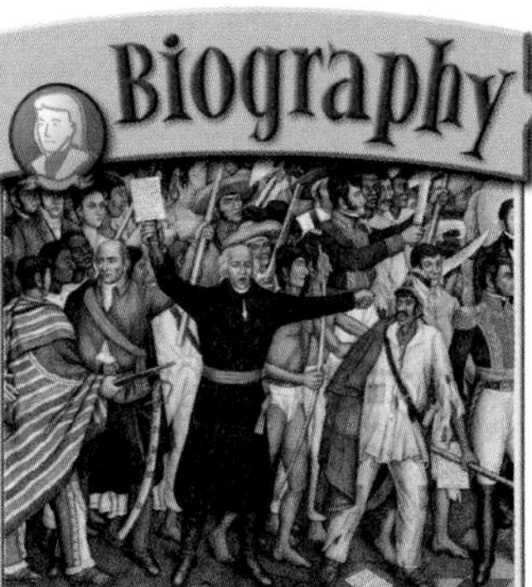

Father Hidalgo Father Miguel Hidalgo y Costilla, shown in the center of the illustration above, was born in 1753. He was a *criollo* who felt great sympathy for the Native Americans and *mestizos*. As a priest in the small village of Dolores, Father Hidalgo joined a secret group that fought for Mexico's independence.

Father Hidalgo is known as the Father of Mexican Independence. Every September 16, Mexicans shout slogans from the *Grito de Dolores* in celebration of their independence and Father Hidalgo.

War with the United States 4

In 1821, the new nation of Mexico was far larger than it is today. You can see on the map on page 178 that northern Mexico included much of what is now the Southwestern United States. Spanish explorers claimed this entire region in the 1500s and 1600s. Spanish and Mexican priests built missions there in the 1600s.

Desert and Distance Much of this land was desert. Travel was slow and communication difficult. Mexico was at war with Native Americans in this region, such as the Apache and the Comanche tribes. For all these reasons, few Mexicans settled there.

To encourage settlement, the Mexican government invited foreigners to move into these northern lands. Most of the newcomers were from the United States and still felt some loyalty to that country. By the 1830s, settlers in Texas from the United States greatly outnumbered those from Mexico.

Place • You can still see this mission, built in 1700, in Arizona today. ▼

Texas Independence In 1835, many settlers in Texas decided to break away from Mexico and rose in revolt. After several fierce battles, the Texans won their independence. They set up the Republic of Texas in 1836.

Most Texans wanted to become part of the United States. In 1845, the United States agreed, but Mexico and the United States could not agree where the boundary between Texas and Mexico should be. Each side claimed land that the other wanted.

The Mexican War In 1846, the dispute grew into a war. During the next two years, U.S. forces won control of northern Mexico; it was made official when Mexico was forced to sign the **Treaty of Guadalupe Hidalgo**.

Biography

Miguel Hidalgo y Costilla attended the university in Mexico City and eventually became a priest. He was sent to the town of Dolores in 1803, where he devoted himself to helping the poor and improving their lives. Hidalgo learned to speak to the people in their native languages, and thus won their trust.

Mexican citizens continue to honor Father Hidalgo in their Independence Day celebration. Each year, the president of Mexico broadcasts the message as might have been spoken by Father Hidalgo in 1810.

CRITICAL THINKING ACTIVITY

Comparing Review the "three guarantees" on which Mexico's independence was based: Mexico would be independent, it would be Catholic, and there would be equality for all Mexicans. Ask students to consider how these guarantees are similar to and different from the rights guaranteed to United States citizens. You might draw a Venn diagram on the chalkboard and record students' responses in the appropriate areas of the diagram.

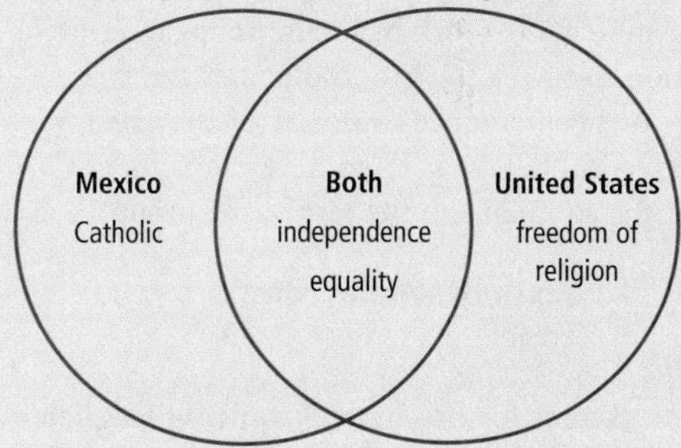

Class Time 15 minutes

Activity Options

Interdisciplinary Link: Speech

Class Time One class period

Task Writing the speech Father Hidalgo might have given, known as the *Grito de Dolores*

Purpose To identify arguments urging Native Americans and *mestizos* to fight for independence

Supplies Needed
- Writing paper
- Pencils or pens
- Tape recorders

Block Scheduling

Activity Recall with students that no one knows the exact words of Father Hidalgo's famous speech. Brainstorm with students some of the ideas that he might have incorporated. Have them write a short speech urging the Native Americans and *mestizos* to revolt against Spanish rule. Encourage students to focus on the key issues and use convincing words. Have them deliver their speeches to the class.

FOCUS ON VISUALS

Interpreting the Map Direct students' attention to the map key. Ask which rivers partially formed the Mexican border in 1819.

Answer Arkansas River, Red River

Extension Ask students to imagine that they were settlers living in San Antonio at the time of the Texas annexation in 1845. Ask them to write letters stating how they felt about Texas's becoming part of the United States.

ASSESS & RETEACH

Reading Social Studies Have students fill in the first three boxes on the graphic organizer on page 172.

Formal Assessment
- Section Quiz, p. 93

RETEACHING ACTIVITY

Ask students to work in small groups to create a section review. Each group member can write a one- or two-sentence summary of the text for one section heading. Then have each student read the summary to the rest of the group.

In-depth Resources: Unit 3
- Reteaching Activity, p. 19

Access for Students Acquiring English
- Reteaching Activity, p. 51

Mexican Territory Lost to the United States as of 1853

GEOGRAPHY SKILLBUILDER: Interpreting a Map

1. **Location** • What river formed the new border between Mexico and the Texas territory?
2. **Region** • About how many U.S. states were eventually formed from what had been Mexican territory?

Geography Skillbuilder Answers

1. Rio Grande
2. seven, and small parts of others

A few years later, in 1853, the **Gadsden Purchase** gave the United States more of Mexico's northern land. The two countries have since made slight adjustments to the border, but they have not fought a war again.

SECTION 1 ASSESSMENT

Terms & Names

1. **Explain the significance of:** (a) *peninsular* (b) *criollo* (c) *mestizo* (d) *encomienda* (e) Father Miguel Hidalgo (f) Treaty of Guadalupe Hidalgo (g) Gadsden Purchase

Using Graphics

2. Use a chart like this one to list the five social groups that made up New Spain and their characteristics.

Social Group	Characteristics

Main Ideas

3. (a) What advantages allowed Cortés's small army to conquer Mexico?
   (b) In what ways did Spanish rule change life in Mexico?
   (c) What were the results of the war between Mexico and the United States?

Critical Thinking

4. **Analyzing Causes** Why did Mexicans decide to fight for independence from Spain?

Think About
- influential events around the world
- reasons for discontent among the *criollos, mestizos,* and Native Americans

ACTIVITY -OPTION- Reread the information about the first meeting of the Spanish and the Aztec. Write a **short story** describing the event from either the Spanish or Aztec viewpoint.

Section 1 Assessment

1. Terms & Names
- a. *peninsular,* p. 175
- b. *criollo,* p. 175
- c. *mestizo,* p. 175
- d. *encomienda,* p. 176
- e. Father Miguel Hidalgo, p. 176
- f. Treaty of Guadalupe Hidalgo, p. 177
- g. Gadsden Purchase, p. 178

2. Using Graphics

Peninsulares	Spanish officials, ruling class
Criollos	Born, Mexico; Spanish parents
Mestizos	European and Native American ancestry
Africans	Enslaved by Europeans
Native Americans	Lowest level of society

3. Main Ideas
- a. Many Native Americans helped Cortés conquer Montezuma; Cortés's army had better weapons and horses.
- b. They brought new animals, new trades, and Christianity.
- c. The United States won control of northern Mexico; Mexico was forced to sign the Treaty of Guadalupe Hidalgo.

4. Critical Thinking

They were tired of the poverty and hardship endured under Spanish rule; people in other countries were gaining freedom.

ACTIVITY OPTION

Integrated Assessment
- Rubric for writing a short story

Government in Mexico: Revolution and Reform

TERMS & NAMES
Benito Juárez
Francisco Madero
hacienda
Emiliano Zapata
ejido
Institutional Revolutionary Party
Vicente Fox

MAIN IDEA

Through periods of reform and revolution, Mexico struggled to establish a strong democratic national government.

WHY IT MATTERS NOW

Other countries, such as the United States, are more willing to work as partners with Mexico because its government is democratic.

DATELINE

QUERÉTARO, MEXICO, 1917—Amid the battles of the Mexican Revolution, politicians today took a step toward reforming the nation. After many weeks of discussion and debate, they produced a new constitution for Mexico. The document presents new approaches to issues such as education, landownership, and religion.

Perhaps this constitution will become a basis for a new and stable Mexican government. For now, though, the constant fighting among sides continues. The potential power of the constitution will be revealed only in time, as the government puts the document into action.

Location • President Venustiano Carranza organized the meetings that resulted in the 1917 constitution. ▲

A Struggle for Power 1

The constitution of 1917 was written as a response to the struggles of the previous century. During that time, Mexico spent many years fighting wars. In 1821, Mexico won its war for independence from Spain. Two decades later, Mexicans entered into and then lost a war with the United States over Texas, California, and other lands. During all these years, still another struggle was going on—a struggle for power within Mexico.

TAKING NOTES
Use your time line to take notes about Mexico.

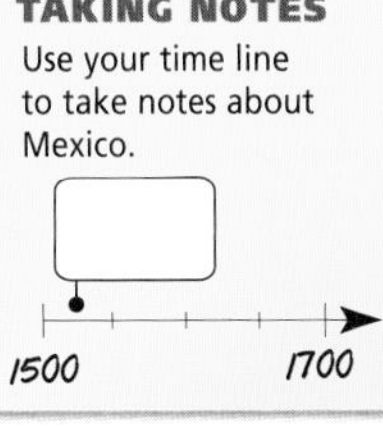

SECTION OBJECTIVES

1. To explain the need for reforms in Mexico and understand what events led to the end of reform
2. To identify the changes brought about by the Mexican Revolution
3. To describe the government of Mexico

SKILLBUILDER
- Interpreting a Map, p. 183

CRITICAL THINKING
- Analyzing Motives, p. 180

FOCUS & MOTIVATE

WARM-UP

Identifying Problems Have students read Dateline and ask these questions.

1. Why might the new approaches to land ownership and religion be controversial?
2. What might delay the implementation of the new constitution?

INSTRUCT: Objective 1

A Struggle for Power/ Benito Juárez Brings Reform

- What were the concerns of the people planning reforms in Mexico? power of national government, landownership, role of Catholic church
- What rights were granted by the constitution of 1857? freedom of speech, equality, end to slavery, reduction of military power

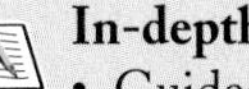

In-depth Resources: Unit 3
- Guided Reading Worksheet, p. 12

Reading Study Guide
(Spanish and English), pp. 50–51

Program Resources

In-depth Resources: Unit 3
- Guided Reading Worksheet, p. 12
- Reteaching Activity, p. 20

Reading Study Guide
(Spanish and English), pp. 50–51

Formal Assessment
- Section Quiz, p. 94

Integrated Assessment
- Rubric for writing a speech

Outline Map Activities

Access for Students Acquiring English
- Guided Reading Worksheet, p. 46

Technology Resources
classzone.com

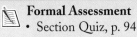

TEST-TAKING RESOURCES
- Strategies for Test Preparation
- Test Practice Transparencies
- Online Test Practice

In the years after independence, army leaders often took over Mexico's government. In some parts of the country, bandits attacked travelers. Elsewhere, Mexicans fought with Spanish landowners. Everywhere, a few people enjoyed great wealth, while many suffered in poverty.

Benito Juárez Brings Reform 1

By the 1850s, many Mexicans were eager for reform. They found a leader in **Benito Juárez,** a man who rose from poverty to become president of Mexico and a hero to his people. He became minister of justice in 1855, and he later became chief justice of the Supreme Court. In 1858, he gained the presidency, giving control of the Mexican government to the reformers.

Culture • Benito Juárez was a Zapotec, one of the many Native American groups in Mexico. He grew up in a mountain village, studied law, and then went into politics. ▲

Response to Reform The reformers wrote a new constitution for Mexico in 1857. For the first time, Mexicans had a bill of rights, promising them freedom of speech and equality under the law. The constitution of 1857 also ended slavery and forced labor. However, the new constitution did not promise freedom of religion. Nor did it make Catholicism Mexico's official religion, as many church leaders had hoped it would. The reformers also cut back the army's power in the government.

These reforms stirred up a storm of controversy. Church leaders, army leaders, and wealthy landowners were outraged. From 1858 to 1860, the War of Reform raged between the reformers and their opponents.

Foreign Intervention in Mexico The War of Reform left Mexico so weak—because of death, debt, and unemployment—that the country was an easy target for foreign takeover. Spain, Britain, and France sent troops into Mexico. In 1863, after more than a year of fighting, the French marched into Mexico City and established themselves in control of the country. They made a European nobleman named Maximilian emperor. Maximilian did not reign long. The Mexicans overthrew Maximilian and executed him in 1867.

Culture • Mexican forces fight their way to victory in an early battle against the French in 1862. ▼

MORE ABOUT...
Benito Juárez

Benito Juárez is sometimes called the "Abraham Lincoln of Mexico." Both Lincoln and Juárez were born to poor families—Juárez in an adobe hut, Lincoln in a log cabin. Both were avid readers and studied law when they were young men. For both, the study of law was a means toward the goal of politics. Most importantly, both men believed that justice should be administered fairly and equally to all people. In their political lives they acted on this belief.

The Juárez Monument, a great white marble semicircle erected to the memory of Benito Juárez, stands in Mexico City. Juárez is honored in his country the way Lincoln is honored in the United States.

CRITICAL THINKING ACTIVITY

Analyzing Motives Review the constitution of 1857, which called for an end to slavery in Mexico. What might have motivated Mexico to take this important step? Lead students to consider that the United States had not yet abolished slavery.

Class Time 15 minutes

Activity Options

Multiple Learning Styles: Interpersonal

Class Time One class period

Task Presenting multiple points of view about the War of the Reform in a debate

Purpose To understand the controversy that arose around the issues of reform from groups protecting different interests

Supplies Needed

- Writing paper
- Pencils or pens

Activity Review with the class the reforms proposed in the constitution of 1857—freedom of speech, the end of slavery, a cutback of the power of the army in government. Assign student groups the following roles: large landowners, army leaders, landless farmers, and reformers. Have groups summarize their points of view concerning these issues. Then have groups take turns debating the issues.

An End to Reform That same year, Benito Juárez and the reformers returned to power. Juárez remained president until his death in 1872. Unfortunately, his successors cared about reform less than he had. Poverty and lack of education remained problems. A few rich families held most of the political and economic power. Not until the 20th century did a new wave of reform begin.

The Mexican Revolution ❷

By 1910, the divisions between rich and poor in Mexico were huge. Just 800 families owned more than 90 percent of the farmland. Of Mexico's 15 million people, 10 million owned no land at all.

A Decade of War Once again, many Mexicans decided to fight for reforms. And once again, the struggle turned bloody. From 1910 to 1920, Mexico endured the Mexican Revolution.

The Revolution was a fight among many armies. Almost every part of Mexico had an army of rebels and reformers with particular goals. One of the first revolutionary leaders was a wealthy rancher named **Francisco Madero,** who became president in 1911. For Madero and his supporters, the key issue was free, honest elections. For others, however, the most important problem was landownership.

Reading Social Studies
A. Possible Answers
poverty, lack of education, land ownership issues, political power

The Problem of Land Poor farmers wanted land of their own. They believed the government should give each farm family a few acres by breaking up the giant *haciendas.* A ***hacienda*** is a big farm or ranch, often as large as 40,000 or 50,000 acres. Much of the *hacienda* land had once belonged to village farmers. But a law passed in 1883 allowed some of the wealthiest ranch owners to easily take away land from the village farmers. During the 1880s and 1890s, the ranch owners took over millions of acres of land owned by village farmers, and that land became part of their *haciendas.*

Reading Social Studies

A. Analyzing Issues What sorts of concerns led Mexicans to fight in the Mexican Revolution?

Region • During the Revolution, Francisco "Pancho" Villa was a famous leader in the north of Mexico. ▼

Region • Like this one, many of Mexico's *haciendas* were situated on huge pieces of land. ▶

INSTRUCT: Objective ❷

The Mexican Revolution/ A Continuing Revolution

- According to revolutionary leaders, what were the important problems facing Mexico? free, honest elections; landownership; farmers' rights
- How did the distribution of land change after the Revolution? land was more equally distributed among the people, especially villagers and small farmers
- How did the most powerful political party in Mexico preserve the idea of the Revolution? by including the word *revolution* in its name

MORE ABOUT...

Pancho Villa

Pancho Villa, now revered as a national hero in Mexico, was at one time a cattle thief and a bandit. He spent most of his youth living off the land and robbing mines. These experiences helped him develop skills as a guerrilla fighter. In 1910, Villa was inspired to support the revolutionary leader Francisco Madero. While fighting bravely, Villa was captured and imprisoned in Mexico City in 1912. He escaped to Texas, where he organized followers and returned to fight in Mexico a year later. He was retired as a general in 1920 and was assassinated in 1923.

Activity Options

Interdisciplinary Links: Speech/Language Arts

Class Time One class period

Task Writing a speech urging people to support revolutionary goals

Purpose To identify arguments that would urge the people of Mexico to support the reforms of the Revolution

Supplies Needed
- Writing paper
- Pencils or pens

Block Scheduling

Activity Review the goals of revolutionary leaders such as Francisco Madero. Brainstorm a list of the most important goals of the movement and record them on the chalkboard. Have each student focus on one goal and write a short speech explaining that goal and urging others to support it. Have students deliver their speeches to the class.

A VOICE FROM MEXICO

Explain to students that a quote such as this one is known as a primary source. Primary sources are the direct words of participants or observers of an event. Zapata's words show that he had strong feelings about others' joining the revolution, as indicated by the words "Rise up with us . . ." Point out that the pronouns *I, me, we,* and *us* let the reader know that Zapata speaks for himself and his people. Prompt a discussion about how resources such as this quote are convincing indicators of history.

Connections to Citizenship

Only seven basic colors are used in most national flags—red, white, blue, green, yellow, black, and orange. These colors were originally used in heraldry, a system of designs that emerged in the Middle Ages. In the Mexican flag, adopted in 1821, red traditionally stands for union, white for religion, and green for independence.

Emiliano Zapata was a legendary fighter for farmers' rights. With his famous motto—"Land and Liberty!"—Zapata gathered an army in the south of Mexico and urged farmers to join him.

A VOICE FROM MEXICO

> Join me. . . . We want a much better president. Rise up with us because we don't like what the rich men pay us. It is not enough for us to eat and dress ourselves. I also want for everyone to have his piece of land so that he can plant and harvest corn, beans, and other crops. What do you say? Are you going to join us?
>
> *Emiliano Zapata*

Connections to Citizenship

The Mexican Flag In the middle of Mexico's flag sits an eagle holding a snake in its mouth—a symbol from an Aztec legend about the founding of their capital, Tenochtitlán. The legend says that in the 1100s, the Aztec sun god told the Aztec to build a city on the spot where they saw an eagle with a snake in its mouth. When they saw just such an eagle in the middle of Lake Texcoco, they knew where to build their capital.

Each of the flag's red, white, and green stripes represents one of the "three guarantees" of the Mexican War of Independence. By flying a flag that combines a symbol from Aztec times with one from the period of independence, Mexico shows how important its roots are to its modern identity.

A Continuing Revolution ❷

Over the course of a decade, dozens of large and small armies fought with one another. In 1913, Madero was murdered. The same fate befell Zapata in 1919. Between 1910 and 1920, more than 1 million Mexicans died in the battles of the Revolution. In 1920, a new government managed to make peace among the many armies. The fighting was over, but the Revolution—the effort to reform Mexico's government and economy—went on.

Answering Demands for Land In 1917, a new constitution was written, and one of its promises was to distribute land more equally among the people. Between 1920 and 1940, the government broke up many of the giant *haciendas.* Millions of acres were divided among small farmers or given to *ejidos.* An ***ejido*** (eh•HEE•daw) is a community farm owned by all the villagers together. Farmers were proud and happy to have their own land once again.

Reading Social Studies
B. Possible Answer
The constitution promised to distribute land more equally among the people.

Reading Social Studies
B. Summarizing How did the 1917 constitution respond to concerns about landownership?

Activity Options

Interdisciplinary Links: Art/History

Block Scheduling

Class Time One class period

Task Creating a banner to symbolize the issues of the Revolution in Mexico

Purpose To illustrate the concerns of the people supporting the Revolution

Supplies Needed
- Poster paper
- Poster paint and brushes

Activity Review the concerns of the people that led to the Revolution in Mexico, such as inequalities, land distribution, and elections. Then have students work in small groups to design banners that represent one or more of these issues. Encourage students to use their imaginations to create colorful and interesting symbols.

The Revolution in Politics The idea of the Revolution was so important and popular among new Mexicans that the most powerful political party called itself the party of the Revolution. Its name changed several times, but the word *revolution* was always part of it. Today, it is called the **Institutional Revolutionary Party** (Partido Revolucionario Institucional, or PRI). This party won every presidential election in Mexico from 1929 until 2000, with power passing peacefully from one president to the next.

Mexico's Government Today ❸

December 1, 2000, was a historic occasion in Mexico. On that day, **Vicente Fox** became Mexico's new president. Fox was the first president in more than 70 years who did not belong to the PRI. Instead, he belonged to the National Action Party (Partido Acción Nacional, or PAN). The election of a president from a party other than the PRI confirmed that Mexico was entering a time of new political possibilities.

National Government Mexico's official name is Estados Unidos Mexicanos, or the United Mexican States. Thirty-one states make up the nation. Mexico is a democracy and a republic. All Mexicans who are 18 or older have the right to vote. The Mexican government has three branches. As in the United States, these branches are the executive, legislative, and judicial.

Geography Skillbuilder Answers

1. Baja California, Sonora, Chihuahua, Coahuila, Nuevo Leon, Tamaulipas
2. Baja California, Baja California Sur

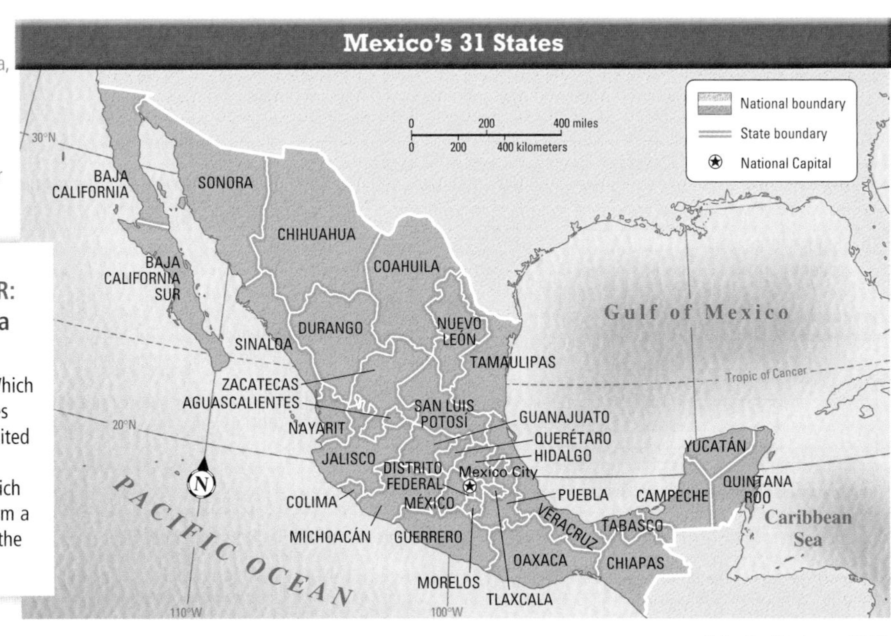

GEOGRAPHY SKILLBUILDER: Interpreting a Map

1. **Location** • Which Mexican states border the United States?
2. **Region** • Which two states form a peninsula on the Pacific coast?

INSTRUCT: Objective ❸

Mexico's Government Today

- What was the historic significance of the election of Vicente Fox in the year 2000? first president not from Institutional Revolutionary Party
- How are the systems of government in the United States and Mexico alike? democratic republic; executive, legislative, and judicial branches; power shared by national and state governments; states have elected governors

FOCUS ON VISUALS

Interpreting the Map Direct students' attention to the bodies of water that border Mexico. Ask students to name them.

Answer Pacific Ocean, Caribbean Sea, Gulf of Mexico

Extension Ask students to use a Unit Atlas map to determine how many countries border Mexico.

Activity Options

Differentiating Instruction: Less Proficient Readers

Some students may be confused about the causes of the Revolution. As a group, reread the sections "The Mexican Revolution" and "A Continuing Revolution." Work with students to create a concept web like the one shown to record the causes of the Revolution.

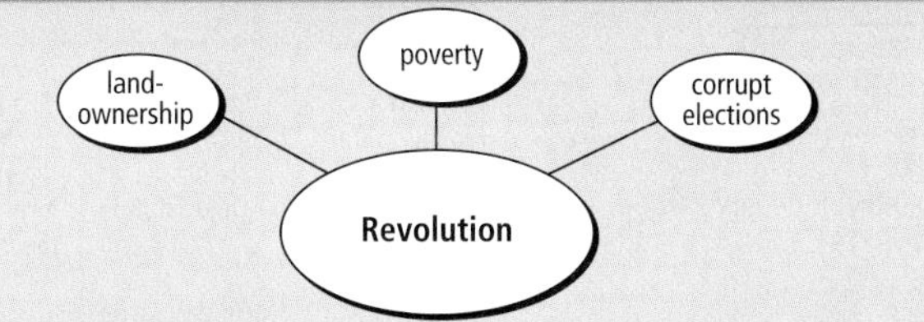

ASSESS & RETEACH

Reading Social Studies Have students continue to take notes on the graphic organizer on page 172.

Formal Assessment
- Section Quiz, p. 94

RETEACHING ACTIVITY

Have students work in teams to prepare questions and answers for a section review. The teams can exchange questions, answer them, and then verify their answers.

In-depth Resources: Unit 3
- Reteaching Activity, p. 20

Access for Students Acquiring English
- Reteaching Activity, p. 52

State Government Like the United States, Mexico has a federal system of government, in which power is shared between the national government and state governments. Voters in each state elect a governor. Each state also has its own legislature that makes laws. However, the national government has some control over the state governments. For example, the president and the national Senate together can remove a state governor from office.

Local Government Most towns and villages depend on money from the national government. Therefore, a local government has less say in its town's affairs than the national government does. However, local governments do provide essential public services, such as maintaining sewer systems and public safety.

Region • The Mexican government issues stamps that celebrate its people and achievements. ▶

SECTION 2 ASSESSMENT

Terms & Names

1. **Explain the significance of:** (a) Benito Juárez (b) Francisco Madero (c) *hacienda* (d) Emiliano Zapata (e) *ejido* (f) Institutional Revolutionary Party (g) Vicente Fox

Using Graphics

2. Use a chart like this one to record details about each level of Mexico's government.

National	State	Local

Main Ideas

3. (a) What changes did reformers such as Benito Juárez help bring about in Mexico?
   (b) Why was the Mexican Revolution fought among many armies?
   (c) How did the Mexican government help farmers gain land of their own?

Critical Thinking

4. **Analyzing Points of View**
   How did different groups view the need for reform and change in Mexico during the years from 1850 to 1940?

Think About
- Juárez and the reformers
- Madero and his supporters
- the owners of the *haciendas*
- Zapata and the small farmers

ACTIVITY -OPTION- Think about a constitution's bill of rights. Write a **speech** explaining why a bill of rights is important to citizens.

Section 2 Assessment

1. Terms & Names
- a. Benito Juárez, p. 180
- b. Francisco Madero, p. 181
- c. *hacienda,* p. 181
- d. Emiliano Zapata, p. 182
- e. *ejido,* p. 182
- f. Institutional Revolutionary Party, p. 183
- g. Vicente Fox, p. 183

2. Using Graphics

National	State	Local
both a democracy and a republic, three branches (executive, legislative, judicial)	31 states; each state elects a governor; each state has its own legislature	towns and villages depend on the national government; local governments provide services

3. Main Idea
- a. Reformers wrote a new constitution granting a bill of rights, the end of slavery, and equality under the law.
- b. Almost every part of Mexico had an army of rebels and reformers with different goals.
- c. The Mexican government broke up many of the *haciendas* and gave the land to small farmers.

4. Critical Thinking

Groups had opposing views concerning equality, land ownership, and the need for reform.

ACTIVITY OPTION

Integrated Assessment
- Rubric for writing a speech

Mexico's Changing Economy

MAIN IDEA

In the mid-1900s, the basis of Mexico's economy changed from farming to industry and tourism.

WHY IT MATTERS NOW

Mexico's successful expansion of its economy has helped the nation to prosper.

TERMS & NAMES
- Carlos Salinas de Gortari
- privatization
- distribution
- *maquiladora*
- nationalize
- PEMEX
- tourism

DATELINE

EXTRA

MEXICO CITY, MEXICO, MARCH 18, 1938

President Lázaro Cárdenas's radio address today established a new course for Mexico's economy. Speaking to the nation, the president announced that foreigners will no longer be allowed to control petroleum companies in Mexico.

The oil industry will now be run by the Mexican government itself. It is hoped that this change will boost both Mexico's economy and its national identity.

Human-Environment Interaction • The Mexican government will now own all oil-producing equipment, such as these oil wells in Veracruz. ▶

Farming in a Time of Change ❶

The 1938 decision that the government would own Mexico's oil industry was made in an effort to expand Mexico's economy. The expansion was necessary because, from ancient times until the mid-1900s, most Mexicans worked in just one industry—farming. Since the 1950s, great numbers of Mexicans have left farming for other kinds of work. However, farming is still important to Mexico's economy.

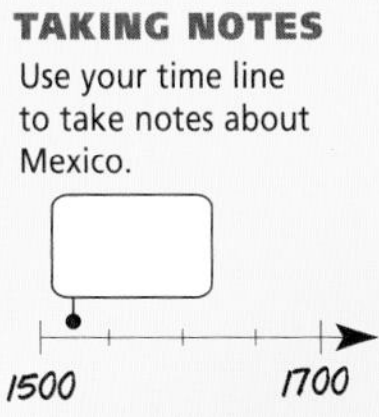

TAKING NOTES
Use your time line to take notes about Mexico.

SECTION OBJECTIVES

1. To describe farming systems in Mexico
2. To identify the steps that the Mexican government has taken to support business
3. To explain the importance of Mexico's natural resources
4. To recognize the importance of tourism

SKILLBUILDER
- Interpreting a Map, p. 188

CRITICAL THINKING
- Comparing, p. 186
- Summarizing, p. 188

FOCUS & MOTIVATE

WARM-UP

Making Inferences Have students read Dateline and discuss these questions.

1. Why might it be important for a country to control its own resources?
2. How might gaining control of the oil industry boost Mexico's economy?

INSTRUCT: Objective ❶

Farming in a Time of Change

- Why did the *ejido* system fail to lift Mexican farmers out of poverty? Farmers did not own land, so they could not use it as loan security.
- How did privatization change farming in Mexico? Farmers could vote to divide their *ejido* into individual farms; families could then sell, rent, or trade their land.

In-depth Resources: Unit 3
- Guided Reading Worksheet, p. 13

Reading Study Guide
(Spanish and English), pp. 52–53

Program Resources

In-depth Resources: Unit 3
- Guided Reading Worksheet, p. 13
- Reteaching Activity, p. 21

Reading Study Guide
(Spanish and English), pp. 52–53

Formal Assessment
- Section Quiz, p. 95

Integrated Assessment
- Rubric for making a flow chart

Outline MapActivities

Access for Students Acquiring English
- Guided Reading Worksheet, p. 47

Technology Resources
classzone.com

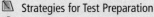

TEST-TAKING RESOURCES
- Strategies for Test Preparation
- Test Practice Transparencies
- Online Test Practice

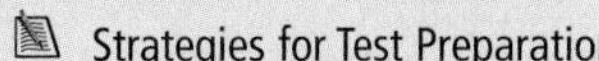

CRITICAL THINKING ACTIVITY

Comparing Ask students to consider the following question: If you were a Mexican farmer, would you rather work on an *ejido* or on a private farm?

Have students work in small groups to brainstorm a list of benefits and drawbacks concerning each system. Then, have each group decide, based on their ideas, which system they would prefer. Have a spokesperson present each group's conclusions to the class.

Class Time 15 minutes

INSTRUCT: Objective 2

The Growth of Business and Industry

- What steps did the Mexican government take to encourage industry in the mid-1900s? built power plants, constructed homes for factory workers, lent money to new businesses, lowered some taxes, privatized many businesses
- What were the effects of government aid on business? encouraged production and new factories, improved distribution
- What is a *maquiladora* and what does it do? a factory along the Mexican/U.S. border usually owned by non-Mexicans; imports parts and exports a finished product
- What goal has NAFTA achieved? the reduction of taxes charged on items traded between Mexico, the United States, and Canada

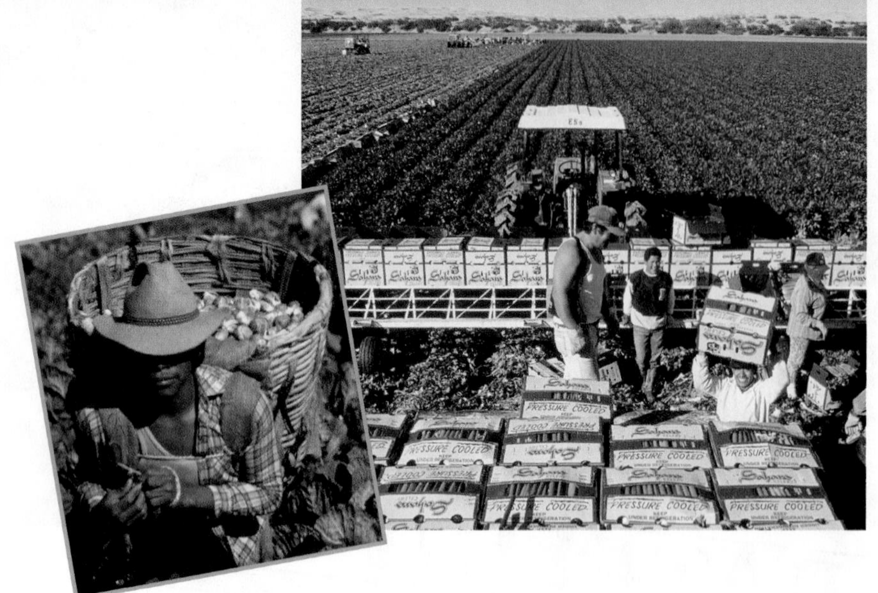

Human-Environment Interaction • At far left, a worker on a small Mexican farm carries crops on his back. At left, a private farm turns out hundreds of crates of produce at once. ◄

The Problems of Farming About one-fourth of Mexican workers are farmers. Many small farmers still work on the *ejidos,* or community farms, that were set up after the Revolution.

Although the *ejido* system gave land to many poor villagers, it did not lift them out of poverty. Farmers could not use the land as security for a bank loan because they did not really own their land—the *ejido* did. Without much money, they could not buy tractors, plows, or fertilizer. They had to continue to farm in the old ways, with hand tools on worn-out soil.

Reading Social Studies

A. Clarifying How does using land on an *ejido* differ from owning one's own land?

Reading Social Studies A. Possible Answer Land in an *ejido* belongs to the community, not to individuals. That meant farmers could not get loans based on land ownership.

A New System In 1991, Mexican president **Carlos Salinas de Gortari** decided it was time to change the *ejido* system. Under Salinas's new laws, farmers could vote to divide their *ejido* into individual farms. Each farm family would have its own piece of land. The family could sell, rent, or trade its land. This process of replacing community ownership with individual, or private, ownership is called **privatization.**

The supporters of privatization hope that private farms will be able to grow more crops. Many of these farms are run like big businesses. Banks lend them money so they can afford to buy and use modern machinery. They can then grow the crops that Mexico sells to other countries, such as cotton, coffee, sugar cane, and strawberries.

Ejidos still make up about half the farmland in Mexico. This is partly because some farmers do not want to divide their *ejidos* into private farms. They worry that privatization might once again put most of the country's land into the hands of a few wealthy people.

Activity Options

Skillbuilder Mini-Lesson: Finding and Summarizing the Main Idea

Explaining the Skill Explain to students that summarizing is restating the main ideas in a logical sequence, using their own words.

Applying the Skill Have students read the part of the text entitled "A New System" and apply the following steps:

1. Look for the main idea in each paragraph. Restate it in your own words. Mexico's president decided to change the *ejido* system. Private farms are run like big businesses. Some farmers worry about privatization.
2. Include key facts such as names and dates. In 1991, President Salinas proposed privatization; *ejidos* still hold about half the farmland.
3. Review your summary and think about whether someone who has not read the section would understand it. Have students reread summaries and delete nonessential details.

Region • In this Mexican factory, workers make car parts. ▲

The Growth of Business and Industry 2

During the mid-1900s, industry became a larger part of Mexico's economy because the Mexican government took steps to encourage its growth. For example, the government built new power plants to supply energy for factories. It also constructed homes for factory workers.

The Mexican government also helped new companies get started by lending them money from the national bank. Sometimes the government lowered taxes on businesses or helped companies pay back money they had borrowed.

Effects of Government Aid The new policies encouraged production. As a result, new factories sprang up that made products such as steel, chemicals, paper, soft drinks, and textiles. The Mexican government also promoted the building of highways, railroads, and airports to aid manufacturers in the distribution of goods. **Distribution** is the process of moving products to their markets.

Privatization of Business In the same way that it was privatizing farms, the Mexican government during the 1990s began to privatize businesses. It raised millions of dollars by selling businesses, such as banks, mines, and steel mills, to private companies. By 2000, only a few key industries—such as the oil industry—were still in government hands.

Foreign-Owned Businesses in Mexico During the 1990s, many of Mexico's fastest-growing factories were situated along its border with the United States. These factories are called *maquiladoras*. In Mexico, a ***maquiladora*** (mah•kee•lah•DAW•rah) is a factory that imports duty-free parts from the United States to make products that it then exports back across the border. The lack of a tax on the parts helps keep operating costs low. Also, most of Mexico's *maquiladoras* are owned by foreigners, who save money because wages are lower in Mexico than in countries such as the United States. Although the wages are not great, *maquiladoras* have provided hundreds of thousands of jobs in Mexico.

Vocabulary

duty-free: free of government-imposed taxes

Na Bolom Since 1951, an organization called Na Bolom (headquarters shown below) has tried to help maintain Mexico's national heritage. Na Bolom works with the Maya living in the southeastern part of Mexico to uphold their traditional ways of life while also developing their economic opportunities. The group also tries to protect the often-threatened resources of the surrounding environment, such as the rain forest.

MORE ABOUT...

Maquiladoras

As of December 2000, there were 3,653 registered *maquiladoras* operating in Mexico within one to 30 miles of the United States border. These factories employed 1.3 million Mexicans. The largest percentage of *maquiladoras* (36%) manufactures electronic components and appliances. The next largest manufacturing industries, at about 20% each, are automobile equipment and clothing.

Citizenship IN ACTION

Native Planet, a volunteer organization similar to Na Bolom, also tries to help maintain the Maya heritage. The goal of people involved with Native Planet is to reinforce cultural pride of indigenous peoples in the world, while also showing them how to survive in the modern era. Native Planet volunteers support local schools in Maya communities, promote health care and humanitarian aid projects, and teach people in these communities how to promote and benefit from ecotourism.

Activity Options

Differentiating Instruction: Gifted and Talented

Block Scheduling

Presenting a Product Challenge advanced students to imagine that they are president of a sporting goods equipment company with a factory in Mexico. Have students prepare a presentation to show a new customer how their product will be distributed in the United States. Students should provide a

- physical description of the product (what it is, how it works, why it will bring in new business)
- plan for packaging the product
- plan for distributing the product, using efficient transportation

Encourage students to use charts and graphs. Provide time for the "president" to make the presentation to a new "customer."

INSTRUCT: Objective 3

Mexico's Rich Resources

- What is Mexico's most important natural resource? petroleum, or oil
- What is the role of PEMEX? runs the oil industry

CRITICAL THINKING ACTIVITY

Summarizing Draw two concept webs on the chalkboard and write the topics "Minerals" and "Oil Industry" in the center circles. Have students copy the webs on their own papers and fill in the surrounding circles with related words and phrases. Invite students to use their completed webs to summarize the information.

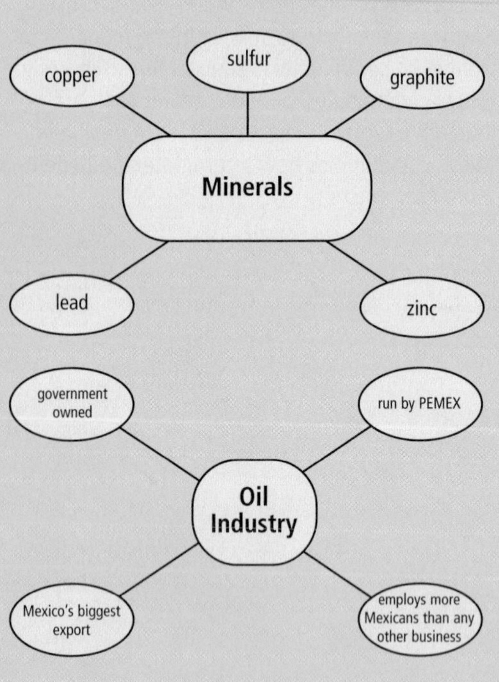

Class Time 10 minutes

NAFTA Giving tax breaks on trade items is also a goal of the North American Free Trade Agreement (NAFTA). NAFTA has reduced taxes on items traded among Mexico, the United States, and Canada. However, some Mexicans are not sure NAFTA is a good idea—they worry that having such close ties to the United States and Canada gives those countries too much influence over Mexico. Despite these concerns, Mexico has nearly doubled its trade with the United States and with Canada since NAFTA was approved in 1992.

Reading Social Studies
B. Possible Answer
Nationalization means to establish government ownership. Privatization means to establish individual, or private, ownership.

Mexico's Rich Resources 3

Just as gold and silver drew the Spanish to Mexico in the 1500s, other minerals that are found there attract worldwide interest today. While Mexico produces more silver than any other country in the world, it also mines lead, zinc, graphite, sulfur, and copper.

The Booming Oil Industry By far the most important of Mexico's natural resources is petroleum, or oil. In 1938, the Mexican government decided to **nationalize,** or establish government control of, the oil industry. Today, when many other businesses have been privatized, the oil industry is still government owned. An agency called **PEMEX** (which stands for Petróleos Mexicanos, or Mexican Petroleum) runs the industry. PEMEX is Mexico's largest and most important company. Oil is Mexico's biggest export, and the United States is its largest buyer.

Reading Social Studies
B. Contrasting How does nationalization differ from privatization?

Geography Skillbuilder Answers
1. southern region, or Gulf Coast
2. cattle, other minerals

GEOGRAPHY SKILLBUILDER: Interpreting a Map
1. **Location** • In what region of Mexico is oil produced?
2. **Region** • What resource is found all over Mexico?

Activity Options

Differentiating Instruction: Students Acquiring English/ESL

Word Usage Point out the word *nationalize* in the section "Mexico's Rich Resources." Students may be familiar with the noun *nation* or the adjective *national* but be confused about the meaning of the verb *nationalize.* Write this word equation to help students determine the meaning of the word.

Then ask: What word means "to bring together under a central government"?

NATIONAL (REFERRING TO A CENTRAL GOVERNMENT) + IZE (TO MAKE) = NATIONALIZE

Tourism Is Big Business 4

Mexico's second-largest business is tourism. **Tourism** is the business of helping people travel on vacations. Tourists come to Mexico to enjoy its warm weather and its sunny beaches. Many people also visit the ancient Native American ruins. Visitors admire art in Mexico City's spectacular museums. They also shop for fine silver jewelry, weavings, wood carvings, and other handicrafts.

Place • Tourists who travel to Santa Catalina Island, off western Mexico, can see rattleless rattlesnakes, which live nowhere else on Earth. ▲

Economic Effects of Tourism A popular tourist place in Mexico is Cancún, on the country's southeastern coast. The story of Cancún shows the effect tourism has had on Mexico's economy.

Until 1970, Cancún was a small Maya village of about 100 people. Its shoreline had white sand beaches and palm trees. The weather was sunny almost every day. In 1970, the Mexican government decided Cancún was an ideal place for a holiday resort. Working together, the government and private businesses built an airport, new roads, and skyscraper hotels.

Today, more than 2.5 million people from all over the world visit Cancún's resorts each year. Because of the tourism boom, the once tiny village has become a city with about 500,000 people.

SECTION 3 ASSESSMENT

Terms & Names

1. Explain the significance of: (a) Carlos Salinas de Gortari (b) privatization (c) distribution (d) *maquiladora* (e) nationalize (f) PEMEX (g) tourism

Using Graphics

2. Use a chart like this one to take notes on four important parts of Mexico's economy.

Mexico's Economy	
Farming	
Industry	
Mining	
Tourism	

Main Ideas

3. (a) How has privatization changed the *ejido* system?
   (b) What part has Mexico's government played in the growth of industry?
   (c) How does tourism contribute to Mexico's economy?

Critical Thinking

4. **Forming and Supporting Opinions**
   Do you think the privatization of farmland in Mexico has been a positive step for Mexico's farmers? Why or why not?

Think About

- the concerns of some *ejido* farmers
- farmers' need to modernize
- Mexico's history of land reform

ACTIVITY -OPTION- Draw a **flow chart** that shows the process by which goods are produced and distributed by a *maquiladora.*

INSTRUCT: Objective 4

Tourism Is Big Business

- What are some tourist attractions in Mexico? beaches, ruins, museums, craft shops
- How did the government and private businesses promote tourism in Cancún? built airport, roads, hotels

MORE ABOUT...

Rattleless Rattlesnakes

A rattlesnake's rattle is made up of rings of thick dry scales that are loosely joined together. These scales make a buzzing sound when the snake shakes its tail. Many harmless snakes are mistaken for rattlers because they vibrate their tails in dry grass or leaves, making a similar sound. The rattlesnake always lifts its tail in the air when it shakes it to scare off predators. The harmless snakes keep their tails down near the grass or leaves.

ASSESS & RETEACH

Reading Social Studies Have students continue to take notes on the graphic organizer on page 172.

Formal Assessment
- Section Quiz, p. 95

RETEACHING ACTIVITY

Have students work in small groups to create outlines of the section. Assign each group one of the following aspects to outline—farming, business, industry, or tourism. Groups can present their outlines to the rest of the class.

In-depth Resources: Unit 3
- Reteaching Activity, p. 21

Access for Students Acquiring English
- Reteaching Activity, p. 53

Section 3 Assessment

1. Terms & Names

a. Carlos Salinas de Gortari, p. 186
b. privatization, p. 186
c. distribution, p. 187
d. *maquiladora,* p. 187
e. nationalize, p. 188
f. PEMEX, p. 188
g. tourism, p. 189

2. Using Graphics

Farming	one-fourth of workers are farmers
Industry	1990s: government privatized businesses
Mining	oil is major resource
Tourism	second-largest business

3. Main Ideas

a. It has put land in the hands of individual farmers.
b. The Mexican government has encouraged the growth of industry by building power plants and aiding manufacturers in the distribution of goods.
c. Tourism is the second-largest industry.

4. Critical Thinking

Answers may vary. Privatization is positive, as it gives ownership to farmers; however, land might fall into the hands of the wealthy.

ACTIVITY OPTION

Integrated Assessment
- Rubric for making a flow chart

Mexico's Culture Today

TERMS & NAMES
Diego Rivera
Frida Kahlo
Octavio Paz
rural
urban
Day of the Dead
fiesta

MAIN IDEA

Modern Mexican culture reflects a blending of Native American and Spanish heritages, as well as many new elements.

WHY IT MATTERS NOW

Cultures change over time, but understanding their histories can help you better understand their characteristics today.

SECTION OBJECTIVES

1. To recognize that Mexico's culture is a blend of many traditions
2. To explain the factors contributing to the growth of cities in Mexico
3. To describe life in the countryside of Mexico
4. To explain the significance of holidays celebrated in Mexico

SKILLBUILDER

- Interpreting a Graph, p. 192

FOCUS & MOTIVATE

WARM-UP

Making Inferences Have students read Dateline, examine the photograph, and discuss the following questions to help them understand a cultural celebration in Mexico.

1. How do people in Mexico celebrate the life of family members who have passed away?
2. In what ways might this celebration help people remember and honor the past?

INSTRUCT: Objective 1

Mexico's Blend of Cultures

- What traditions have blended to form Mexico's culture today? Native American, Spanish, modern Mexican, African American
- What examples of architecture in Mexico City represent the three traditions? Native American: ruins of the Aztec marketplace; Spanish: Catholic church; modern Mexican: skyscrapers

 **In-depth Resources: Unit 3**
- Guided Reading Worksheet, p. 14

 **Reading Study Guide** (Spanish and English), pp. 54–55

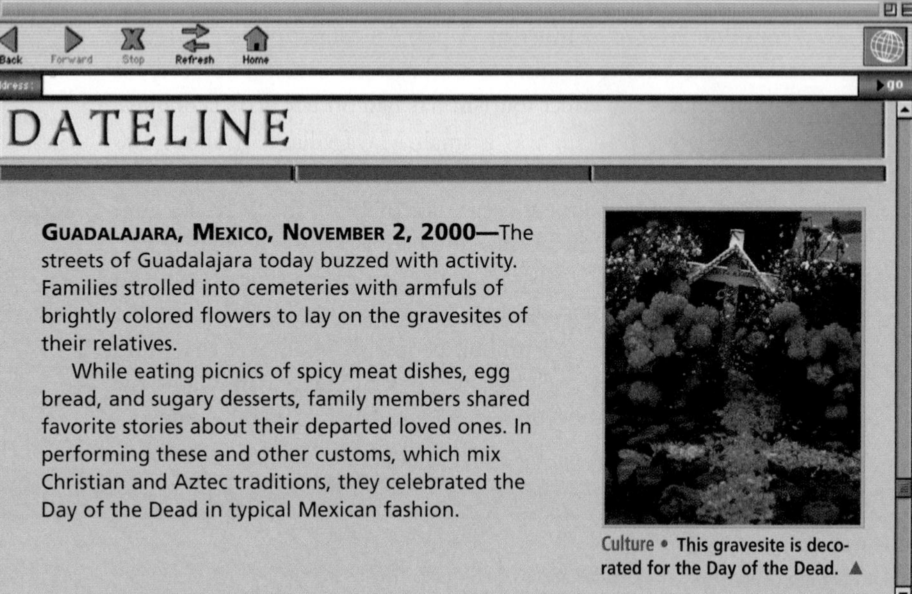

DATELINE

GUADALAJARA, MEXICO, NOVEMBER 2, 2000—The streets of Guadalajara today buzzed with activity. Families strolled into cemeteries with armfuls of brightly colored flowers to lay on the gravesites of their relatives.

While eating picnics of spicy meat dishes, egg bread, and sugary desserts, family members shared favorite stories about their departed loved ones. In performing these and other customs, which mix Christian and Aztec traditions, they celebrated the Day of the Dead in typical Mexican fashion.

Culture • This gravesite is decorated for the Day of the Dead. ▲

Mexico's Blend of Cultures 1

Mexican culture today reflects the same pattern seen in the Day of the Dead celebrations—a mixing of traditions. This mix includes three main cultures. The first two, Native American and Spanish, have long histories. The third, modern Mexican, results from the natural changes Mexicans have gone through over time.

TAKING NOTES
Use your time line to take notes about Mexico.

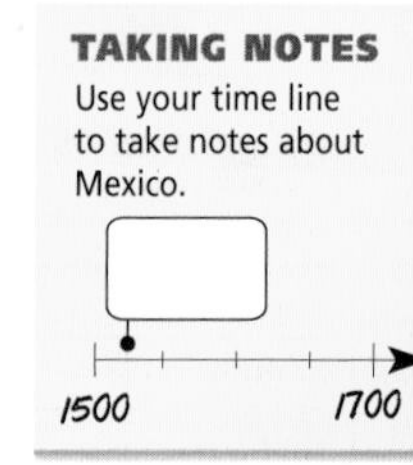

Program Resources

 **In-depth Resources: Unit 3**
- Guided Reading Worksheet, p. 14
- Reteaching Activity, p. 22

Reading Study Guide (Spanish and English), pp. 54–55

 **Formal Assessment**
- Section Quiz, p. 96

 **Integrated Assessment**
- Rubric for writing a journal entry

 **Outline Map Activities**

 **Access for Students Acquiring English**
- Guided Reading Worksheet, p. 48

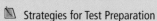

 **Technology Resources classzone.com**

TEST-TAKING RESOURCES

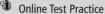

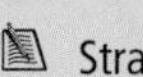 Strategies for Test Preparation
 Test Practice Transparencies
Online Test Practice

Culture in Architecture A plaza near the center of Mexico City symbolizes the traditions that have come together in Mexico. At the center of this Plaza of Three Cultures stand the stone ruins of the Aztec marketplace. Nearby, a Spanish Catholic church borders the plaza. Beyond the plaza, skyscrapers rise against the sky and stand as landmarks of modern Mexico.

Culture in Other Forms of Art Just as Mexico's architecture reveals the multiple layers of its culture, so do other art forms. A series of historical murals decorates the walkways of the National Palace in Mexico City. Painted by **Diego Rivera,** one of Mexico's most celebrated 20th-century artists, these murals depict scenes from the Aztec Empire, New Spain, and the Mexican Revolution. Rivera's wife, **Frida Kahlo** (FREE•duh KAH•loh), is another favorite Mexican painter. After being injured badly in a bus accident, Kahlo painted many famous self-portraits from her bed.

Mexican literature also echoes the country's three cultural traditions. **Octavio Paz,** who won the Nobel Prize in literature in 1990, often writes about the connections between elements of Mexico's past.

BACKGROUND

Each year, a foundation in Sweden gives a Nobel Prize to a world leader in a particular field, such as literature.

Life in the City ❷

About one of every five Mexicans lives in Mexico City. Thousands of people move there each year, hoping to work in factories or attend one of the city's universities or colleges. While Mexico City is by far the largest city in the country, many other cities and towns are growing quickly. Like Mexico City, they offer a blend of opportunities and problems.

Place • **Aztec ruins, Spanish colonial architecture, and modern high-rises share space in the Plaza of Three Cultures.** ▼

MORE ABOUT...
Architecture

Although the Spanish invaders destroyed much of the pre-Columbian art in Mexico, they did teach European building techniques to the Mexican people. The result was a style that combined European and Native American influences. For example, simple mission churches became more ornate and decorative. The Revolution also had an influence on architecture. Many large buildings were constructed and decorated with murals and sculptures reflecting the revolutionary spirit of the times.

MORE ABOUT...
Mexico City

A lot of superlatives are needed to describe Mexico City. Since it was founded as Tenochtitlán in 1325, it has been the longest continually inhabited city in the Western Hemisphere. Mexico City in 2001 had the third largest population in the world. With over 20 million people, it falls behind only Tokyo and New York City. The city also boasts the largest public square in the world, called the *Zócalo.*

INSTRUCT: Objective ❷

Life in the City

- What factors have contributed to the growth of Mexico City? job opportunities, education, and cultural events
- What problems have resulted from this growth? heavy traffic, pollution, large gap between rich and poor citizens

Activity Options

Multiple Learning Styles: Visual/Interpersonal

Class Time One class period

Task Creating small murals that reflect the times in which students live

Purpose To illustrate the life and times of a specific period in the way that Diego Rivera did in Mexico

Supplies Needed

- Poster board
- Poster paint and brushes

Activity Recall with students that Diego Rivera recorded the history of his country in his murals. Have students work with partners to create murals that illustrate life in the United States in the early 21st century. You might want to brainstorm categories that students might consider, such as current events or aspects of popular culture. Provide space for students to display their work.

FOCUS ON VISUALS

Interpreting the Graph Have students compare the percentage of rural population in 1910 with the percentage in 1999. Ask students to consider factors that might have contributed to this decrease. Have them share their ideas about why people leave rural areas to move to the cities.

Possible Responses Families might not be able to earn enough in farming; parents might want better educational opportunities for their children; people might feel isolated in rural areas.

Extension Have students make a graph to predict the percentage of people living in rural areas and cities 20 years from now. Suggest that they consider the effect of overcrowded cities on family life. Have students compare their graphs.

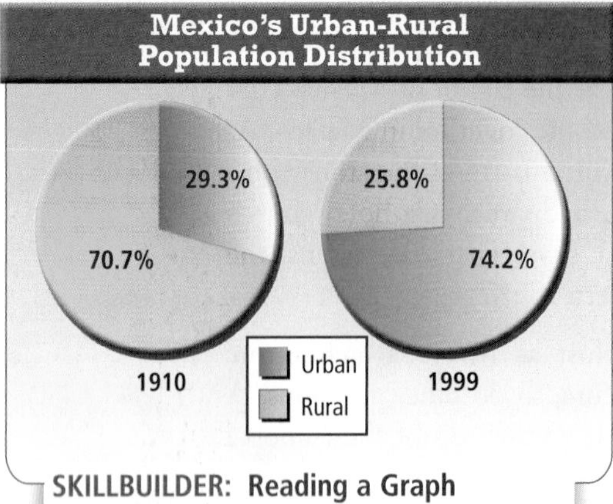

SKILLBUILDER: Reading a Graph

1. What percentage of Mexico's population was urban in 1910? in 1999?
2. What does this tell you about changes that have taken place in Mexican society?

Skillbuilder Answers

1. 29.3% (1910), 74.2% (1999)
2. Urbanization has increased greatly.

A Lively Capital With more than 18 million inhabitants, Mexico City is the second-largest city in the world, after Tokyo, Japan. It is also the cultural center of Mexico. The great marble Palace of Fine Arts houses the national opera, theater, and symphony. It is also home to the famous Ballet Folklórico, a group that performs spectacular dances based on Mexican traditions.

The Cost of Growth Growth has, however, created problems. Streets are jammed with traffic. Car exhaust creates a blanket of smog over the city. The government has responded to the pollution problem in a number of ways. One solution was to free taxi drivers from paying taxes on their vehicles if they drive cars that pollute less.

Activity Options

Interdisciplinary Links: Language Arts/Writing

Class Time One class period

Task Writing a persuasive paragraph urging people in Mexico City not to drive their cars to work

Purpose To identify and present arguments convincing people in Mexico City to help reduce air pollution

Supplies Needed

- Writing paper
- Pencils or pens

Activity Recall with students that the increased population in Mexico City has resulted in an increase in traffic and pollution. Tell students that they are going to write a paragraph persuading citizens of that city to leave their cars at home and find alternative means of transportation. Encourage students to state the argument in the beginning of the paragraph and then support it with details.

Culture • The dancers in Ballet Folklórico wear detailed, bright costumes. ▶

As is true in most large cities, a noticeable gap exists between the lives of Mexico City's rich and poor citizens. While luxurious homes and fine shopping centers line some streets, many people also live in poverty.

Life in the Countryside ❸

In the smaller villages and farming towns, much of Mexico's older way of life still goes on. At the center of each village is a plaza, where people gather to talk and visit with neighbors. Most people speak Spanish, but many also speak Native American languages, such as those of the Aztec and Maya.

Each village sets aside one day a week as market day. People gather in the plaza to buy, sell, or trade food, clothing, and other goods. Farmers come in from surrounding areas, bringing vegetables and handicrafts. The scent of freshly baked *tortillas* and *frijoles,* or beans, fills the air.

Poverty in the Countryside Mexico's **rural** areas—those in the countryside—face the serious problem of poverty. Some homes have only one room and a dirt floor. Farmhouses may lack electricity and running water. These hardships have driven many rural Mexicans to seek jobs in **urban,** or city, settings.

Education is also more limited in the rural areas than in the urban ones. Without education, it is especially hard for people to escape poverty.

Mexican Muralist After the Mexican Revolution, artist Diego Rivera (1886–1957) painted a series of famous murals on the walls of the National Palace in Mexico City. Rivera wanted to use his art to remind Mexicans of the important events in their country's past. This mural shows a typical day in the Aztec capital, Tenochtitlán—now Mexico City.

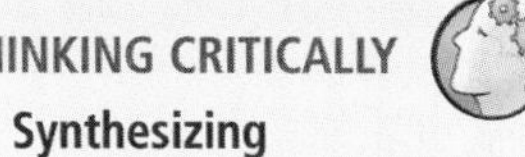

THINKING CRITICALLY

1. **Synthesizing**
   What aspects of Aztec society was Rivera celebrating in this mural?
2. **Recognizing Details**
   Identify three activities depicted in the mural.

For more on Diego Rivera, go to

RESEARCH LINKS CLASSZONE.COM

Thinking Critically Answers

1. This mural shows aspects of Aztec society such as industry, defense, commerce, and communal life.
2. buying and selling; eating; loading and unloading goods; traveling

INSTRUCT: Objective ❸

Life in the Countryside

- In what ways does the plaza serve as the center of life in the countryside? People gather to talk or visit, and to buy, sell, or trade on market day.
- Why is a limited opportunity for education an especially serious problem in rural areas? Without an education, people have a hard time getting jobs that pay well enough for them to escape poverty.

Spotlight on CULTURE

Diego Rivera's murals deal with Mexican life, history, and social problems, and reveal interpretations of these themes. Another of his murals in Mexico City depicts the peaceful life of the Native Americans in contrast to their enslavement by Spanish settlers and eventually to their independence after the Revolution. Ask students why the location of these murals in Mexico City is important for Mexican citizens today.

Activity Options

Differentiating Instruction: Less Proficient Readers

If students have difficulty understanding that there are pros and cons to both city and country life in Mexico, review the text and graph on pages 191–193. Work with students to create a chart like the one shown at the right comparing education, family life, traffic, air quality, and entertainment.

Issues	City	Countryside
Education		
Family Life		
Traffic		
Air Quality		
Entertainment		

INSTRUCT: Objective 4

Holidays

- How is the celebration of Independence Day in Mexico similar to Independence Day celebrations in the United States? fireworks, music
- Why do towns and villages hold fiesta celebrations? to honor a holy person, as big neighborhood parties

Connections to History

While some historians doubt the authenticity of the story of Guadalupe, most Mexicans believe Juan Diego's story. A painting of Mary can be found in most homes. Bus drivers keep a statue of her near the driver's seat, and many stores display her image. Our Lady of Guadalupe is honored by the rich and the poor in Mexico today.

ASSESS & RETEACH

Reading Social Studies Have students complete the time line on page 172.

Formal Assessment
- Section Quiz, p. 96

RETEACHING ACTIVITY

Have students work in groups to summarize one aspect of life in Mexico today—blend of cultures, life in the cities, life in the country, or celebrations. Ask each group to summarize the main points and supporting details for their assigned topic.

In-depth Resources: Unit 3
- Reteaching Activity, p. 22

Access for Students Acquiring English
- Reteaching Activity, p. 54

Connections to History

Our Lady of Guadalupe In 1531, many Mexicans believe, Mary, the mother of Jesus, appeared to Juan Diego, a Native American man, in the Villa de Guadalupe Hidalgo. Mary asked Juan Diego to carry some roses in his cloak to the local bishop. Amazingly, Juan Diego found that roses were blooming even in winter on a harsh, rocky hillside. More astonishingly, when he opened his cloak, Juan Diego saw on it an image of Mary.

Many Mexicans saw these events as miraculous. They began calling Mary "Our Lady of Guadalupe" (shown at right). She was so loved that Father Hidalgo carried a banner with her image to rally support for independence. Mexicans still regard her as their protector.

Holidays 4

September 16 is Mexico's Independence Day. To celebrate, Mexicans reenact Father Hidalgo's call in 1810 to rise up against Spanish rule. Then people watch fireworks, dance, and play music in the streets late into the night.

Another major holiday has a somber name—the **Day of the Dead.** Nevertheless, Mexicans see it as a joyful time. They set aside November 1 and 2 to remember and honor their loved ones who have died. They decorate the graves with candles and flowers. Bakeries sell loaves of bread shaped like bones, and many stores sell small candy skulls. Relatives gather for meals at the cemeteries.

At least once a year, each village or town celebrates a **fiesta.** A fiesta is a holiday with parades, games, and feasts. It usually takes place on a saint's day—a day set aside by the Catholic Church to honor the memory of a holy person. While these days have religious origins, they are also celebrated as big neighborhood parties.

SECTION 4 ASSESSMENT

Terms & Names

1. Explain the significance of: (a) Diego Rivera (b) Frida Kahlo (c) Octavio Paz (d) rural (e) urban (f) Day of the Dead (g) fiesta

Using Graphics

2. Use a chart like this one to list the positive and negative features of living in urban or in rural Mexico.

	Positive Features	Negative Features
Rural Mexico		
Urban Mexico		

Main Ideas

3. (a) What three traditions mix in Mexico's culture today?
(b) Why is Mexico City growing so rapidly? What are the effects of that growth?
(c) Describe some of Mexico's holidays.

Critical Thinking

4. Comparing

Do you see any similarities between the mix of cultural traditions in Mexico and the mix in the United States? What are they?

Think About
- forms of art, music, dance, and architecture
- languages spoken
- holidays celebrated

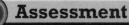

ACTIVITY -OPTION- Imagine that you are traveling through Mexico. Write a **journal entry** describing what you saw and did in its cities and towns.

Section 4 Assessment

1. Terms & Names
- **a.** Diego Rivera, p. 191
- **b.** Frida Kahlo, p. 191
- **c.** Octavio Paz, p. 191
- **d.** rural, p. 193
- **e.** urban, p. 193
- **f.** Day of the Dead, p. 194
- **g.** fiesta, p. 194

2. Using Graphics

	Positive	Negative
Rural Mexico	traditional way of life neighbors	poverty limited education
Urban Mexico	educational opportunities culture	traffic pollution gap between rich and poor

3. Main Ideas
- **a.** Native American, Spanish, and modern Mexican cultures mix.
- **b.** It is growing due to education and employment. Effects include problems with traffic and pollution.
- **c.** Independence Day features fireworks and music. Families honor the dead on the Day of the Dead.

4. Critical Thinking

In both, different languages are spoken, and different cultures influence art and music.

ACTIVITY OPTION

Integrated Assessment
- Rubric for writing a journal entry

194 CHAPTER 7

SKILLBUILDER

Reading a Graph

▶▶ Defining the Skill

Graphs use pictures and symbols, along with words, to show information. There are many different kinds of graphs. Bar graphs, line graphs, and pie graphs are the most common. Bar graphs compare numbers or sets of numbers. The length or height of each bar shows a quantity. It is easy to compare different categories using a bar graph.

▶▶ Applying the Skill

The bar graph to the right shows the number of Mexican state governors who belong to each political party. Use the strategies listed below to help you interpret the graph.

How to Read a Graph

Strategy ❶ Read the title to identify the main idea of the graph.

Strategy ❷ Read the vertical axis (the one that goes up and down) on the left side of the graph. This one shows the number of state governors. Each bar represents the number of Mexican state governors who were members of a particular political party.

Strategy ❸ Read the horizontal axis (the one that runs across the bottom of the graph). This one shows the three political parties of Mexico's governors in 2000.

Strategy ❹ Summarize the information given in each part of the graph. Use the title to help you focus on what information the graph is presenting.

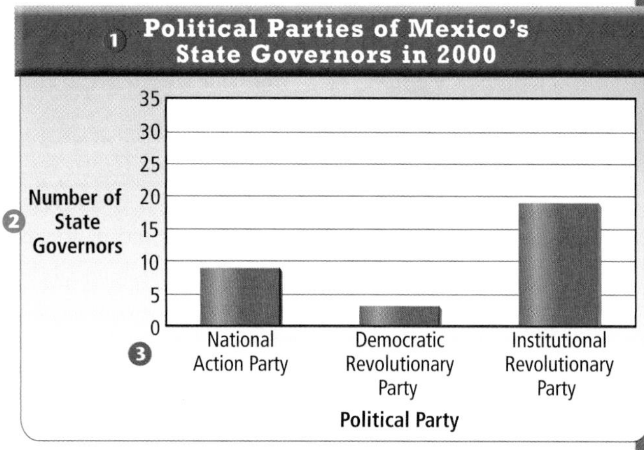

Write a Summary

Writing a summary will help you understand the information in the graph. The paragraph to the right summarizes the information from the bar graph.

❹ In the year 2000, the state governors in Mexico belonged to three different political parties. The majority belonged to the Institutional Revolutionary Party. The next largest group was the National Action Party, and the smallest number of governors belonged to the Democratic Revolutionary Party. This shows that the Institutional Revolutionary Party was probably the most powerful party in Mexico.

▶▶ Practicing the Skill

Turn to page 176 in Chapter 7, Section 1, "The Roots of Modern Mexico." Look at the graph entitled "Population of Mexico in 1810," and write a paragraph summarizing what you learned from it.

SKILLBUILDER

Reading a Graph

Defining the Skill

Display a variety of graphs. Point out that graphs are often used to show clearly what a textbook describes. Knowing how to interpret a graph will help students better understand what they read.

Applying the Skill

How to Read a Graph Guide students through each strategy. Ask volunteers to read the title and the labels on the vertical and horizontal axes. Explain that the title tells the main idea of the graph. Ask a volunteer to tell what this graph shows (political parties of state governors of Mexico as of 2000). Then point out that the numbers along the vertical axis increase by fives. Explain to students that they must estimate the number if a bar is above or below a numbered line.

Write a Summary

Discuss the data on the graph. Ask questions such as, About how many governors belong to the National Action Party? (9) Which political party is represented by the most state governors? (Institutional Revolutionary Party) the fewest governors? (Democratic Revolutionary Party)

Practicing the Skill

Have students turn to the graph on page 176. Explain that the strategy for reading a pie graph is slightly different. Ask a volunteer to read the title aloud. Point out the key and the percents written in each colored section. Ask questions about the graph, such as, What color represents the *peninsulares?* or Which group made up 19.8 percent of the population in 1810? Have students use information from the graph as an outline for their summaries.

In-depth Resources: Unit 3
- Skillbuilder Practice, p. 17

Career Connection: Statistician

Block Scheduling

Encourage students who enjoy analyzing and creating charts and graphs to find out about careers that utilize this skill. For example, students might choose to research the work of statisticians. Tell students that statisticians create charts and graphs, usually using computers, based on the numerical data (statistics) they collect and analyze.

1. Explain that statisticians often work in areas that are of particular interest to them. A person interested in sports might collect statistics for football or basketball. A person interested in population (demographer) collects data about people (where they live, how many of them vote, etc.).
2. Help students find out what aptitudes and education are needed to become a statistician.
3. Ask students to create a large poster showing their findings so they can share their career information with classmates.

OBJECTIVE

Students analyze a Mexican folk tale, drawn from Aztec mythology, that imaginatively explains how music appeared on Earth, stolen from the House of the Sun by the god of the winds, Quetzalcoatl.

FOCUS & MOTIVATE

Making Inferences To help students picture the main character and origins of "How Quetzalcoatl Brought Music to the World," have them study the photograph at the top of page 196 and answer the following questions:

1. How would you describe the character pictured on page 196?
2. What does this artifact indicate about the people who first told this story?

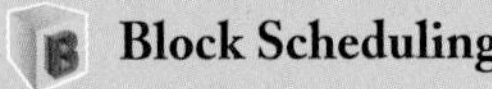

Block Scheduling

MORE ABOUT...

"How Quetzalcoatl Brought Music to the World"

This folk tale is based on an Aztec myth. Myths use supernatural characters and events to answer questions about the world. They express a culture's religious beliefs and social values.

Quetzalcoatl was a primary god in the Aztec pantheon and a main character in Aztec creation myths. He was the god of civilization, learning, and books. *Quetzal* refers to the magnificently plumed quetzal bird of South America, and *coatl* means "snake."

Quetzalcoatl could also appear as a man or as a masked figure. He was associated with the planet Venus.

How Quetzalcoatl Brought Music to the World

THIS MEXICAN FOLK TALE was first written down as a poem in Nahuatl (NAH•WAHT•uhl), the language of the Aztec, in the 1500s. The myth on which it is based may be a thousand years older. It was passed by word of mouth from one storyteller to another for centuries.

When time began, Earth had no music. Brooks did not babble, and birds did not sing. This, however, does not mean that music did not exist. High in the heavens in the Palace of the Sun, musicians of every sort filled the air with dazzling notes.

The god Smoking Mirror[1] envied this music, and he knew that the Sun would never share his prized possession with Earth. Therefore, Smoking Mirror lifted his voice to the air and shouted in every valley and cave for his friend Quetzalcoatl,[2] the feathered serpent who was also Lord of the Winds.

At a distance just beyond the horizon, Quetzalcoatl slumbered in silence. At the sound of Smoking Mirror's call, Quetzalcoatl slowly opened one eye and muttered with a sigh, "What does Smoking Mirror want now? Just as I find a dream, he finds a problem." With that being said, he whirled his serpent body 'round and stormed toward Smoking Mirror's ringing voice.

When Quetzalcoatl reached Smoking Mirror, he asked, "What is the matter now?" Smoking Mirror replied, "This morning, as I walked upon the bright and brilliant Earth that the two of us created, I realized that it is incomplete."

1. A name of Tezcatlipoca (tehs•KAH•tlee•POH•kah), a powerful Aztec god.
2. The quetzal is a bird that lives in the rain forests of southern Mexico and Central America. *Coatl* means "serpent" in Nahuatl.

Activity Options

Differentiating Instruction: Less Proficient Readers

Building Language Skills Group students by threes, matching less proficient readers with more fluent readers. Duplicate copies of the Literature Connection and give each group a copy and two highlighters. Have groups go through the selection and highlight the dialogue, using different colors for the lines of Smoking Mirror and Quetzalcoatl. Tell them to use a pencil or pen to circle and practice words that are unfamiliar or difficult to pronounce. Have them underline descriptions that tell how the characters speak and act. Then have a fluent reader take the part of narrator and other members of the group take the parts of Smoking Mirror and Quetzalcoatl to read the story aloud. Alternatively, have groups retell or act the story, using their own words.

Quetzalcoatl responded doubtfully, "That is not so! There are beautiful creeks and colorful birds in the world! Don't you smell the fragrance of the rose or feel the soft grass beneath your feet?"

"I do," responded Smoking Mirror, "but there is no music! Earth cannot sing its joy when there is no music. That is why I must ask you to go to the Palace of the Sun and return with musicians who are able to spread music across Earth."

"The musicians are faithful servants of the Sun," replied Quetzalcoatl. "They will never leave him." Smoking Mirror remained silent. After a while, the silence began to bother Quetzalcoatl, and he realized the importance of music. So up he went, through the blue smoke of the sky, to the Palace of the Sun.

The Sun saw the Lord of the Winds coming and told his musicians that they must be very quiet, or the feathered serpent would carry them away to the dark and silent Earth. As Quetzalcoatl reached the palace, the glorious music stopped, and the musicians turned their instruments away from the feathered serpent.

But with his command over the wind, Quetzalcoatl brought forth fierce storm clouds that blocked the radiance of the Sun. He then produced his own light as a guide for the frightened musicians. Mistaking the serpent's light for the light of the Sun, the musicians stepped into Quetzalcoatl's embrace. The feathered serpent gently floated the musicians to Earth.

When the musicians saw Smoking Mirror, they knew instantly that they had been tricked. What they saw, however, was not the horrible place described by the Sun. Instead, Earth was full of wondrous colors and activity. It was still within reach of the Sun's warming rays. The musicians then embarked on journeys throughout the world. On their way, they taught the birds to sing, the brooks to babble, the leaves to rustle, and the people to make music of their very own.

Reading THE LITERATURE

What words and phrases give the reader a sense of Quetzalcoatl's personality? How does the author convey Quetzalcoatl's attitude toward the task Smoking Mirror gives him?

Thinking About THE LITERATURE

What does this folk tale suggest about the role of music in Aztec society? Why might this folk tale be popular in Mexico today?

Writing About THE LITERATURE

Write a dialogue involving the musicians, Quetzalcoatl, and the Sun, in which Quetzalcoatl tries to persuade the musicians to come with him to Earth and the Sun urges them to stay. As a class, compare and contrast the arguments each side makes.

Further Reading To learn about more Mexican stories, read *The Tree Is Older Than You Are,* edited by Naomi Shihab Nye.

INSTRUCT

Reading the Literature

Possible Responses

restless, drifted, whispering, pleaded; Quetzalcoatl's anger surged up; "The same old story," he grumbled. . . . "Still, I suppose I must go. . . ."

Thinking About the Literature

Possible Responses

Music was important, a source of pleasure and communication, seen as a gift from the gods. The tale may be popular because it entertains and tells about Mexico's cultural history.

Writing About the Literature

Possible Responses

Student dialogues should include arguments from Quetzalcoatl trying to persuade the musicians to come (e.g., Earth is bright and beautiful; they will see the sun but have freedom; their music will spread) and arguments from the sun urging them to stay (e.g., they don't know what Earth and people are like; they may be unhappy; they can't return).

MORE ABOUT...

Music from Ancient Mexico

Aztec dance troupes perform in the United States. Maya marimba music is available on CD. Instruments with Maya and Inca roots, such as the pan flute and the ocarina, can be heard on recordings.

World Literature

More to Think About

Making Personal Connections Ask students to think about the ways in which music affects their lives. What music do they listen to, and what are its roots? Where in the larger culture do they hear music? How is music important and why? Invite them to give examples of how music connects cultures.

Vocabulary Activity Have students choose words and phrases from the story to compile a poem that conveys the main ideas of the myth. The poem does not have to rhyme.

TERMS & NAMES

1. *criollo,* p. 175
2. *mestizo,* p. 175
3. Father Miguel Hidalgo, p. 176
4. Treaty of Guadalupe Hidalgo, p. 177
5. Benito Juárez, p. 180
6. *hacienda,* p. 181
7. *maquiladora,* p. 187
8. PEMEX, p. 188
9. Diego Rivera, p. 191
10. fiesta, p. 194

REVIEW QUESTIONS

Possible Responses

1. Montezuma's empire consisted of between 5 million and 6 million people. Even with more sophisticated weapons, Cortés needed more people to fight on his side.
2. It occurred at a time when political writers in the United States and Europe publicly stated that people should be free to choose their own governments.
3. Mexicans wanted land to be distributed more equally among the people.
4. For the first time in more than 70 years, the new president was not a member of the PRI.
5. Under the *ejido* system, farmers did not own their land and could not use the land as security for a bank loan to buy machinery or other supplies.
6. Oil is Mexico's most important resource and its biggest export.
7. The rapid growth of Mexico City has caused problems such as traffic, air pollution, and poverty.
8. On September 16, Mexicans celebrate Independence Day.

CHAPTER 7 ASSESSMENT

TERMS & NAMES

Explain the significance of each of the following:

1. *criollo* 2. *mestizo* 3. Father Miguel Hidalgo 4. Treaty of Guadalupe Hidalgo 5. Benito Ju
6. *hacienda* 7. *maquiladora* 8. PEMEX 9. Diego Rivera 10. fiesta

REVIEW QUESTIONS

The Roots of Modern Mexico *(pages 173–178)*

1. Why did Cortés need the help of some Native Americans to defeat Montezuma II?
2. How was Mexico's War of Independence connected to events elsewhere in the world at that time?

Government in Mexico: Revolution and Reform *(pages 179–184)*

3. What concerns did Mexicans in the 1800s and 1900s have about land ownership?
4. What was unique about Mexico's presidential election in 2000?

Mexico's Changing Economy *(pages 185–189)*

5. Why has the *ejido* system been unsuccessful at ending poverty for many villagers?
6. What is the role of oil in Mexico's economy?

Mexico's Culture Today *(pages 190–194)*

7. What are some of the problems caused by Mexico City's rapid growth?
8. What event do Mexicans celebrate every September 16?

CRITICAL THINKING

Sequencing Events

1. Using your completed time line from Reading Social Studies, p. 172, list the points in Mexico's history when the nation's government changed drastically.

Forming and Supporting Opinions

2. Do you think the Mexican government's decision in the 1990s to privatize farming and business was a good one? Explain.

Hypothesizing

3. Modern artists, such as Diego Rivera and Octavio Paz, have used images from both Mexico's past and its present in their works. What do you think this says about the way Mexicans today view their nation's history?

Visual Summary

1 The Roots of Modern Mexico

- From 1810 to 1821, Mexico fought for and gained independence from Spain.

2 Government in Mexico: Revolution and Reform

- The Mexican Revolution (1910–1920) established a new government.
- Today, the government of Mexico is a federal republic.

3 Mexico's Changing Economy

- Farming is still an important part of Mexico's economy, but today its top businesses are oil production and tourism.

4 Mexico's Culture Today

- Mexico's culture combines its Native American and Spanish pasts with new elements.
- Mexico's holidays and arts show the influence of its history.

CRITICAL THINKING: Possible Responses

1. Sequencing Events

1521—Hernán Cortés defeats the Aztecs; Mexico becomes part of the Spanish Empire.

1821—Mexico gains independence from Spain.

1858—Benito Juárez, a reformer, becomes president.

1911—A revolutionary, Francisco Madero, becomes president.

1917—A new constitution is written.

1929—A member of the PRI is elected every term from 1929 to 2000.

2000—Vicente Fox, a member of the National Action Party, is elected president.

2. Forming and Supporting Opinions

Possible Responses Many people believe it is a good decision, because the goal is for farms to be run the way businesses are. Farmers who are able to borrow money can keep up with modern farming methods and produce more exports. However, some people believe that the wealthy might again control most or all of the land.

3. Hypothesizing

Mexico's culture today reflects the value Mexicans have placed on a mix of Native American, Spanish, Mexican, and African influences and cultural elements.

e the map and your knowledge of world cultures and ography to answer questions 1 and 2.

ditional Test Practice, pp. S1–S33

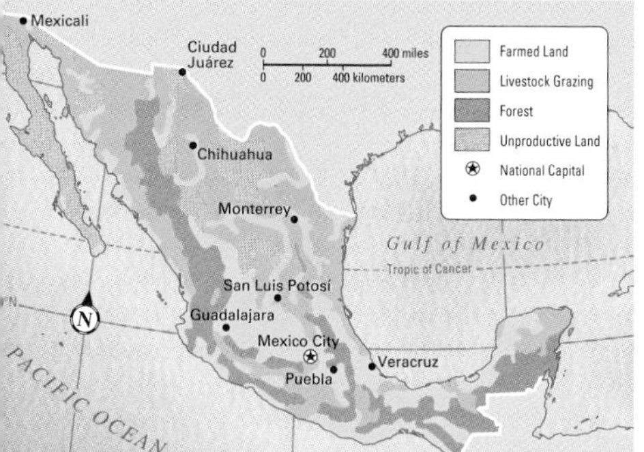

What are the most common uses of land in the northern region of Mexico?

A. Forest, livestock grazing, unproductive
B. Farmed land, forest, livestock grazing
C. Livestock grazing, mining, unproductive
D. Farmed land, livestock grazing, unproductive

Where does more farming occur in Mexico?

A. Along the southern border
B. In the northern half
C. In the southern half
D. On the western peninsula

The following quotation about the painter Diego Rivera is from a news interview. Use the quotation and your knowledge of world cultures and geography to answer question 3.

PRIMARY SOURCE

When people think of Mexico, when people close their eyes and think of Mexico, most of the time, you imagine a world similar to that described by Diego Rivera in his paintings.

GREGORIO LUKE, interview on *NewsHour with Jim Lehrer,* July 15, 1999

3. The passage supports which of the following points of view?

A. Rivera's paintings have little connection to Mexico.
B. Diego Rivera has captured the spirit of Mexico in his paintings.
C. Rivera has painted aspects of Mexico in a realistic style.
D. If you see Diego Rivera's paintings, you don't need to go to Mexico.

STANDARDS-BASED ASSESSMENT

1. **Answer A** is correct because the map shows that a great deal of land in the north is used for livestock grazing, and much is forest or unproductive. Answers B and D are incorrect because little land in the north is farmed; answer C is incorrect because mining is not on the map.
2. **Answer C** is correct because the map shows more farming occurs in the southern half. Answer A is incorrect because the land on the southern border is mostly forest; answer B is incorrect because little in the north is used for farming; answer D is incorrect because the peninsula is mostly unproductive land.
3. **Answer B** is correct because it focuses on Rivera's success in capturing the essence of Mexico; answer A is incorrect because Rivera is famous as a painter of Mexico; answer C is incorrect because the passage says nothing about Rivera's style; answer D is incorrect because seeing Rivera's work doesn't replace a visit to Mexico.

TERNATIVE ASSESSMENT

WRITING ABOUT HISTORY

Mexico City is the second largest city in the world and the cultural center of Mexico. Imagine that you are a travel consultant and create a guide for visitors to Mexico City.

- Use the Internet and the library to learn about the museums and cultural institutions, such as the opera, theater, symphony, and ballet. You might also contact the Mexican tourist bureau for information.
- Design a guide that would inform visitors to the city about these institutions. Write a description of each. Include illustrations and captions.

COOPERATIVE LEARNING

Work with a group of classmates to create a news show about the experience of Mexicans during the Mexican Revolution. Choose a specific year and a specific location as the focus of your news show. Group members can take the roles of anchor, reporter, or interviewee.

INTEGRATED TECHNOLOGY

Doing Internet Research

The varied climates in Mexico affect the way people live and the type of work they do. Use the Internet to research the climate in one region of Mexico.

- Research online and print encyclopedias, atlases, and almanacs to learn about the climate of the region you chose.
- Find out how the climate in that particular region affects the way of life of the people there.
- Include maps or graphs that visually support the information in your report.

For Internet links to support this activity, go to

INTEGRATED TECHNOLOGY

Discuss how students might find information about Mexico's climate by using the Internet. Brainstorm possible key words, Web sites, and search engines. Remind students that climate refers to the kind of weather a place has over a long period of time and that temperature, rainfall, and wind are all factors of climate.

Alternative Assessment

1. Rubric

The guide should

- list several prominent sites and institutions of Mexico City.
- use colorful language and verbal images to entice tourists.
- use correct grammar, spelling, and punctuation, including any necessary diacritical marks for Spanish names.

2. Cooperative Learning

Briefly discuss the style of news broadcasts that students have seen. Divide students into small groups, and have them assign the roles of anchor, reporter, and interviewees. You might want to brainstorm types of questions reporters ask. Discourage students from asking questions that can be answered with yes or no.

PLANNING GUIDE: UNIT 3 ATLAS AND CHAPTER 8

Central America and the Caribbean Islands

	OVERVIEW	COPYMASTERS	INTEGRATED TECHNOLOGY
UNIT ATLAS AND CHAPTER RESOURCES	Students will examine how historical events have affected the political, economic, and cultural development of Central America and the Caribbean Islands.	**In-depth Resources: Unit 3** • Guided Reading Worksheets, pp. 23–26 • Skillbuilder Practice, p. 29 • Unit Atlas Activities, pp. 1–2 • Geography Workshop, pp. 47–48 **Reading Study Guide** (Spanish and English), pp. 58–67 **Outline Map Activities**	• eEdition Plus Online • EasyPlanner Plus Online • eTest Plus Online • eEdition • Power Presentations • EasyPlanner • Electronic Library of Primary Sources • Test Generator • Reading Study Guide • The World's Music • Critical Thinking Transparencies CT15
	KEY IDEAS		
SECTION 1 Establishing Independence pp. 203–207	• Colonization brought many changes to Central American and Caribbean populations. • Colonies fought for independence from European powers. • The United States used military power and influence to effect change in the region.	**In-depth Resources: Unit 3** • Guided Reading Worksheet, p. 23 • Reteaching Activity, p. 31 **Reading Study Guide** (Spanish and English), pp. 58–59	**Critical Thinking Transparencies CT16** **classzone.com** Reading Study Guide
SECTION 2 Building Economies and Cultures pp. 208–213	• Caribbean nations face challenges due to single-product economies. • Central American economies rely on agriculture, tourism, and industry. • Native American, African, and European influences have combined to form Caribbean culture. • Central American culture draws from Native American beliefs and languages.	**In-depth Resources: Unit 3** • Guided Reading Worksheet, p. 24 • Reteaching Activity, p. 32 **Reading Study Guide** (Spanish and English), pp. 60–61	**classzone.com** Reading Study Guide
SECTION 3 Cuba Today pp. 215–220	• Cuba became Communist under Fidel Castro, and allied with the Soviet Union instead of the United States. • The breakup of the Soviet Union devastated Cuba's economy. • Under communism, the Cuban government controls education, health care, food supply, and other basic needs.	**In-depth Resources: Unit 3** • Guided Reading Worksheet, p. 25 • Reteaching Activity, p. 33 **Reading Study Guide** (Spanish and English), pp. 62–63	**classzone.com** Reading Study Guide
SECTION 4 Guatemala Today pp. 221–225	• Jacobo Arbenz Guzmán led a revolution against dictatorial rule. • After years of civil war and military rule, Guatemala is now a democracy. • Agriculture and tourism are important industries in Guatemala. • Urban life differs greatly from rural life in Guatemala.	**In-depth Resources: Unit 3** • Guided Reading Worksheet, p. 26 • Reteaching Activity, p. 34 **Reading Study Guide** (Spanish and English), pp. 64–65	**classzone.com** Reading Study Guide

KEY TO RESOURCES

Audio
CD-ROM
Copymaster
Internet
Overhead Transparency
Pupil's Edition
Teacher's Edition
Video

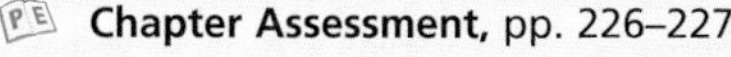

ASSESSMENT OPTIONS

Chapter Assessment, pp. 226–227

Formal Assessment
- Chapter Tests: Forms A, B, C, pp. 113–124

Test Generator

Online Test Practice

Strategies for Test Preparation

Section Assessment, p. 207

Formal Assessment
- Section Quiz, p. 109

Integrated Assessment
- Rubric for writing a biography

Test Generator

Test Practice Transparencies TT21

Section Assessment, p. 213

Formal Assessment
- Section Quiz, p. 110

Integrated Assessment
- Rubric for making a chart

Test Generator

Test Practice Transparencies TT22

Section Assessment, p. 220

Formal Assessment
- Section Quiz, p. 111

Integrated Assessment
- Rubric for making a poster

Test Generator

Test Practice Transparencies TT23

Section Assessment, p. 225

Formal Assessment
- Section Quiz, p. 112

Integrated Assessment
- Rubric for writing a paragraph

Test Generator

Test Practice Transparencies TT24

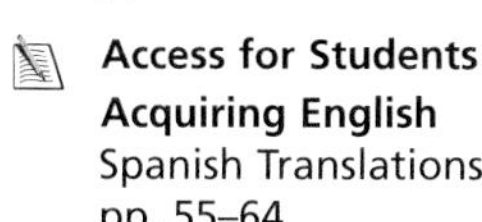

RESOURCES FOR DIFFERENTIATING INSTRUCTION

Students Acquiring English/ESL

Reading Study Guide (Spanish and English), pp. 58–67

Access for Students Acquiring English Spanish Translations, pp. 55–64

TE Activity
- Parts of Speech, p. 211

Modified Lesson Plans for English Learners

Less Proficient Readers

Reading Study Guide (Spanish and English), pp. 58–67

TE Activities
- Sequencing, p. 206
- Summarizing, p. 212
- Taking Notes, p. 223

Gifted and Talented Students

TE Activities
- Researching Exports, p. 210
- Advising the President, p. 222

CROSS-CURRICULAR CONNECTIONS

Literature
Delacre, Lulu. ***Salsa Stories.*** New York: Scholastic Press, 2000. A girl's notebook shares stories and recipes by loved ones from Cuba, Peru, and Guatemala.

Popular Culture
Ancona, George. ***Cuban Kids.*** New York: Marshall Cavendish, 2000. Children at work and play.
Simons, Suzanne. ***Trouble Dolls: A Guatemalan Legend.*** New York: Scholastic, 2000. Traditional Guatemalan good-luck charms.

Language Arts/Literature
Ada, Alma Flor. ***Under the Royal Palms: A Childhood in Cuba.*** New York: Atheneum, 1998. Children's author recalls growing up in Cuba.

Science
Lasky, Kathryn. ***The Most Beautiful Roof in the World: Exploring the Rainforest Canopy.*** San Diego, CA: Harcourt Brace & Co., 1997. Follows scientists studying the canopy in Belize.
Nellis, David. ***Puerto Rico and Virgin Islands Wildlife Viewing Guide.*** Helena, MT: Falcon Publishing, Inc., 1999. Wildlife on the ground, in the air, and under water.

History
Gaines, Ann Graham. ***The Panama Canal in American History.*** Springfield, NJ: Enslow Publishers, 1999. Planning, building, and maintaining the canal.

ENRICHMENT ACTIVITIES

The following activities are especially suitable for classes following block schedules.

Teacher's Edition, pp. 205, 209, 210, 214, 216, 217, 218, 222
Pupil's Edition, pp. 207, 213, 220, 225
Unit Atlas, pp. 142–149
Interdisciplinary Challenge, pp. 236–237
Linking Past and Present, pp. 248–249
Outline Map Activities

INTEGRATED TECHNOLOGY

Go to **classzone.com** for lesson support and activities for Chapter 8.

BLOCK SCHEDULE LESSON PLAN OPTIONS: 90-MINUTE PERIOD

DAY 1

CHAPTER PREVIEW, pp. 200–202
Class Time 20 minutes

- **Discussion** Use the prompts in Before You Read, TE, p. 202 to lead a discussion about what students know and want to learn about Central America and the Caribbean.

SECTION 1, pp. 203–207
Class Time 70 minutes

- **Role-Playing** Organize students into groups. Assign each the role of a Native American, African, or European in 1600, living in the Caribbean and Central America. Have the groups prepare a brief description of the effect of Spanish colonization from the point of view of their assigned role. Have groups share their descriptions, and discuss what their person might predict for his or her descendants.
Class Time 35 minutes
- **Time Line** Work as a class to create a time line showing events leading to the United States' construction of the Panama Canal.
Class Time 15 minutes
- **Peer Teaching** Have students work in pairs to write two questions about the content in From Colonies to Independence PE p. 205. Have pairs exchange questions and answer.
Class Time 20 minutes

DAY 2

SECTION 2, pp. 208–213
Class Time 45 minutes

- **Identifying Details** Assign the Latin America economy to half the class, and the Caribbean economy to the other half. Subdivide each half into working groups. Have each prepare an outline of the past and present economy of their assigned region. Then have students predict the economy's future, based on its past and present. Students can reread Economies of the Caribbean Islands and Economies of Central America PE p. 210. Have the groups present their outlines and discuss their predictions.

SECTION 3, pp. 215–220
Class Time 45 minutes

- **Letter Writing** Have students write a letter to the editor supporting or opposing Fidel Castro's 40 years of rule. Assign pro and con positions randomly. Tell students to support their positions by using specific examples and events from Cuba's fight for independence. When they have finished, discuss as a class.
Class Time 25 minutes
- **Demonstration** Have students reread Cuba's Economy, PE p. 218. Ask for three volunteers to represent the U.S., Cuba, and the Soviet Union. Have them stage a brief scene in front of the class demonstrating the economic relationship over the past 40 years between these countries.
Class Time 20 minutes

DAY 3

SECTION 4, pp. 221–225
Class Time 35 minutes

- **Contrasting** Organize students into pairs, and assign one to be a rich Guatemalan, the other a poor Guatemalan. Have the pairs work together to create a contrasting description, with alternating sentences describing a specific aspect of life. An example: As a wealthy Guatemalan, I live in a city. As a poor Guatemalan, I live in the country. Students can use information from the PE, section 4, and you can also share the information in the More About... in the TE, p. 223. Have the pairs stand and deliver their descriptions by reading their sentences alternately.

CHAPTER 8 REVIEW AND ASSESSMENT, pp. 226–227
Class Time 55 minutes

- **Review** Have students use the charts they created for Reading Social Studies on PE p. 202 to review problems and solutions in the Caribbean and Latin America.
Class Time 20 minutes
- **Assessment** Have students complete the Chapter 8 Assessment.
Class Time 35 minutes

TECHNOLOGY IN THE CLASSROOM

MULTIMEDIA PRESENTATION

With the advent of the World Wide Web and multimedia presentation programs, both of which employ hyperlinks to go from page to page, it has become increasingly common to organize and present information in a nonlinear manner. Students can create this type of presentation in a multimedia software program by designing a primary page that links directly to several other pages that contain details about aspects of the topic introduced on the first page.

ACTIVITY OUTLINE

Objective Students will create a multimedia presentation that highlights the unique characteristics of a few Caribbean nations.

Task Ask students to research the culture, history, and government of several countries. Then have them create multimedia presentations, with hyperlinks from a map on the first page, that provides text and pictures to describe these countries.

Class Time 3–4 class periods

DIRECTIONS

1. Ask students to imagine that their family is planning a trip to the Caribbean.
2. Give each student a blank outline map of the Caribbean (available at the first link at **classzone.com**), and ask them to label the following: Cuba, Puerto Rico, Jamaica, Haiti, Trinidad and Tobago, and the U.S. Virgin Islands.
3. Have students, either individually or in small groups, use library resources and the Web sites at **classzone.com** to gather information about the countries they have mapped. They should take notes to answer these questions:
    - What languages are spoken here?
    - What are some unique things about this country's culture (as food, music, and holidays)?
    - What European country colonized this island or island group?
    - What country controls this island or island group now?
    - What type of government does this island or island group have?

    Students may want to make charts to help organize their notes.
4. Have students use a multimedia presentation program to create pages that contain a large map of the Caribbean. They can scan the maps they have already created or use new ones.
5. Tell students to insert text over the map indicating the location and name of each of the countries they have studied. Then have them create hyperlinks from each country to additional pages in the presentation program.
6. Have students place text and images on each of the pages to show some of the things they have learned about each country. They should be sure to include information about unique features that set each country apart from the others. Ask them to include at least four pieces of information for each country.
7. Have students share their presentations with the class and discuss the differences among the island countries they have studied.

CHAPTER 8 OBJECTIVE

Students will examine how historical events have affected the political, economic, and cultural development of Central America and the Caribbean Islands.

FOCUS ON VISUALS

Interpreting the Photograph Have students look at the photograph of the community market in Guatemala. Ask them to notice the colorful clothing and the items for sale. Point out that this chapter focuses on the history and economy of Central America and the Caribbean Islands. Ask students to explain what part a traditional market like this might play in a country's economy.

Possible Responses A market like this is one layer of the economy—where small farmers and craftspeople bring things to sell. Bigger farms and factories would sell their products in different ways.

Extension Ask students to imagine that they could visit a market such as this one. Have them write a "postcard" home to a friend describing the market.

CRITICAL THINKING ACTIVITY

Hypothesizing Prompt a discussion about how life in Central America today probably contrasts with life during ancient Maya rule. Ask students to think about aspects of life that are better now than they were 1,000 years ago. Then ask them to think of ways in which life may have been better then.

Class Time 10 minutes

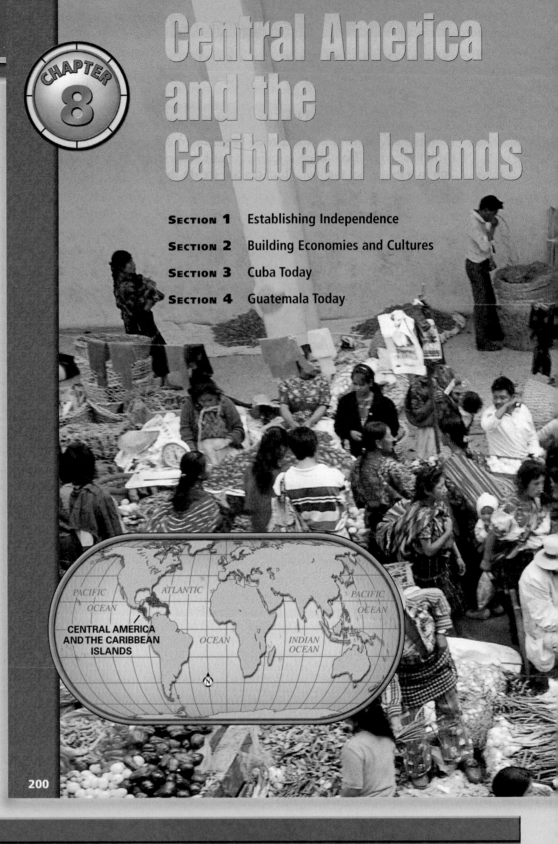

CHAPTER 8

Central America and the Caribbean Islands

PACIFIC OCEAN
ATLANTIC OCEAN
PACIFIC OCEAN
INDIAN OCEAN
CENTRAL AMERICA AND THE CARIBBEAN ISLANDS

200

Recommended Resources

BOOKS FOR THE TEACHER

Coates, Antony, ed. ***Central America: A Natural and Cultural History***. New Haven, CT: Yale University Press, 1998. Geography, flora, fauna.

Ryan, Alan, ed. ***The Reader's Companion to Cuba***. San Diego: Harcourt Brace & Co., 1997. Social life, culture, and attitudes.

Shea, Maureen E. ***Culture and Customs of Guatemala***. Westport, CT: Greenwood Press, 2000. An overview of the culture.

VIDEOS

Spirits of the Rainforest. Discovery Channel Video, 1996. Amazon to rainforest in Peru.

SOFTWARE

Amazon Trail. Learning Company, 1994. 4,000-mile journey through the Latin American rainforest.

INTERNET

For more information about Central America and the Caribbean Islands, visit **classzone.com.**

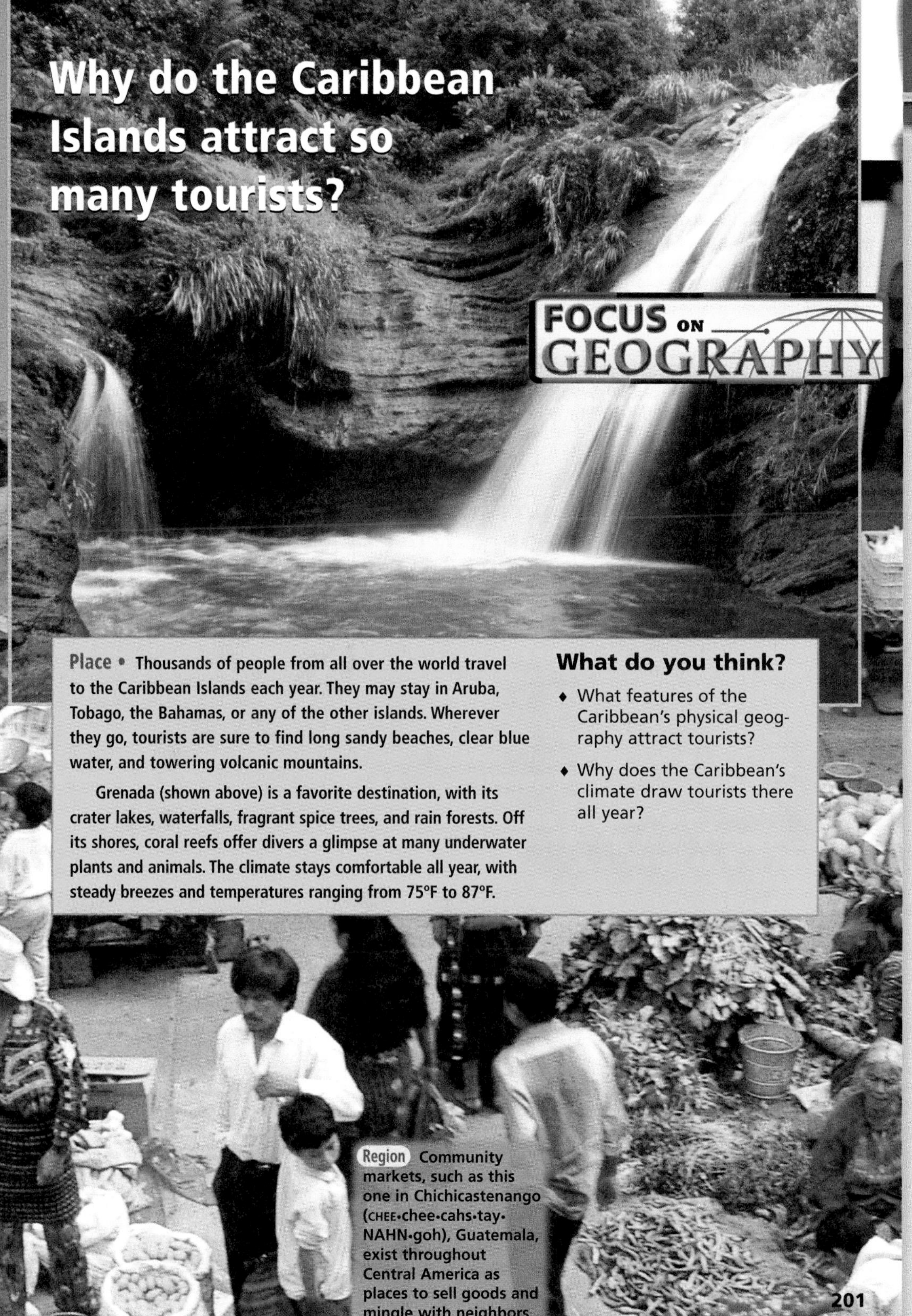

Why do the Caribbean Islands attract so many tourists?

FOCUS ON GEOGRAPHY

Place • Thousands of people from all over the world travel to the Caribbean Islands each year. They may stay in Aruba, Tobago, the Bahamas, or any of the other islands. Wherever they go, tourists are sure to find long sandy beaches, clear blue water, and towering volcanic mountains.

Grenada (shown above) is a favorite destination, with its crater lakes, waterfalls, fragrant spice trees, and rain forests. Off its shores, coral reefs offer divers a glimpse at many underwater plants and animals. The climate stays comfortable all year, with steady breezes and temperatures ranging from 75°F to 87°F.

What do you think?

- What features of the Caribbean's physical geography attract tourists?
- Why does the Caribbean's climate draw tourists there all year?

Region Community markets, such as this one in Chichicastenango (CHEE•chee•cahs•tay•NAHN•goh), Guatemala, exist throughout Central America as places to sell goods and mingle with neighbors.

201

FOCUS ON GEOGRAPHY

Objectives

- To help students describe the climate and identify some landforms in the Caribbean Islands
- To help students recognize how climate, landforms, and location affect the tourist industry

What Do You Think?

1. Ask students to describe the different landforms and regions they think exist in the Caribbean. Ask them what they think attracts tourists to each kind of landform.
2. Have students hypothesize why the climate in the Caribbean makes it more attractive to North Americans and Europeans in the winter than in the summer. Then ask them to think about where tourists might come from who would be particularly interested in visiting the Caribbean in July and August.

Why do the Caribbean Islands attract so many tourists?

In addition to considering climate and geography, encourage students to consider recreational opportunities and historical aspects that might also attract tourists.

MAKING GEOGRAPHIC CONNECTIONS

Ask students to identify other islands that may be similar to the Caribbean Islands in terms of climate and landforms. Have them locate these islands on a world map or a globe. Why do so many North American tourists choose to visit the Caribbean Islands?

Implementing the National Geography Standards

Standard 1 Use data to create a thematic map

Objective To create a map illustrating different economic activities in Central America and the Caribbean

Class Time 30 minutes

Task Have students use a variety of symbols and colors to create a map that illustrates the locations of different economic activities in Central America and the Caribbean. Have students scan the chapter and create a list of regional economic activities. Students will use the list when plotting economic activities on the map.

Evaluation Students should create a legend that explains what their symbols represent.

BEFORE YOU READ

What Do You Know?

Have students brainstorm a list of countries that are located in Central America or the Caribbean. Then poll students to find out how many of them came from, have relatives who came from, or know someone who came from that region. If any of your students are from a Central American or Caribbean country or have visited there, ask them to describe life there.

What Do You Want to Know?

Suggest that students make a chart with the headings *What I Want to Know* and *What I Learned* in their notebooks. Ask them to fill in the first column. As they read the chapter, have them fill in the second column.

READ AND TAKE NOTES

Reading Strategy: Identifying Problems and Solutions Explain to students that they will be reading about major issues that Central American and Caribbean nations are facing or have faced in the past. Read aloud the issues listed in the first column of the chart. Tell the students to look for these events and terms as they read the chapter and to take notes about the problems surrounding these issues and any solutions that were found.

In-depth Resources: Unit 3
- Guided Reading Worksheets, pp. 23–26

READING SOCIAL STUDIES

BEFORE YOU READ

▸▸ What Do You Know?

Before you read the chapter, think about what you already know about Central America and the Caribbean Islands. You may have heard Caribbean music like reggae. Perhaps you have seen colorful woven cloth from Central America. Do you know where the Panama Canal is? Have you heard of Fidel Castro?

Place • This painting shows people working on a Caribbean sugar cane plantation. ▲

▸▸ What Do You Want to Know?

Decide what else you want to know about Central America and the Caribbean Islands. In your notebook, record what you hope to learn from this chapter.

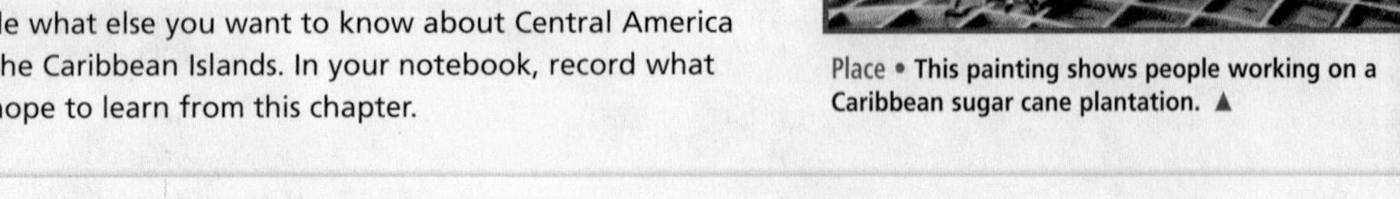

READ AND TAKE NOTES

Reading Strategy: Identifying Problems and Solutions Identifying problems and solutions is a useful strategy for increasing your understanding of what you read. You can make a chart that lists issues. In the chart, identify the problems with those issues and the solutions to the problems to better understand the issues.

- Copy the chart into your notebook.
- As you read the chapter, look for issues related to government and economy.
- Record the problems and solutions involved in each issue on your chart.

Culture • Cubans celebrate a holiday called Carnival. ◄

Issues	Problems	Solutions
Colonization (1492–1800s)	Forced labor	Independence
The Panama Canal (1903–1914)	Acquiring the land	Panama breaks from Colombia
Spanish-American War (1898)	Cuba-Puerto Rico under Spain	U.S. war with Spain
Dictatorships	Unlimited governments	Democracy
Economic Development	Single-product economy	Diversified economy
U.S./Cuba Relations	Cuban Revolution	Diplomacy
Civil War in Guatemala	Land, jobs	Peace agreement (1996)

Teaching Strategy

Reading the Chapter This is a thematic chapter that focuses on how events in the past—including colonization, slavery, and dictatorship—have affected the people, government, economy, and culture of Central America and the Caribbean Islands. Encourage students to think about why the history of this region has been so different from that of the United States.

Integrated Assessment The Chapter Assessment on page 227 describes several activities for integrated assessment. You may wish to have students work on these activities during the course of the chapter and then present them at the end.

Establishing Independence

TERMS & NAMES
West Indies
dependency
mulatto
ladino
dictator

AIN IDEA

entral America and the Caribbean lands have struggled to ecome independent nations ith democratic governments.

WHY IT MATTERS NOW

The quest for democracy continues in Central America and the Caribbean, as it does elsewhere in the world today.

DATELINE

EXTRA

ST. DOMINGUE, CARIBBEAN ISLANDS, 1803

Word has arrived from France that François Dominique Toussaint L'Ouverture died there recently. He was taken to France last year to be imprisoned after leading the slave rebellion here in St. Domingue. The rebellion was so successful that St. Domingue seems just days away from declaring its independence once and for all.

Toussaint was given the nickname L'Ouverture, which means "opening" in French, because he always found an opening in enemy lines. His success in leading the rebellion is staggering when one realizes that his army of former slaves, mostly uneducated, defeated, in turn, the French, the British, and the Spanish.

Culture • Toussaint has become a hero to many. ▲

Central America and the Caribbean ❶

Central America includes seven nations—Belize, Guatemala, Honduras, El Salvador, Nicaragua, Costa Rica, and Panama. Its neighbors—St. Domingue, now called Haiti, and the other islands of the Caribbean—are known as the **West Indies.** They include 13 nations and 11 dependencies. A **dependency** is a place that is governed by or closely connected with another country. For example, Puerto Rico is a dependency of the United States.

TAKING NOTES

Use your chart to take notes about Central America and the Caribbean.

Issues	Problems	Solutions
Colonization		
Panama Canal		

SECTION OBJECTIVES

1. To explain the impact of colonization on Central America and the Caribbean
2. To explain how the countries of the region gained their independence
3. To identify ways in which the United States has shaped the region
4. To describe the persistence of dictatorships

SKILLBUILDER

- Interpreting a Map, p. 204

CRITICAL THINKING

- Recognizing Important Details, p. 204
- Contrasting, p. 205

FOCUS & MOTIVATE

WARM-UP

Making Inferences Have students read Dateline and discuss these questions:

1. Why was the rebellion successful?
2. What effect did the rebellion have on enslaved people in other parts of the region?

INSTRUCT: Objective ❶

Central America and the Caribbean

- Why did European rulers bring Africans to the Caribbean? to replace the Native American workers who had died
- Why did so many Native Americans in Central America survive the encounter with the Spanish? They withdrew to the inland mountains.

In-depth Resources: Unit 3
- Guided Reading Worksheet, p. 23

Reading Study Guide
(Spanish and English), pp. 58–59

Program Resources

In-depth Resources: Unit 3
- Guided Reading Worksheet, p. 23
- Reteaching Activity, p. 31

Reading Study Guide
(Spanish and English), pp. 58–59

Formal Assessment
- Section Quiz, p. 109

Integrated Assessment
- Rubric for writing a short biography

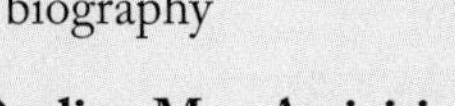

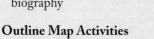

Outline Map Activities

Access for Students Acquiring English
- Guided Reading Worksheet, p. 55

Technology Resources
classzone.com

TEST-TAKING RESOURCES
- Strategies for Test Preparation
- Test Practice Transparencies
- Online Test Practice

FOCUS ON VISUALS

Interpreting the Map Have students point out the features that make this a political map—such as national boundaries, capital cities, and other major urban centers. Ask them to notice the large number of different nations and dependencies in the Caribbean. Direct students' attention to the map key, and ask what symbol stands for the capital of a nation or dependency. What is the capital of Jamaica?

Responses A star inside a circle; Kingston

Extension Ask students to choose one of the nations in the Caribbean as a focus. Have them use an almanac, encyclopedia, or Internet resources to create a short "country profile," including area in square miles, population, climate, and major products.

CRITICAL THINKING ACTIVITY

Recognizing Important Details Have students reread "Central America and the Caribbean." Ask them to explain why the encounter with Europeans proved so deadly for the Native Americans living in the Caribbean Islands, while most of the Native Americans in Central America were able to survive the same encounter.

Class Time 10 minutes

Political Map of Central America and the Caribbean in 2001

National boundary
Capital city
Other city
0 100 200 miles
0 100 200 kilometers
Gulf of Mexico
ATLANTIC OCEAN
Caribbean Sea
PACIFIC OCEAN
SOUTH AMERICA
Tropic of Cancer
90°W 80°W 70°W 20°N 10°N
BAHAMAS
Nassau
Havana
CUBA
DOMINICAN REPUBLIC
Puerto Rico (U.S.)
San Juan
Virgin Is. (U.K.)
Anguilla (U.K.)
ST. KITTS & NEVIS
ANTIGUA & BARBUDA
Guadeloupe (Fr.)
Martinique (Fr.)
BARBADOS
DOMINICA
ST. LUCIA
ST. VINCENT & THE GRENADINES
GRENADA
TRINIDAD & TOBAGO
Virgin Is. (U.S.)
Montserrat (U.K.)
HAITI
Port-au-Prince
Santo Domingo
Kingston
JAMAICA
Netherlands Antilles
Aruba (Neth.)
BELIZE
Belmopan
GUATEMALA
Guatemala City
HONDURAS
Tegucigalpa
NICARAGUA
Managua
San Salvador
EL SALVADOR
COSTA RICA
San José
Panama City
PANAMA

GEOGRAPHY SKILLBUILDER: Reading a Political Map

1. **Location** • Which two nations are located on the same island?
2. **Region** • Which Central American nations border Mexico?

Geography Skillbuilder Answers

1. Haiti and the Dominican Republic island of Hispaniola
2. Belize and Guatemala

Peoples of the Caribbean When Christopher Columbus reached the West Indies in 1492, about 750,000 Native Americans lived there. The Europeans who took over forced Native Americans to work on plantations and in mines. The harsh labor, plus European diseases unknown in the Americas, killed many Native Americans. Others died in battles with the Europeans. Within a few years, nearly all the Native Americans on the islands had died.

Without Native Americans to use for labor, European rulers looked for a new source of workers. By the 1520s, the Spanish began bringing shiploads of enslaved Africans to the West Indies. Other European nations joined in the slave trade. From the 1500s to the mid-1800s, about 10 million enslaved Africans arrived in the West Indies. As a result, many people in the Caribbean today are **mulattos** (mu•LAT•ohz), or people who have African and European ancestry. Few Caribbean people have Native American ancestry.

Reading Social Studies A. Answer
The mountainous te rain gave them a pl to hide and escape Europeans.

BACKGROUND
About 75 million people live in Central America and the Caribbean today.

Region • **Like this man, many Caribbean people are of African descent.** ▼

Central American People In Central America, the situation for the Native Americans was different. When the Spanish arrived in 1501, many Native Americans withdrew to the inland mountains and thus survived what proved to be a deadly encounter with the Spanish elsewhere. Today, one-fifth of Central Americans are Native Americans.

Reading Social Studies

A. Finding Causes How did geography help the Native Americans in Central America survive the arrival of Europeans?

Activity Options

Multiple Learning Styles: Interpersonal

Class Time 25 minutes

Task Writing a newspaper article

Purpose To encourage students to think about what Columbus's landing in the West Indies might have been like for both Europeans and Native Americans who witnessed the event

Supplies Needed
- Paper and pencil

Activity Ask students to imagine that they are reporters covering the landing of Christopher Columbus and his crew in the West Indies in 1492. Have them write a newspaper article vividly describing the landing. Instruct them to try to capture the event from the points of view of both the Europeans and the Native Americans. Encourage them to write creative first-person quotations from people who participated in or witnessed the event.

204 CHAPTER 8

The slave ships also arrived in Central America, bringing Africans to the region. Africans from the Caribbean also migrated to the area. As the years passed, the Spanish, Native Americans, and Africans intermarried. Those people with mixed European and Native American ancestry are called ***ladinos*** (luh•DEE•naws), and they make up about two-thirds of Central America's population today.

Region • Central Americans today reflect the region's rich mix of cultures and people. ▲

From Colonies to Independence 2

From the 1500s to the 1800s, European nations ruled Central America and the West Indies as colonies. These differed from the dependencies of today. Usually, a dependency is free to break off its connection to the other country, but a colony must either win its independence or be granted it.

The Lure of Gold and Sugar Spain was the first European country to colonize this region. Soon, however, the French, Dutch, and English set up their own colonies in the islands. While they found the gold they were looking for, they also found another source of wealth—sugar. Soon many of the islands were home to sugar plantations. Most of the workers who grew and cut the sugar cane were enslaved.

Becoming Independent By the 1800s, the people of Central America and the Caribbean Islands began to demand their independence. In 1804, the French colony of St. Domingue became the first nation in the region to win independence.

BACKGROUND

After independence, St. Domingue took a new name—Haiti (HAY•tee).

Other nations soon followed St. Domingue's example. In 1821, Guatemala, Honduras, Costa Rica, and Nicaragua declared independence from Spain. At first they united, but since 1839, they have existed as separate nations.

Reggae After the Caribbean island of Jamaica gained independence from Britain in 1962, musicians such as Bob Marley (shown below) developed a new style of music called reggae (REHG•ay). Today, reggae is popular around the world. It is known for its political messages.

With their lyrics, the early reggae singers commented on social and economic problems. Reggae is enjoyed both for pleasure and for its ability to let the poor and the oppressed make their voices heard. Music has played a similar role in many cultures throughout history.

Relations with the United States 3

The United States of America is the largest neighbor of Central America and the West Indies. U.S. policies have long played a part in shaping the region's history.

CRITICAL THINKING ACTIVITY

Contrasting Ask students to explain the difference between the ancestry of mulattos and the ancestry of *ladinos*. Then pose questions such as this: If a Caribbean person had an African mother and a European father, would that person be a mulatto or a *ladino?* Then ask students to describe the possible lineage of a *ladino*.

Class Time 10 minutes

INSTRUCT: Objective 2

From Colonies to Independence

- Which European nations established colonies in Central America and the West Indies? Spain, France, Netherlands, England
- What sources of wealth did European nations discover in the West Indies? sugar and gold
- Which was the first nation in the region to gain independence? What is that nation called today? the French colony of St. Domingue; Haiti

The WORLD'S HERITAGE

Bob Marley, a native of Jamaica, introduced reggae to millions of people all over the world. He was more than just a musician, however. For his devotion to world peace, Marley received a medal from the United Nations. When he died of cancer at 36, his funeral in Jamaica was attended by thousands. Decades after his death, his legend lives on through numerous Web sites devoted to his life and work.

Activity Options

Interdisciplinary Link: World History/Government

Class Time One class period

Task Writing a short report to explain why so many Haitians left Haiti in the 1990s

Purpose To understand Haiti's political and economic problems

Supplies Needed
- Reference sources
- Pencil and paper

Block Scheduling

Activity Provide students with resources to help them find out as much as they can about the political and economic problems that faced Haiti in the 1990s. Ask them to research the percentage of the population that lived in poverty during that decade. Also ask them to identify major political events that led Haitians to flee their country, often risking their lives to do so. Have students write short reports to share with the class.

INSTRUCT: Objective 3

Relations with the United States

- How did the United States gain control of the land around the Panama Canal? It urged the people living in the area to break away from Colombia, form their own country, and then sell the land to the United States.
- Why did the United States declare war on Spain in the Spanish-American war? to help the people of Cuba and Puerto Rico gain freedom from Spain and to protect U.S. business interests on the islands
- What were the results of the Spanish-American War? Puerto Rico became a U.S. dependency; Cuba became an independent country.

FOCUS ON VISUALS

Interpreting the Map Have students relate this map of Panama and the Panama Canal to the larger regional map of Central America on page 204. Have them notice the countries that border Panama as well as the route of the canal itself. Ask why the Panama Canal is located where it is.

Possible Responses The canal was built at the narrowest part of the isthmus because it was easier to build there; it cost less to build there; there were several nearby countries that could use it.

Extension Have students use the map to determine the answer to these questions: If a ship is sailing from the Caribbean Sea to the Pacific Ocean, in what direction is it traveling? to the southeast The Caribbean Sea is a part of which larger Ocean? the Atlantic Ocean

The Panama Canal Until the early 1900s, people who wanted to travel from one side of North or South America to the other had two choices. They could make the long, dangerous trip by land or sail all the way around South America. Because the Isthmus of Panama is about 40 miles wide, by 1900, first France, then the United States were eager to build a canal across it to connect the two oceans.

Caribbean Sea
NICARAGUA
Panama Canal
COSTA RICA
Panama City
PANAMA
COLOMBIA
PACIFIC OCEAN
80°W

Human-Environment Interaction • **By using the Panama Canal, ships save 2,000 to 8,000 miles, depending on where they start and end their trips.** ▲

At that time, Panama was part of the South American nation of Colombia. The United States tried to buy from Colombia a strip of land on the isthmus. Colombia refused, and so the United States urged the people of Panama to break away. Soon, Panama did revolt. After establishing its own country in 1903, Panama agreed to lease the United States land to build a canal. The canal opened in 1914 and soon became one of the most important transportation routes in the world. Now ships only had to sail the 50 miles from one end of the canal to the other.

The Spanish-American War Long after Central America broke free of Spanish rule, Spain continued to control Cuba and Puerto Rico. During the late 1800s, the islanders there rebelled many times but were not able to gain independence.

In 1898, the United States declared war on Spain. It did so partly because it wanted to help the people of Cuba and Puerto Rico gain freedom from Spain. However, the United States also wanted to protect the many sugar cane plantations that U.S. businesses owned on the islands. By the end of the war, which lasted less than a year, Spain had lost its last colonies in the Americas. Puerto Rico became a U.S. dependency. Cuba became an independent country, but the U.S. military set up bases there and kept tight control over the country.

Reading Social Studies B. Answers

Spain lost its Caribbean coloni in the Spanish-American War (1 the United States gained influence.

Reading Social Studies

B. Recognizing Effects How did world affairs in the late 1800s affect the Caribbean?

Place • **The United States controlled the Panama Canal and the land around it until December 31, 1999, when control passed to Panama.** ▶

206

Activity Options

Differentiating Instruction: Less Proficient Readers

Sequencing To help students identify and sequence the events that led to the construction of the Panama Canal, work with them to create a flowchart like the one shown here.

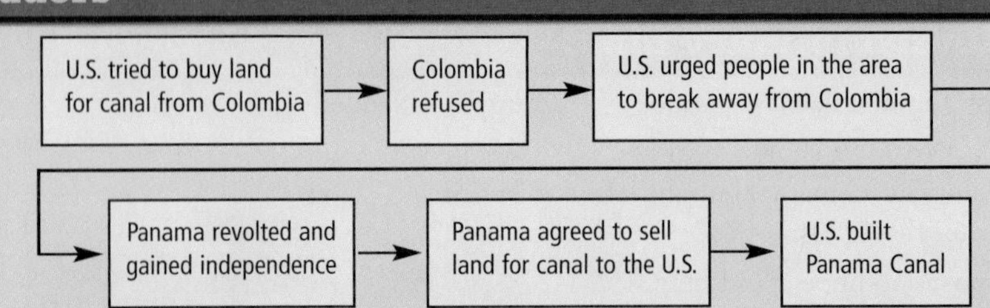
U.S. tried to buy land for canal from Colombia
Colombia refused
U.S. urged people in the area to break away from Colombia
Panama revolted and gained independence
Panama agreed to sell land for canal to the U.S.
U.S. built Panama Canal

Dictatorships and Democracy 4

Independence did not necessarily bring freedom to the people of Central America and the West Indies. Only Costa Rica has been a democracy since the beginning of the 20th century.

Dictatorships Nearly all the other countries have spent many years under the rule of dictators. A **dictator** is a person who has complete control over a country's government. At times, dictators ruled Guatemala, Honduras, Panama, El Salvador, Haiti, and the Dominican Republic. Dictators often used violence to grab and keep power.

Culture • Nicaragua's first female president, Violetta Barrios de Chamorro, governed from 1990 to 1996. ▲

Hopes for Democracy Most countries in Central America and the West Indies now have democratically elected governments. The people of these countries removed most of their dictators, one by one. For example, in 1990, Nicaraguans replaced dictatorial rule by electing Violetta Barrios de Chamorro (vee•oh•LE•tuh BA•ree•ohs day chah•MOH•roh) as president. However, elected leaders sometimes refuse to give up power, thus becoming dictators. Sometimes, elections are not run fairly. Nevertheless, freedom is more widespread than it was 50 years ago.

SECTION 1 ASSESSMENT

Terms & Names

1. Explain the significance of: (a) West Indies (b) dependency (c) mulatto (d) *ladino* (e) dictator

Using Graphics

2. Use a chart like this one to note the effects Europeans had on the Native American populations in Central America and the Caribbean.

In Central America	In the Caribbean

Main Ideas

3. (a) Why are there more Native Americans in Central America today than in the Caribbean?
(b) Describe the difference between dependencies and colonies.
(c) List two reasons why the United States declared war on Spain.

Critical Thinking

4. **Identifying Problems**
Why have some countries in Central America and the Caribbean had trouble establishing democracies?

Think About
- the region's history of being controlled by other countries and dictatorship
- the powers and resources of dictators

ACTIVITY -OPTION- Imagine you lived on the island that became Haiti in 1804. Write a short **biography** of your island's hero, François Dominique Toussaint L'Ouverture.

INSTRUCT: Objective 4

Dictatorships and Democracy

- What is a dictator? a person with complete control over a country's government
- How do elected leaders become dictators? They refuse to give up control.
- What changes are occurring in Central American governments? More leaders are elected democratically, although problems with free elections continue.

ASSESS & RETEACH

Reading Social Studies Have students complete the first four rows of the chart on page 202.

Formal Assessment
- Section Quiz, p. 109

RETEACHING ACTIVITY

Arrange students into groups of four. Have each student in the group write a summary of one of the four headings in the section. Then have students read their summaries aloud to their group and answer two questions from other group members.

In-depth Resources: Unit 3
- Reteaching Activity, p. 31

Access for Students Acquiring English
- Reteaching Activity, p. 61

Section 1 Assessment

1. Terms & Names

a. West Indies, p. 203
b. dependency, p. 203
c. mulatto, p. 204
d. *ladino*, p. 205
e. dictator, p. 207

2. Using Graphics

In Central America	In the Caribbean
Native Americans survived by retreating inland to the mountains	Native Americans killed by harsh labor and disease
Created a population of *ladinos*	Created a population of mulattos

3. Main Ideas

a. Caribbean, died of disease or were killed. Central America, survived by going to mountains.
b. A colony must win or be granted independence. A dependency usually is free to choose.
c. to help the people of Cuba and Puerto Rico gain independence, to protect U.S. business interests

4. Critical Thinking

Possible Responses countries under rule of dictators for a long time; people had little experience with democratic practices

ACTIVITY OPTION

Integrated Assessment
- Rubric for writing a short biography

SECTION 2 Building Economies and Cultures

TERMS & NA
sugar cane
single-produc economy
diversify

SECTION OBJECTIVES

1. To identify economic challenges faced by Caribbean nations
2. To explain how Central American countries are attempting to diversify their economies
3. To describe influences that have affected languages, religions, and music of the Caribbean
4. To explain the formation of a shared Central American culture

SKILLBUILDER
- Interpreting a Map, p. 210

CRITICAL THINKING
- Recognizing Important Details, p. 209
- Comparing, p. 212

FOCUS & MOTIVATE

WARM-UP

Making Inferences Have students read Dateline and discuss these questions:

1. How might the loss of the banana crop affect the people of Honduras?
2. Why might a hurricane be more devastating to Honduras than to the United States?

INSTRUCT: Objective 1

The Economies of the Caribbean Islands

- What was the Caribbean Islands' main industry in the colonial period? growing sugar cane
- What is a single-product economy? one that depends on one product for jobs and income

In-depth Resources: Unit 3
- Guided Reading Worksheet, p. 24

Reading Study Guide (Spanish and English), pp. 60–61

MAIN IDEA

The economies and cultures of this region reflect both the colonial past and efforts to modernize.

WHY IT MATTERS NOW

Though small in size, the region exports its culture and products to the United States and beyond.

DATELINE

TEGUCIGALPA, HONDURAS, 1998— A devastating hurricane hit Central America this week, reminding people that live here of the dangers of having an economy based mainly on agriculture. In addition to about 10,000 deaths, Honduras faces the loss of 70 percent of its crops to the hurricane's ferocious winds.

For example, the banana industry, which brings in $200 million for Hondurans each year, was nearly wiped out. Honduran president Carlos Roberto Flores Facusse described the horrific results of Hurricane Mitch: "In 72 hours, we saw what took as much as 50 years to build... destroyed."

Human-Environment Interaction • Many houses have been destroyed by Hurricane Mitch. ▲

The Economies of the Caribbean Islands 1

In the Caribbean, the economies of most islands have depended on growing one or two crops to sell to other countries. However, the islanders have also worked hard to create new businesses as well as new industries.

TAKING NOTES
Use your chart to take about Central America the Caribbean.

Issues	Problems	Solu
Colonization		
Panama Canal		

Program Resources

In-depth Resources: Unit 3
- Guided Reading Worksheet, p. 24
- Reteaching Activity, p. 32

Reading Study Guide (Spanish and English), pp. 60–61

Formal Assessment
- Section Quiz, p. 110

Integrated Assessment
- Rubric for making a chart

Outline Map Activities

Access for Students Acquiring English
- Guided Reading Worksheet, p. 56

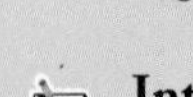

Technology Resources
classzone.com

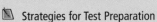

TEST-TAKING RESOURCES
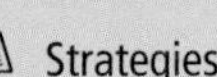
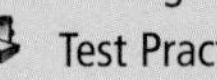
- Strategies for Test Preparation
- Test Practice Transparencies
- Online Test Practice

Human-Environment Interaction • **Many Caribbean islanders spent their days planting and then harvesting sugar cane.** ▲

The Colonial Period For most of the colonial period, the Caribbean Islands focused mainly on one industry—growing **sugar cane.** From the 1600s to the 1800s, most islanders worked on sugar cane plantations. In the early years, the majority of these workers were enslaved. Even after slavery ended by the late 1800s, most workers owned no land of their own. Instead, they planted, tended, and cut sugar cane on plantations owned by the wealthy.

During the colonial period, most of the islands traded only with their ruling countries. Cuba, for example, sold its sugar to Spain and bought goods from Spain in return. But after the colonial period ended, many of the islands traded mostly with the United States.

Reading Social Studies
A. Answer
It forced people to find new ways to make a living, instead of relying on a single product.

Single-Product Economies Sugar was so valuable that plantation owners raised few other crops. Most did not even grow food. Many of the islands had to buy almost all their food from other countries. A country that depends on just one product for almost all its jobs and income has a **single-product economy.**

Reading Social Studies

A. Analyzing Issues Why did production of sugar elsewhere in the world affect the economies of the Caribbean?

Economies Must Change A single-product economy can be unstable—if something happens to that single product, the country's economy will be ruined. By the late 1800s, the sugar business in the West Indies was in trouble. People raising sugar cane in other parts of the world offered the West Indies fierce competition. Steam-powered machines allowed these foreigners to process sugar cane at lower prices. The people of the West Indies had to find new ways to make a living.

The islanders found they needed to diversify their economies. To **diversify** an economy means to invest in a variety of industries. People began to raise other crops, such as pineapples and bananas. Industries, such as textiles, medical supplies, and electronic equipment, also developed in the Caribbean.

FOCUS ON VISUALS

Interpreting the Picture Ask students to point out the different jobs the sugar-cane workers in the picture are doing. Then have them suggest explanations for some details in the painting. Ask what they can tell about the climate and vegetation of the islands. What is the function of the windmill?

Possible Responses Trees and lush vegetation show that the climate is tropical or subtropical. The mill might be used to crush and grind sugar cane into syrup.

Extension Ask students to find out how sugar was grown, harvested, and processed in the 1800s and 1900s, focusing on the work done by people and by machines. They can use encyclopedias and Internet resources for information.

CRITICAL THINKING ACTIVITY

Recognizing Important Details Ask students to identify the numerous ways in which Caribbean islanders began to diversify their economy. You may want to create a concept web on the chalkboard to record their responses.

Class Time 10 minutes

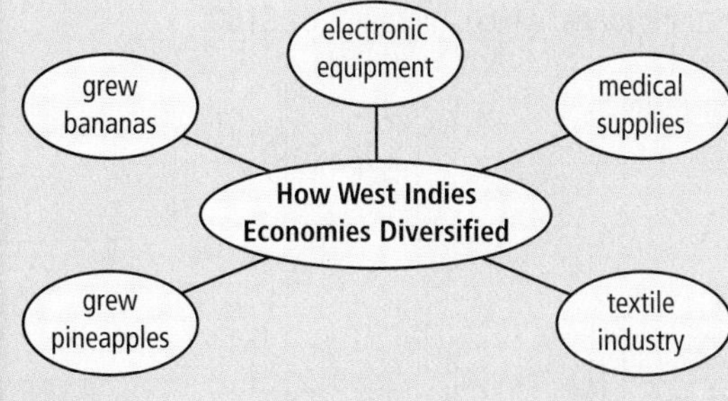

Activity Options

Interdisciplinary Link: Economics/Math

Block Scheduling

Class Time 30 minutes

Task Calculating how the decline in the price of an export affects a country's profits

Purpose To demonstrate how dependence on a single product leaves a country vulnerable to changes in world prices

Supplies Needed

• Paper and pencil

Activity Tell students that the imaginary country of Zabana has a single-product economy based on "Zissle." Zabana exports 200 tons of Zissle each year. Have students work in small groups to create graphs or charts to show Zabana's income during four years when Zissle is exported at four different prices: $50 per ton–1998, $60 per ton–1999, $75 per ton–2000, $100 per ton–2001. Then have them show what happens to Zabana's economy when a disaster reduces Zissle exports, still valued at $100 per ton, by 50 percent in 2002.

INSTRUCT: Objective 2

The Economies of Central America

- What was the first crop that Central American countries produced for export? coffee
- In what ways did United Fruit Company change economies in Central America? It established large banana plantations, creating another important product; it gave the United States a foothold in Central American economies.
- What did countries in the region develop to diversify their economies? tourism and industries
- What do factories in Costa Rica manufacture? machinery, furniture, cloth, medicine

Connections to Citizenship

One way of improving the lives of women in developing countries is to provide them with small loans so that they can start their own small businesses. The first organization in the world to provide what is known as "micro-credit" was the Grameen Bank in Bangladesh. By 2001 Grameen had more than 2.3 million borrowers, 95 percent of whom were women. Almost all of Grameen's customers pay back their loans, which average just $160.

Percentage of Tourism in GDP, 1997–1999

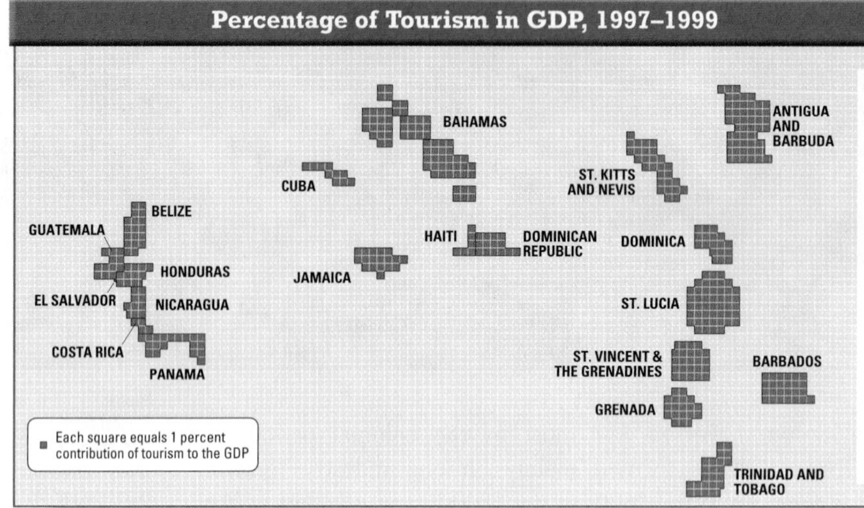

GEOGRAPHY SKILLBUILDER: Interpreting a Map

Human-Environment Interaction • In this cartogram, countries with a greater percentage of GDP coming from tourism are larger than countries with a smaller percentage. In which country does tourism make up the greatest percentage of GDP?

Geography Skillbuilder Answer

1. Antigua, Barbud and St. Lucia are about equal.

María Elena Cuadra Movement

In Central America, poverty and unemployment are widespread. In Nicaragua, many women take jobs in factories, like the one shown here, under poor conditions and for low wages.

Since 1994, the María Elena Cuadra Movement of Working and Unemployed Women (MEC) has been working to improve the position of women at work and in the home. MEC protects workers' rights, trains women in new skills, and lends money to women to start their own small businesses.

Tourism Takes Hold One of the most important industries in the Caribbean Islands is tourism. With warm weather and beautiful beaches, the islands attract tourists from around the world. About 8 million tourists flock there each year. On some of the smaller islands, such as Antigua, tourism is now the major industry.

The Economies of Central America 2

After the Spanish Central American countries became independent in the 1820s, they wanted to increase their trade with other nations. To do so, they needed to develop exports. During the 1800s, several Central American countries began to produce coffee. Soon it became an important export crop for Costa Rica, Guatemala, Honduras, El Salvador, and Nicaragua. In exchange for their exports, Central American countries purchased imports from other parts of the world.

The United States Steps In In the late 1800s, a major new export business got started in Central America. U.S.-based United Fruit Company (UFCO) set up huge banana plantations in the hot, wet lowlands of Central America.

Activity Options

Differentiating Instruction: Gifted and Talented

Block Scheduling

Have students work in a group to research the major exports of all seven Central American countries. Ask them to record their findings on a chart. When the chart is completed, encourage them to use it to answer the following kinds of questions:

- Is there any product that is exported by all seven countries?
- How many of these countries export pineapples?
- What is the major export of Honduras?
- Does any country have a unique export?
- Which country exports the largest number of manufactured goods?

UFCO did a huge amount of business, and bananas became another important Central American product.

Two Crops Are Not Enough The Central American economies grew to depend on bananas and coffee. Whenever the price of these items on the world market fell, Central Americans faced hardship. Like the Caribbean islanders, Central Americans wanted to diversify their economies.

Central American countries have worked to build more factories. Costa Rica, for example, has factories that make machinery, furniture, cloth, and medicine. The countries of the region have also developed tourism as an industry. Visitors arrive by the thousands in Guatemala to see the spectacular ruins of Maya temples. As different businesses grow, Central Americans will not be as dependent on agriculture as they were in the past.

Reading Social Studies

B. Making Inferences What are some factors that would cause the price of a product on the world market to fall?

Reading Social Studies B. Posssible Answers

1. an unusually large crop (oversupply), a drop in world demand, introduction of a similar product

Region • Coffee grows well in the cool highlands of Central America. ▲

Caribbean Cultures ❸

Each country in the Caribbean has its own particular way of life. Native American, African, and European influences blend differently from place to place.

Languages In the Caribbean, people speak a variety of languages. These reflect the area's history. Look at page 145 of the Unit Atlas to see which languages are spoken in which areas.

Religions Roman Catholicism is the most widespread religion in the West Indies. However, many islanders practice religions that have African influences. In Haiti, elements of Catholic and African religious practices combine in the religion known as voodoo.

Strange but TRUE

A Language Lives On When Europeans arrived on the Caribbean island of St. Vincent in 1635, they encountered a startling mystery. The women of St. Vincent spoke one language, and the men spoke another.

Originally, they all spoke the Arawak language. However, when speakers of the Carib language attacked the Arawak-speaking islands, most of the Arawak-speaking men were killed. Only the female Arawak speakers survived.

The Arawakan women continued to speak their language, while the conquering men spoke the Carib language. Today, the language inherited from the Arawakan women is called Garífuna. Like the woman at the right, 30,000 people living on the Caribbean coast of Central America still speak it.

INSTRUCT: Objective ❸

Caribbean Cultures

- What three ethnic influences have combined to form Caribbean cultures? Native American, African, European
- What is the most widespread religion in the West Indies? Catholicism
- What other religions have developed in the Caribbean as a result of African influences? voodoo, Santeria, Shango

In 1635 enslaved Africans escaped after two Spanish slave ships shipwrecked near St. Vincent. The descendants of these people and the Island Caribs (descendants of the Arawak and Carib tribes) are known as the Garifuna. Today, they live in Belize, Guatemala, and Honduras. Their unique culture combines African traditions of music, dance, religious rites, and ceremonies with Native American techniques for raising crops, hunting, and fishing.

The word *Garifuna* means "cassava-eating people." It is probably derived from the word *Kalipuna,* the name of the South American tribe that conquered the Arawak hundreds of years ago.

Activity Options

Differentiating Instruction: Students Acquiring English/ESL

To ensure that students acquiring English understand the vocabulary used in this chapter, discuss the groups of words listed at right. Guide students to understand each word's part of speech. Make sure they understand how the words in each group are related by asking them to take turns using the words in sentences.

- revolt/revolution
- industry/industrial
- export/import
- oppose/opposition
- culture/cultural
- tour/tourist/tourism

CRITICAL THINKING ACTIVITY

Comparing Ask students to compare the different blends of cultures that predominate in the Caribbean and in Central America. Then ask the following questions: Which region is most influenced by African culture? Which is most influenced by Native American culture? In which is Native American influence strong? What historical developments explain these differences?

Class Time 10 minutes

The merengue represents a mixture of African and European culture. In the late 1700s and early 1800s, enslaved Africans watched their French owners dancing minuets in their mansions. At their own parties, they imitated the owners' dances. The enslaved people found these dances boring, however, so they added a drum beat. In other ways, though, the dance was similar to the formal dances of the French. Dancers formed a circle, in which men and women stood across from each other at arm's length.

Thinking Critically Possible Answers

1. They show how closely dancing is tied to Latin American history.
2. Dancing can preserve cultural traditions, songs, and rhythms; it can unite communities and introduce the culture to outsiders.

In Cuba, Yoruba beliefs from Africa combine with Catholic beliefs in the religion called Santeria. In Trinidad, the Shango religion blends Catholic, Baptist, and West African beliefs.

Music Music, too, shows a blending of cultures. From Cuban salsa to Jamaican reggae, much Caribbean music combines African and European styles to make something completely new. The rap music that is popular in the United States also has Caribbean roots. Some of the first rappers to perform in the United States introduced to this country the Jamaican technique of mixing together tracks from different songs.

Central American Cultures 

The countries of Central America share a common history. In their cultures, the Native American and Spanish heritages blend.

Languages Look again at the map on page 145. You can see that in most of Central America, people speak Spanish. Central Americans also speak about 80 Native American languages, nearly half of which are Maya languages.

Merengue Dancing is a popular pastime in Latin America. One of the favorite dances (and the national dance of the Dominican Republic) is the merengue (*shown below*). This dance form originated in the neighboring nations of Haiti and the Dominican Republic. It is known for its unique dance step, called a sliding step, in which the dancer always rests his or her weight on the same foot. People commonly explain this step with one of two legends. One legend says that enslaved people developed the step while chained to one another at the leg. Another legend has it that a Dominican war hero, wounded in the leg during a revolt, designed the step. Whatever the true story behind the dance, the merengue did arise as a folk dance in rural areas. Later it became a favorite of ballroom dancers.

THINKING CRITICALLY 

1. **Drawing Conclusions**
   What do the two legends about the origins of the merengue's sliding step tell you about the significance of dancing in Latin American culture?
2. **Synthesizing**
   What different roles do you think dancing can play in a culture?

For more on the merengue, go to

Activity Options

Differentiating Instruction: Less Proficient Readers

Summarizing Pair a less proficient reader with a more skilled reader. Have them read each main heading, up to this point in the section, and formulate a question about each one. Then have the students take turns reading aloud the information under each heading. Instruct them to stop reading after each one and discuss whether the question they asked was answered. Then have them summarize the information in each of the paragraphs.

For example:

1. Historically, on what was the economy of the Caribbean Islands based?
2. What was the overall problem of economies in Central American countries?
3. What influences blended to form Caribbean cultures?

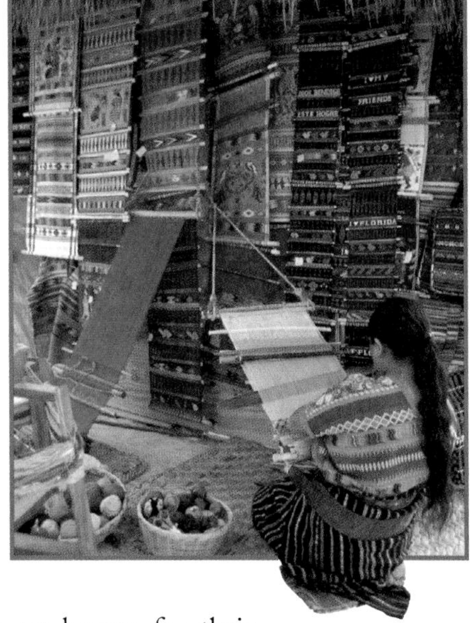

Culture • By wearing the belt around her waist, the weaver stretches the threads tight to weave on them. ▲

Religions Catholicism is the most widespread religion in Central America. In recent years, however, millions of Central Americans have become Protestants and Mormons. Also, some ancient Maya religious beliefs still thrive. An example is the companion spirit. Maya people today believe that when a person is born, so is an animal. That animal, the companion spirit, lives through the same experiences as the person does. A person usually learns about his or her companion spirit in dreams.

Crafts Many Central American towns are known for their crafts, such as weaving, embroidery, pottery, silversmithing, and basketmaking. Many of the styles and methods used originated with ancient Native Americans. For example, many weavers use a backstrap loom—a 2,000-year-old device consisting of threads that are attached to a fixed post or tree on one end and a belt on the other end.

SECTION 2 ASSESSMENT

Terms & Names

1. Explain the significance of: (a) sugar cane (b) single-product economy (c) diversify

Using Graphics

2. Use a chart like this one to check off the languages spoken in each region, country, or island mentioned in the text.

	Guatemala	Haiti
Native American		
Spanish		
French		
English		

Main Ideas

3. (a) List some ways in which sugar production affected the lives of Caribbean people.
(b) How have Central American countries diversified their economies?
(c) How do the religions in Central America and the Caribbean reflect the regions' histories?

Critical Thinking

4. **Forming and Supporting Opinions** How do you think the success of a foreign-owned company affects a country?

Think About
- the importance of UFCO's product to Central American economies
- the jobs created by a successful company
- who receives the company's profits

ACTIVITY -OPTION- A country with a single-product economy risks losing a great deal if that product fails. Make a **chart** of the factors that put the country in this risky position but that it cannot control.

INSTRUCT: Objective 4

Central American Cultures

- Approximately how many Native American languages are spoken in Central America? 80
- How do languages in Central America reflect the region's history? Most Central Americans speak Spanish, but there are also about 80 Native American languages spoken in Central America, mainly by the Maya.
- What religious beliefs are important in Central America? mainly Roman Catholicism, as well as some Protestant groups
- What is the Maya people's belief about companion spirits? When a person is born, an animal is born. That animal lives through the same experiences as the person.

ASSESS & RETEACH

Reading Social Studies Have students fill in the fifth row of the chart on page 202.

Formal Assessment
- Section Quiz, p. 110

RETEACHING ACTIVITY

Assign each one of the four section headings to small groups of students. Ask students to work together to create concept webs that review the material in their assigned sections. Then have a student from each group copy the web onto the board and review the information.

In-depth Resources: Unit 3
- Reteaching Activity, p. 32

Access for Students Acquiring English
- Reteaching Activity, p. 62

Section 2 Assessment

1. Terms & Names

a. sugar cane, p. 209
b. single-product economy, p. 209
c. diversify, p. 209

2. Using Graphics

	Guatemala	Haiti
Native American	✔	
Spanish	✔	
French		✔
English		

3. Main Ideas

a. It increased demand for enslaved workers, created a single-product economy, and made people dependent on imports.
b. planted crops other than sugar, built factories, and developed tourism
c. The religions reflect colonial powers (Catholicism), slavery (African religions), and ancient Native American beliefs.

4. Critical Thinking

Possible Response A foreign-owned company creates jobs for local people; however, it uses natural resources to make profits that go back to the parent company, not into the national economy.

ACTIVITY OPTION

Integrated Assessment
- Rubric for making a chart

SKILLBUILDER

Reading a Political Map

Defining the Skill

Display an encyclopedia entry for your state and show students a political map. Ask students to identify the map features they see. Responses may include boundaries, capital city, other cities, bordering states, highways, airports, and so on. Discuss situations in which students might use a political map, such as locating major cities and estimating the distance between them.

Applying the Skill

How to Read a Political Map Discuss the strategies for reading a political map with students. Point out that the title identifies the region shown on the map, and features such as the key and scale will help them use the map.

Make a Chart

You might want to create a chart like the one shown at right on the chalkboard. Ask students questions such as the following to guide them in reading the map and completing the chart.

- What countries are located in Central America?
- What is the capital of Guatemala?

Practicing the Skill

Have students turn to the political map on page 204. Ask them to make a chart in which they list the countries of the Caribbean, the capital city for each, and other cities. If students need further practice with this skill, have them use a political map of a region of the United States to identify states and capitals.

In-depth Resources: Unit 3
- Skillbuilder Practice, p. 29

SKILLBUILDER

Reading a Political Map

▶▶ Defining the Skill

Political maps show the boundaries of nations and other political areas, such as dependencies. Lines show these boundaries. Often political maps also show capitals and other cities.

▶▶ Applying the Skill

The political map at right shows the nations of Central America. Use the strategies listed below to identify the information shown on the map.

How to Read a Political Map

Strategy ❶ Read the title. It tells you which region's political areas are being represented.

Strategy ❷ Read the key. It tells you what each symbol stands for. This key shows boundaries between nations, national capitals, and other cities.

Strategy ❸ Read the scale. It tells you how many miles or kilometers each inch represents.

1 Political Map of Central America in 2001

Make a Chart

A chart can help you organize information given on maps. The chart below organizes information about the map you just studied.

Central America		
Countries	Capitals	Other Cities
Guatemala	Guatemala City	Quetzaltenango
Belize	Belmopan	Belize City
Honduras	Tegucigalpa	San Pedro Sula
El Salvador	San Salvador	Santa Ana
Nicaragua	Managua	
Costa Rica	San José	
Panama	Panama City	Colón

▶▶ Practicing the Skill

Turn to Chapter 8, Section 1, "Establishing Independence." Study the political map of Central America and the Caribbean on page 204. Make a chart listing the nations of the Caribbean and the capitals that are shown on the map.

214 CHAPTER 8

Career Connection: Cartographer

Block Scheduling

Encourage students who enjoy reading political maps to find out more about the people who create them. Tell them that mapmakers are called cartographers and that they use data from land surveyors, photographs, satellites, and other sources to determine the location and names of places and other features on a political map.

1. Suggest that students look for specific information about how cartographers do their work. Help them explore the impact of technology, particularly the Global Positioning System (GPS) and the Geographic Information Systems (GIS), on the field.
2. Does a person need to know geography well to be a cartographer? Help students find out what aptitudes and education are needed to become a cartographer.
3. Have students organize their findings in an outline or other visual aid.

SECTION 3

Cuba Today

TERMS & NAMES
José Martí
Fidel Castro
Communism
malnutrition
Carnival

AIN IDEA

'ter Cuba became independent, e country was ruled by a series of ctators. Since 1959, revolutionary ader Fidel Castro has ruled e nation.

WHY IT MATTERS NOW

At the turn of the century, Cuba was the only Communist country in the Western Hemisphere.

DATELINE

SIERRA MAESTRA, CUBA, 1956—Word has leaked out that revolutionary leader Fidel Castro has returned to Cuba. Three years ago, he received a 15-year prison sentence for leading an attack on the government. He was released early, a year ago, at which time he fled to Mexico to organize a group of revolutionaries.

Sources say that he and about 80 of his followers have returned to Cuba, though many of them were killed or captured upon arrival. The rest are hiding out in the Sierra Maestra, mountains in southeastern Cuba. People have been speculating about what kind of revolution may develop now that Castro is back in Cuba.

Place • While hiding deep in the Sierra Maestra, Castro is organizing his followers for revolution. ▲

Independence and Revolution ❶

Long after Mexico and Central America gained independence, Cuba, the largest island in the Caribbean, still suffered Spanish rule. In 1895, Cubans led by **José Martí** continued fighting for the nation's independence. Three years later, the Spanish-American War reached the island. By the end of the war, Cuba had gained its independence from Spain. However, the United States maintained great influence over the nation.

TAKING NOTES

Use your chart to take notes about Central America and the Caribbean.

Issues	Problems	Solutions
Colonization		
Panama Canal		

SECTION OBJECTIVES

1. To describe Cuba's history and its relationship with the United States
2. To identify the effects of Communism on Cuba's economy
3. To describe life in Cuba since the revolution

SKILLBUILDER

- Interpreting a Map, p. 219

CRITICAL THINKING

- Hypothesizing, p. 217
- Synthesizing, p. 220

FOCUS & MOTIVATE

WARM-UP

Analyzing Motives Have students read Dateline and answer these questions:

1. Why do you think Fidel Castro led an attack on the Cuban government?
2. Why has he returned to Cuba, and what do you predict will happen?

INSTRUCT: Objective ❶

Independence and Revolution

- How did the Spanish-American War affect Cuba? ended Spanish rule but brought in a strong U.S. influence
- Why did Fidel Castro begin a revolution in the 1950s? as a protest against dictators
- How did the Soviet Union win Cuba as an ally? provided weapons, food, and machinery; helped trade by buying Cuban sugar
- In what way did Castro fail to keep his promise to the people? became a dictator

 **In-depth Resources: Unit 3**
- Guided Reading Worksheet, p. 25

 **Reading Study Guide** (Spanish and English), pp. 62–63

Program Resources

 **In-depth Resources: Unit 3**
- Guided Reading Worksheet, p. 25
- Reteaching Activity, p. 33

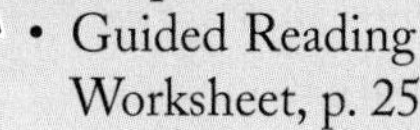

 **Reading Study Guide** (Spanish and English), pp. 62–63

 **Formal Assessment**
- Section Quiz, p. 111

  **Integrated Assessment**
- Rubric for making a poster

 **Outline Map Activities**

 **Access for Students Acquiring English**
- Guided Reading Worksheet, p. 57

 **Technology Resources**
classzone.com

TEST-TAKING RESOURCES

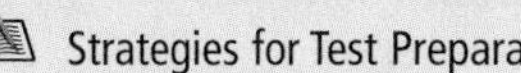
Strategies for Test Preparation

Test Practice Transparencies
Online Test Practice

Connections to Science

Malaria is a serious infectious disease spread by mosquitoes. Symptoms can include fever, chills, headache, muscle ache, and fatigue. To protect themselves against contracting malaria, people take antimalarial drugs and take measures to avoid being bitten by mosquitoes. The risk of contracting malaria exists in rural areas of most countries in Central America and the Dominican Republic. It is not a problem in the rest of the Caribbean.

Yellow fever is a viral disease that takes its name from the jaundice that affects some patients. Symptoms include fever, muscle pain, headache, shivers, loss of appetite, nausea, and vomiting. Central America is not a risk area for yellow fever, although several Caribbean Islands are risk areas.

FOCUS ON VISUALS

Interpreting the Photographs Have students look at the side-by-side photographs on this page. Ask them to compare the photos and describe in their own words the contrasts they observe. Encourage them to discuss whether they think an ordinary Cuban like the man pictured would support Castro or not.

Possible Responses A poor person might have supported Castro's revolution at first, but might now be disillusioned.

Extension Although tourism from the United States is still restricted, Cuba is trying to increase its tourist industry. Have students collect and display ads and travel brochures about Cuban tourism.

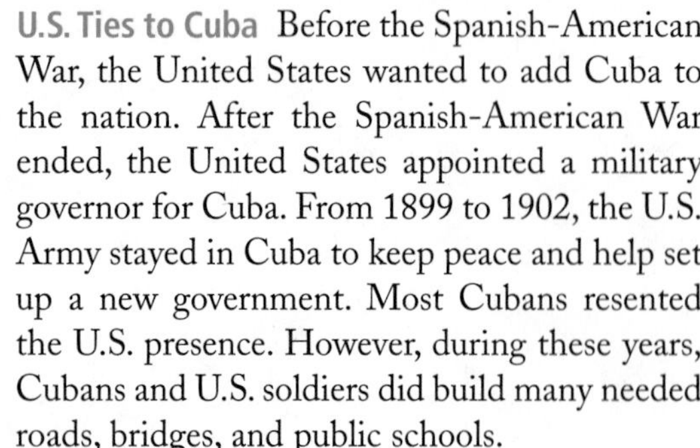

Partners Against Disease

A partnership between the United States and Cuba helped save Cubans from serious diseases, especially malaria and yellow fever. These illnesses killed or weakened many people each year.

In 1900, Dr. Carlos Finlay from Cuba and Dr. Walter Reed from the United States worked together to find the cause of yellow fever. They learned that mosquitoes like the one shown below spread the disease, just as they do malaria. To help put an end to these diseases, the U.S. Army sprayed chemicals that killed off the mosquitoes.

U.S. Ties to Cuba Before the Spanish-American War, the United States wanted to add Cuba to the nation. After the Spanish-American War ended, the United States appointed a military governor for Cuba. From 1899 to 1902, the U.S. Army stayed in Cuba to keep peace and help set up a new government. Most Cubans resented the U.S. presence. However, during these years, Cubans and U.S. soldiers did build many needed roads, bridges, and public schools.

In 1902, the U.S. Army withdrew from Cuba, which then became independent. However, the United States insisted that it still had the right to send soldiers to Cuba any time. The U.S. Navy also kept a large base for its ships at Guantánamo Bay.

Time of Dictatorship After independence, Cuba had a series of leaders. Some were elected, and some took power by force. Most governed as dictators. Cuba's dictators were careful to stay friendly with the United States. They welcomed U.S. businesses and tourists. Havana, Cuba's capital, offered tourists luxury hotels and casinos. Most Cubans, however, remained poor.

Revolution Takes Hold Time and again, Cubans protested against the dictators. In the 1950s, Cubans who were angry at the government found a leader for their cause. He was a young lawyer named **Fidel Castro.** Born in 1927 to a wealthy family, Castro was known for being a dynamic speaker. In college, he developed a deep interest in politics. By late 1956, Castro and a few followers had established headquarters for their revolution in the mountains of southeastern Cuba.

Reading Social Studies

A. U.S. businesses and tourists helped Cuba's economy, which benefited the dictators and other wealthy Cubans

Reading Social Studies

A. Drawing Conclusions If Cubans resented the U.S. presence in Cuba, why did their dictators stay friendly with the United States?

Culture • The two extremes of life in Cuba are apparent in this contrast of one of Havana's fancy hotels with one Cuban's simple dwelling. ▼

Activity Options

Interdisciplinary Link: Science/Health

Class Time One class period

Task Preparing an oral report on the eradication of yellow fever in the Panama Canal Zone in the early twentieth century

Purpose To learn about an important medical advance developed as part of the construction of the Panama Canal

Supplies Needed

- Reference materials
- Internet access, if available

Block Scheduling

Activity Ask pairs of students to research yellow fever to find out how the commission headed by Walter Reed discovered the cause of the disease. Encourage them to explain why it was necessary to eliminate the disease before the Panama Canal could be built. When they have completed their research, ask them to prepare an oral presentation to share with the group.

A few at a time, Cubans began to join Castro's small army. The rebel army won several battles against government troops. As the revolution grew stronger, Cuba's dictator, Fulgencio Batista (fool•HEHN•see•oh buh•TEES•tuh), fled the country on January 1, 1959. On January 8, 1959, Castro and his followers marched triumphantly into Havana. More than half a million Cubans greeted them joyfully. In a speech to the crowds, Castro promised that Cuba would have no more dictators.

The revolution had succeeded. Castro became the new commander-in-chief of Cuba's army. By July 1959, he had taken full control of Cuba's government.

> ***A VOICE FROM CUBA***
>
> We cannot ever become dictators. Those who do not have the people with them must resort to being dictators. We have the love of the people, and because of that love, we will never turn away from our principles.
>
> *Fidel Castro*

Cuba in the Cold War Castro took power in Cuba during the Cold War—a period of conflict between the United States and the Soviet Union. Castro needed the friendship of a powerful country. The Soviet Union was eager to have Cuba as an ally. It proved its interest in the smaller country by engaging in large-scale trade with Cuba as well as providing Cuba with weapons. This attention helped Castro choose to side with the Soviet Union in the Cold War.

Cuba Becomes a Communist Country The Soviet Union practiced an economic and political system known as **Communism**. Under this system, the government plans and controls a country's economy—in effect, the government owns the country's farms, factories, and businesses. Soon, Castro began to adopt Communist policies for Cuba's economy. His government took over the big sugar cane plantations, many of which had been owned by U.S. companies. His government then took over U.S. banks, oil refineries, and other businesses on the island.

Human-Environment Interaction • One of Castro's Communist policies was to take government control of farms, such as this one. ▼

MORE ABOUT...

Castro's Speeches

Fidel Castro has given thousands of speeches in his lifetime, some of them lasting many hours. In 1960 he addressed the United Nations for four and a half hours—the longest speech ever given before the General Assembly. His longest speeches have run more than seven hours, exhausting most of his listeners.

A VOICE FROM CUBA

Read aloud the excerpt from Castro's speech. Have students identify the pledge made by Castro to the people. (not to become a dictator) Then ask why Castro believed he could keep this promise. (He had the "love," or devotion of the people, and therefore did not need to resort to force.)

CRITICAL THINKING ACTIVITY

Hypothesizing Have students reread the excerpt from Castro's speech on page 217. Point out that Castro did become a dictator. Ask students to hypothesize about why he may have changed his mind about the kind of regime he would maintain, and about how Cuba might be different today had Castro stuck to his promise never to become a dictator.

Class Time 15 minutes

Activity Options

Interdisciplinary Link: History

Class Time One class period

Task Advising the President of the United States during the 1962 Cuban Missile Crisis

Purpose To inform students about an important historical event and give them an appreciation of the difficult decisions world leaders must face

Supplies Needed

- Fact sheets describing the Cuban Missile Crisis

Block Scheduling

Activity In advance, prepare fact sheets explaining the main points of the Cuban Missile Crisis. Distribute the fact sheets and discuss the event with students. Make sure they understand how great the risk of war was. Then ask volunteers to play the roles of President Kennedy and his advisors. Have some students advise the president to allow the missiles to remain in Cuba and others urge him to force the Soviet Union to remove the missiles.

INSTRUCT: Objective 2

Cuba's Economy

- What is Cuba's most important product? sugar
- How did the breakup of the Soviet Union affect Cuba's economy? increased prices, decreased food supplies, hurt trade, affected supplies of fuel oil for transportation and production

MORE ABOUT...

Cuba's Economy

Government control over the economy is stronger in Cuba than almost anywhere else in the world. A military enterprise controls tourist services, a general controls the sugar industry, and the government owns factories, hotels, and restaurants.

In return, the United States cut off all trade with Cuba. Cubans could no longer sell their huge sugar crop to the United States. Instead, the Soviet Union bought Cuba's sugar. The Soviet government also sent weapons, farm machinery, food, and money to Cuba. In 1961, Castro declared that he and Cuba were Communist.

Castro as Dictator While the poor usually supported Castro's policies, many wealthier Cubans did not. They were particularly upset when he redistributed land so that no family or farm owned more than a certain amount. Castro also imprisoned people who spoke out against him. As a result, Cubans who opposed Castro began to flee to the United States and other countries. Over the years, hundreds of thousands of people left Cuba.

Place • In recent years, Cubans like these have continued to protest Castro's policies. ▲

Castro has kept a tight hold on power for more than 40 years. Without ever being elected, he has remained head of state. His government has controlled all newspapers and radio and television stations. No one has been allowed to criticize his actions or the government. Despite his 1959 promise, Castro became a dictator.

Reading Social Studies

B. Clarifying What actions did Castro take that went against his 1959 promise?

Reading Social Studies B. Possible Answers

did not allow criticism or opposition, took power as a dictator

Cuba's Economy 2

Since the Cuban revolution, the Cuban economy has changed in many ways. Since the collapse of the Soviet Union in 1991, Cuba has struggled to maintain its Communist way of life without Soviet aid.

Sugar and the Economy Sugar is Cuba's most important product in the world economy. The yearly sugar cane harvest is a key event for Cuba's economy. Once, workers with machetes cut the tough canes by hand. Then, with Soviet help, Castro's government bought huge machines to cut most of the cane. The cane harvest grew to record size.

Vocabulary

machete: large, wide-bladed knife

While the Soviet Union was powerful, it traded oil, grain, and machinery with Cuba for sugar. Most of these products were worth more than sugar, so the Soviet Union was, in large part, supporting Cuba's economy.

Living in Cuba 3

As with the economy, ways of life in Cuba have been greatly affected by Communism. Both education and health care reflect these changes.

Activity Options

Skillbuilder Mini-Lesson: Reading a Graph

Block Scheduling

Explaining the Skill Review with students that when they read a graph, they should first identify the graph's main idea and what each axis measures, then examine the data.

Applying the Skill Display the graph shown. Have students identify the title and the labels on the horizontal and vertical axes. Have them summarize the data.

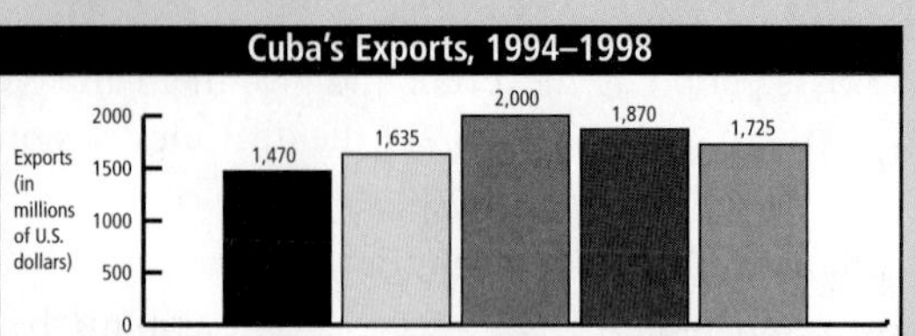

Culture • These Cuban children sing their national anthem together at school. ▲

Education After the revolution, Cuba's government set up many new schools. In the 1960s, teachers and even schoolchildren went into small villages to teach those who could not read or write. Many older people who had never had the chance to go to school learned to read for the first time.

Today, Cuban children must go to public school from ages 6 to 12. They can choose whether to continue their education after that. All schools are free, including college. Besides their academic subjects, students must take classes that teach Communist beliefs.

Health Care Like education, all health care in Cuba is paid for by the government. Cuba's health care system is probably the largest in Latin America. Every small village has a clinic.

However, Cuba had economic problems in the 1990s that affected both education and health care. Lack of fuel for buses and cars prevented some children from getting to school.

Cuba and Its Capital

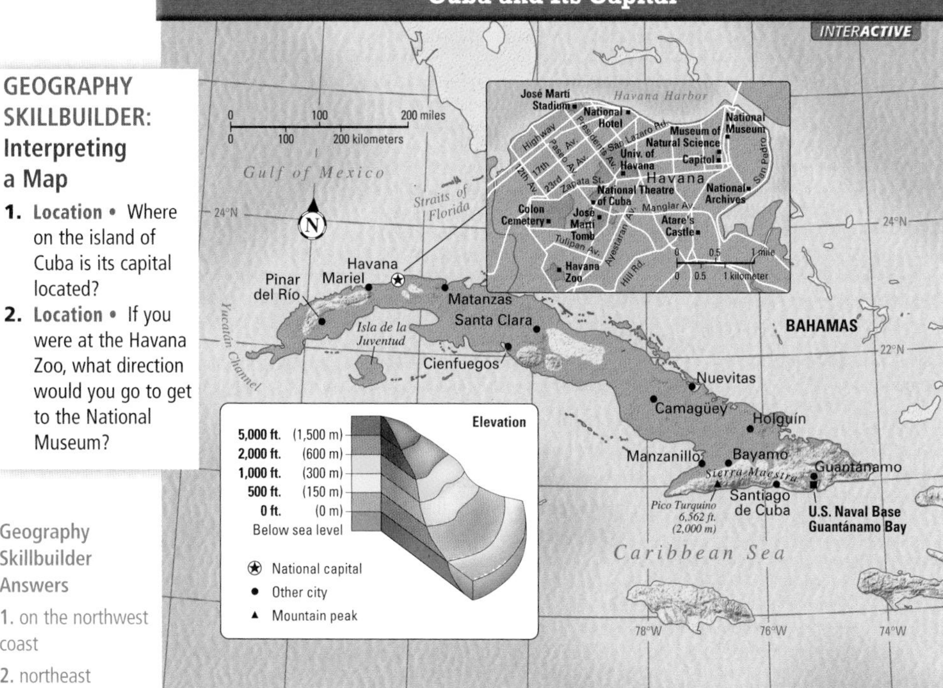

GEOGRAPHY SKILLBUILDER: Interpreting a Map

1. **Location** • Where on the island of Cuba is its capital located?
2. **Location** • If you were at the Havana Zoo, what direction would you go to get to the National Museum?

Geography Skillbuilder Answers

1. on the northwest coast
2. northeast

INSTRUCT: Objective 3

Living in Cuba

- How has Communism affected the education system in Cuba? more people learned to read and write, laws require children ages 6 to 12 to go to school, all schools are free
- Who pays for health care in Cuba? the government
- What kinds of entertainment do Cubans enjoy? music; sports, especially baseball
- What is the significance of Carnival? celebrates the end of sugar harvest

FOCUS ON VISUALS

Interpreting the Map Direct students to notice the special features of the map, including the key. Be sure they understand the function of the inset map, which shows downtown Havana. Ask them to use the elevation scale to describe the terrain of the island.

Possible Responses Most of the island is low-lying, less than 500 feet (150 m) in elevation. There are mountains at the southeastern end of the island and on nearby offshore islands. There are also some highland areas in the center and just west of Havana.

Extension Ask students to use the inset map of Havana to give directions for the following trips: (a) from José Martí Stadium to the University of Havana; (b) from the National Hotel to the National Museum.

Activity Options

Multiple Learning Styles: Intrapersonal

Class Time 30 minutes

Task Writing a diary entry by a Cuban child

Purpose To make students aware of some of the differences between life in Cuba and life in the United States

Supplies Needed

- Paper
- Pencils or pens

Activity Ask students to imagine they are sixth graders in Cuba. Then have them write a diary entry using what they learned about Cuba in this section. Suggest that students might include details about their education, health care, housing, favorite activities, or celebrations.

CRITICAL THINKING ACTIVITY

Synthesizing Ask students to think about what they have learned about the Castro regime. Then ask them to describe all of the ways in which they think Castro has improved life for the Cuban people and all of the ways he has failed to improve their lives. If necessary, remind them of the restrictions on free speech and protests. Finally, ask them to compare the ways in which Cuban citizens and American citizens can influence their respective governments and economies.

Class Time 15 minutes

ASSESS & RETEACH

Reading Social Studies Have students add details to the "Dictatorships," "Economic Development," and "U.S./Cuba Relations" sections of the chart on page 202.

Formal Assessment
- Section Quiz, p. 111

RETEACHING ACTIVITY

Ask students to create time lines for the events in Cuba's history during the twentieth century, based on the material presented in this section. Have students share and compare their time lines.

In-depth Resources: Unit 3
- Reteaching Activity, p. 33

Access for Students Acquiring English
- Reteaching Activity, p. 63

Culture • During Carnival, the air fills with the lively music played by musicians such as these drummers. ▲

Food shortages caused malnutrition. **Malnutrition** is poor health due to a lack of eating the right kinds of food. It often causes sickness. From the 1960s to the 1980s, Cuba had almost wiped out malnutrition. Since 1990, malnutrition has again become a problem in Cuba.

Arts, Sports, and Holidays

Both Castro's government and earlier dictators have placed strict limits on what artists, writers, and filmmakers may say in their work. As a result, some Cubans have fled to other countries in search of greater freedom of expression.

However, some art forms thrive in Cuba. Music is part of everyday life there. The unique sound of Cuban music combines African drums and Spanish guitars. Each different style and rhythm has its own name—son, mambo, cha-cha, rumba, salsa.

Sports are also popular in Cuba. The nation's favorite sport is baseball. Cuba's baseball team won a gold medal at the Olympics in 1996.

A favorite Cuban holiday, **Carnival,** takes place each year at the end of July. Cities and villages celebrate the end of the sugar harvest with festivals filled with music and dancing.

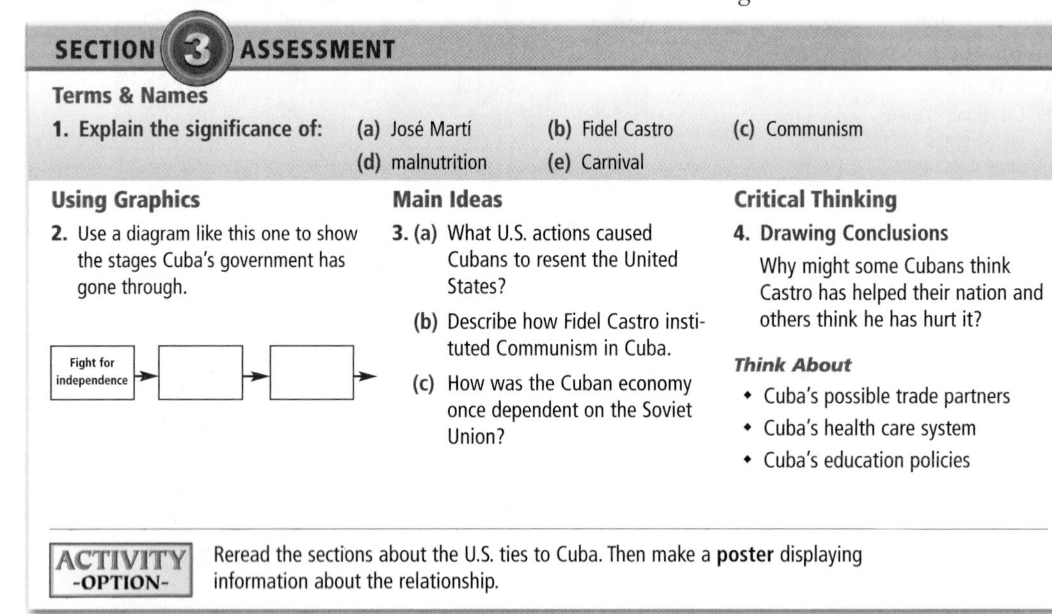

SECTION 3 ASSESSMENT

Terms & Names

1. **Explain the significance of:** (a) José Martí (b) Fidel Castro (c) Communism (d) malnutrition (e) Carnival

Using Graphics

2. Use a diagram like this one to show the stages Cuba's government has gone through.

Main Ideas

3. (a) What U.S. actions caused Cubans to resent the United States?
   (b) Describe how Fidel Castro instituted Communism in Cuba.
   (c) How was the Cuban economy once dependent on the Soviet Union?

Critical Thinking

4. **Drawing Conclusions** Why might some Cubans think Castro has helped their nation and others think he has hurt it?

Think About
- Cuba's possible trade partners
- Cuba's health care system
- Cuba's education policies

ACTIVITY -OPTION- Reread the sections about the U.S. ties to Cuba. Then make a **poster** displaying information about the relationship.

Section 3 Assessment

1. Terms & Names
- **a.** José Martí, p. 215
- **b.** Fidel Castro, p. 216
- **c.** Communism, p. 217
- **d.** malnutrition, p. 220
- **e.** Carnival, p. 220

2. Using Graphics

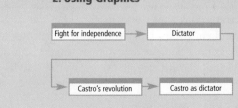

Fight for independence → Dictator → Castro's revolution → Castro as dictator

3. Main Ideas
- **a.** remaining in Cuba after the Spanish-American War; maintaining a naval base at Guantanamo Bay; supporting Cuba's dictators; holding vast assets in Cuba's banks and oil refineries
- **b.** He nationalized sugar-cane plantations and took over foreign banks, oil refineries, and other businesses.
- **c.** The Soviet Union provided food, machinery, military aid, and support for the sugar trade.

4. Critical Thinking

Possible Responses Health care and education are free, illiteracy has fallen, and health status of the poor has improved. However, there is a lack of political freedom, a low standard of living, and a lack of trade with the United States.

ACTIVITY OPTION

Integrated Assessment
- Rubric for making a poster

Guatemala Today

TERMS & NAMES
Rafael Carrera
Jacobo Arbenz Guzmán
departamento

AIN IDEA

he establishment of a stable overnment in Guatemala has been struggle, causing much suffering long the way.

WHY IT MATTERS NOW

Guatemala's increased stability has improved its relationship with its neighbors, such as the United States.

DATELINE

GUATEMALA CITY, GUATEMALA, DECEMBER 28, 1996

Tonight, Guatemala City is celebrating. However, the festivities don't mark a holiday or a party. They are, instead, signs of the historic action that will take place here tomorrow.

Guatemala's leaders will join their opponents in signing a peace agreement that will end a civil war that killed more than 100,000 Guatemalans over the past four decades. Finally, hopes for peace have arrived in the country that has suffered through Latin America's longest-lasting and most deadly civil war.

Place • **Many Guatemalans expressed relief and appreciation when they heard about the peace agreement.** ▶

History of Government ❶

In 1821, along with three other Central American states, Guatemala gained independence from Spain. It broke from the other states in 1839 to become the nation we know today. Between 1821 and 1839, peasants in the mountains had staged revolts against the government. In 1837, an uneducated farmer, **Rafael Carrera** (rah•fy•EHL kuh•REHR•uh), led a revolt and emerged as a new leader for Guatemala. In 1854, he took over the presidency, which he held until his death in 1865.

TAKING NOTES

Use your chart to take notes about Central America and the Caribbean.

Issues	Problems	Solutions
Colonization		
Panama Canal		

SECTION OBJECTIVES

1. To identify key events and figures in Guatemala's history
2. To describe Guatemala's government today
3. To identify important aspects of Guatemala's economy
4. To describe life in Guatemala today

CRITICAL THINKING

• Comparing and Contrasting, p. 223

FOCUS & MOTIVATE

WARM-UP

Making Inferences Ask students to read Dateline and discuss the following questions:

1. In what ways do you think the long civil war affected Guatemalans?
2. What effects might the war have had on industry and trade?

INSTRUCT: Objective ❶

History of Government

• Who was Jacobo Arbenz Guzmán? a revolutionary, president of Guatemala

• What significant events took place in 1944, 1954, and 1996? 1944: revolutionaries won control of Guatemala; 1954: military takeover, civil war began; 1996: civil war ended

• How did the United States influence Guatemalan politics in the 1950s? accused government of supporting Communism in land redistribution, supported military takeover of Arbenz government

In-depth Resources: Unit 3
• Guided Reading Worksheet, p. 26

Reading Study Guide
(Spanish and English), pp. 64–65

Program Resources

In-depth Resources: Unit 3
• Guided Reading Worksheet, p. 26
• Reteaching Activity, p. 34

Reading Study Guide
(Spanish and English), pp. 64–65

Formal Assessment
• Section Quiz, p. 112

Integrated Assessment
• Rubric for writing a paragraph

Outline Map Activities

Access for Students Acquiring English
• Guided Reading Worksheet, p. 58

Technology Resources
classzone.com

TEST-TAKING RESOURCES

Strategies for Test Preparation
Test Practice Transparencies
Online Test Practice

MORE ABOUT...

Jacobo Arbenz Guzmán

Throughout the long civil war, Jacobo Arbenz Guzmán remained a hero, particularly to poor people and left-wing political parties. He was ousted in 1954 after the United States intervened, calling him a Communist. Guzmán lived in exile, mostly in Mexico, until his death in 1971. In 1995 his ashes were buried in Guatemala in a ceremony attended by more than 100,000 people.

FOCUS ON VISUALS

Interpreting the Photograph Direct students to study the photograph of the soldiers and read the caption. Ask them to suggest reasons why people of Native American heritage—*ladinos* or Maya—might have opposed the military government and supported the rebels.

Possible Responses They are the people who would probably have benefited the most from the land policies of the Guzmán reform government; they probably gained little under the military rulers.

Extension Ask students to research and report on the events and negotiations that led to the 1996 peace agreement in Guatemala.

From Dictatorial Rule to Reforms After Carrera died, a steady flow of dictators filled Guatemala's presidency. In 1944, a set of military officers revolted and won control of the nation. One of these officers, **Jacobo Arbenz Guzmán** (YAH•koh•boh AHR•bayns gooz•MAHN), saw the need for social reforms in Guatemala. When he became president in 1951, Guzmán decided to develop a market economy and raise Guatemala's standard of living. Guzmán also redistributed 1.5 million acres of land to 100,000 families. As in Mexico, the goal of redistributing land was to give many more people access to land that they could farm.

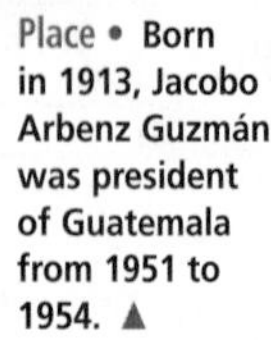

Place • Born in 1913, Jacobo Arbenz Guzmán was president of Guatemala from 1951 to 1954. ▲

U.S. Ties to Guatemala Serious opposition to Guzmán's redistribution program arose in the United States. Both United Fruit Company and the U.S. government owned much land in Guatemala. Guzmán established a policy of giving farmers any land that was not already being used. Eighty-five percent of UFCO's land in Guatemala was unused. Thus, UFCO was at great risk of losing that land.

Reading Social Studies
A. Possible Answ
Probably not, because the count was torn by civil war, and the regim was not likely to favor reform.

The United States Steps In The United States took action in 1954. Accusing Guzmán of supporting Communism—a political system that the United States believed threatened national safety—the U.S. Central Intelligence Agency (CIA) supported an invasion of Guatemala's capital, Guatemala City. A Guatemalan colonel, Carlos Castillo Armas (CAR•lohs kah•STEE•yoh AR•mahs), led the attack. A frightened Guzmán quickly gave in. A new government, backed by the United States, took control of Guatemala.

Place • Rebel forces in Guatemala have included thousands of *ladinos* and Maya people. ▼

Civil War Takes Hold After 1954, Guatemala's government was ruled mainly by military officers. During much of this era, a civil war raged between government forces and rebels who opposed the government. Many people who expressed disagreement with government policy were murdered. More than 100,000 Guatemalans were killed or kidnapped before a peace agreement was reached in 1996.

Reading Social Studies

A. Synthesizing Given the opposition to the new government after Guzmán's overthrow, do you think it continued social reforms?

Activity Options

Differentiating Instruction: Gifted and Talented

Block Scheduling

Advising the President Challenge advanced students to act as policy advisors to the president and cabinet of an actual contemporary Central American country. They will make recommendations for reforms in three areas: economic growth, free elections, and education. Individual students (or small groups) should focus on one issue. To develop a policy, students should do the following:

- Investigate the current situation in the country chosen—for example, the percentage of illiteracy
- Identify problems in the current situation
- Suggest solutions in a written memo to the president and cabinet

Students should then present their policy ideas to the class.

Guatemala's Government Today ❷

Guatemala's current constitution was written in 1985. It established the nation's government as a democratic republic with three branches. They are executive, legislative, and judicial.

Government's Three Branches A president heads Guatemala's executive branch. He or she is elected by the people every four years and may not be reelected. The president appoints a cabinet, or a group of advisers, to carry out the government's work.

Guatemala's legislative branch is called Congress. It has 113 members who are elected to four-year terms. Guatemala's Congress is unicameral, or has one chamber. Members of Congress may be reelected.

Location • Since the 1930s, Guatemala City's National Palace has housed the government's offices. ▲

Guatemala's judicial branch has different levels of courts, somewhat like the United States. Unlike U.S. Supreme Court justices, whom the President appoints to serve for as long as they choose, Guatemala's Supreme Court judges are elected to five-year terms.

State and Local Governments Guatemala is a federal republic, so the national government shares power with state and local governments. Governors head Guatemala's 22 states, called ***departamentos*** (deh•pahr•tah•MEHN•taws). The president appoints each of the governors. Mayors elected by popular vote oversee the city governments.

Guatemala's Economy ❸

At the turn of the century, Guatemala had the largest gross domestic product (GDP) in Central America. It also had the fastest growing GDP in the region.

Agriculture Guatemala's dominant industry is agriculture, which employs more than half of its work force. The nation's economy largely relies on the export of agricultural products. Since 1870, coffee has been Guatemala's leading export. Other agricultural exports include sugar, a spice called cardamom, and bananas.

INSTRUCT: Objective ❷

Guatemala's Government Today

- What type of government does Guatemala have today? democratic republic
- How does the office of the presidency differ from that of the United States? cannot be reelected after a four-year term; appoints governors of states

CRITICAL THINKING ACTIVITY

Comparing and Contrasting Read aloud the paragraph describing the judicial system in Guatemala. Ask a volunteer to explain how the system differs from that in the United States. Then have students discuss the benefits and drawbacks of appointing judges for life as opposed to electing them for five-year terms.

Class Time 15 minutes

INSTRUCT: Objective ❸

Guatemala's Economy

- What is the dominant industry in Guatemala? agriculture
- What agricultural products are important exports for Guatemala? coffee, followed by sugar, cardamom, bananas
- What other industries support Guatemala's economy? textiles, clothing, tourism

Activity Options

Differentiating Instruction: Less Proficient Readers

Taking Notes To help students understand how the government in Guatemala is organized, work with them to reread "Guatemala's Government Today." Then have them use a graphic organizer like the one shown to identify the three branches of government and the responsibilities of each.

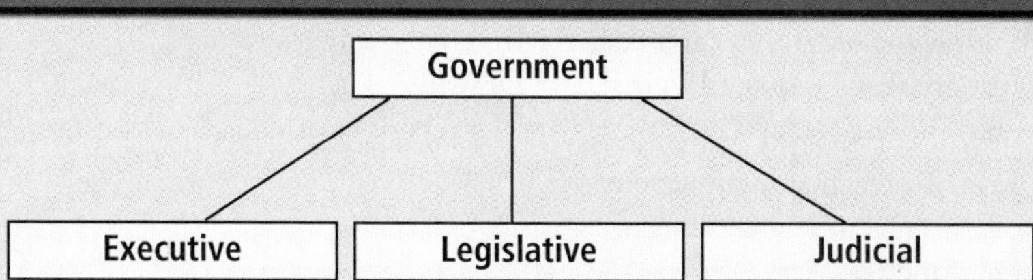

MORE ABOUT...

Poverty in Guatemala

Guatemala is a very poor country. Three-quarters of the population live in poverty, and two-thirds live in extreme poverty. Income distribution is highly skewed, with the wealthiest 10 percent of the population receiving about half of the total income and the top 20 percent receiving almost two-thirds of the total income. GNP per person is estimated to be about $1,600 a year—only about 5 percent of the GNP per person in the United States. The rates of infant mortality and illiteracy are among the highest in the Western Hemisphere.

INSTRUCT: Objective 4

Living in Guatemala

- How are *ladinos* different from the Maya in Guatemala? either mixed Spanish and Maya ancestry or do not practice Maya ways of life or speak Maya languages
- How does life in rural areas differ from life in urban areas in Guatemala? fewer children attend school, homes are small and often have no electricity or running water

Human-Environment Interaction • Bananas are produced in mass quantity on Guatemala's many banana plantations. ▲

Banana production began in the early 1900s, when U.S. companies built banana plantations in Guatemala. These fruit companies also developed railroads, ports, and communication systems in order to transport the bananas to foreign markets.

Other Parts of the Economy Guatemala also relies on manufacturing to bring in money. Food, beverages, and clothing are among its manufactured goods. These goods are sold both within Guatemala and as exports to other countries. While other Central American nations purchase many of Guatemala's manufactured goods, the United States purchases more of Guatemala's exports than any other country does.

In past years, Guatemala's economy has boomed with the sale of both textiles and clothing. Also, new nontraditional agricultural products, such as cut flowers and winter fruits, are selling quite well on the international market. Tourism is also a strong industry in Guatemala, which is home to many ancient Maya ruins.

Reading Social Studies
B. Possible Answe
No, because it exports manufactured goods as wel as coffee and bananas.

Reading Social Studies
B. Clarifying Does Guatemala have a single-product economy? Why or why not?

Living in Guatemala 4

More than half of Guatemala's people are Maya. The rest are *ladinos.* In Guatemala, *ladinos* are either of mixed Maya and Spanish ancestry, or they are of Maya ancestry but no longer practice Maya ways or speak Maya languages. Like the ancient Maya, most of Guatemala's Maya today work in agriculture and live in small rural villages. They speak Maya languages, though many of them also speak Spanish. They wear traditional clothing, much of which they weave by hand.

Education Guatemalan children are required to attend school from the age of 7 through 13. However, about one-third do not. Most of these children live in rural areas that have no schools. Only 15 percent of Guatemalans attend high school.

Place • Standing nearly 150 feet high, this Maya temple at Tikal is one of Guatemala's major tourist attractions. ▼

Activity Options

Multiple Learning Styles: Spatial

Class Time One class period

Task Drawing a diagram of Tikal, a Maya city that flourished in the area that is present-day Guatemala

Purpose To teach students about one of the ancient civilizations of Central America

Supplies Needed
- Reference sources
- Drawing paper
- Markers or crayons

Activity Have students conduct research to learn about the ancient city of Tikal. Then ask them to draw a diagram showing the position of temples, homes, ball courts, and other structures within the city. Invite them to decorate the sides of their posters with examples of Maya culture, such as masks and calendars.

Culture • This Guatemalan woman sells many traditional Maya handwoven cloths. ▲

Two Sides of Guatemala Daily life in Guatemala is a matter of extremes. On the one hand, rural Guatemalans have few of the comforts that North Americans take for granted, such as indoor bathrooms, running water, and electricity. Outside the cities, most homes are very small, and many have dirt floors.

On the other hand, urban Guatemalans live in modern homes, attend schools and universities, and go to theaters, museums, and restaurants. Many of the cultural influences in the cities, such as movies, restaurant chains, clothing styles, magazines, cars, and television programs, come from foreign countries. The cultural influences in the rural areas are much more local in origin.

SECTION 4 ASSESSMENT

Terms & Names

1. Explain the significance of: **(a)** Rafael Carrera **(b)** Jacobo Arbenz Guzmán **(c)** *departamento*

Using Graphics

2. Use a table like this one to keep track of the sequence of events in Guatemala's history.

Year	Event
1837	
1839	
1944	
1951	
1954	
1985	
1996	

Main Ideas

3. (a) Why did the United States feel threatened when Jacobo Arbenz Guzmán established his policy of land redistribution?

(b) After 1954, how did the Guatemalan government treat people who expressed disagreement with its policies?

(c) What is Guatemala's leading export?

Critical Thinking

4. Synthesizing

What is the impact of having a powerful nation, such as the United States, be the largest purchaser of Guatemala's exports?

Think About

- the role of exports in Guatemala's economy
- U.S. interventions in Guatemala

ACTIVITY -OPTION- Write a **paragraph** explaining three facts about Guatemala that you did not know before you read this section.

ASSESS & RETEACH

Reading Social Studies Have students complete the "Civil War in Guatemala" section of the chart on page 202.

Formal Assessment
- Section Quiz, p. 112

RETEACHING ACTIVITY

Divide the class into four groups and assign each group one section heading. Then ask each group to write a summary of the material included under that heading. Call on a volunteer from each group to read the summary aloud.

In-depth Resources: Unit 3
- Reteaching Activity, p. 34

Access for Students Acquiring English
- Reteaching Activity, p. 64

Section 4 Assessment

1. Terms & Names

a. Rafael Carrera, p. 221
b. Jacobo Arbenz Guzmán, p. 222
c. *departamento,* p. 223

2. Using Graphics

<1837>	Carrera leads a revolt
<1839>	Guatemala is independent
<1944>	Military takeover
<1951>	Guzmán becomes President
<1954>	CIA supports invasion; civil war begins
<1985>	New constitution written
<1996>	Civil war ends

3. Main Ideas

a. Both UFCO and the U.S. government owned much land that would have been redistributed.
b. It had many of them killed.
c. coffee

4. Critical Thinking

Guatemala is vulnerable to economic and political changes in the United States; the United States may have excessive political power in Guatemala's internal affairs.

ACTIVITY OPTION

Integrated Assessment
- Rubric for writing a paragraph

ASSESSMENT

TERMS & NAMES

1. West Indies, p. 203
2. dependency, p. 203
3. mulatto, p. 204
4. *ladino,* p. 205
5. dictator, p. 207
6. single-product economy, p. 209
7. diversify, p. 209
8. Fidel Castro, p. 216
9. Communism, p. 217
10. Rafael Carrera, p. 221

REVIEW QUESTIONS

Possible Responses

1. The canal provides a water route between the Atlantic Ocean and the Pacific Ocean, eliminating a long trip around South America.
2. The Spanish-American War freed the islands from Spain.
3. If something happens to the product, or the market price drops, the economy can be ruined.
4. They depended on sugar, then on bananas and coffee.
5. Castro ruled as a dictator, adopted Communist policies, redistributed land, and suppressed dissent.
6. It made Cuba dependent on the Soviet Union, so that the collapse of the Soviet Union caused hardship.
7. They did not want to lose the land owned by the U.S. government and UFCO.
8. *Ladinos* are either of mixed Maya and Spanish ancestry or are of Maya ancestry but no longer practice Maya ways or speak Maya languages.

CHAPTER 8 ASSESSMENT

TERMS & NAMES

Explain the significance of each of the following:

1. West Indies
2. dependency
3. mulatto
4. *ladino*
5. dictator
6. single-product economy
7. diversify
8. Fidel Castro
9. Communism
10. Rafael Carrera

REVIEW QUESTIONS

Establishing Independence *(pages 203–207)*

1. Why did the United States want to build the Panama Canal?
2. What event freed the Caribbean Islands from Spanish rule?

Building Economies and Cultures *(pages 208–213)*

3. What risk did Caribbean islanders face by having single-product economies?
4. What two crops did Central American economies depend on before the countries diversified their economies?

Cuba Today *(pages 215–220)*

5. Why did hundreds of thousands of people flee from Cuba to the United States after Castro took over?
6. How did Cuba's relationship with the Soviet Union affect Cuba?

Guatemala Today *(pages 221–225)*

7. Why did Jacobo Arbenz Guzmán's program of land redistribution upset people in the United States?
8. What are the cultural backgrounds of Guatemala's *ladinos?*

CRITICAL THINKING

Identifying Problems and Solutions

1. Using your completed chart from Reading Social Studies, p. 202, choose one issue, such as dictatorship or colonization, and summarize the problems with it. Then summarize the solutions to the problems.

Drawing Conclusions

2. How do you think Fidel Castro's control of newspapers and television and radio stations has helped keep him in power?

Contrasting

3. Contrast the fates of Native Americans in the Caribbean Islands and in Central America after Europeans took over.

Visual Summary

1 Establishing Independence

- People in Central America and the Caribbean fought first for independence and then for democracy.

- Central America's people are mostly *ladinos* and Native Americans, while the Caribbean's are largely of African or European descent.

2 Building Economies and Cultures

- The countries of the region had to diversify their economies so they would be more stable.
- The region's languages, religions, and music were influenced by the cultures of the colonizers, the enslaved Africans, and the Native Americans.

3 Cuba Today

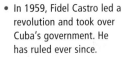

- In 1959, Fidel Castro led a revolution and took over Cuba's government. He has ruled ever since.
- Throughout the 1900s, both the United States and the Soviet Union influenced events in Cuba.

4 Guatemala Today

- After decades of struggle, Guatemala's government has become more stable.
- Guatemala's economy relies heavily on agriculture, especially coffee production.
- More than half of Guatemala's people are Maya. The rest are *ladinos.*

CRITICAL THINKING: Possible Responses

1. Identifying Problems and Solutions

Sample answers: Students may say that some countries, having fought for independence from colonizers, found that their leaders had become dictators. Over time, they overthrew their dictators and organized democratic elections.

2. Drawing Conclusions

Control over the media has enabled Castro to control what people know and what they think; people who oppose him are unable to air their views.

3. Contrasting

Most Native Americans living in the West Indies at the time of the European conquest died from disease or were killed by harsh labor or in battle. In contrast, the Native American peoples of Central America retreated inland to mountainous areas and survived.

the map and your knowledge of world cultures and geography to answer questions 1 and 2.

dditional Test Practice, pp. S1–S33

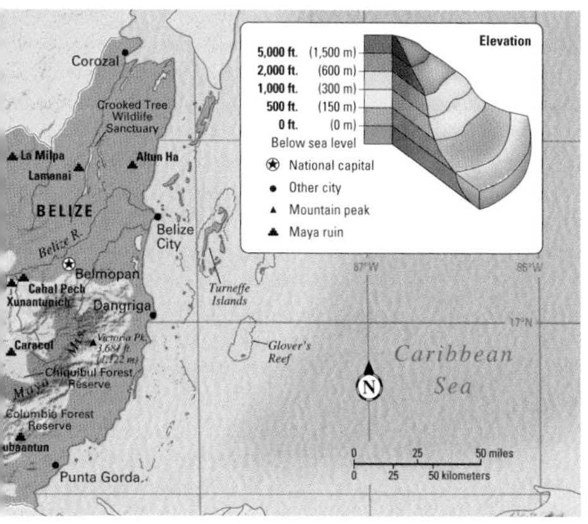

What natural features most likely attract tourists to Belize?

A. Mountains
B. Cities
C. Lakes
D. Hot springs

What human-made features might attract tourists to Belize?

A. Mountains
B. Maya ruins
C. Coral reefs
D. Rivers

The following passage is from the acceptance speech that President Oscar Arias of Costa Rica gave when he received the Nobel Peace Prize. Use the quotation and your knowledge of world cultures and geography to answer question 3.

PRIMARY SOURCE

Peace is not a matter of prizes or trophies. It is not the product of a victory or command. It has no finishing line, no final deadline, no fixed definition of achievement.

Peace is a never-ending process, the work of many decisions by many people in many countries. It is an attitude, a way of life, a way of solving problems and resolving conflicts. . . . It requires us to live and work together.

OSCAR ARIAS, acceptance speech, December 10, 1987

3. Which of the following statements most accurately restates Arias's belief about peace?

A. Leaders like Arias are the only people who can help to bring about world peace.
B. Now that Arias has received the Nobel Prize, he can stop working for peace.
C. Peace has been achieved whenever a country is not presently fighting a war.
D. Individual people can help create peace by the choices they make every day.

TERNATIVE ASSESSMENT

WRITING ABOUT HISTORY

Imagine that you have traveled to one of the Caribbean islands, such as Grenada or Jamaica. Write a letter to a friend describing what you saw.

Using the Internet or library resources, research the island. You might find books about the island in the travel section of the library.

In your letter, describe scenery and the sights you visited, any special events you attended, and what you enjoyed most about the island. Include photographs or illustrations of special sights or experiences.

COOPERATIVE LEARNING

With a group of classmates, create a bulletin board about one of the countries in the Caribbean or Central America. Describe the climate, list the natural resources, and include information about the cities, politics, and the economy. Include maps, charts, and illustrations. Share the tasks of researching, writing, and creating the display with the members of your group.

INTEGRATED TECHNOLOGY

Doing Internet Research

The cultures of Central America and the Caribbean are rich and varied. Use the Internet to research the culture of one country, such as Belize or Cuba, to prepare a presentation.

- Use the Internet or resources in the library to learn about the culture of your chosen country.
- Learn about the art, music, religion, and other traditions of that country. Find out how its people have preserved cultural traditions of the past.
- Include copies of photographs or create your own drawings to include in your presentation.

For Internet links to support this activity, go to

STANDARDS-BASED ASSESSMENT

1. **Answer A** is the correct answer because mountains are physical features that often attract tourists; Answer B is incorrect because cities are not physical features; Answer C is incorrect because Belize has few lakes; Answer D is incorrect because there are no hot springs noted on the map.
2. **Answer B** is the correct answer because Maya ruins are interesting human-made features; Answers A, C, and D are incorrect because they name natural rather than human-made features.
3. **Answer D** is the correct answer because it echoes Arias's statement that peace is a way of life and the work of many decisions by many people. Answer A is incorrect because it contradicts that peace is the work of many. Answers B and C are incorrect because they contradict that point that peace has no finishing line or final achievement and that it is a never-ending process.

INTEGRATED TECHNOLOGY

Before students begin, discuss aspects of cultural traditions observed by their own families. Then, explore ways to find information about other cultural traditions using the Internet. Brainstorm possible keywords, Web sites, and search engines students might explore.

Alternative Assessment

1. Rubric

The letter should

- accurately convey information about Jamaica.
- use colorful language and verbal images.
- reflect the thoughts and experiences of a traveler.
- use correct grammar, spelling, and punctuation.

2. Cooperative Learning

Write the names of countries in the Caribbean and Central America on index cards. Place the cards in a bag, and have a student from each group choose a card to determine the country they will research. Remind students that a bulletin board display should be interesting, detailed, and colorful.

PLANNING GUIDE: UNIT 3 ATLAS AND CHAPTER 9

South America

	OVERVIEW	COPYMASTERS	INTEGRATED TECHNOLOGY
UNIT ATLAS AND CHAPTER RESOURCES	The students will explore the history and geography of South America and the effects of both on the government, economy, and culture today.	**In-depth Resources: Unit 3** • Guided Reading Worksheets, pp. 35–38 • Skillbuilder Practice, p. 41 • Unit Atlas Activities, pp. 1–2 • Geography Workshop, pp. 47–48 **Reading Study Guide** (Spanish and English), pp. 68–75 **Outline Map Activities**	• eEdition Plus Online • EasyPlanner Plus Online • eTest Plus Online • eEdition • Power Presentations • EasyPlanner • Electronic Library of Primary Sources • Test Generator • Reading Study Guide • The World's Music • Critical Thinking Transparencies CT17
SECTION 1 Establishing Independence pp. 231–235	**KEY IDEAS** • European settlement in South America had mostly negative effects. • Simón Bolívar and José San Martin led wars for independence. • South America's population includes immigrants from many countries.	**In-depth Resources: Unit 3** • Guided Reading Worksheet, p. 35 • Reteaching Activity, p. 43 **Reading Study Guide** (Spanish and English), pp. 68–69	**Map Transparencies MT18** classzone.com Reading Study Guide
SECTION 2 Building Economies and Cultures pp. 238–242	• Transportation barriers pose challenges to South American economies. • Brazil, Chile, and Argentina are industrial leaders in the region. • The arts flourish in South America, and there are stark differences between rural and urban life.	**In-depth Resources: Unit 3** • Guided Reading Worksheet, p. 36 • Reteaching Activity, p. 44 **Reading Study Guide** (Spanish and English), pp. 70–71	classzone.com Reading Study Guide
SECTION 3 Brazil Today pp. 243–247	• Brazil is the largest country in South America and has the strongest economy. • The Brazilian government supports agriculture and other industries. • Brazil's historic ties to Portugal have influenced its culture today.	**In-depth Resources: Unit 3** • Guided Reading Worksheet, p. 37 • Reteaching Activity, p. 45 **Reading Study Guide** (Spanish and English), pp. 72–73	**Critical Thinking Transparencies CT18** classzone.com Reading Study Guide
SECTION 4 Peru Today pp. 250–254	• Peru's major landforms include mountains, rain forests, and deserts. • Peru has struggled to achieve economic and political stability. • Peru's culture draws heavily from Native American heritage, though most Peruvians today live in cities.	**In-depth Resources: Unit 3** • Guided Reading Worksheet, p. 38 • Reteaching Activity, p. 46 **Reading Study Guide** (Spanish and English), pp. 74–75	classzone.com Reading Study Guide

KEY TO RESOURCES

 Audio
 CD-ROM
  Copymaster
 Internet
 Overhead Transparency
Pupil's Edition
 Teacher's Edition
 Video

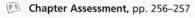

ASSESSMENT OPTIONS

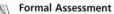

Chapter Assessment, pp. 256–257

Formal Assessment
- Chapter Tests: Forms A, B, C, pp. 129–140

Test Generator

Online Test Practice

Strategies for Test Preparation

Section Assessment, p. 235

Formal Assessment
- Section Quiz, p. 125

Integrated Assessment
- Rubric for making a poster

Test Generator

Test Practice Transparencies TT25

Section Assessment, p. 242

Formal Assessment
- Section Quiz, p. 126

Integrated Assessment
- Rubric for writing a description

Test Generator

Test Practice Transparencies TT26

Section Assessment, p. 247

Formal Assessment
- Section Quiz, p. 127

Integrated Assessment
- Rubric for writing a speech

Test Generator

Test Practice Transparencies TT27

Section Assessment, p. 254

Formal Assessment
- Section Quiz, p. 128

Integrated Assessment
- Rubric for writing a short story

Test Generator

Test Practice Transparencies TT28

RESOURCES FOR DIFFERENTIATING INSTRUCTION

Students Acquiring English/ESL

Reading Study Guide (Spanish and English), pp. 68–75

Access for Students Acquiring English Spanish Translations, pp. 65–74

TE Activity
- Understanding Word Meaning, p. 245

Modified Lesson Plans for English Learners

Less Proficient Readers

Reading Study Guide (Spanish and English), pp. 68–75

TE Activities
- Categorizing, p. 240
- Interpreting a Graph, p. 245

Gifted and Talented Students

 **TE Activity**
- Identifying Problems and Solutions, p. 246

CROSS-CURRICULAR CONNECTIONS

Humanities
Merrill, Yvonne. ***Hands-on Latin America: Art Activities for All Ages.*** Salt Lake City, UT: Kits Publishing, 1998.
A sophisticated craft book with strong cultural connections.

Literature
Loya, Olga. ***Momentos Mágicos/Magic Moments.*** Little Rock, AR: August House, 1997. Bilingual story collection.

Health
Pascoe, Elaine. ***Mysteries of the Rain Forest: 20th Century Medicine Man.*** Woodbridge, CT: Blackbirch Press, 1998. Modern doctor studies medicinal benefits of rain forest vegetation.

Language Arts/Literature
de Jesus, Carolina Maria. ***Bitita's Diary: The Childhood Memoirs of Carolina Maria de Jesus.*** Armonk, NY: M. E. Sharpe, 1998. Child growing up in a slum in 1920s Brazil.

Science
Arnold, Caroline. ***South American Animals.*** New York: Morrow Junior Books, 1999. Birds, mammals, reptiles, and amphibians.

Welsbacher, Anne. ***Jaguars.*** Edina, MN: ABDO publishing, 2000. Cats indigenous to South America.

ENRICHMENT ACTIVITIES

The following activities are especially suitable for classes following block schedules.

Teacher's Edition, pp. 233, 234, 241, 246, 251, 253, 255
Pupil's Edition, pp. 235, 242, 247, 254
Unit Atlas, pp. 142–149
Interdisciplinary Challenge, pp. 236–237
Linking Past and Present, pp. 248–249
Outline Map Activities

INTEGRATED TECHNOLOGY

Go to **classzone.com** for lesson support and activities for Chapter 9.

BLOCK SCHEDULE LESSON PLAN OPTIONS: 90-MINUTE PERIOD

DAY 1

CHAPTER PREVIEW, pp. 228–229
Class Time 45 minutes

- **Discussion** Discuss the objective for Chapter 9 on TE p. 228.
  Class Time 10 minutes
- **Political Map** Divide the class into pairs and use the Skillbuilder: Interpreting a Map on TE p. 232 to introduce students to the countries and geography of South America.
  Class Time 35 minutes

SECTION 1, pp. 231–235
Class Time 45 minutes

- **Skit** Divide the class into two groups. Have each group create and perform a skit to illustrate the history of South America.

DAY 2

SECTION 2, pp. 238–242
Class Time 45 minutes

- **Categorizing** Divide students into pairs and use the Activity Option on TE p. 240 to help students review the information under the head Products and Industries of South America. Reconvene the class to go over the charts.
  Class Time 25 minutes
- **Internet** Extend students' background knowledge of the physical geography of South America by visiting **classzone.com.**
  Class Time 20 minutes

SECTION 3, pp. 243–247
Class Time 45 minutes

- **Peer Teaching** Have pairs of students review the Main Idea for Section 3 on PE p. 243 and find three details to support it. Then have each pair list two additional important ideas and trade lists with another group to find details.
  Class Time 15 minutes
- **Skillbuilder** Use the questions in Skillbuilder: Interpreting a Graph on PE p. 244 to guide a discussion about Latin American economies.
  Class Time 15 minutes
- **Distinguishing Fact from Opinion** Have students read the ads they wrote for the Critical Thinking Activity on TE p. 247. Ask other students to identify statements of fact and opinion in each advertisement.
  Class Time 15 minutes

DAY 3

SECTION 4, pp. 250–254
Class Time 35 minutes

- **Small Groups** Divide the class into four groups and have each group do the Skillbuilder Mini-Lesson: Reading a Political Map on TE p. 251. After each group has completed its map, have the entire class share their maps and explain why they chose their respective routes.
  Class Time 25 minutes
- **Summarizing** Read The Economy of Peru on PE pp. 251–252 aloud as a class. On the board, lead the class in listing the factors that affect Peru's struggle for a stable economy.
  Class Time 10 minutes

CHAPTER 9 REVIEW AND ASSESSMENT, pp. 256–257
Class Time 55 minutes

- **Review** Have students use the charts they created for Reading Social Studies on PE p. 230 to review generalizations and supporting facts about South America.
  Class Time 20 minutes
- **Assessment** Have students complete the Chapter 9 Assessment.
  Class Time 35 minutes

TECHNOLOGY IN THE CLASSROOM

COMPARING NUMBERS AND STATISTICS ON THE INTERNET

The Internet offers a wealth of numerical data that can be helpful for student research and supplement topics students are studying. Some Web sites even allow users to specify how they would like to see the data presented. When students use this type of site, they not only practice statistical analysis skills, but they also learn about ways that the Internet can be used to effectively organize and present numerical information.

ACTIVITY OUTLINE

Objective Students will use the interactive forms on a Web site to compare and contrast economic and social indicators for seven South American countries. They will generate charts of economic and social statistics by inputting their requests into interactive forms.

Task Have students compare and contrast such indicators as GDP per capita, life expectancy, and unemployment rates by using an Internet program specifically designed to do this type of analysis. Ask them to discuss their interpretation of the data, then input the data on a computer.

Class Time 1–2 class periods

DIRECTIONS

1. Ask students to go to the Web site at **classzone.com.** Explain that they are going to examine data for South American countries that will tell them about the countries' economies and standards of living.
2. Have students click the boxes in the right-hand frame next to the following countries: Argentina, Brazil, Chile, Columbia, Guyana, and Venezuela. Then have them click "Data Menu."
3. Ask students to select the following data fields: GDP per capita, unemployment, life expectancy (women/men), and illiteracy rate (total).
4. Discuss what each of these data fields measures. Ask students to explain what they think each type of data reveals about a country. For example, what does it mean for one country to have a much lower life expectancy or per capita income than another?
5. Have students click "View Info" to view the data. Ask them to look carefully at the chart and compare the data for the different countries. Which country appears to be strong and which appears to have the most problems?
6. (optional step) Have students return to the data field page, and discuss the meanings of some of the other indicators, such as population density, population under 15 years old, infant mortality, and spending on education. Based on what they have learned about the seven countries so far, how do they predict the countries will compare on these indicators? Have students find out by selecting these data fields and clicking "View Info."
7. Have pairs of students discuss why they think that the data numbers for different countries vary. Ask them to list their ideas, and then hold a class discussion on what students have hypothesized.
8. Have students use a word processing program to type sentences or paragraphs that describe how the countries compare to one another and hypothesize why this might be the case.

CHAPTER 9 OBJECTIVE

Students will explore the history and geography of South America and the effects of both on its present-day governments, economies, and cultures.

FOCUS ON VISUALS

Interpreting the Photograph Have students look at the photograph of the Amazon River and the surrounding rain forest. Ask them to suggest words to describe the landscape. Then have them think of words to describe the urban scene in the photograph on the facing page. Point out that this chapter focuses on the geography and history of South America. Ask them to explain how these contrasting photographs might be a symbol for present-day South America.

Possible Response Modern South America is a place of great contrasts, with both busy cities and vast undeveloped wilderness.

Extension Have students research a topic related to the Amazon and write a background report to share with the class. For example, they might focus on the river, towns in the Amazon Basin, indigenous peoples, wildlife, or the rain forest.

CRITICAL THINKING ACTIVITY

Hypothesizing Encourage students to discuss the ways in which a landscape such as this one would affect history, economy, and lifestyles in a region. Would the river be a barrier or a help to transportation? What kind of settlements would be likely in this landscape? Help students recall the connections among land, climate, and lifestyles.

Class Time 15 minutes

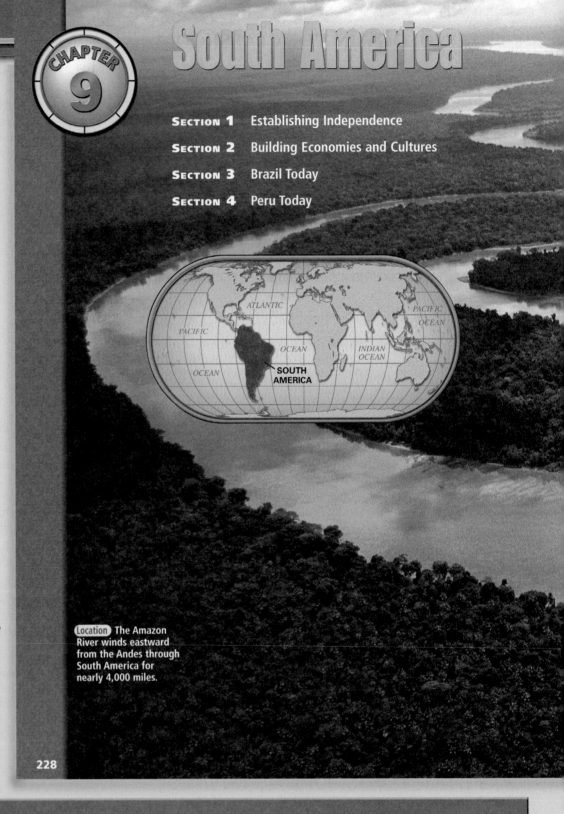

CHAPTER 9

South America

ATLANTIC OCEAN

PACIFIC OCEAN

PACIFIC OCEAN

INDIAN OCEAN

SOUTH AMERICA

Location The Amazon River winds eastward from the Andes through South America for nearly 4,000 miles.

228

Recommended Resources

BOOKS FOR THE TEACHER

Early, Edwin. ***The History Atlas of South America.*** New York: Macmillan, 1998. Comprehensive coverage.

Rocha, Jan. ***Brazil: A Guide to the People, Politics and Culture.*** New York: Interlink Books, 2000. A travel guide with extensive cultural essays.

VIDEOS

Warriors of the Amazon. NOVA Boston: WGBH Educational Foundation, 1996. Endangered Yanomamo Indians of Brazil and Venezuela.

SOFTWARE

A Field Trip to the Rainforest. Pleasantville, NY: Sunburst, 1998. Animals, plants, and geography.

INTERNET

For more information about South America, visit **classzone.com.**

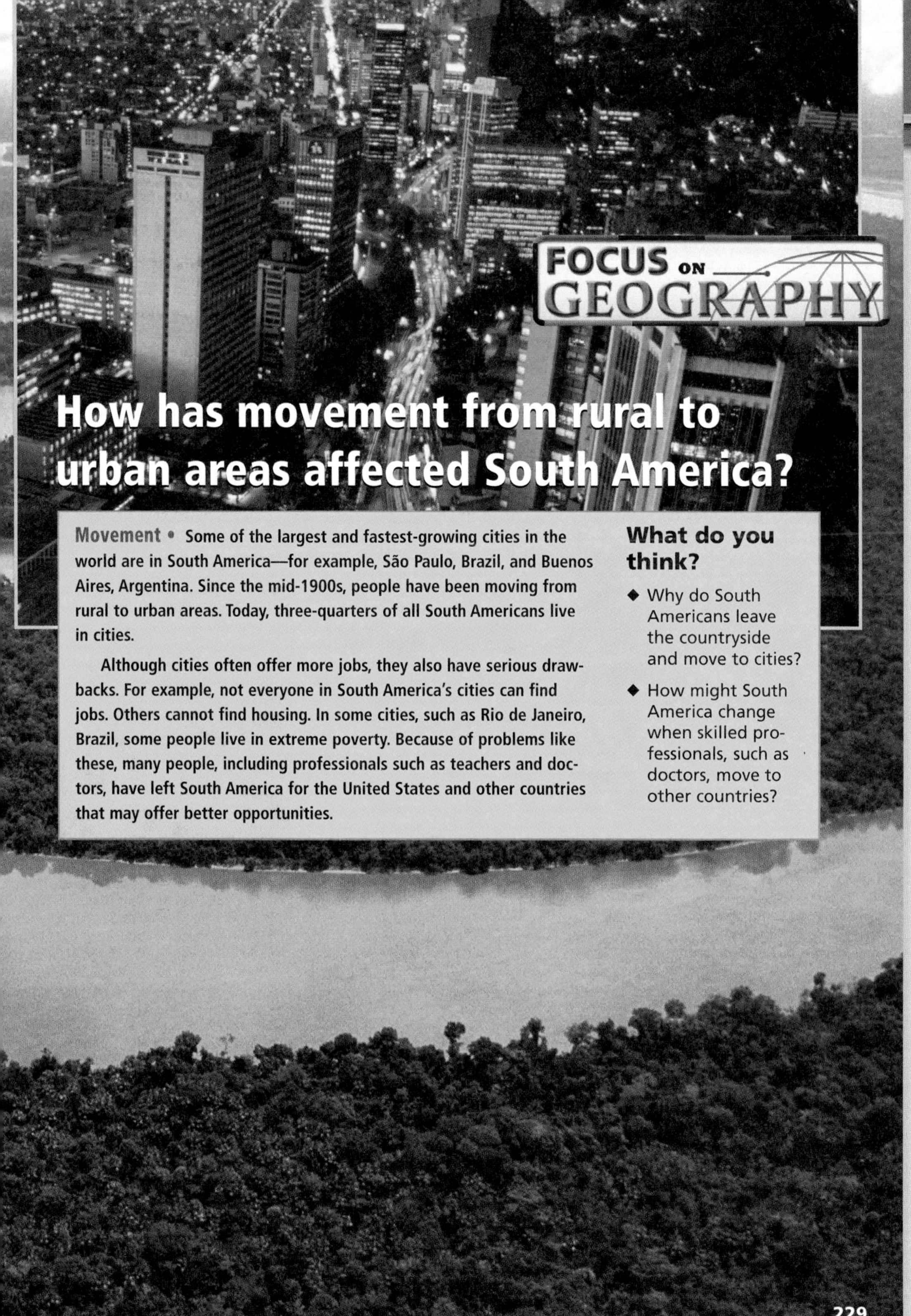

How has movement from rural to urban areas affected South America?

Movement • Some of the largest and fastest-growing cities in the world are in South America—for example, São Paulo, Brazil, and Buenos Aires, Argentina. Since the mid-1900s, people have been moving from rural to urban areas. Today, three-quarters of all South Americans live in cities.

Although cities often offer more jobs, they also have serious drawbacks. For example, not everyone in South America's cities can find jobs. Others cannot find housing. In some cities, such as Rio de Janeiro, Brazil, some people live in extreme poverty. Because of problems like these, many people, including professionals such as teachers and doctors, have left South America for the United States and other countries that may offer better opportunities.

What do you think?

- Why do South Americans leave the countryside and move to cities?
- How might South America change when skilled professionals, such as doctors, move to other countries?

229

FOCUS ON GEOGRAPHY

Objectives

- To help students identify movement of people as an important issue in South America today.
- To help students explain the impact of the rural-to-urban population shift.

What Do You Think?

1. Ask students what opportunities might prompt someone to move to a new place. Guide them to consider how people's basic needs for food, shelter, education, and health services factor into the decision.
2. Ask students to describe what would happen to health care in their community if all but one doctor moved away. Then have students consider the effect on an entire country of losing skilled professionals.

How has South America been affected by the movement of people?

Have students speculate about the changes in South America caused by the rapid growth of cities. Encourage them to consider the effects of the movement of people on housing, food, employment, and services in both urban and rural environments.

MAKING GEOGRAPHIC CONNECTIONS

Ask students to think about population changes in their school, community, or state over the past ten years. Has the population increased or decreased? Do they notice any pattern of movement? What has been the impact of these changes?

Implementing the National Geography Standards

Standard 18 Develop innovative plans to improve the quality of environments in large cities

Objective To create a transportation plan that reduces pollution for a South American city

Class Time 40 minutes

Task Organize students into groups. Have each group develop a transportation plan that will reduce pollution from cars and buses in a South American city. Encourage students to consider improvements such as public transportation, pedestrian walkways, bicycle lanes, and "green" space.

Evaluation Have groups share their ideas. Each group should describe specifically how each idea will improve the city's environment while providing transportation around the city.

BEFORE YOU READ

What Do You Know?

Ask students to think about things they know about or associate with South America. These might be related to sports, food, music, clothing, historical events, or people. Provide time for students to write as many ideas as they can. Invite volunteers to share their lists. If any students, or their ancestors, have come from South America, invite these students to share what they remember or have learned.

What Do You Want to Know?

Ask each student to develop two questions about present-day South America for which they would like to find answers. They should write these questions in their notebooks. Encourage students to write answers as they read the chapter.

READ AND TAKE NOTES

Reading Strategy: Making Generalizations
Point out the two headings on the chart: Generalizations and Facts. Explain that the generalizations, or conclusions, can be supported by facts from the chapter. Tell students that when the chart is completed, they will be able to see the relationships between the facts and generalizations.

In-depth Resources: Unit 3
- Guided Reading Worksheets, pp. 35–38

READING SOCIAL STUDIES

BEFORE YOU READ

▸▸ What Do You Know?

Before you read the chapter, consider what you already know about South America. What do you know about the region's history? What do you know about its role in the world's economy and culture? In particular, what do you know about Brazil and Peru? You may know that Brazil is famous for its successes in soccer. Think about what else you have read or seen on the news or in sports reports. Also consider connections to your own life: Have you ever heard music from South America, such as salsa?

▸▸ What Do You Want to Know?

Decide what else you want to know about South America. In your notebook, record what you hope to learn from this chapter.

Region • This Andean girl plays the region's music, called *huayno.* ▲

READ AND TAKE NOTES

Reading Strategy: Making Generalizations
Making generalizations is a useful strategy for understanding themes in social studies. A generalization is a conclusion supported by facts. Use the chart of generalizations below to better understand events and situations in South America.

- Copy the chart into your notebook.
- As you read the chapter, look for facts that support each generalization. Examples may be found in more than one section of the chapter.
- Beside each generalization, record the facts that support it.

Place • People crowd the streets of São Paulo in Brazil. ▲

Generalizations	Facts
Independence led to different types of governments in South America.	Some South American countries sought to establish representative governments, but many others were ruled by military leaders.
South American countries are trying to cooperate with one another and with the United States.	OAS and the Free Trade Agreement of 1994 are examples of cooperation among South American countries.
Geography and politics affect the economies of South America.	The Amazon rain forest and the Andes are natural barriers, and political agreements, such as the Free Trade Zone, help economies.
Urban growth presents serious problems for some South American countries.	The large numbers of poor people who crowd into major cities often lack decent housing and basic utilities.
South America's different cultures contribute to world culture.	Three South American poets have been awarded the Nobel Prize in Literature.

Teaching Strategy

Reading the Chapter This is a thematic chapter focusing on the history, geography, and current issues of South America. As they read the chapter, encourage students to look for aspects of the history and geography of South American nations that have affected their governments, economies, and citizens' lives today.

Integrated Assessment The Chapter Assessment on page 257 describes several activities for integrated assessment. You may wish to have students work on these activities during the course of the chapter and then present them at the end.

Establishing Independence

SECTION 1

TERMS & NAMES
Simón Bolívar
José de San Martín
Pan-American
Organization of American States

AIN IDEA

ter 300 years of rule by Spain
d Portugal, South Americans won
dependence and established their
wn nations.

WHY IT MATTERS NOW

Because the United States has close economic and political ties with South America, it is important to understand the history of the region.

DATELINE

EXTRA

CAJAMARCA, THE ANDES, 1533

Word has just arrived that the Inca ruler, Atahualpa, was executed by a small group of Spanish adventurers. A few weeks ago, Atahualpa's messengers informed him that the Spanish were coming. Atahualpa commanded 30,000 warriors and saw no reason to worry.

However, the Spanish, with cannons and guns, killed around 4,000 of Atahualpa's guards and then captured the Inca king. Atahualpa tried to buy his freedom, offering the Spanish a stack of gold 9 feet high in a room measuring 17 feet by 22 feet. The Spanish accepted the gold—and then executed Atahualpa anyway. Now many fear for the future of the Inca Empire.

Culture • The portrait on the left shows Atahualpa, while the one on the right shows Pizarro. ▲

Europeans Arrive in South America ❶

In 1531, Spanish explorer Francisco Pizarro landed on the coast of what is now the South American country of Peru. He had with him horses, guns, cannons, and about 200 soldiers. His forces began the long climb up into the Andes Mountains, following the Inca road that led to the city of Cajamarca (KAH•hah•MAHR•kah).

TAKING NOTES

Use your chart to take notes about South America.

Generalizations	Facts
Independence led to different types of governments in South America	
South American countries are trying to cooperate...	

SECTION OBJECTIVES

1. To examine the impact of European settlement on South America
2. To identify key events in the fight for independence
3. To describe governments and population in South America today

SKILLBUILDER
- Interpreting a Map, p. 232
- Reading a Time Line, p. 233

CRITICAL THINKING
- Distinguishing Fact from Opinion, p. 232
- Evaluating Decisions, p. 234

FOCUS & MOTIVATE

WARM-UP

Making Inferences Have students read Dateline and discuss the confrontation between the Inca and the Spanish army.

1. Why do you think Atahualpa believed that he could buy his freedom?
2. What did the Spanish actions suggest about the future of the Inca Empire?

INSTRUCT: Objective ❶

Europeans Arrive in South America

- What were some effects of Spanish and Portuguese settlement in South America? death of many Native Americans, importation of African slaves, great wealth for Spain

In-depth Resources: Unit 3
- Guided Reading Worksheet, p. 35

Reading Study Guide
(Spanish and English), pp. 68–69

Program Resources

In-depth Resources: Unit 3
- Guided Reading Worksheet, p. 35
- Reteaching Activity, p. 43

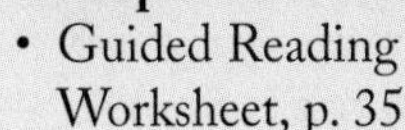

Reading Study Guide
(Spanish and English), pp. 68–69

Formal Assessment
- Section Quiz, p. 125

Integrated Assessment
- Rubric for making a poster

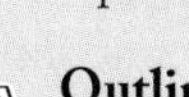

Outline Map Activities

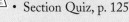

Access for Students Acquiring English
- Guided Reading Worksheet, p. 65

Technology Resources
classzone.com

TEST-TAKING RESOURCES

Strategies for Test Preparation

Test Practice Transparencies

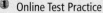

Online Test Practice

CRITICAL THINKING ACTIVITY

Distinguishing Fact from Opinion Have students reread "Europeans Establish Control" on page 232. Ask each student to write one fact from the text and one opinion about it. Collect their responses and randomly read the statements aloud. Have students identify each as a fact or an opinion. Then discuss features that students can look for to distinguish a fact from an opinion.

Class Time 15 minutes

FOCUS ON VISUALS

Interpreting the Map Ask students to point out how the colors of the map key relate to the map. What features besides European claims does the map include? present-day country borders, rivers Point out that claiming a large land area, such as Brazil, did not necessarily mean that the colonial power had explored or settled the entire territory. Ask students to suggest reasons why France, Britain, and the Netherlands wanted to have small colonies in a continent dominated by Spain and Portugal.

Possible Responses access to natural resources, land for plantations

Extension Have students use the map to determine the answers to these questions: What countries in South America are landlocked? Bolivia, Paraguay What countries does the Equator cross? Ecuador, Colombia, Brazil

Europeans Establish Control When Pizarro first encountered the Inca, he found a kingdom weakened by a bitter civil war. Pizarro quickly captured and executed the Inca ruler, Atahualpa. The Inca Empire soon fell under Spanish control.

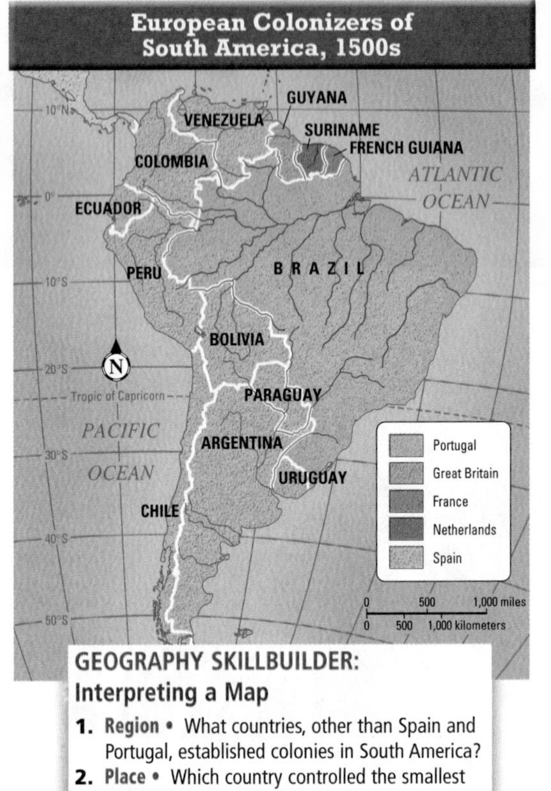

GEOGRAPHY SKILLBUILDER: Interpreting a Map

1. **Region** • What countries, other than Spain and Portugal, established colonies in South America?
2. **Place** • Which country controlled the smallest area of South America?

Geography Skillbuilder Answers

1. Great Britain, France, Netherlands
2. France

Meanwhile, Portugal had claimed what is now Brazil, and so the Portuguese began to settle the region. However, dense rain forests prevented much exploration of the region's interior. The Portuguese therefore built most of their settlements along the Atlantic Coast.

Colonial South America Many Spanish and Portuguese settlers soon made their way to South America. As happened throughout the New World, the arrival of the Europeans led to the deaths of many Native Americans. Millions died from disease or overwork. As the Native American population shrank, the Europeans imported enslaved Africans to work mainly on the large sugar cane plantations in Brazil.

For nearly 300 years, Europeans ruled much of South America. Spain and Portugal between them claimed most of the land. Ships loaded with South American silver, gold, and sugar regularly sailed to these two countries. Both, especially Spain, grew enormously wealthy from their South American colonies.

Independence ❷

In the early 1800s, Spain and Portugal were still taking most of the wealth out of the South American colonies. People of Spanish or Portuguese descent born in South America wanted to share in the political and economic power. They were encouraged by the American Revolution in 1776 and the French Revolution

Activity Options

Interdisciplinary Links: Language Arts/Writing

Class Time One class period

Task Writing a letter about Pizarro's capture of the Inca leader

Purpose To compare and contrast points of view about a historical event

Supplies Needed

- Textbook
- Writing paper
- Pens or pencils

Activity Review the events in Pizarro's capture of Atahualpa. Assign half of the students the role of a Spanish soldier in Pizarro's army and the other half the role of a Native American in Atahualpa's army. Ask each student to write a letter describing the event from the point of view of his or her assigned character. Encourage students to try to imagine their character's feelings and opinions about the event and its effects. Have them share their letters, and discuss the contrasting points of view.

Key Independence Days in South America

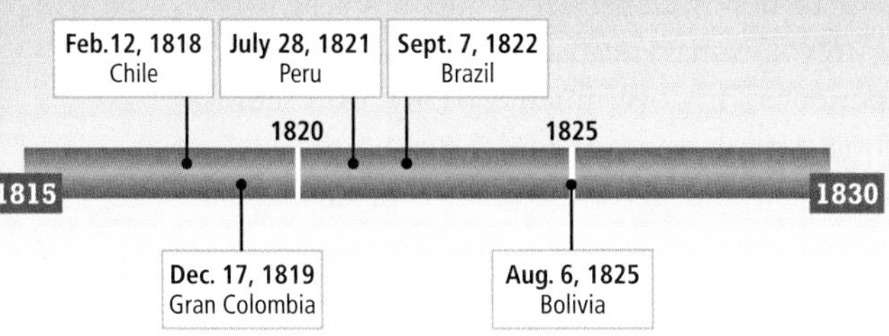

SKILLBUILDER: Reading a Time Line

1. For each country on the time line, what event occurred on the date given?
2. On what day did Brazil achieve independence?

Skillbuilder Answers

1. The country gained independence.
2. September 7, 1822

in 1789. At the same time, the *mestizos* and mulattos wanted to bring about change because they were often treated no better than slaves. South Americans soon decided to fight for independence.

Gaining Independence Beginning in 1810, two generals led a series of wars for independence. One was **Simón Bolívar** (see•MOHN boh•LEE•var), whose leadership freed the northern parts of South America. The other was **José de San Martín** (san mahr•TEEN). He was responsible for defeating Spanish forces in the south. By 1825, nearly all of Spanish South America was independent.

Reading Social Studies

A. Recognizing Important Details How did Brazil gain its independence differently from other South American countries?

Reading Social Studies A. Answer

It gained independence peacefully, choosing a Portuguese prince as its ruler.

Meanwhile, Brazil gained its independence without a major war. When the French general Napoleon Bonaparte invaded Portugal in 1807, the Portuguese royal family fled to Brazil. After Napoleon's later defeat, the Portuguese king returned to Portugal in 1821. He left his son Pedro to be regent of Brazil. When the Brazilians demanded their freedom in 1822, Pedro agreed. Brazil then named Pedro its emperor.

Biography

Simón Bolívar Simón Bolívar (shown at right) was born in Caracas, Venezuela, in 1783. As a teenager, Bolívar lived in Spain. He was influenced by the European Enlightenment and the philosophers Voltaire and Rousseau. A dream of freedom and independence for Hispanic America stirred Bolívar's soul, and he returned to South America.

Bolívar became a leader of the revolution in Venezuela in 1810. His clever military tactics led to victory over the Spanish and the creation of the republic of Gran Colombia, which included what are now Colombia, Panama, Venezuela, and Ecuador. Bolívar became president of Gran Colombia and continued fighting farther south. Victorious there, he soon became the president of Peru and Bolivia, which was named for him. Because of his role in gaining South American independence from Spain, Bolívar is often called "the Liberator." He died in Colombia in 1830.

INSTRUCT: Objective 2

Independence

- What factors encouraged the fight for independence? hope for a share in political and economic power, revolutions in the United States and France
- Who led the wars for independence? Simón Bolívar, José de San Martín

FOCUS ON VISUALS

Interpreting the Time Line Ask students to recall the steps to follow in reading a time line. Have one volunteer read aloud the title of the time line and another point out the time span it covers. Then ask students to summarize the information shown on the time line.

Possible Response Most South American countries gained their independence in the period from 1818 to 1825.

Extension Have students research how various South American countries celebrate their Independence Days. Ask them to describe the public ceremonies, special events such as fireworks or parades, and other traditional ways of marking the day. How do these celebrations compare with the American Fourth of July?

Biography

Simón Bolívar's inherited wealth allowed him to travel in Europe, and he hoped to establish a relationship with Great Britain. He requested aid from Great Britain for his first military efforts, but only received promises of neutrality. By 1830 his hopes for a united South America with close ties to Great Britain had faded.

Activity Options

Skillbuilder Mini-Lesson: Reading a Political Map

Block Scheduling

Explaining the Skill Review the features of political maps with students and discuss situations in which they might use their map-reading skills with a political map. Be sure to include features such as boundaries, country names, capitals, other cities, and rivers in your discussion.

Applying the Skill Provide students with copies of a political map of South America. Have pairs work together to list the names of the countries and their capitals. Then have them use the map key to identify and list other features. You might want to ask specific questions such as: In what direction would you travel if you went from Buenos Aires to Santiago? What is the name of one river in Brazil?

INSTRUCT: Objective 3

Governments of South America/The People of South America

- Who supported unlimited governments? Why? wealthy people and former Spanish officials; to keep their property and power
- How did governments in South America change in the 1990s? Most became more democratic.
- What is the purpose of the OAS? to promote economic cooperation, social justice, and equality of all people
- Where did immigrants to South America come from after the 1880s? Britain, Switzerland, other parts of Europe and Asia

CRITICAL THINKING ACTIVITY

Evaluating Decisions Briefly review features of Augusto Pinochet's regime for students. Point out that after ruling Chile as a dictator for almost 20 years, Pinochet was voted out of office. Tell students that Pinochet allowed his office to come up for election. Ask them to evaluate Pinochet's decision to allow the election. Encourage them to consider his assumption that he would win, the pressure from others to look like a legitimately elected leader, and whether he might have lost power anyway if he refused to allow the election. Ask students what might prompt a dictator to expose himself to the risk of an election.

Class Time 15 minutes

Governments of South America 3

However, South America's new independence did not lead to a stable, fair society, as revolutionary leaders had hoped. These leaders had little experience in government. Many of them wanted to establish constitutions that set limits on the powers of government. Doing this would allow citizens to participate in government.

Place • General Augusto Pinochet (pee•noh•CHEHT) ruled Chile as a military dictator from 1974 to 1990. ▲

Unlimited Governments However, wealthy citizens and former Spanish officials in South America wanted to keep their property and power. They did not want all citizens to have a say in government. To maintain order and to protect their interests, the powerful often gave control of the government to the military. This frequently resulted in unlimited governments, in which one person or one group held total power. In other cases, military leaders used their armies to take over limited governments in South America. By the 1990s, however, the majority of governments in South America were democratic.

BACKGROUND

A small group of people, usually military, who join together to seize government power is called a *junta* (HUN•tuh).

South American Cooperation Simón Bolívar tried to create a united South America—a nation of states like the United States. Although he was not successful, in the late 1800s the U.S. government began encouraging Pan-American unity. *Pan* means "all," so **Pan-American** means "all of the Americas."

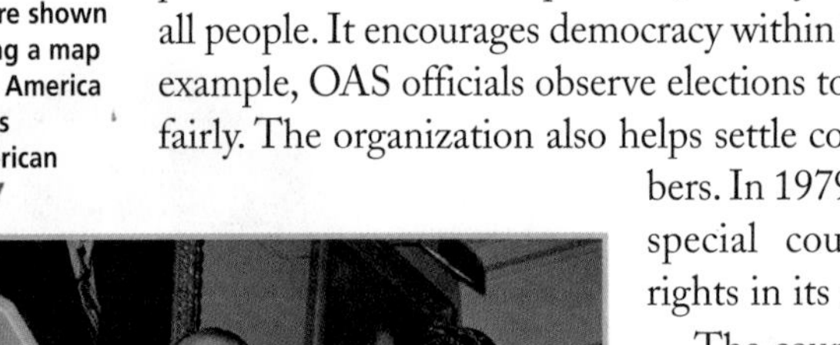

In 1948, Latin American nations joined with the United States to form the **Organization of American States** (OAS). The OAS promotes economic cooperation, social justice, and the equality of all people. It encourages democracy within its member nations. For example, OAS officials observe elections to make sure they are run fairly. The organization also helps settle conflicts among its members. In 1979, the OAS established a special court to protect human rights in its member countries.

The countries of South America will probably never join together in the way that Simón Bolívar envisioned. They are, however, working together to achieve justice and a better life for all their people.

Region • An OAS official and a U.S. government official are shown here using a map of South America to discuss Pan-American issues. ▼

Reading Social Studies

B. Analyzing Motives Why do you think the United States wanted to establish Pan-American ties?

Reading Social Studies B. Possible Answers

to encourage democratic governments, to promote economic ties and trade

Activity Options

Interdisciplinary Link: Speech

Class Time One class period

Task Making a political speech opposing unlimited government

Purpose To identify drawbacks of unlimited government and deliver a reasoned and compelling oral argument

Supplies Needed
- Writing paper
- Pens or pencils
- Tape recorders, if desired

B Block Scheduling

Activity Point out that many South Americans have struggled against military dictatorships for a long time. Have each student develop a brief speech opposing unlimited, undemocratic government. Tell students to state clearly their reasons for opposition and to include details that support their arguments. Invite volunteers to deliver their speeches and allow time for discussion after each speech.

The People of South America 3

Until the 1800s, immigrants to South America came mostly from Spain and Portugal. During the 1880s, South America attracted many more European immigrants.

Immigrants Influence Society The new immigrants helped build South America's economy by establishing a variety of industries. Also, as all immigrants do, they brought their own customs to their new home. For example, the British introduced the game of football (which people in the United States call soccer), and it quickly became a popular sport across South America.

Region • Welsh settlers started sheep ranches in South America, while Swiss immigrants in Brazil developed the cheese industry. ▼

South America's Population Today There are many ethnic groups within South America's population—Native Americans, descendants of Europeans or of enslaved Africans, and people of mixed ancestry. Some of these groups live in particular regions, while others are more widespread. For example, many Africans live in the tropical lowlands where their enslaved ancestors worked. Native Americans make up a large part of the population in the Andean nations of Peru, Bolivia, and Ecuador. The majority of South Americans, however, are either *mestizos* or mulattos.

SECTION 1 ASSESSMENT

Terms & Names

1. **Explain the significance of:** (a) Simón Bolívar (b) José de San Martín (c) Pan-American (d) Organization of American States

Using Graphics

2. Use a chart like this one to list the contributions to South America's history made by the four main groups from which the people of South America are descended.

Group	Contributions
Native Americans	
European Colonists	
Enslaved Africans	
Other Immigrants	

Main Ideas

3. (a) From what South American products did European nations profit?
   (b) How has South America's population changed since the 1500s?
   (c) What challenges have many South American governments faced in recent years?

Critical Thinking

4. **Synthesizing**
   What challenges do you think South Americans faced in their fights for independence?

   Think About
   - differences in wealth between the colonial rulers and the general population
   - who held political power
   - living conditions

ACTIVITY -OPTION- Imagine you work for the OAS. Make a **poster** that highlights the benefits the organization offers its members.

ASSESS & RETEACH

Reading Social Studies Have students add facts to the first two rows of the chart on page 230.

Formal Assessment
- Section Quiz, p. 125

RETEACHING ACTIVITY

Ask pairs of students to write two important ideas about each of the following aspects of South America: independence, governments, European influence, and people. Have pairs read aloud their ideas to the group.

In-depth Resources: Unit 3
- Reteaching Activity, p. 43

Access for Students Acquiring English
- Reteaching Activity, p. 71

Section 1 Assessment

1. Terms & Names

a. Simón Bolívar, p. 233
b. José de San Martín, p. 233
c. Pan-American, p. 234
d. Organization of American States, p. 234

2. Using Graphics

Group	Contributions
Native Americans	languages, traditional culture
European Colonists	skills, ideas of independence
Enslaved Africans	cultural traditions, economic development
Other Immigrants	new industries, sports

3. Main Ideas

a. silver, gold, and sugar
b. The population includes people from Britain, Switzerland, Africa, and other nations of Europe and Asia.
c. Governments have faced military takeovers, unfair elections, and human rights issues.

4. Critical Thinking

Most people were poor and uneducated and with little power, while colonial rulers had political and military strength.

ACTIVITY OPTION

Integrated Assessment
- Rubric for making a poster

OBJECTIVE

People throughout Latin America recognize the historical and cultural value of Aztec, Inca, and Maya ruins. In this activity, students describe through art, words, and diagrams the wonders of the Maya city Chichén Itzá.

 **Block Scheduling**

PROCEDURE

Provide supplies such as poster board and colored markers. Students who choose the first option should form groups of three or four and divide the work among themselves. Students who choose the second option should work alone first, then find a partner with whom they can share their journal.

LANGUAGE ARTS CHALLENGE

Class Time 50 minutes

Students should find books and magazine articles about Chichén Itzá and the Maya. They might display artwork from these sources or incorporate music into the advertisement.

Possible Solutions

Students' advertisements should explain the purpose of the "light and sound" show at Chichén Itzá in an attractive, appealing way. Images that might be included: a cenote, a carving or other representation of a Maya god, El Castillo, or the surrounding jungle.

Students' journal entries should describe "light and sound" clearly and vividly, conveying the students' emotional and critical responses to it.

Interdisciplinary Challenge

Explore the Mysteries of Chichén Itzá

You are a movie director making a documentary film called *Mysteries of the Maya*. You are shooting your film at the ruins of Chichén Itzá, a city of stone deep in the jungle of Mexico's Yucatán Peninsula. Many things about the Maya culture are still a mystery. You want to inform your audience about Chichén Itzá, but you also want to make them feel the mood of the place—awesome and mysterious.

COOPERATIVE LEARNING On these pages you will find challenges that you and your crew will face in making the documentary. Working with your crew, decide which one of these problems you will solve. Divide the work among crew members. Look for helpful information in the Data File. Keep in mind that you will present your solution to the class.

LANGUAGE ARTS CHALLENGE

"As darkness falls, colored lights play on the buildings . . ."

While your film crew visits Chichén Itzá, you are awed by the nighttime "light and sound show." As darkness falls, colored lights play on the buildings, and a narrator tells the Maya's story. How can you include this scene in your film? How can you convey the mood of this site to a TV audience? Choose one of these options. Use the Data File for help.

ACTIVITIES

1. Write a brief (60-second) promotional advertisement to be shown during TV station breaks. Include sketches or descriptions of photographs or art you will use in the commercial. If possible, think of background music, too.
2. As a member of the film crew, write a journal entry describing your experiences at the light and sound show.

Standards for Evaluation

LANGUAGE ARTS CHALLENGE

Option 1 Advertisements should
- emphasize visual highlights of Chichén Itzá.
- outline historical importance of Chichén Itzá.

Option 2 Journal entries should
- describe "light and sound" in detail.
- include personal reactions to the show.

SCIENCE CHALLENGE

Option 1 Diagram and speech should
- depict El Castillo's appearance accurately.
- utilize Data File information.

Option 2 Storyboard should
- connect El Castillo, calendar, and spring equinox.
- depict the serpent of the spring equinox.

ART CHALLENGE

Option 1 The drawing and speech should
- show Chac and his physical characteristics.
- capture the sacred quality of Chac.

Option 2 The design should
- incorporate symbols of the cenote.
- demonstrate understanding of the cenote.

CIENCE CHALLENGE

El Castillo, the pyramid of the serpent god"

rom the top of El Castillo, the pyramid of the serpent god Kukulcan, you an see the stone ruins and the jungle surrounding them. This great temple-yramid is the heart of Chichén Itzá. Maya priests, who were also scientists, tudied the skies and made an accurate calendar. This pyramid reflects their nowledge of astronomy. How will your film explain this knowledge? Use ne of these options to present information. Look in the Data File for help.

CTIVITIES

. Draw a diagram of El Castillo. Write a speech in which the film's narrator uses the diagram to explain the significance of the pyramid's structure and location.

. Draw a storyboard to show what happens at El Castillo during the spring equinox.

ctivity Wrap-Up

s a group, review the way you solved the challenge you hose. Organize your solution and present it to the class.

DATA FILE

CHICHÉN ITZÁ

- **City was founded by Maya** about sixth century A.D.
- The name, which means **"mouth of the wells of Itzá,"** refers to the site's two deep natural wells, or cenotes. The Itzá were a Maya group.
- One **cenote** supplied water. The other, about 200 feet across, was sacred to the rain god. Human sacrifices, mostly young people, were thrown into it, along with gold and jade ornaments.
- The Maya had many gods. **Chac** was the rain god. Kukulcan was pictured as a feathered serpent.
- Major buildings—**Pyramid of Kukulcan** (El Castillo), **Temple of the Warriors, Great Ball Court**—were built about A.D. 900–1200.
- City was abandoned about A.D. 1450.

TEMPLE OF KUKULCAN/ EL CASTILLO ("THE CASTLE")

- Four-sided pyramid represents the **Maya calendar** in several ways.
- Four steep stairways, with 91 steps each, climb each side of the pyramid. Including the top platform, the **steps total 365.** There are 18 flat platforms—the number of months in the Maya calendar.
- At the **spring and fall equinoxes,** sunlight falls on one staircase in a pattern that looks like a serpent creeping down the pyramid.

To learn more about Chichén Itzá, visit

SCIENCE CHALLENGE

Class Time 50 minutes

Provide students with poster board, rulers, markers, and tape. Have students form groups of three or four, and ask each group to outline the solution to the challenge. Then have them assign tasks to each member of the group.

Possible Solution

Students should use the film and Data File to make an accurate diagram. (The number of steps and platforms corresponds to the calendar.) The pattern on the staircase at the spring equinox looks like a serpent.

ALTERNATIVE CHALLENGE... ART CHALLENGE

There are almost no lakes or rivers on the Yucatán Peninsula. Finding a cenote meant the difference between survival and death. At the sacred cenote at Chichén Itzá the people made sacrifices to the rain god Chac, the god of water and harvest. He had a tangle of knots for hair, two fangs, and tears that poured from his eyes. The frog was associated with Chac because the song of the frog told people that rain was coming. The Maya included carvings and paintings of Chac in their cities. Why was he so important to them?

- Make a drawing of Chac and write a speech that tells about the importance of Chac and the symbolism of his features.
- Design an enclosure for the opening of the sacred cenote at Chichén Itzá, including symbols representing the cenote.

Activity Wrap-Up

To help students evaluate the creativity of their challenge solutions, have them make a grid with criteria like the one shown. Then have them rate each solution on a scale from 1 to 5.

Originality	1	2	3	4	5
Creativity	1	2	3	4	5
Accuracy	1	2	3	4	5
Overall effectiveness	1	2	3	4	5

Building Economies and Cultures

TERMS & NAMES
free-trade zone
economic indicator
urbanization

MAIN IDEA

While South America's countries work to overcome challenges to their economies, their cultures flourish.

WHY IT MATTERS NOW

Cultural elements from South America, such as music, literature, and dance, are popular around the world.

SECTION OBJECTIVES

1. To explain how geography affects the economy of a region
2. To identify South America's natural resources and industries
3. To describe daily life and the arts in South America

SKILLBUILDER
- Interpreting a Map, p. 240
- Interpreting a Chart, p. 241

CRITICAL THINKING
- Recognizing Effects, p. 239

FOCUS & MOTIVATE

WARM-UP

Making Inferences Have students read Dateline and consider the importance of the meeting in Brazil.

1. Why do you think this type of meeting was held for the first time?
2. What outcome do people hope for?

INSTRUCT: Objective 1

Geography and Trade in South America

- How does a transportation barrier differ from a transportation corridor? A barrier makes transportation difficult; a corridor is a pathway along which goods and people move.
- What geographic features of South America are barriers to trade? dense rain forest, Andes

In-depth Resources: Unit 3
- Guided Reading Worksheet, p. 36

Reading Study Guide
(Spanish and English), pp. 70–71

DATELINE

Bogotá, Colombia, September 9, 2000— All over Bogotá, people are talking about last week's meeting in Brazil's capital, Brasília. It marked the first-ever meeting of the presidents of South America's 12 nations.

Many people here say that holding this meeting in Brazil's capital indicates that Brazil is going to take a leadership role in the region. As one Colombian official in Bogotá said: "We have serious economic and political problems here. Brazil, the biggest and richest nation in South America, needs to exercise leadership." Indeed, it seems that Brazil has decided to do just that.

Region • **Brazil's president, Fernando Henrique Cardoso, sponsored the 12-nation meeting.** ▲

Geography and Trade in South America 1

Many South American nations have found that working together results in greater economic opportunity. Partly this is because they face similar challenges and possibilities. A common factor influencing many of the region's economies is geography. South America's physical geography presents the region with both transportation barriers and transportation corridors.

TAKING NOTES
Use your chart to take notes about South America.

Generalizations	Facts
Independence led to different types of governments in South America.	
South American countries are trying to cooperate . . .	

Program Resources

In-depth Resources: Unit 3
- Guided Reading Worksheet, p. 36
- Reteaching Activity, p. 44

Reading Study Guide
(Spanish and English), pp. 70–71

Formal Assessment
- Section Quiz, p. 126

Integrated Assessment
- Rubric for writing a description

Outline Map Activities

Access for Students Acquiring English
- Guided Reading Worksheet, p. 66

Technology Resources
classzone.com

TEST-TAKING RESOURCES

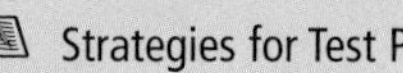
Strategies for Test Preparation

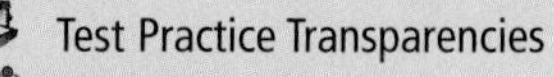
Test Practice Transparencies

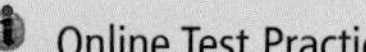
Online Test Practice

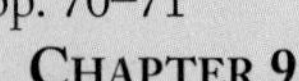

hinking Critically
nswers
. Mistral celebrates
e beauty, gifts,
nd value of the
nd.
. Students may say
uch phrases as
the land most
reen with
rchards" appealed
South Americans'
eelings.

Barriers South America's transportation barriers have interfered with trade and contacts with other cultures. For example, Portuguese explorers had trouble penetrating the dense Amazon rain forest. Because of this, they built their settlements along the coastline. Today, rain forests and rugged regions such as the Andes still prevent easy travel across the continent.

Corridors South America also has transportation corridors, such as the Amazon River system. Before the Europeans arrived, Native Americans canoed along the Amazon and its tributaries. Today, oceangoing vessels enter the Amazon system on Brazil's north coast. They carry goods such as food, clothing, and tools. They bring back lumber, rubber, animal skins, Brazil nuts, and other raw materials for shipment overseas.

Movement • Ships like this one can carry goods inland on the Amazon for more than 2,000 miles. ▲

Spotlight on CULTURE

Gabriela Mistral Chilean poet Gabriela Mistral (1889–1957; shown at right) used poetry to express her deep feelings for the people and land of South America, as in this poem, "Chilean Earth." She particularly loved children, and their rhymes and lullabies influenced her writing.

South Americans and people around the world greatly admire Mistral's work. As the first South American to win the Nobel Prize in Literature, she became a symbol of the hopes and dreams of a whole continent.

THINKING CRITICALLY

1. **Drawing Conclusions**
   What about Chile does Mistral celebrate in this poem?
2. **Forming and Supporting Opinions**
   Why do you think Mistral's poetry inspired South Americans?

For more on Gabriela Mistral, go to

RESEARCH LINKS
CLASSZONE.COM

Chilean Earth

We dance on Chilean earth
more beautiful than Lia and Raquel:
the earth that kneads men,
their lips and hearts without bitterness.

The land most green with orchards,
the land most blond with grain,
the land most red with grapevines,
how sweetly it brushes our feet!

Its dust molded our cheeks,
Its rivers, our laughter,
and it kisses our feet with a melody
that makes my mother sigh.
For the sake of its beauty,
we want to light up the fields with song.
It is free,
and for freedom we want
to bathe its face in music.

Tomorrow we will open its rocks;
we will create vineyards and orchards;
tomorrow we will exalt its people.
Today we need only to dance!

CRITICAL THINKING ACTIVITY

Recognizing Effects Review what students know about the value and condition of the Amazon rain forest. Discuss the effects of industry on the rain forest. Point out that much industry is made possible by the Amazon River system, which enables large ships to travel more than 2,000 miles inland. Then have students speculate on how travel on the Amazon River has affected the rain forest. Encourage them to recognize that there have been both positive and negative effects.

Class Time 15 minutes

Gabriela Mistral was actually a pen name for Lucila Godoy y Alcayaga. She began writing poetry while working as a schoolteacher, and she continued teaching until her poetry was widely recognized. She played an active role in the educational systems of Mexico and Chile. Mistral received honorary degrees from the Universities of Florence and Guatemala, and she taught Spanish literature in the United States at Columbia University, Middlebury College, and Vassar College.

Activity Options

Multiple Learning Styles: Linguistic

Class Time One class period

Task Writing a poem about an important place

Purpose To use language to describe a place and express feelings about it

Supplies Needed
- Writing paper
- Pencils or pens

Activity Invite a volunteer to read aloud Gabriela Mistral's poem on page 239. Point out that she uses poetry to describe her homeland and express her feelings about it. Ask students to choose a special place to write about. Have them write a poem in which they describe the place and communicate their feelings about it. Suggest that after students read their poems aloud, they can compile a class poetry collection.

INSTRUCT: Objective 2

Products and Industries of South America

- What resources are important to South American export economies? minerals, large farms producing agricultural products such as beef, wool, and coffee
- Which countries are the industrial leaders in South America? Venezuela, Chile, Argentina, Brazil
- How are South American nations trying to improve their economies? by cooperating to create a Free-Trade Zone and reduce trade barriers

FOCUS ON VISUALS

Interpreting the Map Explain that this is a thematic map showing the agricultural products of South America. Point out the map key and the lines of latitude. Ask students to note which products thrive in the northern part of the country, which thrive in the southern part, and which thrive in both parts. Have them consider factors that influence where products are grown in South America.

Possible Responses Students should notice that products such as cacao, fruit, and coffee are grown in the northern part, sheep are raised mainly in the southern part, and timber, wheat, and cattle are found in both parts. Students might say that climate influences where agricultural products thrive.

Extension Ask students to refer to the map and make a generalization about important exports.

South American Agriculture, 2001

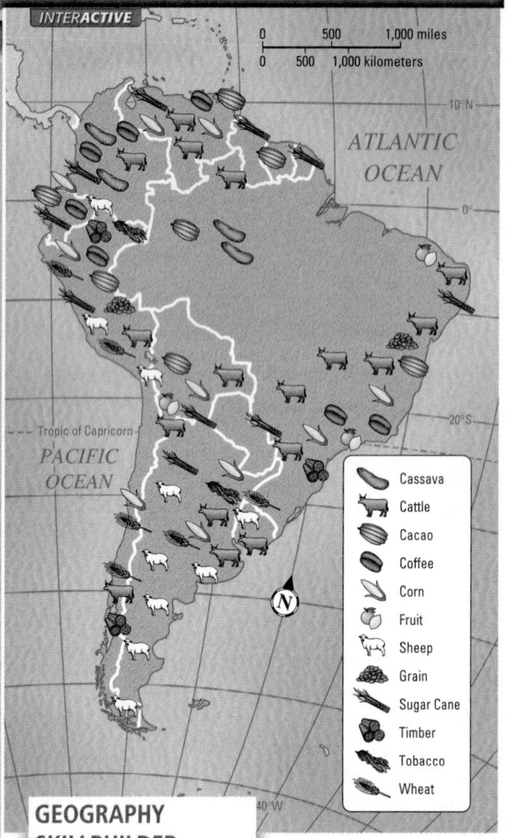

INTERACTIVE

0 500 1,000 miles
0 500 1,000 kilometers

ATLANTIC OCEAN

PACIFIC OCEAN

Tropic of Capricorn

10°N 0° 20°S 40°W

Cassava
Cattle
Cacao
Coffee
Corn
Fruit
Sheep
Grain
Sugar Cane
Timber
Tobacco
Wheat

GEOGRAPHY SKILLBUILDER: Interpreting a Map

1. **Place** • Name an agricultural product from southern South America.
2. **Region** • Name an agricultural product that is produced in many parts of South America.

Geography Skillbuilder Answers

1. sheep, wheat, cattle, corn, timber
2. sugar cane, cattle, sheep, timber

Products and Industries of South America 2

The rich natural resources of South America include abundant minerals and fertile land. However, few South American countries have fully developed their natural resources.

Mineral Resources Under the surface of South America's land lie many precious minerals—including gold, iron ore, lead, petroleum, tin, and copper. Many South American countries mine these minerals for export. For example, Chile mostly mines copper, Bolivia has a great amount of tin, and Colombia supplies the world with emeralds.

Agricultural Products South America is not only rich in mineral resources, but it also boasts some of the largest farms in the world. These farms produce goods for export, such as beef, grain, sugar, wool, bananas, and coffee. However, most of South America's farms are small. On these farms, individual farmers struggle to grow even enough food to feed their families. Many poor farmers have given up and moved to cities, hoping to find jobs there.

Reading Social Studies

A. Contrasting How do South America's small farms and large farms differ?

Reading Social Studies A. Answer

Large farms produce goods for export, while small ones struggle for subsistence.

Manufacturing The most important South American industrial countries are Venezuela, Chile, Argentina, and Brazil. In fact, Brazil is one of the most important industrial nations in the world. It manufactures enough cars and trucks to supply the entire continent. Brazil also manufactures computers, televisions, and airplanes. In other South American countries, manufactured goods include shoes, furniture, beverages, and textiles.

Economic Cooperation Lack of funding prevents many South American countries from developing manufacturing. To improve their economies, countries may cooperate economically. For example, in 1994, the heads of 34 North and South American countries met in Miami, Florida, at the first Summit of the Americas.

240 CHAPTER 9

Activity Options

Differentiating Instruction: Less Proficient Readers

Categorizing Using a chart to categorize aspects of the South American economy may help less proficient readers understand and remember what they read. Create a chart like the one shown, including the headings. Have students reread aloud "Products and Industries of South America." Then have them add items to each column in the chart.

Mineral Resources	Agricultural Products	Manufacturing
gold, iron ore, lead, petroleum, tin, copper, emeralds	beef, grain, sugar, wool, bananas, coffee	cars, computers, televisions, airplanes, shoes, furniture, beverages, textiles

BACKGROUND

The countries continued to develop the idea of a Free-Trade Zone of the Americas at additional Summit of the Americas meetings in 1998 and in 2001.

There, they agreed to create the huge Free-Trade Zone of the Americas, which would include almost every country in North and South America by the year 2005. In a **free-trade zone,** people and goods move across borders without being taxed. Many South Americans are confident that the Free-Trade Zone of the Americas will lead to greater prosperity in the region.

Country	Literacy Rate	Life Expectancy
Argentina	96%	75 years
Bolivia	83%	64 years
Brazil	85%	63 years
Chile	95%	76 years
Colombia	91%	70 years
Ecuador	90%	71 years
Guyana	98%	64 years
Paraguay	92%	74 years
Peru	89%	70 years
Suriname	93%	71 years
Uruguay	97%	75 years
Venezuela	91%	73 years

SKILLBUILDER: Interpreting a Chart

1. Which nation has the highest literacy rate? Which has the lowest?
2. What is the life expectancy in Brazil?

Skillbuilder Answers

1. Guyana (highest); Bolivia (lowest)
2. 63 years

Economic Indicators Economic cooperation among nations can be challenging if some economies are strong and others are weak. Differences can be measured by **economic indicators,** statistics that show how a country's economy is doing. The literacy rate shows the percentage of a country's people who can read and write at an elementary school level. Life expectancy, or the average age to which people in a country live, gives clues about a country's health care and nutrition.

Reading Social Studies

B. Making Inferences How do you think a high literacy rate can help boost a country's economy?

Reading Social Studies B. Possible Answer

Literate people can hold better jobs and help the country prosper.

Daily Life in South America ❸

As with all regions, South America is home to both urban and rural areas. Many of the urban areas are enormous. City populations include some very wealthy people and many middle-class people who work in government or business. Millions more, however, live in extreme poverty. Nevertheless, South Americans enjoy a proud tradition of music and literature.

The Urban Setting For the past 50 years or so, South America has experienced major **urbanization,** meaning that many people have moved from the countryside to cities. Multiple factors caused this movement to occur. For example, the growth of manufacturing created more jobs in cities. At the same time, many rural people lived in poverty, without enough land to support their families. The promise of jobs, schools, and health services drew them to cities.

Today, several South American cities rank among the largest in the world. In 2000, São Paulo (sown POW•loh) in Brazil was home to nearly 18 million people, and Buenos Aires (BWAY•nos AIR•ays) in Argentina had nearly 13 million people.

FOCUS ON VISUALS

Interpreting the Chart Discuss what the terms *literacy rate* and *life expectancy* mean. Point out that the countries are listed in alphabetical order. Explain that different observations can be made when the information is presented differently. Have students reorder the countries based on life expectancy, from longest to shortest. What do they notice about the literacy rates?

Possible Responses Students should notice that, with the exception of Guyana, higher literacy rates tend to correspond with longer life expectancy.

Extension Have students consider why a higher literacy rate might correspond with a higher life expectancy rate.

INSTRUCT: Objective ❸

Daily Life in South America

- Why have many South Americans moved from rural to urban areas? jobs, schools, and health services; lack of land to farm
- What have been the effects of urbanization? overcrowding of cities, shortage of housing, slums
- What are some of the literary accomplishments of South American writers? Nobel Prizes, creation of style of magical realism

Activity Options

Interdisciplinary Link: Economics

Block Scheduling

Class Time One class period

Task Simulating free-trade and non–free-trade exchanges using classroom items

Purpose To explore the economic differences between a free-trade and a non–free-trade zone

Supplies Needed

- Play money
- Classroom items for trade
- Writing paper
- Calculators
- Pencils

Activity Divide the class into a free-trade zone, in which tax isn't charged on goods traded, and a non–free-trade zone, in which it is. (Determine in advance the tax rate and who collects it.) Give each student an equal amount of money and a classroom item as a product to sell. Allow time for students within each zone to exchange items. Have groups evaluate the number of goods traded and the amount spent. Then have them discuss the effect of taxes on the total amounts.

MORE ABOUT…

Gabriel García Márquez

Gabriel García Márquez, a well-known writer of short stories and novels, was born in 1928 in Aracataca, Colombia. Márquez won the Nobel Prize in Literature in 1982 for his novel *One Hundred Years of Solitude*. Set in the fictional town of Macondo, the novel is the story of several generations of a Colombian family and the decay of their town.

ASSESS & RETEACH

Reading Social Studies Have students continue adding facts to the chart on page 230.

Formal Assessment
- Section Quiz, p. 126

RETEACHING ACTIVITY

Have students work in small groups to list aspects of South America's geography and natural resources that support its economy. Ask groups to share their lists with others.

In-depth Resources: Unit 3
- Reteaching Activity, p. 44

Access for Students Acquiring English
- Reteaching Activity, p. 72

Region • Peruvians play *huayno* on flutes like the one this girl is using. ▲

Houses cannot be built quickly enough to keep up with the growing number of people. Large slums surround South America's biggest cities. In these areas, people live in shacks of cardboard, wood scraps, or tin. They often have no electricity or running water.

The Arts Despite the terrible poverty of millions, the arts in South America have thrived. The region's literature is admired throughout the world. The Nobel Prize in Literature has been given to three South Americans—Chilean poets Gabriela Mistral (mih•STRAHL) and Pablo Neruda (neh•ROO•duh), and Colombian novelist Gabriel García Márquez (gar•SEE•uh MAR•kez). Many South American poets and writers express their unique cultural heritage in their work. South American novelists also founded a literary style, magical realism, in which everyday reality mixes with fantasy.

Music, too, is an important part of the culture of South America. The traditional music of the Andean regions, called *huayno,* is played on flutes and drums. Some forms of music, such as salsa, have African roots. As more and more South Americans move to urban areas, musicians there are combining the traditional musical styles with rock and other popular types of music from Europe and North America.

SECTION 2 ASSESSMENT

Terms & Names

1. **Explain the significance of:** (a) free-trade zone (b) economic indicator (c) urbanization

Using Graphics

2. Use a spider map like this one to show six factors that have influenced South America's economic progress.

Main Ideas

3. (a) Name one geographic feature that is a transportation barrier and one that is a transportation corridor.
   (b) List five of South America's important natural resources.
   (c) Name the three South Americans who have won the Nobel Prize in Literature.

Critical Thinking

4. **Identifying Problems**
   What challenges do South American countries face in building strong economies?

   Think About
   - physical geography
   - the needs of a rapidly growing population
   - the effects of urbanization

ACTIVITY -OPTION- Look at the physical map of South America on page 142. Write a short **description** of a way to overcome one of South America's transportation barriers.

Section 2 Assessment

1. Terms & Names
- a. free-trade zone, p. 241
- b. economic indicator, p. 241
- c. urbanization, p. 241

2. Using Graphics

South America's Economic Progress: free-trade zone; agricultural products; mineral resources; transportation barriers; transportation corridors; lack of money

3. Main Ideas
- a. The Andes Mountains are a transportation barrier; the Amazon River is a transportation corridor.
- b. Answers may include gold, iron ore, lead, petroleum, tin, copper, beef, grain, sugar, wool, bananas, or coffee.
- c. Gabriela Mistral, Pablo Neruda, and Gabriel García Márquez have each won the Nobel Prize in Literature.

4. Critical Thinking

Challenges include a physically impassable interior, a rapidly growing population, cooperation among countries, and overcrowded cities.

ACTIVITY OPTION

Integrated Assessment
- Rubric for writing a description

Brazil Today

TERMS & NAMES
inflation
São Paulo
Rio de Janeiro
Brasília
Carnival

MAIN IDEA

As the largest country in South America, Brazil has achieved economic success while facing challenges such as unemployment.

WHY IT MATTERS NOW

Brazil's huge land area, population, and economic success enable it to influence its neighbors in South America and North America.

DATELINE

RIO DE JANEIRO, BRAZIL, DECEMBER 30, 2001—Preparations are almost complete for the New Year's Eve celebration tomorrow. Shop owners are sold out of the small blue and white boats that worshipers of the sea goddess Iemanjá will set adrift on the sea. They will fill the boats with offerings to the goddess in hopes that she will bless them with a good fishing season.

Millions of people will also throw flowers into the water. If the flowers drift back to shore, Iemanjá is not happy with the offering. If the flowers drift out to sea, the people can expect to have a good year.

Culture • These residents of Rio de Janeiro get ready to set their boat off to greet Iemanjá. ▲

Brazil: Regional Leader 1

Almost anything that happens in Brazil is newsworthy—because Brazil is the largest country in South America, covering almost half the continent. Its population of 172 million is close in size to the combined population of all the other South American countries. Brazil's gross domestic product is larger than that of any other South American country.

TAKING NOTES
Use your chart to take notes about South America.

Generalizations	Facts
Independence led to different types of governments in South America.	
South American countries are trying to cooperate...	

SECTION OBJECTIVES

1. To identify the regional importance of Brazil's size and its government
2. To examine Brazil's economy
3. To describe Brazil's people and culture

SKILLBUILDER
- Interpreting a Graph, p. 244
- Interpreting a Map, p. 246

CRITICAL THINKING
- Hypothesizing, p. 245
- Distinguishing Fact from Opinion, p. 247

FOCUS & MOTIVATE

WARM-UP

Making Predictions Ask students to read Dateline and consider the significance of the New Year's celebration.

1. What types of things do you think the worshippers will offer to Iemanja?
2. How do you think Brazilians would react to the fate of the flowers?

INSTRUCT: Objective 1

Brazil: Regional Leader/ The Government of Brazil

- How does Brazil compare in size and population to the rest of South America? largest country in size, population nearly equals combined populations of all other South American countries

In-depth Resources: Unit 3
- Guided Reading Worksheet, p. 37

Reading Study Guide
(Spanish and English), pp. 72–73

Program Resources

In-depth Resources: Unit 3
- Guided Reading Worksheet, p. 37
- Reteaching Activity, p. 45

Reading Study Guide
(Spanish and English), pp. 72–73

Formal Assessment
- Section Quiz, p. 127

Integrated Assessment
- Rubric for writing a speech

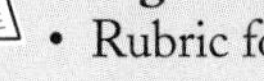
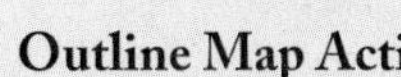
Outline Map Activities

Access for Students Acquiring English
- Guided Reading Worksheet, p. 67

Technology Resources
classzone.com

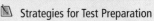

TEST-TAKING RESOURCES
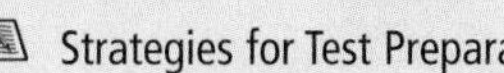
Strategies for Test Preparation
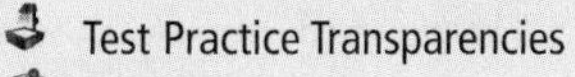
Test Practice Transparencies
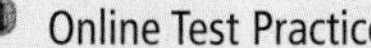
Online Test Practice

FOCUS ON VISUALS

Interpreting the Graph Point out the title and the labels on the horizontal and vertical axes of the graph. Explain that this chart shows Brazil as the economic leader of South America. Ask students how many other countries' GDPs would need to be combined to equal or exceed the GDP of Brazil.

Possible Response Students should note that when all of the other countries' GDPs are combined, the total still does not equal Brazil's.

Extension Ask students to determine how many times greater Brazil's GDP is than Venezuela's GDP.

INSTRUCT: Objective 2

The Economy of Brazil

- How has Brazil's government helped the economy grow? built automobile factories, thus reducing imports of cars and encouraging exports
- Brazil is a leading producer of what crops? coffee, oranges, bananas, corn
- What are some causes of unemployment? not enough jobs, workers do not have the education or training needed for available jobs

The Government of Brazil 1

After gaining independence from Portugal in 1822, Brazil was ruled by a series of emperors. In 1889, Brazil became a constitutional republic. Beginning in 1930, a series of dictators and military leaders ruled Brazil. Democratic government was restored in 1985. Power today is shared by a president, an elected congress, and a court system.

The Economy of Brazil 2

Brazil has the largest economy in South America. The country's gross domestic product is nearly twice that of Argentina, which is the next largest economy in South America.

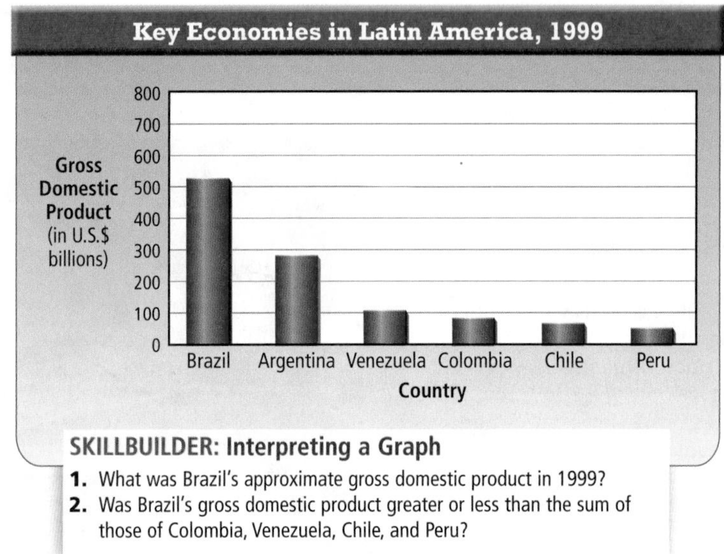

SKILLBUILDER: Interpreting a Graph

1. What was Brazil's approximate gross domestic product in 1999?
2. Was Brazil's gross domestic product greater or less than the sum of those of Colombia, Venezuela, Chile, and Peru?

Skillbuilder Answers

1. almost 800 billion U.S. dollars
2. greater than

The Growth of Industry In Brazil, the government controls or influences certain industries to help the economy grow. For example, in the 1950s, the government promoted the building of automobile factories to cut down on the number of cars imported into Brazil. By the late 1980s, Brazil was building more than 1 million vehicles a year, enough to export some to other countries.

Agricultural Production In the world, Brazil is second only to the United States in exporting crops. When it comes to coffee, Brazil produces more than any other country in the world. Brazil is also a leading producer of oranges, bananas, and corn.

Activity Options

Differentiating Instruction: Less Proficient Readers

Interpreting a Graph To help less proficient readers read and understand the graph "Key Economies in Latin America," work with students to create a list of the countries and their gross domestic products, like the one shown. Have students use the data in the list to answer questions about the economies of South American nations.

Brazil	$525 billion
Argentina	$300 billion
Venezuela	$100 billion
Columbia	$95 billion
Chile	$75 billion
Peru	$50 billion

Unemployment and Inflation Even though Brazil's economy is the strongest in Latin America, many Brazilians are unemployed. Unemployment results when not enough businesses hire workers, or when workers do not have the education or training they need for the jobs available. Even those who have jobs may face hard times, especially due to inflation. **Inflation** is a general increase in the price of goods or services. It occurs when goods or services are in great demand, allowing producers to charge higher prices for them. The combination of unemployment and inflation has led to much poverty in Brazil.

Reading Social Studies
A. Answer
Inflation makes goods and services more expensive.

Place • **Automobile plants supply Brazil with many cars, like these parked in a plant lot, for export each year.** ▲

Reading Social Studies

A. Synthesizing How would inflation pose challenges even to people who do have jobs?

The People of Brazil 3

When explorers from Portugal arrived in Brazil in 1500, as many as 5 million Native Americans lived there. During the 1500s, the Portuguese established large sugar cane plantations in northeastern Brazil. At first they enslaved Native Americans to work on the plantations. Soon, however, many Native Americans died of disease. The plantation owners then turned to Africa for labor. Eventually, Brazil brought over more enslaved Africans than any other North or South American country.

BACKGROUND

Of the 10 million to 15 million enslaved Africans brought to North and South America, more than 3.5 million ended up in Brazil.

Today, Native Americans make up less than 1 percent of Brazil's population. In northeastern Brazil, most people have African ancestors, while many people in Brazil have both European and African ancestors.

Strange but TRUE

Can Cars Run on Sugar? Brazil's first major export was sugar made from sugar cane (shown at right). However, the plentiful sugar cane isn't just used to make sugar. Today, Brazilians use sugar cane to produce ethanol. Like gasoline, ethanol is a fuel used to run cars.

Nearly half of the cars sold in Brazil run on a gasoline-ethanol mix. Thousands run on ethanol alone. Producing ethanol from sugar cane makes good use of an abundant resource and creates more jobs in the sugar cane industry. Also, its production is less harmful to the environment than is gasoline's.

CRITICAL THINKING ACTIVITY

Hypothesizing Ask students to imagine that they are part of a Brazilian government task force developing a plan to end inflation. Have them suggest steps the government might take to slow inflation. Encourage students to discuss why they think their plans might or might not be successful.

Class Time 15 minutes

Strange but TRUE

Although Brazil leads the world in sugar cane production today, sugar cane was not native to the Americas. Sugar cane was first cultivated on South Pacific islands and in India more than 8,000 years ago. Alexander the Great mentions finding sugar cane in 326 B.C. in what is now Pakistan.

From India sugar cane traveled to China, and it reached Europe in the A.D. 600s. By the 1400s Europeans had transported sugar cane to Northern Africa and to islands in the Atlantic Ocean. Christopher Columbus brought sugar cane to the Caribbean Islands in 1493, and the Portuguese first planted it in Brazil.

Activity Options

Differentiating Instruction: Students Acquiring English

Understanding Word Meaning Write *inflate* and *inflation* on the chalkboard. Discuss things we inflate to make them bigger, such as balloons, tires, or balls. Make sure students understand that when something inflates, it increases in size. Explain that the word *inflation* is used in this section to describe an economic process. Point out that when products are in high demand, vendors can charge more for them. Use a diagram like the one shown to discuss the cycle of increased cost, increased wages, and so on.

High demand for goods, services → Higher prices → Higher wages → Higher prices

INSTRUCT: Objective 3

The People of Brazil/The Culture of Brazil

- How have different ethnic groups influenced Brazilian culture? Native American influence on Portuguese language, African religious beliefs, African music and dance at Carnival
- What is Brazil's official language? Portuguese
- What is the predominant religion in Brazil? Catholicism
- What is the most popular sport in Brazil? football (American soccer)

FOCUS ON VISUALS

Interpreting the Map Ask a student volunteer to explain the concept of population density. Then direct students' attention to the different types of information shown in the key: density per square mile, density per square kilometer, and population in metropolitan areas. Ask them to use the map and key to make a generalization about population distribution in Brazil.

Possible Response Most people of Brazil live in a narrow strip along the coast. Of all the metropolitan areas larger than one million people, only two are in the interior.

Extension Have students research population densities in various areas of the United States, including their own home state, county, city, and neighborhood. (Local information may be available from county or municipal offices.) Have them compare these figures with those of Brazil.

Population Density of Brazil, 2000

GEOGRAPHY SKILLBUILDER: Interpreting a Map

1. **Place** • How many cities in Brazil have populations greater than 5 million?
2. **Region** • Which parts of Brazil are nearly unpopulated?

Place • **The overcrowding of Brazil's cities, such as São Paulo, has put many stresses on the nation.** ▲

Geography Skillbuilder Answers
1. two
2. the interior

City Populations Today, four out of five Brazilians live in cities. Brazil's two largest cities, **São Paulo** and **Rio de Janeiro,** are growing quickly. In 2000, São Paulo's population was close to 18 million. At the rate it is growing, the population will be more than 20 million in 2015. The national population is also increasing rapidly. In 1999, Brazil's population was almost 172 million. If current trends continue, by 2025 it could reach 210 million.

Because of much crowding along Brazil's Atlantic coast, the government wanted people to move into Brazil's vast interior. In 1956, it decided to create a new capital, **Brasília,** 600 miles inland. Now, like every other city in Brazil, Brasília has problems with overcrowding.

Reading Social Studies
B. Hypothesizing What problems can rapid population growth cause?

Reading Social Studies B. Possible Answers
overcrowding, housing shortages, lack of city services

The Culture of Brazil 3

Brazil's lively culture is a blend of influences from the many cultural groups that have come to Brazil over the centuries. Brazil's music, foods, and religious practices reflect that blend.

A Rich Mix Brazilian languages, religions, and musical traditions all reflect the multiple roots of Brazil's culture. For example, Brazil's official language is Portuguese. Included in Brazilian Portuguese, however, are many words from Tupi-Guarani (TOO•pee-GWAH•ruh•NEE), the language of Native Americans from the interior of northern Brazil.

Activity Options

Differentiating Instruction: Gifted and Talented

Block Scheduling

Identifying Problems and Solutions Have students work as a group to identify problems of overcrowded cities in South America. Then have them propose ways cities might tackle the problems caused by urbanization. Suggest that students then divide into smaller groups to focus on specific factors such as jobs, housing, health care, education, and city services like electricity and transportation. Encourage them to think about how one solution might lead to a new problem. Then have the groups come together to present their proposals and discuss how they interact, complement each other, or conflict.

As for religion, most Brazilians are Catholic, the religion brought to Brazil by the Portuguese. However, the number of non-Catholics is increasing. In 1940, only 5 percent of Brazilians were not Catholic. In 2000, non-Catholics had risen to 20 percent of the population because immigrants and missionaries had brought other religions to Brazil. Even so, more Catholics live in Brazil than in any other country in the world.

In addition, African religions still thrive in Brazil. For example, many people worship the African sea goddess Iemanjá. African influences can also be heard in Brazilian music, such as samba, which is based on African rhythms.

Culture • Top: Players on Brazil's women's soccer team move in for the win. Bottom: Brazilian soccer hero Pelé (PAY•lay) smiles for the crowd. ▲

Holidays A Brazilian holiday called **Carnival** highlights the country's cultural diversity. This famous festival occurs during the four days before Lent. Carnival includes huge parades and street parties. In Rio de Janeiro, groups of African Brazilians perform samba dances. The dancers wear elaborate costumes of feathers and brightly colored, sparkling cloth.

BACKGROUND

In Christianity, Lent is the 40-day period before Easter.

Sports Brazilian football, called soccer in the United States, is a sport that most of the country gets excited about. Brazil is often a finalist in the World Cup, the sport's world championship competition. Brazilians enjoy watching professional football; millions of them also enjoy playing the game.

SECTION 3 ASSESSMENT

Terms & Names

1. Explain the significance of: (a) inflation (b) São Paulo (c) Rio de Janeiro (d) Brasília (e) Carnival

Using Graphics

2. Use a chart like this one to record key facts about Brazil's geography, history, government, economy, and culture.

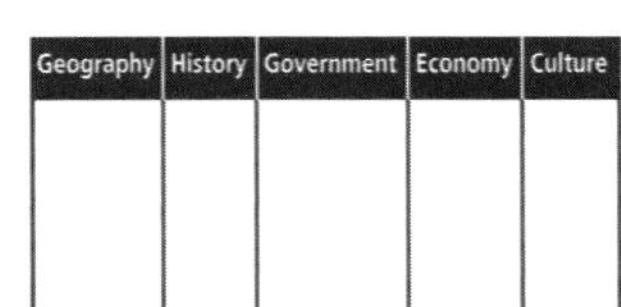

Geography	History	Government	Economy	Culture

Main Ideas

3. (a) Why can Brazil be described as an "economic giant"?

(b) Where do most people in Brazil live today?

(c) Why did the Brazilian government move the capital inland to Brasília?

Critical Thinking

4. Drawing Conclusions

Why do you think Brazil's economy is the most successful in South America?

Think About

- the government's role in the economy
- natural resources
- population size

ACTIVITY -OPTION- Write a **speech** about the different groups that have influenced Brazilian culture. Include descriptions of each group's contributions.

CRITICAL THINKING ACTIVITY

Distinguishing Fact from Opinion Ask students to write a three- to four-sentence radio advertisement designed to motivate people to move to the new inland city of Brasília. When they have finished, have students read the ads aloud. Ask other students to identify statements of fact and opinion in each advertisement. Discuss the effectiveness of the facts and the opinions.

Class Time 15 minutes

ASSESS & RETEACH

Reading Social Studies Have students complete the chart on page 230.

Formal Assessment

- Section Quiz, p. 127

RETEACHING ACTIVITY

Divide students into three groups and assign each group one of these topics: Brazilian government, economy, or culture. Have groups summarize the main ideas for their topics and then share their summaries with the others.

In-depth Resources: Unit 3

- Reteaching Activity, p. 45

Access for Students Acquiring English

- Reteaching Activity, p. 73

Section 3 Assessment

1. Terms & Names

a. inflation, p. 245
b. São Paulo, p. 246
c. Rio de Janeiro, p. 246
d. Brasília, p. 246
e. Carnival, p. 247

2. Using Graphics

Geography	History	Government	Economy	Culture
largest country cities on coast	independence from Portugal military dictatorships democratic government	emperors dictators military leaders democracy	strongest economy high unemployment inflation coffee producer	blend Native American, European, and African Catholic, African religions Carnival, football

3. Main Ideas

a. Its gross domestic product is nearly twice that of the next strongest economy in South America.
b. Most Brazilians live in cities.
c. Brasília was created to encourage people to move inland, away from the crowded coast.

4. Critical Thinking

The government actively manages industries to encourage growth, there are vast lands with many natural resources, and there is a large labor pool.

ACTIVITY OPTION

Integrated Assessment

- Rubric for writing a speech

OBJECTIVE

Students learn about the natural resources found by early Latin Americans and evaluate the legacy and benefits these resources have brought the world.

FOCUS & MOTIVATE

Drawing Conclusions Ask students to study the pictures and read each paragraph. Then have them answer the following questions:

1. What are some uses of the natural resources found in Latin America?
2. How have the plants found in Latin America benefited or improved parts of your life?

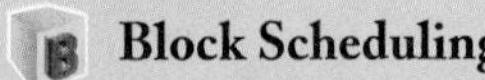
Block Scheduling

MORE ABOUT...

Food

Sapodilla is a fruit that grows on trees native to Central and South America. The chicle from the trunk of the tree was once a primary ingredient in chewing gum. Today the sapodilla tree is planted mainly for its sweet, light brown fruit that may be eaten chilled or dried. Sapodilla is a popular fruit in many restaurants.

Natural Rubber

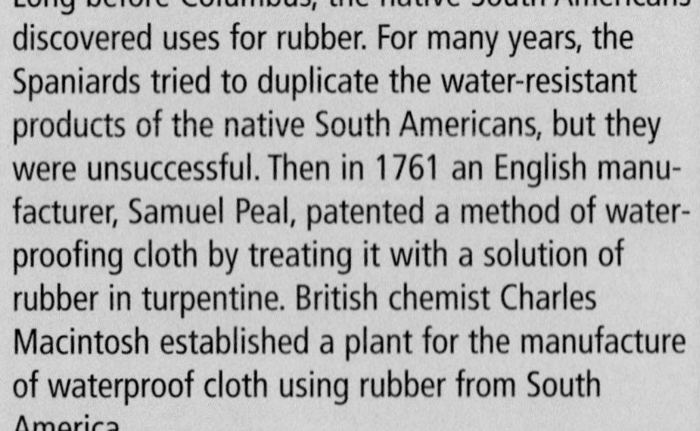

Long before Columbus, the native South Americans discovered uses for rubber. For many years, the Spaniards tried to duplicate the water-resistant products of the native South Americans, but they were unsuccessful. Then in 1761 an English manufacturer, Samuel Peal, patented a method of waterproofing cloth by treating it with a solution of rubber in turpentine. British chemist Charles Macintosh established a plant for the manufacture of waterproof cloth using rubber from South America.

The Legacy of Latin America

Food

People the world over have developed a taste for foods that originated in Latin America. Chile peppers, tomatoes, sweet potatoes, corn, and chocolate are just a few of the Latin American foods regularly found in kitchens and restaurants around the globe.

Natural Rubber

Ancient Latin Americans were the first to make use of natural rubber. Harvested from trees of the genus *Hevea,* native to Brazil, natural rubber is used around the world for everything from erasers to tires for racecars, airplanes, and trucks. The Maya invented chewing gum more than 1,000 years ago. They chewed a rubberlike substance called chicle (CHIHK•uhl), which is made from the sap of the sapodilla tree. Natural chicle was used to make chewing gum in the United States until the 1940s, when it was replaced with artificial ingredients.

Activity Options

Interdisciplinary Link: Music

Class Time One class period

Task Using music to inform others of the wealth of natural resources found in Latin American countries

Supplies Needed

- Encyclopedia and other research materials containing information about Latin American natural resources
- Musical instruments

Activity Have students work in pairs and choose one Latin American country for which they will prepare a song. Suggest they do research on resources, such as minerals and plants, found in that country. Ask them to describe, in song, those resources and how they benefit people throughout the world. Have students perform for the class.

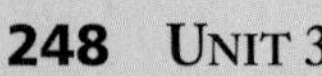

Cowboys

Pioneers in the southwestern United States first learned many of the skills of cattle ranching from *vaqueros* (vah•KEH•rohs), Mexican cowboys working on Texas cattle ranches. The *vaqueros* showed the newcomers how to use lariats, saddles, spurs, and branding irons.

Language

Some of the words we use every day come from Spanish or from Native American languages. Examples are *coyote, patio, tomato, cocoa, cafeteria, canyon, corral, chile, lariat, lasso, rodeo,* and *stampede.*

stam·pede (stăm-pēd′) *n.* **1.** A sudden frenzied rush of panic-stricken animals. **2.** A sudden headlong rush or flight of a crowd of people. **3.** A mass impulsive action: *a stampede of support for the candidate. v.* **-ped·ed, -ped·ing, -pedes** —*tr.* **1.** To cause (a herd of animals) to flee in panic. **2.** To cause (a crowd of people) to act on mass impulse. —*intr.* **1.** To flee in a headlong rush. **2.** To act on mass impulse. [Spanish *estampida,* uproar, stampede, from Provençal, from *estampir,* to stamp, of Germanic origin.] —**stam·ped′er** *n.*

Life-Saving Medicines

Native peoples in Latin America, especially in the Amazon rain forest, have been sharing their knowledge of medicinal plants for centuries—to the world's benefit. Two important contributions to modern medicine have been quinine, used to cure malaria, and curare, used to fight serious nerve diseases, such as multiple sclerosis.

Quinine ▲

Find Out More About It!

Study the text and photos on these pages to learn about inventions, creations, and contributions that have come from Latin America. Then choose the item that interests you the most and do research in the library or on the Internet to learn more about it. Use the information you gather to create a poster that celebrates the contribution.

INSTRUCT

- What are two medicinal uses of plants found in the Amazon rain forest?
- Where did the pioneers of the Southwest first learn the skills needed for ranching?
- Which ancient people invented chewing gum more than 1,000 years ago?

MORE ABOUT... Cowboys

Strong horsemen were an integral part of the Spanish culture. As early as the Middle Ages, horsemen called *caballeros* were herding horses, cattle, burros, and sheep on the Spanish plains. In Latin America, the Spanish cowboy was called the *llanero* in Venezuela, the *gaucho* in Argentina, and the *vaquero* in Mexico. From the Mexican *vaquero* evolved the well-known American cowboy.

Life-Saving Medicines

Quinine is a chemical found in the cinchona tree native to the Andes Mountains in South America. The ancient Inca used quinine as a remedy for fever. In 1633, Calancha, a monk living in Peru, described how the Indians ground the cinchona bark into a fine powder and used it to treat fevers. Quinine is still used today in the treatment of malaria.

More to Think About

Making Personal Connections Ask students to think about foods from Latin American countries that are part of their everyday lives, such as bananas. Have them consider the place of origin of fruits, vegetables, and meats that their families purchase in the supermarket. Encourage students to examine the labels of the foods their families buy and ask them to bring in labels from foods grown in Latin American countries.

Vocabulary Activity Ask students to use a dictionary to find the definitions for some of the words listed in the language section. Have students write a sentence for each word. Tell them that each sentence should be written so that the meaning of the word can be determined by the context of the sentence.

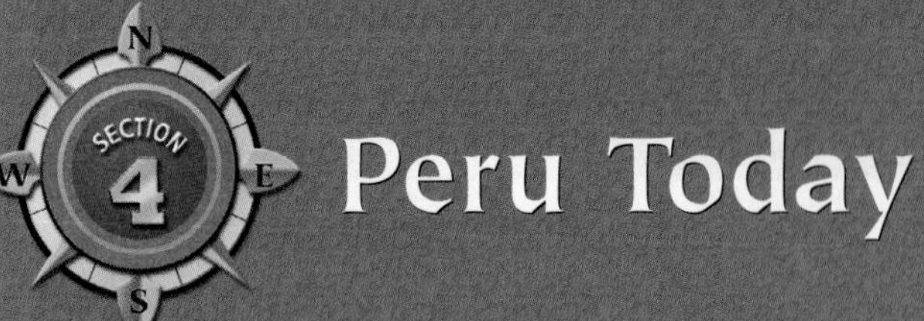

Peru Today

TERMS & NAMES
oasis
guerrilla warfare
Alberto Fujimori
Quechua

MAIN IDEA	WHY IT MATTERS NOW
Peru's physical features and its unstable governments have posed challenges to the development of its economy and its people.	Peru's economic problems also affect its trading partners, including the United States.

SECTION OBJECTIVES

1. To identify Peru's major landforms
2. To examine Peru's struggle for a stable economy and government
3. To describe Peru's people, daily life, and culture

SKILLBUILDER
- Interpreting a Map, p. 251

CRITICAL THINKING
- Comparing, p. 251
- Summarizing, p. 252

FOCUS & MOTIVATE

WARM-UP

Hypothesizing Have students read Dateline and discuss the impact of guano as an income source.

1. How might increased demand for guano affect Peru's environment and people?
2. What might happen if other guano sources were discovered?

INSTRUCT: Objective 1

The Land of Peru

- What are the major landforms in Peru? mountains, rain forest, desert
- How do these landforms affect movement? They act as transportation barriers.

In-depth Resources: Unit 3
- Guided Reading Worksheet, p. 38

Reading Study Guide
(Spanish and English), pp. 74–75

DATELINE

CHINCHA ISLANDS, PERU, 1842— Officials recently announced a major new source of prosperity for Peru. Dried sea-bird droppings, called guano, are now being exported around the world to be used as plant fertilizer.

The idea of using guano as fertilizer came from ancient Inca farmers. Recently, scientific research has confirmed that guano is rich in phosphates and nitrates, which help plants to grow. Fertilizer companies in Britain and other countries are now paying huge sums of money for the right to dig guano from Peru.

Human-Environment Interaction • Guano comes from a number of sea birds, including cormorants like this one. ▲

The Land of Peru 1

Though Peru is rich in resources, such as guano, variations in its physical geography also present problems. Three types of landforms exist in Peru: mountains, rain forest, and desert. Each type has its own special characteristics, but all three are transportation barriers rather than transportation corridors. Traveling from one part of Peru to another is not easy.

TAKING NOTES
Use your chart to take notes about South America.

Generalizations	Facts
Independence led to different types of governments in South America.	
South American countries are trying to cooperate ...	

Program Resources

In-depth Resources: Unit 3
- Guided Reading Worksheet, p. 38
- Reteaching Activity, p. 46

Reading Study Guide
(Spanish and English), pp. 74–75

Formal Assessment
- Section Quiz, p. 128

Integrated Assessment
- Rubric for writing a story

Outline Map Activities

Access for Students Acquiring English
- Guided Reading Worksheet, p. 68

Technology Resources
classzone.com

TEST-TAKING RESOURCES

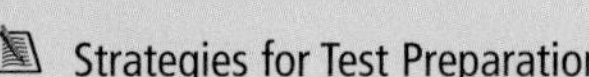
Strategies for Test Preparation

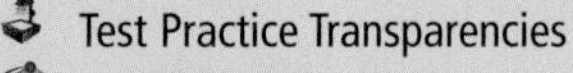
Test Practice Transparencies

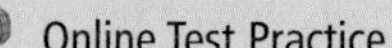
Online Test Practice

Physical Features of Peru

PACIFIC OCEAN
Gulf of Guayaquil
selva
Amazon R.
ANDES MOUNTAINS
Lake Titicaca
0°
4°S
N

Elevation
13,100 ft. (4,000 m)
6,600 ft. (2,000 m)
1,600 ft. (500 m)
650 ft. (200 m)
0 ft. (0 m)
Below sea level

0 250 500 miles
0 250 500 kilometers

GEOGRAPHY SKILLBUILDER: Interpreting a Map

1. **Region** • Through which of Peru's landforms does the Amazon River run?
2. **Place** • What is the highest elevation shown on this map?

Geography Skillbuilder Answers

1. mountains, rain forest *(selva)*
2. above 13,100 feet

Three Types of Landforms Look at the map above. You can see that the Andes Mountains run the entire length of Peru, dividing the country in two. In places, the mountains are so steep that they are practically impassable. Notice that off the eastern slopes of the Andes stretches rain forest, which is called selva in Spanish. Now look to the west of the Andes, along the Pacific coast. Here the northern stretches of Chile's Atacama Desert reach into Peru. Most of Peru's cities, large farms, and factories are located in the desert, in or near oases. An **oasis** is a fertile region in a desert that formed around a river or spring.

Reading Social Studies
A. Possible Answer
Mountains are too cold and rugged; rain forest is impassable.

The Economy of Peru ❷

Peru has many resources, but it also has many problems. The country's harsh geography affects its economy. For example, the cold, rocky highlands and the cool, dry desert cover so much area that there is not enough arable land to feed the growing population.

Reading **Social Studies**

A. Clarifying Why do you think Peru's cities are located in the desert rather than in other environments?

Place • Because of cold currents in the Pacific Ocean, Peru's desert has an average summer temperature of only 73°F. ▶

FOCUS ON VISUALS

Interpreting the Map Have students point out the areas representing Peru's three types of landforms—mountains, rain forest (selva), and desert. Point out the apparent contradiction of a desert area that is located along a seacoast. Ask students to find the source of the Amazon River and relate the river's course to Peru's landforms.

Response The Amazon begins with streams that flow eastward out of the Andes. The selva, along the river, is part of the rain forest of the Amazon Basin.

Extension Have students look up facts and figures about the Atacama Desert, one of the world's driest places. Ask them to look for an explanation of why a desert area would form along a seacoast.

CRITICAL THINKING ACTIVITY

Comparing Ask students to explain why the landforms of Peru are transportation barriers and then compare them with landforms in other countries that serve as transportation corridors. Encourage them to use the map of Peru to demonstrate why certain landforms, such as rivers, may be transportation corridors in some places but not in others.

Class Time 10 minutes

Activity Options

Skillbuilder Mini-Lesson: Reading a Political Map

Block Scheduling

Explaining the Skill Review the features of physical and political maps. Tell students that they can use physical and political maps of Peru to help them understand how geography affects movement, the economy, and daily life.

Applying the Skill Organize students into teams, and give each team a physical map of Peru. Have teams use the map to plan a Pan-Peru highway. Tell students that their highway must connect points on the Colombia border, the Pacific Coast, the border of Bolivia, and the border of Chile. Then have teams compare their maps with a political map of Peru. How do the locations of their highways compare with the locations of real highways? How did the landforms affect both their plans and the actual highways?

INSTRUCT: Objective 2

The Economy of Peru/ The Government of Peru

- Why is it difficult for Peru to grow enough food to feed its people? Much of the land is not arable.
- Why are Peru's mineral resources hard to develop? located in dense rain forests and high mountains
- What Peruvian resources are exported? fishmeal, sugar cane, coffee, cotton, silver, copper
- What problems has the Peruvian government faced since independence in 1821? civil wars, guerrilla warfare, government corruption

CRITICAL THINKING ACTIVITY

Summarizing Have students review "The Economy of Peru" and summarize the factors affecting the struggle to implement a stable economy. Encourage them to include aspects of the agriculture, mining, and fishing industries.

Class Time 10 minutes

Place • At left, a freight train transports copper ingots through Peru. At center, cotton plants grow in Ica, Peru. ▲

Place • Many crates of anchovies are sold in Peru each year. ▲

Agriculture Like many other countries, Peru must import certain foods. These include grains, vegetable oils, and some meats, many of which come from the United States. However, Peruvians do grow sugar cane, cotton, and coffee for export. Also, southern Peru has a large dairy industry that serves markets both in Peru and beyond. Meats from cattle, sheep, alpaca, and goats are also processed and distributed within the country.

Fishing The cold waters along the Pacific Coast are fine fishing grounds. Sardines and anchovies are the most important fish in the Peruvian catch. They are dried and made into fishmeal, which is sold as feed for livestock throughout the world.

Mining Peru is an important supplier of metals such as silver, copper, and bismuth. It also contains oil and gold deposits. However, the richest deposits of minerals in the country exist in dense rain forests and at elevations of over 12,000 feet. Because it is difficult to mine in these locations, Peru's mineral resources have not brought the country the great wealth that they could.

The Government of Peru 2

Peru declared itself independent of Spain in 1821. The nation was not completely free, however, until December of 1824, when Simón Bolívar finally drove out the Spanish. Following independence, Peru's military leaders began fighting one another. Struggles between military and civilian leaders continued until late in the 20th century.

Guerrilla Warfare Perhaps the greatest struggle in Peru's modern history arose in the early 1980s. At that time, Communist groups rose up to fight against the democracy that they felt was failing Peru. The most powerful of these groups was Sendero Luminoso (sen•DAIR•oh loo•mih•NOH•soh), or Shining Path.

BACKGROUND

Although in Spanish *guerrilla* is pronounced geh-REE-yuh, in English it is pronounced like the word *gorilla*. The word is Spanish for "small-scale war."

Activity Options

Multiple Learning Styles: Spatial/ Visual

Class Time One class period

Task Creating a time line of changes in Peru's government

Purpose To express information visually

Supplies Needed
- Writing paper
- Pens or pencils
- Ruler

Activity Review "The Government of Peru" and point out that achieving independence in 1821 was the beginning of a long struggle. Ask students to create their own time line of important dates related to Peru's government from 1821 to 2002. Tell students to label each date and event clearly. Students might want to illustrate their time lines. Have students share and compare their time lines.

Reading Social Studies
B. Possible Answers
disrupted daily life, caused fear

Sendero Luminoso fought for changes using **guerrilla warfare,** or nontraditional military tactics characterized by small groups using surprise attacks. The military responded, and many citizens died in the crossfire. Until Sendero Luminoso's leader, Abimael Guzmán Reynoso (ah•bee•mah•EHL gooz•MAHN ray•NAW•saw), was imprisoned in 1992, the civil war continued.

Reading Social Studies

B. Drawing Conclusions How do you think the violence and corruption in government affected people's daily lives in Peru?

Government in Crisis The 1990s did not bring better times to Peru. From 1990 to 2000, **Alberto Fujimori,** the son of Japanese immigrants, was president. At first, many of Peru's poor rural people supported him. By May 2000, however, he and his officials were accused of corruption. Many resignations followed, and in November, Fujimori abandoned the presidency and fled to Japan. The new president, Alejandro Toledo (al•eh•HAHN•droh toh•LAY•doh), faced the challenge of trying to win back the trust of Peruvians after the government scandals.

Citizenship IN ACTION

Committee to Protect Journalists In a democracy, freedom of the press helps to prevent wrongdoing by the government. In Peru, for example, many administrators resigned in disgrace after journalists exposed widespread corruption in Alberto Fujimori's administration. Elsewhere, however, many governments have jailed journalists like the one shown below for their reports.

The Committee to Protect Journalists (CPJ) was formed in 1981 to promote freedom of the press. CPJ tracks abuses against the press all over the world, makes the abuses public, and organizes protests.

3

Peruvian People and Culture

Today, more Native Americans live in Peru than in any other South American country. Forty-five percent of Peru's people are Native Americans—the descendants of the Inca. Many of these people are **Quechua** (KEHCH•wuh), people who live in the Andes highlands and speak the Inca language Quechua. Along with Spanish, Quechua is one of Peru's official languages. Many people in the highlands speak the language of another Native American group, the Aymara (EYE•mah•RAH). The Inca conquered the Aymara in the 15th century, but the language lived on.

After Native Americans, *mestizos* are Peru's next largest group. Peru's population also includes people with European, African, and Asian ancestors.

Urban and Rural Life Most of Peru's people live in cities or towns. Lima is Peru's capital and its biggest city, with about 7 million people. Lima has grown very quickly, which poses some severe problems. At the dawn of the 21st century, many neighborhoods lacked basic city services, such as electricity, running water, and public transportation.

Citizenship IN ACTION

While journalists in the United States are protected by law, journalists in many other parts of the world are not. The Committee to Protect Journalists (CPJ) was formed by a group of U.S. journalists in response to brutal and threatening treatment of fellow journalists in nations around the world.

The organization receives no government funding. Instead, it relies on contributions from foundations, corporations, and individuals. In one year alone, the CPJ documented the deaths of 24 journalists.

INSTRUCT: Objective 3

Peruvian People and Culture

- What is the largest ethnic group in Peru? Native Americans (descendants of the Inca)
- What is unusual about Peru's official languages? include both Spanish and the Inca language Quechua
- Where do most of Peru's people live? Why? in cities or towns; to have access to jobs and a better life
- Why do rural farmers have a hard time finding work in Peru's cities? little education, do not speak Spanish
- How does Peru's culture reflect its ethnic diversity? Inca beliefs mixed with Catholicism, traditional themes in literature, many speakers of Native American languages

Activity Options

Interdisciplinary Link: Current Events

Class Time One class period

Task Creating a picture-essay about Native Americans living in Peru today

Purpose To explore the beliefs, practices, customs, and issues of a Native American population in South America today

Supplies Needed
- Reference materials about Native Americans in Peru
- Paper
- Glue

Block Scheduling

Activity Recall with students that Native Americans make up almost half of Peru's population. Have students research Peruvian Native American life today to learn about their lifestyle, customs, occupations, religious practices, and so on. Have students create a picture-essay in which they include photos with captions. Students might make photocopies of illustrations or create original drawings. Display completed picture-essays.

ASSESS & RETEACH

Reading Social Studies Have students complete the "Generalizations/Facts" chart on page 230.

Formal Assessment
- Section Quiz, p. 128

RETEACHING ACTIVITY

Have students reread "The Economy of Peru" and make a list of Peru's economic advantages and disadvantages. Then have each student write a generalization about Peru's economy.

In-depth Resources: Unit 3
- Reteaching Activity, p. 46

Access for Students Acquiring English
- Reteaching Activity, p. 74

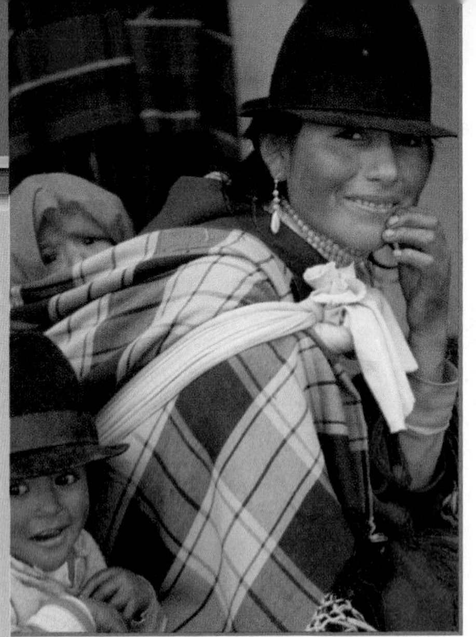

Region • Like this mother and daughter, many Quechua live in the Andes of Ecuador, Peru, and Bolivia. ▲

Many of Peru's rural farmers are very poor. They farm such small plots of land that often they cannot grow enough to feed their families. These rural people, who are mainly Native Americans, often move to the cities in search of a better life. However, many have little education and cannot speak Spanish, making it hard to find work.

Religion Peru's religions reflect multiple cultural traditions. Catholicism is the national religion of Peru, and more than 90 percent of Peruvians are Catholic. However, many Inca religious practices also still exist. At times, the Inca and Catholic customs mix. For example, some villages honor Catholic saints with traditional Inca festivals.

Literature The literature of Peru reveals modern themes as well as traditional ones. Peru's most famous living novelist, Mario Vargas Llosa (MAHR•yoh VAHR•guhs YOH•suh), is known for his belief that a novel should represent life to the fullest. César Vallejo (SAY•sar vuh•YAY•hoh), a *mestizo,* is Peru's most famous poet and is considered one of the world's best Spanish-language poets. His poetry describes what life is like for Peru's Native Americans and tells about suffering and struggles that all people may face.

SECTION 4 ASSESSMENT

Terms & Names

1. **Explain the significance of:** (a) oasis (b) guerrilla warfare (c) Alberto Fujimori (d) Quechua

Using Graphics

2. Use a chart like this one to list the products generated by each of Peru's industries.

Industry	Products
Agriculture	
Fishing	
Mining	

Main Ideas

3. (a) Describe the three types of landforms in Peru.
   (b) Describe the challenges facing Peru's economy.
   (c) Which group makes up the largest part of Peru's population?

Critical Thinking

4. **Forming and Supporting Opinions**
   Why did Peru's guerrillas use violence to bring about change? Were they right to do so? Explain.

Think About
- reasons for discontent in Peru
- the history of democracy in Peru
- other possible methods of demanding change

ACTIVITY -OPTION- Imagine you live in Peru. Write a **short story** describing your daily life, including what you do for work and for recreation.

Section 4 Assessment

1. Terms & Names
- **a.** oasis, p. 251
- **b.** guerrilla warfare, p. 253
- **c.** Alberto Fujimori, p. 253
- **d.** Quechua, p. 253

2. Using Graphics

Agriculture	sugar cane, cotton, coffee, beef, sheep, alpaca, goat meat
Fishing	sardines, anchovies
Mining	silver, copper, bismuth, oil, gold

3. Main Ideas
- **a.** mountains, rain forest, and desert
- **b.** Peru cannot produce enough food for its citizens. Mineral deposits are located in places difficult to mine.
- **c.** Native Americans, descendants of the Inca

4. Critical Thinking

Possible Response people with little power could not think of other ways; they should back up opinion with historical facts

ACTIVITY OPTION

Integrated Assessment
- Rubric for writing a story

SKILLBUILDER

Reading a Time Line

▶▶ Defining the Skill

A time line is a visual list of dates and events shown in the order in which they occurred. Time lines can be horizontal or vertical. On horizontal time lines, the earliest date is on the left. On vertical time lines, the earliest date is usually at the top.

▶▶ Applying the Skill

The time line below shows the dates of expeditions to explore the Amazon River. Use the strategies listed below to help you read the time line.

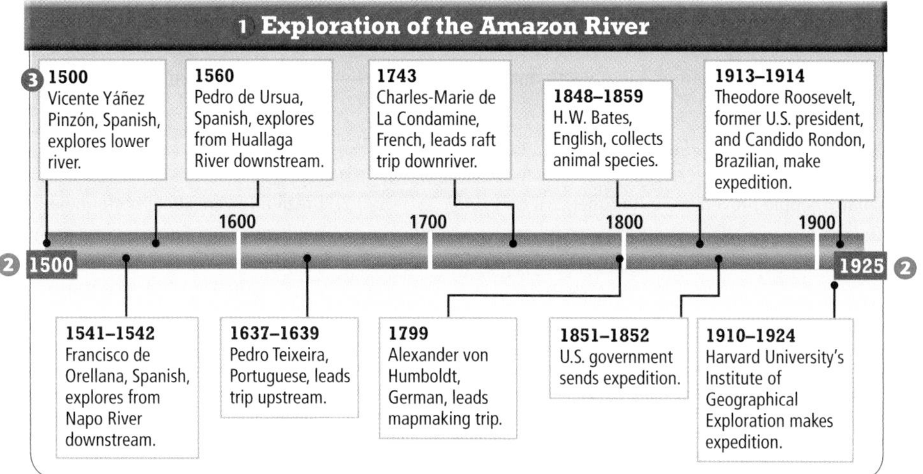

1 Exploration of the Amazon River
3 1500 Vicente Yáñez Pinzón, Spanish, explores lower river.
1560 Pedro de Ursua, Spanish, explores from Huallaga River downstream.
1743 Charles-Marie de La Condamine, French, leads raft trip downriver.
1848–1859 H.W. Bates, English, collects animal species.
1913–1914 Theodore Roosevelt, former U.S. president, and Candido Rondon, Brazilian, make expedition.
1600
1700
1800
1900
2 1500
1925 2
1541–1542 Francisco de Orellana, Spanish, explores from Napo River downstream.
1637–1639 Pedro Teixeira, Portuguese, leads trip upstream.
1799 Alexander von Humboldt, German, leads mapmaking trip.
1851–1852 U.S. government sends expedition.
1910–1924 Harvard University's Institute of Geographical Exploration makes expedition.

How to Read a Time Line

Strategy 1 Read the title. It will tell you the main idea of the time line.

Strategy 2 Read the dates at the beginning and the end of the time line. These will show the period of time that the time line covers.

Strategy 3 Read the dates and events in order, beginning with the earliest one. Think about how each event may have influenced later events. Take note of which nations were involved in each expedition.

Strategy 4 Summarize the main idea of the time line. Remember that the title will help you focus on the main idea.

Write a Summary

Writing a summary can help you understand the information shown on a time line. The summary to the right states the time period covered and the main idea of the time line.

4 The time line covers the period between 1500 and 1925. During that period of time, people from Europe and the United States explored the Amazon River. The time line shows that on their expeditions, people explored the river, made maps, and collected animal species.

▶▶ Practicing the Skill

Turn to page 233 in Chapter 9, Section 1. Look at the time line entitled "Key Independence Days in South America," and write a paragraph summarizing what you learned from it.

SKILLBUILDER

Reading a Time Line

Defining the Skill

Point out that time lines help organize dates and events in a visual way, which can make understanding and seeing connections between these events easier. Discuss time lines students have seen or made, and have students tell what information was included in the time lines. Brainstorm topics for which a time line could provide a useful way of organizing information, such as scientific discoveries or inventions.

Applying the Skill

How to Read a Time Line Explain to students that they need to work through each strategy one at a time to get the most information out of a time line. Encourage students to notice the amount of time between events and to recognize that some events happen over a period of time, shown by a span of years.

Write a Summary

Guide students as they identify the information that will go into their summaries. You might want to ask questions such as: What is the title of the time line? What is the period of time covered? What special features are noted on the time line?

Practicing the Skill

Have students turn to the time line on page 233 and apply the same set of strategies, in order. After students have completed their summaries, have volunteers read them aloud. If students still need additional practice, have them create a time line for their own lives in which they include at least six events.

In-depth Resources: Unit 3
- Skillbuilder Practice, p. 41

Career Connection: Paleontologist

Encourage students who enjoy reading the time line to find out about careers for which time lines are useful tools. For example, a paleontologist is a scientist who studies fossils of life forms that existed many thousands or even millions of years ago. Because this time span is so broad, creating and using time lines is a very helpful practice for paleontologists.

1. Suggest that students look for information about specific tasks that paleontologists undertake. Some research and organize museum exhibits; others work for oil companies, locating deposits of coal and petroleum.
2. Help students find out what subject areas they should study if they want to be paleontologists.
3. Ask students to create a collage or other visual aid to show what they have learned.

B Block Scheduling

ASSESSMENT

TERMS & NAMES

1. Simón Bolívar, p. 233
2. Pan-American, p. 234
3. free-trade zone, p. 241
4. urbanization, p. 241
5. São Paulo, p. 246
6. Brasília, p. 246
7. Carnival, p. 247
8. oasis, p. 251
9. guerrilla warfare, p. 253
10. Quechua, p. 253

REVIEW QUESTIONS

1. Native Americans died from disease or being overworked in mines and on plantations.
2. Wealthy South Americans and former Spanish officials worked to preserve their property, power, and interests by giving control of the government to the military.
3. South America's natural resources include fertile land, fishing grounds, and minerals.
4. Many people have moved to cities for jobs, education, and health services.
5. The causes of unemployment include not enough businesses hiring and workers lacking necessary education or training.
6. The government of Brazil built the city of Brasília to relieve crowding along the coast and to encourage settlement inland.
7. Landforms that make up Peru are mountains, rain forests, and deserts.
8. Peru has the largest population of Native Americans in all of South America. Forty-five percent of all Peruvians are Native American.

CHAPTER 9 ASSESSMENT

TERMS & NAMES

Explain the significance of each of the following:

1. Simón Bolívar
2. Pan-American
3. free-trade zone
4. urbanization
5. São Paulo
6. Brasília
7. Carnival
8. oasis
9. guerrilla warfare
10. Quechua

REVIEW QUESTIONS

Establishing Independence *(pages 231–235)*

1. What caused the death of so many Native Americans after the arrival of the Europeans in South America?
2. How did many South American nations end up with unlimited governments after independence?

Building Economies and Cultures *(pages 238–242)*

3. List three of South America's major natural resources.
4. Why have so many people moved to South America's cities in the past 50 years?

Brazil Today *(pages 243–247)*

5. What are two causes of unemployment in Brazil?
6. Why did the government of Brazil build the city of Brasília?

Peru Today *(pages 250–254)*

7. Describe the three types of landforms that make up Peru.
8. How does the Native American population in Peru compare in size with the Native American populations elsewhere in South America?

CRITICAL THINKING

Making Generalizations

1. Using your completed chart from Reading Social Studies, p. 230, write a paragraph explaining the facts that support one generalization about South America.

Contrasting

2. Brazil and Venezuela both gained independence from European countries. How did the road to freedom for Brazil contrast with that of Venezuela?

Synthesizing

3. When South American nations form trade blocs, how is that different from uniting politically?

Visual Summary

1 Establishing Independence

- After 300 years of colonial rule, the countries of South America gained independence.
- South America's population today reflects the arrival of immigrants from Europe, Africa, and Asia.

2 Building Economies and Cultures

- Geographic features and growing populations have challenged economic progress in South America.
- The countries of the region work together to benefit all of their economies.

3 Brazil Today

- Brazil is South America's largest and most economically successful country.
- Despite economic success, Brazil still faces unemployment and inflation.

4 Peru Today

- Peru's physical geography hinders the development of its economy.
- Peru has suffered many political problems, such as guerrilla warfare and corrupt government.

CRITICAL THINKING: Possible Responses

1. Making Generalizations
Students should state one of the generalizations in the chart on page 230, then cite relevant facts from the text. For example, for the second statement, a student might cite the organization of the OAS and the creation of a free-trade zone.

2. Contrasting
Students may say that Venezuela, under Simón Bolívar, had to fight the Spanish for independence, while Brazil achieved independence from Portugal without fighting and chose a Portuguese prince as ruler.

3. Synthesizing
Students may say that trade blocs are economic agreements that allow free trade among member nations and restrict trade with other nations. The nations remain politically independent.

the map and your knowledge of world cultures and graphy to answer questions 1 and 2.

ditional Test Practice, pp. S1–S33

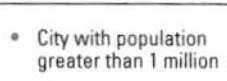

ow many South American cities had populations over 1 illion in 1950?

. 0
5
. 9
12

ow many more South American cities had populations over million in 2000 than in 1950?

. 5
. 16
. 21
. 26

In the following passage, Daniel Munduruku describes learning the ancient myths of his people. Use the quotation and your knowledge of world cultures and geography to answer question 3.

PRIMARY SOURCE

When I was a five-year-old boy, my grandfather would sit me on his lap and. . . tell the stories that explained the origins of our people and the vision of the universe. . . . . Now . . . I realize that those ancient myths say what cannot be said. They are pure poetry, and through them I see how an identity can be created in the oral tradition.

DANIEL MUNDURUKU, *Tales of the Amazon, How the Munduruku Indians Live*

3. According to the passage, in what way is storytelling important to the Munduruku culture?

A. The stories preserve the beliefs and identity of the Munduruku.

B. Telling stories is a way for grandparents to be close to young children.

C. The stories are an example of the art of the Munduruku culture.

D. The stories are important to the Munduruku if they are written down.

ERNATIVE ASSESSMENT

WRITING ABOUT HISTORY

you were a passenger in an expedition flying over Peru, you ould be able to see the long coastline, mountains, rain forests, d desert. Imagine that you are such traveler, and write a urnal entry about what you see from your flight.

Use the Internet or reference sources in the library to research the geography of Peru.

In your journal entry, describe the varied landforms you see. Include descriptive details, as well as your personal impressions of the views.

OOPERATIVE LEARNING

ith a group of classmates, decide on a problem that your untry faces, such as overcrowding. Then, plan a new city to lp solve the problem. Describe how the city would be ganized, what services would be provided for the people, and here to locate homes, businesses, and parks. Create a map of ur new city with illustrations of important buildings and a list key services that the city provides.

INTEGRATED TECHNOLOGY

Doing Internet Research

Many useful products are harvested from the rain forest. Use the Internet to research one product, such as a medicinal plant, and find out how it is used in different parts of the world.

- Using the Internet or books and periodicals in the library, find out about your product.
- Focus your research on how the product is harvested, transported, and used worldwide.
- Create a presentation of your findings. Include visuals such as maps, charts, graphs, or illustrations to support your findings.

For Internet links to support this activity, go to

STANDARDS-BASED ASSESSMENT

1. **Answer B** is the correct answer because the map for 1950 shows five red dots, each indicating a city of more than 1 million people; answer A is incorrect because it gives the amount for the year 1900; answers C and D are incorrect because the amounts given are inaccurate.
2. **Answer C** is the correct answer because there were 26 cities in 2000 and 5 in 1950, so the difference is 21; answer A is incorrect because it indicates the difference from 1900 to 1950; answers B and D are incorrect because the amounts given are inaccurate.
3. **Answer A** is correct because it focuses on the importance of oral tradition in preserving Munduruku identity; answers B and C are incorrect because they do not mention that storytelling preserves the Munduruku culture; answer D is incorrect because the passage refers to stories passed down in the oral tradition.

INTEGRATED TECHNOLOGY

Discuss possible keywords students might use to start their research about rain forest products. Have students focus their research on one product and look for information about how it is used in different parts of the world. Students' reports should clearly identify the product, its use, and its origin. Students should include a list of Web sites used.

Alternative Assessment

1. Rubric

The journal entry should
- accurately describe the physical geography of Peru as seen from the air.
- reflect the thoughts and experiences of a traveler.
- use descriptive, colorful language.
- use correct grammar, spelling, and punctuation.

2. Cooperative Learning

Explain to students that their task is to try to create a city that will help solve problems that have been identified. Suggest that students clearly state the reason for building the new city, describe the key services the city will provide, and show a map of the new city, with important streets, buildings, and other points of interest clearly labeled.

Europe, Russia, and the Independent Republics

Before You Read

Previewing Unit 4

Unit 4 explores European culture and society from its roots in ancient Greece and Rome to the many challenges facing the region today. It begins by identifying the major physical features of Europe that influenced human development. It then highlights the developments in European history that help explain the modern Europe we know today and concludes with information about Europe's role in the world.

UNIT 4

Place Completed in A.D. 80, the Colosseum in Rome, Italy, held 50,000 spectators. There they watched battles between gladiators, among other contests. The Colosseum is the largest structure that survives from the Roman Empire.

258

Unit Level Activities

1. Create a Time Line

Hang a large sheet of mural paper on the wall. Draw a horizontal line along the bottom of the paper. Show students how to divide the time line by adding small vertical lines marking every hundred years between 800 B.C. and A.D. 2000. Have students write in the years. As students read the chapter, ask volunteers to write events on the time line and to illustrate the events.

2. Identify Locations on a Map

Hang a large map of Europe on the wall. Before you begin Chapter 10, ask students to identify any European countries they have visited. Have them use red pins to mark on the map all the countries they have visited. Then ask if any students have ancestors who came from Europe, and have them use blue pins to mark those countries on the map. As you read each chapter, ask volunteers to identify the region being studied and to put a yellow pin on the map.

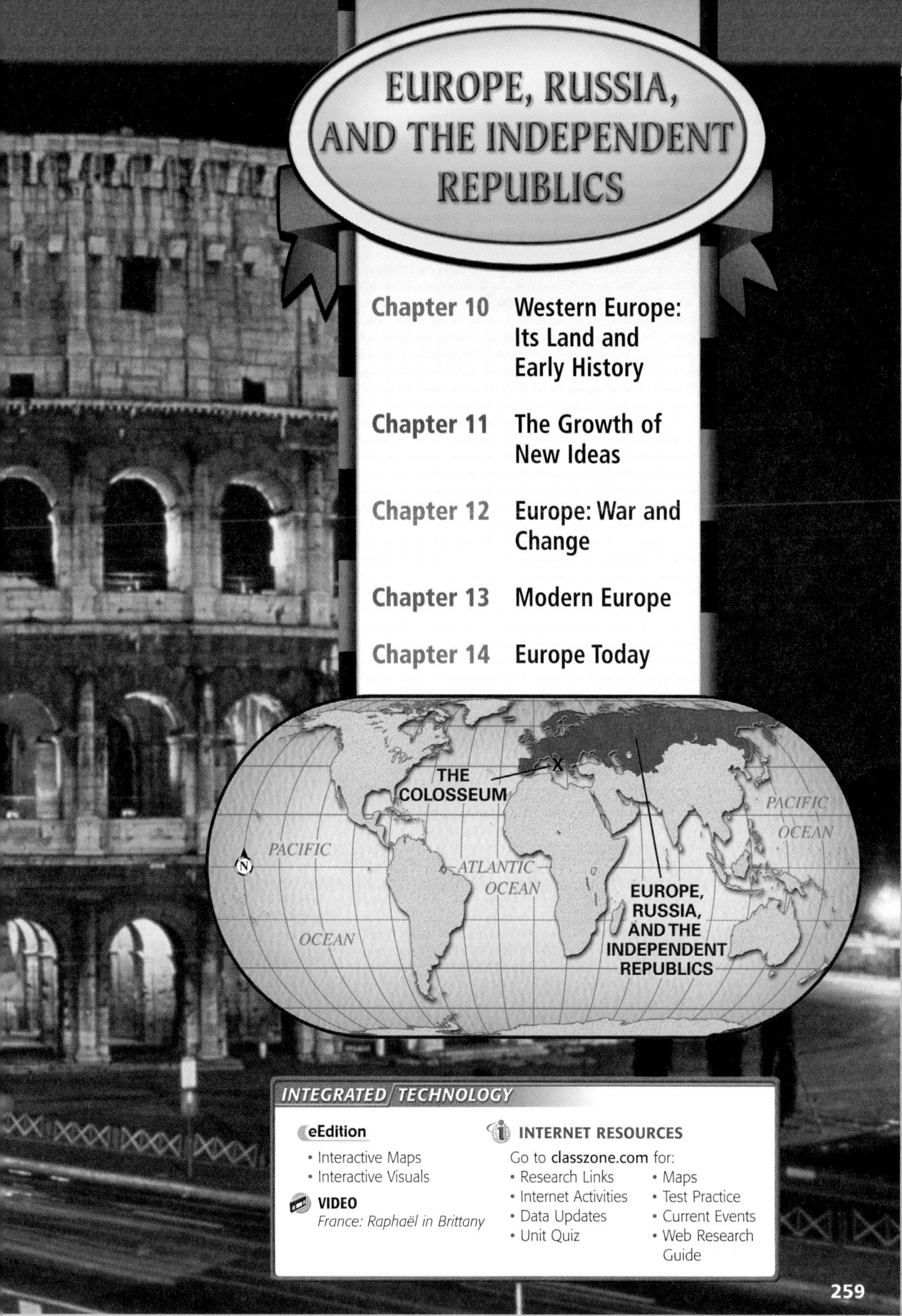

INTEGRATED TECHNOLOGY

eEdition
- Interactive Maps
- Interactive Visuals

VIDEO
France: Raphaël in Brittany

INTERNET RESOURCES
Go to **classzone.com** for:
- Research Links
- Internet Activities
- Data Updates
- Unit Quiz
- Maps
- Test Practice
- Current Events
- Web Research Guide

FOCUS ON VISUALS

Interpreting the Photograph Direct students' attention to the photograph of the Colosseum and read the caption aloud. Ask students to use their imaginations to picture what the inside of this large building must have looked like when it was completed in A.D. 80. Then ask them to list some of the events they think might have been held there. Have them compare the structure and style of the Colosseum with buildings of similar size and function today. What style of building would they be likely to see today? What use would it be put to?

Possible Responses A building of this size today would be more sleek and modern; it would be used for sports events or concerts.

Extension Ask students to draw a plan of what they think the inside of the Colosseum looked like. Have them indicate seating, playing field, entrances, and exits.

Implementing the National Geography Standards

Standard 18 Describe the immigrant experience

Objective To role-play being immigrants who are adjusting to life in their new country

Class Time 20 minutes

Task Have students role-play being immigrants to a country in Europe. Encourage them to describe how it feels to be in that situation. Ask students to explore the immigrants' perceptions of the new nation and of how to adjust to life in a different environment in order to appreciate the significance of people's beliefs, attitudes, and values in environmental adaptation.

Evaluation Students should mention the following cultural differences: language, religion, social customs, and cuisine.

ATLAS Europe, Russia, and the Independent Republics

ATLAS OBJECTIVES

1. Describe and locate physical features of Europe, Russia, and the Independent Republics
2. Examine maps and data concerning climate, landmass, and population of the region
3. Identify political features of Europe, Russia, and the Independent Republics
4. Analyze the population and examine a road map of Europe
5. Compare data on the countries of Europe, Russia, and the Independent Republics

FOCUS & MOTIVATE

Ask students how this physical map can help them understand the geography of Europe, Russia, and the Independent Republics of the former Soviet Union. Answers will vary, but students should understand the use of a physical map and the key that explains it.

INSTRUCT: Objective 1

Physical Map of Europe, Russia, and the Independent Republics

- What mountains separate France and Spain? Pyrenees
- The Aral Sea lies in what two countries? Kazakhstan and Uzbekistan
- The Ural River drains into what body of water? Caspian Sea

In-depth Resources: Unit 4
- Unit Atlas Activities, p. 1

UNIT 4 Atlas

Physical Geography

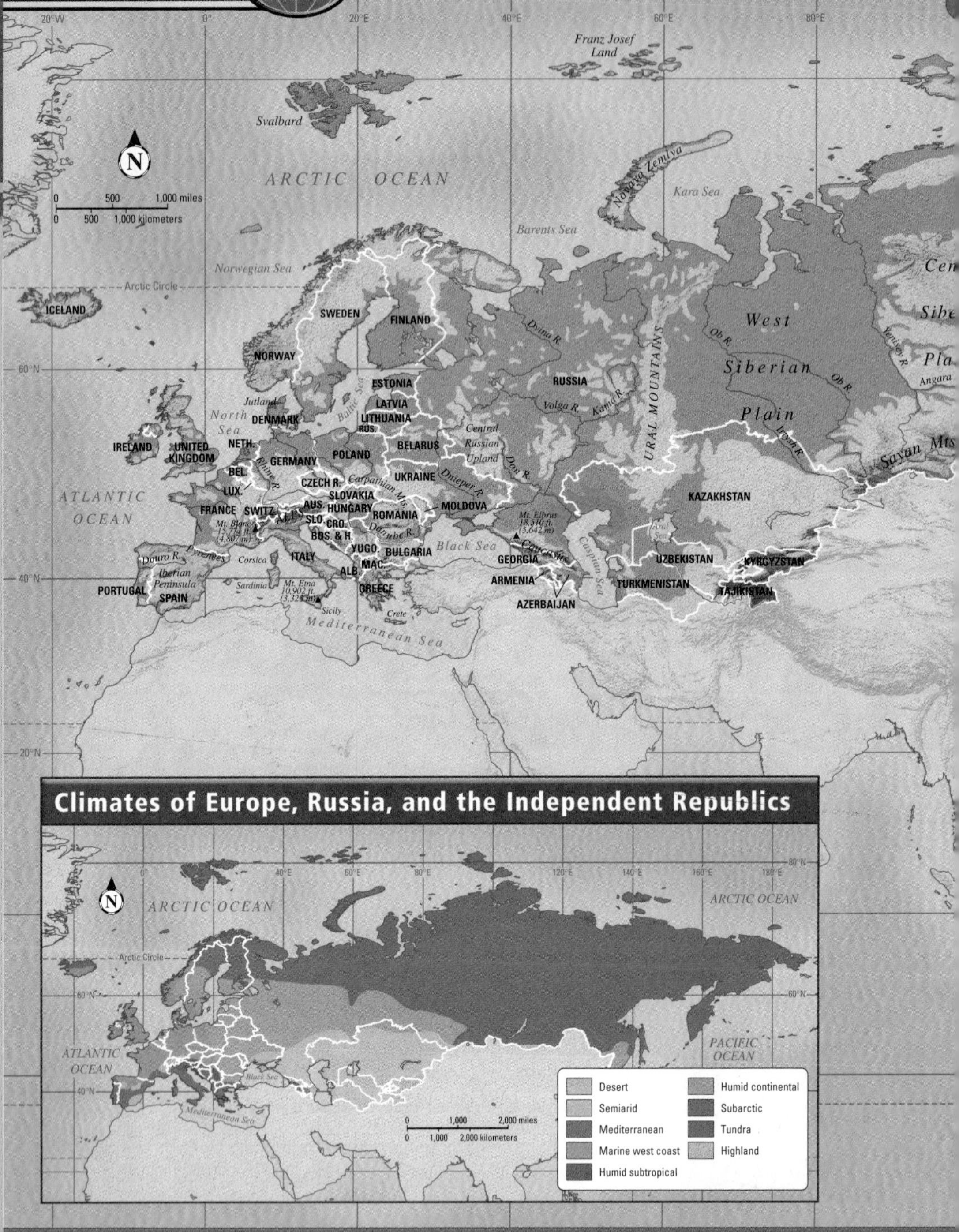

ARCTIC OCEAN

Climates of Europe, Russia, and the Independent Republics

Desert	Humid continental
Semiarid	Subarctic
Mediterranean	Tundra
Marine west coast	Highland
Humid subtropical	

260 UNIT 4

Activity Options

Interdisciplinary Link: Math

Block Scheduling

Explaining the Skill Students will demonstrate an understanding of elevation and other features shown on a physical map, as well as the use of a compass rose and scale of miles.

Applying the Skill Ask students to plan a trip that will take them through any three high mountain ranges of Europe, Russia, and the Independent Republics. Beginning with one mountain range, they should write a description of the route they take, the direction and distance they travel, and their elevation gains and losses as they travel from one range to the next. They should tell when they pass from one country to another and identify any major rivers they cross.

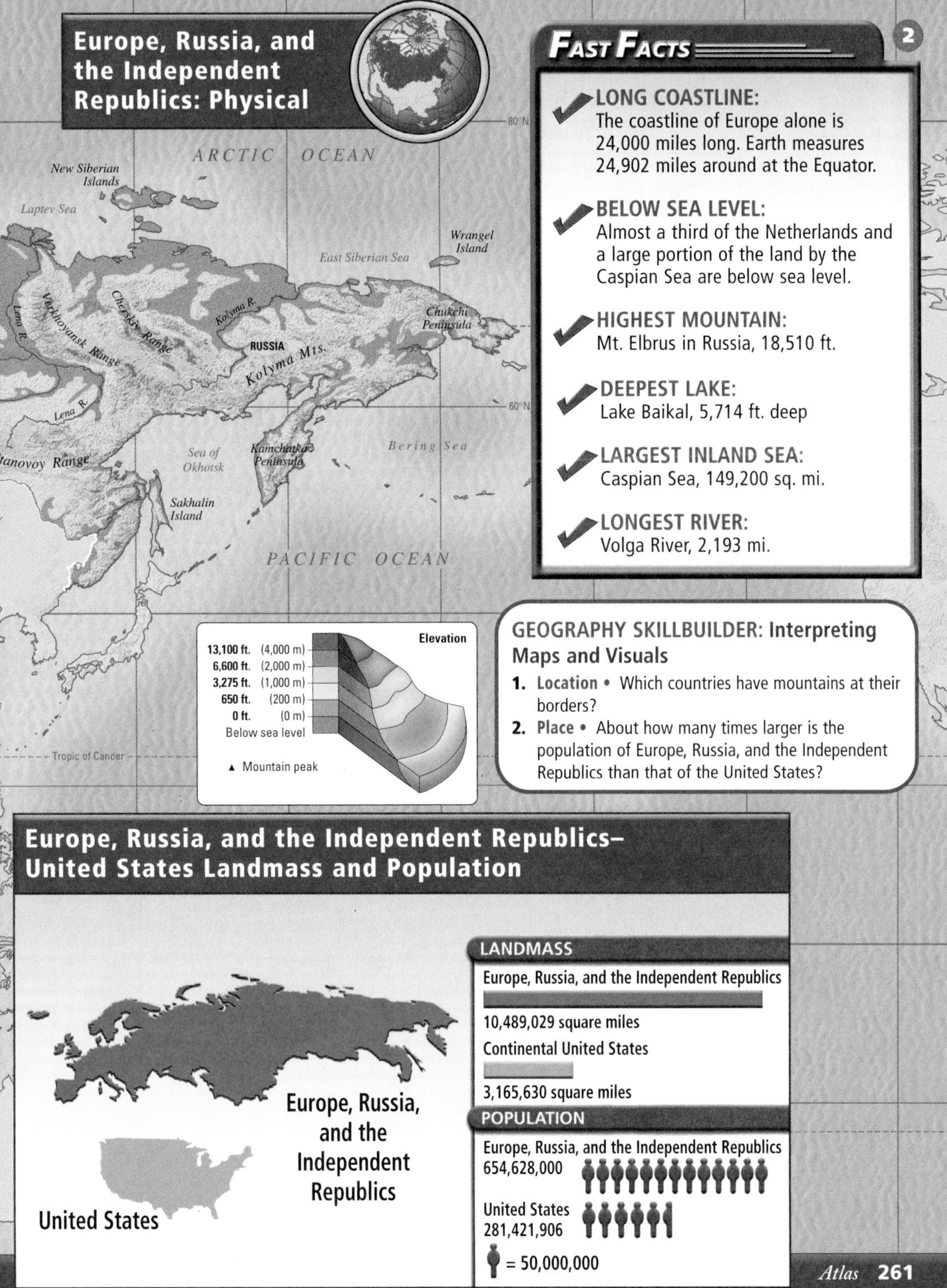

INSTRUCT: Objective 2

Climate of Europe, Russia, and the Independent Republics

- What kind of climate would you find east of the Caspian Sea? desert

Landmass

- About how many times larger are Europe, Russia, and the Independent Republics than the United States? about three and one-third times as large

Population

- What is the population of Europe, Russia, and the Independent Republics? 654,628,000 people

Fast Facts

- Is the coastline of Europe longer or shorter than the equator? By how much? shorter by 902 miles

MORE ABOUT...

Lake Baikal

At more than a mile deep, Lake Baikal is the deepest lake in the world. It is so deep that it holds more than 20 percent of the world's supply of fresh, unfrozen water. It is also an extremely old lake, dating back 25 million years. Many kinds of fish and other wildlife are found only in and around Lake Baikal.

Fast Facts

Urge students to begin their own Fast Facts file. They should add to it as they read the unit and by looking in sources such as almanacs and encyclopedias.

Geography Skillbuilder

Answers

1. Spain, France, Italy, Switzerland, Austria, Liechtenstein, Germany, Slovenia, Croatia, Bosnia and Herzegovina, Greece, Albania, Bulgaria, Romania, Russia, Kyrgyzstan
2. about two and one-third times

Country Profiles

The Netherlands The Netherlands is one of the world's smaller nations. It would be much smaller were it not for the hard work of generations of Dutch people. The North Sea, swamps, and lakes once covered more than 40 percent of the land. But the Netherlands is a crowded place, and the country has always needed more land for farms and people. The solution was simple: take the land from the sea. The Dutch would first build a dike around a section of sea or swamp and then drain it with pumps and canals. These new lands, called polders, are below the level of the sea. The Dutch must constantly battle with the sea to protect their land. Pumps run constantly to keep the polders dry. Once the pumps were powered by windmills. Today, electric pumps do the work. These polders are rich farmland. Dairy farming and dairy products are key components of the Netherlands economy.

INSTRUCT: Objective 3

Political Map of Europe, Russia, and the Independent Republics

- What European countries border the Baltic Sea? Sweden, Finland, Russia, Estonia, Latvia, Lithuania, Poland, Germany, Denmark
- Approximately how far is Russia from its eastern to its most western border? about 5,000 miles
- What is the capital of Belarus? Minsk

MORE ABOUT...

The Caspian Sea

The Caspian Sea is the largest inland body of water in the world. It covers 149,200 square miles, an area larger than Japan and nearly five times the area of Lake Superior, the largest of the Great Lakes. Unlike Lake Superior, the Caspian Sea contains saltwater. In fact, it was once connected to the oceans by the Sea of Azov, the Black Sea, and the Mediterranean Sea. Because it is fed by freshwater rivers and is now isolated from the ocean, its salinity is only one-third that of the ocean.

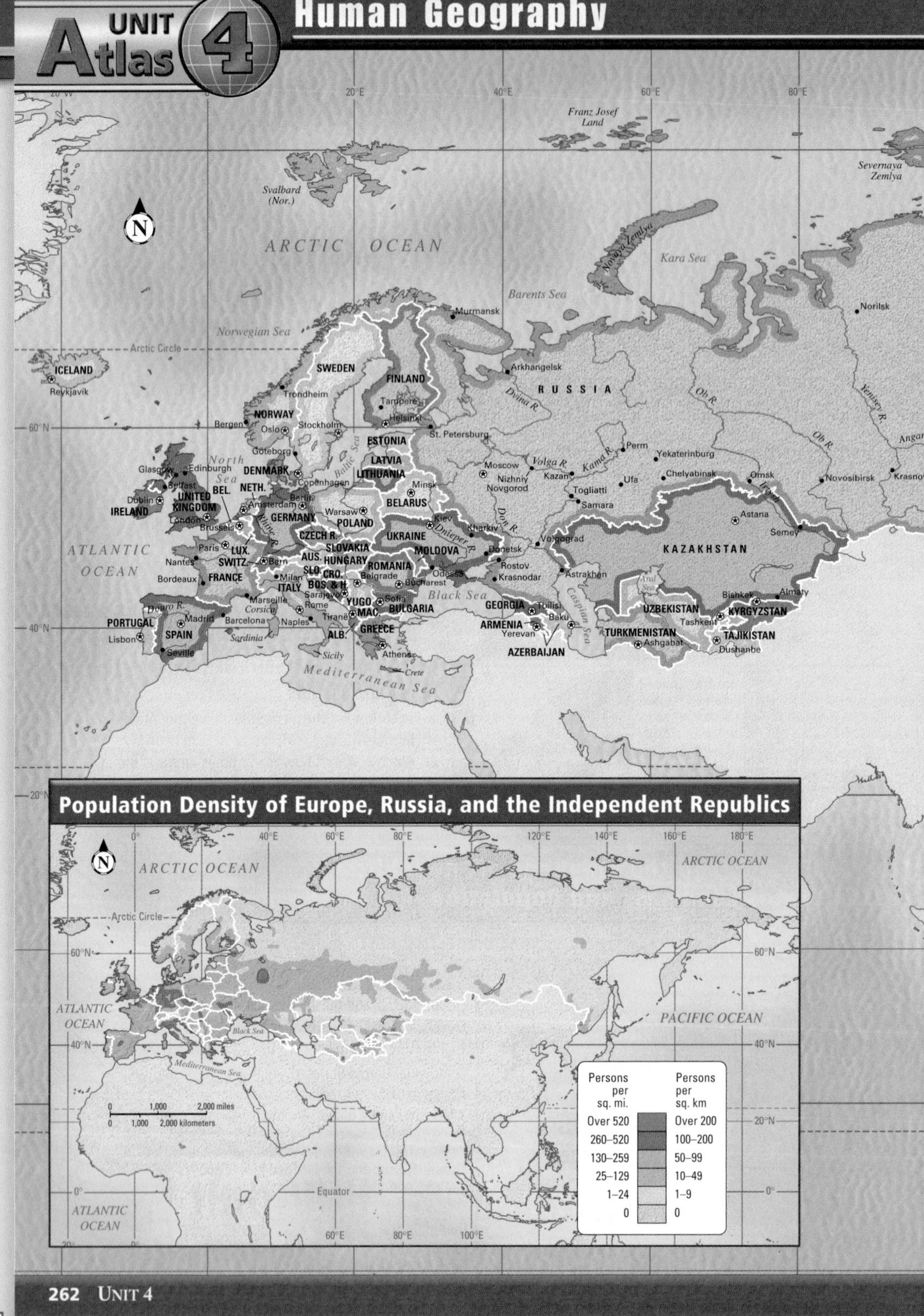

Activity Options

Interdisciplinary Link: Popular Culture

Class Time 20 minutes

Task Using a map scale and map to determine the actual size of countries

Purpose To use a map scale and map to learn about the size of Europe, Russia, and the Independent Republics

Supplies Needed

- Paper and pencil
- Ruler

Activity Tell students that they represent a travel company and have been asked to test a new train route for tourists. They will begin in Madrid and visit many of the capitals of Europe, Russia, and the Independent Republics. From Madrid, they will go to Paris, Berlin, Vienna, Sarajevo, and Sofia. The train will swing north to Kiev and Moscow before turning south to Ashgabat in Turkmenistan. Have students calculate the number of miles they will travel. If the train travels 60 miles per hour, how long will the journey take?

Europe, Russia, and the Independent Republics: Political

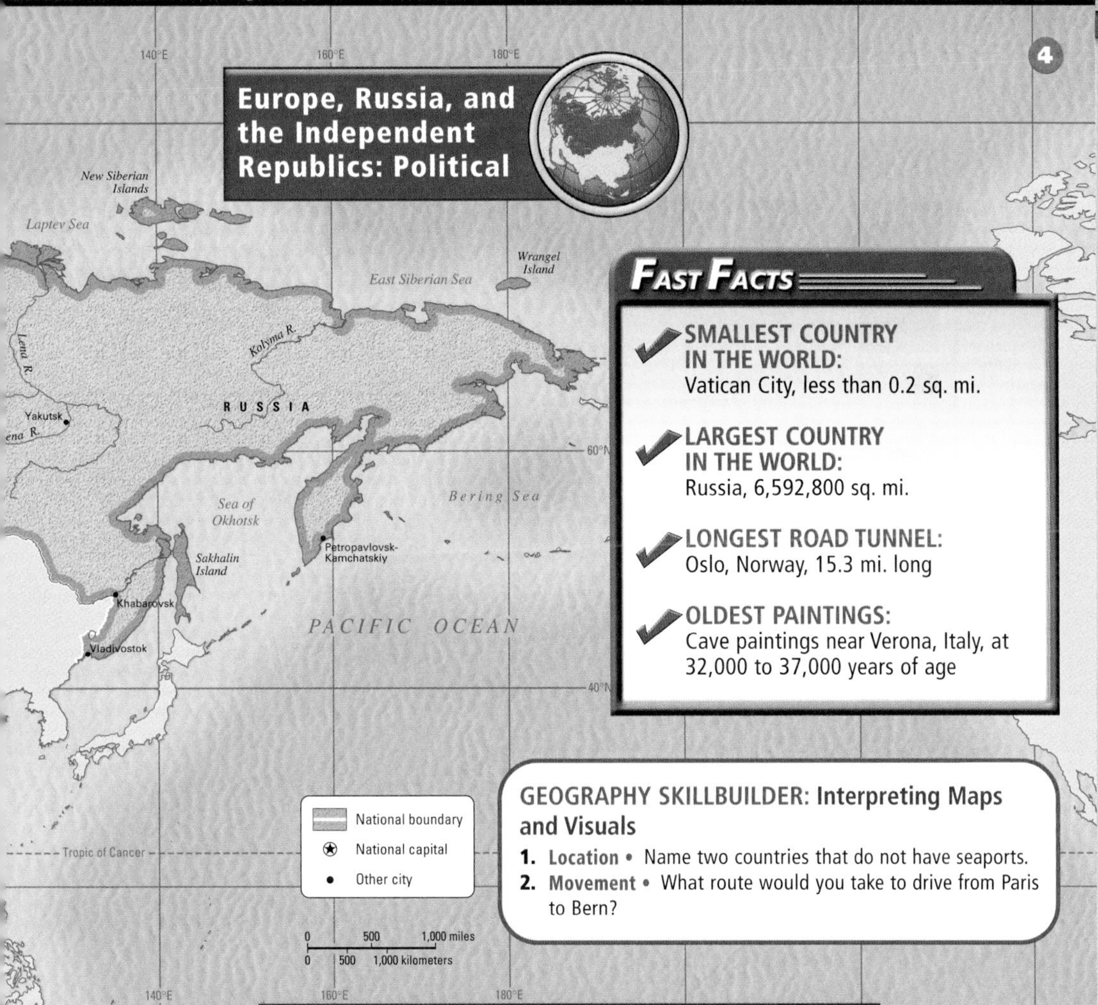

National boundary
National capital
Other city

FAST FACTS

SMALLEST COUNTRY IN THE WORLD:
Vatican City, less than 0.2 sq. mi.

LARGEST COUNTRY IN THE WORLD:
Russia, 6,592,800 sq. mi.

LONGEST ROAD TUNNEL:
Oslo, Norway, 15.3 mi. long

OLDEST PAINTINGS:
Cave paintings near Verona, Italy, at 32,000 to 37,000 years of age

GEOGRAPHY SKILLBUILDER: Interpreting Maps and Visuals

1. **Location** • Name two countries that do not have seaports.
2. **Movement** • What route would you take to drive from Paris to Bern?

Road Map of Selected European Countries

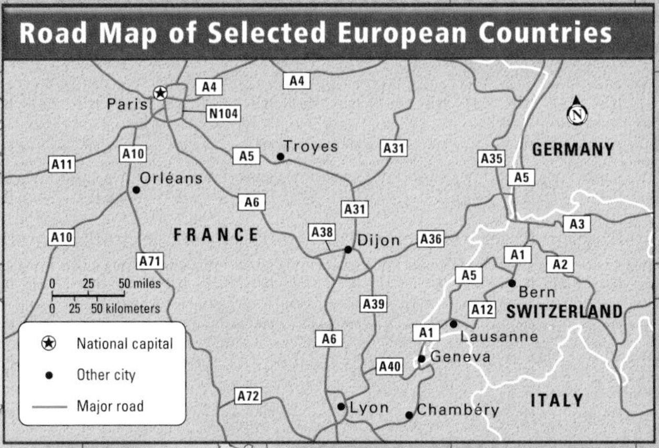

National capital
Other city
Major road

INSTRUCT: Objective 4

Population Density and Road Map of Europe

- What is the population density near London, England? over 520 people per square mile
- What is the population density around Murmansk, Russia? between 1 and 24 people per square mile
- What would be the shortest route to take if you were driving from Troyes to Lyon, France? A31 to Dijon and N79 to Lyon

MORE ABOUT...

The Oldest Cave Paintings

Archaeologists have known about the Fumane Cave on a hill near Verona, Italy, for many years. Only recently, however, have they discovered that this cave may contain what may be the oldest cave paintings in the world. The paintings date back between 32,000 and 37,000 years. The images found are somewhat mysterious. They are painted in red on rock. One picture is of an animal, which may be a weasel or cat, with five legs. The other picture is of a human wearing a mask. The mask has horns, and archaeologists speculate that the picture may represent a wizard.

Fast Facts

Encourage students to continue adding details to their Fast Facts files as they do research and continue their reading.

Geography Skillbuilder

Possible Answers

1. include Switzerland, Czech Republic, Austria, Liechtenstein, Slovakia, Hungary, Macedonia, Belarus, Azerbaijan, Kazakhstan, Uzbekistan, Turkmenistan, Tajikistan, Kyrgyzstan.
2. A6 from Paris, A38 to Dijon, A36 to Basel, and 4 to Bern

Country Profiles

Vatican City Located on just 109 acres, Vatican City is the smallest independent nation in the world. It is the center of the Roman Catholic Church, and the Pope's home is located there. Vatican City is located in the middle of Rome on the bank of the Tiber River. In most ways, the city is completely independent. It has its own radio station, telephone and banking system, post office, gardens, and pharmacy. However, many essential supplies must be imported. These include food, water, gas, and electricity. Although Vatican City is a peaceful nation, it does require protection for the Pope and for its irreplaceable works of art. The need for protection is provided by Swiss Guards who have been on duty in the Vatican since 1506.

Data File

For updates on these statistics, go

DATA UPDATE
CLASSZONE.COM

Country Flag	Country/Capital	Currency	Population (2001 estimate)	Life Expectancy (years)	Birthrate (per 1,000 pop.) (2000)
	Albania Tiranë	Lek	3,510,000	71	19
	Andorra Andorra la Vella	French Franc	68,000	83	11
	Armenia Yerevan	Dram	3,336,000	75	10
	Austria Vienna	Euro*	8,151,000	78	10
	Azerbaijan Baku	Manat	7,771,000	70	15
	Belarus Minsk	Ruble	10,350,000	68	9
	Belgium Brussels	Euro*	10,259,000	78	11
	Bosnia-Herzegovina Sarajevo	Conv. Mark	3,922,000	73	13
	Bulgaria Sofia	Lev	7,707,000	71	8
	Croatia Zagreb	Kuna	4,334,000	73	11
	Czech Republic Prague	Koruna	10,264,000	75	9
	Denmark Copenhagen	Danish Krone	5,353,000	77	12
	Estonia Tallinn	Kroon	1,423,000	70	8
	Finland Helsinki	Euro*	5,176,000	78	11
	France Paris	Euro*	59,551,000	79	13
	Georgia Tbilisi	Lavi	4,989,000	73	9
	Germany Berlin	Euro*	83,029,000	77	9

*On January 1, 2002, the euro became the common currency for 12 of the member nations of the European Union.

DATA FILE OBJECTIVE

1. Examine and compare data on Europe, Russia, and the Independent Republics

FOCUS & MOTIVATE

Which nation of Europe, Russia, and the Independent Republics has the highest rate of car ownership? Which has the highest birthrate? Tell students to scan the Data File to find the answers. car ownership: San Marino; highest birthrate: Uzbekistan Have students find these nations on the map on page 262.

INSTRUCT: Objective 5

Data File

- Which country has the fewest doctors per 100,000 people? Liechtenstein
- What is the official currency of Russia? ruble
- How does the life expectancy in Denmark compare with that of Uzbekistan? Denmark: 77 years; Uzbekistan: 69 years

In-depth Resources: Unit 4
- Data File Activities, p. 2

Activity Options

Multiple Learning Styles: Logical

Class Time 45 minutes

Task Determining relationships among data

Purpose To draw conclusions about countries in different parts of the region based on data

Activity Ask students to work in pairs and choose two countries from Europe, two Independent Republics formed from the former Soviet Union, and Russia. Have them create their own chart in which they compare life expectancy, infant mortality, numbers of doctors, and passenger cars per 1,000 people. Have students look for a relationship among these data. Then guide a class discussion. What conclusions have students drawn? What is the basis and reasoning for their conclusions?

5

DATA FILE

Infant Mortality (per 1,000 live births) (2000)	Doctors (per 100,000 pop.) (1990–1998)	Literacy Rate (percentage) (1991–1998)	Passenger Cars (per 1,000 pop.) (1996–1997)	Total Area (square miles)	Map (not to scale)
41.3	129	83	10 (1990)	11,100	
6.4	253	100	552	174	
41.0	316	98	2	11,506	
4.9	302	100	468	32,378	
83.0	360	99	36	33,436	
15.0	443	100	111	80,154	
5.6	395	99	434	11,787	
25.2	143	86	23	19,741	
14.9	345	98	202	42,822	
8.2	229	98	160	21,830	
4.6	303	99	428	30,448	
4.7	290	100	339	16,637	
13.0	297	99	294	17,413	
4.2	299	100	378	130,560	
4.8	303	99	437	212,934	
53.0	436	99	80	26,911	
4.7	350	100	504	137,830	

MORE ABOUT...

Quality of Life

The Soviet Union collapsed in 1991, and the wall dividing East and West Berlin was opened in 1989. Since then, data show that significant changes have occurred in the quality of life among people of the region. Record the following data on the chalkboard, and ask students to draw conclusions about changes that have occurred between Eastern and Western Europe following the breakup of the Soviet Union.

Germany

Life expectancy before 1991: 72

Life expectancy in 2000: 77

Cars per 1,000 people before 1991: 479

Cars per 1,000 people in 1997: 504

Poland

Life expectancy before 1991: 66

Life expectancy in 2000: 74

Cars per 1,000 people before 1991: 138

Cars per 1,000 people in 1997: 195

Slovenia

Life expectancy before 1991: 73

Life expectancy in 2000: 75

Cars per 1,000 people before 1991: 289

Cars per 1,000 people in 1997: 343

Activity Options

Differentiating Instruction: Gifted and Talented

Class Time Two class periods

Task Preparing a time line to depict the history of one republic

Purpose To learn about the history of the Independent Republics formed after the breakup of the Soviet Union

Supplies Needed

- Library and Internet resources
- Art supplies
- Computer

Activity Tell students that the Independent Republics formed after the collapse of the Soviet Union—nations such as Kazakhstan, Kyrgyzstan, Uzbekistan, and Tajikistan—have histories dating back thousands of years. Have students choose one country and explore its history. Ask them to prepare a time line to show major events. Then have them tell the story of these countries to the class.

MORE ABOUT...
Flags

Flags have special meanings to the people of the countries they fly over. Each color and symbol has a special meaning. The thirteen stripes of the U.S. flag stand for the original states. The color white stands for innocence; red stands for valor; blue for vigilance and justice. Many Christian nations, such as Denmark and Switzerland, include a cross. Many Muslim nations, such as Turkmenistan, Azerbaijan, and Uzbekistan, include a crescent moon and star to symbolize life and peace. They may also use the colors black, green, red, and white to represent unity. The Star of David appears on Israel's flag.

UNIT 4 Atlas

Data File

For updates on these statistics, go

5 

Country Flag	Country/Capital	Currency	Population (2001 estimate)	Life Expectancy (years)	Birthrate (per 1,000 pop.) (2000)
	Greece Athens	Euro*	10,624,000	78	10
	Hungary Budapest	Forint	10,106,000	71	9
	Iceland Reykjavik	Krona	278,000	80	15
	Ireland Dublin	Euro*	3,841,000	76	15
	Italy Rome	Euro*	57,680,000	78	9
	Kazakhstan Astana	Tenge	16,731,000	65	14
	Kyrgyzstan Bishkek	Som	4,753,000	67	22
	Latvia Riga	Lat	2,385,000	70	8
	Liechtenstein Vaduz	Swiss Franc	33,000	73	14
	Lithuania Vilnius	Litas	3,611,000	72	10
	Luxembourg Luxembourg	Euro*	443,000	77	13
	Macedonia Skopje	Denar	2,046,000	73	15
	Malta Valletta	Lira	395,000	77	12
	Moldova Chisinau	Leu	4,432,000	67	11
	Monaco Monaco	French Franc	32,000	79	20
	Netherlands Amsterdam	Euro*	15,981,000	78	13
	Norway Oslo	Krone	4,503,000	79	13
	Poland Warsaw	Zloty	38,634,000	74	10

*On January 1, 2002, the euro became the common currency for 12 of the member nations of the European Union.

Activity Options

Differentiating Instruction: Citizenship Activities

Class Time One class period

Task Creating a flag to represent a school or community

Purpose To think about the symbolism of flags

Supplies Needed
- Colored markers, pencils, or crayons
- Art paper

Activity Explain to students that they have been asked to create a flag to represent their school or community. Have them think about the most important ideas or qualities of their school or community. What makes it special? Why are they proud to live in their community? What do they hope their community can become? Then have them create symbols and assemble them into a flag that will represent their school or community. Have them draw the flags on art paper and display them in the room. Provide opportunities for them to explain the meaning of their flags.

5

DATA FILE

Infant Mortality (per 1,000 live births) (2000)	Doctors (per 100,000 pop.) (1990–1998)	Literacy Rate (percentage) (1991–1998)	Passenger Cars (per 1,000 pop.) (1996–1997)	Total Area (square miles)	Map (not to scale)
6.7	392	97	223	50,950	
8.9	357	99	222	35,919	
4.0	326	100	489	39,768	
6.2	219	100	292	27,135	
5.5	554	98	540	116,320	
59.0	353	99	61	1,048,300	
77.0	301	97	32	76,641	
16.0	282	100	174	24,595	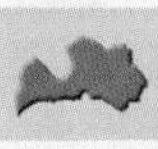
5.1	100	100	592 (1993)	62	
15.0	395	100	242	25,174	
5.0	272	100	515	999	
16.3	204	89	132	9,927	
5.3	261	91	321	124	
43.0	400 (1995)	99	46	13,012	
5.9	664	100	548	0.6	
5.0	251	100	372	16,033	
4.0	413	100	399	125,050	
8.9	236	99	195	124,807	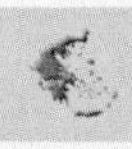

MORE ABOUT...

Currency

Explain that some cultures did not have money. Instead, they traded for the things they wanted. Ask students to consider how convenient it might be to trade a cow for some cloth, especially if the owner of the cloth did not want a cow. People began using money because it has a definite value. If the cow is worth 50 dollars and the cloth is worth 20, what would you do? If you pay in money, the exchange is easy. You can also save the money; it will be worth as much next year as it is today. Of course, money itself is of no real value. The paper it is printed on or the metals it is made from have little real value. Their value comes from the agreement among people of a society that each coin or piece of currency has a certain value.

Activity Options

Multiple Learning Styles: Logical / Mathematical

Class Time 30 minutes

Task Creating a table to aid in estimating population density of six countries

Purpose To compare the population density of different countries

Activity Working independently, have students choose six countries from the Data File that have high populations. Then ask them to create a table with three columns and six rows. They should label the first column *country*, the second column *population*, and the third column *population density*. Have them fill out the first two columns with information from the Data File. Then have them turn to the population density map on page 262. Based on the map and on the key provided, have them estimate the population density of the six countries. Are they surprised at the results?

UNIT 4 Atlas — Data File

For updates on these statistics, go

5 

MORE ABOUT...
Census

Nations have been taking censuses for thousands of years. Ancient Rome was among the first. Its interest in people and property had mainly to do with taxation and military service. Modern census-taking for the purpose of learning about how people live, their longevity, how much they earn, and what they do, for example, began in the 17th century. The first census of an entire nation took place in Canada in 1665. Sweden followed in 1749, Italy in 1770, and the United States in 1790. Today, most industrialized nations conduct a census every five or ten years. Census-taking remains an uncertain science, however. People move around and are hard to find, so many do not get counted. In South America, the government sends helicopters to survey the rain forest, looking for homes. In some cities, thousands of people live on the street and may never be counted.

Country Flag	Country/Capital	Currency	Population (2001 estimate)	Life Expectancy (years)	Birthrate (per 1,000 pop.) (2000)
	Portugal Lisbon	Euro*	10,066,000	76	11
	Romania Bucharest	Leu	22,364,000	70	11
	Russia Moscow	Ruble	145,470,000	67	8
	San Marino San Marino	Italian Lira	27,000	80	11
	Slovakia Bratislava	Koruna	5,415,000	73	11
	Slovenia Ljubljana	Tolar	1,930,000	75	9
	Spain Madrid	Euro*	40,038,000	78	9
	Sweden Stockholm	Krona	8,875,000	80	10
	Switzerland Bern	Franc	7,283,000	80	11
	Tajikistan Dushanbe	Ruble	6,579,000	68	21
	Turkmenistan Ashgabat	Manat	4,603,000	66	21
	Ukraine Kiev	Hryvnya	48,760,000	68	8
	United Kingdom London	Pound	59,648,000	77	12
	Uzbekistan Tashkent	Som	25,155,000	69	23
	Vatican City Vatican City	Vatican Lira/Italian Lira	870 (2000)	N/A	N/A
	Yugoslavia Belgrade	New Dinar	10,677,000	73	11
	United States Washington, D.C.	Dollar	281,422,000	77	15

*On January 1, 2002, the euro became the common currency for 12 of the member nations of the European Union.

268 UNIT 4

Activity Options
Multiple Learning Styles: Visual

Class Time One class period

Task Drawing an important building or monument

Purpose To learn about a capital city in Europe, Russia, or the Independent Republics

Supplies Needed
- Encyclopedia, Internet, or other information sources
- Art supplies

Activity Point out to students that a country's capital is always special. It has unique buildings for its leaders and monuments to recognize accomplishments of the people. Have students select one of the capital cities listed in the Data File and do research to learn about one important building or monument that makes the city special. Ask students to make a drawing of the building or monument. Have them share their drawings and tell what is important about the building or monument.

DATA FILE

5

Infant Mortality (per 1,000 live births) (2000)	Doctors (per 100,000 pop.) (1990–1998)	Literacy Rate (percentage) (1991–1998)	Passenger Cars (per 1,000 pop.) (1996–1997)	Total Area (square miles)	Map (not to scale)
6.0	312	91	295	35,514	
20.5	184	98	106	92,042	
20.0	421	100	120	6,592,812	
8.8	252	99	955	23	
8.8	353	100	185	18,923	
5.2	228	99	343	7,819	
5.7	424	97	384	195,363	
3.5	311	100	417	173,730	
4.8	323	100	460	15,942	
117.0	201	99	31	55,251	
73.0	300 (1997)	98	N/A	188,455	
22.0	299	100	97	233,089	
5.7	164	100	434	94,548	
72.0	309	88	37	173,591	
N/A	N/A	100	N/A	0.17	
10.4	203	98	173	39,448	
7.0	251	97	489	3,787,319	

GEOGRAPHY SKILLBUILDER: Interpreting a Chart

1. Place • Which country in the region has the highest life expectancy?
2. Place • How many fewer cars per thousand people does Greece have than Germany?

MORE ABOUT...

Life Expectancy

Life expectancy is a means of measuring how long people of a particular age can expect to live. Students should be aware that life expectancy is based on many factors, including health care, nutrition, physical condition, heredity, and occupation. Typically, more highly industrialized nations have higher life expectancies than poor countries. This may be because of a higher standard of living that permits better and more frequent health care, better food, and safer working conditions. It is also true that in most countries, women have a longer life expectancy than men.

Geography Skillbuilder

Answers

1. Andorra with a life expectancy of 83 years
2. 281 fewer cars per thousand people

Activity Options

Interdisciplinary Links: Science/Health

Block Scheduling

Class Time One class period

Task Researching and preparing a line graph showing changes in life expectancy

Purpose To learn how life expectancy has changed

Supplies Needed

- Library or Internet resources
- Pencil or pen
- Graph paper

Activity Have students do research to learn about life expectancy since 1900. Students might research life expectancy in the United States or in one of the other nations listed in the Data File. Have students plot their findings on a line graph. Invite them to share what they learn with the class.

Western Europe: Its Land and Early History

	OVERVIEW	COPYMASTERS	INTEGRATED TECHNOLOGY
UNIT ATLAS AND CHAPTER RESOURCES	The students will explore the geography of Europe, the lasting achievements of the ancient Greek and Roman civilizations, and developments in Europe during the medieval period.	**In-depth Resources: Unit 4** • Guided Reading Worksheets, pp. 3–6 • Skillbuilder Practice, p. 9 • Unit Atlas Activities, pp. 1–2 • Geography Workshop, pp. 61–62 **Reading Study Guide** (Spanish and English), pp. 78–87 **Outline Map Activities**	• eEdition Plus Online • EasyPlanner Plus Online • eTest Plus Online • eEdition • Power Presentations • EasyPlanner • Electronic Library of Primary Sources • Test Generator • Reading Study Guide • Critical Thinking Transparencies CT19
	KEY IDEAS		
SECTION 1 A Land of Varied Riches pp. 273–277	• Geographical features of Europe contributed to the development of different cultures. • Europe's temperate climate benefits its agricultural and tourist industries. • The natural resources of Europe affect what it produces today.	**In-depth Resources: Unit 4** • Guided Reading Worksheet, p. 3 • Reteaching Activity, p. 11 **Reading Study Guide** (Spanish and English), pp. 78–79	**Map Transparencies MT24, 25** classzone.com Reading Study Guide
SECTION 2 Ancient Greece pp. 278–282	• Greek city-states differed in forms of government, but shared common beliefs and a way of life. • Democracy was first practiced in sixth-century Greece. • Greek culture spread across Europe through military conquests.	**In-depth Resources: Unit 4** • Guided Reading Worksheet, p. 4 • Reteaching Activity, p. 12 **Reading Study Guide** (Spanish and English), pp. 80–81	classzone.com Reading Study Guide
SECTION 3 Ancient Rome pp. 284–289	• Rome became a republic in 509 B.C., but the lives of patricians and plebeians were very different. • The Roman Empire expanded under Julius Caesar and Augustus. • After initial persecution by the Romans, Christianity became the official religion of the Roman Empire.	**In-depth Resources: Unit 4** • Guided Reading Worksheet, p. 5 • Reteaching Activity, p. 13 **Reading Study Guide** (Spanish and English), pp. 82–83	**Critical Thinking Transparencies CT20** classzone.com Reading Study Guide
SECTION 4 Time of Change: The Middle Ages pp. 290–295	• The fall of the Roman Empire brought rapid decline to most of Europe. • The Roman Catholic Church was a central part of life in the Middle Ages. • Nobles and peasants benefited from feudalism and manorialism differently.	**In-depth Resources: Unit 4** • Guided Reading Worksheet, p. 6 • Reteaching Activity, p. 14 **Reading Study Guide** (Spanish and English), pp. 84–85	classzone.com Reading Study Guide

KEY TO RESOURCES

Audio · CD-ROM · Copymaster · Internet · Overhead Transparency · Pupil's Edition · Teacher's Edition · Video

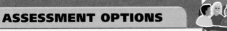

ASSESSMENT OPTIONS

Chapter Assessment, pp. 296–297

Formal Assessment
- Chapter Tests: Forms A, B, C, pp. 145–156

Test Generator

Online Test Practice

Strategies for Test Preparation

Section Assessment, p. 277

Formal Assessment
- Section Quiz, p. 141

Integrated Assessment
- Rubric for writing a short story

Test Generator

Test Practice Transparencies TT29

Section Assessment, p. 282

Formal Assessment
- Section Quiz, p. 142

Integrated Assessment
- Rubric for presenting an oral report

Test Generator

Test Practice Transparencies TT30

Section Assessment, p. 289

Formal Assessment
- Section Quiz, p. 143

Integrated Assessment
- Rubric for creating a comparison chart

Test Generator

Test Practice Transparencies TT31

Section Assessment, p. 295

Formal Assessment
- Section Quiz, p. 144

Integrated Assessment
- Rubric for writing journal entries

Test Generator

Test Practice Transparencies TT32

RESOURCES FOR DIFFERENTIATING INSTRUCTION

Students Acquiring English/ESL

Reading Study Guide (Spanish and English), pp. 78–87

Access for Students Acquiring English Spanish Translations, pp. 75–84

Modified Lesson Plans for English Learners

Less Proficient Readers

Reading Study Guide (Spanish and English), pp. 78–87

TE Activities
- Rereading, p. 274
- Understanding Key Concepts, p. 292

Gifted and Talented Students

TE Activity
- Researching a Historical Figure, p. 281

CROSS-CURRICULAR CONNECTIONS

Humanities
Hart, Avery, and Paul Mantell. ***Knights and Castles: 50 Hands-On Activities to Experience the Middle Ages.*** Charlotte, VT: Williamson Publishing, 1998. Games, celebrations, food, and customs.

Stroud, Jonathan. ***Ancient Rome: A Guide to the Glory of Imperial Rome.*** New York: Kingfisher, 2000. Ancient Rome in a travel guide format.

Literature
Spires, Elizabeth. ***I Am Arachne: Fifteen Greek and Roman Myths.*** New York: Farrar, Straus & Giroux, 2001. First-person retellings.

Popular Culture
Connolly, Peter, and Hazel Dodge. ***The Ancient City: Life in Classical Athens and Rome.*** London: Oxford University Press, 2000. Detailed illustrations and descriptions.

Copeland, Tim. ***Ancient Greece.*** Cambridge, UK: Cambridge University Press, 1998. Uses the Olympic Games as a method of exploring the ancient world.

Science/Math
Nardo, Don. ***Roman Roads and Aqueducts.*** San Diego, CA: Lucent Books, 2001. Engineering feats of ancient Rome.

Language Arts/Literature
Cushman, Karen. ***Catherine, Called Birdy.*** New York: Harper, 1995. Spirited fictional diary of a young noblewoman in medieval England.

ENRICHMENT ACTIVITIES

The following activities are especially suitable for classes following block schedules.

Teacher's Edition, pp. 276, 281, 283, 287, 288, 293

Pupil's Edition, pp. 277, 282, 289, 295

Unit Atlas, pp. 260–269

Outline Map Activities

INTEGRATED TECHNOLOGY

Go to **classzone.com** for lesson support and activities for Chapter 10.

BLOCK SCHEDULE LESSON PLAN OPTIONS: 90-MINUTE PERIOD

DAY 1

UNIT PREVIEW, pp. 258–259
Class Time 20 minutes

- **Discussion** Discuss the objective for Chapter 10 on TE p. 270.

UNIT ATLAS, pp. 260–269
Class Time 20 minutes

- **Small Groups** Divide the class into five groups and have each group answer the questions designated for each Unit Atlas objective.

SECTION 1, pp. 273–277
Class Time 50 minutes

- **Small Groups** Divide the class into three groups. Assign each group one of the following headings: "The Geography of Europe," "Climate," and "Natural Resources." Have students turn their headings into a question, then read to find out the answer to their question. Reconvene as a whole class to discuss the answers.
Class Time 30 minutes
- **Understanding Climate** Use the maps from the Activity Option on TE p. 275 to help students understand how landforms affect climate. Have students identify the location of the Alps and the Pyrenees and draw them on their maps. Ask students to use arrows to show the direction of the wind that blows southward from the Arctic Circle. Discuss with students the effects of the mountains on the region's climate.
Class Time 20 minutes

DAY 2

SECTION 2, pp. 278–282
Class Time 50 minutes

- **Small Groups** Divide students into groups of three to four. Have them skim the section for information on the achievements of the ancient Greeks and fill in the chart in the Section Assessment on PE p. 282. Review the chart with the class.
Class Time 25 minutes
- **Peer Teaching** After reading aloud Strange but True on TE p. 280, ask a panel of four students to debate the following question: Are Spartan virtues important in today's world? Have the rest of the class act as debate judges, deciding which speakers make the most convincing arguments.
Class Time 25 minutes

SECTION 3, pp. 284–289
Class Time 40 minutes

- **Creating a Time Line** Lead the class in creating a time line on the board that traces the development of ancient Rome from its early beginnings.
Class Time 20 minutes
- **Summarizing** Have students use the Terms & Names to write a summary of the section.
Class Time 20 minutes

DAY 3

SECTION 4, pp. 290–295
Class Time 35 minutes

- **Hypothesizing** Use the Critical Thinking Activity on TE p. 291 as a starting point to discuss why the church played such an important role in medieval life.
Class Time 10 minutes
- **Role-Playing** Have pairs of students create a pantomine depicting a feudal relationship. Ask for volunteers to perform their pantomine in front of the class. Have the rest of the class guess what social groups are being portrayed.
Class Time 25 minutes

CHAPTER 10 REVIEW AND ASSESSMENT, pp. 296–297
Class Time 55 minutes

- **Review** Divide students into pairs. Using the Visual Summary on PE p. 296, have students provide two supporting details for each statement.
Class Time 20 minutes
- **Assessment** Have students complete the Chapter 10 Assessment.
Class Time 35 minutes

TECHNOLOGY IN THE CLASSROOM

INTERNET RESEARCH

The Web lends itself to student research because of the availability of Web sites with valuable, up-to-date information, including text and pictures. However, the Web also contains many sites that are not up-to-date or that have irrelevant, erroneous, or inappropriate information. One way to ensure that students use appropriate Web sites is to have them look through a limited set of preselected sites. Students should be given a list of sites sponsored by organizations that are generally considered reliable, such as major news sources, television networks, nonprofit organizations, or museums. This will save a good deal of classroom time, and students will be less likely to contact sites that are inappropriate.

ACTIVITY OUTLINE

Objective Students will use the Internet to research daily life in ancient Greece and Rome. They will use a word processing program to create paragraphs describing what they have learned.

Task Ask students to visit Web sites to gather information about ancient Greek and Roman housing, clothing, food, family life, recreation, education, business and work, holidays, and religion. Have them discuss their findings and write paragraphs from the point of view of ancient Greeks and Romans. Suggest that they write additional paragraphs describing some of the things the Romans borrowed from the Greeks.

Class Time 2–3 class periods

DIRECTIONS

1. Ask students what they think it would have been like to live in ancient Greece and Rome. List their ideas, and discuss why the Romans might have borrowed some ideas from Greek culture.
2. Have students make charts with two columns, one labeled "Ancient Greece" and the other labeled "Ancient Rome."
3. Have students visit the Web sites at **classzone.com** to find out what life was like in ancient Greece and Rome. As they explore the sites, have them take notes on features of daily life such as housing, clothing, food, family life, recreation, education, business and work, holidays, and religion.
4. As a class discuss these questions:
   - What were the major similarities and differences between daily life in ancient Greece and Rome?
   - What did the Romans borrow from the Greeks?
5. Have students choose one aspect of daily life to focus on. Ask each student to assume the role of a person from ancient Greece, and use a word processing program to write a one-paragraph story describing this aspect of their life. Then have them write an additional one-paragraph story describing the same aspect of life from the point of view of a person from ancient Rome. If time permits, encourage students to illustrate their stories.
6. Finally, ask students to write a final paragraph in which they name an item, custom, or skill the Romans borrowed from the Greeks. They should include what this meant to each civilization.

CHAPTER 10 OBJECTIVE

Students will explore the geography of Europe, the lasting achievements of the ancient Greek and Roman civilizations, and developments in Europe during the medieval period.

FOCUS ON VISUALS

Interpreting the Photograph Have students look at the photograph of Segovia, Spain, and identify the castle and arches of the Roman aqueduct. Point out that the aqueduct probably was built in about A.D. 100, at least 1,000 years before the castle was built. This chapter focuses on the early history of Western Europe. What do these two structures show about that history?

Possible Responses They show that the history of Europe extends many centuries back in time, with different cultures following one another.

Extension Have students look up the word *aqueduct* in an encyclopedia to discover the original purpose of aqueducts and how they are used today.

CRITICAL THINKING ACTIVITY

Hypothesizing Prompt a discussion about why the history of Europe is so important to people in our country. Guide students to consider where many of our ancestors came from, where our language and many of our customs originated, how the political systems in our country and Europe are similar, and other ways in which our country is connected culturally and politically to Europe.

Class Time 10 minutes

Western Europe: Its Land and Early History

Region Many European cities show their history in their architecture. In Segovia, Spain, an ancient Roman aqueduct lies below the walls of a castle built in the Middle Ages.

Recommended Resources

BOOKS FOR THE TEACHER

Gies, Frances and Joseph. ***Life in a Medieval City.*** New York: HarperCollins, 1981. Authoritative resource with a companion volume, ***Life in a Medieval Village.***

Rees, Rosemary. ***The Ancient Romans.*** Des Plaines, IL: Heinemann Library, 1999. Examination of social, economic, political, and cultural life.

VIDEOS

Ancient Greece. Wynnewood, PA: Schlessinger Media, 1998. Explore the mysteries of ancient Greece, including a visit to the Acropolis.

SOFTWARE

The Romans. Library Video Company, 1998. Construct a mosaic, view clothing, explore mythology, and more.

INTERNET

For more information about Europe, visit **classzone.com.**

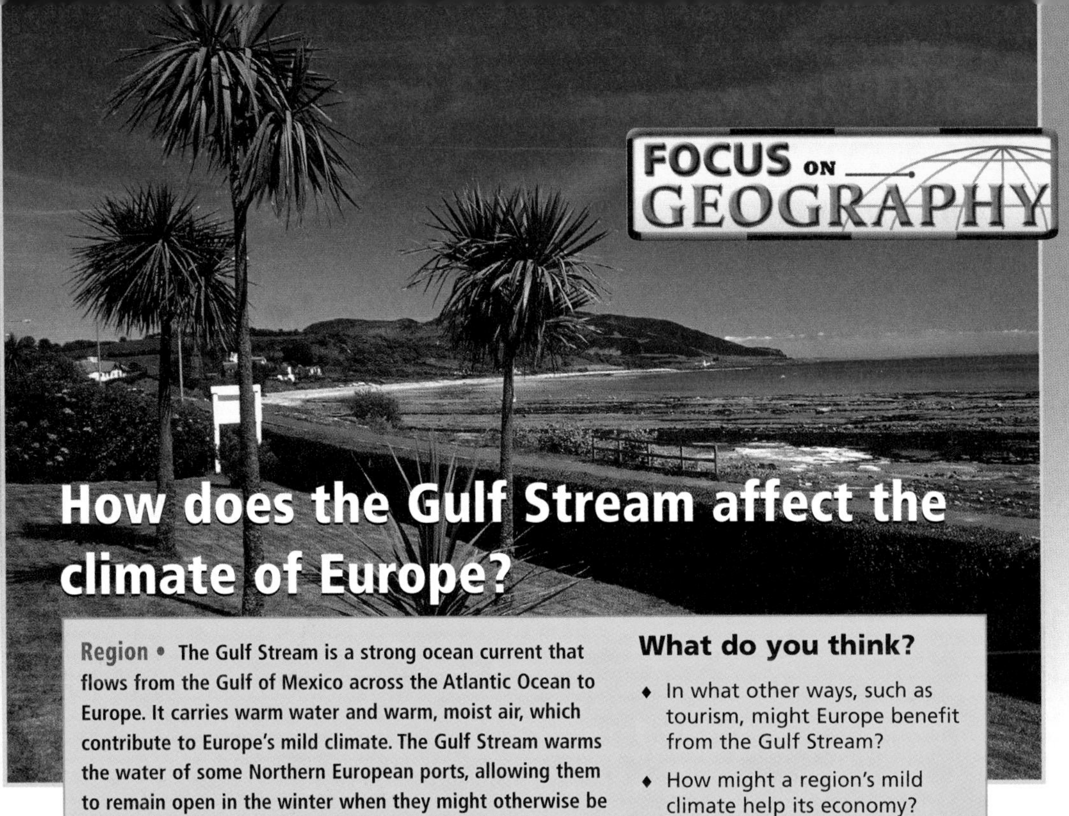

How does the Gulf Stream affect the climate of Europe?

Region • The Gulf Stream is a strong ocean current that flows from the Gulf of Mexico across the Atlantic Ocean to Europe. It carries warm water and warm, moist air, which contribute to Europe's mild climate. The Gulf Stream warms the water of some Northern European ports, allowing them to remain open in the winter when they might otherwise be frozen. Palm trees even grow in Scotland, which is as far north as southern Alaska!

What do you think?

- In what other ways, such as tourism, might Europe benefit from the Gulf Stream?
- How might a region's mild climate help its economy?

FOCUS ON GEOGRAPHY

Objectives

- To help students understand how geography affects climate in Europe
- To show a relationship between climate and the economy

What Do You Think?

1. Point out that a moderate climate extends the season when people want to visit a country, sightsee, or take part in outdoor sports. Have them speculate whether it is easier for cultures to develop and prosper in a moderate climate rather than in one that is either extremely hot or frigid.
2. Students should understand that economy refers to both the goods and services produced in a region and how those goods are produced. Help students to recognize how climate affects the ability to raise crops and the ease with which people and goods move from one region to another.

How does the Gulf Stream affect the climate of Europe?

Help students understand that the Gulf Stream brings warm water from the Gulf of Mexico across the ocean, making Europe's climate milder than that of other places at the same latitude.

MAKING GEOGRAPHIC CONNECTIONS

Ask students to think about how climate affects people in Africa, Latin America, and Antarctica. Discuss how the lives of people living in these places might differ from those of people living in Europe because of climate.

Implementing the National Geography Standards

Standard 3 Describe and analyze the spatial arrangement of urban land-use patterns

Objective To build a scale model of the students' local community

Class Time 40 minutes

Task Invite students to work together to build a scale model of their community, indicating the commercial, residential, and industrial areas. After the model is built, have each student refer to the model to write an explanation of the distribution of these three types of land use in the community.

Evaluation In their explanations of the three types of land use, students should mention local landforms and transportation routes.

BEFORE YOU READ

What Do You Know?

Poll students to see how many of them have ever been to Europe, have relatives from Europe, or are themselves European-born. Have them describe what they know about the Continent from visits, relatives, or personal experience. Ask whether they have seen any sports events on television that took place in Europe or read any books set in Europe. Have them describe their impressions of Europe from these sources. How might Europe differ from the United States?

What Do You Want to Know?

Suggest that students make a three-column chart with the headings *Ancient Greece, Ancient Rome,* and *Middle Ages.* Have them work in pairs to write what they want to know about each topic. After the students have taken notes, have volunteers read their notes aloud. Ask students to note on their charts any additional items that interest them about each topic. They can add to their charts as they read each section.

READ AND TAKE NOTES

Reading Strategy: Categorizing Categorizing facts and details in a meaningful way can help students identify similarities and differences across cultures and time periods. Point out that they might use this strategy to categorize other topics in the chapter.

In-depth Resources: Unit 4
- Guided Reading Worksheets, pp. 3–6

CHAPTER 10 READING SOCIAL STUDIES

BEFORE YOU READ

▸▸ What Do You Know?

Before you read the chapter, think about what you already know about Europe. What are some of its geographical features? What do you know about its early history? Have you ever read myths from ancient Greece or ancient Rome? Have you ever heard of Julius Caesar or Hercules? What do you know about knights and castles from the Middle Ages?

Region • Ancient Greece made important contributions in literature, philosophy, and architecture. ▼

▸▸ What Do You Want to Know?

Decide what you want to know about these early periods of European history. Record your questions in your notebook before you read this chapter.

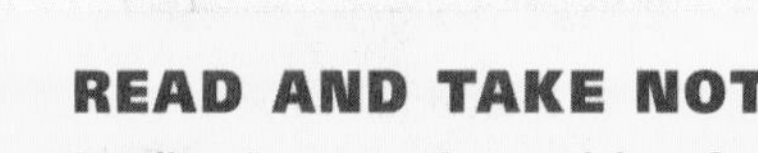

READ AND TAKE NOTES

Reading Strategy: Categorizing One way to make sense of what you read is to categorize, or sort, information. Making a chart to categorize the information in this chapter will help you to understand the contributions made by early European cultures.

- Copy the chart below into your notebook.
- As you read, look for information relating to the categories of social structure, architecture, religion, and arts and sciences.
- Write your notes under the appropriate headings.

Region • The Middle Ages saw the rise of the Catholic Church and the growth of a middle class. ▲

Region • Ancient Rome made its mark in government, law, and engineering. ▲

Time Period	Social Structure	Architecture	Religion	Arts and Sciences
Ancient Greece	city-states, oligarchy, democracy, government participation limited to free, adult males	built temples atop the Acropolis, the Parthenon	shared religious beliefs within city-states, honored gods and goddesses through literature	myths, poems, plays, philosophy, architecture
Ancient Rome	republic, senate, patricians, plebeians, empire	public buildings, lighthouses	different religious beliefs, rise of Christianity	great literature, Aeneid
Middle Ages	no central government, feudalism, manorialism, growth of towns, guilds	castles	Catholicism, church became center of community, clash between government and church	the Bayeux Tapestry

Teaching Strategy

Reading the Chapter This is a thematic chapter that focuses on how the development of Europe's diverse cultures has been affected by geographic features and the contributions of the ancient Greeks and Romans, as well as how the events of the Middle Ages transformed Europe into a modern society. As students read about each period, encourage them to think about the system of government in place and about how much economic freedom individuals had.

Integrated Assessment The Chapter Assessment on page 297 describes several activities that can be used for integrated assessment. You may wish to have students work on these activities during the course of the chapter and then present them at the end.

A Land of Varied Riches

TERMS & NAMES
Mediterranean Sea
peninsula
fjord
Ural Mountains
plain

MAIN IDEA

Europe is a continent with varied geographic features, abundant natural resources, and a climate that can support agriculture.

WHY IT MATTERS NOW

The development of Europe's diverse cultures has been shaped by the continent's diverse geography.

DATELINE

EXTRA

LONDON, ENGLAND, MAY 6, 1994

Rough waters have always made the English Channel, which separates England and France, difficult to cross. Now, however, you can travel under the water! Today, a tunnel nicknamed "the Chunnel" opens, allowing high-speed trains to travel between London and Paris in about three hours. The Chunnel—short for Channel Tunnel—was carved through chalky earth under the sea floor and took seven years to build. It is the largest European construction project of the 20th century.

Movement • Eurostar trains make the 31-mile trip under the English Channel in only 20 minutes. ▲

Chunnel
English Channel
France

The Channel connects England ... ▲

The Geography of Europe ❶

Today, cars, airplanes, and trains are common forms of high-speed transportation across Europe. Before the 19th century, however, the fastest form of transportation was to travel by water—on top of it, rather than under it.

TAKING NOTES
Use your chart to take notes about Western Europe.

Time Period	Social Structure	Architecture
Ancient Greece		
Ancient Rome		

SECTION OBJECTIVES

1. To describe Europe's geography
2. To describe factors affecting Europe's climate
3. To explain how Europe's natural resources affect what it produces today

SKILLBUILDER

• Interpreting a Map, p. 277

CRITICAL THINKING

• Making Inferences, p. 274
• Generalizing, p. 275

FOCUS & MOTIVATE

WARM-UP

Making Inferences Have students read Dateline and encourage them to think about how transportation has changed since ancient times.

1. What are some advantages of the Chunnel?
2. How might improvements in transportation affect economic development?

INSTRUCT: Objective ❶

The Geography of Europe

• What are the distinctive features of Europe's geography? oceans, rivers, peninsulas, mountain ranges, central plain
• What geographical feature of Europe influenced the development of different cultures? Mountain ranges separated groups of people from one another as they settled.
• What is an important feature of the Great European Plain? rich farmland

In-depth Resources: Unit 4
• Guided Reading Worksheet, p. 3

Reading Study Guide (Spanish and English), pp. 78–79

Program Resources

In-depth Resources: Unit 4
• Guided Reading Worksheet, p. 3
• Reteaching Activity, p. 11

Reading Study Guide (Spanish and English), pp. 78–79

Formal Assessment
• Section Quiz, p. 141

Integrated Assessment
• Rubric for writing a story

Outline Map Activities

Access for Students Acquiring English
• Guided Reading Worksheet, p. 75

Technology Resources
classzone.com

TEST-TAKING RESOURCES
Strategies for Test Preparation
Test Practice Transparencies
Online Test Practice

CRITICAL THINKING ACTIVITY

Making Inferences Using the map on page 277 or in the Unit Atlas, draw students' attention to the lengthy coastlines of many European countries. Ask them to infer how this access to water, plus the number of great inland rivers, might affect the history of nations and civilizations. Encourage them to consider such factors as trade, warfare and defense, ways of making a living, and the spread of skills and ideas.

Class Time 15 minutes

MORE ABOUT...
Peninsulas

Europe has sometimes been described as "a peninsula of peninsulas." Other peninsulas jutting into the Mediterranean Sea are the Italian Peninsula, the Balkan or Greek Peninsula, and Asia Minor (Turkey). The strategically located Crimean Peninsula, which extends into the Black Sea from southern Ukraine, is also historically important. The word *peninsula* comes from the Latin words for "almost" *(paene)* and "island" *(insula)*.

Reading Social Studies
A. Possible Answer
Before the 1800s, it was easier and quicker to travel by water than over land.

Waterways Look at the map of Europe on page 277. Water surrounds the continent to the north, south, and west. The southern coast of Europe borders the warm waters of the **Mediterranean Sea.** Europe also has many rivers. The highly traveled Rhine and Danube rivers are two of the most important. The Volga, which flows nearly 2,200 miles through western Russia, is the continent's longest. For hundreds of years, these and other waterways have been home to boats and barges carrying people and goods inland across great distances.

A. Clarifying Why were waterways important for the movement of people and goods?

Landforms Several large **peninsulas,** or bodies of land surrounded by water on three sides, form the European continent. In Northern Europe, the Scandinavian Peninsula is home to Norway and Sweden. Along the jagged shoreline of this peninsula are beautiful fjords (fyawrdz). A **fjord** is a long, narrow, deep inlet of the sea located between steep cliffs. In Western Europe, the Iberian Peninsula includes Portugal and Spain. The Iberian Peninsula is separated from the rest of the continent by a mountain range called the Pyrenees (PEER•uh•NEEZ). The entire continent of Europe, itself surrounded by water on three sides, is a giant peninsula.

BACKGROUND

Europe can be divided into four areas: Western Europe, Northern Europe, Eastern Europe, and Russia and its neighboring countries.

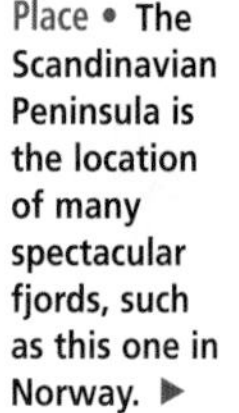

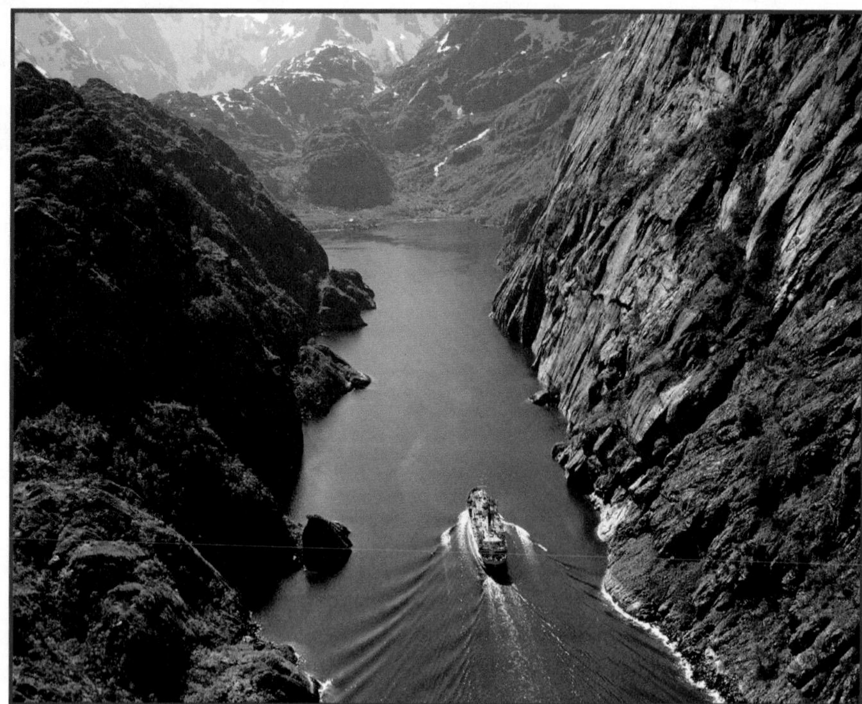

Place • The Scandinavian Peninsula is the location of many spectacular fjords, such as this one in Norway. ▶

Activity Options

Differentiating Instruction: Less Proficient Readers

Rereading To ensure that less proficient readers are able to pick out important details from the text, ask them to reread the section on the geography of Europe. Then have them write answers to questions such as those in the next column.

- What sea forms the southern border of Europe?
- What is a fjord?
- What is a peninsula?
- What countries in Northern Europe are located on the Scandinavian Peninsula?

Place • The Alps remain snowcapped year-round. ▶

Mountain ranges, including the towering Alps, also stretch across much of the continent. Along Europe's eastern border, the **Ural Mountains** (YUR•uhl) divide the continent from Asia. The many mountain ranges of Europe separated groups of people from one another as they settled the land thousands of years ago. This is one of the reasons why different cultures developed across the continent.

Reading Social Studies

B. Clarifying What natural landform separates Europe from Asia?

Reading Social Studies B. Answer the Ural Mountains

The Great European Plain Not all of Europe is mountainous. A vast region called the Great European Plain stretches from the coast of France to the Ural Mountains. A **plain** is a large, flat area of land, usually without many trees. The Great European Plain is the location of some of the world's richest farmland. Ancient trading centers attracted many people to this area, which today includes some of the largest cities in Europe—Paris, Berlin, Warsaw, and Moscow.

Climate ❷

Vocabulary

Gulf Stream: a warm ocean current that flows northeast from the Gulf of Mexico through the Atlantic Ocean.

Although the Gulf Stream brings warm air and water to Europe, the winters are still severe in the mountains and in the far north. In some of these areas, cold winds blow southward from the Arctic Circle and make the average temperature fall below 0°F in January. The Alps and the Pyrenees, however, protect the European countries along the Mediterranean Sea from these chilling winds. In these warmer parts of southern Europe, the average temperature in January stays above 50°F.

INSTRUCT: Objective ❷

Climate

- What are the average January and July temperatures in southern Europe? January: above 50°F; July: about 80°F
- Why is the Mediterranean coast a popular vacation spot? has hot, dry summers

CRITICAL THINKING ACTIVITY

Generalizing Provide students with the average summer and winter temperatures for five cities in North America and five cities located at about the same latitudes in Europe. Then ask students to make generalizations concerning climate and latitude based on the data about the climates on the two continents.

Class Time 10 minutes

MORE ABOUT...

The Ural Mountains

Although the Urals mark the traditional boundary between Europe and Asia, they are relatively low-lying mountains. The rugged range stretches about 1,500 miles north-south across Russia, from Arctic tundra in the north to deserts north of the Caspian Sea in Kazakhstan. The highest peaks are in the north. The foothills of the range slope westward toward the Volga River. In the east, the land drops to the lowlands of western Siberia.

Activity Options

Multiple Learning Styles: Visual

Class Time 30 minutes

Task Identifying access to major oceans and seas

Purpose To explore the importance of a country's location in relation to a major sea or ocean

Supplies Needed

- Photocopies of a map of Europe showing current national boundaries and country names
- Colored pencils

Activity Distribute copies of the map of Europe. Have students use pencils of one color to identify countries on the Mediterranean Sea and pencils of another color to identify countries on the North Sea. Then have students use a third color to identify all remaining countries on the Atlantic Ocean. When they have finished, ask students to identify the countries that do not have ocean access.

INSTRUCT: Objective 3

Natural Resources

- What are some of Europe's mineral resources? coal and iron ore
- What factors make Europe a good place to grow crops? rich soil, plentiful rainfall, moderate temperatures

FOCUS ON VISUALS

Interpreting the Photographs Have students examine the photographs and point out details that demonstrate the variety of resources and landforms in different regions of Western Europe. Ask them to identify the natural resources shown and explain how humans are using these different environments and interacting with them.

Possible Responses Forests, mountains, oceans, grazing land. The sheep are grazing on hillsides that are probably too steep to farm; factories use natural resources and energy.

Extension Ask students to collect photographs illustrating the geographical variety of Western Europe, using travel magazines, tour brochures, or posters. Have them write captions for the pictures, then arrange them as a bulletin board display.

The summers in the south are usually hot and dry, with an average July temperature around 80°F. This makes the Mediterranean coast a popular vacation spot. Elsewhere in Europe, in all but the coldest areas of the mountains and the far north, the average July temperature ranges from 50°F to 70°F.

Natural Resources 3

Europe has a large variety of natural resources, including minerals. The rich coal deposits of Germany's Ruhr (rur) Valley region have helped to make that area one of the world's major industrial centers. Russia and Ukraine have large deposits of iron ore, which is used to make iron for automobiles and countless other products.

Region • Western Europe benefits from a varied landscape rich in natural resources. ▼

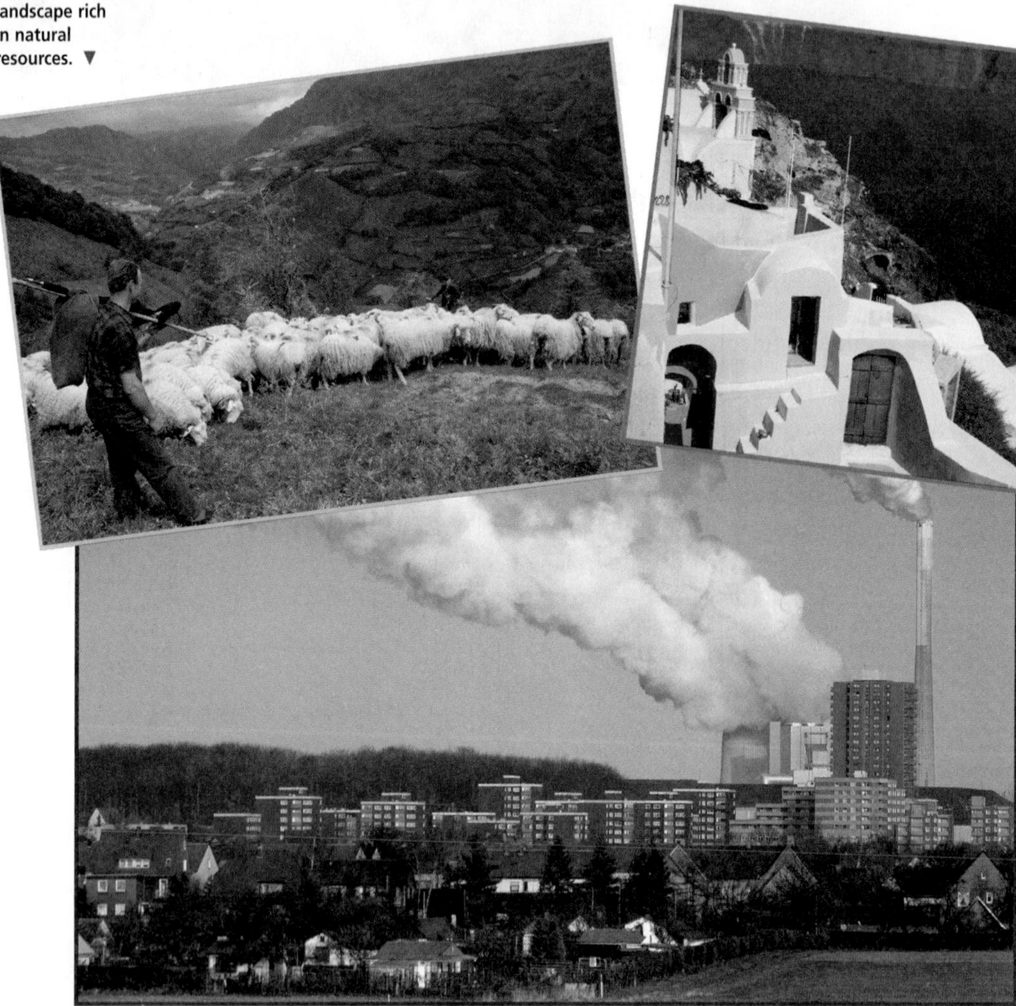

Activity Options

Interdisciplinary Link: Current Events

Block Scheduling

Class Time Two 30-minute sessions

Task Holding a roundtable discussion

Purpose To identify and report on current economic issues in Europe

Supplies Needed
- Writing paper
- Pencils or pens
- Current newspapers and newsmagazines
- Internet access

Activity Divide the class into six to eight small groups and ask each group to report on a different European nation. In the first session, have students research newspapers, newsmagazines, or the Internet for current news about their chosen country, focusing on resource-related issues, and write a short news item (one to two paragraphs). Next, ask one member from each group to present the news item in a roundtable discussion on "Europe Today." Have the other students discuss the presentations.

Land Use in Europe Today

GEOGRAPHY SKILLBUILDER: Interpreting a Map

1. **Place** • What are the three most common uses of land in Europe?
2. **Location** • Where is the majority of unused land?

Vocabulary

precipitation: moisture, including rain, snow, and hail, that falls to the ground

Europe also has rich soil and plentiful rainfall. The average precipitation for the Great European Plain, for example, is between 20 and 40 inches per year. The map above shows the agricultural uses of the land, highlighting the major crops. Notice that few parts of the continent are too cold or too hot and dry to support some form of agriculture. These characteristics have made Europe a world leader in crop production.

Geography Skillbuilder Answers

1. forests, growing wheat, growing rye and potatoes
2. in the far north and the mountains of Scandinavia

SECTION 1 ASSESSMENT

Terms & Names

1. Explain the significance of: (a) Mediterranean Sea (b) peninsula (c) fjord (d) Ural Mountains (e) plain

Using Graphics

2. Use a spider map like this one to list the different geographic features of Europe, and give a few specific examples of each.

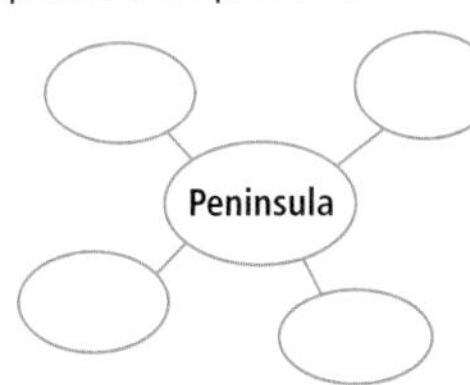

Main Ideas

3. (a) How does the Gulf Stream affect the climate of Europe?
   (b) What separates Europe from Asia?
   (c) How do waterways, such as rivers and seas, strengthen trade in Europe?

Critical Thinking

4. **Recognizing Effects**
   How did Europe's many mountain ranges affect its development?

Think About

- climate
- trade and travel
- the separation of groups of people

ACTIVITY -OPTION- Reread the information about the Chunnel. Write a **short story** in which you imagine what it might have been like to work on the Chunnel's construction.

FOCUS ON VISUALS

Interpreting the Map Review with students the importance of using the key to get information from a thematic map like this one. Make sure they recognize the difference between the gray color that indicates unused land and the neutral shade indicating nosubject areas. Ask them to use the key to determine the most widely grown crop in this part of Europe.

Response wheat

Extension Have students use the map to determine the answers to these questions: Where are most vineyards located? on the Mediterranean coast What land use dominates Scandinavia? forests

ASSESS & RETEACH

Reading Social Studies Have students list several facts about landforms, waterways, climate, and natural resources of Europe.

Formal Assessment
- Section Quiz, p. 141

RETEACHING ACTIVITIES

Divide the class into groups and assign each group one of the section topics. Ask each group to write a short summary of the topic. Have one member from each group read the summary aloud.

In-depth Resources: Unit 4
- Reteaching Activity, p. 11

Access for Students Acquiring English
- Reteaching Activity, p. 81

Section 1 Assessment

1. Terms & Names

a. Mediterranean Sea, p. 274
b. peninsula, p. 274
c. fjord, p. 274
d. Ural Mountains, p. 275
e. plain, p. 275

2. Using Graphics

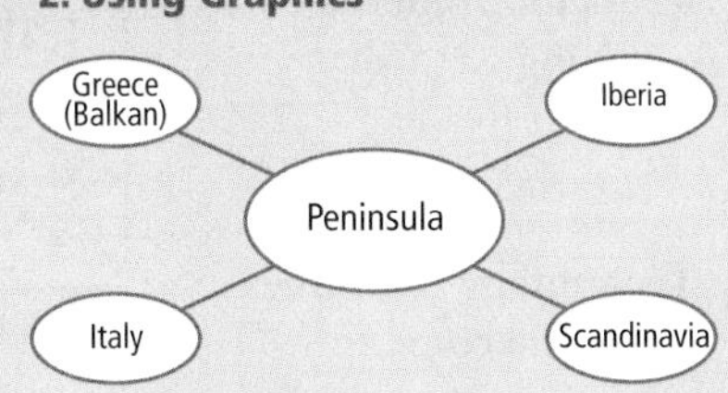

3. Main Ideas

a. It carries warm water and warm, moist air across Europe, resulting in a relatively mild climate.
b. the Ural Mountains
c. They carry people and goods across great distances.

4. Critical Thinking

Mountains separated groups of people when they settled thousands of years ago, which partly explains why different cultures developed in different areas.

ACTIVITY OPTION

Integrated Assessment
- Rubric for writing a story

Ancient Greece

TERMS & NAMES
city-state
polis
Aegean Sea
oligarchy
Athens
philosopher
Aristotle
Alexander the Great

MAIN IDEA

The ancient Greeks developed a complex society, with remarkable achievements in the arts, sciences, and government.

WHY IT MATTERS NOW

The achievements of the ancient Greeks continue to influence culture, science, and politics in the world today.

SECTION OBJECTIVES

1. To describe the geography of Greece, the development of Greek city-states, and the birth of democracy in Greece
2. To describe the achievements of Greek culture and explain its spread and influence

SKILLBUILDER
• Interpreting a Map, p. 279

CRITICAL THINKING
• Analyzing Motives, p. 281

FOCUS & MOTIVATE

WARM-UP

Making Inferences Have students read Dateline and ask them to think about excavations in their own region.

1. What artifacts might be discovered underground in your region?
2. Why do you think these artifacts would be different from those found in Athens?

INSTRUCT: Objective 1

The Land and Early History of Greece/Athens and Sparta

- How did geography influence the development of ancient Greece? Mountains separated city-states; people became skilled sailors and established colonies overseas.
- How were city-states alike and different? alike: common language, religious beliefs, way of life; different: laws, forms of government
- What kind of government did Athens have by the end of the sixth century? a democracy

In-depth Resources: Unit 4
• Guided Reading Worksheet, p. 4

Reading Study Guide
(Spanish and English), pp. 80–81

DATELINE

Athens, Greece, February 2, 1997— Five years after construction workers began building the new Athens subway, artifacts from ancient Greek civilization are still being discovered. When completed, the new subway will reduce traffic and air pollution in the capital. Historians and archaeologists, however, have been the first to benefit from this massive public works project.

Workers have discovered statues, coins, jewelry, and gravesites from ancient Greece. Recently, workers digging the foundation for a downtown Athens station found an ancient dog collar decorated with gemstones. Local officials have promised to create permanent displays of some artifacts in stations throughout the new subway system.

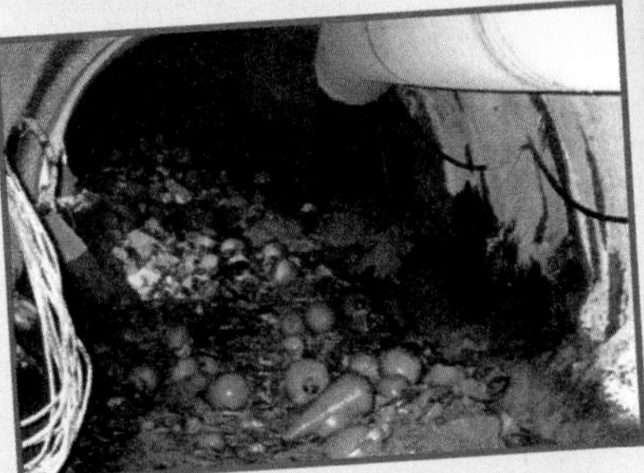
Place • Building the subway in Athens led to spectacular discoveries of ancient artifacts. ▲

1 The Land and Early History of Greece

The Greek Peninsula is mountainous, which made travel by land difficult for early settlers. Most of the rocky land also contains poor soil and few large trees, but settlers were able to cultivate the soil to grow olives and grapes. The greatest natural resource of the peninsula is the water that surrounds it. The ancient Greeks depended on these seas for fishing and trade, and they became excellent sailors.

TAKING NOTES
Use your chart to take notes about Western Europe.

Time Period	Social Structure	Architecture
Ancient Greece		
Ancient Rome		

Program Resources

In-depth Resources: Unit 4
• Guided Reading Worksheet, p. 4
• Reteaching Activity, p. 12

Reading Study Guide
(Spanish and English), pp. 80–81

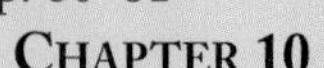

Formal Assessment
• Section Quiz, p. 142

Integrated Assessment
• Rubric for presenting an oral report

Outline Map Activities

Access for Students Acquiring English
• Guided Reading Worksheet, p. 76

Technology Resources
classzone.com

TEST-TAKING RESOURCES
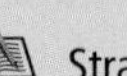 Strategies for Test Preparation
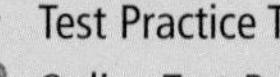 Test Practice Transparencies
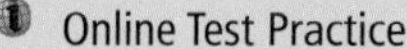 Online Test Practice

The Formation of City-States As the ancient Greek population grew, people created city-states. A **city-state** included a central city, called a **polis,** and surrounding villages. Each ancient Greek city-state had its own laws and form of government. The city-states were united by a common language, shared religious beliefs, and a similar way of life.

The Growth of Colonies By the mid-eighth century B.C., the Greeks were leaving the peninsula in search of better land and greater opportunities for trade. During the next 200 years, they built dozens of communities on the islands and coastline of the **Aegean Sea** (ih•JEE•uhn). Some Greeks settled as far away as modern-day Spain and North Africa.

Once established, these distant Greek communities traded with each other and with those communities on the Greek Peninsula. This made a great variety of goods available to the ancient Greeks, including wheat for bread, timber for building boats, and iron ore for making strong tools and weapons.

Geography Skillbuilder Answers
1. its location on the waterway into the Black Sea
2. Crete

Metropolis When ancient Greeks moved away from a large polis to a distant community, they referred to their former city-state as their metropolis. In Greek, this means "mother-city." Today, we use the word *metropolis* to mean any large urban area, such as Los Angeles, London, Tokyo, or Athens *(shown below).*

Connections to Language

Many English words come from Greek. These include words that end in the suffix *-logy,* meaning "study of," and words that end in the suffix *-phobia,* meaning "fear of." Common English words using these suffixes include biology, geology, psychology, claustrophobia, xenophobia, and hydrophobia.

FOCUS ON VISUALS

Interpreting the Map Have students relate this area map to the map of Europe on page 277 or in the Unit Atlas. Have them point out the area of the Mediterranean Sea in which Greece is located. Ask them to recall what kind of landform these two land areas (Greece and Asia Minor) represent.

Response peninsula

Extension Have students use the map scale to estimate these distances: Athens to Sparta about 200 miles; Ionia to Ephesus about 80 miles; Athens to Crete about 380 miles.

Greek Colonization, 800 B.C.

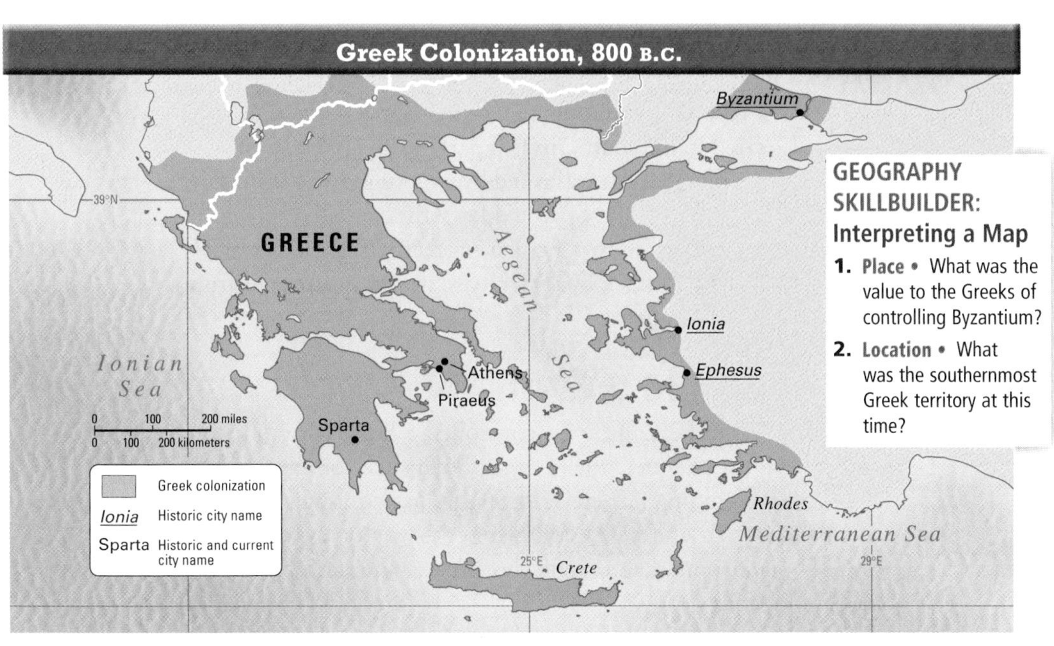

GEOGRAPHY SKILLBUILDER: Interpreting a Map
1. **Place** • What was the value to the Greeks of controlling Byzantium?
2. **Location** • What was the southernmost Greek territory at this time?

Activity Options
Multiple Learning Styles: Visual/Linguistic

Class Time One hour

Task Creating museum brochures for Athens and Sparta

Purpose To reinforce the differences between Sparta and Athens through a visual learning experience

Supplies Needed
- Poster board or paper
- Markers or crayons
- Books with pictures of Athens and Sparta

Activity Divide the class into groups of five students. Assign Athens to half the groups and Sparta to the other half. Ask each group to design a museum brochure to inform visitors about the city-state. Encourage students to include information about the location of each state, the kind of government it has, and other details they learn through their research. Remind students that the brochures must be historically accurate. Allow time for brief presentations of the brochures to the class.

Strange but TRUE

From the minute he was born, a Spartan boy belonged to the state. At the age of seven, boys were assigned to military companies made up of 15 boys. Discipline in these units was strict. The boys took their meals in a public dining hall, where they ate with their unit. The bravest boy became the captain of the unit and was allowed to command and punish the other unit members. The young soldiers in training slept on beds made out of reeds, which they gathered themselves. Living conditions were tough, because Spartans valued endurance, and the rejection of luxuries denoted strength—all still considered "Spartan" virtues today.

INSTRUCT: Objective 2

Learning and the Arts

- In what cultural fields did the ancient Greeks excel? literature, philosophy, architecture
- Who were usually the major characters in Greek poetry and tragic plays? gods and goddesses
- What were some of the topics that interested Socrates and other philosophers? friendship, knowledge, justice, government, human behavior
- How did Greek culture spread beyond Greece? through the expansion of the Greek empire under Alexander the Great, a military leader who conquered the Mediterranean and lands as far east as India during the fourth century B.C.

Spartan Soldiers Sparta was the only city-state with a permanent army. At age seven, Spartan boys were sent by their families for military training. They had to remain in the army until they were 30 years old.

Individual Forms of Government Some ancient Greek city-states were oligarchies (AHL•ih•GAHR•kees). An **oligarchy** is a system in which a few powerful, wealthy individuals rule. The word *oligarchy* comes from an ancient Greek word meaning "rule by the few." Other city-states were ruled by a tyrant, a single person who took control of the government against the wishes of the community. Still other ancient Greek city-states developed an early form of democracy. The word *democracy* comes from an ancient Greek word meaning "rule by the people." In a democracy, citizens take part in the government.

Reading Social Studies

A. Comparing Compare the three forms of government most common in ancient Greek city-states.

Reading Social Studies A. Answer: A small group of wealthy men ruled an oligarchy; one ruler became a tyrant; citizens took part in ruling a democracy.

Athens and Sparta 1

Athens, centrally located on the Greek Peninsula, was one of the largest and most important ancient Greek city-states. By the end of the sixth century B.C., Athens had developed a democratic form of government. Athenian citizens took part in political debates and voted on laws, but not everyone who lived in Athens enjoyed these rights. Participation in government was limited to free, adult males whose fathers had been citizens of Athens. Women, slaves, and foreign residents could not take part in government.

Athens's chief rival among the other Greek city-states was Sparta. Located in the southernmost part of the Greek Peninsula, Sparta was an oligarchy. It was ruled by two kings, who were supported by other officials. Sparta, like Athens, had a powerful army. Each city-state's army helped protect it from slave rebellions, guard against attack by rival city-states, and defend it from possible foreign invaders.

BACKGROUND

After the defeat of Persia, Athens became the most powerful Greek city-state. The most important Athenian leader of the time was Pericles (PEHR•ih•KLEEZ), who lived from c. 495 to 429 B.C.

Learning and the Arts 2

In 480 B.C., the Persians, who controlled a large empire to the east, tried to conquer the Greek Peninsula. Several Greek city-states, including Athens and Sparta, joined forces to defeat the Persians. In the years following this victory, the ancient Greeks made remarkable achievements in literature, learning, and architecture.

Activity Options

Differentiating Instruction: Students Acquiring English/ESL

Using Section Vocabulary To ensure that students acquiring English understand the vocabulary terms used in this section, discuss as a group the words listed to the right. Discuss where and when students have seen these words in print sources or heard them on radio or television. Then ask students to write each word in a sentence.

- community
- united
- rival
- victory
- architecture

Literature To honor their gods and goddesses, the ancient Greeks created myths and wrote poems and plays. Some of the greatest Greek plays were written during the fifth century B.C. During that time, the playwrights Aeschylus (EHS•kuh•luhs), Sophocles (SAHF•uh•KLEEZ), and Euripides (yu•RIHP•ih•DEEZ) wrote tragedies, which are serious plays that end unhappily. Many of these stories have been the basis for modern films and operas.

In addition to using the gods as characters, ancient Greek playwrights sometimes poked fun at important citizens, including generals and politicians. Aristophanes (ar•ih•STAHF•uh•NEEZ) was a popular writer of comedies of this type.

Philosophy Ancient Greece was the birthplace of some of the finest thinkers of the ancient world. Socrates (SAHK•ruh•TEEZ) was an important philosopher of the fifth century B.C. A **philosopher** studies and thinks about why the world is the way it is. Socrates studied and taught about friendship, knowledge, and justice. Another great philosopher, Plato (PLAY•toh), was a student of Socrates who studied and taught about human behavior, government, mathematics, and astronomy.

Reading Social Studies

B. Making Inferences Why do you think philosophers felt the need to teach?

Reading Social Studies
B. Possible Answer
They wanted to pass on their ideas, to share them.

The ancient Greek philosopher Heraclitus (HEHR•uh•KLY•tuhs) wrote the following lines.

A VOICE FROM ANCIENT GREECE

One cannot step twice into the same river, for the water into which you first stepped has flowed on.

Heraclitus

Many people continue to study and write about the same philosophical questions that these, and other, ancient Greek philosophers explored.

The World's Heritage

Ancient Greek Architecture
Ancient Greek builders created some of the world's most impressive works of architecture. They built several beautiful temples atop the Acropolis (uh•KRAHP•uh•lihs) in Athens, shown at right. The most famous of the temples is the Parthenon (PAHR•thuh•nahn).

In the United States and elsewhere, government buildings, such as courthouses and post offices, have been built similar in style to the Parthenon. This use of ancient architecture echoes the democratic ideals of ancient Greece.

The World's Heritage

The main part of the Parthenon was destroyed in 1687, after the Venetians attempted to conquer Athens. The 2,000-year-old building, which was being used as a powder house, exploded.

By the 19th century, Greece had become part of the Turkish Empire. In 1801, the Turkish government granted the British ambassador in Constantinople, Lord Elgin, permission to remove some of the most beautiful marble statues from the Acropolis. These statues included many sculptures from the Parthenon created under the direction of Phidias, the greatest Greek sculptor. Lord Elgin sold these statues, known as the Elgin Marbles, to the British government in 1816. Despite attempts by the Greek government to have the statues returned to Greece, they remain in the British Museum, in London.

A VOICE FROM ANCIENT GREECE

Ask a volunteer to read aloud the words of the philosopher Heraclitus. Explain that the message contains the philosopher's view that things in life continually change. Ask students to relate his message to situations in their own lives.

CRITICAL THINKING ACTIVITY

Analyzing Motives After students have discussed the quotation from Heraclitus and its meaning for them, ask them to take a broader look at the Greek philosophers and their influence. Ask them to suggest reasons that thinkers 2,500 years later are still puzzling over the same questions about the world and human behavior.

Class Time 15 minutes

Activity Options

Differentiating Instruction: Gifted and Talented

Block Scheduling

Researching a Historical Figure Have students use library resources and the Internet to find information about famous Greek figures, such as Plato, Sophocles, Herodotus, Pericles, Hippocrates, or Alexander the Great. Encourage students not to attempt a complete biographical sketch but to focus on just one or two aspects of the historical figures they have chosen—for example, Herodotus's approach to history or Alexander's conquests. Ask students to write a two-page report focusing on those aspects and on how the person's achievements have influenced contemporary literature, history, or science.

Biography

Aristotle may have been the first philosopher to conduct organized scientific research. He analyzed the functions of parts of animals and wrote several treatises on the subject. He also wrote about literature, analyzing the plots, characters, and themes of classic Greek dramas.

ASSESS & RETEACH

Reading Social Studies Have students fill in the column on Greece in the chart on page 272.

Formal Assessment
- Section Quiz, p. 142

RETEACHING ACTIVITIES

Ask students to work in groups of three to write a review of this section. Have one student summarize each heading. Then have students combine their work in a complete summary.

In-depth Resources: Unit 4
- Reteaching Activity, p. 12

Access for Students Acquiring English
- Reteaching Activity, p. 82

Aristotle At the age of 17, Aristotle (384–322 B.C.) began studying philosophy with Plato. After Plato died, Aristotle received his most important assignment—to teach Alexander, the teenage son of King Philip II of Macedonia.

After teaching Alexander, Aristotle returned to Athens. There he taught and wrote about poetry, government, and astronomy. He started a famous school called the Lyceum (ly•SEE•uhm). Aristotle also collected and studied plants and animals. The work of this brilliant philosopher continues to greatly influence scientists and philosophers today.

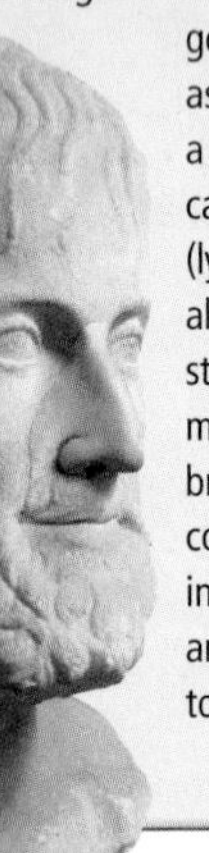

The Spread of Greek Culture The city-states of ancient Greece were constantly at war with one another. By the fourth century B.C., this fighting had weakened their ability to defend themselves against foreign invaders. In 338 B.C., King Philip II of Macedonia conquered the land. After Philip died, his son, Alexander—who had been taught by **Aristotle**—took control.

Region • In this mosaic Alexander the Great is shown riding into battle on his beloved horse, Bucephalus (byoo•SEHF•ah•luhs).

Alexander the Great was an excellent military leader, and his armies conquered vast new territories. As Alexander's empire expanded, Greek culture, language, and ideas were spread throughout the Mediterranean region and as far east as modern-day India. Upon Alexander's death, however, his leading generals fought for control of his territory and divided it among themselves. This marked the end of one of the great empires of the ancient world.

SECTION 2 ASSESSMENT

Terms & Names

1. **Explain the significance of:** (a) city-state (b) polis (c) Aegean Sea (d) oligarchy (e) Athens (f) philosopher (g) Aristotle (h) Alexander the Great

Using Graphics

2. Use a chart like this one to list and describe the ancient Greek achievements in government, literature, and architecture.

Government	Literature	Architecture

Main Ideas

3. (a) Why were the surrounding areas of water an important natural resource of the Greek Peninsula?
   (b) Which people were allowed to participate in the government of ancient Athens?
   (c) How did Alexander the Great help to spread Greek culture?

Critical Thinking

4. **Summarizing**
   Why was the fifth century B.C. a remarkable time in ancient Greek history?

Think About
- warfare
- leaders
- literature and philosophy

ACTIVITY -OPTION- Reread the information about the individual forms of government common in ancient Greece. Present an **oral report** to the class that compares and contrasts two of the forms.

Section 2 Assessment

1. Terms & Names

a. city-state, p. 279
b. polis, p. 279
c. Aegean Sea, p. 279
d. oligarchy, p. 280
e. Athens, p. 280
f. philosopher, p. 281
g. Aristotle, p. 282
h. Alexander the Great, p. 282

2. Using Graphics

Government	Literature	Architecture
formation of city-states, each with its own laws and form of government	tragedies (Aeschylus, Sophocles, and Euripides)	temples (Acropolis)
creation of democracy in Athens at the end of the sixth century B.C.	comedies (Aristophanes)	
	philosophy (Socrates, Plato, Aristotle, and Heraclitus)	

3. Main Ideas

a. The oceans provided fish and facilitated trade.
b. Free, adult males whose fathers had been citizens could participate.
c. Alexander conquered vast territories, spreading Greek language and ideas.

4. Critical Thinking

Democracy first developed at that time; many great playwrights, philosophers, architects, and artists lived during the fifth century B.C.

ACTIVITY OPTION

Integrated Assessment
- Rubric for presenting an oral report

Making a Generalization

▶▶ Defining the Skill

To make generalizations means to make broad judgments based on information. When you make generalizations, you should gather information from several sources.

▶▶ Applying the Skill

The following three passages contain different information on the government of ancient Athens. Use the strategies listed below to make a generalization about Athenian government based on the passages.

How to Make a Generalization

Strategy ❶ Look for all the information that the sources have in common. These three sources all explain about Athenian government.

Strategy ❷ Form a generalization that describes ancient Athenian government in a way that all three sources would support. State your generalization in a sentence.

❶ Athenian citizens took part in political debates and voted on laws, but not everyone who lived in Athens enjoyed these rights. Participation in government was limited to free, adult males whose fathers had been citizens in Athens.

—*World Cultures and Geography*

In return for playing their parts as soldiers or sailors, ❶ ordinary Athenians insisted on controlling the government.

—*Encyclopaedia Britannica*

Unlike representative democracies or republics, in which one man is elected to speak for many, Athens was a true ❶ democracy: every citizen spoke for himself.

—*Classical Greece*

Make a Chart

Using a chart can help you make generalizations. The chart below shows how the information you just read can be used to generalize about the government of ancient Athens.

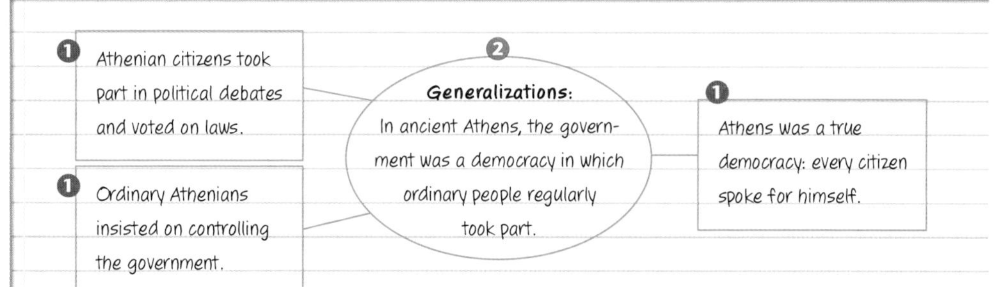
❶ Athenian citizens took part in political debates and voted on laws.
❶ Ordinary Athenians insisted on controlling the government.
❷ Generalizations:
In ancient Athens, the government was a democracy in which ordinary people regularly took part.
❶ Athens was a true democracy: every citizen spoke for himself.

▶▶ Practicing the Skill

Turn to Chapter 10, Section 2, "Ancient Greece." Read the sections on literature and philosophy. Also read about ancient Greek writings in an encyclopedia, a library book, or on the Internet. Then make a chart like the one above to form a generalization about the importance of knowledge and learning to the ancient Greeks.

SKILLBUILDER

Making a Generalization

Defining the Skill

Explain to students that when they generalize, they draw conclusions based on information gathered from different sources. Ask students to think about real-life situations where being able to generalize might help them, such as estimating how much money they might need for a movie or deciding what clothing to take on a trip.

Applying the Skill

How to Make a Generalization Point out the two strategies for making a generalization and help students work through each. For example, ask students to read the three passages, and have a volunteer tell what information they all have in common. Ask students to think of a generalization about the Athenian government that would be supported by all three sources.

Make a Chart

Discuss the three passages and the point being made by each. Ask questions such as: How did Athenian citizens participate in government? What did ordinary Athenians expect in return for being soldiers or sailors? In what way was Athens a true democracy? Have volunteers read the completed chart aloud.

Practicing the Skill

Find information about Greek writers in several sources, and distribute copies to students. Also have them reread the suggested sections on literature and philosophy. Suggest that students make notes of main points on a chart like the one shown. Then ask them to make a generalization.

In-depth Resources: Unit 4
- Skillbuilder Practice, p. 9

Career Connection: Weather Forecaster

Encourage students who enjoy making generalizations to learn about careers that utilize this skill. For example, students might choose to research the work of meteorologists, scientists who study the atmosphere. Many meteorologists are weather forecasters who bring together and analyze scientific data from many sources to make generalizations about what the weather will be like.

Block Scheduling

1. Help students find information about the many sources weather forecasters use: ground-weather stations, weather balloons, high-tech aircraft, radar, satellites, and computer models.
2. Help students determine the education required to work in this field.
3. Have students create a chart to explain the process of becoming a weather forecaster.

SECTION OBJECTIVES

1. To describe the Roman Republic and the spread of Rome's power
2. To explain the establishment and influence of the Roman Empire
3. To identify reasons for the rise of Christianity

SKILLBUILDER
- Interpreting a Map, pp. 286, 287

CRITICAL THINKING
- Contrasting, p. 285
- Finding Causes, p. 288

FOCUS & MOTIVATE

WARM-UP

Making Inferences Have students read Dateline, examine the illustration, and discuss building roadways 2,000 years ago.

1. Why did Rome need a roadway system?
2. What types of traffic were common?

INSTRUCT: Objective 1

The Beginnings of Ancient Rome/The Expansion of the Roman World

- In what way did the government of Rome change in 509 B.C.? It became a republic.
- What were the two groups of Roman citizens? wealthy patricians, workers called plebeians
- How did Rome win control of the western Mediterranean? by defeating Carthage

In-depth Resources: Unit 4
- Guided Reading Worksheet, p. 5

Reading Study Guide
(Spanish and English), pp. 82–83

SECTION 3

Ancient Rome

MAIN IDEA

The ancient Romans made important contributions to government, law, and engineering.

WHY IT MATTERS NOW

The cultural achievements of the Romans continue to influence the art, architecture, and literature of today.

TERMS & NAMES
republic
Senate
patrician
plebeian
Julius Caesar
empire
Augustus
Constantine

DATELINE

EXTRA

ROME, 295 B.C.

Yet another Roman road was completed today! Rome is famous for its vast network of roadways. Repairing old roads and adding new ones keeps Roman engineers busy. Construction is time-consuming because the lengthy roads, which are paved with large stones, must be carefully planned. However, the benefits are worth the effort.

The roads connect the great city to distant lands under Roman rule. These roadways also enable the army to move quickly. These days, it seems that almost all roads lead to Rome. In fact, when this massive undertaking is finished, Roman roads will stretch for tens of thousands of miles across the land.

Location • All roads lead to Rome—including the Via Appia (VEE•uh APP•ee•uh) shown here. ▲

The Beginnings of Ancient Rome 1

Ancient Rome began as a group of villages located along the banks of the Tiber River in what is now Italy. There, early settlers herded sheep and grew wheat, olives, and grapes. Around 750 B.C., these villages united to form the city of Rome.

TAKING NOTES
Use your chart to take notes about Western Europe.

Time Period	Social Structure	Architecture
Ancient Greece		
Ancient Rome		

Program Resources

In-depth Resources: Unit 4
- Guided Reading Worksheet, p. 5
- Reteaching Activity, p. 13

Reading Study Guide
(Spanish and English), pp. 82–83

Formal Assessment
- Section Quiz, p. 143

Integrated Assessment
- Rubric for creating a chart

Outline Map Activities

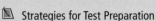

Access for Students Acquiring English
- Guided Reading Worksheet, p. 77

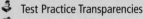

Technology Resources
classzone.com

TEST-TAKING RESOURCES

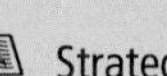
Strategies for Test Preparation

Test Practice Transparencies

Online Test Practice

The Formation of the Roman Republic For more than 200 years, kings ruled Rome. Then, in 509 B.C., Rome became a republic. A **republic** is a nation in which power belongs to the citizens, who govern themselves through elected representatives.

BACKGROUND

In ancient Rome, a senator held his position for life.

The Senate The Roman **Senate** was an assembly of elected representatives. It was the single most powerful ruling body of the Roman Republic. Each year, the Senate selected two leaders, called consuls, to head the government and the military.

Patricians At first, most of the people elected to the Senate were patricians (puh•TRIHSH•uhns). In ancient Rome, a **patrician** was a member of a wealthy, landowning family who claimed to be able to trace its roots back to the founding of Rome. The patricians also controlled the law, since they were the only citizens who were allowed to be judges.

Plebeians An ordinary, working male citizen of ancient Rome—such as a farmer or craftsperson—was called a **plebeian** (plih•BEE•uhn). Plebeians had the right to vote, but they could not hold public office until 287 B.C., when they gained equality with patricians.

Roman Law It may be hard to believe, but in the early Roman Republic, laws were not written down. Only the patrician judges knew what the laws were. This meant that judges usually ruled in favor of fellow patricians and against plebeians.

The plebeians grew tired of unfair treatment and demanded that the judges create a written code of laws that applied to all Roman citizens. This code, called the Law of the Twelve Tables, was written around 450 B.C. It formed the foundation of Roman law.

The Expansion of the Roman World 1

Over hundreds of years, Rome grew into a mighty city. By the third century B.C., Rome ruled most of the Italian Peninsula. This gave Rome control of the central Mediterranean.

The city-state of Carthage, which ruled North Africa and southern Spain, controlled the western Mediterranean. To take control over this area as well, Rome fought Carthage and eventually won.

As Rome's population grew, its army also expanded in size and strength. Under the leadership of ambitious generals, Rome's highly trained soldiers set out to conquer new territories one by one.

Strange but TRUE

The Twelve Tables refers to 12 sets of laws covering issues from treason to intermarriage between classes. These laws are important because they established the principle of a written legal code for Roman law. After the laws were passed, justice was no longer based solely on interpretation by judges.

By today's standards, the punishments set by the Twelve Tables seem harsh. Debtors could be bound in chains. Slanderers could be clubbed to death. Arsonists could be burned at the stake. Judges who accepted bribes could also be put to death.

CRITICAL THINKING ACTIVITY

Contrasting Work with students to identify ways in which the Roman Senate and the United States Senate differ. Ask them to think about who was eligible to hold office in each senate, how much power each body held, and how each body was related to the head of government and to the head of the military.

Ask a volunteer to record the comparison points in a two-column chart on the chalkboard. Then extend the discussion by asking students to compare the rights and responsibilities of citizens in ancient Rome with those of American citizens today. Ask them to explain the difference between a republic and a democracy and to suggest modern nations that are republics but not democracies.

Class Time 10 minutes

Activity Options

Interdisciplinary Link: Language Arts

Class Time 30 minutes

Task Identifying words with Latin roots

Purpose To demonstrate that many English words come from Latin

Supplies Needed
- Writing paper
- Pencils or pens
- Dictionaries

Activity Write the following Latin roots on the chalkboard: *aqua* (water), *audi* (hearing), *cent* (hundred), *creat* (make), *mare* (sea), *oct* (eight), *ped* (foot), *scrib/script* (write), *uni* (one), *vis* (see). Have students work in small groups to list as many words as they can for each root. After about 15 minutes, ask volunteers from each group to read the lists aloud.

INSTRUCT: Objective 2

From Republic to Empire

- Who was Julius Caesar? a general who became dictator of Rome in the first century B.C.
- In what way did Augustus carry on the work of his great-uncle Julius Caesar? by continuing to expand the Roman Empire
- What were some of the activities that took place during the Augustan Age? The empire grew, public buildings and lighthouses were built, trade increased, and famous works of literature were written.

FOCUS ON VISUALS

Interpreting the Map Review the use of a map key with students, making sure they can relate the colors to the regions shown. Check their understanding by asking what areas Rome controlled by 241 B.C.

Response most of the Italian Peninsula, island of Sicily

Extension Point out that the names on this map are modern. The Romans sometimes used different names for these provinces. For example, Romans called Italy *Italia.* Ask students to suggest a modern English term that derives from the Roman name for Spain, *Hispania.* Hispanic Have interested students research the names of other provinces in the Roman Empire such as *Britannia* and show how they relate to modern European nations.

Extent of Roman Control, 509 B.C. to 146 B.C.

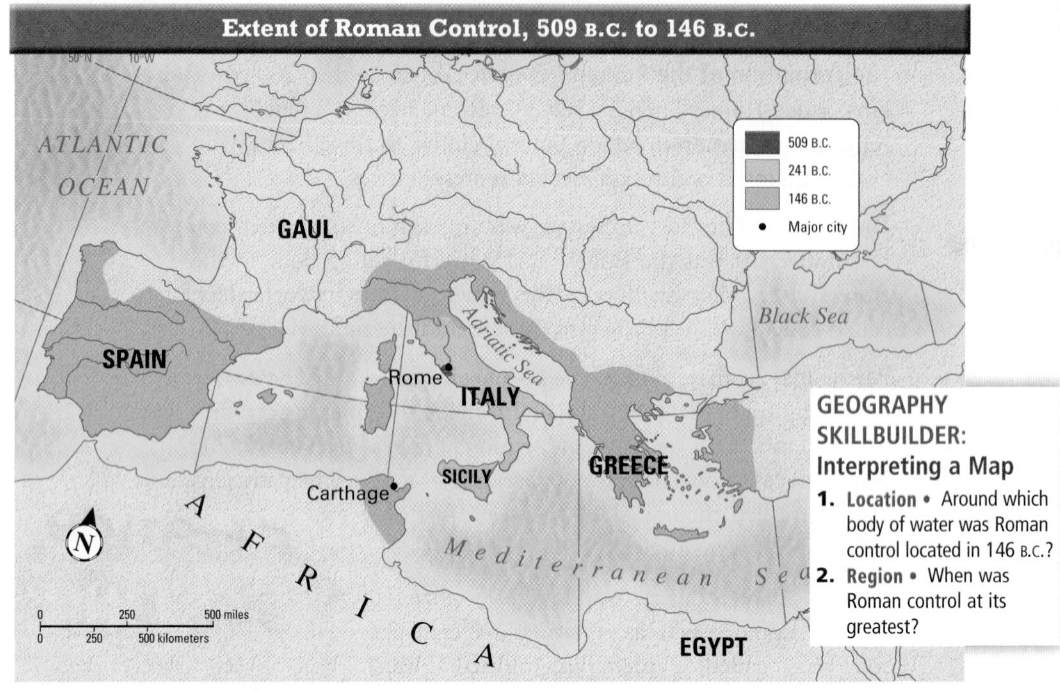

GEOGRAPHY SKILLBUILDER: Interpreting a Map

1. **Location** • Around which body of water was Roman control located in 146 B.C.?
2. **Region** • When was Roman control at its greatest?

As Rome's control over its neighbors expanded, its culture and language continued to spread into Spain and Greece. By the end of the second century B.C., the Romans ruled most of the land surrounding the Mediterranean Sea. The ancient Romans even called the Mediterranean *mare nostrum* (MAH•ray NOH•struhm), which means "our sea."

Geography Skillbuilder Answers
1. Mediterranean Sea
2. in 146 B.C.

From Republic to Empire 2

Region • Once in power, Julius Caesar had his likeness stamped on coins such as this one. ▼

As the Roman Republic grew, its citizens became a more and more diverse group of people. Many Romans practiced different religions and followed different customs, but they were united by a common system of government and law. In the middle of the first century B.C., however, Rome's form of government changed.

The End of the Roman Republic **Julius Caesar,** a successful Roman general and famous speaker, was the governor of the territory called Gaul. By conquering nearby territories to expand the land under his control, he increased both his power and his reputation. The Roman Senate feared that Caesar might become too powerful, and they ordered him to resign. Caesar, however, had other ideas.

BACKGROUND
Ancient Gaul included the lands that are modern-day France, Belgium, and parts of northern Italy.

Activity Options

Multiple Learning Styles: Linguistic

Class Time 15 minutes

Task Identifying the origin of the terms *empire* and *emperor*

Purpose To trace the Latin roots of basic words used in the section

Supplies Needed
- Paper
- Pencils or pens
- Dictionaries

Activity Point out the heading "From Republic to Empire" on page 286 and the explanation of *empire* on page 287. Then ask students to use their dictionaries to determine the origins of the words *empire* and *emperor.* Suggest that they locate the dictionary entries and look in the first line of each entry to find information about the word's origin. When they have finished, ask volunteers to read their results aloud. Be sure that students understand that empire comes from the Latin word *imperium,* and emperor comes from the Latin word *imperator.*

Rather than resign, Caesar fought a long, fierce battle for control of the Roman Republic. In 45 B.C., he finally triumphed and returned to Rome. Caesar eventually became dictator of the Roman world. A dictator is a person who holds total control over a government. Caesar's rule marked the end of the Roman Republic.

The Beginning of the Roman Empire Julius Caesar had great plans to reorganize the way ancient Rome was governed, but his rule was cut short. On March 15, 44 B.C., a group of senators, angered by Caesar's plans and power, stabbed him to death on the floor of the Roman Senate. A civil war then erupted that lasted for several years.

In 27 B.C., Caesar's adopted son, Octavian, was named the first emperor of Rome. This marks the official beginning of the Roman Empire. An **empire** is a nation or group of territories ruled by a single, powerful leader, or emperor. As emperor, Octavian took the name **Augustus.**

The Augustan Age Augustus ruled the Roman Empire for more than 40 years. During this time, called the Augustan Age, the empire continued to expand. To help protect the enormous amount of land under his control, Augustus sent military forces along its borders, which now extended northward to the Rhine and Danube rivers.

Reading **Social Studies**

A. Recognizing Important Details How many years separated the rules of Julius Caesar and Augustus?

Reading Social Studies
A. Answer
17 years (44–27 B.C.)

Region • Sculptures of Augustus were sent all over the Roman Empire to let people know what their leader looked like. ▲

Geography Skillbuilder Answers
1. Africa, Europe
2. Syria

The Roman Empire, A.D. 14

GEOGRAPHY SKILLBUILDER: Interpreting a Map

1. **Location •** Name two continents on which the Roman Empire was located.
2. **Location •** What was the easternmost territory of the Roman Empire in A.D. 14?

FOCUS ON VISUALS

Interpreting the Sculpture Ask students what details they see in the sculpture of Emperor Augustus. What character traits did the sculptor try to convey in this statue? Why did the Roman Empire erect statues like this one throughout its territories?

Possible Responses Students are likely to say that the statue conveys strength, power, fairness, dignity, or intelligence. Erecting statues like this one may have helped create a sense of the greatness of Augustus throughout the empire.

Extension Ask students to think about how political leaders today make sure that their images are widely recognized.

FOCUS ON VISUALS

Interpreting the Map Direct students to compare this map with the one on page 286. Have them calculate the time difference between this map (A.D. 14) and the latest date shown on the previous one (146 B.C.). Ask them to describe where the greatest territorial expansion had taken place in the interim.

Response 160 years; into central and northern Europe, to the Rhine and Danube rivers (Gaul)

Extension Have students compare this map with a map of modern Europe and list the modern nations that were all or partly within the Roman Empire in A.D. 14.

Activity Options

Interdisciplinary Link: Literature

B Block Scheduling

Class Time One class period

Task Listening to a reading of part of the murder scene from Shakespeare's *Julius Caesar*

Purpose To demonstrate how writers depict historical events and give students a sense of the emotions behind events

Supplies Needed

- Photocopies of Act III, Scene 1, "O mighty Caesar . . . ," *Julius Caesar*

Activity Set the scene by explaining that Shakespeare wrote a play about the assassination of Julius Caesar. In this scene, Marc Antony, Brutus, and Cassius appear and react to the murder. Read the characters' speeches aloud to the group, and encourage students to follow along. Discuss the material to help students understand the characters' reactions—how Brutus places blame and how Antony uses sarcasm.

Spotlight on CULTURE

To supply water to the city of Rome, engineers built 11 major aqueducts in the period between 312 B.C. and A.D. 226. Aqueducts brought water from springs, lakes, and rivers in the hills around the city. Most of the Roman water system actually ran through underground channels, with only a relatively few miles carried across valleys on the spectacular stone arches that we now think of as "aqueducts."

Water flowed from the hills into distribution tanks. From there it traveled through pipes (of lead, tile, or stone) to different parts of the city. The best water was used for drinking and cooking, while some was only clean enough for watering gardens. Some very rich citizens had water piped directly into their private villas. Most people used water from public fountains that were located at 100-meter intervals throughout the city. Huge quantities of water also supplied the enormous bath complexes, such as the Baths of Caracalla.

CRITICAL THINKING ACTIVITY

Finding Causes Have students review the different stages in Rome's development to make sure that they understand the difference between the Roman Republic and the Roman Empire. Then ask them to identify the events that caused the end of the Roman Republic.

Class Time 10 minutes

Region • A diver holds an artifact from an ancient Roman shipwreck in the Mediterranean Sea. ▲

While the Roman army kept peace, architects and engineers built many new public buildings. Trade increased, with olive oil, wine, pottery, marble, and grain being shipped all across the Mediterranean. Lighthouses were constructed, too, to help ships find their way into port.

The Augustan Age was also a time of great Roman literature. One of the most famous works of the age is the *Aeneid* (ih•NEE•ud). This long poem tells the story of Rome's founding. Augustus himself asked the famous poet Virgil to write it. This period of peace and cultural growth that Augustus created in the Roman Empire was called the "Pax Romana" (pahks roh•MAH•nah). The Pax Romana, or Roman Peace, lasted for 200 years.

Vocabulary

marble: a hard, smooth stone, often white in color

Thinking Critical[ly] Answers

1. to give their citi[zens] a good water supp[ly]

2. Possible respon[se:] No, because cities could not have flourished, and pe[o]ple would not hav[e] been as healthy.

Architecture Various inventions helped the Roman Empire grow and prosper. In addition to buildings and roads, Roman architects and engineers constructed water systems called aqueducts. Ancient aqueducts were raised tunnels that carried fresh water over long distances.

Built throughout the empire, aqueducts poured millions of gallons of water into Rome and other cities every day. They supplied clean water to private homes, fountains, and public baths. Today, some ancient Roman aqueducts still stand in France, Spain, and even on the outskirts of Rome itself.

THINKING CRITICALLY 

1. **Analyzing Motives**
   Why did Romans want a way to transport water?
2. **Hypothesizing**
   Do you think the Roman Empire would have grown so large and prosperous without the aqueducts?

For more on Roman architecture, go to 

288 CHAPTER 10

Activity Options

Interdisciplinary Link: Art

Class Time One class period

Task Drawing a Roman structure

Purpose To familiarize students with the architecture of ancient Rome

Supplies Needed
- Drawing paper
- Colored pencils
- Illustrations/models of Roman architecture

Block Scheduling

Activity Display examples of Roman architecture and ask students to comment on what they see. Ask them where they think the materials came from, how the materials were transported to the site, who built the structures, and what they were used for. Then ask them to choose a Roman structure, such as the Colosseum, and draw it. Display the drawings. Ask students if they can identify any structures in the United States that have elements of Roman architecture.

The Rise of Christianity ❸

Reading **Social Studies**

B. Making Inferences How do you think the Roman Empire indirectly helped the spread of Christianity?

In the years following the death of Augustus in A.D. 14, a new religion from the Middle East began to take hold in the rest of the Mediterranean world: Christianity. At first, this religion became popular mainly in the eastern half of the Roman Empire. Many followers there preached about its teachings. Christianity spread along the transportation network constructed by the Romans. By the third century A.D., this religion had spread throughout the empire.

Reading Social Studies
B. Possible Answer
Roman roads and the stability of Roman laws let ideas spread throughout the empire.

Most earlier Roman leaders had tolerated the different religions practiced throughout the empire. Christians, however, were viewed with suspicion and suffered persecution as early as A.D. 64. Roman leaders and people of other religions even blamed the Christians for natural disasters. Many Christians during this time were punished or killed for their beliefs.

Region • Constantine (died A.D. 337) was the first Christian emperor of Rome. ▼

The First Christian Emperor ❸

Things changed when **Constantine** became emperor of Rome in A.D. 306. In A.D. 312, before a battle, Constantine claimed to have had a vision of a cross in the sky. The emperor promised that if he won the battle, he would become a Christian. Constantine was victorious, and the next year he fulfilled his promise. Christianity became the official religion of the Roman Empire. Today, Christianity has nearly two billion followers worldwide.

SECTION 3 ASSESSMENT

Terms & Names

1. **Explain the significance of:** (a) republic (b) Senate (c) patrician (d) plebeian (e) Julius Caesar (f) empire (g) Augustus (h) Constantine

Using Graphics

2. Use a chart like this one to outline the achievements of ancient Rome's Augustan Age.

Achievement	Effects

Main Ideas

3. (a) On what waterway is the city of Rome located?
   (b) What helped to unite the many different citizens of the Roman Republic?
   (c) How did Christianity spread throughout the Roman Empire?

Critical Thinking

4. **Drawing Conclusions**
   Why was ancient Rome able to control most of the land surrounding the Mediterranean Sea?

Think About
- the location of the Italian Peninsula
- Rome's army
- Rome's wars with Carthage

ACTIVITY -OPTION- Review the information about the beginnings of ancient Rome. Create a **chart** that compares the two important classes of Roman society: patricians and plebeians.

INSTRUCT: Objective ❸

The Rise of Christianity/ The First Christian Emperor

- How did Roman attitudes toward Christians change over time? Romans at first were tolerant, then became suspicious of and persecuted Christians.
- How did Christianity come to be the official religion of the Roman Empire? Constantine, the Roman emperor, converted to Christianity.

ASSESS & RETEACH

Reading Social Studies Have students fill in the column on Rome in the chart on page 272.

Formal Assessment
- Section Quiz, p. 143

RETEACHING ACTIVITIES

With students, brainstorm ideas for newspaper articles based on events they learned about in this section. Have students work in groups to write a newspaper article about one of the events.

In-depth Resources: Unit 4
- Reteaching Activity, p. 13

Access for Students Acquiring English
- Reteaching Activity, p. 83

Section 3 Assessment

1. Terms & Names

- a. republic, p. 285
- b. Senate, p. 285
- c. patrician, p. 285
- d. plebeian, p. 285
- e. Julius Caesar, p. 286
- f. empire, p. 287
- g. Augustus, p. 287
- h. Constantine, p. 289

2. Using Graphics

Achievement	Effects
Literature	Virgil's *Aeneid*
Growth of empire	peace, cultural growth
Trade	olive oil, wine, pottery, grain shipped across Mediterranean
Engineering	lighthouses, aqueducts

3. Main Ideas

- a. the Tiber River
- b. a common system of government and law
- c. At first, Christianity spread by means of the Roman transportation network; it spread more rapidly after Constantine converted in A.D. 312 and made it the official religion of Rome.

4. Critical Thinking

Rome was centrally located; its large, powerful army defeated its only regional rival for power, Carthage, in 146 B.C.

ACTIVITY OPTION

Integrated Assessment
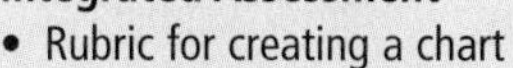
- Rubric for creating a chart

SECTION 4

Time of Change: The Middle Ages

TERMS & NAMES
medieval
Charlemagne
feudalism
manorialism
guild
Magna Carta

MAIN IDEA	WHY IT MATTERS NOW
The Middle Ages was a time of great change in Western Europe.	Some developments that occurred during the Middle Ages continue to affect life in Europe today.

SECTION OBJECTIVES

1. To describe Europe after the fall of the Roman Empire
2. To identify Charlemagne and describe the Church's role in the Middle Ages
3. To explain the relationships among people under feudalism and manorialism
4. To explain the changes that took place as towns developed and grew

CRITICAL THINKING

- Forming and Supporting Opinions, p. 291
- Recognizing Important Details, p. 293

FOCUS & MOTIVATE

WARM-UP

Drawing Conclusions Have students read Dateline and discuss the invasion of Rome.

1. How might the Visigoths have been able to overthrow the Roman government?
2. Why did invaders not attack the city earlier?

INSTRUCT: Objective 1

Western Europe in Collapse

- What were the effects of the collapse of the Roman Empire? no central government, towns abandoned, unsafe travel, decreased trade
- What is the period between the end of ancient times and the beginning of the modern world called? Middle Ages or medieval era

In-depth Resources: Unit 4
- Guided Reading Worksheet, p. 6

Reading Study Guide
(Spanish and English), pp. 84–85

DATELINE

Rome, A.D. 476—A Germanic tribe called the Visigoths has attacked our city of Rome and overthrown the emperor, Romulus Augustulus. The Roman army—no longer as large or as well organized as it was during the height of the empire—was unable to fight off the invaders.

After looting the great city, fierce bands of warriors and bandits have continued raiding towns and villages throughout Western Europe. They are stealing jewels and money, killing both people and animals, and even seizing control of entire territories. The Roman Empire seems to have breathed its last breath.

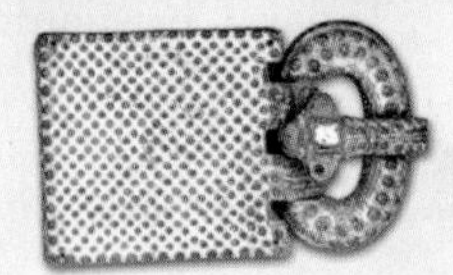

Region • Visigoth artifacts, like these saddle buckles, were found near Rome. ▲

Western Europe in Collapse 1

As the Roman Empire collapsed in the fifth century, more and more people fled to the countryside to escape invaders from the north and east. Eventually, there was no central government to maintain roads, public buildings, or water systems. Most towns and cities in Western Europe shrank or were totally abandoned. Long-distance travel became unsafe, and trade less common.

TAKING NOTES

Use your chart to take notes about Western Europe.

Time Period	Social Structure	Architecture
Ancient Greece		
Ancient Rome		

Program Resources

In-depth Resources: Unit 4
- Guided Reading Worksheet, p. 6
- Reteaching Activity, p. 14

Reading Study Guide
(Spanish and English), pp. 84–85

Formal Assessment
- Section Quiz, p. 144

Integrated Assessment
- Rubric for writing a journal

Outline Map Activities

Access for Students Acquiring English
- Guided Reading Worksheet, p. 78

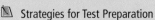

Technology Resources
classzone.com

TEST-TAKING RESOURCES

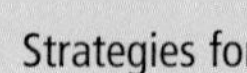

Strategies for Test Preparation

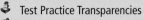

Test Practice Transparencies

Online Test Practice

Reading Social Studies

A. Clarifying Who provided leadership during the Middle Ages?

Reading Social Studies
A. Answers
1. military leaders and the Roman Catholic Church

The Beginning of the Medieval Era The period of history between the fall of the Roman Empire and the beginning of the modern world is called the Middle Ages, or **medieval** (MEE•dee•EE•vuhl) era. During this time, many of the advances and inventions of the ancient world were lost. Without a strong central government, many Europeans turned to military leaders and the Roman Catholic Church for leadership and support.

Charlemagne and the Christian Church ❷

Among the most famous military leaders was the Germanic King Charlemagne (SHAHR•luh•mayn). In the late 700s, **Charlemagne,** or Charles the Great, worked to bring political order to the northwestern fringes of what had been the Roman Empire. This great warrior not only fought to increase the size of his kingdom, he also worked to improve life for those who lived there.

A New Roman Emperor Eventually, news of Charlemagne's accomplishments spread to Rome. Although the old empire was gone, Rome was now the center of the Catholic Church. The Pope recognized that joining forces with Charlemagne might bring greater power to the Church.

In 800, the Pope crowned Charlemagne as the new Holy Roman Emperor. During Charlemagne's rule, education improved, the government became stronger, and Catholicism spread. But after Charlemagne's death, Western Europe was once again without a strong political leader.

Region • Charlemagne established order and supported education and culture for a brief period in the early Middle Ages. ▲

The Role of the Church ❷

Throughout Western Europe in medieval times, each community was centered around a church. The church offered religious services, established orphanages, and helped care for the poor, sick, and elderly. They also hosted feasts, festivals, and other celebrations. As communities grew, their members often donated money and labor to build new and larger churches.

Monks and Nuns Some people chose to dedicate their lives to serving God and the Church. These religious people were called monks and nuns. Monks were men who devoted their time to praying, studying, and copying and decorating holy books by hand. Monks lived in communities called monasteries. Many monasteries became important centers of learning in medieval society.

INSTRUCT: Objective ❷

Charlemagne and the Christian Church/The Role of the Church

- Why did Rome remain important after the fall of the Roman Empire? because it was the center of Roman Catholicism
- What changes took place under the rule of Charlemagne? improved education, stronger government, spread of Roman Catholicism
- What were the functions of the church in medieval times? offered religious services, cared for the sick and needy, established orphanages, hosted feasts and celebrations, preserved books and learning

CRITICAL THINKING ACTIVITY

Forming and Supporting Opinions Ask students to think about how life in the early Middle Ages differed from life in ancient Greece and Rome. Then ask them to form an opinion as to whether the fall of Rome was a positive or negative event. Have them support their opinions by comparing standards of living, political systems, cultural achievements, and peace and stability in the two eras.

Class Time 15 minutes

Activity Options

Multiple Learning Styles: Intrapersonal

Class Time 20 minutes

Task Writing a diary entry from the perspective of a medieval monk or nun

Purpose To help students understand what life was like in a medieval monastery

Supplies Needed
- Writing paper
- Pencils or pens

Activity Review the section "Monks and Nuns" beginning on page 291. Explain to students that monks and nuns took religious vows and were subject to rules governing their daily lives. Explain that during the medieval period, monasteries were centers of knowledge and education. Monasteries maintained schools and libraries. Ask students to use what they know about monastic life in order to write a fictional diary entry of a day in a monk's or nun's life.

INSTRUCT: Objective 3

Two Medieval Systems/ Medieval Ways of Life

- Who owned most of the land in Europe by the beginning of the Middle Ages? powerful nobles—lords, kings, high church officials
- Under feudalism, how were land and loyalty related? Nobles or rulers granted land to lesser nobles in exchange for political loyalty and help in war.
- What were the benefits of the manor system for the nobles and the peasants? nobles: food and labor; peasants: protection

Connections to History

Today we think of tapestries as works of art. During the Middle Ages, however, tapestries were used to make castle rooms easier to heat. The tapestries were woven to fit particular walls, on which they were hung using large rods. If the tapestry was moved to a different wall, it was cut to fit a smaller wall or to allow access to a door. This explains why tapestries seen in museums today may have different sections of different lengths.

Location • Convents and monasteries often were located in hard-to-reach areas. ▲

Women who served the Church were called nuns. In the Middle Ages, it was common for a woman to become a nun after her husband died. Nuns prayed, sewed, taught young girls, cared for the poor, and also copied and decorated books. They lived in secluded communities called convents.

Vocabulary

secluded: to be separate or hidden away

Two Medieval Systems 3

During the Middle Ages, almost all the land was owned by powerful nobles—lords, kings, and high church officials. The central government was not very strong. The nobles sometimes even controlled the king and constantly fought among themselves. To protect their lands and position, nobles developed a system known as feudalism.

The Feudal System **Feudalism** was a system of political ties in which the nobles, such as kings, gave out land to less powerful nobles, such as knights. In return for the land, the noble, called a vassal, made a vow to provide various services to the lord. The most important was to furnish his lord with knights, foot soldiers, and arms for battle.

The parcel of land granted to a vassal by his lord was called a fief (feef). The center of the lord's fief was the manor, which consisted of a large house or castle, surrounding farmland, villages, and a church. A fief might also include several other manors or castles belonging to the fief-owner's vassals.

The Bayeux Tapestry This famous work of art depicts the invasion of England by William the Conqueror in 1066. The Bayeux (by•YOO) Tapestry is a series of scenes from the point of view of the invaders, who came from Normandy. Normandy is a part of what is now France. The work is an important source of information about not only the conquest of England, but also medieval armor, clothing, and other aspects of culture.

Although called a tapestry, the work is really an embroidered strip of linen about 230 feet long. It includes captions in Latin. The Bayeux Tapestry was probably made by nuns in England about 1092.

Activity Options

Differentiating Instruction: Less Proficient Readers

Understanding Key Concepts To help students understand the obligations and rewards of feudalism and manorialism, work with them to create a diagram like the one shown. As you create the diagram, be sure students understand that each group of people had responsibilities to other groups, and that each group benefited from the systems.

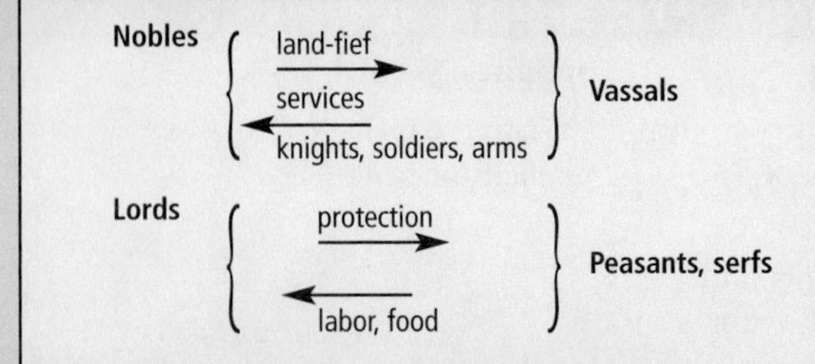

292 CHAPTER 10

Place • Although castles were large, they were built for defense. Castles were usually located on high ground with a series of walls and towers. ▲

Manorialism On the manor, peasants lived and farmed, but they usually did not own the land they lived on. In exchange for their lord's protection, the peasants contributed their labor and a certain amount of the food they raised. Some peasants, known as serfs, actually belonged to the fief on which they lived. They were not slaves, but they were not free to leave the land without the permission of the lord. This system, in which the lord received food and work in exchange for his protection, is known as **manorialism.**

Medieval Ways of Life

Medieval nobles had more power than the peasants. However, the difference in the standard of living between the very rich and the very poor was not as great as the difference today.

Castle Life The manor houses or castles may have been large, but they were built more for defense than for comfort. Thick stone walls and few windows made the rooms cold, damp, and dark. Fires added warmth but made the air smoky. Medieval noble families may have slept on feather mattresses, but lice and other pests were a constant annoyance. Most castles did not have indoor plumbing.

CRITICAL THINKING ACTIVITY

Recognizing Important Details Ask students to read the second sentence after the heading "Medieval Ways of Life." Then ask them to find details in the paragraphs that follow to show living conditions were difficult for both the rich and the poor in medieval Europe.

Class Time 5 minutes

MORE ABOUT...
Castles

Early medieval castles were built around a central mound (or motte), on which stood a wooden citadel. Within the castle's wooden walls was an open space, or bailey, where local people could gather for safety during wars. A moat around the walls gave still more protection. Later in the Middle Ages, castles were built of stone. As the picture on page 293 shows, the walls around the inner bailey contained a series of towers linked by a walkway along the top of the walls. Other buildings, including the central keep, stood within the walls. Also within the castle walls were wells to supply water.

Activity Options

Multiple Learning Styles: Linguistic

Class Time 30 minutes

Task Writing a dialogue between a lord and his serf

Purpose To understand the relationship between a lord and a serf and the responsibilities of and benefits for each

Supplies Needed
- Writing paper
- Pencils or pens

Block Scheduling

Activity Recall with students that under the manor system, lords protected peasants in return for labor and the food that the peasants raised. Invite students to work in pairs to write a dialogue between one lord and his serf. Suggest that they might focus on a discussion regarding problems or living conditions in the manor, or future plans. Remind them to keep in mind the social class gap between the two. Ask volunteers to read their dialogues aloud.

INSTRUCT: Objective 4

The Growth of Medieval Towns/The Late Middle Ages

- Why did people begin to move back into towns in the 11th century? Fewer people were needed on farms.
- What were the benefits of belonging to a guild? protected workers' rights, set wages and prices, settled disputes
- What were the conditions of the Magna Carta? limited power of king, gave nobles greater say

Connections to Economics

The rise of a middle class of merchants, skilled craftsworkers, and others began in the later Middle Ages, along with the growth of towns and trade. After the Industrial Revolution, in the 1800s and 1900s, the middle class grew rapidly in Western Europe. Today, most Western Europeans belong to the middle class. Nevertheless, a small group of very rich people controls an enormous percentage of the world's wealth. According to *Forbes* magazine, the world's 538 billionaires owned $1.72 trillion worth of assets in 2000—approximately 10 percent of total world production that year.

Peasant Life Peasants lived outside the castle walls in small dwellings, often with dirt floors and straw roofs. They owned little furniture and slept on straw mattresses. It was common for peasant families to keep their farm animals inside their homes.

Peasants often worked two or three days a week for their lord, harvesting crops and repairing roads and bridges. The rest of the week they farmed their own small plots. Many days were religious festivals during which no one worked.

The Middle Class In the early Middle Ages, only a small percentage of people in Western Europe were wealthy landowners. Most people worked on manor lands or at some sort of craft. However, those workers who found jobs in towns often were able to save money and build businesses. Eventually, their improved status led to the rise of a middle class.

Unlike nobles, the members of this new middle class did not live off the land they owned. They had to continuously earn money, as most people do today.

The Growth of Medieval Towns 4

By the middle of the 11th century, life was improving for many people in Western Europe. New farming methods increased the supply of food and shortened the time it took to harvest crops. Fewer farmers were needed, and workers began to leave the countryside in search of other opportunities. People moved back into towns or formed new ones that grew into booming centers of trade. The population increased, and more and more people owned property or started businesses.

Reading Social Studies
B. Possible Answe
to protect their businesses and crafts b joining with others

Guilds As competition among local businesspeople grew, tradespeople and craftspeople created their own guilds, or business associations. Similar to modern trade unions, a **guild** protected workers' rights, set wages and prices, and settled disputes. Membership in a guild was also a common requirement for citizens who sought one of the few elective public offices.

B. Analyzing Motives Why did people create guilds?

The Late Middle Ages 4

Over time, the towns of the late Middle Ages grew in size, power, and wealth. The citizens of these towns began to establish local governments and to elect leaders.

Activity Options

Skillbuilder Mini-Lesson: Reading a Time Line

Explaining the Skill Remind students that in a time line, dates and events are arranged in chronological order. Seeing events on a time line helps a reader see them in relation to one another as well as in order.

Applying the Skill Have students review the section on the Middle Ages and jot down important dates. Encourage them to include approximate dates, such as the 11th century, as well as exact dates, such as A.D. 476. Have students create time lines that begin with the fall of Rome and end with the signing of the Magna Carta. Suggest that students illustrate the more interesting events listed on their time lines.

Extension Ask students to choose an event on their time line that they consider important. Then ask them to research the event and write a paragraph about it.

Governments Challenge the Church The Pope insisted that he had supreme authority over all the Christian lands. Kings and other government leaders, however, did not agree that the Pope was more powerful than they were. This is an issue that continues to be discussed today.

The Magna Carta The rulers of Western Europe also struggled for power with members of the nobility. In England, nobles rebelled against King John. In 1215, the nobles forced the English king to sign a document called the **Magna Carta** (MAG•nuh KAHR•tuh), or Great Charter. This document limited the king's power and gave the nobles a larger role in the government.

Region • High taxes and failures on the battlefield made King John one of the most hated kings of England. ▼

Region • The Magna Carta influenced the creators of the U.S. Constitution. ▲

SECTION 4 ASSESSMENT

Terms & Names

1. **Explain the significance of:** (a) medieval (b) Charlemagne (c) feudalism (d) manorialism (e) guild (f) Magna Carta

Using Graphics

2. Use a flow chart like this one to show how Europe changed over four time periods: A.D. 476, the 800s, the mid-1000s, and the 1200s.

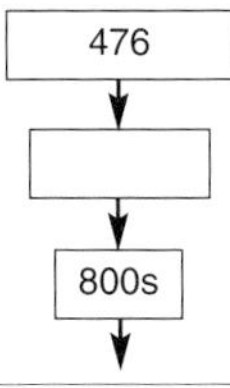

Main Ideas

3. (a) Why is this era of European history called the Middle Ages?
   (b) Describe the role of the Church in medieval society.
   (c) How did manorialism help both nobles and peasants?

Critical Thinking

4. **Contrasting**
   How did life differ for nobles and peasants under feudalism?

Think About
- where they lived
- what they ate
- how they did their work

ACTIVITY -OPTION- Review the information about serfs. Write a series of short **journal entries** describing what a week in the life of a serf might have been like during the Middle Ages.

MORE ABOUT...
The Magna Carta

The rights and protections of the Magna Carta originally benefited only the barons who forced King John to sign the document. They included trial by a jury of one's peers and the right to due process of law—which would prevent the king's arbitrarily seizing property or imprisoning a noble. Such legal limitations on a monarch were unprecedented. Over time, these rights were also extended to ordinary English people.

ASSESS & RETEACH

Reading Social Studies Have students complete the chart on page 272.

Formal Assessment
- Section Quiz, p. 144

RETEACHING ACTIVITIES

Have students work in small groups to identify the most important themes in this section. Then ask them to write a two- to three-sentence summary of each theme. Have a volunteer from each group read the group's summary.

In-depth Resources: Unit 4
- Reteaching Activity, p. 14

Access for Students Acquiring English
- Reteaching Activity, p. 84

Section 4 Assessment

1. Terms & Names

a. medieval, p. 291
b. Charlemagne, p. 291
c. feudalism, p. 292
d. manorialism, p. 293
e. guild, p. 294
f. Magna Carta, p. 295

2. Using Graphics

476—collapse of Roman Empire

800s—Charlemagne was emperor; government was stronger; Roman Catholicism spread.

mid-1000s—Farms prospered; towns and cities grew; guilds protected workers.

1200s—Rulers struggled for power; nobles rebelled; Magna Carta limited the king's power.

3. Main Ideas

a. It falls between ancient times and the beginning of the modern era.
b. Communities were built around a church, which held religious services and festivals. Monks and nuns taught children, cared for the sick and needy, and preserved books and learning.
c. Nobles protected the peasants and were supported by the food and labor they provided.

4. Critical Thinking

peasants: few rights, owned little, depended on lord and worked hard to supply him with food and labor; nobles: housed and fed better, ran estates, often went to war

ACTIVITY OPTION

Integrated Assessment
- Rubric for writing a journal

ASSESSMENT

TERMS & NAMES

1. peninsula, p. 274
2. plain, p. 275
3. city-state, p. 279
4. Aegean Sea, p. 279
5. Athens, p. 280
6. republic, p. 285
7. empire, p. 287
8. Constantine, p. 289
9. feudalism, p. 292
10. Magna Carta, p. 295

REVIEW QUESTIONS

Possible Responses

1. Europe has a variety of waterways and landforms, particularly peninsulas, mountains, and a central plain. It has a varied climate and abundant resources.
2. The Great European Plain has some of the richest farmland; trade with farmers attracted many people.
3. a common language, shared religious beliefs, similar way of life
4. They united to defeat the Persians.
5. Rome controlled most of the land surrounding the Mediterranean.
6. They feared he was too powerful, a threat to the republic.
7. There was no strong central government; the Church gave people a sense of purpose.
8. Technological advances in farming methods meant that fewer people were needed on farms. Many people left farms, moving to towns and starting businesses.

CHAPTER 10 ASSESSMENT

TERMS & NAMES

Explain the significance of each of the following:

1. peninsula
2. plain
3. city-state
4. Aegean Sea
5. Athens
6. republic
7. empire
8. Constantine
9. feudalism
10. Magna Carta

REVIEW QUESTIONS

A Land of Varied Riches *(pages 273–277)*

1. What is special about Europe's physical environment?
2. Why is the Great European Plain an important region?

Ancient Greece *(pages 278–282)*

3. What helped to unite the separate city-states of ancient Greece?
4. What caused the people of Athens to join forces with their rival city-state, Sparta, in 480 B.C.?

Ancient Rome *(pages 284–289)*

5. Why did the ancient Romans call the Mediterranean Sea "our sea"?
6. Why did the Roman Senate ask Julius Caesar to resign?

Time of Change: The Middle Ages *(pages 290–295)*

7. Why did the people turn to the Roman Catholic Church for leadership and support during the Middle Ages?
8. What contributed to the growth of towns during the Middle Ages?

CRITICAL THINKING

Recognizing Effects

1. Using your completed chart from Reading Social Studies, p. 272, identify changes in art, culture, religion, and social structure from the time of ancient Rome to the Middle Ages.

Hypothesizing

2. Many myths and plays of ancient Greece have been the basis for modern films and dramas. What does this indicate about these ancient stories and characters?

Analyzing Causes

3. How did the long, peaceful reign of Augustus help to promote architecture, literature, and art in the Roman Empire?

Visual Summary

1 A Land of Varied Riches

- The rich natural resources and varied geography of Europe helped to shape its development.

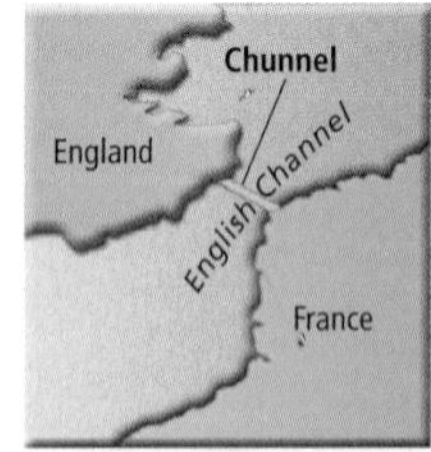

2 Ancient Greece

- The ancient Greeks developed a complex society and system of government.
- The achievements of the ancient Greeks in architecture, literature, and philosophy had a lasting impact on the world.

3 Ancient Rome

- Under strong leadership, ancient Rome experienced a time of great growth.
- Ancient Rome's contributions to government, engineering, and literature influenced Western culture.

4 Time of Change: The Middle Ages

- The social order and government of the Middle Ages transformed Europe into a modern society.

CRITICAL THINKING: Possible Responses

1. Recognizing Effects

Depending on their notes, students may trace the social/political changes from republic to empire to feudalism to the rise of the middle class; the change from worship of many gods to the supremacy of the Catholic Church; the decline in science, arts, and learning after the fall of Rome.

2. Hypothesizing

They dealt with universal themes that continue to appeal to people today.

3. Analyzing Causes

Augustus's long reign brought peace to the region, allowing creativity to flourish. Architects and engineers built many new buildings. The age was also a time of great literature.

ANDARDS-BASED ASSESSMENT

the map and your knowledge of world cultures and graphy to answer questions 1 and 2.

ditional Test Practice, pp. S1–S33

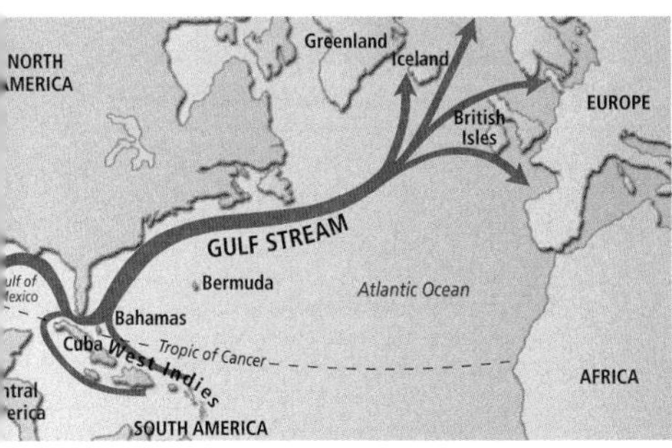

Vhat bodies of water does the Gulf Stream flow through on he way to Europe?

A. Atlantic Ocean and Gulf of Mexico
B. Atlantic Ocean and Tropic of Cancer
C. Atlantic Ocean and West Indies
D. Gulf of Mexico and Tropic of Cancer

Which of the following regions do you think is most directly ffected by the Gulf Stream?

A. Africa
B. Central America
C. Europe
D. South America

The following passage is from a biography of Julius Caesar. Use the quotation and your knowledge of world cultures and geography to answer question 3.

PRIMARY SOURCE

The conspirators who murdered Julius Caesar . . . failed miserably to destroy him. Though they killed his body, they could not extinguish his spirit, his vision for Rome's future. . . . Many plebes [common people] . . . and soldiers saw him as their champion. And his bloody death unleashed the pent-up distrust and anger they felt for the . . . ruling class.

DON NARDO, *Julius Caesar*

3. The passage supports which of the following observations?

A. Caesar's assassination was celebrated by the masses in Rome.
B. Caesar's energy and his hopes for Rome lived on in spite of his death.
C. The soldiers joined with the conspirators to remove Caesar from power.
D. The death of Caesar signaled the end to his vision for Rome.

LTERNATIVE ASSESSMENT

WRITING ABOUT HISTORY

The Pope crowned Charlemagne as the Holy Roman Emperor n 800. Research this event and write a biography of Charlemagne in which you focus on his reign as emperor.

- Use the Internet or library resources to research Charlemagne and his impact on political, economic, and religious life in Europe.
- As a biographer, focus on the facts and important details you found in your research. Also, include your impressions of his reign.

COOPERATIVE LEARNING

Work with a small group of classmates to design and create a playbill for a Greek tragedy. Your playbill, or program for the audience, should include a cover illustration, a list of the characters, a summary of the story, and information about the playwright. Group members can share the responsibilities of researching, writing, and illustrating the playbill.

INTEGRATED TECHNOLOGY

Doing Internet Research

Different regions of Europe contain different natural resources. Focus on one European region, such as the Ruhr Valley, the Mediterranean Sea, or the independent republics of Eastern Europe. Use the Internet to research the region and its natural resources.

- Use the Internet, as well print encyclopedias, atlases, and other reference books to learn about the region.
- Focus your research on the natural resources found there, how they contributed to the region's development, and how the region has developed in modern times.
- Organize your findings into a report. Include a map that shows the region and its resources.

For Internet links to support this activity, go to

STANDARDS-BASED ASSESSMENT

1. **Answer A** is the correct answer because according to the map, the Gulf Stream moves through the Gulf of Mexico and then across the Atlantic Ocean; answers B, C, and D are incorrect because the Tropic of Cancer and West Indies are not bodies of water.
2. **Answer C** is the correct answer because the Gulf Stream flows to Europe and affects its Western coast; answers A, B, and D are incorrect because the Gulf Stream does not flow toward the regions named.
3. **Answer B** is the correct answer because it focuses on the point that Caesar was a hero to the plebes and soldiers who believed in his vision for Rome; answer A is incorrect because many mourned Caesar's death; answer C is incorrect because the soldiers saw Caesar as their champion; answer D is incorrect because Caesar's vision lived on after his death.

INTEGRATED TECHNOLOGY

Discuss how students might find information about the natural resources of Europe using the Internet. Brainstorm ideas for keywords, Web sites, and search engines. Suggest that students make a copy of their chosen map and include symbols and a key to indentify the location of various types of natural resources.

Alternative Assessment

1. Rubric

The biography should

- accurately portray information about Charlemagne's life.
- be organized chronologically and include references to dates.
- be written in an interesting style.
- use correct grammar, spelling, and punctuation.

2. Cooperative Learning

Discuss the features of a playbill, a guide to a play. Point out that a playbill for a Greek tragedy should include a cover illustration that reflects the way actors dressed in ancient Greece, a list of characters, biographical information about the playwright, and a brief summary of the story.

The Growth of the Western World

	OVERVIEW	COPYMASTERS	INTEGRATED TECHNOLOGY
UNIT ATLAS AND CHAPTER RESOURCES	The students will explore the political and social changes that brought Europe from the Renaissance to the beginning of the modern era.	**In-depth Resources: Unit 4** • Guided Reading Worksheets, pp. 15–18 • Skillbuilder Practice, p. 21 • Unit Atlas Activities, pp. 1–2 • Geography Workshop, pp. 61–62 **Reading Study Guide** (Spanish and English), pp. 88–97 **Outline Map Activities**	• eEdition Plus Online • EasyPlanner Plus Online • eTest Plus Online • eEdition • Power Presentations • EasyPlanner • Electronic Library of Primary Sources • Test Generator • Reading Study Guide • Critical Thinking Transparencies CT21
	KEY IDEAS		
SECTION 1 Renaissance Connections pp. 301–306	• The Renaissance was a new era of creativity and learning in Western Europe. • The Protestant Reformation led to changes in church practices and an increase in education for all people.	**In-depth Resources: Unit 4** • Guided Reading Worksheet, p. 15 • Reteaching Activity, p. 23 **Reading Study Guide** (Spanish and English), pp. 88–89	classzone.com Reading Study Guide
SECTION 2 Traders, Explorers, and Colonists pp. 307–312	• Europeans sought shorter trade routes to bring spices from Asia. • Explorers such as da Gama and Magellan discovered new routes and brought European influence to new lands. • European exploration led to imperialism, which brought hardship to African and Native American peoples.	**In-depth Resources: Unit 4** • Guided Reading Worksheet, p. 16 • Reteaching Activity, p. 24 **Reading Study Guide** (Spanish and English), pp. 90–91	classzone.com Reading Study Guide
SECTION 3 The Age of Revolution pp. 313–317	• The Industrial Revolution changed the way goods were produced and brought about the growth of cities and factories. • Capitalism is the private ownership of factories and businesses. • The French Revolution claimed thousands of lives before Napoleon gained control of France.	**In-depth Resources: Unit 4** • Guided Reading Worksheet, p. 17 • Reteaching Activity, p. 25 **Reading Study Guide** (Spanish and English), pp. 92–93	Critical Thinking Transparencies CT22 classzone.com Reading Study Guide
SECTION 4 The Russian Empire pp. 318–322	• The early czars held unlimited power and sought to minimize the influence of nobles. • Peter the Great and Catherine the Great helped make Russia a strong empire. • The Russian Revolution in 1917 overthrew the monarchy.	**In-depth Resources: Unit 4** • Guided Reading Worksheet, p. 18 • Reteaching Activity, p. 26 **Reading Study Guide** (Spanish and English), pp. 94–95	classzone.com Reading Study Guide

KEY TO RESOURCES

- Audio
- CD-ROM
- Copymaster
- Internet
- Overhead Transparency
- Pupil's Edition
- Teacher's Edition
-  Video

ASSESSMENT OPTIONS

Chapter Assessment, pp. 324–325

Formal Assessment
- Chapter Tests: Forms A, B, C, pp. 161–172

Test Generator

Online Test Practice

Strategies for Test Preparation

Section Assessment, p. 306

Formal Assessment
- Section Quiz, p. 157

Integrated Assessment
- Rubric for writing a letter

Test Generator

Test Practice Transparencies TT33

Section Assessment, p. 311

Formal Assessment
- Section Quiz, p. 158

Integrated Assessment
- Rubric for writing a journal entry

Test Generator

Test Practice Transparencies TT34

Section Assessment, p. 317

Formal Assessment
- Section Quiz, p. 159

Integrated Assessment
- Rubric for writing a poem or lyrics

Test Generator

Test Practice Transparencies TT35

Section Assessment, p. 322

Formal Assessment
- Section Quiz, p. 160

Integrated Assessment
- Rubric for writing a summary

Test Generator

Test Practice Transparencies TT36

RESOURCES FOR DIFFERENTIATING INSTRUCTION

Students Acquiring English/ESL

Reading Study Guide (Spanish and English), pp. 88–97

Access for Students Acquiring English Spanish Translations, pp. 85–94

TE Activities
- Prefixes, p. 302
- Understanding Multiple Meanings, p. 314

Modified Lesson Plans for English Learners

Less Proficient Readers

Reading Study Guide (Spanish and English), pp. 88–97

TE Activity
- Determining Cause and Effect, p. 321

Gifted and Talented Students

TE Activity
- Researching a Topic, p. 315

CROSS-CURRICULAR CONNECTIONS

Humanities
Lassieur, Allison. ***Leonardo da Vinci and the Renaissance in World History.*** Berkeley Heights, NJ: Enslow, 2000. Places da Vinci in the context of his world.

Murrell, Kathleen Berton. ***Eyewitness: Russia.*** New York: DK Publishing, 2000. From empire to today's federation.

Literature
Maynard, Christopher. ***The History News: Revolution.*** Cambridge, MA: Candlewick Press, 1999. Presents in fictional newspaper format facts relating to the French, American, Russian, and Chinese revolutions.

Science/Math
Sharth, Sharon. ***Way to Go!: Finding Your Way With a Compass.*** Pleasantville, NY: Reader's Digest, 2000. History, science, and activities using a compass.

History
Aaseng, Nathan. ***You Are the Explorer.*** Minneapolis: Oliver Press, 2000. Readers face the same dilemmas as explorers and must solve the problems.

Macaulay, David. ***Ship.*** Boston: Houghton Mifflin, 1993. Underwater archaeology uncovers caravels from the 15th century.

ENRICHMENT ACTIVITIES

The following activities are especially suitable for classes following block schedules.

Teacher's Edition, pp. 303, 304, 305, 309, 310, 312, 315, 319, 320
Pupil's Edition, pp. 306, 311, 317, 322

Unit Atlas, pp. 260–269

Technology: 1781, p. 323
Outline Map Activities

INTEGRATED TECHNOLOGY

Go to **classzone.com** for lesson support and activities for Chapter 11.

BLOCK SCHEDULE LESSON PLAN OPTIONS: 90-MINUTE PERIOD

DAY 1

CHAPTER PREVIEW, pp. 298–299
Class Time 20 minutes

- **Hypothesize** Use the "What do you think?" questions in Focus on Geography on PE p. 299 to help students hypothesize about the connection between trade and disease.

SECTION 1, pp. 301–306
Class Time 70 minutes

- **Identifying Cause and Effect** Working on their own, have students create a matching game of causes and effects of the Renaissance. Have students trade games and compete, then review and discuss as a class.
  Class Time 30 minutes
- **Discussion** Lead a discussion reviewing the reasons wealthy Renaissance patrons supported artists and scholars. Invite students to analyze the motives behind such support.
  Class Time 10 minutes
- **Forming and Supporting Opinions** Have students imagine they are living in Germany in 1530. Have each student create a logical written argument first in support of, and then in opposition to, translating the Latin Bible into German. Have students share their arguments when they finish.
  Class Time 30 minutes

DAY 2

SECTION 2, pp. 307–311
Class Time 50 minutes

- **Analyzing Motives** Have five students sit in a line across the classroom, and give the first student a pencil, saying it costs $.50. Tell students to "buy" and "sell" the pencil to the next in line, adding a $.20 profit for the seller each time. When the last student has "bought" the pencil, have students compare the first price ($.50) with the last price ($1.30). Point out that this is why European spice traders wanted to buy directly from Asia.
  Class Time 25 minutes
- **Calculating Mileage** Tell students that the Portuguese sailed east, around Africa, looking for the shortest route to Asia, while the Spanish sailed west, across the Atlantic. Have students predict which route they think is shorter. Ask students to use the scale on the globe to calculate mileage for each water route. Then have them use their calculations to verify their predictions.
  Class Time 25 minutes

SECTION 3, pp. 313–317
Class Time 40 minutes

- **Peer Competition** Divide the class into pairs. Assign each pair one of the Terms & Names for this section. Have pairs make up three questions that can be answered with the term or name. Have pairs take turns asking the class their questions.

DAY 3

SECTION 4, pp. 318–322
Class Time 35 minutes

- **Time Line** Have each student choose one important event or aspect of Russian history covered in this section and write it in large letters on a piece of paper. Holding their papers, have students stand up and organize themselves into a human time line. Discuss each event in order, identifying any missing events.

CHAPTER 11 REVIEW AND ASSESSMENT, pp. 324–325
Class Time 55 minutes

- **Review** Have students prepare a summary of the chapter by reviewing the Main Idea and Why It Matters Now features of each section in Chapter 11.
  Class Time 20 minutes
- **Assessment** Have students complete the Chapter 11 Assessment.
  Class Time 35 minutes

TECHNOLOGY IN THE CLASSROOM

VIEWING ART ON THE WEB

One advantage of the Internet is that it offers a variety of images that supplement textbook material. For example, the Internet offers excellent opportunities to view works of art from many periods and places, and in many styles. Students can view some of the world's most famous artworks from their classroom or computer lab.

ACTIVITY OUTLINE

Objective Students will view examples of pre-Renaissance and Renaissance art and compare and contrast these artworks in a document on the computer.

Task Ask students to visit Web sites to explore the differences between pre-Renaissance and Renaissance painting styles. Have them copy and paste one painting from each period into a document on the computer and write one-page reports comparing the two artworks.

Class Time Two class periods

S

1. Have students reread Chapter 11. As a class or individually, have students list words that characterize the Renaissance.
2. Have students go through the first two Web sites at **classzone.com** to learn about artistic techniques that became common during the Renaissance. Ask them to take notes on the methods of painting that characterized the Renaissance and on how these methods differed from pre-Renaissance art. As an option, they may want to sketch some examples of linear perspective and other elements of Renaissance art.
3. Discuss students' findings as a class, making sure they understand that Renaissance art was characterized by attention to linear perspective, proportion, and careful observation of the subject.
4. Have students go to the third Web site at **classzone.com** to compare pre-Renaissance (Gothic and Byzantine) paintings with Renaissance paintings. Ask them to choose two paintings, one pre-Renaissance and one Renaissance, to compare and contrast.
5. Tell students to copy both paintings from the Web and paste them into a document on the computer. Ask them to label each painting with the URL where they found the painting, the name of the artist, the date, and the country.
6. Have students use a word processing program to write one-page reports comparing and contrasting these paintings. Encourage them to focus on the techniques employed in the Renaissance movement and to tell why the pre-Renaissance painting would have been considered outdated during the Renaissance.

CHAPTER 11 OBJECTIVE

Students will explore the political and social changes that brought Europe from the Renaissance to the beginning of the modern era.

FOCUS ON VISUALS

Interpreting the Photograph Have students look closely at the photograph. Ask them to identify the most distinguishing feature of the Cape of Saint Vincent. Then ask why they think Prince Henry's School of Navigation was located in this apparently desolate place.

Possible Responses There is water on both sides of the Cape of Saint Vincent. Its location made it a perfect spot from which to launch voyages of trade and exploration to the south.

Extension Have students make a list of the subjects that might have been taught at the School of Navigation.

CRITICAL THINKING ACTIVITY

Analyzing Motives Remind students that Portugal is a small country bordering Spain to the north and east. The Atlantic Ocean lies to the west and south. Display a world map and a map of Western Europe, and invite students to speculate about why Portugal became one of the greatest seafaring nations of the 15th and 16th centuries.

Class Time 15 minutes

CHAPTER 11

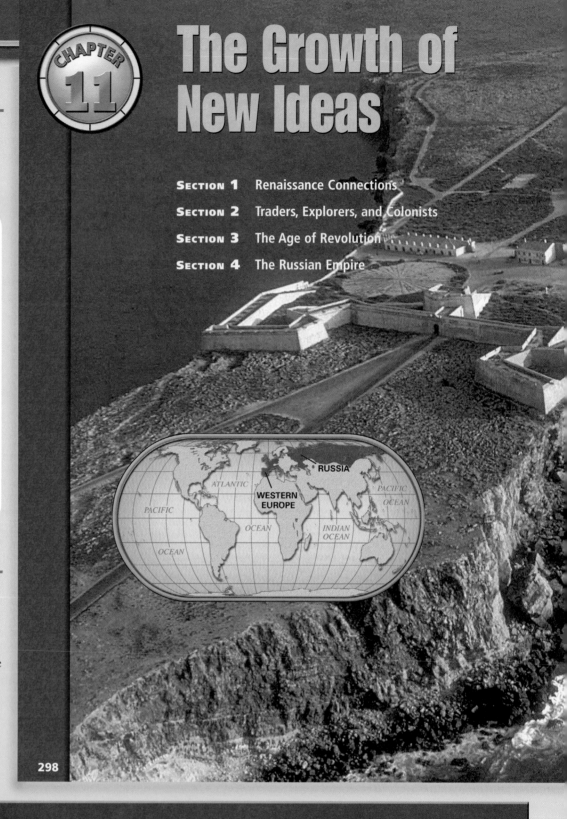

The Growth of New Ideas

298

Recommended Resources

BOOKS FOR THE TEACHER

Corrick, James A. ***The Industrial Revolution.*** San Diego, CA: Lucent Books, 1998. Overview of European and American Industrial Revolutions.

Fritz, Jean. ***Around the World in a Hundred Years: From Henry the Navigator to Magellan.*** New York: Putnam & Grosset, 1998. Chronicles 15th-century European exploration.

Hosking, Geoffrey A. ***Russia and the Russians: A History.*** Cambridge, MA: Harvard University Press, 2001. Cultural history, ancient to modern.

SOFTWARE

Explorers of the New World. Cambridge, MA: Softkey. Journeys, findings, and results of exploration.

INTERNET

For more information about Europe and Russia, visit **classzone.com.**

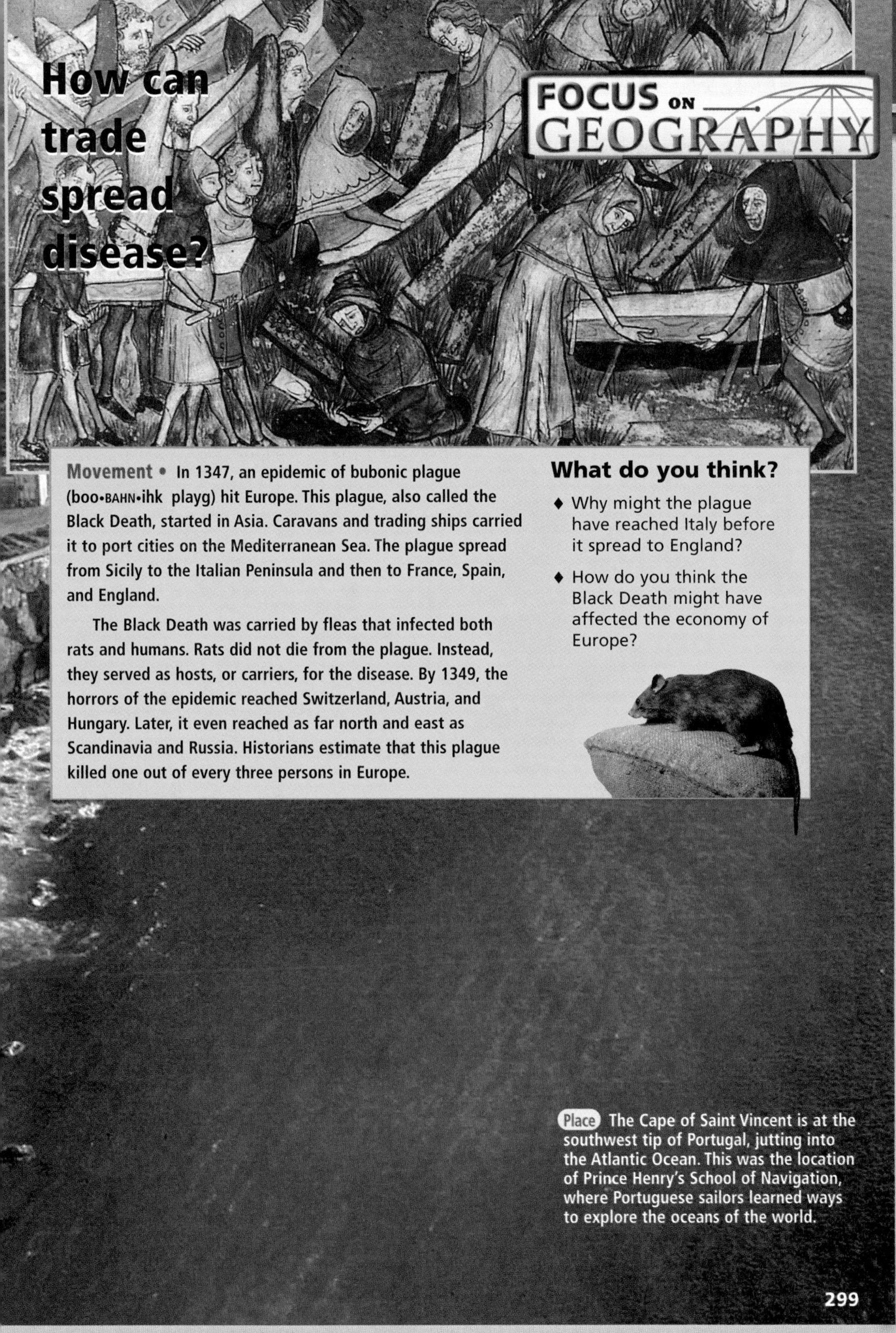

How can trade spread disease?

FOCUS ON GEOGRAPHY

Movement • In 1347, an epidemic of bubonic plague (boo•BAHN•ihk playg) hit Europe. This plague, also called the Black Death, started in Asia. Caravans and trading ships carried it to port cities on the Mediterranean Sea. The plague spread from Sicily to the Italian Peninsula and then to France, Spain, and England.

The Black Death was carried by fleas that infected both rats and humans. Rats did not die from the plague. Instead, they served as hosts, or carriers, for the disease. By 1349, the horrors of the epidemic reached Switzerland, Austria, and Hungary. Later, it even reached as far north and east as Scandinavia and Russia. Historians estimate that this plague killed one out of every three persons in Europe.

What do you think?

- Why might the plague have reached Italy before it spread to England?
- How do you think the Black Death might have affected the economy of Europe?

Place The Cape of Saint Vincent is at the southwest tip of Portugal, jutting into the Atlantic Ocean. This was the location of Prince Henry's School of Navigation, where Portuguese sailors learned ways to explore the oceans of the world.

FOCUS ON GEOGRAPHY

Objectives

- To help students understand the relationships among geography, trade, and economics
- To identify the physical and social impact of trade on communities

What Do You Think?

1. Make sure students understand the location of Sicily relative to the Italian mainland and England. Encourage them to compare the distances and to consider which route was likely to have more frequent trade contacts.
2. Ask students to think about what happens to an economy when there are fewer people to make and buy goods.

How can trade spread disease?

Ask students to develop a list of the different activities in trading. Lead them to understand that trade enables people to share and exchange food, goods, clothing, plants, animals, and ideas. Have students consider how these activities might transmit disease.

MAKING GEOGRAPHIC CONNECTIONS

Have students consider the recent spread of the West Nile virus. The virus, which came from Africa, recently has been found in birds along the East Coast of the United States. Ask students to describe how they think the speed of travel and the extent of human trade affect the spread of disease among people and animals.

Implementing the National Geography Standards

Standard 13 Explain why peoples sometimes engage in conflict to control Earth's surface

Objective To identify and list factors that contribute to conflicts between countries

Class Time 20 minutes

Task Ask each student to list two pairs of countries that have adjoining borders. Beneath each pair, students should indicate factors that might contribute to conflict between the countries, such as boundary disputes, dissimilar languages, competition for natural resources, and so on.

Evaluation Students should list at least three reasons for possible conflict between each pair of countries.

BEFORE YOU READ

What Do You Know?

Ask students to identify classroom items that are made in a factory, such as clothing, books, desks, and clocks. Lead students to conclude that almost all of the items we use in our everyday lives are mass-produced in a factory. Ask students to discuss what they know about factories and mass production, including why mass-produced items are less costly to make.

What Do You Want to Know?

Ask each student to copy the topic names from the Influences column in the chart to the right into their notebook. Have each student develop two questions about each topic that they would like answered. As students read the chapter, have them look for answers to these questions.

READ AND TAKE NOTES

Reading Strategy: Categorizing Explain to students that this chart will help them organize the main ideas of the chapter. Point out that there were many changes in this time period. Tell them that the completed chart will be an organized way to look at how those changes are connected. They may also find answers to many of their questions.

In-depth Resources: Unit 4
- Guided Reading Worksheets, pp. 15–18

READING SOCIAL STUDIES

BEFORE YOU READ

▸▸ What Do You Know?

Do you know who first sailed around the world? Do you know that Leonardo da Vinci drew plans for a helicopter 400 years before it was actually built? Think of other discoveries, inventions, events, and famous people. What do you think life was like for common people during this time? Think about movies you have seen, books you have read, and what you have learned in other classes about the Renaissance, the Industrial Revolution, and political revolutions in France, Russia, and the United States.

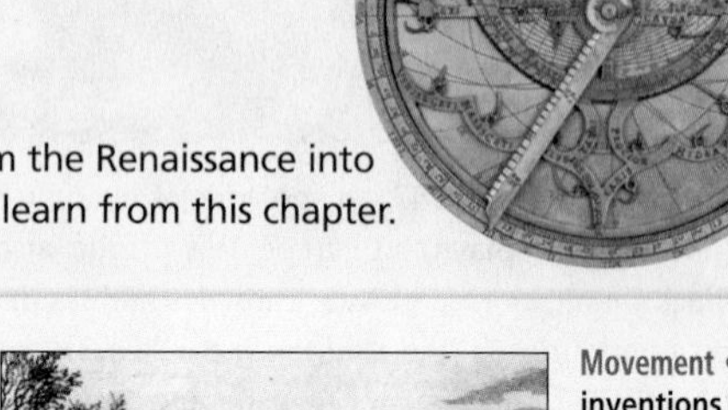

▸▸ What Do You Want to Know?

Decide what you know about changes in the West from the Renaissance into the 1800s. In your notebook, record what you hope to learn from this chapter.

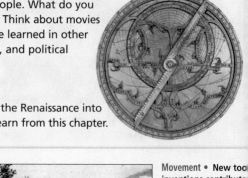

READ AND TAKE NOTES

Reading Strategy: Categorizing One way to make sense of what you read is to categorize ideas. Categorizing means sorting information by certain traits, ideas, or characteristics. Use the chart below to categorize details about the topics covered in this chapter.

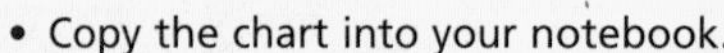

- Copy the chart into your notebook.

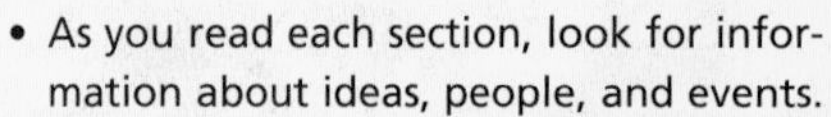

- As you read each section, look for information about ideas, people, and events.
- Record key details in each category.

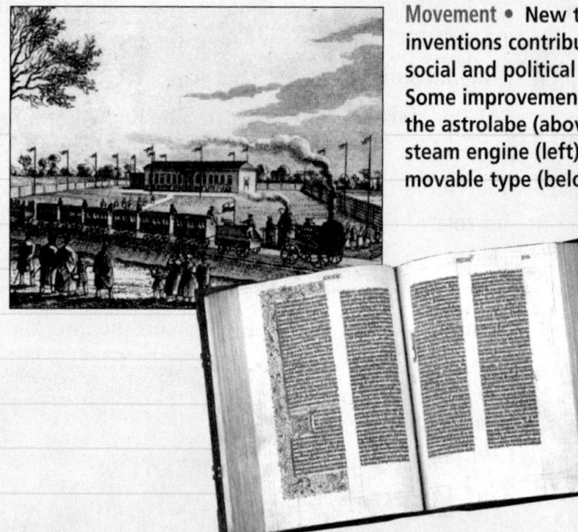

Movement • New too inventions contribute social and political ch Some improvements the astrolabe (above), steam engine (left), a movable type (below)

Influences	New Ideas	People/Achievements	Events/Effects
The Renaissance	emphasis on education, arts	trade, da Vinci, Shakespeare, printing press	new class of aristocrats, patrons
European Exploration and Conquest	desire to control trade routes	navigation, da Gama, Columbus, Magellan	imperialism, colonialism, slavery
Scientific and Industrial Revolutions	capitalism	locomotive, Galileo, telescope	factories, disease, pollution
Political Revolutions	political rights, equality, nationalism	French constitution, Napoleon	French Revolution, Reign of Terror
The Russian Empire	revolution, Western ideas	Ivan IV, Peter the Great, Catherine the Great, Nicholas II	expansion, Bloody Sunday, revolution

Teaching Strategy

Reading the Chapter This is a chronological chapter that covers periods in Europe from the 1400s to the early 1900s. These include the Renaissance, European exploration and trade, the Industrial Revolution, and the French and Russian Revolutions. The chapter identifies the ways in which these events are connected and explores the causes and effects of each. Encourage students to identify specific relationships among events.

Integrated Assessment The Chapter Assessment on pages 324–325 describes several activities that may be used for integrated assessment. You may wish to have students work on these activities during the course of the chapter and then present them at the end.

Renaissance Connections

MAIN IDEA

The rebirth of art, literature, and ideas during the Renaissance changed European society.

WHY IT MATTERS NOW

Many accomplishments of the Renaissance are high points of Western culture and continue to inspire artists, writers, and thinkers of today.

TERMS & NAMES
- Crusades
- Renaissance
- Florence
- Leonardo da Vinci
- William Shakespeare
- Reformation
- Martin Luther
- Protestant

DATELINE

EXTRA

PARIS, FRANCE, 1269

Paris is buzzing with activity as thousands of European soldiers assemble here. This is the starting-off point for the eighth Crusade, which has nearly a thousand miles to travel. King Louis IX of France, who is in command, is confident that his armies can restore European power over the Holy Land.

Since the Crusades began in 1096, the Christians have fought against the Muslims and founded four states in the eastern Mediterranean. European power has weakened since then. However, King Louis's army looks ready to recapture the lost territory for Christianity.

Movement • Crusaders will make their way toward the Holy Land. ▲

Europeans Encounter New Cultures ❶

The **Crusades**—a series of expeditions from the 11th to the 13th centuries by Western European Christians to capture the Holy Lands from Muslims—greatly changed life in Western Europe. The Crusades opened up trade routes, linking Western Europe with southwestern Asia and North Africa. They also helped Europeans rediscover the ideas of ancient Greece and Rome.

TAKING NOTES

Use your chart to take notes about people and ideas.

Influences	New Ideas	People/ Achievements
The Renaissance		
European Exploration		

SECTION OBJECTIVES

1. To explain the scope of Europe's Renaissance
2. To describe the growth of arts and learning in Europe during the Renaissance
3. To identify the religious conflicts that led to the Protestant Reformation
4. To describe the Reformation and Luther's work

SKILLBUILDER
- Interpreting a Map, p. 302

CRITICAL THINKING
- Summarizing, p. 302
- Analyzing Motives, p. 304
- Hypothesizing, p. 305

FOCUS & MOTIVATE

WARM-UP

Making Inferences Have students read Dateline and discuss these questions.

1. Why is this Crusade important to King Louis IX?
2. What long-term effect does the king hope to have on the lands around the Mediterranean?

INSTRUCT: Objective ❶

Europeans Encounter New Cultures/The Rebirth of Europe

- What was the Renaissance? When did it exist? a new era of creativity and learning in Western Europe; from the 14th to the 16th century
- What kinds of great works emerged during the Renaissance? paintings, large sculptures, impressive architecture, literature

In-depth Resources: Unit 4
- Guided Reading Worksheet, p. 15

Reading Study Guide
(Spanish and English), pp. 88–89

Program Resources

In-depth Resources: Unit 4
- Guided Reading Worksheet, p. 15
- Reteaching Activity, p. 23

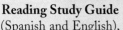

Reading Study Guide
(Spanish and English), pp. 88–89

Formal Assessment
- Section Quiz, p. 157

Integrated Assessment
- Rubric for writing a letter

Outline Map Activities

Access for Students Acquiring English
- Guided Reading Worksheet, p. 85

Technology Resources
classzone.com

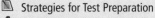

TEST-TAKING RESOURCES
- Strategies for Test Preparation
- Test Practice Transparencies
- Online Test Practice

CRITICAL THINKING ACTIVITY

Summarizing Have students write a summary about what they have learned so far about the Renaissance. Remind students that a summary includes only the most important ideas about a topic. Have students work together to create a concept web on the chalkboard with the word *Renaissance* in the center. As they suggest main ideas, record their suggestions in the outer circles. Then ask each student to write a summary about the Renaissance, using the ideas in the web.

Class Time 15 minutes

FOCUS ON VISUALS

Analyzing the Photograph Point out the Duomo, the cathedral in the center of the photograph, and tell students that the construction of the cathedral's dome marks the beginning of Renaissance architecture. Studying in Rome, the architect Filippo Brunelleschi learned about the construction of columns and arches. He studied ancient Roman ruins and used the principles he discovered when designing the dome. The dome's diameter is 130 feet. Ask students why they think it might be difficult to build a dome of this size without internal or external support. Why is it unlikely that such a project would have succeeded or even been attempted before the Renaissance?

Possible Response The diameter of the dome is too large to be supported with columns. The design required a sophisticated understanding of mathematics and engineering—neither of which flourished during medieval times.

Extension Have students write a letter from Filippo Brunelleschi to a fellow architect about his design for the Duomo.

Italian City-States, c. 1350

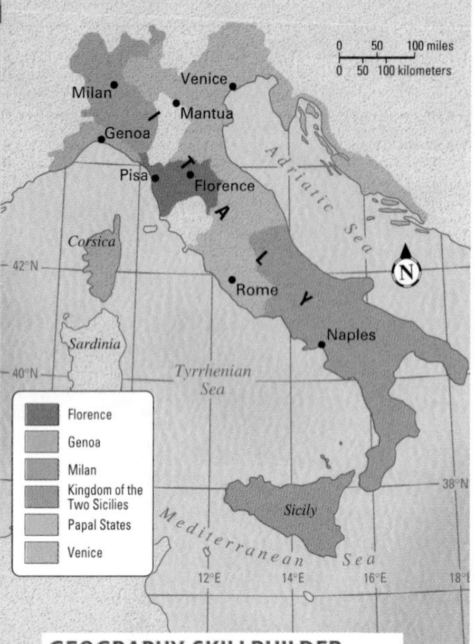

GEOGRAPHY SKILLBUILDER: Interpreting a Map

1. **Location** • Which city-state does not have access to water?
2. **Location** • Which city-state was in the best position to trade by land and sea with Asia?

Over time, this interest in the ancient world sparked a new era of creativity and learning in Western Europe. This cultural era, which lasted from the 14th to the 16th century, is called the **Renaissance**.

BACKGROUND

After Rome fell, many achievements of ancient times were lost. Many books and manuscripts, however, were preserved by Muslim and Christian scholars.

The Rebirth of Europe ❶

The Renaissance began on the Italian Peninsula in the mid-14th century. During this time, many artists, architects, writers, and scholars created works of great importance. These included beautiful paintings, large sculptures, impressive buildings, and thought-provoking literature. As new ideas and achievements spread across the continent of Europe, they changed the way people viewed themselves and the world.

Geography Skillbuilder Answers

1. Milan
2. Venice

The Italian City-States In the 14th century, the Italian Peninsula was divided into many independent city-states. Some of these city-states, such as **Florence**, were bustling centers of banking, trade, and manufacturing.

Region • Florence, once a wealthy city-state, remains an important economic and cultural center. The Duomo, shown here, is a symbol of the city's Renaissance past. ▼

Activity Options

Differentiating Instruction: Students Acquiring English/ESL

Prefixes Students acquiring English may be confused about the meaning of the words *rebirth* and *rediscover*. Explain that the prefix *re-* means "again." Write the word equation *re-* + discover = rediscover *(discover again)* on the chalkboard. Then read aloud the heading "The Rebirth of Europe" on page 302. Ask students to explain in their own words what that heading means, focusing particularly on the word *rebirth*.

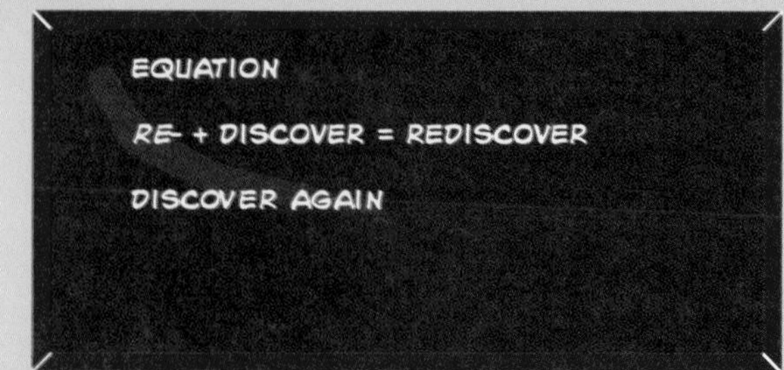

Region • The wealthy merchants in Italy built large palaces, called palazzos, such as Florence's Palazzo Medici shown here. ▶

The wealthy businesspeople who lived in these city-states were members of a new class of aristocrats. Unlike the nobles of the feudal system, these aristocrats lived in cities, and their wealth came from money and goods rather than from the lands they owned.

A Changing View of the World Religion was important to people's daily life during the Renaissance, but many wealthy Europeans began to turn increased attention to the material comforts of life.

New wealth allowed aristocratic families to build large homes for themselves in the city centers, decorating them with luxurious objects. They ate expensive food and dressed in fine clothes and jewels, often acquired as a result of the expanded trade routes. Aristocrats also placed increased emphasis on education and the arts.

Learning and the Arts Flourish ❷

Wealthy citizens were proud of their city-states and often became generous patrons. A patron gave artists and scholars money and, sometimes, a place to live and work. They hired architects and designers to improve local churches, to design grand new buildings, and to create public sculptures and fountains. As one Italian city-state made additions and improvements, others competed to outdo it.

BACKGROUND

Some Renaissance architects, such as Filippo Brunelleschi (BROO•nuh•LEHS•kee), studied the ruins of Roman buildings and modeled their new buildings after ancient designs.

The Medici Family Among the most famous patrons of the Renaissance were the Medici (MEHD•uh•chee). They were a wealthy family of bankers and merchants. In fact, they were the most powerful leaders of Florence from the early 1400s until the 18th century.

Along with Lorenzo, pictured below, the Medici family included famous princes and dukes, two queens, and four popes. Throughout the 15th and 16th centuries, the Medici supported many artists, including Botticelli, Michelangelo, and Raphael. Today, Florence is still filled with important works of art made possible by the Medici.

Biography

Among the many famous members of the Medici family, three are especially noteworthy. Lorenzo Medici, grandson of the family founder, ruled Florence from 1469 until 1492. Under his leadership, Florence became one of Italy's most powerful cities.

Catherine Medici (1519–1589) was married to one French king, Henry II, and was the mother of three other French kings.

Marie Medici (1573–1642) was married to French king Henry IV. When he was assassinated in 1610, her eight-year-old son, Louis XIII, inherited the throne. Because of his youth, Marie ruled France as queen regent for seven years.

INSTRUCT: Objective ❷

Learning and the Arts Flourish/ The Northern Renaissance

- What did patrons do to improve their city-states? hired artists to beautify cities and scholars to increase their status
- How did art subjects change in the Renaissance? from solely religious subjects to a wider variety of subjects
- In addition to being an artist, what else was Leonardo da Vinci? He was an engineer, a scientist, and an inventor.
- What effect did Renaissance ideas have on Northern Europe? inspired artists and writers

Activity Options

Interdisciplinary Link: Art

Class Time One class period

Task Drawing an object, showing a light source

Purpose To understand one of the important changes in art that took place during the Renaissance

Supplies Needed
- Drawing paper
- Pencils
- Small objects to draw
- Information about Renaissance artist Giotto and copies of his drawings and paintings

 **Block Scheduling**

Activity Explain that Giotto was among the first artists to show how light shines on an object in nature. He illuminated one side of an object and painted the opposite side in shadow. Place several small objects in a place where they will cast a shadow. Ask students to draw one object, showing the direction of the light source and the resulting shadows. Suggest students make realistic drawings.

CRITICAL THINKING ACTIVITY

Analyzing Motives Remind students that wealthy Renaissance patrons supported artists and scholars to improve and beautify their city-states. Invite students to analyze the motives behind such support. Lead students to consider various ways in which improving the appearance and status of their city-states would ultimately benefit these wealthy patrons, both socially and economically.

Class Time 15 minutes

Connections to Math

One Renaissance artist interested in perspective was Jacopo Bellini. He lived in Venice in the 1400s, and he used perspective to paint three-dimensional scenes on flat wooden panels. Most of his paintings, famous in his own time, have been lost. Today he is best known for his drawings, which range from simple sketches to more detailed compositions, all of which display his interest and skill in creating the illusion of space.

MORE ABOUT...

William Shakespeare

Many people consider William Shakespeare to be the greatest playwright who ever lived. In Shakespeare's time, attending plays in the afternoon was a popular form of entertainment for English people from all walks of life. Shakespeare wrote more than two dozen plays, including both comedies and tragedies. He also wrote many sonnets and other poems. Shakespeare's mastery of words, images, and rhythm was so great that his work is still among the most widely read in English, even though the language has changed a great deal in the nearly four centuries since his death. Today Shakespeare's plays are staged more often, and in more countries, than ever before.

Culture • Leonardo da Vinci completed the painting *La Belle Ferronnière* in 1495. ▲

As part of the competition to improve the appearance and status of their individual city-states, patrons wanted to attract the brightest and best-known scholars and poets of the time. Patrons believed that the contributions of these individuals would, in turn, add to the greatness of their city-states and attract more wealth.

The Visual Arts: New Subjects and Methods Most medieval art was based on religious subjects. Painters and sculptors of the early Renaissance created religious art too, but they also began to depict other subjects. Some made portraits for wealthy patrons. Others created works showing historical scenes or mythological stories.

Reading Social Studies

A. Contrasting How did the subject matter of Renaissance art differ from medieval art?

Reading Social Studies
A. Possible Answer
It depicted more than religious subjects.

Leonardo da Vinci One of the most famous artists and scientists of the Renaissance was **Leonardo da Vinci** (lee•uh•NAHR•doh duh VIHN•chee) (1452–1519). Among his best-known paintings are the *Mona Lisa,* a portrait of a young woman with a mysterious smile, and *The Last Supper.* Da Vinci was more than just a talented painter, however.

Throughout his life, da Vinci observed the world around him. He studied the flow of water, the flight of birds, and the workings of the human body. Da Vinci, who became a skilled engineer, scientist, and inventor, filled notebooks with thousands of sketches of his discoveries and inventions. He even drew ideas for flying machines, parachutes, and submarines—hundreds of years before they were built.

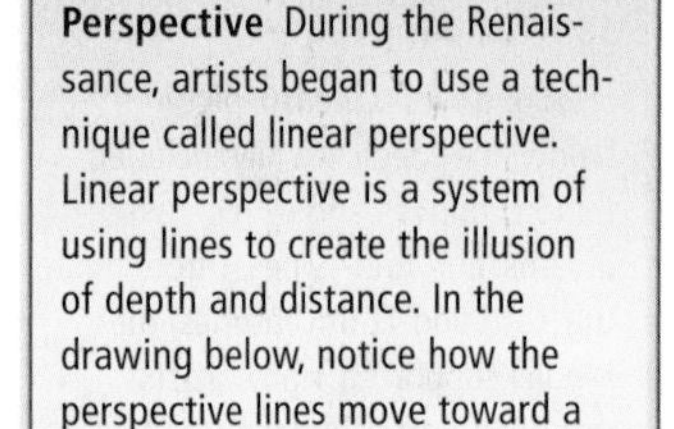

Connections to Math

Perspective During the Renaissance, artists began to use a technique called linear perspective. Linear perspective is a system of using lines to create the illusion of depth and distance. In the drawing below, notice how the perspective lines move toward a single point in the distance, giving the picture depth.

The Northern Renaissance ❷

As the new Renaissance ideas about religion and art spread to Northern Europe, they inspired artists and writers working there. The Dutch scholar and philosopher Desiderius Erasmus (ih•RAS•muhs) (1466–1536), for example, criticized the church for its wealth and poked fun at its officials. During the late 16th and early 17th centuries, another writer—the Englishman **William Shakespeare**—wrote a series of popular stage plays. Many of his works, including *Romeo and Juliet* and *Macbeth,* are still read and performed around the world.

Activity Options

Skillbuilder Mini-Lesson: Making a Generalization — **Block Scheduling**

Explaining the Skill Remind students that making generalizations means to make broad judgments based on information gathered from several sources.

Applying the Skill Have students reread "Leonardo da Vinci" and use these strategies to begin to develop a generalization. Students will need to consult additional sources to complete the generalization.

1. Make a list of things Leonardo da Vinci studied. water, birds, body
2. Make a second list of the subjects he drew. people, buildings, machines, ideas for inventions
3. Use this information to write a generalization about da Vinci. Leonardo da Vinci had a wide range of interests.

The Reformation 3

BACKGROUND

In 1516, the English writer Thomas More published a famous book called *Utopia.* It describes the author's idea of a perfect society. Today, the word "utopia" is used to describe any ideal place.

Roman Catholicism was still the most powerful religion in Western Europe. Some of the views of the northern Renaissance writers and scholars, however, were in conflict with the Roman Catholic Church. These new ideas would eventually lead to the **Reformation,** a 16th-century movement to change church practices.

Martin Luther The German monk **Martin Luther** (1483–1546) was one of the most important critics of the church. The wealth and corruption of many church officials disturbed him. Luther also spoke out against the church's policy of selling indulgences—the practice of forgiving sins in exchange for money.

In 1517, Luther wrote 95 theses, or statements of belief, attacking the sale of indulgences and other church practices. Copies were printed and handed out throughout Western Europe. After this, Luther was excommunicated, or cast out and no longer recognized as a member of a church, and went into hiding. While in hiding, he translated the Bible from Latin into German so that all literate, German-speaking people could read it. Under Luther's leadership, many Europeans began to challenge the practices of the Roman Catholic Church.

Thinking Critically Possible Answers
1. multiple copies, books made faster and more cheaply
2. monks or nuns

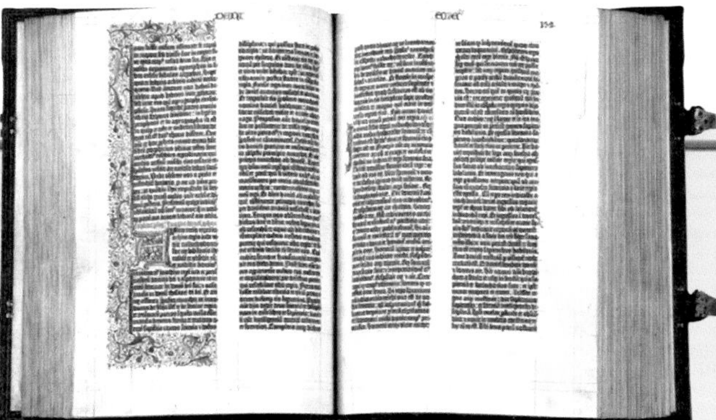

The Printing Press Until the Renaissance, each copy of a book had to be written by hand—usually by monks or nuns. A Renaissance invention, however, changed that forever. Around 1450, a German printer named Johann Gutenberg (Yoh•HAHN GOO•tuhn•BERG) began to use a method of printing with movable type. This meant that multiple copies of books, such as this Bible, could be printed quickly and less expensively.

Although many Renaissance books dealt with religious subjects, printers also published plays, poetry, works of philosophy and science, and tales of travel and adventure. As greater numbers of books were published, more and more Europeans learned to read.

For more on the printing press, go to

THINKING CRITICALLY 

1. **Recognizing Effects**
   What were three effects of the invention of Gutenberg's printing press?
2. **Synthesizing**
   Before the printing press, who produced the books?

INSTRUCT: Objective 3

The Reformation

- What was the Reformation? a 16th-century movement to change church practices
- What were Martin Luther's criticisms of the Roman Catholic Church? wealth, corruption, selling indulgences

CRITICAL THINKING ACTIVITY

Hypothesizing Discuss with students why they think Martin Luther translated the Bible from Latin into German. Explain that at that time, Latin was a language used by priests but not understood by most people. Translating the Bible into German, the language of the people, meant that more people would have direct access to the Bible's ideas and would thus be less dependent on the authority of the church.

Class Time 15 minutes

Spotlight on CULTURE

Soon after Gutenberg created his first printing press, he began to improve his invention. The first method, which involved pressing by hand, did not produce evenly printed pages. Adding a large screw to the printing press solved the problem, and soon Gutenberg could print 300 pages a day. Within 50 years of Gutenberg's first efforts, there were more than 1,000 printing shops in Europe, and several million books had been printed.

Activity Options

Multiple Learning Styles: Logical

Class Time One class period

Task Using logical reasoning to support two opposing points of view

Purpose To create logical arguments to support and oppose translating the Latin Bible into other languages

Supplies Needed
- Writing paper
- Pencils or pens

Block Scheduling

Activity Have students imagine they are living in Germany in 1530. They have learned that Martin Luther has translated the Latin Bible into German. Ask each student to create a logical written argument first to support, and then to oppose, this accomplishment. Remind students that their reasons on each side must make sense and support the main argument. Have students share their arguments when they finish.

INSTRUCT: Objective 4

A Conflict over Religious Beliefs

- What were Martin Luther's followers called and why? Protestants; they protested against the church
- Why did Protestants push to expand education? so more people could read the Bible

ASSESS & RETEACH

Reading Social Studies Have students fill in the top row of the chart on page 300, "The Renaissance."

Formal Assessment
- Section Quiz, p. 157

RETEACHING ACTIVITY

Have students work in small groups to create a five-question quiz about the section. Encourage students to write questions about the most important ideas and their supporting details. Then have groups exchange quizzes, answer questions, and return the quizzes to check their answers.

In-depth Resources: Unit 4
- Reteaching Activity, p. 23

Access for Students Acquiring English
- Reteaching Activity, p. 91

A Conflict over Religious Beliefs 4

Reading Social Studies

B. Clarifying How did Protestants get their name?

Reading Social Studies
B. Possible Answer
They protested the intolerance of the church.

Luther's followers were called **Protestants** because they protested events at an assembly that ended the church's tolerance of their beliefs. Many people in Western Europe still supported the church, however. This conflict led to religious wars that ended in 1555. At that time, the Peace of Augsburg declared that German rulers could decide the official religion of their own state.

The Spread of Protestant Ideas By 1600, Protestantism had spread to England and the Scandinavian Peninsula. Protestants pushed to expand education for more Europeans. They did this because being able to read meant being able to study the Bible. They also encouraged translation of the Bible into the native language of each country.

The Counter Reformation The Roman Catholic Church responded to Protestantism by launching its own movement in the mid-16th century. As part of this movement, called the Counter Reformation, the church stopped selling indulgences. It also created a new religious order called the Society of Jesus, or the Jesuits. Jesuit missionaries and scholars worked to spread Catholic ideas across Europe, to Asia, and to the lands of the "new world" across the Atlantic Ocean.

Region • Martin Luther's writings and actions changed Christianity forever. ▲

SECTION 1 ASSESSMENT

Terms & Names

1. Explain the significance of:

(a) Crusades	(b) Renaissance	(c) Florence	(d) Leonardo da Vinci
(e) William Shakespeare	(f) Reformation	(g) Martin Luther	(h) Protestants

Using Graphics

2. Use a spider map like this one to chart the characteristics and accomplishments of the Renaissance.

Main Ideas

3. (a) Where and when did the Renaissance begin?

(b) In what ways were the wealthy Europeans of the Renaissance different from the wealthy Europeans of feudal times?

(c) What was the Counter Reformation?

Critical Thinking

4. Hypothesizing
Why do you think Protestantism spread so quickly in Northern Europe?

Think About
- new methods of printing
- the ideas of the northern Renaissance
- the work of Martin Luther

ACTIVITY -OPTION- Write a **letter** to an imagined patron asking for support to create a project—such as a public sculpture, park, fountain, or building—to beautify your community.

Section 1 Assessment

1. Terms & Names
- **a.** Crusades, p. 301
- **b.** Renaissance, p. 302
- **c.** Florence, p. 302
- **d.** Leonardo da Vinci, p. 304
- **e.** William Shakespeare, p. 304
- **f.** Reformation, p. 305
- **g.** Martin Luther, p. 305
- **h.** Protestant, p. 306

2. Using Graphics

Renaissance Accomplishment: printing press, art, architecture, literature, patrons, Reformation

3. Main Ideas
- **a.** The Renaissance began on the Italian Peninsula in the mid-14th century.
- **b.** The wealth of Renaissance Europeans came from money and goods. The wealth of feudal Europeans came from land.
- **c.** It was the Roman Catholic Church's response to its Protestant critics.

4. Critical Thinking

Possible Response The Protestants' support of widespread education and the translation of the Bible might have had great appeal to Northern Europeans.

ACTIVITY OPTION

Integrated Assessment
- Rubric for writing a letter

Traders, Explorers, and Colonists

TERMS & NAMES
Prince Henry the Navigator
Christopher Columbus
Ferdinand Magellan
circumnavigate
imperialism

[M]AIN IDEA

[Eur]opean trade and exploration [ch]anged the lives of many people [on] both sides of the Atlantic.

WHY IT MATTERS NOW

Today, citizens of the Americas continue to feel the effects of European exploration and colonization.

DATELINE

SAGRES, PORTUGAL, 1421— Portugal's Prince Henry may not have journeyed to sea, but he has earned a well-deserved nickname: "The Navigator." He has organized expeditions of sailors to explore the west coast of Africa. Five years ago, Henry also founded a School of Navigation. It is here in Sagres, at Portugal's southwestern tip, which juts into the Atlantic Ocean.

Astronomers, geographers, and mathematicians gather here to study and teach new methods of traveling across the seas. They plan expeditions using the latest maps, tools, and information about the winds and currents of the Atlantic Ocean. Sometimes the scholars add to their knowledge by talking with sea captains about their voyages.

Movement • Prince Henry of Portugal founded the School of Navigation. ▲

Trade Between Europe and Asia ❶

For centuries before the Renaissance, European traders traveled back and forth across the Mediterranean. Merchants commonly journeyed from southern Europe to North Africa and to the eastern Mediterranean. Spices were one of the most important items traded at this time.

TAKING NOTES
Use your chart to take notes about people and ideas.

Influences	New Ideas	People/ Achievements
The Renaissance		
European Exploration		

SECTION OBJECTIVES

1. To recognize the importance of trade routes between Europe and Asia
2. To identify Portuguese explorers and routes
3. To identify Spanish and English explorers and routes
4. To describe the effect of European exploration on Africa and North and South America

SKILLBUILDER
• Interpreting a Map, pp. 308, 311

CRITICAL THINKING
• Summarizing, p. 310

FOCUS & MOTIVATE

WARM-UP

Drawing Conclusions Have students read Dateline and discuss these questions.

1. Why do you think Prince Henry invited people with different kinds of knowledge to his school?
2. What could you say about the maps, tools, and information available to early navigators?

INSTRUCT: Objective ❶

Trade Between Europe and Asia

• What item was in great demand in Europe? Why? spices; to preserve food and improve flavor

• Why did Europeans want a new route to Asia? land route was long, made spices expensive

In-depth Resources: Unit 4
• Guided Reading Worksheet, p. 16

Reading Study Guide
(Spanish and English), pp. 90–91

Program Resources

In-depth Resources: Unit 4
• Guided Reading Worksheet, p. 16
• Reteaching Activity, p. 24

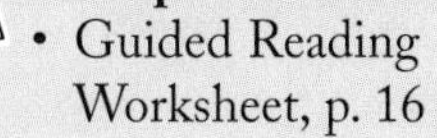

Reading Study Guide
(Spanish and English), pp. 90–91

Formal Assessment
• Section Quiz, p. 158

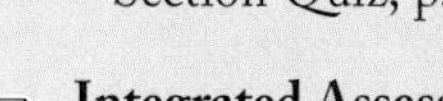

Integrated Assessment
• Rubric for writing a journal entry

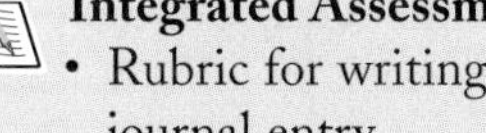

Outline Map Activities

Access for Students Acquiring English
• Guided Reading Worksheet, p. 86

Technology Resources
classzone.com

TEST-TAKING RESOURCES
Strategies for Test Preparation
Test Practice Transparencies
Online Test Practice

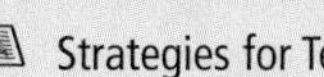

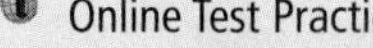

FOCUS ON VISUALS

Interpreting the Map Have students trace the routes of the Portuguese explorers shown on the map. Ask them why the explorers could not sail directly east through the Mediterranean Sea and into the Indian Ocean but instead had to take the longer, more hazardous route around the southern tip of Africa.

Possible Responses There was a small land barrier between the Mediterranean Sea and the Red Sea. The later construction of the Suez Canal made it possible to travel between the Mediterranean and Red seas.

Extension Have students imagine they are crew members on one of the Portuguese ships and write a letter home describing their experiences.

INSTRUCT: Objective 2

Leaders in Exploration

- Why did Prince Henry send explorers down the coast of Africa? to find a shortcut to Asia
- Which Portuguese explorer first rounded the southern tip of Africa? Bartolomeu Dias
- What was the result of Vasco da Gama's expedition to India? Portugal controlled the sea route to Asia.

The Spice Trade Spices were in great demand by Europeans. Before refrigeration, meat and fish spoiled quickly. To help preserve food and to improve its flavor, people used spices such as pepper, cinnamon, nutmeg, and cloves. These spices came from Asia.

For centuries, Italian merchants from Genoa and Venice controlled the spice trade. They sailed to ports in the eastern Mediterranean, where they would purchase spices and other goods from traders who had traveled across Asia. The Italian merchants would then bring these goods back to Europe.

BACKGROUND
In addition to spices, European countries traded for precious metals, which they used to make coins. Metals such as gold and silver were scarce in Europe.

The Possibility of Great Wealth Transporting goods across these great distances was costly. Everyone along the way had to be paid and wanted to earn a profit. By the time the spices reached Europe, they had to be sold at extremely high prices.

European merchants knew that if they could trade directly with people in Asia, they could make enormous profits. In the 15th century, Europeans began to search for a new route to Asia.

Leaders in Exploration 2

The small country of Portugal is at the westernmost part of the European continent. Portuguese sailors had navigated the waters of the Atlantic Ocean for centuries. As shown on the map below, they traveled down the west coast of Africa and as far west into the Atlantic as Madeira, the Azores, and the Canary Islands.

Exploring the African Coast In the early 1400s, Portugal's **Prince Henry the Navigator** decided to send explorers farther down the coast of Africa. He believed that if explorers could find a way around Africa, it might be a shortcut to Asia. Portuguese explorers returned home from these expeditions with gold dust, ivory, and more knowledge of navigation. By the time Henry died in 1460, the Portuguese had ventured around the great bulge of western Africa to present-day Sierra Leone.

Portuguese Explorers, 1400s

GEOGRAPHY SKILLBUILDER: Interpreting a Map

1. **Movement** • Which explorer reached Asia?
2. **Location** • Which continent was most explored by the Portuguese?

Geography Skillbuilder Answers
1. da Gama
2. Africa

Activity Options

Multiple Learning Styles: Interpersonal

Class Time 25 minutes

Task Demonstrating transactions along an "in-class" trade route

Purpose To understand why transporting goods across great distances makes them more costly

Supplies Needed
- Pencil

Activity Have five students sit in a line across the classroom. Give the first student a pencil and state that it costs $0.50. Instruct students to "buy" and "sell" the pencil to the next person in line, each time adding a $0.20 profit for the seller. Continue until the fifth and final student buys the pencil. Keep track of each transaction. Compare the first price ($0.50) and the last price ($1.30). Point out that this is the same process that drove up the cost of spices.

The Race Around Africa Bold Portuguese explorers continued to push farther down the African coast. Finally, in 1488, Bartolomeu Dias (BAHR•too•loo•MAY•oo DEE•uhsh) rounded the southern tip of Africa. The Portuguese named the tip the Cape of Good Hope.

Less than ten years later, Vasco da Gama (vas•KOH deh GAH•muh) led a sea expedition all the way to Asia. Da Gama and his crew traveled for 317 days and 13,500 miles before reaching the coast of India. They were the first Europeans to discover a sea route to Asia. Now, the riches of Asia could be brought directly to Europe. After setting up trading posts along the coast of the Indian Ocean, Portugal ruled these waterways.

Connections to Science

New Ships In the early 15th century, Portuguese shipbuilders designed a sturdy ship called a caravel, pictured below. Built for exploration and trade, the caravel was small and had a narrow body. This helped the ship to cut through waves and to travel in shallow water.

The caravel also used a combination of square and triangular sails. These made sailing easier against strong, shifting winds.

Europe Enters a New Age 3

Portugal was not the only European country to understand that whoever controlled trade with Asia would have great power and wealth. Spain and England quickly entered the race to find a direct sea route of their own.

Christopher Columbus Some explorers believed that the shortest way to Asia was to sail west across the Atlantic Ocean. Queen Isabella of Spain agreed to fund an expedition across the Atlantic.

In August 1492, an Italian named **Christopher Columbus** and 90 crew members left Spain aboard three ships—the *Santa Maria,* the *Pinta,* and the *Niña.* The Atlantic Ocean proved to be wider than maps of the time suggested. On October 12, after weeks at sea, the crew spotted land. Although Columbus thought he had found Asia, they were off the coast of an island in the Caribbean. This was still a great distance from their spice-rich destination.

Reading Social Studies

A. Recognizing Important Details What continent did Columbus reach, and where did he think he was?

Reading Social Studies A. Answer He reached the Caribbean, but he thought he had found Asia.

Ferdinand Magellan In 1519, Spain funded an expedition for the Portuguese explorer **Ferdinand Magellan** (muh•JEHL•uhn). Magellan left Spain with five ships and more than 200 sailors. As they traveled west, the crew battled violent storms and rough seas. Food was in short supply, and starving sailors ate rats and sawdust. Some died of disease.

Connections to Science

Caravels were used by the Portuguese and Spanish for both coastal and ocean journeys. The caravel was smaller and easier to maneuver than the other ship of the time, the galleon. Caravels varied in size and weight, depending on their purpose. For example, caravels used for coastal trading weighed approximately 10.2 metric tons, while caravels used for ocean expeditions weighed approximately 51 metric tons. Two of the three ships that Christopher Columbus used in his 1492 expedition were caravels—the *Niña* and the *Pinta.*

INSTRUCT: Objective 3

Europe Enters a New Age

- What lands did Christopher Columbus discover while searching for a route to Asia? the Caribbean Islands
- What did members of Magellan's crew accomplish that no one had ever done before? They circumnavigated the globe.
- What did the explorer John Cabot accomplish? In 1497, he sailed west from England, hoping to find Asia. Instead, he landed in present-day Newfoundland in Canada.
- Why did King Henry VII fund Cabot's voyage? He wanted a share in the riches of Asia, which Portugal and Spain had monopolized.

Activity Options

Interdisciplinary Links: Geography/Mathematics

Block Scheduling

Class Time 30 minutes

Task Calculating the mileage of two routes

Purpose To compare the distances of an eastward and westward sailing route from Spain to Asia

Supplies Needed
- World map or globe
- Calculators
- Paper
- Pens or pencils

Activity Point out to students that the Portuguese sailed east around Africa looking for the shortest route to Asia, while the Spanish sailed west, across the Atlantic and around South America. Have students study the map or globe and decide which route they think is shorter. Then have them use the scale on the map or globe to calculate two water routes from Spain to Asia: one traveling east, one traveling west.

MORE ABOUT...

Navigation

The first compasses were simple devices: pieces of magnetic iron, floating on straw or cork in a bowl of water. They were first used by Chinese and Mediterranean sailors around the year 1000 A.D. In the 1300s, compass makers began dividing the compass card into 32 different points of direction. Other improvements and continued experience enabled sailors to use compasses with increasing accuracy over time.

CRITICAL THINKING ACTIVITY

Summarizing At great expense of lives and time, Ferdinand Magellan's expedition was the first time human beings ever sailed all the way around the world. Have students summarize this undertaking by listing the obstacles the explorers faced. violent storms, rough seas, lack of food and fresh water, disease, hostile peoples, years at sea If students have trouble recalling the information, have them reread the information on pages 309–310. Ask them to imagine they are reporters in Spain who witnessed the return of the surviving members of Magellan's crew. Have them write and orally present a two-sentence "breaking news report" that summarizes the historic voyage.

Class Time 15 minutes

INSTRUCT: Objective 4

The Outcomes of Exploration

- What is imperialism? the practice of one country controlling the government and economy of another country
- What were some unexpected outcomes of European explorations for the people of Africa and North and South America? diseases, religious conversion, slavery

Movement • **Sailors figured their ship's position with the astrolabe. It measured the position of the sun and stars in relation to the horizon.** ▼

By the time Magellan and his ships reached the Philippines in Asia, the sailors had spent 18 long months at sea. Then, during a battle there, Magellan and several crew members were killed. The expedition returned to Spain after a three-year journey. Only one boat and 18 crew members succeeded. They had to **circumnavigate,** or sail completely around, the world.

Reading Social Studies

B. Identifying Problems What were the main problems faced by Magellan and his crew?

Reading Social Studies
B. Possible Answers
storms, food shortages, disease

John Cabot King Henry VII of England did not want Portugal and Spain to claim all the riches of Asia. He funded a voyage by Italian-born Giovanni Caboto, called John Cabot by the English, who believed that a northern route across the Atlantic Ocean might be a shortcut to Asia.

Aboard one small ship, Cabot and 18 crew members sailed west from England in May 1497. When they reached land the following month, Cabot thought they had found Asia. Most likely, they landed in present-day Newfoundland in Canada.

The Outcomes of Exploration 4

The kings and queens of Europe sent explorers in search of a direct trade route to Asia. These expeditions, however, turned out to have unexpected results.

A Clash of Cultures European countries founded many new colonies along the coastal areas of Africa and North and South America. This practice of one country controlling the government and economy of another country or territory is called **imperialism.** These conquered lands were already home to large, self-ruling populations. They had their own cultural traditions. After the arrival of the Europeans, the lives of these indigenous peoples would never be the same.

Vocabulary

indigenous: born and living in a place, rather than having come from somewhere else

Religious Conversion The European monarchs were Christians. They had strong religious beliefs, and they sent missionaries and other religious officials to help convert conquered peoples to Christianity. The European rulers also hoped that these new converts would help Christianity overcome other powerful religions, especially Islam.

The Spread of Diseases Without knowing it, the European explorers and colonists carried diseases with them, including smallpox, malaria, and measles. These diseases were unknown in the Americas, and killed tens of thousands of people there.

Activity Options

Interdisciplinary Link: Geography

Class Time One class period

Task Developing a board game based on alternative northern routes to Asia that John Cabot could have used

Purpose To develop creative ways to organize and express information about sailing routes from England to Asia

Supplies Needed
- World map or globe
- Paper
- Pens or pencils
- Blank map outlines

Block Scheduling

Activity Have students find two or three alternative water routes that John Cabot could have used to reach Asia. Working in small groups, have students develop board games using those routes to get players from *Start* in England to *Finish* in Asia. Tell students that they must use actual waterways. They must decide where and how players move, create delays due to weather conditions, and determine rules for playing and winning. Have groups play each other's games.

Columns, Cabot, and Magellan, 1492–1522

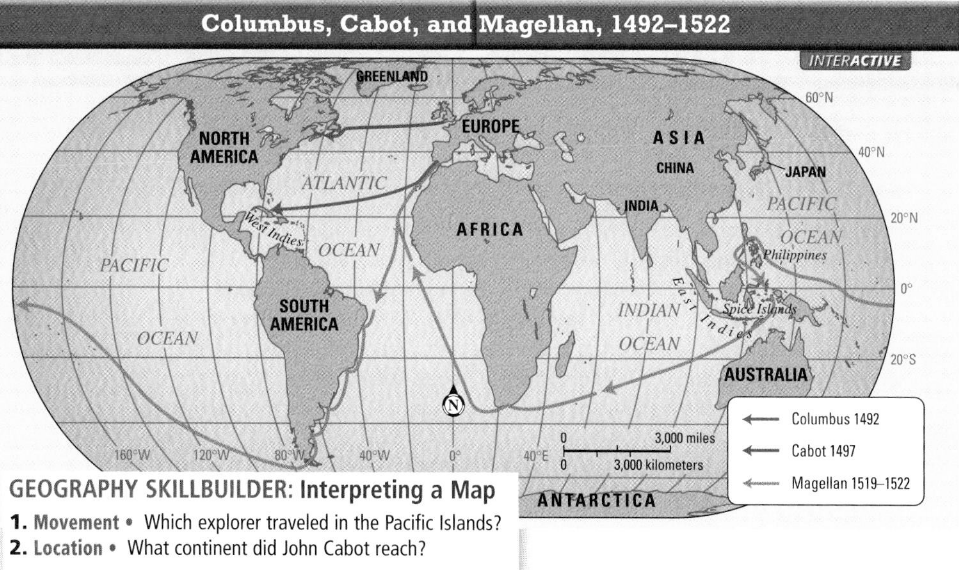

GEOGRAPHY SKILLBUILDER: Interpreting a Map

1. **Movement** • Which explorer traveled in the Pacific Islands?
2. **Location** • What continent did John Cabot reach?

Slavery European explorations also led to an expanding slave trade. The Portuguese purchased West Coast African people to work as slaves back in Portugal, where the work force had been reduced by plague. In other colonized areas, such as Mexico and parts of South America, Europeans forced conquered peoples to work the land where they lived. For hundreds of years, Africans and conquered peoples of the Americas would be forced to work under horrible conditions.

Geography Skillbuilder Answers
1. Magellan
2. North America

ECTION 2 ASSESSMENT

rms & Names

Explain the significance of: (a) Prince Henry the Navigator (b) Christopher Columbus (c) Ferdinand Magellan (d) circumnavigate (e) imperialism

ing Graphics

Use a chart like this one to compare characteristics of the voyages of Christopher Columbus and Vasco da Gama.

lumbus's Voyage	Da Gama's Voyage

Main Ideas

3. (a) Why were spices so important to Europeans?
   (b) Why did Europeans want to find a new route to Asia?
   (c) Name three ways in which European exploration affected the indigenous peoples of North and South America.

Critical Thinking

4. **Making Inferences**
   Why do you think the Portuguese became leaders of European exploration?

Think About
- the location of Portugal
- early Portuguese voyages
- Prince Henry and his School of Navigation

CTIVITY -OPTION- Reread the information about Magellan's voyage around the world. Write a **journal entry** describing the events of the voyage from the point of view of a crew member.

ASSESS & RETEACH

Reading Social Studies Have students fill in the second row of the chart on page 300, "European Exploration and Conquest."

Formal Assessment
- Section Quiz, p. 158

RETEACHING ACTIVITY

Assign students different headings of the section to review. Ask each student to draw a picture or a cartoon that illustrates a main idea of the assigned reading. Tell them to write captions for their drawings. Arrange the drawings sequentially and use them to review the entire section.

In-depth Resources: Unit 4
- Reteaching Activity, p. 24

Access for Students Acquiring English
- Reteaching Activity, p. 92

Section 2 Assessment

1. Terms & Names

a. Prince Henry the Navigator, p. 308
b. Christopher Columbus, p. 309
c. Ferdinand Magellan, p. 309
d. circumnavigate, p. 310
e. imperialism, p. 310

2. Using Graphics

Columbus's Voyage	da Gama's Voyage
Left in August 1492 Three ships Sailed west Spotted land after weeks at sea Made it to Caribbean Islands	Left about 5 years after Columbus Sailed east Spotted land after 317 days at sea Made it to the coast of India

3. Main Ideas

a. Before refrigeration, spices preserved food and improved its flavor.
b. European merchants would make more money buying the spices directly.
c. European explorers brought new diseases, slavery, and religion.

4. Critical Thinking

Possible Response The Portuguese had been sailors and navigators for centuries. Also, Prince Henry's School of Navigation probably increased their chances of success.

ACTIVITY OPTION

Integrated Assessment
- Rubric for writing a journal entry

SKILLBUILDER

Researching Topics on the Internet

Defining the Skill

Take a poll to discover how many students have used the Internet. Have students share the kinds of information they research on the Internet. Brainstorm real-life situations in which the Internet could provide useful information, such as planning a trip, finding directions, or learning information about a subject.

Applying the Skill

How to Research Topics on the Internet

Point out the three strategies for doing research on the Internet. Explain to students that they need to work through each strategy in order to fully understand and successfully use the Internet for research. Remind students that the Internet can provide access to a range of material not easily accessed in other ways. Impress upon them the importance of being a critical consumer.

Practicing the Skill

Have students work with partners to create a list of interesting topics about the Renaissance that they might want to explore further. Then they can come up with a list of keywords that they could use in their research. If students need additional practice, suggest a real-life topic they might research on the Internet. Examples include an interest or hobby, a current event, or directions to a given destination.

In-depth Resources: Unit 4
- Skillbuilder Practice, p. 21

SKILLBUILDER

Researching Topics on the Internet

▸▸ Defining the Skill

The Internet is a computer network that connects libraries, museums, universities, government agencies, businesses, news organizations, and private individuals all over the world. Each location on the Internet has a home page with its own address, or URL (universal resource locator). With a computer connected to the Internet, you can reach the home pages of many organizations and services. The international collection of home pages, known as the World Wide Web, is an excellent source of up-to-date information about the regions and countries of the world.

▸▸ Applying the Skill

The Web page shown below is the European Reading Room at the Library of Congress Web site. Use the strategies listed below to help you understand how to research topics on the Internet.

How to Research Topics on the Internet

Strategy ❶ Once on the Internet, go directly to the Web page. For example, type http://www.loc.gov/rr/european/extlinks.html in the box at the top of the Web browser and press ENTER. The Web page will appear on your screen.

Strategy ❷ Explore the European Reading Room links. Click any of the links to find more information about a subject. These links take you to other Web sites.

Strategy ❸ Always confirm information you have found on the Internet. The Web sites of universities, government agencies, museums, and trustworthy news organizations are more reliable than others. You can often find information about a site's creator by looking for copyright information or reviewing the home page.

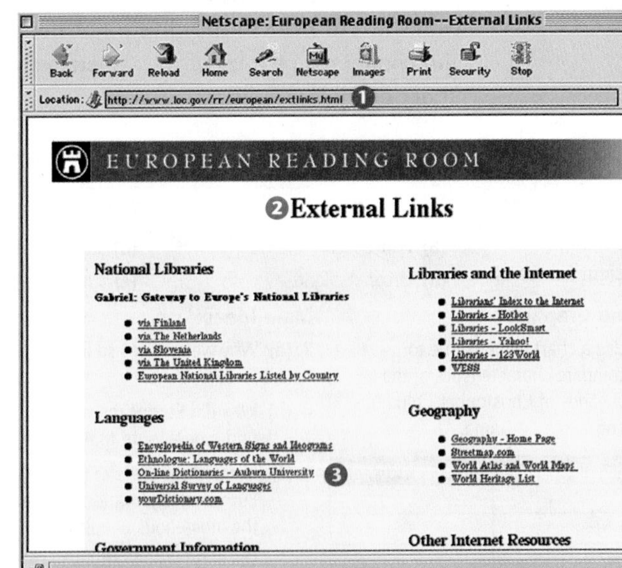

▸▸ Practicing the Skill

Turn to Chapter 11, Section 1, "Renaissance Connections." Reread the section and make a list of topics you would like to research.

For Internet links to support this activity, go to

312 CHAPTER 11

Career Connection: Reference Librarian

Encourage students who enjoy researching topics on the Internet to find out about careers that use this skill. Tell students that a reference librarian, for example, helps people locate information about a vast range of subjects.

1. Locate information about the typical tasks a reference librarian performs. Elicit from students that reference librarians must be able to find specific information on the Internet as well as in traditional reference books.
2. Help students find out what qualities and training a person needs in order to become a reference librarian.
3. Have students create a poster or other visual aid that summarizes their findings.

Block Scheduling

The Age of Revolution

MAIN IDEA

Scientific, industrial, and political revolutions transformed European society.

WHY IT MATTERS NOW

European revolutions in science, technology, and politics helped to create modern societies throughout the world.

TERMS & NAMES
- Scientific Revolution
- Industrial Revolution
- labor force
- capitalism
- French Revolution
- Reign of Terror
- Napoleon Bonaparte

DATELINE

EXTRA

LEIPZIG, GERMANY, APRIL 1839

A new era in German history has begun. The Leipzig-Dresden railway is open for business. Although short rail lines have been in service for a few years, this is the first long-distance railway in this part of Europe.

The steam locomotive that powers the German train was made in England. It is the latest improvement to George Stephenson's "Rocket" train, which set a speed record of 30 mph in 1829. Already, this new form of transportation is changing Europe. The railroads are attracting many passengers and are also ideal for hauling goods. It seems that wherever new train stations are built, growth and prosperity soon follow.

Region • Leipzig's railway will now take passengers all the way to Dresden. ▲

Changes in Science and Industry ①

The steam-powered locomotive was only one in a long line of technological improvements made in Europe since the 1600s. In fact, scientists and inventors made so many discoveries during these years that Europe experienced both a scientific and an industrial revolution. These periods of great change would help to create modern societies.

TAKING NOTES

Use your chart to take notes about people and ideas.

Influences	New Ideas	People/ Achievements
The Renaissance		
European Exploration		

SECTION OBJECTIVES

1. To describe the changes that occurred during the Scientific and Industrial Revolutions
2. To explain the effect of the Industrial Revolution on Europe's labor force
3. To identify the growth in citizens' rights and changes in government in France and England

SKILLBUILDER
- Interpreting a Chart, p. 315

CRITICAL THINKING
- Analyzing Motives, p. 317

FOCUS & MOTIVATE

WARM-UP

Making Inferences Have students read Dateline and discuss the following questions to help them understand how transportation changes affect an economy.

1. What might a long-rail service offer that a short-line service does not?
2. What other kinds of services and supplies will be needed around the new train stations?

INSTRUCT: Objective ①

Changes in Science and Industry

- What were some inventions of the Scientific Revolution? telescope, microscope, classification systems
- What was the Industrial Revolution, and in what ways did it change workers' lives? a change in the way goods were produced; workers moved to cities to work in factories

In-depth Resources: Unit 4
- Guided Reading Worksheet, p. 17

Reading Study Guide
(Spanish and English), pp. 92–93

Program Resources

In-depth Resources: Unit 4
- Guided Reading Worksheet, p. 17
- Reteaching Activity, p. 25

Reading Study Guide
(Spanish and English), pp. 92–93

Formal Assessment
- Section Quiz, p. 159

Integrated Assessment
- Rubric for writing a poem

Outline Map Activities

Access for Students Acquiring English
- Guided Reading Worksheet, p. 87

Technology Resources
classzone.com

TEST-TAKING RESOURCES

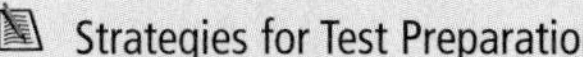

Strategies for Test Preparation

Test Practice Transparencies

Online Test Practice

One of the leaders of the Renaissance's Scientific Revolution was Galileo. He emphasized two ideas: the importance of controlled experiments and the value of observation. He used these ideas in his own scientific research. For example, he tested the long-held belief, first established by Aristotle, that heavier objects fall to the ground faster than lighter ones. After dropping heavy objects and light objects from the same height, he observed that they landed at the same time. Controlled experimentation and careful observation had disproved an accepted belief and laid the groundwork for understanding basic physics.

INSTRUCT: Objective 2

The Workshop of the World

- What is capitalism? a system in which factories and businesses are privately owned
- How were cities affected by industrialization? became dirtier, more crowded, polluted, and full of disease

MORE ABOUT...

The Industrial Revolution

The Industrial Revolution marked the beginning of modern technology, but it could not have taken place without coal miners, whose only tools were picks and shovels. The machines in the new factories required much more power than the traditional horses or water wheels could provide. One important new power source was the steam engine, which burned coal to heat water. Steam engines and many factory machines were made of iron, and smelting—the process of separating iron from ore—also required coal.

The Scientific Revolution In the 16th and 17th centuries, scientific discoveries changed the way Europeans looked at the world. This led to the **Scientific Revolution.**

In Italy, Galileo Galilei (GAL•uh•LEE•oh GAL•uh•LAY) (1564–1642) studied the stars and planets using a new invention called the telescope. Later in Holland, Antoni van Leeuwenhoek (LAY•vuhn•huk) (1632–1723) used a microscope to explore an unknown world found in a drop of water. The Swedish botanist Carolus Linnaeus (lih•NEE•uhs) (1707–1778) even developed a system to name and classify all living things on Earth.

Culture • In 1610, Galileo used his telescope to observe that Jupiter had moons. ▲

The Scientific Method During the Scientific Revolution, scientists began doing research in a new way, called the scientific method. This scientific method is still used by scientists today.

First, scientists identify a problem. Next, they collect data about the problem. Using this data, they develop an explanation for the problem and test the explanation by performing experiments. Finally, they reach a conclusion.

The Industrial Revolution Many inventions of the Scientific Revolution began to change the way people worked all across Europe. Machines performed jobs that once had been done by humans and animals. This brought about such great change that it led to a revolution in the way goods were produced: the **Industrial Revolution.**

Machines were grouped together to make products in large factories. Early factories were built in the countryside near streams and rivers so that they could be powered by water. By the late 1700s, however, new steam engines were used to power the machinery. More and more factories could now be built in cities. People, in turn, moved from the countryside to the cities in search of work.

A. Finding Causes How did the Scientific Revolution lead to the Industrial Revolution?

Reading Social Studies
A. Possible Answe[r]
New inventions changed the way people worked.

The Workshop of the World 2

The Industrial Revolution began in England in the late 1700s. The first English factories made textiles, or cloth. The steam-powered machines of the textile industry produced large amounts of goods quickly and cheaply. So many factories were built in England that the country earned the nickname "The Workshop of the World."

Activity Options

Differentiating Instruction: Students Acquiring English/ESL

Understanding Multiple Meanings Point out the word *revolution* in the headings "The Scientific Revolution" and "The Industrial Revolution." Explain that the word *revolution* has several meanings. In this context, it means "major changes in ideas, materials, and methods." Tell students that later in this section they will read about the French Revolution. Explain that *revolution* can also mean "the overthrow and replacement of a government." Ask students to use *revolution* in sentences, using both meanings of the word.

BACKGROUND

In the 1850s, laws were finally passed to help protect women and children from long hours and harsh working conditions.

Hard Work for Low Pay The Industrial Revolution created a need for workers, or a **labor force,** in cities. The workers who ran the textile machines made up part of this labor force. Most workers could earn more income in cities than on farms, but life could be hard. Factory laborers worked long hours and received low pay. In fact, many families often sent their children to work to help create more income.

In 1838, women and children made up more than 75 percent of all textile factory workers. Children as young as seven were forced to work 12 hours a day, six days a week.

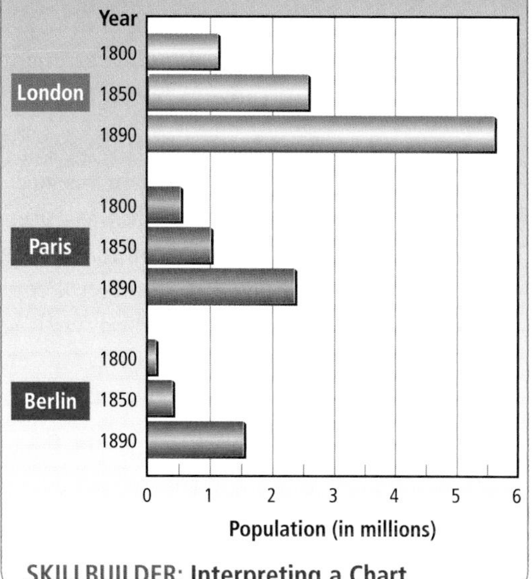

SKILLBUILDER: Interpreting a Chart

1. Which city had the largest growth in population?
2. What was the population of Paris in 1890?

Skillbuilder Answers
1. London
2. about 2.3 million people

The Spread of Industrialization The textile industry in 18th-century England was one step in the development of an economic system called **capitalism.** In this system, factories and other businesses that make and sell goods are privately owned. Private business owners make decisions about what goods to produce. They sell these goods at a price that will earn a profit.

Industrialization spread from England to other countries, including Germany, France, Belgium, and the United States. Cities in these countries grew rapidly and became more crowded and dirtier. Diseases, such as cholera (KAHL•uhr•uh) and typhoid (TY•foyd) fever, spread. Smoke from factories blackened city skies, and pollution fouled the rivers.

Place • Factories, like this one in Sheffield, England, were found throughout Western Europe by the mid-19th century. ▶

FOCUS ON VISUALS

Interpreting the Chart Discuss the chart with students, asking them to make a generalization based on the facts. What happened between 1800 and 1890 to cause the population explosion in London, Paris, and Berlin? How much did the population grow during these years?

Possible Responses With the development of the steam engine, more factories were built in cities, and people moved to the cities in search of work; London—from about 1.2 million to 5.7 million; Paris—from 0.5 million to 2.3 million; Berlin—from about 0.2 million to 1.6 million.

Extension Have students research population growth in selected American cities during the same period. Encourage them to create a graph displaying their results.

MORE ABOUT...
Child Labor

The exploitation of very young children in the early factories led to an international movement to control the use of child labor. The movement began in Great Britain in 1802, with a law to control the use of children from poor families as apprentices in cotton mills. The International Association for Labor Legislation was established in 1900 to promote regulations governing work by children. Today, hardly any children in Europe, North America, Australia, or New Zealand hold regular jobs before age 15. In other parts of the world, however, children as young as 7 still work in factories, fields, mines, and quarries, and the governments there do little to control the working conditions. In some countries in the Middle East, more than 10 percent of all workers are children. Child labor is also common in parts of Latin America and Asia.

Activity Options

Differentiating Instruction: Gifted and Talented

Researching a Topic Have students work in pairs to research machines and other labor-saving devices that were invented during the Industrial Revolution, such as the spinning jenny, the flying shuttle, and the steam engine. Then ask students to create time lines that display the information they discovered. Next, have students elaborate on their time lines, adding information about significant events, new social ideas, or scientific developments that occurred during the Industrial Revolution, both in Europe and the United States. The time lines could be annotated to make clear connections between inventions and ensuing developments. Allow time for students to compare their findings.

INSTRUCT: Objective 3

The French Revolution

- What were some of the causes of the French Revolution? food shortages, hunger, heavy taxes, uncaring government
- What happened during the Reign of Terror? 17,000 people were executed for disagreeing with revolutionary leaders
- Who restored order to France? Napoleon Bonaparte

MORE ABOUT...

Marie Antoinette

Marie Antoinette was only a teenager when she became queen of France in 1774. She had little formal education and was bored by the affairs of the government. She soon became very unpopular with the French people, as she was quite unconcerned about their suffering and France's financial crisis. A popular but unsubstantiated story about her claims that she once asked a government official why the people were so angry. When he replied, "Because they have no bread," she answered, "Then let them eat cake." She did not understand that a large number of her subjects were too poor to afford either.

MORE ABOUT...

The Bastille

The Bastille held only seven prisoners when a mob stormed it on July 14, 1789. The rebels wanted the man in charge of the prison to turn over the weapons and ammunition that were stored there. His refusal to answer the demand so enraged the mob that it stormed and captured the prison. After they took over the government, the revolutionaries demolished the Bastille. July 14—Bastille Day—is a national holiday that the people of France celebrate with parades and fireworks.

The French Revolution 3

Along with changes in science, technology, and the economy came new ideas about government. In the late 18th century, many ordinary citizens began to fight for more political rights.

Ripe for Political Change By the 1780s, the French government was deeply in debt because of bad investments and the costs of waging wars. Life was miserable for the common working people. Poor harvests combined with increased population had led to food shortages and hunger. People were forced to pay heavy taxes. At the same time, the French king, Louis XVI, and his queen, Marie Antoinette, continued to enjoy an expensive life at court, entertaining themselves and the French nobility.

Reading Social Studies

B. Analyzing Motives Why did the French citizens demand a new government?

Reading Social Studies
B. Possible Answer
Their lives were very hard, and their rulers were insensitive to their suffering.

Storming the Bastille The citizens of France demanded changes in the government, without success. Then, on July 14, 1789, angry mobs stormed a Paris prison called the Bastille (ba•STEEL). The attack on this prison, which reflected the royal family's power, became symbolic of the **French Revolution.**

Revolts spread from Paris to the countryside, and poor and angry workers burned the homes of the nobility. By 1791, France had a new constitution that made all French citizens equal under the law.

Region • The storming of the Bastille remains a symbol of the French Revolution. ▼

Activity Options

Interdisciplinary Link: Government

Class Time 20 minutes

Task Creating a list of questions that French citizens might have used to design a new government

Purpose To understand the role of government

Supplies Needed
- Paper
- Pens and pencils

Activity Point out to students that when the people of France overthrew the monarchy, they were faced with the task of creating a new government. Lead a discussion about the kinds of decisions a new government must make, such as how to raise money, how to make laws, and how to resolve conflicts. Have each student write a list of questions to be addressed by the framers of a new constitution.

BACKGROUND

Until the French Revolution, the guillotine was only used to execute nobles. It was considered the most humane type of execution.

The French Republic In 1792, France became a republic. King Louis XVI was found guilty of treason, or betraying one's country. In 1793, he and Marie Antoinette were sentenced to death. They were beheaded on the guillotine (GIHL•uh•teen).

Still, France was not at peace. The new revolutionary leaders refused to tolerate any disagreement. Between 1793 and 1794, these new leaders executed 17,000 people. This period of bloodshed became known as the **Reign of Terror**.

Napoleon French leaders continued to struggle for power until 1799, when General **Napoleon Bonaparte** (nuh•POH•lee•uhn BOH•nuh•PAHRT) took control. The French Revolution and the disorder that followed were finally over.

However, the new sense of equality brought about by the Revolution stirred feelings of nationalism among the French. Nationalism is pride in and loyalty to one's nation. Soon, the citizens of other European nations began to fight for more political power. Slowly, they, too, won more rights.

Region • Napoleon Bonaparte crowned himself emperor of France in 1804. He led France to victory in what became known as the Napoleonic Wars. ▲

SECTION 3 ASSESSMENT

Terms & Names

1. **Explain the significance of:** (a) Scientific Revolution (b) Industrial Revolution (c) labor force (d) capitalism (e) French Revolution (f) Reign of Terror (g) Napoleon Bonaparte

Using Graphics

2. Use a chart like this one to list some of the scientific, industrial, and political changes that occurred during the Age of Revolution.

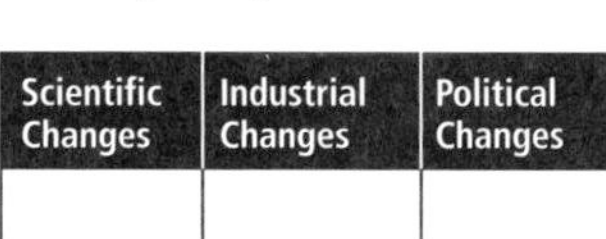

Scientific Changes	Industrial Changes	Political Changes

Main Ideas

3. (a) Describe at least three inventions or discoveries of the Scientific Revolution.
   (b) How did the Industrial Revolution change the way people in Europe worked?
   (c) What changes occurred in France after the French Revolution?

Critical Thinking

4. **Recognizing Effects** How did industrialization change the cities to which it spread?

Think About
- population
- diseases
- the environment

ACTIVITY -OPTION- Reread the section about the French Revolution. Write a **poem** or **lyrics** for a folk song that describe the events from the point of view of a common citizen or a member of the royal family.

CRITICAL THINKING ACTIVITY

Analyzing Motives Remind students that the French Revolution was a dramatic break with the past. For the first time, a monarch was no longer the primary authority; the people were in charge. How, then, could Napoleon Bonaparte rise to power a mere ten years after the storming of the Bastille? Ask students who they think Napoleon's supporters might have been and why they might have welcomed an emperor. Discuss other circumstances where new rulers and/or governments have been welcomed by the same people who sought to overthrow such rulers and/or governments in the past.

Class Time 15 minutes

ASSESS & RETEACH

Reading Social Studies Have students fill in the third and fourth rows of the chart on page 300, "Scientific and Industrial Revolution" and "Political Revolution."

Formal Assessment
- Section Quiz, p. 159

RETEACHING ACTIVITY

Ask students to write a three-minute radio news report about the French Revolution. Tell students to imagine that they are writing their report in 1794. Tell them to include past events that led up to the Revolution. When they are finished, have students present their reports to the class.

In-depth Resources: Unit 4
- Reteaching Activity, p. 25

Access for Students Acquiring English
- Reteaching Activity, p. 93

Section 3 Assessment

1. Terms & Names

a. Scientific Revolution, p. 314
b. Industrial Revolution, p. 314
c. labor force, p. 315
d. capitalism, p. 315
e. French Revolution, p. 316
f. Reign of Terror, p. 317
g. Napoleon Bonaparte, p. 317

2. Using Graphics

Scientific Changes	Industrial Changes	Political Changes
invented telescope, microscope	factories, labor force	French Revolution

3. Main Ideas

a. Three inventions are telescopes, microscopes, and scientific method.
b. More and more people worked in factories in cities.
c. The French monarchy was overthrown and France became a republic with a new constitution.

4. Critical Thinking

Possible Response Cities became crowded, polluted, and riddled with disease.

ACTIVITY OPTION

Integrated Assessment
- Rubric for writing a poem

The Russian Empire

TERMS & NAMES
czar
Ivan the Terrible
Peter the Great
Catherine the Great
Russian Revolution

SECTION OBJECTIVES

1. To describe the impact of the early czars on the internal and external affairs of Russia
2. To explain the growth of Russia into a large empire under strong rulers
3. To explain problems of Russia's expansion
4. To identify causes of the Russian Revolution

SKILLBUILDER
- Interpreting a Map, p. 320

CRITICAL THINKING
- Recognizing Important Details, p. 320

FOCUS & MOTIVATE

WARM-UP

Drawing Conclusions Have students read Dateline and discuss the following questions to help them explore how the cathedral expresses Russia's growth and power.

1. What do the size and decorations of the Cathedral of St. Basil say about Russia's economy?
2. What does the addition of the Tatars' lands mean for Russia's future?

INSTRUCT: Objective 1

Russia Rules Itself

- Who was the first czar of Russia after it broke free of the Mongols? Ivan IV, or Ivan the Terrible
- What is an unlimited government? a government in which a single ruler holds all the power
- Why were the early czars and nobles in conflict? Czars saw nobles as a threat to their control.

In-depth Resources: Unit 4
- Guided Reading Worksheet, p. 18

Reading Study Guide
(Spanish and English), pp. 94–95

MAIN IDEA

Strong leaders built Russia into a large empire, but the country's citizens had few rights and struggled with poverty.

WHY IT MATTERS NOW

Russia has had a great influence on world politics and is experiencing a period of great change.

DATELINE

Moscow, Russia, 1560—Today, the most magnificent church in Moscow opened with a grand celebration. The Cathedral of St. Basil has ten domes—each one unique. The massive structure, built of bricks and white stone, is decorated with brilliant colors.

Ivan IV built this cathedral to celebrate his victory eight years ago over the Tatars (TAH•tuhrz). These Turkish people who live in Central Asia have long threatened Russia's security.

The victory also added the lands of the Tatars, including their capital at Kazan, to our growing empire. Russians everywhere should be proud of Moscow's new church and of the victory it symbolizes.

Place • Ivan IV has honored a Russian victory over the Tatars with the construction of St. Basil's Cathedral. ▲

Russia Rules Itself 1

Russia, geographically the world's largest nation, is located in both Europe and Asia. It takes up large parts of both continents, and both continents have helped shape its history.

Mongols from eastern Asia conquered Russia in the 13th century and ruled it for about 200 years. During the 15th century, Russia broke free of Mongol rule. At this time, the most important Russian city was Moscow, located in the west.

TAKING NOTES
Use your chart to take notes about people and ideas.

Influences	New Ideas	People/ Achievements
The Renaissance		
European Exploration		

Program Resources

In-depth Resources: Unit 4
- Guided Reading Worksheet, p. 18
- Reteaching Activity, p. 26

Reading Study Guide
(Spanish and English), pp. 94–95

Formal Assessment
- Section Quiz, p. 160

Integrated Assessment
- Rubric for writing a summary

Outline Map Activities

Access for Students Acquiring English
- Guided Reading Worksheet, p. 88

Technology Resources
classzone.com

TEST-TAKING RESOURCES
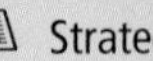
Strategies for Test Preparation
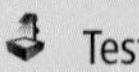
Test Practice Transparencies
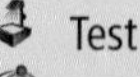
Online Test Practice

Reading Social Studies

A. Clarifying Why did the Russian people give Ivan IV the nickname Ivan the Terrible?

Reading Social Studies A. Answer

They named him Ivan the Terrible because he was so cruel.

The First Czars of Russia In 1547, a 16-year-old leader in Moscow was crowned the first **czar** (zahr), or emperor, of modern Russia. His official title was Ivan IV, but the people nicknamed him **Ivan the Terrible.** Ivan was known for his cruelty, especially toward those he viewed as Russia's enemies. During his rule of 37 years, the country was constantly at war.

During the reigns of Ivan the Terrible and the czars who followed him, Russia had an unlimited government. This is a form of government in which a single ruler holds all the power. The people have no say in how the country is run.

Region • Ivan the Terrible is said to have worn this fur-trimmed crown at his coronation in 1547. ▲

Conflicts at Home The first Russian czars were often in conflict with the Russian nobles, who possessed much land and wealth. The czars viewed the nobles as a threat to their control over the people. Ivan the Terrible ordered his soldiers to murder Russian nobles and church leaders who opposed him.

The poor farmers, or peasants, of Russia also suffered under the first czars. New laws forced the peasants to become serfs, who had to remain on the farms where they worked.

The Expansion of Russia ❷

In addition to strengthening their control over the Russian people, the czars wanted to gain new territory. Throughout the 17th and 18th centuries, rulers such as Peter the Great and Catherine the Great conquered neighboring lands.

BACKGROUND

In 1721, Peter the Great changed his title from *czar* to *emperor,* a title that he thought sounded more European.

A Window on the West An intelligent man with big ideas for his country, **Peter the Great** ruled Russia from 1682 to 1725. After defeating Sweden in war and winning land along the Baltic Sea, Peter built a port city called St. Petersburg. This city, which Peter saw as Russia's "window on the west," became the new capital.

One of Peter's goals was to have closer ties with Western Europe. He hoped to use the ideas and inventions of the Scientific Revolution to modernize and strengthen Russia. During his rule, Peter reformed the army and the government and built new schools. He even ordered Russians to dress like Europeans and to shave off their beards. Peter's reforms made Russia stronger, but they did not improve life for Russian peasants.

Movement • Peter the Great brought to Russia many of the improvements of the Scientific and Industrial Revolutions. ▲

MORE ABOUT...
Ivan the Terrible

Ivan Vasilyevich, who became Czar Ivan IV and was nicknamed Ivan the Terrible, set out to establish a Christian state based on the principles of justice. One of his many reforms was to limit the power of aristocrats who inherited their land and titles. He started a war against Sweden, Poland, and Lithuania in 1558 by invading a territory called Livonia. The war lasted 24 years, and when it began to go badly for Russia, the disappointed czar ordered the execution of several advisers he suspected of treason. This began a reign of terror in which many thousands of Russians were killed by the czar's troops.

INSTRUCT: Objective ❷
The Expansion of Russia

- In what ways did Peter the Great modernize and strengthen Russia? He brought ideas and inventions from Western Europe, reformed army and government, and so on.
- What ideas did Catherine the Great borrow from Western Europe? new schools, new towns, expanded trade, encouraged art, science, literature
- Which group of citizens suffered most under Russia's new strength? the peasants

Activity Options
Multiple Learning Styles: Logical

Class Time One class period

Task Writing an advertisement to encourage Russians to move to the new city of St. Petersburg

Purpose To identify the aspects of St. Petersburg that made it different from other Russian cities

Supplies Needed
- Photos and information about St. Petersburg
- Writing paper
- Pens and pencils

Block Scheduling

Activity Explain that Peter the Great invited architects to design buildings, streets, and public areas of St. Petersburg to imitate aspects of London, Paris, and Vienna. Have students study photos and information and then create a brochure that might have persuaded people to move to the city in the early 1700s. Suggest that they describe its appearance and growth opportunities and the advantages of living in a city supported by the czar.

CRITICAL THINKING ACTIVITY

Recognizing Important Details Help students identify important details about the reigns of Peter the Great and Catherine the Great. Work with students to create a spider map for each, recording important details around each ruler's name.

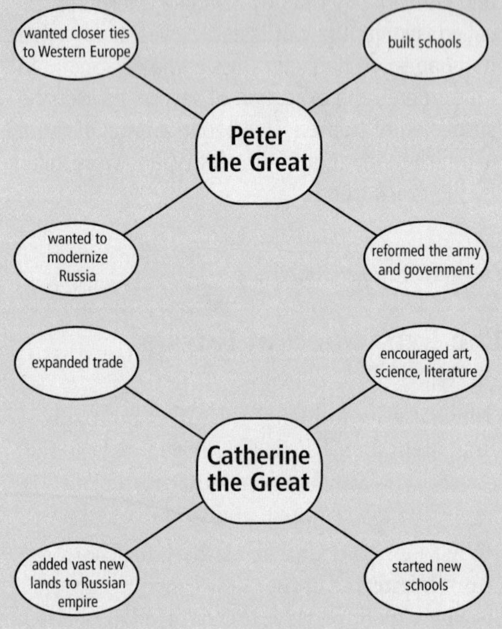

INSTRUCT: Objective 3

A Divided Russia

- Russia was divided into what two major classes in the 19th century? nobles, serfs
- Why did Alexander II's efforts to end serfdom fail? heavy taxes, poor land for farming
- What happened on Bloody Sunday? government troops shot protesting workers

The Expansion of Russia, 1584–1796

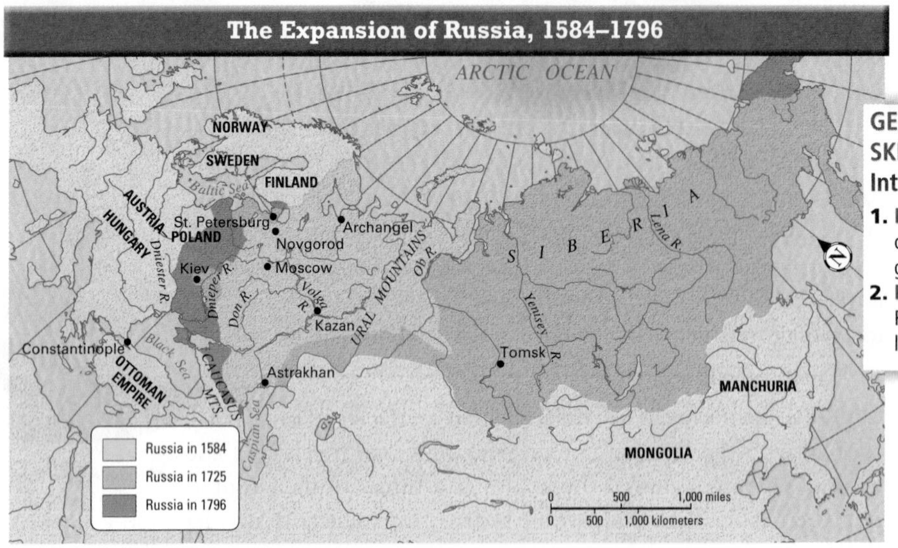

GEOGRAPHY SKILLBUILDER: Interpreting a Map

1. **Location** • What body of water did Russia gain access to in 1796?
2. **Place** • When did Russia gain the most land?

Geography Skillbuilder Answers
1. Black Sea
2. between 1584 and 1725

A Great Empress **Catherine the Great** took control of Russia in 1762 and ruled until her death in 1796. Catherine added vast new lands to the empire, including the present-day countries of Ukraine (yoo•KRAYN) and Belarus (behl•uh•ROOS). Like Peter the Great, Catherine borrowed many ideas from Western Europe. She started new schools and encouraged art, science, and literature. Catherine also built new towns and expanded trade.

During Catherine's reign, Russia became one of Europe's most powerful nations. The lives of the peasants, however, remained miserable. Catherine thought about freeing them, but she knew the nobles would oppose her. When the peasants rebelled in the 1770s, Catherine crushed their uprising.

Movement • Catherine the Great continued Peter the Great's practice of bringing the ideas of Western Europe to Russia. ▼

A Divided Russia 3

In the 19th century, Russia remained a divided nation. Most people were poor peasants, and most of the wealth belonged to the nobles. This division would lead to conflict and eventually to a political revolution.

The Nobles Many Russian nobles sent their children to be educated in Germany and France. In fact, many noble families spoke French at home, speaking Russian only to their servants. The Western Europeans introduced many new ideas to the Russian nobles, among them the idea that a nation's government should reflect the wishes of its citizens.

BACKGROUND

Catherine the Great was born in Germany. She came to Russia at 15 to marry the heir to the throne, Peter III. He was a weak ruler, however, and Catherine, supported by the army and the people, overthrew him.

Activity Options

Skillbuilder Mini-Lesson: Making a Generalization

Block Scheduling

Explaining the Skill Remind students that to make a generalization, they must examine many sources of information about a topic and then draw a conclusion based on all the facts and evidence they have found.

Applying the Skill Have students reread the section entitled "A Great Empress" and use these strategies to support a given generalization. Write the following generalization on the board: "The lives of the Russian nobles were very different from the lives of the peasants." Ask students to reread pages 320–321 and to make a list of information that specifically supports this generalization. Nobles owned land; sent their children to be educated in the west; did not speak Russian at home; were army officers or government officials. When they finish, brainstorm other reference sources where they might locate information to support this generalization.

Many Russian nobles were army officers or government officials. Most supported the czar and were proud of Russia's growing power. In 1825, one group of nobles tried to replace the government. Their attempt to gain more power failed.

The Serfs In the 19th century, the Russian serfs still had no land or money of their own. They worked on farms owned by others and received little help from the Russian government.

In 1861, Alexander II decided to end serfdom in Russia. He hoped that freeing the serfs would help his country compete with Western Europe. The serfs had to pay a heavy tax, though, and the land they were given was often not good for farming. Most former serfs felt that they had gained very little.

BACKGROUND

In the 1850s, Russia fought the Crimean War against Turkey. Two of Turkey's allies were Britain and France. When Russia lost, Alexander II thought this proved that his country was still far less advanced than Western European nations.

Bloody Sunday The serfs were not the only unhappy Russians. Many university students, artists, and writers believed that the government's treatment of the serfs was unfair. Some joined groups that tried to overthrow the government. In addition, workers in Russia's cities complained about low pay and poor working conditions.

In 1905, a group of workers marched to the royal palace in St. Petersburg with a list of demands. Government troops shot many of them. News of the events of this "Bloody Sunday" spread across Russia, making people even angrier with the government and czar.

Thinking Critically Possible Answers

1. They admired the culture of Western Europe and hoped it would strengthen and modernize Russia.
2. They hoped the public would admire and/or emulate the values and customs represented in the works of art.

The Hermitage is truly enormous. Its buildings run almost half a mile along the bank of the Neva River in the center of St. Petersburg, with almost 400 rooms and 3 million pieces of art.

The Hermitage was originally a royal residence known as the Winter Palace. When Catherine the Great moved into the Winter Palace in 1762, she brought her art collection with her. Two years later, she had a long, narrow building added to house her extensive collection. This addition was called the "Little Hermitage." Over the next hundred years, several other buildings were added.

Spotlight on CULTURE

The Hermitage Museum One of the world's largest art museums is the Hermitage in St. Petersburg. It contains many works of art, including French, Spanish, and British paintings. Part of the collection is in the Winter Palace, a former royal residence.

Both Peter the Great and Catherine the Great collected European art. On a trip to Amsterdam in 1716, Peter bought paintings by the famous Dutch artist Rembrandt. About 50 years later, Catherine bought more than 200 works of art when she visited Germany. These royal collections became part of the Hermitage when it opened as a public museum in 1852.

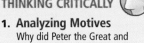

THINKING CRITICALLY

1. **Analyzing Motives**
   Why did Peter the Great and Catherine the Great collect art from Western Europe?
2. **Making Inferences**
   Why do you think the works of art were displayed in a museum?

For more on the Hermitage Museum, go to

Activity Options

Differentiating Instruction: Less Proficient Readers

Determining Cause and Effect Less proficient readers may need to review the reasons why the lives of Russian serfs did not improve after they gained their freedom. Have students create one cause-and-effect graphic organizer labeled "Problems Before Freedom" and one labeled "Problems After Freedom." Have students reread "The Serfs" on page 321 and fill in the organizer.

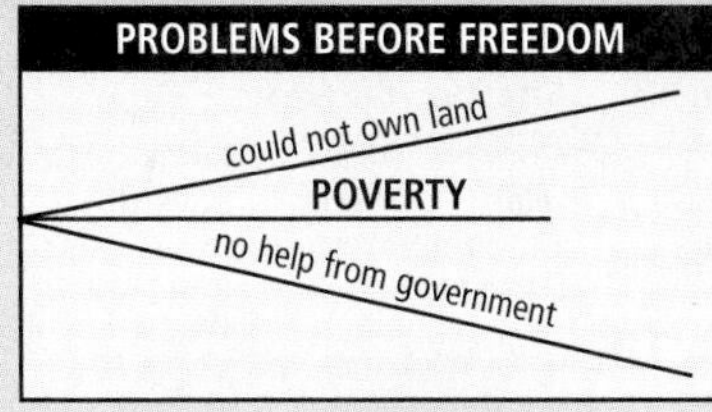

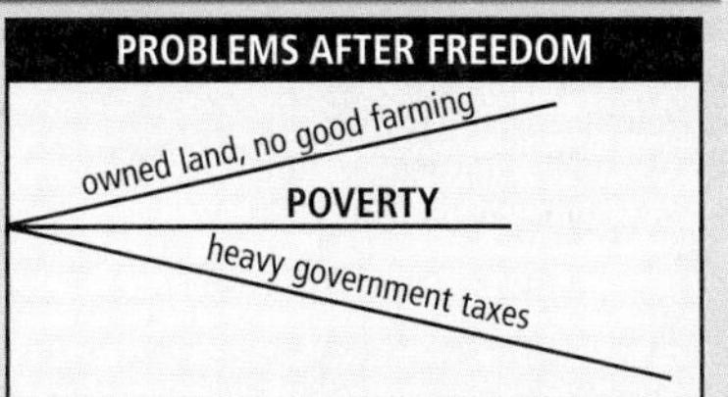

INSTRUCT: Objective 4

The End of the Russian Empire

- What problems led to the Russian Revolution in 1917? food shortages and worker strikes
- What happened to the Russian monarchy? It was overthrown.

ASSESS & RETEACH

Reading Social Studies Have students fill in the bottom line of the chart on page 300, "The Russian Empire."

Formal Assessment
- Section Quiz, p. 160

RETEACHING ACTIVITY

Ask each student to choose one important event or aspect of Russian history that was discussed in this section and write it in large letters on a piece of paper. Holding their papers, have students stand up and organize themselves into a human time line. Have the class review and discuss, in order, each event represented on the time line. If necessary, have students identify missing events.

In-depth Resources: Unit 4
- Reteaching Activity, p. 26

Access for Students Acquiring English
- Reteaching Activity, p. 94

The End of the Russian Empire 4

In 1914, World War I began. Nicholas II—a quiet, shy man who did not want war—ruled Russia, but he failed to keep his country out of the battle. Russia, whose allies included the United Kingdom and France, suffered terrible losses fighting Germany and its allies.

During World War I, there were food shortages in the cities and workers went on strike. Russian revolutionaries organized the workers against the czar. Even the Russian army turned against their ruler, and in 1917, Nicholas was forced to give up power. This overturning of the Russian monarchy is known as the **Russian Revolution.**

Nicholas II and the royal family (the Romanovs) were imprisoned by the revolutionaries. On July 17, 1918, they were all shot to death. This execution ended more than 300 years of rule by the Romanov family and nearly 400 years of czarist rule.

Reading Social Studies

B. Analyzing Motives Why did Russian workers strike?

Reading Social Studies
B. Possible Answer
There were food shortages in the cities.

Rasputin One of the most influential people at the court of Czar Nicholas II was Rasputin. He came from Siberia in eastern Russia and was a self-styled holy man. Crown prince Alexis suffered from the disease hemophilia, and no doctor in Russia could cure him. Rasputin seemed to mysteriously heal the boy, gaining favor with Nicholas's wife, Czarina Alexandra. However, in 1916, Russian nobles killed Rasputin out of fear of the considerable power and influence the monk had.

SECTION 4 ASSESSMENT

Terms & Names

1. **Explain the significance of:** (a) czar (b) Ivan the Terrible (c) Peter the Great (d) Catherine the Great (e) Russian Revolution

Using Graphics

2. Use a chart like this one to describe three characteristics of czars of Russia.

Ivan the Terrible	Peter the Great	Catherine the Great	Nicholas II

Main Ideas

3. (a) What effects did an unlimited government have on Russian peasants?
   (b) How did Peter the Great help reform Russia?
   (c) Alexander II ended serfdom in 1861, but this did little to help the serfs. Why?

Critical Thinking

4. **Finding Causes**
   What events led to the Russian Revolution?

Think About
- the life of the serfs
- Bloody Sunday
- the events of World War I

ACTIVITY -OPTION- Look at the map on page 320 that shows the expansion of Russia. Write a brief **summary** to describe how the Russian nation grew from the 1500s to 1800.

Section 4 Assessment

1. Terms & Names
- **a.** czar, p. 319
- **b.** Ivan the Terrible, p. 319
- **c.** Peter the Great, p. 319
- **d.** Catherine the Great, p. 320
- **e.** Russian Revolution, p. 322

2. Using Graphics

Ivan the Terrible	Peter the Great	Catherine the Great	Nicholas II
Russia constantly at war	Closer ties with Western Europe	Added new lands to empire	Russia suffered terrible losses in WWI

3. Main Ideas
- **a.** They had no say in how their country was run.
- **b.** Peter the Great reformed the government and developed close ties to Western Europe.
- **c.** They were heavily taxed and given poor land to farm.

4. Critical Thinking

Possible Response The unfair treatment of serfs, Bloody Sunday, and food shortages brought about by World War I all led to the Russian Revolution.

ACTIVITY OPTION

Integrated Assessment
- Rubric for writing a summary

INTERACTIVE

James Watt's Double-Action Steam Engine

Amid the excitement of the Industrial Revolution, James Watt (1736–1819), a Scottish inventor, patented a new steam engine. Steam power had been used for many years, but Watt's invention was an improved, double-action steam engine. This system, in which the steam pushes from both sides of the piston rather than from just one, enhanced efficiency and increased power. Watt's invention helped to advance manufacturing and transportation and influenced later inventions. Watt's double-action steam engine was one of the most important inventions of the Industrial Revolution.

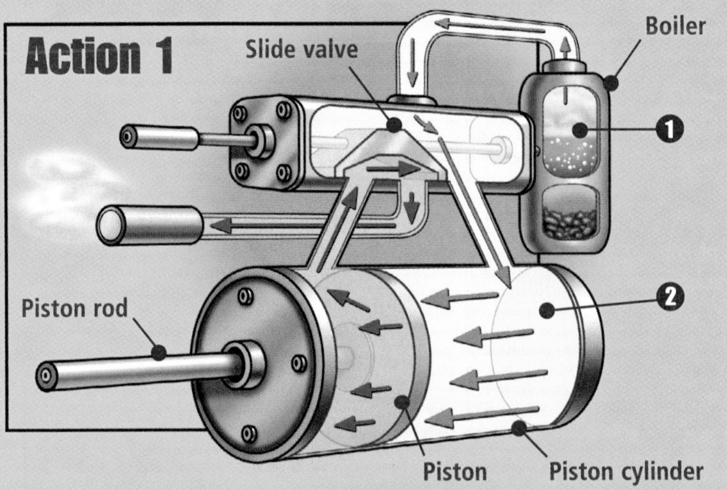

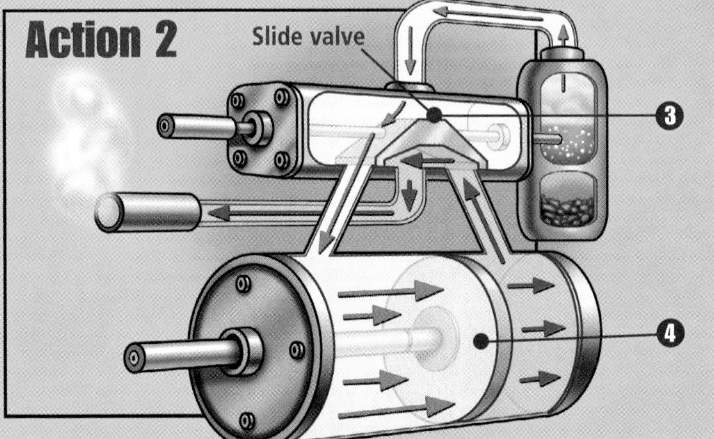

How the Engine Works

Steam from the **boiler** enters the **piston cylinder**. The pressure of the steam pushes the **piston** to one side, moving the **piston rod**. When the piston reaches the end of the stroke, the **slide valve** shifts the steam to the other side of the piston, forcing it back and releasing the steam it compresses as exhaust.

1. As water is converted to steam, its volume increases 1,600 percent.
2. When the steam enters the piston cylinder, it forces the piston rod to one side.
3. As the piston reaches the end of its stroke, the slide valve channels steam to the other side of the piston.
4. The piston rod is pushed back, forcing the "old" steam out as exhaust.

Thinking Critically

1. **Drawing Conclusions** How did the steam engine help power the Industrial Revolution?
2. **Recognizing Effects** How did Watt's steam engine change the lives of working people?

OBJECTIVES

1. To illustrate how the double-action steam engine works
2. To explain the great influence of the steam engine on industrial growth

INSTRUCT

- How did Watt's engine improve on others?
- What is the function of the boiler?

Block Scheduling

MORE ABOUT...

Rapid Industrial Growth

In 1783 Richard Arkwright adapted one of Watt's engines for a textile mill. Within 20 years, England and Scotland had 2,400 steam-powered looms. By 1857 some 250,000 were in operation. Imports of raw cotton increased six-fold from 1775 to 1790. Fabrics once spun by workers at home were mass produced in factories, greatly reducing costs. The price of cotton yarn in 1830 was about one-twentieth of its price in 1760.

Connect to History

Strategic Thinking How does the steam engine's history show the influence inventors have on each other?

Connect to Today

Strategic Thinking Why was the steam engine eventually replaced by electric and internal combustion engines?

Thinking Critically

1. **Drawing Conclusions Possible Response** The steam engine powered large, complex machines that made the Industrial Revolution possible. It also influenced later inventions that impacted on the revolution.

2. **Recognizing Effects Possible Responses** Steam power changed the way people worked. Early textile mills were powered by water from rivers and were located in the country. Many tasks, such as weaving, were done by hand. With the invention of the steam engine, people living in the country moved to towns and cities to find work in factories. Watt's invention also led to the development of railroads, which made transportation much faster and easier.

CHAPTER 11 ASSESSMENT

TERMS & NAMES

1. Renaissance, p. 302
2. Leonardo da Vinci, p. 304
3. Reformation, p. 305
4. Ferdinand Magellan, p. 309
5. circumnavigate, p. 310
6. imperialism, p. 310
7. Industrial Revolution, p. 314
8. Napoleon Bonaparte, p. 317
9. Peter the Great, p. 319
10. Russian Revolution, p. 322

REVIEW QUESTIONS

Possible Responses

1. Before the Renaissance artists painted religious subjects; after the Renaissance they began to paint portraits, historical scenes, and mythological stories.
2. They were called Protestants because they protested against the Catholic Church.
3. Spices from Asia were transported over long distances. Each merchant along the trade route needed to make a profit.
4. They brought back gold dust, ivory, and an increased knowledge of navigation.
5. The Industrial Revolution began in England in the late 1700s.
6. National debt, food shortages, and heavy taxes led to the French Revolution.
7. Ivan the Terrible was known for his cruelty, especially against his enemies in other countries.
8. Catherine the Great started schools; encouraged art, science, and literature; built towns; and expanded trade.

CHAPTER 11 ASSESSMENT

TERMS & NAMES

Explain the significance of each of the following:

1. Renaissance
2. Leonardo da Vinci
3. Reformation
4. Ferdinand Magellan
5. circumnavigate
6. imperialism
7. Industrial Revolution
8. Napoleon Bonaparte
9. Peter the Great
10. Russian Revoluti

REVIEW QUESTIONS

Renaissance Connections *(pages 301–306)*

1. How did the subjects chosen by artists change during the Renaissance?
2. Why were the followers of Martin Luther called Protestants?

Traders, Explorers, and Colonists *(pages 307–311)*

3. Why were spices from Asia so expensive when sold in Europe?
4. What did Portuguese explorers bring back from their expeditions to western Africa?

The Age of Revolutions *(pages 313–317)*

5. When and where did the Industrial Revolution begin?
6. What conditions in France during the 1780s led to the French Revolution?

The Russian Empire *(pages 318–322)*

7. How did Ivan the Terrible earn his nickname?
8. What ideas did Catherine the Great borrow from Western Europe?

CRITICAL THINKING

Finding Causes

1. Using your completed chart from Reading Social Studies, p. 300, list the events that led to the growth of cities during the Industrial Revolution.

Recognizing Effects

2. What were the effects of the Crusades on life in Western Europe?

Analyzing Causes

3. In 19th-century Russia, the lives of poor citizens were very different from those of wealthy citizens. How do you think this division led to political revolution?

Visual Summary

1 Renaissance Connections

- European society was transformed by the art, literature, and ideas of the Renaissance.
- The accomplishments of this period are an important part of Western culture.

2 Traders, Explorers, and Colonists

- People on both sides of the Atlantic were changed by the voyages of the European explorers.
- European exploration led to colonization and to the slave trade.

3 The Age of Revolution

- The Age of Revolution resulted in great changes in European society, industry, and politics.
- These changes were felt around the world.

4 The Russian Empire

- Many citizens of the Russian Empire were deprived of their rights.
- Today, Russia is a large nation experiencing great change.

CRITICAL THINKING: Possible Responses

1. Finding Causes
Early factories were built in the countryside near rivers that were used for power. During the Industrial Revolution, steam engines powered machines, so factories could be built in cities. People moved to cities in search of work.

2. Recognizing Effects
The Crusaders opened up trade routes that connected Western Europe with Asia and North Africa. This increased trade brought new ideas and products, and even new people, to Western Europe.

3. Analyzing Causes
The peasants were poor, and the wealth of the country belonged to the nobles. The nobles had opportunities for education and thus wielded power, while the peasants had no hope of bettering themselves. This inequality resulted in violence and the overthrow of the government.

STANDARDS-BASED ASSESSMENT

the map and your knowledge of geography to answer questions 1 and 2.

Additional Test Practice, pp. S1–S33

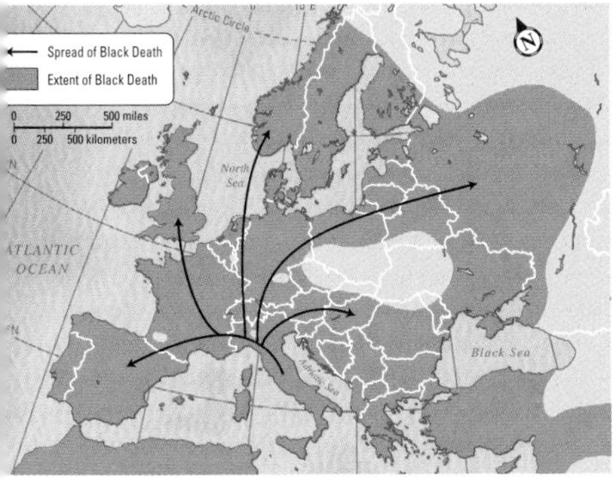

1. Judging from the arrows on the map, how might the plague have spread?
   - **A.** mostly from north to south
   - **B.** through the center of Eastern Europe
   - **C.** along trade routes out of Italy
   - **D.** in a circular pattern

2. According to the map, which body of water did the plague cross as it spread through Europe?
   - **A.** Adriatic Sea
   - **B.** Atlantic Ocean
   - **C.** Black Sea
   - **D.** North Sea

The following passage is from a biography of Leonardo da Vinci. Use the quotation and your knowledge of world cultures and geography to answer question 3.

PRIMARY SOURCE

Although we have called Leonardo a scientist, it is not a title he would have understood. The word "scientist" was not used before 1840. Leonardo did, however, use the word "science." To him it meant knowledge proved true by experience. He contrasted it with "speculation," by which he meant guesswork not proved by experience.

STEWART ROSS, *Leonardo da Vinci*

3. What did Leonardo da Vinci consider the basis of science?
   - **A.** guesses
   - **B.** experience
   - **C.** dedication
   - **D.** speculations

ALTERNATIVE ASSESSMENT

WRITING ABOUT HISTORY

Imagine that you are a film maker. Write a proposal for a documentary film about Leonardo da Vinci and one of his inventions. In your proposal, describe the invention, tell how it works, and explain how da Vinci came up with the idea. Remember that the purpose of your documentary is to inform and entertain the viewer.

- Research the Internet or library books about Leonardo da Vinci to find an interesting invention.
- If possible, include a copy of da Vinci's sketch or a photograph of the invention.

COOPERATIVE LEARNING

With a group of three or four classmates, research the voyage of Ferdinand Magellan, the first explorer to lead an expedition that sailed around the world. Then prepare a presentation for your class. Group members can share the responsibilities of finding out who funded the trip, what hardships the sailors faced, and what eventually happened to Magellan. One member can create a map of Magellan's route.

INTEGRATED TECHNOLOGY

Doing Internet Research

The Scientific Revolution changed the way people viewed the world. Use the Internet to research one discovery or invention of a great scientist of the time, such as Galileo Galilei.

- Using the Internet, as well as other library resources, find out about the discovery or invention. You might find information in a biography of the scientist.
- Another source of information might be a science museum.
- Create a poster to explain your findings. Include a diagram or other illustration of the scientific discovery. Explain how the discovery or invention has affected life in modern times.

For Internet links to support this activity, go to

STANDARDS-BASED ASSESSMENT

1. **Answer C** is the correct answer because, according to the map, the plague originated in Italy and fanned outward. Answers A and D are incorrect because the map does not show arrows going north to south or in a circular pattern; answer B is incorrect because the center of Eastern Europe did not suffer the plague.
2. **Answer D** is the correct answer because the map shows an arrow crossing the North Sea; answers A, B, and C are incorrect because no arrows on the map cross those bodies of water.
3. **Answer B** is the correct answer because it focuses on the importance of experience to scientific inquiry; answers A and D are incorrect because da Vinci thought that guesses and speculation were not part of science; answer C is incorrect because the passage does not mention dedication.

INTEGRATED TECHNOLOGY

Brainstorm possible keywords, Web sites, and search engines that students might use to find information about European scientists. To help them focus their research, you may want to write the framework for an outline on the chalkboard and have each student copy and complete it. Then have students write brief reports using the facts in their outlines.

Alternative Assessment

1. Rubric

The documentary film proposal should
- present information about Leonardo da Vinci and his invention in a persuasive way.
- be organized logically.
- cover the topic comprehensively.
- use correct grammar, spelling, and punctuation.

2. Cooperative Learning

In advance, collect any materials that students might need, such as reference books, a globe, or an atlas. You might want to assign students the roles of researchers, illustrators, and presenters. Brainstorm interesting presentations, such as making a videotape, re-creating a ship's log, or writing and performing a play.

PLANNING GUIDE: UNIT 4 ATLAS AND CHAPTER 12

Europe: War and Change

	OVERVIEW	COPYMASTERS	INTEGRATED TECHNOLOGY
UNIT ATLAS AND CHAPTER RESOURCES	Students will examine how the development of nationalism in Europe led to two world wars and a cold war during the 20th century.	**In-depth Resources: Unit 4** • Guided Reading Worksheets, pp. 27–29 • Skillbuilder Practice, p. 32 • Unit Atlas Activities, pp. 1–2 • Geography Workshop, pp. 61–62 **Reading Study Guide** (Spanish and English), pp. 98–105 **Outline Map Activities**	• eEdition Plus Online • EasyPlanner Plus Online • eTest Plus Online • eEdition • Power Presentations • EasyPlanner • Electronic Library of Primary Sources • Test Generator • Reading Study Guide • Critical Thinking Transparencies CT23
	KEY IDEAS		
SECTION 1 European Empires pp. 329–332	• Nationalism, colonialism, and empire building led to conflicts among European nations.	**In-depth Resources: Unit 4** • Guided Reading Worksheet, p. 27 • Reteaching Activity, p. 34 **Reading Study Guide** (Spanish and English), pp. 98–99	classzone.com Reading Study Guide
SECTION 2 Europe at War pp. 333–338	• Countries join alliances to unite for a common cause, and for defense. • World War I brought many changes to the political makeup of Europe. • Hitler's aggression, including the invasion of Poland, led to World War II. • The Allies helped Western European nations rebuild after World War II.	**In-depth Resources: Unit 4** • Guided Reading Worksheet, p. 28 • Reteaching Activity, p. 35 **Reading Study Guide** (Spanish and English), pp. 100–101	**Critical Thinking Transparencies CT24** classzone.com Reading Study Guide
SECTION 3 The Soviet Union pp. 342–347	• The Iron Curtain served as a barrier between Eastern and Western Europe. • Joseph Stalin wielded dictatorial power over the Soviet Union and many Eastern European nations. • NATO and the Warsaw Pact nations would not trade or cooperate with each other during the Cold War.	**In-depth Resources: Unit 4** • Guided Reading Worksheet, p. 29 • Reteaching Activity, p. 36 **Reading Study Guide** (Spanish and English), pp. 102–103	classzone.com Reading Study Guide

KEY TO RESOURCES

- Audio
- CD-ROM
- Copymaster
- Internet
- Overhead Transparency
- Pupil's Edition
- Teacher's Edition
- Video

ASSESSMENT OPTIONS

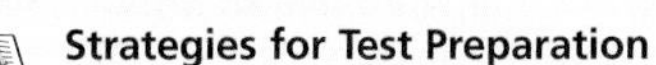

Chapter Assessment, pp. 348–349

Formal Assessment
- Chapter Tests: Forms A, B, C, pp. 176–187

Test Generator

Online Test Practice

Strategies for Test Preparation

Section Assessment, p. 332

Formal Assessment
- Section Quiz, p. 173

Integrated Assessment
- Rubric for making an outline

Test Generator

Test Practice Transparencies TT37

Section Assessment, p. 338

Formal Assessment
- Section Quiz, p. 174

Integrated Assessment
- Rubric for writing a letter

Test Generator

Test Practice Transparencies TT38

Section Assessment, p. 347

Formal Assessment
- Section Quiz, p. 175

Integrated Assessment
- Rubric for writing a scene

Test Generator

Test Practice Transparencies TT39

RESOURCES FOR DIFFERENTIATING INSTRUCTION

Students Acquiring English/ESL

Reading Study Guide (Spanish and English), pp. 98–105

Access for Students Acquiring English Spanish Translations, pp. 95–102

TE Activity
- Base Words and Suffixes, p. 334

Modified Lesson Plans for English Learners

Less Proficient Readers

Reading Study Guide (Spanish and English), pp. 98–105

TE Activity
- Main Ideas and Details, p. 337

Gifted and Talented Students

TE Activities
- Debating, p. 336
- Reviewing a Movie, p. 346

CROSS-CURRICULAR CONNECTIONS

Humanities

Belloli, Andrea. ***Exploring World Art.*** Los Angeles: J. Paul Getty Museum, 1999. Places Western European art in a broad global context.

Moore, Reavis. ***Native Artists of Europe.*** Santa Fe, NM: Publishers Group West, 1994. Singer, painter, wood carver, potter, and musician from Western Europe.

Popular Culture

CultureGrams. ***Volume I, The Americas and Europe: The Nations Around Us.*** Salt Lake City: Millennial Star Network, 2000. Reports on seventy-four countries.

Steele, Philip. ***Houses Through the Ages.*** Mahwah, NJ: Troll Associates, 1994. From the caves of Stone Age hunters to modern apartment houses in Western Europe.

Science

Millard, Anne. ***Street Through Time.*** New York: DK Publishing, 1998. A European street from Stone Age path to 20th-century city street.

History

Burgan, Michael. ***Belgium.*** New York: Children's Press, 2000. Geography, plants, animals, history, economy, language, religions, culture, sports, arts, and people.

Costain, Meredith. ***Welcome to the United Kingdom.*** Broomall, PA: Chelsea House, 2000. Full overview with suggested activities.

ENRICHMENT ACTIVITIES

The following activities are especially suitable for classes following block schedules.

Teacher's Edition, pp. 330, 331, 335, 336, 339, 343, 345, 346
Pupil's Edition, pp. 332, 338, 347

Unit Atlas, pp. 260–269
Literature Connections, pp. 340–341

Outline Map Activities

INTEGRATED TECHNOLOGY

Go to **classzone.com** for lesson support and activities for Chapter 12.

BLOCK SCHEDULE LESSON PLAN OPTIONS: 90-MINUTE PERIOD

DAY 1

CHAPTER PREVIEW, pp. 326–327
Class Time 20 minutes

- **Small Groups** In small groups have students share the photo albums they made for Implementing the National Geography Standards on PE p. 327. Ask them to explain the significance of the images they chose. Exhibit student albums in the classroom.

SECTION 1, pp. 329–332
Class Time 70 minutes

- **Peer Teaching** Assign each student a partner and have the pairs do the Reteaching Activity on TE p. 332. Reconvene as a class and ask students to share their sentences.
Class Time 25 minutes
- **Map Reading** Use the map and the Geography Skillbuilder questions on PE p. 330 to lead students in identifying which Western European countries controlled which colonies.
Class Time 10 minutes
- **Cause and Effect** Use the Skillbuilder Mini-Lesson on TE p. 331 to guide a discussion about the causes and effects of colonialism.
Class Time 15 minutes
- **Internet** Extend students' background knowledge of the physical geography of Europe by visiting **classzone.com.**
Class Time 20 minutes

DAY 2

SECTION 2, pp. 333–338
Class Time 65 minutes

- **Peer Competition** Divide the class into pairs. Assign each pair one of the Terms & Names for this section. Have pairs make up five questions that can be answered with the term or name. Have groups take turns asking the class their questions. Give points for correct answers.
Class Time 20 minutes
- **Brainstorming** Lead the class to consider why a country might choose to remain neutral during a time of war. Record students' suggestions on the chalkboard.
Class Time 10 minutes
- **Interview** Have the class reread the biography of Anne Frank on PE p. 337. Ask students to imagine that they are newspaper reporters who have been assigned to interview Anne Frank while she is in hiding in Amsterdam. Have each student make a list of questions to ask about her life. Conduct the interview as a whole class.
Class Time 35 minutes

SECTION 3, pp. 342–347
Class Time 25 minutes

- **Discussion** Discuss the objectives for the section on TE p. 342.
Class Time 5 minutes
- **Peer Teaching** Have pairs of students review the Main Idea for the section on PE p. 342 and find three details to support it. Then have each pair list two additional important ideas and trade lists with another group to find details.
Class Time 10 minutes
- **Analyzing Issues** Use the Critical Thinking Activity on TE p. 343 to guide students in a discussion about how the Soviet Union became the strongest nation in Europe.
Class Time 10 minutes

DAY 3

SECTION 3, continued
Class Time 35 minutes

- **Political Cartoon** Review with students what they learned about political cartoons on PE p. 339. Have students reread "Joseph Stalin" in their texts. Then ask them to do Activity Options: Multiple Learning Styles on TE p. 344. When they have finished their cartoons, invite volunteers to share them with the rest of the class.

CHAPTER 12 REVIEW AND ASSESSMENT, pp. 348–349
Class Time 55 minutes

- **Review** Have students prepare a summary of the chapter, using the Terms & Names listed on the first page of each section.
Class Time 20 minutes
- **Assessment** Have students complete the Chapter 12 Assessment.
Class Time 35 minutes

TECHNOLOGY IN THE CLASSROOM

USING AND CREATING SOUNDS ON THE COMPUTER

Although we frequently think of computers and the Internet as visual media, they can also be very useful in storing and transmitting sounds. Students may already be accustomed to listening to music on the computer, but they may not have used computers to listen to other people speak or to record their own voices. The recent versions of most Internet browsers support audio, and most computers allow the user to record sounds, provided they have a microphone.

ACTIVITY OUTLINE

In this activity, students will listen to sound files in Real Audio format that describe the role of radio in England during World War II. Although recent versions of their Internet browsers should allow them to play the files automatically, it would be a good idea to check this feature before having students start this activity. Students will also need to use a microphone attached to the computer to create their own radio broadcasts.

Objective Students will listen to an on-line audio file and create their own audio files that will be organized into a multimedia presentation.

Task Have students listen to a description of the role radio played in England during World War II. Then have them create radio broadcasts that might have been heard at that time, and save them on the computer. Have them put their broadcasts together as a class multimedia presentation.

Class Time Three class periods

DIRECTIONS

1. Direct students to the Web site at **classzone.com** and select "Radio." Have them link to "Wartime Radio" and listen to British people discussing the role of radio in World War II. Students will need to listen carefully, and may even need to play the sound file several times. They can use the sliding navigator button in the Real Audio window to "rewind" the sound file so they can listen to all or part of it again.
2. As a class, discuss the role of radio during World War II. Discuss issues such as why radio was important, how radio was used, and why people liked the radio.
3. Divide the class into small groups and ask each group to pretend they lived in England during World War II. Have them prepare short (15–30 second) sound clips that might have been heard on the radio during that period.
4. Have students record their sound clips with a microphone, saving the sound files on the computer. Some students may want to try speaking with a British accent.
5. Have several students compile a page in a multimedia presentation program that will serve as the introductory page for students' sound files.
6. Groups can transfer their sound files to the folder containing the multimedia presentation so they can easily link from the introductory page to the sound files.
7. Have members of each group take turns creating links from the introductory page to their sound file. Play the presentation for the class so students can listen to other groups' broadcasts.

CHAPTER 12 OBJECTIVE

Students will examine how the development of nationalism in Europe led to two world wars and a cold war during the 20th century.

FOCUS ON VISUALS

Interpreting the Photograph Have students look at the nighttime photograph of Berlin and notice details that tell them something about the city today. Draw their attention to the contrasts between the ruined church, now a monument, and the modern skyscrapers. Ask them to point out specific features in the photograph that demonstrate Germany's recovery and prosperity.

Possible Responses brightly lit buildings, signs, and cafes, skyscraper office buildings, people on the sidewalk, busy traffic, construction scaffolding

Extension Have students write and illustrate a short guide to the sights of Berlin, using information from encyclopedias, tourist guides, and the Internet.

CRITICAL THINKING ACTIVITY

Hypothesizing Encourage a discussion on why a city might have been bombed even though it was not strictly a military target. Point out that such attacks are still the subject of debate many years after the war. Direct students' attention to the map of Europe and its major cities. Have students consider the long-term effects of such destruction.

Class Time 15 minutes

CHAPTER 12

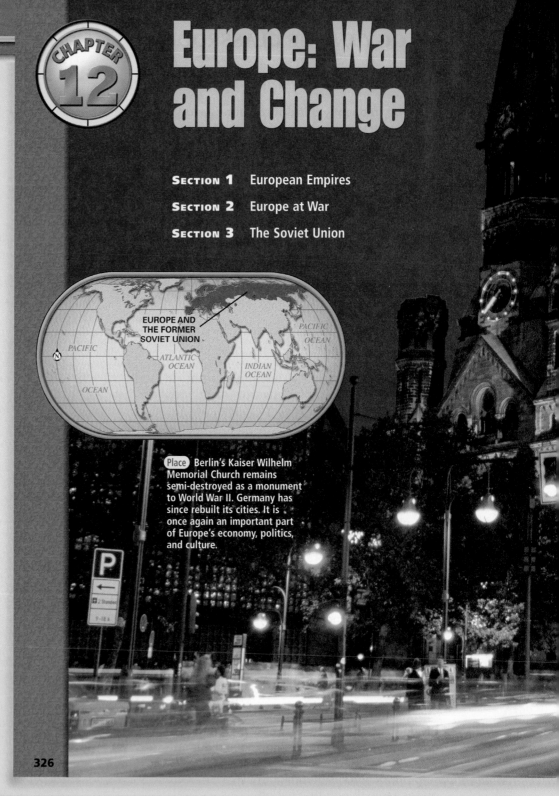

Europe: War and Change

EUROPE AND THE FORMER SOVIET UNION

PACIFIC OCEAN

ATLANTIC OCEAN

INDIAN OCEAN

PACIFIC OCEAN

Place Berlin's Kaiser Wilhelm Memorial Church remains semi-destroyed as a monument to World War II. Germany has since rebuilt its cities. It is once again an important part of Europe's economy, politics, and culture.

326

Recommended Resources

BOOKS FOR THE TEACHER

Map Packs Continents Series: Europe. Washington, D.C.: National Geographic Society, 1998. Sixty-five map transparencies with guide and suggested projects.

Mazower, Mark. ***Dark Continent: Europe's Twentieth Century.*** New York: Knopf, 1999. From World War I to the European Union.

Williams, Roger. ***Insight Guide Continental Europe.*** Maspeth, NY: Langenscheidt, 2000. History and culture of continental Europe.

VIDEOS

Europe: The Road to Unity. Washington, D.C.: National Geographic Society, 1993. The challenges in uniting citizens of diverse cultures and languages.

SOFTWARE

Great Museums of the World. Renton, WA: CounterTop Software, 1998. Electronic tours of the great European art museums.

INTERNET

For more information about Europe and the EU, visit **classzone.com.**

How has Europe's small landmass affected its history?

Place • The continent of Europe is home to more than 40 countries. Yet, it is approximately the same size as the United States. Since many European nations share borders with several other countries, Europeans often speak three or more languages. Across Europe, approximately 50 languages are spoken.

Europe is densely populated. In fact, the continent has almost three times as many people as the United States. So many people, living so close together, has sometimes led to competition and warfare over land and resources.

What do you think?

- How might the differences among Europeans cause conflict?
- How might the closeness of so many countries help to unite Europe?

FOCUS ON GEOGRAPHY

Objectives

- To help students recognize the diversity of languages and cultures in Europe
- To identify shared borders and densely populated areas as factors in numerous European wars

What Do You Think?

1. Guide students to understand that the desire for independence and autonomy, set against the desire to expand and control, might cause conflict among nations.
2. Students may note that people in bordering nations often speak each other's languages, which might promote an open exchange between them. Geographic closeness makes trade and travel easier. Also, the desire for protection against other countries might contribute to unity.

How has Europe's small landmass affected its history?

Have students consider how the desire to expand and control larger areas might have led to invasions, colonies, conflict, and sometimes war. On the other hand, it is easy to travel from one country to another and share language and cultural traits.

MAKING GEOGRAPHIC CONNECTIONS

Have students compare Europe's small landmass with the larger one of the United States. Ask them to speculate on how and why our history might have been different if instead of 50 states there had been 50 separate countries.

Implementing the National Geography Standards

Standard 6 Compile a series of photographs that show structures that have come to represent or symbolize a particular city

Objective To compile a photo album of famous structures in Europe

Class Time 30 minutes

Task Students use newspapers or travel magazines to create a photo album containing images of buildings, structures, or statues that have come to represent or symbolize a European city. Some examples are the Tower Bridge in London, the Eiffel Tower in Paris, and the Colosseum in Rome. Ask students to write captions describing each location.

Evaluation The photo albums should contain at least five examples of buildings, structures, or statues.

BEFORE YOU READ

What Do You Know?

Ask students to describe movies they have seen or books they have read about Europe. Invite them to describe the topics of these movies or books. Then encourage students who have learned about World War I or World War II from these sources, or from museum or family visits, to share what they know about these events.

What Do You Want to Know?

Ask small groups of students to work together to make up questions about Europe in the 1900s and Europe today. Suggest that students record the questions in their notebooks, leaving space to write answers as they read the chapter.

READ AND TAKE NOTES

Reading Strategy: Analyzing Cause and Effect Explain to students that many historians believe that if people understand the causes of certain events, it may be possible to prevent similar events from occurring again. Point out that when students have completed the causes/event/effects charts, they will be able to see the connections between causes and effects more clearly.

In-depth Resources: Unit 4
- Guided Reading Worksheets, pp. 27–29

READING SOCIAL STUDIES

BEFORE YOU READ

▸▸ What Do You Know?

Do you know that during World War I, armies trained dogs to guard supplies and assist soldiers? What do you know about World War I and World War II? Have you ever seen a movie or read a book about either conflict? What do you hear in the news about current events in Europe? Think about how events in Europe in the past century might have contributed to life there today.

▸▸ What Do You Want to Know?

Decide what you know about Europe's history in the 1900s and what it is like there today. In your notebook, record what you hope to learn from this chapter.

Region • The image of the hammer and sickle became th[e] symbol of the Soviet Union. ▲

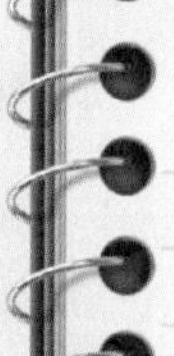

READ AND TAKE NOTES

Reading Strategy: Analyzing Causes and Effects Analyzing causes and effects is an essential skill for understanding what you read in social studies, because events are caused by other events or situations. This sequence is called a chain of events. Understanding which causes lead to which events is essential in understanding history and other areas of social studies. Use the chart below to show causes and effects discussed in Chapter 12.

- Copy the chart into your notebook.
- As you read, record causes and effects for each event.

Region • During World War I, armies trained do[gs to] assist them. ▲

Causes	Event	Effects
domino effect of alliances	World War I	many deaths, poverty, homelessness, unemployment, shift in many European political boundaries
election of Hitler, fascism, Germany invades Poland	World War II	free governments established in Western Europe, NATO, Soviet Union occupies Eastern Europe
defeat of Germany, puppet governments throughout Eastern Europe	Growth of Soviet Union	Cold War, Stalin's five-year plans, collective farming, secret police

Teaching Strategy

Reading the Chapter This is a chronological chapter focusing on the causes and effects of wars that involved much of Europe during the 20th century. Encourage students to note the factors that led nations to war, as well as the events that happened as a result of war.

Integrated Assessment The Chapter Assessment on page 348 describes several activities for integrated assessment. You may wish to have students work on these activities during the course of the chapter and then present them at the end.

European Empires

TERMS & NAMES
- nationalism
- colonialism
- Austria-Hungary
- dual monarchy

MAIN IDEA

The beginning of the 20th century was a time of change in Europe, as feelings of nationalism began to take hold.

WHY IT MATTERS NOW

Feelings of nationalism continue to lead to conflicts that change the map of Europe.

DATELINE

EXTRA

NORWAY, SEPTEMBER 1905

It could have been war in the Scandinavian Peninsula. The armies of Norway and Sweden had begun preparations.

Instead, Sweden ended the crisis peacefully by granting Norway independence. Norway had been under Swedish control since 1814. Although Norway ran its own affairs within the country, Sweden set foreign policy and controlled Norway's international shipping and trade.

Prince Charles of Denmark has been invited to become king of Norway. The Norwegians will vote to approve their new leader. If chosen, he will become King Haakon VII.

The king's role will be largely ceremonial. His chief task will be to help unite the newly independent people of Norway.

Region • Prince Charles of Denmark, pictured here with his family, hopes to become King Haakon VII of Norway.

The Spread of Nationalism ❶

Norway's independence from Sweden was a sign of new ideas that were sweeping across Europe at the time. During the late 19th and early 20th centuries, **nationalism,** or strong pride in one's nation or ethnic group, influenced the feelings of many Europeans. An ethnic group includes people with similar languages and traditions, but who are not necessarily ruled by a common government.

TAKING NOTES

Use your chart to take notes about war and change in Europe.

Causes	Event	Effects
	World War I	
	World War II	

SECTION OBJECTIVES

1. To explain how nationalism and colonialism led to conflicts among European nations in the early 20th century
2. To identify the dual monarchy of Austria-Hungary

SKILLBUILDER
- Interpreting a Map, pp. 330, 332

CRITICAL THINKING
- Recognizing Effects, p. 330
- Clarifying, p. 332

FOCUS & MOTIVATE

WARM-UP

Making Inferences Have students read Dateline and discuss how these two countries resolved a question that could have led to war.

1. Why do you think Swedish officials decided to grant Norway its independence peacefully?
2. What might this decision signify for future relations between the two countries?

INSTRUCT: Objective ❶

The Spread of Nationalism

- What is nationalism? strong pride in one's nation or ethnic group
- Why were European powers interested in gaining colonies in Asia and Africa? They could obtain raw materials cheaply and turn them into manufactured goods.

In-depth Resources: Unit 4
- Guided Reading Worksheet, p. 27

Reading Study Guide
(Spanish and English), pp. 98–99

Program Resources

In-depth Resources: Unit 4
- Guided Reading Worksheet, p. 27
- Reteaching Activity, p. 34

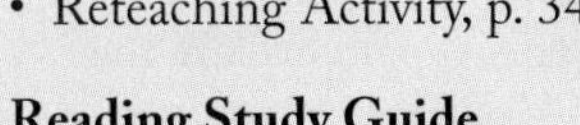

Reading Study Guide
(Spanish and English), pp. 98–99

Formal Assessment
- Section Quiz, p. 173

Integrated Assessment
- Rubric for making an outline

Outline Map Activity

Access for Students Acquiring English
- Guided Reading Worksheet, p. 95

Technology Resources
classzone.com

TEST-TAKING RESOURCES

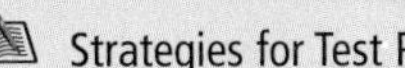

Strategies for Test Preparation

Test Practice Transparencies

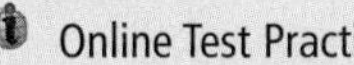

Online Test Practice

CRITICAL THINKING ACTIVITY

Recognizing Effects Ask students to consider the relationship between the freedoms and rights that citizens have and the responsibilities they are willing to take on for their country. Encourage them to discuss how a medieval peasant might have felt when ordered to go to war by his lord, then compare that with the feelings of a citizen who has elected government leaders. If it seems appropriate, have students discuss their own feelings about nationalism, patriotism, and citizens' responsibilities.

Class Time 15 minutes

FOCUS ON VISUALS

Interpreting the Map Ask students to explain the map key by pointing out the colors that represent smaller colonial powers, such as Belgium and Denmark, and then locate those countries' possessions on the map. Have them discuss which countries were the dominant colonial powers in Asia and Africa in 1914, and why this might be so.

Possible Responses Great Britain (U.K.) and France in Asia and Africa; also in Africa, Germany. Those nations were the leading European powers.

Extension Remind students that some European nations also had possessions in South America and the Caribbean. Ask them to make a map showing those colonies. They should use different colors to represent each colonial nation's possessions and include a map key.

Constitutional Monarchies In part, the spread of nationalism was fueled by the fact that more Europeans than ever before could vote. For centuries, many monarchs had unlimited power. In country after country, however, citizens demanded the right to elect lawmakers who would limit their monarch's authority. This kind of government is called a constitutional monarchy. A constitutional monarchy not only has a king or queen, but also a ruling body of elected officials. The United Kingdom is one example of a constitutional monarchy.

By 1900, many countries in Western Europe had become constitutional monarchies. Citizens of these countries strongly supported the governments that they helped to elect. When one country threatened another, most citizens were willing to go to war to defend their homeland.

Reading Social Studies

A. Contrasting How does a constitutional monarchy differ from a democracy?

Reading Social Studies A. Answer A constitutional monarchy has a king or queen with limited power, and may or may not be a democracy.

The Defense of Colonial Empires At the beginning of the 20th century, many Western European countries—including France, Italy, the United Kingdom, Germany, and even tiny Belgium—had colonies in Asia and Africa. Colonies supplied the raw materials that the ruling countries needed to produce goods in their factories back home. Asian and African colonies, sometimes larger than the ruling country, were also important markets for manufactured goods.

Geography Skillbuilder Answers

1. United Kingdom
2. Africa

European Colonial Possessions, 1914

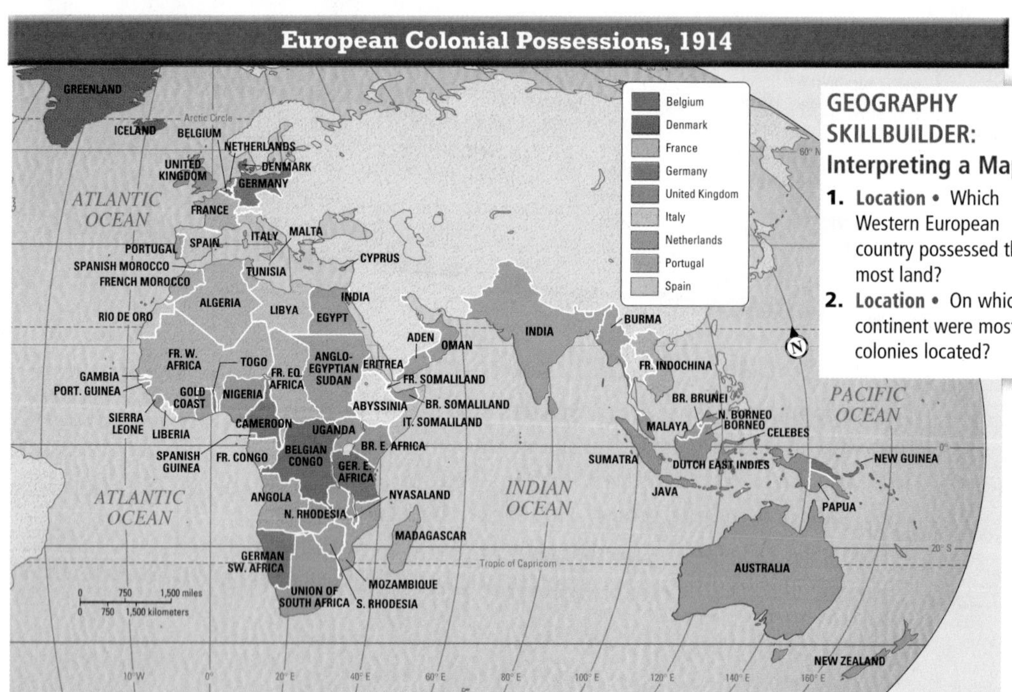

GEOGRAPHY SKILLBUILDER: Interpreting a Map

1. **Location** • Which Western European country possessed the most land?
2. **Location** • On which continent were most colonies located?

Activity Options

Interdisciplinary Link: Music

Class Time One class period

Task Listening to music that expresses nationalistic themes

Purpose To show how nationalism influenced the arts, including music, in the 19th and early 20th centuries

Supplies Needed

- CD or cassette player
- CDs or cassettes of music by composers such as Grieg, Dvorak, Smetana, Bartok, Liszt, Borodin, or Tchaikovsky
- Reference sources, including music encyclopedias

Block Scheduling

Activity Explain that nationalism influenced many European composers to use folksongs and folk themes in their music (for example, Liszt's "Hungarian Rhapsodies"). Have students with an interest or background in classical music find short pieces by some of the composers listed and present them in class. Students should explain the composer's national background and the folk tradition or story that each piece represents.

Location • **In 1914, the United Kingdom could truthfully state that the sun never set on the British Empire.** ◄

During this period of **colonialism,** Western European nations spent much of their wealth on building strong armies and navies. Their military forces helped to defend borders at home as well as colonies in other parts of the world. Colonies were so important that the ruling countries sometimes fought one another for control of them. They also struggled to extend their territories.

Thinking Critically Answers

1. It gave artists from many countries a chance to work together and share traditions and experience.

2. It served as a center for developing and spreading new artistic ideas.

FOCUS ON VISUALS

Interpreting the Illustration Ask students to study the details of this illustration praising the British Empire. Point out that the background represents the Union Jack, the flag of the United Kingdom. Ask them to identify the four territories shown and explain why the artist might have chosen them.

Possible Responses India, Australia, New Zealand, South Africa. They were the largest and most valuable British possessions.

Extension Have students use the map on page 330 to point out British possessions on a globe. Ask them to use the globe to demonstrate that "the sun never set on the British Empire."

The Ballets Russes Begun in Paris, France, in 1909, the Ballets Russes (ba•LAY ROOS) was a dance company under the direction of the Russian producer Sergey Diaghilev (dee•AH•guh•LEHF). It was a critical and commercial success, and it spread artistic ideas.

Talented dancers and choreographers, such as Nijinsky, worked for Diaghilev. Famous composers—including Claude Debussy (duh•BYOO•see) and Igor Stravinsky—wrote music for performances. Pablo Picasso, Marc Chagall, and other great artists designed the sets. The Ballets Russes continued until Diaghilev's death in 1929.

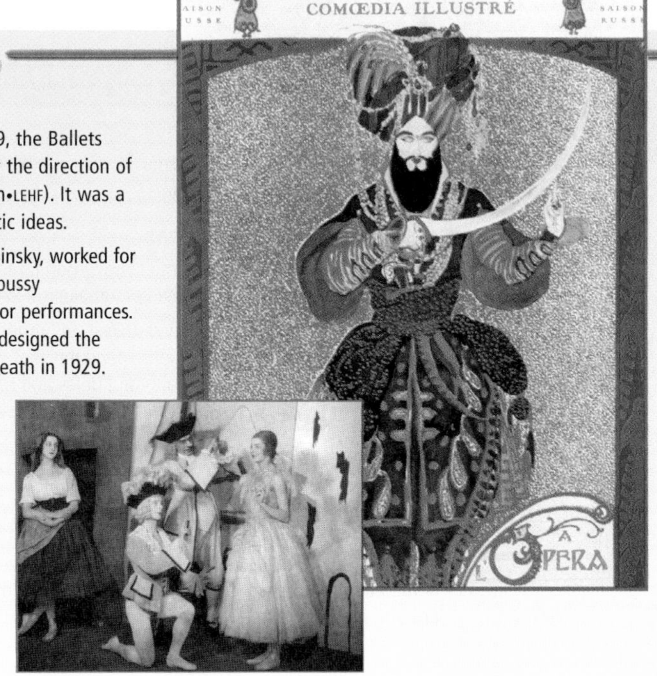

THINKING CRITICALLY

1. **Synthesizing**
   How did the Ballets Russes benefit the European art and theater communities?
2. **Clarifying**
   How was the Ballets Russes more than a collection of dancers, musicians, and artists?

For more on the Ballets Russes, go to

Spotlight on CULTURE

The first choreographer of the Ballets Russes was Michel Fokine. Formerly with the St. Petersburg Ballet company, Fokine believed that a dancer could express a character's emotion and tell a story through his or her movements. Male dancers, such as Vaslav Nijinsky, were important soloists in Fokine's ballets. Among Fokine's well-known ballets are *Prince Igor* (1909), *The Firebird* (1910), and *Petrushka* (1911). He also created the famous solo *The Dying Swan* (1905) for the ballerina Anna Pavlova. Fokine left Russia in 1918 and settled in New York City.

Activity Options

Skillbuilder Mini-Lesson: Identifying Causes and Effects

Block Scheduling

Explaining the Skill Discuss with students the importance of understanding what caused certain events in history. Explain that it is helpful when studying the history of Europe to understand how certain events led to other, often undesirable, events.

Applying the Skill Have students reread "The Defense of Colonial Empires," pages 330–331. Ask them to identify the causes and effects of colonialism. Causes: Colonies provided raw materials for production and were markets for manufactured goods. Effects: European nations built up armies and navies and fought one another for control of colonies.

FOCUS ON VISUALS

Interpreting the Map Have students notice the many countries crowded in this part of Europe. How might this cause problems?

Possible Response Countries would want more land for people and agriculture.

Extension Have students compare this map with a current atlas and list present-day countries once in Austria-Hungary.

CRITICAL THINKING ACTIVITY

Clarifying Ask students to consider why the countries of Eastern Europe were dependent on the countries of Western and Northern Europe. Guide them to recognize that these agricultural countries needed the manufactured goods produced in Western and Northern Europe.

Class Time 10 minutes

ASSESS & RETEACH

Reading Social Studies Have students create a cause-and-effect chart like the one on page 328 to show the causes and effects of nationalism and colonialism in Europe.

Formal Assessment
- Section Quiz, p. 173

RETEACHING ACTIVITY

Have students work in pairs to review the section and then write three sentences about the spread of nationalism in Europe.

In-depth Resources: Unit 4
- Reteaching Activity, p. 34

Access for Students Acquiring English
- Reteaching Activity, p. 100

Austria-Hungary, 1900

UNITED KINGDOM, NETH., Berlin, Amsterdam, GERMANY, London, BELGIUM, Brussels, LUX., Paris, FRANCE, SWITZ., Vienna, AUSTRIA-HUNGARY, RUSSIA, ROMANIA, Bucharest, Belgrade, BOSNIA, SERBIA, BULGARIA, Sofia, ITALY, Adriatic Sea, MONTENEGRO, Constantinople, OTTOMAN EMPIRE (TURKEY), Black Sea, SPAIN, Madrid, Rome, GREECE; scale 0–250–500 miles, 0–250–500 kilometers

GEOGRAPHY SKILLBUILDER: Interpreting a Map
1. **Location** • Name three countries that bordered Austria-Hungary.
2. **Region** • What was the capital of Austria-Hungary?

Geography Skillbuilder Answers
1. (any three) Russia, Germany, Switzerland, Italy, Serbia, Romania
2. Vienna

Austria-Hungary By the end of the 19th century, most nations of Western and Northern Europe had become industrialized. The majority of Eastern Europe, including Russia, remained agricultural. These Eastern European countries imported most of their manufactured goods from Western and Northern Europe.

The largest empire in Eastern Europe in 1900 was **Austria-Hungary.** The empire was a **dual monarchy**, in which one ruler governs two nations. As you can see in the map above, Austria-Hungary also included parts of many other present-day countries, including Romania, the Czech Republic, and portions of Poland.

Reading Social Studies
B. Making Inferences Why do you think governing a dual monarchy was difficult?

Reading Social Studies B. Possible Answers
Two nations would have different needs; they would include conflicting ethnic groups.

SECTION 1 ASSESSMENT

Terms & Names
1. **Explain the significance of:** (a) nationalism (b) colonialism (c) Austria-Hungary (d) dual monarchy

Using Graphics
2. Look at the map on page 330 that shows European colonial territories. Use a chart like the one below to list the major colonial powers and their colonies.

Nation	Locations of Colonies

Main Ideas
3. (a) Identify one reason for the spread of nationalism in Europe.
(b) Why did Western European nations spend much of their wealth on armies and navies?
(c) How did the nations of Eastern Europe differ from those of Western and Northern Europe at the end of the 19th century?

Critical Thinking
4. **Drawing Conclusions** Why were their colonies so important to European nations?

Think About
- land and people
- competition among nations
- the production and sale of goods

ACTIVITY -OPTION- Reread the information about the Ballets Russes. Write an **outline** of a story or book that might be a good choice for a ballet. Explain your choice.

Section 1 Assessment

1. Terms & Names
a. nationalism, p. 329
b. colonialism, p. 331
c. Austria-Hungary, p. 332
d. dual monarchy, p. 332

2. Using Graphics

Nation	Locations of Colonies
United Kingdom	Africa, India, Australia, New Zealand
France	West Africa, French Indochina
Germany	Africa
Belgium	Belgian Congo
Netherlands	Dutch East Indies

3. Main Ideas
a. More Europeans could vote than in the past.
b. They wanted to defend their borders at home, as well as their colonies.
c. Western and Northern European nations were industrialized; most Eastern European nations remained agricultural.

4. Critical Thinking
Colonies provided raw materials and a market for manufactured goods; colonies helped make nations wealthier and stronger.

ACTIVITY OPTION

Integrated Assessment
- Rubric for making an outline

Europe at War

TERMS & NAMES
World War I
alliance
Adolf Hitler
fascism
Holocaust
World War II
NATO

MAIN IDEA

During the first half of the 20th century, European countries fought each other over land, wealth, and ideals.

WHY IT MATTERS NOW

The changes brought about by the two world wars continue to affect Europe today.

DATELINE

SARAJEVO, BOSNIA-HERZEGOVINA, JUNE 28, 1914—Today, Archduke of Austria-Hungary Franz Ferdinand and his wife, Duchess Sophie, were murdered as they drove through Sarajevo. A nineteen-year-old Serb, Gavrilo Princip, jumped on the Archduke's automobile and fired two shots. The first killed the Duchess. The second killed the Archduke, who was next in line to be emperor of Austria-Hungary.

The Serbians have protested against Austria-Hungary since 1908, when the empire took over Bosnia and Herzegovina (BAHZ•nee•uh HEHRT•suh•GOH•VEE•nuh). Princip has been arrested.

Region • Archduke Franz Ferdinand and his wife, Duchess Sophie, were fatally shot in Sarajevo. ▲

The World at War 1

Because of the murder of Archduke Franz Ferdinand in 1914, the emperor of Austria-Hungary declared war on Serbia. When Russia sent troops to defend Serbia, Germany declared war on Russia. Russia supported Serbia because both Russians and Serbians share a similar ethnic background—they are both Slavic peoples. This was the beginning of **World War I.**

TAKING NOTES
Use your chart to take notes about war and change in Europe.

Causes	Event	Effects
	World War I	
	World War II	

SECTION OBJECTIVES

1. To identify the issues that led to World War I
2. To describe Europe after World War I
3. To identify Germany's actions that led to World War II
4. To describe Europe after World War II

SKILLBUILDER
- Interpreting a Political Cartoon, p. 334
- Interpreting a Map, p. 335

FOCUS & MOTIVATE

WARM-UP

Making Inferences Have students read Dateline and think about causes and effects of the assassination.

1. What part might nationalism have played in inspiring the murder?
2. How do you think the nations of Europe reacted to this event?

INSTRUCT: Objective 1

The World at War

- Why did Germany declare war on Russia? to support Austria after Russia sent help to Serbia
- Why would a country join an alliance? to unite for a common cause; to receive support if attacked

In-depth Resources: Unit 4
- Guided Reading Worksheet, p. 28

Reading Study Guide (Spanish and English), pp. 100–101

Program Resources

In-depth Resources: Unit 4
- Guided Reading Worksheet, p. 28
- Reteaching Activity, p. 35

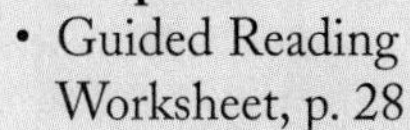

Reading Study Guide (Spanish and English), pp. 100–101

Formal Assessment
- Section Quiz, p. 174

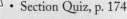

Integrated Assessment
- Rubric for writing a letter

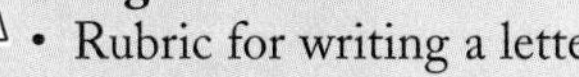

Outline Map Activities

Access for Students Acquiring English
- Guided Reading Worksheet, p. 96

Technology Resources classzone.com

TEST-TAKING RESOURCES
- Strategies for Test Preparation
- Test Practice Transparencies
- Online Test Practice

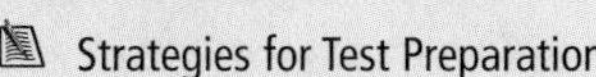

FOCUS ON VISUALS

Interpreting the Political Cartoon Ask students to give their first impressions of what this cartoon represents. Have them read the inscription on the pedestal and notice the weapons the figure is carrying or wearing. Then have them present their interpretations.

Possible Responses The figure looks not quite human but is probably a soldier. He is carrying a weapon with a bayonet and wearing a gas mask. Those new weapons are the kind of progress that the war represents.

Extension Ask students to sketch a cartoon that represents their opinion or attitude about a current political event or issue.

MORE ABOUT...

World War I Weaponry

Trench warfare was a terrifying way to fight, made worse by new weapons. Machine guns made it deadly to try to cross no man's land, the area between opposing lines of trenches. Other new weapons included long-range guns and poison gas. Later in the war, daring pilots took to the skies in small planes to take part in aerial dogfights.

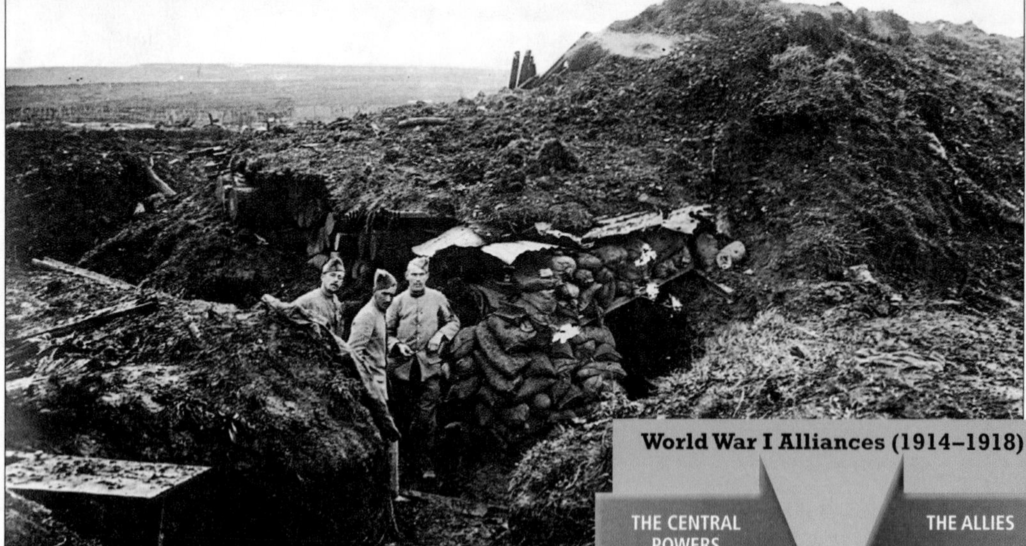

Place • World War I was primarily fought in trenches, which were dug by the armies for better defense. ▲

World War I Alliances (1914–1918)

THE CENTRAL POWERS	THE ALLIES
Austria-Hungary	Russia (dropped out in 1917)
Germany	France
Turkey (Ottoman Empire)	United Kingdom
Bulgaria	Italy (joined 1915)
	United States (joined 1917)

World War I Alliances European rulers wanted other leaders to think twice before declaring war on their countries. To help defend themselves, several countries joined alliances (uh•LY•uhn•sez). An **alliance** is an agreement among people or nations to unite for a common cause. Each member of an alliance agrees to help the other members in case one of them is attacked.

When Germany joined the war to support Austria-Hungary, France came in on the side of Russia. Germany then invaded Belgium, which was neutral, to attack France. Because Great Britain had promised to protect Belgium, it, too, declared war on Germany. After German submarines sank four American merchant ships, the United States joined the side of Russia, France, and Great Britain.

The chart above shows the major powers on both sides of World War I. Italy had originally been allied with Germany and Austria-Hungary but joined the Allies after the war began. Russia dropped out of the war completely after the revolution in that country in 1917.

Reading Social Studies

A. Recognizing Important Details Why did Great Britain enter World War I?

Reading Social Studies
A. Answer
because Germany had invaded Belgium, an ally of Britain

SKILLBUILDER: Interpreting a Political Cartoon

1. What does the artist mean by naming the figure "Progress"?
2. Why is the man wearing a gas mask?

Skillbuilder Answers

1. The artist is using "Progress" sarcastically, referring to the soldier's new weapons.
2. Poison gas was one of the weapons introduced in World War I.

Activity Options

Differentiating Instruction: Students Acquiring English/ESL

Base words and suffixes Write the words *ally* and *alliance* on the chalkboard. Explain that the word *alliance* was formed by adding the suffix *-ance* to the base word *ally*. Explain that an ally is "a person, group, or nation united with another." Review that an alliance is "an agreement to unite for a common cause." Have students practice using the words in sentences. You might want to provide models, such as "Great Britain was an ally of Belgium. Great Britain and Belgium formed an alliance."

ALLY

ALLIANCE = ALLY + ANCE

World War I was costly in terms of human life. When it was over, nearly 22 million civilians and soldiers on both sides were dead. The Allies had won, and Europe had been devastated.

War Dogs During World War I, dogs were trained to guard ammunition, to detect mines, and to carry messages. Dogs even helped to search for the wounded.

War dogs saved many lives. They were especially helpful in forested areas and at night. These dogs are wearing protective masks to keep them safe from poison gas attacks.

Europe After World War I ❷

More people were killed during World War I than during all the wars of the 19th century combined. Afterward, people in many countries on both sides of the costly war—and even those not directly involved—were poor, homeless, and without work.

The Allies blamed Germany for much of the killing and damage during the war. In 1919, Germany and the Allies signed the Treaty of Versailles (vuhr•SY).

aphy
ilder
rs
atic Sea
three)
, Romania,
, Hungary,
ny

Europe After World War I

GRAPHY
BUILDER:
preting a Map
cation • What body water does the ast of Yugoslavia ach?
cation • Name ee countries that rder Czechoslovakia.

INSTRUCT: Objective ❷

Europe After World War I

- What did the Treaty of Versailles require of Germany? to pay for damage done to Allied countries; to give up valuable territory
- How did the division of Austria-Hungary affect some ethnic groups in Eastern Europe? They became independent nations.

FOCUS ON VISUALS

Interpreting the Map Remind students that World War I brought the breakup of Austria-Hungary and the formation of several new nations. Have them look back at the map on page 332, then list the new nations that appear on this map of Europe.

Possible Responses Czechoslovakia, Yugoslavia, Poland, Albania. Austria and Hungary were separated. Romania and Italy gained territory.

Extension Have students draw a map of Yugoslavia after World War I, showing and labeling the smaller nations that were combined to create the new country.

Strange but TRUE

Animals other than dogs supported the war effort. The U.S. Army Signal Corps in World War I used more than 600 carrier pigeons in France. One of the bravest was called Cher Ami (SHER ah•MEE), meaning "dear friend." Cher Ami delivered 12 important messages in France. Even after being fatally wounded by enemy fire, he made it back to his own lines.

Activity Options

Interdisciplinary Link: Art

Class Time One class period

Task Designing a medal for heroic dogs

Purpose To combine form and function in the creation of a military-style decoration

Supplies Needed
- Drawing paper
- Pencils and markers
- Reference sources, including an encyclopedia

Block Scheduling

Activity Explain that the carrier pigeon Cher Ami earned a major French medal for service. Find and display illustrations of medals. Then ask students to design a medal to honor the service of a war dog that saved human lives. Encourage students to find examples of other designs on medals that they might incorporate into their own. Provide a bulletin board or other area for students to display their work.

INSTRUCT: Objective 3

World War II

- Why did the majority of Germans support Hitler? They thought he would help Germany recover from World War I.
- What ideas do fascists support? strong central government, military dictatorship, racism, extreme nationalism
- What action by Germany brought about World War II? its invasion of Poland
- Which countries remained neutral? Denmark, Norway, Sweden, Switzerland

MORE ABOUT...

The Holocaust

Today, the German government continues to try to repair the damages and loss that so many families suffered. Throughout Europe, many sites connected to the Holocaust—including some of the former concentration camps—are permanent memorials to remind future generations of what happened.

Memorials to the Holocaust have also been built in many cities in the United States as well. The United States Holocaust Memorial Museum is located in Washington, D.C. The museum contains a three-floor exhibition that presents the history of the Holocaust through artifacts, photographs, films, and eyewitness reports.

The Treaty of Versailles demanded that Germany be punished by being forced to pay for the damage done to the Allied countries. Germany was also made to give up valuable territory.

A New Map of Europe Additional treaties during the following year also altered the political boundaries of many European countries. As the map on page 335 shows, Austria-Hungary was divided as a result of the war, becoming two separate countries. This allowed several Eastern European ethnic groups that had been part of Austria-Hungary to gain their independence.

Reading Social Studies
B. Answers
economic problems, resentment at paying reparations

World War II 3

By the 1930s, Germany was still paying for the damage done to the Allied countries during World War I. The German economy was in ruins, and the Germans greatly wished to rebuild their own country. In 1933, citizens elected **Adolf Hitler** and the National Socialist, or Nazi, Party. The Nazi Party believed in fascism. **Fascism** (FASH•IHZ•uhm) is a philosophy that supports a strong, central government controlled by the military and led by a powerful dictator. People believed that this new leader would help Germany recover.

World War II Alliances (1939–1945)

THE AXIS POWERS	THE ALLIES
Germany	United Kingdom
Italy	France (until June 1940)
Japan	Soviet Union (formerly "Russia")
	United States (joined in 1941)

Reading Social Studies

B. Finding Causes What conditions led Germans to find hope in Adolf Hitler?

BACKGROUND

Like Germany, Italy was also ruled by a fascist dictator after World War I: Benito Mussolini (1883–1945).

Hitler and the Nazi Party Fascists practiced an extreme form of patriotism and nationalism. Fascists also had racist beliefs.

In the 1930s, Hitler unjustly blamed the Jewish citizens of Germany, among other specific groups, for the country's problems. His Nazi followers seized Jewish property and began to send Jews, along with disabled people, political opponents, and others, to concentration camps. During this **Holocaust,** millions of people were deliberately killed, and others starved or died from disease.

In 1934, Hitler took command of the armed forces. Then, in 1939, Hitler's army invaded Poland. **World War II** had begun. By June 1940, Hitler's army had swept through Western Europe, conquering Belgium, the Netherlands, Luxembourg, France, Denmark, and Norway. A year later, Germany invaded the Soviet Union.

Activity Options

Differentiating Instruction: Gifted and Talented

Block Scheduling

Debating Examine the issue of neutrality in times of war. Divide students into groups, one group supporting neutrality and the other opposing it. Suggest that students examine issues such as a country's motives for remaining neutral; what is required from countries choosing neutrality; and/or whether other countries always truly recognize and observe this neutrality. Allow time for students to prepare their arguments.

WWII Alliances The chart on page 336 shows the major powers on both sides of World War II. As in World War I, the United States at first tried to stay out of the conflict but entered the war after Japan bombed U.S. military bases at Pearl Harbor in Hawaii on December 7, 1941.

Europe After World War II 4

World War II turned much of Europe into a battleground. By the end of the war, the United States, France, and the United Kingdom occupied Western Europe. The Soviet Union occupied Eastern Europe, including the eastern part of Germany.

Once peace was established, the western allies helped to set up free governments in Western Europe. In 1949, the countries of Western Europe joined Canada and the United States to form a defense alliance called **NATO** (NAY•toh). The members of this alliance, whose name stands for North Atlantic Treaty Organization, agreed to defend one another if they were attacked by the Soviet Union or any other country. Without a common enemy, political differences quickly separated the Soviet Union from Western Europe and the United States.

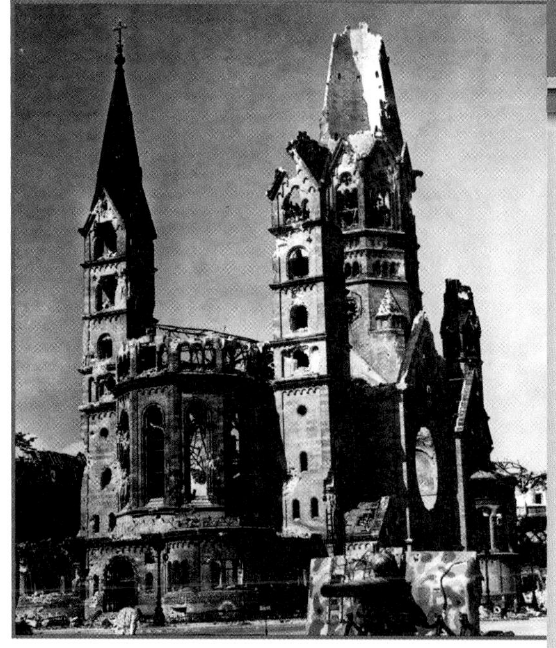

Place • The Kaiser Wilhelm Memorial Church in Berlin was nearly destroyed by Allied bombs. The ruins still stand today as a World War II monument. See pages 326–327. ▲

Biography

Anne Frank In July 1942, during World War II, Anne Frank and her family went into hiding in Amsterdam—a city in the Netherlands. The Frank family were Jewish and were afraid they would be sent to a concentration camp. Anne was only thirteen.

For two years, Anne, her father, mother, sister, and four other people lived in rooms in an attic. Their rooms were sealed off from the rest of the building. While in hiding, Anne kept a diary. Although the family was discovered and Anne died in a concentration camp, her diary was eventually published. Today, this famous book—translated into many languages and the basis for a play and a film—lives on.

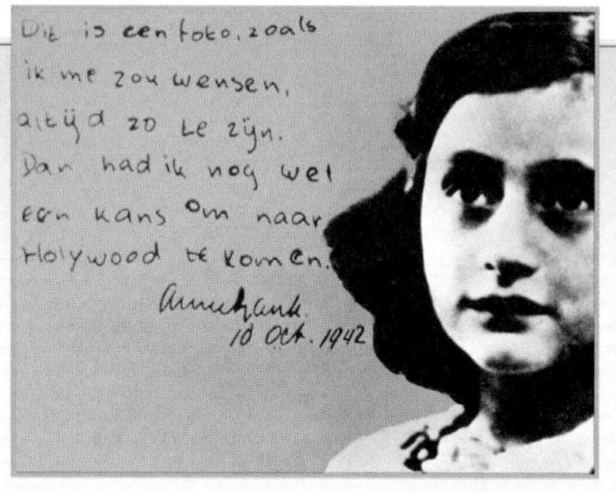

INSTRUCT: Objective 4

Europe After World War II

- How did the Allies help Western European nations after the war? helped them set up free governments
- Why was NATO formed? to provide defense against attack by the Soviet Union or other countries

CRITICAL THINKING ACTIVITY

Recognizing Effects Explain to students that an occupied area is one in which a victorious army keeps its troops. For example, the United States and the United Kingdom occupied Western European countries after World War II. Have students brainstorm the positive and negative effects such a situation might have on the occupied country. Suggest that they consider freedom of movement within the country, economic factors, and national pride.

Class Time 15 minutes

Biography

Approximately 500,000 visitors each year line up at the entrance to the Anne Frank House, located in the center of Amsterdam. After watching an introductory video, visitors enter the "Secret Annex" through a revolving bookshelf that hides the entrance. There they can see the rooms occupied by Anne's family and the Van Daans for two years. Anne's original diary and a model of the annex during the occupation are on display in the house. The house serves as a reminder that nearly all of the Jewish people of Amsterdam were killed during World War II, and shows what life was like for those in hiding.

Activity Options

Differentiating Instruction: Less Proficient Readers

Class Time 30 minutes

Task Highlighting main ideas and details in text

Purpose To understand the relationship of main ideas and details

Supplies Needed
- Photocopies of pp. 336–337
- Colored highlighters

Activity Reread aloud the text under "World War II" and "Europe After World War II." Stop at the end of each subsection and work with students to identify the main idea. Tell them to highlight the main idea. Then work together to identify the important details that support the main ideas and highlight them in color. Continue the process until each subsection has been highlighted in a different color. Encourage students to use their colored copies as study guides.

FOCUS ON VISUALS

Interpreting the Map Have students compare the map of Europe after World War II with that on page 335 of the Continent after World War I. What major change do they notice?

Possible Response division of Germany into East and West

Extension Have students use the map to name the countries that border East and West Germany.

ASSESS & RETEACH

Reading Social Studies Have students fill in the first two rows of the chart on page 328.

Formal Assessment
- Section Quiz, p. 174

RETEACHING ACTIVITY

Divide students into four groups. Ask each group to summarize the information under one of the headings in the section. When they have finished, have groups share their summaries with the other groups.

In-depth Resources: Unit 4
- Reteaching Activity, p. 35

Access for Students Acquiring English
- Reteaching Activity, p. 101

Europe After World War II

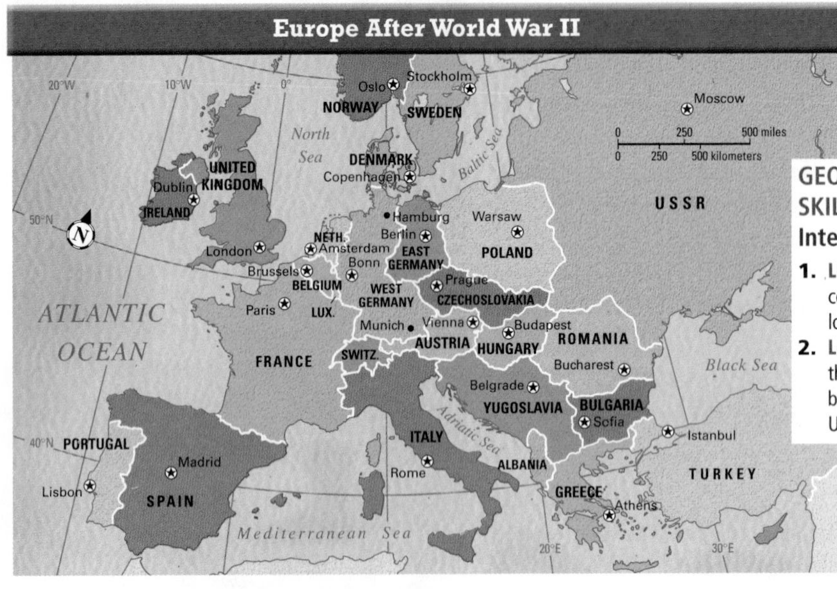

GEOGRAPHY SKILLBUILDER: Interpreting a Map
1. **Location** • In what country is Berlin located?
2. **Location** • Name three countries that border the Soviet Union.

Geography Skillbuilder Answers
1. East Germany
2. Poland, Czechoslovakia, Hungary, Romania

The Marshall Plan United States Secretary of State George C. Marshall created the Economic Cooperation Act of 1948, also known as the Marshall Plan. This plan provided U.S. aid—agricultural, industrial, and financial—to countries of Western Europe. The Marshall Plan greatly benefited war-torn Europe. It may also have prevented economic depression or political instability.

SECTION 2 ASSESSMENT

Terms & Names

1. **Explain the significance of:** (a) World War I (b) alliance (c) Adolf Hitler (d) fascism (e) Holocaust (f) World War II (g) NATO

Using Graphics

2. Use a Venn diagram like this one to compare the countries that were involved in World War I and World War II.

Involved in WWI | Involved in Both | Involved in WWII

Main Ideas

3. (a) What event set off World War I?
   (b) When did World War II begin and end? Which countries won?
   (c) What happened at the end of World War II?

Critical Thinking

4. **Making Inferences** How did World War I change Europe?

Think About
- the destruction and many deaths
- the Treaty of Versailles
- Austria-Hungary

ACTIVITY -OPTION- Look at the photographs in this section. Write a **letter** in which you describe what it might have been like to visit Europe just after World War I or World War II.

Section 2 Assessment

1. Terms & Names
- a. World War I, p. 333
- b. alliance, p. 334
- c. Adolf Hitler, p. 336
- d. fascism, p. 336
- e. Holocaust, p. 336
- f. World War II, p. 336
- g. NATO, p. 337

2. Using Graphics

Involved in WWI:	Involved in Both World Wars:	Involved in WWII:
Austria-Hungary, Turkey, Bulgaria	Germany, Russia, France, United Kingdom, Italy, United States	Japan

3. Main Ideas
- a. The assassination of the archduke and duchess of Austria set off WWI.
- b. began: 1939; ended: 1945. The Allies [Great Britain (U.K.), United States, France, Soviet Union] won.
- c. Western Allies occupied Western Europe; Soviet Union took control of Eastern Europe; Western powers formed NATO; Marshall Plan helped European economies rebuild.

4. Critical Thinking

Austria-Hungary was broken up; Eastern European ethnic groups became independent; Germany gave up territory.

ACTIVITY OPTION

Integrated Assessment
- Rubric for writing a letter

SKILLBUILDER

Reading a Political Cartoon

▶▶ Defining the Skill

Political cartoons—also known as editorial cartoons—express an opinion about a serious subject. A political cartoonist uses symbols, familiar objects, and people to make his or her point quickly and visually. Sometimes the caption and words in the cartoon help to clarify the meaning. Although a cartoonist may use humor to make a point, political cartoons are not always funny.

▶▶ Applying the Skill

This political cartoon was created in the period between World War I and World War II. However, Europeans were already concerned about developments in Germany.

How to Read a Political Cartoon

Strategy 1 Read the cartoon's title and any other words. For example, some cartoons have labels, captions, and thought balloons. Then study the cartoon as a whole.

Strategy 2 If the cartoon has people in it, are they famous? Sometimes the cartoonist wants to comment on a famous person, such as a world leader. Look for symbols or details in the cartoon. For example, in this cartoon a German soldier is climbing out of the Versailles Treaty. Think about the relationships between the words and the images.

Strategy 3 Summarize the cartoonist's message. What is the cartoonist's point of view about the subject? What does this cartoonist think was the cause of Hitler's rise to power?

Make a Chart

A chart can help you to analyze the information in a political cartoon. Once you understand the cartoon's elements, you can summarize its meaning. Use a chart such as this one to help you organize the information.

Important Words	Hitler Party; Versailles Treaty
Important Symbols/Images	German soldier with "Hitler Party" on his helmet crawling out of the Versailles Treaty that officially ended World War I.
Summary 3	The terms of the Versailles Treaty led to the rise of Hitler's party in Germany; Hitler's party, symbolized by a soldier, is war-like and threatens Europe.

▶▶ Practicing the Skill

Study the political cartoon in Chapter 12, Section 2, on page 334. Make a chart similar to the one above in which you list the important parts of the cartoon and write a summary of the cartoon's message.

SKILLBUILDER

Reading a Political Cartoon

Defining the Skill

Display several political cartoons or have students turn to the example in their text on page 334. Ask students to look for common features, such as the use of recognizable people, objects or symbols, and captions. Point out that political cartoons assume a certain knowledge about important events on the part of readers. Explain that they are also called editorial cartoons, meaning they express a point of view in the same way an editorial does.

Applying the Skill

How to Read a Political Cartoon Point out the strategies for reading a political cartoon and guide students in working through each one. Suggest that students note any significant words and identify any people or symbols they recognize. Remind students to look at the cartoon (text and drawing) as a whole.

Make a Chart

Guide students in completing a chart. Ask questions such as, What important words do you see? What does the appearance of the German soldier suggest? What might the cartoonist's purpose have been in drawing this cartoon?

Practicing the Skill

Have students work in pairs to complete a chart for the cartoon on page 334. If students need more practice with this skill, work with them to analyze a political cartoon from a current newspaper or newsmagazine.

In-depth Resources: Unit 4
- Skillbuilder Practice, p. 32

Activity Options

Career Connection: Political Cartoonist

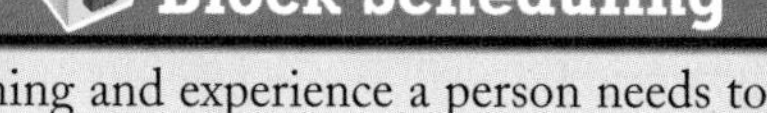

Block Scheduling

Encourage students who enjoy reading political cartoons to find out about people who create them for a living. Tell students that most political cartoonists work for a single newspaper or magazine.

1. Suggest that students look for interviews with actual political cartoonists and articles about how they rose to their positions.
2. Help students learn what training and experience a person needs to become a political cartoonist.
3. Invite students to summarize in cartoon form what they have learned and to share their cartoon with the class.

LITERATURE CONNECTIONS

OBJECTIVE

Students analyze an Irish folk tale in which Bláithín, the wife of folk hero Fíonn Mac Cumhail, saves her husband from a Scottish giant.

FOCUS & MOTIVATE

Drawing Conclusions To help students predict elements of this story, have them study the illustrations on pages 340–341 and answer the following questions:

1. What can you tell about the characters in this tale from the illustrations?
2. What might be special about the round cakes that are pictured?

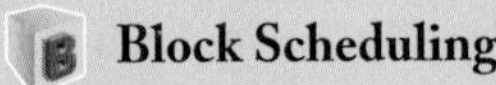

MORE ABOUT...

"Fíonn Mac Cumhail and the Giant's Causeway"

Celtic culture arose in Europe around 700 B.C. and flourished throughout Europe until the Romans conquered much of the region between 300 B.C. and A.D. 100. The only place where Celtic culture was preserved was Ireland, Scotland, Wales, and parts of England and France. The early Celts did not have a written language. However, during the Middle Ages the Celts adopted the Roman alphabet and became prolific writers. Many of their tales and legends have survived.

Fionn Mac Cumhail and the Giant's Causeway[1]

FIONN MAC CUMHAIL, more commonly known as Finn MacCool, is a familiar figure in Irish folk tales. He first appears in the ancient Celtic tales known as the Fenian cycle. In the following story, retold by Una Leavy, Fionn is portrayed as a clever giant, hard at work with the Fianna, his band of Irish warriors. They begin to build a bridge from Ireland to Scotland, because, as the boastful Fionn says, "There are giants over there that I'm longing to conquer." Plans suddenly change, however, and Fionn must go home.

Fionn Mac Cumhail and the Fianna worked quickly on the bridge, splitting stones into splendid pillars and columns. Further and further they stretched out into the ocean. From time to time, there came a distant rumble. "Is it thunder?" asked the Fianna, but they went on working. Then one of their spies came ashore. "I've just been to Scotland!" he said. "There's a huge giant there called Fathach Mór. He's doing long jumps—you can hear the thumping. He has a magic little finger with the strength of ten men! He's in training for the long jump to Antrim."[2]

Fionn's face paled. "The strength of ten men!" he thought. "I'll never fight him. He'll squash me into a pancake." But he could not admit that he was nervous, so he said to the Fianna, "I've just had a message from Bláithín, my wife. I must go home at once—you can all take a holiday."

He set off by himself and never did a man travel faster. Bláithín was surprised to see him. "And is the great causeway finished already?" she asked.

"No indeed," replied Fionn.

"What's the matter?" Bláithín asked. So Fionn told her.

"What will I do, Bláithín?" he asked. "There's the strength of ten men in his magic little finger. He'll squish me into a jelly!"

Bláithín laughed. "Just leave him to me. Stoke up the fire and fetch me the sack of flour. Then go outside and find nine flat stones." Fionn did as he was told. Bláithín worked all night making ten oatcakes. In each she put a large flat stone, all except the last. This one she marked with her thumbprint. "Go and cut down some wood," she said. "You must make an enormous cradle."

1. The Giant's Causeway, which takes its name from this legend, is a striking natural rock formation on the coast of Ireland.

2. The region of Ireland where Fionn and the Fianna are building their bridge to Scotland. It is the location of the actual Giant's Causeway.

Activity Options

Differentiating Instruction: Less Proficient Readers

Building Language Skills Have students work in small groups, each including less proficient readers and more fluent readers. Duplicate copies of the Literature Connection and give each student a copy of the folk tale. Assign character parts to the more proficient readers and sections of narrative text to less proficient readers. Suggest that all students highlight their speaking parts with a marker. Allow time for students to practice their parts and then have groups present the folk tale to the class.

ionn worked all morning. The cradle was just finished when there was mighty rumble and the dishes shook.

"It's him," squealed Fionn.

"Don't worry!" said Bláithín. "Put on this bonnet. Now into the radle and leave me to do the talking."

"Does Fionn Mac Cumhail live here?" boomed a great voice bove her.

"He does," said Bláithín, "though he's away at the moment. He's one to capture the giant, Fathach Mór."

"I'm Fathach Mór!" bellowed the giant. "I've been searching for ionn everywhere."

"Did you ever see Fionn?" she asked. "Sure you're only a baby comared with him. He'll be home shortly and you can see for yourself. But ow that you're here, would you do me a favor? The well has run dry nd Fionn was supposed to lift up the mountain this morning. There's pring water underneath it. Do you think you could get me some?"

"Of course," shouted the giant as he scooped out a hole in the nountain, the size of a crater.

Fionn shook with fear in the cradle and even Bláithín turned pale. 3ut she thanked the giant and invited him in. "Though you and Fionn re enemies, you are still a guest," she said. "Have some fresh bread." nd she put the oatcakes before him. Fathach Mór began to eat. lmost at once he gave a piercing yell and spat out two teeth.

"What kind of bread is this?" he screeched. "I've broken my teeth n it."

"How can you say such a thing?" asked Bláithín. "Even the child n the cradle eats them!" And she gave Fionn the cake with the humbprint. Fathach looked at the cradle. "Whose child is that?" he sked in wonder.

"That's Fionn's son," said Bláithín.

"And how old is he?" he asked then.

"Just ten months," replied Bláithín.

"Can he talk?" asked the giant.

"Not yet, but you should hear him roar!" At once, Fionn began o yell.

"Quick, quick," cried Bláithín. "Let him suck your little finger. If ionn comes home and hears him, he'll be in such a temper. With n anxious glance at the door, the giant gave Fionn his finger. Fionn it off the giant's magic little finger. Screeching, the iant bolted from the house. Fionn leaped from he cradle in bib and bonnet and danced his 3láithín round the kitchen.

Reading THE LITERATURE

Before reading this story, how did you expect Fionn Mac Cumhail to act? Did you expect him to be the hero of the story? Who is? How does the character solve the problem in the story? What skills are used to solve it?

Thinking About THE LITERATURE

In many European myths and legends, the heroes are powerful and fearless. How does Fionn act in this story? What words does the author use to make clear Fionn's attitude toward the danger he faces? How does he differ from other legendary figures that you have read about?

Writing About THE LITERATURE

Often, myths are created in order to answer questions about or explain mysteries in the world. This legend explains why the causeway was never finished. How might the story about Fionn be different if the causeway had been finished?

About the Author

Una Leavy, the author of *Irish Fairy Tales & Legends,* is an Irish writer who lives with her husband and children in County Mayo, Ireland.

Further Reading *The Names upon the Harp* by Marie Heaney recounts myths and legends of early Irish literature, including the stories about Fionn Mac Cumhail that make up the Fenian cycle.

INSTRUCT

Reading the Literature

Possible Responses

The title hints that Fíonn will be the hero, but when he gets scared, he goes to his wife, Bláithín, for help. Her cleverness and domestic skills enable them to outwit Fathach Mór by convincing him that Fíonn is their baby. When Fathach Mór lets Fíonn suck his magic finger, Fíonn bites it off.

Thinking About the Literature

Possible Responses

Fíonn's face paled; "He'll squash me . . ."; he was nervous; "What will I do, Bláithín? . . . "; he squealed; he shook with fear. Most legendary figures are strong and brave, but Fíonn is frightened and nervous.

Writing About the Literature

Possible Responses

Students' responses should explain how the causeway was finished. They might say that Fathach Mór finished it to get back to Scotland, or that Bláithín finished it herself.

MORE ABOUT...

Celtic Monuments

Made up of 40,000 separate stone columns, the Giant's Causeway juts out from the coast into the channel separating Scotland and Ireland. Some of its pillars are 82 feet high. It was formed 50 to 60 million years ago by inland lava flows that cooled when they reached the sea.

Though the Giant's Causeway is natural, the Celts left monuments throughout Ireland and Europe, including circular earthworks, terraced hillsides, burial mounds, and a huge horse carved into the ground in England.

 **World Literature**

More to Think About

Making Personal Connections Brainstorm a list or web of folk tale characters students know. Include Fíonn Mac Cumhail and add characters from American folk tales, Native American tales, and folk tales from other lands. Then ask: Why do people tell these stories, even today? What do folk tales explain about a culture?

Vocabulary Activity Have students write lines from the story on index cards or slips of paper. Read or display the lines and see who can correctly match each line with its speaker.

SECTION OBJECTIVES

1. To explain the postwar division between the countries of Eastern and Western Europe
2. To describe the effects of the dictatorship of Joseph Stalin on the Soviet Union
3. To describe the climate of the Cold War

SKILLBUILDER
- Interpreting a Map, p. 344

CRITICAL THINKING
- Analyzing Issues, p. 343

FOCUS & MOTIVATE

WARM-UP

Recognizing Effects Have students read Dateline and discuss the Warsaw Pact.

1. Why would countries located between the Soviet Union and Western Europe be willing to have Soviet troops stationed there?
2. Why do you think Yugoslavia did not agree to sign the Warsaw Pact?

INSTRUCT: Objective 1

East Against West

- In what way did the Iron Curtain serve as a barrier? kept people from traveling between the East and the West
- How did the Soviet Union control the Eastern European countries? through puppet governments, through force; people could be jailed for expressing anti-Soviet views

In-depth Resources: Unit 4
- Guided Reading Worksheet, p. 29

Reading Study Guide (Spanish and English), pp. 102–103

The Soviet Union

TERMS & NAMES
Iron Curtain
puppet government
one-party system
Joseph Stalin
collective farm
Warsaw Pact
Cold War

MAIN IDEA	WHY IT MATTERS NOW
After World War II, the Soviet Union was the most powerful country in Europe, but life for most Soviet citizens was difficult.	Russia, the former Soviet Union, remains powerful and is currently experiencing great change.

DATELINE EXTRA

WARSAW, POLAND, MAY 14, 1955

Today, the Soviet Union and most Eastern European countries announced that they have signed the Warsaw Treaty of Friendship, Cooperation, and Mutual Assistance. The members of this alliance agree to offer military defense to one another for a period of 20 years.

Yugoslavia is the only country in Eastern Europe that did not sign the agreement.

The new treaty, also called the Warsaw Pact, allows the Soviet Union to keep troops in the countries that are located between the Soviet Union and Western Europe. The Warsaw Pact is a response to the formation of NATO, an alliance that Western European countries joined six years ago.

Region • Warsaw hosted Eastern European officials who signed a military alliance here in the Palace of Culture. ▲

East Against West 1

After World War II, political differences divided the Soviet-controlled countries of Eastern Europe from those of Western Europe. These differences gave rise to an invisible wall known as the **Iron Curtain.** While there was no actual curtain, people of the East were restricted from traveling outside of their countries. Westerners who wished to visit the East also faced restrictions.

TAKING NOTES
Use your chart to take notes about war and change in Europe.

Causes	Event	Effects
	World War I	
	World War II	

Program Resources

In-depth Resources: Unit 4
- Guided Reading Worksheet, p. 29
- Reteaching Activity, p. 36

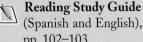

Reading Study Guide (Spanish and English), pp. 102–103

Formal Assessment
- Section Quiz, p. 175

Integrated Assessment

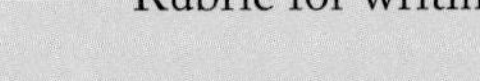

- Rubric for writing a scene

Outline Map Activities

Access for Students Acquiring English
- Guided Reading Worksheet, p. 97

Technology Resources
classzone.com

TEST-TAKING RESOURCES

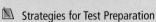

Strategies for Test Preparation

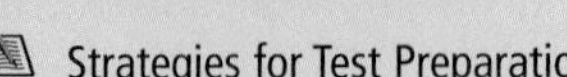

Test Practice Transparencies

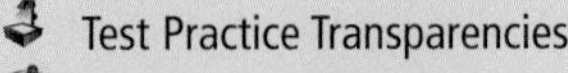

Online Test Practice

The Strongest Nation in Europe The Union of Soviet Socialist Republics, or USSR, was the official name of the Soviet Union. It included 15 republics, of which Russia was the largest. The Soviet Union entered World War II in 1941, when Germany invaded its borders. German troops destroyed much of the western Soviet Union and killed millions of people. This invasion brought the Soviet Union close to collapse. However, with the defeat of Germany, the Soviet Union rose to become the strongest nation in Europe.

Region • The hammer and sickle became the symbol of Soviet Communism. The tools represent the unity of the peasants (sickle) with the workers (hammer). ▲

Vocabulary
establish: set up; create

Communism After World War II, the Soviet Union established Communist governments in Eastern Europe. The Soviets made sure—either by politics or by force—that these new Eastern European governments were loyal to the Soviet Union.

Reading Social Studies
A. Possible Answers
through fear and military force; by controlling elections

Soviet Control of Eastern Europe The Soviet Union controlled the countries of Eastern Europe through puppet governments. A **puppet government** is one that does what it is told by an outside force. In this case, the Eastern European governments followed orders from Soviet leaders in Moscow.

Most Eastern Europeans did have the chance to vote, but they had only one political party to choose from: the Communist Party. All other parties were outlawed. This meant that there was only one candidate to choose from for each government position. This is an example of a **one-party system.** Soviet citizens could not complain about the government. In fact, they could be jailed for expressing any view that the Soviet leaders did not like.

Reading Social Studies
A. Making Inferences
How do you think the Soviet Union enforced a one-party system in Eastern Europe?

Movement • The government-controlled factories in the Soviet Union did not produce enough of certain items. When goods that were often in short supply—such as bread and shoes—finally became available, people had to wait in long lines to buy them. ◀

MORE ABOUT...
The Iron Curtain
It was Britain's great wartime leader, Winston Churchill, who coined the term "Iron Curtain." On a visit to the United States in March 1946, he delivered an address in Fulton, Missouri. In that address, he warned Americans that "an iron curtain has descended across the Continent." Churchill was referring to the isolationism of the Soviet Union, which cut off the Soviet bloc—the Soviet Union and its Eastern European satellite countries—from the rest of the world.

CRITICAL THINKING ACTIVITY
Analyzing Issues Point out to students that at one point during World War II, the Soviet Union was on the verge of collapsing. However, at the end of the war, it was the strongest nation in Europe. To explain how this was possible, guide students to consider the size of the Soviet Union and its influence over Eastern European countries as well as the devastation suffered in the rest of the Continent.

Class Time 10 minutes

Activity Options
Multiple Learning Styles: Visual
Class Time One class period
Task Creating a political symbol
Purpose To understand how symbols are used to convey political messages and stir emotions

Supplies Needed
- Encyclopedia
- Paper
- Pencils, pens, markers

Block Scheduling
Activity Recall with students the political symbols they have seen in the text, such as the Soviet hammer and sickle. Point out that symbols can be a form of propaganda. They stir people's emotions. Have small groups of students create a political symbol to represent a current political concept in our country. Provide space for groups to display and explain their symbols and brief written descriptions.

INSTRUCT: Objective 2

Joseph Stalin

- Who was Joseph Stalin? Communist leader of the Soviet Union during World War II
- How did Stalin control agriculture in the Soviet Union? through collective farms
- In what way were the secret police important to Stalin's government? arrested people whom Stalin did not trust and who did not support the government; sent them to Siberia

FOCUS ON VISUALS

Interpreting the Map Have students use the key to point out the area of the map showing the Soviet Union. Point out the vast extent of the country, especially the remote area of Siberia. Ask students to identify the oceans and seas that border Soviet territory on the north, south, east, and west. How might this access to the sea have influenced Russian and Soviet history?

Possible Responses North: Arctic Ocean; South: Black Sea; East: Pacific Ocean; West: Baltic Sea. Compared with its size, the Soviet Union had little seacoast in its European territory; this may have limited trade and economic development and prompted Russian and Soviet leaders to try to annex coastal territories.

Extension Have students compare this map with one of present-day Europe. Invite volunteers to point out differences and changes in national borders, both in the former Soviet Union and in the rest of Europe.

Region • Joseph Stalin ruled the Soviet Union from 1928 to 1953. ▲

Joseph Stalin 2

Joseph Stalin (STAH•lihn) (1879–1953) ruled the Soviet Union during World War II. Stalin took power after the death of Vladimir Lenin. Lenin was a Communist leader who had helped overthrow the czar and ruled the Soviet Union from 1917 until his death in 1924. The name Stalin is related to the Russian word for "steel." Stalin was greatly feared, and his rule was indeed as tough as steel. He controlled the government until his death.

The Five-Year Plans Under Stalin, the government controlled every aspect of Soviet life. Stalin hoped to strengthen the country with his five-year plans, which were sets of economic goals. For example, Stalin ordered many new factories to be built. The Soviet government decided where and what types of factories to build, how many goods to produce, and how to distribute them. These decisions were based on the Communist theory that this would benefit the most people.

Geography Skillbuilder Answers

1. East Germany, Poland, Czechoslovakia, Hungary, Romania, Bulgaria, Albania
2. East Germany

Region • The Soviet government managed the factories while citizens provided the actual labor. ◄

GEOGRAPHY SKILLBUILDER: Interpreting a Map

1. **Region •** Which countries were behind the Iron Curtain but not in the Soviet Union?
2. **Location •** What was the western most country in the Warsaw Pact?

Activity Options

Multiple Learning Styles: Visual

Class Time One hour

Task Creating a political cartoon about Stalin

Purpose To examine Stalin's policies and consider how people in democratic countries might have viewed them

Supplies Needed
- Textbook
- Drawing paper
- Pencils, pens, markers

Activity Review with students what they learned about political, or editorial, cartoons. Then have them reread the section "Joseph Stalin" in their textbooks. Ask students to create a political cartoon about Stalin and his policies that might have appeared in a newspaper in the United States. When they have finished their cartoons, invite volunteers to share them with the rest of the class.

The Iron Curtain and the Warsaw Pact Nations, 1955

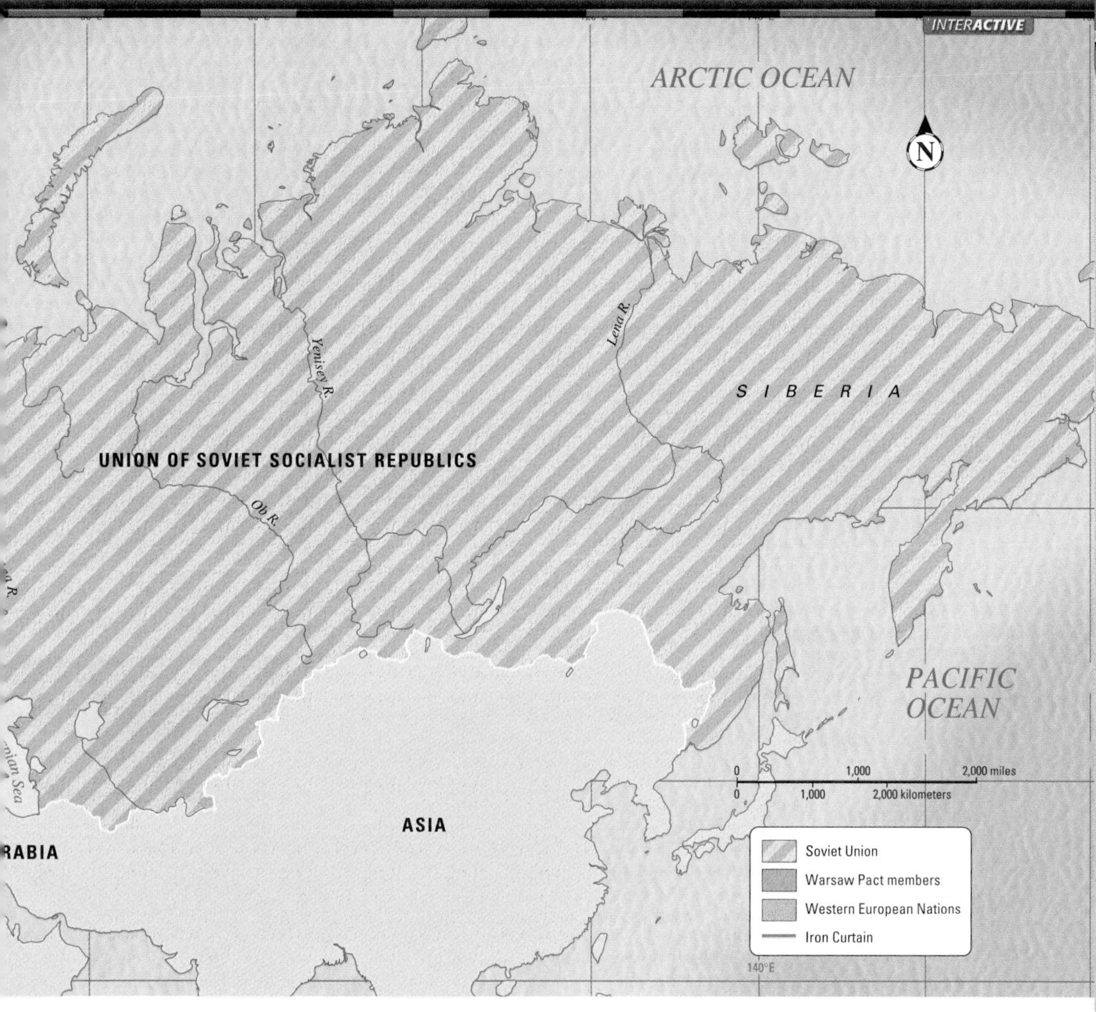

Soviet Agriculture Stalin also hoped to strengthen the Soviet Union by controlling the country's agriculture. During the 1930s, peasants were forced to move to collective farms. A **collective farm** was government-owned and employed large numbers of workers. All the crops produced by the collective farms were distributed by the government. Sometimes farm workers did not receive enough food to feed themselves and their families.

Region • Similar to urban factory workers, Russian peasants labored on government-controlled collective farms. ▼

MORE ABOUT...

The Soviet Union

Some facts about the Soviet Union:

- It was often just called Russia, because that was the dominant republic.
- It was the largest country in the world, extending about 6,800 miles (10,940 km) east to west and crossing 11 time zones.
- It was organized in 1922, mainly from Russia, Ukraine, and Belarus, eventually consisting of 15 republics.
- It was dissolved relatively peacefully in December 1991 and replaced by 15 independent states.

FOCUS ON VISUALS

Interpreting the Photographs Ask students to study the photographs of Soviet farm and factory workers and describe their impressions of the workers and their working conditions. Ask them to compare the farm workers with their ideas of an American farm in the 1930s or 1940s.

Possible Responses The farm workers are doing all the work by hand; they don't seem to have any farm machines. The workers in both pictures look very young.

Extension Have students look in history textbooks, popular histories, and other sources to find other photographs illustrating life in the Soviet Union from the 1930s through the 1950s.

Activity Options

Interdisciplinary Links: Art/Language Arts

Class Time One class period

Task Using persuasive writing techniques to create a propaganda poster

Purpose To understand how the employment of propaganda techniques can be used to manipulate the public

Supplies Needed

- Drawing paper
- Pens, markers

Block Scheduling

Activity Remind students that Joseph Stalin's government forced agricultural workers to join collective farms. Explain that the Communist government often employed propaganda posters to persuade workers that what they were doing was for the good of the country. Have students work in small groups to create propaganda posters. Brainstorm examples of persuasive words, phrases, and images that students might include in their posters.

INSTRUCT: Objective 3

The Cold War

- What goal united the Soviet Union, the United States, and the United Kingdom from 1941 to 1945? the goal of defeating the Axis powers
- How was the Cold War different from World Wars I and II? NATO and the Warsaw Pact nations did not fight, but they would not trade or cooperate with each other either.
- What did the United States and its European allies fear? Soviet and Communist influence on other countries
- What did the Soviet Union fear from Western Europe? another invasion

Spotlight on CULTURE

Sergei Mikhailovich Eisenstein was born on January 23, 1898, in Riga, Latvia. In 1918, he joined the Red Army, where he worked as a poster designer. Shortly after, Eisenstein moved to Moscow and studied Japanese culture and language. It was this experience that influenced his new method of film editing known as montage. This method involves the grouping of unlike objects in such a way that they suggest a new meaning.

Thinking Critically Answers

1. Possible responses: He was responding to world events; he wanted to glorify Russian history and encourage patriotism.
2. He wanted to show that Germany was an ancient enemy of Russia but had been defeated before.

The Secret Police Stalin used his secret police to get rid of citizens he did not trust. The secret police arrested those who did not support the Soviet government. Suspects were transported to slave-labor camps in Siberia. Millions of men and women were sent to this remote and bitterly cold region of northeastern Russia. Many never returned home.

The Cold War 

From 1941 to 1945, the United Kingdom, the United States, and the Soviet Union shared a goal: to defeat the Axis Powers. They became allies to make that happen. Once the war ended, however, these countries no longer had a common enemy—and had little reason to work together. Most Western European countries were constitutional monarchies or democracies, and most Eastern European countries had Communist, largely Soviet-controlled, governments.

Spotlight on CULTURE

Soviet Film The Russian director Sergey Eisenstein (EYE•zen•stine) (1898–1948), bottom right, made only six movies, but they are among the most important works in film history. The silent film *Battleship Potemkin* (1925), whose poster is to the right, is one of Eisenstein's most famous. It is about a mutiny at sea. The director's use of close-ups and his method of combining short scenes changed the way films were made all over the world.

Just before the start of World War II, Eisenstein made the film *Alexander Nevsky* (1938). It tells the story of a historic battle that the Russians won against German-speaking invaders in the 1200s. This film became very popular during World War II, which it seemed to foreshadow.

THINKING CRITICALLY 

1. **Clarifying**
   What influenced Eisenstein to direct war films?
2. **Synthesizing**
   What did Eisenstein want to show about the relationship between Russians and Germans?

For more on Sergey Eisenstein, go to 

Activity Options

Differentiating Instruction: Gifted and Talented

 **Block Scheduling**

Class Time One class period

Task Writing a review of a movie based on a historical event

Purpose To identify elements of a motion picture that appeal to feelings of patriotism and national pride

Supplies Needed
- Writing paper
- Pencils or pens

Activity Discuss any motion pictures that students have seen that had a historical background. Explain that most movie reviews assess films on such qualities as plot development, quality of performances, and special effects. Ask students to write a review of a movie that was based on an event in history. Encourage them to consider the features discussed, as well as the movie's appeal to feelings of patriotism and national pride. If possible, provide time for students to share their reviews and discuss their opinions.

Region • The Brandenburg Gate was a part of the Berlin Wall that separated East Berlin from West Berlin. ◄

Reading Social Studies
B. Possible Answers
Each side distrusted the other's political systems; the Soviet Union also wanted protection against future invasion.

The members of NATO and the nations in the **Warsaw Pact**—the alliance of Eastern European countries behind the Iron Curtain—refused to trade or cooperate with each other. The countries never actually fought, so this period of political noncooperation is called the **Cold War.** Both sides in the Cold War were hesitant to start a war that would involve the use of newly developed nuclear weapons, which could cause destruction on a global scale.

Reading Social Studies
B. Comparing Compare the Soviet Union's fears in the Cold War with those of the United States and Western Europe.

The United States and Western Europe feared that the Soviet Union would influence other countries to become Communist. At the same time, the Soviet Union wanted to protect itself against invasion. This led the countries on either side of the Iron Curtain to view and treat each other as possible threats. The tense international situation caused by the Cold War would continue for almost 40 years.

SECTION 3 ASSESSMENT

Terms & Names

1. **Explain the significance of:** (a) Iron Curtain (b) puppet government (c) one-party system (d) Joseph Stalin (e) collective farm (f) Warsaw Pact (g) Cold War

Using Graphics

2. Use a chart like this one to describe three elements of Joseph Stalin's rule of the Soviet Union.

Five-Year Plans	Agriculture	Secret Police

Main Ideas

3. (a) What happened to the Soviet Union during World War II?
(b) How did the governments of most Western and Eastern European countries differ?
(c) How did Joseph Stalin rule the Soviet Union?

Critical Thinking

4. **Analyzing Motives**
Why do you think the Soviet Union wanted to control the countries of Eastern Europe?

Think About
- the events of World War II
- the location of the Eastern European countries
- the governments of the Soviet Union and Western Europe

ACTIVITY -OPTION- Reread the information about the secret police. Write a dramatic **scene** in which the main character is sent to a labor camp in Siberia.

MORE ABOUT...
The Brandenburg Gate

Built around 1790, the Brandenburg Gate was the symbol of the city of Berlin. It stands at the western end of the famous avenue, Unter den Linden. The gate was modeled after a famous Greek triumphal arch in Athens. The statue on the top represents a chariot drawn by four horses. The gate was badly damaged in World War II but restored in the late 1950s. After the Berlin Wall cut it off from West Berlin, the gate became a symbol of German reunification.

ASSESS & RETEACH

Reading Social Studies Have students complete the graphic organizer on page 328.

Formal Assessment
- Section Quiz, p. 175

RETEACHING ACTIVITY

Have students work in groups to create five-question quizzes on the section. When they have finished, have groups exchange quizzes and answer the questions.

In-depth Resources: Unit 4
- Reteaching Activity, p. 36

Access for Students Acquiring English
- Reteaching Activity, p. 102

Section 3 Assessment

1. Terms & Names

a. Iron Curtain, p. 342
b. puppet government, p. 343
c. one-party system, p. 343
d. Joseph Stalin, p. 344
e. collective farm, p. 345
f. Warsaw Pact, p. 347
g. Cold War, p. 347

2. Using Graphics

Five-Year Plans	Agriculture	Secret Police
Government-set economic goals	Controlled by government	Eliminated citizens Stalin didn't trust
Where to build factories	Peasants live on collective farms	Arrested those who did not support the government
What types of factories to build	Crops distributed by government	Sent suspects to labor camps
How many goods to produce	Workers lack food	
How to distribute goods		

3. Main Ideas

a. invaded by Germany; widespread, severe destruction and millions of civilians killed
b. Western Europe, mainly constitutional monarchies or democracies; Eastern Europe, Communist dictatorships
c. used terror and repression to control people; controlled industry and agriculture to develop the economy

4. Critical Thinking

Soviets wanted a line of "buffer states" between themselves and Western Europe, to protect themselves against invasion.

ACTIVITY OPTION

Integrated Assessment
- Rubric for writing a scene

ASSESSMENT

TERMS & NAMES

1. nationalism, p. 329
2. colonialism, p. 331
3. World War I, p. 333
4. alliance, p. 334
5. World War II, p. 336
6. NATO, p. 337
7. Adolf Hitler, p. 336
8. Warsaw Pact, p. 347
9. Iron Curtain, p. 342
10. Cold War, p. 347

REVIEW QUESTIONS

Possible Reponses

1. The largest empire was Austria-Hungary.
2. Expanded armies protected their countries' borders and colonies.
3. Alliances provided help against outside attack.
4. It required Germany to pay for damages and give up valuable territory.
5. Poland
6. The Soviet Union installed puppet Communist governments that allowed only one, Communist Party candidate in elections.
7. It prevented travel and communication between Western and Eastern Europe.
8. Both had nuclear weapons that could have caused worldwide destruction.

CHAPTER 12 ASSESSMENT

TERMS & NAMES

Explain the significance of each of the following:

1. nationalism
2. colonialism
3. World War I
4. alliance
5. World War II
6. NATO
7. Adolf Hitler
8. Warsaw Pact
9. Iron Curtain
10. Cold War

REVIEW QUESTIONS

European Empires *(pages 329–332)*

1. What was the largest empire in Eastern Europe in 1900?
2. What is one reason why European nations built up their military?

Europe at War *(pages 333–338)*

3. Why did European countries join alliances?
4. What did the Treaty of Versailles require Germany to do?
5. What country did Germany invade to begin World War II?

The Soviet Union *(pages 342–347)*

6. Why did most Eastern European voters have only one political party to choose from?
7. How did the Iron Curtain affect the lives of Eastern Europeans?
8. Why were both sides in the Cold War hesitant to start a war?

CRITICAL THINKING

Identifying Problems

1. Using your completed chart from Reading Social Studies, p. 328, list some of the causes and effects of World Wars I and II.

Making Inferences

2. Why might a citizen who has helped elect a government be more willing to fight to defend it?

Hypothesizing

3. How do you think Soviet peasants felt about collective farms? Why?

Visual Summary

European Empires 1

- In early-20th-century Europe, feelings of nationalism arose.
- Western European nations ruled colonial empires.

Europe at War 2

- Due to a complex set of alliances, most of Europe was drawn into World War I.
- The Treaty of Versailles set the stage for an even more widespread conflict—World War II.

The Soviet Union 3

- After World War II, the Soviet Union was very powerful.
- However, life was difficult for many Soviet citizens.

CRITICAL THINKING: Possible Responses

1. Identifying Problems

Causes of World War I: increasing nationalism; protests against power of Austria-Hungary; assassination of archduke of Austria; alliances bring in Russia and Germany, then other nations

Effects of World War I: destruction left many homeless; breakup of Austria-Hungary; independence for many ethnic groups

Causes of World War II: weakness of German economy; dictatorship of Adolf Hitler; German invasion of Poland

Effects of World War II: millions dead; occupation of Western Europe by Allies; occupation of Eastern Europe by Soviet Union

2. Making Inferences

A citizen might want to defend a government he or she voted for.

3. Hypothesizing

Farmers resented losing their own lands and animals, having to move to collectives, feeding people in cities while not having enough food for themselves.

ANDARDS-BASED ASSESSMENT

the map and your knowledge of geography to ver questions 1 and 2.

ditional Test Practice, pp. S1–S33

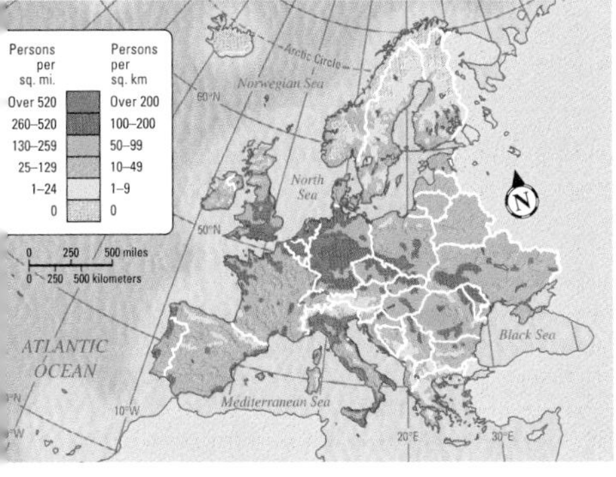

hich areas of Europe are the least densely populated?

The area just south of the North Sea

. Areas on the Atlantic Coast

. Areas along the Mediterranean Sea

Areas near the Arctic Circle

dging from this map, why do you think some areas are ore densely populated than others?

. Many people tend to move to mountainous regions.

. Many people tend to settle in cities near coasts or rivers.

. Many people tend to avoid living on peninsulas.

. Many people like to live on islands because of the fishing.

The excerpt is from a famous speech that Winston Churchill gave during the Cold War. Use the quotation and your knowledge of world cultures and geography to answer question 3.

PRIMARY SOURCE

From Stettin in the Baltic to Trieste in the Adriatic, an iron curtain has descended across the Continent. Behind that line lie all the capitals of the ancient states of Central and Eastern Europe. . . . All these famous cities and the populations around them lie in what I might call the Soviet sphere, and all are subject, in one form or another, not only to Soviet influence but to a very high and in some cases increasing measure of control from Moscow.

WINSTON CHURCHILL,
"Iron Curtain" speech, Fulton, Missouri

3. According to the passage, what threat did the iron curtain provide?
   - **A.** The Communist Soviet Union controlled the region behind the iron curtain.
   - **B.** Famine and disease threatened countries in Central and Eastern Europe.
   - **C.** Hitler and his forces wanted to reconquer the region behind the iron curtain.
   - **D.** A barrier prevented people from visiting the cities of Central and Eastern Europe

TERNATIVE ASSESSMENT

WRITING ABOUT HISTORY

nagine that you are a government official from a European ation living in one of the nation's colonies in Asia or Africa. rite a letter to your family telling them what life in the colony like in the year 1900.

Use the Internet or the library to research one of the European colonies in Asia or Africa.

In your letter, identify the colony and describe its location. Include information about the colony's climate, geography, culture, and people. Describe your country's plans for the colony.

OOPERATIVE LEARNING

ith a small group, design a monument to commemorate n event in 20th-century Europe, such as a particular battle om World War I, the Holocaust, or World War II. Share the sponsibilities of researching the event, designing the monu- ent, and presenting it to your class. In your presentation, ovide some background information and suggest a ermanent location for your monument.

INTEGRATED TECHNOLOGY

Doing Internet Research

Countries such as Switzerland and Sweden were not directly involved in World War I or World War II. Use the Internet to research one of the neutral countries and include your findings in a report.

- Use the Internet and other library resources to identify a country that wasn't directly involved in the wars.
- Another source of information might be a historical museum or archive.
- Look for information about the country's government and economy in the first half of the 20th century, how the country was affected by the wars, and what life was like for its citizens.

For Internet links to support this activity, go to

STANDARDS-BASED ASSESSMENT

1. **Answer D** is the correct answer because the least densely populated areas are in the north, near the Arctic Circle; answers A, B, and C are incorrect because the areas south of the North Sea and around the Atlantic and Mediterranean coasts are densely populated.
2. **Answer B** is the correct answer because the areas near coasts and rivers tend to be more populated; answer A is incorrect because people may find it difficult to settle in mountainous regions; answer C is incorrect because some peninsulas have areas of high density; answer D is incorrect because many islands are sparsely populated.
3. **Answer A** is the correct answer because it focuses on the Soviet control highlighted in the passage; answers B and C are incorrect because neither famine, disease, nor Hitler are mentioned in the passage; answer D is incorrect because the iron curtain was a figure of speech, not a real barrier.

INTEGRATED TECHNOLOGY

Discuss how students might find information about a European country that was not directly involved in the world wars. Have them make lists of what they might want to find out about the countries in addition to the items suggested on page 349. Emphasize that as students do their research, they need to keep a log listing the information they get from each Web site.

Alternative Assessment

1. Rubric

The letter should
- cover important information about the colony.
- accurately reflect the viewpoint of a European colonial official.
- reflect the student's understanding of colonialism.
- use correct grammar, spelling, and punctuation.

2. Cooperative Learning

Display photographs of a variety of monuments for students to observe, such as the Vietnam Veterans' Memorial in Washington, D.C. Groups should choose an event to commemorate, agree on a location for their monument, and then create a design.

PLANNING GUIDE: UNIT 4 ATLAS AND CHAPTER 13

Modern Europe

	OVERVIEW	COPYMASTERS	INTEGRATED TECHNOLOGY
UNIT ATLAS AND CHAPTER RESOURCES	Students will examine recent changes in modern Europe, including political, economic, and cultural aspects of life in Eastern Europe before and after the breakup of the Soviet Union. They will also look at the European Union and its activities.	**In-depth Resources: Unit 4** • Guided Reading Worksheets, pp. 37–39 • Skillbuilder Practice, p. 42 • Unit Atlas Activities, pp. 1–2 • Geography Workshop, pp. 61–62 **Reading Study Guide** 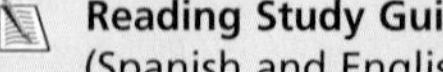(Spanish and English), pp. 106–113 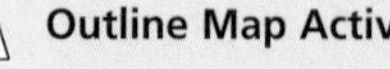**Outline Map Activities**	 • eEdition Plus Online • EasyPlanner Plus Online • eTest Plus Online • eEdition • Power Presentations • EasyPlanner • Electronic Library of Primary Sources • Test Generator • Reading Study Guide • The World's Music • Critical Thinking Transparencies CT25
SECTION 1 Eastern Europe Under Communism pp. 353–358	**KEY IDEAS** • The Soviet Union attempted to create a national identity by controlling cultural and artistic expression. • The Soviets controlled factories, railroads, businesses, and other aspects of the economy. • Some Eastern European nations tried to change their Soviet-dominated governments and economies.	**In-depth Resources: Unit 4** • Guided Reading Worksheet, p. 37 • Reteaching Activity, p. 44 **Reading Study Guide** (Spanish and English), pp. 106–107	 classzone.com Reading Study Guide
SECTION 2 Eastern Europe and Russia pp. 360–366	• After the collapse of the Soviet Union in 1991, many Eastern European nations declared their independence. • Conflicts among ethnic groups led to war in the Balkan states. • Modern Russia enjoys many freedoms, but the free-market economy also poses difficulties.	 **In-depth Resources: Unit 4** • Guided Reading Worksheet, p. 38 • Reteaching Activity, p. 45  **Reading Study Guide** (Spanish and English), pp. 108–109	 classzone.com Reading Study Guide
SECTION 3 The European Union pp. 367–371	• Many countries want to join the European Union to gain economic and political aid. However, they must first commit to certain improvements. • The European Union works to improve trade among member nations. • The European Union provides job training and protects citizens' rights in its culturally diverse member nations.	 **In-depth Resources: Unit 4** • Guided Reading Worksheet, p. 39 • Reteaching Activity, p. 46 **Reading Study Guide** (Spanish and English), pp. 110–111	 **Critical Thinking Transparencies CT26**  classzone.com  Reading Study Guide

KEY TO RESOURCES

Audio

CD-ROM
Copymaster

Internet

Overhead Transparency
Pupil's Edition

Teacher's Edition

Video

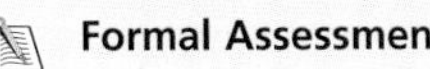

ASSESSMENT OPTIONS

Chapter Assessment, pp. 374–375

Formal Assessment
- Chapter Tests: Forms A, B, C, pp. 191–202

Test Generator

Online Test Practice

Strategies for Test Preparation

Section Assessment, p. 358

Formal Assessment
- Section Quiz, p. 188

Integrated Assessment
- Rubric for writing a speech

Test Generator

Test Practice Transparencies TT40

Section Assessment, p. 366

Formal Assessment
- Section Quiz, p. 189

Integrated Assessment
- Rubric for writing a personal essay

Test Generator

Test Practice Transparencies TT41

Section Assessment, p. 371

Formal Assessment
- Section Quiz, p. 190

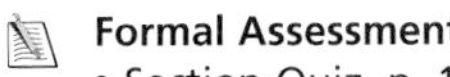

Integrated Assessment
- Rubric for writing a postcard or e-mail

Test Generator

Test Practice Transparencies TT42

RESOURCES FOR DIFFERENTIATING INSTRUCTION

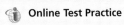

Students Acquiring English/ESL

Reading Study Guide
(Spanish and English), pp. 106–113

Access for Students Acquiring English
Spanish Translations, pp. 103–110

TE Activity
- Borrowed Words, p. 361

Modified Lesson Plans for English Learners

Less Proficient Readers

Reading Study Guide
(Spanish and English), pp. 106–113

TE Activities
- Identifying Main Ideas and Details, p. 354
- Identifying Sequence, p. 363
- Setting a Purpose, p. 368

Gifted and Talented Students

TE Activities
- Comparing and Contrasting, p. 355
- Creating Crossword Puzzles, p. 364

CROSS-CURRICULAR CONNECTIONS

Humanities
Adler, Naomi. ***Play Me a Story: Nine Tales About Musical Instruments.*** Brookfield, CT: Millbrook Press, 1998. Folk tales linked to native instruments, including the Russian balalaika.

Literature
Riordan, James. ***Russian Folk-Tales.*** New York: Oxford University Press, 2000. Ten traditional folk tales.

Popular Culture
Carona, Laurel. ***Life in Moscow.*** San Diego: Lucent, 2001. Focuses on life after the fall of communism.
Toht, Patricia. ***Daily Life in Ancient and Modern Moscow.*** Minneapolis: Runestone Press, 2001. Nice overview of a tumultuous city history.

History
Burgan, Michael. ***Cold War: The Collapse.*** Austin, TX: Raintree Steck-Vaughn, 2001. Based on CNN's Cold War series, chronicles the final years of communism.

Economics
Thompson, Clifford, ed. ***Russia & Eastern Europe.*** New York: H. W. Wilson Co., 1998. Articles about post–Cold War economic conditions.

Science
Dommermuth-Costa, Carol. ***Nikola Tesla: A Spark of Genius.*** Minneapolis: Lerner Publications, 1994. Serbian-born scientist who pioneered work with alternating-current electricity.

ENRICHMENT ACTIVITIES

The following activities are especially suitable for classes following block schedules.

Teacher's Edition, pp. 356, 357, 359, 362, 364
Pupil's Edition, pp. 358, 366, 371

Unit Atlas, pp. 258–269
Literature Connections, pp. 372–373

Outline Map Activities

INTEGRATED TECHNOLOGY

Go to **classzone.com** for lesson support and activities for Chapter 13.

BLOCK SCHEDULE LESSON PLAN OPTIONS: 90-MINUTE PERIOD

DAY 1

CHAPTER PREVIEW, pp. 350–351
Class Time 30 minutes

- **Hypothesize** Use the "What do you think?" questions in Focus on Geography on PE p. 351 to help students hypothesize about the relationship between a country's economy and its natural resources.

SECTION 1, pp. 353–358
Class Time 60 minutes

- **Small Groups** On the chalkboard, list the section objectives from TE p. 353. Divide the class into three groups. Have each group select one objective and prepare a summary of the section based on it. Remind students that when they summarize, they should include the main ideas and the most important details in their own words. Reconvene as a whole class and have each group share its summary.
  Class Time 35 minutes
- **Comparing** Use the Critical Thinking Activity on TE p. 354 to guide a discussion about the techniques used by the Soviet Union and the United States to create a national identity among their citizenry.
  Class Time 15 minutes
- **Word Web** Have students work in pairs to organize important information about the Soviet economy using a word web.
  Class Time 10 minutes

DAY 2

SECTION 2, pp. 360–366
Class Time 65 minutes

- **Internet** Extend students' background knowledge of the economies of Eastern Europe by visiting **classzone.com.**
  Class Time 25 minutes
- **Venn Diagram** Divide the class into pairs. Use the Critical Thinking Activity on TE p. 364 to have groups create a Venn diagram comparing Soviet life before and after the breakup of the Soviet Union.
  Class Time 25 minutes
- **Geography Skillbuilder** Use the questions in the Geography Skillbuilder on PE p. 365 to guide a discussion about Russia's natural resources.
  Class Time 15 minutes

SECTION 3, pp. 367–371
Class Time 25 minutes

- **Analyzing Motives** Have students read Dateline on PE p. 367. Use the questions in Focus & Motivate on TE p. 367 to help students analyze the motives behind the creation of the European Union.
  Class Time 10 minutes
- **Word Problems** Use the Activity Option on TE p. 369 to teach students about the relative value of the euro compared to the U.S. dollar. After you have created two or three word problems with the class, divide students into small groups. Have the groups create three work problems, and have the class solve them.
  Class Time 15 minute

DAY 3

SECTION 3, continued
Class Time 35 minutes

- **Locating Places on a Map** Use the Activity Options on TE p. 370 to have students locate countries and cities in Western Europe.
  Class Time 20 minutes
- **Summary** Have students write a short summary of the section using the Terms & Names on PE p. 371.
  Class Time 15 minutes

CHAPTER 13 REVIEW AND ASSESSMENT, pp. 374–375
Class Time 55 minutes

- **Chapter Review** Divide students into pairs. Using the Visual Summary on PE p. 374, have students provide two supporting details for each statement.
  Class Time 20 minutes
- **Assessment** Have students complete the Chapter 13 Assessment.
  Class Time 35 minutes

TECHNOLOGY IN THE CLASSROOM

INTERNET RESEARCH

The Web is a valuable research tool, providing valuable, up-to-date information, including text and pictures. However, some sites should be screened before students do research on the Web, as many sites are not up to date or have erroneous or inappropriate information. Students should be given a list of sites sponsored by organizations that are generally considered reliable, such as major news sources, television networks, nonprofit organizations, or museums. This procedure will save class time and steer students in the right direction.

ACTIVITY OUTLINE

Objective Students will go to Web sites to find out about the Berlin Wall. Then they will use a word processing program to write several paragraphs about the significance of the wall.

Task Have students go to the Web sites and take notes on the Berlin Wall and its significance. Ask them to write paragraphs explaining the significance of the dismantling of the wall in the 1980s.

Class Time Two class periods

DIRECTIONS

1. As a class, discuss what students know about the Berlin Wall. Encourage volunteers to identify the location of Berlin and share what they know about why the wall was constructed.
2. Tell students that the Berlin Wall was torn down in 1989. This event is considered to be one of the most significant happenings of the 1980s. Ask them to find out more about the events surrounding the building and dismantling of the wall by going to the Web sites at **classzone.com.** Help students focus their research by suggesting that they create an outline with the following headings:
   - Why the wall was built
   - What the wall represented
   - What life was like in the East
   - What life was like in the West
   - Why the wall was taken down

   They should then take notes by adding important facts and details under each heading.
3. Have students use the information in their outlines to write one- or two-paragraph reports. Provide an opportunity for students to share what they have learned.

CHAPTER 13 OBJECTIVE

Students will examine recent changes in Europe, including political, economic, and cultural aspects of life in Eastern Europe before and after the breakup of the Soviet Union. They will also look at the European Union.

FOCUS ON VISUALS

Interpreting the Photograph Direct students' attention to the photograph of the locomotive and have them describe it. Have they seen trains like this before? If so, where? Discuss how it may be different from other trains they have seen. What conclusions might students draw about Europe based on this photograph?

Possible Responses Students' descriptions should indicate that the locomotive looks very modern. They may conclude that because the train is modern, the economies and technology of Eastern Europe are probably advanced as well.

Extension Ask students to research modern bullet trains. Challenge them to find out how fast bullet trains travel, where they are found, and why they are not more common.

CRITICAL THINKING ACTIVITY

Comparing Ask students to compare this passenger train with the trains they see where they live. Explain that people travel by train more frequently in Europe than in the United States. Ask students why this might be the case.

Class Time 5 minutes

CHAPTER 13

Modern Europe

SECTION 1 Eastern Europe Under Communism

SECTION 2 Eastern Europe and Russia

SECTION 3 The European Union

EUROPE AND RUSSIA
ATLANTIC OCEAN
PACIFIC OCEAN
PACIFIC OCEAN
INDIAN OCEAN

118
SNCF

Movement High-speed trains, like this one in France, make travel in Europe very convenient.

350

Recommended Resources

BOOKS FOR THE TEACHER

Kort, G. Michael. ***The Handbook of the New Eastern Europe.*** Brookfield, CT: Twenty-First Century Books, 2001.

Pinder, John. ***The European Union: A Very Short Introduction.*** New York: Oxford University Press, 2001. Compact and useful introduction.

Winters, Paul A., ed. ***The Collapse of the Soviet Union.*** San Diego: Greenhaven Press, 1999. Essays, articles, and interviews.

VIDEOS

People Power: The End of Soviet-Style Communism. Cambridge, MA: WGBH Public Television, 1999. Extensive coverage of Gorbachev and Yeltsin.

INTERNET

For more information about Eastern Europe and the European Union, visit **classzone.com.**

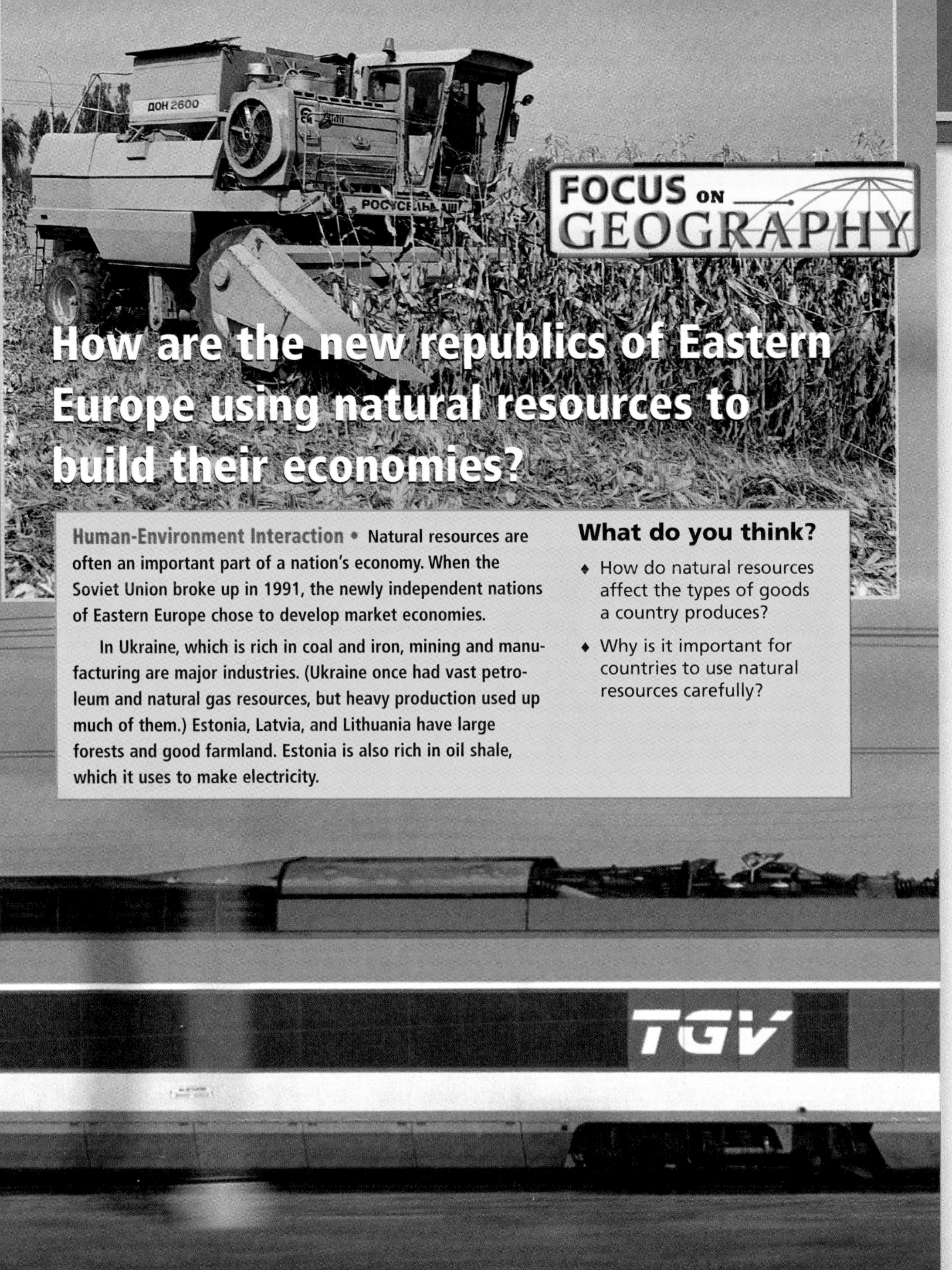

How are the new republics of Eastern Europe using natural resources to build their economies?

Human-Environment Interaction • Natural resources are often an important part of a nation's economy. When the Soviet Union broke up in 1991, the newly independent nations of Eastern Europe chose to develop market economies.

In Ukraine, which is rich in coal and iron, mining and manufacturing are major industries. (Ukraine once had vast petroleum and natural gas resources, but heavy production used up much of them.) Estonia, Latvia, and Lithuania have large forests and good farmland. Estonia is also rich in oil shale, which it uses to make electricity.

What do you think?

- How do natural resources affect the types of goods a country produces?
- Why is it important for countries to use natural resources carefully?

351

FOCUS ON GEOGRAPHY

Objectives

- To identify the relationship between natural resources and economic development
- To recognize the importance of the wise use of natural resources

What Do You Think?

1. Remind students that natural resources are anything that occurs naturally in the environment, such as iron, coal, and trees. Ask students to discuss how different types of natural resources would affect the kinds of goods a nation produces.
2. Challenge students to discuss how depleting a natural resource would affect an economy based on that resource.

How are the new republics of Eastern Europe using natural resources to build their economies?

Have students consider how the people of Eastern Europe turned to the natural resources in their countries to develop individual economies. Do they think the availability of natural resources was taken into account when the Soviet Union divided into independent nations?

MAKING GEOGRAPHIC CONNECTIONS

Have students list natural resources that exist in their area. Then discuss how local industries and businesses use these resources. Invite them to speculate on how the local economy might be different if other natural resources were available.

Implementing the National Geography Standards

Standard 16 Develop and implement a personal plan to conserve water and recycle materials

Objective To create an environmental diary

Class Time 15 minutes

Task Have each student create an environmental diary by listing activities he or she can do to conserve water and recycle materials. Ask students to record the activities they were able to complete for at least a week. At the end of the week, have them speculate about how and why the list of activities might change within the next ten years.

Evaluation Students should list at least ten different activities. Students should complete at least five of the activities.

BEFORE YOU READ

What Do You Know?

Have students look at the Unit Atlas map and name the countries of Eastern Europe. Then ask them to recall information they have heard about any of these countries in the news or in a sports context. Ask students whether anyone in their family—themselves, family members, or ancestors—came from one of the countries in Eastern Europe or the former Soviet Union. Encourage them to share family stories about the culture, daily life, or history of the region.

What Do You Want to Know?

Suggest that students scan the chapter, looking at photographs and maps and reading heads and captions. As they preview the material, have them write questions in their notebooks that they would like answered.

READ AND TAKE NOTES

Reading Strategy: Comparing Explain to students that in this chapter they will be reading about various countries before and after a major change in their histories—the breakup of the Soviet Union. Tell students that their completed charts will provide a good tool for comparing the ways in which this event changed different aspects of life.

In-depth Resources: Unit 4
- Guided Reading Worksheets, pp. 37–39

CHAPTER 13 READING SOCIAL STUDIES

BEFORE YOU READ

▶▶ What Do You Know?

Before you read the chapter, think about what you already know about Europe. Do you have family, friends, or neighbors who were born in Europe? Have you read books, such as the Harry Potter series, that take place in Europe? Think about what you have seen or heard about Italy, England, France, or Germany in the news, during sporting events, and in your other classes.

Region • Euros are the most visible symbol of economic unity in Europe. ▲

▶▶ What Do You Want to Know?

Decide what you know about Europe today. Then, in your notebook, record what you hope to learn from this chapter.

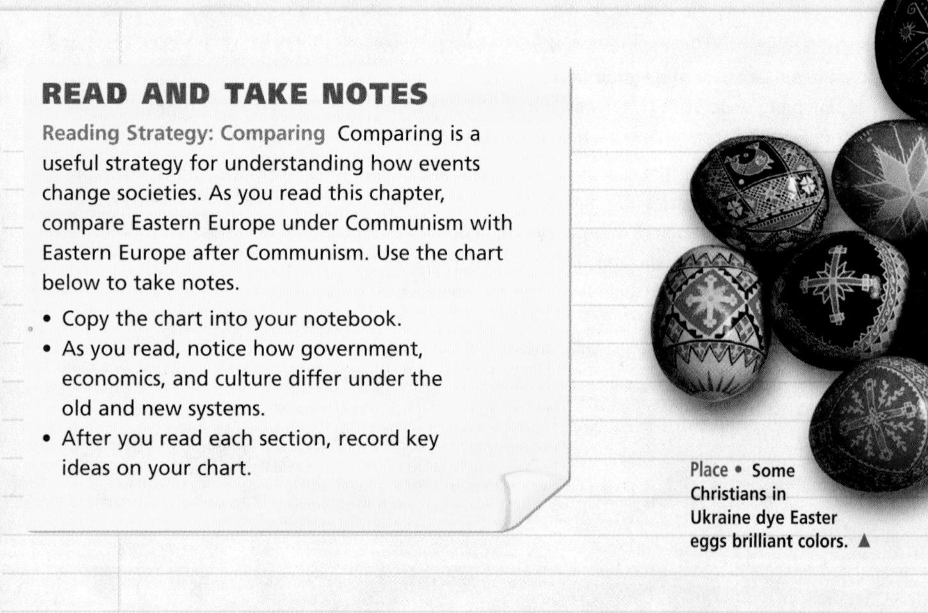

READ AND TAKE NOTES

Reading Strategy: Comparing Comparing is a useful strategy for understanding how events change societies. As you read this chapter, compare Eastern Europe under Communism with Eastern Europe after Communism. Use the chart below to take notes.

- Copy the chart into your notebook.
- As you read, notice how government, economics, and culture differ under the old and new systems.
- After you read each section, record key ideas on your chart.

Place • Some Christians in Ukraine dye Easter eggs brilliant colors. ▲

Aspect	Under Communism	After Communism
Government	government distributed propaganda, controlled economy, restricted cultural activities	greater freedom for citizens, democracy in many Eastern European countries
Economy	controlled by government, widespread poverty, private property seized	change to free-market economies in Eastern Europe, inflation, unemployment
Culture	many cultural celebrations outlawed, art censored, sports and space programs well funded	writers given greater freedom, freedom of religion, increased cultural freedom

Teaching Strategy

Reading the Chapter This is a thematic chapter that examines the political, social, and economic events that have occurred in Eastern Europe since the 1950s. It then analyzes changes that have occurred following the breakup of the Soviet Union. Encourage students to pay particular attention to the effects of Soviet dominance on the countries of Eastern Europe and to the impact of those countries' sudden independence.

Integrated Assessment The Chapter Assessment on page 375 describes several activities for integrated assessment. You may wish to have students work on these activities during the course of the chapter and then present them at the end.

SECTION 1

Eastern Europe Under Communism

TERMS & NAMES
- propaganda
- private property rights
- Nikita Khrushchev
- deposed
- détente

AIN IDEA

e Communist government of e Soviet Union controlled the es of its citizens.

WHY IT MATTERS NOW

Today, many republics of the former Soviet Union have become independent nations.

DATELINE

EXTRA

THE KREMLIN, MOSCOW, APRIL 12, 1961

A 27-year-old Soviet pilot has become the first person to travel into space. Soviet officials proudly announced today that cosmonaut Yuri Gagarin had orbited Earth in 1 hour and 29 minutes.

His 4.75-ton spacecraft, *Vostok I*, flew at a maximum altitude of 187 miles above the planet. Its top speed was 18,000 miles per hour.

Gagarin graduated from the Soviet Air Force cadet school just four years ago. He is the son of a carpenter and began to study flying while in college. Gagarin's space flight puts the Soviet Union a giant step ahead of the United States in the space race.

Movement • Yuri Gagarin becomes the first human in space. ▶

Soviet Culture ❶

The Soviet space program of the 1950s and 1960s brought international attention to that country. Daily life for citizens of the Soviet Union and of the Eastern European countries under its control, however, was difficult. Most people were poor and had little, if any, say in their government.

TAKING NOTES

Use your chart to take notes about modern Europe.

Aspect	Under Communism	After Communism
Government		
Economy		

SECTION OBJECTIVES

1. To describe the degree of control the Soviet Union had over its citizens
2. To describe the Soviet economy
3. To explain attempts to change Soviet-dominated governments and economies

SKILLBUILDER
- Interpreting a Map, p. 355

CRITICAL THINKING
- Comparing, p. 354
- Drawing Conclusions, p. 357

FOCUS & MOTIVATE

WARM-UP

Analyzing Motives Have students read Dateline and answer these questions.

1. Why did the United States and the Soviet Union compete to put people into space?
2. What might the Soviet Union have achieved by putting the first person into space?

INSTRUCT: Objective ❶

Soviet Culture

- Why did the Soviet government want to create a sense of national identity? to discourage ethnic groups from seeking independence
- What actions did the Soviet government take to try to establish a national identity? outlawed cultural celebrations, destroyed churches, prohibited use of native languages, controlled media and artistic expression

In-depth Resources: Unit 4
- Guided Reading Worksheet, p. 37

Reading Study Guide
(Spanish and English), pp. 106–107

Program Resources

In-depth Resources: Unit 4
- Guided Reading Worksheet, p. 37
- Reteaching Activity, p. 44

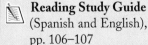

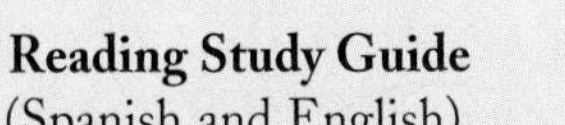

Reading Study Guide
(Spanish and English), pp. 106–107

Formal Assessment
- Section Quiz, p. 188

Integrated Assessment
- Rubric for writing a speech

Outline Map Activities

Access for Students Acquiring English
- Guided Reading Worksheet, p. 103

Technology Resources
classzone.com

TEST-TAKING RESOURCES

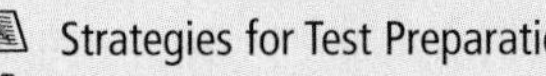

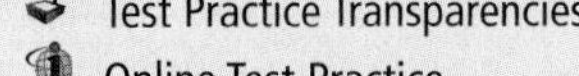

- Strategies for Test Preparation
- Test Practice Transparencies
- Online Test Practice

Strange but TRUE

Laika, a female dog, was part Siberian husky. She weighed about 13 pounds. The area that Laika traveled in was padded, and she had enough space to lie down and stand. She wore a harness and life support system. Before being recruited for the space program, Laika was a stray living on the streets of Moscow. American newspapers dubbed her "Muttnik."

Belka and Strelka were not alone on their journey into space. With them went 40 mice and 2 rats. Belka and Strelka returned to Earth after one day in orbit. Later, Strelka had six puppies, one of which was given to President John F. Kennedy.

CRITICAL THINKING ACTIVITY

Comparing Ask students to reread the two paragraphs under "Creating a National Identity." Discuss the ways the Soviet government went about creating a national identity. Then prompt a discussion about whether the government of the United States tries to do the same. Have students compare and contrast the ways each government tries or has tried to achieve this goal.

Class Time 15 minutes

Space Dogs Four years before Yuri Gagarin blasted into space, a Russian dog orbited the planet. Her name was Laika (LY•kuh), which means "Barker." Laika, pictured below, was launched into space on *Sputnik 2* in November 1957. The Soviets did not then have the ability to bring a spacecraft down safely, and Laika lived in space for only a few days.

In August 1960, however, the Russians sent two other dogs into space. Named Belka and Strelka, they were the first living creatures to go into space and return safely to Earth.

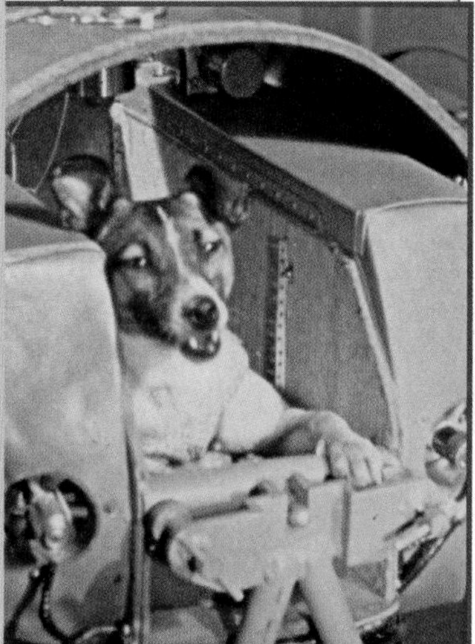

Creating a National Identity The Soviet government was fearful that some ethnic groups might want to break away from the Soviet Union. To keep this from happening, Soviet leaders tried to create a strong national identity. They wanted people in the republic of Latvia, for example, to think of themselves as Soviets, not as Latvians.

To help achieve its goals, the Soviet government created and distributed **propaganda** (PRAHP•uh•GAN•duh), or material designed to spread certain beliefs. Soviet propaganda included pamphlets, posters, artwork, statues, songs, and films. It praised the Soviet Union, its leaders, and Communism.

Soviet Control of Daily Life To prevent different ethnic groups from identifying with their individual cultures rather than with the Soviet Union, the Soviet government outlawed many cultural celebrations. It destroyed churches and other religious buildings and killed thousands of religious leaders. The members of many ethnic groups were not allowed to speak their native languages or celebrate certain holidays.

The Soviet government also controlled communications media, such as newspapers, books, and radio. This meant that most Soviet citizens could not learn much about other nations around the world.

Region • This statue, a form of propaganda, displays the Soviet belief in the unity of the worker (hammer and the farmer (sickle). ▼

Literature and the Arts The works of many writers, poets, and other artists who lived during the Soviet era often were banned or censored. Soviet artists were forced to join government-run unions. These unions told artists what kinds of works they could create. Artists who disobeyed were punished. Some were imprisoned or even killed.

Activity Options

Differentiating Instruction: Less Proficient Readers

Identifying Main Ideas and Details Point out the headings "Creating a National Identity" and "Soviet Control of Daily Life." Explain to students that headings such as these identify main ideas. The paragraphs that follow provide details about the main ideas. On the chalkboard, draw a cluster diagram like the one shown here and guide students in completing it for "Soviet Control of Daily Life." Then have them work with a partner to develop a diagram for "Literature and the Arts."

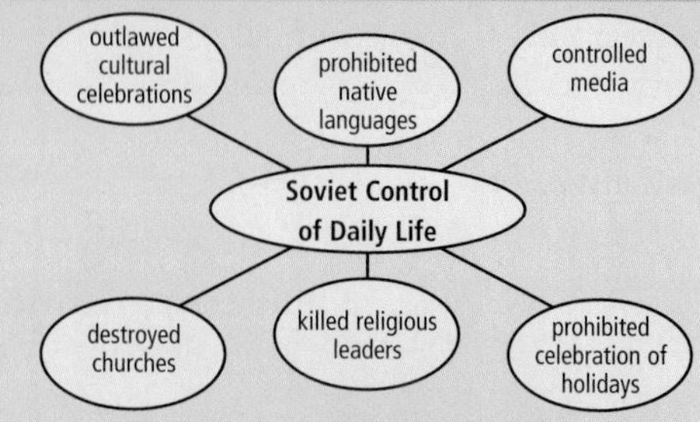

GEOGRAPHY SKILLBUILDER: Interpreting a Map

1. **Place •** Where in the Soviet Union do most Uralic and Altaic people live?
2. **Region •** What is the most common ethnicity of the Soviet Union?

Sports The leaders of the Soviet Union wanted their country to be seen as equal to, if not better than, other powerful nations. One way to achieve this goal was to become a strong competitor in the Olympics and in other international sports competitions.

The Soviet government supported its top athletes and provided for all their basic needs. It even hired and paid for the coaches and paid for all training. The hockey teams and gymnasts of the Soviet Union were among the best in the world.

Region • Romanian gymnasts, like Nadia Comaneci, won medals at the Olympics. ▼

The Soviet Economy

In addition to controlling the governments of the Soviet Union and of those Eastern European countries under its influence, Soviet leaders also ran the economy. When the Soviets installed Communist governments in Eastern Europe after World War II, they promised to improve industry and to bring new wealth to be shared among all citizens. This did not happen.

Government Control Communism in the Soviet Union did not support **private property rights,** or the right of individuals to own land or an industry. The Soviets wanted all major industries to be owned by the government rather than by private citizens. So the government took over factories, railroads, and businesses.

INSTRUCT: Objective 2

The Soviet Economy

- What major industries did the Soviet government take over? factories, railroads, businesses
- What was the Soviet government's role in production? decided what would be produced, how it would be produced, and who would get what was produced

FOCUS ON VISUALS

Interpreting the Map Have students read the title of the map and note that its time frame is about 1950. To avoid any confusion about the term "Caucasian," explain that in the context of ethnic groups, it refers only to certain peoples of the region called the Caucasus, between the Black and Caspian seas. Also ask students to notice the relationship between the ethnic term "Uralic" and the Ural Mountains, the traditional boundary between Europe and Asia, where some of these peoples still live. Have students point out the different ethnic groups that occupy the Caucasus. What might be the consequences of this mix?

Possible Responses Caucasian, Indo-European, Uralic, and Altaic. Many groups in a small area might result in ethnic conflicts.

Extension Suggest that interested students choose one of the regions shown on the map and learn more about its people, languages, geography, and history. They can prepare a short oral report for the class.

Activity Options

Differentiating Instruction: Gifted and Talented

Comparing and Contrasting Have students work in groups to find out how many medals the Soviet Union won in the winter and summer Olympic Games held from 1952 to 1992. Help them think of keywords to find this information on the Internet; results are also available in print sources. Explain that in the first years after the breakup of the Soviet Union, athletes from the former Soviet republics competed together as the Unified Team. (Later, each new nation sent its own athletes.) Then have each group create a chart that compares the final medal standings of the United States and the Soviet Union for each of the Olympic years in which both countries competed. (There were boycotts in 1980 and 1984.) Have groups share their charts with the class.

INSTRUCT: Objective 3

Attempts at Change

- What characterized the era known as "The Thaw"? Citizens had greater freedom; Khrushchev visited the United States.
- What was the "Prague Spring"? a period of improvement when the Soviet government loosened control over Czechoslovakia
- After détente, what did the Soviet Union continue to spend most of its money on? armed forces and nuclear weapons

Spotlight on CULTURE

Aleksandr Solzhenitsyn wrote several books about his experiences in the labor camps. The first was *One Day in the Life of Ivan Denisovich,* published in the Soviet Union in 1962. Nikita Khrushchev, who was promoting anti-Stalinist feelings, intervened personally to allow the book to be published. When Khrushchev was removed from office, Solzhenitsyn's writings were once again banned. *The Gulag Archipelago* was published in the West in 1973. It describes the cruel repression of the Soviet labor camps. Despite his treatment at home, Solzhenitsyn received recognition in other parts of the world. In 1970, he was awarded the Nobel Prize in Literature. However, he declined it because of threats that he would not be able to return to the Soviet Union if he went to Sweden to accept it.

Thinking Critically Answers

1. The government wanted citizens to think that things were wonderful in their country and that everyone was happy with the system.
2. Possible response: The U.S. government does not censor literature, but some local governments and libraries do.

The Soviet government decided what would be produced, how it would be produced, and who would get what was produced. These choices were made based on Soviet interests, not on the interests of the republics or of individuals. Communist countries of Eastern Europe were often unable to meet the needs—including bread, meat, and clothing—of their citizens.

Reading Social Studies

A. Clarifying Who benefited most from Soviet industry?

Reading Social Studies A. Answer The Soviet government, not its citizens, benefited.

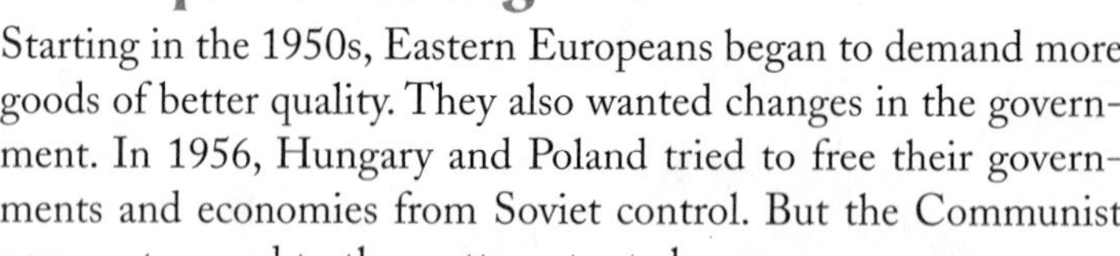

Attempts at Change 3

Starting in the 1950s, Eastern Europeans began to demand more goods of better quality. They also wanted changes in the government. In 1956, Hungary and Poland tried to free their governments and economies from Soviet control. But the Communist army put an end to these attempts at change.

Khrushchev From 1958 until 1964, **Nikita Khrushchev** (KRUSH•chehf) ruled the Soviet Union. During this period, called "The Thaw," writers and other citizens began to have greater freedoms. Khrushchev even visited the United States in 1959, but the thaw in the Cold War did not last. In 1964, with the Soviet economy growing weaker, Khrushchev was **deposed,** or removed from power.

Spotlight on CULTURE

Solzhenitsyn In 1945, army officer Aleksandr Solzhenitsyn (SOHL•zhuh•NEET•sihn), far right, called the Soviet leader Joseph Stalin "the boss." For this, he was sentenced to eight years in slave-labor camps. Later, Solzhenitsyn wrote books about his experiences in those camps. He also wrote a letter against censorship. The government called him a traitor, and in 1969 it forced Solzhenitsyn to leave the writers' union. Five years later, Solzhenitsyn left the country.

Although Solzhenitsyn's works were banned, many Soviet citizens read them in secret. Copies of his and other banned books were passed from person to person across the nation. Through such writings, Soviet citizens learned many things that the government had tried to hide from them.

THINKING CRITICALLY

1. **Analyzing Motives** Why would the Soviet government stop people from reading Solzhenitsyn's books?
2. **Comparing** Compare the censorship of literature in the Soviet Union with censorship in the United States.

For more on Aleksandar Solzhenitsyn, go to RESEARCH LINKS CLASSZONE.COM

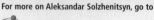

Activity Options

Skillbuilder Mini-Lesson: Reading a Political Cartoon

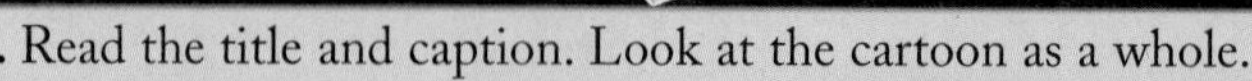

Block Scheduling

Explaining the Skill Remind students that political cartoons are created to express an opinion about a serious subject. They often use humor as well as symbols to make their point. They may have a title or caption.

Applying the Skill Challenge students to create political cartoons that express an opinion about the Soviet economy as seen through the eyes of Eastern Europeans during the 1950s and 1960s. Then have students use the following strategies as they share their cartoons in small groups.

1. Read the title and caption. Look at the cartoon as a whole.
2. Try to identify people in the cartoon. Look at how features or characteristics of the people may have been exaggerated. Look for details and symbols.
3. Summarize the cartoonist's message and point of view.

Reading Social Studies

B. Recognizing Important Details How did the Soviet Union maintain control over other Eastern European nations?

Reading Social Studies B. Answer They used military force and economic pressure.

The Prague Spring In January 1968 in Czechoslovakia, Alexander Dubček (DOOB•chek) became the First Secretary of the Czechoslovak Communist Party. His attempts to lessen the Soviet Union's control over Czechoslovakia led to a period of improvement called the "Prague Spring." Czech citizens enjoyed greater freedoms, including more contact with Western Europe. In August of that year, however, the Soviet Union sent troops to force a return to strict Communist control. Dubček was later replaced, and Soviet controls were back in place.

Place • Nikita Khrushchev, the son of a miner and grandson of a peasant, lessened government control of Soviet citizens. ▲

Détente The member nations of NATO, which were concerned about starting an all-out war with the Soviet Union, were unable to stop the Soviet control of Eastern Europe. In the 1970s, however, leaders of the Soviet Union and the United States began to have more contact with each other. This led to a period of **détente** (day•TAHNT), or lessening tension, between the members of NATO and the Warsaw Pact nations.

Region • Citizens of Czechoslovakia protested Soviet control in 1968. ◄

MORE ABOUT...
Nikita Khrushchev

Joseph Stalin's death on March 5, 1953, left a void in the Communist leadership of the Soviet Union. At first, a group of leaders shared power. Gradually, Nikita Khrushchev (1894–1971) gained influence. In 1956, he delivered an address to the 20th Party Congress attacking Stalin and his policies. The "destalinization" of the Soviet Union had already begun, but now Khrushchev led it. Many legal procedures were revived, labor camps were closed, and the secret police became less powerful. By 1958, Khrushchev had gained complete power in the Soviet Union. He was a shrewd leader who sought peaceful coexistence with the West rather than confrontation.

CRITICAL THINKING ACTIVITY

Drawing Conclusions Have students study the photograph at the bottom of the page, which shows the violence that ended the "Prague Spring." Have them describe what they see happening in the photograph. Encourage them to identify details and emotions that the photo captures. Then ask them to discuss what message they think the photographer intended to convey about this event.

Class Time 10 minutes

Activity Options

Interdisciplinary Links: Language Arts/Writing

B Block Scheduling

Class Time One class period

Task Writing a journal entry about life under Soviet rule

Purpose To develop an understanding of the points of view of citizens in Eastern Europe

Supplies Needed
- Writing paper
- Pens or pencils

Activity Encourage students to imagine that they are young people living in one of the nations under Soviet control in the 1960s, 1970s, or 1980s. Have them write a journal entry describing either the restrictions and problems of an ordinary day or the events of an upheaval such as the violent ending to the "Prague Spring."

ASSESS & RETEACH

Reading Social Studies Have students add notes to the first column of the chart on page 352.

Formal Assessment
- Section Quiz, p. 188

RETEACHING ACTIVITY

Divide students into three groups and assign each group a heading in the section. Tell students to review the material under their assigned heading and then work together to create a short quiz. Have students orally quiz other groups.

In-depth Resources: Unit 4
- Reteaching Activity, p. 44

Access for Students Acquiring English
- Reteaching Activity, p. 108

Place • The old city of Dubrovnik is in Croatia, a part of the former Yugoslavia, which was a Communist country in Eastern Europe. ▶

Economic Crisis By the 1980s, economic conditions in the Soviet Union and in those countries under its control had still not improved. Even after détente, the Soviet government continued to spend most of its money on the armed forces and nuclear weapons. In addition, people who lived in the non-Russian republics of the Soviet Union now wanted more control over their own affairs. Many citizens began to reject the Soviet economic system, but the Soviet leaders refused to give up any of their power or control.

SECTION 1 ASSESSMENT

Terms & Names

1. **Explain the significance of:** (a) propaganda (b) private property rights (c) Nikita Khrushchev (d) deposed (e) détente

Using Graphics

2. Use a chart like this one to list and describe major aspects of Soviet culture.

Aspects of Soviet Culture

Main Ideas

3. (a) Why did Soviet leaders try to create a strong national identity?
   (b) What began to happen in Eastern Europe in the 1950s?
   (c) Describe the significance of the "Prague Spring."

Critical Thinking

4. **Analyzing Motives**
   Why do you think the works of many writers, poets, and artists were banned or censored during the Soviet era?

Think About
- what Soviet citizens learned from Solzhenitsyn's works
- the government's use of propaganda
- what life was like for most Soviet citizens

ACTIVITY -OPTION- Reread the information under "Literature and the Arts" and the Spotlight on Culture feature. Write a **speech** for or against censorship in the arts.

Section 1 Assessment

1. Terms & Names
- a. propaganda, p. 354
- b. private property rights, p. 355
- c. Nikita Khrushchev, p. 356
- d. deposed, p. 356
- e. détente, p. 357

2. Using Graphics

Most people were poor.
Citizens had little say in government.
Officials suppressed ethnic identity.
Soviet propaganda praised Soviet Union.

3. Main Ideas
- a. They feared that ethnic groups might want to break away from the Soviet Union.
- b. People began trying to regain their independence from the Soviet Union.
- c. The Soviet Union loosened control over Czechoslovakia.

4. Critical Thinking

Possible Response The Soviet government feared that the truth about events and conditions in the Soviet Union might incite people to revolt.

ACTIVITY OPTION

Integrated Assessment
- Rubric for writing a speech

SKILLBUILDER

Using an Electronic Card Catalog

▶▶ Defining the Skill

To find books, magazines, or other sources of information in a library, you may use an electronic card catalog. This catalog is a computerized search program on the Internet that lists every book, periodical, or other resource found in the library. You can search for resources in the catalog in four ways: by title, by author, by subject, and by keyword. Once you have typed in your search information, the catalog will give you a list of every resource that matches it. This is called bibliographic information. You can use an electronic card catalog to build a bibliography, or a list of books, on the topic you are researching.

▶▶ Applying the Skill

The screen below shows the results of an electronic search for information about the Danube River. To use the information on the screen, follow the strategies listed below.

How to Use an Electronic Card Catalog

Strategy ❶ To begin your search, choose Subject, Title, Author, or Keyword. The student doing this search chose "Subject" and then typed in "Danube River."

Strategy ❷ Based on your search, the catalog will give you a list of records that match that subject. You must then select one of the records to view the details about the resource. The catalog will then give you a screen like the one to the right. This detailed record lists the author, title, and information about where and when the resource was published, and by whom.

Strategy ❸ Locate the call number for the book. The call number indicates the section in the library where you will find the book. You can also find out if the book is available in the library you are using. If not, it may be available in another library in the network.

SEARCH REQUEST: Danube River
1 Subject Title Author Keyword
Find Options Locations Backup Startover Help
2 Lessner, Erwin Christian. The Danube; the dramatic history of the great river and the people touched by its flow. Westport, Conn.: Greenwood Press, 1961.
AUTHOR: Lessner, Erwin Christian
2 TITLE: The Danube; the dramatic history of the great river and the people touched by its flow
2 PUBLISHED: Westport, Conn., Greenwood Press, 1961
PAGING: 529 p.
NOTES: Includes maps, bibliography
3 CALL NO: 914.9603 L Book Available

▶▶ Practicing the Skill

Review the text in Chapter 13, Section 1 to find a topic that interests you, such as Yuri Gagarin. Use the Subject search on an electronic card catalog to find information about your topic. Make a bibliography about the subject. Organize your bibliography alphabetically by author. For each book you list, also include the title, city, publisher, and date of publication.

SKILLBUILDER

Using an Electronic Card Catalog

Defining the Skill

Tell students that many libraries today use an electronic card catalog, a computer program that keeps track of every book and other resource in the collection of a single library or a larger library system. (Some students may also be familiar with the older card catalog system, which consists of file drawers containing index cards.) Like older systems, an electronic card catalog indexes resources in four ways: Subject, Title, Author, and Keyword.

Applying the Skill

How to Use an Electronic Card Catalog If possible, use an actual electronic card catalog to take students through the strategies step by step. Remind them to follow these steps in order whenever they use an electronic card catalog. Library software systems may vary in the way they work. Students should follow any on-screen directions.

Write a Bibliography

Discuss the information shown on the screen, and help students identify which items they should include in a bibliography entry. Remind them to list entries alphabetically by the author's last name.

Practicing the Skill

Have students use the Subject search as suggested to create a bibliography about a topic in Section 1. If students need additional practice, suggest that they use an Author search to create a list of works by an author whose books they enjoy reading.

In-depth Resources: Unit 4
- Skillbuilder Practice, p. 42

Career Connection: Database Programmer

Encourage students who enjoy using an electronic card catalog to research careers that involve creating and maintaining databases. Tell students that a database is a large collection of data that can be retrieved and searched quickly, usually using a computer. An electronic card catalog is one type of database.

Block Scheduling

1. Suggest that students find out what other kinds of databases exist and what a database programmer does.
2. Help students learn about the aptitudes and education needed to become a database programmer.
3. Invite students to create a graphic organizer such as a flow chart to show what they have learned.

SECTION OBJECTIVES

1. To identify changes in Eastern Europe and Russia after the breakup of the Soviet Union
2. To describe the war in the Balkans
3. To describe modern Russia's culture, government, resources, industry, and economies

SKILLBUILDER
- Interpreting a Map, pp. 361, 363, 365

CRITICAL THINKING
- Summarizing, p. 362
- Comparing, p. 365

FOCUS & MOTIVATE

WARM-UP

Making Inferences Have students read Dateline and discuss these questions.

1. How do you think people reacted when Gorbachev began to remove troops from Eastern Europe?
2. What do you think were the effects of Gorbachev's changes?

INSTRUCT: Objective 1

The Breakup of the Soviet Union/Modern Eastern Europe

- What major change occurred in Eastern Europe during 1991? Soviet republics and nations under Soviet control declared independence and set up new governments.
- How are the economies of Eastern European countries changing? from command economies to free-market economies

In-depth Resources: Unit 4
- Guided Reading Worksheet, p. 38

Reading Study Guide (Spanish and English), pp. 108–109

Eastern Europe and Russia

SECTION 2

TERMS & NAMES
Mikhail Gorbachev
parliamentary republic
coalition government
ethnic cleansing
Duma

MAIN IDEA
After the breakup of the Soviet Union, many former Soviet republics and countries of Eastern Europe became independent.

WHY IT MATTERS NOW
Nations once under Soviet rule are taking steps toward new economies and democratic governments.

DATELINE

THE KREMLIN, MOSCOW, 1988—To reduce military spending, the Soviet Union has begun removing large numbers of troops and arms from Eastern Europe. This latest news is just one of many changes in the Soviet government since Mikhail Gorbachev (GAWR•buh•chawf) came to power three years ago.

Although Gorbachev believes in the ideals of the Soviet system, he thinks that change is necessary to help solve the country's economic and political problems. Since 1985 Gorbachev has reduced Cold War tensions with the United States. At home in the Soviet Union, he has allowed more political and economic freedom.

Region • Mikhail Gorbachev leads the Soviet Union toward a freer society. ▲

The Breakup of the Soviet Union 1

Mikhail Gorbachev's reforms did not solve the problems of the Soviet Union. The economy continued to get worse. When Gorbachev did not force the countries of Eastern Europe to remain Communist, this further displeased many Communists.

TAKING NOTES
Use your chart to take notes about modern Europe.

Aspect	Under Communism	After Communism
Government		
Economy		

Program Resources

In-depth Resources: Unit 4
- Guided Reading Worksheet, p. 38
- Reteaching Activity, p. 45

Reading Study Guide (Spanish and English), pp. 108–109

Formal Assessment
- Section Quiz, p. 189

Integrated Assessment
- Rubric for writing a personal essay

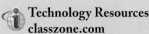

Outline Map Activities

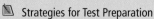

Access for Students Acquiring English
- Guided Reading Worksheet, p. 104

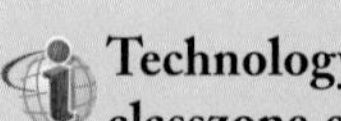
Technology Resources
classzone.com

TEST-TAKING RESOURCES
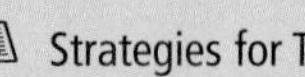
Strategies for Test Preparation
Test Practice Transparencies
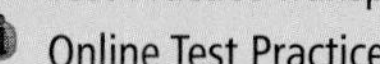
Online Test Practice

Vocabulary

coup d'état: the overthrow of a government, usually by a small group in a position of power; often shortened to "coup"

In 1991, a group of more traditional Soviet leaders tried to take over the Soviet government. Thousands of people opposed this coup d'état (KOO•day•TAH), and the coup failed. Then, one by one, the Soviet republics declared independence. The Warsaw Pact was dissolved. By the end of 1991, the Soviet Union no longer existed. The huge country had become 15 different nations.

Modern Eastern Europe 

Each former Soviet republic set up its own non-Communist government. The countries of Eastern Europe that had been under Soviet control held democratic elections, and many wrote or revised their constitutions.

BACKGROUND

The Central Asian Soviet republics were mostly Muslim. These republics are now the countries of Kazakhstan, Turkmenistan, Uzbekistan, Kyrgyzstan, and Tajikistan.

In some countries, such as the Czech Republic, former Communists were banned from important government posts. In other countries, such as Bulgaria, the former Communists reorganized themselves into a new political party and have won elections. Many different ethnic groups also tried to create new states within a nation or to reestablish old states that had not existed in many years.

Parliamentary Republics Today, most of the countries of Eastern Europe are parliamentary republics. A **parliamentary republic** is a form of government led by the head of the political party with the most members in parliament. The head of government, usually a prime minister, proposes the programs that the government will undertake. Most of these countries also have a president who has ceremonial, rather than political, duties.

Former Soviet Republics and Warsaw Pact Members, 2001

GEOGRAPHY SKILLBUILDER: Interpreting a Map

1. **Location** • Which former Soviet republics and Warsaw Pact members border Russia?
2. **Region** • On which continent are most of these countries located?

Geography Skillbuilder Answers

1. Estonia, Latvia, Belarus, Ukraine, Georgia, Azerbaijan, Kazakhstan; Poland and Lithuania border enclave on Baltic Sea
2. Slightly more than half are in Europe; the rest are in Central Asia.

FOCUS ON VISUALS

Interpreting the Map Explain to students that this map shows both the former republics that made up the Soviet Union and the Eastern European nations that were more or less under Soviet control. Have students trace the border of the former Soviet Union to make this distinction. Then ask students to name the former Soviet republics that do not border other European nations.

Possible Responses Georgia, Armenia, Azerbaijan, Kazakhstan, Uzbekistan, Turkmenistan, Kyrgyzstan, Tajikistan

Extension Have students choose one of the former Soviet republics and prepare a short illustrated report on its history and government since 1991, including a map and photographs if possible. Ask students to present their reports in class.

Activity Options

Differentiating Instruction: Students Acquiring English/ESL

Borrowed Words Use the terms *détente* and *coup d'état* to introduce the idea that everyday English uses a number of words and terms taken directly from other languages. Encourage students to suggest words or terms from their own languages that they think would be useful to native English speakers. Then draw up a chart like this one and ask them to use their dictionaries to find out how the term is used in English.

esprit (French)	
savoir-faire (French)	
chic (French)	
gumbo (Bantu)	
machismo (Spanish)	
cliché (French)	

CRITICAL THINKING ACTIVITY

Summarizing Ask students to summarize the economic and social problems many Eastern European countries faced after they gained independence. As students identify information that should be included in a summary, have them record their ideas on a concept web. Then have students use the web to write a summary.

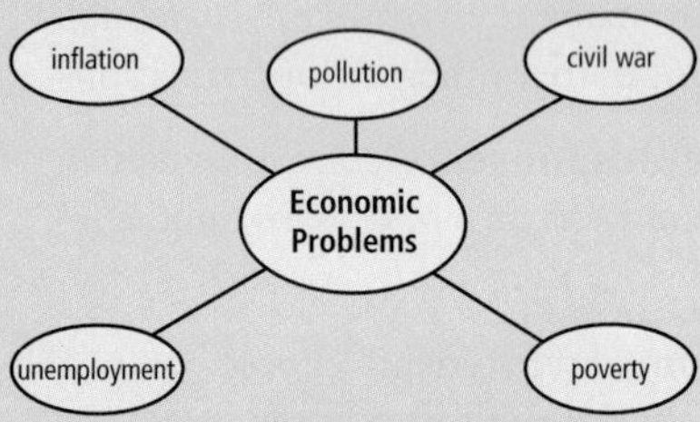

Class Time 15 minutes

Spotlight on CULTURE

In Ukraine, decorating Easter eggs is an art. The Easter eggs, which are called *pysanky,* are decorated using wax and dyes. It can take 15 or more hours to decorate a single egg. Every color and design has symbolic meaning. Red, for example, symbolizes love, and for centuries was the most important color for *pysanky*. Brown symbolizes Earth, and green represents spring, renewal, and freedom. The sun design, a circle with lines running out from the edges, symbolizes good fortune, while the flower stands for love and charity.

Thinking Critically Answers

1. The Soviet government discouraged religious and cultural traditions.
2. Students' accounts will vary, but most will probably mention traditional foods, comparable to the Ukrainian Easter eggs.

In some countries, small political parties have joined forces to work together to form a government. This is called a **coalition government.**

New Economies Under Soviet rule, Eastern Europe struggled economically and its people's freedoms were severely restricted. Although Eastern Europeans gained their freedom, they also faced problems such as inflation and unemployment.

Eastern Europe's countries are changing from command economies to free-market economies. Some countries, such as Slovakia, made this change slowly. Others, such as Poland, reformed their economic system and achieved economic success.

Many former Soviet republics, which did not quickly reform their economic systems, are in bad economic shape. Some of these nations are terribly poor. Struggles for power have led to violence and sometimes civil war. Pollution from the Soviet era threatens people's health. Still, some republics, including Ukraine, Latvia, Lithuania, and Estonia, are making progress as independent nations.

Defense After the breakup of the Soviet Union, Eastern European nations no longer looked to the Soviet government to defend them. Many wanted to become members of NATO. Belonging to NATO would help assure them of protection in case of invasion.

A. Comparing Compare a command economy with a free-market economy.

Reading Social Studies A. Answer In a command economy, the government owns industries, decides what and how much to produce, and how to distribute it. A free-market economy operates primarily through supply and demand.

Easter in Ukraine In Ukraine, most Christians belong to the Orthodox Church. These Ukrainians are known for the special way in which they celebrate the Easter holiday. They create beautiful Easter eggs, which are dyed bright colors and covered with intricate designs. These eggs are so beautiful that people around the world collect them.

Ukrainians also bake a special bread for Easter. They decorate it with designs made from pieces of dough. Families bring the bread and other foods to church to be blessed on Easter. Then they eat the foods for the holiday feast.

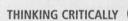

THINKING CRITICALLY

1. **Analyzing Issues**
   Why were Ukrainian Easter eggs not common during the Soviet era?
2. **Comparing**
   How do your family's holiday customs compare with Ukrainian customs?

For more on Easter in Ukraine, go to

Activity Options

Multiple Learning Styles: Visual/Kinesthetic

Class Time One class period

Task Creating a poster promoting travel to an Eastern European country

Purpose To learn more about the countries of Eastern Europe

Supplies Needed
- Library or Internet resources
- Poster board
- Markers or crayons

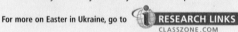

Block Scheduling

Activity Explain that the countries of Eastern Europe are building modern tourist facilities and encouraging tourists to visit. Ask each student to choose a country from this region and learn more about it. Students should take notes on the country's physical geography, climate, and attractions. Then have them create posters promoting travel to their chosen countries. Tell them to create titles and slogans for their posters. Have them share their posters and what they have learned with their classmates.

In 1999 three new members joined NATO: Poland, Hungary, and the Czech Republic. In 2001 Bulgaria, Romania, Slovakia, Slovenia, and the Baltic states were also working to become NATO members.

Vocabulary

Baltic states: Estonia, Latvia, and Lithuania—former Soviet republics that are on the Baltic Sea

War in the Balkan Peninsula ❷

Since the late 1980s, much of Eastern Europe has been a place of turmoil and struggle. Yugoslavia, one of the countries located on Europe's Balkan Peninsula, has experienced terrible wars, extreme hardships, and great change.

Under Tito After World War II, Yugoslavia came under Marshal Tito's (TEE•toh) dictatorship. Tito controlled all the country's many different ethnic groups, which included Serbs, Croats, and Muslims. His rule continued until his death in 1980. Slobodan Milošević (sloh•boh•DON muh•LAW•shuh•vich) became Yugoslavia's president in 1989, after years of political turmoil.

BACKGROUND

By 1991 Croatia, Slovenia, Macedonia, and Bosnia-Herzegovina had gained independence from Yugoslavia. Only Serbia and Montenegro were still part of the Yugoslavian federation.

Milosevic Slobodan Milošević, a Serb, wanted the Serbs to rule Yugoslavia. The Serbs in Bosnia began fighting the Croats and Muslims living there. The Bosnian Serbs murdered many Muslims so that Serbs would be in the majority. The Serbs called these killings of members of minority ethnic groups **ethnic cleansing.** Finally, NATO attacked the Bosnian Serbs and ended the war.

Connections to Science

Pollution Soviet leaders thought that industry would improve life for everyone. Developing industry was so important that the Soviet government did not worry about pollution. Few laws were passed to protect the environment.

In the 1970s and 1980s there was not enough money to modernize industry or to reduce pollution. Some areas also could not afford proper sewage systems or recycling plants. Today, Eastern Europe has some of the worst pollution problems on the continent.

The Balkan States, 1991 and 2001

AUSTRIA · HUNGARY · Slovenia · Croatia · Vojvodina · YUGOSLAVIA · Belgrade · ROMANIA · Bosnia and Herzegovina · Serbia · SAN MARINO · Adriatic Sea · ITALY · Montenegro · Kosovo · BULGARIA · Macedonia · ALBANIA · GREECE

National boundaries, 2001 · Yugoslavia, 1991 · Autonomous province boundaries, 2001 · National capital

0 100 200 miles · 0 100 200 kilometers · 12°E · 18°E

Geography Skillbuilder Answers

1. Macedonia
2. Slovenia, Croatia, Bosnia-Herzegovina, Macedonia, and two autonomous provinces plus two countries in federation

GEOGRAPHY SKILLBUILDER: Interpreting a Map

1. **Location** • Which Balkan state borders Greece?
2. **Region** • How many countries developed from Yugoslavia?

INSTRUCT: Objective ❷

War in the Balkan Peninsula

- What were the three major ethnic groups in Yugoslavia when Tito was dictator? Serbs, Croats, Muslims
- Why did Milosevic order ethnic cleansing? to kill minority ethnic groups so the Bosnian Serbs would be in the majority

Connections to Science

Since the fall of communism, environmental awareness has increased in Eastern Europe. Environmental organizations began to form in the 1980s. In 1990 the Regional Environmental Center for Central and Eastern Europe was formed. It is a not-for-profit organization devoted to solving environmental problems and encouraging public participation in environmental issues.

FOCUS ON VISUALS

Interpreting the Map Explain to students that the nation of Yugoslavia was formed after World War I as a federation of republics that brought together people of diverse ethnic and religious groups. Ask students to identify the countries that border the former Yugoslavia.

Response Austria, Hungary, Romania, Bulgaria, Greece, Albania, small part of Italy

Extension Have students research and report on the pre–World War I backgrounds of the republics of the former Yugoslavia.

Activity Options

Differentiating Instruction: Less Proficient Readers

Identifying Sequence Some students may have difficulty following the sequence of events in Yugoslavia that began with Tito's rise to power after World War II and ended with Milosevic's arrest for war crimes. To clarify these events, help students create a sequence chart like the one shown here. Then have students use the chart to review "War in the Balkan Peninsula."

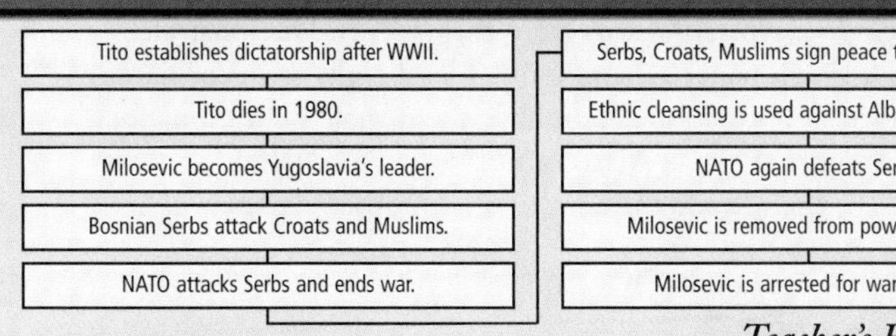

Tito establishes dictatorship after WWII.
Tito dies in 1980.
Milosevic becomes Yugoslavia's leader.
Bosnian Serbs attack Croats and Muslims.
NATO attacks Serbs and ends war.
Serbs, Croats, Muslims sign peace treaty in 1995.
Ethnic cleansing is used against Albanians in 1999.
NATO again defeats Serbs.
Milosevic is removed from power in 2000.
Milosevic is arrested for war crimes.

INSTRUCT: Objective 3

Modern Russia

- What new freedoms do Russians have since the breakup of the Soviet Union? freedom to elect leaders and own property, freedom of speech and religion, end of censorship
- How has Russia's economy changed since the Soviet Union collapsed? developing a free-market economy; private property ownership; new businesses
- What difficulties have resulted from the free-market economy? uncontrolled high prices; economic inequality; crime

Before Communist rule, religion was very important in many Russians' lives. Most homes, even peasant cottages, had a special spot called the "beautiful corner" *(krasny ugol)* for an icon. One aftereffect of the breakup of the Soviet Union was a religious revival.

In 1995 the Serbs, Croats, and Muslims of Bosnia signed a peace treaty. In 1999 Milošević began using ethnic cleansing against the Albanians in Kosovo, a region of Serbia. NATO launched an air war against Yugoslavia that ended with the defeat of the Serbs. In 2000, public protests led to Milošević's removal. He was subsequently arrested and tried for war crimes by the United Nations.

Modern Russia 3

Life in Russia has improved since the breakup of the Soviet Union. Russian citizens can elect their own leaders. They enjoy more freedom of speech. New businesses have sprung up, and some Russians have become wealthy.

Unfortunately, Russia still faces serious problems. Many leaders are dishonest. The nation has been slow to reform its economic system. Most of the nation's new wealth has gone to a small number of people, so that many Russians remain poor. The crime rate has grown tremendously. The government has also fought a war against Chechnya (CHECH•nee•yah), a region of Russia that wants to become independent.

Russian Culture The fall of communism helped most Russians to follow their cultural practices more freely. Russians gained the freedom to practice the religion of their choice. They can also buy and read the great works of Russian literature that once were banned. At the beginning of the 21st century, writers and other artists also have far more freedom to express themselves.

New magazines and newspapers are being published. Even new history books are being written. For the first time in decades, these publications are telling more of the truth about the Soviet Union.

Russia's Government Russia has a democratic form of government. The president is elected by the people. The people also elect members of the **Duma** (DOO•muh), which is part of the legislature.

The World's Heritage

Russian Icons A special feature of Russian Orthodox churches is their beautiful religious paintings called icons (EYE•kahns). Russian icons usually depict biblical figures and scenes. They often decorate every corner of a church.

The greatest Russian icon painter was Andrei Rublev (AHN•dray ruhb•LYAWF). He worked in the late 1300s and early 1400s. Rublev's paintings, one of which is shown below, are brightly colored and highlighted in gold. His work influenced many later painters, and today he is considered one of the world's great religious artists.

Reading Social Studies

B. Identifying Problems What are the main problems that face Russia today?

Reading Social Studies
B. Possible Answers
crime; dishonest officials; economic problems; small group of newly rich people, but many still poor

BACKGROUND

One of the most popular pastimes in Russia is the game of chess. In fact, many of the world's greatest chess players, such as Boris Spassky, have been Russian.

Activity Options

Differentiating Instruction: Gifted and Talented

Crossword Puzzle Have students work in pairs to create crossword puzzles that include important terms from this section. Demonstrate how they can use graph paper to configure at least eight terms or names in a crossword-puzzle formation. Then have them write definitions for each term and number them as Across and Down clues. When students have finished, have them trade their puzzles with other classmates.

Block Scheduling

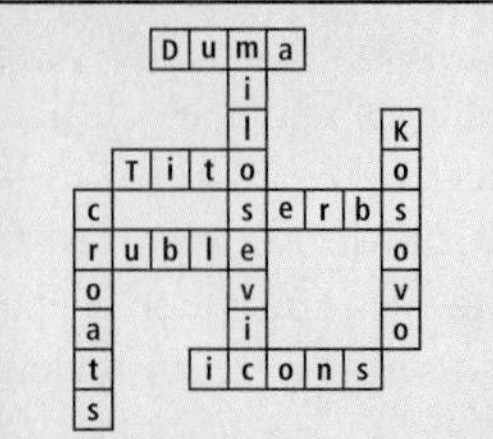

Russia's Natural Resources Today

Geography Skillbuilder Answers
1. oil, coal, natural gas
2. forest

GEOGRAPHY SKILLBUILDER: Interpreting a Map

1. **Human-Environment Interaction** • Name three of Russia's more common natural resources.
2. **Place** • What is the most common type of land in Russia?

BACKGROUND

Russian highways are in poor condition. Also, many rivers and major ports are closed by ice in the winter. As a result, most Russian goods are transported by railroad.

Democracy is still new to the Russian people. Some citizens are working to improve the system to reduce corruption and to ensure that everyone receives fair treatment. Even the thought of changing the government is new to most Russians. Under the Soviets, people had to accept things the way they were.

Resources and Industry The map above shows Russia's major natural resources. The country is one of the world's largest producers of oil. Russia also contains the world's largest forests. Its trees are made into lumber, paper, and other wood products.

Russian factories produce steel from iron ore. Other factories use that steel to make tractors and other large machines. Since Russian ships can reach both the Pacific and Atlantic oceans, Russia also has a large fishing industry.

Economics Following the lead of Eastern European countries, Russia has been moving toward a free-market economy. Citizens can own land, and foreign companies are encouraged to do business in Russia. These changes have given many Russians more opportunities, but they have also brought difficulties.

Connections to Language

The Russian Language More than 150 million people speak Russian. It is related to other Slavic languages of Eastern Europe, including Polish, Serbian, and Bulgarian.

Russian is written using the Cyrillic (suh•RIHL•ihk) alphabet, which has 33 characters.

Many of the newly independent republics are now returning to the Latin alphabet, used to write English and most other languages of the Western world. The major powers in the world economy base their languages on the Latin alphabet, making communication easier with other countries.

FOCUS ON VISUALS

Interpreting the Map Go over the map key with students. Make sure that they notice the dotted areas showing "farmland" overlaying both forest and grassland areas. Ask them to point out the oceans in which there is fishing.

Response Arctic Ocean, Sea of Okhotsk

Extension Ask students to find and display photographs or paintings of Russian forests and steppe lands.

CRITICAL THINKING ACTIVITY

Comparing Ask students to compare the quality of life of Soviet citizens before and after the breakup of the Soviet Union. Create a Venn diagram to help make comparisons.

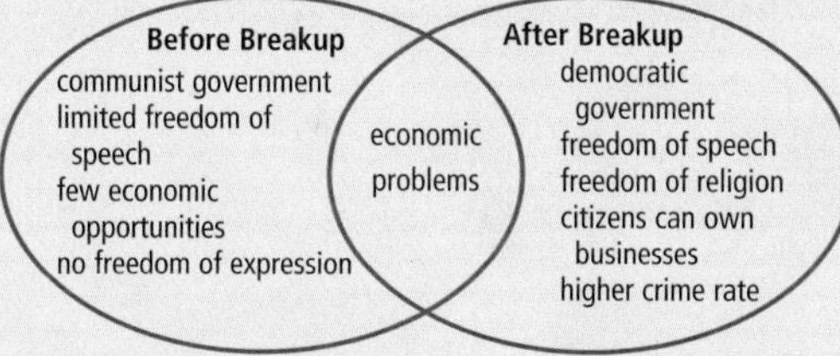

Class Time 15 minutes

Connections to Language

The Cyrillic alphabet was invented in the early 800s by followers of two missionaries, St. Cyril and St. Methodius, who traveled to Eastern Europe to convert the Slavs. The alphabet is based on Greek, but because the Slavic languages have more sounds than Greek, more letters were added. The first book written in Cyrillic was a translation of the Bible.

Activity Options

Interdisciplinary Link: Geography

Class Time One class period

Task Creating natural resource maps of Eastern European countries

Purpose To identify natural resources found in Eastern Europe

Supplies Needed
- Library or Internet resources
- Drawing paper
- Pencils or pens

Activity Have students study the map of Russia's natural resources. Then assign groups of students to create natural resource maps of the other independent countries of Eastern Europe. Suggest that groups delegate the responsibilities of researching and illustrating. Display the completed maps, and use them to compare and contrast the natural resources of countries throughout Eastern Europe.

ASSESS & RETEACH

Reading Social Studies Have students add notes to the second column of the chart on page 352.

Formal Assessment
- Section Quiz, p. 189

RETEACHING ACTIVITY

Divide the class into groups of four, and assign a major heading from the section to each group. Ask students to work together to prepare an outline about the information presented in their assigned material. Then have a person from each group write the outline on the board. Use students' outlines to review the section.

In-depth Resources: Unit 4
- Reteaching Activity, p. 45

Access for Students Acquiring English
- Reteaching Activity, p. 109

Place • **Forestry is a major industry in Russia. These harvested logs are being floated downriver to be processed.** ▶

Prices are no longer controlled by the government. This means that companies can charge a price that is high enough for them to make a profit. At the beginning of the 21st century, however, people's wages have not risen as fast as prices. Many people cannot afford to buy new products.

Some Russians have done well in the new economy. On the other hand, people with less education and less access to power have not done as well. Also, today most new businesses and jobs are in the cities, which means that people in small towns have fewer job opportunities.

BACKGROUND

The Russian government is unable to enforce tax laws. Many people don't pay their taxes. Without that money, the government cannot provide basic services, such as health care.

SECTION 2 ASSESSMENT

Terms & Names

1. Explain the significance of: (a) Mikhail Gorbachev (b) parliamentary republic (c) coalition government (d) ethnic cleansing (e) Duma

Using Graphics

2. Use a flow chart like this one to outline the changes in Eastern Europe and Russia from 1988 through 2000.

1988:
↓
↓

Main Ideas

3. (a) What happened to the governments of the former Soviet republics after independence?

(b) How have the economies of Eastern European countries changed now that those countries are free?

(c) In what ways has life in Russia improved since the breakup of the Soviet Union?

Critical Thinking

4. Making Inferences

Why do you think many Eastern European countries would like to join NATO?

Think About
- what happened to the Warsaw Pact
- the economies of Eastern Europe
- the relationship between Eastern Europe and Russia

ACTIVITY -OPTION- Reread the information in the Spotlight on Culture feature. Write a short, personal **essay** that describes a special family, school, neighborhood, or holiday celebration in which you participated.

Section 2 Assessment

1. Terms & Names

a. Mikhail Gorbachev, p. 360
b. parliamentary republic, p. 361
c. coalition government, p. 362
d. ethnic cleansing, p. 363
e. Duma, p. 364

2. Using Graphics

1988	Soviet Union removes many troops and arms from Eastern Europe.
1991	Soviet republics declare independence; Yeltsin becomes first freely elected president of Russia.
1999	Some Eastern European countries join NATO.

3. Main Ideas

a. They set up their own non-Communist governments, mostly parliamentary republics, and held elections.
b. They changed from command economies to free-market economies.
c. People can elect their leaders; they have freedom of speech and religion; they can own land and businesses and travel freely; writers and artists have more freedom.

4. Critical Thinking

Possible Response They might want to join NATO because they are no longer defended by the Soviet Union. Many have weak economies and cannot afford a large military for self-defense.

ACTIVITY OPTION

Integrated Assessment
- Rubric for writing a personal essay

The European Union

TERMS & NAMES
European Union
currency
euro
tariff
standard of living
Court of Human Rights

AIN IDEA

ropeans want to maintain a high ality of life for all citizens while eserving their unique cultures.

WHY IT MATTERS NOW

A prosperous and culturally diverse Europe provides goods and markets for the rest of the world.

DATELINE

WESTERN EUROPE, DECEMBER 2001—Starting next month, people in many Western European nations will begin trading their old bills and coins for euros—the new money of the European Union (EU). The design of the bills, below, is the same for all EU members.

The design of the euro coins, however, will be different. Individual countries are minting their own. As shown here, one side has a standard euro design. The other side has national symbols that relate to each country. In 1996, artists and sculptors from all over Europe entered a contest to design the coins. The winner was Luc Luycx (lewk lowx) from Belgium.

Region • Euros reached the European market in January 2002. ▲

Western Europe Today 1

Today, in Western Europe, all national leaders share their power with elected lawmakers. Citizens take part in government by voting and through membership in a variety of political parties. The Unit Atlas on pages 260–269 shows modern Europe.

TAKING NOTES
Use your chart to take notes about modern Europe.

Aspect	Under Communism	After Communism
Government		
Economy		

SECTION OBJECTIVES

1. To identify the benefits of membership in the European Union
2. To describe the economies of European Union members
3. To describe the cultures of European Union members

CRITICAL THINKING
- Analyzing Issues, p. 368

FOCUS & MOTIVATE

WARM-UP

Analyzing Motives Have students read Dateline and answer questions about the euro.

1. Why might Western European nations decide to switch to a single currency?
2. Why do you think each country wants to mint its own coins?

INSTRUCT: Objective 1

Western Europe Today

- What was the original goal of the European Union? to encourage trade
- Why do many former Communist countries want to join the European Union? to gain economic and political advantages
- What changes do Eastern European nations need to make to join the EU? economic, legal, and environmental improvements

In-depth Resources: Unit 4
- Guided Reading Worksheet, p. 39

Reading Study Guide
(Spanish and English), pp. 110–111

Program Resources

In-depth Resources: Unit 4
- Guided Reading Worksheet, p. 39
- Reteaching Activity, p. 46

Reading Study Guide
(Spanish and English), pp. 110–111

Formal Assessment
- Section Quiz, p. 190

Integrated Assessment
- Rubric for writing a postcard or e-mail

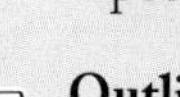

Outline Map Activities

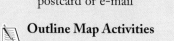

Access for Students Acquiring English
- Guided Reading Worksheet, p. 105

Technology Resources
classzone.com

TEST-TAKING RESOURCES

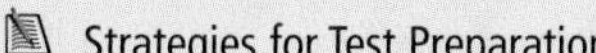

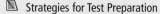

Strategies for Test Preparation
Test Practice Transparencies
Online Test Practice

CRITICAL THINKING ACTIVITY

Analyzing Issues Point out to students that the current members of the EU want Eastern European countries to make economic, legal, and environmental improvements before admitting them to membership in the EU. Use the following questions to help students analyze this issue.

- What might happen if Eastern European countries joined the EU without improving their economies?
- How might environmental problems affect a nation's economy, as well as that of its neighbors?
- Why would it be helpful for all members of the EU to have similar legal systems?

Class Time 20 minutes

FOCUS ON VISUALS

Interpreting the Photograph Have individual students identify the countries on the Euro plaque, identifying each country by shape and by comparing its flag to the chart of EU members. Ask them to identify the EU members not included on the plaque and think of a reason for their absence.

Response Denmark, Greece, Sweden, United Kingdom; have not adopted the euro

Extension Have students research the flags of the five countries that have won initial EU approval (see Background, page 368).

Members of the European Union, 2001
Austria
Belgium
Denmark
Finland
France
Germany
Greece
Ireland
Italy
Luxembourg
Netherlands
Portugal
Spain
Sweden
United Kingdom

The European Union Many countries of Western Europe belong to a group called the **European Union** (EU). At first, countries joined the EU to encourage trade. This economic group, however, is becoming a loose political union.

Many former Communist countries of Eastern Europe want to join the Union too. They know that membership will help them economically and politically. Eastern European countries, however, cannot automatically join the EU. Many must first make legal, economic, and environmental improvements. The EU has agreed to include them over time. With a possible membership of more than 20 nations by 2003, the EU may be the best hope for European peace and prosperity.

BACKGROUND

In 2001, the EU gave initial approval to the Czech Republic, Estonia, Hungary, Poland, and Slovenia to join in the near future.

Regional Governments In Western Europe, each nation also has regional governments, similar to those of individual states in the United States. Regional governments are demanding—and receiving—greater power. As a result, many people in Western Europe enjoy increased self-rule and participation in the political process.

Region • The headquarters of the European Central Bank is located in Frankfurt, Germany. ◄

Activity Options

Differentiating Instruction: Less Proficient Readers

Setting a Purpose To help students set a purpose for reading about the European Union, prepare a set of questions for students to use as a guide as they read the paragraphs under "The European Union" (page 368). Write the questions on the chalkboard, and read them aloud to students. When students have finished reading and answering the questions, review and discuss their answers as a group.

You might want to ask questions such as the following:

- Why did countries originally join the European Union?
- Why do former Communist countries want to join the EU?
- What must Eastern European countries do before they can join the EU?

EU Economies ❷

BACKGROUND

Some EU nations, including the United Kingdom and Denmark, have not agreed to give up their existing currency.

Traditionally, each European nation has had its own **currency,** or system of money. The EU is meant to make international trade much simpler. With more Europeans using the **euro,** the currency of the EU, currency no longer has to be exchanged every time a payment crosses a border.

Improved Trade To encourage trade, members have also done away with tariffs on the goods they trade with one another. A **tariff** is a duty or fee that must be paid on imported or exported goods, making them more expensive. EU members have lifted border controls as well. This means that goods, services, and people flow freely among these member nations.

Another goal of the EU is to achieve economic equality among its members. To reach this goal, EU members are sharing their wealth. Poorer countries such as Ireland receive money to help them build businesses.

Reading Social Studies

A. Clarifying How would improved trade raise the standard of living?

Reading Social Studies
A. Possible Answer Trade can make more goods available and may create new jobs.

A Higher Standard of Living Member nations hope that increased trade and shared wealth will help give all citizens of the EU a high standard of living. A person's **standard of living,** or quality of life, is based on the availability of goods and services.

People who have a high standard of living have enough food and housing, good transportation and communications, and access to schools and health care. They also have a high rate of literacy, meaning that most adults are able to read.

Additional Benefits The members of the EU are helping the countries of Eastern Europe to raise their environmental standards. They are willing to pay up to 75 percent of the cost for a new waste treatment system in Romania, for example. The program includes recycling centers for paper, glass, and plastics. It will clean up and close old dumping grounds, which were leaking pollution into the ground water.

Connections to Economics

Tourism For many European nations, tourism is an important part of the economy. In fact, the continent represents about 60 percent of the world's tourist market. Visitors come to enjoy Europe's climate, historic sites, museums, and food.

Popular destinations include Spain, Italy, Austria, and the United Kingdom. France, below, is the most visited country in the world. In 1999 it hosted more than 73 million tourists.

INSTRUCT: Objective ❷

EU Economies

- What have members of the EU done to improve trade? done away with tariffs, lifted border controls, made development loans
- What steps did the EU take to improve the environment in Romania? new waste treatment system, including recycling centers and cleaning up old dumping grounds
- How is the EU improving the lives of people? tries to raise standards of living, provides job training, allows people to work in any member country, allows people to vote in local elections wherever they live, protects people's rights through a Court of Justice

Connections to Economics

Within the EU, travel and tourism account for about $1.4 billion in economic activity each year. This amounts to about 4.8 percent of the gross domestic product of the EU as a whole, and experts have predicted that tourism will increase by approximately 3.7 percent each year. Currently, travel and tourism provide more than 9 million jobs—nearly one of every 17 jobs in Europe.

Activity Options

Interdisciplinary Link: Math

Class Time 15 minutes

Task Converting euros to U.S. dollars

Purpose To learn about the relative value of the euro compared with the United States dollar

Supplies Needed
- Writing paper
- Pencils

Activity Explain to students that when Americans travel to Europe, they exchange, or trade, their U.S. money for euros. Explain that the exchange rate varies, but one euro is generally worth a little less than one U.S. dollar. Create word problems such as this one for students to solve.

- A European is traveling to America. She has 1,200 euros. If one euro is worth 90 cents, how many U.S. dollars will she get in exchange? ($1,080)

INSTRUCT: Objective 3

Cultural Diversity

- What are some examples of cultural diversity among EU nations? different languages, unique foods, certain ways of doing business, special games and celebrations
- What conveniences do most major European cities offer? excellent public transportation, sidewalk cafés

MORE ABOUT...

European Languages

Much of Europe's rich cultural diversity is a result of its many languages—there are about 50 in all. Until recently, the use of many of these languages has been discouraged. The French Constitution, for example, declares that "the language of the country is French." Breton, Basque, Catalan, and other languages spoken in France have been ignored, and the specific dialect of French spoken in Paris has been recognized as the official language for the whole country. Other countries have likewise discouraged the use of minority languages. This is changing, however. In 1992, the Council of Europe adopted the European Charter for Regional or Minority Languages. The charter acknowledges the cultural value of Europe's regional languages and seeks to preserve them.

The EU also runs programs that train people for jobs. As citizens of a member nation, people are not limited to a job in their own country. They may work in any part of the EU. They can even vote in local elections wherever they live. In addition, the Council of Europe's **Court of Human Rights** protects the rights of all its citizens in whichever member country they live.

Cultural Diversity 3

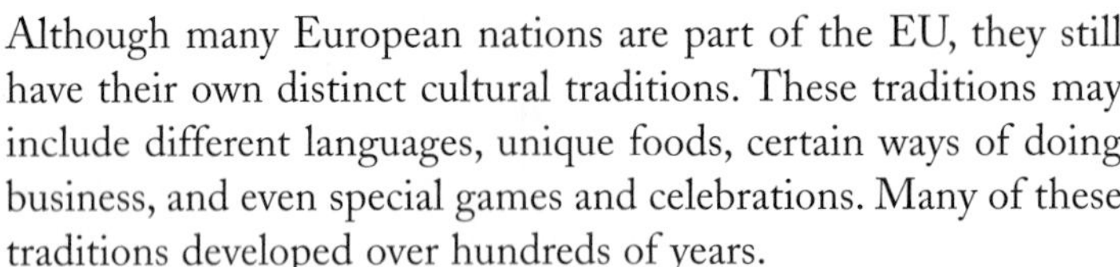

Although many European nations are part of the EU, they still have their own distinct cultural traditions. These traditions may include different languages, unique foods, certain ways of doing business, and even special games and celebrations. Many of these traditions developed over hundreds of years.

Some nations are a mix of several cultures. In Belgium, for example, Flemings live in the north and speak Dutch. Another major group, the Walloons, lives in the south. They speak French. A third group of German-speaking Belgians lives in the eastern part of the country. Many Belgian cities include people from all three groups.

Reading Social Studies
B. Possible Answers
conflicting cultural traditions, national pride, economic inequality among members

Reading **Social Studies**

B. Identifying Problems What are the main problems facing the European Union?

City Life Many of the world's famous and exciting cities are located in Western Europe. London, Madrid, Paris, Amsterdam, and Rome are just a few of the major centers for the arts, business, and learning. These cities are centuries old, and Europeans work hard to preserve them.

Europeans also take pride in the conveniences that their cities offer. Most major urban areas have excellent public transportation, including subways, buses, and trains. Sidewalk cafés are also popular, where people come to meet friends, eat, and relax.

Region • Many Europeans center their social lives around urban sidewalk cafés, such as this one in Italy. ▼

Activity Options

Multiple Learning Styles: Visual/Spatial

Class Time 20 minutes

Task Locating countries and cities of Western Europe on a map

Purpose To learn where Western European countries and cities are located in relation to one another

Supplies Needed
- Map of Europe

Activity Read aloud the paragraphs under "Cultural Diversity." As you read about the various languages spoken in Belgium, ask students to locate the areas on the map. Then, as you read about London, Madrid, Paris, Amsterdam, and Rome, point out the cities and have students identify the country in which each city is located.

Region • Quaint small European villages are popular tourist attractions. ▶

BACKGROUND

Many European families cannot make a living on a small farm. The government may offer support to such families, to help preserve the nation's rural culture.

Country Life European cities have much to offer, but the countryside is also popular—especially for vacationers. The Italian region of Tuscany (TUHS•kuh•nee) and the French region of Provence (pruh•VAHNS) are two of the best-known examples of the many beautiful rural areas.

Small European villages may have only a café, a grocery store, a post office, a town square, and a collection of houses. Many families who live in such areas have been farming or raising animals on the same land for generations. Some even live in houses that their families have owned for hundreds of years.

SECTION 3 ASSESSMENT

Terms & Names

1. Explain the significance of: (a) European Union (EU) (b) currency (c) euro (d) tariff (e) standard of living (f) Court of Human Rights

Using Graphics

2. Use a chart like this one to compare aspects of city life and country life in Europe.

City Life	Country Life

Main Ideas

3. (a) Describe the importance of the new shared currency that is based on the euro.
   (b) Can any European country automatically join the EU? Why or why not?
   (c) List at least two benefits, other than a shared currency, for countries that are members of the EU.

Critical Thinking

4. **Synthesizing**
   Why may the EU be the best hope for European peace and prosperity?

Think About

- the number of member countries
- the goals of the EU
- modern European conflicts

ACTIVITY -OPTION- Choose one photograph from this section that shows a place in Europe. Write a **postcard** or **e-mail** to a friend or family member as if you were there. What sights and sounds will you describe?

ASSESS & RETEACH

Reading Social Studies Have students add notes about EU membership to the second column of the chart on page 352, noting which countries are on their way to acceptance.

Formal Assessment
- Section Quiz, p. 190

RETEACHING ACTIVITY

Assign a part of the section to each student. Ask each to create a cluster diagram to show the main ideas and supporting details for their assigned text. Then have students meet in groups and share their clusters. Encourage them to add details they may have overlooked as they study their classmates' work.

In-depth Resources: Unit 4
- Reteaching Activity, p. 46

Access for Students Acquiring English
- Reteaching Activity, p. 110

Section 3 Assessment

1. Terms & Names

a. European Union (EU), p. 368
b. currency, p. 369
c. euro, p. 369
d. tariff, p. 369
e. standard of living, p. 369
f. Court of Justice, p. 370

2. Using Graphics

City Life	Country Life
many cafés, stores	beautiful scenery
excellent public transportation	many families farmed same land for generations

3. Main Ideas

a. The euro makes trade among EU countries easier.
b. No; they must meet certain economic, legal, and environmental conditions.
c. economic assistance, improved trading opportunities, improved standard of living

4. Critical Thinking

Possible Response The EU creates a common economic bond and goals to reduce conflicts among members.

ACTIVITY OPTION

Integrated Assessment
- Rubric for writing a postcard or e-mail

OBJECTIVE

Students learn about the contributions of Europe to the world's languages, ideas, and inventions.

FOCUS & MOTIVATE

Making Inferences Ask students to study the pictures and read each paragraph. Then have them answer the following questions:

1. How might the printing press have changed other European technologies and inventions?
2. What do the stylistic similarities between the Agora and the White House suggest about the influence of Greek art on Roman art?

 **Block Scheduling**

MORE ABOUT...

Nitroglycerin

Along with his scientific and technological interests, Alfred Nobel wrote plays, novels, and poetry in his free time. He also traveled widely and spoke five languages. He became very wealthy from his work with explosives and decided that profits should reward human ingenuity. Nobel set up a fund of about $9 million in his will, and the interest from this fund is used for the Nobel Prizes—annual awards to those whose work benefits humanity. Today, they are the most honored prizes in the world.

Architecture

Roman architecture had a great deal of stylistic unity. For example, the Romans were the first to fully utilize the arch and the vault. The use of these reduced or eliminated the need for columns, and allowed the roof to rest completely on the outer walls. The Romans still used columns, but more as sculptural decoration.

The Legacy of Europe

Movable Type

Before Johann Gutenberg (1400–1468) invented the printing press in Germany, European monks copied books by hand. Movable type made it possible to print multiple copies of books quickly, allowing people access to them. The printing process advanced greatly in the 1930s. By the mid-1940s, printed works included complex illustrations and color. Today, people create text and images and print them directly from their computers.

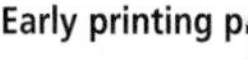

Early printing p

Nitroglycerin

For almost 20 years, Alfred Nobel (1833–1896), shown at lef worked on developing a way to safely contain and ignite nitroglycerin, a powerful explosive. Eventually, chemists and doctors realized that nitroglycerin widens blood vessels and can be used to treat patients with heart conditions. Given ir tablet, patch, or oral-spray form, nitroglycerin has saved countless lives.

Architecture

The White House in Washington, D.C., is one of many buildings in the United States that has been influenced by Roman architecture. Roman buildings often featured vaulted domes, columns, and large interior spaces. This type of architecture has influenced other buildings in the United States, including many banks and courthouses.

Activity Option

Interdisciplinary Link: Government

Class Time 60 minutes

Task Preparing arguments, pro or con, about the effects of the printing press on government

Supplies Needed

- Information on the number of books available in Europe before, and after, the invention of the printing press

Activity Have students work in groups. Ask each group to imagine that they are advisors to a European king in the early 1400s. The king is considering the effects of a new invention, the printing press, on his rule. Each group should prepare a recommendation for the king, either to allow individual entrepreneurs to establish printing presses, or to outlaw this new invention as a dangerous idea. Remind each group to clearly state their reasons, pro or con, and to predict the effects the printing press will have on the king's government and his power.

372 UNIT 4

·mocracy

·und 500 B.C., several ·ek city-states established ·ocracies, replacing their ·le-ruler governments. · word *democracy* derives ·n two Greek words: ·*os,* meaning "people," · *kratos,* meaning "power." ·ay, the idea of rule by the ·ple is found around the ·ld, from the United States ·rance to India.

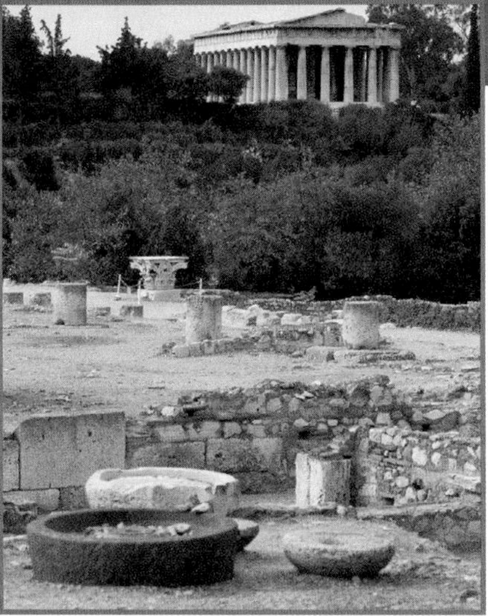

The agora, or marketplace, of ancient Athens was often the scene of political activities. ▲

Find Out More About It!

Study the text and photos on these pages to learn about inventions, creations, and contributions that have come from Europe. Then choose the item that interests you the most and use the library or the Internet to learn more about it. Use the information you gather to create a poster celebrating the contribution.

Yacht ▼

Robot ▶

Flamingo ▶

·ropean ·nguages

·ny words of different ·opean languages have ·de their way into English. · example, the word *dinner* ·ctually French in origin. As ·lish people traveled and settled ·und the world, they borrowed words ·n such European languages as German, Spanish, and ·wegian to use in their everyday communication. Examples ·ome of these words we use today are *kindergarten* ·rman), *dinner* (French), *yacht* (Dutch), *corridor* (Italian), ·*illa* (Spanish), *flamingo* (Portuguese), *robot* (Czech), · *ski* (Norwegian).

INSTRUCT

- What architectural elements do the Agora and the White House share?
- Where was democracy first practiced in Europe?
- When was the printing press invented?

MORE ABOUT...

Democracy

Athens was the most powerful Greek city-state with a democratic government. All government officials in Athens were citizens chosen by lot. An Athenian citizen had to volunteer to participate in the political process. If the citizen chose to participate, he was eligible to be a member of either a jury or a council that ran the daily government.

MORE ABOUT...

European Languages

Today, English is the most widely spoken language in the world. This was not always true, however. In the 1500s, English was spoken by fewer than two million people, and all of them lived in the British Isles. Since then, because of various historical events, English has spread throughout the world. Today, English is the native language for roughly 400 million people. Another 100 million people speak English along with their native language. Additionally, some English is used by 200 million people.

With all these speakers, English is constantly growing and incorporating words from other languages. Today, the English vocabulary is larger than that of any other language. Examples of words incorporated from other languages include algebra from Arabic, fashion from French, piano from Italian, and canyon from Spanish.

More to Think About

Making Personal Connections Ask students to reread the section called "Architecture" and review the elements of Roman style: columns, domes, pediments, and arches. Ask them to identify any local buildings which show these elements of Roman influence. Have them bring in other examples from magazines and newspapers and create a display.

Vocabulary Activity Have students draw a visual expression of the title of each section. One example might be a drawing of the outline of Europe with several "heads" shown talking to represent "European Languages." Encourage students to be creative, but not deliberately tricky. Have them share their completed pictures with the class and discuss.

TERMS & NAMES

1. propaganda, p. 354
2. Nikita Khrushchev, p. 356
3. détente, p. 357
4. Mikhail Gorbachev, p. 360
5. ethnic cleansing, p. 363
6. Duma, p. 364
7. currency, p. 369
8. euro, p. 369
9. tariff, p. 369
10. standard of living, p. 369

REVIEW QUESTIONS

Possible Responses

1. The Soviets wanted to discourage ethnic identity so minorities would not seek independence.
2. The Soviet government controlled publications and the work of writers and artists, preventing people from learning about other nations.
3. through military force and economic control
4. A group of Soviet officials tried to undo Gorbachev's reforms, but public protest stopped them; republics in the Soviet Union declared their independence; the Soviet Union came to an end.
5. oil, coal, iron, lead, gold, natural gas
6. The EU unites 15 European countries in an economic and, increasingly, political union. It improves trade and reduces the possibility of conflicts among member countries.
7. no tariffs; free flow of goods, services, and people; political freedoms; job training; economic aid to poorer members; protection of rights by a Court of Justice
8. Paris, London, Madrid, Rome, Amsterdam

CHAPTER 13 ASSESSMENT

TERMS & NAMES

Explain the significance of each of the following:

1. propaganda
2. Nikita Khrushchev
3. détente
4. Mikhail Gorbachev
5. ethnic cleansing
6. Duma
7. currency
8. euro
9. tariff
10. standard of living

REVIEW QUESTIONS

Eastern Europe Under Communism *(pages 353–358)*

1. Why did the Soviet government outlaw many cultural celebrations?
2. Explain why most Soviet citizens learned little about other nations around the world.
3. How did the Soviet Union maintain control over Eastern European countries?

Eastern Europe and Russia *(pages 360–366)*

4. How did the Soviet Union change during 1991?
5. List at least three of Russia's major natural resources.

The European Union *(pages 367–371)*

6. What is the importance of the European Union (EU)?
7. Name one benefit of being a member of the EU.
8. Identify at least three of Europe's major centers of the arts, business, and learning.

CRITICAL THINKING

Comparing

1. Using your completed chart from Reading Social Studies, p. 352, compare Eastern Europe under Communism with Eastern Europe after Communism.

Summarizing

2. Outline the changes to the Russian economy since the breakup of the Soviet Union.

Recognizing Important Details

3. What types of changes must Eastern European countries make in order to join the EU?

Visual Summary

1 Eastern Europe Under Communism

- The Soviet Union's communist government controlled the lives of its citizens.
- Under Nikita Khrushchev, citizens began to have greater freedom.

2 Eastern Europe and Russia

- Today, independent nations once under Soviet rule are taking steps toward new economies and greater freedom.

3 The European Union

- Many European countries are members of an economic and political alliance called the European Union (EU).

CRITICAL THINKING: Possible Responses

1. Comparing

Under the Soviet Union, the governments of Eastern European countries were Communist; citizens' rights were strictly limited. The governments created in the 1990s were democratically elected and organized as parliamentary republics that protected the rights of citizens. Countries moved toward free-market economies.

2. Summarizing

The economy has changed from a command economy to a free-market economy. Now people can own land and have the right to start businesses. Foreign companies are encouraged to do business in Russia. The government no longer controls prices. Some people have grown wealthy; many remain poor, but standards of living have improved for many.

3. Recognizing Important Details

They must make environmental, legal, and economic improvements.

ANDARDS-BASED ASSESSMENT

the map and your knowledge of geography to wer questions 1 and 2.

ditional Test Practice, pp. S1–S33

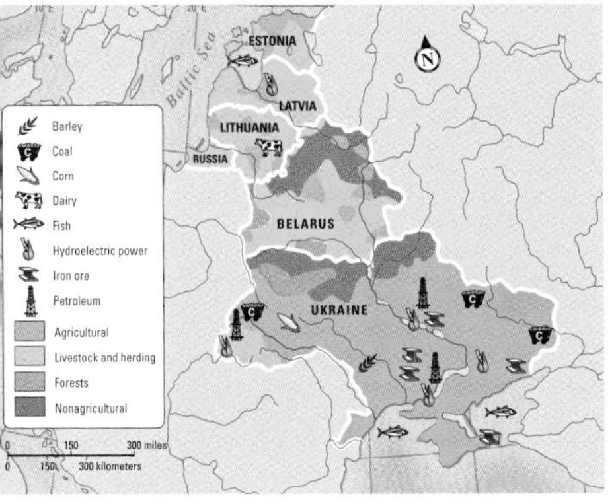

Which of the following countries has the greatest variety of ower resources?

. Estonia
. Latvia
. Lithuania
. Ukraine

you were a dairy farmer, in which country would you robably live?

. Belarus
. Lithuania
. Russia
. Ukraine

In the following passage, Mikhail Gorbachev explains his reasons for reforming the Soviet economy. Use the quotation and your knowledge of world cultures and geography to answer question 3.

PRIMARY SOURCE

The country began to lose momentum. Economic failures became more frequent. Difficulties began to accumulate and deteriorate, and unresolved problems to multiply. . . . Analyzing the situation, we first discovered a slowing economic growth. In the last fifteen years the national income growth rates had declined by more than a half.

MIKHAIL GORBACHEV, *Perestroika*

3. According to Gorbachev, what difficulty did the Soviet economy face?

A. Jobs were being eliminated.
B. Prices were rapidly rising.
C. Growth rates were declining.
D. Factories were being closed.

TERNATIVE ASSESSMENT

WRITING ABOUT HISTORY

During "The Thaw" in the Soviet Union (1958–1964), people gained many freedoms. Research these freedoms and find out ow these changes affected the lives of average citizens in the Soviet Union. Then write a journal entry that might have been written by a citizen after he or she experienced one of these reedoms for the first time. Share your entry with the class.

COOPERATIVE LEARNING

When the Soviet Union broke apart into 15 nations, each country had to set up a new government. In a group of three to ve classmates, create a government and constitution. Assign each group member a role in the new government, such as a epresentative or president. Outline the responsibilities of each member. Work together to write a brief constitution for your government including sections that outline basic rights, reedoms, and responsibilities of citizens.

INTEGRATED TECHNOLOGY

Doing Internet Research

Use the Internet and library resources to research the economy of any one of the Balkan states. Write a short report of your findings. List the Web sites that you used to prepare your report.

- Specifically look for information about the types of industry that exist in that state as well as any economic problems the state might face.
- Include graphs or charts that show the state's major imports and exports.

For Internet links to support this activity, go to

STANDARDS-BASED ASSESSMENT

1. **Answer D** is the correct answer because Ukraine shows the symbols for coal, hydroelectric power, and petroleum. Answers A and C are incorrect because Estonia and Lithuania do not show any symbols for power resources; answer B is incorrect because Latvia shows only the symbol for hydroelectric power.
2. **Answer B** is the correct answer because Lithuania shows the symbol for dairies. Answers A, C, and D are incorrect because Belarus, Russia, and Ukraine do not show the symbol for dairies.
3. **Answer C** is the correct answer because it restates Gorbachev's assertion that growth rates had declined. Answer A is incorrect because the passage does not mention jobs; answer B is incorrect because the passage does not mention prices; answer D is incorrect because the passage does not mention factories.

INTEGRATED TECHNOLOGY

Discuss how students might locate information about the Balkan states and their economies. Brainstorm possible keywords, Web sites, and search engines that students can use. You might have students create charts to help them organize and present the information they find.

Alternative Assessment

1. Rubric

The journal entry should

- accurately reflect the thoughts and experiences of a Soviet citizen.
- discuss the topic of increased freedom in the Soviet Union.
- reflect the student's understanding of the time period 1958–1964.
- use correct grammar, spelling, and punctuation.

2. Cooperative Learning

Before students begin writing their constitutions, review what they have learned about the newly formed governments of Eastern European countries after the breakup of the Soviet Union. List problems these new countries faced. Have students refer to their lists so their constitutions can address or avoid these problems.

PLANNING GUIDE: UNIT 4 ATLAS AND CHAPTER 14

Europe Today

	OVERVIEW	COPYMASTERS	INTEGRATED TECHNOLOGY
UNIT ATLAS AND CHAPTER RESOURCES	The students will learn about the government, economy, and culture of the United Kingdom, Sweden, France, Germany, and Poland.	**In-depth Resources: Unit 4** • Guided Reading Worksheets, pp. 47–51 • Skillbuilder Practice, p. 54 • Unit Atlas Activities, pp. 1–2 • Geography Workshop, pp. 61–62 **Reading Study Guide** (Spanish and English), pp. 114–125 **Outline Map Activities**	• eEdition Plus Online • EasyPlanner Plus Online • eTest Plus Online • eEdition • Power Presentations • EasyPlanner • Electronic Library of Primary Sources • Test Generator • Reading Study Guide • The World's Music • Critical Thinking Transparencies CT27, 28 • There Is No Food Like My Food
	KEY IDEAS		
SECTION 1 The United Kingdom pp. 379–383	• England, Scotland, Wales, and Northern Ireland make up the United Kingdom. • Trade is important to the British economy.	**In-depth Resources: Unit 4** • Guided Reading Worksheet, p. 47 • Reteaching Activity, p. 56 **Reading Study Guide** (Spanish and English), pp. 114–115	classzone.com Reading Study Guide
SECTION 2 Sweden pp. 384–387	• Sweden's legislature, the Riksdag, makes the country's laws. • Outdoor recreation and the arts are popular in Sweden.	**In-depth Resources: Unit 4** • Guided Reading Worksheet, p. 48 • Reteaching Activity, p. 57 **Reading Study Guide** (Spanish and English), pp. 116–117	classzone.com Reading Study Guide
SECTION 3 France pp. 390–393	• France's government and economy quickly recovered after World War II. • Nuclear power is the source of 75% of France's energy. • French culture centers around the capital city of Paris.	**In-depth Resources: Unit 4** • Guided Reading Worksheet, p. 49 • Reteaching Activity, p. 58 **Reading Study Guide** (Spanish and English), pp. 118–119	classzone.com Reading Study Guide
SECTION 4 Germany pp. 394–397	• Both the division and the reunification of Germany presented many challenges. • Germany has a rich cultural heritage, especially in literature and music.	**In-depth Resources: Unit 4** • Guided Reading Worksheet, p. 50 • Reteaching Activity, p. 59 **Reading Study Guide** (Spanish and English), pp. 120–121	classzone.com Reading Study Guide
SECTION 5 Poland pp. 400–403	• Solidarity, a workers' party, led the fight for political change in Poland. • Poland's constitution protects individual freedoms, including freedom of speech. • The Polish economy has struggled with the shift to a free-market system.	**In-depth Resources: Unit 4** • Guided Reading Worksheet, p. 51 • Reteaching Activity, p. 60 **Reading Study Guide** (Spanish and English), pp. 122–123	classzone.com Reading Study Guide

KEY TO RESOURCES

- Audio
- CD-ROM
- Copymaster
- Internet
- Overhead Transparency
- Pupil's Edition
- Teacher's Edition
- Video

ASSESSMENT OPTIONS

Chapter Assessment, pp. 404–405

Formal Assessment
- Chapter Tests: Forms A, B, C, pp. 208–219

Test Generator

Online Test Practice

Strategies for Test Preparation

Section Assessment, p. 383

Formal Assessment
- Section Quiz, p. 203

Integrated Assessment
- Rubric for writing a description

Test Generator

Test Practice Transparencies TT43

Section Assessment, p. 387

Formal Assessment
- Section Quiz, p. 204

Integrated Assessment
- Rubric for writing a description

Test Generator

Test Practice Transparencies TT44

Section Assessment, p. 393

Formal Assessment
- Section Quiz, p. 205

Integrated Assessment
- Rubric for drawing a portrait

Test Generator

Test Practice Transparencies TT45

Section Assessment, p. 397

Formal Assessment
- Section Quiz, p. 206

Integrated Assessment
- Rubric for writing a short story

Test Generator

Test Practice Transparencies TT46

Section Assessment, p. 403

Formal Assessment
- Section Quiz, p. 207

Integrated Assessment
- Rubric for writing a speech

Test Generator

Test Practice Transparencies TT47

RESOURCES FOR DIFFERENTIATING INSTRUCTION

Students Acquiring English/ESL

Reading Study Guide (Spanish and English), pp. 114–125

Access for Students Acquiring English Spanish Translations, pp. 111–122

TE Activities
- Prefixes, p. 396
- Identifying Sequence of Events, p. 401

Modified Lesson Plans for English Learners

Less Proficient Readers

Reading Study Guide (Spanish and English), pp. 114–125

TE Activity
- Identifying Sequence of Events, p. 401

Gifted and Talented Students

TE Activity
- Forming and Supporting Opinions, p. 402

CROSS-CURRICULAR CONNECTIONS

Humanities

Lisandrelli, Elaine Slivinski. ***Ignacy Jan Paderewski: Polish Pianist and Patriot.*** Greensboro, NC: Morgan Reynolds, 1999. Portrait of a famous Polish hero.

Literature

Kuniczak, W. S., ed., ***The Glass Mountain: Twenty-Eight Ancient Polish Folktales and Fables.*** New York: Hippocrene Books, 1997. Literate anthology of rarely anthologized tales.

Banks, Lynne Reid. ***Melusine: A Mystery.*** New York: Avon Books, 1997. A teenage boy uncovers a mystery in the south of France.

Science

Linder, Greg. ***Marie Curie: A Photo-Illustrated Biography.*** Mankato, MN: Bridgestone Books, 1999. Polish scientist who discovered radium and won two Nobel Prizes.

History

Ayer, Eleanor H. ***Poland: A Troubled Past, A New Start.*** Tarrytown, NY: Benchmark Books, 1996. History, geography, daily life, culture, and customs.

Epler, Doris. ***The Berlin Wall: How It Rose and Why It Fell.*** Brookfield, CT: Millbrook Press, 1992. A history of the Wall, from its construction during the Cold War to its fall in 1989.

Lobel, Anita. ***No Pretty Pictures: A Child of War.*** New York: Avon Books, 2000. Life as a Polish Jew during WWII, and in Sweden for years after.

Popular Culture

Lalley, Linda. ***The Volkswagen Beetle.*** Mankato, MN: Riverfront Books, 1999. Traces the history, development, and design of this popular car.

Aronson, Marc. ***Art Attack: A Short Cultural History of the Avant-Garde.*** New York: Clarion Books, 1998. Traces the story of bohemians, radicals, hipsters, and hippies of the avant-garde movement begun in Paris.

ENRICHMENT ACTIVITIES

The following activities are especially suitable for classes following block schedules.

Teacher's Edition, pp. 380, 382, 385, 386, 392, 395, 398, 402
Pupil's Edition, pp. 383, 387, 393, 397, 403

Unit Atlas, pp. 260–269
Interdisciplinary Challenge, pp. 388–389

Outline Map Activities

INTEGRATED TECHNOLOGY

Go to **classzone.com** for lesson support and activities for Chapter 14.

BLOCK SCHEDULE LESSON PLAN OPTIONS: 90-MINUTE PERIOD

DAY 1

CHAPTER PREVIEW, pp. 376–377
Class Time 20 minutes

- **Hypothesize** Use the "What do you think?" questions in Focus On Geography on PE p. 377 to help students hypothesize about the challenges presented by environmental pollution.

SECTION 1, pp. 379–383
Class Time 40 minutes

- **Peer Competition** Divide the class into pairs. Assign each pair one of the Terms & Names for this section. Have pairs make up five questions that can be answered with the Term or Name. Have groups take turns asking the class their questions.
Class Time 20 minutes
- **Geography Skills** Help students use the map on PE p. 380 to identify the location of Great Britain, using latitude and longitude. Then divide the class into small groups. List the names of five cities on the map on the board and have groups identify their locations, using latitude and longitude.
Class Time 10 minutes
- **Brainstorming** Give students five minutes to list as many products as they can think of that the United States imports from the United Kingdom. Together as a class make a list on the board.
Class Time 10 minutes

SECTION 2, pp. 384–387
Class Time 30 minutes

- **Peer Teaching** Have pairs of students review the Main Idea for the section and find three details to support it. Then have each pair list two additional important ideas and trade lists with another group to find details.
Class Time 10 minutes
- **Section Assessment** Have students do the Section 2 Assessment.
Class Time 20 minutes

DAY 2

SECTION 3, pp. 390–393
Class Time 40 minutes

- **Small Groups** Divide the class into small groups. Assign each group one section objective on TE p. 390 to help them prepare a summary of this section. Remind students that when they summarize, they should include the main ideas and most important details in their own words. Reconvene as a whole class for discussion.

SECTION 4, pp. 394–397
Class Time 40 minutes

- **Decorating a Memorial** Use Focus & Motivate: Analyzing Motives on TE p. 394 to help the class imagine how Germans felt on the event of their reunification. Then divide the class into groups of two to three students and tell each group to plan and draw a sketch for a painting to decorate the Berlin Wall memorial. Have the class vote on which sketch they like best.

SECTION 5, pp. 399–403
Class Time 10 minutes

- **Comparing** Extend the Critical Thinking Activity on TE p. 401 by leading students in a discussion about the effects of guaranteeing even small ethnic groups a voice in government by reserving seats in parliament for them. Ask students if they think the United States would benefit by a similar arrangement.

DAY 3

SECTION 5, continued
Class Time 35 minutes

- **Creating a Chart** Divide the class into small groups. Have each group create a chart contrasting life in Poland before and after communism. Reconvene the class and have groups share their information for you to include in a master chart on the board.

CHAPTER 14 REVIEW AND ASSESSMENT, pp. 404–405
Class Time 55 minutes

- **Review** Have pairs of students review the information in the Visual Summary on p. 404 and find three details to support each statement.
Class Time 20 minutes
- **Assessment** Have students complete the Chapter 14 Assessment.
Class Time 35 minutes

TECHNOLOGY IN THE CLASSROOM

CHARTING AND GRAPHING DATA WITH A SPREADSHEET PROGRAM

Spreadsheet programs can be invaluable tools for analyzing and comparing data. Most spreadsheet programs allow the user to input numbers into rows and columns and to then create charts and graphs based on those numbers. These charts and graphs present the data in a visual manner that makes it easier to compare the data.

ACTIVITY OUTLINE

Objective Students will use a spreadsheet program to graph economic and social indicator numbers for Bosnia and Herzegovina, France, Germany, Poland, Sweden, the United Kingdom, and the United States. They will use a word processing program to write an analysis of the data.

Task Have students use an interactive Web site to generate economic and social data for these countries. Tell them to input this data into a spreadsheet and create column, bar, or line graphs to display the data. Have them conclude by writing paragraphs analyzing and explaining the graphs.

Class Time Two class periods

DIRECTIONS

1. Ask students to go to **classzone.com.** Have them click the boxes in the right-hand frame next to Bosnia and Herzegovina, France, Germany, Poland, Sweden, the United Kingdom, and the United States. Then have them click "Data Menu."
2. Ask students to select the following data fields: GDP per capita, life expectancy (women/men), illiteracy rate (total), and telephones.
3. Discuss the meanings of these four data fields, making sure students understand what each one measures. Ask them to explain what they think each type of data reveals about a country. For example, what does it mean for one country to have a lower life expectancy or GDP per capita than another country? Do students predict most of these countries will have figures that are higher than, lower than, or about the same as the United States? Why?
4. Have students click "View Info" to view the data. Ask them to look carefully at the chart and compare the data for the different countries. Why do they think Poland's figures are noticeably different? Explain that Bosnia and Herzegovina have recently been involved in a war. How do students think this fact might affect the data?
5. Assign each pair of students one of the categories: GDP per capita, life expectancy (women/men), or telephones. Have them enter the numbers for each country into a spreadsheet. The first (A) column should list the countries and the second (B) column should list the numbers.
6. Have students use the spreadsheet program's charting or graphing feature to create column, bar, or line graphs for the data they have entered. The charts should show the countries along the horizontal axis (x-axis) and the economic or social indicators up the vertical axis (y-axis).
7. Have students write paragraphs analyzing the data and hypothesizing reasons to explain it. In particular, they should explain why they think the data for several of the countries is similar to that of the United States. They should then explain why the data for two of the countries is significantly different.

CHAPTER 14 OBJECTIVE

Students will learn about the government, economy, and culture of the United Kingdom, Sweden, France, Germany, and Poland.

FOCUS ON VISUALS

Interpreting the Photograph Direct students' attention to the photograph of London's Piccadilly Circus. Ask students to identify both modern and historical elements in the scene. Tell students that the chapter describes Europe today, which is built on its long past. Which elements in the photograph show this connection?

Possible Responses Modern elements include shopping district and subway entrance. Historical elements include the old buildings and streets. Point out to students that the modern subway runs beneath the historical streets and the modern stores are located in the first floor of the old buildings.

Extension Have students work in groups, and assign each group a photograph of Stockholm, Paris, Berlin, or Warsaw. Ask the groups to identify and list both modern and historical elements.

CRITICAL THINKING ACTIVITY

Identifying Problems What problems may develop in a city trying to balance modern needs and methods with historical places, ideas, and customs? Direct students to look at the photograph again, and ask them to imagine what specific problems might have come up when the subway was being built in Piccadilly Circus.

Class Time 15 minutes

Europe Today

CHAPTER 14

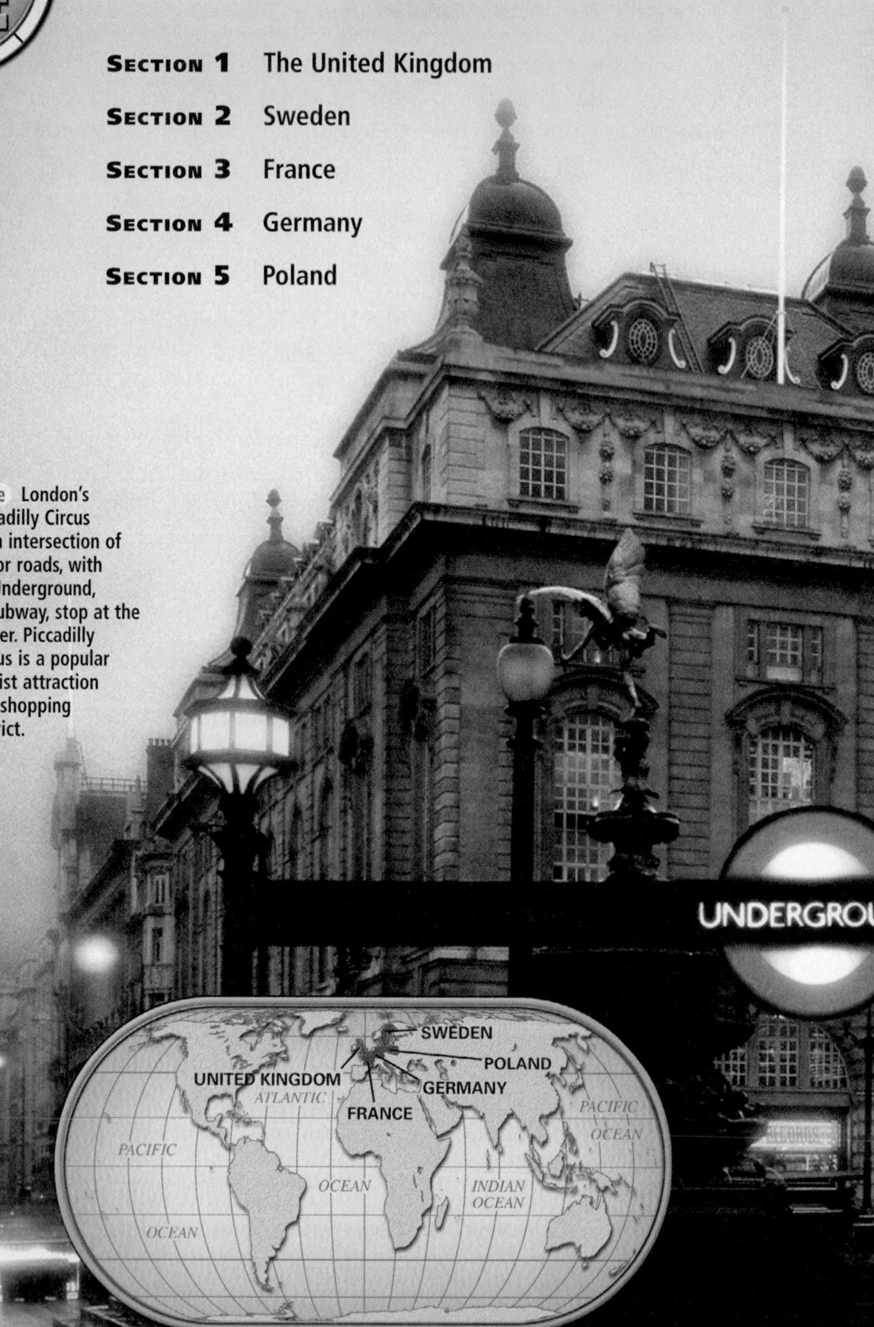

Place London's Piccadilly Circus is an intersection of major roads, with an Underground, or subway, stop at the center. Piccadilly Circus is a popular tourist attraction and shopping district.

376

Recommended Resources

BOOKS FOR THE TEACHER

Knab, Sophie Hodorowicz. ***Polish Customs, Traditions, and Folklore.*** New York: Hippocrene Books, 1993. Folklore resources are particularly good.

Lord, Richard. ***Germany.*** Milwaukee: G. Stevens, 1999. Geography, history, and economics of unified Germany.

Zickgraf, Ralph. ***Sweden (Major World Nations).*** Philadelphia: Chelsea House Publishing, 1997. Overview of Sweden's history, government, and economy.

VIDEOS

Wondrous Kingdom: England, Scotland, and Ireland. Questar, 1999. One-hour guided tour.

INTERNET

For more information about modern Europe, visit **classzone.com.**

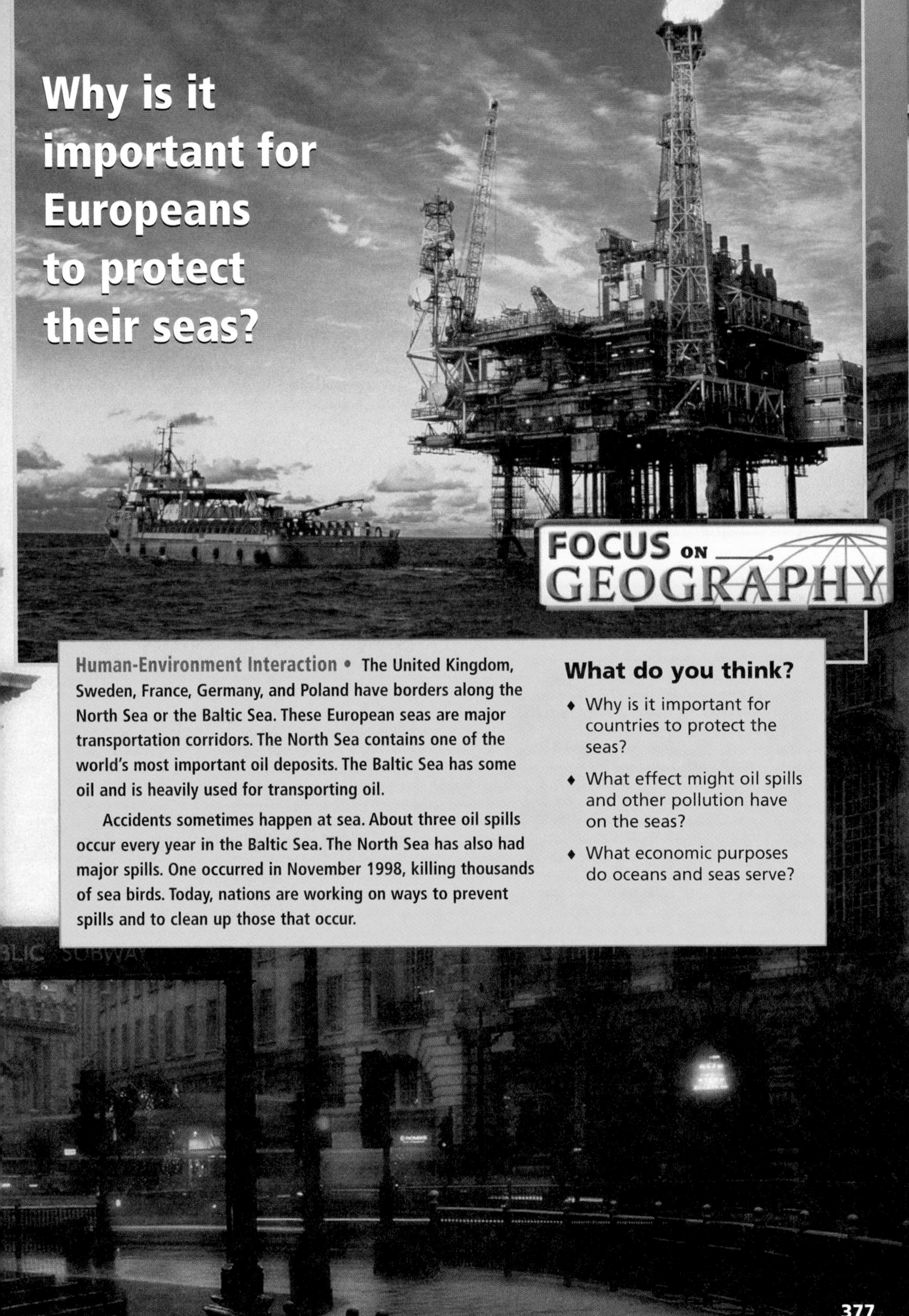

Why is it important for Europeans to protect their seas?

FOCUS ON GEOGRAPHY

Human-Environment Interaction • The United Kingdom, Sweden, France, Germany, and Poland have borders along the North Sea or the Baltic Sea. These European seas are major transportation corridors. The North Sea contains one of the world's most important oil deposits. The Baltic Sea has some oil and is heavily used for transporting oil.

Accidents sometimes happen at sea. About three oil spills occur every year in the Baltic Sea. The North Sea has also had major spills. One occurred in November 1998, killing thousands of sea birds. Today, nations are working on ways to prevent spills and to clean up those that occur.

What do you think?

- Why is it important for countries to protect the seas?
- What effect might oil spills and other pollution have on the seas?
- What economic purposes do oceans and seas serve?

377

FOCUS ON GEOGRAPHY

Objectives

- To help students recognize the risks to the environment caused by economic development
- To help students understand the relationship between economics and geography

What Do You Think?

1. Invite students to discuss what they like about seas and oceans. Ask why these bodies of water are important.
2. Guide a discussion of the effects of pollution on the sea. Ask how oil spills might affect humans and other animals.
3. Have students list the economic benefits countries get from the seas, such as food, recreation, and trade.

Why is it important for Europeans to protect their seas?

Have students consider the many things Europeans get from the sea. Guide them to understand that the seas provide many economic benefits, including oil reserves and transportation corridors.

MAKING GEOGRAPHIC CONNECTIONS

Have students discuss the importance of oceans and seas, as well as rivers, lakes, and streams, to the United States. Ask them why it is also important for us to protect our water resources.

Implementing the National Geography Standards

Standard 11 Draw some general conclusions about how transportation and communication innovations affect patterns of economic interaction

Objective To give a presentation about how a transportation or communication innovation will affect the economy

Class Time 40 minutes

Task Ask students to put together a presentation to explain to prospective clients how a new service will affect patterns of economic interaction. Have students pretend that they are developing one of the following services: refrigerated railroad cars, airfreight services, telephone services, fax transmission services, or satellite-based communication systems.

Evaluation Students should use at least two drawings or charts to convey their ideas.

BEFORE YOU READ

What Do You Know?

Ask students to brainstorm one thing they associate with the European countries of Great Britain, Sweden, France, Germany, and Poland. Encourage them to think of products, ideas, people, or major events. Write their ideas on the board, and ask students to compare the items listed for each country. Are there more items for one country than others? How do the items themselves compare? Ask students to notice which country has the fewest items. Have students speculate about why the class knows the least about that country.

What Do You Want to Know?

Have each student make a chart with the headings What I Want to Know and What I Learned in their notebooks. Then ask students to work in pairs to develop questions they hope to have answered about the European countries covered in this chapter. They can record their questions in the first column of their chart, and, as they read the chapter, record the facts and ideas they learn in the second column.

READ AND TAKE NOTES

Reading Strategy: Comparing Explain to students that this chart compares five European countries in several ways. Point out that when it is completed, they will have a useful tool for understanding the differences and similarities of these countries.

In-depth Resources: Unit 4
- Guided Reading Worksheets, pp. 47–51

READING SOCIAL STUDIES

BEFORE YOU READ

▸▸ What Do You Know?

Did you know that from the end of WW II until 1990, Germany was two separate countries and Poland was controlled by the Soviet Union? Do you have relatives or friends who come from the United Kingdom, Sweden, France, Germany, or Poland? Have you ever seen the Queen of England or the Pope, who is from Poland, on television? Have you heard of the Nobel Prize, which is awarded in Sweden? Think about what you have learned in other classes, what you have read, and what you have heard or seen in the news about these countries.

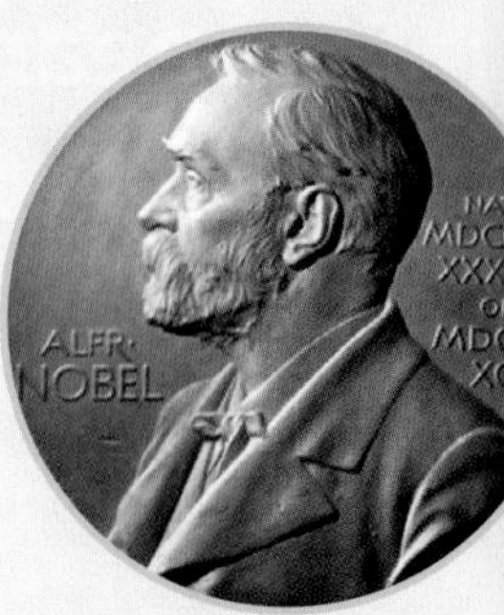

Region • Sweden's Nobel Priz honors great achievements worldwide. ▲

▸▸ What Do You Want to Know?

Consider what you know about the countries covered in Chapter 14. In your notebook, record what you hope to learn from this chapter.

READ AND TAKE NOTES

Reading Strategy: Comparing Comparing is a useful strategy for evaluating two or more similar subjects. Making comparisons also helps you to better understand what you have learned. Use the chart below to compare information about the United Kingdom, Sweden, France, Germany, and Poland.

- Copy the chart into your notebook.
- As you read, look for information for each category.
- Record details under the appropriate headings.

Movement • The German-made Volkswagen Beetle is the best-selling car ever. ▲

Country	Physical Geography	Government	Economy	Culture	Interesting Facts
United Kingdom	islands	constitutional monarchy	manufacturing, trade	music, literature	home of J. K. Rowling
Sweden	Scandinavian Peninsula	constitutional monarchy	engineering, communications	Nobel Prize	Workers get long vacations.
France	agricultural land	parliamentary republic	tourism	museums, literature	home to many artists
Germany	North and Baltic seas	democratic republic	rebuilt eastern part	music, literature	complex machinery
Poland	coast along Baltic Sea	parliamentary republic	free market	literature	publications sold tax-free

Teaching Strategy

Reading the Chapter This is a thematic chapter focusing on the government structures, economies, unique cultures, and recent history of the United Kingdom, Sweden, France, Germany, and Poland. Ask students to compare and contrast each of these countries as they read. Encourage them to consider how history and geography help explain the similarities and differences.

Integrated Assessment The Chapter Assessment on page 405 describes several activities for integrated assessment. You may wish to have students work on these activities during the course of the chapter and then present them at the end.

The United Kingdom

TERMS & NAMES
London
secede
Good Friday Accord
Charles Dickens

MAIN IDEA

The United Kingdom is a small nation in Western Europe with a history of colonization.

WHY IT MATTERS NOW

British economic, political, and cultural traditions have influenced nations around the world.

DATELINE

EXTRA

BARCELONA, SPAIN, MAY 30, 1999

They did it! Manchester United won the triple crown of soccer. The British fans here are going wild, and their excitement is easy to understand. Manchester United is only the third soccer team to win its league championship, cup titles, and the Champions League final.

Football, or "soccer" as Americans call it, is the world's most popular sport. The British invented a form of the game, called "mob football," in the 1300s. Back then, the playing field was the size of a small town, and there might have been as many as 500 players. A set of rules for the game was developed in 1863. Today, football is the national pastime in the United Kingdom.

Region • Manchester United celebrates after winning soccer's triple crown. ▲

SECTION OBJECTIVES

1. To identify the four regions of the United Kingdom
2. To describe the cultural heritage of the United Kingdom
3. To explain the British economy

SKILLBUILDER
- Interpreting a Map, p. 380

CRITICAL THINKING
- Identifying Problems, p. 381

FOCUS & MOTIVATE

WARM-UP

Making Inferences Have students read Dateline and discuss British sports.

1. How are the British and Americans alike and different in their interest in soccer?
2. How would you describe British fans?

INSTRUCT: Objective 1

A Kingdom of Four Political Regions

- What four regions make up the United Kingdom? Scotland, England, Wales, Northern Ireland
- Who actually leads the government in the United Kingdom? the prime minister

In-depth Resources: Unit 4
- Guided Reading Worksheet, p. 47

Reading Study Guide
(Spanish and English), pp. 114–115

A Kingdom of Four Political Regions 1

The United Kingdom is a small island nation of Western Europe. Its culture has had an enormous impact on the world. The nation's official name is the United Kingdom of Great Britain and Northern Ireland. **London,** located in southeastern England, is the capital.

TAKING NOTES
Use your chart to take notes about Europe today.

Country	Physical Geography	Government
United Kingdom		
Sweden		

Program Resources

In-depth Resources: Unit 4
- Guided Reading Worksheet, p. 47
- Reteaching Activity, p. 56

Reading Study Guide
(Spanish and English), pp. 114–115

Formal Assessment
- Section Quiz, p. 203

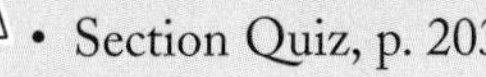

Integrated Assessment
- Rubric for writing a description

Outline Map Activities

Access for Students Acquiring English
- Guided Reading Worksheet, p. 111

Technology Resources
classzone.com

TEST-TAKING RESOURCES

Strategies for Test Preparation

Test Practice Transparencies

Online Test Practice

FOCUS ON VISUALS

Interpreting the Map Instruct students to look at the map. Ask them what two bodies of water separate England from France. What main waterway lies between England and Ireland? What ocean lies to the north of the United Kingdom?

Responses English Channel, Strait of Dover; Irish Sea; North Atlantic Ocean

Extension Ask students to compare the size of the United Kingdom with their own state. They can use the scale of miles on this map to estimate the length and width of Britain. Have them turn to page 260 of the Unit Atlas and use the scale of miles to find the length and width of their state.

Four different political regions make up the United Kingdom: Scotland, England, Wales, and Northern Ireland (see the map below). The British monarchy has ruled over the four regions for hundreds of years.

Vocabulary
monarchy: government by king or queen

National Government Today, the government of the United Kingdom is a constitutional monarchy. The British monarch is a symbol of power rather than an actual ruler. The power to govern belongs to Parliament, which is the national lawmaking body.

The British Parliament has two parts. The House of Lords is made up of nobles. Elected representatives make up the House of Commons. The House of Commons is the more powerful of the two houses.

The prime minister leads the government. He or she is usually the leader of the political party that wins the most seats in the House of Commons. The other political parties go into "opposition," which means their role is to question government policies.

Reading Social Studies
A. Clarifying Who is the head of government in the United Kingdom?

Reading Social Studies A. Possible Answer The prime minister is the head of government.

Regional Government in Great Britain Recently, the national government of the United Kingdom has returned some self-rule to some regions of Great Britain. In the late 1990s, voters in Wales approved plans for their own assembly, or body of lawmakers. Also at this time, the Scots voted to create their own parliament. Both Wales's assembly and Scotland's parliament met for the first time in 1999.

The United Kingdom Today

INTERACTIVE

GEOGRAPHY SKILLBUILDER: Interpreting a Map

1. **Location** • Which region of the United Kingdom is on a different island?
2. **Location** • Which body of water separates the United Kingdom from France?

Geography Skillbuilder Answers
1. Northern Ireland
2. English Channel

Activity Options

Multiple Learning Styles: Logical

Class Time 30 minutes

Task Creating a chart comparing and contrasting the governments of the United Kingdom and the United States

Purpose To understand similarities and differences in the structures of governments

Supplies Needed
- Writing paper
- Pencils or pens
- Ruler

Block Scheduling

Activity Have students create a chart comparing the governments of the United States and the United Kingdom. Suggest that they compare the heads of government and the legislative bodies for both nations. They might add Sweden, France, Germany, and Poland as they read this chapter. Have students share their charts in class.

BACKGROUND

The roots of division in Northern Ireland go back to at least the 1600s, when English and Scottish colonists settled there. These Protestant settlers took over the lands of Irish Catholics.

Governing Northern Ireland Throughout the 20th century, there were conflicts in Northern Ireland between Irish Catholic nationalists and Irish Protestants who supported the government of the United Kingdom. In fact, during the 1960s, many Irish Catholics wanted Northern Ireland to **secede,** or withdraw from, the United Kingdom. They hoped to unite Northern Ireland with the Republic of Ireland. Irish Protestants—a majority in the region—generally wanted to remain part of the United Kingdom.

In 1969, riots broke out, and the British government sent in troops to stop them. Violence between groups of Protestants and Catholics continued for almost 30 years. In 1998, representatives from both sides signed the **Good Friday Accord**. This agreement set up the Northern Ireland Assembly, which represents both Catholic and Protestant voters. For this government to succeed in Northern Ireland, however, the former enemies will need to work together.

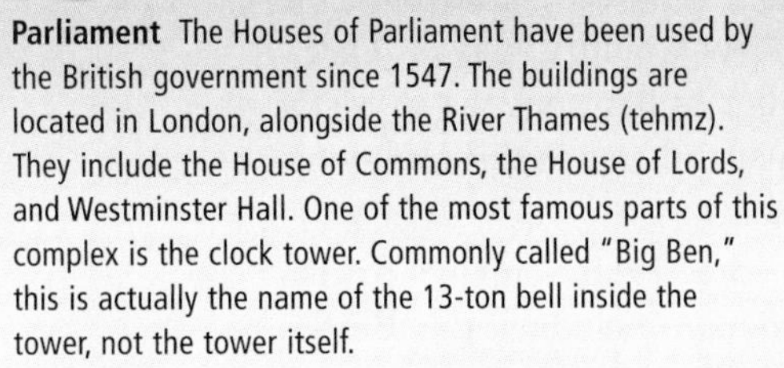

Parliament The Houses of Parliament have been used by the British government since 1547. The buildings are located in London, alongside the River Thames (tehmz). They include the House of Commons, the House of Lords, and Westminster Hall. One of the most famous parts of this complex is the clock tower. Commonly called "Big Ben," this is actually the name of the 13-ton bell inside the tower, not the tower itself.

In 1834, a fire destroyed much of the original buildings. The reconstruction by architect Sir Charles Barry was completed in 1860. The Houses of Parliament are visited and photographed by tourists from around the world.

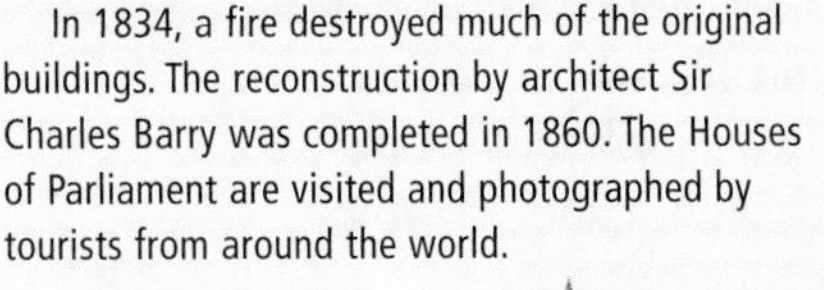

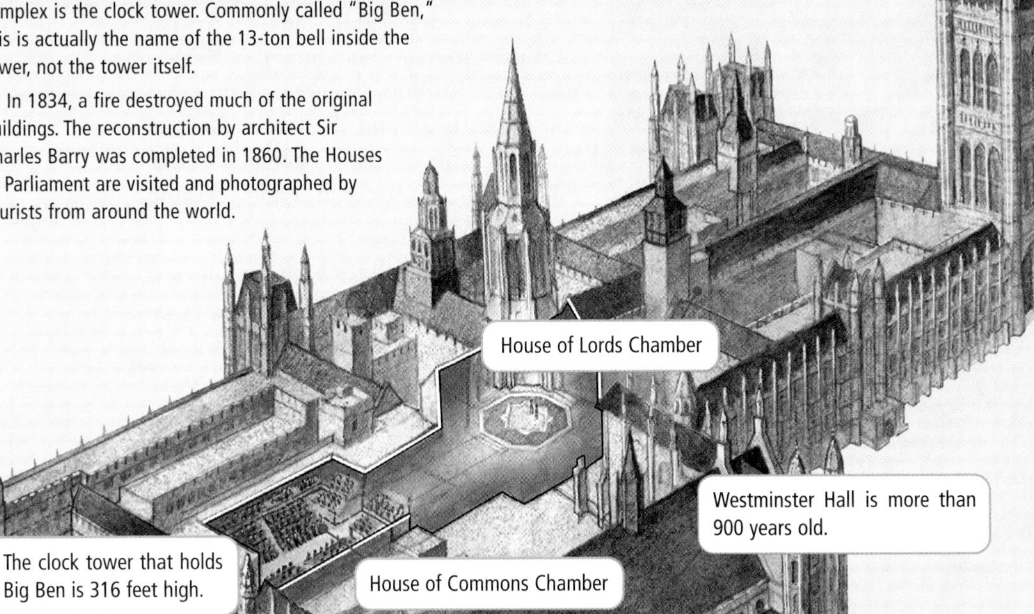
When Parliament is in session, the Union Jack flies from Victoria Tower.
House of Lords Chamber
Westminster Hall is more than 900 years old.
The clock tower that holds Big Ben is 316 feet high.
House of Commons Chamber

CRITICAL THINKING ACTIVITY

Identifying Problems Have students work individually to list problems they identify in the discussion of Northern Ireland, including the information given as Background. Then ask them to discuss the following question: Why is it so difficult to govern Northern Ireland? List their responses on the chalkboard.

Class Time 10 minutes

The Houses of Parliament rest on the site of Westminster Palace, which was built in the 11th century by William II, the son of William the Conqueror. Only Westminster Hall, which was completed in 1097, survives from the original buildings. It has been in continuous use for more than 900 years. The clock, in contrast, is relatively new. It was installed in 1859. There are four clock faces, each 23 feet square. Each number is two feet high and each minute hand is 14 feet long. The clock is known for its extreme accuracy. The bell, Big Ben, came from the original Westminster Palace and was refurbished for installation with the clock.

Activity Options

Interdisciplinary Link: Government

Class Time One class period

Task Writing arguments in support of and opposition to regional self-rule

Purpose To compare advantages and disadvantages of regional self-rule

Supplies Needed
- Writing paper
- Pens or pencils

Activity Guide a discussion to help students explore some advantages and disadvantages of regional self-rule. Encourage them to make comparisons with federal and state government in the United States. List their ideas on the board. Assign half the class to write a paragraph in support of regional self-rule, and assign the other half to write in opposition. Invite volunteers to read their paragraphs in class. Discuss.

INSTRUCT: Objective 2

Cultural Heritage

- Why did the United Kingdom's culture spread to so many parts of the world? The United Kingdom was an imperial power with colonies around the world.
- Who are some important modern British musicians? Beatles, Rolling Stones, Elton John, Sting, Dido
- Who are some important British writers? William Shakespeare, Mary Shelley, Sir Arthur Conan Doyle, Charles Dickens, Virginia Woolf, George Orwell, C. S. Lewis, J. K. Rowling

MORE ABOUT...

Charles Dickens

Charles Dickens (1812–1870) is widely recognized as a major figure in English literature. As a child, he enjoyed reading adventure stories, fairy tales, and novels. In the 1820s, he became a newspaper reporter, covering Parliament debates and writing feature stories. Dickens's keen observation skills, understanding of humanity, and sense of humor contributed to his success as a writer. Some books by Dickens include *David Copperfield* and *Great Expectations.*

Region • London's New Globe Theater is a replica of the 17th-century playhouse that originally hosted William Shakespeare's works. ◄

Cultural Heritage 2

The United Kingdom has a rich cultural heritage that includes the great Renaissance playwright William Shakespeare. With a long history as an imperial power, the nation has been exporting its culture around the world for hundreds of years. For example, India, Canada, and other former British colonies modeled their governments on the British parliamentary system. British culture has also set trends in sports, music, and literature.

Music British music influenced the early music of Canada and the United States, both former British colonies. One British tune long familiar to people in the United States is "God Save the Queen." You probably know it as "My Country, 'Tis of Thee." Several countries have put the words of their national anthems to this traditional British melody.

Region • In the early 1960s, the Beatles became wildly popular, not only in the United Kingdom, but also around the world. ▼

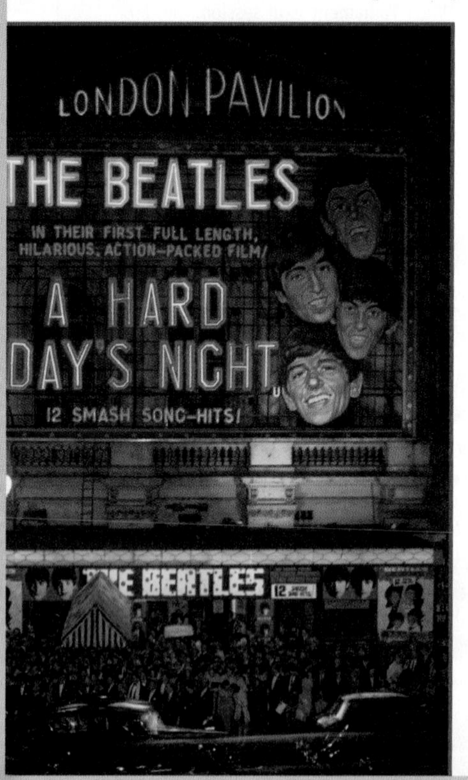

During the 1960s, many British musical groups—including the Beatles and the Rolling Stones—dominated music charts around the world. In later decades, other British singers, including Elton John, Sting, and Dido, became popular favorites.

Literature The best-known cultural export of the United Kingdom, aside from the English language itself, may be literature. In the 19th century, Mary Shelley dreamed up Frankenstein's monster, and Sir Arthur Conan Doyle first wrote about Sherlock Holmes. Another popular author of the time was **Charles Dickens** (1812–1870), who wrote *Oliver Twist* and *A Christmas Carol.*

Activity Options

Skillbuilder Mini-Lesson: Using an Electronic Card Catalog

Block Scheduling

Explaining the Skill Remind students that books, magazines, and other library resources are listed in the library's electronic card catalog. These resources can be found by looking for the title, the author's name, or the subject.

Applying the Skill Assign individual students topics from the discussion of Britain's cultural heritage. Ask each to find library resources for his or her topic, using the title, author's name, or subject catalogs. For example, one student might be assigned to find works about William Shakespeare. Another might be asked to find books written by Shakespeare. Have students make lists of their findings. In class discussion, ask them to summarize their discoveries about the use of the electronic card catalog.

Reading Social Studies

B. Recognizing Important Details Why was the United Kingdom able to spread British culture across the world?

Reading Social Studies
B. Possible Answer
It has a long history as an imperial power.

Two gifted British writers of the 20th century are Virginia Woolf and George Orwell. Modern British authors have also given the world many popular stories for young people. They include C. S. Lewis, who wrote *The Chronicles of Narnia,* and J. K. Rowling, who created the Harry Potter books.

Region • J. K. Rowling's Harry Potter books have captured the imaginations of children worldwide. ▲

The British Economy ❸

The United Kingdom is an important trading and financial center. Many British citizens also make their living in mining and manufacturing. Factories in the United Kingdom turn out a variety of products ranging from china to sports cars. The nation has plenty of coal, natural gas, and oil to fuel its factories, but it has few other natural resources.

The need for imported goods makes trade another major industry of the United Kingdom. The nation imports many raw materials used in manufacturing. It also imports food, because the farms of this nation produce only enough to feed about two-thirds of its large population.

SECTION 1 ASSESSMENT

Terms & Names

1. **Explain the significance of:** (a) London (b) secede (c) Good Friday Accord (d) Charles Dickens

Using Graphics

2. Use a chart like this one to describe the major aspects of the United Kingdom's modern government, economy, and culture.

Modern United Kingdom	
Government	
Economy	
Culture	

Main Ideas

3. (a) Identify the four regions that make up the United Kingdom.
   (b) What role does the British monarch play in the government of the modern United Kingdom?
   (c) What impact has the culture of the United Kingdom had on its colonies and on other parts of the world?

Critical Thinking

4. **Analyzing Issues**
   Why do you think the conflict in Northern Ireland is so difficult to resolve?

Think About

- differences in religious beliefs
- recent changes in British regional governments
- the long period of continued violence

ACTIVITY -OPTION- Reread the information about British football from the "Dateline" feature that opens the section. Write a **description** of a sport that interests you.

INSTRUCT: Objective ❸

The British Economy

- What natural resources are found in the United Kingdom? coal, natural gas, oil
- Why is trade an important industry of the United Kingdom? The country must import food and raw materials for manufacturing.

ASSESS & RETEACH

Reading Social Studies Have students add details from Section 1 to the chart on page 378.

Formal Assessment
- Section Quiz, p. 203

RETEACHING ACTIVITY

Have pairs of students write one main-idea sentence for the discussion following each section heading. Then have students meet in groups to compare sentences and to decide on the best statement of each main idea.

In-depth Resources: Unit 4
- Reteaching Activity, p. 56

Access for Students Acquiring English
- Reteaching Activity, p. 118

Section 1 Assessment

1. Terms & Names

a. London, p. 379
b. secede, p. 381
c. Good Friday Accord, p. 381
d. Charles Dickens, p. 382

2. Using Graphics

Modern United Kingdom	
Government	constitutional monarchy; Parliament governs; prime minister leads government
Economy	based on trade and finance; few natural resources; imports raw materials/food
Culture	cultural exports: English language, government, music, literature

3. Main Ideas

a. England, Wales, Scotland, Northern Ireland
b. The monarch is a symbol of power.
c. It has influenced language, government, music, and literature throughout the world.

4. Critical Thinking

People fear loss of religious freedom, distrust one another, and cannot forget past experiences.

ACTIVITY OPTION

Integrated Assessment
- Rubric for writing a description

Sweden

TERMS & NAMES
- Riksdag
- ombudsman
- armed neutrality
- hydroelectricity
- acid rain
- skerry

SECTION OBJECTIVES

1. To identify governmental, economic, and environmental issues of Sweden
2. To describe Sweden's culture

SKILLBUILDER
- Interpreting a Map, p. 385

CRITICAL THINKING
- Analyzing Issues, p. 385
- Analyzing Causes, p. 386

FOCUS & MOTIVATE

WARM-UP

Drawing Conclusions After students read Dateline, discuss the Nobel Prize.

1. Why do you think Alfred Nobel established the Nobel Prize?
2. Why do you think the Nobel Prize has become so important?

INSTRUCT: Objective 1

Sweden's Government/The Economy and the Environment

- What legislative body makes the laws of Sweden? Riksdag
- What environmental problems does Sweden face? safe sources for electrical power, acid rain

In-depth Resources: Unit 4
- Guided Reading Worksheet, p. 48

Reading Study Guide (Spanish and English), pp. 116–117

MAIN IDEA	WHY IT MATTERS NOW
Sweden offers its people a high standard of living, although it also faces environmental problems.	Modern Sweden is dealing with environmental issues that affect many countries around the world.

DATELINE

STOCKHOLM, SWEDEN, DECEMBER 4, 2001—This week in Stockholm, Sweden's capital, hundreds of past winners of the Nobel Prize gather to celebrate the centennial, or 100th anniversary, of this award. Concerts, lectures, and banquets lead up to the award ceremony in Stockholm City Hall on December 10.

The first Nobel Prize ceremony was held in Stockholm in 1901. Since then, awards in physics, chemistry, economics, medicine, literature, and peace have gone to more than 700 people, representing every inhabited continent. Besides achieving worldwide honor and fame, the winners receive a medal and a cash prize. The award was established at the request of Alfred Nobel (1833–1896), a Swedish chemist and millionaire who invented dynamite.

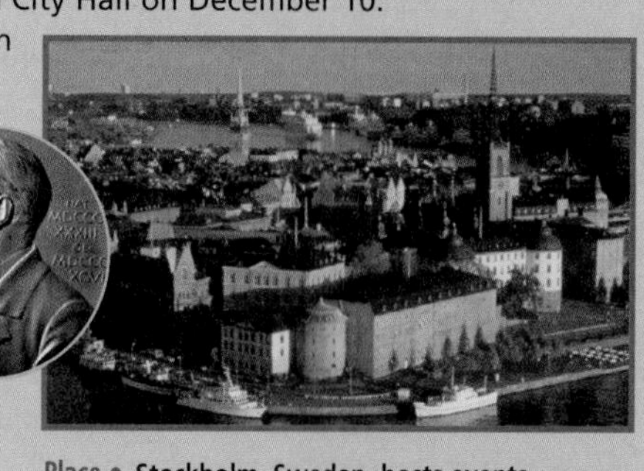

Place • Stockholm, Sweden, hosts events celebrating the 100th anniversary of the Nobel Prize. ▲

Sweden's Government 1

Home of the Nobel Prize, Sweden shares the Scandinavian Peninsula with Norway in Northern Europe (see the map on page 385). The country is a constitutional monarchy; the Swedish monarch has only ceremonial powers and cannot make laws. Instead, the people elect representatives to four-year terms in the Swedish parliament, called the **Riksdag** (REEKS•DAHG).

TAKING NOTES
Use your chart to take notes about Europe today.

Country	Physical Geography	Government
United Kingdom		
Sweden		

Program Resources

In-depth Resources: Unit 4
- Guided Reading Worksheet, p. 48
- Reteaching Activity, p. 57

Reading Study Guide (Spanish and English), pp. 116–117

Formal Assessment
- Section Quiz, p. 204

Integrated Assessment
- Rubric for writing a description

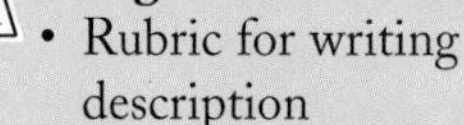

Outline Map Activities

Access for Students Acquiring English
- Guided Reading Worksheet, p. 112

Technology Resources
classzone.com

TEST-TAKING RESOURCES

Strategies for Test Preparation

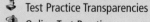

Test Practice Transparencies

Online Test Practice

384 CHAPTER 14

The Riksdag The 349 members of the Riksdag nominate Sweden's prime minister. They also appoint ombudsmen. **Ombudsmen** are officials who protect citizens' rights and make sure that the Swedish courts and civil service follow the law.

Region • Women are active in Swedish government. ▶

Swedish citizens vote to determine how many members of each political party serve in the Riksdag. Before 1976, the Social Democratic Labour Party had been in power for nearly 44 years. Today, the Swedish government includes four other parties.

Reading Social Studies
A. Possible Answer
Sweden tries not to form military alliances. It is a strong supporter of the United Nations.

Foreign Policy Since World War I, Sweden's foreign policy has been one of **armed neutrality.** This means that in times of war, the country has its own military forces but does not take sides in other nations' conflicts.

Even during peacetime, the Swedish government tries not to form military alliances. Unless Sweden is directly attacked, it will not become involved in war. The country is a strong supporter of the United Nations.

Reading Social Studies
A. Synthesizing How does Sweden's neutrality affect its foreign relations?

The Economy and the Environment ❶

Privately owned businesses and international trade are important to Sweden's economy. It exports many goods, including metals, minerals, and wood. Engineering and communications are major industries. The automobile industry also provides many jobs.

Sweden Today

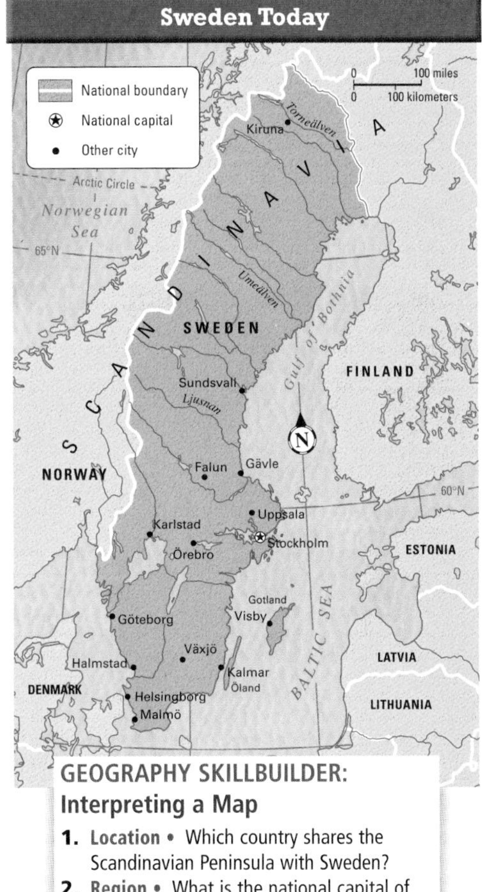

GEOGRAPHY SKILLBUILDER: Interpreting a Map
1. **Location •** Which country shares the Scandinavian Peninsula with Sweden?
2. **Region •** What is the national capital of Sweden?

Geography Skillbuilder Answers
1. Norway
2. Stockholm

CRITICAL THINKING ACTIVITY

Analyzing Issues Prompt a discussion of Sweden's policy of armed neutrality by asking students what some advantages of this position might be; then ask them to consider what the disadvantages might be. Record student answers in a chart to save for future discussions.

Class Time 10 minutes

FOCUS ON VISUALS

Interpreting the Map Direct students' attention to the map. Ask them which countries share a border with Sweden.

Response Norway, Finland

Extension Point out the Arctic Circle on the map and ask students what they know about it. Then have them look at a map of North America and locate the Arctic Circle. Ask which state in the United States it passes through. Discuss the kinds of weather they would expect in places near the Arctic Circle.

Activity Options

Multiple Learning Styles: Interpersonal

Class Time One class period

Task Writing a travel itinerary

Purpose To learn about the geography and culture of Sweden

Supplies Needed
- Library or Internet resources
- Writing paper
- Pencils or pens

Block Scheduling

Activity Invite students to imagine that they are traveling to Sweden with a family member and to write an itinerary for the trip. Explain that an itinerary lists places they will visit each day. Suggest that students look through the section for topics and study the map on this page to identify cities or geographic features they would visit. They should then research the topics and write an itinerary for a one-week visit.

CRITICAL THINKING ACTIVITY

Analyzing Causes Guide a discussion of Sweden's sources of power. Ask students why the Swedish government is looking for power sources other than nuclear power. Discuss why solar- and wind-powered energy sources are being investigated. Students might suggest that such alternative sources of energy may be safer for the environment.

Class Time 10 minutes

INSTRUCT: Objective 2

Daily Life and Culture

- In what way are Swedes a homogeneous people? Most are Caucasian, 90 percent are Lutherans, and most speak Swedish.
- What types of recreation are popular in Sweden? cross-country and downhill skiing, skating, ice hockey, ice fishing, hiking, camping, tennis, soccer, and outdoor performances
- In what areas has Sweden contributed to world culture? drama, literature, film

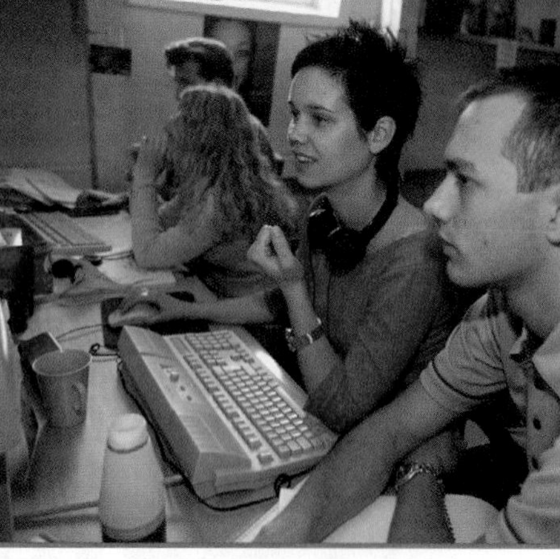

Place • Many in Sweden's highly educated labor force work in the high-tech and engineering industries. ▲

The Swedish Labor Force After World War II, many Swedes left their towns and villages to find work in the large cities in the south. Today, more than 80 percent of the population lives in these urban areas. Much of Sweden's labor force is highly educated and enjoys a high standard of living.

Power Sources **Hydroelectricity,** or power generated by water, is the main source of electrical power in Sweden. Nuclear power is also widely used. The Swedish government is looking into other, safer sources of energy, which include solar- and wind-powered energy.

Acid Rain Sweden and its neighboring countries share similar environmental problems. One of the most severe problems is **acid rain.** Acid rain occurs when air pollutants come back to Earth in the form of precipitation. These pollutants may soon poison many trees throughout the region. Sweden and neighboring countries are working to clean up the environment by trying to control air pollutants produced by cars and factories.

Reading Social Studies
B. Clarifying What causes acid rain?

Region • December 13 is St. Lucia's Day, one of Sweden's most important Christian holidays. ▼

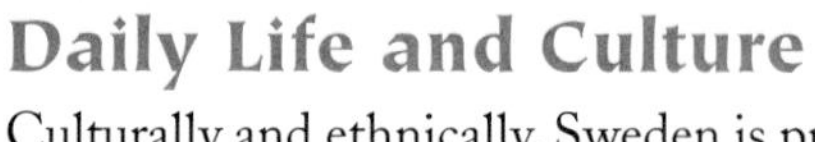

Daily Life and Culture 2

Culturally and ethnically, Sweden is primarily a homogeneous country. Ninety percent of the population are native to Sweden and are members of the Lutheran Church of Sweden. The majority of people speak Swedish.

Since World War II, immigrants from Turkey, Greece, and other countries have brought some cultural diversity to Sweden's population. Today, about one in nine people living in Sweden is an immigrant or the child of an immigrant.

Vocabulary
homogeneous: the same throughout

Reading Social Studies
B. Possible Answer
air pollutants returning to Earth in the form of rain

Recreation Workers in Sweden have many benefits, including long vacations. The Swedes love taking time to enjoy both winter and summer sports. Sweden, with its cold weather and many hills and mountains, is a great place for cross-country and downhill skiing. Skating, hockey, and ice fishing are also popular.

Activity Options

Multiple Learning Styles: Visual

Block Scheduling

Class Time One class period

Task Creating a bulletin board display

Purpose To describe popular sports in Sweden

Supplies Needed
- Library or Internet resources
- Writing and drawing paper
- Pencils, pens, markers

Activity Point out to students that the people of Sweden enjoy many sports throughout the year. Have students work in small groups and create a bulletin board display that describes these sports. Students should first do research. They should then plan a display that explains the sport, including equipment, rules, and so on. Their display should include diagrams, pictures, and other graphic organizers, as well as brief descriptions.

Place • Sweden's cold winters have made downhill and cross-country skiing popular. ▶

Many small islands, called **skerries,** dot the Swedish coast. In the summer, many people visit these islands to hike, camp, and fish. Tennis, soccer, and outdoor performances such as concerts are popular as well.

Contributions to World Culture Sweden is well-known for its contributions to drama, literature, and film. The late 19th-century and early 20th-century plays of August Strindberg are produced all over the world. Astrid Lindgren's children's books, including *Pippi Longstocking* (1945), still delight readers everywhere. Ingmar Bergman is famous for the many great films he directed.

SECTION 2 ASSESSMENT

Terms & Names

1. **Explain the significance of:** (a) Riksdag (b) ombudsman (c) armed neutrality (d) hydroelectricity (e) acid rain (f) skerry

Using Graphics

2. Use a spider map like this one to outline the major aspects of Sweden's government, economy, and culture.

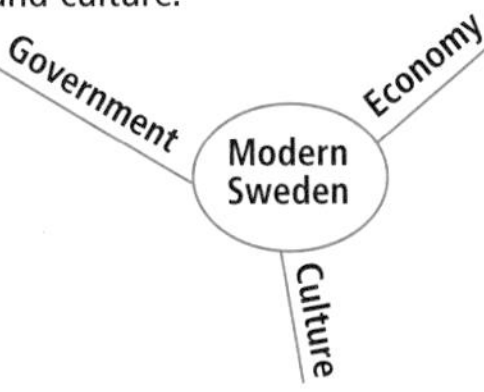

Main Ideas

3. (a) On which European peninsula is Sweden located? What other country shares this peninsula?
   (b) What happened to the Swedish labor force after World War II?
   (c) How has immigration since World War II changed the population of Sweden?

Critical Thinking

4. **Evaluating Decisions**
   What do you think might be the advantages and disadvantages of armed neutrality for Sweden?

Think About

- Sweden's location
- the damage and expense of war
- the benefits of alliances

ACTIVITY -OPTION- Reread the "Dateline" feature at the beginning of this section. Write a short **description** of which category you would like to earn a Nobel Prize in and why.

ASSESS & RETEACH

Reading Social Studies Have students add details from Section 2 to the chart on page 378.

Formal Assessment
- Section Quiz, p. 204

RETEACHING ACTIVITY

Assign each student one page of Section 2 to review. Each student should select one important or interesting idea and illustrate it. Then have students meet in groups and share their illustrations. They should explain their drawings and tell why the topic is important.

In-depth Resources: Unit 4
- Reteaching Activity, p. 57

Access for Students Acquiring English
- Reteaching Activity, p. 119

Section 2 Assessment

1. Terms & Names

a. Riksdag, p. 384
b. ombudsman, p. 385
c. armed neutrality, p. 385
d. hydroelectricity, p. 386
e. acid rain, p. 386
f. skerry, p. 387

2. Using Graphics

Modern Sweden
- Government: constitutional monarchy, Riksdag, ombudsmen, armed neutrality
- Culture: homogeneous population; outdoor recreation; drama, literature, film
- Economy: international trade, engineering, communications, automobile industry, labor force highly educated

3. Main Ideas

a. Sweden is located on the Scandinavian peninsula; Norway
b. Many Swedes moved from the countryside to large cities.
c. The population is less homogeneous.

4. Critical Thinking

Advantages—fewer deaths, less destruction, lower costs of war. Disadvantages—no allies in case of war.

ACTIVITY OPTION

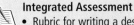

Integrated Assessment
- Rubric for writing a description

OBJECTIVE

Students work cooperatively to explore Florence at the time of the Renaissance.

PROCEDURE

In addition to the Data File, provide resources such as biographies, art books, documentary films, and magazine articles. Encourage students to divide the responsibilities of researching, writing, and presenting monologues. Some students may prefer to write journal entries individually. Give these students a chance to exchange solutions and offer constructive criticism.

 **Block Scheduling**

LANGUAGE ARTS CHALLENGE

Class Time 50 minutes

Suggest that students explore biographies and histories of the period. For the second option, journals or histories of daily life might be helpful. Invite volunteers to read their monologues and journal entries aloud.

Possible Solutions

Monologues should capture the key events in the figure's life and emphasize the figure's contributions. Presenters should bring the character to life. Journal entries should cover topics, such as guilds, the role of the apprentice, or the job description of the master, and include details of daily life.

Interdisciplinary Challenge

Spend a Day in Renaissance Florence

You are a traveler visiting Florence, Italy, in the year 1505. It is exciting to be here now. All over Europe, people have heard about the Renaissance, or cultural rebirth, that is taking place in this beautiful city. Artists, architects, writers, and scientists are turning out brilliant work. In the day you spend here, you want to learn about this new cultural movement. You want to be able to tell people at home about Renaissance Florence.

COOPERATIVE LEARNING On these pages are challenges you will encounter as you tour Renaissance Florence. Working with a small group, choose one of these challenges to solve. Divide the work among group members. Look for helpful information in the Data File. Keep in mind that you will present your solution to the class.

LANGUAGE ARTS CHALLENGE

"Florence is home to brilliant artists and writers."

Why did the Renaissance start here? Florence is home to brilliant artists and writers. Successful merchants and craftworkers, along with several powerful families, have made the city rich. Many wealthy people are patrons, or sponsors, of artists' work. You are curious about the people of Florence. Who are the leading figures? What is life like here? Choose one of these options to discover the answers. Use the Data File for help.

ACTIVITIES

1. Choose one major figure who lived in Florence during the Renaissance and research his or her life. Then write a short first-person monologue in which, speaking as that person, you describe your life and work.
2. Imagine you are an ordinary young Florentine living in 1505—for example, a goldsmith's apprentice. Write journal entries for a week in your life.

Standards for Evaluation

LANGUAGE ARTS CHALLENGE

Option 1 Monologues should
- highlight major events in the figure's life.
- explain the figure's significance.

Option 2 Journal entries should
- describe the roles of apprentice and master.
- capture details of life in Renaissance times.

SCIENCE CHALLENGE

Option 1 Diagrams should
- accurately reproduce the Duomo.
- show use of vaults and supporting walls.

Option 2 Interview questions should
- reflect Brunelleschi's contributions to architecture and Florence.

ARTS CHALLENGE

Option 1 Slide shows should
- include 5–6 major works of art.
- explain why they reflect the Renaissance.

Option 2 Maps should
- include major art sites in Florence.
- highlight works of art to be seen at those sites.

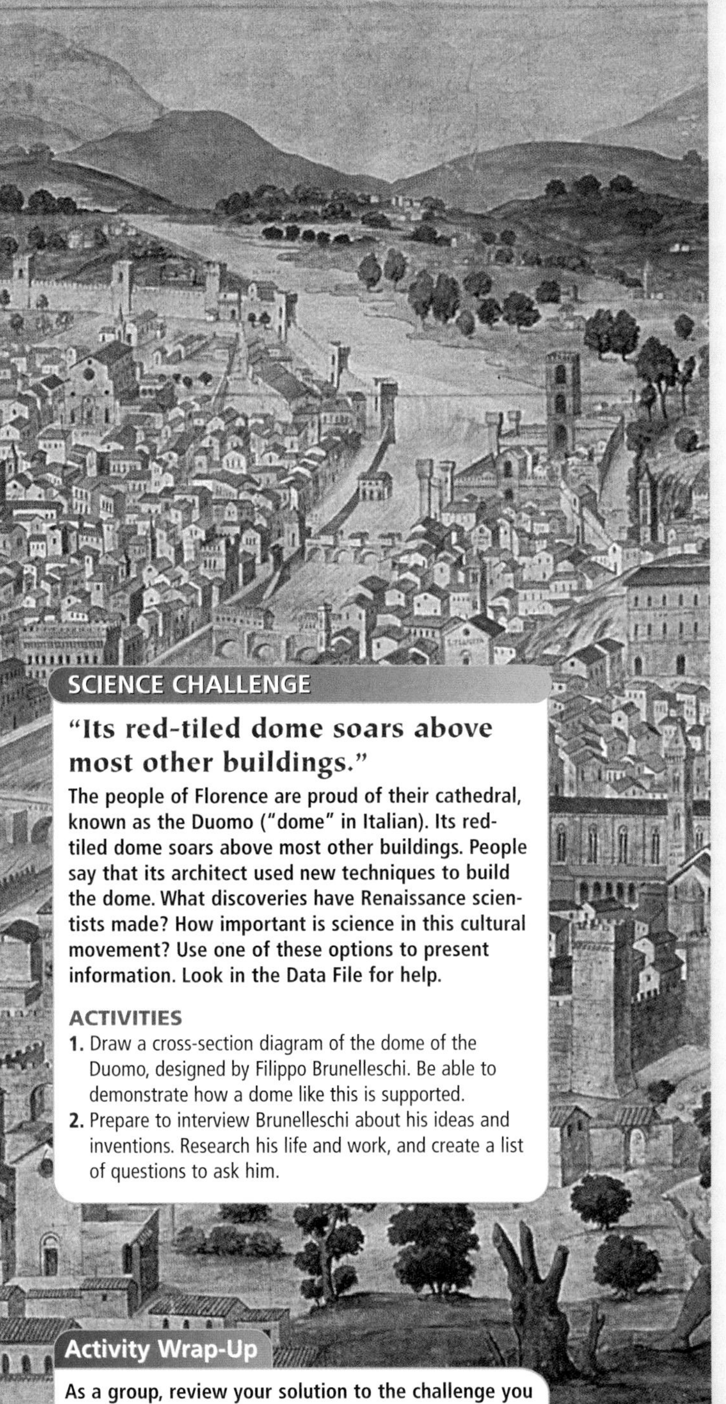

SCIENCE CHALLENGE

"Its red-tiled dome soars above most other buildings."

The people of Florence are proud of their cathedral, known as the Duomo ("dome" in Italian). Its red-tiled dome soars above most other buildings. People say that its architect used new techniques to build the dome. What discoveries have Renaissance scientists made? How important is science in this cultural movement? Use one of these options to present information. Look in the Data File for help.

ACTIVITIES

1. Draw a cross-section diagram of the dome of the Duomo, designed by Filippo Brunelleschi. Be able to demonstrate how a dome like this is supported.
2. Prepare to interview Brunelleschi about his ideas and inventions. Research his life and work, and create a list of questions to ask him.

Activity Wrap-Up

As a group, review your solution to the challenge you selected. Then present your solution to the class.

DATA FILE

LANDMARKS OF RENAISSANCE FLORENCE

- Florence is built on both sides of the **Arno River**. Its population during the Renaissance was about 100,000. Most of the famous buildings are on the right bank. Besides its artists, Renaissance Florence was known for its craftworkers, such as goldsmiths and leatherworkers.
- The **Duomo** stands on the Piazza del Duomo, an open square. In 1418, Filippo Brunelleschi won a contest to build a dome over the unfinished church. He invented new methods and machines to build it. As in earlier domes, vaults or pointed arches support the dome. Brunelleschi added a circular support wall, called a drum, to build it higher.
- The **Ponte Vecchio** ("Old Bridge"), built in 1345, is one of several bridges across the Arno River. Shops, especially those of goldsmiths, line both sides of the bridge.
- The **Pitti Palace**, built in 1458, is on the left bank of the river.

MAJOR FIGURES OF THE RENAISSANCE

- **Filippo Brunelleschi** (1377–1446), architect of the Duomo and the Pitti Palace.
- **Dante** (1265–1321), poet, author of *Divine Comedy.* Dante pioneered the usage of everyday language, instead of Latin, in literature.
- **Isabella d'Este** (1474–1539), noblewoman and patron of many artists.
- **Leonardo da Vinci** (1452–1519), painter, sculptor, engineer, scientist.
- **Michelangelo** (1475–1564), sculptor, painter, architect; sculptor of *David* (1504).
- **Raphael** (1483–1520), painter and architect.

To learn more about Renaissance Florence, go to

SCIENCE CHALLENGE

Class Time 50 minutes

For both options, students will need outside reference sources. For the first option, provide students with poster board and rulers. Encourage students to display their diagrams.

Possible Solutions

The diagrams should emphasize the use of vaults and supporting walls. Interview questions should reflect an understanding of Brunelleschi's life and works and his contributions to the field of architecture.

ALTERNATIVE CHALLENGE...

ARTS CHALLENGE

Churches, palaces, and museums in Florence contain Renaissance paintings and sculptures. For example, the Uffizi Gallery houses da Vinci's *Adoration of the Magi* and Botticelli's *The Birth of Venus.* In the Accademia Gallery you can see Michelangelo's well-known sculpture *David.* In the Pitti Palace are paintings by Raphael and Rubens.

- Design a slide show of 5–6 slides of Renaissance works of art. Prepare a brief explanation of each work and its significance.
- Create a map for an "art tour" of Florence. Show key buildings and museums and include highlights of each visit.

Activity Wrap-Up

To help students evaluate their challenge solutions, have them make a grid with criteria like the one shown. Then have them rate each solution on a scale from 1 to 5, with 1 representing the lowest score and 5 the highest.

Originality	1	2	3	4	5
Creativity	1	2	3	4	5
Accuracy	1	2	3	4	5
Overall effectiveness	1	2	3	4	5

France

TERMS & NAMES
Charles de Gaulle
French Resistance
Jean Monnet
socialism
European Community
impressionism

SECTION OBJECTIVES

1. To describe the government of France's Fifth Republic
2. To describe France's economy
3. To explain France's contribution to world culture

SKILLBUILDER
- Interpreting a Map, p. 391
- Interpreting a Graph, p. 392

FOCUS & MOTIVATE

WARM-UP

Making Inferences Have students read Dateline and ask the following questions.

1. Why do you think the French Resistance was formed?
2. How do you think Parisians felt when they saw the French army enter the city?

INSTRUCT: Objective 1

The Fifth Republic

- What contributions did Charles de Gaulle make to France during the war? led military as general and led the French in exile
- How is the government of the Fifth Republic organized? power split between president and parliament; parliament divided into Senate and National Assembly, with prime minister as head

In-depth Resources: Unit 4
- Guided Reading Worksheet, p. 49

Reading Study Guide
(Spanish and English), pp. 118–119

MAIN IDEA

France was ruined politically and economically by World War II but has since made a full recovery.

WHY IT MATTERS NOW

France is an important member of the European Union and continues to influence the world's economy and cultures.

DATELINE

EXTRA

PARIS, FRANCE, AUGUST 26, 1944

Paris is free! The church bells are still ringing from yesterday's celebrations. After four long years of German control, Paris finally has been liberated.

General Charles de Gaulle returned from the United Kingdom yesterday and celebrated the liberation by leading a parade from the Arc de Triomphe to Notre Dame Cathedral.

The liberation of Paris is the result of a two-and-a-half-month advance of Allied forces from the beaches of Normandy in northern France. The French Resistance in Paris began disrupting the German occupiers on August 19, and yesterday, the French army entered Paris.

Region • The liberation of Paris is a significant symbolic victory for the Allies. ▲

The Fifth Republic 1

During World War II, **Charles de Gaulle** (1890–1970) was a general in the French army. After Germany conquered France in 1940, de Gaulle fled to the United Kingdom. There, he became the leader of the French in exile and stayed in contact with the French Resistance. The **French Resistance** established communications for the Allied war effort, spied on German activity, and sometimes assassinated high-ranking German officers.

TAKING NOTES
Use your chart to take notes about Europe today.

Country	Physical Geography	Government
United Kingdom		
Sweden		

Program Resources

In-depth Resources: Unit 4
- Guided Reading Worksheet, p. 49
- Reteaching Activity, p. 58

Reading Study Guide
(Spanish and English), pp. 118–119

Formal Assessment
- Section Quiz, p. 205

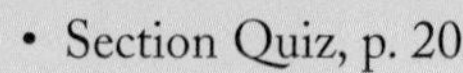

Integrated Assessment
- Rubric for drawing a portrait

Outline Map Activities

Access for Students Acquiring English
- Guided Reading Worksheet, p. 113

Technology Resources
classzone.com

TEST-TAKING RESOURCES

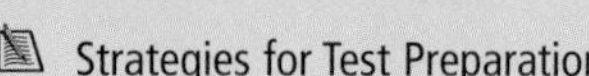

Strategies for Test Preparation
Test Practice Transparencies

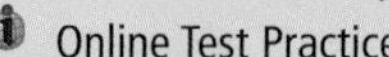

Online Test Practice

Geography Skillbuilder Possible Answers

1. Luxembourg, Monaco, Italy, Switzerland, Spain, Belgium, Germany

2. English Channel, Bay of Biscay, Mediterranean Sea

France Today

National capital · Other city

GEOGRAPHY SKILLBUILDER: Interpreting a Map

1. **Location** • Name three countries that border France.
2. **Location** • Which bodies of water does France have access to?

On December 21, 1958, Charles de Gaulle was elected president of France. He reorganized the French constitution and instituted the Fifth Republic of France.

The Government of the Fifth Republic France is a parliamentary republic. Governmental power is split between the president and parliament. The president is elected by the public to a seven-year term; beginning in 2002, the president will serve a five-year term. The president's primary responsibilities are to act as guardian of the constitution and to ensure proper functioning of other authorities.

Reading Social Studies

A. Comparing Compare the term and role of the president of France with the president of the United States.

Parliament has two parts: the Senate and the National Assembly. The president chooses a prime minister, who heads parliament and is largely responsible for the internal workings of the government. The French government is very active in the country's economy.

Reading Social Studies

A. Possible Answer United States: four-year terms, two-term maximum; France: seven-year terms, unlimited terms

A Centralized Economy 2

World War II left France poor and in need of rebuilding. The National Planning Board, established by Jean Monnet (moh•NAY) in 1946, launched a series of five-year plans to modernize France and set economic goals for the country.

Connections to History

Lascaux Cave Paintings On September 12, 1940, while hiking in the hills of Lascaux (lah•SKOH) near the town of Montignac in southern France, four teenage boys discovered ancient cave paintings. They found the caves after their dog fell in a hole in the ground.

Henri Breuil (broy), one of the first archaeologists on the scene, counted more than 600 images of horses (shown here), deer, and bison. The cave paintings are about 17,000 years old, making them some of the oldest works of art yet discovered.

INSTRUCT: Objective 2

A Centralized Economy

- What was the result of France's five-year plan following the war? a mixed economy with both private ownership and the nationalization of some businesses and industries
- What was France's main source of power following the war, and what is it today? coal, oil, gas; nuclear energy

FOCUS ON VISUALS

Interpreting the Map France has at various times been at war with both Germany and Great Britain. Ask students to identify one physical feature between France and Great Britain that is not between France and Germany. How might this have affected each conflict?

Possible Responses A body of water, the English Channel, lies between France and Great Britain. This may have helped France defend against Great Britain more easily.

Extension Following World War I, France built a line of defense against German invasion along its eastern border, called the Maginot Line. Have students find out more about the Maginot Line.

Connections to History

When Lascaux was first discovered, the paintings were in perfect condition. The cave was first opened to public viewing in 1948. However, as many as 100,000 visitors a year and artificial lighting took its toll on the artwork, and the cave was closed to the public in 1963. A partial replica of the cave was opened in 1983. About 300,000 visit it annually.

Activity Options

Interdisciplinary Links: Government/History

Class Time One class period

Task Creating a chart on historical French governments

Purpose To learn about the various types of governments in France's history

Supplies Needed
- Reference materials related to French history
- Pencils or pens
- Paper

Activity Have students work in small groups to research the forms of government France has had in its history, beginning in the 15th century. Ask each group to create a chart that includes brief descriptions of each form of government and the time period in which each existed.

FOCUS ON VISUALS

Interpreting the Graph Ask students to determine if most countries on the graph get more than 50 percent of their power from nuclear energy or less than 50 percent. How many countries generate 20 percent or less of their power by nuclear energy? What percentage of the United States energy is generated by nuclear energy?

Possible Response Most countries get less than 50 percent of their power from nuclear energy; twelve countries; 20 percent

Extension Explain that controversy has surrounded the use of nuclear energy in the United States. Have students find out more about nuclear energy and list some of the advantages and disadvantages of its use.

INSTRUCT: Objective 3

The Culture of Paris

- How has Paris contributed to world art? museums, home to leading school of fine arts, center of artistic activities and movements, such as impressionism
- Who are some important French literary figures? Marcel Proust, Albert Camus, Simone de Beauvoir

The result of these plans was a mixed economy, with both public and private sectors. The French government nationalized, or took over, major banks; insurance companies; the electric, coal, and steel industries; schools; universities; hospitals; railroads; airlines; and even an automobile company.

This nationalization of industry is a form of socialism. **Socialism** is an economic system in which some businesses and industries are controlled by the government. The government also provides many health and welfare benefits, such as health care, housing, and unemployment insurance. However, today the French government is slowly placing more of the economy under the control of private companies.

Nuclear Energy Generation, 1999

France
Lithuania
Belgium
Bulgaria
Slovakia
Sweden
Ukraine
South Korea
Hungary
Slovenia
Armenia
Switzerland
Japan
Finland
Germany
Spain
United Kingdom
Czech Republic
United States
Russia
Canada
Romania
Argentina
South Africa
Mexico
Netherlands
India
Brazil
Pakistan
China

0 10 20 30 40 50 60 70 80
Percent of Power Generated

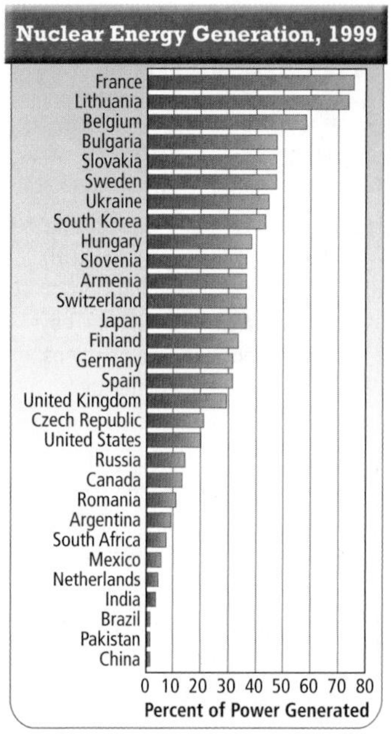

Energy The French economy grew rapidly after 1946, and the country's industry was powered mainly by coal, oil, and gas. When worldwide oil prices rose in the 1970s, the French economy suffered. In the 1980s, France turned to nuclear power so that its economy would be less dependent on oil. Today France draws 75 percent of its power from nuclear energy, a higher percentage than any other nation in the world.

Most famous for its wines, France also exports grains, automobiles, electrical machinery, and chemicals. Although only about 7 percent of the labor force works on farms, France exports more agricultural products than any other nation in the European Community.

The **European Community** is an association developed after World War II to promote economic unity among the countries of Western Europe. Its success gave rise to greater unity, both politically and economically, in the European Union.

Region • Nuclear power plants are a common sight in the French countryside. ▼

The Culture of Paris 3

Paris, the capital city of France, is famous for its contributions to world culture, most especially in the arts. Nicknamed "City of Light," Paris has long been an intellectual and artistic center.

Reading Social Studies
B. Possible Answer France wants to lessen its dependence upon oil.

Reading Social Studies
B. Analyzing Motives Why does France produce a large amount of nuclear energy?

BACKGROUND
Tourism is a major industry in France. The country hosts more than 70 million visitors annually, making it the most visited country in the world.

Activity Options

Interdisciplinary Link: Art

Block Scheduling

Class Time One class period

Task Creating an Impressionist-style visual

Purpose To help students gain an understanding of Impressionist-style art

Supplies Needed
- Art books with examples of Impressionist paintings
- Colored pencils, pastels, or watercolor paints
- Drawing paper

Activity Display the copies of Impressionist art and discuss the style with students. Then invite students to select a subject from this chapter and create an Impressionist-style visual of it. Encourage students to use the medium with which they are most comfortable.

Edouard Manet (muh•NAY) (1832–1883) helped influence one of the most important art movements of modern times, impressionism. **Impressionism** is an art style that uses light to create an impression of a scene rather than a strictly realistic picture. Manet inspired such artists as Claude Monet (moh•NAY), Pierre Renoir (ruhn•WAHR), and Paul Cézanne (say•ZAHN). This group of artists worked together in Paris and shared their thoughts and opinions of art.

Region • Monet and his family often modeled for Manet, as in this 1874 painting, *Monet Working on His Boat in Argenteuil.* ▲

Paris's Musée d'Orsay and the Louvre (loove) house two of the greatest collections of fine art in the world. The School of Fine Arts leads a tradition of education and art instruction that has produced artists such as Pierre Bonnard (baw•NAHR) (1867–1947) and Balthus (1908–2001).

Literature France has a rich tradition of literature as well. Marcel Proust, who wrote *Remembrance of Things Past,* was an influential writer in the early 20th century. Other significant writers include Albert Camus (kah•MOO), who wrote *The Stranger,* and Simone de Beauvoir (boh•VWAHR), author of *The Mandarins.*

SECTION 3 ASSESSMENT

Terms & Names

1. Explain the significance of: (a) Charles de Gaulle (b) French Resistance (c) Jean Monnet (d) socialism (e) European Community (f) impressionism

Using Graphics

2. Use a chart like this one to list the major aspects of French government, economy, and culture.

	Major Aspects
Government	
Economy	
Culture	

Main Ideas

3. (a) What role does the French government play in the country's economy?

(b) What is France's primary source of power?

(c) Name three contributions of French culture to the world.

Critical Thinking

4. Clarifying

How was the liberation of Paris a symbolic victory?

Think About

- the actions of the French Resistance
- the cultural life of Paris

ACTIVITY -OPTION- Reread the text on Manet. Draw an impressionist **portrait** of a classmate, friend, or family member.

ASSESS & RETEACH

Reading Social Studies Have students add details from Section 3 to the chart on page 378.

Formal Assessment
- Section Quiz, p. 205

RETEACHING ACTIVITY

Work with students to develop an outline of this section on the chalkboard. Use the section heads as the main-idea headings; then ask students to identify details related to each one.

In-depth Resources: Unit 4
- Reteaching Activity, p. 58

Access for Students Acquiring English
- Reteaching Activity, p. 120

Section 3 Assessment

1. Terms and Names

a. Charles de Gaulle, p. 390
b. French Resistance, p. 390
c. Jean Monnet, p. 391
d. socialism, p. 392
e. European Community, p. 392
f. impressionism, p. 393

2. Using Graphics

	Major Aspects
Government	parliamentary republic; power split between president and parliament; parliament has two parts, Senate and National Assembly; prime minister heads parliament
Economy	mixed, with both public and private sectors; some industries nationalized
Culture	important center of painting and literature, birthplace of impressionism

3. Main Ideas

a. The French government controls the major banks; insurance companies; electric, coal, and steel industries; schools; universities; hospitals; railroads; airlines; and an automobile company.
b. Today France's primary source of power is nuclear energy.
c. French culture has contributed artists, museums, and writers.

4. Critical Thinking

Paris was the center of French cultural life, and freeing it from German control was a symbol of freeing France itself.

ACTIVITY OPTION

Integrated Assessment
- Rubric for drawing a portrait

SECTION OBJECTIVES

1. To describe Germany as a divided, and then as a reunified, nation
2. To describe Germany's culture

SKILLBUILDER

- Interpreting a Map, p. 395

CRITICAL THINKING

- Hypothesizing, p. 395
- Recognizing Effects, p. 396

FOCUS & MOTIVATE

WARM-UP

Analyzing Motives Have students read Dateline and respond to these questions.

1. How would you feel if you were German, living in Berlin in October 1990?
2. Why was it important to save a part of the wall as a memorial?

INSTRUCT: Objective 1

A Divided Germany/Reunified Germany

- Why was West Germany able to achieve an economic miracle? U.S. loans helped revitalize industry and the general economy.

In-depth Resources: Unit 4
- Guided Reading Worksheet, p. 50

Reading Study Guide
(Spanish and English), pp. 120–121

Germany

SECTION 4

TERMS & NAMES
- Berlin Wall
- reunification
- Ludwig van Beethoven
- Rainer Maria Rilke

MAIN IDEA	WHY IT MATTERS NOW
Germany has overcome many obstacles to become both a unified and a modern nation.	Germany has helped to shape recent European history and contemporary Western culture.

DATELINE

BERLIN, GERMANY, OCTOBER 3, 1990— It is just past midnight. Church bells are ringing, fireworks are exploding, bands are playing, and the streets are filled with celebrating Germans. At midnight, the treaty to reunite East and West Germany became official. Germany is whole once more!

Just a year ago, the Berlin Wall—a 103-mile-long barrier of concrete and barbed wire—still separated East and West Berlin. Constructed in 1961, the Wall kept East Germans from escaping from their Communist government to democratic West Germany. Then, in 1989, as the Communist government weakened, the Wall came down.

Location • Germans celebrate unification in front of the Reichstag, the seat of the federal government. ▲

A Divided Germany 1

Today, the reunified nation of Germany is one of the largest countries in Europe. When World War II ended in 1945, however, Germany was divided. United States, French, and British soldiers occupied the new West German nation, and Soviet soldiers occupied the new East Germany.

TAKING NOTES
Use your chart to take notes about Europe today.

Country	Physical Geography	Government
United Kingdom		
Sweden		

Program Resources

In-depth Resources: Unit 4
- Guided Reading Worksheet, p. 50
- Reteaching Activity, p. 59

Reading Study Guide
(Spanish and English), pp. 120–121

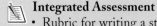

Formal Assessment
- Section Quiz, p. 206

Integrated Assessment
- Rubric for writing a story

Outline Map Activities

Access for Students Acquiring English
- Guided Reading Worksheet, p. 114

Technology Resources
classzone.com

TEST-TAKING RESOURCES

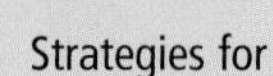

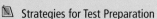

Strategies for Test Preparation

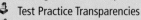

Test Practice Transparencies

Online Test Practice

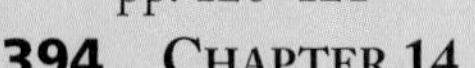

West Germany The United States helped West Germany set up a democratic government. In part, the United States supported the new nation because it was located between the Communist countries of Eastern Europe and the rest of Western Europe.

With the help of U.S. loans, West Germany experienced a so-called economic miracle. In 20 years, it rebuilt its factories and became one of the world's richest nations. Its economy later became the driving force behind the European Union.

East Germany In contrast to West Germany, East Germany remained poor. Most East Germans saw West Germany, and Western Europe in general, as a place where people had better lives. East Germany's Communist government, however, discouraged contact between east and west.

By 1989, the Soviet Union's control of Eastern Europe was weakening. Hungary, a Soviet ally, relaxed control over its borders with Western Europe. East Germans began crossing the Hungarian border into Austria and eventually made their way into West Germany. After the **Berlin Wall** came down in 1989, more East Germans fled to West Germany.

BACKGROUND

Although Berlin was inside East Germany, it had a large free zone defended by the Allies. The Berlin Wall separated the two parts of the city: East Berlin and West Berlin.

Geography Skillbuilder Answers
1. Poland
2. West Germany

Germany Today

GEOGRAPHY SKILLBUILDER: Interpreting a Map

1. **Location** • Which country is nearest to Germany's national capital?
2. **Place** • Which was larger, East or West Germany?

CRITICAL THINKING ACTIVITY

Hypothesizing Ask a volunteer to summarize the relationship that existed between East and West Germany from the end of World War II to 1990. Then have the class discuss why the East German government tried to discourage contact between the people of East and West Germany.

Class Time 15 minutes

FOCUS ON VISUALS

Interpreting the Map Ask students to study the map. Have students compare the size of the former East Germany and West Germany. About how far is Berlin from the border of West Germany?

Possible Responses West Germany was about twice the size of East Germany; about 100 miles

Extension Have students meet in groups and discuss the following question: How has the reunification of Germany affected its relations with neighboring countries? Have volunteers in each group share their conclusions with the class.

Activity Options

Interdisciplinary Links: Geography/Math

Class Time 30 minutes

Task Comparing the length and height of the Berlin Wall with places familiar to students

Purpose To appreciate the size of the Berlin Wall

Supplies Needed
- State and local maps
- Paper
- Pencils

Block Scheduling

Activity Tell students that the Berlin Wall was approximately 100 miles long. Much of it was made of concrete barricades that averaged nearly 12 feet high. Have students compare the length and height of the Berlin Wall to places or things with which they are familiar. They might compare it to the distance around their city or the height of a basketball hoop. Have volunteers share their comparisons with the class.

INSTRUCT: Objective 2

German Culture

- Who are three of Germany's best-known composers? Johann Sebastian Bach, George Frederick Handel, Ludwig van Beethoven
- What two German writers have been awarded the Nobel Prize in Literature? Günter Grass, Thomas Mann

CRITICAL THINKING ACTIVITY

Recognizing Effects Remind students that culture is the literature, music, and art that a group of people enjoy. It is also the way the people do things, how they live, and what they value as a group. Ask students to name some of the cultural traditions of the United States. Then guide a discussion of why these traditions are important to people. Finally, talk about how cultural traditions may help to unite the people of East Germany and West Germany.

Class Time 15 minutes

Biography

Ludwig van Beethoven (1770–1827) began studying music as a child. By the time he was 11, he had become an assistant court organist. He published his own compositions at age 12. In his twenties, Beethoven had become renowned as a great pianist. He stopped performing in 1808, but he continued composing. When he died, 10,000 people attended his funeral.

Region • The new Volkswagen Beetle is typical of German car design, known for its simplicity and style. The first Volkswagen was designed by Ferdinand Porsche (POOR•sheh) in 1934. ▲

Reunified Germany 1

Since the 1990 **reunification,** or the reuniting of East and West Germany, the German government has spent billions of dollars rebuilding the eastern part of the country. The effort has included roads, factories, housing, and hospitals. The city of Berlin, once again the nation's capital, was also rebuilt. The newly reunified nation also restored the Reichstag (RYK•shtahg), where the Federal Assembly meets.

However, reunification has also caused tensions between "Ossies" (OSS•eez) and "Wessies" (VEHSS•eez). Many Ossies complain about the lack of jobs and the cost of housing. Many Wessies complain about paying taxes to rebuild the nation and to help support the former East Germans.

Reading Social Studies

Clarifying Why did East Germany need to be rebuilt and not West Germany?

Vocabulary

Ossies: former East Germans

Wessies: former West Germans

German Culture 2

Germany's rich cultural traditions may help to unite its people, who are especially proud of their music and literature. Germans are also famous for designing high-quality products, such as cars, electronic appliances, and other complex machinery.

Music Three of Germany's best-known composers are Johann Sebastian Bach (bahck) (1685–1750), George Frederick Handel (HAHN•duhl) (1685–1759), and **Ludwig van Beethoven** (LOOD•vig vahn BAY•TOH•vuhn) (1770–1827). Their music is still performed and recorded all over the world. German composer Richard Wagner (VAHG•nuhr) (1813–1883) wrote many operas, including a series based on German myths and legends known as the Ring Cycle.

Reading Social Studies Possible Answer

Unlike West Germany, East Germany's economy was too poor to maintain roads and buildings.

Biography

Beethoven Perhaps the best-loved German composer is Ludwig van Beethoven. Beethoven began to lose his hearing when he was in his 20s. By the time he was 50, he was almost deaf.

Beethoven refused to let his deafness stop him from creating music, however. "I will grapple with Fate, it shall not overcome me," he wrote. In 1824, he finished his Ninth Symphony, which ends with a section containing the well-known "Ode to Joy." An orchestra played this same symphony at an open-air concert during the destruction of the Berlin Wall.

Activity Options

Differentiating Instruction: Students Acquiring English/ESL

Prefixes Point out the word *reunified* in the first head on this page. Explain that this word is formed by combining the prefix *re-* with the base word *unified.* Remind students that they can often figure out the meaning of unfamiliar words if they know what the word parts mean. Here, *re-* means "again" and *unified* means "brought together." *Reunified,* therefore, means "brought together again." Have students find other words on this page that are formed from the prefix *re-* (*reunification, reuniting, rebuilding, rebuilt, rebuild*). Have students define the words and restate the sentences in their own words.

Place • Half-timber architecture, shown here, is common throughout Germany. ▲

BACKGROUND

More than 100 million people around the world speak German.

Literature One of the greatest writers in the German language was **Rainer Maria Rilke** (RIHL•kuh) (1875–1926). His poems, which are still admired and studied today, were a way for Rilke to communicate his feelings and experiences.

Other important 20th-century German authors include Günter Grass (grahs) (b. 1927) and Thomas Mann (man) (1875–1955). Grass has written about the horrors of World War II, the setting for his novel *The Tin Drum.* Both writers were awarded the Nobel Prize in Literature—Mann in 1929 and Grass in 1999.

SECTION 4 ASSESSMENT

Terms & Names

1. Explain the significance of: (a) Berlin Wall (b) reunification (c) Ludwig van Beethoven (d) Rainer Maria Rilke

Using Graphics

2. Use a chart like this one to compare aspects of Germany before and after reunification.

Before Reunification	After Reunification

Main Ideas

3. (a) Describe the economic miracle that occurred in West Germany.
   (b) Why has there been tension between the Ossies and the Wessies?
   (c) On what projects has Germany spent billions of dollars since 1990?

Critical Thinking

4. **Synthesizing** What makes Germany an important European country?

Think About

- its location
- its size
- its role in modern history

ACTIVITY -OPTION- Reread the "Dateline" feature at the beginning of the section. Write a **short story** describing what it might have been like to celebrate the reunification of Germany in 1990.

ASSESS & RETEACH

Reading Social Studies Have students add details to line 4 of the chart on page 378.

Formal Assessment
- Section Quiz, p. 206

RETEACHING ACTIVITY

Have small groups of students collaborate in writing a three- or four-sentence summary of the discussion under one of the section headings. Have groups share their summaries in class.

In-depth Resources: Unit 4
- Reteaching Activity, p. 59

Access for Students Acquiring English
- Reteaching Activity, p. 121

Section 4 Assessment

1. Terms & Names

a. Berlin Wall, p. 395
b. reunification, p. 396
c. Ludwig van Beethoven, p. 396
d. Rainer Maria Rilke, p. 397

2. Using Graphics

Before Reunification	After Reunification
Germany divided Occupation: West Germany by U.S.; East Germany by Soviet Union West Germany economic miracle; East Germany poor Berlin divided by wall People cannot move freely	Germany unified No occupation East Germany being rebuilt Wall removed People can move freely

3. Main Ideas

a. West Germany rebuilt its economy and became one of the world's richest nations.
b. Ossies need jobs and housing; Wessies dislike supporting Ossies.
c. Billions have been spent on rebuilding eastern Germany's roads, factories, housing, and hospitals.

4. Critical Thinking

Germany is centrally located, large, and economically powerful, and is a political leader in Europe.

ACTIVITY OPTION

Integrated Assessment
- Rubric for writing a story

SKILLBUILDER

Making an Outline

Defining the Skill

Explain that making an outline is essential for writing a research report and is also a useful study skill. After studying the strategies, students might, for example, outline Chapter 14 in order to have a concise guide to main ideas and details that could be used for review.

Applying the Skill

How to Make an Outline Ask a volunteer to read the first strategy. Point out the Roman numerals used to designate each main idea. Continue to work through each strategy. Reinforce the need to follow each strategy.

Make an Outline

Review the outline of Marie Curie's biography. Point out the relationship between main ideas and supporting ideas. Ask questions such as, How does item C support the first main idea: Who was she? Then discuss details that add to the supporting ideas. Ask how they relate to the main ideas.

Practicing the Skill

Brainstorm topics from the chapter that students might outline. Have students who chose similar topics compare their outlines. If students need more practice, have them turn to the discussion of the divided Germany. Work together to create an outline of this material.

In-depth Resources: Unit 4
- Skillbuilder Practice, p. 54

SKILLBUILDER

Making an Outline

▸▸ Defining the Skill

Before writing a research report, you must decide on your topic and then gather information about it. When you have all of the information you need, then you begin to organize it. One way of organizing your information before writing the report is to make an outline. An outline lists the main ideas in the order in which they will appear in the report. It also organizes the main ideas and supporting details according to their importance. The form of every outline is the same. Main ideas are listed on the left and labeled with capital Roman numerals. Supporting ideas are indented and labeled with capital letters. Supporting details are indented farther and labeled with numerals.

▸▸ Applying the Skill

The outline to the right is for a biography of Marie Curie, one of the great physicists of all time. Use the strategies listed below to help you learn how to make an outline.

How to Make an Outline

Strategy ❶ Read the main ideas of this report. They are labeled with capital Roman numerals. Each main idea will need at least one paragraph.

Strategy ❷ Read the supporting ideas for each main idea. These are labeled with capital letters. Notice that some of the main ideas require more supporting ideas than others.

Strategy ❸ Read the supporting details that are included in this outline. These are labeled with numerals. The writer of this outline did not include the supporting details for some of the supporting ideas. It is not necessary to include every piece of information that you have. An outline is intended merely as a guide for you to follow as you write the report.

Strategy ❹ A report can be organized in different ways. This biography is organized chronologically, that is, according to time. It starts with Curie's birth and ends with her legacy after death. The outline follows the order of events in her life. A report can be organized in other ways, such as comparing and contrasting or according to advantages and disadvantages. The outline should clearly reflect the way the report is organized.

❶ I. Who Was She?
- A. Polish-born physicist
- ❷ B. Birth and early life
- C. Schooling
  - 1. In secret in Poland (women were not allowed to e
    ❸ higher education)
  - 2. In France at the Sorbonne
    - ❹ a. license of physical sciences, 1893
    - b. license of mathematical sciences, 1894

II. The Physicist
- A. Life and work with husband, Pierre Curie
  - 1. Discoveries
    - a. polonium, summer 1898
    - b. radium, fall 1898
  - 2. Nobel Prize in Physics, 1903
    - a. shared with Henri Becquerel
    - b. Marie was the first woman to ever be awa
      a Nobel Prize
- B. Her own accomplishments
  - 1. Became the first female professor at the Sor
    - a. took over Pierre's position after his death,
  - 2. Her research on radioactivity was published, 1
  - 3. Nobel Prize in Chemistry, 1911

▸▸ Practicing the Skill

Look through Chapter 14 and find a topic that interests you. Gather information about that topic, and then write an outline for a report about that topic. Be sure to use the correct outline form.

Career Connection: Biographer

Encourage students who enjoy making an outline to learn about careers in writing that use this skill. For example, biographers write about other people. Because much research goes into a biography, making an outline is important for a biographer.

1. Point out to students that a biographer usually chooses a subject who is interesting to her or him.

2. Suggest that students find out how a biographer researches a subject and organizes the findings.

3. Ask students to research the education and experience an aspiring biographer should obtain. Have students prepare an outline of their findings.

B Block Scheduling

398 CHAPTER 14

Section 5 Poland

TERMS & NAMES
Solidarity
Lech Walesa
Czeslaw Milosz
censorship
dissident

MAIN IDEA

Poland has gone through the difficulties of establishing a new democratic government and a new economic system.

WHY IT MATTERS NOW

Poland is an excellent example of the success that has been achieved by the newly independent Eastern European nations.

DATELINE

EXTRA

GDAŃSK, POLAND, 1980

In response to recent increases in food prices, many strikes have broken out across Poland. Today's strikes are much larger than the strikes that occurred in 1976. The shipyards in Gdańsk have 17,000 striking workers. One of the strikers' demands is the right to form labor unions. The recent strikes are yet another sign of the country's weakening economy, which has continued to decline over the past decade. Poland's attempts to improve its economic health by borrowing money from other nations have not helped, as the government is unable to repay those loans.

Place • Polish workers protest poor conditions under the Communist government. ▲

Political and Economic Struggles ❶

The strikes and riots of the 1970s and 1980s were not the first actions Polish citizens took against their government. In 1956, Polish workers had rioted to protest their low wages.

In fact, there have been political and economic struggles in Poland since World War II ended in 1945. At that time, Communists took over the government and set strict wage and price controls.

TAKING NOTES
Use your chart to take notes about Europe today.

Country	Physical Geography	Government
United Kingdom		
Sweden		

SECTION OBJECTIVES

1. To explain Poland's political and economic struggles
2. To describe Poland's government and economy
3. To describe Poland's culture

SKILLBUILDER
- Interpreting a Map, p. 400

CRITICAL THINKING
- Drawing Conclusions, p. 400
- Comparing, p. 401

FOCUS & MOTIVATE

WARM-UP

Analyzing Motives After students read Dateline, discuss unions and strikes.

1. What conditions might prompt workers to strike?
2. What happens during a labor strike?

INSTRUCT: Objective ❶

Political and Economic Struggles

- Who led Poland's efforts toward political and economic change? workers, Solidarity movement

In-depth Resources: Unit 4
- Guided Reading Worksheet, p. 51

Reading Study Guide
(Spanish and English), pp. 122–123

Program Resources

In-depth Resources: Unit 4
- Guided Reading Worksheet, p. 51
- Reteaching Activity, p. 60

Reading Study Guide
(Spanish and English), pp. 122–123

Formal Assessment
- Section Quiz, p. 207

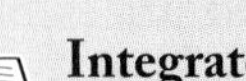

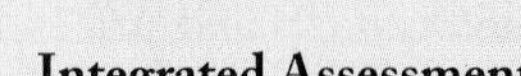

Integrated Assessment
- Rubric for writing a speech

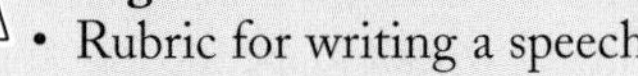

Outline Map Activities

Access for Students Acquiring English
- Guided Reading Worksheet, p. 115

Technology Resources
classzone.com

TEST-TAKING RESOURCES
- Strategies for Test Preparation
- Test Practice Transparencies
- Online Test Practice

FOCUS ON VISUALS

Interpreting the Map Ask students to look at the map and identify the capital of Poland. Where is the city of Gdańsk? What river flows into the Gulf of Gdańsk?

Responses Warsaw; Gdańsk is in north central Poland on the Gulf of Gdańsk; Vistula River

Extension Have students look at the cities shown on the map and at the scale of miles. Then ask them to create a chart showing the distance that each of the cities is from Poland's capital.

CRITICAL THINKING ACTIVITY

Drawing Conclusions Remind students that the Communist Party was originally organized to protect workers. Ask them why the Communist leaders of Poland now opposed Solidarity. What conclusions can they draw about the Communist Party's success?

Class Time 10 minutes

GEOGRAPHY SKILLBUILDER: Interpreting a Map

1. **Location** • Name a port city in Poland.
2. **Location** • How many different countries border Poland?

Geography Skillbuilder Answers

1. Gdańsk
2. seven

Solidarity In 1980, labor unions throughout Poland joined an organization called **Solidarity.** This trade union was led by **Lech Walesa** (LEK wah•LEHN•suh), an electrical worker from the shipyards of Gdańsk (guh•DAHNSK).

In the beginning, Solidarity's goals were to increase pay and improve working conditions. Before long, however, the organization set its sights on bigger goals. In late 1981, members of Solidarity were calling for free elections and an end to Communist rule. Even though Solidarity had about 10 million members, the government fought back. It suspended the organization, cracked down on protesters, and arrested thousands of members, including Walesa.

BACKGROUND

Poland's capital and largest city is Warsaw. The nation's citizens are called Poles.

Region • In 1980, Solidarity leader Lech Walesa gained the support of labor unions. Ten years later, he became Poland's president. ◄

Activity Options

Interdisciplinary Link: World History

Class Time 30 minutes

Task Creating a placard for a Solidarity strike

Purpose To identify and express ideas that motivated the Solidarity movement

Supplies Needed

- Art paper or poster board
- Markers

Activity Review the Solidarity movement with students, discussing its original goals and how those goals changed. Explain to students that during demonstrations and strikes, members of Solidarity would make placards, or posters, stating their demands or expressing their views. Ask students to work independently to create placards that Solidarity members might have displayed. Have students post their placards in the classroom.

Region • Poland's senate helps ensure that all the country's citizens have representation. ◄

A Free Poland ❷

In the late 1980s, economic conditions continued to worsen in Poland. The government asked Solidarity leaders to help them solve the country's economic difficulties. Finally, the Communists agreed to Solidarity's demand for free elections.

When the elections were held in 1989, many Solidarity candidates were elected, and the Communists lost power. In 1990, Lech Walesa became the president of a free Poland.

Reading Social Studies

A. Recognizing Important Details What led to free elections in Poland?

Reading Social Studies
A. Possible Answer
The Communist government needed Solidarity's help to solve economic problems, and Solidarity demanded free elections.

A New Constitution Today, Poland is a parliamentary republic. The country approved a new constitution in 1997. This constitution guarantees civil rights such as free speech. It also helps to balance the powers held by the president, the prime minister, and parliament.

BACKGROUND

The Polish president, who is elected every five years, is the head of state.

Parliament Poland's parliament is made up of two houses. The upper house, or senate, has 100 members. The lower house, which has 460 members, chooses the prime minister. Usually, as in the United Kingdom, the prime minister is a member of the largest party or alliance of parties within parliament.

A number of seats in parliament are reserved for representatives of the small German and Ukrainian ethnic groups in Poland. In this way, all Polish citizens are ensured a voice in their government.

A Changing Economy ❷

Besides a new government, the Poles have also had to deal with a changing economy. In 1990, Poland's new democratic government quickly switched from a command economy to a free market economy. Prices were no longer controlled by the government, and trade suddenly faced international competition.

INSTRUCT: Objective ❷

A Free Poland/A Changing Economy

- Why did Poland's government finally ask Solidarity for help? The economy had continued to worsen.
- How does Poland's new constitution help protect peoples' freedom? guarantees freedom of speech; balances powers of president, prime minister, and parliament so no one has too much power
- Why did Poland experience high unemployment following its switch to a free-market economy? Polish companies could not compete with high-quality foreign goods, and industries closed, so workers were laid off.

CRITICAL THINKING ACTIVITY

Comparing Review the discussion of Poland's new parliament. Then ask students to work independently to create Venn diagrams comparing Poland's new parliament with the United States Congress. Create a class diagram and have volunteers complete it with ideas from their own diagrams.

Class Time 15 minutes

Activity Options

Differentiating Instruction: Less Proficient Readers

Identifying Sequence of Events Discuss the events that occurred when Poland changed from a command economy to a free-market economy. Then have students work in small groups to create a sequence-of-events chart like the one shown. Tell them to add as many boxes as necessary to show events. Have groups share their charts in class.

Change: command to free-market economy	
Prices not controlled; international trade permitted	
More goods for sale	
Prices increase	

INSTRUCT: Objective 3

Poland's Culture

- Why is Poland's literature filled with stories of national independence and kingdoms won and lost? It reflects Poland's history.
- How have the Polish government's actions toward the arts changed since Communist rule ended? Before communism, the government exercised censorship; now it supports and encourages the arts.

MORE ABOUT...

Czeslaw Milosz

Like that of many Polish writers, the work of Milosz remained unpublished in Poland for many years. When he won the Nobel Prize in 1980, however, the government suddenly recognized him. "My name was defrozen," he said. "The Polish public at large had not heard my name because of censorship." When Milosz visited Warsaw in 1981, he was a hero. People saw him as a symbol of freedom.

Inflation Although Polish shops were able to sell goods that had not been available before, prices rose quickly—by almost 80 percent. With this inflation, or a continual rise in prices, people's wages could not keep up with the cost of goods.

Many Polish companies, which could not compete with high-quality foreign goods, went out of business. This, in turn, resulted in high unemployment. As more and more people lost their jobs, Poland's overall standard of living fell.

Region • With Poland's economy on the rise, unemployment has decreased. ▲

An Improving Economy In time, new Polish businesses found success, giving more people work. Inflation started to drop. By 1999, inflation was down to around 7 percent. By 2000, Poland no longer needed the economic aid it had been receiving from the United States.

One way to measure the strength of a country's economy is to look at consumer spending. Between 1995 and 2000, Poles bought new cars at a high rate of half a million each year. Today, Poland has 2 million small and medium-sized businesses. The success of these small businesses is another sign of Poland's healthy economy.

Poland's Culture 

The history of Poland has been one of ups and downs. In the 1500s and 1600s, Poland was a large and powerful kingdom. By 1795, Russia, Prussia, and Austria had taken control of its land, and Poland ceased to exist as an independent country. Poland did not become a republic until 1918, after World War I. Throughout the centuries, however, Poland has had a rich culture.

BACKGROUND

A famous Polish general and patriot, Thaddeus Kosciusko (KAHS•ee•UHS•koh), fought on the side of the colonists during the American Revolution.

Literature Polish literature is full of accounts of struggles for national independence and stories about glorious kingdoms won and lost by heroic patriots.

One of Poland's best-known writers of recent times is **Czeslaw Milosz** (CHEH•slawv MEE•LAWSH) (b. 1911). Milosz published his first book of poems in the 1930s. After World War II, he worked as a diplomat in the United States and then France.

Activity Options

Differentiating Instruction: Gifted and Talented

Block Scheduling

Forming and Supporting Opinions Tell students that many Polish writers resented the Communist government's censorship of their writing. Have students write speeches expressing their opinion of censorship. Tell them to consider these questions:

- Why is censorship a big issue for writers?
- How does censorship affect the citizens of a country?
- Why do some governments think it is necessary to control what people read?

Have students practice their speeches and deliver them in class.

402 CHAPTER 14

Milosz, who became a professor at the University of California at Berkeley, won the Nobel Prize in Literature in 1980.

Place • In 1978, Poland's pride was greatly boosted when Polish-born John Paul II was elected pope. He was the first non-Italian to be elected pope in 456 years. ▲

Reading Social Studies

B. Analyzing Motives Why did the Communist government control the media?

Reading Social Studies
B. Possible Answer
It was trying to censor any criticism of government actions.

Censorship Under Communist rule, the Polish media were controlled by the government. The government decided what the media could and could not say. It outlawed any information that did not support and praise the accomplishments of Communism. As a result of this **censorship,** many writers could not publish their works. Some of them became dissidents. A **dissident** is a person who openly disagrees with a government's policies.

Supporting the Arts In order to help Polish writers, the government now allows publications printed in Poland to be sold tax-free. To help Polish actors, screenwriters, and directors, movie theaters are repaid their costs for showing Polish movies. Public-sponsored television stations are supported not only by free-market advertising but also by fees the public pays to own television sets.

SECTION 5 ASSESSMENT

Terms & Names

1. **Explain the significance of:** (a) Solidarity (b) Lech Walesa (c) Czeslaw Milosz (d) censorship (e) dissident

Using Graphics

2. Use a chart like this one to compare and contrast one aspect of Poland with the same aspect of the United Kingdom, Sweden, France, or Germany.

Poland	Other Country

Main Ideas

3. (a) How did the Polish government respond to Solidarity's goals?
(b) What was the outcome of Poland's free election in the late 1980s?
(c) Describe the recent changes in the economy of Poland.

Critical Thinking

4. **Summarizing**
How would you describe what life was like in Poland before the changes of 1990?

Think About
- strikes and riots
- the Communist government
- censorship

ACTIVITY -OPTION- Reread the information about Solidarity. Write a short **speech** that might have been given to gain support for the organization in the 1980s.

ASSESS & RETEACH

Reading Social Studies Have students complete the chart on page 378.

Formal Assessment
- Section Quiz, p. 207

RETEACHING ACTIVITY

Ask each student to write one question about the content that follows each heading in the section. Then have students meet in small groups and take turns asking their questions and discussing the answers.

In-depth Resources: Unit 4
- Reteaching Activity, p. 60

Access for Students Acquiring English
- Reteaching Activity, p. 122

Section 5 Assessment

1. Terms & Names

a. Solidarity, p. 400
b. Lech Walesa, p. 400
c. Czeslaw Milosz, p. 402
d. censorship, p. 403
e. dissident, p. 403

2. Using Graphics

Poland	Other Country
Formerly Communist Parliamentary republic Free-market economy Becoming economically prosperous	Sweden: never Communist Sweden: constitutional monarchy Sweden: free-market economy Sweden: prosperous economy

3. Main Ideas

a. The government opposed Solidarity at first, and then asked for its help.
b. Solidarity won; Communists lost power.
c. Poland became a free-market economy, which resulted in inflation and unemployment. The economy is now improving.

4. Critical Thinking

Workers protested low wages; Solidarity organized strikes; the Communist government censored writers and media.

ACTIVITY OPTION

Integrated Assessment
- Rubric for writing a speech

CHAPTER 14 ASSESSMENT

TERMS & NAMES

1. London, p. 379
2. Good Friday Accord, p. 381
3. armed neutrality, p. 385
4. hydroelectricity, p. 386
5. Charles de Gaulle, p. 390
6. socialism, p. 392
7. Berlin Wall, p. 395
8. reunification, p. 396
9. censorship, p. 403
10. dissident, p. 403

REVIEW QUESTIONS

1. The House of Lords and the House of Commons form Parliament; the House of Commons is more powerful.
2. It has many people, little land, and few natural resources.
3. They make sure that the courts and civil service follow the law, and protect citizens' rights.
4. It has hills, mountains, and cold weather.
5. France's main source of energy is nuclear power.
6. Manet, Monet, Renoir, and Cézanne are Impressionist painters.
7. The U.S. government provided loans for rebuilding industry.
8. Bach, Handel, Beethoven, and Wagner are German composers.
9. He led Solidarity, which helped win Poland its freedom. He became free Poland's first president.
10. It protects civil rights and ensures a balance of power among the president, prime minister, and parliament.

CHAPTER 14 ASSESSMENT

TERMS & NAMES

Explain the significance of each of the following:

1. London
2. Good Friday Accord
3. armed neutrality
4. hydroelectricity
5. Charles de Gaulle
6. socialism
7. Berlin Wall
8. reunification
9. censorship
10. dissident

REVIEW QUESTIONS

The United Kingdom *(pages 379–383)*

1. What are the two houses that form the British Parliament? Which is more powerful?
2. Why is it necessary for the United Kingdom to import foods and other goods?

Sweden *(pages 384–387)*

3. What role do ombudsmen play in the Swedish government?
4. Why is Sweden an excellent place for skiing?

France *(pages 390–393)*

5. What is France's main source of energy?
6. Identify at least three Impressionist painters.

Germany *(pages 394–397)*

7. What role did the United States government play in West Germany after World War II?
8. Identify at least three famous German composers.

Poland *(pages 399–403)*

9. Why is Lech Walesa important to modern Poland?
10. Describe the new constitution that Poland approved in 1997.

CRITICAL THINKING

Comparing

1. Using your completed chart from Reading Social Studies, p. 378, compare the governments and economies of the United Kingdom, Sweden, France, Germany, and Poland.

Hypothesizing

2. One of Sweden's severe environmental problems is acid rain. Why might it be difficult for a country to solve this problem?

Making Inferences

3. Why do you think it was important to the United States that West Germany have a democratic government?

Visual Summary

The United Kingdom 1

- British economic, political, and cultural traditions have influenced nations around the world.

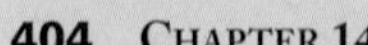

2 Sweden

- Sweden offers its people a high standard of living.
- Sweden is dealing with environmental issues such as nuclear power and acid rain.

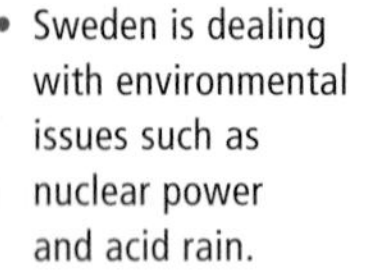

France

- France has made a speedy recovery from World War II.

3

Germany 4

- Germany has overcome many obstacles to become a unified and modern nation.

Poland 5

- Poland is an example of the success made possible by the recent independence of Eastern European nations.

CRITICAL THINKING: Possible Responses

1. Comparing

Great Britain and Sweden are constitutional monarchies. France and Poland are parliamentary republics. Germany, once split into Communist and democratic parts, is today democratic.

Great Britain and Sweden are free-trade economies. France has a mixed economy with public and private sectors. Poland has changed from a command economy to a free-market one.

2. Hypothesizing

Possible Answers Acid rain is caused by pollution from many countries. It's difficult to get all countries responsible to agree to reduce their polluting emissions. Also, the industries responsible may resist regulations that would reduce pollution.

3. Making Inferences

West Germany was a large, populous country that was historically highly industrialized. The United States realized West Germany would be powerful again and wanted it to be a democratic ally rather than a member of the Soviet-dominated East European bloc of nations.

ANDARDS-BASED ASSESSMENT

the map and your knowledge of geography to wer questions 1 and 2.

ditional Test Practice, pp. S1–S33

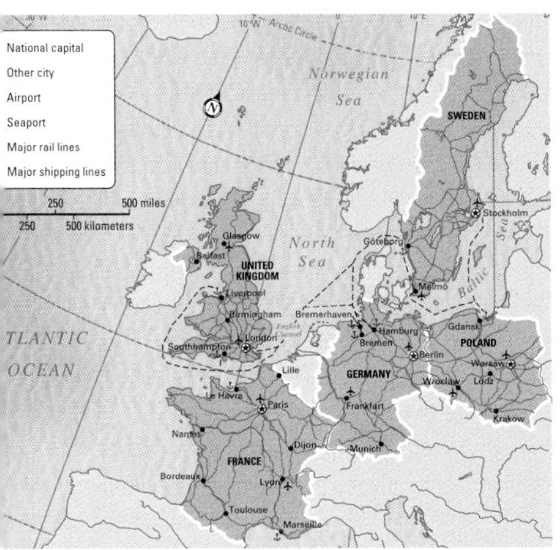

/hich of these cities has an airport but is not a capital?

. Birmingham
. Paris
. Stockholm
. Wroclaw

a ship is traveling along a major shipping line between verpool and Göteborg, what seaport would it pass?

. Gdansk
. Lille
. Southhampton
. Stockholm

The following excerpt is from a speech about Lech Walesa when he won the Nobel Peace Prize in 1983. Use the quotation and your knowledge of world cultures and geography to answer question 3.

PRIMARY SOURCE

He was faced with overwhelming difficulties; the choice of strategy was not easy. The goal was clear enough; the workers' right to organize and the right to negotiate with the country's officials on the workers' social and economic situation. But which of the many available paths would lead him to this goal? . . . Walesa's chosen strategy was that of peace and negotiation.

EGIL AARVIK, excerpt of presentation speech

3. What peaceful strategy did Walesa use to achieve his goal?

A. opposing Communists
B. negotiating with officials
C. facing overwhelming difficulties
D. choosing a clear path

TERNATIVE ASSESSMENT

WRITING ABOUT HISTORY

)n November 9, 1989, the Berlin Wall was torn down, and East nd West Germany were reunited. Write a headline and a eature article that might have appeared in a newspaper the ext day. Include historic information about the wall. Conduct esearch to find quotes from people who were present at the vent and include several of them in your article. Share your rticle with the class.

OOPERATIVE LEARNING

lany festivals and holidays are celebrated in the United ingdom, Sweden, France, Germany, and Poland. Working in a mall group, create a presentation about one of these. Work ogether to research the meaning of the holiday or festival you hoose, when it is celebrated, and the foods and activities ssociated with it. Decide what form your presentation will take nd divide the tasks needed to complete the project.

INTEGRATED TECHNOLOGY

Doing Internet Research

Use the Internet or other library resources to research the life and work of a well-known artist, author, or poet from the United Kingdom, Sweden, France, Germany, or Poland. Prepare a presentation of your findings.

- You might research biographies of the person and collections of his or her work.
- In your research, find out what was historically or socially significant about your subject's writing or artwork.
- Include a copy of the person's artwork or an excerpt from his or her writing.

For Internet links to support this activity, go to

STANDARDS-BASED ASSESSMENT

1. **Answer D** is the correct answer. The map indicates that Wroclaw has an airport, but it is not the capital of Poland. Answer A is incorrect because Birmingham has no airport; answers B and C are incorrect because, while Paris and Stockholm each have an airport, they are the capitals of France and Sweden.
2. **Answer C** is the correct answer because the map indicates a route between the two ports that passes by Southhampton. Answers A and D are incorrect because both Gdansk and Stockholm lie farther away from Liverpool than Göteborg does. Answer B is incorrect because Lille is not a seaport.
3. **Answer B** is the correct answer because it restates that Walesa's strategy was peace and negotiation. Answer A is incorrect because opposing Communists may be done peacefully or violently; answer C is incorrect because facing difficulties was Walesa's problem, not his solution; answer D is incorrect because it restates that Walesa made a choice but not what that choice was.

INTEGRATED TECHNOLOGY

Reports should be clearly and logically organized. They should begin with a statement that attracts the reader's interest. Paragraphs should be organized around main ideas that are supported by specific details. If possible, students should include photographs and other visuals that they download from the Internet.

Alternative Assessment

1. Rubric

The feature article should:

- use a journalistic style and be written in an unbiased way.
- have a catchy headline and portray the fall of the Berlin Wall accurately.
- include quotations by eyewitnesses.
- use correct grammar, spelling, and punctuation.

2. Cooperative Learning

Encourage students in each group to assign roles of researcher, recorder, and presenter. Presentations should respond to the questions from the text; include specific, accurate information; and include photographs, drawings, models, or other visuals.

The Attack: September 11, 2001

Terrorism is the use of violence against people or property to force changes in societies or governments. Acts of terrorism are not new. Throughout history, individuals and groups have used terror tactics to achieve political or social goals.

In recent decades, however, terrorist groups have carried out increasingly destructive and high-profile attacks. The growing threat of terrorism has caused many people to feel vulnerable and afraid. However, it also has prompted action from many nations, including the United States.

Many of the terrorist activities of the late 20th century occurred far from U.S. soil. As a result, most Americans felt safe from such violence. All that changed, however, on the morning of September 11, 2001.

A Surprise Strike

As the nation began another workday, 19 terrorists hijacked four airplanes heading from East Coast airports to California. The hijackers crashed two of the jets into the twin towers of the World Trade Center in New York City. They slammed a third plane into the Pentagon outside Washington, D.C. The fourth plane crashed into an empty field in Pennsylvania after passengers apparently fought the hijackers.

The attacks destroyed the World Trade Center and badly damaged a section of the Pentagon. In all, some 3,000 people died. Life for Americans would never be the same after that day. Before, most U.S. citizens viewed terrorism as something that happened in other countries. Now they knew it could happen on their soil as well.

Officials soon learned that those responsible for the attacks were part of a largely Islamic terrorist network known as al-Qaeda. Observers, including many Muslims, accuse al-Qaeda of preaching a false and extreme form of Islam. Its members believe, among other things, that the United States and other Western nations are evil.

U.S. president George W. Bush vowed to hunt down all those responsible for the attacks. In addition, he called for a greater international effort to combat global terrorism. "This battle will take time and resolve," the president declared. "But make no mistake about it: we will win."

Securing the Nation

As the Bush Administration began its campaign against terrorism, it

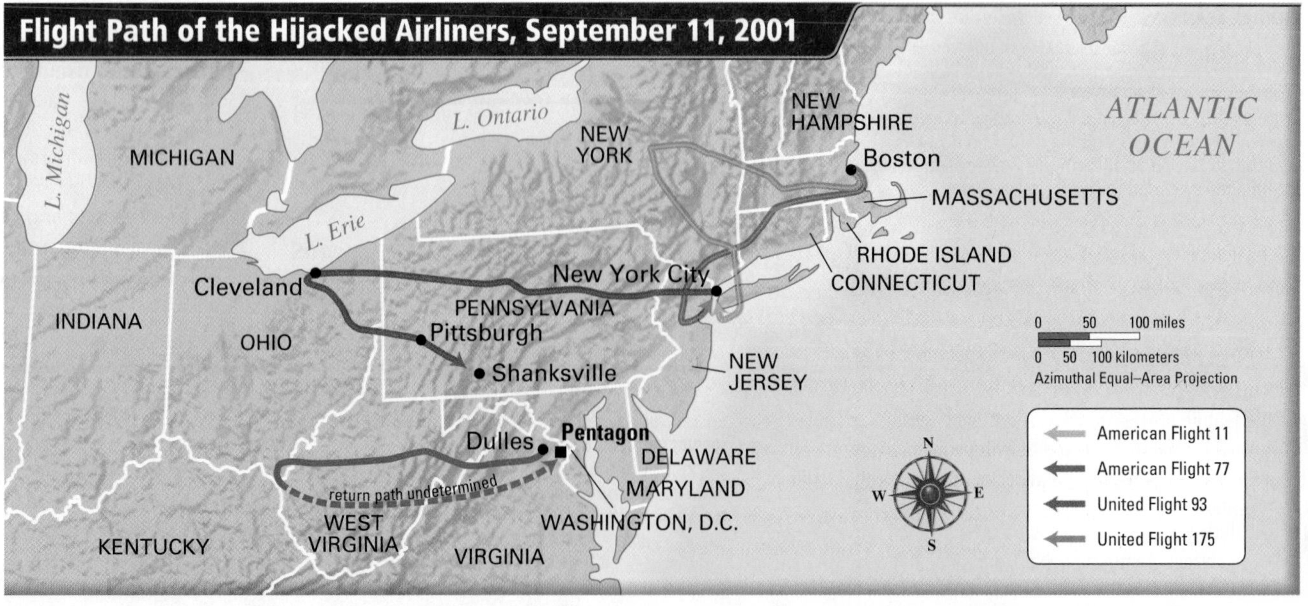

406

also sought to prevent any further attacks on America. In October 2001, the president signed into law the USA Patriot Act. The law gave the federal government a broad range of new powers to strengthen national security.

The new law enabled officials to detain foreigners suspected of terrorism for up to seven days without charging them with a crime. Officials could also monitor all phone and Internet use by suspects, and prosecute terrorist crimes without any time restrictions or limitations.

In addition, the government created a new cabinet position, the Department of Homeland Security, to coordinate national efforts against terrorism. President Bush named former Pennsylvania governor Tom Ridge as the first Secretary of Homeland Security.

Underneath a U.S. flag posted amid the rubble of the World Trade Center, rescue workers search for survivors of the attack.

Some critics charged that a number of the government's new anti-terrorism measures violated people's civil rights. Supporters countered that occasionally limiting some civil liberties was justified in the name of greater national security.

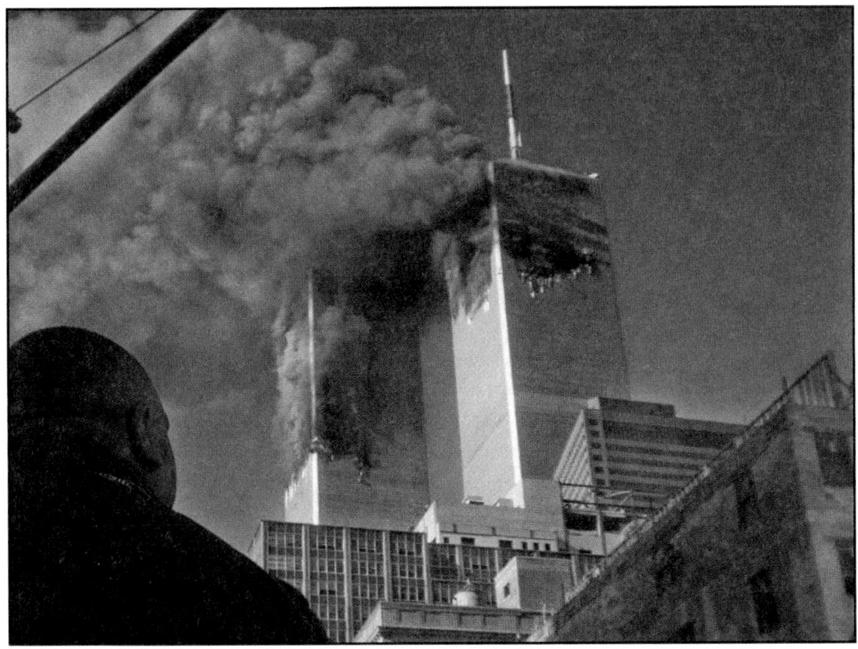

Stunned bystanders look on as smoke billows from the twin towers of the World Trade Center moments after an airplane slammed into each one.

The federal government also stepped in to ensure greater security at the nation's airports. The September 11 attacks had originated at several airports, with four hijackings occurring at nearly the same time. In November 2001, President Bush signed the Aviation and Transportation Security Act into law. The law put the federal government in charge of airport security. Before, individual airports had been responsible for security. The new law created a federal security force to inspect passengers and carry-on bags. It also required the screening of checked baggage.

While the September 11 attacks shook the United States, they also strengthened the nation's unity and resolve. In 2003, officials approved plans to rebuild on the World Trade Center site and construct a memorial. Meanwhile, the country has grown more unified as Americans recognize the need to stand together against terrorism.

407

Fighting Back

The attack against the United States on September 11, 2001, represented the single most deadly act of terrorism in modern history. By that time, however, few regions of the world had been spared from terrorist attacks. Today, America and other nations are responding to terrorism in a variety of ways.

The Rise of Terrorism

The problem of modern international terrorism first gained world attention during the 1972 Summer Olympic Games in Munich, Germany. Members of a Palestinian terrorist group killed two Israeli athletes and took nine others hostage. Five of the terrorists, all the hostages, and a police officer were later killed in a bloody gun battle.

Since then, terrorist activities have occurred across the globe. In Europe, the Irish Republican Army (IRA) used terrorist tactics for decades against Britain. The IRA has long opposed British control of Northern Ireland. Since 1998, the two sides have been working toward a peaceful solution to their conflict. In South America, a group known as the Shining Path terrorized the residents of Peru throughout the late 20th century. The group sought to overthrow the government and establish a Communist state.

Africa, too, has seen its share of terrorism. Groups belonging to the al-Qaeda terrorist organization operated in many African countries. Indeed, officials have linked several major attacks against U.S. facilities in Africa to al-Qaeda. In 1998, for example, bombings at the U.S. embassies in Kenya and Tanzania left more than 200 dead and 5,000 injured.

Most terrorists work in a similar way: targeting high profile events or crowded places where people normally feel safe. They include such places as subway stations, bus stops, restaurants, or shopping malls. Terrorists choose these spots carefully in order to gain the most attention and to achieve the highest level of intimidation.

Terrorists use bullets and bombs as their main weapons. In recent years, however, some terrorist groups have used biological and chemical agents in their attacks. These actions involve the release of bacteria or poisonous gas into the air. Gas was the weapon of choice for a radical Japanese religious cult, Aum Shinrikyo. In 1995, cult members released sarin, a deadly nerve gas, in subway stations in Tokyo. Twelve people were killed and more than 5,700 injured. The possibility of this type of terrorism is particularly worrisome, because biochemical agents are relatively easy to acquire.

Terrorism: A Global Problem

PLACE	YEAR	EVENT
Munich, Germany	1972	Palestinians take Israeli hostages at Summer Olympics; hostages and terrorists die in gun battle with police
Beirut, Lebanon	1983	Terrorists detonate truck bomb at U.S. marine barracks, killing 241
Tokyo, Japan	1995	Religious extremists release lethal gas into subway stations, killing 12 and injuring thousands
Omagh, Northern Ireland	1998	Faction of Irish Republican Army sets off car bomb, killing 29
Moscow, Russia	2002	Rebels from Chechnya seize a crowded theater; rescue effort leaves more than 100 hostages and all the terrorists dead

408

Hunting Down Terrorists

Most governments have adopted an aggressive approach to tracking down and punishing terrorist groups. This approach includes spying on the groups to gather information on membership and future plans. It also includes striking back harshly after a terrorist attack, even to the point of assassinating known terrorist leaders.

Another approach that governments use is to make it more difficult for terrorists to act. This involves eliminating a terrorist group's source of funding. President Bush issued an executive order freezing the U.S. assets of alleged terrorist organizations as well as various groups accused of supporting terrorism. President Bush asked other nations to freeze such assets as well. By the spring of 2002, the White House reported, the United States and other countries had blocked nearly $80 million in alleged terrorist assets.

Battling al-Qaeda

In one of the more aggressive responses to terrorism, the United States quickly took military action against those it held responsible for the September 11 attacks.

U.S. officials had determined that members of the al-Qaeda terrorist group had carried out the assault under the direction of the group's leader, Osama bin Laden. Bin Laden was a Saudi Arabian millionaire who lived in Afghanistan. He directed his terrorist activities under the protection of the country's extreme Islamic government, known as the Taliban.

The United States demanded that the Taliban turn over bin Laden. The Taliban refused. In October 2001, U.S. forces began bombing Taliban air defenses, airfields, and command centers. They also struck numerous al-Qaeda training camps. On the ground, the United States provided assistance to rebel groups opposed to the Taliban. By December, the United States had driven the Taliban from power and severely weakened the al-Qaeda network. However, as of 2003, Osama bin Laden was still believed to be at large.

Osama bin Laden delivers a videotaped message from a hidden location shortly after the U.S.-led strikes against Afghanistan began.

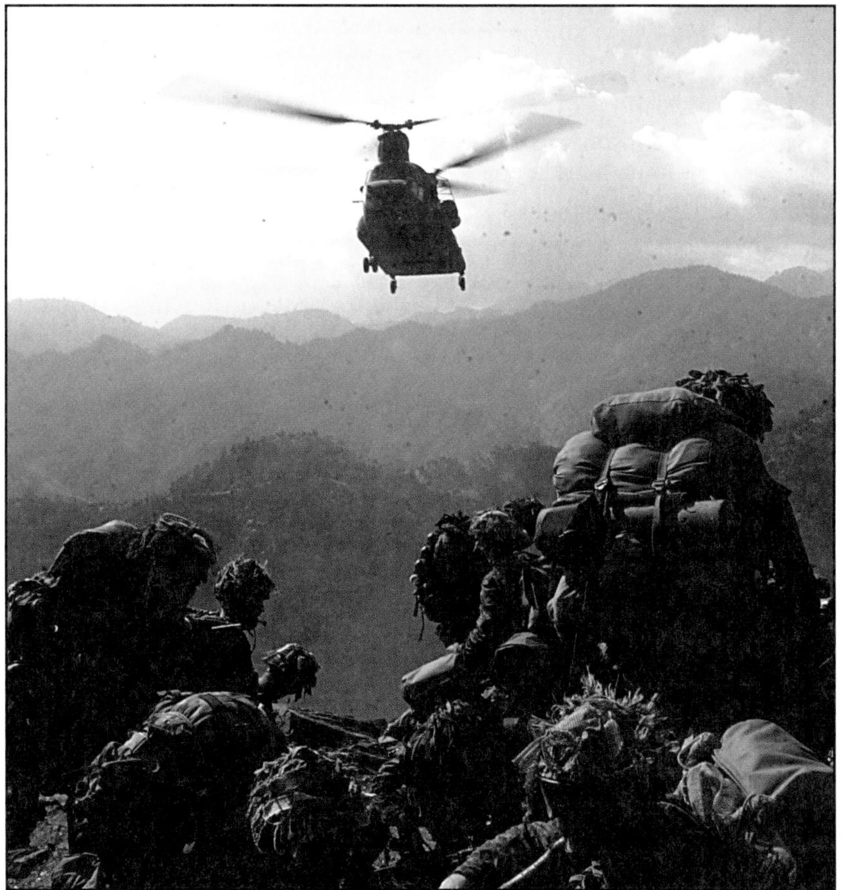

Troops battling the Taliban in Afghanistan await transport by helicopter.

409

The War in Iraq

In the ongoing battle against terrorism, the United States confronted the leader of Iraq, Saddam Hussein. The longtime dictator had concerned the world community for years. During the 1980s, Hussein had used chemical weapons to put down a rebellion in his own country. In 1990, he had invaded neighboring Kuwait—only to be pushed back by a U.S.-led military effort. In light of such history, many viewed Hussein as an increasing threat to peace and stability in the world. As a result, the Bush Administration led an effort in early 2003 to remove Hussein from power.

The Path to War

One of the main concerns about Saddam Hussein was his possible development of so-called weapons of mass destruction. These are weapons that can kill large numbers of people. They include chemical and biological agents as well as nuclear devices.

Bowing to world pressure, Hussein allowed inspectors from the United Nations to search Iraq for such outlawed weapons. Some investigators, however, insisted that the Iraqis were not fully cooperating with the inspections.

U.S. and British officials soon threatened to use to force to disarm Iraq. During his State of the Union address in January 2003, President Bush declared Hussein too great a threat to ignore in an age of increased terrorism. Reminding Americans of the September 11 attacks, Bush stated, "Imagine those 19 hijackers with other weapons and other plans—this time armed by Saddam Hussein. It would take one vial, one canister, one crate slipped into this country to bring a day of horror like none we have ever known. We will do everything in our power to make sure that day never comes."

Operation Iraqi Freedom

In the months that followed, the UN Security Council debated what action to take. Some countries, such as France and Germany, called for letting the inspectors continue searching for weapons. British prime minister Tony Blair, however, accused the Iraqis of "deception and evasion" and insisted inspections would never work.

On March 17, President Bush gave Saddam Hussein and his top aides 48 hours to leave the country or face a military strike. The Iraqi leader refused. On March 19, a coalition led by the United States and Britain launched air strikes in and around the Iraqi capital, Baghdad. The next day, coalition forces marched into Iraq though Kuwait. The invasion of Iraq to remove Saddam Hussein, known as Operation Iraqi Freedom, had begun.

The military operation met with strong opposition from numerous countries. Russian president Vladimir Putin claimed the invasion could "in no way be justified." He and others criticized the policy of attacking a nation to prevent it from future misdeeds. U.S. and British officials, however, argued that they would not wait for Hussein to strike first.

As coalition troops marched north to Baghdad, they met pockets of stiff resistance and engaged in fierce fighting in several southern cities. Meanwhile, coalition forces parachuted into northern Iraq and began moving south toward the capital city. By early April, Baghdad had fallen and the regime of Saddam Hussein had collapsed. After less than four weeks of fighting, the coalition had won the war.

U.S. Army Specialist Shoshana Johnson was one of several Americans held prisoner and eventually released during the war in Iraq.

410

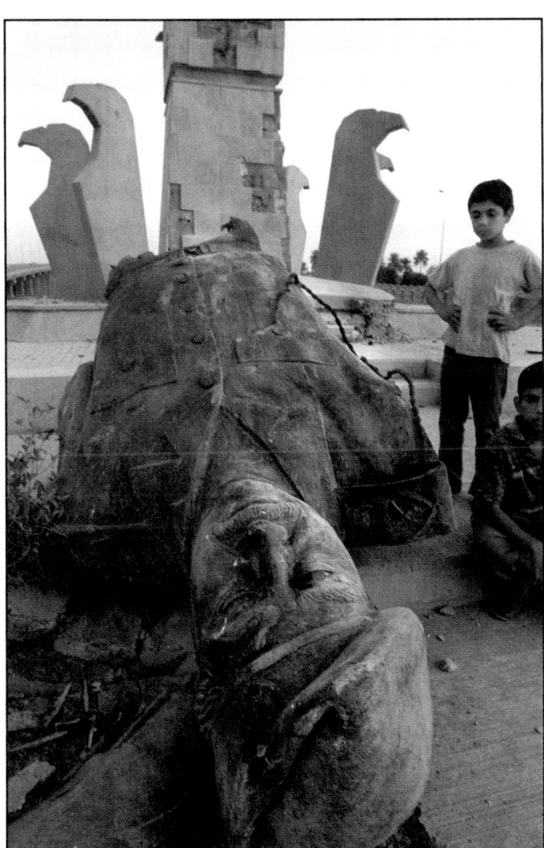

As the regime of Saddam Hussein collapsed, statues of the dictator toppled.

The Struggle Continues

Despite the coalition victory, much work remained in Iraq. The United States installed a civil administrator, retired diplomat L. Paul Bremer, to help oversee the rebuilding of the nation. With the help of Bremer and others, the Iraqis established their own interim government several months after the war. The new governing body went to work creating a constitution and planning democratic elections.

Meanwhile, numerous U.S. troops had to remain behind to help maintain order and battle pockets of fighters loyal to Saddam Hussein. As for the defeated dictator, intelligence officials searched for clues of his whereabouts. The former Iraqi leader disappeared toward the end of the war, and it was unclear whether he had died or escaped.

Finally, the United States and Britain came under increasing fire for failing to find any weapons of mass destruction in the months after the conflict ended. U.S. and British officials insisted that it would be only a matter of time before they found Hussein's deadly arsenal.

Despite the unresolved issues, coalition leaders declared the defeat of Saddam Hussein to be a victory for global security. In a post-war speech to U.S. troops aboard the aircraft carrier *USS Abraham Lincoln*, President Bush urged the world community to keep moving forward in its battle against terrorism. "We do not know the day of final victory, but we have seen the turning of the tide," declared the president. "No act of the terrorists will change our purpose, or weaken our resolve, or alter their fate. Their cause is lost. Free nations will press on to victory."

President George W. Bush and British prime minister Tony Blair stood together throughout the war.

Special Report Assessment

1. Main Ideas

a. What steps did the U.S. government take to make the nation more secure after the attacks on September 11, 2001?

b. Why did the United States take military action against the Taliban in Afghanistan?

c. What was the result of Operation Iraqi Freedom?

2. Critical Thinking

Analyzing Issues Is it important for the U.S. government to respect people's civil rights as it wages a war against terrorism? Why or why not?

THINK ABOUT

- what steps are necessary to protect the nation
- a government that grows too powerful

Answers 1.a. passed the USA Patriot Act, which gave the government increased security powers; created the cabinet position, the Department of Homeland Security; enacted the Transportation Security Act to make airports safer. b. The Taliban protected al-Qaeda and refused to hand over its leader Osama bin Laden, who is believed to have masterminded the attacks of September 11, 2001.
c. The regime of Saddam Hussein fell and an interim body was established to create a new government.
2. Answers will vary. Some students may say maintaining civil rights is very important, or else the government may come to resemble a dictatorship; others might say that maintaining civil rights is not as important as keeping society safe from terrorists attacks.

411

World Cultures AND GEOGRAPHY

Skillbuilder Handbook

Table of Contents

R1

1.1 Summarizing

Defining the Skill

When you **summarize,** you restate a paragraph, passage, or chapter in fewer words. You include only the main ideas and most important details. It is important to use your own words when summarizing.

Applying the Skill

The passage below describes the origins of several state names in the United States. Use the strategies listed below to help you summarize the passage.

How to Summarize

Strategy (1) Look for topic sentences stating the main idea or ideas. These are often at the beginning of a section or paragraph. Briefly restate each main idea—in your own words.

Strategy (2) Include key facts and any names, dates, numbers, amounts, or percentages from the text.

Strategy (3) After writing your summary, review it to see that you have included only the most important details.

STATES' NAMES

(1) The name of a state often comes from that state's geography. For example, (2) the name for Montana comes from a Latin word that means "mountainous."

(1) Other states are named after people who were in power at the time the area was explored or settled. Present-day Louisiana was explored by a Frenchman named La Salle. (2) He named the area Louisiana after the French king at the time, Louis XIV. The state of (2) Georgia was named after King George II of England, who granted the right to start the colony.

(1) Still other states get their names from Native American tribes living in the area when Europeans arrived. (2) Arkansas, Alabama, and Massachusetts were named for the Native American tribes living there.

Write a Summary

You should be able to write your summary in a short paragraph. The paragraph at right summarizes the passage you just read.

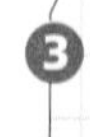

The names of states in the United States often came from geographical features, the names of people in power, or the names of Native American tribes.

Practicing the Skill

Turn to Chapter 4, Section 2, "A Constitutional Democracy." Read "Three Branches of Government" and write a paragraph summarizing the passage.

1.2 Taking Notes

Defining the Skill

When you **take notes,** you write down the important ideas and details of a paragraph, passage, or chapter. A chart or an outline can help you organize your notes to use in the future.

Applying the Skill

The following passage describes the vegetation regions of North and South America. Use the strategies listed below to help you take notes on the passage.

How to Take and Organize Notes

Strategy ① Look at the title to find the main topic of the passage.

Strategy ② Identify the main ideas and details of the passage. Then summarize the main idea and details in your notes.

Strategy ③ Identify key terms and define them. The term *vegetation region* is shown in boldface type and underlined; both techniques signal that it is a key term

Strategy ④ In your notes, use abbreviations to save time and space. You can abbreviate words such as *north (n.), south (s.),* and *United States (U.S.)* to save time and space.

① **VEGETATION OF NORTH AND SOUTH AMERICA**

② North and South America have similar ③ **vegetation regions.** Some of these are are ② deserts, tropical rain forests, forests, and grasslands. Both North and South America have deserts, which are very dry. ② The Painted Desert in the United States and the Patagonia area of Argentina are examples of deserts.

A large percentage of the world's tropical rain forests lie in South America and southern North America. ② The Amazon region in Brazil has the largest tropical rain forest in the world.

② North America has many more forests than South America does. Both continents have grasslands. ② The central part of North America is a type of grassland called prairie.

Make a Chart

Making a chart can help you take notes on a passage. The chart below contains notes from the passage you just read.

② Item	Notes
1. ③ vegetation region	an area of land that has similar plants
a. desert	Painted in ④ N. America; Patagonia in ④ S. America
b. tropical rain forest	Amazon region in Brazil largest in world
c. forest	④ N. America has more than ④ S. America
d. grassland	central ④ N. America type of grassland called prairie

Practicing the Skill

Turn to Chapter 1, Section 1, "The World at Your Fingertips." Read "History and Geography" and use a chart to take notes on the passage.

R3

1.3 Sequencing Events

Defining the Skill

Sequence is the order in which events follow one another. By being able to follow the sequence of events through history, you can get an accurate sense of the relationship among events.

Applying the Skill

The following passage shows the sequence of events that led to the American Civil War. Use the strategies listed below to help you follow the sequence of events.

How to Find the Sequence of Events

Strategy ① Look for specific dates provided in the text. The dates may not always read from earliest to latest, so be sure to match an event with the date.

Strategy ② Look for clues about time that allow you to order events according to sequence. Words and phrases such as *day, week, month,* or *year* may help to sequence the events.

THE CIVIL WAR BEGINS

On ① December 20, 1860, South Carolina became the first state to secede, or leave the Union. Many other states followed. In ② February of the following year, these states united to form the Confederate States of America, also known as the Confederacy. Soon afterward, Confederate soldiers began seizing federal forts. By the time that Lincoln was sworn in as president on ① March 4, 1861, only four southern forts remained in Union hands. The most important was Fort Sumter, on an island in Charleston harbor.

On the morning of ① April 12, Confederate cannons began firing on Fort Sumter. The war between North and South had begun.

Make a Time Line

Making a time line can help you sequence events. The time line below shows the sequence of events in the passage you just read.

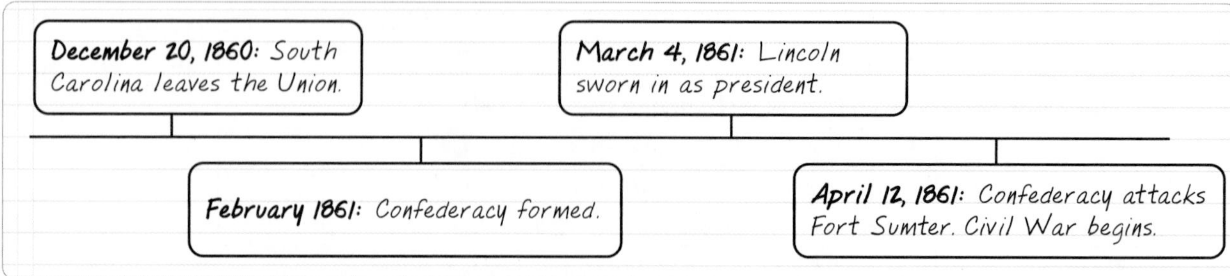

December 20, 1860: South Carolina leaves the Union.
February 1861: Confederacy formed.
March 4, 1861: Lincoln sworn in as president.
April 12, 1861: Confederacy attacks Fort Sumter. Civil War begins.

Practicing the Skill

Turn to Chapter 11, Section 2, "Traders, Explorers, and Colonists." Read "Leaders in Exploration" and "Europe Enters a New Age," then make a time line showing the sequence of events in those two passages.

1.4 Finding Main Ideas

Defining the Skill

The **main idea** is a statement that summarizes the subject of a speech, an article, a section of a book, or a paragraph. Main ideas can be stated or unstated. The main idea of a paragraph is often stated in the first or last sentence. If it is the first sentence, it is followed by sentences that support that main idea. If it is the last sentence, the details build up to the main idea. To find an unstated idea, you must use the details of the paragraph as clues.

Applying the Skill

The following paragraph describes the Taino, some of the first Native Americans to see European explorers. Use the strategies listed below to help you identify the main idea.

How to Find the Main Idea

Strategy ① Identify what you think may be the stated main idea. Check the first and last sentences of the paragraph to see if either could be the stated main idea.

Strategy ② Identify details that support the main idea. Some details explain that idea. Others give examples of what is stated in the main idea.

DECLINE OF THE TAINO

① The Taino were some of the first people in the western hemisphere to see European explorers, and the meeting nearly destroyed them. Christopher Columbus landed in what is now the West Indies in 1492. He thought he had found India, so he called the Taino people he found there "Indians." The Europeans were looking for gold. ② They enslaved the Taino, forcing them to work long hours in mines or in other dangerous jobs. ② Many Taino died because of this treatment. ② Others died from diseases the Europeans unknowingly brought with them. ② By the mid-1500s, almost all of the Taino had been destroyed. However, some survived. The Taino still exist today.

Make a Chart

Making a chart can help you identify the main idea and details in a passage or paragraph. The chart below identifies the main idea and details in the paragraph you just read.

Main Idea: *Meeting Europeans led to the near destruction of the Taino.*

Detail: *The Europeans enslaved the Taino and made them mine for gold.*
Detail: *Many Taino died from the hard work.*
Detail: *Other Taino died from diseases brought by the Europeans.*
Detail: *Almost all the Taino were destroyed by the mid-1500s.*

Practicing the Skill

Turn to Chapter 6, Section 2, "Ancient Latin America." Read "The Spanish in Latin America" and create a chart that identifies the main idea and the supporting details.

R5

1.5 Categorizing

Defining the Skill

To **categorize** is to sort people, objects, ideas, or other information into groups, called categories. Historians categorize information to help them identify and understand patterns in historical events.

Applying the Skill

The following passage discusses the involvement of various countries in World War II. Use the strategies listed below to help you categorize information.

How to Categorize

Strategy ❶ First, decide what kind of information needs to be categorized. Decide what the passage is about and how that information can be sorted into categories. For example, look at the different ways countries reacted to World War II.

Strategy ❷ Then find out what the categories will be. To find how countries reacted to the war, look for clue words such as *in response, some, other,* and *both.*

Strategy ❸ Once you have chosen the categories, sort information into them. Which countries were Axis Powers? Which were Allies? What about the ones who were conquered and those who never fought at all?

WORLD WAR II ALLIANCES

❶During World War II, most countries around the world had to choose which side to take in the conflict. Some countries were conquered so quickly that they could not join either side. ❷ *Others* chose to remain neutral. ❷ Italy and Japan joined with Germany to form the Axis Powers. In 1939, Germany began the war by conquering Poland. ❷ *In response,* Great Britain and France (the first two Allied Powers) declared war on Germany. Germany quickly conquered many countries, including Denmark, Norway, the Netherlands, Belgium, and France. ❷ During 1941, *both* the Soviet Union and the United States entered the war on the Allied side. ❷ Sweden and Switzerland remained neutral during the war.

Make a Chart

Making a chart can help you categorize information. You should have one more column than you have categories. The chart below shows how the information from the passage you just read can be categorized.

❸

Name of Alliance	Axis	Allied Powers	Neutral
Countries	• Germany • Japan • Italy	• Great Britain • France • Soviet Union • United States	• Sweden • Switzerland

Practicing the Skill

Turn to Chapter 14, Section 4, "Germany." Read "A Divided Germany" and make a chart in which you categorize the changes that took place in Germany after World War II.

1.6 Making Public Speeches

Defining the Skill

A speech is a talk given in public to an audience. Some speeches are given to persuade the audience to think or act in a certain way, or to support a cause. You can learn how to **make public speeches** effectively by analyzing great speeches in history.

Applying the Skill

The following is a part of the Gettysburg Address given by Abraham Lincoln in 1863 to dedicate a cemetery on the Civil War battlefield at Gettysburg. Use the strategies listed below to help you analyze Lincoln's speech and prepare a speech of your own.

How to Analyze and Prepare a Speech

Strategy 1 Choose one central idea or theme and organize your speech to support it. Lincoln organized his speech around the idea of giving meaning to the deaths of American soldiers.

Strategy 2 Use words or images that will win over your audience. Lincoln said the dead soldiers had given the living a task—to help the nation "have a new birth of freedom."

Strategy 3 Repeat words or images to drive home your main point—as if it is the "hook" of a pop song. Lincoln repeats the word *here,* to emphasize the idea that the new freedom will begin here, in a place of sacrifice and death. It all starts "here."

GETTYSBURG ADDRESS

The world will little note, nor long remember what we say (3) here, but it can never forget what they did (3) here. (1) It is for us the living, rather, to be dedicated (3) here to the unfinished work which they who fought (3) here have thus far so nobly advanced. It is rather for us to be (3) here dedicated to the great task remaining before us—that from these honored dead we take increased devotion to that cause for which they gave the last full measure of devotion—that we (3) here highly resolve that these dead shall not have died in vain— (2) that this nation, under God, shall have a new birth of freedom—and that government of the people, by the people, for the people, shall not perish from the earth.

Make an Outline

Making an outline like the one to the right will help you make an effective public speech.

Title: Gettysburg Address

I. **Introduce Theme:** The living will give meaning to these soldiers' deaths.
   A. Those who fought will never be forgotten.
   B. It is job of the living to complete the task for which the soldiers died—preserving the government and spreading freedom.

II. **Repeat theme:** Here we dedicate ourselves to the task
   A. We take up the cause from those who gave their lives for it.
   B. We resolve that the soldiers will not have died in vain.

III. **Conclude:** Here there will be a new birth of freedom.

Practicing the Skill

Turn to Chapter 3, Section 2, "A Rich Diversity in Climate and Resources." Read the section, especially the quote from John F. Kennedy on page 79, and use the quote as the subject for your speech. First, make an outline like the one to the right to organize your ideas. Then write your speech. Next, practice giving your speech. Make it a three-minute speech.

R7

2.1 Analyzing Points of View

Defining the Skill

Analyzing points of view means looking closely at a person's arguments to understand the reasons behind that person's beliefs. The goal of analyzing a point of view is to understand different thoughts, opinions, and beliefs about a topic.

Applying the Skill

The following passage describes the difference between Native American and European attitudes about land use. Use the strategies below to help you analyze the points of view.

How to Analyze Points of View

Strategy ❶ Look for statements that show you a particular point of view on an issue. For example, Native Americans believed land should be preserved for the future. European colonists believed land could be owned and changed as desired.

Strategy ❷ Think about why different people or groups held a particular point of view. Ask yourself what they valued. What were they trying to gain or to protect? What were they willing to sacrifice?

Strategy ❸ Write a summary that explains why different groups of people might have taken different positions on this issue.

LAND USE

Native Americans and Europeans had many conflicts because of differing ideas about land use. ❶ Most Native Americans believed that land must be preserved for future generations. They believed the present generation had the right to use land for hunting and farming. However, no one had the right to buy or sell land. Also, no one should ever damage or destroy land.

In contrast, Europeans had a long history of taming wilderness and owning land. As a result, ❶ Europeans believed that they could buy land, sell it, and alter it. For example, if they wished to mine for gold underground, they could destroy landscape as they dug the mine. ❶ Europeans used land to make money.

Make a Diagram

Using a diagram can help you analyze points of view. The diagram below analyzes the different points of view of the Native Americans and the European colonists in the passage you just read.

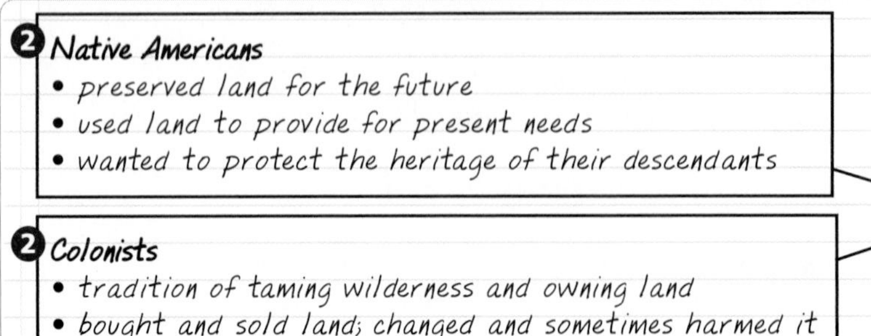

❷ Native Americans
• preserved land for the future
• used land to provide for present needs
• wanted to protect the heritage of their descendants
❷ Colonists
• tradition of taming wilderness and owning land
• bought and sold land; changed and sometimes harmed it
• wanted to make money
❸ Native Americans wanted to preserve the land. They did so because they valued their heritage. The colonists bought land, sold it, and changed it. They did so because they valued making money

Practicing the Skill

Turn to Chapter 5, Section 2, "A Constitutional Monarchy." Read "Many Cultures, Many Needs" about the disagreement between separatists and nationalists regarding Quebec's independence. Make a diagram to analyze their different points of view.

2.2 Comparing and Contrasting

Defining the Skill

Comparing means looking at the similarities and differences between two or more things. **Contrasting** means examining only the differences between them. Historians compare and contrast events, personalities, behaviors, beliefs, and situations in order to understand them.

Applying the Skill

The following passage describes the states of Alaska and Texas. Use the strategies below to help you compare and contrast these two states.

How to Compare and Contrast

Strategy ① Look for two aspects of the subject that can be compared and contrasted. This passage compares Alaska and Texas, the two largest states in the United States.

Strategy ② To contrast, look for clue words that show how two things differ. Clue words include *however, but, on the other hand, while,* and *yet.*

Strategy ③ To find similarities, look for clue words indicating that two things are alike. Clue words include *both, together, alike,* and *similarly.*

TWO BIG STATES

① Texas and Alaska are ③ *alike* in being the two largest states in the United States, ② *but* they have little else in common. Alaska is more than twice as large as Texas. ② *However,* Texas has about 30 times the population of Alaska. Because of their size, ③ *both* states have a variety of climate zones, ② *yet* Texas is generally warmer than Alaska. One reason for this is that Alaska lies much further north. More than 30 percent of Alaska's land is within the Arctic Circle. Texas has very fertile soil and more farms than any other state, ② *while* agriculture is a small part of Alaska's economy. Finally, Texas became a state more than 100 years before Alaska did.

Make a Venn Diagram

Making a Venn diagram will help you identify similarities and differences between two things. In the overlapping area, list characteristics shared by both subjects. Then, in the separate ovals, list the characteristics of each subject not shared by the other. This Venn diagram compares and contrasts Texas and Alaska.

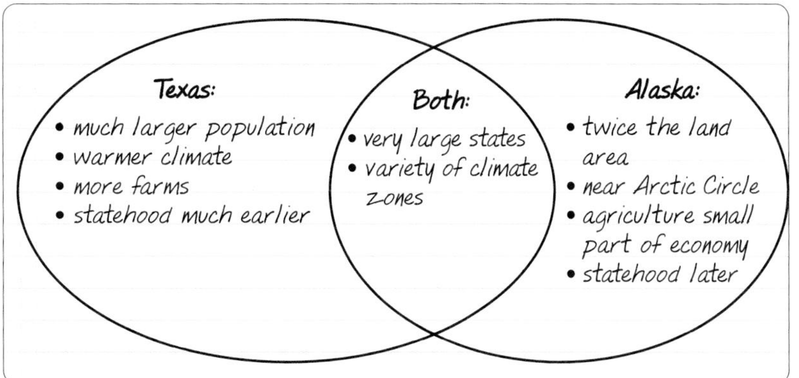

Practicing the Skill

Turn to Chapter 7, Section 2, "Government in Mexico: Revolution and Reform." Read "Mexico's Government Today" and make a Venn diagram showing the similarities and differences between Mexico's government and that of the United States.

R9

2.3 Analyzing Causes; Recognizing Effects

Defining the Skill

A **cause** is an action in history that makes something happen. An **effect** is the historical event that is the result of the cause. A single event may have several causes. It is also possible for one cause to result in several effects. Historians identify cause-and-effect relationships to help them understand why historical events took place.

Applying the Skill

The following paragraph describes events that caused changes in the way of life of the ancient Maya people of Central America. Use the strategies below to help you identify the cause-and-effect relationships.

How to Analyze Causes and Recognize Effects

Strategy ① Ask why an action took place. Ask yourself a question about the title and topic sentence, such as, "What caused Maya civilization to decline?"

Strategy ② Look for effects. Ask yourself, "What happened?" (the effect). Then ask, "Why did it happen?" (the cause). For example, "What caused the Maya to abandon their cities?"

Strategy ③ Look for clue words that signal causes, such as *cause, contributed,* and *led to.*

① DECLINE OF MAYA CIVILIZATION

① The civilization of the Maya went into a mysterious decline around A.D. 900. ② Maya cities in the southern lowlands were abandoned, trade ceased, and the huge stone pyramids of the Maya fell into ruin. No one really understands what happened to the Maya, but there are many theories.

③ Some believe that a change in climate *caused* the decline of Maya civilization. Three long droughts between 810 and 910 meant that there was not enough water for Maya crops. ② As a result, the Maya abandoned their cities. ③ Other researchers believe that additional problems *contributed* to the crisis. They include overpopulation and warfare among the Maya nobility.

Make a Diagram

Using a diagram can help you understand causes and effects. The diagram below shows two causes and an effect for the passage you just read.

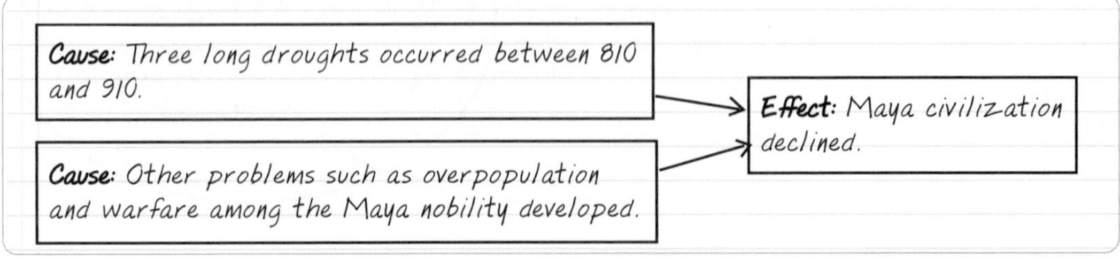

Cause: Three long droughts occurred between 810 and 910.
Cause: Other problems such as overpopulation and warfare among the Maya nobility developed.
Effect: Maya civilization declined.

Practicing the Skill

Turn to Chapter 3, Section 2, "A Rich Diversity in Climate and Resources." Read Connections to History, "The Dust Bowl Disaster" and make a diagram about the causes and effects of the drought on the Great Plains during the 1930s.

2.4 Making Inferences

Defining the Skill

Inferences are ideas that the author has not directly stated. **Making inferences** involves reading between the lines to interpret the information you read. You can make inferences by studying what is stated and using your common sense and previous knowledge.

Applying the Skill

The passage below explains what we know about the early Native Americans who built gigantic mounds of earth for ceremonial purposes. Use the strategies below to help you make inferences from the passage.

How to Make Inferences

Strategy ❶ Read to find statements of facts and ideas. Knowing the facts will give you a good basis for making inferences.

Strategy ❷ Use your knowledge, logic, and common sense to make inferences that are based on facts. Ask yourself, "What does the author want me to understand?" For example, from the facts about how the earth for the mounds was transported, you can make the inference that building the mounds was incredibly hard work and took a very long time. See other inferences in the chart below.

THE MOUND BUILDERS

Early Native American people built gigantic mounds of earth for ceremonial purposes. ❶ The people who built the mounds did not have wheeled vehicles or pack animals to transport the earth. The mounds are made of hundreds of tons of earth. Some mounds appear to have been used as burial plots. ❶ Many Indian artifacts were found buried in those mounds. Other giant mounds served as platforms for temples. ❶ Still other mounds seem to have been symbolic. For example, the Great Serpent Mound in Ohio is shaped like a giant snake.

Make a Chart

Making a chart will help you organize information and make logical inferences. The chart below organizes information from the passage you just read.

❶ *Stated Facts and Ideas*	❷ *Inferences*
The Mound Builders did not have wheeled vehicles or pack animals to transport the earth. The mounds contain hundreds of tons of earth.	*Building the mounds was incredibly difficult and time-consuming work, as the builders had to carry the earth by hand.*
Artifacts were found buried in the mounds.	*Native Americans buried objects with their dead.*
Some mounds were symbolic, such as the Great Serpent Mound.	*The serpent may have been important in their religion.*

Practicing the Skill

Turn to Chapter 3, Section 1, "From Coast to Coast." Read Strange but True, "Quakes Shake Central U.S. Lands" and use a chart like the one above to make inferences about earthquakes in the United States.

R11

2.5 Making Decisions

Defining the Skill

Making decisions involves choosing between two or more options, or courses of action. In most cases, decisions have consequences, or results. Sometimes decisions may lead to new problems. By understanding how historical figures made decisions, you can learn how to improve your own decision-making skills.

Applying the Skill

The passage below explains the decision British Prime Minister Chamberlain faced when Germany threatened aggression in 1938. Use the strategies below to analyze his decision.

How to Make Decisions

Strategy ❶ Identify a decision that needs to be made. Think about what factors make the decision difficult.

Strategy ❷ Identify possible consequences of the decision. Remember that there can be more than one consequence to a decision.

Strategy ❸ Identify the decision that was made.

Strategy ❹ Identify actual consequences that resulted from the decision.

PEACE OR WAR?

In 1938, German leader Adolf Hitler demanded a part of Czechoslovakia where mostly Germans lived. ❶ British Prime Minister Neville Chamberlain had to decide how to respond to that aggression. ❷ He could threaten to go to war, but he feared that Britain was not ready. ❷ If he gave Germany the region, he might avoid war, but he would be setting a bad example by giving in to a dictator. ❸ Along with the French leader, Chamberlain decided to give Germany what it demanded. In exchange, Hitler promised not to take any more land in Europe. ❹ Six months later, Germany took the rest of Czechoslovakia and later invaded Poland.

Make a Flow Chart

A flow chart can help you identify the process of making a decision. The flow chart below shows the decision-making process in the passage you just read.

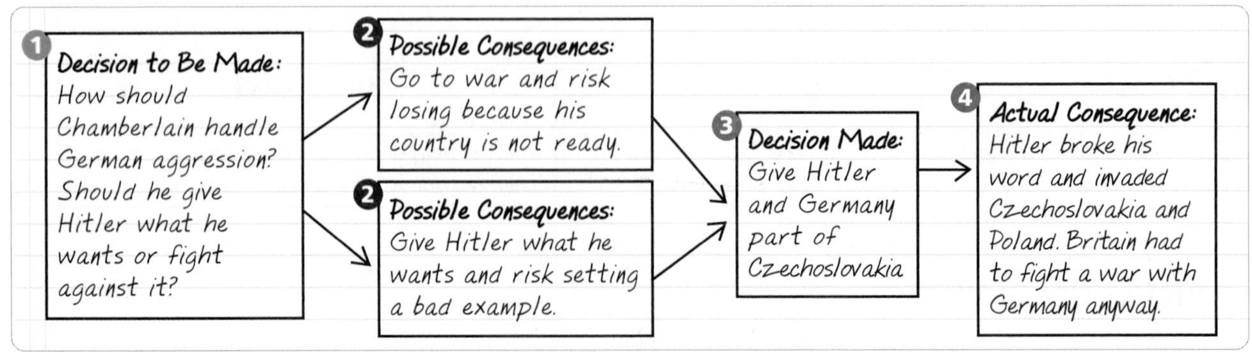

❶ Decision to Be Made: How should Chamberlain handle German aggression? Should he give Hitler what he wants or fight against it?
❷ Possible Consequences: Go to war and risk losing because his country is not ready.
❷ Possible Consequences: Give Hitler what he wants and risk setting a bad example.
❸ Decision Made: Give Hitler and Germany part of Czechoslovakia
❹ Actual Consequence: Hitler broke his word and invaded Czechoslovakia and Poland. Britain had to fight a war with Germany anyway.

Practicing the Skill

Turn to Chapter 13, Section 2, "Eastern Europe and Russia." Read "The Breakup of the Soviet Union" and make a flow chart to identify Gorbachev's decision and its consequences as described in the section.

R12

2.6 Recognizing Propaganda

Defining the Skill

Propaganda is communication that aims to influence people's opinions, emotions, or actions. Propaganda is not always factual. Rather, it uses one-sided language or striking symbols to sway people's emotions. Modern advertising often uses propaganda. By thinking critically, you will avoid being swayed by propaganda.

Applying the Skill

The following political cartoon from World War I shows an immigrant being exposed as an "enemy alien." Use the strategies listed below to help you understand how the cartoon works as propaganda.

How to Recognize Propaganda

Strategy 1 Identify the aim, or purpose, of the cartoon. Point out the subject and explain the point of view.

Strategy 2 Identify those images in the cartoon that viewers might respond to emotionally and identify the emotions.

Strategy 3 Think critically about the cartoon. What facts has the cartoon ignored?

Make a Chart

Making a chart will help you think critically about a piece of propaganda. The chart below summarizes the information about the anti-immigrant cartoon.

1 *Identify Purpose*	*The cartoon shows "an enemy alien" being exposed by the U.S. government.*
2 *Identify Emotions*	*The cartoonist uses the man's posture to suggest he is sneaky and has something to hide. The cartoonist also shows him trying to hide under a "cloak" of patriotism—the U.S. flag.*
3 *Think Critically*	*Some viewers may conclude that all recent immigrants are enemy aliens, when probably only a tiny minority were helping the enemy during World War I.*

Practicing the Skill

Turn to Chapter 13, Section 1, "Eastern Europe Under Communism." Read "Soviet Culture," paying special attention to the photograph on page 354. Use a chart like the one above to think critically about the statue shown in the photograph as an example of propaganda.

R13

2.7 Identifying Facts and Opinions

Defining the Skill

Facts are events, dates, statistics, or statements that can be proved to be true. Opinions are judgments, beliefs, and feelings. By identifying facts and opinions, you will be able to think critically when a person is trying to influence your own opinion.

Applying the Skill

The following passage tells about a man named John Henry, who became a legend in American folklore. Use the strategies listed below to distinguish facts from opinions.

How to Recognize Facts and Opinions

Strategy ❶ Look for specific information that can be proved or checked for accuracy.

Strategy ❷ Look for assertions, claims, and judgments that express opinions. In this case the author gives a direct opinion about the stories told of John Henry.

Strategy ❸ Think about whether statements can be checked for accuracy. Then, identify the facts and opinions in a chart.

JOHN HENRY: MAN AND MYTH

❶ John Henry was an African American who worked for the railroads in the late 1800s. His job was to lay down track for the trains. Many famous stories have been written about John Henry, and the most famous is the one about his race against the steam drill. ❶ The first written version of the story dates from 1900.

According to the tale, John was helping dig out the Big Bend Tunnel, when someone started using a steam-powered drill. John Henry was the strongest man on the job, so he raced the drill to see whether he could dig faster than a machine. John Henry won the race, but he died soon after. ❷ This legend may be the most interesting story in U.S. history.

Make a Chart

The chart below analyzes the facts and opinions from the passage above.

Statement	❸ *Can It Be Proved?*	❸ *Fact or Opinion*
John Henry worked for the railroads in the 1800s.	*Yes. Check newspapers and other historical documents.*	*Fact*
The first version of the story dates from 1900.	*Yes. Check the date of the actual written version of the story.*	*Fact*
The legend of John Henry's race against the steam-powered drill may be the most interesting story in American history.	*No. This is the opinion of the author; other stories may be just as interesting.*	*Opinion*

Practicing the Skill

Turn to Chapter 3, Section 2, "A Rich Diversity in Climate and Resources," and read the section entitled "Neighbors and Leaders." Make a chart in which you analyze key statements to determine whether they are facts or opinions.

2.8 Forming and Supporting Opinions

Defining the Skill

When you form **opinions,** you interpret and judge the importance of events and people in history. You should always **support your opinions** with facts, examples, and quotes.

Applying the Skill

The following passage describes the unification of Europe. Use the strategies listed below to form and support your opinions about the topic.

How to Recognize Facts and Opinions

Strategy ❶ Look for important information about the subject. Information can include facts, quotations, and examples.

Strategy ❷ Form an opinion about the subject by asking yourself questions about the information. For example, how important was the subject? How does it relate to similar subjects in your own experience?

Strategy ❸ Support your opinions with facts, quotations, and examples. If the facts do not support the opinion, then rewrite your opinion so that it is supported by the facts.

UNIFYING EUROPE

In 1951, six nations—Belgium, France, Italy, Luxembourg, the Netherlands, and West Germany—placed their coal and steel industries under the control of one group. Six years later, the same countries formed an organization called the European Economic Community, or Common Market. ❶ This made trade easier among member countries. By 1995, 15 European countries were members of what is now called the European Union (EU).

❶ The European Union has passed laws to clean up the environment. Through the EU, nations have shared scientific information and new technology. ❶ They have also combined police forces to fight crime.

Make a Chart

Making a chart can help you organize your opinions and supporting facts. The following chart summarizes one possible opinion about the unification of Europe.

❷ *Opinion*	*The establishment of the European Union has been good for Europe.*
❸ *Facts*	*The Common Market made trade easier.*
	The EU passed laws to clean up the environment.
	EU nations have shared scientific information and new technology.
	EU nations work together to fight crime.

Practicing the Skill

Turn to Chapter 8, Section 3, "Cuba Today." Read "Living in Cuba," and form your own opinion about the Cuban way of life. Make a chart like the one above to summarize your opinion and the supporting facts and examples.

R15

2.9 Identifying and Solving Problems

Defining the Skill

Identifying problems means finding and understanding the difficulties faced by a particular group of people during a certain time. **Solving problems** means understanding how people tried to remedy those problems. By studying how people solved problems in the past, you can learn ways to solve problems today.

Applying the Skill

The following paragraph describes the problems that resulted when women tried to gain equality. Use the strategies listed below to learn how women responded.

How to Identify Problems and Solutions

Strategy 1 Look for the difficulties or problems caused by the situation.

Strategy 2 Consider how the problem affected people or groups with different points of view. For example, the main problem described here is how to increase job opportunities for women.

Strategy 3 Look for the solutions that people or groups employed to deal with the problem. Think about whether the solution was a good one for people or groups with differing points of view.

JOB OPPORTUNITIES FOR WOMEN

1 In the mid-20th century women had limited job opportunities. In 1950, only 30 percent of women worked outside the home. By 1960, that number had risen to about 40 percent. Women working outside the home faced several obstacles.

2 For example, the jobs traditionally open to women were in areas like clerical work. Such jobs paid poorly and offered few chances to advance. 2 Many men were unwilling to share power with women. 3 In the 1970s, women demanded more rights. They began to move into "blue collar" jobs, such as construction work. 3 Other women gained the education to move into professions like law. By 2000, women were working in every field.

Make a Chart

Making a chart will help you identify and organize information about problems and solutions. The chart below shows problems and solutions included in the passage you just read.

1 Problem	2 Differing Points of View	3 Solution
In the mid-20th century, there were limited job opportunities for women.	Women wanted jobs with better pay and opportunity to advance. Men did not want to share power.	Women demanded rights. Women gained education. In time, women were working in every field.

Practicing the Skill

Turn to Chapter 12, Section 1, "European Empires." Read "The Spread of Nationalism." Then make a chart that summarizes the problems faced by the monarchies of Europe and their citizens and the solutions they agreed on.

2.10 Evaluating

Defining the Skill

To **evaluate** is to make a judgment about something. Historians evaluate the actions of people in history. One way to do this is to examine both the positives and the negatives of a historical action, then decide which is stronger—the positive or the negative.

Applying the Skill

The following passage describes President Franklin Roosevelt's attempts to pull the United States out of the Great Depression. Use the strategies listed below to evaluate how successful he was.

How to Evaluate

Strategy 1 Before you evaluate a person's actions, first determine what that person was trying to do. In this case, think about what Roosevelt wanted to accomplish.

Strategy 2 Look for statements that show the positive, or successful, results of his actions. For example, did he achieve his goals?

Strategy 3 Also look for statements that show the negative, or unsuccessful, results of his actions. Did he fail to achieve something he tried to do?

Strategy 4 Write an overall evaluation of the person's actions.

THE NEW DEAL

1 As U.S. president, Franklin Roosevelt wanted to lift the country out of the Great Depression. He called his policies the New Deal. 2 Roosevelt used government resources to create jobs. He said, "People who are hungry and out of a job are the stuff of which dictatorships are made." 2 He created the Public Works Administration to build roads and bridges and the Civilian Conservation Corps to create parks. The idea was to put people back to work, so they would spend money and help the economy. 3 Roosevelt's New Deal did not end the Depression, but it gave people hope—and the people re-elected Roosevelt three times.

Make a Diagram for Evaluating

Using a diagram can help you evaluate. List the positives and negatives of the historical person's actions and decisions. Then make an overall judgment. The diagram below shows how the information from the passage you just read can be diagrammed.

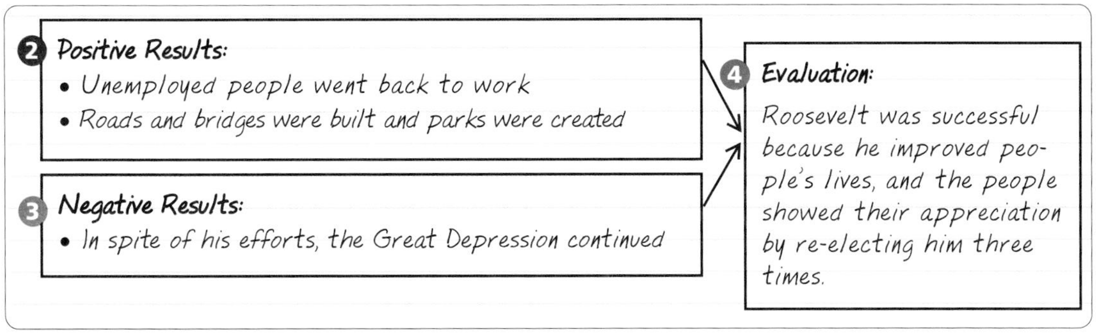

2 *Positive Results:*
- *Unemployed people went back to work*
- *Roads and bridges were built and parks were created*

3 *Negative Results:*
- *In spite of his efforts, the Great Depression continued*

4 *Evaluation:*
Roosevelt was successful because he improved people's lives, and the people showed their appreciation by re-electing him three times.

Practicing the Skill

Turn to Chapter 2, Section 2, "A Rich Diversity in Climate and Resources." Read "Neighbors and Leaders," and make a diagram in which you evaluate U.S. and Canadian cooperation.

2.11 Making Generalizations

Defining the Skill

To **make generalizations** means to make broad judgments based on information. When you make generalizations, you should gather information from several sources.

Applying the Skill

The following three passages contain ideas on America as a melting pot of different cultures. Use the strategies listed below to make a generalization about these ideas.

How to Make Generalizations

Strategy 1 Look for information that the sources have in common. These three sources all look at the relationship of immigrants to the United States.

Strategy 2 Form a generalization that describes this relationship in a way that all three sources would agree with. State your generalization in a sentence.

THE MELTING POT

At the turn of the century, 1 many native-born Americans thought of their country as a melting pot, a mixture of people of different cultures and races who blended together by abandoning their native languages and customs.

—*The Americans*

1 A nation, like a tree, does not thrive well till it is engrafted with a foreign stock.

—*Journals* (Emerson, 1823)

1 The United States has often been called a *melting pot.* But in other ways, U.S. society is an example of *cultural pluralism.* . . . That is, large numbers of its people have retained features of the cultures of their ancestors.

—*World Book Encyclopedia*

Make a Chart

Using a chart can help you make generalizations. The chart below shows how the information you just read can be used to generalize about the contributions of immigrants.

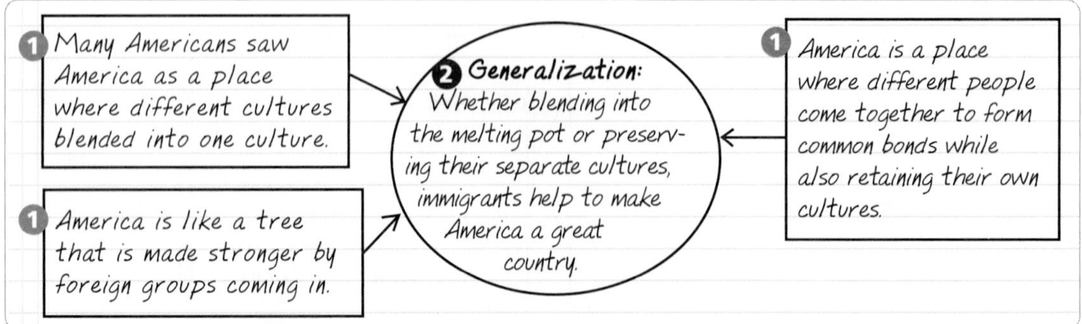

1 Many Americans saw America as a place where different cultures blended into one culture.
1 America is like a tree that is made stronger by foreign groups coming in.
2 Generalization: Whether blending into the melting pot or preserving their separate cultures, immigrants help to make America a great country.
1 America is a place where different people come together to form common bonds while also retaining their own cultures.

Practicing the Skill

Turn to Chapter 7, Section 1, "The Roots of Modern Mexico." Read "Life in New Spain," and study the graph on page 176. Also read Biography, "Father Hidalgo" on page 177. Use a chart like the one above to make a generalization about the class system in Mexico.

3.1 Interpreting Time Lines

Defining the Skill

A **time line** is a visual list of events and dates shown in the order in which they occurred. Time lines can be horizontal or vertical. On horizontal time lines, the earliest date is on the left. On vertical time lines, the earliest date is often at the top.

Applying the Skill

The time line below lists dates and events associated with the Civil Rights Movement in the United States. Use the strategies listed below to help you interpret the information.

How to Read a Time Line

Strategy 1 Read the dates at the beginning and end of the time line. These will show the period of history that is covered. The time line to the right is a vertical time line. It shows the sequence of events from top to bottom instead of from left to right.

Strategy 2 Read the dates and events in sequential order, beginning with the earliest one. Pay particular attention to how the entries relate to each other. Think about whether earlier events influenced later events.

Strategy 3 Summarize the focus, or main idea, of the time line. Try to write a main idea sentence that describes the time line.

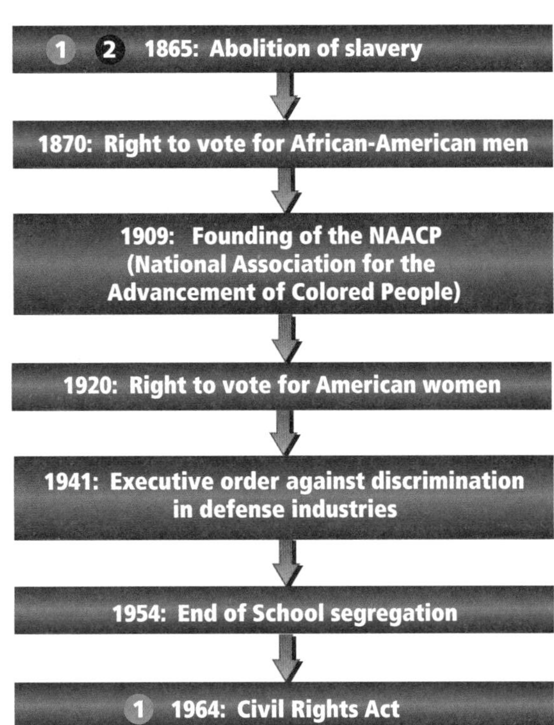

1 2 1865: Abolition of slavery
1870: Right to vote for African-American men
1909: Founding of the NAACP (National Association for the Advancement of Colored People)
1920: Right to vote for American women
1941: Executive order against discrimination in defense industries
1954: End of School segregation
1 1964: Civil Rights Act

Write a Summary

Writing a summary can help you understand information shown on a time line. The summary to the right states the main idea of the time line and tells how the events are related.

3 *Although slavery was abolished in 1865, gaining civil rights was a long and gradual process for African Americans. It involved the passage of amendments granting them the vote, an executive order against discrimination during World War II, and the Civil Rights Act of 1964.*

Practicing the Skill

Turn to Chapter 9, page 233, and write a summary of the information shown on the time line.

R19

4.1 Using an Electronic Card Catalog

Defining the Skill

An **electronic card catalog** is a library's computerized search program that will help you locate books and other materials in the library. You can search the catalog by entering a book title, an author's name, or a subject of interest to you. The electronic card catalog also provides basic information about each book (author, title, publisher, and date of publication). You can use an electronic card catalog to create a bibliography (a list of books) on any topic that interests you.

Applying the Skill

The screen shown below is from an electronic search for information about the Amazon River. Use the strategies listed below to help you use the information on the screen.

How to Use an Electronic Card Catalog

Strategy 1 Begin searching by choosing either subject, title, or author, depending on the topic of your search. For this search, the user chose "Subject" and typed in the words "Amazon River."

Strategy 2 Once you have selected a book from the results of your search, identify the author, title, city, publisher, and date of publication.

Strategy 3 Look for any special features in the book. This book is illustrated, and it includes maps, bibliographical references, and an index.

Strategy 4 Locate the call number for the book. The call number indicates the section in the library where you will find the book. You can also find out if the book is available in the library you are using. If not, it may be in another library in the network.

```
Search Request:
1 Subject          Title          Author

Find  Options  Locations  Backup  Startover  Help

2 Barter, James. The Amazon. San Diego: Lucent
Books, ©2003.
    2 AUTHOR:  Barter, James
     2 TITLE:  The Amazon/James Barter.
 2 PUBLISHED:  San Diego: Lucent Books, ©2003.
    3 PAGING:  112p.: ill., maps; 24 cm.
      SERIES:  Rivers of the World
     3 NOTES:  Includes bibliographical
               references (p. 100-104) and index.
4 CALL NUMBER: J 981.1 BA
```

Practicing the Skill

Turn to Chapter 5, "Canada Today," and find a topic that interests you, such as the First Nations of northern Canada, the cultural conflicts in Quebec, the history of the St. Lawrence Seaway, or the design of Canada's flag. Use the SUBJECT search on an electronic card catalog to find information about your topic. Make a bibliography of books about the subject. Be sure to include the author, title, city, publisher, and date of publication for all the books in your bibliography.

4.2 Creating a Database

Defining the Skill

A **database** is a collection of data, or information, that is organized so that you can find and retrieve information on a specific topic quickly and easily. Once a computerized database is set up, you can search it to find specific information without going through the entire database. The database will provide a list of all information in the database related to your topic. Learning how to use a database will help you learn how to create one.

Applying the Skill

The chart below is a database for famous mountains in the Western Hemisphere. Use the strategies listed below to help you understand and use the database.

How to Create a Database

Strategy 1 Identify the topic of the database. The keywords, or most important words, in this title are "Mountains" and "Western Hemisphere." These words were used to begin the research for this database.

Strategy 2 Identify the kind of data you need to enter in your database. These will be the column headings of your database. The key words "Mountain," "Location," "Height," and "Interesting Facts" were chosen to focus the research.

Strategy 3 Identify the entries included under each heading.

Strategy 4 Use the database to help you find information quickly. For example, in this database you could search for "Volcanoes" to find a list of famous mountains that are also volcanoes.

1 FAMOUS MOUNTAINS OF THE WESTERN HEMISPHERE

2 MOUNTAIN	LOCATION	HEIGHT ABOVE SEA LEVEL (FEET)	INTERESTING FACTS
3 Aconcagua	Argentina	22,834	Tallest mountain in Western Hemisphere
Cotopaxi	Ecuador	19,347	4 One of tallest active volcanoes
Logan	Canada	19,524	Tallest mountain in Canada
Mauna Kea	Hawaii	13,796	Tallest mountain in world from ocean floor to peak
Mauna Loa	Hawaii	13,677	4 Largest volcano in the world
McKinley	Alaska	20,320	Tallest mountain in North America
Orizaba	Mexico	18,700	Tallest mountain in Mexico
Shasta	California	14,162	Well-known twin peaks

Practicing the Skill

Create a database of Central and South American countries that shows the name of each country, its location, its land area, and its population. Use the information on pages 146–149 "Data File" to provide the data. Use a format like the one above for your database.

R21

4.3 Using the Internet

Defining the Skill

The **Internet** is a computer network that connects to universities, libraries, news organizations, government agencies, businesses, and private individuals throughout the world. Each location on the Internet has a home page with its own address, or URL (Universal Resource Locator). With a computer connected to the Internet, you can reach the home pages of many organizations and services. The international collection of home pages, known as the World Wide Web, is a good source of up-to-date information about current events as well as research on subjects in geography.

Applying the Skill

The Web page below shows helpful links for Unit 1 of *World Cultures and Geography.* Use the strategies listed below to help you understand how to use the Web page.

How to Use the Internet

Strategy ❶ Go directly to a Web page. For example, type http://www.classzone.com/books/wc_survey/ in the box at the top of the screen and press ENTER (or RETURN). The Web page will appear on your screen. Then click on the link for Unit 1. A new web page will appear. Find the link for research links. Click on it and you will go to a third web page. Finally, click on the link for *Web sites on the world* and it will take you to the screen shown here.

Strategy ❷ Explore the links on the right side of the screen. Click on any one of the links to find out more about a specific subject. These links take you to other pages on this Web site. Some pages include links to related information that can be found at other places on the Internet.

Strategy ❸ When using the Internet for research, you should confirm the information you find. Web sites set up by universities, government agencies, and reputable news sources are more reliable than other sources. You can often find information about the creator of a site by looking for copyright information.

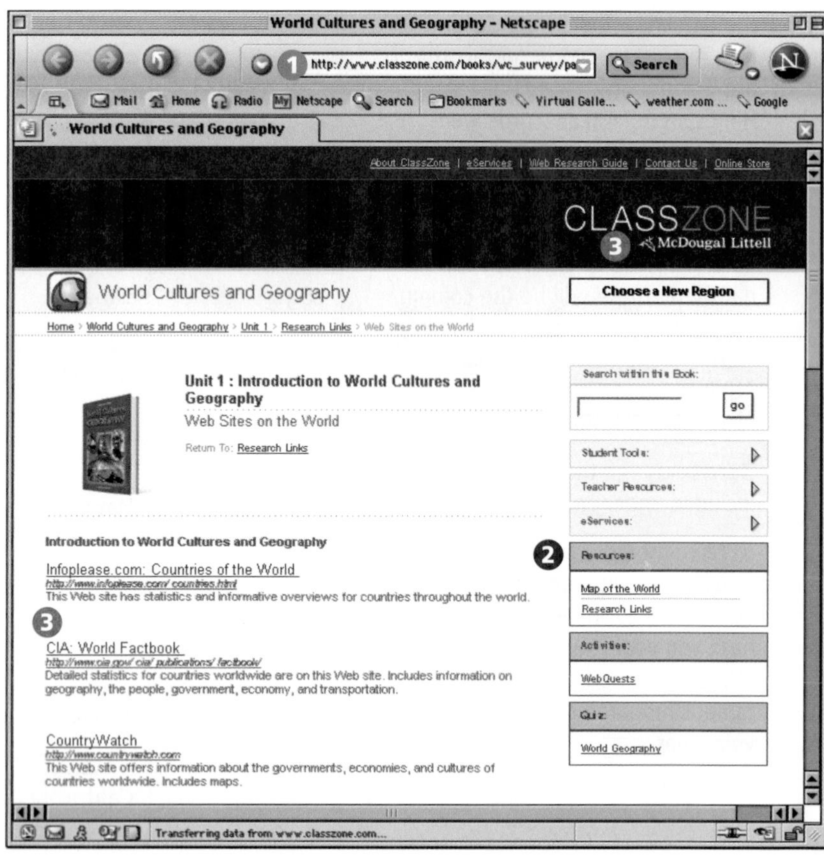

Practicing the Skill

Turn to Chapter 23, Section 1, "Physical Geography." Read the section and make a list of thematic maps you would like to research. If you have Internet access, go to classzone.com. There you will find links that provide more information about the topics in the section.

4.4 Creating a Multimedia Presentation

Defining the Skill

Movies, CD-ROMs, television, and computer software are different kinds of media. To **create a multimedia presentation,** you need to collect information in different media and organize them into one presentation.

Applying the Skill

The illustration below shows students using computers to create a multimedia presentation. Use the strategies listed below to help you create your own multimedia presentation.

How to Create a Multimedia Presentation

Strategy 1 Identify the topic of your presentation and decide which media are best for an effective presentation. For example, you may want to use video or photographic images to show the impressive architecture of medieval castles Or, you may want to use CDs or audiotapes to provide music or to make sounds that go with your presentations, like the sound of a drawbridge closing.

Strategy 2 Research the topic in a variety of sources. Images, text, props, and background music should reflect the region and the historical period of your topic.

Strategy 3 Write the script for the presentation and then record it using a microphone and audiotape. You could use a narrator and characters' voices to tell the story. Primary sources are an excellent source for script material.

Strategy 4 Videotape the presentation or create it on your computer. Having the presentation as a file on your computer will preserve it for future viewing and allow you to show it to different groups of people.

Practicing the Skill

Turn to Chapter 10, "Western Europe: Its Land and Early History." Choose a topic from the chapter and use the strategies listed above to create a multimedia presentation about it.

R23

Glossary

A

absolute location *n.* the exact spot on Earth where a place is found. (p. 36)

acid rain *n.* rain or snow that carries air pollutants to Earth. (p. 386)

Aegean (ih•JEE•uhn) **Sea** *n.* a branch of the Mediterranean Sea that is located between Greece and Turkey. (p. 279)

alliance (uh•LY•uhns) *n.* an agreement among people or nations to unite for a common cause and to help any alliance member that is attacked. (p. 334)

Anasazi *n.* a group of Native Americans who developed a complex civilization in the U.S. Southwest. (p. 88)

armed neutrality *n.* a policy by which a country maintains military forces but does not take sides in the conflicts of other nations. (p. 385)

Athens *n.* the capital of Greece and once one of the most important ancient Greek city-states. (p. 280)

Austria-Hungary *n.* in the 1900s, a dual monarchy in which the Hapsburg emperor ruled both Austria and Hungary. (p. 332)

B

Berlin Wall *n.* a wire-and-concrete wall that divided Germany's East Berlin and West Berlin from 1961 to 1989. (p. 395)

bilingual *adj.* able to speak two languages. (p. 133)

Bill of Rights *n.* ten amendments to the U.S. Constitution that list specific freedoms guaranteed to every U.S. citizen. (p. 96)

Brasília *n.* the capital of Brazil. (p. 246)

C

capitalism *n.* an economic system in which the factories and businesses that make and sell goods are privately owned and the owners make the decisions about what goods to produce. (p. 315)

Carnival *n.* a Cuban holiday that celebrates the end of the harvest each year in July. (p. 220); *n.* a Brazilian holiday that occurs during the four days before Lent. (p. 247)

cartographer *n.* a person who makes maps. (p. 45)

censorship *n.* the outlawing of materials that contain certain information. (p. 403)

chinampa (chee•NAHM•pah) *n.* a floating garden on which the Aztec grew crops. (p. 162)

circumnavigate *v.* to sail completely around. (p. 310)

citizen *n.* a legal member of a country. (p. 20)

citizenship *n.* the status of a citizen, which includes certain duties and rights. (p. 91)

city-state *n.* a central city and its surrounding villages, which together follow the same law, have one form of government, and share language, religious beliefs, and ways of life. (p. 279)

climate *n.* the typical weather of a region over a long time. (p. 75)

coalition government *n.* a government formed by political parties joining together. (p. 362)

Cold War *n.* after World War II, a period of political noncooperation between the members of NATO and the Warsaw Pact nations, during which these countries refused to trade or cooperate with each other. (p. 347)

R24

collective farm *n.* a government-owned farm that employs large numbers of workers, often in Communist countries. (p. 345)

colonialism *n.* a system by which a country maintains colonies outside its borders. (p. 331)

Columbian Exchange *n.* the exchange of goods between Europe and its colonies in North and South America. (p. 166)

Communism *n.* an economic and political system in which property is owned collectively and labor is organized in a way that is supposed to benefit all people. (p. 217)

competition *n.* the rivalry among businesses to sell the most goods to consumers and make the greatest profit. (p. 106)

constitutional amendment *n.* a formal change or addition to the U.S. Constitution. (p. 96)

constitutional monarchy *n.* a government ruled by a king or queen whose power is determined by the nation's constitution and laws. (p. 125)

consumer *n.* a person who uses goods or services. (p. 105)

continent *n.* a landmass above water on earth. (p. 35)

Court of Human Rights *n.* the Council of Europe's court that protects the rights of all citizens in whichever of its member countries they live. (p. 370)

criollo (kree•AW•yaw) *n.* a person born in Mexico whose parents were born in Spain. (p. 175)

Crusades *n.* a series of military expeditions led by Western European Christians in the 11th, 12th, and 13th centuries to reclaim control of the Holy Lands from the Muslims. (p. 301)

culture *n.* the beliefs, customs, laws, art, and ways of living that a group of people share. (p. 21)

culture region *n.* an area of the world in which many people share similar beliefs, history, and languages. (p. 24)

culture trait *n.* the food, clothing, technology, beliefs, language, and tools that the people of a culture share. (p. 21)

currency *n.* money used as a form of exchange. (p. 369)

czar (zahr) *n.* in Russia, an emperor. (p. 319)

D

Day of the Dead *n.* a Mexican holiday, on November 1 and 2, for honoring loved ones who have died. (p. 194)

deforestation *n.* the process of cutting and clearing away trees from a forest. (p. 157)

democracy *n.* a government that receives its power from the people. (p. 91)

departamento (deh•pahr•tah•MEHN•taw) *n.* a Guatemalan state. (p. 223)

dependency *n.* a place governed by or closely connected with a country that it is not officially part of. (p. 203)

deposed *v.* removed from power. (p. 356)

détente (day•TAHNT) *n.* a relaxing of tensions between nations. (p. 357)

dictator *n.* a person who has complete control over a country's government. (p. 207)

dissident *n.* a person who openly disagrees with a government's policies. (p. 403)

distribution *n.* the process of moving products to their markets. (p. 187)

diversify *v.* to conduct business activities in a variety of industries. (p. 209)

dual monarchy *n.* a form of government in which one ruler governs two nations. (p. 332)

Duma (DOO•muh) *n.* one of the two houses of the Russian legislature. (p. 364)

R25

E

economic indicator *n.* a measure that shows how a country's economy is doing. (p. 241)

economics *n.* the study of how resources are managed in the production, exchange, and use of goods and services. (p. 20)

economy *n.* the system by which business owners in a region use productive resources to provide goods and services that satisfy people's wants. (p. 79)

ejido (eh•HEE•daw) *n.* in Mexico, a community farm owned by the people of a village together. (p. 182)

El Niño (ehl NEE•nyaw) *n.* a current in the Pacific Ocean that results from changes in air pressure and that causes changes in weather patterns. (p. 159)

empire *n.* a nation or group of territories ruled by an emperor. (p. 287)

encomienda (ehn•kaw•MYEHN•dah) *n.* a system that the rulers of Spain established in Mexico, under which Spanish men received a Native American village to oversee and gain tribute from. (p. 176)

equal opportunity *n.* a guarantee that government and private institutions will not discriminate against people on the basis of factors such as race, religion, age, gender, or disability. (p. 90)

erosion *n.* the process by which environmental factors, such as wind, rivers, and rain, wear away soil and stone. (p. 74)

ethnic cleansing *n.* the organized killing of members of an ethnic group or groups. (p. 363)

euro *n.* the common unit of currency used by European Union countries. (p. 369)

European Community *n.* an association developed after World War II to promote economic unity among the countries of Western Europe. (p. 392)

European Union (EU) *n.* an economic and political grouping of countries in Western Europe. (p. 368)

export *n.* a product traded with or sold to another country. (p. 129)

F

factor of production *n.* one of the elements needed for production of goods or services to occur. (p. 103)

fascism (FASH•IHZ•uhm) *n.* a political philosophy that promotes a strong, central government controlled by the military and led by a powerful dictator. (p. 336)

federal government *n.* a national government. (p. 96)

feudalism *n.* in medieval Europe, a political and economic system in which lords gave land to less powerful nobles, called vassals, in return for which the vassals agreed to provide various services to the lords. (p. 292)

fiesta *n.* a holiday celebrated by a village or town, with events such as parades, games, and feasts. (p. 194)

First Nations *n.* in Canada, a group of descendants of the first settlers of North America, who came from Asia. (p. 120)

fjord (fyawrd) *n.* a long, narrow, deep inlet of the sea located between steep cliffs. (p. 274)

Florence *n.* a city in Italy that was a bustling center of banking, trade, and manufacturing during the 14th century. (p. 302)

Francophone *n.* a French-speaking person. (p. 133)

free enterprise/market economy *n.* an economy that allows business owners to compete in the market with little government interference. (p. 105)

R26

free-trade zone *n.* an area in which goods can move across borders without being taxed. (p. 241)

French Resistance *n.* an anti-German movement in France during World War II. (p. 390)

French Revolution *n.* a revolution that began on July 14, 1789, and that led to France's becoming a republic. (p. 316)

G

Gadsden Purchase *n.* the 1853 purchase by the United States of a piece of land from northern Mexico. (p. 178)

GDP *n.* gross domestic product, or the total value of the goods and services produced in a country during a given time period. (p. 104)

geography *n.* the study of people, places, and the environment. (p. 18)

glacier *n.* a thick sheet of ice that moves slowly across land. (p. 73)

globalization *n.* the spreading of an idea, product, or technology around the world. (p. 111)

Good Friday Accord *n.* an agreement signed in 1998 by Ireland's Protestants and Catholics that established the Northern Ireland Assembly to represent voters from both groups. (p. 381)

government *n.* the people and groups within a society that have the authority to make laws, to make sure they are carried out, and to settle disagreements about them. (p. 19)

guerrilla warfare *n.* nontraditional military tactics by small groups involving surprise attacks. (p. 253)

guild *n.* a business association created by people working in the same industry to protect their common interests and maintain standards within the industry. (p. 294)

H

hacienda *n.* in Spanish-speaking countries, a big farm or ranch. (p. 181)

hieroglyph *n.* a picture or symbol used in hieroglyphics. (p. 161)

history *n.* a record of the past. (p. 18)

Holocaust *n.* the organized killing of European Jews and others by the Nazis during World War II. (p. 336)

hydroelectricity *n.* electrical power generated by water. (p. 386)

I

immigrant *n.* a person who comes to a country to take up residence. (p. 88)

imperialism *n.* the practice of one country's controlling the government and economy of another country or territory. (p. 310)

import *n.* a product brought into a country through trade or sale. (p. 130)

impressionism *n.* a style of art that creates an impression of a scene rather than a strictly realistic picture. (p. 393)

Industrial Revolution *n.* a period of change beginning in the late 18th century, during which goods began to be manufactured by power-driven machines. (p. 314)

industry *n.* any area of economic activity. (p. 129)

inflation *n.* a continuing increase in the price of goods and services, or a continuing decrease in the capability of money to buy goods and services. (p. 245)

Institutional Revolutionary Party *n.* the most powerful political party in Mexico from the 1920s to 2000, which won every presidential election during that time. (p. 183)

interdependence *n.* the economic, political, and social dependence of culture regions on one another. (p. 26)

R27

Iron Curtain *n.* a political barrier that isolated the peoples of Eastern Europe after World War II, restricting their ability to travel outside the region. (p. 342)

L

labor force *n.* a pool of available workers. (p. 315)

ladino (lah•DEE•noh) *n.* a person of mixed European and Native American ancestry. (p. 205)

landform *n.* a feature of Earth's surface, such as a mountain, valley, or plateau. (p. 73)

latitude *n.* a measure of distance north or south of the equator. (p. 36)

limited government *n.* government in which the powers of the leaders are limited. (p. 95)

London *n.* the capital of England. (p. 379)

longitude *n.* a measure of distance east or west of a line called the prime meridian. (p. 36)

M

Machu Picchu (MAH•choo PEEK•choo) *n.* an ancient stone city built by the Inca. (p. 165)

Magna Carta (MAG•nuh KAHR•tuh) *n.* a document signed by England's King John in 1215 that guaranteed English people basic rights. (p. 295)

malnutrition *n.* poor nutrition, usually from not eating the right foods, which can result in poor health. (p. 220)

manorialism *n.* a social system in which peasants worked on a lord's land and supplied him with food in exchange for his protection of them. (p. 293)

map projection *n.* one of the different ways of showing Earth's curved surface on a flat map. (p. 47)

maquiladora (mah•kee•lah•DAW•rah) *n.* in Mexico, a factory that imports duty-free parts from the United States to make products that it then exports back across the border. (p. 187)

market economy See **free enterprise/ market economy.**

medieval (MEE•dee•EE•vuhl) *adj.* relating to the period of history between the fall of the Roman Empire and the beginning of the modern world, often dated from 476 to 1453. (p. 291)

Mediterranean Sea *n.* an inland sea that borders Europe, Southwest Asia, and Africa. (p. 274)

mestizo (mehs•TEE•saw) *n.* a person of mixed European and Native American ancestry. (p. 175)

migrate *v.* to move from one area in order to settle in another. (p. 38)

mulatto (mu•LAT•oh) *n.* a person of mixed African and European ancestry. (p. 204)

multiculturalism *n.* an acceptance of many cultures instead of just one. (p. 121)

N

national identity *n.* a sense of belonging to a nation. (p. 132)

nationalism *n.* strong pride in one's nation or ethnic group. (p. 329)

nationalize *v.* to establish government control of a service or industry. (p. 188)

NATO (NAY•toh) *n.* the North Atlantic Treaty Organization, a defense alliance formed in 1949, with the countries of Western Europe, Canada, and the United States agreeing to defend one another if attacked. (p. 337)

O

oasis *n.* a region in a desert that is fertile because it is near a river or spring. (p. 251)

R28

oligarchy (AHL•ih•GAHR•kee) *n.* a government in which a few powerful individuals rule. (p. 280)

ombudsman *n.* a Swedish official who protects citizens' rights and ensures that the courts and civil service follow the law. (p. 385)

one-party system *n.* a system in which there is only one political party and only one candidate to choose from for each government position. (p. 343)

Organization of American States (OAS) *n.* an organization of the nations of the Americas that encourages democracy and promotes economic cooperation, social justice, and the equality of all people. (p. 234)

P

Pan-American *adj.* relating to all of the Americas. (p. 234)

Parliament *n.* Canada's national lawmaking body. (p. 125)

parliamentary republic *n.* a republic whose head of government, usually a prime minister, is the leader of the political party that has the most members in the parliament. (p. 361)

patrician (puh•TRIHSH•uhn) *n.* in ancient Rome, a member of a wealthy, landowning family that claimed to be able to trace its roots back to the founding of Rome. (p. 285)

patriotism *n.* love for one's country. (p. 91)

PEMEX *n. Petróleos Mexicanos,* or "Mexican Petroleum," a government agency that runs the oil industry in Mexico. (p. 188)

peninsula *n.* a body of land surrounded by water on three sides. (p. 274)

peninsular (peh•neen•soo•LAHR) *n.* a person who was born in Spain but who lived in Mexico after the Spanish took control. (p. 175)

philosopher *n.* a person who studies and thinks about why the world is the way it is. (p. 281)

plain *n.* a large flat area of land that usually does not have many trees. (p. 275)

plebeian (plih•BEE•uhn) *n.* a common citizen of ancient Rome. (p. 285)

polis *n.* the central city of a city-state. (p. 279)

political process *n.* legal activities through which a citizen influences public policy. (p. 91)

precipitation *n.* moisture that falls to Earth, such as rain or snow. (p. 75)

prime minister *n.* in a parliamentary democracy, the leader of the cabinet and often also of the executive branch. (p. 125)

private property rights *n.* the right of individuals to own land or industry. (p. 355)

privatization *n.* the process of replacing community ownership with individual, or private, ownership. (p. 186)

profit *n.* the money that remains after the costs of producing a product are paid. (p. 105)

propaganda (PRAHP•uh•GAN•duh) *n.* material designed to spread certain beliefs. (p. 354)

Protestant *n.* a member of a Christian church based on the principles of the Reformation. (p. 306)

puppet government *n.* a government that is controlled by an outside force. (p. 343)

Q

Quechua (KEHCH•wuh) *n.* people who live in the Andes highlands and speak the Inca language Quechua. (p. 253)

R29

R

Reformation *n.* a 16th-century movement to change practices within the Roman Catholic Church. (p. 305)

refugee *n.* a person who flees a country because of war, disaster, or persecution. (p. 122)

Reign of Terror *n.* the period between 1793 and 1794 during which France's new leaders executed thousands of its citizens. (p. 317)

relative location *n.* the location of one place in relation to other places. (p. 37)

Renaissance *n.* an era of creativity and learning in Western Europe from the 14th century to the 16th century. (p. 302)

republic *n.* a form of government in which people rule through elected representatives. (p. 285)

reunification *n.* the uniting again of parts. (p. 396)

Riksdag (RIHKS•DAHG) *n.* Sweden's parliament. (p. 384)

Rio de Janeiro *n.* a city in Brazil. (p. 246)

river system *n.* a network that includes a major river and its tributaries. (p. 74)

rural *adj.* of the countryside. (p. 193)

Russian Revolution *n.* the 1917 revolution that removed the Russian monarchy from power after it had ruled for 400 years. (p. 322)

S

São Paulo *n.* a city in Brazil. (p. 246)

scarcity *n.* a word economists use to describe the conflict between people's desires and limited resources. (p. 20)

Scientific Revolution *n.* a period of great scientific change and discovery during the 16th and 17th centuries. (p. 314)

secede *v.* to withdraw from a political union, such as a nation. (p. 381)

Senate *n.* the assembly of elected representatives that was the most powerful ruling body of the Roman Republic. (p. 285)

separatist *n.* a person who wants to separate from a body to which he or she belongs, such as a church or nation. (p. 127)

single-product economy *n.* an economy that depends on just one product for the majority of its income. (p. 209)

skerry *n.* a small island. (p. 387)

socialism *n.* an economic system in which businesses and industries are owned collectively or by the government. (p. 392)

Solidarity *n.* a trade union in Poland that originally aimed to increase pay and improve working conditions and that later opposed Communism. (p. 400)

standard of living *n.* a measure of quality of life. (p. 369)

sugar cane *n.* a plant from which sugar is made. (p. 209)

T

tariff *n.* a fee imposed by a government on imported or exported goods. (p. 369)

technology *n.* tools and equipment made through scientific discoveries. (p. 112)

thematic map *n.* a map that focuses on a specific idea or theme. (p. 46)

tourism *n.* the business of helping people travel on vacations. (p. 189)

transportation barrier *n.* a geographic feature that prevents or slows transportation. (p. 131)

transportation corridor *n.* a path that makes transportation easier. (p. 131)

R30

Treaty of Guadalupe Hidalgo *n.* a treaty that the United States forced Mexico to sign in 1848, giving Mexico's northern lands to the United States. (p. 177)

tributary *n.* a stream or river that flows into a larger river. (p. 157)

Tropical Zone *n.* the region of the world that lies between the latitudes 23°27' north and 23°27' south. (p. 158)

U

United States Constitution *n.* the document that is the foundation for all U.S. laws and the framework for the U.S. government. (p. 94)

unlimited government *n.* a government in which the leaders have almost absolute power. (p. 95)

Ural (YUR•uhl) **Mountains** *n.* a mountain range that divides Europe from Asia. (p. 275)

urban *adj.* of the city. (p. 193)

urbanization *n.* the movement of people from the countryside to cities. (p. 241)

V

value *n.* a principle or ideal by which people live. (p. 110)

vegetation *n.* plant life, such as trees, plants, and grasses. (p. 75)

W

Warsaw Pact *n.* a treaty signed in 1955 that established an alliance among the Soviet Union, Albania, Bulgaria, Czechoslovakia, East Germany, Hungary, Poland, and Romania. (p. 347)

weather *n.* the state of Earth's atmosphere at a given time and place. (p. 75)

West Indies *n.* the Caribbean Islands. (p. 203)

World War I *n.* a war fought from 1914 to 1918 between the Allies (Russia, France, the United Kingdom, Italy, and the United States) and the Central Powers (Austria-Hungary, Germany, Turkey, and Bulgaria). (p. 333)

World War II *n.* a war fought from 1939 to 1945 between the Axis powers (Germany Italy, and Japan) and the Allies (the United Kingdom, France, the Soviet Union, and the United States). (p. 336)

R31

Spanish Glossary

A

absolute location [ubicación absoluta] *s.* lugar exacto donde se halla un lugar en la Tierra. (pág. 36)

acid rain [lluvia ácida] *s.* lluvia o nieve que lleva sustancias contaminantes a la Tierra. (pág. 386)

Aegean (ih•JEE•uhn) **Sea** [mar Egeo] *s.* parte del Mar Mediterráneo ubicada entre Grecia y Turquía. (pág. 279)

alliance (uh•LY•uhns) [alianza] *s.* acuerdo de unión entre pueblos o naciones por una causa común y de ayuda mutua en caso de que uno sea atacado. (pág. 334)

Anasazi [anasazi] *s.* grupo de indígenas que desarrollaron una cultura compleja en el sudoeste de Estados Unidos. (pág. 88)

armed neutrality [neutralidad armada] *s.* política mediante la cual un país mantiene fuerzas armadas pero no participa en conflictos de otras naciones. (pág. 385)

Athens [Atenas] *s.* capital de Grecia y una de las ciudades-estado más importantes de la antigua Grecia. (pág. 280)

Austria-Hungary [Austria-Hungría] *s.* monarquía dual mediante la cual a comienzos del siglo XX el emperador de la dinastía de los Hasburgo gobernó Austria y Hungría. (pág. 332)

B

Berlin Wall [Muro de Berlín] *s.* pared de cemento y alambre que desde 1961 a 1989 dividía la parte este de Berlín de la parte oeste. (pág. 395)

bilingual [bilingüe] *adj.* que puede hablar dos idiomas. (pág. 133)

Bill of Rights [Declaración de Derechos] *s.* las diez enmiendas a la Constitución de los Estados Unidos que enumeran libertades específicas garantizadas a todos los ciudadanos estadounidenses. (pág. 96)

C

capitalism [capitalismo] *s.* sistema económico en el cual las empresas y comercios que fabrican y venden productos y mercancías son de propiedad privada; los dueños de dichas empresas y comercios deciden lo que desean producir y vender. (pág. 315)

Carnival [carnaval] *s.* feriado cubano que celebra cada año el fin de la cosecha en el mes de julio. (pág. 220); *s.* feriado brasileño de cuatro días anterior al comienzo de la cuaresma. (pág. 247)

cartographer [cartógrafo] *s.* persona que hace mapas. (pág. 45)

censorship [censura] *s.* prohibición de materiales que contienen cierta información. (pág. 403)

chinampa (chee•NAHM•pah) *s.* huerta flotante en la que los aztecas sembraban sus cultivos. (pág. 162)

circumnavigate [circunnavegar] *v.* dar la vuelta alrededor de algo en una nave. (pág. 310)

citizen [ciudadano] *s.* habitante legal de un país. (pág. 20)

citizenship [ciudadanía] *s.* estado de un ciudadano que incluye derechos y obligaciones. (pág. 91)

city-state [ciudad estado] *s.* ciudad central y sus aldeas aledañas que acatan las mismas leyes, tienen una sola forma de gobierno y comparten una lengua, creencias religiosas y estilos de vida. (pág. 279)

climate [clima] *s.* condición típica de la atmósfera en determinada región. (pág. 75)

R32

coalition government [gobierno de coalición] *s.* gobierno formado por la unión de partidos políticos. (pág. 362)

Cold War [Guerra Fría] *s.* período político posterior a la Segunda Guerra Mundial, caracterizado por la falta de cooperación y relaciones comerciales entre los países miembros de la OTAN y las naciones del Pacto de Varsovia. (pág. 347)

collective farm [granja colectiva] *s.* granja que pertenece al gobierno, que emplea a gran número de trabajadores generalmente en países comunistas. (pág. 345)

colonialism [colonialismo] *s.* sistema mediante el cual un país mantiene colonias en otras partes del mundo. (pág. 331)

Columbian Exchange [intercambio colombino] *s.* intercambio de mercancías entre Europa y sus colonias en América. (pág. 166)

Communism [comunismo] *s.* sistema político y económico en el cual la propiedad es colectiva y la actividad laboral es organizada de manera de beneficiar a todos los individuos. (pág. 217)

competition [competencia] *s.* rivalidad entre empresas por vender la mayor cantidad de productos y mercaderías a los consumidores y por obtener el mayor beneficio. (pág. 106)

constitutional amendment [enmienda constitucional] *s.* cambio formal a la constitución de los Estados Unidos. (pág. 96)

constitutional monarchy [monarquía constitucional] *s.* gobierno encabezado por un rey o reina, cuyo poder está determinado por la constitución y leyes de la nación. (pág. 125)

consumer [consumidor] *s.* persona que usa productos o servicios. (pág. 105)

continent [continente] *s.* masa continental sobre agua en la Tierra. (pág. 35)

Court of Human Rights [Corte de Derechos Humanos] *s.* corte que protege los derechos de ciudadanos que habiten en países miembros. (pág. 370)

criollo (kree•AW•yaw) *s.* persona nacida en México cuyos padres provienen de España. (pág. 175)

Crusades [las cruzadas] *s.* serie de expediciones militares dirigidas por cristianos de Europa occidental en los siglos XI, XII y XIII, para apoderarse de nuevo de las Tierras Santas, en poder de los musulmanes. (pág. 301)

culture [cultura] *s.* conjunto de creencias, costumbres, leyes, formas artísticas y de vida compartidas por un grupo de personas. (pág. 21)

culture region [región cultural] *s.* territorio donde muchas personas comparten creencias, historia y lenguas similares. (pág. 24)

culture trait [característica cultural] *s.* alimento, vestimenta, tecnología, creencia, lengua u otro elemento compartido por un pueblo o cultura. (pág. 21)

currency [moneda] *s.* sistema que sirve para medir el valor de las cosas que se intercambian. (pág. 369)

czar (zahr) [zar] *s.* emperador ruso. (pág. 319)

D

Day of the Dead [Día de los Muertos] *s.* feriado mexicano del 1 al 2 de noviembre en honor a los seres queridos que han muerto. (pág. 194)

deforestation [deforestación] *s.* proceso mediante el cual se cortan árboles y se van eliminando bosques. (pág. 157)

democracy [democracia] *s.* sistema de gobierno mediante el cual los gobernantes reciben el poder del pueblo. (pág. 91)

departamento (deh•pahr•tah•MEHN•taw) *s.* nombre que reciben los estados en Guatemala. (pág. 223)

dependency [territorio dependiente] *s.* lugar gobernado o que está estrechamente conectado con un país del cual no forma parte. (pág. 203)

deposed [depuesto] *v.* removido del poder. (pág. 356)

détente (day•TAHNT) [distensión] *s.* disminución de la tensión entre países. (pág. 357)

dictator [dictador] *s.* persona que tiene el control absoluto del gobierno de un país. (pág. 207)

R33

dissident [disidente] *s.* persona que abiertamente muestra desacuerdo con la política de un gobierno. (pág. 403)

distribution [distribución] *s.* proceso mediante el cual se transporta la mercancía a los mercados. (pág. 187)

diversify [diversificar] *v.* llevar a cabo actividades comerciales en una variedad de industrias. (pág. 209)

dual monarchy [monarquía dual] *s.* gobierno en que un solo jefe gobierna dos naciones. (pág. 332)

Duma (DOO•muh) [Duma] *s.* una de las dos cámaras de la legislatura rusa. (pág. 364)

E

economic indicator [indicador económico] *s.* medida que muestra el estado de la economía de un país. (pág. 241)

economics [economía] *s.* estudio del uso de los recursos naturales y del modo de producción, intercambio y utilización de los productos, mercaderías y servicios. (pág. 20)

economy [economía] *s.* sistema de administrar recursos de producción en una región con el fin de proveer productos y servicios que satisfagan las necesidades humanas. (pág. 79)

ejido *s.* granja comunitaria en México que pertenece colectivamente a los habitantes de un pueblo o localidad. (pág. 182)

El Niño *s.* corriente del océano Pacífico que surge de los cambios en la presión atmosférica y que causa cambios meteorológicos. (pág. 159)

empire [imperio] *s.* nación o conjunto de territorios gobernados por un emperador. (pág. 287)

encomienda *s.* sistema establecido en México por los gobernantes españoles mediante el cual los españoles controlaban comunidades indígenas y recaudaban tributo. (pág. 176)

equal opportunity [igualdad de oportunidades] *s.* garantía de que el gobierno e instituciones privadas no discriminarán en contra de ciertas personas por motivos de raza, religión, edad o sexo. (pág. 90)

erosion [erosión] *s.* desgaste del terreno y suelo producido por factores ambientales tales como el viento, la lluvia y los ríos. (pág. 74)

ethnic cleansing [limpieza étnica] *s.* matanza sistemática (genocidio) de uno o varios grupos étnicos que conforman una minoría. (pág. 363)

euro [euro] *s.* unidad monetaria de los países miembros de la Unión Europea. (pág. 369)

European Community [Comunidad Europea] *s.* asociación creada después de la Segunda Guerra Mundial para promover la unidad económica entre los países de Europa occidental. (pág. 392)

European Union [Unión Europea] *s.* asociación económica y política de países de Europa occidental. (pág. 368)

export [exportación] *s.* mercadería que se vende a otro país. (pág. 129)

F

factor of production [factor de producción] *s.* elemento necesario para producir mercaderías y ofrecer servicios. (pág. 103)

fascism (FASH•IHZ•uhm) [fascismo] *s.* filosofía que promueve un gobierno centralista fuerte, controlado por el ejército y dirigido por un dictador poderoso. (pág. 336)

federal government [gobierno federal] *s.* gobierno nacional. (pág. 96)

feudalism [feudalismo] *s.* sistema político y económico de la Europa medieval en el que los señores feudales repartían tierras a miembros de la nobleza menos poderosos, llamados vasallos, quienes, a cambio de éstas, se comprometían a brindar varios servicios a los señores feudales. (pág. 292)

R34

fiesta *s.* feriado celebrado por una comunidad o pueblo, con desfiles, juegos y banquetes. (pág. 194)

First Nation [Primera Nación] *s.* nombre que reciben en Canadá los descendientes de los primeros pobladores en Norteamérica que provenían de Asia. (pág. 120)

fjord (fyawrd) [fiordo] *s.* entrada larga y estrecha del mar formada entre acantilados abruptos. (pág. 274)

Florence [Florencia] *s.* ciudad italiana que durante el siglo XIV mantuvo una dinámica actividad bancaria, comercial y manufacturera. (pág. 302)

Francophone [francófono] *s.* persona que habla francés. (pág. 133)

free enterprise/market economy [librecambismo/economía de libre mercado] *s.* economía que permite a empresarios y comerciantes competir en el mercado con poca interferencia del gobierno. (pág. 105)

free-trade zone [zona de libre comercio] *s.* área en la que tanto personas como mercaderías pueden circular más allá de las fronteras sin que esta actividad sea gravada con impuestos. (pág. 241)

French Resistance [Resistencia francesa] *s.* en Francia, un movimiento antialemán durante la Segunda Guerra Mundial. (pág. 390)

French Revolution [Revolución francesa] *s.* revolución que comenzó el 14 de julio de 1789 y tuvo como resultado la conversión de Francia en una república. (pág. 316)

G

Gadsden Purchase [Compra de Gadsden] *s.* compra mediante la cual los Estados Unidos adquirieron en 1853 parte del territorio del norte de México. (pág. 178)

GDP [PIB, Producto Interno Bruto] *s.* valor total de los productos y servicios producidos en un país durante un período determinado. (pág. 104)

geography [geografía] *s.* estudio de los pueblos, lugares y el medio ambiente. (pág. 18)

glacier [glaciar] *s.* espesas formaciones de hielo que se mueven lentamente por la tierra. (pág. 73)

globalization [globalización] *s.* difusión de una idea, producto o tecnología por todo el mundo. (pág. 111)

Good Friday Accord [Acuerdo del Viernes Santo] *s.* acuerdo firmado por los protestantes y católicos de Irlanda del Norte que estableció la Asamblea de Irlanda del Norte, asamblea esta que representa a los votantes de ambos grupos. (pág. 381)

government [gobierno] *s.* los individuos y grupos en una sociedad que tienen la autoridad de crear leyes y hacerlas cumplir, y de resolver desacuerdos que puedan surgir con respecto a ellas. (pág. 19)

guerrilla warfare [guerrilla] *s.* tácticas militares no tradicionales llevadas a cabo por grupos pequeños que realizan ataques sorpresivos. (pág. 253)

guild [gremio] *s.* asociación creada por personas que trabajan en una misma industria, con el fin de proteger sus intereses comunes y mantener ciertos criterios y principios aplicables a la industria. (pág. 294)

H

hacienda *s.* en países hispanos, finca o granja de gran tamaño. (pág. 181)

hieroglyph [jeroglífico] *s.* dibujo y símbolo de la escritura jeroglífica. (pág. 161)

history [historia] *s.* un registro de los acontecimientos del pasado. (pág. 18)

Holocuast [Holocausto] *s.* matanza sistemática (genocidio) de los judíos europeos y otros por el partido nazi durante la Segunda Guerra Mundial. (pág. 336)

hydroelectricity [electricidad hidráulica] *s.* energía eléctrica producida por el agua. (pág. 386)

R35

I

immigrant [inmigrante] *s.* persona que llega a un país y allí se establece. (pág. 88)

imperialism [imperialismo] *s.* práctica mediante la cual un país controla el gobierno y la economía de otro país o territorio. (pág. 310)

import [importación] *s.* mercadería y producto que se compra del extranjero. (pág. 130)

impressionism [impresionismo] *s.* estilo de arte que crea una impresión de algo en lugar de una obra con características concretas. (pág. 393)

Industrial Revolution [Revolución industrial] *s.* período de cambio en el siglo XVIII que dio lugar a la fabricación de productos por máquinas. (pág. 314)

industry [industria] *s.* sector o rama de actividad económica. (pág. 129)

inflation [inflación] *s.* aumento continuo del precio de mercaderías y servicios, o disminución continua de la capacidad de la moneda de comprar mercaderías y servicios. (pág. 245)

Institutional Revolutionary Party [PRI, Partido Revolucionario Institucional] *s.* partido político mexicano más poderoso desde 1920 hasta 2000, período en que ganó todas las elecciones presidenciales. (pág. 183)

interdependence [interdependencia] *s.* dependencia económica, política y social que mantienen las sociedades de diversas regiones culturales. (pág. 26)

Iron Curtain [Cortina de Hierro] *s.* barrera política que aisló los países de Europa del Este luego de la Segunda Guerra Mundial, limitando la capacidad de movimiento y tránsito fuera de esta región. (pág. 342)

L

labor force [fuerza laboral] *s.* trabajadores disponibles. (pág. 315)

ladino *s.* persona de descendencia mixta, con sangre europea e indígena. (pág. 205)

landform [accidente geográfico] *s.* característica del suelo o terreno, como montañas, valles y mesetas. (pág. 73)

latitude [latitud] *s.* distancia norte-sur con relación al ecuador, de la superficie terrestre. (pág. 36)

limited government [gobierno limitado] *s.* gobierno en que los gobernantes tienen poderes limitados. (pág. 95)

London [Londres] *s.* capital de Inglaterra. (pág. 379)

longitude [longitud] *s.* distancia este-oeste de un punto de la Tierra, a partir de la línea inicial llamada primer meridiano (meridiano de Greenwich). (pág. 36)

M

Machu Picchu (MAH•choo PEEK•choo) *s.* antigua ciudad de piedra construida por los incas. (pág. 165)

Magna Carta (MAG•nuh KAHR•tuh) [Carta Magna] *s.* documento firmado por el rey Juan de Inglaterra en 1215 que garantizó los derechos básicos de las personas en ese país. (pág. 295)

malnutrition [desnutrición] *s.* mala nutrición debida a la falta de alimentos nutritivos que puede causar mala salud. (pág. 220)

manorialism [régimen señorial] *s.* sistema social en el que campesinos trabajan las tierras de un señor, a cambio de protección y seguridad. (pág. 293)

map projection [proyección cartografía] *s.* una de las diversas maneras de mostrar la curvatura de la Tierra en una superficie plana. (pág. 47)

maquiladora *s.* fábrica en México que importa partes sin arancel aduanero de los Estados Unidos para fabricar productos que luego envía de vuelta a través de la frontera. (pág. 187)

R36

market economy Ver **free enterprise.**

medieval (MEE•dee•EE•vuhl) [medieval] *adj.* que pertenece al período de la historia comprendido entre la caída del Imperio romano y el comienzo del mundo moderno, aproximadamente desde 476 a 1453. (pág. 291)

Mediterranean Sea [mar Mediterráneo] *s.* mar interno que bordea Europa, el sudoeste de Asia y África. (pág. 274)

mestizo *s.* persona de descendencia mixta, con sangre europea e indígena. (pág. 175)

migrate [migrar] *v.* irse de un área para establecerse en otra. (pág. 38)

mulatto [mulato] *s.* persona con descendencia mixta, de sangre europea y africana. (pág. 204)

multiculturalism [multiculturalismo] *s.* la aceptación de muchas culturas en vez de una solamente. (pág. 121)

N

national identity [identidad nacional] *s.* sentimiento de pertenencia a una nación. (pág. 132)

nationalism [nacionalismo] *s.* intenso orgullo por el país o grupo étnico propio. (pág. 329)

nationalize [nacionalizar] *v.* pasar al control gubernamental un servicio o industria. (pág. 188)

NATO (NAY•toh) [OTAN, Organización del Tratado del Atlántico Norte] *s.* alianza de defensa que agrupa a los países de Europa occidental, Canadá y Estados Unidos, que acuerdan la defensa común en caso de ataque. (pág. 337)

O

oasis [oasis] *s.* región fértil en un desierto que se formó alrededor de un río o manantial. (pág. 251)

oligarchy (AHL•ih•GAHR•kee) [oligarquía] *s.* gobierno de sólo unos pocos individuos poderosos. (pág. 280)

ombudsman [defensor del pueblo] *s.* funcionario del gobierno sueco que protege los derechos de los ciudadanos y asegura que los tribunales y la administración pública cumplan con la ley. (pág. 385)

one-party system [sistema monopartidista] *s.* sistema donde sólo se puede votar por un partido político y por un candidato para cada puesto de gobierno. (pág. 343)

Organization of American States (OAS) [OEA, Organización de los Estados Americanos] *s.* organización de todas las naciones americanas, que promueve la democracia y la cooperación económica, la justicia social y la igualdad entre los pueblos. (pág. 234)

P

Pan-American [panamericano] *adj.* relativo a todas las Américas. (pág. 234)

Parliament [Parlamento] *s.* cuerpo legislativo nacional de Canadá. (pág. 125)

parliamentary republic [república parlamentaria] *s.* república cuyo jefe de estado, en general un primer ministro, es el líder del partido político que tiene la mayoría de representantes en el parlamento. (pág. 361)

patrician (puh•TRIHSH•uhn) [patricio] *s.* miembro de familia adinerada y hacendada en la antigua Roma, que afirmaba que sus orígenes se remontan a la época de la fundación de Roma. (pág. 285)

patriotism [patriotismo] *s.* amor por el país propio. (pág. 91)

PEMEX *s.* Petróleos Mexicanos, agencia gubernamental que administra la industria petrolera en México. (pág. 188)

peninsula [península] *s.* territorio rodeado de agua en tres de sus lados. (pág. 274)

peninsular *s.* persona que nació en España pero vivió en México luego de la conquista de los españoles. (pág. 175)

philosopher [filósofo] *s.* persona que estudia y piensa sobre el mundo y su naturaleza. (pág. 281)

R37

plain [llanura] *s.* superficie extensa y plana que suele no tener muchos árboles. (pág. 275)

plebeian (plih•BEE•uhn) [plebeyo] *s.* ciudadano corriente (sin título de nobleza) en la antigua Roma. (pág. 285)

polis [polis] *s.* ciudad central de una ciudad estado. (pág. 279)

political process [proceso político] *s.* actividades permitidas por la ley mediante las cuales el ciudadano influye en las políticas públicas. (pág. 91)

precipitation [precipitación] *s.* humedad como la lluvia o la nieve que cae a la Tierra. (pág. 75)

prime minister [primer ministro] *s.* en una democracia parlamentaria, líder del gabinete y frecuentemente de la administración ejecutiva. (pág. 125)

private property rights [derechos de propiedad privada] *s.* derechos individuales de ser propietario de bienes raíces, campos o industrias. (pág. 355)

privatization [privatización] *s.* proceso mediante el cual la propiedad que pertenece a la comunidad se convierte en propiedad privada o individual. (pág. 186)

profit [ganancia] *s.* dinero que sobra luego de pagar el costo de producir un producto. (pág. 105)

propaganda (PRAHP•uh•GAN•duh) [propaganda] *s.* material cuyo objetivo es difundir ciertas creencias. (pág. 354)

Protestant [protestante] *s.* miembro de una iglesia cristiana fundada de acuerdo a los principis de la Reforma. (pág. 306)

puppet government [gobierno títere] *s.* gobierno que hace lo que le indica un poder exterior. (pág. 343)

Q

Quechua (KEHCH•wuh) *s.* habitante de los Andes que habla una lengua inca del mismo nombre. (pág. 253)

R

Reformation [Reforma] *s.* movimiento del siglo XVI que se propuso cambiar las prácticas de la Iglesia Católica. (pág. 305)

refugee [refugiado] *s.* persona que huye de un país a raíz de una guerra, catástrofe, o porque es objeto de persecuciones. (pág. 122)

Reign of Terror [reino del Terror] *s.* período comprendido entre 1793 y 1794 durante el cual las nuevas autoridades en Francia ejecutaron miles de ciudadanos. (pág. 317)

relative location [ubicación relativa] *s.* ubicación de un lugar en relación con otros. (pág. 37)

Renaissance *s.* período de creatividad y de aprendizaje en Europa occidental entre los siglos XIV y XVI. (pág. 302)

republic [república] *s.* forma de gobierno controlado por sus cuidadanos a través de representantes elegidos por los cuidadanos. (pág. 285)

reunification [reunificación] *s.* acción de unificar nuevamente las partes. (pág. 396)

Riksdag (RIHKS•DAHG) *s.* parlamento sueco. (pág. 384)

Rio de Janeiro *s.* ciudad brasileña. (pág. 246)

river system [sistema fluvial] *s.* que incluye ríos principales y sus tributarios. (pág. 74)

rural [rural] *adj.* que pertenece al campo. (pág. 193)

Russian Revolution [Revolución rusa] *s.* revolución de 1917 que eliminó la monarquía rusa del poder luego de 400 años de vigencia. (pág. 322)

S

São Paulo [San Pablo] *s.* ciudad brasileña. (pág. 246)

scarcity [escasez] *s.* palabra usada por los economistas para describir el conflicto que existe entre el deseo de los seres humanos y los recursos limitados para satisfacerlo. (pág. 20)

R38

Scientific Revolution [Revolución científica] *s.* período de grandes cambios científicos y descubrimientos durante los siglos XVI a XVII. (pág. 314)

secede [separarse] *v.* independizarse de una unidad política, como una nación. (pág. 381)

Senate [Senado] *s.* asamblea más poderosa de la República romana, cuyos representantes eran elegidos. (pág. 285)

separatist [separatista] *s.* persona que desea separarse del cuerpo al que pertenece, como la iglesia o la nación. (pág. 127)

single-product economy [economía de un solo producto] *s.* economía cuya mayor parte de los ingresos depende de un solo producto. (pág. 209)

skerry [arrecife] *s.* islote. (pág. 387)

socialism [socialismo] *s.* sistema económico en donde algunos negocios e industrias le pertenecen a una cooperativa o al gobierno. (pág. 392)

Solidarity [Solidaridad] *s.* sindicato polaco cuya finalidad inicial fue aumentar el salario, mejorar las condiciones laborales de los trabajadores y luego oponerse al Comunismo. (pág. 400)

standard of living [nivel de vida] *s.* forma de medir la calidad de vida. (pág. 369)

sugar cane [caña de azúcar] *s.* planta de la que se extrae el azúcar. (pág. 209)

T

tariff [arancel aduanero] *s.* tarifa o suma de dinero impuesto por el gobierno en productos que se importan o exportan. (pág. 369)

technology [tecnología] *s.* herramientas o equipos que se crean a partir de ciertos descubrimientos. (pág. 112)

thematic map [mapa temático] *s.* mapa que se centra en una idea o tema particular. (pág. 46)

tourism [turismo] *s.* industria que estimula a la gente a viajar por placer. (pág. 189)

transportation barrier [barrera al transporte] *s.* obstáculo que impede el transporte o lo disminuye. (pág. 131)

transportation corridor [vía de transporte] *s.* camino que facilita el transporte. (pág. 131)

Treaty of Guadalupe Hidalgo [Tratado de Guadalupe Hidalgo] *s.* tratado que Estados Unidos exigió a México que firmara en 1848, mediante el cual México cedió a Estados Unidos su territorio en el norte. (pág. 177)

tributary [tributario] *s.* arroyo o río que confluye en otro río de mayor caudal. (pág. 157)

Tropical Zone [Zona Tropical] *s.* región del mundo comprendida entre las latitudes 23°27′ al norte y 23°27′ al sur. (pág. 158)

U

United States Constitution [Constitución de Estados Unidos] *s.* documento que fundamenta todas las leyes del gobierno de Estados Unidos y constituye su marco jurídico. (pág. 94)

unlimited government [gobierno ilimitado] *s.* gobierno en el que las autoridades ejercen la mayor parte del poder. (pág. 95)

Ural (YUR•uhl) **Mountains** [Montes Urales] *s.* cadena montañosa que divide Europa de Asia. (pág. 275)

urban [urbano] *adj.* que pertenece a la ciudad. (pág. 193)

urbanization [urbanización] *s.* movimiento de personas del campo a la ciudad. (pág. 241)

V

value [valor] *s.* principio e ideal básico de las personas. (pág. 110)

vegetation [vegetación] *s.* conjunto de plantas, arbustos, árboles y hierbas. (pág. 75)

R39

W

Warsaw Pact [Pacto de Varsovia] *s.* tratado firmado en 1955 que estableció una alianza entre la Unión Soviética, Albania, Bulgaria, Checoslovaquia, Alemania Oriental, Hungría, Polonia y Rumania. (pág. 347)

weather [tiempo] *s.* estado atmosférico cercano a la Tierra en un momento y lugar determinados. (pág. 75)

West Indies [Indias Occidentales/las Antillas] *s.* islas del Caribe. (pág. 203)

World War I [Primera Guerra Mundial] *s.* guerra de 1914 a 1918 entre los aliados (Rusia, Francia, el Reino Unido, Italia y los Estados Unidos) y las potencias centrales (Imperio autro-húngaro, Alemania, Turquía y Bulgaria). (pág. 333)

World War II [Segunda Guerra Mundial] *s.* guerra de 1939 a 1945 entre las potencias del Eje (Alemania, Italia y Japón) y los aliados (el Reino Unido, Francia, la Unión Soviética y los Estados Unidos). (pág. 336)

R40

Index

An *i* preceding a page reference in italics indicates that there is an illustration, and usually text information as well, on that page. An *m* or a *c* preceding an italic page reference indicates a map or a chart, as well as text information on that page.

R41

C

D

E

R43

F

G

R44

R45

L

M

N

R47

O

P

Q

R

S

R49

T

U

V

W

R51

Y

Z

R52

Acknowledgments

Text Credits

Chapter 1, pages 28–29: "How Thunder and Earthquake Made Ocean," from *Keepers of the Earth: Native American Stories and Environmental Activities for Children* by Michael J. Caduto and Joseph Bruchac. Copyright © 1988, 1989 Michael J. Caduto and Joseph Bruchac. Reprinted by permission of Fulcrum Publishing.

Chapter 3, page 83: Excerpts from *The Dust Bowl: Men, Dust, and the Depression* by Paul Bonnifield. Copyright © 1979 by the University of New Mexico Press. Reprinted by permission of the University of New Mexico Press.

Chapter 4, page 100: "Coney" from *Subway Swinger* by Virginia Schonborg, copyright © 1970 by Virginia Schonborg. Used by permission of HarperCollins Publishers.

Chapter 4, page 100: "Knoxville, Tennessee" from *Black Feeling, Black Talk, Black Judgment* by Nikki Giovanni. Copyright 1968, 1970 by Nikki Giovanni. Reprinted by permission of HarperCollins Publishers Inc.

page 101: "Scenic" from *Collected Poems, 1953–1993* by John Updike, copyright © 1993 by John Updike. Used by permission of Alfred A. Knopf, a division of Random House, Inc.

Chapter 4, page 115: Excerpt from "Diversity: The American Journey" by Frank A. Blethen, *Seattle Times* Publisher and CEO. Reprinted by permission of the author.

Chapter 5, page 139: Excerpt from "Canadian Bakin': Nation ends 50 years of hockey frustrations" by Michael Hunt, *Milwaukee Journal Sentinel.* Reprinted by permission of the Milwaukee Journal Sentinel.

Chapter 7, page 199: Excerpt from interview from the *NewsHour with Jim Lehrer,* July 15, 1999. Copyright © 1999 by NewsHour. Reprinted by permission of NewsHour.

Chapter 8, page 227: Excerpt from "Nobel Accceptance Speech" by Oscar Arias Sánchez. Copyright © The Nobel Foundation. Reprinted by permission of The Nobel Foundation.

Chapter 9, page 239: "Chilean Earth" from *A Gabriela Mistral Reader,* translation copyright 1993 by Maria Giachetti. Reprinted by permission of White Pine Press.

page 249: Copyright © 2000 by Houghton Mifflin Company, Reproduced by permission from *The American Heritage Dictionary of the English Language, Fourth Edition.*

Chapter 10, page 281: Quote by Herakleitos, translated by Guy Davenport, from *7 Greeks,* copyright © 1995 by Guy Davenport. Reprinted by Sales Territory: U.S./Canadian rights only.

Chapter 12, page 340: "The Giant's Causeway" from *Irish Fairy Tales and Legends* retold by Una Leavy. Copyright © 1996 by The Watts Publishing Group Ltd.

Chapter 12, page 349: Excerpt from *Blood, Toil, Tears and Sweat* by Winston Churchill. Reproduced with permission of Curtis Brown Ltd., London, on behalf of the Estate of Sir Winston S. Churchill. Copyright © the Estate of Sir Winston S. Churchill.

Chapter 14, page 405: Excerpt from "Presentation Speech" by Egil Aarvik. Copyright © The Nobel Foundation. Reprinted by permission of The Nobel Foundation.

Art Credits

Beverly Doyle 28; John Edwards Studio 81; Ken Goldammer 12; Nenad Jakesevic 163, 164, 168, 381, 404; Rich McMahon 44, 323; Gary Overacre 340–341; Matthew Pippin xiv, 236–237. All other artwork created by Publicom, Inc.

Map Credits

This product contains proprietary property of MAPQUEST.COM. Unauthorized use, including copying, of this product is expressly prohibited.

Photography Credits

Cover NASA; (See page 14 for full credits.); **vii** *left* Terry Wild Studio; *bottom* Copyright © Panoramic Images; *top* D. Robert and Lorri Franz/Corbis; *right* Copyright © SuperStock; **viii** *top* Michel Zabe/Art Resource, New York; *center* AFP/Corbis; *bottom* Staffan Widstand/Corbis; **ix** *top left* Erich Lessing/Art Resource, New York; *bottom left* Dave Bartruff/Corbis; *bottom right* Hulton|Archive/Getty Images; *top* Erich Lessing/Art Resource, New York; *top right* Reunion des Musées Nationaux/Art Resource, New York; *center right* Scott Gilchrist/Archivision.com; **x** *top left* S. Bavister/Robert Harding Picture Library; *bottom left* John Noble/Corbis; *bottom* John Launois/Black Star Publishing/PictureQuest; *bottom right* AFP/Corbis; *top* Copyright © Brannhage/Premium/Panoramic Images; **S12** Reprinted with the permission of the *St. Louis Post Dispatch,* 2002; **S13** The Granger Collection, New York; **S26** Kevin Schafer/Corbis; **S29** The Granger Collection, New York; **S30–31** Stephanie Maze/Corbis; **S32** Mary Evans Picture Library.

R53

UNIT ONE

2–3 NASA; **4** *bottom right* NOAA; *left* © Roger Ressmeyer/Corbis; *top right* Science Museum/Science and Society Picture Library, London.

Chapter 1

14 *top left* Brian A. Vikander/Corbis; *top center* Owen Franken/Corbis; *bottom left* Tim Thompson/Corbis; *center left* Kevin Schafer/Corbis; *bottom center* Maria Taglienti/The Image Bank/GettyImages; *center right* James A. Sugar/Corbis; *bottom right* Nicholas deVore III/Photographers Aspen/PictureQuest; *top right* Dean Conger/Corbis; **15** *top* John Callahan/Stone/GettyImages; *bottom left* The Purcell Team/Corbis; *center left* Helen Norman/Corbis; *center* Nik Wheeler/Corbis; *bottom center* Dennis Degnan/Corbis; *center right* Neil Rabinowitz/Corbis; *bottom right* Martin Rogers/Corbis; **16** *top* Copyright © Jim West; *bottom* K. Gilham/Robert Harding Picture Library; **17** *top* The Image Bank/Getty Images; *bottom* The Military Picture Library/Corbis; **18** *bottom* © Roger Ressmeyer/Corbis; *top* Skjold Photographs; **19** *left* © Franz-Marc Frei/Corbis; *right* © Peter Turnley/Corbis; *center* © Owen Franken/Corbis; **20** *right* Copyright © Jim West; *left* © Reuters NewMedia Inc./Corbis; **21** © Ed Reinke/AP/Wide World Photos; **22** Joseph Sohm/ Visions of America, LLC/PictureQuest; **23** *bottom* Joe Sohm, Chromosohm/Stock Connection/ PictureQuest; *top* Scott Teven/Stock Connection/ PictureQuest; **24** *right* The Granger Collection, New York; **26** *left* © Neil Rabinowitz/Corbis; *right* © Dave G. Houser/Corbis; *center* © Roman Soumar/Corbis; **28–29** Illustration © Bill Cigliano **30** *bottom left* Picture Finders/eStock Photography/PictureQuest.

Chapter 2

32–33 © David Hardy/Photo Researchers, Inc.; **33** Dagli Orti/Biblioteca Nacionale Marciana Venice/The Art Archive; **34** *top* © Stuart Franklin/Magnum Photos; *bottom* © Francis E. Caldwell, Affordable Photo; **35** *left* The Granger Collection, New York; **36** Schafer and Hill/Stone/GettyImages; **37** *top* © Francis E. Caldwell, Affordable Photo; *bottom* Digital Image © 1996 Corbis; Original image courtesy of NASA/CORBIS; **38** © Thonig/Premium Stock/PictureQuest; **39** David Muench/ Stone/GettyImages; **42** *left* © Bettmann/Corbis; *top right* © Victoria & Albert Museum, London/Art Resource, NY; *bottom right* The Art Archive/India Office Library; **43** *bottom* Austrian Archives/Corbis; *top right* The Granger Collection, New York; **45** *right* The Granger Collection, New York; **46** Illustration © Bill Cigliano; **47** Library of Congress, Prints and Photographs Division; **50** David Muench/Stone/GettyImages; **51** NASA.

UNIT TWO

52–53 Copyright © Panoramic Images.

Chapter 3

66–67 Paul A. Souders/Corbis; **67** *top* James Randklev/Corbis; **68** *bottom* Michael Melford/The Image Bank/GettyImages; *top* Raymond Gehman/Corbis; **69** *right* Corbis; *left* Pat O'Hara/Corbis; **70** Michael Melford/The Image Bank/GettyImages; **71** *top* D. Robert and Lorri Franz/Corbis; *center* Bettmann/Corbis; *bottom* Bettmann/Corbis; **72** *right* Paul A. Souders/Corbis; *left* Copyright © Didier Dorval/Masterfile; **73** M. L. Fuller and the U. S. Geological Society; **74** Paul A. Souders/Corbis; **75** Copyright © F. Hoffman/The Image Works; **76** Bettmann/Corbis; **78** *top* Raymond Gehman/Corbis; *bottom* Phillip Gould/Corbis; **79** David Reed/AP/Wide World Photos; **82** *right* David Reed/AP/Wide World Photos; *left* Corbis.

Chapter 4

84–85 Copyright © Tom Jelen/Panoramic Images; **85** *bottom right* Dennis Brack/Black Star Publishing/PictureQuest; *top left* Reuters NewMedia Inc./Corbis; *bottom left* Ed Kashi/Corbis; *top center* Copyright © Marianne Barcellona/Time Pix; *center* Reuters NewMedia Inc./Corbis; *bottom center* Copyright © David Lassman/The Image Works; *top right* AFP/Corbis; **86** *top* The Granger Collection, New York; *bottom* Reuters NewMedia Inc./Corbis; **87** The Granger Collection, New York; **88** *bottom* Copyright © Nancy Richmond/The Image Works; *top* The Granger Collection, New York; **89** The Granger Collection, New York; **90** Reuters NewMedia Inc./Corbis; **92** *top left* NASA; *bottom right* European Space Agency; **93** *bottom right* European Space Agency; **93–94** NASA The Everett Collection; **94** The Everett Collection; **95** Art Resource, New York; **96** *bottom* Copyright © Topham/The Image Works; *top* The Granger Collection, New York; **100** *top* Tony Freeman; *bottom* Copyright © Rafael Macia/Photo Researchers; **101** *bottom* Copyright © Jim Corwin/Photo Researchers; **102** Copyright © Norbert Schwerine/The Image Works; **103** Teri Bloom; **104** *bottom* U.S. Treasury Department; **105** Kevin R. Morris/Bohemian Nomad PictureMakers/Corbis; **106** Copyright © Monika Graff/The Image Works; **108** *top* Corbis; *center* Spencer Grant III/Stock Boston/PictureQuest; *bottom* Paul A. Souders/Corbis; **109** *bottom right* Culver Pictures; *top* Archive Photos/PictureQuest; *bottom left* Farrell Greham/Corbis; *center* Michael S. Yamashita/Corbis; **110** Tom Vano/Index Stock Imagery/PictureQuest; **111** *right* Terry Wild Studio; *left* William Folsom/Words and Pictures/PictureQuest; **114** *left* The Granger Collection, New York; *right* Terry Wild Studio; *center* The Everett Collection.

R54

Chapter 5

116–117 Copyright © First Light/Panoramic Images; **117** *top* Canadian Museum of Civilization, image number S93–2826; **118** *top* Kevin R. Morris/Corbis; *bottom* Hockey Hall of Fame; **119** *all* The Granger Collection, New York; **120** Kevin R. Morris/Corbis; **121** Private Collection/Phillips, Fine Art Auctioneers, New York/The Bridgeman Art Library; **123** *top* Copyright © James Schwabel/Panoramic Images; **124** Copyright © SuperStock; **125** Copyright © Bill Brooks/Masterfile; **126** Ron Poling/Canadian Press/AP/Wide World Photos; **127** Yves Marcoux/Stone/GettyImages; **128** Bettmann/Corbis; **131** The Granger Collection, New York; **132** Hockey Hall of Fame; **133** Copyright © J. A. Kraulis/Masterfile; **134** Lee Snider/Corbis; **135** Archivision.com; **138** *top left* Copyright © James Schwabel/Panoramic Images; *bottom left* Copyright © Bill Brooks/Masterfile; *top right* The Granger Collection, New York; *bottom right* Copyright © J. A. Kraulis/Masterfile.

UNIT THREE

140–141 Bill Pogue/Stone/GettyImages.

Chapter 6

150–151 Cosmo Condina/Stone/GettyImages; **151** *top* Kevin Schafer/Corbis; **152** *right* Richard A. Cooke/Corbis; *left* Archivo Iconografico, S. A./Corbis; **153** AFP/Corbis; **154** Danny Lehman/Corbis; **156** *top* Guido A. Rossi/The Image Bank/GettyImages; *bottom* Stephen Frink/Corbis; **157** Galen Rowell/Corbis; **160** *left* Lara Fox; *right* After Hansen 1990: II. Drawn by TW Rutledge. Used by permission; **161** *top* Richard A. Cooke/Corbis; *bottom* Copyright © Craig Lovell/Painet; **162** *top* Mireille Vautier/Woodfin Camp/PictureQuest; *bottom* Michel Zabe/Art Resource, New York; **165** *bottom* Edwin Sulca Lagos; *top* Werner Forman Archive/Art Resource, New York; **166** Archivo Iconografico, S. A./Corbis; **168** *center left* Kevin Schafer/Corbis; *top left* Guido A. Rossi/The Image Bank/GettyImages; *bottom right* Werner Forman Archive/Art Resource, New York.

Chapter 7

170–171 Robert Frerck/Stone/GettyImages; **171** *top* R.G.K. Photography/Stone/GettyImages; **172** *top* Erich Lessing/Art Resource, New York; *center Cabacete* [Helmet] (c. 1480). Private collection. Photograph courtesy of the Cleveland (Ohio) Museum of Art; *bottom* Danny Lehman/Corbis; **173** *right* The Granger Collection, New York; *left* Richard T. Nowitz/Corbis; **174** *bottom center* Copyright © SPL/Custom Medical Stock Photography; *bottom left* The Granger Collection, New York; *top left* Erich Lessing/Art Resource, New York; *top right Cabacete* [Helmet] (c. 1480). Private collection. Photograph courtesy of the Cleveland (Ohio) Museum of Art; **175** David Hiser/Stone/GettyImages; **177** *top* Schalkwijk/Art Resource, New York; *bottom* Greg Probst/Corbis; **179** *left* PhotoDisc/GettyImages; *right* Bettmann/Corbis; **180** *top* Bettmann/Corbis; *bottom* The Granger Collection, New York; **181** *center* Bettmann/Corbis; *bottom* Tom Bean/Corbis; **184** Latin Focus.com; **185** J. P. Courau/D. Donne Bryant Stock Photography; **186** *left* Annie GriffithsBelt/Corbis; *right* Robert Frerck/Stone/GettyImages; **187** *top* Danny Lehman; *bottom* Jose Welbers/Latin Focus.com; **189** ZSSD/Latin Focus.com; **190** *right* Kal Muller/Woodfin Camp/PictureQuest; *left* Copyright © Erwin Nielsen/Painet; **191** Peter Menzel/Stock Boston/PictureQuest; **192** Schalkwijk/Art Resource, New York; **193** Danny Lehman/Corbis; **194** Charles and Josette Lenars/Corbis; **196** *top* Cosmo Condina/Stone/GettyImages; *bottom* Copyright © Bernard Neuwirtch/Painet; **197** Gary S. Withey/Bruce Coleman/Picture Quest; **198** *top right* Annie Griffiths Belt/Corbis; *bottom right* Danny Lehman/Corbis; *bottom left* Schalkwijk/Art Resource, New York.

Chapter 8

200–201 Copyright © Michel Frigan/Film Works Inc.; **201** *top* Dave G. Houser/Corbis; **202** *bottom* Reuters NewMedia Inc./Corbis; *top right* Copyright © Topham/The Image Works; **203** Copyright © The Image Works; **204** James Strachan/Stone/GettyImages; **205** *center Ladino* Reuters NewMedia Inc./Corbis; *bottom Ladino* Reuters NewMedia Inc./Corbis; *top Ladino* Martin Rogers/Corbis; *bottom right* Archive Photos/PictureQuest; **206** Jeff Greenberg/eStock Photography/PictureQuest; **207** Bill Gentile/Corbis; **208** AFP/Corbis; **209** Copyright © Topham/The Image Works; **210** AFP/Corbis; **211** *top* Bill Ross/Corbis; *bottom* Copyright © Rob Crandall/StockBoston; **212** Suzanne Murphy-Larronde; **213** Nik Wheeler/Corbis; **215** Sovfoto/Eastfoto/PictureQuest; **216** *center left* Bettmann/Corbis; *bottom left* Richard Bickel/Corbis; *bottom right* Latin Focus.com; **217** Latin Focus.com; **218** AFP/Corbis; **219** Copyright © Marge George/Spectrum Stock; **220** Reuters NewMedia Inc./Corbis; **221** Scott Sandy/AP/Wide World Photos; **222** *top* Latin Focus.com; *bottom* Cindy Karp/Black Star Publishing/PictureQuest; **223** Jan Butchofsky-Houser/Corbis; **224** *top* Martin Rogers/Corbis; *bottom* World Pictures/eStock Photography/PictureQuest; **225** Jan Butchofsky-Houser/Corbis; **226** *bottom left* Copyright © The Image Works; *top* Suzanne Murphy-Larronde; *bottom center* Sovfoto/Eastfoto/PictureQuest; *bottom right* Jan Butchofsky-Houser/Corbis.

R55

Chapter 9

228–229 Layne Kennedy/Corbis; **229** *top* Eye Ubiquitous/Corbis; **230** *bottom* David Wolf; *top* Steve Vidler/eStock Photography/PictureQuest; **231** *all* The Granger Collection, New York; **233** Christie's Images/Corbis; **234** *top* AFP/Corbis; *bottom* Bettmann/Corbis; **235** *bottom* Latin Focus.com; *top* Richard Hamilton Smith/Corbis; **238** Reuters NewMedia Inc./Corbis; **239** *top* Victor Englebert; *bottom* Corbis; **242** Steve Vidler/eStock Photography/PictureQuest; **243** Copyright © Larry Luxner/Luxner News; **245** *bottom* Tim Page/Corbis; *top* Copyright © Les Stone/The Image Works; **246** David Wolf; **247** *top* Reuters NewMedia Inc./Corbis; *bottom* Bettmann/Corbis; **248** *top* Macduff Everton/Corbis; *bottom* AFP/Corbis; **249** *top* Dagli Orti/National History Museum Mexico City/The Art Archive; *bottom* Garden Picture Library; **250** Latin Focus.com; **251** Cory Langley; **252** *right* Richard Hamilton Smith/Corbis; *left* Victor Engelbert; *center* Charles and Josette Lenars/Corbis; **253** Victor Engelbert; **254** Staffan Widstrand/Corbis; **256** *bottom left* Christie's Images/Corbis; *bottom center* Victor Engelbert; *top* David Wolf; *bottom right* Staffan Widstrand/Corbis.

UNIT FOUR

258–259 Stuart Dee/The Image Bank/GettyImages.

Chapter 10

270–271 Copyright © James L. Stanfield/National Geographic Society Image Collection; **271** *top* Robert Harding Picture Library; **272** *top* Sef/Art Resource, New York; *bottom left, bottom right* Erich Lessing/Art Resource, New York; **273** Bill Ross/Corbis; **274** Arnulf Husmo/Stone/GettyImages; **275** Walter Bibikow/Index Stock Imagery/PictureQuest; **276** *top left* Jonathan Blair/Corbis; *top right* Eye Ubiquitous/Corbis; *bottom* Johan Elzenga/Stone/GettyImages; **278** *all* Greek Culture Ministry/AP/Wide World Photos; **279** HorreeZirkzee Produk/Corbis; **280** Foto Marburg/Art Resource, New York; **281** Sef/Art Resource, New York; **282** *left* Nimatallah/Art Resource, New York; *right* Scala/Art Resource, New York; **284** Copyright © Macduff Everton/The Image Works; **285** Erich Lessing/Art Resource, New York; **286** Giraudon/Art Resource, New York; **287** Erich Lessing/Art Resource, New York; **288** *top left* Jeff Rotman; *center* Scala/Art Resource, New York; *bottom right* O. Alamany and E. Vicens/Corbis; *bottom center* Bettmann/Corbis; **289** Erich Lessing/Art Resource, New York; **290** Art Resource, New York; **291** Reunion des Musées Nationaux/Art Resource, New York; **292** Catherine Karnow/Corbis; *spread* Musée de la Tapisserre, Bayoux, France/The Bridgeman Art Library; **293** *top* Jose Fuste Raga/eStockPhotography/PictureQuest; **294** Erich Lessing/Art Resource, New York; **295** *right* Dept. of the Environment, London/The Bridgeman Art Library; *left* The Granger Collection, New York; **296** *top left* Sef/Art Resource, New York; *bottom* Erich Lessing/Art Resource, New York; *top right* Jose Fuste Raga/eStockPhotography/PictureQuest.

Chapter 11

298–299 Bruno Barbey/Magnum/PictureQuest; **299** *top* The Granger Collection, New York; *center* Leonard L. T. Phodes/Animals Animals; **300** *center* Mary Evans Picture Library; *top* Reunion des Musées Nationaux/Art Resource, New York; *bottom* The Pierpont Morgan Library/Art Resource, New York; **301** The Granger Collection, New York; **302** Alinari/Art Resource, New York; **303** *top* Scott Gilchrist/Archivision.com; *bottom* Palazzo Medici-Riccardi, Florence, Italy/The Bridgeman Art Library; **304** *bottom* Scala/Art Resource, New York; *top* Reunion des Musées Nationaux/Art Resource, New York; **305** The Pierpont Morgan Library/Art Resource, New York; **306** Corbis; **307** Giraudon/Art Resource, New York; **309** North Wind Pictures; **310** Reunion des Musées Nationaux/Art Resource, New York; **313** AKG London; **314** *center* Scala/Art Resource, New York; *top* NASA; *bottom* Copyright © Will & Deni McIntyre/Photo Researchers; **315** The Granger Collection, New York; **316** Hulton-Deutsch Collection/Corbis; **317** Victoria & Albert Museum, London/Art Resource, New York; **318** *right* © Courtesy of the Estate of Ruskin Spear/Private Collection/Phillips, Fine Art Auctioneers, New York/The Bridgeman Art Library; *left* Giraudon/Art Resource, New York; **319** *top* Scala/Art Resource, New York; *bottom* Roger Tidman/Corbis; **320** Erich Lessing/Art Resource, New York; **321** Chuck Nacke/Woodfin Camp/PictureQuest; **322** Hulton-Deutsch Collection/Corbis; **324** *top left* North Wind Pictures; *bottom left* Alinari/Art Resource, New York; *bottom right* Hulton-Deutsch Collection/Corbis; *top right* © Courtesy of the Estate of Ruskin Spear/Private Collection/Phillips, Fine Art Auctioneers, New York/The Bridgeman Art Library; **325** The Library of Congress Website.

Chapter 12

326–327 Michael S. Yamashita/Corbis; **327** *top* Owen Franken/Corbis; **328** *top* Ralph White/Corbis; *bottom* Hulton|Archive/Getty Images; **329** Hulton|Archive/Getty Images; **331** *bottom* Hulton|Archive/Getty Images; *center* Gianni Dagli Orti/Corbis; *top* Mark Rykoff/Rykoff Collection/Corbis; **333** *right* Bettmann/Corbis; *left* Hulton|Archive/Getty Images; **334** *top* Hulton|Archive/Getty Images; *bottom* Art Young; **335** Hulton|Archive/Getty Images; **337** *bottom* Hulton|Archive/Getty Images; *top* Hulton-Deutsch Collection/Corbis; **339** Reprinted with the permission of the *St. Louis Post Dispatch,* 2002; **342** Paul Almasy/Corbis; **343** *bottom* Sovfoto/Eastfoto/PictureQuest; *top* Ralph White/Corbis; **344** *bottom* Hulton|Archive/Getty Images; *top* Sovfoto/Eastfoto/PictureQuest; **345** Culver Pictures; **346** *bottom* Hulton|Archive/Getty Images; *center* The Kobal Collection; **347** Dave Bartruff/Corbis; **348** *right* Sovfoto/

R56

Eastfoto/PictureQuest; *left* Hulton|Archive/Getty Images; *center* Hulton-Deutsch Collection/Corbis; **349** Mandeville Special Collections at UCSD.

Chapter 13

350–351 Copyright © Brannhage/Premium/Panoramic Images; **351** *top* Sovfoto/Eastfoto; **352** *top* Premium Stock/Corbis; *bottom* Craig Aurness/Corbis; **353** Mark Rykoff/Corbis; **354** *left* NASA/AP/Wide World Photos; *right* Sovfoto/Eastfoto; **355** Copyright © Giuliano Bevilacqua/TimePix; **356** Bryn Colton/Corbis; **357** *top* Bettmann/Corbis; *bottom* AP/Wide World Photos; **358** Bojan Brecelj/Corbis; **360** David and Peter Turnley/Corbis; **362** Craig Aurness/Corbis; **363** Copyright © Bios (F. Gilson)/Peter Arnold; **364** Scala/Art Resource, New York; **366** Sovfoto/Eastfoto/PictureQuest; **367** *bottom* Hoa Qui/Index Stock Imagery/PictureQuest; *top* Premium Stock/Corbis; **368** AFP/Corbis; **369** Mike Mazzaschi/Stock Boston/PictureQuest; **370** S. Bavister/Robert Harding Picture Library; **371** Copyright © Malcolm S. Kirk/Peter Arnold; **372** *bottom* John Neubauer/Photo Edit/PictureQuest; *top right* Underwood & Underwood/Corbis; *top left* Bettmann/Corbis; **373** *top* Wolfgang Kaehler/Corbis; *left* Paul A. Souders/Corbis; *center* Roger Ressmeyer/Corbis; *bottom right* Academy of Natural Sciences of Philadelphia/Corbis; **374** *left* Bettmann/Corbis; *right* Premium Stock/Corbis; *center* David and Peter Turnley/Corbis.

Chapter 14

376–377 Alan Thornton/Stone/GettyImages; **377** *top* Mark A. Leman/Stone/GettyImages; **378** *top* Ted Spiegel/Corbis; *bottom* www.carpix.net; **379** *bottom right* Michael Neveux/Corbis; *center* AFP/Corbis; **382** *bottom* John Launois/Black Star Publishing/PictureQuest; *top* Copyright © Julian Nieman/Collections; **383** AFP/Corbis; **384** *right* Nik Wheeler/Corbis; *left* Ted Spiegel/Corbis; **385** Hans T. Dahlskog/Pressens Bild; **386** *bottom* AFP/Corbis; *top* Alex Farnsworth/The Image Works; **387** John Noble/Corbis; **388–389** Museo de Firenze Com'era, Florence, Italy/The Bridgeman Art Library; **390** *right* Bettmann/Corbis; *left* Corbis; **391** Bettmann/Corbis; **392** Robert Estall/Corbis; **393** Art Resource, New York; **394** *right* AFP/Corbis; *left* Thomas Hoepker/Magnum/PictureQuest; **396** *bottom center* Erich Lessing/Art Resource, New York; *top* www.carpix.net; *bottom right* Josef Karl Stieler/Archivo Iconografico, S. A./Corbis; **397** Carmen Redondo/Corbis; **399** Chuck Fishman/Contact Press Images/PictureQuest; **400** Bettmann/Corbis; **401** Dennis Chamberlain/Black Star Publishing/PictureQuest; **402** Steven Weinberg/Stone/GettyImages; **403** Vittoriano Rastelli/Corbis; **404** *top left* Copyright © Alex Farnsworth/The Image Works; *top right* Thomas Hoepker/Magnum/PictureQuest; *bottom right* Bettmann/Corbis; *center* Corbis.

Special Report

407 *top* AP/Wide World Photos; *bottom* AP/Wide World Photos; **409** *top* AP/Wide World Photos; *bottom* AP/Wide World Photos; **410** AP/Wide World Photos; **411** *left* © Mario Tama/Getty Images; *right* © Reuters NewMedia Inc./Corbis.

McDougal Littell Inc. has made every effort to locate the copyright holders for the images used in this book and to make full acknowledgment for their use. Omissions brought to our attention will be corrected in subsequent editions.

R57